Mathematics MCQ

Asit Das Gupta, PhD

Bharati Bhawan
PUBLISHERS & DISTRIBUTORS

Published by

BHARATI BHAWAN (Publishers & Distributors)

www.bharatibhawan.in   email: editorial@bbpd.in

4271/3 Ansari Road, Daryaganj, NEW DELHI 110 002, Phone: 23286557
A-61 B/2 Sector 63, NOIDA 201 307, Phone: 4757400
Thakurbari Road, PATNA 800 003, Phone: 2670325
10 Raja Subodh Mallick Square, KOLKATA 700 013, Phone: 22250651
No. 98 Sirsi Circle, Mysore Road, BENGALURU 560 018, Phone: 26740560

First edition 1999
Seventh edition 2008
Revised print 2010
2023 print

Mathematics MCQ
Printed at Jai Hind Press, Patna-800 020

Preface

This book is meant for helping students prepare for JEE-IIT and other technical institutes. It will also be useful for preparing for those examinations in which multiple-choice questions (MCQs) are asked.

The salient features of this book are as follows.

1. All the concepts of a chapter are covered by the questions, which are graded and arranged topic-wise.
2. Most of the questions are divided into two types—Type 1 and Type 2. Type 1 questions have only one option correct, whereas Type 2 questions may have multiple options correct.
3. In the examinations, the questions may range from simple to difficult, with emphasis on the student's ability to do maximum number of questions correctly. Therefore, questions of different difficulty levels have been included in this book.
4. At the end of each chapter, hints to a large number of questions have been given to help students in case of difficulty. Also, a number of difficult problems have been fully solved.
5. The number of questions in each chapter has been kept sufficiently large to provide rigorous practice.
6. One separate part is devoted to miscellaneous questions covering all forms of objective questions. Miscellaneous Questions 1 and 3 contains multiple-choice question with *one correct option.* Miscellaneous Questions 2 contains multiple-choice question with *one or more correct options.* Miscellaneous Questions 4 contains questions on comprehension. Miscellaneous Questions 5 contains questions on matching. Miscellaneous Questions 6 contains questions on fill in/completion. Miscellaneous Questions 7 contains questions on assertion-reason. This is followed by practice worksheets which should help students practice answering such questions in a fixed period of time.

I would like to thank the members of my family and the editorial and production people at Bharati Bhawan for their cooperation and efforts.

Author

To the Students

In order to do well in a test of multiple-choice questions (MCQs) you need to sharpen a few skills, know how to overcome potential problems and understand the different ways in which an MCQ can be framed.

- MCQs can be based just to test the examinee's knowledge of a particular concept, formula, theorem, etc. These should be relatively simple to answer, provided you have studied all the topics well.
- To score well in an MCQ test, speed and accuracy of calculations is an asset. You have to develop the ability to do a number of steps mentally. Practise solving a number of problems in a short interval, say about 40 questions in an hour.
- Read a question carefully and decide whether you can arrive at the answer by the process of elimination. This is an important method and it is illustrated below.

EXAMPLE The lines whose equations are $x+y=1$, $2x-y=3$, $2x+3y=6$ and $3x+4y=5$.

(a) form a trapezium (b) form a rectangle

(c) are concurrent (d) form a quadrilateral

Since no two pair of lines are parallel, the first two options are eliminated. Now, if the point of intersection of a pair of lines is the same as the point of intersection of the other two lines then the four lines are concurrent, otherwise the lines form a quadrilateral. In this case you will find that the option (d) is the most appropriate.

- An MCQ may have more than one option correct. You need to develop the ability to recognise problems which may have multiple correct options.

EXAMPLE The line $y=mx+2$ will cut the pair of lines $2x^2-3xy+y^2-x+y=0$ at only one real point if is equal to

(a) 1 (b) 2 (c) –2 (d) –1

The figures ahead indicate that the given line can cut the pair of lines at only one real point under three circumstances.

The separate equations of the lines of the pair are $2x-y-1=0$ and $x-y=0$.

Case I The line $y = mx + 2$ passes through the point of intersection of the pair of lines.

The point of intersection of $2x - y - 1 = 0$ and $x - y = 0$ is (1, 1). It is on $y = mx + 2$ when $1 = m + 2$, i.e., $m = -1$. Thus, option (d) is correct.

Case II The line $y = mx + 2$ is parallel to the line $x - y = 0$ of the pair.

The slope of $x - y = 0$ is 1. $\therefore\ m = 1$.

Therefore, option (a) is also correct.

Case III The line $y = mx + 2$ is parallel to the line $2x - y - 1 = 0$ of the pair.

The slope of $2x - y - 1 = 0$ is 2. $\therefore\ m = 2$.

Therefore, option (b) is also correct.

EXAMPLE There are 10 people sitting in a row. The number of ways of selecting 4 people from among them so that no two consecutive people are selected is

(a) ${}^7C_3$ (b) 45 (c) ${}^7C_4$ (d) none of these

The number of ways of doing a work is always unique. But it is quite possible that two or more options in the question give the same value.

A selection of four persons according to the restriction corresponds to an integral solution of the equation

$$x_1 + x_2 + x_3 + x_4 + x_5 = 6\ (= 10 - 4),$$

where $x_1 \geq 0, x_2 \geq 1, x_3 \geq 1, x_4 \geq 1, x_5 \geq 0$.

$\therefore$ the required number of selections of 4 persons the number of integral solutions of the above equation coefficient of x^6 in $(x^0 + x^1 + x^2 + \dots)(x + x^2 + \dots)(x + x^2 + \dots)(x + x^2 + \dots)(x^0 + x^1 + \dots)$ coefficient of x^3 in $(1 - x)^{-5} = {}^7C_3$.

Thus, option (a) is correct.

But ${}^nC_r = {}^nC_{n-r} \Rightarrow {}^7C_3 = {}^7C_4$, so, option (c) is also correct.

An MCQ may appear to be beyond the bounds of the +2 syllabus. However, a little application will allow you to solve such a question.

EXAMPLE The number of integral solutions (x, y) of the equations $\sqrt{y} + y\sqrt{x} = 20$, $x\sqrt{x} + y\sqrt{y} = 65$ is

(a) 0 (b) 1 (c) 2 (d) none of these

Here the removal of radicals will turn the equations into higher degree polynomial equations. But the method of solving such equations is not included in the +2 syllabus. However, their integral solutions can be obtained by using the following ordinary mathematical argument.

As $\sqrt{x}$, $\sqrt{y}$ appear and 20, 65 are integers, clearly x and y must be positive integers which are perfect squares. So, the possible values of x, y must belong to the set {1, 4, 9, 16, ...}. By trial we find that $x = 1, y = 16$ and $x = 16, y = 1$ satisfy both the equations. Hence, there are exactly two integral solutions.

If you keep in mind the points made above and solve the questions given in this book, you should be able to tackle most MCQs quickly.

Author

□

Contents

□

Part 1

Algebra

1

Progressions, Related Inequalities and Series

• *Type 1* •

Choose the most appropriate option (a, b, c or d).

1. If $a_1, a_2, a_3, \ldots$ are in AP then a_p, a_q, a_r are in AP if p, q, r are in

(a) AP (b) GP

(c) HP (d) none of these

2. Let t_r denote the rth term of an AP. If $t_m = \frac{1}{n}$ and $t_n = \frac{1}{m}$ then t_{mn} equals

(a) $\frac{1}{mn}$ (b) $\frac{1}{m} + \frac{1}{n}$

(c) 1 (d) 0

3. If $p, q, r, s \in N$ and they are four consecutive terms of an AP then the pth, qth, rth, sth terms of a GP are in

(a) AP (b) GP

(c) HP (d) none of these

4. If in a progression $a_1, a_2, a_3, \ldots$, etc., $(a_r - a_{r+1})$ bears a constant ratio with $a_r \cdot a_{r+1}$ then the terms of the progression are in

(a) AP (b) GP

(c) HP (d) none of these

5. If $\frac{a_2a_3}{a_1a_4} = \frac{a_2 + a_3}{a_1 + a_4} = 3\left(\frac{a_2 - a_3}{a_1 - a_4}\right)$ then a_1, a_2, a_3, a_4 are in

(a) AP (b) GP

(c) HP (d) none of these

6. Let x, y, z be three positive prime numbers. The progression in which $\sqrt{x}, \sqrt{y}, \sqrt{z}$ can be three terms (not necessarily consecutive) is

(a) AP (b) GP

(c) HP (d) none of these

7. Let $f(x) = 2x + 1$. Then the number of real values of x for which the three unequal numbers $f(x), f(2x), f(4x)$ are in GP is

(a) 1 (b) 2

(c) 0 (d) none of these

8. If $a_r > 0, r \in N$ and $a_1, a_2, a_3, \dots, a_{2n}$ are in AP then

$$\frac{a_1 + a_{2n}}{\sqrt{a_1} + \sqrt{a_2}} + \frac{a_2 + a_{2n-1}}{\sqrt{a_2} + \sqrt{a_3}} + \frac{a_3 + a_{2n-2}}{\sqrt{a_3} + \sqrt{a_4}} + \dots + \frac{a_n + a_{n+1}}{\sqrt{a_n} + \sqrt{a_{n+1}}}$$

is equal to

(a) $n - 1$ (b) $\dfrac{n(a_1 + a_{2n})}{\sqrt{a_1} + \sqrt{a_{n+1}}}$

(c) $\dfrac{n-1}{\sqrt{a_1} + \sqrt{a_{n+1}}}$ (d) none of these

9. If $a_1, a_2, a_3, \dots, a_{2n+1}$ are in AP then

$$\frac{a_{2n+1} - a_1}{a_{2n+1} + a_1} + \frac{a_{2n} - a_2}{a_{2n} + a_2} + \dots + \frac{a_{n+2} - a_n}{a_{n+2} + a_n}$$

is equal to

(a) $\dfrac{n(n+1)}{2} \cdot \dfrac{a_2 - a_1}{a_{n+1}}$ (b) $\dfrac{n(n+1)}{2}$

(c) $(n+1)(a_2 - a_1)$ (d) none of these

10. Let $a_1, a_2, a_3, \dots$ be in AP and a_p, a_q, a_r be in GP. Then $a_q : a_p$ is equal to

(a) $\dfrac{r-p}{q-p}$ (b) $\dfrac{q-p}{r-q}$

(c) $\dfrac{r-q}{q-p}$ (d) none of these

11. If a, b, c are in GP then $a + b, 2b, b + c$ are in

(a) AP (b) GP

(c) HP (d) none of these

12. If a, b, c, d are nonzero real numbers such that

$$(a^2+b^2+c^2)(b^2+c^2+d^2) \le (ab+bc+cd)^2$$

then a, b, c, d are in

(a) AP (b) GP
(c) HP (d) none of these

13. If $4a^2+9b^2+16c^2 = 2(3ab+6bc+4ca)$, where a, b, c are nonzero numbers, then a, b, c are in

(a) AP (b) GP
(c) HP (d) none of these

14. If a, b, c are in AP then $\frac{a}{bc}, \frac{1}{c}, \frac{2}{b}$ are in

(a) AP (b) GP
(c) HP (d) none of these

15. If in an AP, $t_1 = \log_{10} a$, $t_{n+1} = \log_{10} b$ and $t_{2n+1} = \log_{10} c$ then a, b, c are in

(a) AP (b) GP
(c) HP (d) none of these

16. If $n!$, $3 \times n!$ and $(n+1)!$ are in GP then $n!$, $5 \times n!$ and $(n+1)!$ are in

(a) AP (b) GP
(c) HP (d) none of these

17. In an AP, the pth term is q and the $(p+q)$th term is 0. Then the qth term is

(a) $-p$ (b) p
(c) $p+q$ (d) $p-q$

18. In a sequence of $(4n+1)$ terms the first $(2n+1)$ terms are in AP whose common difference is 2, and the last $(2n+1)$ terms are in GP whose common ratio is 0.5. If the middle terms of the AP and GP are equal then the middle term of the sequence is

(a) $\dfrac{n \cdot 2^{n+1}}{2^n - 1}$ (b) $\dfrac{n \cdot 2^{n+1}}{2^{2n} - 1}$
(c) $n \cdot 2^n$ (d) none of these

19. If $x^2+9y^2+25z^2 = xyz\left(\frac{15}{x}+\frac{5}{y}+\frac{3}{z}\right)$ then x, y, z are in

(a) AP (b) GP
(c) HP (d) none of these

20. If a, b, c, d and p are distinct real numbers such that

$$(a^2+b^2+c^2)p^2-2(ab+bc+cd)p+(b^2+c^2+d^2)\le 0$$

then a, b, c, d are in

(a) AP (b) GP

(c) HP (d) none of these

21. The largest term common to the sequences 1, 11, 21, 31, ... to 100 terms and 31, 36, 41, 46, ... to 100 terms is

(a) 381 (b) 471

(c) 281 (d) none of these

22. The interior angles of a convex polygon are in AP, the common difference being 5°. If the smallest angle is $2\pi/3$ then the number of sides is

(a) 9 (b) 16

(c) 7 (d) none of these

23. The minimum number of terms of $1+3+5+7+\ldots$ that add up to a number exceeding 1357 is

(a) 15 (b) 37

(c) 35 (d) 17

24. In the value of 100 ! the number of zeros at the end is

(a) 11 (b) 22

(c) 23 (d) 24

25. The sum of all the proper divisors of 9900 is

(a) 33851 (b) 23952

(c) 23951 (d) none of these

26. The sum of all odd proper divisors of 360 is

(a) 77 (b) 78

(c) 81 (d) none of these

27. In the sequence 1, 2, 2, 3, 3, 3, 4, 4, 4, 4, ... , where n consecutive terms have the value n, the 150th term is

(a) 17 (b) 16

(c) 18 (d) none of these

28. In the sequence 1, 2, 2, 4, 4, 4, 4, 8, 8, 8, 8, 8, 8, 8, 8, ... , where n consecutive terms have the value n, the 1025th term is

(a) 2^9 (b) 2^{10}

(c) 2^{11} (d) 2^8

29. Let $\{t_n\}$ be a sequence of integers in GP in which $t_4 : t_6 = 1 : 4$ and $t_2 + t_5 = 216$. Then t_1 is

(a) 12 (b) 14
(c) 16 (d) none of these

30. If $\log\left(\frac{5c}{a}\right)$, $\log\left(\frac{3b}{5c}\right)$ and $\log\left(\frac{a}{3b}\right)$ are in AP, where a, b, c are in GP, then a, b, c are the lengths of sides of

(a) an isosceles triangle (b) an equilateral triangle
(c) a scalene triangle (d) none of these

31. Let S be the sum, P be the product and R be the sum of the reciprocals of n terms of a GP. Then $P^2R^n : S^n$ is equal to

(a) $1 : 1$
(b) $(\text{common ratio})^n : 1$
(c) $(\text{first term})^2 : (\text{common ratio})^n$
(d) none of these

32. If the pth, qth and rth terms of an AP are in GP then the common ratio of the GP is

(a) $\dfrac{p+q}{r+q}$ (b) $\dfrac{r-q}{q-p}$
(c) $\dfrac{p-r}{p-q}$ (d) none of these

33. The number of terms common between the series $1 + 2 + 4 + 8 + \dots$ to 100 terms and $1 + 4 + 7 + 10 + \dots$ to 100 terms is

(a) 6 (b) 4
(c) 5 (d) none of these

34. The 10th common term between the series $3 + 7 + 11 + \dots$ and $1 + 6 + 11 + \dots$ is

(a) 191 (b) 193
(c) 211 (d) none of these

35. Three consecutive terms of a progression are 30, 24, 20. The next term of the progression is

(a) 18 (b) $17\frac{1}{7}$
(c) 16 (d) none of these

36. If three numbers are in GP then the numbers obtained by adding the middle number to each of the three numbers are in

(a) AP (b) GP
(c) HP (d) none of these

37. If a_1, a_2, a_3 are in AP, a_2, a_3, a_4 are in GP and a_3, a_4, a_5 are in HP then a_1, a_3, a_5 are in

(a) AP (b) GP
(c) HP (d) none of these

38. If a, b, c, d are four numbers such that the first three are in AP while the last three are in HP then

(a) $bc = ad$ (b) $ac = bd$
(c) $ab = cd$ (d) none of these

39. If the first two terms of an HP be 2/5 and 12/23 then the largest positive term of the progression is the

(a) 6th term (b) 7th term
(c) 5th term (d) 8th term

40. If $x, 2y, 3z$ are in AP, where the distinct numbers x, y, z are in GP, then the common ratio of the GP is

(a) 3 (b) $\frac{1}{3}$
(c) 2 (d) $\frac{1}{2}$

41. If $x > 1, y > 1, z > 1$ are three numbers in GP then

$$\frac{1}{1+\ln x}, \frac{1}{1+\ln y}, \frac{1}{1+\ln z}$$

are in

(a) AP (b) HP
(c) GP (d) none of these

42. If $a, a_1, a_2, a_3, \ldots a_{2n-1}, b$ are in AP, $a, b_1, b_2, b_3, \ldots, b_{2n-1}, b$ are in GP and $a, c_1, c_2, c_3, \ldots, c_{2n-1}, b$ are in HP, where a, b are positive, then the equation $a_n x^2 - b_n x + c_n = 0$ has its roots

(a) real and unequal (b) real and equal
(c) imaginary (d) none of these

43. If a, x, b are in AP, a, y, b are in GP and a, z, b are in HP such that $x = 9z$ and $a > 0, b > 0$ then

(a) $|y| = 3z$ (b) $x = 3|y|$

(c) $2y = x + z$ (d) none of these

44. If three numbers are in HP then the numbers obtained by subtracting half of the middle number from each of them are in

(a) AP (b) GP

(c) HP (d) none of these

45. a, b, c, d, e are five numbers in which the first three are in AP and the last three are in HP. If the three numbers in the middle are in GP then the numbers in the odd places are in

(a) AP (b) GP

(c) HP (d) none of these

46. Let $a_1, a_2, a_3, \ldots, a_{10}$ be in AP and $h_1, h_2, h_3, \ldots, h_{10}$ be in HP. If $a_1 = h_1 = 2$ and $a_{10} = h_{10} = 3$ then $a_4 h_7$ is

(a) 2 (b) 3

(c) 5 (d) 6

47. If in an AP, $S_n = p \cdot n^2$ and $S_m = p \cdot m^2$, where S_r denotes the sum of r terms of the AP, then S_p is equal to

(a) $\frac{1}{2}p^3$ (b) mnp

(c) p^3 (d) $(m+n)p^2$

48. If S_r denotes the sum of the first r terms of an AP then $\dfrac{S_{3r} - S_{r-1}}{S_{2r} - S_{2r-1}}$ is equal to

(a) $2r - 1$ (b) $2r + 1$

(c) $4r + 1$ (d) $2r + 3$

49. S_r denotes the sum of the first r terms of a GP. Then S_n, $S_{2n} - S_n$, $S_{3n} - S_{2n}$ are in

(a) AP (b) GP

(c) HP (d) none of these

50. If $(1-p)(1 + 3x + 9x^2 + 27x^3 + 81x^4 + 243x^5) = 1 - p^6, p \neq 1$ then the value of $\frac{p}{x}$ is

(a) $\frac{1}{3}$ (b) 3

(c) $\frac{1}{2}$ (d) 2

51. If the sum of the series $1+\frac{2}{x}+\frac{4}{x^2}+\frac{8}{x^3}+\ldots$ to ∞ is a finite number then

(a) $x<2$ (b) $x>\frac{1}{2}$

(c) $x>-2$ (d) $x<-2$ or $x>2$

52. Let S_n denote the sum of the first n terms of an AP. If $S_{2n}=3S_n$ then $S_{3n}:S_n$ is equal to

(a) 4 (b) 6

(c) 8 (d) 10

53. In a GP of even number of terms, the sum of all terms is 5 times the sum of the odd terms. The common ratio of the GP is

(a) $\frac{-4}{5}$ (b) $\frac{1}{5}$

(c) 4 (d) none of these

54. In an AP, $S_p=q$, $S_q=p$ and S_r denotes the sum of the first r terms. Then S_{p+q} is equal to

(a) 0 (b) $-(p+q)$

(c) $p+q$ (d) pq

55. The coefficient of x^{15} in the product

$$(1-x)(1-2x)(1-2^2\cdot x)(1-2^3\cdot x)\ldots(1-2^{15}\cdot x)$$

is equal to

(a) $2^{105}-2^{121}$ (b) $2^{121}-2^{105}$

(c) $2^{120}-2^{104}$ (d) none of these

56. The coefficient of x^{49} in the product $(x-1)(x-3)\ldots(x-99)$ is

(a) -99^2 (b) 1

(c) $-2\,500$ (d) none of these

57. If a, b, c are in AP then $a+\frac{1}{bc}, b+\frac{1}{ca}, c+\frac{1}{ab}$ are in

(a) AP (b) GP
(c) HP (d) none of these

58. The AM of two given positive numbers is 2. If the larger number is increased by 1, the GM of the numbers becomes equal to the AM of the given numbers. Then the HM of the given numbers is

(a) $\frac{3}{2}$ (b) $\frac{2}{3}$
(c) $\frac{1}{2}$ (d) none of these

59. Let a, b be two positive numbers, where $a > b$ and $4 \times GM = 5 \times HM$ for the numbers. Then a is

(a) $4b$ (b) $\frac{1}{4}b$
(c) $2b$ (d) b

60. If $a, a_1, a_2, a_3, \ldots, a_{2n}, b$ are in AP and $a, g_1, g_2, g_3, \ldots, g_{2n}, b$ are in GP and h is the HM of a and b then

$$\frac{a_1 + a_{2n}}{g_1 g_{2n}} + \frac{a_2 + a_{2n-1}}{g_2 g_{2n-1}} + \ldots + \frac{a_n + a_{n+1}}{g_n g_{n+1}}$$

is equal to

(a) $\frac{2n}{h}$ (b) $2nh$
(c) nh (d) $\frac{n}{h}$

61. Let $a_1 = 0$ and $a_1, a_2, a_3, \ldots, a_n$ be real numbers such that $|a_i| = |a_{i-1} + 1|$ for all i then the AM of the numbers $a_1, a_2, a_3, \ldots, a_n$ has the value A where

(a) $A < -\frac{1}{2}$ (b) $A < -1$
(c) $A \geq -\frac{1}{2}$ (d) $A = -\frac{1}{2}$

62. Let there be a GP whose first term is a and the common ratio is r. If A and H are the arithmetic mean and the harmonic mean respectively for the first n terms of the GP, $A \cdot H$ is equal to

(a) $a^2 r^{n-1}$ (b) ar^n
(c) $a^2 r^n$ (d) none of these

63. If the first and the $(2n-1)$th terms of an AP, a GP and an HP are equal and their nth terms are a, b and c respectively then

(a) $a=b=c$ (b) $a \geq b \geq c$

(c) $a+c=b$ (d) $ac-b^2=0$

64. $\dfrac{a^n+b^n}{a^{n-1}+b^{n-1}}$ is the HM between a and b if n is

(a) 0 (b) $\dfrac{1}{2}$

(c) $-\dfrac{1}{2}$ (d) 1

65. If the harmonic mean between P and Q be H then $H\left(\dfrac{1}{P}+\dfrac{1}{Q}\right)$ is equal to

(a) 2 (b) $\dfrac{PQ}{P+Q}$

(c) $\dfrac{P+Q}{PQ}$ (d) $\dfrac{1}{2}$

66. Let x be the AM and y, z be two GMs between two positive numbers. Then $\dfrac{y^3+z^3}{xyz}$ is equal to

(a) 1 (b) 2

(c) $\dfrac{1}{2}$ (d) none of these

67. If HM : GM = 4 : 5 for two positive numbers then the ratio of the numbers is

(a) 4 : 1 (b) 3 : 2

(c) 3 : 4 (d) 2 : 3

68. In a GP of alternately positive and negative terms, any term is the AM of the next two terms. Then the common ratio is

(a) –1 (b) –3

(c) –2 (d) $-\dfrac{1}{2}$

69. If a, b, c are in AP, and p, p' are the AM and GM respectively between a and b, while q, q' are the AM and GM respectively between b and c, then

(a) $p^2 + q^2 = p'^2 + q'^2$ (b) $pq = p'q'$
(c) $p^2 - q^2 = p'^2 - q'^2$ (d) none of these

70. If $-\frac{\pi}{2} < \theta < \frac{\pi}{2}$ then the minimum value of $\cos^3\theta + \sec^3\theta$ is
(a) 1 (b) 2
(c) 0 (d) none of these

71. If $a > 1, b > 1$ then the minimum value of $\log_b a + \log_a b$ is
(a) 0 (b) 1
(c) 2 (d) none of these

72. The minimum value of $4^x + 4^{1-x}, x \in R$, is
(a) 2 (b) 4
(c) 1 (d) none of these

73. If $x = \log_5 3 + \log_7 5 + \log_9 7$ then
(a) $x \geq \frac{3}{2}$ (b) $x \geq \frac{1}{\sqrt[3]{2}}$
(c) $x \geq \frac{3}{\sqrt[3]{2}}$ (d) none of these

74. If $a_n > 1$ for all $n \in N$ then

$$\log_{a_2} a_1 + \log_{a_3} a_2 + \ldots + \log_{a_n} a_{n-1} + \log_{a_1} a_n$$

has the minimum value
(a) 1 (b) 2
(c) 0 (d) none of these

75. The product of n positive numbers is 1. Their sum is
(a) a positive integer (b) divisible by n
(c) equal to $n + \frac{1}{n}$ (d) greater than or equal to n

76. If x, y, z are three real numbers of the same sign then the value of $\frac{x}{y} + \frac{y}{z} + \frac{z}{x}$ lies in the interval
(a) $[2, +\infty)$ (b) $[3, +\infty)$
(c) $(3, +\infty)$ (d) $(-\infty, 3)$

77. The least value of $2\log_{100} a - \log_a 0.0001, a > 1$ is

(a) 2 (b) 3
(c) 4 (d) none of these

78. If $0 < x < \pi/2$ then the minimum value of $(\sin x + \cos x + \operatorname{cosec} 2x)^3$ is

(a) 27 (b) 13.5
(c) 6.75 (d) none of these

79. If x, y, z are positive then the minimum value of

$$x^{\log y - \log z} + y^{\log z - \log x} + z^{\log x - \log y} \text{ is}$$

(a) 3 (b) 1
(c) 9 (d) 16

80. a, b, c are three positive numbers and abc^2 has the greatest value $\frac{1}{64}$. Then

(a) $a = b = \frac{1}{2}, c = \frac{1}{4}$ (b) $a = b = \frac{1}{4}, c = \frac{1}{2}$
(c) $a = b = c = \frac{1}{3}$ (d) none of these

81. If $a > 0, b > 0, c > 0$ and the minimum value of

$$a(b^2 + c^2) + b(c^2 + a^2) + c(a^2 + b^2)$$

is λabc then λ is

(a) 2 (b) 1
(c) 6 (d) 3

82. The value of $\sum_{n=1}^{10} \int_0^n x\,dx$ is

(a) an even integer (b) an odd integer
(c) a rational number (d) an irrational number

83. The sum of $0.2 + 0.004 + 0.00006 + 0.0000008 + \ldots$ to ∞ is

(a) $\frac{200}{891}$ (b) $\frac{2000}{9801}$
(c) $\frac{1000}{9801}$ (d) none of these

84. If $(2n + r)r, n \in N, r \in N$ is expressed as the sum of k consecutive odd natural numbers then k is equal to

(a) r (b) n
(c) $r + 1$ (d) $n + 1$

85. $\sum_{r=1}^{n} r^2 - \sum_{m=1}^{n} \sum_{r=1}^{m} r$ is equal to

(a) 0 (b) $\frac{1}{2}\left(\sum_{r=1}^{n} r^2 + \sum_{r=1}^{n} r\right)$

(c) $\frac{1}{2}\left(\sum_{r=1}^{n} r^2 - \sum_{r=1}^{n} r\right)$ (d) none of these

86. If $(1+x)(1+x^2)(1+x^4) \dots (1+x^{128}) = \sum_{r=0}^{n} x^r$ then n is

(a) 255 (b) 127

(c) 63 (d) none of these

87. The value of $\sum_{n=1}^{m} \log \frac{a^{2n-1}}{b^{m-1}}$ $(a \neq 0, 1; b \neq 0, 1)$ is

(a) $m \log \frac{a^{2m}}{b^{m-1}}$ (b) $\log \frac{a^{2m}}{b^{m-1}}$

(c) $\frac{m}{2} \log \frac{a^{2m}}{b^{2m-2}}$ (d) $\frac{m}{2} \log \frac{a^{2m}}{b^{m+1}}$

88. The sum of the products of the ten numbers ±1, ±2, ±3, ±4, ±5 taking two at a time is

(a) 165 (b) –55

(c) 55 (d) none of these

89. The sum of the series $\frac{1}{\log_2 4} + \frac{1}{\log_4 4} + \frac{1}{\log_8 4} + \dots + \frac{1}{\log_{2^n} 4}$ is

(a) $\frac{n(n+1)}{2}$ (b) $\frac{n(n+1)(2n+1)}{12}$

(c) $\frac{1}{n(n+1)}$ (d) $\frac{n(n+1)}{4}$

90. If $\sum_{n=1}^{n} n$, $\frac{\sqrt{10}}{3} \cdot \sum_{n=1}^{n} n^2$, $\sum_{n=1}^{n} n^3$ are in GP then the value of n is

(a) 2 (b) 3

(c) 4 (d) nonexistent

91. The value of $\sum_{r=1}^{n}\left\{(2r-1)a+\frac{1}{b^r}\right\}$ is equal to

(a) $an^2+\frac{b^{n-1}-1}{b^{n-1}(b-1)}$ (b) $an^2+\frac{b^n-1}{b^n(b-1)}$

(c) $an^3+\frac{b^{n-1}-1}{b^n(b-1)}$ (d) none of these

92. If $s_n=\sum_{n=1}^{n}\frac{1+2+2^2+\ldots \text{ to } n \text{ terms}}{2^n}$ then s_n is equal to

(a) $2^n-(n+1)$ (b) $1-\frac{1}{2^n}$

(c) $n-1+\frac{1}{2^n}$ (d) 2^n-1

93. Let S_n denote the sum of the cubes of the first n natural numbers and s_n denote the sum of the first n natural numbers. Then $\sum_{r=1}^{n}\frac{S_r}{s_r}$ is equal to

(a) $\frac{n(n+1)(n+2)}{6}$ (b) $\frac{n(n+1)}{2}$

(c) $\frac{n^2+3n+2}{6}$ (d) none of these

94. It is known that $\sum_{r=1}^{\infty}\frac{1}{(2r-1)^2}=\frac{\pi^2}{8}$. Then $\sum_{r=1}^{\infty}\frac{1}{r^2}$ is equal to

(a) $\frac{\pi^2}{24}$ (b) $\frac{\pi^2}{3}$

(c) $\frac{\pi^2}{6}$ (d) none of these

95. It is given that $\frac{1}{1^4}+\frac{1}{2^4}+\frac{1}{3^4}+\ldots$ to $\infty=\frac{\pi^4}{90}$. Then $\frac{1}{1^4}+\frac{1}{3^4}+\frac{1}{5^4}+\ldots$ to ∞ is equal to

(a) $\frac{\pi^4}{96}$ (b) $\frac{\pi^4}{45}$

(c) $\frac{89\pi^4}{90}$ (d) none of these

96. If in a series $t_n = \frac{n}{(n+1)!}$ then $\sum_{n=1}^{20} t_n$ is equal to

(a) $\frac{20!-1}{20!}$ (b) $\frac{21!-1}{21!}$

(c) $\frac{1}{2(n-1)!}$ (d) none of these

97. If t_n denotes the nth term of the series $2+3+6+11+18+\dots$ then t_{50} is

(a) 49^2-1 (b) 49^2

(c) 50^2+1 (d) 49^2+2

98. $2^{1/4} \cdot 4^{1/8} \cdot 8^{1/16} \cdot \dots$ to ∞ is equal to

(a) 1 (b) 2

(c) $\frac{3}{2}$ (d) none of these

99. The sum of n terms of the series

$$1^2 + 2.2^2 + 3^2 + 2.4^2 + 5^2 + 2.6^2 + \dots$$

is $\frac{n(n+1)^2}{2}$ when n is even. When n is odd, the sum is

(a) $\frac{n^2(n+1)}{2}$ (b) $\frac{n(n^2-1)}{2}$

(c) $2(n+1)^2 \cdot (2n+1)$ (d) none of these

100. If n is an odd integer greater than or equal to 1 then the value of $n^3-(n-1)^3+(n-2)^3-\dots+(-1)^{n-1}\cdot 1^3$ is

(a) $\frac{(n+1)^2 \cdot (2n-1)}{4}$ (b) $\frac{(n-1)^2 \cdot (2n-1)}{4}$

(c) $\frac{(n+1)^2 \cdot (2n+1)}{4}$ (d) none of these

101. Observe that

$$1^3 = 1, 2^3 = 3 + 5, 3^3 = 7 + 9 + 11, 4^3 = 13 + 15 + 17 + 19.$$

Then n^3 as a similar series is

(a) $\left[2\left\{\frac{n(n-1)}{2}+1\right\}-1\right]+\left[2\left\{\frac{(n+1)n}{2}+1\right\}+1\right] + \ldots + \left[2\left\{\frac{(n+1)n}{2}+1\right\}+2n-3\right]$

(b) $(n^2+n+1)+(n^2+n+3)+(n^2+n+5)+\ldots+(n^2+3n-1)$

(c) $(n^2-n+1)+(n^2-n+3)+(n^2-n+5)+\ldots+(n^2+n-1)$

(d) none of these

102. Let $t_r = 2^{r/2} + 2^{-r/2}$. Then $\sum_{r=1}^{10} t_r^2$ is equal to

(a) $\frac{2^{21}-1}{2^{10}}+20$ (b) $\frac{2^{21}-1}{2^{10}}+19$

(c) $\frac{2^{21}-1}{2^{20}}-1$ (d) none of these

103. Let $S_k = \lim_{n\to\infty} \sum_{i=0}^{n} \frac{1}{(k+1)^i}$. Then $\sum_{k=1}^{n} kS_k$ equals

(a) $\frac{n(n+1)}{2}$ (b) $\frac{n(n-1)}{2}$

(c) $\frac{n(n+2)}{2}$ (d) $\frac{n(n+3)}{2}$

104. Let $t_n = n \cdot (n\,!)$. Then $\sum_{n=1}^{15} t_n$ is equal to

(a) $15\,! - 1$ (b) $15\,! + 1$

(c) $16\,! - 1$ (d) none of these

105. The sum of $\frac{3}{1\cdot 2}\cdot\frac{1}{2}+\frac{4}{2\cdot 3}\cdot\left(\frac{1}{2}\right)^2+\frac{5}{3\cdot 4}\cdot\left(\frac{1}{2}\right)^3+\ldots$ to n terms is equal to

(a) $1-\frac{1}{(n+1)2^n}$ (b) $1-\frac{1}{n\cdot 2^{n-1}}$

(c) $1+\dfrac{1}{(n+1)2^n}$ (d) none of these

106. Let $f(n)=\left[\dfrac{1}{2}+\dfrac{n}{100}\right]$ where $[x]$ denotes the integral part of x. Then the value of $\sum_{n=1}^{100} f(n)$ is

(a) 50 (b) 51

(c) 1 (d) none of these

107. $A_r; r=1,2,3,\ldots,n$ are n points on the parabola $y^2=4x$ in the first quadrant. If $A_r=(x_r, y_r)$, where $x_1, x_2, x_3, \ldots, x_n$ are in GP and $x_1=1, x_2=2$, then y_n is equal to

(a) $-2^{\frac{n+1}{2}}$ (b) 2^{n+1}

(c) $(\sqrt{2})^{n+1}$ (d) $2^{\frac{n}{2}}$

108. In the given square, a diagonal is drawn, and parallel line segments joining points on the adjacent sides are drawn on both sides of the diagonal. The length of the diagonal is $n\sqrt{2}$ cm. If the distance between consecutive line segments be $1/\sqrt{2}$ cm then the sum of the lengths of all possible line segments and the diagonal is

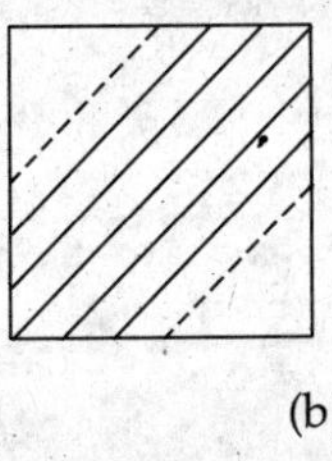

(a) $n(n+1)\sqrt{2}$ cm (b) n^2 cm

(c) $n(n+2)$ cm (d) $n^2\sqrt{2}$ cm

109. ABCD is a square of length $a, a \in N, a>1$. Let $L_1, L_2, L_3, \ldots$ be points on BC such that $BL_1=L_1L_2=L_2L_3=\ldots=1$ and $M_1, M_2, M_3, \ldots$ be points on CD such that $CM_1=M_1M_2=M_2M_3=\ldots=1$. Then $\sum_{n=1}^{a-1}(AL_n^2+L_nM_n^2)$ is equal to

(a) $\dfrac{1}{2}a(a-1)^2$ (b) $\dfrac{1}{2}a(a-1)(4a-1)$

(c) $\frac{1}{2}(a-1)(2a-1)(4a-1)$ (d) none of these

110. The sum of infinite terms of a decreasing GP is equal to the greatest value of the function $f(x)=x^3+3x-9$ in the interval [–2, 3] and the difference between the first two terms is $f'(0)$. Then the common ratio of the GP is

(a) $-\frac{2}{3}$ (b) $\frac{4}{3}$

(c) $\frac{2}{3}$ (d) $-\frac{4}{3}$

111. The lengths of three unequal edges of a rectangular solid block are in GP. The volume of the block is 216 cm^3 and the total surface area is 252 cm^2. The length of the longest edge is

(a) 12 cm (b) 6 cm

(c) 18 cm (d) 3 cm

112. ABC is a right-angled triangle in which $\angle B=90°$ and $BC=a$. If n points $L_1, L_2, \dots, L_n$ on AB are such that AB is divided in $n+1$ equal parts and $L_1M_1, L_2M_2, \dots, L_nM_n$ are line segments parallel to BC and $M_1, M_2, \dots, M_n$ are on AC then the sum of the lengths of $L_1M_1, L_2M_2, \dots, L_nM_n$ is

(a) $\frac{a(n+1)}{2}$

(b) $\frac{a(n-1)}{2}$

(c) $\frac{an}{2}$

(d) impossible to find from the given data

• *Type 2* •

Choose the correct options. One or more options may be correct.

113. If AM of the numbers 5^{1+x} and 5^{1-x} is 13 then the set of possible real values of x is

(a) $\left\{5, \frac{1}{5}\right\}$ (b) {1, –1}

(c) $\{x \mid x^2-1=0, x \in R\}$ (d) none of these

114. If the AM of two positive numbers be three times their geometric mean then the ratio of the numbers is

(a) $3 \pm 2\sqrt{2}$ (b) $\sqrt{2} \pm 1$

(c) $17 + 12\sqrt{2}$ (d) $(3 - 2\sqrt{2})^{-2}$

115. If a, b, c are in HP then $\frac{1}{b-a} + \frac{1}{b-c}$ is equal to

(a) $\frac{2}{b}$ (b) $\frac{2}{a+c}$

(c) $\frac{1}{a} + \frac{1}{c}$ (d) none of these

116. S_r denotes the sum of the first r terms of an AP. Then $S_{3n} : (S_{2n} - S_n)$ is

(a) n (b) $3n$

(c) 3 (d) independent of n

117. If $a^x = b^y = c^z$ and x, y, z are in GP then $\log_c b$ is equal to

(a) $\log_b a$ (b) $\log_a b$

(c) $\frac{z}{y}$ (d) none of these

118. The value of $\sum_{r=1}^{n} \frac{1}{\sqrt{a+rx} + \sqrt{a+(r-1)x}}$ is

(a) $\frac{n}{\sqrt{a} + \sqrt{a+nx}}$ (b) $\frac{\sqrt{a+nx} - \sqrt{a}}{x}$

(c) $\frac{n(\sqrt{a+nx} - a)}{x}$ (d) none of these

119. Let $\sum_{n=1}^{n} r^4 = f(n)$. Then $\sum_{r=1}^{n} (2r-1)^4$ is equal to

(a) $f(2n) - 16f(n)$ for all $n \in N$

(b) $f(n) - 16f\left(\frac{n-1}{2}\right)$ when n is odd

(c) $f(n) - 16f\left(\frac{n}{2}\right)$ when n is even

(d) none of these

120. If $2 \cdot {}^nP_1, {}^nP_2, {}^nP_3$ are three consecutive terms of an AP then they are

(a) in GP (b) in HP

(c) equal (d) none of these

121. In a GP the product of the first four terms is 4 and the second term is the reciprocal of the fourth term. The sum of the GP up to infinite terms is

(a) 8 (b) –8

(c) $\frac{8}{3}$ (d) $-\frac{8}{3}$

122. If $\sum_{k=1}^{n}\left(\sum_{m=1}^{k} m^2\right) = an^4 + bn^3 + cn^2 + dn + e$ then

(a) $a = \frac{1}{12}$ (b) $b = \frac{1}{6}$

(c) $d = \frac{1}{6}$ (d) $e = 0$

123. If a, b, c, d are four positive numbers then

(a) $\left(\frac{a}{b}+\frac{b}{c}\right)\left(\frac{c}{d}+\frac{d}{e}\right) \geq 4 \cdot \sqrt{\frac{a}{e}}$ (b) $\left(\frac{a}{b}+\frac{c}{d}\right)\left(\frac{b}{c}+\frac{d}{e}\right) \geq 4 \cdot \sqrt{\frac{a}{e}}$

(c) $\frac{a}{b}+\frac{b}{c}+\frac{c}{d}+\frac{d}{e}+\frac{e}{a} \geq 5$ (d) $\frac{b}{a}+\frac{c}{b}+\frac{d}{c}+\frac{e}{d}+\frac{a}{e} \geq \frac{1}{5}$

124. Let $f(x) = \frac{1-x^{n+1}}{1-x}$ and $g(x) = 1 - \frac{2}{x} + \frac{3}{x^2} - \ldots + (-1)^n \frac{n+1}{x^n}$. Then the constant term in $f'(x) \times g(x)$ is equal to

(a) $\frac{n(n^2-1)}{6}$ when n is even

(b) $\frac{n(n+1)}{2}$ when n is odd

(c) $-\frac{n}{2}(n+1)$ when n is even

(d) $-\frac{n(n-1)}{2}$ when n is odd

125. Let a_n = product of the first n natural numbers. Then for all $n \in N$

(a) $n^n \geq a_n$ (b) $\left(\frac{n+1}{2}\right)^n \geq n!$

(c) $n^n \geq a_{n+1}$ (d) none of these

126. Let the sets $A = \{2, 4, 6, 8, \ldots\}$ and $B = \{3, 6, 9, 12, \ldots\}$, and $n(A) = 200$, $n(B) = 250$. Then

(a) $n(A \cap B) = 67$ (b) $n(A \cup B) = 450$

(c) $n(A \cap B) = 66$ (d) $n(A \cup B) = 384$

127. Let a, x, b be in AP; a, y, b be in GP and a, z, b be in HP. If $x = y + 2$ and $a = 5z$ then

(a) $y^2 = xz$ (b) $x > y > z$

(c) $a = 9, b = 1$ (d) $a = \frac{1}{4}, b = \frac{9}{4}$

128. Let $S_1, S_2, S_3, \ldots$ be squares such that for each $n \geq 1$, the length of a side of S_n equals the length of a diagonal of S_{n+1}. If the length of a side of S_1 is 10 cm then for which of the following values of n is the area of S_n less than 1 cm^2?

(a) 7 (b) 8

(c) 9 (d) 10

129. Three positive numbers form a GP. If the middle number is increased by 8, the three numbers form an AP. If the last number is also increased by 64 along with the previous increase in the middle number, the resulting numbers form a GP again. Then

(a) common ratio = 3 (b) first number = $\frac{4}{9}$

(c) common ratio = –5 (d) first number = 4

130. If a, b, c are in GP and a, p, q are in AP such that $2a, b + p, c + q$ are in GP then the common difference of the AP is

(a) $\sqrt{2}a$ (b) $(\sqrt{2} + 1)(a - b)$

(c) $\sqrt{2}(a + b)$ (d) $(\sqrt{2} - 1)(b - a)$

131. If x, y, z are positive numbers in AP then

(a) $y^2 \geq xz$

(b) $y \geq 2\sqrt{xz}$

(c) $\frac{x+y}{2y-x}+\frac{y+z}{2y-z}$ has the minimum value 2

(d) $\frac{x+y}{2y-x}+\frac{y+z}{2y-z}\geq 4$

132. Between two unequal numbers, if a_1, a_2 are two AMs; g_1, g_2 are two GMs and h_1, h_2 are two HMs then $g_1 \cdot g_2$ is equal to

(a) a_1h_1 (b) a_1h_2

(c) a_2h_2 (d) a_2h_1

133. The numbers 1, 4, 16 can be three terms (not necessarily consecutive) of

(a) no AP (b) only one GP

(c) infinite number of APs (d) infinite number of GPs

Answers

1. a	**2.** c	**3.** b	**4.** c	**5.** c
6. d	**7.** c	**8.** b	**9.** a	**10.** c
11. c	**12.** b	**13.** c	**14.** d	**15.** b
16. a	**17.** b	**18.** a	**19.** c	**20.** b
21. d	**22.** a	**23.** b	**24.** d	**25.** c
26. a	**27.** a	**28.** b	**29.** a	**30.** d
31. a	**32.** b	**33.** c	**34.** a	**35.** b
36. c	**37.** b	**38.** a	**39.** c	**40.** b
41. b	**42.** c	**43.** b	**44.** b	**45.** b
46. d	**47.** c	**48.** b	**49.** b	**50.** b
51. d	**52.** b	**53.** c	**54.** b	**55.** a
56. c	**57.** a	**58.** a	**59.** a	**60.** a
61. c	**62.** a	**63.** d	**64.** a	**65.** a
66. b	**67.** a	**68.** c	**69.** c	**70.** b
71. c	**72.** b	**73.** c	**74.** d	**75.** d
76. b	**77.** c	**78.** b	**79.** a	**80.** b
81. c	**82.** c	**83.** b	**84.** a	**85.** c
86. a	**87.** c	**88.** b	**89.** d	**90.** c
91. b	**92.** c	**93.** a	**94.** c	**95.** a
96. b	**97.** d	**98.** a	**99.** a	**100.** a
101. c	**102.** b	**103.** d	**104.** c	**105.** a
106. b	**107.** c	**108.** d	**109.** b	**110.** c
111. a	**112.** c	**113.** b, c	**114.** c, d	**115.** a, c
116. c, d	**117.** a, c	**118.** a, b	**119.** a	**120.** a, b,
121. a, b, c, d	**122.** a, c	**123.** a, b, c	**124.** b, c	**125.** a,
126. c, d	**127.** a, c	**128.** b, c, d	**129.** a, d	**130.** b
131. a, d	**132.** b, d	**133.** c, d		

Hints and Solutions to Selected Questions

1. If p, q, r are in AP then in an AP or a GP or an HP $a_1, a_2, a_3, \ldots$, etc., the terms a_p, a_q, a_r are in AP, GP or HP respectively.

3. See the hint to Q. No. 1.

5. $\dfrac{a_1+a_4}{a_1a_4}=\dfrac{a_2+a_3}{a_2a_3}$; so $\dfrac{1}{a_4}+\dfrac{1}{a_1}=\dfrac{1}{a_3}+\dfrac{1}{a_2}$ or $\dfrac{1}{a_4}-\dfrac{1}{a_3}=\dfrac{1}{a_2}-\dfrac{1}{a_1}$ ... (1)

Also $\dfrac{3(a_2-a_3)}{a_2a_3}=\dfrac{a_1-a_4}{a_1a_4}$; so $3\left(\dfrac{1}{a_3}-\dfrac{1}{a_2}\right)=\dfrac{1}{a_4}-\dfrac{1}{a_1}$ ... (2)

Clearly, (1) and (2) $\Rightarrow \dfrac{1}{a_2}-\dfrac{1}{a_1}=\dfrac{1}{a_3}-\dfrac{1}{a_2}=\dfrac{1}{a_4}-\dfrac{1}{a_3}$; so $\dfrac{1}{a_1}, \dfrac{1}{a_2}, \dfrac{1}{a_3}$ are in AP.

6. If in AP, $\sqrt{y}=\sqrt{x}+(n-1)d$ and $\sqrt{z}=\sqrt{x}+(m-1)d$

$\therefore \dfrac{\sqrt{y}-\sqrt{x}}{\sqrt{z}-\sqrt{x}}=\dfrac{n-1}{m-1}$, a rational number. As x, y, z are prime, $\dfrac{\sqrt{y}-\sqrt{x}}{\sqrt{z}-\sqrt{x}}$ is irrational.

$\therefore$ irrational = rational (absurd). So $\sqrt{x}, \sqrt{y}, \sqrt{z}$ are not in AP.
Similarly, they are not in GP or HP.

7. $2x+1, 4x+1, 8x+1$ are in GP $\Rightarrow (4x+1)^2=(2x+1)(8x+1)$
$\Rightarrow x=0$ and for this value $f(x), f(2x), f(4x)$ are equal.

8. $a_1+a_{2n}=a_2+a_{2n-1}=\ldots=a_n+a_{n+1}=k$ (say).

Expression $=k\left\{\dfrac{\sqrt{a_1}-\sqrt{a_2}}{a_1-a_2}+\ldots+\dfrac{\sqrt{a_n}-\sqrt{a_{n+1}}}{a_n-a_{n+1}}\right\}=\dfrac{k}{-d}(\sqrt{a_1}-\sqrt{a_{n+1}})$,

where d = common difference

$$=\frac{k}{-d}\cdot\frac{a_1-a_{n+1}}{\sqrt{a_1}+\sqrt{a_{n+1}}}=(a_1+a_{2n})\cdot\frac{-nd}{-d(\sqrt{a_1}+\sqrt{a_{n+1}})}.$$

. Use $a_r=a_1+(r-1)d$, where $d=a_2-a_1$.

$a_1+(p-1)d, a_1+(q-1)d, a_1+(r-1)d$ are in GP.

$$\therefore \quad \frac{a_q}{a_p} = \frac{a_1 + (q-1)d}{a_1 + (p-1)d} = \frac{a_1 + (r-1)d}{a_1 + (q-1)d} = \frac{r-p}{q-p}.$$

12. On simplification, $(b^2 - ac)^2 + (c^2 - bd)^2 + (ad - bc)^2 \le 0$

$$\Rightarrow \quad b^2 = ac, c^2 = bd, ad = bc \quad \Rightarrow \quad \frac{b}{a} = \frac{c}{b} = \frac{d}{c}.$$

13. $(2a)^2 + (3b)^2 + (4c)^2 - (2a)(3b) - (3b)(4c) - (4c)(2a) = 0$

$$\Rightarrow \quad (2a - 3b)^2 + (3b - 4c)^2 + (4c - 2a)^2 = 0 \quad \Rightarrow \quad 2a = 3b = 4c$$

$$\Rightarrow \quad \frac{a}{\frac{1}{2}} = \frac{b}{\frac{1}{3}} = \frac{c}{\frac{1}{4}}.$$

19. See the hint to Q. No. 13.

20. $(ap - b)^2 + (bp - c)^2 + (cp - d)^2 \le 0 \Rightarrow ap - b = 0, bp - c = 0, cp - d = 0$

$$\therefore \quad p = \frac{b}{a} = \frac{c}{b} = \frac{d}{e}.$$

24. 100 ! contains 5, 10, 15, ... , 100 —— 20 in number

25, 50, 75, 100 —— 4 in number

$\therefore$ number of zeros = 20 + 4

($\because$ the numbers in the 2nd line have 5×5 as a factor).

25. $9900 = 3^2 \times 2^2 \times 5^2 \times 11$

$\therefore$ sum of all divisors

$$= (3^0 + 3^1 + 3^2)(2^0 + 2^1 + 2^2)(5^0 + 5^1 + 5^2)(11^0 + 11^1).$$

$\therefore$ sum of all proper divisors

$$= 13 \times 7 \times 31 \times 12 - 1 - 9900.$$

26. $360 = 2^3 \times 3^2 \times 5$.

Sum of all odd proper divisors $= (3^0 + 3^1 + 3^2)(5^0 + 5^1) - 1$.

27. Let the 150th term $= n$. Then

$$1 + 2 + 3 + \ldots + (n-1) < 150 < 1 + 2 + 3 + \ldots + n$$

$$\therefore \quad n(n-1) < 300 < n(n+1) \quad \Rightarrow \quad n = 17.$$

28. Let the 1025th term $= 2^n$. Then

$$1 + 2 + 4 + 8 + \ldots + 2^{n-1} < 1025 \le 1 + 2 + 4 + 8 + \ldots + 2^n$$

$\therefore \quad 2^n - 1 < 1025 < 2^{n+1} - 1$ or $2^n < 1026 < 2^{n+1} \Rightarrow n = 10.$

30. $2\log\dfrac{3b}{5c} = \log\dfrac{5c}{a} + \log\dfrac{a}{3b} \quad\Rightarrow\quad \left(\dfrac{3b}{5c}\right)^2 = \dfrac{5c}{a}\cdot\dfrac{a}{3b} \quad\Rightarrow\quad 3b = 5c.$

Also $b^2 = ac$. So, $9ac = 25c^2$ or $9a = 25c$.

$\therefore \quad \dfrac{9a}{5} = 5c = 3b \quad\Rightarrow\quad \dfrac{a}{5} = \dfrac{b}{3} = \dfrac{c}{\frac{9}{5}} \quad\Rightarrow\quad b + c < a.$

33. For GP, $t_n = 2^{n-1}$; for AP, $T_m = 1 + (m-1)3$. They are common if

$$2^{n-1} = 3m - 2 \text{ or } 2^{n-2} + 1 = \frac{3m}{2} \le 150 \quad\Rightarrow\quad n \le 9, m \le 100.$$

By trial, $n = 1, m = 1; n = 3, m = 2; n = 5, m = 6; n = 7, m = 22; n = 9, m = 86.$

35. $\dfrac{1}{24} - \dfrac{1}{30} = \dfrac{1}{20} - \dfrac{1}{24}$ and so in HP.

42. Clearly a_n, b_n, c_n are the middle terms of the given AP, GP, HP respectively. So, a_n is the AM of a, b; b_n is the GM of a, b and c_n is the HM of a, b. Also a_n, b_n, c_n are positive because a, b are positive.

$\therefore\ a_n, b_n, c_n$ are in GP; so discriminant $= b_n^2 - 4a_nc_n = -3a_nc_n < 0.$

55. Product

$$= (-1)(-2)(-2^2)\ldots(-2^{15})(x-1)\left(x-\frac{1}{2}\right)\left(x-\frac{1}{2^2}\right)\left(x-\frac{1}{2^3}\right)\ldots\left(x-\frac{1}{2^{15}}\right)$$

$$= 2^{1+2+3+\ldots+15}(x-1)\left(x-\frac{1}{2}\right)\left(x-\frac{1}{2^2}\right)\ldots\left(x-\frac{1}{2^{15}}\right)$$

$\therefore$ coefficient of $x^{15} = 2^{1+2+3+\ldots 15}\cdot\left(-1-\dfrac{1}{2}-\dfrac{1}{2^2}-\ldots-\dfrac{1}{2^{15}}\right).$

56. There are 50 factors. So, the coefficient of $x^{49} = -1 - 3 - 5 - \ldots - 99.$

60. Here, $a + b = a_1 + a_{2n} = a_2 + a_{2n-1} = \ldots = a_n + a_{n+1}$

and $ab = g_1 \cdot g_{2n} = g_2 \cdot g_{2n-1} = \ldots = g_n \cdot g_{n+1}$ and $h = \dfrac{2ab}{a+b}.$

61. Here, $a_i^2 = (a_{i-1}+1)^2 \quad\Rightarrow\quad a_i^2 - a_{i-1}^2 = 2a_{i-1} + 1.$

Putting $i = 2, 3, \ldots, n+1$ and adding,

$$a_{n+1}^2 - a_1^2 = 2(a_1 + a_2 + \ldots + a_n) + n$$

$$\therefore \quad a_{n+1}^2 - 0 = 2nA + n; \quad \therefore \quad A = (a_{n+1}^2 - n)\frac{1}{2n} = \frac{a_{n+1}^2}{2n} - \frac{1}{2} \geq -\frac{1}{2}.$$

62. $A = \dfrac{a + ar + ar^2 + \ldots + ar^{n-1}}{n} = \dfrac{a(1-r^n)}{n(1-r)};$

$$H = \frac{1}{\dfrac{\dfrac{1}{a} + \dfrac{1}{ar} + \ldots + \dfrac{1}{ar^{n-1}}}{n}} = \frac{n(r-1)r^n}{\dfrac{r}{a}(r^n - 1)}.$$

63. nth term is the middle term in each case. So a, b, c are the AM, GM, HM respectively of the same two numbers. For any two numbers AM, GM, HM are in GP.

64. $\text{HM} = \dfrac{2ab}{a+b} = \dfrac{2}{a^{-1} + b^{-1}} = \dfrac{a^n + b^n}{a^{n-1} + b^{n-1}}$, where $n = 0$.

72. AM ≥ GM for positive numbers. So, $\dfrac{4^x + \dfrac{4}{4^x}}{2} \geq \sqrt{4^x \cdot \dfrac{4}{4^x}} = 2.$

73. AM ≥ GM for positive numbers. So,

$$\frac{x}{3} \geq \sqrt[3]{\log_5 3 \cdot \log_7 5 \cdot \log_9 7} = \sqrt[3]{\log_9 3}$$

$$\therefore \quad \frac{x}{3} \geq \sqrt[3]{\frac{1}{\log_3 9}} = \sqrt[3]{\frac{1}{2}}.$$

74. $A \geq G \Rightarrow \dfrac{\log_{a_2} a_1 + \log_{a_3} a_2 + \ldots + \log_{a_1} a_n}{n}$

$$\geq \sqrt[n]{\log_{a_2} a_1 \cdot \log_{a_3} a_2 \cdot \ldots \log_{a_1} a_n} = 1.$$

75. $A = \dfrac{a_1 + a_2 + \ldots + a_n}{n}$, $G = \sqrt[n]{a_1 a_2 \ldots a_n} = 1;$

$A \geq G \Rightarrow a_1 + a_2 + \ldots + a_n \geq n.$

76. $\dfrac{x}{y}$, etc., are positive. $A \geq G \Rightarrow \dfrac{\dfrac{x}{y} + \dfrac{y}{z} + \dfrac{z}{x}}{3} \geq \sqrt[3]{\dfrac{x}{y} \cdot \dfrac{y}{z} \cdot \dfrac{z}{x}} = 1.$

77. $2\log_{100} a - \log_a \left(\dfrac{1}{100}\right)^2 = 2\{\log_{100} a + \log_a 100\}.$

Also, $\dfrac{\log_{100} a + \log_a 100}{2} \geq \sqrt{\log_{100} a \times \log_a 100} = 1.$

78. $\dfrac{\sin x + \cos x + \text{cosec } 2x}{3} \geq \sqrt[3]{\sin x \cdot \cos x \cdot \text{cosec } 2x} = \sqrt[3]{\dfrac{1}{2}}\cdot$

79. $\dfrac{x^{\log y - \log z} + y^{\log z - \log x} + z^{\log x - \log y}}{3}$

$$\geq \sqrt[3]{x^{\log y - \log z} \cdot y^{\log z - \log x} \cdot z^{\log x - \log y}} = 1$$

because $\log \{x^{\log y - \log z} \cdot y^{\log z - \log x} \cdot z^{\log x - \log y}\} = 0$ (on simplification).

80. $\dfrac{a + b + \frac{c}{2} + \frac{c}{2}}{4} \geq \sqrt[4]{a \cdot b \cdot \frac{c}{2} \cdot \frac{c}{2}}$ or $\dfrac{a+b+c}{4} \geq \sqrt[4]{\dfrac{abc^2}{4}}$

$\therefore \dfrac{(a+b+c)^4}{4^4} \geq \dfrac{abc^2}{4}$; or $abc^2 \leq \dfrac{1}{64}(a+b+c)^4$.

$\therefore$ the greatest value of $abc^2 = \dfrac{1}{64}(a+b+c)^4$.

Also for the greatest value of abc^2 the numbers have to be equal, i.e., $a = b = \dfrac{c}{2}\cdot$

Also, given the greatest value $= \dfrac{1}{64}\cdot$ So $a + b + c = 1$.

81. Use $A \geq G$ for the numbers $ab^2, ac^2, bc^2, ba^2, ca^2, cb^2$.

83. Sum $= \dfrac{2}{10} + \dfrac{4}{10^3} + \dfrac{6}{10^5} + \dfrac{8}{10^7} + \ldots$ to ∞ which is an arithmetico-geometric series.

84. $(2n + r)r = (n + r)^2 - n^2 = \{1 + 3 + 5 + \ldots$ to $(n + r)$ terms$\}$
$- \{1 + 3 + 5 + \ldots$ to n terms$\}$.

86. $n =$ index of the highest power of $x = 1 + 2 + 4 + \ldots + 128$.

88. $(1 - 1 + 2 - 2 + \ldots + 5 - 5)^2 = 1^2 + 1^2 + 2^2 + 2^2 + \ldots + 5^2 + 5^2 + 2S$,
where S is the required sum

or $0 = 2(1^2 + 2^2 + 3^2 + 4^2 + 5^2) + 2S$.

94. Here $\frac{1}{1^2}+\frac{1}{3^2}+\frac{1}{5^2}+\dots$ to $\infty=\frac{\pi^2}{8}\cdot$ Let $\frac{1}{1^2}+\frac{1}{2^2}+\frac{1}{3^2}+\dots$ to $\infty=x$.

Then $x=\frac{1}{1^2}+\frac{1}{2^2}+\frac{1}{3^2}+\dots$ to ∞

$$=\left(\frac{1}{1^2}+\frac{1}{3^2}+\frac{1}{5^2}+\dots \text{ to } \infty\right)+\left(\frac{1}{2^2}+\frac{1}{4^2}+\frac{1}{6^2}+\dots \text{ to } \infty\right)$$

$$=\frac{\pi^2}{8}+\frac{1}{4}\left(\frac{1}{1^2}+\frac{1}{2^2}+\frac{1}{3^2}+\dots \text{ to } \infty\right)=\frac{\pi^2}{8}+\frac{1}{4}x.$$

95. $x=\frac{1}{1^4}+\frac{1}{3^4}+\frac{1}{5^4}+\dots$ to ∞

$$=\left(\frac{1}{1^4}+\frac{1}{2^4}+\frac{1}{3^4}+\dots \text{ to } \infty\right)-\left(\frac{1}{2^4}+\frac{1}{4^4}+\dots \text{ to } \infty\right)$$

$$=\frac{\pi^4}{90}-\frac{1}{16}\left(\frac{1}{1^4}+\frac{1}{2^4}+\frac{1}{3^4}+\dots \text{ to } \infty\right)=\frac{\pi^4}{90}-\frac{1}{16}\cdot\frac{x^4}{90}\cdot$$

96. $t_n=\frac{(n+1)-1}{(n+1)!}=\frac{1}{n!}-\frac{1}{(n+1)!}$.

97. $t_n=2+1+3+5+7+\dots$ to n terms $=2+(n-1)^2$.

98. Value $=2^{\frac{1}{4}+\frac{2}{8}+\frac{3}{16}+\dots \text{ to } \infty}$.

99. When n is even, let $n=2m$. From the question,

$1^2+2.2^2+3^2+2.4^2+5^2+2.6^2+\dots$ to $2m$ terms $=\frac{2m(2m+1)^2}{2}$.

When n is odd, let $n=2m+1$.

$\therefore\quad (1^2+2\cdot 2^2+3^2+2\cdot 4^2+5^2+2\cdot 6^2+\dots+$ to $2m$ terms$)+(2m+1)^2$

$$=\frac{2m(2m+1)^2}{2}+(2m+1)^2=\frac{(n-1)n^2}{2}+n^2.$$

100. As n is odd, the value of the given expression

$$=1^3-2^3+3^3-\dots+n^3$$

$$=(1^3+2^3+3^3+\dots+n^3)-2\{2^3+4^3+\dots(n-1)^3\}$$

$$=\left\{\frac{n(n+1)}{2}\right\}^2-16\left\{1^3+2^3+\dots+\left(\frac{n-1}{2}\right)^3\right\}$$

$$= \frac{n^2(n+1)^2}{4} - 16 \cdot \left\{ \frac{\frac{n-1}{2} \cdot \frac{n+1}{2}}{2} \right\}^2.$$

101. $1^3 = 1 \cdot (1-1) + 1, 2^3 = (2 \cdot 1 + 1) + 5, 3^3 = (3 \cdot 2 + 1) + 9 + 11,$

$4^3 = (4 \cdot 3 + 1) + 15 + 17 + 19$, etc.

$\therefore \quad n^3 = \{n \cdot (n-1) + 1\} + \ldots$, next term being 2 more than the previous

$$= (n^2 - n + 1) + (n^2 - n + 3) + \ldots$$

103. $S_k = 1 + \dfrac{1}{k+1} + \dfrac{1}{(k+1)^2} + \ldots$ to $\infty = \dfrac{1}{1 - \dfrac{1}{k+1}} = \dfrac{k+1}{k}$

$\therefore \quad kS_k = k + 1$

$\therefore \quad \sum\limits_{k=1}^{n} kS_k = \sum\limits_{k=1}^{n} (k+1) = 2 + 3 + \ldots + (n+1).$

104. $t_n = n(n\,!) = \{(n+1) - 1\}n\,! = (n+1)\,! - n\,!.$

105. $t_n = \dfrac{n+2}{n(n+1)} \cdot \left(\dfrac{1}{2}\right)^n = \dfrac{2(n+1) - n}{n(n+1)} \cdot \left(\dfrac{1}{2}\right)^n = \dfrac{1}{n} \cdot \left(\dfrac{1}{2}\right)^{n-1} - \dfrac{1}{n+1} \cdot \left(\dfrac{1}{2}\right)^n.$

106. $f(1) = \left[\dfrac{1}{2} + \dfrac{1}{100}\right] = 0, \ldots, f(49) = \left[\dfrac{1}{2} + \dfrac{49}{100}\right] = \left[\dfrac{99}{100}\right] = 0$

$f(50) = \left[\dfrac{1}{2} + \dfrac{50}{100}\right] = 1, f(51) = \left[\dfrac{1}{2} + \dfrac{51}{100}\right] = 1, \ldots, f(100) = \left[\dfrac{1}{2} + \dfrac{100}{100}\right] = 1.$

107. $y = 2\sqrt{x}$, being in the first quadrant.

The sequence of x-coordinates: 1, 2, 4, 8, …

$\therefore$ the sequence of y-coordinates: $2, 2\sqrt{2}, 2\sqrt{4}, 2\sqrt{8}, \ldots$, where the common ratio is $\sqrt{2}$.

$\therefore \quad y_n = 2 \cdot (\sqrt{2})^{n-1}.$

108. Lengths of line segments on one side of the diagonal are

$$\sqrt{2}, 2\sqrt{2}, 3\sqrt{2}, \ldots, (n-1)\sqrt{2}.$$

So, the required sum $= 2\{\sqrt{2} + 2\sqrt{2} + 3\sqrt{2} + \ldots + (n-1)\sqrt{2}\} + n\sqrt{2}$

$$= 2\sqrt{2}\{1 + 2 + 3 + \ldots + (n-1)\} + n\sqrt{2}.$$

109.

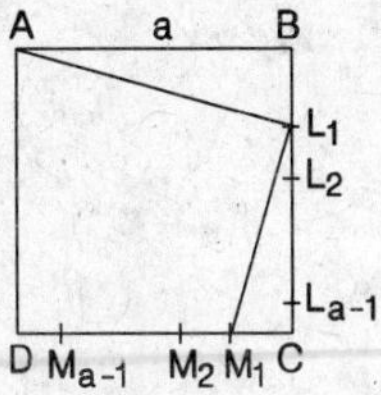

$$AL_1^2 + L_1M_1^2 = (a^2 + 1^2) + \{(a-1)^2 + 1^2\}$$
$$AL_2^2 + L_2M_2^2 = (a^2 + 2^2) + \{(a-2)^2 + 2^2\}$$
$$\dots\dots\dots\dots\dots\dots\dots\dots\dots\dots\dots$$
$$AL_{a-1}^2 + L_{a-1}M_{a-1}^2 = a^2 + (a-1)^2 + \{1^2 + (a-1)^2\}.$$

$\therefore$ the required sum

$$= (a-1)a^2 + \{1^2 + 2^2 + \dots + (a-1)^2\} + 2\{1^2 + 2^2 + \dots + (a-1)^2\}$$
$$= (a-1)a^2 + 3 \cdot \frac{(a-1)a(2a-1)}{6} = a(a-1)\left\{a + \frac{2a-1}{2}\right\}.$$

110. Let the GP be $a, ar, ar^2, \dots$ $(0 < r < 1)$. From the question,

$$\frac{a}{1-r} = 3^3 + 3.3 - 9$$

$\{\because f'(x) = 3x^2 + 3 > 0$; so, $f(x)$ is monotonically increasing;

$\therefore f(3)$ is the greatest value in $[-2, 3]$.$\}$

Also, $f'(0) = 3$. So, $a - ar = 3$.

Solving, $a = 27(1-r)$ and $a(1-r) = 3$ we get $r = \frac{2}{3}, \frac{4}{3} \cdot$ But $r < 1$.

111. Let the edges be $\frac{a}{r}, a, ar$, where $r > 1$.

From the question, $\frac{a}{r} \cdot a \cdot ar = 216$, i.e., $a^3 = 216$, i.e., $a = 6$ and

$$2\left(\frac{a}{r} \cdot a + a \cdot ar + \frac{a}{r} \cdot ar\right) = 252; \quad \therefore \frac{1}{r} + r + 1 = \frac{7}{2} \Rightarrow r = \frac{1}{2}, 2.$$

$\therefore \quad a = 6, r = 2$; so the longest side $= ar = 12$.

112.

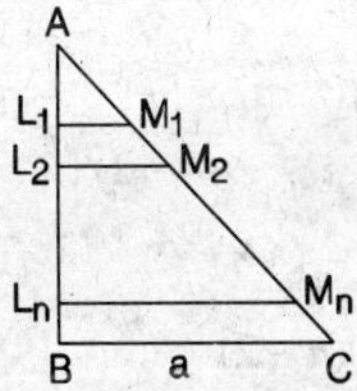

$$\frac{AL_1}{AB} = \frac{L_1M_1}{BC}; \quad \therefore \frac{1}{n+1} = \frac{L_1M_1}{a}; \quad \therefore L_1M_1 = \frac{a}{n+1}$$
$$\frac{AL_2}{AB} = \frac{L_2M_2}{BC}; \quad \therefore \frac{2}{n+1} = \frac{L_2M_2}{a}; \quad \therefore L_2M_2 = \frac{2a}{n+1}, \text{ etc.}$$

$\therefore$ the required sum $= \frac{a}{n+1} + \frac{2a}{n+1} + \frac{3a}{n+1} + \dots + \frac{na}{n+1}$

$$= \frac{a}{n+1}(1+2+3+\ldots+n) = \frac{a}{n+1} \cdot \frac{n(n+1)}{2}.$$

118. $$\frac{1}{\sqrt{a+rx}+\sqrt{a+(r-1)x}} = \frac{\sqrt{a+rx}-\sqrt{a+(r-1)x}}{a+rx-a-(r-1)x}$$

$$= \frac{1}{x}[\sqrt{a+rx}-\sqrt{a+(r-1)x}]$$

$$\therefore \quad t_1+t_2+\ldots+t_n = \frac{1}{x}[\{\sqrt{a+x}-\sqrt{a}\}+\{\sqrt{a+2x}-\sqrt{a+x}\}$$

$$+\ldots+\{\sqrt{a+nx}-\sqrt{a+(n-1)x}\}]$$

$$= \frac{1}{x}[\sqrt{a+nx}-\sqrt{a}] = \frac{a+nx-a}{x(\sqrt{a+nx}+\sqrt{a})}.$$

119. $$\sum_{r=1}^{n}(2r-1)^4 = 1^4+3^4+5^4+\ldots+(2n-1)^4$$

$$= 1^4+2^4+3^4+\ldots+(2n)^4-\{2^4+4^4+\ldots+(2n)^4\}$$

$$= f(2n)-16\{1^4+2^4+\ldots+n^4\} = f(2n)-16f(n).$$

If $n = 2m$, then $\sum_{r=1}^{n}(2r-1)^4 = 1^4+3^4+5^4+\ldots+(4m-1)^4$

$$= 1^4+2^4+3^4+\ldots+(2m)^4+(2m+1)^4+\ldots$$

$$+(4m-1)^4+(4m)^4-\{2^4+4^4+\ldots+(4m)^4\}$$

$$= f(4m)-16\{1^4+2^4+\ldots+(2m)^4\}$$

$$= f(2n)-16f(n), \text{ etc.}$$

120. $2 \cdot {}^nP_2 = 2 \cdot {}^nP_1 + {}^nP_3$; or $\quad 2 \cdot \frac{n!}{(n-2)!} = 2 \cdot \frac{n!}{(n-1)!} + \frac{n!}{(n-3)!}$

or $\quad 2n(n-1) = 2n + n(n-1)(n-2)$

or $\quad 2(n-1) = 2+(n-1)(n-2) \quad$ or $\quad n = 2, 3.$

Clearly, $n \neq 2$. So, $n = 3$.

Then the numbers are $2 \cdot {}^3P_1, {}^3P_2, {}^3P_3$, i.e., 6, 6, 6.

122. $$\sum_{k=1}^{n}\left(\sum_{m=1}^{k} m^2\right) = \sum_{k=1}^{n} \frac{k(k+1)(2k+1)}{6} = \frac{1}{6}\sum_{k=1}^{n}(2k^3+3k^2+k)$$

$$=\frac{1}{3}\cdot\left\{\frac{n(n+1)}{2}\right\}^2+\frac{1}{2}\frac{n(n+1)(2n+1)}{6}+\frac{1}{6}\frac{n(n+1)}{2}.$$

$a=$ coefficient of $n^4=\frac{1}{3}\cdot\frac{1}{4}$; $b=$ coefficient of $n^3=\frac{1}{3}\cdot\frac{1}{2}+\frac{1}{6}$, etc.

123. As AM $\geq$ GM, $\dfrac{\frac{a}{b}+\frac{b}{c}}{2}\geq\sqrt{\frac{a}{b}\cdot\frac{b}{c}}$ and $\dfrac{\frac{c}{d}+\frac{d}{e}}{2}\geq\sqrt{\frac{c}{d}\cdot\frac{d}{e}}$.

$\therefore$ $\left(\frac{a}{b}+\frac{b}{c}\right)\left(\frac{c}{d}+\frac{d}{e}\right)\geq 4\sqrt{\frac{a}{c}}\cdot\sqrt{\frac{c}{e}}$.

Also $\dfrac{\frac{a}{b}+\frac{b}{c}+\frac{c}{d}+\frac{d}{e}+\frac{e}{a}}{5}\geq\sqrt[5]{\frac{a}{b}\cdot\frac{b}{c}\cdot\frac{c}{d}\cdot\frac{d}{e}\cdot\frac{e}{a}}=1.$

124. $f(x)=1+x+x^2+\ldots+x^n$

$\therefore$ $f'(x)\cdot g(x)=(1+2x+3x^2+\ldots+nx^{n-1})$

$$\times\left\{1-\frac{2}{x}+\frac{3}{x^2}-\ldots+(-1)^n\cdot\frac{n+1}{x^n}\right\}$$

$\therefore$ the required constant term $=1-2^2+3^2-\ldots+(-1)^{n-1}n^2$.

Separate the case of n odd or even.

125. AM > GM $\Rightarrow$ $\dfrac{1+2+3+\ldots+n}{n}\geq\sqrt[n]{1\cdot2\cdot3\cdot\ldots n}$,

equality holding when $n=1$

$\Rightarrow$ $\dfrac{n+1}{2}\geq\sqrt[n]{a_n}$ $\Rightarrow$ $\left(\dfrac{n+1}{2}\right)^n\geq a_n=n!$.

Also $n\geq1, n\geq2, \ldots, n=n$ $\Rightarrow$ $n^n\geq1\cdot2\cdot3\cdot\ldots n=a_n$.

126. $n(A\cap B)=$ number of common terms in 2, 4, 6, 8, ... to 200 terms and 3, 6, 9, 12, ... to 250 terms.

$n(A\cup B)=n(A)+n(B)-n(A\cap B)=200+250-n(A\cap B)$.

128. Lengths of sides of $S_1, S_2, S_3, \ldots$ are $10, \frac{10}{\sqrt{2}}, \frac{10}{(\sqrt{2})^2}, \ldots$ respectively.

$\therefore$ S_n has its side $=10\cdot\left(\frac{1}{\sqrt{2}}\right)^{n-1}$

$\therefore$ area of $S_n = 100 \cdot \left(\frac{1}{2}\right)^{n-1} = \frac{200}{2^n} < 1$ (from the question)

$\therefore$ $2^n > 200.$

131. AM of $x, z = y$, GM of $x, z = \sqrt{xz}$ and AM $\geq$ GM. So, $y^2 \geq xz$.

$$\frac{x+y}{2y-x} = \frac{x+y}{x+z-x} = \frac{x+y}{z}, \quad \frac{y+z}{2y-z} = \frac{y+z}{x}$$

$$\therefore \quad \frac{\frac{x+y}{2y-x} + \frac{y+z}{2y-z}}{2} \geq \sqrt{\frac{x+y}{z} \cdot \frac{y+z}{x}}$$

$$= \sqrt{1 + \frac{y(x+y+z)}{xz}} = \sqrt{1 + \frac{3y^2}{xz}} \qquad \{\because \ x+z = 2y\}$$

$$\therefore \quad \frac{x+y}{2y-x} + \frac{y+z}{2y-z} \geq 2\sqrt{1 + 3 \cdot \frac{y^2}{xz}} \geq 4 \qquad (\because \ y^2 \geq xz)$$

133. $4 = 1 + (n-1)d,\ 16 = 1 + (m-1)d \quad \Rightarrow \quad \frac{15}{3} = \frac{m-1}{n-1}$

or $\quad \frac{n-1}{1} = \frac{m-1}{5} = p =$ positive integer

$\therefore$ $n = p + 1$, $m = 5p + 1$. So n, m have infinite pairs of values.

Also, $4 = 1 \cdot r^n,\ 16 = 1 \cdot r^m \quad \Rightarrow \quad r^{m-n} = 4 = r^n$. So, $m - n = n$.

$\therefore$ $\frac{m}{2} = \frac{n}{1} = q =$ positive integer. So m, n have infinite pairs of values.

□

2

Equation, Inequation and Expression

• Type 1 •

Choose the most appropriate option (a, b, c or d).

1. If x is a real number such that $x(x^2+1)$, $(1/2)x^2$, 6 are three consecutive terms of an AP then the next two consecutive term of the AP are

(a) 14, 6 (b) $-2, -10$

(c) 14, 22 (d) none of these

2. The number of real solutions of $x - \frac{1}{x^2-4} = 2 - \frac{1}{x^2-4}$ is

(a) 0 (b) 1

(c) 2 (d) infinite

3. The number of values of a for which

$$(a^2-3a+2)x^2+(a^2-5a+6)x+a^2-4=0$$

is an identity in x is

(a) 0 (b) 2

(c) 1 (d) 3

4. The number of values of the pair (a, b) for which

$$a(x+1)^2+b(x^2-3x-2)+x+1=0$$

is an identity in x is

(a) 0 (b) 1

(c) 2 (d) infinite

5. The number of values of the triplet (a, b, c) for which

$$a\cos 2x + b\sin^2 x + c = 0$$

is satisfied by all real x is

(a) 0 (b) 2

(c) 3 (d) infinite

6. The polynomial $(ax^2 + bx + c)(ax^2 - dx - c)$, $ac \neq 0$, has

(a) four real zeros (b) at least two real zeros

(c) at most two real zeros (d) no real zeros

7. Let $f(x) = ax^3 + 5x^2 - bx + 1$. If $f(x)$ when divided by $2x + 1$ leaves 5 as remainder, and $f'(x)$ is divisible by $3x - 1$ then

(a) $a = 26, b = 10$ (b) $a = 24, b = 12$

(c) $a = 26, b = 12$ (d) none of these

8. $x^{3^n} + y^{3^n}$ is divisible by $x + y$ if

(a) n is any integer ≥ 0

(b) n is an odd positive integer

(c) n is an even positive integer

(d) n is a rational number

9. If x, y are rational numbers such that

$$x + y + (x - 2y)\sqrt{2} = 2x - y + (x - y - 1)\sqrt{6}$$

then

(a) x and y cannot be determined

(b) $x = 2, y = 1$

(c) $x = 5, y = 1$

(d) none of these

10. The number of real solutions of the equation

$$2^{x/2} + (\sqrt{2} + 1)^x = (5 + 2\sqrt{2})^{x/2}$$

is

(a) one (b) two

(c) four (d) infinite

11. The number of real solutions of the equation $e^x = x$ is

(a) 1 (b) 2

(c) 0 (d) none of these

12. The sum of the real roots of the equation $x^2 + |x| - 6 = 0$ is

(a) 4 (b) 0

(c) −1 (d) none of these

13. The solutions of the equation $2x - 2[x] = 1$, where $[x]$ = the greatest integer less than or equal to x, are

(a) $x = n + \frac{1}{2}, n \in N$ (b) $x = n - \frac{1}{2}, n \in N$

(c) $x = n + \frac{1}{2}, n \in Z$ (d) $n < x < n + 1, n \in Z$

14. The number of real solutions of the equation $\sin(e^x) = 5^x + 5^{-x}$ is

(a) 0 (b) 1

(c) 2 (d) infinitely many

15. The number of real solutions of $1 + |e^x - 1| = e^x(e^x - 2)$ is

(a) 0 (b) 1

(c) 2 (d) 4

16. The equation $2\sin^2\frac{x}{2} \cdot \cos^2 x = x + \frac{1}{x}, 0 < x \le \frac{\pi}{2}$ has

(a) one real solution

(b) no real solution

(c) infinitely many real solutions

(d) none of these

17. If $y \neq 0$ then the number of values of the pair (x, y) such that

$$x + y + \frac{x}{y} = \frac{1}{2} \text{ and } (x + y)\frac{x}{y} = -\frac{1}{2}, \text{ is}$$

(a) 1 (b) 2

(c) 0 (d) none of these

18. The number of real solutions of the equation $\log_{0.5} x = |x|$ is

(a) 1 (b) 2

(c) 0 (d) none of these

19. The equation $\sqrt{x+1} - \sqrt{x-1} = \sqrt{4x-1}$ has

(a) no solution (b) one solution

(c) two solutions (d) more than two solutions

20. The number of solutions of the equation $|x| = \cos x$ is

(a) one (b) two

(c) three (d) zero

21. The product of all the solutions of the equation

$$(x-2)^2 - 3|x-2| + 2 = 0$$

is

(a) 2 (b) –4

(c) 0 (d) none of these

22. If $0 < x < 1000$ and $\left[\frac{x}{2}\right] + \left[\frac{x}{3}\right] + \left[\frac{x}{5}\right] = \frac{31}{30}x$, where $[x]$ is the greatest integer less than or equal to x, the number of possible values of x is

(a) 34 (b) 32

(c) 33 (d) none of these

23. The solution set of $(x)^2 + (x+1)^2 = 25$, where (x) is the least integer greater than or equal to x, is

(a) $(2, 4)$ (b) $(-5, -4] \cup (2, 3]$

(c) $[-4, -3) \cup [3, 4)$ (d) none of these

24. If $3^{x+1} = 6^{\log_2 3}$ then x is

(a) 3 (b) 2

(c) $\log_3 2$ (d) $\log_2 3$

25. If $(\sqrt{2})^x + (\sqrt{3})^x = (\sqrt{13})^{x/2}$ then the number of values of x is

(a) 2 (b) 4

(c) 1 (d) none of these

26. The number of real solutions of the equation $\frac{6-x}{x^2-4} = 2 + \frac{x}{x+2}$ is

(a) two (b) one

(c) zero (d) none of these

27. The number of real solutions of

$$\sqrt{x^2 - 4x + 3} + \sqrt{x^2 - 9} = \sqrt{4x^2 - 14x + 6} \text{ is}$$

(a) one (b) two

(c) three (d) none of these

28. If $[x]$ = the greatest integer less than or equal to x, and (x) = the least integer greater than or equal to x, and $[x]^2 + (x)^2 > 25$ then x belongs to

(a) $[3, 4]$ (b) $(-\infty, -4]$

(c) $[4, +\infty)$ (d) $(-\infty, -4] \cup [4, +\infty)$

29. Let R = the set of real numbers, Z = the set of integers, N = the set of natural numbers. If S be the solution set of the equation $(x)^2 + [x]^2 = (x-1)^2 + [x+1]^2$, where (x) = the least integer greater than or equal to x and $[x]$ = the greatest integer less than or equal to x, then

(a) $S = R$ (b) $S = R - Z$

(c) $S = R - N$ (d) none of these

30. If $[x]^2 = [x+2]$, where $[x]$ = the greatest integer less than or equal to x, then x must be such that

(a) $x = 2, -1$ (b) $x \in [2, 3)$

(c) $x \in [-1, 0)$ (d) none of these

31. The solution set of $\left|\frac{x+1}{x}\right| + |x+1| = \frac{(x+1)^2}{|x|}$ is

(a) $\{x \mid x \geq 0\}$ (b) $\{x \mid x > 0\} \cup \{-1\}$

(c) $\{-1, 1\}$ (d) $\{x \mid x \geq 1 \text{ or } x \leq -1\}$

32. The number of solutions of $|[x] - 2x| = 4$, where $[x]$ is the greatest integer $\leq x$, is

(a) 2 (b) 4

(c) 1 (d) infinite

33. The set of real values of x satisfying $|x-1| \leq 3$ and $|x-1| \geq 1$ is

(a) $[2, 4]$ (b) $(-\infty, 2] \cup [4, +\infty)$

(c) $[-2, 0] \cup [2, 4]$ (d) none of these

34. The set of real values of x satisfying $\big||x-1| - 1\big| \leq 1$ is

(a) $[-1, 3]$ (b) $[0, 2]$

(c) $[-1, 1]$ (d) none of these

35. If $x \in Z$ (the set of integers) such that $x^2 - 3x < 4$ then the number of possible values of x is

(a) 3 (b) 4

(c) 6 (d) none of these

36. If x is an integer satisfying $x^2 - 6x + 5 \leq 0$ and $x^2 - 2x > 0$ then the number of possible values of x is

(a) 3 (b) 4

(c) 2 (d) infinite

37. The solution set of the inequation $\log_{1/3}(x^2+x+1)+1>0$ is

(a) $(-\infty,-2)\cup(1,+\infty)$ (b) $[-1,2]$

(c) $(-2,1)$ (d) $(-\infty,+\infty)$

38. If $5^x+(2\sqrt{3})^{2x}\geq 13^x$ then the solution set for x is

(a) $[2,+\infty)$ (b) $\{2\}$

(c) $(-\infty,2]$ (d) $[0,2]$

39. If $3^{x/2}+2^x>25$ then the solution set is

(a) R (b) $(2,+\infty)$

(c) $(4,+\infty)$ (d) none of these

40. If $\sin^x\alpha+\cos^x\alpha\geq 1, 0<\alpha<\frac{\pi}{2}$, then

(a) $x\in[2,+\infty)$ (b) $x\in(-\infty,2]$

(c) $x\in[-1,1]$ (d) none of these

41. The solution set of $x^2+2\leq 3x\leq 2x^2-5$ is

(a) ϕ (b) $[1,2]$

(c) $(-\infty,-1]\cup[5/2,+\infty)$ (d) none of these

42. The solution set of $\frac{x^2-3x+4}{x+1}>1, x\in R$, is

(a) $(3,+\infty)$ (b) $(-1,1)\cup(3,+\infty)$

(c) $[-1,1]\cup[3,+\infty)$ (d) none of these

43. The number of integral solutions of $\frac{x+2}{x^2+1}>\frac{1}{2}$ is

(a) 4 (b) 5

(c) 3 (d) none of these

44. If a, b, c are nonzero, unequal rational numbers then the roots of the equation $abc^2x^2+(3a^2+b^2)cx-6a^2-ab+2b^2=0$ are

(a) rational (b) imaginary

(c) irrational (d) none of these

45. If l, m are real and $l\neq m$ then the roots of the equation

$$(l-m)x^2-5(l+m)x-2(l-m)=0 \text{ are}$$

(a) real and equal (b) nonreal complex
(c) real and unequal (d) none of these

46. If a, b, c, d are four consecutive terms of an increasing AP then the roots of the equation $(x-a)(x-c)+2(x-b)(x-d)=0$ are

(a) real and distinct (b) nonreal complex
(c) real and equal (d) integers

47. If a, b, c are three distinct positive real numbers then the number of real roots of $ax^2+2b\,|x|-c=0$ is

(a) 4 (b) 2
(c) 0 (d) none of these

48. The equation $x^2-6x+8+\lambda(x^2-4x+3)=0, \lambda \in R$, has

(a) real and unequal roots for all λ
(b) real roots for $\lambda < 0$ only
(c) real roots for $\lambda > 0$ only
(d) real and unequal roots for $\lambda = 0$ only

49. If $\cos\theta, \sin\phi, \sin\theta$ are in GP then roots of $x^2+2\cot\phi \cdot x+1=0$ are always

(a) equal (b) real
(c) imaginary (d) greater than 1

50. The roots of $ax^2+bx+c=0$, where $a \neq 0$ and coefficients are real, are nonreal complex and $a+c<b$. Then

(a) $4a+c>2b$ (b) $4a+c<2b$
(c) $4a+c=2b$ (d) none of these

51. The equation $(a+2)x^2+(a-3)x=2a-1, a \neq -2$ has roots rational for

(a) all rational values of a except $a=-2$
(b) all real values of a except $a=-2$
(c) rational values of $a > \frac{1}{2}$
(d) none of these

52. If $a \cdot 3^{\tan x}+a \cdot 3^{-\tan x}-2=0$ has real solutions, $x \neq \frac{\pi}{2}, 0 \leq x \leq \pi$, then the set of possible values of the parameter a is

(a) $[-1, 1]$ (b) $[-1, 0)$
(c) $(0, 1]$ (d) $(0, +\infty)$

53. If $a > 1$, roots of the equation $(1-a)x^2 + 3ax - 1 = 0$ are
(a) one positive and one negative
(b) both negative
(c) both positive
(d) both nonreal complex

54. If $a \in R, b \in R$ then the equation $x^2 - abx - a^2 = 0$ has
(a) one positive root and one negative root
(b) both roots positive
(c) both roots negative
(d) nonreal roots

55. If the roots of the equation $x^2 - 2ax + a^2 + a - 3 = 0$ are less than 3 then
(a) $a < 2$ (b) $2 \le a \le 3$
(c) $3 < a \le 4$ (d) $a > 4$

56. If α, β are the roots of $x^2 - 3x + a = 0$, $a \in R$ and $\alpha < 1 < \beta$ then
(a) $a \in (-\infty, 2)$ (b) $a \in \left(-\infty, \frac{9}{4}\right]$
(c) $a \in \left(2, \frac{9}{4}\right]$ (d) none of these

57. If α, β be the roots of $4x^2 - 16x + \lambda = 0$, $\lambda \in R$ such that $1 < \alpha < 2$ and $2 < \beta < 3$ then the number of integral solutions of λ is
(a) 5 (b) 6
(c) 2 (d) 3

58. The number of integral values of a for which $x^2 - (a-1)x + 3 = 0$ has both roots positive and $x^2 + 3x + 6 - a = 0$ has both roots negative is
(a) 0 (b) 1
(c) 2 (d) infinite

59. If X denotes the set of real numbers p for which the equation $x^2 = p(x + p)$ has its roots greater than p then X is equal to
(a) $\left(-2, -\frac{1}{2}\right)$ (b) $\left(-\frac{1}{2}, \frac{1}{4}\right)$

(c) null set ϕ (d) $(-\infty, 0)$

60. If $\cos^4 x + \sin^2 x - p = 0, p \in R$ has real solutions then

(a) $p \leq 1$ (b) $\frac{3}{4} \leq p \leq 1$

(c) $p \geq \frac{3}{4}$ (d) none of these

61. If one root of the equation $(k^2 + 1)x^2 + 13x + 4k = 0$ is reciprocal of the other then k has the value

(a) $-2 + \sqrt{3}$ (b) $2 - \sqrt{3}$

(c) 1 (d) none of these

62. If the ratio of the roots of $\lambda x^2 + \mu x + \nu = 0$ is equal to the ratio of the roots of $x^2 + x + 1 = 0$ then λ, μ, ν are in

(a) AP (b) GP

(c) HP (d) none of these

63. p, q, r and s are integers. If the AM of the roots of $x^2 - px + q^2 = 0$ and GM of the roots of $x^2 - rx + s^2 = 0$ are equal then

(a) q is an odd integer (b) r is an even integer

(c) p is an even integer (d) s is an odd integer

64. If α, β are roots of the equation $(x - a)(x - b) = c, c \neq 0$, then the roots of the equation $(x - \alpha)(x - \beta) + c = 0$ are

(a) a, c (b) b, c

(c) a, b (d) $a + c, b + c$

65. If the roots of $4x^2 + 5k = (5k + 1)x$ differ by unity then the negative value of k is

(a) -3 (b) $-\frac{1}{5}$

(c) $-\frac{3}{5}$ (d) none of these

66. The harmonic mean of the roots of the equation

$$(5 + \sqrt{2})x^2 - (4 + \sqrt{5})x + 8 + 2\sqrt{5} = 0 \text{ is}$$

(a) 2 (b) 4

(c) 6 (d) 8

67. If the product of the roots of the equation $x^2 - 5x + 4^{\log_2 \lambda} = 0$ is 8 then λ is

(a) $\pm 2\sqrt{2}$ (b) $2\sqrt{2}$

(c) 3 (d) none of these

68. If the roots of $a_1 x^2 + b_1 x + c_1 = 0$ are α_1, β_1, and those of

$a_2 x^2 + b_2 x + c_2 = 0$ are α_2, β_2 such that $\alpha_1\alpha_2 = \beta_1\beta_2 = 1$

then

(a) $\dfrac{a_1}{a_2} = \dfrac{b_1}{b_2} = \dfrac{c_1}{c_2}$ (b) $\dfrac{a_1}{c_2} = \dfrac{b_1}{b_2} = \dfrac{c_1}{a_2}$

(c) $a_1 a_2 = b_1 b_2 = c_1 c_2$ (d) none of these

69. If α, β are the roots of $ax^2 + c = bx$ then the equation $(a + cy)^2 = b^2 y$ in y has the roots

(a) α^{-1}, β^{-1} (b) α^2, β^2

(c) $\alpha\beta^{-1}, \alpha^{-1}\beta$ (d) α^{-2}, β^{-2}

70. If the roots of $ax^2 - bx - c = 0$ change by the same quantity then the expression in a, b, c that does not change is

(a) $\dfrac{b^2 - 4ac}{a^2}$ (b) $\dfrac{b - 4c}{a}$

(c) $\dfrac{b^2 + 4ac}{a^2}$ (d) none of these

71. If α, β are the roots of $x^2 - px + q = 0$ then the product of the roots of the quadratic equation whose roots are $\alpha^2 - \beta^2$ and $\alpha^3 - \beta^3$ is

(a) $p(p^2 - q)^2$ (b) $p(p^2 - q)(p^2 - 4q)$

(c) $p(p^2 - 4q)(p^2 + q)$ (d) none of these

72. If the sum of the roots of the quadratic equation $ax^2 + bx + c = 0$ is equal to the sum of the squares of their reciprocals then $\dfrac{b^2}{ac} + \dfrac{bc}{a^2}$ is equal to

(a) 2 (b) −2

(c) 1 (d) −1

73. If the absolute value of the difference of roots of the equation $x^2 + px + 1 = 0$ exceeds $\sqrt{3p}$ then

(a) $p < -1$ or $p > 4$ (b) $p > 4$

(c) $-1 < p < 4$ (d) $0 \le p < 4$

74. If α, β are roots of $x^2 + px + q = 0$ and γ, δ are the roots of $x^2 + px - r = 0$ then $(\alpha - \gamma)(\alpha - \delta)$ is equal to

(a) $q + r$ (b) $q - r$

(c) $-(q + r)$ (d) $-(p + q + r)$

75. If α, β are roots of $375x^2 - 25x - 2 = 0$ and $s_n = \alpha^n + \beta^n$ then $\lim_{n \to \infty} \sum_{r=1}^{n} s_r$ is

(a) $\frac{7}{116}$ (b) $\frac{1}{12}$

(c) $\frac{29}{358}$ (d) none of these

76. The quadratic equation whose roots are the AM and HM of the roots of the equation $x^2 + 7x - 1 = 0$ is

(a) $14x^2 + 14x - 45 = 0$ (b) $45x^2 - 14x + 14 = 0$

(c) $14x^2 + 45x - 14 = 0$ (d) none of these

77. Let $\alpha \neq \beta$ and $\alpha^2 + 3 = 5\alpha$ while $\beta^2 = 5\beta - 3$. The quadratic equation whose roots are $\frac{\alpha}{\beta}$ and $\frac{\beta}{\alpha}$ is

(a) $3x^2 - 31x + 3 = 0$ (b) $3x^2 - 19x + 3 = 0$

(c) $3x^2 + 19x + 3 = 0$ (d) none of these

78. If a and b are rational and b is not a perfect square then the quadratic equation with rational coefficients whose one root is $\frac{1}{a + \sqrt{b}}$ is

(a) $x^2 - 2ax + (a^2 - b) = 0$ (b) $(a^2 - b)x^2 - 2ax + 1 = 0$

(c) $(a^2 - b)x^2 - 2bx + 1 = 0$ (d) none of these

79. If $\frac{1}{4 - 3i}$ is a root of $ax^2 + bx + 1 = 0$, where a, b are real, then

(a) $a = 25, b = -8$ (b) $a = 25, b = 8$

(c) $a = 5, b = 4$ (d) none of these

80. If α, β, γ be the roots of the equation $x(1+x^2)+x^2(6+x)+2=0$ then the value of $\alpha^{-1}+\beta^{-1}+\gamma^{-1}$ is

(a) -3
(b) $\frac{1}{2}$
(c) $-\frac{1}{2}$
(d) none of these

81. If the roots of $x^3-12x^2+39x-28=0$ are in AP then their common difference is

(a) ± 1
(b) ± 2
(c) ± 3
(d) ± 4

82. The roots of the equation $x^3+14x^2-84x-216=0$ are in

(a) AP
(b) GP
(c) HP
(d) none of these

83. If $z_0=\alpha+i\beta$, $i=\sqrt{-1}$, then the roots of the cubic equation

$$x^3-2(1+\alpha)x^2+(4\alpha+\alpha^2+\beta^2)x-2(\alpha^2+\beta^2)=0 \text{ are}$$

(a) $2, z_0, \bar{z}_0$
(b) $1, z_0, -z_0$
(c) $2, z_0, -\bar{z}_0$
(d) $2, -z_0, \bar{z}_0$

84. If 3 and $1+\sqrt{2}$ are two roots of a cubic equation with rational coefficients then the equation is

(a) $x^2-5x^2+9x-9=0$
(b) $x^3-3x^2-4x+12=0$
(c) $x^3-5x^2+7x+3=0$
(d) none of these

85. Let a, b, c be real numbers and $a \neq 0$. If α is a root of $a^2x^2+bx+c=0$, β is a root of $a^2x^2-bx-c=0$, and $0<\alpha<\beta$ then the equation $a^2x^2+2bx+2c=0$ has a root γ that always satisfies

(a) $\gamma=\frac{1}{2}(\alpha+\beta)$
(b) $\gamma=\alpha+\frac{\beta}{2}$
(c) $\gamma=\alpha$
(d) $\alpha<\gamma<\beta$

86. Let a, b, c be three real numbers such that $2a+3b+6c=0$. Then the quadratic equation $ax^2+bx+c=0$ has

(a) imaginary roots
(b) at least one root in $(0, 1)$
(c) at least one root in $(-1, 0)$
(d) both roots in $(1, 2)$

87. If the equations $2x^2 - 7x + 1 = 0$ and $ax^2 + bx + 2 = 0$ have a common root then

(a) $a = 2, b = -7$ (b) $a = -\frac{7}{2}, b = 1$

(c) $a = 4, b = -14$ (d) none of these

88. The quadratic equations $x^2 + (a^2 - 2)x - 2a^2 = 0$ and $x^2 - 3x + 2 = 0$ have

(a) no common root for all $a \in R$

(b) exactly one common root for all $a \in R$

(c) two common roots for some $a \in R$

(d) none of these

89. If the equations $ax^2 + bx + c = 0$ and $cx^2 + bx + a = 0$, $a \neq c$ have a negative common root then the value of $a - b + c$ is

(a) 0 (b) 2

(c) 1 (d) none of these

90. If the equations $x^2 + ix + a = 0$, $x^2 - 2x + ia = 0$, $a \neq 0$ have a common root then

(a) a is real (b) $a = \frac{1}{2} + i$

(c) $a = \frac{1}{2} - i$ (d) the other root is also common

91. If $x^2 - 2r \cdot p_r x + r = 0$; $r = 1, 2, 3$ are three quadratic equations of which each pair has exactly one root common then the number of solutions of the triplet (p_1, p_2, p_3) is

(a) 2 (b) 1

(c) 9 (d) 27

92. If $(\lambda^2 + \lambda - 2)x^2 + (\lambda + 2)x < 1$ for all $x \in R$ then λ belongs to the interval

(a) $(-2, 1)$ (b) $\left(-2, \frac{2}{5}\right)$

(c) $\left(\frac{2}{5}, 1\right)$ (d) none of these

93. The least integral value of k for which $(k - 2)x^2 + 8x + k + 4 > 0$ for all $x \in R$, is

(a) 5 (b) 4
(c) 3 (d) none of these

94. The set of possible values of x such that $5^x + (2\sqrt{3})^{2x} - 169$ is always positive is

(a) $[3, +\infty)$ (b) $[2, +\infty)$
(c) $(2, +\infty)$ (d) none of these

95. If all real values of x obtained from the equation
$$4^x - (a-3)2^x + a - 4 = 0$$
are nonpositive then

(a) $a \in (4, 5]$ (b) $a \in (0, 4)$
(c) $a \in (4, +\infty)$ (d) none of these

96. The set of possible values of λ for which
$$x^2 - (\lambda^2 - 5\lambda + 5)x + (2\lambda^2 - 3\lambda - 4) = 0$$
has roots whose sum and product are both less than 1 is

(a) $\left(-1, \frac{5}{2}\right)$ (b) $(1, 4)$
(c) $\left[1, \frac{5}{2}\right]$ (d) $\left(1, \frac{5}{2}\right)$

97. If $\log_{10} x + \log_{10} y \geq 2$ then the smallest possible value of $x + y$ is

(a) 10 (b) 30
(c) 20 (d) none of these

98. If $f(x) = \dfrac{x^2 - 1}{x^2 + 1}$ for every real number x then the minimum value of f

(a) does not exist because f is unbounded
(b) is not attained even though f is bounded
(c) is equal to 1
(d) is equal to -1

99. If $ax^2 + bx + 6 = 0$ does not have two distinct real roots, where $a \in R$, $b \in R$, then the least value of $3a + b$ is

(a) 4 (b) -1
(c) 1 (d) -2

100. If $ab = 2a + 3b$, $a > 0$, $b > 0$ then the minimum value of ab is

(a) 12 (b) 24

(c) $\frac{1}{4}$ (d) none of these

101. If $x^2 + px + 1$ is a factor of the expression $ax^3 + bx + c$ then

(a) $a^2 + c^2 = -ab$ (b) $a^2 - c^2 = -ab$

(c) $a^2 - c^2 = ab$ (d) none of these

102. If $x^2 - 1$ is a factor of $x^4 + ax^3 + 3x - b$ then

(a) $a = 3, b = -1$ (b) $a = -3, b = 1$

(c) $a = 3, b = 1$ (d) none of these

103. The number of values of k for which

$$\{x^2 - (k-2)x + k^2\}\{x^2 + kx + (2k-1)\}$$

is a perfect square is

(a) 1 (b) 2

(c) 0 (d) none of these

104. If $x + \lambda y - 2$ and $x - \mu y + 1$ are factors of the expression

$$6x^2 - xy - y^2 - 6x + 8y - 12$$

then

(a) $\lambda = \frac{1}{3}, \mu = \frac{1}{2}$ (b) $\lambda = 2, \mu = 3$

(c) $\lambda = \frac{1}{3}, \mu = -\frac{1}{2}$ (d) none of these

105. If $x - y$ and $y - 2x$ are two factors of the expression

$$x^3 - 3x^2y + \lambda xy^2 + \mu y^3$$

then

(a) $\lambda = 11, \mu = -3$ (b) $\lambda = 3, \mu = -11$

(c) $\lambda = \frac{11}{4}, \mu = -\frac{3}{4}$ (d) none of these

106. If $x + y$ and $y + 3x$ are two factors of the expression

$$\lambda x^3 - \mu x^2 y + xy^2 + y^3$$

then the third factor is

(a) $y + 3x$ (b) $y - 3x$

(c) $y - x$ (d) none of these

107. If x, y, z are real and distinct then

$$f(x, y) = x^2 + 4y^2 + 9z^2 - 6yz - 3zx - 2xy$$

is always

(a) non-negative (b) nonpositive
(c) zero (d) none of these

108. If $x^2 + y^2 + z^2 = 1$ then the value of $xy + yz + zx$ lies in the interval

(a) $\left[\frac{1}{2}, 2\right]$ (b) $[-1, 2]$
(c) $\left[-\frac{1}{2}, 1\right]$ (d) $\left[-1, \frac{1}{2}\right]$

109. If $a \in R, b \in R$ then the factors of the expression $a(x^2 - y^2) - bxy$ are

(a) real and different (b) real and identical
(c) complex (d) none of these

110. If a, b, c are in HP then the expression

$$a(b-c)x^2 + b(c-a)x + c(a-b)$$

(a) has real and distinct factors (b) is a perfect square
(c) has no real factor (d) none of these

111. The number of positive integral values of k for which

$$(16x^2 + 12x + 39) + k(9x^2 - 2x + 11)$$

is a perfect square is

(a) two (b) zero
(c) one (d) none of these

112. If $(x-1)^3$ is a factor of $x^4 + ax^3 + bx^2 + cx - 1$ then the other factor is

(a) $x - 3$ (b) $x + 1$
(c) $x + 2$ (d) none of these

• Type 2 •

Choose the correct options. One or more options may be correct.

113. If $x^2 - bx + c = 0$ has equal integral roots then

(a) b and c are integers
(b) b and c are even integers

(c) b is an even integer and c is a perfect square of a positive integer

(d) none of these

114. Let A, G and H be the AM, GM and HM of two positive numbers a and b. The quadratic equation whose roots are A and H is

(a) $Ax^2 - (A^2 + G^2)x + AG^2 - 0$

(b) $Ax^2 - (A^2 + H^2)x + AH^2 = 0$

(c) $Hx^2 - (H^2 + G^2)x + HG^2 = 0$

(d) none of these

115. Let A, G and H are the AM, GM and HM respectively of two unequal positive integers. Then the equation $Ax^2 - |G|x - H = 0$ has

(a) both roots as fractions

(b) at least one root which is a negative fraction

(c) exactly one positive root

(d) at least one root which is an integer

116. Let $x^2 - px + q = 0$, where $p \in R$, $q \in R$, have the roots α, β such that $\alpha + 2\beta = 0$ then

(a) $2p^2 + q = 0$ (b) $2q^2 + p = 0$

(c) $q < 0$ (d) none of these

117. The cubic equation whose roots are the AM, GM and HM of the roots of $x^2 - 2px + q^2 = 0$ is

(a) $(x - p)(x - q)(x - p - q) = 0$

(b) $(x - p)(x - |q|)(px - q^2) = 0$

(c) $x^3 - \left(p + |q| + \frac{q^2}{p}\right)x^2 + \left(p|q| + q^2 + \frac{|q|^3}{p}\right)x - |q|^3 = 0$

(d) none of these

118. If $x^2 + ax + b = 0$ and $x^2 + bx + a = 0$, $a \neq b$, have a common root α then

(a) $a + b = 1$ (b) $\alpha + 1 = 0$

(c) $\alpha = 1$ (d) $a + b + 1 = 0$

119. The line $y + 14 = 0$ cuts the curve whose equation is $x(x^2 + x + 1) + y = 0$ at

(a) three real points (b) one real point

(c) at least one real point (d) no real point

120. If a, b, c are in GP, where a, c are positive, then the equation $ax^2 + bx + c = 0$ has

(a) real roots

(b) imaginary roots

(c) ratio of roots $= 1 : w$ where w is a nonreal cube root of unity

(d) ratio of roots $= b : ac$

121. If α, β are the roots of the equation $x^2 + x + 3 = 0$ then the equation $3x^2 + 5x + 3 = 0$ has a root

(a) $\frac{\alpha}{\beta}$ (b) $\frac{\beta}{\alpha}$

(c) $\frac{\alpha}{\beta} + \frac{\beta}{\alpha}$ (d) none of these

122. If α, β are the roots of $x^2 - 2ax + b^2 = 0$ and γ, δ are the roots of $x^2 - 2bx + a^2 = 0$ then

(a) AM of $\alpha, \beta =$ GM of γ, δ (b) GM of $\alpha, \beta =$ AM of γ, δ

(c) $\alpha, \beta, \gamma, \delta$ are in AP (d) $\alpha, \beta, \gamma, \delta$ are in GP

123. If the roots of the equation $ax^2 - 4x + a^2 = 0$ are imaginary and the sum of the roots is equal to their product then a is

(a) -2 (b) 4

(c) 2 (d) none of these

124. If x, y, z are three consecutive terms of a GP, where $x > 0$ and the common ratio is r, then the inequality $z + 3x > 4y$ holds for

(a) $r \in (-\infty, 1)$ (b) $r = \frac{24}{5}$

(c) $r \in (3, +\infty)$ (d) $r = \frac{1}{2}$

125. The equation $||x-1| + a| = 4$ can have real solutions for x if a belongs to the interval

(a) $(-\infty, 4]$ (b) $(-\infty, -4]$

(c) $(4, +\infty)$ (d) $[-4, 4]$

126. The equation $|x+1||x-1| = a^2 - 2a - 3$ can have real solutions for x if a belongs to

(a) $(-\infty, -1] \cup [3, +\infty)$ (b) $[1-\sqrt{5}, 1+\sqrt{5}]$

(c) $[1-\sqrt{5}, -1] \cup [3, 1+\sqrt{5}]$ (d) none of these

127. The common roots of the equations $x^3 + 2x^2 + 2x + 1 = 0$ and $1 + x^{130} + x^{1988} = 0$ are (where ω is a nonreal cube root of unity)

(a) ω (b) ω^2

(c) -1 (d) $\omega - \omega^2$

128. If α is a root of the equation $2x(2x+1) = 1$ then the other root is

(a) $3\alpha^3 - 4\alpha$ (b) $-2\alpha(\alpha + 1)$

(c) $4\alpha^3 - 3\alpha$ (d) none of these

129. For the equation $2x^2 + 6\sqrt{2}x + 1 = 0$

(a) roots are rational

(b) if one root is $p + \sqrt{q}$ then the other is $-p + \sqrt{q}$

(c) roots are irrational

(d) if one root is $p + \sqrt{q}$ then the other is $p - \sqrt{q}$

130. If α, β are the real roots of $x^2 + px + q = 0$ and α^4, β^4 are the roots of $x^2 - rx + s = 0$ then the equation $x^2 - 4qx + 2q^2 - r = 0$ has always

(a) two real roots

(b) two negative roots

(c) two positive roots

(d) one positive root and one negative root

131. The equation $x^{3/4(\log_2 x)^2 + \log_2 x - 5/4} = \sqrt{2}$ has

(a) at least one negative solution

(b) exactly one irrational solution

(c) exactly three real solutions

(d) two nonreal complex roots

132. If a, b, c are rational and no two of them are equal then the equations

$$(b-c)x^2 + (c-a)x + a - b = 0$$

and $$a(b-c)x^2 + b(c-a)x + c(a-b) = 0$$

(a) have rational roots

(b) will be such that at least one has rational roots

(c) have exactly one root common

(d) have at least one root common

133. The equations $x^2 + b^2 = 1 - 2bx$ and $x^2 + a^2 = 1 - 2ax$ have one and only one root common. Then

(a) $a - b = 2$ (b) $a - b + 2 = 0$

(c) $|a - b| = 2$ (d) none of these

134. If $px^2 + qx + r = 0$ has no real roots and p, q, r are real such that $p + r > 0$ then

(a) $p - q + r < 0$ (b) $p - q + r > 0$

(c) $p + r = q$ (d) all of these

135. Let p and q be roots of the equation $x^2 - 2x + A = 0$, and let r and s be the roots of the equation $x^2 - 18x + B = 0$. If $p < q < r < s$ are in arithmetic progression then

(a) $A = -83, B = -3$ (b) $A = -3, B = 77$

(c) $q = 3, r = 7$ (d) $p + q + r + s = 20$

136. The quadratic equation $x^2 - 2x - \lambda = 0, \lambda \neq 0,$

(a) cannot have a real root if $\lambda < -1$

(b) can have a rational root if λ is a perfect square

(c) cannot have an integral root if $n^2 - 1 < \lambda < n^2 + 2n$ where $n = 0, 1, 2, 3, \ldots$

(d) none of these

137. A quadratic equation whose roots are $\left(\frac{\gamma}{\alpha}\right)^2$ and $\left(\frac{\beta}{\alpha}\right)^2$, where α, β, γ are the roots of $x^3 + 27 = 0$, is

(a) $x^2 - x + 1 = 0$ (b) $x^2 + 3x + 9 = 0$

(c) $x^2 + x + 1 = 0$ (d) $x^2 - 3x + 9 = 0$

138. The graph of the curve $x^2 = 3x - y - 2$ is

(a) between the lines $x = 1$ and $x = \frac{3}{2}$

(b) between the lines $x = 1$ and $x = 2$

(c) strictly below the line $4y = 1$

(d) none of these

139. $a(x^2 - y^2) + \lambda\{x(y+1)+1\}$ can be resolved into linear rational factors. Then

(a) $\lambda = 1$ (b) $\lambda = \frac{4a^2}{a-1}, a \neq 1$

(c) $\lambda = 0, a = 1$ (d) none of these

140. $x^2 - 4$ is a factor of $f(x) = (a_1x^2 + b_1x + c_1) \cdot (a_2x^2 + b_2x + c_2)$ if

(a) $b_1 = 0, c_1 + 4a_1 = 0$

(b) $b_2 = 0, c_2 + 4a_2 = 0$

(c) $4a_1 + 2b_1 + c_1 = 0, 4a_2 + c_2 = 2b_2$

(d) $4a_1 + c_1 = 2b_1, 4a_2 + 2b_2 + c_2 = 0$

141. $ax^2 + by^2 + cz^2 + 2ayz + 2bzx + 2cxy$ can be resolved into linear factors if a, b, c are such that

(a) $a = b = c$ (b) $ab + bc + ca = 1$

(c) $a + b + c = 0$ (d) none of these

142. If a, b are the real roots of $x^2 + px + 1 = 0$ and c, d are the real roots of $x^2 + qx + 1 = 0$ then $(a-c)(b-c)(a+d)(b+d)$ is divisible by

(a) $a + b + c + d$ (b) $a + b - c - d$

(c) $a - b + c - d$ (d) $a - b - c - d$

143. If $x \in [2, 4]$ then for thc expression $x^2 - 6x + 5$

(a) the least value = –4 (b) the greatest value = 4

(c) the least value = 3 (d) the greatest value = –3

144. If $0 < a < 5$, $0 < b < 5$ and $\frac{x^2+5}{2} = x - 2\cos(a + bx)$ is satisfied for at least one real x then the greatest value of $a + b$ is

(a) π (b) $\frac{\pi}{2}$

(c) 3π (d) 4π

145. Let $f(x) = x^2(x+2) + x + 3$. Then

(a) $f(-3-k) < 0$ and $f(-2+k) > 0$ for all $k > 0$

(b) $f(-3-k) > 0$ and $f(-2+k) < 0$ for all $k > 0$

(c) $f(x) = 0$ has a root α such that $[\alpha] + 3 = 0$, where $[\alpha]$ is the greatest integer less than or equal to α

(d) $f(x) = 0$ has exactly one root α such that $(\alpha) + 2 = 0$, where (α) is the smallest integer greater than or equal to α

Answers

1. c	**2.** a	**3.** c	**4.** a	**5.** d
6. b	**7.** c	**8.** a	**9.** b	**10.** a
11. c.	**12.** b	**13.** c	**14.** a	**15.** b
16. b	**17.** b	**18.** a	**19.** a	**20.** b
21. c	**22.** c	**23.** b	**24.** d	**25.** c
26. b	**27.** a	**28.** d	**29.** b	**30.** d
31. b	**32.** b	**33.** c	**34.** a	**35.** b
36. a	**37.** c	**38.** c	**39.** c	**40.** b
41. a	**42.** b	**43.** c	**44.** a	**45.** c
46. a	**47.** b	**48.** a	**49.** b	**50.** b
51. a	**52.** c	**53.** c	**54.** a	**55.** a
56. a	**57.** d	**58.** b	**59.** c	**60.** b
61. b	**62.** b	**63.** c	**64.** c	**65.** b
66. b	**67.** b	**68.** b	**69.** d	**70.** c
71. b	**72.** a	**73.** b	**74.** c	**75.** b
76. c	**77.** b	**78.** b	**79.** a	**80.** c
81. c	**82.** b	**83.** a	**84.** d	**85.** d
86. b	**87.** c	**88.** b	**89.** a	**90.** c
91. a	**92.** b	**93.** a	**94.** c	**95.** a
96. d	**97.** c	**98.** d	**99.** d	**100.** b
101. c	**102.** b	**103.** a	**104.** a	**105.** c
106. b	**107.** a	**108.** c	**109.** a	**110.** b
111. c	**112.** b	**113.** a, c	**114.** a, c	**115.** b, c
116. a, c	**117.** b, c	**118.** c, d	**119.** b	**120.** b, c
121. a, b	**122.** a, b	**123.** c	**124.** a, b, c, d	**125.** a, b
126. a, c	**127.** a, b	**128.** b, c	**129.** b, c	**130.** a, d
131. b, c	**132.** a, c	**133.** a, b, c	**134.** b	**135.** b, c, d
136. a, c	**137.** c	**138.** c	**139.** b, c	**140.** a, b, c, d
141. a, c	**142.** a, b	**143.** a, d	**144.** c	**145.** a, c, d

Hints and Solutions to Selected Questions

1. Using $2\left(-\frac{1}{2}x^2\right) = x(x^2+1)+6$, we get $x^3+x^2+x+6=0$. By trial $x=-2$ satisfies it.

$\therefore \quad (x+2)(x^2-x+3)=0 \quad \Rightarrow \quad x=-2$ because $x^2-x+3=0$ has $D<0$.

$\therefore$ the given terms are $-10, -2, 6$, where common difference $= 8$.

2. $x^2-4 \neq 0 \Rightarrow x \neq 2$. But the given equation implies that if $x \neq 2$ then $x=2$. This is a contradiction, so x has no value.

3. It is an identity in x if $a^2-3a+2=0$, $a^2-5a+6=0$, $a^2-4=0$ hold at the same time.

5. $a(1-2\sin^2 x)+b\sin^2 x+c=0$, i.e., $(b-2a)\sin^2 x+(a+c)=0$.

It is an identity if $b-2a=0$, $a+c=0$. So $\frac{a}{1}=\frac{b}{2}=\frac{c}{-1}$.

6. $D=b^2-4ac$, $D'=d^2+4ac \quad \Rightarrow \quad D+D'=b^2+d^2>0$

$\therefore$ at least one of D, D' is positive.

7. Use $f\left(-\frac{1}{2}\right)=5$ and $f'\left(\frac{1}{3}\right)=0$.

8. $x^{3^n}+(-x)^{3^n}=0$ if $n \geq 0$ and n is an integer.

9. Can hold if $x+y=2x-y$, $x-2y=0$, $x-y-1=0$.

10. $\left(\frac{\sqrt{2}}{\sqrt{5+2\sqrt{2}}}\right)^x+\left(\frac{\sqrt{2}+1}{\sqrt{5+2\sqrt{2}}}\right)^x=1$ which is of the form

$\cos^x\alpha+\sin^x\alpha=1$; $\therefore$ $x=2$.

11.

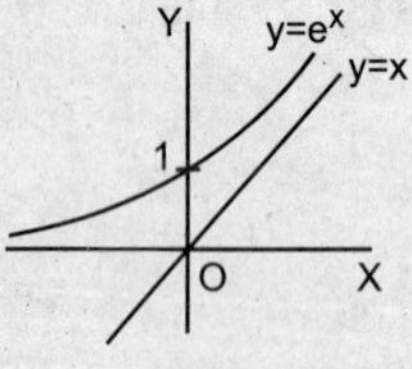

Clearly, $e^x > x$ for all x.

13. If $x = n \in \mathbb{Z}$, the equation is $2n - 2[n] = 1$ or $2n - 2n = 1$ (impossible).

If $x = n + k,\ n \in \mathbb{Z},\ 0 < k < 1$ then the equation is

$$2(n+k) - 2[n+k] = 1$$

or $\quad 2n + 2k - 2n = 1 \quad$ or $\quad k = \frac{1}{2}$ and $n \in \mathbb{Z}$.

14. $5^x + 5^{-x} \geq 2$ ($\because$ AM $\geq$ GM); but $\sin e^x \leq 1$.

15. $2 + |e^x - 1| = (e^x)^2 - 2e^x + 1 = |e^x - 1|^2$

$\therefore \quad |e^x - 1|^2 - |e^x - 1| - 2 = 0$

or $\quad |e^x - 1| = 2, -1 \quad \Rightarrow \quad |e^x - 1| = 2$

$\Rightarrow \quad e^x - 1 = 2, -2 \quad \Rightarrow \quad e^x = 3, -1 \quad \Rightarrow \quad e^x = 3.$

16. $\sin^2 \frac{x}{2} \cdot \cos^2 x = \dfrac{x + \frac{1}{x}}{2} \geq \sqrt{x \cdot \frac{1}{x}} = 1$, equality will hold when $x = \frac{1}{x}$.

$\therefore\ \sin^2 \frac{x}{2} \cdot \cos^2 x = 1$ and then $x = \frac{1}{x}$, i.e., $x = 1$ for which the equation is not satisfied.

17. $xy + y^2 + x = \frac{y}{2},\ x^2 + xy = -\frac{y}{2}$.

Adding, $(x+y)^2 + x = 0$, and subtracting $y^2 - x^2 = y - x$.

Solving these equations, $x = -1,\ y = 2$ and $x = -\frac{1}{4} = y$.

18.

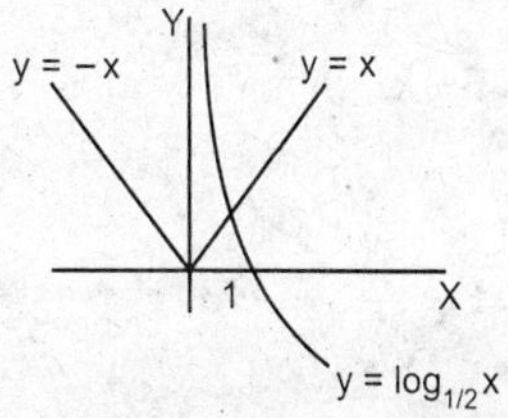

It is clear from the graph that there is only one point of intersection of the curves $y = |x|$ and $y = \log_{1/2} x$.

So, there is only one real solution.

20.

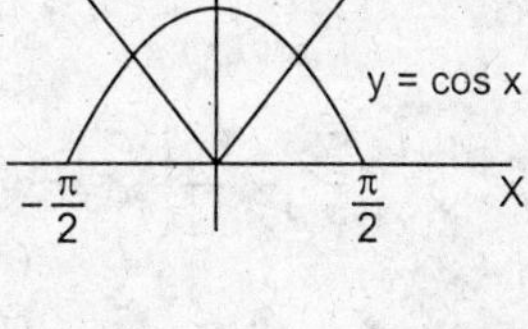

Clearly, there are two points of intersection of $y = |x|$ and $y = \cos x$.

Hence, there are two real solutions.

22. $\left[\frac{x}{2}\right]+\left[\frac{x}{3}\right]+\left[\frac{x}{5}\right]=\frac{31}{30}x=\frac{x}{2}+\frac{x}{3}+\frac{x}{5}$

$\Rightarrow \quad \frac{x}{2}, \frac{x}{3}, \frac{x}{5}$ are all integers. So, x = multiple of the LCM of 2, 3, 5

$\therefore \quad x = 30\times 1, 30\times 2, 30\times 3, \dots, 30\times 33.$

23. If $x=n\in Z, n^2+(n+1)^2=25$ or $2n^2+2n-24=0$

or $n^2+n-12=0 \quad \therefore\ n=3,-4 \quad \therefore\ x=3,-4.$

If $x=n+k, n\in Z, 0<k<1$ then $(n+1)^2+(n+2)^2=25$

or $2n^2+6n-20=0$ or $n^2+3n-10=0 \quad \therefore\ n=2,-5$

$\therefore \quad x=2+k, -5+k,$ where $0<k<1.$

24. $3^{x+1}=2^{\log_2 3}\times 3^{\log_2 3}=3\times 3^{\log_2 3} \quad \therefore\ 3^x=3^{\log_2 3}.$

25. $2^{x/2}+3^{x/2}=(\sqrt{13})^{x/2} \quad \Rightarrow \quad \left(\frac{2}{\sqrt{13}}\right)^{x/2}+\left(\frac{3}{\sqrt{13}}\right)^{x/2}=1$

which is of the form $\cos^{x/2}\alpha+\sin^{x/2}\alpha=1. \quad \therefore\ \frac{x}{2}=2.$

28. If $x=n\in Z, n^2+n^2>25$. So, $n^2>\frac{25}{2}$

$\therefore \quad x=n=4, 5, 6, \dots$ or $-4, -5, -6, \dots$

If $x=n+k, n\in Z, 0<k<1$ then $n^2+(n+1)^2>25$

or $n^2+n-12>0$

$\therefore \quad n<-4$ or $n>3 \quad \therefore\ x<-4+k$ or $x>3+k$, where $0<k<1$

$\therefore \quad x\le -4$ or $x\ge 4.$

29. $\{(x-1)\}^2=\{(x)-1\}^2=(x)^2-2(x)+1$

$\{[x+1]\}^2=\{[x]+1\}^2=[x]^2+2[x]+1$

$\therefore$ equation gives $[x]-(x)+1=0.$

If $x=n\in Z, n-n+1=0$ (absurd).

If $x=n+k, n\in Z, 0<k<1$ then $n-\{n+1\}+1=0.$

This is true for all n and k.

31. $\frac{|x+1|}{|x|}+|x+1|=\frac{|x+1|^2}{|x|} \quad \Rightarrow \quad |x+1|\left\{\frac{1}{|x|}+1-\frac{|x+1|}{|x|}\right\}=0$

$\therefore \quad |x+1|=0$ or $1+|x|-|x+1|=0.$

$|x+1|=0 \Rightarrow x=-1$.

If $x<-1, 1+|x|-|x+1|=0 \Rightarrow 1-x+x+1=0 \Rightarrow 2=0$ (absurd).

If $-1 \le x<0, 1+|x|-|x+1|=0 \Rightarrow 1-x-(x+1)=0 \Rightarrow x=0$ (not possible).

If $x \ge 0, 1+x-(x+1)=0 \Rightarrow 0=0 \Rightarrow x$ can have any value in the interval.

$\therefore \quad x=-1, x>0 \ (\because \ x \ne 0)$.

32. If $x=n \in \mathbb{Z}, |n-2n|=4 \quad \therefore \ n=\pm 4$.

If $x=n+k, n \in \mathbb{Z}, 0<k<1$ then $|n-2(n+k)|=4$.

$\therefore \quad |-n-2k|=4$. It is possible if $k=\frac{1}{2}$.

Then $|-n-1|=4$, i.e., $n+1=\pm 4$

$\therefore \quad n=3, -5$.

34. $||x-1|-1| \le 1 \Rightarrow 0 \le |x-1| \le 2 \Rightarrow |x-1| \le 2 \Rightarrow -1 \le x \le 3$.

37. $\log_{1/3}(x^2+x+1)>-1=\log_{1/3}\left(\frac{1}{3}\right)^3 \Rightarrow x^2+x+1<\left(\frac{1}{3}\right)^{-1}$

$\Rightarrow \quad x^2+x-2<0$. Use sign scheme.

38. $\left(\frac{5}{13}\right)^x+\left(\frac{12}{13}\right)^x \ge 1 \quad \therefore \ \cos^x\alpha+\sin^x\alpha \ge 1$, where $\cos\alpha=\frac{5}{13}$.

Equality holds for $x=2$. If $x<2$, both $\cos\alpha$ and $\sin\alpha$ increase (being positive fractions). So, $\cos^x\alpha+\sin^x\alpha>1$ if $x<2$. Thus $x \le 2$.

39. By trial, $3^{x/2}+2^x \le 25$ for $x=1, 2, 3, 4$. But $3^{x/2}+2^x>25$ for $x>4$.

40. See the hint to Q. No. 38.

41. Solve the inequations $x^2-3x+2 \le 0$ and $2x^2-3x-5 \ge 0$. If S_1, S_2 are the respective solution sets then $S_1 \cap S_2$ is the required solution set.

42. $\frac{x^2-3x+4}{x+1}>1 \Rightarrow \frac{x^2-4x+3}{x+1}>0 \Rightarrow (x-1)(x-3)(x+1)>0, \ x \ne -1$

{multiplying by $(x+1)^2$}

$\therefore \quad$ from general sign scheme:

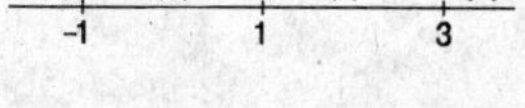

$\{\because$ for $x=0$, expression $>0\}$.

43. $\dfrac{x+2}{x^2+1} > \dfrac{1}{2} \Rightarrow \dfrac{-x^2+2x+3}{x^2+1} > 0 \Rightarrow -x^2+2x+3 > 0$

$\{\because\ x^2+1$ is always positive$\}$.

By sign scheme, $-1 < x < 3$. As x is integer, $x = 0, 1, 2$.

44. $D = c^2(3a^2+b^2)^2 - 4abc^2(-6a^2 - ab + 2b^2) = c^2(3a^2 - b^2 + 4ab)^2$.

46. If $k(> 0)$ be the common difference then the equation is

$$3x^2 - (6a+10k)x + a(a+2k) + 2(a+k)(a+3k) = 0$$

$\therefore \quad D = (6a+10k)^2 - 4\cdot 3\cdot\{a^2+2ak+2a^2+8ak+6k^2\} = 28k^2 > 0.$

47. $a|x|^2 + 2b|x| - c = 0 \quad \therefore\ |x| = \dfrac{-2b \pm \sqrt{4b^2+4ac}}{2} = -b \pm \sqrt{b^2+ac}$.

a, b, c are positive. So, $|x| = -b + \sqrt{b^2+ac}$.

$\therefore \quad x$ has two real values.

48. $(1+\lambda)x^2 - (6+4\lambda)x + 8 + 3\lambda = 0$

$\therefore \quad D = (6+4\lambda)^2 - 4(1+\lambda)(8+3\lambda) = \lambda^2 + \lambda + 1 > 0.$

49. $\sin^2\phi = \cos\theta\cdot\sin\theta \quad \therefore\ \cos 2\phi = 1 - \sin 2\theta = (\cos\theta - \sin\theta)^2$.

$D = 4\cot^2\phi - 4 = 4\cdot\dfrac{\cos 2\phi}{\sin^2\phi} = 4\cdot\left(\dfrac{\cos\theta - \sin\theta}{\sin\phi}\right)^2 > 0$ because

$\cos\theta = \sin\theta \quad \Rightarrow \quad$ three numbers are equal which is a special case.

50. $b^2 - 4ac < 0$ and $a + c < b$.

$b^2 - 4ac < 0 \quad \Rightarrow \quad a, c$ must be of the same sign.

If a, c are both positive, $(a+c)^2 < b^2$.

$\therefore\ b^2 - 4ac < 0 \quad \Rightarrow \quad (a+c)^2 - 4ac < 0$

$\Rightarrow \quad (a-c)^2 < 0$ which is absurd.

So a, c will both be negative. For example, $-x^2 + 4x - 5 = 0$.

It satisfies, $D < 0$ and $a + c < b$. Also $4a + c = -4 - 5 = -9 < 8 = 2b$.

Clearly, when a, c are negative, b must be positive. Otherwise, the equation becomes one where coefficients are all positive.

51. The sum of the coefficients $= (a+2) + (a-3) - (2a-1) = 0$. So, $x = 1$ is a root.

$\therefore$ other root $= -\dfrac{2a-1}{a+2}$ = rational for all a, $a \neq -2$.

52. Let $3^{\tan x} = y$. Then $ay + \dfrac{a}{y} - 2 = 0$ or $ay^2 - 2y + a = 0$.

$D \geq 0 \Rightarrow 4 - 4a^2 \geq 0$. Also roots are positive as $y = 3^{\tan x} > 0$.

$\therefore$ sum of the roots $= \dfrac{2}{a} > 0 \Rightarrow a > 0$.

53. $\alpha + \beta = \dfrac{3a}{a-1}$, $\alpha\beta = \dfrac{1}{a-1}$. As $a > 1$, $\alpha + \beta > 0$ and $\alpha\beta > 0$.

55. $\alpha - 3 < 0$, $\beta - 3 < 0$ and $D \geq 0$

$\Rightarrow \alpha + \beta - 6 < 0$, $\alpha\beta - 3(\alpha + \beta) + 9 > 0$ and $4a^2 - 4(a^2 + a - 3) \geq 0$

$\Rightarrow 2a - 6 < 0$, $a^2 + a - 3 - 3 \cdot 2a + 9 > 0$ and $3 - a \geq 0$.

Solve these and find the common value.

56. $\alpha - 1 < 0$, $\beta - 1 > 0$ and $D > 0$

$\Rightarrow (\alpha - 1)(\beta - 1) < 0$ and $D > 0 \Rightarrow \alpha\beta - (\alpha + \beta) + 1 < 0$ and $D > 0$.

57. $\alpha - 1 > 0, \beta - 1 > 0 \Rightarrow \alpha + \beta - 2 > 0$ and $\alpha\beta - (\alpha + \beta) + 1 > 0$.

$\alpha - 2 < 0, \beta - 2 > 0 \Rightarrow \alpha\beta - 2(\alpha + \beta) + 4 < 0$.

$\alpha - 3 < 0, \beta - 3 < 0 \Rightarrow \alpha + \beta - 6 < 0$ and $\alpha\beta - 3(\alpha + \beta) + 9 > 0$.

Also $D > 0$. Thus obtain six inequations and solve them together.

58. $\alpha > 0, \beta > 0, D_1 \geq 0 \Rightarrow \alpha + \beta > 0, \alpha\beta > 0, D_1 \geq 0$.

$\gamma < 0, \delta < 0, D_2 \geq 0 \Rightarrow \gamma + \delta < 0, \gamma\delta > 0, D_2 \geq 0$.

$\therefore a - 1 > 0, 3 > 0, (a-1)^2 - 12 \geq 0$

and $-3 < 0, 6 - a > 0, 9 - 4(6 - a) \geq 0$.

Solve the inequations together.

59. $\alpha - p > 0$, $\beta - p > 0$ and $D \geq 0$

$\Rightarrow \alpha + \beta - 2p > 0$, $\alpha\beta - p(\alpha + \beta) + p^2 > 0$ and $p^2 + 4p^2 \geq 0$

$\therefore p - 2p > 0, -p^2 - p \cdot p + p^2 > 0 \Rightarrow p < 0, p^2 < 0$ (absurd).

60. $\cos^4 x - \cos^2 x + 1 - p = 0$; as $0 \leq \cos^2 x \leq 1$, the roots of $y^2 - y + 1 - p = 0$ lie between 0 and 1.

$\therefore \alpha \geq 0, \beta \geq 0, \alpha - 1 \leq 0, \beta - 1 \leq 0$ and $D \geq 0$

$\Rightarrow \quad \alpha+\beta \geq 0, \alpha\beta \geq 0, \alpha+\beta-2 \leq 0, \alpha\beta-(\alpha+\beta)+1 \geq 0$ and $D \geq 0$

$\therefore \quad 1 \geq 0, 1-p \geq 0, 1-2 \leq 0, 1-p-1+1 \geq 0$ and $1-4(1-p) \geq 0.$

67. $4^{\log_2 \lambda} = 8 \quad \Rightarrow \quad 2^{2\log_2 \lambda} = 2^3 \quad \Rightarrow \quad 2\log_2 \lambda = 3$

$\Rightarrow \quad \lambda^2 = 2^3 \quad \Rightarrow \quad \lambda = 2\sqrt{2} \quad (\because \lambda > 0).2 \quad 2$

68. Roots of the second equation are reciprocal of those of the first.

$\therefore$ $c_1x^2 + b_1x + a_1 = 0$ and $a_2x^2 + b_2x + c_2 = 0$ have both roots common.

70. Let α, β be the roots of $ax^2 - bx - c = 0$ and let α', β' be the roots of $a'x^2 - b'x - c' = 0$ such that $|\alpha - \beta| = |\alpha' - \beta'|$, i.e.,

$$(\alpha+\beta)^2 - 4\alpha\beta = (\alpha'+\beta')^2 - 4\alpha'\beta'$$

$$\Rightarrow \quad \frac{b^2+4ac}{a^2} = \frac{b'^2+4a'c'}{a'^2}.$$

Hence, the expression $\dfrac{b^2+4ac}{a^2}$ does not vary in value.

75. $\sum_{r=1}^{n} s_r = (\alpha+\beta) + (\alpha^2+\beta^2) + \ldots + (\alpha^n + \beta^n)$

$$= (\alpha + \alpha^2 + \ldots + \alpha^n) + (\beta + \beta^2 + \ldots + \beta^n)$$

$$\therefore \quad \lim_{n\to\infty} \sum_{r=1}^{n} s_r = (\alpha + \alpha^2 + \alpha^3 + \ldots \text{ to } \infty) + (\beta + \beta^2 + \beta^3 + \ldots \text{ to } \infty)$$

$$= \frac{\alpha}{1-\alpha} + \frac{\beta}{1-\beta} \text{ (clearly } |\alpha| < 1, |\beta| < 1).$$

77. Clearly α, β are the roots of the equation $x^2 - 5x + 3 = 0$. Use $\alpha + \beta = 5$, $\alpha\beta = 3$.

80. $2x^3 + 6x^2 + x + 2 = 0$ has roots α, β, γ.

So, $2x^3 + x^2 + 6x + 2 = 0$ has roots $\alpha^{-1}, \beta^{-1}, \gamma^{-1}$ (writing coefficients in reverse order).

$$\therefore \quad \alpha^{-1} + \beta^{-1} + \gamma^{-1} = -\frac{1}{2}.$$

81. If roots are $\alpha - k, \alpha, \alpha + k$ then

$$\alpha - k + \alpha + \alpha + k = 12 \quad \Rightarrow \quad \alpha = 4,$$

$$(\alpha - k)\alpha(\alpha + k) = 28 \quad \Rightarrow \quad (4^2 - k^2)4 = 28 \quad \Rightarrow \quad k^2 = 9.$$

82. Put $x = -2$; then $(-2)^3 + 14(-2)^2 - 84(-2) - 216 = 0$

$\therefore$ the equation is $(x+2)(x^2 + 12x - 108) = 0$

or $(x+2)(x-6)(x+18) = 0.$

Roots are $-2, 6, -18$.

85. Let $f(x) = a^2x^2 + 2bx + 2c$. From the question, $a^2\alpha^2 + b\alpha + c = 0$ and $a^2\beta^2 - b\beta - c = 0$.

Now, $f(\alpha) = a^2\alpha^2 + 2b\alpha + 2c = b\alpha + c = -a^2\alpha^2$

$$f(\beta) = a^2\beta^2 + 2b\beta + 2c = 3b\beta + 3c = 3(b\beta + c) = 3a^2\beta^2.$$

Clearly, $0 < \alpha < \beta \quad \Rightarrow \quad \alpha, \beta$ are real. So $f(\alpha) < 0, f(\beta) > 0$.

So, $f(\gamma) = 0$ where $\alpha < \gamma < \beta$.

86. Take the equation $2ax^3 + 3bx^2 + 6cx = 0$. Clearly, $x = 0, 1$ are two roots. Observe that $ax^2 + bx + c = 0$ is the derived equation. So, one root lies between 0 and 1.

89. As the coefficients are in reverse order, the roots of $ax^2 + bx + c = 0$ are α, β, while the roots of $cx^2 + bx + a = 0$ are $\frac{1}{\alpha}, \frac{1}{\beta}$.

One negative root is common

$\Rightarrow$ (i) $\alpha = \frac{1}{\alpha} < 0$; so, $\alpha = -1$

or (ii) $\alpha = \frac{1}{\beta} < 0 \quad \Rightarrow \quad \alpha\beta = 1 \quad \Rightarrow \quad \frac{c}{a} = 1 \quad \Rightarrow \quad c = a$ (not possible).

90. As all the coefficients are not real, one common root does not imply that the other root is also common.

Let α be the common root. Then $\alpha^2 + i\alpha + a = 0, \alpha^2 - 2\alpha + ia = 0$

$$\Rightarrow \quad \frac{\alpha^2}{a} = \frac{\alpha}{a(1-i)} = \frac{1}{-2-i} \quad \Rightarrow \quad a^2(1-i)^2 = a(-2-i).$$

91. $x^2 - 2p_1x + 1 = 0$ has roots α, β $\qquad x^2 - 4p_2x + 2 = 0$ has roots α, γ

$x^2 - 6p_3x + 3 = 0$ has roots β, γ

$\Rightarrow \quad \alpha + \beta = 2p_1, \alpha + \gamma = 4p_2, \beta + \gamma = 6p_3, \alpha\beta = 1, \alpha\gamma = 2, \beta\gamma = 3.$

$\therefore \quad \gamma - \beta = 4p_2 - 2p_1$ and $\alpha(\gamma - \beta) = 1 \quad \Rightarrow \quad \alpha = \dfrac{1}{4p_2 - 2p_1}$.

Also $\dfrac{\beta}{\gamma} = \dfrac{1}{2} \quad \therefore \ 2\beta = \gamma \quad \therefore \ \beta \cdot 2\beta = 3 \quad \therefore \ \beta = \pm\sqrt{\dfrac{3}{2}}$

$$\therefore \quad \frac{1}{4p_2 - 2p_1}\left(\pm\sqrt{\frac{3}{2}}\right) = 1 \qquad \ldots (1)$$

Again $\dfrac{1}{\alpha} + \dfrac{1}{\beta} = 2p_1, \dfrac{1}{\alpha} + \dfrac{1}{\gamma} = 2p_2, \dfrac{1}{\beta} + \dfrac{1}{\gamma} = 2p_3$

$\therefore \quad \dfrac{1}{\alpha} + \dfrac{1}{\beta} + \dfrac{1}{\gamma} = p_1 + p_2 + p_3$

$\therefore \quad \dfrac{1}{\alpha} = p_1 + p_2 - p_3, \dfrac{1}{\beta} = p_1 - p_2 + p_3, \dfrac{1}{\gamma} = -p_1 + p_2 + p_3.$

So, $\dfrac{\gamma}{\beta} = 2 = \dfrac{p_1 - p_2 + p_3}{-p_1 + p_2 + p_3}$ and $\dfrac{\gamma}{\alpha} = 3 = \dfrac{-p_1 + p_2 + p_3}{p_1 + p_2 - p_3}$.

Thus, we get $\dfrac{p_1}{7} = \dfrac{p_2}{4} = \dfrac{p_3}{9} = t$. Using this in (1), we get two values of t.

92. $(\lambda^2 + \lambda - 2)x^2 + (\lambda + 2)x - 1 < 0$ for all $x \in R$ if

$D < 0$ and $a < 0$, i.e., $(\lambda + 2)^2 + 4(\lambda^2 + \lambda - 2) < 0$ and $\lambda^2 + \lambda - 2 < 0$.

94. $5^x + (2\sqrt{3})^{2x} > 13^2 \quad \Rightarrow \quad 5^x + 12^x > 13^2$.

We know $5^2 + 12^2 = 13^2$. So for $x > 2, 5^x + 12^x > 5^2 + 12^2 = 13^2$.

95. $(2^x)^2 - (a-4)2^x - 2^x + (a-4) = 0 \Rightarrow (2^x - a + 4)(2^x - 1) = 0$

$\therefore \quad 2^x = 1, a - 4$. As $x \le 0, 0 < a - 4 \le 1$.

97. Here $xy \ge 100$. Now, $(x+y)^2 = (x-y)^2 + 4xy \ge 400 + (x-y)^2 \ge 400$

$\therefore$ the smallest possible value of $x + y$ is 20 $\quad (\because x > 0, y > 0)$.

98. $f(x) = \dfrac{x^2 + 1 - 2}{x^2 + 1} = 1 - \dfrac{2}{x^2 + 1}$; $f(x)$ is minimum when $\dfrac{2}{x^2 + 1}$ is maximum, i.e., 2.

99. Here $b^2 - 24a \le 0$. Let $3a + b = \lambda$.

Then $b^2 - 8(\lambda - b) \le 0$ or $b^2 + 8b - 8\lambda \le 0$.

Since b is real, therefore, $64 + 32\lambda \ge 0 \quad \therefore \ \lambda \ge -2$.

100. $\frac{2a+3b}{2} \geq \sqrt{2a \cdot 3b} \quad \Rightarrow \quad \frac{ab}{2} \geq \sqrt{6ab}$

$\Rightarrow \quad \sqrt{ab} \geq 2\sqrt{6} \quad \Rightarrow \quad ab \geq 24.$

101. It is a factor if $ax^3 + bx + c = (x^2 + px + 1)(ax + c)$

$= ax^3 + (c + ap)x^2 + (pc + a)x + c$

$\Rightarrow \quad c + ap = 0, pc + a = b$

$\Rightarrow \quad p = \frac{-c}{a} = \frac{b-a}{c}.$

102. Let $f(x) = x^4 + ax^3 + 3x - b$. Then $f(1) = 0, f(-1) = 0$.

103. $x^2 - (k-2)x + k^2 = 0$ and $x^2 + kx + 2k - 1 = 0$ should have both roots common or each should have equal roots.

$\therefore \quad$ (i) $\frac{1}{1} = \frac{-(k-2)}{k} = \frac{k^2}{2k-1}$

or (ii) $(k-2)^2 - 4k^2 = 0$ and $k^2 - 4(2k-1) = 0$.

(i) $\Rightarrow k = -k + 2, 2k - 1 = k^2$. So $k = 1$

(ii) $\Rightarrow (3k-2)(-k-2) = 0, k^2 - 8k + 4 = 0$. So, no value of k is possible.

104. $6x^2 - xy - y^2 - 6x + 8y - 12 = 6(x + \lambda y - 2)(x - \mu y + 1)$.

Equate coefficients and solve for λ, μ.

105. $x - y$ is a factor $\Rightarrow$ putting $y = x$, expression $= 0$ which gives $1 - 3 + \lambda + \mu = 0$.

$y - 2x$ is a factor $\Rightarrow$ putting $y = 2x$, expression $= 0$ which gives $1 - 6 + 4\lambda + 8\mu = 0$.

106. As it is a third-degree homogeneous expression in x, y, we have $y^3 + y^2x - \mu yx^2 + \lambda x^3$

$= (y + x)(y + 3x)(y + mx)$

$= y^3 + (m + 3 + 1)y^2x + (3 + m + 3m)yx^2 + 3mx^3$

$\Rightarrow \quad 1 = m + 4, -\mu = 3 + 4m, \lambda = 3m.$

107. $f(x, y) = \frac{1}{2}[(x-2y)^2 + (2y-3z)^2 + (3z-x)^2] \geq 0.$

108. Let $xy + yz + zx = \lambda$. Then

$$x^2 + y^2 + z^2 - \lambda = \frac{1}{2}[(x-y)^2 + (y-z)^2 + (z-x)^2] \geq 0 \quad \Rightarrow \quad 1 - \lambda \geq 0.$$

Again, $(x+y+z)^2 = x^2 + y^2 + z^2 + 2\lambda = 1 + 2\lambda \quad \Rightarrow \quad 1 + 2\lambda \geq 0$

109. For $ax^2 - bxy - ay^2$, $D = b^2 + 4a^2 > 0$.

110. As $a(b-c) + b(c-a) + c(a-b) = 0$, $x = 1$ is a root of the corresponding equation. The other root of the equation

$$= \frac{c(a-b)}{a(b-c)} = 1 \text{ because } a, b, c \text{ in HP implies } \frac{1}{b} - \frac{1}{a} = \frac{1}{c} - \frac{1}{b}, \text{ i.e., } \frac{a-b}{a} = \frac{b-c}{c}$$

$\therefore$ $x = 1, 1$ are the roots of the corresponding equation. So, $(x-1)^2$ is a factor.

111. Let $f(x) = x^4 + ax^3 + bx^2 + cx - 1$.

$(x-1)^3$ is a factor $\quad \Rightarrow \quad f(1) = 0, f'(1) = 0, f''(1) = 0$.

This gives a, b, c.

113. Let α, α be equal integral roots. So $\alpha + \alpha = b, \alpha \cdot \alpha = c$

$\therefore \quad b = 2\alpha, c = \alpha^2$. Also $\alpha = \sqrt{c} = \text{real} \quad \Rightarrow \quad c > 0$.

115. Let α, β be the roots. $\alpha + \beta = \frac{|G|}{A}$, $\alpha\beta = \frac{-H}{A}$.

As $A > |G| > H$,

$\alpha + \beta =$ positive fraction ... (1)

and $\alpha\beta =$ negative fraction ... (2)

Also $AH = |G|^2$, $D = |G|^2 + 4AH > 0 \quad \Rightarrow$ roots are real and unequal.

From (2), at least one root is a fraction and of the two roots one root is positive while the other is negative. Also, (1) $\Rightarrow$ the positive root must be bigger in magnitude. So the equation has a negative fractional root.

117. $\alpha + \beta = 2p$, $\alpha\beta = q^2$.

$\therefore$ AM of roots $= p$, GM of roots $= |q|$, HM of roots $= \frac{2\alpha\beta}{\alpha + \beta} = \frac{2q^2}{2p}$

$\therefore$ cubic equation is $(x-p)(x-|q|)\left(x - \frac{q^2}{p}\right) = 0$.

119. Solving the equations, $x^3 + x^2 + x - 14 = 0$

$\Rightarrow \quad (x-2)(x^2+3x+7) = 0 \qquad \therefore \; x = 2, \dfrac{-3 \pm \sqrt{9-28}}{2}.$

124. $y = xr, z = xr^2$ and $z + 3x > 4y \quad \Rightarrow \quad xr^2 + 3x > 4xr$

$\Rightarrow \quad r^2 - 4r + 3 > 0 \, (\because \; x > 0) \qquad \therefore \; r < 1 \;$ or $\; r > 3.$

125. $|x-1| + a = \pm 4 \quad \Rightarrow \quad |x-1| = 4 - a$ or $\; -4 - a.$

$\therefore \quad$ for real x, $4 - a \geq 0$ or $\; -4 - a \geq 0.$

126. $|x^2 - 1| = a^2 - 2a - 3$

$\Rightarrow \quad x^2 - 1 = \pm (a^2 - 2a - 3)$ provided $a^2 - 2a - 3 \geq 0$ ... (1)

$\Rightarrow \quad x^2 = a^2 - 2a - 2, -a^2 + 2a + 4.$

For real solution, $a^2 - 2a - 2 \geq 0$ ... (2)

or $\quad -a^2 + 2\alpha + 4 \geq 0$ ... (3)

If S_1, S_2, S_3 are solution sets of (1), (2), (3) respectively then for real solution, $a \in S_1 \cap S_2$ or $a \in S_1 \cap S_3$.

127. $(x^3 + 1) + 2x(x+1) = 0$ or $\; (x+1)\{x^2 + x + 1\} = 0 \; \Rightarrow \; x = -1, \omega, \omega^2.$

Of these $x = \omega, \omega^2$ satisfy the equation $1 + x^{130} + x^{1988} = 0.$

128. If β be the other root then $\alpha + \beta = -\dfrac{1}{2}, \alpha\beta = -\dfrac{1}{4}\cdot$

Also, $4\alpha^2 + 2\alpha - 1 = 0$ ... (1)

So, $\beta = -\dfrac{1}{2} - \alpha = -\dfrac{1 + 2\alpha}{2}$

$= -\dfrac{4\alpha^2 + 2\alpha + 2\alpha}{2} = -2\alpha(\alpha + 1)$

$= -2\alpha^2 - 2\alpha = \alpha \cdot (4\alpha^2 - 1) - 2\alpha$, from (1).

129. As the coefficients are not rational, irrational roots need not appear in conjugate pair.

Here $\alpha + \beta = -3\sqrt{2}, \alpha\beta = \dfrac{1}{2}\cdot$ Let $\alpha = p + \sqrt{q}$. Then prove that

$\beta = -p + \sqrt{q}.$

130. $\alpha + \beta = -p, \alpha\beta = q, \alpha^4 + \beta^4 = r, \alpha^4\beta^4 = s.$

Then $D = 16q^2 - 4(2q^2 - r) = 8q^2 + 4r$

$$= 8\alpha^2\beta^2 + 4(\alpha^4 + \beta^4) = 4(\alpha^2 + \beta^2)^2 > 0.$$

Also, the product of the roots

$$= 2q^2 - r = 2\alpha^2\beta^2 - (\alpha^4 + \beta^4) = -(\alpha^2 - \beta^2)^2 < 0.$$

131. Taking logarithm, $\left(\frac{3}{4}y^2 + y - \frac{5}{4}\right)y = \frac{1}{2}$ where $y = \log_2 x$

or $\quad 3y^3 + 4y^2 - 5y - 2 = 0 \quad$ or $\quad (y-1)(3y^2 + 7y + 2) = 0$

or $\quad (y-1)(3y+1)(y+2) = 0 \quad \Rightarrow \quad \log_2 x = 1, -\frac{1}{3}, -2$

$\Rightarrow \quad x = 2, \sqrt[3]{\frac{1}{2}}, \frac{1}{4}.$

133. $(x+b)^2 = 1 \quad \Rightarrow \quad x = -b \pm 1$

$(x+a)^2 = 1 \quad \Rightarrow \quad x = -a \pm 1$

Clearly, $a \neq b$. So, one and only one root common means,

$-b + 1 = -a - 1$ or $-b - 1 = -a + 1$.

134. Let $\alpha + i\beta, \alpha - i\beta$ be the roots. Then $\alpha^2 + \beta^2 = \frac{r}{p} > 0$. So p, r are of the same sign. Also $p + r > 0$. So p, r are both positive.

If $q < 0, p - q + r > 0$.

If $q > 0, (p+r)^2 - (p-r)^2 = 4pr \geq q^2$ ($\because$ roots are nonreal)

$\therefore \quad (p+r)^2 \geq q^2 + (p-r)^2 \geq q^2 \quad \therefore \quad p + r > q.$

136. $D = 4 + 4\lambda, \quad D < 0 \quad \Rightarrow \quad 4 + 4\lambda < 0 \quad \Rightarrow \quad \lambda < -1.$

$D = 4(1+\lambda) = 4(1+p^2) \neq$ perfect square.

If α is an integral root, $\alpha^2 - 2\alpha - \lambda = 0$

or $\quad \lambda = \alpha^2 - 2\alpha = (\alpha - 1)^2 - 1$

$\therefore \quad \lambda = (\text{integer})^2 - 1$. But if $n^2 - 1 < \lambda < n^2 + 2n$ then

$n^2 - 1 < \lambda < (n+1)^2 - 1$

$\therefore \quad \lambda \neq (\text{integer})^2 - 1.$

137. $x^3 + 27 = 0 \quad \Rightarrow \quad x = (-27)^{1/3} = -3, -3\omega, -3\omega^2$

$\therefore \quad \frac{\gamma}{\alpha} = \omega^2, \frac{\beta}{\alpha} = \omega \quad$ or $\quad \frac{\gamma}{\alpha} = \frac{1}{\omega}, \frac{\beta}{\alpha} = \omega \quad$ or $\quad \frac{\gamma}{\alpha} = \frac{1}{\omega^2}, \frac{\beta}{\alpha} = \frac{1}{\omega}.$

In all the cases, $\frac{\gamma}{\alpha}, \frac{\beta}{\alpha}$ are ω or ω^2.

$\therefore$ the equation is $x^2 - (\omega + \omega^2)x + \omega \cdot \omega^2 = 0$.

138. $y = -x^2 + 3x - 2 = \frac{9}{4} - \left(x^2 - 3x + \frac{9}{4}\right) - 2 = \frac{1}{4} - \left(x - \frac{3}{2}\right)^2 \le \frac{1}{4}$.

140. (i) $x - 2$ is a factor of $a_1x^2 + b_1x + c_1$ and $x + 2$ is a factor of $a_2x^2 + b_2x + c_2$

or (ii) $x + 2$ is a factor of $a_1x^2 + b_1x + c_1$ and $x - 2$ is a factor of $a_2x^2 + b_2x + c_2$.

(i) $\Rightarrow 4a_1 + 2b_1 + c_1 = 0$ and $4a_2 - 2b_2 + c_2 = 0$.

(ii) $\Rightarrow 4a_1 - 2b_1 + c_1 = 0$ and $4a_2 + 2b_2 + c_2 = 0$.

144. $2\cos(a + bx) = x - \frac{x^2 + 5}{2} = \frac{-x^2 + 2x - 5}{2} = \frac{1}{2}[-4 - (x^2 - 2x + 1)]$

$$= \frac{1}{2}\{-4 - (x - 1)^2\} = -2 - \frac{1}{2}(x - 1)^2 \le -2$$

$\Rightarrow$ $\cos(a + bx) = -1$ and it takes place when $a + bx = 2n\pi - \pi$ and $x = 1$.

$\therefore$ $a + b = \pi, 3\pi, \ldots$; but $a + b$ is less than 10.

145. $f(-3 + k) = -(-3 + k)^2 \cdot (1 + k) - k < 0$ if $k > 0$; also $f(-3) < 0$.

$f(-2 + k) = -(k - 2)^2 \cdot k + k + 1 > 0$ if $k > 0$; also $f(-2) > 0$.

$\therefore$ $f(x) = 0$ has a root α lying between -3 and -2.

$\therefore$ $[\alpha] = -3$ and $(\alpha) = -2$.

□

3

Complex Numbers

• Type 1 •

Choose the most appropriate option (a, b, c or d).

1. If $a < 0, b > 0$ then $\sqrt{a} \cdot \sqrt{b}$ is equal to

(a) $-\sqrt{|a| \cdot b}$

(b) $\sqrt{|a| \cdot b} \cdot i$

(c) $\sqrt{|a| b}$

(d) none of these

2. The value of the sum $\sum_{n=1}^{13} (i^n + i^{n+1})$, where $i = \sqrt{-1}$, is

(a) i

(b) $i - 1$

(c) $-i$

(d) 0

3. If n_1, n_2 are positive integers then

$$(1+i)^{n_1} + (1+i^3)^{n_1} + (1+i^5)^{n_2} + (1+i^7)^{n_2}$$

is a real number if and only if

(a) $n_1 = n_2 + 1$

(b) $n_1 + 1 = n_2$

(c) $n_1 = n_2$

(d) n_1, n_2 are any two positive integers

4. The complex number $\dfrac{2^n}{(1+i)^{2n}} + \dfrac{(1+i)^{2n}}{2^n}$, $n \in Z$, is equal to

(a) 0 (b) 2

(c) $\{1+(-1)^n\}\cdot i^n$ (d) none of these

5. The smallest positive integral value of n for which $\left(\frac{1-i}{1+i}\right)^n$ is purely imaginary with positive imaginary part, is

(a) 1 (b) 3

(c) 5 (d) none of these

6. If $(a+ib)^5=\alpha+i\beta$ then $(b+ia)^5$ is equal to

(a) $\beta+i\alpha$ (b) $\alpha-i\beta$

(c) $\beta-i\alpha$ (d) $-\alpha-i\beta$

7. If $i=\sqrt{-1}$, the number of values of i^n+i^{-n} for different $n\in Z$ is

(a) 3 (b) 2

(c) 4 (d) 1

8. Im(z) is equal to

(a) $\frac{1}{2}(z+\bar{z})i$ (b) $\frac{1}{2}(z-\bar{z})$

(c) $\frac{1}{2}(\bar{z}-z)i$ (d) none of these

9. The value of $(1+i)^3+(1-i)^6$ is

(a) i (b) $2(-1+5i)$

(c) $1-5i$ (d) none of these

10. Taking the value of a square root with positive real part only, the value of $\sqrt{-3-4i}+\sqrt{3+4i}$ is

(a) $1+i$ (b) $1-3i$

(c) $1+3i$ (d) none of these

11. $\sin^{-1}\left\{\frac{1}{i}(z-1)\right\}$, where z is nonreal, can be the angle of a triangle if

(a) Re(z) = 1, Im(z) = 2 (b) Re(z) = 1, $-1\le$ Im(z) ≤ 1

(c) Re(z) + Im(z) = 0 (d) none of these

12. If n is an odd integer, $i=\sqrt{-1}$ then $(1+i)^{6n}+(1-i)^{6n}$ is equal to

(a) 0 (b) 2

(c) −2 (d) none of these

13. If $z_1 = 9y^2 - 4 - 10ix$, $z_2 = 8y^2 - 20i$, where $z_1 = \overline{z}_2$, then $z = x + iy$ is equal to

(a) $-2 + 2i$ (b) $-2 \pm 2i$

(c) $-2 \pm i$ (d) none of these

14. The complex numbers $\sin x - i\cos 2x$ and $\cos x - i\sin 2x$ are conjugate to each other for

(a) $x = n\pi$ (b) $x = 0$

(c) $x = (2n+1)\frac{\pi}{2}$ (d) no value of x

15. If $z = 1 + i\tan\alpha$, where $\pi < \alpha < \frac{3\pi}{2}$, then $|z|$ is equal to

(a) $\sec\alpha$ (b) $-\sec\alpha$

(c) $\operatorname{cosec}\alpha$ (d) none of these

16. If z is a complex number satisfying the relation $|z+1| = z + 2(1+i)$ then z is

(a) $\frac{1}{2}(1+4i)$ (b) $\frac{1}{2}(3+4i)$

(c) $\frac{1}{2}(1-4i)$ (d) $\frac{1}{2}(3-4i)$

17. If $(1+i)z = (1-i)\overline{z}$ then z is

(a) $t(1-i), t \in R$ (b) $t(1+i), t \in R$

(c) $\frac{t}{1+i}, t \in R^+$ (d) none of these

18. If z_1, z_2 are two nonzero complex numbers such that

$$|z_1 + z_2| = |z_1| + |z_2| \text{ then amp } \frac{z_1}{z_2} \text{ is equal to}$$

(a) π (b) $-\pi$

(c) 0 (d) $\frac{\pi}{2}$

19. The complex number z is purely imaginary if

(a) $z\overline{z}$ is real (b) $z = \overline{z}$

(c) $z+\bar{z}=0$ (d) none of these

20. If $z=x+iy$ such that $|z+1|=|z-1|$ and amp $\frac{z-1}{z+1}=\frac{\pi}{4}$ then

(a) $x=\sqrt{2}+1, y=0$ (b) $x=0, y=\sqrt{2}+1$

(c) $x=0, y=\sqrt{2}-1$ (d) $x=\sqrt{2}-1, y=0$

21. Let $z=\frac{\cos\theta+i\sin\theta}{\cos\theta-i\sin\theta}, \frac{\pi}{4}<\theta<\frac{\pi}{2}$. Then arg z is

(a) 2θ (b) $2\theta-\pi$

(c) $\pi+2\theta$ (d) none of these

22. If $z=\frac{\sqrt{3}+i}{\sqrt{3}-i}$ then the fundamental amplitude of z is

(a) $-\frac{\pi}{3}$ (b) $\frac{\pi}{3}$

(c) $\frac{\pi}{6}$ (d) none of these

23. If $\frac{1+2i}{2+i}=r(\cos\theta+i\sin\theta)$ then

(a) $r=1, \theta=\tan^{-1}\frac{3}{4}$ (b) $r=\sqrt{5}, \theta=\tan^{-1}\frac{4}{3}$

(c) $r=1, \theta=\tan^{-1}\frac{4}{3}$ (d) none of these

24. If $z=x+iy$ satisfies amp $(z-1)$ = amp $(z+3i)$ then the value of $(x-1):y$ is equal to

(a) 2 : 1 (b) 1 : 3

(c) –1 : 3 (d) none of these

25. Let z be a complex number of constant modulus such that z^2 is purely imaginary then the number of possible values of z is

(a) 2 (b) 1

(c) 4 (d) infinite

26. If ω is an imaginary cube root of unity then $(1+\omega-\omega^2)^7$ equals

(a) 128ω (b) -128ω

(c) $128\omega^2$ (d) $-128\omega^2$

27. If ω is a nonreal cube root of unity then the expression

$(1-\omega)(1-\omega^2)(1+\omega^4)(1+\omega^8)$ is equal to

(a) 0 (b) 3
(c) 1 (d) 2

28. If $3^{49}(x+iy) = \left(\frac{3}{2}+\frac{\sqrt{3}}{2}i\right)^{100}$ and $x = ky$ then k is

(a) $-\frac{1}{3}$ (b) $\sqrt{3}$
(c) $-\sqrt{3}$ (d) $-\frac{1}{\sqrt{3}}$

29. $x^{3m}+x^{3n-1}+x^{3r-2}$, where $m, n, r \in N$, is divisible by

(a) x^2-x+1 (b) x^2+x+1
(c) x^2+x-1 (d) x^2-x-1

30. If $x^2-x+1=0$ then the value of $\sum_{n=1}^{5}\left(x^n+\frac{1}{x^n}\right)^2$ is

(a) 8 (b) 10
(c) 12 (d) none of these

31. If $1+x^2=\sqrt{3}x$ then $\sum_{n=1}^{24}\left(x^n-\frac{1}{x^n}\right)^2$ is equal to

(a) 48 (b) -48
(c) $\pm 48(\omega-\omega^2)$ (d) none of these

32. The smallest positive integral value of n for which $(1+\sqrt{3}i)^{n/2}$ is real is

(a) 3 (b) 6
(c) 12 (d) 0

33. If $i=\sqrt{-1}$, ω = nonreal cube root of unity then

$$\frac{(1+i)^{2n}-(1-i)^{2n}}{(1+\omega^4-\omega^2)(1-\omega^4+\omega^2)}$$ is equal to

(a) 0 if n is even (b) 0 for all $n \in Z$
(c) $2^{n-1}\cdot i$ for all $n \in N$ (d) none of these

34. If $z^2-z+1=0$ then z^n-z^{-n}, where n is a multiple of 3, is

(a) $2(-1)^n$ (b) 0

(c) $(-1)^{n+1}$ (d) none of these

35. If ω is a nonreal cube root of unity then

$$\frac{1+2\omega+3\omega^2}{2+3\omega+\omega^2}+\frac{2+3\omega+\omega^2}{3+\omega+2\omega^2} \text{ is equal to}$$

(a) -1 (b) 2ω

(c) 0 (d) -2ω

36. If $(x-1)^4-16=0$ then the sum of nonreal complex values of x is

(a) 2 (b) 0

(c) 4 (d) none of these

37. If $z_r=\cos\frac{2r\pi}{5}+i\sin\frac{2r\pi}{5}$, $r=0, 1, 2, 3, 4, \ldots$ then $z_1z_2z_3z_4z_5$ is equal to

(a) -1 (b) 0

(c) 1 (d) none of these

38. If $e^{i\theta}=\cos\theta+i\sin\theta$ then for the $\triangle ABC$, $e^{iA}\cdot e^{iB}\cdot e^{iC}$ is

(a) $-i$ (b) 1

(c) -1 (d) none of these

39. If $(\sqrt{3}+i)^n=(\sqrt{3}-i)^n$, $n\in N$ then the least value of n is

(a) 3 (b) 4

(c) 6 (d) none of these

40. If the fourth roots of unity are z_1, z_2, z_3, z_4 then $z_1^2+z_2^2+z_3^2+z_4^2$ is equal to

(a) 1 (b) 0

(c) i (d) none of these

41. If $x^3-1=0$ has the nonreal complex roots α, β then the value of $(1+2\alpha+\beta)^3-(3+3\alpha+5\beta)^3$ is

(a) -7 (b) 6

(c) -5 (d) 0

42. If $i=\sqrt{-1}$ then $4+5\left(-\frac{1}{2}+i\frac{\sqrt{3}}{2}\right)^{334}-3\left(\frac{1}{2}+i\frac{\sqrt{3}}{2}\right)^{365}$ is equal to

(a) $1 - i\sqrt{3}$ (b) $-1 + i\sqrt{3}$

(c) $4\sqrt{3}i$ (d) $-i\sqrt{3}$

43. If $(\sqrt{3} - i)^n = 2^n, n \in Z,$ the set of integers, then n is a multiple of

(a) 6 (b) 10

(c) 9 (d) 12

44. If $z(2 - i2\sqrt{3})^2 = i(\sqrt{3} + i)^4$ then amplitude of z is

(a) $\frac{5\pi}{6}$ (b) $-\frac{\pi}{6}$

(c) $\frac{\pi}{6}$ (d) $\frac{7\pi}{6}$

45. If z is a nonreal root of $\sqrt[7]{-1}$ then $z^{86} + z^{175} + z^{289}$ is equal to

(a) 0 (b) -1

(c) 3 (d) 1

46. If α is nonreal and $\alpha = \sqrt[5]{1}$ then the value of $2^{|1 + \alpha + \alpha^2 + \alpha^{-2} - \alpha^{-1}|}$ is equal to

(a) 4 (b) 2

(c) 1 (d) none of these

47. The value of amp $(i\omega)$ + amp $(i\omega^2)$, where $i = \sqrt{-1}$ and $\omega = \sqrt[3]{1}$ = nonreal, is

(a) 0 (b) $\frac{\pi}{2}$

(c) π (d) none of these

48. If α, β be two complex numbers then $|\alpha|^2 + |\beta|^2$ is equal to

(a) $\frac{1}{2}(|\alpha + \beta|^2 - |\alpha - \beta|^2)$ (b) $\frac{1}{2}(|\alpha + \beta|^2 + |\alpha - \beta|^2)$

(c) $|\alpha + \beta|^2 + |\alpha - \beta|^2$ (d) none of these

49. The set of values of $a \in R$ for which $x^2 + i(a - 1)x + 5 = 0$ will have a pair of conjugate complex roots is

(a) R (b) $\{1\}$

(c) $\{a \mid a^2 - 2a + 21 > 0\}$ (d) none of these

50. Nonreal complex numbers z satisfying the equation $z^3 + 2z^2 + 3z + 2 = 0$ are

(a) $\frac{-1 \pm \sqrt{-7}}{2}$ (b) $\frac{1+\sqrt{7}i}{2}, \frac{1-\sqrt{7}i}{2}$

(c) $-i, \frac{-1+\sqrt{7}i}{2}, \frac{-1-\sqrt{7}i}{2}$ (d) none of these

51. For a complex number z, the minimum value of $|z| + |z-2|$ is

(a) 1 (b) 2

(c) 3 (d) none of these

52. If $|z| = 1$ then $\frac{1+z}{1+\bar{z}}$ is equal to

(a) z (b) $\bar{z}$

(c) $z + \bar{z}$ (d) none of these

53. If α is a nonreal cube root of unity then $|\alpha^n|, n \in \mathbb{Z}$, is equal to

(a) 1 (b) 3

(c) 0 (d) none of these

54. If z be a complex number satisfying $z^4 + z^3 + 2z^2 + z + 1 = 0$ then $|z|$ is equal to

(a) $\frac{1}{2}$ (b) $\frac{3}{4}$

(c) 1 (d) none of these

55. Let $z_1 = a + ib, z_2 = p + iq$ be two unimodular complex numbers such that $\text{Im}(z_1\bar{z}_2) = 1$. If $\omega_1 = a + ip, \omega_2 = b + iq$ then

(a) $\text{Re}(\omega_1\omega_2) = 1$ (b) $\text{Im}(\omega_1\omega_2) = 1$

(c) $\text{Re}(\omega_1\omega_2) = 0$ (d) $\text{Im}(\omega_1\bar{\omega}_2) = 1$

56. If $|z_1 - 1| < 1, |z_2 - 2| < 2, |z_3 - 3| < 3$ then $|z_1 + z_2 + z_3|$

(a) is less than 6 (b) is more than 3

(c) is less than 12 (d) lies between 6 and 12

57. If $|z - i| \le 2$ and $z_0 = 5 + 3i$ then the maximum value of $|iz + z_0|$ is

(a) $2 + \sqrt{31}$ (b) 7

(c) $\sqrt{31} - 2$ (d) none of these

58. If $|z| = \max\{|z-1|, |z+1|\}$ then

(a) $|z+\bar{z}|=\frac{1}{2}$ (b) $z+\bar{z}=1$

(c) $|z+\bar{z}|=1$ (d) none of these

59. $|z-4|<|z-2|$ represents the region given by

(a) $\text{Re}(z)>0$ (b) $\text{Re}(z)<0$

(c) $\text{Re}(z)>2$ (d) none of these

60. If $\log_{1/2}\frac{|z|^2+2|z|+4}{2|z|^2+1}<0$ then the region traced by z is

(a) $|z|<3$ (b) $1<|z|<3$

(c) $|z|>1$ (d) $|z|<2$

61. $\left|\frac{z-1}{z+1}\right|=1$ represents

(a) a circle (b) an ellipse

(c) a straight line (d) none of these

62. If $2z_1-3z_2+z_3=0$ then z_1, z_2, z_3 are represented by

(a) three vertices of a triangle

(b) three collinear points

(c) three vertices of a rhombus

(d) none of these

63. If A, B, C are three points in the Argand plane representing the complex numbers z_1, z_2, z_3 such that $z_1=\frac{\lambda z_2+z_3}{\lambda+1}$, where $\lambda \in R$, then the distance of A from the line BC is

(a) λ (b) $\frac{\lambda}{\lambda+1}$

(c) 1 (d) 0

64. The roots of the equation $1+z+z^3+z^4=0$ are represented by the vertices of

(a) a square (b) an equilateral triangle

(c) a rhombus (d) none of these

65. If $\text{Re}\left(\frac{z+4}{2z-i}\right)=\frac{1}{2}$ then z is represented by a point lying on

(a) a circle (b) an ellipse

(c) a straight line (d) none of these

66. The angle that the vector representing the complex number $\dfrac{1}{(\sqrt{3}-i)^{25}}$ makes with the positive direction of the real axis is

(a) $\dfrac{2\pi}{3}$ (b) $-\dfrac{\pi}{6}$

(c) $\dfrac{5\pi}{6}$ (d) $\dfrac{\pi}{6}$

67. If P, P′ represent the complex number z_1 and its additive inverse respectively then the complex equation of the circle with PP′ as a diameter is

(a) $\dfrac{z}{z_1} = \left(\dfrac{\overline{z}_1}{z}\right)$ (b) $z\overline{z} + z_1\overline{z}_1 = 0$

(c) $z\overline{z}_1 + \overline{z}z_1 = 0$ (d) none of these

68. If $|z_1| = |z_2| = |z_3| = |z_4|$ then the points representing z_1, z_2, z_3, z_4 are

(a) concyclic (b) vertices of a square

(c) vertices of a rhombus (d) none of these

69. Suppose z_1, z_2, z_3 are the vertices of an equilateral triangle inscribed in the circle $|z| = 2$. If $z_1 = 1 + \sqrt{3}i$ and z_1, z_2, z_3 are in the clockwise sense then

(a) $z_2 = 1 - \sqrt{3}i, z_3 = -2$ (b) $z_2 = 2, z_3 = 1 - \sqrt{3}i$

(c) $z_2 = -1 + \sqrt{3}i, z_3 = -2$ (d) none of these

70. Suppose z_1, z_2, z_3 are the vertices of an equilateral triangle circumscribing the circle $|z| = 1$. If $z_1 = 1 + \sqrt{3}i$ and z_1, z_2, z_3 are in the anticlockwise sense then z_2 is

(a) $1 - \sqrt{3}i$ (b) 2

(c) $\dfrac{1}{2}(1 - \sqrt{3}i)$ (d) none of these

71. If amp $\dfrac{z-1}{z+1} = \dfrac{\pi}{3}$ then z represents a point on

(a) a straight line (b) a circle

(c) a pair of lines (d) none of these

72. If the roots of $z^3 + iz^2 + 2i = 0$ represent the vertices of a ΔABC in the Argand plane then the area of the triangle is

(a) $\frac{3\sqrt{7}}{2}$ (b) $\frac{3\sqrt{7}}{4}$

(c) 2 (d) none of these

73. The equation $z\bar{z} + (4-3i)z + (4+3i)\bar{z} + 5 = 0$ represents a circle whose radius is

(a) 5 (b) $2\sqrt{5}$

(c) $\frac{5}{2}$ (d) none of these

74. If z is a complex number such that $\left|\frac{z-3i}{z+3i}\right| = 1$ then z lies on

(a) the real axis (b) the line $\text{Im}(z) = 3$

(c) a circle (d) none of these

75. Let z_1 and z_2 be two nonreal complex cube roots of unity and $|z - z_1|^2 + |z - z_2|^2 = \lambda$ be the equation of a circle with z_1, z_2 as ends of a diameter then the value of λ is

(a) 4 (b) 3

(c) 2 (d) $\sqrt{2}$

76. Let $\lambda \in R$. If the origin and the nonreal roots of $2z^2 + 2z + \lambda = 0$ form the three vertices of an equilateral triangle in the Argand plane then λ is

(a) 1 (b) $\frac{2}{3}$

(c) 2 (d) -1

77. The equation $|z - i| + |z + i| = k, k > 0$, can represent an ellipse if k is

(a) 1 (b) 2

(c) 4 (d) none of these

78. The equation $|z + i| - |z - i| = k$ represents a hyperbola if

(a) $-2 < k < 2$ (b) $k > 2$

(c) $0 < k < 2$ (d) none of these

79. Let $OP \cdot OQ = 1$ and let O, P, Q be three collinear points. If O and Q represent the complex numbers 0 and z then P represents

(a) $\frac{1}{\bar{z}}$ (b) $\bar{z}$

(c) $\frac{1}{\bar{z}}$ (d) none of these

80. Let $z = 1 - t + i\sqrt{t^2 + t + 2}$, where t is a real parameter. The locus of z in the Argand plane is

(a) a hyperbola (b) an ellipse

(c) a straight line (d) none of these

81. The area of the triangle whose vertices are i, α, β, where $i = \sqrt{-1}$ and α, β are the nonreal cube roots of unity, is

(a) $\frac{3\sqrt{3}}{2}$ (b) $\frac{3\sqrt{3}}{4}$

(c) 0 (d) $\frac{\sqrt{3}}{4}$

• Type 2 •

Choose the correct options. One or more options may be correct.

82. The nonzero real value of x for which $\frac{(1+ix)(1+2ix)}{1-ix}$ is purely real is

(a) $\sqrt{2}$ (b) 1

(c) $-\sqrt{2}$ (d) none of these

83. If $z_1 = \frac{1}{a+i}$, $a \neq 0$ and $z_2 = \frac{1}{1+bi}$, $b \neq 0$ such that $z_1 = \bar{z}_2$ then

(a) $a = 1, b = 1$ (b) $a = -1, b = 1$

(c) $a = 1, b = -1$ (d) none of these

84. If z_1, z_2, z_3, z_4 are roots of the equation

$$a_0z^4 + a_1z^3 + a_2z^2 + a_3z + a_4 = 0,$$

where a_0, a_1, a_2, a_3 and a_4 are real, then

(a) $\bar{z}_1, \bar{z}_2, \bar{z}_3, \bar{z}_4$ are also roots of the equation

(b) z_1 is equal to at least one of $\bar{z}_1, \bar{z}_2, \bar{z}_3, \bar{z}_4$

(c) $-\bar{z}_1, -\bar{z}_2, -\bar{z}_3, -\bar{z}_4$ are also roots of the equation

(d) none of these

85. If α is a complex constant such that $\alpha z^2 + z + \overline{\alpha} = 0$ has a real root then

(a) $\alpha + \overline{\alpha} = 1$

(b) $\alpha + \overline{\alpha} = 0$

(c) $\alpha + \overline{\alpha} = -1$

(d) the absolute value of the real root is 1

86. If amp $(z_1 z_2) = 0$ and $|z_1| = |z_2| = 1$ then

(a) $z_1 + z_2 = 0$ (b) $z_1 z_2 = 1$

(c) $z_1 = \overline{z}_2$ (d) none of these

87. If z is a nonzero complex number then $\left|\frac{|\overline{z}|^2}{z\overline{z}}\right|$ is equal to

(a) $\left|\frac{\overline{z}}{z}\right|$ (b) 1

(c) $|\overline{z}|$ (d) none of these

88. If ω is a nonreal cube root of unity then the value of

$$1 \cdot (2-\omega)(2-\omega^2) + 2 \cdot (3-\omega)(3-\omega^2) + \ldots + (n-1)(n-\omega)(n-\omega^2)$$

is

(a) real (b) $\frac{n^2(n-1)^2}{4} - n + 1$

(c) $\left\{\frac{n(n+1)}{2}\right\}^2 - n$ (d) not real

89. If z is a complex number satisfying $z + z^{-1} = 1$ then $z^n + z^{-n}$, $n \in N$, has the value

(a) $2(-1)^n$ when n is a multiple of 3

(b) $(-1)^{n-1}$ when n is not a multiple of 3

(c) $(-1)^{n+1}$ when n is a multiple of 3

(d) 0 when n is not a multiple of 3

90. The value of $\alpha^{-n} + \alpha^{-2n}$, $n \in N$ and α is a nonreal cube root of unity, is

(a) 3 if n is a multiple of 3 (b) -1 if n is not a multiple of 3

(c) 2 if n is a multiple of 3 (d) none of these

91. The value of $\alpha^{4n-1} + \alpha^{4n-2} + \alpha^{4n-3}$, $n \in N$ and α is a nonreal fourth root of unity, is

(a) 0 (b) –1

(c) 3 (d) none of these

92. Let x be a nonreal complex number satisfying $(x-1)^3+8=0$ then x is

(a) $1+2\omega$ (b) $1-2\omega$

(c) $1-2\omega^2$ (d) none of these

93. If $z=\dfrac{1+3i}{1+i}$ then

(a) $\text{Re}(z)=2\text{Im}(z)$ (b) $\text{Re}(z)+2\text{Im}(z)=0$

(c) $|z|=\sqrt{5}$ (d) $\text{amp } z=\tan^{-1}2$

94. If z is different from $\pm i$ and $|z|=1$ then $\dfrac{z+i}{z-i}$ is

(a) purely real

(b) nonreal, whose real and imaginary parts are equal

(c) purely imaginary

(d) none of these

95. If z_1, z_2 are two complex numbers then

(a) $|z_1+z_2|\le|z_1|+|z_2|$ (b) $|z_1-z_2|\ge|z_1|-|z_2|$

(c) $|z_1+z_2|\ge|z_1\cdot z_2|$ (d) $|z_1-z_2|\le|z_1+z_2|$

96. Let z_1, z_2 be two complex numbers represented by points on the circle $|z|=1$ and $|z|=2$ respectively then

(a) $\max|2z_1+z_2|=4$ (b) $\min|z_1-z_2|=1$

(c) $\left|z_2+\dfrac{1}{z_1}\right|\le 3$ (d) none of these

97. ABCD is a square, vertices being taken in the anticlockwise sense. If A represents the complex number z and the intersection of the diagonals is the origin then

(a) B represents the complex number iz

(b) D represents the complex number $i\bar{z}$

(c) B represents the complex number $i\bar{z}$

(d) D represents the complex number $-iz$

98. If $z(\overline{z+\alpha})+\bar{z}(z+\alpha)=0$, where α is a complex constant, then z is represented by a point on

(a) a straight line (b) a circle
(c) a parabola (d) none of these

99. If z_1, z_2, z_3, z_4 are the four complex numbers represented by the vertices of a quadrilateral taken in order such that $z_1 - z_4 = z_2 - z_3$ and $\text{amp}\,\frac{z_4 - z_1}{z_2 - z_1} = \frac{\pi}{2}$ then the quadrilateral is a

(a) rhombus (b) square
(c) rectangle (d) a cyclic quadrilateral

100. If z_0, z_1 represent points P, Q on the locus $|z-1| = 1$ and the line segment PQ subtends an angle $\pi/2$ at the point $z = 1$ then z_1 is equal to

(a) $1 + i(z_0 - 1)$ (b) $\frac{i}{z_0 - 1}$
(c) $1 - i(z_0 - 1)$ (d) $i(z_0 - 1)$

101. If $|z_1| = |z_2| = |z_3| = 1$ and z_1, z_2, z_3 are represented by the vertices of an equilateral triangle then

(a) $z_1 + z_2 + z_3 = 0$ (b) $z_1 z_2 z_3 = 1$
(c) $z_1 z_2 + z_2 z_3 + z_3 z_1 = 0$ (d) none of these

102. Let A, B, C be three collinear points which are such that $AB \cdot AC = 1$ and the points are represented in the Argand plane by the the complex numbers $0, z_1, z_2$ respectively. Then

(a) $z_1 z_2 = 1$ (b) $z_1 \bar{z}_2 = 1$
(c) $|z_1||z_2| = 1$ (d) none of these

103. If z_1, z_2, z_3, z_4 are represented by the vertices of a rhombus taken in the anticlockwise order then

(a) $z_1 - z_2 + z_3 - z_4 = 0$ (b) $z_1 + z_2 = z_3 + z_4$
(c) $\text{amp}\,\frac{z_2 - z_4}{z_1 - z_3} = \frac{\pi}{2}$ (d) $\text{amp}\,\frac{z_1 - z_2}{z_3 - z_4} = \frac{\pi}{2}$

104. If $\text{amp}\,\frac{z-2}{2z+3i} = 0$ and $z_0 = 3 + 4i$ then

(a) $z_0 \bar{z} + \bar{z}_0 z = 12$ (b) $z_0 z + \bar{z}_0 \bar{z} = 12$
(c) $z_0 \bar{z} + \bar{z}_0 z = 0$ (d) none of these

105. If $z_1 \neq -z_2$ and $|z_1 + z_2| = \left|\frac{1}{z_1} + \frac{1}{z_2}\right|$ then

(a) at least one of z_1, z_2 is unimodular

(b) both z_1, z_2 are unimodular

(c) $z_1 \cdot z_2$ is unimodular

(d) none of these

106. Let $z_1 = \frac{(\sqrt{3}+i)^2 \cdot (1-\sqrt{3}i)}{1+i}$, $z_2 = \frac{(1+\sqrt{3}i)^2 \cdot (\sqrt{3}-i)}{1-i}$. Then

(a) $|z_1| = |z_2|$

(b) $\text{amp } z_1 + \text{amp } z_2 = 0$

(c) $3|z_1| = |z_2|$

(d) $3\text{amp } z_1 + \text{amp } z_2 = 0$

107. If $|z_1 + z_2| = |z_1 - z_2|$ then

(a) $|\text{amp } z_1 - \text{amp } z_2| = \frac{\pi}{2}$

(b) $|\text{amp } z_1 - \text{amp } z_2| - \pi$

(c) $\frac{z_1}{z_2}$ is purely real

(d) $\frac{z_1}{z_2}$ is purely imaginary

108. If $|z_1 + z_2|^2 = |z_1|^2 + |z_2|^2$ then

(a) $\frac{z_1}{z_2}$ is purely real

(b) $\frac{z_1}{z_2}$ is purely imaginary

(c) $z_1\bar{z}_2 + z_2\bar{z}_1 = 0$

(d) $\text{amp } \frac{z_1}{z_2} = \frac{\pi}{2}$

Answers

1. b	**2.** b	**3.** d	**4.** c	**5.** b
6. a	**7.** a	**8.** c	**9.** b	**10.** d
11. b	**12.** a	**13.** b	**14.** d	**15.** b
16. c	**17.** a	**18.** c	**19.** c	**20.** b
21. a	**22.** b	**23.** a	**24.** b	**25.** c
26. d	**27.** b	**28.** d	**29.** b	**30.** a
31. b	**32.** b	**33.** a	**34.** b	**35.** b
36. a	**37.** c	**38.** c	**39.** c	**40.** b
41. a	**42.** c	**43.** d	**44.** b	**45.** b
46. a	**47.** c	**48.** b	**49.** b	**50.** a
51. b	**52.** a	**53.** a	**54.** c	**55.** d
56. c	**57.** b	**58.** c	**59.** d	**60.** a
61. c	**62.** b	**63.** d	**64.** b	**65.** c
66. d	**67** a	**68.** a	**69.** a	**70.** d
71. b	**72.** c	**73.** b	**74.** a	**75.** b
76. b	**77.** c	**78.** a	**79.** c	**80.** a
81. d	**82.** a, c	**83.** c	**84.** a, b	**85.** a, c, d
86. b, c	**87.** a, b	**88.** a, b	**89.** a, b	**90.** b, c
91. b	**92.** b, c	**93.** a, c	**94.** c	**95.** a, b
96. a, b, c	**97.** a, d	**98.** b	**99.** c, d	**100.** a, c
101. a, b	**102.** b, c	**103.** a, c	**104.** b	**105.** c
106. a, d	**107.** a, d	**108.** b, c, d		

Hints and Solutions to Selected Questions

3. Expression $= (1+i)^{n_1} + (1-i)^{n_1} + (1+i)^{n_2} + (1-i)^{n_2}$

$$= 2^{n_1/2}\left(\frac{1}{\sqrt{2}} + i\frac{1}{\sqrt{2}}\right)^{n_1} + 2^{n_1/2}\left(\frac{1}{\sqrt{2}} - i\frac{1}{\sqrt{2}}\right)^{n_1}$$

$$+ 2^{n_2/2}\left(\frac{1}{\sqrt{2}} + i\frac{1}{\sqrt{2}}\right)^{n_2} + 2^{n_2/2}\left(\frac{1}{\sqrt{2}} - i\frac{1}{\sqrt{2}}\right)^{n_2}$$

$$= 2^{n_1/2}\left\{\left(\cos\frac{\pi}{4} + i\sin\frac{\pi}{4}\right)^{n_1} + \left(\cos\frac{\pi}{4} - i\sin\frac{\pi}{4}\right)^{n_1}\right\}$$

$$+ 2^{n_2/2}\left\{\left(\cos\frac{\pi}{4} + i\sin\frac{\pi}{4}\right)^{n_2} + \left(\cos\frac{\pi}{4} - i\sin\frac{\pi}{4}\right)^{n_2}\right\}$$

$$= 2^{n_1/2}\left\{\cos\frac{n_1\pi}{4} + i\sin\frac{n_1\pi}{4} + \cos\frac{n_1\pi}{4} - i\sin\frac{n_1\pi}{4}\right\}$$

$$+ 2^{n_2/2}\left\{\cos\frac{n_2\pi}{4} + i\sin\frac{n_2\pi}{4} + \cos\frac{n_2\pi}{4} - i\sin\frac{n_2\pi}{4}\right\},$$

if n_1, n_2 are integers

$$= 2^{n_1/2}\cdot 2\cos\frac{n_1\pi}{4} + 2^{n_2/2}\cdot 2\cos\frac{n_2\pi}{4} = \text{real.}$$

5. $\dfrac{1-i}{1+i} = \dfrac{(1-i)^2}{1-i^2} = \dfrac{-2i}{2} = -i$

$\therefore \quad \left(\dfrac{1-i}{1+i}\right)^n = (-i)^n = \text{imaginary} \quad \Rightarrow \quad n = 1, 3, 5, \ldots.$

6. $(b+ia)^5 = i^5(a-ib)^5 = i(\alpha - i\beta)$.

7. Take $n = 4m, 4m+1, 4m+2, 4m+3$ and verify.

9. Expression $= (1+i)2i + (-2i)^3$.

11. $\dfrac{z-1}{i}$ must be real. Now, $\dfrac{z-1}{i} = \dfrac{x-1+iy}{i} = y - (x-1)i$

$\therefore \quad x - 1 = 0$. Then $\sin^{-1}\left\{\dfrac{1}{i}(z-1)\right\} = \sin^{-1} y$. So, $-1 \le y \le 1$.

16. If $z = x + iy$, $\sqrt{(x+1)^2 + y^2} = x + 2$ and $0 = y + 2$.

17. $(1+i)(x+iy) = (1-i)(x-iy) \quad \Rightarrow \quad (x-y) + i(x+y) = (x-y) - i(x+y)$

$\therefore \quad x + y = 0 \quad \therefore \quad z = x - ix = x(1-i).$

18. $|z_1 + z_2| = |z_1| + |z_2|$ can hold when $0, z_1, z_2$ are collinear with 0 at one end.

20. $\left|\dfrac{z-1}{z+1}\right| = 1$. So, $\dfrac{z-1}{z+1} = 1 \cdot \left(\cos\dfrac{\pi}{4} + i\sin\dfrac{\pi}{4}\right) = \dfrac{1+i}{\sqrt{2}}$.

$$\therefore \quad z = \frac{\sqrt{2}+1+i}{\sqrt{2}-1-i} = \frac{(\sqrt{2}+1+i)(\sqrt{2}-1+i)}{(\sqrt{2}-1)^2+1}, \text{ etc.}$$

21. $z = (\cos\theta + i\sin\theta)^2 = \cos 2\theta + i\sin 2\theta$, where $\dfrac{\pi}{2} < 2\theta < \pi$.

$\therefore$ z is a complex number in the second quadrant. So, $\dfrac{\pi}{2} < \arg z < \pi$.

$\therefore$ $\arg z = \tan^{-1}(\tan 2\theta) = 2\theta$.

24. $\tan^{-1}\dfrac{y}{x-1} = \tan^{-1}\dfrac{y+3}{x} \quad \therefore \quad xy = (x-1)(y+3) \quad \Rightarrow \quad 3(x-1) = y.$

25. $x^2 + y^2 = 2k^2$, $\text{Re}\{(x+iy)^2\} = 0$ i.e., $x^2 - y^2 = 0$.

Solving these, $x = \pm k$, $y = \pm k$.

29. Roots of $x^2 + x + 1 = 0$ are $x = \omega, \omega^2$. For both these values

$$x^{3m} + x^{3n-1} + x^{3r-2} = 0.$$

30. $x = \dfrac{1 \pm \sqrt{3}i}{2} = \cos\dfrac{\pi}{3} \pm i\sin\dfrac{\pi}{3} \quad \therefore \quad x^{2n} = \cos\dfrac{2n\pi}{3} \pm i\sin\dfrac{2n\pi}{3}$

$$\therefore \quad \left(x^n + \frac{1}{x^n}\right)^2 = x^{2n} + x^{-2n} + 2$$

$$= \cos\frac{2n\pi}{3} \pm i\sin\frac{2n\pi}{3} + \cos\frac{2n\pi}{3} \mp i\sin\frac{2n\pi}{3} + 2$$

$$= 2 + 2\cos\frac{2n\pi}{3}.$$

$$\therefore \text{ the value} = 10 + 2\left\{\cos\frac{2\pi}{3} + \cos\frac{4\pi}{3} + \cos\frac{6\pi}{3} + \cos\frac{8\pi}{3} + \cos\frac{10\pi}{3}\right\}.$$

31. Here $x = \frac{\sqrt{3} \pm \sqrt{3-4}}{2} = \frac{\sqrt{3} \pm i}{2} = \cos\frac{\pi}{6} \pm i\sin\frac{\pi}{6}\cdot$ Now, see the hint to Q. No. 30.

32. $(1+\sqrt{3}i)^{n/2} = \left\{2\left(\frac{1}{2}+\frac{\sqrt{3}}{2}i\right)\right\}^{n/2}$

$$= 2^{n/2}\left(\cos\frac{\pi}{3}+i\sin\frac{\pi}{3}\right)^{n/2}$$

$$= 2^{n/2}\left(\cos\frac{n\pi}{3}+i\sin\frac{n\pi}{3}\right)^{1/2}.$$

Clearly, the least positive integral value of n for which $\cos\frac{n\pi}{3}+i\sin\frac{n\pi}{3}$ is positive real is 6.

35. Expression $= \frac{\omega(2+3\omega+\omega^2)}{2+3\omega+\omega^2} + \frac{\omega(3+\omega+2\omega^2)}{3+\omega+2\omega^2} = \omega+\omega.$

38. $e^{iA}\cdot e^{iB}\cdot e^{iC} = (\cos A + i\sin A)(\cos B + i\sin B)(\cos C + i\sin C)$

$$= \cos(A+B+C) + i\sin(A+B+C) = \cos\pi + i\sin\pi.$$

39. $\left(\frac{\sqrt{3}+i}{\sqrt{3}-i}\right)^n = \left\{\frac{1}{i}\cdot\frac{-1+\sqrt{3}i}{-1-\sqrt{3}i}\cdot(-i)\right\}^n$

$$= \left\{-\frac{\frac{-1+\sqrt{3}i}{2}}{\frac{-1-\sqrt{3}i}{2}}\right\}^n = \left(-\frac{1}{\omega}\right)^n = 1, \text{ when } n = 0, 6, \ldots .$$

40. $1^{1/4} = (\cos 2\pi r + i\sin 2\pi r)^{1/4} = \cos\frac{\pi r}{2} + i\sin\frac{\pi r}{2}; r = 0, 1, 2, 3$

$\therefore \quad 1^{1/4} = 1, i, -1, -i.$

$\therefore \quad \text{value} = 1^2 + i^2 + (-1)^2 + (-i)^2.$

44. $z = \frac{i(\sqrt{3}+i)^4}{4(1-\sqrt{3}i)^2} = \frac{i}{4}\cdot\frac{(2+2\sqrt{3}i)^2}{-2-2\sqrt{3}i} = \frac{i(-2+2\sqrt{3}i)}{2(-1-\sqrt{3}i)} = \frac{\sqrt{3}+i}{1+\sqrt{3}i}$

$$= \frac{(\sqrt{3}+i)(1-\sqrt{3}i)}{1+3} = \frac{2\sqrt{3}-2i}{4} = \frac{\sqrt{3}}{2} - \frac{i}{2}$$

$\therefore$ amp $z = -\tan^{-1}\frac{1}{\sqrt{3}}\cdot$

45. $z^7 = -1 \quad \therefore$ expression $= (z^7)^{12} \cdot z^2 + (z^7)^{25} + (z^7)^{41} \cdot z^2$

$$= (-1)^{12} \cdot z^2 + (-1)^{25} + (-1)^{41} \cdot z^2$$

$$= z^2 - 1 - z^2 = -1.$$

46. $\alpha^5 = 1 \quad \therefore$ index $= |1 + \alpha + \alpha^2 + \alpha^3 - \alpha^4|$

$$= |1 + \alpha + \alpha^2 + \alpha^3 + \alpha^4 - 2\alpha^4|$$

$$= \left|\frac{1-\alpha^5}{1-\alpha} - 2\alpha^4\right| = |-2\alpha^4| = 2|\alpha|^4 = 2 \times 1 = 2.$$

48. $|\alpha + \beta|^2 = (\alpha + \beta)\overline{(\alpha + \beta)} = (\alpha + \beta)(\bar{\alpha} + \bar{\beta}) = \alpha\bar{\alpha} + \beta\bar{\beta} + \alpha\bar{\beta} + \bar{\alpha}\beta$

$$= |\alpha|^2 + |\beta|^2 + \alpha\bar{\beta} + \bar{\alpha}\beta$$

$$|\alpha - \beta|^2 = (\alpha - \beta)(\bar{\alpha} - \bar{\beta}) = \alpha\bar{\alpha} + \beta\bar{\beta} - \alpha\bar{\beta} - \bar{\alpha}\beta$$

$$= |\alpha|^2 + |\beta|^2 - \alpha\bar{\beta} - \bar{\alpha}\beta.$$

49. A quadratic equation $px^2 + qx + r = 0$ can have a pair of conjugate complex roots if all coefficients are real and $D < 0$.

Here $p = 1$, $r = 5$. So, $i(a - 1)$ must be real. Hence, $a = 1$ and then $D < 0$ also.

50. $(z + 1)(z^2 + z + 2) = 0$; nonreal complex roots are found from $z^2 + z + 2 = 0$.

51.

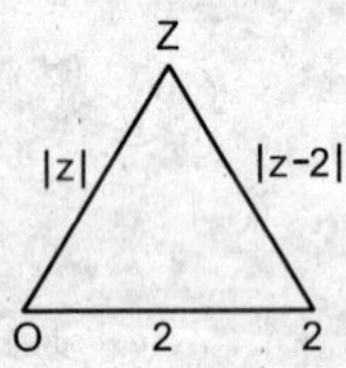

$|z| + |z - 2| \geq 2$

52. $\dfrac{1+z}{1+\bar{z}} = \dfrac{z\bar{z} + z}{1+\bar{z}} \quad (\because \ |z| = 1 \Rightarrow |z|^2 = 1 \Rightarrow z\bar{z} = 1)$

$$= \frac{z(\bar{z}+1)}{1+\bar{z}} = z.$$

54. $z^4 + z^3 + z^2 + z^2 + z + 1 = 0$

or $\quad z^2(z^2 + z + 1) + (z^2 + z + 1) = 0$

or $\quad (z^2 + z + 1)(z^2 + 1) = 0$

$\therefore \quad z = i, -i, \omega, \omega^2$. For each, $|z| = 1$.

55. $\text{Im}(z_1\bar{z}_2) = 1 \quad \Rightarrow \quad bp - aq = 1$

$\omega_1\bar{\omega}_2 = (a + ip)(b - iq) = (ab + pq) + i(bp - aq)$

$\therefore \quad \text{Im}(\omega_1\bar{\omega}_2) = 1.$

56. $|z_1 + z_2 + z_3| = |(z_1 - 1) + (z_2 - 2) + (z_3 - 3) + 6|$

$\leq |z_1 - 1| + |z_2 - 2| + |z_3 - 3| + 6 < 1 + 2 + 3 + 6.$

57. $|iz + z_0| = |i(z - i) - 1 + 5 + 3i| = |i(z - i) + 4 + 3i|$

$\leq |i|\,|z - i| + |4 + 3i| \leq 1 \cdot 2 + 5 = 7.$

58. $|z| = |z - 1| \quad \Rightarrow \quad |z|^2 = |z - 1|^2 \quad \Rightarrow \quad z\bar{z} = (z - 1)(\bar{z} - 1)$

$\Rightarrow \quad z\bar{z} = z\bar{z} - z - \bar{z} + 1$

$\therefore \quad z + \bar{z} = 1$

$|z| = |z + 1| \quad \Rightarrow \quad z + \bar{z} = -1$ (similarly)

$\therefore \quad |z + \bar{z}| = 1.$

60. $\log_{1/2} \dfrac{|z|^2 + 2|z| + 4}{2|z|^2 + 1} < 0 = \log_{1/2} 1 \quad \Rightarrow \quad \dfrac{|z|^2 + 2|z| + 4}{2|z|^2 + 1} > 1$

or $\quad |z|^2 + 2|z| + 4 > 2|z|^2 + 1$

or $\quad |z|^2 - 2|z| - 3 < 0$

or $\quad (|z| + 1)(|z| - 3) < 0$

$\therefore \quad |z| - 3 < 0.$

61. $|z - 1| = |z + 1| \quad \Rightarrow \quad (x - 1)^2 + y^2 = (x + 1)^2 + y^2$

$\Rightarrow \quad x = 0$, which represents a straight line.

Note $\left|\dfrac{z - \alpha}{z - \beta}\right| = k$, where α, β are complex constants, represents a straight line if $k = 1$.

62. $z_2 = \dfrac{2z_1 + z_3}{2 + 1}$

$\Rightarrow$ z_2 divides the line segment joining z_1, z_3 in the ratio 2 : 1 internally.

Note If z_1, z_2, z_3 are related by $az_1 + bz_2 + cz_3 = 0$, where $a + b \pm c = 0$, then z_1, z_2, z_3 will be collinear points.

64.

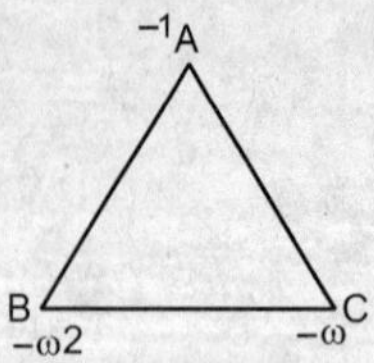

Here $(1+z)(1+z^3)=0$

or $(1+z)^2(z^2-z+1)=0 \quad \Rightarrow \quad z=-1,-1,\dfrac{1\pm\sqrt{3}i}{2}$.

The distinct points corresponding to these are

$$z=-1,\frac{1+\sqrt{3}i}{2},\frac{1-\sqrt{3}i}{2}=-1,-\omega^2,-\omega$$

$$AB = |1-\omega^2| = |\omega^2(\omega-1)|$$

$$= |\omega^2|\,|\omega-1| = |\omega-1|$$

$$BC = |\omega^2-\omega| = |\omega|\,|\omega-1| = |\omega-1|$$

$$CA = |\omega-1|$$

$\therefore \quad AB = BC = CA.$

65. $\dfrac{z+4}{2z-i}=\dfrac{(x+4)+iy}{2x+i(2y-1)}=\dfrac{\{(x+4)+iy\}\{2x-i(2y-1)\}}{\{2x+i(2y-1)\}\{2x-i(2y-1)\}}$

$$=\frac{2x(x+4)+y(2y-1)+i\{2xy-(x+4)(2y-1)\}}{4x^2+(2y-1)^2}.$$

$$\text{Re}\left(\frac{z+4}{2z-i}\right)=\frac{1}{2}$$

$$\Rightarrow \quad \frac{2x(x+4)+y(2y-1)}{4x^2+(2y-1)^2}=\frac{1}{2}$$

$\Rightarrow \quad 16x+2y=1$, which represents a straight line.

66. The required angle = amp $\dfrac{1}{(\sqrt{3}-i)^{25}}$.

Now, $(\sqrt{3}-i)^{25}=\left(\dfrac{-1-\sqrt{3}i}{2}\cdot\dfrac{2}{-i}\right)^{25}=\left(\dfrac{2\omega^2}{-i}\right)^{25}=2^{25}\cdot\dfrac{\omega^2}{-i}$

$$=2^{25}\cdot(\omega^2\cdot i)$$

$\therefore$ the required angle

$$=\text{amp}\,\frac{1}{2^{25}\omega^2\cdot i}=\text{amp}\left(\frac{-\omega i}{2^{25}}\right)$$

$$=\text{amp}\,\frac{-i\cdot\dfrac{-1+\sqrt{3}i}{2}}{2^{25}}$$

$$= \text{amp}\,\frac{\sqrt{3}+i}{2^{26}} = \tan^{-1}\frac{1}{\sqrt{3}} = \frac{\pi}{6}.$$

67. Clearly, $|z| = |z_1|$ $\quad \therefore \quad z\bar{z} = z_1\bar{z}_1$

or $\quad \dfrac{z}{z_1} = \dfrac{\bar{z}_1}{\bar{z}} = \overline{\left(\dfrac{z_1}{z}\right)}.$

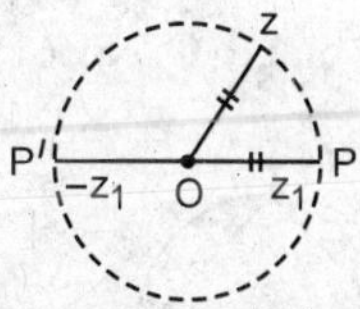

68. $|z_1|$ = the distance of the point representing z_1 from the origin. So, the distances of the four points from the origin are equal.

69.

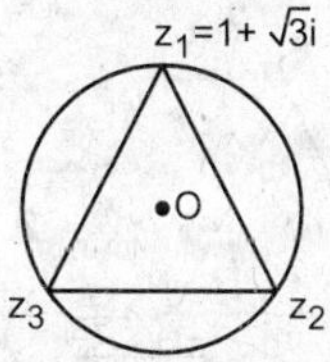

$$\text{amp}\, z_1 = \text{amp}(1+\sqrt{3}i) = \tan^{-1}\sqrt{3} = \frac{\pi}{3}.$$

$$\text{amp}\, z_2 = \frac{\pi}{3} - \frac{2\pi}{3} = -\frac{\pi}{3} \quad \text{and} \quad |z_2| = 2$$

$$\therefore \quad z_2 = 2\left\{\cos\left(-\frac{\pi}{3}\right) + i\sin\left(-\frac{\pi}{3}\right)\right\}.$$

$$\text{amp}\, z_3 = \frac{\pi}{3} + \frac{2\pi}{3} = \pi \quad \text{and} \quad |z_3| = 2 \quad \therefore \quad z_3 = 2\{\cos\pi + i\sin\pi\}.$$

71. If $\left|\dfrac{z-1}{z+1}\right| = r$ then $\dfrac{z-1}{z+1} = r\left(\cos\dfrac{\pi}{3} + i\sin\dfrac{\pi}{3}\right) = r\left(\dfrac{1}{2} + i\dfrac{\sqrt{3}}{2}\right)$

or $$\frac{(x-1)+iy}{(x+1)+iy} = \frac{r}{2} + i\frac{r\sqrt{3}}{2}$$

or $$(x-1)+iy = \frac{r}{2}(x+1) - \frac{yr\sqrt{3}}{2} + i\left\{\frac{ry}{2} + \frac{r\sqrt{3}}{2}(x+1)\right\}$$

$$\Rightarrow \left.\begin{aligned} x-1 &= \frac{r}{2}(x+1) - \frac{yr\sqrt{3}}{2} \\ y &= \frac{ry}{2} + \frac{r\sqrt{3}}{2}(x+1) \end{aligned}\right\} \Rightarrow \frac{x-1}{y} = \frac{x+1-y\sqrt{3}}{y+\sqrt{3}(x+1)}.$$

On simplification, $\sqrt{3}(x^2+y^2) - 2y - \sqrt{3} = 0$, which is a circle.

72. $(z-i)(z^2+2iz-2) = 0 \quad \Rightarrow \quad z = i, \dfrac{-2i \pm \sqrt{4i^2+8}}{2} = i, 1-i, -1-i.$

Find the area of the triangle whose vertices are

$$(0, 1), (1, -1), (-1, -1).$$

73. $x^2 + y^2 + (4-3i)(x+iy) + (4+3i)(x-iy) + 5 = 0$

or $\quad x^2+y^2+8x+6y+5=0$

$\therefore \quad \text{radius} = \sqrt{4^2+3^2-5} = \sqrt{20} = 2\sqrt{5}.$

74. See the hint to Q. No. 61.

75.

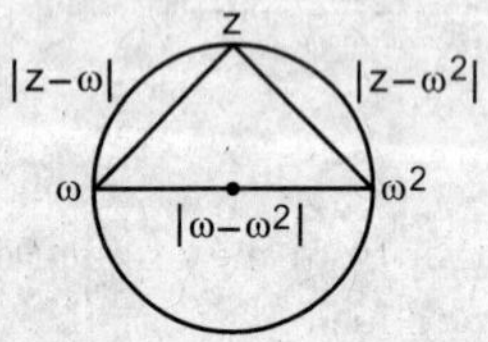

$|z-\omega|^2+|z-\omega^2|^2=\lambda$

$\Rightarrow \quad \lambda = |\omega-\omega^2|^2$

76. $2z^2+2z+\lambda=0 \quad \Rightarrow \quad z=\dfrac{-2\pm\sqrt{4-8\lambda}}{4}.$

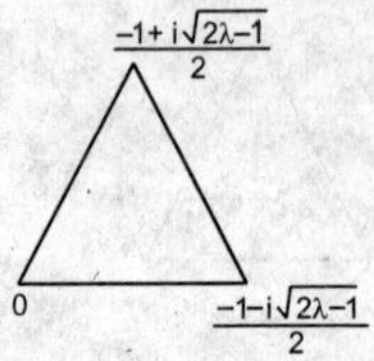

For nonreal roots $4-8\lambda<0$, i.e., $\lambda>\dfrac{1}{2}\cdot$

For equilateral triangle,

$$\left|0-\frac{-1-i\sqrt{2\lambda-1}}{2}\right| = \left|0-\frac{-1+i\sqrt{2\lambda-1}}{2}\right|$$

$$= \left|\frac{-1+i\sqrt{2\lambda-1}}{2}-\frac{-1-i\sqrt{2\lambda-1}}{2}\right|$$

or $\quad \dfrac{1}{4}\{1+2\lambda-1\}=\{\sqrt{2\lambda-1}\}^2$

$\therefore \quad \dfrac{\lambda}{2}=2\lambda-1 \quad \therefore \ \lambda=\dfrac{2}{3}\cdot$

77. $\sqrt{x^2+(y-1)^2}+\sqrt{x^2+(y+1)^2}=k \qquad \ldots (1)$

or $\quad x^2+(y-1)^2-x^2-(y+1)^2=k\left\{\sqrt{x^2+(y-1)^2}-\sqrt{x^2+(y+1)^2}\right\}$

$\therefore \quad \sqrt{x^2+(y-1)^2}-\sqrt{x^2+(y+1)^2}=\dfrac{-4y}{k} \qquad \ldots (2)$

From (1) and (2), $2\sqrt{x^2+(y-1)^2}=k-\dfrac{4y}{k}$

$\Rightarrow \quad 4x^2+\left(4-\dfrac{16}{k^2}\right)y^2=k^2-4.$

For an ellipse, $4-\dfrac{16}{k^2}>0, k^2-4>0.$

For $k = 1, 2$ the coefficient of y^2 is negative or zero.

78. See the hint to Q. No. 77.

It is a hyperbola if $4 - \frac{16}{k^2} < 0$, i.e., $k^2 - 4 < 0$

$\therefore \quad -2 < k < 2.$

79. 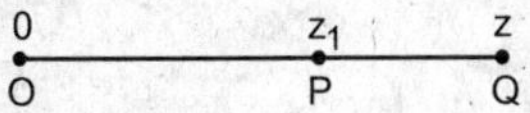

Here $|0 - z_1||0 - z| = 1.$

$\therefore \quad |z_1||z| = 1 \quad \therefore \quad |z_1| = \frac{1}{|z|}.$

Also, $\text{amp}\,\frac{z_1 - 0}{z - 0} = 0 \quad \therefore \quad \text{amp}\,\frac{z_1}{z} = 0 \quad \therefore \quad \text{amp}\, z_1 = \text{amp}\, z.$

$$\therefore \quad z_1 = \frac{1}{|z|}\{\cos(\text{amp}\, z_1) + i\sin(\text{amp}\, z_1)\}$$

$$= \frac{1}{|z|^2}\cdot |z|\,\{\cos(\text{amp}\, z) + i\sin(\text{amp}\, z)\}$$

$$= \frac{1}{|z|^2}\cdot z = \frac{z}{z\bar{z}} = \frac{1}{\bar{z}}.$$

80. $x + iy = 1 - t + i\sqrt{t^2 + t + 2} \quad \Rightarrow \quad x = 1 - t,\ y = \sqrt{t^2 + t + 2}.$

Eliminating t, $y^2 = t^2 + t + 2 = (1 - x)^2 + 1 - x + 2 = \left(x - \frac{3}{2}\right)^2 + \frac{7}{4}$

or $\quad y^2 - \left(x - \frac{3}{2}\right)^2 = \frac{7}{4}$, which is a hyperbola.

84. $a_0 z^4 + a_1 z^3 + a_2 z^2 + a_3 z + a_4 = 0$

$\Rightarrow \quad a_0\bar{z}^4 + a_1\bar{z}^3 + a_2\bar{z}^2 + a_3\bar{z} + a_4 = 0$ (taking modulus of both sides)

$\Rightarrow \quad a_0(\bar{z})^4 + a_1(\bar{z})^3 + a_2(\bar{z})^2 + a_3\bar{z} + a_4 = 0$

$\therefore \quad \bar{z}$ is a root of the equation if z is a root. So, option (a) is correct.

Also if z_1 is real, $z_1 = \bar{z}_1$.

If z_1 is nonreal complex then $\bar{z}_1$ is also a root because imaginary roots occur in conjugate pairs. So, option (b) is correct.

85. Let $z = c$ be a real root.

Then $\alpha c^2 + c + \bar{\alpha} = 0$ ... (1)

Let $\alpha = p + iq$.

Then $(p+iq)c^2+c+p-iq=0$

$\Rightarrow \quad pc^2+c+p=0$ and $qc^2-q=0 \quad \Rightarrow \quad c=\pm 1 \quad (\because q\neq 0)$

$\therefore \quad (1) \Rightarrow \alpha \pm 1+\bar{\alpha}=0.$

Also $|c|=1$.

86. $\text{amp}\,(z_1z_2)=0 \quad \Rightarrow \quad \text{amp}\,z_1+\text{amp}\,z_2=0$

$\therefore \quad \text{amp}\,z_1=-\text{amp}\,z_2=\text{amp}\,\bar{z}_2$

$\because \quad |z_1|=|z_2|$ we get $|z_1|=|\bar{z}_2|$. So, $z_1=\bar{z}_2$.

Also $z_1z_2=\bar{z}_2z_2=|z_2|^2=1$ because $|z_2|=1$.

88. $(n-1)(n-\omega)(n-\omega^2)=(n-1)\{n^2-(\omega+\omega^2)n+\omega^3\}$

$$=(n-1)(n^2+n+1)=n^3-1, \text{ etc.}$$

$$\therefore \quad \text{series}=\sum_{n=1}^{n-1}(n^3-1)=\{1^3+2^3+3^3+\ldots+(n-1)^3\}-(n-1).$$

89. $z+z^{-1}=1 \quad \Rightarrow \quad z^2-z+1=0$

$$\therefore \quad z=\frac{1\pm\sqrt{3}i}{2}=-\omega,-\omega^2$$

$$\therefore \quad z^n+z^{-n}=(-\omega)^n+(-\omega)^{-n} \text{ or } (-\omega^2)^n+(-\omega^2)^{-n}$$

$$=(-1)^n\omega^n+(-1)^{-n}\frac{1}{\omega^n} \text{ or } (-1)^n\omega^{2n}+(-1)^{-n}\cdot\frac{1}{\omega^{2n}}$$

$$=(-1)^n\left(\omega^n+\frac{1}{\omega^n}\right) \text{ or } (-1)^n\left(\omega^{2n}+\frac{1}{\omega^{2n}}\right)$$

$$=(-1)^n\left(\omega^n+\frac{1}{\omega^n}\right), \text{ because } \omega^{3n}=1$$

$$=(-1)^n\cdot(\omega^n+\omega^{2n})$$

$$=(-1)^n\cdot(1+1) \text{ or } (-1)^n\cdot(-1)$$

depending on whether n is a multiple of 3 or not.

91. $\alpha^4=1 \quad \therefore \quad \alpha^{4n-1}+\alpha^{4n-2}+\alpha^{4n-3}=\alpha^{-1}+\alpha^{-2}+\alpha^{-3}$

$$=\alpha^3+\alpha^2+\alpha \quad (\because \alpha^4=1)$$

$$=\frac{\alpha(1-\alpha^3)}{1-\alpha}=\frac{\alpha-\alpha^4}{1-\alpha}=\frac{\alpha-1}{1-\alpha}=-1.$$

92. $(x-1)^3 = -8 = (-2)^3$

$\therefore \quad x - 1 = -2 \times \sqrt[3]{1} = -2, -2\omega, -2\omega^2.$

96. $|2z_1 + z_2| \le |2z_1| + |z_2| = 2|z_1| + |z_2| = 2 \times 1 + 2 = 4.$

From the figure, $|z_1 - z_2|$ is the least when $0, z_1, z_2$ are collinear.

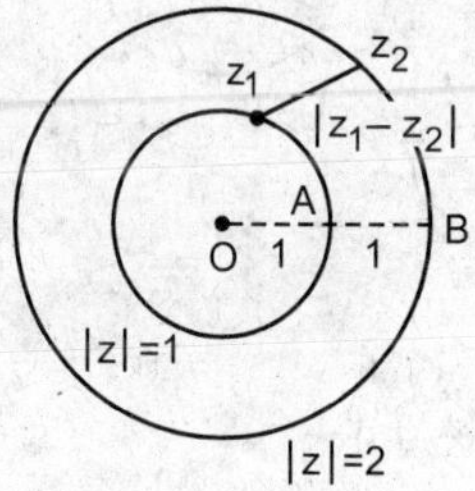

Then $|z_1 - z_2| = 1.$

Again, $\left|z_2 + \frac{1}{z_1}\right| \le |z_2| + \left|\frac{1}{z_1}\right| = 2 + \frac{1}{|z_1|}$

$= 2 + \frac{1}{1} = 3.$

97.

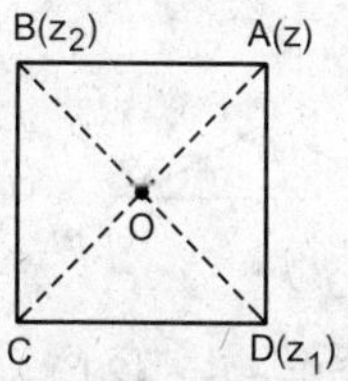

$OB = OD = OA = |z|.$

If amp $z = \theta$ then amp $z_1 = \theta - \frac{\pi}{2}$,

amp $z_2 = \theta + \frac{\pi}{2}$ and $z = |z| \{\cos\theta + i\sin\theta\}$

$\therefore \quad z_1 = |z| \left\{\cos\left(\theta - \frac{\pi}{2}\right) + i\sin\left(\theta - \frac{\pi}{2}\right)\right\}$

$= |z| \{\sin\theta - i\cos\theta\}$

$= |z| (-i)(\cos\theta + i\sin\theta) = -iz$

$z_2 = |z| \left\{\cos\left(\theta + \frac{\pi}{2}\right) + i\sin\left(\theta + \frac{\pi}{2}\right)\right\} = iz.$

98. $z(\bar{z} + \bar{\alpha}) + \bar{z}z + \alpha\bar{z} = 0$

or $\quad z\bar{z} + \frac{\bar{\alpha}}{2} z + \frac{\alpha}{2} \bar{z} = 0$ which is a circle.

99. $z_1 - z_4 = z_2 - z_3$ or $\frac{z_1 + z_3}{2} = \frac{z_2 + z_4}{2}$ i.e., the diagonals bisect each other.

Thus it is a parallelogram. Also, amp $\frac{z_4 - z_1}{z_2 - z_1} = \frac{\pi}{2}$

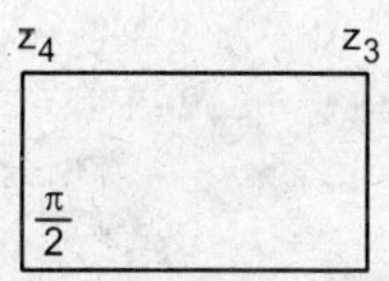

$\Rightarrow$ angle at z_1 is a right angle.

$\therefore$ it is a rectangle and hence a cyclic quadrilateral.

100.

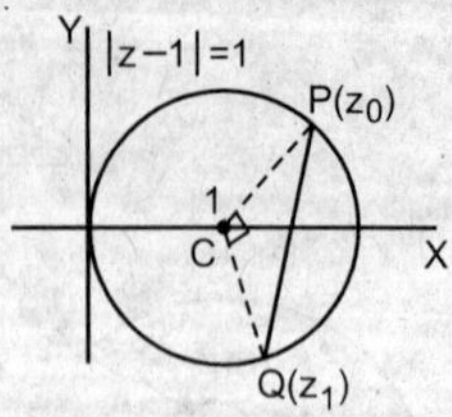

$|z_1 - 1| = 1,\ |z_0 - 1| = 1 \quad \therefore \quad \left|\dfrac{z_1 - 1}{z_0 - 1}\right| = 1$

$$\angle QCP = \frac{\pi}{2} \quad \Rightarrow \quad \text{amp}\,\frac{z_1 - 1}{z_0 - 1} = \frac{\pi}{2}, -\frac{\pi}{2}$$

$$\therefore \quad \frac{z_1 - 1}{z_0 - 1} = 1 \cdot \left(\cos\frac{\pm\pi}{2} + i\sin\frac{\pm\pi}{2}\right) = i, -i$$

$$\therefore \quad z_1 - 1 = (z_0 - 1)i,\ z_1 - 1 = -(z_0 - 1)i.$$

101. $|z_1| = |z_2| = |z_3| \Rightarrow$ the vertices are at equal distances from the origin $z = 0$.

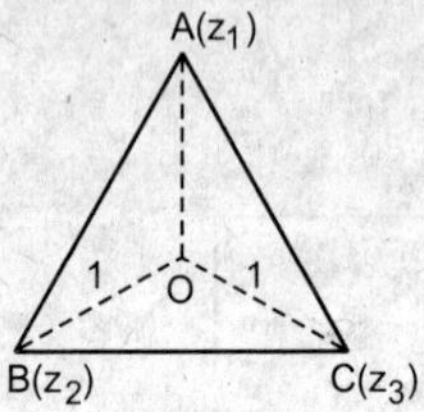

$\therefore$ the origin is at the centroid of the equilateral triangle.

$$\therefore \quad \frac{z_1 + z_2 + z_3}{3} = 0$$

$$\therefore \quad z_1 + z_2 + z_3 = 0$$

With OA as real axis,

$$z_1 = 1,\ z_2 = 1(\cos 120^\circ + i\sin 120^\circ),\ z_3 = 1(\cos 120^\circ - i\sin 120^\circ)$$

$$\therefore \quad z_1 z_2 z_3 = 1 \cdot \left(-\frac{1}{2} + i\frac{\sqrt{3}}{2}\right)\left(-\frac{1}{2} - i\frac{\sqrt{3}}{2}\right) = \left(-\frac{1}{2}\right)^2 + \left(\frac{\sqrt{3}}{2}\right)^2 = 1.$$

102. See the hint to Q. No. 79.

103. Clearly, $\dfrac{z_1 + z_3}{2} = \dfrac{z_2 + z_4}{2} = z_0$

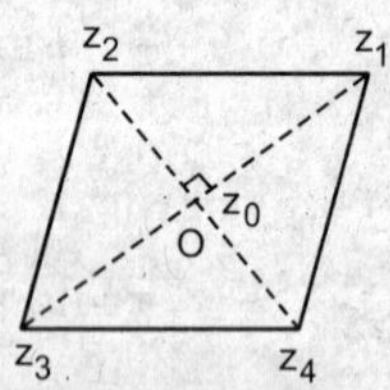

$$\therefore \quad z_1 - z_2 + z_3 - z_4 = 0.$$

Also, $\text{amp}\,\dfrac{z_2 - z_0}{z_1 - z_0} = \dfrac{\pi}{2}$

($\because$ angle at z_0 is a right angle)

$$\text{or} \quad \text{amp}\,\frac{z_2 - \dfrac{z_2 + z_4}{2}}{z_1 - \dfrac{z_1 + z_3}{2}} = \frac{\pi}{2}.$$

104. $\dfrac{z - 2}{2z + 3i} = \dfrac{(x - 2) + iy}{2x + i(2y + 3)}$

$$\therefore \quad \text{amp}\,\frac{z-2}{2z+3i} = \text{amp}\,\{(x-2)+iy\} - \text{amp}\,\{2x+i(2y+3)\}$$

$$\text{or} \quad 0 = \tan^{-1}\frac{y}{x-2} - \tan^{-1}\frac{2y+3}{2x}$$

$$\Rightarrow \quad \frac{y}{x-2} = \frac{2y+3}{2x} \quad \Rightarrow \quad 3x - 4y - 6.$$

Now, $z_0 z = (3+4i)(x+iy) = (3x-4y) + i(4x+3y)$

$$z_0\bar{z} = (3+4i)(x-iy) = (3x+4y) + i(4x-3y)$$

$$\bar{z}_0 z = (3-4i)(x+iy) = (3x+4y) + i(-4x+3y)$$

$$\bar{z}_0\bar{z} = (3-4i)(x-iy) = (3x-4y) + i(-4x-3y).$$

Check the correct answer.

105. $|z_1+z_2| = \left|\frac{z_1+z_2}{z_1z_2}\right|$

$$\text{or} \quad |z_1+z_2|\left(1-\frac{1}{|z_1z_2|}\right) = 0 \quad \therefore \quad |z_1z_2| = 1.$$

107. $|z_1+z_2|^2 = |z_1-z_2|^2$

$$\therefore \quad (z_1+z_2)(\bar{z}_1+\bar{z}_2) = (z_1-z_2)(\bar{z}_1-\bar{z}_2)$$

$$\text{or} \quad z_1\bar{z}_1 + z_1\bar{z}_2 + z_2\bar{z}_1 + z_2\bar{z}_2 = z_1\bar{z}_1 - z_1\bar{z}_2 - z_2\bar{z}_1 + z_2\bar{z}_2$$

$$\text{or} \quad z_1\bar{z}_2 + \bar{z}_1 z_2 = 0$$

$$\text{or} \quad \frac{z_1}{z_2} + \frac{\bar{z}_1}{\bar{z}_2} = 0 \quad \text{or} \quad \frac{z_1}{z_2} + \overline{\left(\frac{z_1}{z_2}\right)} = 0$$

$$\therefore \quad \frac{z_1}{z_2} \text{ is purely imaginary} \qquad \{\because\; z+\bar{z} = 2\text{Re}(z)\}$$

$$\therefore \quad \left|\text{amp}\,\frac{z_1}{z_2}\right| = \frac{\pi}{2}.$$

108. $(z_1+z_2)(\bar{z}_1+\bar{z}_2) = z_1\bar{z}_1 + z_2\bar{z}_2$

or $\quad z_1\bar{z}_2 + z_2\bar{z}_1 = 0$. Now see the hint to Q. No. 107.

□

4

Permutation and Combination

• Type 1 •

Choose the most appropriate option (a, b, c or d).

1. If ${}^nC_{r-1} = 56$, ${}^nC_r = 28$ and ${}^nC_{r+1} = 8$ then r is equal to

(a) 8 (b) 6

(c) 5 (d) none of these

2. The value of ${}^{40}C_{31} + \sum_{j=0}^{10} {}^{40+j}C_{10+j}$ is equal to

(a) ${}^{51}C_{20}$ (b) $2 \cdot {}^{50}C_{20}$

(c) $2 \cdot {}^{45}C_{15}$ (d) none of these

3. In a group of boys, the number of arrangements of 4 boys is 12 times the number of arrangements of 2 boys. The number of boys in the group is

(a) 10 (b) 8

(c) 6 (d) none of these

4. The value of $\sum_{r=1}^{10} r \cdot {}^rP_r$ is

(a) ${}^{11}P_{11}$ (b) ${}^{11}P_{11} - 1$

(c) ${}^{11}P_{11} + 1$ (d) none of these

5. From a group of persons the number of ways of selecting 5 persons is equal to that of 8 persons. The number of persons in the group is

(a) 13 (b) 40
(c) 18 (d) 21

6. The number of distinct rational numbers x such that $0 < x < 1$ and $x = \frac{p}{q}$, where $p, q \in \{1, 2, 3, 4, 5, 6\}$, is
(a) 15 (b) 13
(c) 12 (d) 11

7. The total number of 9-digit numbers of different digits is
(a) 10 (9 !) (b) 8 (9 !)
(c) 9 (9 !) (d) none of these

8. The number of 6-digit numbers that can be made with the digits 0,1, 2, 3, 4 and 5 so that even digits occupy odd places, is
(a) 24 (b) 36
(c) 48 (d) none of these

9. The number of ways in which 6 men can be arranged in a row so that three particular men are consecutive, is
(a) ${}^4P_4$ (b) ${}^4P_4 \times {}^3P_3$
(c) ${}^3P_3 \times {}^3P_3$ (d) none of these

10. Seven different lecturers are to deliver lectures in seven periods of a class on a particular day. A, B and C are three of the lecturers. The number of ways in which a routine for the day can be made such that A delivers his lecture before B, and B before C, is
(a) 420 (b) 120
(c) 210 (d) none of these

11. The total number of 5-digit numbers of different digits in which the digit in the middle is the largest is
(a) $\sum_{n=4}^{9} {}^nP_4$ (b) 33 (3 !)
(c) 30 (3 !) (d) none of these

12. A 5-digit number divisible by 3 is to be formed using the digits 0, 1, 2, 3, 4 and 5 without repetition. The total number of ways in which this can be done is
(a) 216 (b) 600
(c) 240 (d) 3125

13. Let $A = \{x \mid x$ is a prime number and $x < 30\}$. The number of different rational numbers whose numerator and denominator belong to A is

(a) 90 (b) 180

(c) 91 (d) none of these

14. The total number of ways in which six '+' and four '–' signs can be arranged in a line such that no two '–' signs occur together is

(a) $\frac{7!}{3!}$ (b) $6! \times \frac{7!}{3!}$

(c) 35 (d) none of these

15. The total number of words that can be made by writing the letters of the word PARAMETER so that no vowel is between two consonants is

(a) 1440 (b) 1800

(c) 2160 (d) none of these

16. The number of numbers of four different digits that can be formed from the digits of the number 12 356 such that the numbers are divisible by 4, is

(a) 36 (b) 48

(c) 12 (d) 24

17. Let S be the set of all functions from the set A to the set A. If $n(A) = k$ then $n(S)$ is

(a) $k!$ (b) k^k

(c) $2^k - 1$ (d) 2^k

18. Let A be the set of 4-digit numbers $a_1a_2a_3a_4$ where $a_1 > a_2 > a_3 > a_4$ then $n(A)$ is equal to

(a) 126 (b) 84

(c) 210 (d) none of these

19. The number of numbers divisible by 3 that can be formed by four different even digits is

(a) 18 (b) 36

(c) 0 (d) none of these

20. The number of 5-digit even numbers that can be made with the digits 0, 1, 2 and 3 is

(a) 384 (b) 192

(c) 768 (d) none of these

21. The number of 4-digit numbers that can be made with the digits 1, 2, 3, 4 and 5 in which at least two digits are identical, is

(a) $4^5 - 5!$ (b) 505
(c) 600 (d) none of these

22. The number of words that can be made by rearranging the letters of the word APURBA so that vowels and consonants alternate is

(a) 18 (b) 35
(c) 36 (d) none of these

23. The number of words that can be made by writing down the letters of the word CALCULATE such that each word starts and ends with a consonant, is

(a) $\frac{5(7!)}{2}$ (b) $\frac{3(7!)}{2}$
(c) $2(7!)$ (d) none of these

24. The number of numbers of 9 different nonzero digits such that all the digits in the first four places are less than the digit in the middle and all the digits in the last four places are greater than that in the middle is

(a) $2(4!)$ (b) $(4!)^2$
(c) $8!$ (d) none of these

25. In the decimal system of numeration the number of 6-digit numbers in which the digit in any place is greater than the digit to the left of it is

(a) 210 (b) 84
(c) 126 (d) none of these

26. The number of 5-digit numbers in which no two consecutive digits are identical is

(a) $9^2 \times 8^3$ (b) 9×8^4
(c) 9^5 (d) none of these

27. In the decimal system of numeration the number of 6-digit numbers in which the sum of the digits is divisible by 5 is

(a) 180000 (b) 540000
(c) 5×10^5 (d) none of these

28. The sum of all the numbers of four different digits that can be made by using the digits 0, 1, 2 and 3 is

(a) 26664 (b) 39996
(c) 38664 (d) none of these

29. A teacher takes 3 children from her class to the zoo at a time as often as she can, but she does not take the same three children to the zoo more than once. She finds that she goes to the zoo 84 times more than a particular child goes to the zoo. The number of children in her class is

(a) 12 (b) 10

(c) 60 (d) none of these

30. ABCD is a convex quadrilateral. 3, 4, 5 and 6 points are marked on the sides AB, BC, CD and DA respectively. The number of triangles with vertices on different sides is

(a) 270 (b) 220

(c) 282 (d) none of these

31. There are 10 points in a plane of which no three points are collinear and 4 points are concyclic. The number of different circles that can be drawn through at least 3 points of these points is

(a) 116 (b) 120

(c) 117 (d) none of these

32. In a polygon the number of diagonals is 54. The number of sides of the polygon is

(a) 10 (b) 12

(c) 9 (d) none of these

33. In a polygon no three diagonals are concurrent. If the total number of points of intersection of diagonals interior to the polygon be 70 then the number of diagonals of the polygon is

(a) 20 (b) 28

(c) 8 (d) none of these

34. n lines are drawn in a plane such that no two of them are parallel and no three of them are concurrent. The number of different points at which these lines will cut is

(a) $\sum_{k=1}^{n-1} k$ (b) $n(n-1)$

(c) n^2 (d) none of these

35. The number of triangles that can be formed with 10 points as vertices, n of them being collinear, is 110. Then n is

(a) 3 (b) 4

(c) 5 (d) 6

36. There are three coplanar parallel lines. If any p points are taken on each of the lines, the maximum number of triangles with vertices at these points is

(a) $3p^2(p-1)+1$ (b) $3p^2(p-1)$

(c) $p^2(4p-3)$ (d) none of these

37. Two teams are to play a series of 5 matches between them. A match ends in a win or loss or draw for a team. A number of people forecast the result of each match and no two people make the same forecast for the series of matches. The smallest group of people in which one person forecasts correctly for all the matches will contain n people, where n is

(a) 81 (b) 243

(c) 486 (d) none of these

38. A bag contains 3 black, 4 white and 2 red balls, all the balls being different. The number of selections of at most 6 balls containing balls of all the colours is

(a) $42\,(4\,!)$ (b) $2^6 \times 4\,!$

(c) $(2^6-1)(4\,!)$ (d) none of these

39. In a room there are 12 bulbs of the same wattage, each having a separate switch. The number of ways to light the room with different amounts of illumination is

(a) 12^2-1 (b) 2^{12}

(c) $2^{12}-1$ (d) none of these

40. In an examination of 9 papers a candidate has to pass in more papers than the number of papers in which he fails in order to be successful. The number of ways in which he can be unsuccessful is

(a) 255 (b) 256

(c) 193 (d) 319

41. The number of 5-digit numbers that can be made using the digits 1 and 2 and in which at least one digit is different, is

(a) 30 (b) 31

(c) 32 (d) none of these

42. In a club election the number contestants is one more than the number of maximum candidates for which a voter can vote. If the total number of ways in which a voter can vote be 62 then the number of candidates is

(a) 7 (b) 5

(c) 6 (d) none of these

43. The total number of selections of at most n things from $(2n+1)$ different things is 63. Then the value of n is

(a) 3 (b) 2
(c) 4 (d) none of these

44. Let $1 \le m < n \le p$. The number of subsets of the set $A = \{1, 2, 3, \ldots, p\}$ having m, n as the least and the greatest elements respectively, is

(a) $2^{n-m-1} - 1$ (b) 2^{n-m-1}
(c) 2^{n-m} (d) none of these

45. The number of ways in which n different prizes can be distributed among $m (< n)$ persons if each is entitled to receive at most $n-1$ prizes, is

(a) $n^m - n$ (b) m^n
(c) mn (d) none of these

46. The number of possible outcomes in a throw of n ordinary dice in which at least one of the dice shows an odd number is

(a) $6^n - 1$ (b) $3^n - 1$
(c) $6^n - 3^n$ (d) none of these

47. The number of different 6-digit numbers that can be formed using the three digits 0, 1 and 2 is

(a) 3^6 (b) 2×3^5
(c) 3^5 (d) none of these

48. The number of different matrices that can be formed with elements 0, 1, 2, or 3, each matrix having 4 elements, is

(a) 3×2^4 (b) 2×4^4
(c) 3×4^4 (d) none of these

49. Let A be a set of $n (\ge 3)$ distinct elements. The number of triplets (x, y, z) of the elements of A in which at least two coordinates are equal is

(a) ${}^nP_3$ (b) $n^3 - {}^nP_3$
(c) $3n^2 - 2n$ (d) $3n^2(n-1)$

50. The number of different pairs of words (□□□□, □□□) that can be made with the letters of the word STATICS is

(a) 828 (b) 1260
(c) 396 (d) none of these

51. Total number of 6-digit numbers in which all the odd digits and only odd digits appear, is

(a) $\frac{5}{2}(6\,!)$ (b) $6\,!$

(c) $\frac{1}{2}(6\,!)$ (d) none of these

52. The number of divisors of the form $4n+2$ $(n \geq 0)$ of the integer 240 is

(a) 4 (b) 8

(c) 10 (d) 3

53. In the next World Cup of cricket there will be 12 teams, divided equally in two groups. Teams of each group will play a match against each other. From each group 3 top teams will qualify for the next round. In this round each team will play against others once. Four top teams of this round will qualify for the semifinal round, where each team will play against the others once. Two top teams of this round will go to the final round, where they will play the best of three matches. The minimum number of matches in the next World Cup will be

(a) 54 (b) 53

(c) 38 (d) none of these

54. The number of different ways in which 8 persons can stand in a row so that between two particular persons A and B there are always two persons, is

(a) $60(5\,!)$ (b) $15(4\,!) \times (5\,!)$

(c) $4\,! \times 5\,!$ (d) none of these

55. Four couples (husband and wife) decide to form a committee of four members. The number of different committees that can be formed in which no couple finds a place is

(a) 10 (b) 12

(c) 14 (d) 16

56. From 4 gentlemen and 6 ladies a committee of five is to be selected. The number of ways in which the committee can be formed so that gentlemen are in majority is

(a) 66 (b) 156

(c) 60 (d) none of these

57. There are 20 questions in a question paper. If no two students solve the same combination of questions but solve equal number of questions then the maximum number of students who appeared in the examination is

(a) ${}^{20}C_9$ (b) ${}^{20}C_{11}$

(c) ${}^{20}C_{10}$ (d) none of these

58. Nine hundred distinct n-digit positive numbers are to be formed using only the digits 2, 5 and 7. The smallest value of n for which this is possible is

(a) 6 (b) 7

(c) 8 (d) 9

59. The total number of integral solutions for (x, y, z) such that $xyz = 24$ is

(a) 36 (b) 90

(c) 120 (d) none of these

60. The number of ways in which the letters of the word ARTICLE can be rearranged so that the even places are always occupied by consonants is

(a) 576 (b) ${}^4C_3 \times (4\,!)$

(c) $2(4\,!)$ (d) none of these

61. A cabinet of ministers consists of 11 ministers, one minister being the chief minister. A meeting is to be held in a room having a round table and 11 chairs round it, one of them being meant for the chairman. The number of ways in which the ministers can take their chairs, the chief minister occupying the chairman's place, is

(a) $\frac{1}{2}(10\,!)$ (b) $9\,!$

(c) $10\,!$ (d) none of these

62. The number of ways in which a couple can sit around a table with 6 guests if the couple take consecutive seats is

(a) 1440 (b) 720

(c) 5040 (d) none of these

63. The number of ways in which 20 different pearls of two colours can be set alternately on a necklace, there being 10 pearls of each colour, is

(a) $9\,! \times 10\,!$ (b) $5(9\,!)^2$

(c) $(9\,!)^2$ (d) none of these

64. If $r > p > q$, the number of different selections of $p + q$ things taking r at a time, where p things are identical and q other things are identical, is

(a) $p + q - r$ (b) $p + q - r + 1$

(c) $r - p - q + 1$ (d) none of these

65. There are 4 mangoes, 3 apples, 2 oranges and 1 each of 3 other varieties of fruits. The number of ways of selecting at least one fruit of each kind is

(a) 10 !   (b) 9 !

(c) 4 !   (d) none of these

66. The number of proper divisors of $2^p \cdot 6^q \cdot 15^r$ is

(a) $(p+q+1)(q+r+1)(r+1)$

(b) $(p+q+1)(q+r+1)(r+1)-2$

(c) $(p+q)(q+r)r-2$

(d) none of these

67. The number of proper divisors of 1 800 which are also divisible by 10, is

(a) 18   (b) 34

(c) 27   (d) none of these

68. The number of odd proper divisors of $3^p \cdot 6^m \cdot 21^n$ is

(a) $(p+1)(m+1)(n+1)-2$ (b) $(p+m+n+1)(n+1)-1$

(c) $(p+1)(m+1)(n+1)-1$ (d) none of these

69. The number of even proper divisors of 1 008 is

(a) 23   (b) 24

(c) 22   (d) none of these

70. In a test there were n questions. In the test 2^{n-i} students gave wrong answers to i questions where $i = 1, 2, 3, \dots, n$. If the total number of wrong answers given is 2 047 then n is

(a) 12   (b) 11

(c) 10   (d) none of these

71. The number of ways to give 16 different things to three persons A, B, C so that B gets 1 more than A and C gets 2 more than B, is

(a) $\dfrac{16!}{4!5!7!}$ (b) $4!5!7!$

(c) $\dfrac{16!}{3!5!8!}$ (d) none of these

72. The number of ways to distribute 32 different things equally among 4 persons is

(a) $\frac{32\,!}{(8\,!)^3}$ (b) $\frac{32\,!}{(8\,!)^4}$

(c) $\frac{1}{4}(32\,!)$ (d) none of these

73. If $3n$ different things can be equally distributed among 3 persons in k ways then the number of ways to divide the $3n$ things in 3 equal groups is

(a) $k \times 3\,!$ (b) $\frac{k}{3\,!}$

(c) $(3\,!)^k$ (d) none of these

74. In a packet there are m different books, n different pens and p different pencils. The number of selections of at least one article of each type from the packet is

(a) $2^{m+n+p} - 1$ (b) $(m+1)(n+1)(p+1) - 1$

(c) 2^{m+n+p} (d) none of these

75. The number of 6-digit numbers that can be made with the digits 1, 2, 3 and 4 and having exactly two pairs of digits is

(a) 480 (b) 540

(c) 1 080 (d) none of these

76. The number of words of four letters containing equal number of vowels and consonants, repetition being allowed, is

(a) 105^2 (b) 210×243

(c) 105×243 (d) none of these

77. The number of ways in which 6 different balls can be put in two boxes of different sizes so that no box remains empty is

(a) 62 (b) 64

(c) 36 (d) none of these

78. A shopkeeper sells three varieties of perfumes and he has a large number of bottles of the same size of each variety in his stock. There are 5 places in a row in his showcase. The number of different ways of displaying the three varieties of perfumes in the showcase is

(a) 6 (b) 50

(c) 150 (d) none of these

79. The number of arrangements of the letters of the word BHARAT taking 3 at a time is

(a) 72 (b) 120

(c) 14 (d) none of these

80. The number of ways to fill each of the four cells of the table with a distinct natural number such that the sum of the numbers is 10 and the sums of the numbers placed diagonally are equal, is

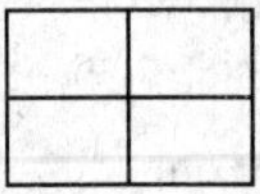

(a) $2! \times 2!$ (b) $4!$

(c) $2(4!)$ (d) none of these

81. In the figure, two 4-digit numbers are to be formed by filling the places with digits. The number of different ways in which the places can be filled by digits so that the sum of the numbers formed is also a 4-digit number and in no place the addition is with carrying, is

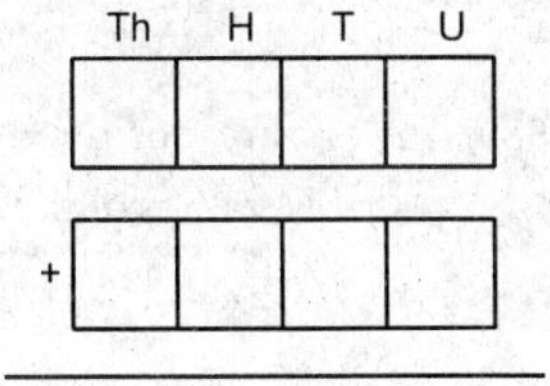

(a) 55^4 (b) 220

(c) 45^4 (d) none of these

82. The number of positive integral solutions of $x + y + z = n, n \in N, n \geq 3$, is

(a) ${}^{n-1}C_2$ (b) ${}^{n-1}P_2$

(c) $n(n-1)$ (d) none of these

83. The number of non-negative integral solutions of $a + b + c + d = n$, $n \in N$, is

(a) ${}^{n+3}P_2$ (b) $\dfrac{(n+1)(n+2)(n+3)}{6}$

(c) ${}^{n-1}C_{n-4}$ (d) none of these

84. The number of points (x, y, z) in space, whose each coordinate is a negative integer such that $x + y + z + 12 = 0$, is

(a) 385 (b) 55

(c) 110 (d) none of these

85. If a, b, c are three natural numbers in AP and $a + b + c = 21$ then the possible number of values of the ordered triplet (a, b, c) is

(a) 15 (b) 14

(c) 13 (d) none of these

86. If a, b, c, d are odd natural numbers such that $a+b+c+d=20$ then the number of values of the ordered quadruplet (a, b, c, d) is

(a) 165 (b) 455

(c) 310 (d) none of these

87. If x, y, z are integers and $x \geq 0, y \geq 1, z \geq 2, x+y+z=15$ then the number of values of the ordered triplet (x, y, z) is

(a) 91 (b) 455

(c) ${}^{17}C_{15}$ (d) none of these

88. If a, b, c are positive integers such that $a+b+c \leq 8$ then the number of possible values of the ordered triplet (a, b, c) is

(a) 84 (b) 56

(c) 83 (d) none of these

89. The number of different ways of distributing 10 marks among 3 questions, each question carrying at least 1 mark, is

(a) 72 (b) 71

(c) 36 (d) none of these

90. The number of ways to give away 20 apples to 3 boys, each boy receiving at least 4 apples, is

(a) ${}^{10}C_8$ (b) 90

(c) ${}^{22}C_{20}$ (d) none of these

91. The position vector of a point P is $\vec{r} = x\vec{i} + y\vec{j} + z\vec{k}$, where $x \in N$, $y \in N$, $z \in N$ and $\vec{a} = \vec{i} + \vec{j} + \vec{k}$. If $\vec{r} \cdot \vec{a} = 10$, the number of possible positions of P is

(a) 36 (b) 72

(c) 66 (d) none of these

• Type 2 •

Choose the correct options. One or more options may be correct.

92. If $P = n(n^2 - 1^2)(n^2 - 2^2)(n^2 - 3^2) \ldots (n^2 - r^2), n > r, n \in N$, then P is divisible by

(a) $(2r+2)!$ (b) $(2r-1)!$

(c) $(2r+1)!$ (d) none of these

93. If $^{n+5}P_{n+1} = \frac{11(n-1)}{2} \cdot {}^{n+3}P_n$ then the value of n is

(a) 7 (b) 8

(c) 6 (d) 5

94. If $^nC_4$, $^nC_5$ and $^nC_6$ are in AP then n is

(a) 8 (b) 9

(c) 14 (d) 7

95. The product of r consecutive integers is divisible by

(a) r (b) $\sum_{k=1}^{r-1} k$

(c) $r!$ (d) none of these

96. There are 10 bags $B_1, B_2, B_3, \ldots, B_{10}$, which contain 21, 22, ... , 30 different articles respectively. The total number of ways to bring out 10 articles from a bag is

(a) $^{31}C_{20} - {}^{21}C_{10}$ (b) $^{31}C_{21}$

(c) $^{31}C_{20}$ (d) none of these

97. If the number of arrangements of $n-1$ things taken from n different things is k times the number of arrangements of $n-1$ things taken from n things in which two things are identical then the value of k is

(a) $\frac{1}{2}$ (b) 2

(c) 4 (d) none of these

98. Kanchan has 10 friends among whom two are married to each other. She wishes to invite 5 of them for a party. If the married couple refuse to attend separately then the number of different ways in which she can invite five friends is

(a) $^8C_5$ (b) $2 \times {}^8C_3$

(c) $^{10}C_5 - 2 \times {}^8C_4$ (d) none of these

99. In a plane there are two families of lines $y = x + r$, $y = -x + r$, where $r \in \{0, 1, 2, 3, 4\}$. The number of squares of diagonals of the length 2 formed by the lines is

(a) 9 (b) 16

(c) 25 (d) none of these

100. There are n seats round a table numbered 1, 2, 3, ..., n. The number of ways in which $m(\leq n)$ persons can take seats is

(a) ${}^{n}P_{m}$ (b) ${}^{n}C_{m} \times (m-1)!$

(c) ${}^{n-1}P_{m-1}$ (d) ${}^{n}C_{m} \times m!$

101. Let $\vec{a} = \vec{i} + \vec{j} + \vec{k}$ and let $\vec{r}$ be a variable vector such that $\vec{r} \cdot \vec{i}$, $\vec{r} \cdot \vec{j}$ and $\vec{r} \cdot \vec{k}$ are positive integers. If $\vec{r} \cdot \vec{a} \leq 12$ then the number of values of $\vec{r}$ is

(a) ${}^{12}C_{9} - 1$ (b) ${}^{12}C_{3}$

(c) ${}^{12}C_{9}$ (d) none of these

102. The total number of ways in which a beggar can be given at least one rupee from four 25-paisa coins, three 50-paisa coins and 2 one-rupee coins, is

(a) 54 (b) 53

(c) 51 (d) none of these

103. For the equation $x + y + z + w = 19$, the number of positive integral solutions is equal to

(a) the number of ways in which 15 identical things can be distributed among 4 persons

(b) the number of ways in which 19 identical things can be distributed among 4 persons

(c) coefficient of x^{19} in $(x^0 + x^1 + x^2 + \ldots + x^{19})^4$

(d) coefficient of x^{19} in $(x + x^2 + x^3 + \ldots + x^{19})^4$

Answers

1. b	**2.** a	**3.** c	**4.** b	**5.** a
6. d	**7.** c	**8.** a	**9.** b	**10.** d
11. d	**12.** a	**13.** c	**14.** c	**15.** b
16. a	**17.** b	**18.** c	**19.** b	**20.** a
21. b	**22.** c	**23.** a	**24.** b	**25.** b
26. c	**27.** a	**28.** c	**29.** b	**30.** d
31. c	**32.** b	**33.** a	**34.** a	**35.** c
36. c	**37.** b	**38.** a	**39.** c	**40.** b
41. a	**42.** c	**43.** a	**44.** b	**45.** d
46. c	**47.** b	**48.** c	**49.** c	**50.** b
51. a	**52.** a	**53.** b	**54.** a	**55.** d
56. a	**57.** c	**58.** b	**59.** c	**60.** a
61. c	**62.** a	**63.** b	**64.** b	**65.** c
66. b	**67.** a	**68.** b	**69.** a	**70.** b
71. a	**72.** b	**73.** b	**74.** a	**75.** c
76. b	**77.** a	**78.** c	**79.** a	**80.** d
81. d	**82.** a	**83.** b	**84.** b	**85.** c
86. a	**87.** a	**88.** b	**89.** c	**90.** a
91. a	**92.** b, c	**93.** a, c	**94.** c, d	**95.** a, b, c
96. a	**97.** b	**98.** b, c	**99.** a	**100.** a, d
101. b, c	**102.** a	**103.** a, d		

Hints and Solutions to Selected Questions

2. Value $= ({}^{40}C_9 + {}^{40}C_{10}) + {}^{41}C_{11} + \ldots + {}^{50}C_{20}$

$= ({}^{41}C_{10} + {}^{41}C_{11}) + {}^{42}C_{12} + \ldots + {}^{50}C_{20}$

$= \ldots = {}^{50}C_{19} + {}^{50}C_{20} = {}^{51}C_{20}$.

4. $r \cdot {}^rP_r = (r+1-1) \times r\,! = (r+1)\,! - r\,!$

$$\therefore \quad \sum_{r=1}^{10} r \cdot {}^rP_r = \sum_{r=1}^{10} \{(r+1)\,! - r\,!\}$$

$$= (2\,! - 1\,!) + (3\,! - 2\,!) + \ldots + (11\,! - 10\,!) = 11\,! - 1.$$

6. As $0 < x < 1$, we have $p < q$.

The number of rational numbers $= 5 + 4 + 3 + 2 + 1 = 15$.

When p, q have a common factor, we get some rational numbers which are not different from those already counted. There are 4 such numbers: $\frac{2}{4}, \frac{2}{6}, \frac{3}{6}, \frac{4}{6}$.

$\therefore$ the required number of rational numbers $= 15 - 4 = 11$.

7. The first place from the left can be filled in 9 ways (any one except 0).

The other eight places can be filled by the remaining 9 digits in ${}^9P_8$ ways.

$\therefore$ the number of 9-digit numbers $= 9 \times {}^9P_8$.

8. $\times | \times | \times |$ Crosses can be filled in ${}^3P_3 - {}^2P_2$ ways

($\because$ 0 cannot go in the first place from the left).

The remaining places can be filled in $3\,!$ ways.

$\therefore$ the required number of numbers $= ({}^3P_3 - {}^2P_2) \times 3\,!$.

10. As the order of A, B, C is not to change they are to be treated identical in arrangement. So, the required number of ways $= \dfrac{7\,!}{3\,!}$.

11. The number of numbers with 4 in the middle $= {}^4P_4 - {}^3P_3$

($\because$ the other four places are to be filled by 0, 1, 2 and 3, and a number cannot begin with 0).

Similarly, the number of numbers with 5 in the middle $= {}^5P_4 - {}^4P_3$, etc.

$\therefore$ the required number of numbers

$$= ({}^4P_4 - {}^3P_3) + ({}^5P_4 - {}^4P_3) + ({}^6P_4 - {}^5P_3) + \dots + ({}^9P_4 - {}^8P_3).$$

12. The numbers are made of 1, 2, 3, 4, 5 or 0, 1, 2, 4, 5.

$\therefore$ the required number of 5-digit numbers $= 5\,! + ({}^5P_5 - {}^4P_4)$.

13. $A = \{2, 3, 5, 7, 11, 13, 17, 19, 23, 29\}$. A rational number is made by taking any two in any order.

$\therefore$ the required number of rational numbers $= {}^{10}P_2 + 1$ (including 1).

14. '–' signs will be put between two '+' signs or at the two ends.

There are 7 places for four '–' signs. So, the required number of ways

$$= {}^7C_4 = \frac{7\,!}{4\,!\,3\,!}$$

(there being no arrangement as the '+' signs are identical as well as '–' signs are identical).

16. $\times\times\left|\begin{matrix}1, 3, 5\\ \times\end{matrix}\right|\begin{matrix}2, 6\\ \times\end{matrix}$

The units place can be filled in 2 ways.

The tens place can be filled in 3 ways.

The remaining two places can be filled by any two of the remaining three digits. So, the required number of numbers $= 2 \times 3 \times {}^3P_2$.

17. Each element of the set A can be given the image in the set A in k ways.

$\therefore$ the required number of functions, i.e., $n(S) = k \times k \times \dots$ (k times) $= k^k$.

18. Any selection of four digits from the ten digits 0, 1, 2, 3, ... , 9 gives one number. So, the required number of numbers $= {}^{10}C_4$.

19. The numbers will be made by 0, 2, 4, 6 or 0, 4, 6, 8.

$\therefore$ the required number of numbers $= ({}^4P_4 - {}^3P_3) + ({}^4P_4 - {}^3P_3)$.

21. The number of numbers when repetition is allowed $= 5^4$.

The number of numbers when digits cannot be repeated $= {}^5P_5$.

$\therefore$ the required number of numbers $= 5^4 - 5\,!$.

23. The number of words like C—C $= \dfrac{7!}{2!2!} =$ the number of words like L—L.

The number of words beginning or ending with C, L $= 2 \times \dfrac{7!}{2!}$.

The number of words beginning or ending with (C or L), $T = 2(2!) \times \dfrac{7!}{2!2!}$.

$\therefore$ the required number of words

$$= \frac{7!}{2!2!} + \frac{7!}{2!2!} + 2 \times \frac{7!}{2!} + 2(2!) \times \frac{7!}{2!2!}$$

$$= \left(\frac{1}{4} + \frac{1}{4} + 1 + 1\right)(7!).$$

24. $\underset{\times\times\times\times}{\underline{1, 2, 3, 4}} \quad 5 \quad \underset{\times\times\times\times}{\underline{6, 7, 8, 9}}$ The required number of numbers $= {}^4P_4 \times {}^4P_4$.

25. Any selection of six digits from 9 digits (excluding 0) will give a number of the required variety.

$\therefore$ the required number of numbers $= {}^9C_6$.

26.

	×	×	×	×	×
Ways:	9	9	9	9	9

0 cannot be placed in the first place. In the next place any digit except the one used in the first place can be used, etc.

27.

	×	×	×	×	×	×
Ways:	9	10	10	10	10	6

0 cannot go in the first place. In other four places any digit can go. After filling the first five places, the last place can be filled by 0 or 5, 1 or 6, 2 or 7, 3 or 8, 4 or 9 depending upon whether the sum of five digits filled is of the form $5m$, $5m+4$, $5m+3$, $5m+2$ or $5m+1$ respectively.

$\therefore$ the required number of numbers $= 9 \times 10^4 \times 2$.

28. The number of numbers with 0 in the units place $= 3! = 6$.

The number of numbers with 1 or 2 or 3 in the units place $= 3! - 2! = 4$.

$\therefore$ the sum of the digits in the units place $= 6 \times 0 + 4 \times 1 + 4 \times 2 + 4 \times 3 = 24$.

Similarly for the tens and the hundreds places.

The number of numbers with 1 or 2 or 3 in the thousands place $= 3!$.

$\therefore$ the sum of the digits in the thousands place

$$= 6 \times 1 + 6 \times 2 + 6 \times 3 = 36.$$

$\therefore$ the required sum $= 36 \times 1000 + 24 \times 100 + 24 \times 10 + 24$.

29. The number of times the teacher goes to the zoo $= {}^nC_3$.

The number of times a particular child goes to the zoo $= {}^{n-1}C_2$.

From the question, ${}^nC_3 - {}^{n-1}C_2 = 84$

or $(n-1)(n-2)(n-3) = 6 \times 84 = 9 \times 8 \times 7 \Rightarrow n - 1 = 9$.

30. 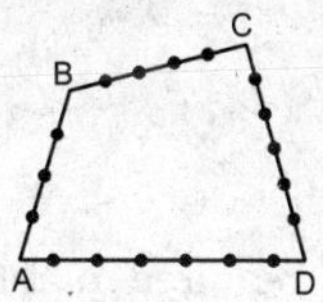

The number of triangles with vertices on sides AB, BC, CD $= {}^3C_1 \times {}^4C_1 \times {}^5C_1$.

Similarly for other cases.

$\therefore$ the total number of triangles

$$= {}^3C_1 \times {}^4C_1 \times {}^5C_1 + {}^3C_1 \times {}^4C_1 \times {}^6C_1 + {}^3C_1 \times {}^5C_1 \times {}^6C_1 + {}^4C_1 \times {}^5C_1 \times {}^6C_1 = 342.$$

31. The number of circles $= ({}^{10}C_3 - {}^4C_3) + 1$.

33. A selection of four vertices of the polygon gives an interior intersection.

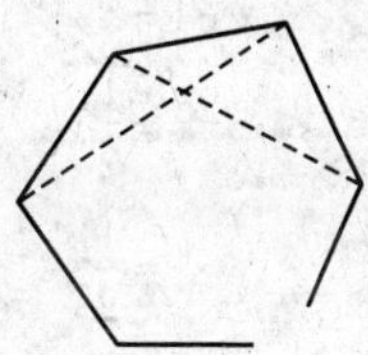

$\therefore$ the number of sides $= n \Rightarrow {}^nC_4 = 70$

$\Rightarrow n(n-1)(n-2)(n-3) = 24 \times 70 = 8 \times 7 \times 6 \times 5$

$\therefore n = 8$

$\therefore$ the number of diagonals $= {}^8C_2 - 8$.

36. The number of triangles with vertices on different lines $= {}^pC_1 \times {}^pC_1 \times {}^pC_1 = p^3$.

The number of triangles with 2 vertices on one line and the third vertex on any one of the other two lines

$$= {}^3C_1\{{}^pC_2 \times {}^{2p}C_1\} = 6p \cdot \frac{p(p-1)}{2}$$

$\therefore$ the required number of triangles $= p^3 + 3p^2(p-1)$.

Note The word "maximum" ensures that no selection of points from each of the three lines are collinear.

37. The smallest number of people

= total number of possible forecasts

= total number of possible results

$= 3 \times 3 \times 3 \times 3 \times 3$.

38. The required number of selections

$$= {}^3C_1 \times {}^4C_1 \times {}^2C_1({}^6C_3 + {}^6C_2 + {}^6C_1 + {}^6C_0).$$

40. The candidate is unsuccessful if he fails in 9 or 8 or 7 or 6 or 5 papers.

$\therefore$ the number of ways to be unsuccessful

$$= {}^9C_9 + {}^9C_8 + {}^9C_7 + {}^9C_6 + {}^9C_5$$

$$= {}^9C_0 + {}^9C_1 + {}^9C_2 + {}^9C_3 + {}^9C_4$$

$$= \frac{1}{2}({}^9C_0 + {}^9C_1 + \ldots + {}^9C_9) = \frac{1}{2} \cdot 2^9 = 2^8.$$

41. Total number of numbers without restriction $= 2^5$.

Two numbers have all the digits equal. So, the required number of numbers $= 2^5 - 2$.

42. $62 = {}^nC_1 + {}^nC_2 + {}^nC_3 + \ldots + {}^nC_{n-1} = 2^n - {}^0C_0 - {}^nC_n = 2^n - 2$

$\therefore \quad 2^n = 64 = 2^6.$

43. $63 = {}^{2n+1}C_1 + {}^{2n+1}C_2 + \ldots + {}^{2n+1}C_n$

$\therefore \quad 64 = {}^{2n+1}C_0 + {}^{2n+1}C_1 + \ldots + {}^{2n+1}C_n$

$$= \frac{1}{2}({}^{2n+1}C_0 + {}^{2n+1}C_1 + \ldots + {}^{2n+1}C_{2n+1})$$

$$= \frac{1}{2} \cdot 2^{2n+1}.$$

44. Total number of subsets

= the number of selections of at least two elements including m, n and natural numbers lying between m and n

= total number of selections from $n - m - 1$ different things

$= 2^{n-m-1}$.

45. Total number of ways $= m \times m \times \ldots$ to n times $= m^n$.

The number of ways in which one gets all the prizes $= m$.

$\therefore$ the required number of ways $= m^n - m$.

46. Total number of ways $= 6 \times 6 \times \ldots$ to n times $= 6^n$.

Total number of ways to show only even numbers $= 3 \times 3 \times \ldots$ to n times $= 3^n$.

$\therefore$ the required number of ways $= 6^n - 3^n$.

47.

	×	×	×	×	×	×
Ways:	2	3	3	3	3	3

0 cannot go in the units place.

$\therefore$ the required number of numbers $= 2 \times 3^5$.

48. The matrix will be of the order 4×1 or 1×4 or 2×2.

For each order, the number of different matrices

= the number of ways to fill four places by 0, 1, 2, 3

$= 4 \times 4 \times 4 \times 4$.

49. Total number of triplets without restriction $= n \times n \times n$.

The number of triplets with all different coordinates $= {}^nP_3$.

$\therefore$ the required number of triplets $= n^3 - n(n-1)(n-2)$.

50. From every arrangement of 7 letters we get a pair by putting a sign (,) after the four letters from the left.

$\therefore$ the required number of pairs $= \dfrac{7!}{2!\,2!}$.

51. Clearly, one of the odd digits 1, 3, 5, 7, 9 will be repeated.

The number of selections of the sixth digit $= {}^5C_1 = 5$.

$\therefore$ the required number of numbers $= 5 \times \dfrac{6!}{2}$.

52. $240 = 2^4 \times 3 \times 5$

$4n + 2 = 2(2n+1) = 2 \times \text{odd}$

$\therefore$ the required number of divisors

= the number of selections of one 2 from four 2's, any number of 3's from one 3 and any number of 5's from one 5.

$= 1 \times 2 \times 2 = 4$.

53. The number of matches in the first round $= {}^6C_2 + {}^6C_2$.

The number of matches in the next round $= {}^6C_2$.

The number of matches in the semifinal round $= {}^4C_2$.

$\therefore$ the required number of matches $= {}^6C_2 + {}^6C_2 + {}^6C_2 + {}^4C_2 + 2$.

Note For "best of three" at least two matches are played.

54. The number of selections of 4 persons including A, B $= {}^6C_2$.

Considering these four as a group, number of arrangements with the other four $= 5\,!$.

But in each group the number of arrangements $= 2\,! \times 2\,!$

$\therefore$ the required number of ways $= {}^6C_2 \times 5\,! \times 2\,! \times 2\,!$.

55. The number of committees of 4 gentlemen $= {}^4C_4 = 1$.

The number of committees of 3 gentlemen, 1 wife $= {}^4C_3 \times {}^1C_1$

($\because$ after selecting 3 gentlemen only 1 wife is left who can be included).

The number of committees of 2 gentlemen, 2 wives $= {}^4C_2 \times {}^2C_2$.

The number of committees of 1 gentleman, 3 wives $= {}^4C_1 \times {}^3C_3$.

The number of committees of 4 wives = 1.

$\therefore$ the required number of committees $= 1 + 4 + 6 + 4 + 1 = 16$.

56. The committee will consist of 4 gentlemen and 1 lady or 3 gentlemen and 2 ladies.

$\therefore$ the number of committees $= {}^4C_4 \times {}^6C_1 + {}^4C_3 \times {}^6C_2$.

57. If r questions are solved by each student then the number of possible selections of questions is ${}^{20}C_r$.

$\therefore$ the number of students $= {}^{20}C_r$.

($\because$ each student has solved different combinations of questions)

$\therefore$ the maximum number of students = maximum value of ${}^{20}C_r = {}^{20}C_{10}$, because ${}^{20}C_{10}$ is the largest among ${}^{20}C_0, {}^{20}C_1 \ldots, {}^{20}C_{20}$—being the middle one.

58. Each place can be filled in 3 ways.

So, n places can be filled in $3 \times 3 \times \ldots$ to n times $= 3^n$.

$3^6 = 729, 3^7 = 2187$. So, for $3^n \geq 900$ the smallest integral n is 7.

59. $24 = 2 \cdot 3 \cdot 4, 2 \cdot 2 \cdot 6, 1 \cdot 6 \cdot 4, 1 \cdot 3 \cdot 8, 1 \cdot 2 \cdot 12, 1 \cdot 1 \cdot 24$

(as product of three positive integers)

$\therefore$ the total number of positive integral solutions of $xyz = 24$ is equal to

$$3\,! + \frac{3\,!}{2\,!} + 3\,! + 3\,! + 3\,! + \frac{3\,!}{2\,!}, \text{ i.e., } 30.$$

Any two of the factors in each factorization may be negative.

$\therefore$ the number of ways to associate negative sign in each case is ${}^3C_2$, i.e., 3.

$\therefore$ the total number of integral solutions $= 30 + 3 \times 30 = 120$.

60. The number of ways to fill the three even places by 4 consonants $= {}^4P_3$.

After filling the even places, remaining places can be filled in ${}^4P_4$ ways.

So, the required number of words $= {}^4P_3 \times {}^4P_4$.

62. A couple and 6 guests can be arranged in $(7-1)!$ ways. But the two people forming the couple can be arranged among themselves in $2!$ ways.

$\therefore$ the required number of ways $= 6! \times 2!$.

63. Ten pearls of one colour can be arranged in $\frac{1}{2} \cdot (10-1)!$ ways. The number of arrangements of 10 pearls of the other colour in 10 places between the pearls of the first colour $= 10!$.

$\therefore$ the required number of ways $= \dfrac{1}{2} \times 9! \times 10!$.

64. The number of selections of p things from p identical things and $r-p$ things from q identical things $= 1 \times 1$.

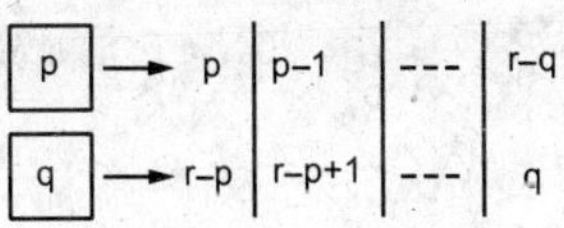

Similarly in all other cases.

$\therefore$ the total number of ways

$= p-(r-q)+1$ or $q-(r-p)+1$

$= p+q-r+1$.

66. $2^p \cdot 6^q \cdot 15^r = 2^{p+q} \cdot 3^{q+r} \cdot 5^r$

$\therefore$ the number of proper divisors

= {total number of selections from $(p+q)$ twos, $(q+r)$ threes and r fives} $-$ 2

$= (p+q+1)(q+r+1)(r+1) - 2$.

67. $1\,800 = 2^3 \times 3^2 \times 5^2$

$\therefore$ the required number of proper divisors

= total number of selections of at least one 2 and one 5 from 2, 2, 2, 3, 3, 5, 5

$= 3 \times (2+1) \times 2 = 18.$

68. $3^p \cdot 6^m \cdot 21^n = 2^m \cdot 3^{p+m+n} \cdot 7^n$

$\therefore$ the required number of proper divisors

= total number of selections of zero 2 and any number of 3's and 7's

$= (p+m+n+1)(n+1) - 1.$

69. $1\,008 = 2^4 \times 3^2 \times 7$

$\therefore$ the required number of even proper divisors

= total number of selections of at least one 2 and any number of 3's or 7's

$= 4 \times (2+1) \times (1+1) - 1.$

70. The number of students giving exactly one wrong answer

$= 2^{n-1} - 2^{n-2}.$

The number of students giving exactly two wrong answers

$= 2^{n-2} - 2^{n-3}$, etc.

$\therefore$ the total number of wrong answers

$= 1 \cdot (2^{n-1} - 2^{n-2}) + 2 \cdot (2^{n-2} - 2^{n-3}) + \ldots$

$\ldots + (n-1)(2 - 2^0) + n \cdot 1 = 2047$ (from question)

$\therefore \quad 2^{n-1} + 2^{n-2} + \ldots + 2 - (n-1) + n = 2\,047$

or $\quad \dfrac{2(2^{n-1} - 1)}{2-1} = 4046$ or $2^n = 4048 = 2^{11}.$

73. (The number of ways of dividing in 3 equal groups) $\times$ (3 !)
= the number of ways to distribute equally among 3 persons.

74. The required number of ways
= total number of ways of selecting any number of books from m different books, any number of pens from n different pens and any number of pencils from p different pencils -1

$= ({}^mC_0 + {}^mC_1 + \ldots + {}^mC_m)({}^nC_0 + {}^nC_1 + \ldots + {}^nC_n)$

$\times ({}^pC_0 + {}^pC_1 + \ldots + {}^pC_p) - 1$

$= 2^m \times 2^n \times 2^p - 1.$

75. The number will have 2 pairs and 2 different digits.

The number of selections $= {}^4C_2 \times {}^2C_2$, and for each selection, number of arrangements $= \dfrac{6!}{2!2!}$ · Therefore, the required number of numbers

$= {}^4C_2 \times {}^2C_2 \times \dfrac{6!}{2!\,2!}$

76. The number of selections of 1 pair of vowels and 1 pair of consonants

$= {}^5C_1 \times {}^{21}C_1$.

The number of selections of 2 different vowels and 2 different consonants

$= {}^5C_2 \times {}^{21}C_2$.

∴ the required number of words

$= {}^5C_1 \times {}^{21}C_1 \times \dfrac{4!}{2!2!} + {}^5C_2 \times {}^{21}C_2 \times 4!$.

77. Each ball can be put in 2 ways (either in one box or the other).

∴ 6 balls can be put in $2 \times 2 \times \ldots$ to six times, i.e., 2^6 ways. But in two of the ways one box is empty. So, the required number of ways $= 2^6 - 2$.

78.

Possibilities	*Selections*	*Arrangements*
One triplet, two different	${}^3C_1 \times {}^2C_2$	${}^3C_1 \times {}^2C_2 \times \dfrac{5!}{3!} = 60$
Two pairs, one different	${}^3C_2 \times {}^1C_1$	${}^3C_2 \times {}^1C_1 \times \dfrac{5!}{2!2!} = 90$

∴ the required number of ways = 150.

79.

Possibilities	*Selections*	*Arrangements*
One pair, one different	${}^1C_1 \times {}^4C_1$	${}^1C_1 \times {}^4C_1 \times \dfrac{3!}{2!} = 12$
Three different	${}^5C_3$	${}^5C_3 \times 3! = 60$

∴ the required number of ways = 72

80. The natural numbers are 1, 2, 3, 4.

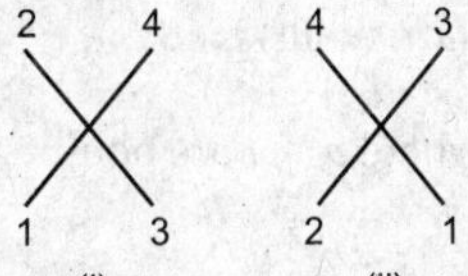

(i) (ii)

Clearly, in one diagonal we have to place 1, 4 and in the other 2, 3.

The number of ways in (i) $= 2! \times 2! = 4$.

The number of ways in (ii) $= 2! \times 2! = 4$.

∴ the total number of ways = 8.

81. If 0 is placed in the units place of the upper number then the units place of the lower number can be filled in 10 ways (filling by any one of 0, 1, 2, … , 9).

If 1 is placed in the units place of the upper number then the units place of the lower number can be filled in 9 ways (filling by any one of 0, 1, 2, … , 8), etc.

$\therefore$ the units column can be filled in $10+9+8+\ldots+1$, i.e., 55 ways. Similarly for the second and the third columns. The number of ways for the fourth column $= 8+7+\ldots+1 = 36$

$\therefore$ the required number of ways $= 55 \times 55 \times 55 \times 36$.

82. The number of positive integral solutions

$$= \text{coefficient of } x^n \text{ in } (x+x^2+x^3+\ldots)^3$$

$$= \text{coefficient of } x^{n-3} \text{ in } (1+x+x^2+\ldots)^3$$

$$= \text{coefficient of } x^{n-3} \text{ in } (1-x)^{-3}$$

$$= \text{coefficient of } x^{n-3} \text{ in } \{{}^2C_0 + {}^3C_1 x + {}^4C_2 x^2 + \ldots\}$$

$$= {}^{n-1}C_{n-3} = \frac{(n-1)!}{(n-3)!\,2!} = \frac{(n-1)(n-2)}{2}.$$

83. The number of non-negative integral solutions

$$= \text{coefficient of } x^n \text{ in } (x^0+x^1+x^2+\ldots)^4$$

$$= \text{coefficient of } x^n \text{ in } (1-x)^{-4}$$

$$= \text{coefficient of } x^n \text{ in } \{{}^3C_0 + {}^4C_1 x + {}^5C_2 x^2 + \ldots\}$$

$$= {}^{n+3}C_n = \frac{(n+3)!}{n!\,3!} = \frac{(n+3)(n+2)(n+1)}{6}.$$

84. The required number of solutions

= the number of positive integral solution of $(a+b+c=12)$

$= \dfrac{(12-1)(12-2)}{2}$. (See the hint to Q. No. 82.)

85. $a+a+d+a+2d=21$ or $a+d=7$

$\therefore \quad a+c=14$ and $b=7$.

The number of positive integral solutions of $(a+c=14)$ is 13.

86. Let $a=2p+1, b=2q+1, c=2r+1, d=2s+1$ where p, q, r are non-negative integers.

$\therefore \quad 2p+1+2q+1+2r+1+2s+1=20$

or $\quad p+q+r+s=8.$

The required number of solutions

$= $ the number of non-negative integral solutions of $(p+q+r+s=8)$

$=\dfrac{(8+3)(8+2)(8+1)}{6}.$ (See the hint to Q. No. 83.)

87. Let $y=p+1, z=q+2$. Then $x\geq 0, p\geq 0, q\geq 0$ and $x+y+z=15$ implies $x+p+q=12$.

$\therefore$ the required number of values of (x, y, z) and hence of (x, p, q)

$=$ the number of non-negative integral solutions of $(x+p+q=12)$

$=$ coefficient of x^{12} in $(x^0+x^1+x^2+\ldots)^3$

$=$ coefficient of x^{12} in $(1-x)^{-3}$

$=$ coefficient of x^{12} in $\{{}^2C_0+{}^3C_1x+{}^4C_2x^2+\ldots\}$

$={}^{14}C_{12}=\dfrac{14\,!}{12\,!\,2\,!}=\dfrac{14\times 13}{2}=91.$

88. Here $a+b+c=n$, where $n=3, 4, 5, 6, 7, 8$ because a, b, c are all positive integers.

The number of positive integral solutions of $(a+b+c=n)$

$={}^{n-1}C_2.$ (See the hint to Q. No. 82.)

$\therefore$ the required number of solutions $=\displaystyle\sum_{n=3}^{8}{}^{n-1}C_2$

$={}^2C_2+{}^3C_2+{}^4C_2+{}^5C_2+{}^6C_2+{}^7C_2$

$={}^3C_0+{}^3C_1+{}^4C_2+{}^5C_3+{}^6C_4+{}^7C_5 \quad (\because\ {}^nC_r={}^nC_{n-r})$

$={}^4C_1+{}^4C_2+{}^5C_3+{}^6C_4+{}^7C_5=\ldots={}^8C_5.$

89. After giving 1 mark to each question, 7 marks are to be distributed among 3 questions.

$\therefore$ the required number of ways

$=$ the number of non-negative integral solutions of $(x+y+z=7)$

$={}^9C_7={}^9C_2.$ (See the hint to Q. No. 87.)

90. After giving away 4 apples to each boy, 8 apples are left to be distributed among 3 boys.

$\therefore$ the required number of ways

$= $ the number of non-negative integral solutions of $(x+y+z=8)$

$= {}^{10}C_8$. (See the hint to Q. No. 87.)

91. $\vec{r}\cdot\vec{a}=10 \quad\Rightarrow\quad x+y+z=10$. Also x, y, z are positive integers.

The required number of positions

$= $ the number of positive integral solutions of $(x+y+z=10)$

$= {}^9C_7 = {}^9C_2$. (See the hint to Q. No. 82.)

92. $P = n(n+1)(n-1)(n+2)(n-2)\ \dots\ (n+r)(n-r)$

$= (n-r)(n-\overline{r-1})\ \dots\ (n-1)n(n+1)(n+2)\ \dots\ (n+r)$

$=$ product of $(2r+1)$ consecutive integers.

$\therefore$ P is divisible by $(2r+1)!$ and so by $(2r-1)!$ also.

96. The required number of ways $= {}^{21}C_{10} + {}^{22}C_{10} + {}^{23}C_{10} + \dots + {}^{30}C_{10}$

$= ({}^{21}C_{10} + {}^{21}C_{11} + {}^{22}C_{12} + {}^{23}C_{13} + \dots + {}^{30}C_{20}) - {}^{21}C_{10} \quad (\because\ {}^nC_r = {}^nC_{n-r})$

$= {}^{22}C_{11} + {}^{22}C_{12} + {}^{23}C_{13} + \dots + {}^{30}C_{20} - {}^{21}C_{10} \quad (\because\ {}^nC_{r-1} + {}^nC_r = {}^{n+1}C_r)$

....

$= {}^{31}C_{20} - {}^{21}C_{10}$.

97. From the question,

$${}^nP_{n-1} = k\cdot\left\{{}^1C_1 \times {}^{n-2}C_{n-3} \times \frac{(n-1)!}{2!} + {}^{n-1}P_{n-1}\right\}$$

$$\Rightarrow \quad n! = k\left\{\frac{(n-2)\times(n-1)!}{2} + (n-1)!\right\}$$

$$\text{or} \quad n = k\left\{\frac{n-2}{2} + 1\right\}.$$

98. The number of ways of inviting, with the couple not included $= {}^8C_5$.

The number of ways of inviting, with the couple included $= {}^8C_3$.

$\therefore$ the required number of ways

$= {}^8C_5 + {}^8C_3 = {}^8C_3 + {}^8C_3 \quad (\because\ {}^8C_5 = {}^8C_3)$

Also ${}^{10}C_5 - 2\times{}^8C_4 = \dfrac{10!}{5!5!} - 2\times\dfrac{8!}{4!4!}$

$$= \frac{10\cdot 9\cdot 8\cdot 7\cdot 6}{120} - 2\times\frac{8\cdot 7\cdot 6\cdot 5}{24}$$

$$= 9 \times 4 \times 7 - 140 = 112 = 2 \times \frac{8!}{3!5!}.$$

99. There are two sets of five parallel lines at equal distances. Clearly, lines like l_1, l_3, m_1, m_3 form a square whose diagonal's length is 2.

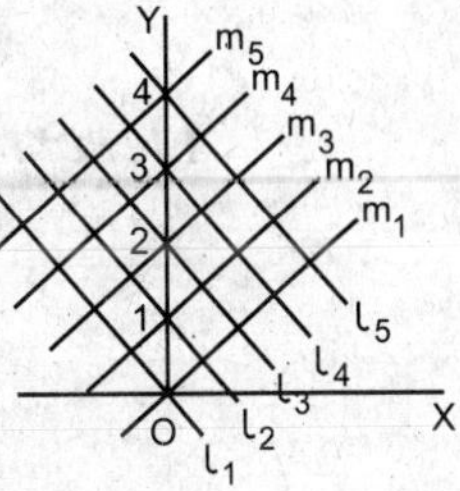

$\therefore$ the number of required squares

$= 3 \times 3$

{$\because$ choices are (l_1, l_3), (l_2, l_4), (l_3, l_5) for one set, etc.}.

100. As the seats are numbered, the arrangement is not circular.

$\therefore$ the required number of ways

= the number of arrangements of n things taking m at a time

$= {}^nP_m = {}^nC_m \times m!$.

101. If $\vec{r} = x\vec{i} + y\vec{j} + z\vec{k}$ then from the question x, y, z are positive integers.

Also $\vec{r} \cdot \vec{a} \le 12 \quad \Rightarrow \quad x + y + z \le 12$.

$\therefore$ the number of values of $\vec{r}$

= the number of positive integral solutions of $(x + y + z \le 12)$

$$= \sum_{n=3}^{12} {}^{n-1}C_2 = {}^2C_2 + {}^3C_2 + \dots + {}^{11}C_2$$

$$= {}^3C_0 + {}^3C_1 + {}^4C_2 + \dots + {}^{11}C_9 \quad (\because \ {}^nC_r = {}^nC_{n-r})$$

$$= {}^4C_1 + {}^4C_2 + \dots + {}^{11}C_9 = \dots = {}^{12}C_9.$$

102.

Possibilities	*Selection*
At least one rupee coin, any number of 50-paisa coins, any number of 25-paisa coins	$2 \times 4 \times 5$
At least two 50-paisa coins, any number of 25-paisa coins	2×5
One 50-paisa coin, at least two 25-paisa coins	1×3
Four 25-paisa coins	1

$\therefore$ the required number of ways = 54

□

5

Determinants and Cramer's Rule

• *Type 1* •

Choose the most appropriate option (a, b, c or d).

1. If $\begin{vmatrix} a+x & a & x \\ a-x & a & x \\ a-x & a & -x \end{vmatrix} = 0$ then x is

(a) 0 (b) a

(c) 3 (d) $2a$

2. $\begin{vmatrix} 0 & p-q & p-r \\ q-p & 0 & q-r \\ r-p & r-q & 0 \end{vmatrix}$ is equal to

(a) $p+q+r$ (b) 0

(c) $p-q-r$ (d) $-p+q+r$

3. If $a \neq b \neq c$ such that $\begin{vmatrix} a^3-1 & b^3-1 & c^3-1 \\ a & b & c \\ a^2 & b^2 & c^2 \end{vmatrix} = 0$ then

(a) $ab+bc+ca=0$ (b) $a+b+c=0$

(c) $abc=1$ (d) $a+b+c=1$

4. $\begin{vmatrix} 1+x & 1 & 1 \\ 1 & 1+x & 1 \\ 1 & 1 & 1+x \end{vmatrix}$ is equal to

(a) $x^2(x+3)$ (b) $3x^3$

(c) 0 (d) x^3

5. If $\begin{vmatrix} 6i & -3i & 1 \\ 4 & 3i & -1 \\ 20 & 3 & i \end{vmatrix} = x + iy$ then

(a) $x = 3, y = 1$ (b) $x = 1, y = 3$

(c) $x = 0, y = 3$ (d) $x = 0, y = 0$

6. The determinant $\begin{vmatrix} xp + y & x & y \\ yp + z & y & z \\ 0 & xp + y & yp + z \end{vmatrix} = 0$ for all $p \in R$ if

(a) x, y, z are in AP (b) x, y, z are in GP

(c) x, y, z are in HP (d) xy, yz, zx are in AP

7. The determinant $\begin{vmatrix} a & a + d & a + 2d \\ a^2 & (a + d)^2 & (a + 2d)^2 \\ 2a + 3d & 2(a + d) & 2a + d \end{vmatrix} = 0$. Then

(a) $d = 0$ (b) $a + d = 0$

(c) $d = 0$ or $a + d = 0$ (d) none of these

8. The value of the determinant $\begin{vmatrix} bc & ca & ab \\ p & q & r \\ 1 & 1 & 1 \end{vmatrix}$, where a, b, c are the pth, qth and rth terms of a HP, is

(a) $ap + bq + cr$ (b) $(a + b + c)(p + q + r)$

(c) 0 (d) none of these

9. The sum of two nonintegral roots of $\begin{vmatrix} x & 2 & 5 \\ 3 & x & 3 \\ 5 & 4 & x \end{vmatrix} = 0$ is

(a) 5 (b) –5

(c) –18 (d) none of these

10. If x, y, z are integers in AP, lying between 1 and 9, and $x51$, $y41$ and $z31$ are three-digit numbers then the value of

$$\begin{vmatrix} 5 & 4 & 3 \\ x51 & y41 & z31 \\ x & y & z \end{vmatrix}$$ is

(a) $x + y + z$ (b) $x - y + z$

(c) 0 (d) none of these

11. If $\Delta_1 = \begin{vmatrix} 1 & 1 & 1 \\ a & b & c \\ a^2 & b^2 & c^2 \end{vmatrix}$, $\Delta_2 = \begin{vmatrix} 1 & bc & a \\ 1 & ca & b \\ 1 & ab & c \end{vmatrix}$ then

(a) $\Delta_1 + \Delta_2 = 0$ (b) $\Delta_1 + 2\Delta_2 = 0$

(c) $\Delta_1 = \Delta_2$ (d) none of these

12. Two nonzero distinct numbers a, b are used as elements to make determinants of the third order. The number of determinants whose value is zero for all a, b is

(a) 24 (b) 32

(c) $a+b$ (d) none of these

13. The value of $\begin{vmatrix} a_1x+b_1y & a_2x+b_2y & a_3x+b_3y \\ b_1x+a_1y & b_2x+a_2y & b_3x+a_3y \\ b_1x+a_1 & b_2x+a_2 & b_3x+a_3 \end{vmatrix}$ is equal to

(a) x^2+y^2 (b) 0

(c) $a_1a_2a_3x^2+b_1b_2b_3y^2$ (d) none of these

14. If $\begin{vmatrix} x_1 & y_1 & 1 \\ x_2 & y_2 & 1 \\ x_3 & y_3 & 1 \end{vmatrix} = \begin{vmatrix} 1 & 1 & 1 \\ b_1 & b_2 & b_3 \\ a_1 & a_2 & a_3 \end{vmatrix}$ then the two triangles whose vertices are $(x_1, y_1), (x_2, y_2), (x_3, y_3)$ and $(a_1, b_1), (a_2, b_2), (a_3, b_3)$, are

(a) congruent (b) similar

(c) equal in area (d) none of these

15. If α, β are nonreal numbers satisfying $x^3-1=0$ then the value of

$\begin{vmatrix} \lambda+1 & \alpha & \beta \\ \alpha & \lambda+\beta & 1 \\ \beta & 1 & \lambda+\alpha \end{vmatrix}$ is equal to

(a) 0 (b) λ^3

(c) λ^3+1 (d) none of these

16. The value of $\begin{vmatrix} {}^{10}C_4 & {}^{10}C_5 & {}^{11}C_m \\ {}^{11}C_6 & {}^{11}C_7 & {}^{12}C_{m+2} \\ {}^{12}C_8 & {}^{12}C_9 & {}^{13}C_{m+4} \end{vmatrix}$ is equal to zero when m is

(a) 6 (b) 4

(c) 5 (d) none of these

17. If $x>0$ and $\neq 1$, $y>0$ and $\neq 1$, $z>0$ and $\neq 1$ then the value of

$\begin{vmatrix} 1 & \log_x y & \log_x z \\ \log_y x & 1 & \log_y z \\ \log_z x & \log_z y & 1 \end{vmatrix}$ is

(a) 0 (b) 1

(c) –1 (d) none of these

18. The value of $\begin{vmatrix} 1 & 1 & 1 \\ (2^x+2^{-x})^2 & (3^x+3^{-x})^2 & (5^x+5^{-x})^2 \\ (2^x-2^{-x})^2 & (3^x-3^{-x})^2 & (5^x-5^{-x})^2 \end{vmatrix}$ is

(a) 0 (b) 30^x

(c) 30^{-x} (d) none of these

19. The value of the determinant $\begin{vmatrix} {}^5C_0 & {}^5C_3 & 14 \\ {}^5C_1 & {}^5C_4 & 1 \\ {}^5C_2 & {}^5C_5 & 1 \end{vmatrix}$ is

(a) 0 (b) –(6 !)

(c) 80 (d) none of these

20. $\begin{vmatrix} \cos C & \tan A & 0 \\ \sin B & 0 & -\tan A \\ 0 & \sin B & \cos C \end{vmatrix}$ has the value

(a) 0 (b) 1

(c) $\sin A \sin B \cos C$ (d) none of these

21. The value of $\begin{vmatrix} x & x^2-yz & 1 \\ y & y^2-zx & 1 \\ z & z^2-xy & 1 \end{vmatrix}$ is

(a) 1 (b) –1

(c) 0 (d) $-xyz$

22. If $\sqrt{-1}=i$, and ω is a nonreal cube root of unity then the value of

$$\begin{vmatrix} 1 & \omega^2 & 1+i+\omega^2 \\ -i & -1 & -1-i+\omega \\ 1-i & \omega^2-1 & -1 \end{vmatrix}$$ is equal to

(a) 1 (b) i

(c) ω (d) 0

23. If $f(x)=\begin{vmatrix} 1 & x & x+1 \\ 2x & x(x-1) & x(x+1) \\ 3x(x-1) & x(x-1)(x-2) & x(x^2-1) \end{vmatrix}$ then $f(100)$ is equal to

(a) 0 (b) 1

(c) 100 (d) –100

24. The value of $\begin{vmatrix} i^m & i^{m+1} & i^{m+2} \\ i^{m+5} & i^{m+4} & i^{m+3} \\ i^{m+6} & i^{m+7} & i^{m+8} \end{vmatrix}$, where $i=\sqrt{-1}$, is

(a) 1 if m is a multiple of 4 (b) 0 for all real m

(c) $-i$ if m is a multiple of 3 (d) none of these

25. If $\Delta_1 = \begin{vmatrix} 7 & x & 2 \\ -5 & x+1 & 3 \\ 4 & x & 7 \end{vmatrix}$, $\Delta_2 = \begin{vmatrix} x & 2 & 7 \\ x+1 & 3 & -5 \\ x & 7 & 4 \end{vmatrix}$ then $\Delta_1 - \Delta_2 = 0$ for

(a) $x=2$ (b) all real x

(c) $x=0$ (d) none of these

26. If $\Delta_1 = \begin{vmatrix} 10 & 4 & 3 \\ 17 & 7 & 4 \\ 4 & -5 & 7 \end{vmatrix}$, $\Delta_2 = \begin{vmatrix} 4 & x+5 & 3 \\ 7 & x+12 & 4 \\ -5 & x-1 & 7 \end{vmatrix}$

such that $\Delta_1 + \Delta_2 = 0$ then

(a) $x=5$ (b) x has no real value

(c) $x=0$ (d) none of these

27. Let $\begin{vmatrix} \lambda^2+3\lambda & \lambda-1 & \lambda+3 \\ \lambda+1 & -2\lambda & \lambda-4 \\ \lambda-3 & \lambda+4 & 3\lambda \end{vmatrix} = p\lambda^4 + q\lambda^3 + r\lambda^2 + s\lambda + t$

be an identity in λ, where p, q, r, s, t are independent of λ. Then the value of t is

(a) 4 (b) 0

(c) 1 (d) none of these

28. Let $\begin{vmatrix} 1+x & x & x^2 \\ x & 1+x & x^2 \\ x^2 & x & 1+x \end{vmatrix} = ax^5 + bx^4 + cx^3 + dx^2 + \lambda x + \mu$

be an identity in x, where a, b, c, d, λ, μ are independent of x. Then the value of λ is

(a) 3 (b) 2

(c) 4 (d) none of these

29. Using the factor theorem it is found that $b+c$, $c+a$ and $a+b$ are three factors of the determinant $\begin{vmatrix} -2a & a+b & a+c \\ b+a & -2b & b+c \\ c+a & c+b & -2c \end{vmatrix}$. The other factor in the value of the determinant is

(a) 4 (b) 2

(c) $a+b+c$ (d) none of these

30. If the determinant $\begin{vmatrix} \cos 2x & \sin^2 x & \cos 4x \\ \sin^2 x & \cos 2x & \cos^2 x \\ \cos 4x & \cos^2 x & \cos 2x \end{vmatrix}$ is expanded in powers of $\sin x$ then the constant term in the expansion is

(a) 1 (b) 2

(c) –1 (d) none of these

31. If $\Delta(x) = \begin{vmatrix} 1 & \cos x & 1-\cos x \\ 1+\sin x & \cos x & 1+\sin x-\cos x \\ \sin x & \sin x & 1 \end{vmatrix}$ then $\int_0^{\pi/2} \Delta(x)\, dx$ is equal to

(a) $\frac{1}{4}$ (b) $\frac{1}{2}$

(c) 0 (d) $-\frac{1}{2}$

32. If $i=\sqrt{-1}$ and $\sqrt[4]{1}=\alpha, \beta, \gamma, \delta$ then $\begin{vmatrix} \alpha & \beta & \gamma & \delta \\ \beta & \gamma & \delta & \alpha \\ \gamma & \delta & \alpha & \beta \\ \delta & \alpha & \beta & \gamma \end{vmatrix}$ is equal to

(a) i (b) $-i$

(c) 1 (d) 0

33. The roots of $\begin{vmatrix} x & a & b & 1 \\ \lambda & x & b & 1 \\ \lambda & \mu & x & 1 \\ \lambda & \mu & \nu & 1 \end{vmatrix} = 0$ are independent of

(a) λ, μ, ν (b) a, b

(c) λ, μ, ν, a, b (d) none of these

34. The value of $\begin{vmatrix} 1 & 0 & 0 & 0 & 0 \\ 2 & 2 & 0 & 0 & 0 \\ 4 & 4 & 3 & 0 & 0 \\ 5 & 5 & 5 & 4 & 0 \\ 6 & 6 & 6 & 6 & 5 \end{vmatrix}$ is

(a) $6!$ (b) $5!$

(c) $1 \cdot 2^2 \cdot 3 \cdot 4^3 \cdot 5^4 \cdot 6^4$ (d) none of these

35. If $\begin{vmatrix} b^2+c^2 & ab & ac \\ ba & c^2+a^2 & bc \\ ca & cb & a^2+b^2 \end{vmatrix}$ = square of a determinant Δ of the third order then Δ is equal to

(a) $\begin{vmatrix} 0 & c & b \\ c & 0 & a \\ b & a & 0 \end{vmatrix}$ (b) $\begin{vmatrix} a & b & c \\ b & c & a \\ c & a & b \end{vmatrix}$

(c) $\begin{vmatrix} 0 & -c & b \\ c & 0 & -a \\ -b & -a & 0 \end{vmatrix}$ (d) none of these

36. The system of equations $ax+4y+z=0$, $bx+3y+z=0$, $cx+2y+z=0$ has nontrivial solutions if a, b, c are in

(a) AP (b) GP

(c) HP (d) none of these

37. If the equations $a(y+z)=x$, $b(z+x)=y$ and $c(x+y)=z$, where $a \neq -1$, $b \neq -1$, $c \neq -1$, admit of nontrivial solutions then

$$(1+a)^{-1}+(1+b)^{-1}+(1+c)^{-1} \text{ is}$$

(a) 2 (b) 1

(c) $\frac{1}{2}$ (d) none of these

38. The system of equations

$$2x-y+z=0$$
$$x-2y+z=0$$
$$\lambda x-y+2z=0$$

has infinite number of nontrivial solutions for

(a) $\lambda=1$ (b) $\lambda=5$

(c) $\lambda=-5$ (d) no real value of λ

39. The equations $x+y+z=6$, $x+2y+3z=10$, $x+2y+mz=n$ give infinite number of values of the triplet (x, y, z) if

(a) $m=3, n \in R$ (b) $m=3, n \neq 10$

(c) $m=3, n=10$ (d) none of these

40. The system of equations $2x+3y=8, 7x-5y+3=0, 4x-6y+\lambda=0$ is solvable if λ is

(a) 6 (b) 8

(c) -8 (d) -6

41. If the system of equations

$$ax + by + c = 0$$
$$bx + cy + a = 0$$
$$cx + ay + b = 0$$

has a solution then the system of equations

$$(b+c)x + (c+a)y + (a+b)z = 0$$
$$(c+a)x + (a+b)y + (b+c)z = 0$$
$$(a+b)x + (b+c)y + (c+a)z = 0$$

has

(a) only one solution
(b) no solution
(c) infinite number of solutions
(d) none of these

• Type 2 •

Choose the correct options. One or more options may be correct.

42. Let $\{\Delta_1, \Delta_2, \Delta_3, \ldots, \Delta_k\}$ be the set of third-order determinants that can be made with the distinct nonzero real numbers $a_1, a_2, a_3, \ldots, a_9$. Then

(a) $k = 9!$ (b) $\sum_{i=1}^{k} \Delta_i - 0$

(c) at least one $\Delta_i = 0$ (d) none of these

43. $\begin{vmatrix} x^2 & (y+z)^2 & yz \\ y^2 & (z+x)^2 & zx \\ z^2 & (x+y)^2 & xy \end{vmatrix}$ is divisible by

(a) $x^2 + y^2 + z^2$ (b) $x - y$

(c) $x - y - z$ (d) $x + y + z$

44. The equation $\begin{vmatrix} 1 & x & x^2 \\ x^2 & 1 & x \\ x & x^2 & 1 \end{vmatrix} = 0$ has

(a) exactly two distinct roots (b) one pair of equal real roots
(c) modulus of each root 1 (d) three pairs of equal roots

45. Let $f(n) = \begin{vmatrix} n & n+1 & n+2 \\ {}^nP_n & {}^{n+1}P_{n+1} & {}^{n+2}P_{n+2} \\ {}^nC_n & {}^{n+1}C_{n+1} & {}^{n+2}C_{n+2} \end{vmatrix}$, where the symbols have their usual meanings. The $f(n)$ is divisible by

(a) n^2+n+1 (b) $(n+1)!$
(c) $n!$ (d) none of these

46. Let $x \neq -1$ and let a, b, c be nonzero real numbers. Then the determinant $\begin{vmatrix} a(1+x) & b & c \\ a & b(1+x) & c \\ a & b & c(1+x) \end{vmatrix}$ is divisible by

(a) $abcx$ (b) $(1+x)^2$
(c) $(1+x)^3$ (d) $x(1+x)^2$

47. The arbitrary constant on which the value of the determinant

$$\begin{vmatrix} 1 & \alpha & \alpha^2 \\ \cos(p-d)a & \cos pa & \cos(p-d)a \\ \sin(p-d)a & \sin pa & \sin(p-d)a \end{vmatrix}$$

does not depend is

(a) α (b) p
(c) d (d) a

48. Let $\Delta(x) = \begin{vmatrix} x+a & x+b & x+a-c \\ x+b & x+c & x-1 \\ x+c & x+d & x-b+d \end{vmatrix}$ and $\int_0^2 \Delta(x)\,dx = -16$, where a, b, c, d are in AP, then the common difference of the AP is

(a) 1 (b) 2
(c) –2 (d) none of these

49. If $A+B+C=\pi$, $e^{i\theta} = \cos\theta + i\sin\theta$ and $z = \begin{vmatrix} e^{2iA} & e^{-iC} & e^{-iB} \\ e^{-iC} & e^{2iB} & e^{-iA} \\ e^{-iB} & e^{-iA} & e^{2iC} \end{vmatrix}$ then

(a) $\text{Re}(z) = 4$ (b) $\text{Im}(z) = 0$
(c) $\text{Re}(z) = -4$ (d) $\text{Im}(z) = -1$

50. If $\begin{vmatrix} a+x & a-x & a-x \\ a-x & a+x & a-x \\ a-x & a-x & a+x \end{vmatrix} = 0$ then x is

(a) 0 (b) a

(c) $3a$ (d) $2a$

51. A value of c for which the system of equations

$$x + y = 1$$
$$(c+2)x + (c+4)y = 6$$
$$(c+2)^2 x + (c+4)^2 y = 36$$

is solvable (consistent) is

(a) 1 (b) 2

(c) 4 (d) none of these

52. Eliminating a, b, c from $x = \dfrac{a}{b-c}$, $y = \dfrac{b}{c-a}$, $z = \dfrac{c}{a-b}$ we get

(a) $\begin{vmatrix} 1 & -x & x \\ 1 & -y & y \\ 1 & -z & z \end{vmatrix} = 0$ (b) $\begin{vmatrix} 1 & -x & x \\ 1 & 1 & -y \\ 1 & z & 1 \end{vmatrix} = 0$

(c) $\begin{vmatrix} 1 & -x & x \\ y & 1 & -y \\ -z & z & 1 \end{vmatrix} = 0$ (d) none of these

53. The system of equations

$$6x + 5y + \lambda z = 0$$
$$3x - y + 4z = 0$$
$$x + 2y - 3z = 0$$

has

(a) only a trivial solution for $\lambda \in R$

(b) exactly one nontrivial solution for some real λ

(c) infinite number of nontrivial solutions for one value of λ

(d) only one solution for $\lambda \neq -5$

Answers

1. a	**2.** b	**3.** c	**4.** a	**5.** d
6. b	**7.** c	**8.** c	**9.** b	**10.** c
11. a	**12.** b	**13.** b	**14.** c	**15.** b
16. c	**17.** a	**18.** a	**19.** b	**20.** a
21. c	**22.** d	**23.** a	**24.** b	**25.** b
26. a	**27.** b	**28.** a	**29.** a	**30.** c
31. d	**32.** d	**33.** b	**34.** b	**35.** a
36. a	**37.** a	**38.** c	**39.** c	**40.** b
41. c	**42.** a, b	**43.** a, b, d	**44.** b, c, d	**45.** a, c
46. a, b, d	**47.** b	**48.** b, c	**49.** b, c	**50.** a, c
51. b, c	**52.** b, c	**53.** c, d		

Hints and Solutions to Selected Questions

2. $\Delta = \begin{vmatrix} 0 & q-p & r-p \\ p-q & 0 & r-q \\ p-r & q-r & 0 \end{vmatrix} = (-1)^3 \begin{vmatrix} 0 & p-q & p-r \\ q-p & 0 & q-r \\ r-p & r-q & 0 \end{vmatrix}$,

$(R \rightleftarrows C)$ taking (-1) common from all the rows

$= (-1)^3 \Delta \quad \therefore \quad \Delta = -\Delta \Rightarrow \Delta = 0.$

3. $\Delta = \begin{vmatrix} a^3 & b^3 & c^3 \\ a & b & c \\ a^2 & b^2 & c^2 \end{vmatrix} + \begin{vmatrix} -1 & -1 & -1 \\ a & b & c \\ a^2 & b^2 & c^2 \end{vmatrix}$

$= abc \begin{vmatrix} a^2 & b^2 & c^2 \\ 1 & 1 & 1 \\ a & b & c \end{vmatrix} - 1 \begin{vmatrix} 1 & 1 & 1 \\ a & b & c \\ a^2 & b^2 & c^2 \end{vmatrix}$

$= (abc - 1) \begin{vmatrix} 1 & 1 & 1 \\ a & b & c \\ a^2 & b^2 & c^2 \end{vmatrix}$

$= (abc-1)(a-b)(b-c)(c-a)$. But $a \neq b \neq c$.

5. $\Delta = \begin{vmatrix} 4+6i & 0 & 0 \\ 4-20i & 0 & 0 \\ 20 & 3 & i \end{vmatrix} = 0 = 0 + i \cdot 0.$

$(R_1 \to R_1 + R_2, R_2 \to R_2 - iR_3)$

6. Using $C_1 \to C_1 - p \times C_2 - C_3$, we get $(xp^2 - z)(xz - y^2) = 0$.

For all p, $xp^2 - z \neq 0$. So, $y^2 = xz$.

7. Use $C_3 \to C_3 - C_2$, $C_2 \to C_2 - C_1$.

8. $\Delta = abc \begin{vmatrix} \frac{1}{a} & \frac{1}{b} & \frac{1}{c} \\ p & q & r \\ 1 & 1 & 1 \end{vmatrix} = abc \begin{vmatrix} x+(p-1)d & x+(q-1)d & x+(r-1)d \\ p-1 & q-1 & r-1 \\ 1 & 1 & 1 \end{vmatrix}$

$= abc \begin{vmatrix} x & x & x \\ p-1 & q-1 & r-1 \\ 1 & 1 & 1 \end{vmatrix} = abcx \begin{vmatrix} 1 & 1 & 1 \\ p-1 & q-1 & r-1 \\ 1 & 1 & 1 \end{vmatrix} = 0.$

$(R_1 \to R_1 - d \times R_2)$

9. Here $\begin{vmatrix} x-5 & 2 & 5 \\ 0 & x & 3 \\ 5-x & 4 & x \end{vmatrix} = 0$ or $(x-5)\begin{vmatrix} 1 & 2 & 5 \\ 0 & x & 3 \\ -1 & 4 & x \end{vmatrix} = 0$

or $(x-5)\begin{vmatrix} 0 & 6 & 5+x \\ 0 & x & 3 \\ -1 & 4 & x \end{vmatrix} = 0$ or $(x-5)\{18-x(5+x)\} = 0$

$\therefore$ the nonintegral roots are the roots of $x^2 + 5x - 18 = 0$.

10. $\Delta = \begin{vmatrix} 5 & 4 & 3 \\ 100x+50+1 & 100y+40+1 & 100z+30+1 \\ x & y & z \end{vmatrix} = \begin{vmatrix} 5 & 4 & 3 \\ 1 & 1 & 1 \\ x & y & z \end{vmatrix}$

$(R_2 \to R_2 - 100R_3 - 10R_1)$

$= x - 2y + z = 0$ $\quad$ ($\because$ x, y, z are in AP.)

11. $\Delta_2 = \dfrac{1}{abc}\begin{vmatrix} a & abc & a^2 \\ b & cab & b^2 \\ c & abc & c^2 \end{vmatrix}$

$= \begin{vmatrix} a & 1 & a^2 \\ b & 1 & b^2 \\ c & 1 & c^2 \end{vmatrix} = -\begin{vmatrix} 1 & a & a^2 \\ 1 & b & b^2 \\ 1 & c & c^2 \end{vmatrix} = -\Delta_1.$

12. As the determinant has to vanish for all a, b, we have at least two rows or two columns identical.

After filling the first column in $2 \times 2 \times 2$ ways and filling another column likewise, the remaining column can be filled in $2 \times 2 \times 2$ ways. So, the number of ways $= 2^3 \times 2^3 = 2^6$.

But each of the two ways give the same determinant.

$\therefore$ the required number of determinants $= 32$.

13. $\Delta = \begin{vmatrix} a_1x+b_1y & a_2x+b_2y & a_3x+b_3y \\ a_1(y-1) & a_2(y-1) & a_3(y-1) \\ b_1x+a_1 & b_2x+a_2 & b_3x+a_3 \end{vmatrix}$

$(R_2 \to R_2 - R_3)$

$= (y-1)\begin{vmatrix} b_1y & b_2y & b_3y \\ a_1 & a_2 & a_3 \\ b_1x+a_1 & b_2x+a_2 & b_3x+a_3 \end{vmatrix}$

$(R_1 \to R_1 - x \times R_2)$

$= (y-1)y\begin{vmatrix} b_1 & b_2 & b_3 \\ a_1 & a_2 & a_3 \\ 0 & 0 & 0 \end{vmatrix}$ (using $R_3 \to R_3 - x \times R_1 - R_2$).

15. $\Delta=(\lambda+1+\alpha+\beta)\begin{vmatrix}1 & 1 & 1\\ \alpha & \lambda+\beta & 1\\ \beta & 1 & \lambda+\alpha\end{vmatrix}$ (using $R_1\to R_1+R_2+R_3$)

$$=\lambda\begin{vmatrix}1 & 0 & 0\\ \alpha & \lambda+\beta-\alpha & 1-\alpha\\ \beta & 1-\beta & \lambda+\alpha-\beta\end{vmatrix}=\lambda\begin{vmatrix}\lambda+\beta-\alpha & 1-\alpha\\ 1-\beta & \lambda+\alpha-\beta\end{vmatrix}$$

$$=\lambda\{\lambda^2-(\alpha-\beta)^2-(1-\alpha)(1-\beta)\}$$

$$=\lambda^3-\lambda\{\alpha^2+\beta^2-2\alpha\beta+1-\alpha-\beta+\alpha\beta\}=0$$

because $\alpha^2=\beta,\ \beta^2=\alpha,\ \alpha\beta=1$.

16. $C_2\to C_2+C_1$ gives $\Delta=\begin{vmatrix}{}^{10}C_4 & {}^{11}C_5 & {}^{11}C_m\\ {}^{11}C_6 & {}^{12}C_7 & {}^{12}C_{m+2}\\ {}^{12}C_8 & {}^{13}C_9 & {}^{13}C_{m+4}\end{vmatrix}$.

For $m=5$, $C_2\equiv C_3$.

17. $\Delta=\dfrac{1}{\log x}\cdot\dfrac{1}{\log y}\cdot\dfrac{1}{\log z}\begin{vmatrix}\log x & \log y & \log z\\ \log x & \log y & \log z\\ \log x & \log y & \log z\end{vmatrix}=0.$

18. Using $R_2\to R_2-R_3$,

$$\Delta=\begin{vmatrix}1 & 1 & 1\\ 2\cdot 2^x\cdot 2\cdot 2^{-x} & 2\cdot 3^x\cdot 2\cdot 3^{-x} & 2\cdot 5^x\cdot 2\cdot 5^{-x}\\ (2^x-2^{-x})^2 & (3^x-3^{-x})^2 & (5^x-5^{-x})^2\end{vmatrix}$$

$$=4\begin{vmatrix}1 & 1 & 1\\ 1 & 1 & 1\\ (2^x-2^{-x})^2 & (3^x-3^{-x})^2 & (5^x-5^{-x})^2\end{vmatrix}=0.$$

19. ${}^5C_0+{}^5C_1+{}^5C_2={}^5C_5+{}^5C_4+{}^5C_3=2^{5-1}=16.$

Use $R_1\to R_1+R_2+R_3$.

21. $\Delta=\begin{vmatrix}x & x^2 & 1\\ y & y^2 & 1\\ z & z^2 & 1\end{vmatrix}+\begin{vmatrix}x & -yz & 1\\ y & -zx & 1\\ z & -xy & 1\end{vmatrix}$

$$=\begin{vmatrix}x & x^2 & 1\\ y & y^2 & 1\\ z & z^2 & 1\end{vmatrix}-\frac{1}{xyz}\begin{vmatrix}x^2 & xyz & x\\ y^2 & zxy & y\\ z^2 & xyz & z\end{vmatrix},\ \text{etc.}$$

22. Use $R_1\to R_1+R_2-R_3$.

23. $f(x) = x \cdot x \cdot (x-1)\begin{vmatrix} 1 & x & x+1 \\ 2 & x-1 & x+1 \\ 3 & x-2 & x+1 \end{vmatrix}$

$$= x^2(x-1)\begin{vmatrix} 1 & x & 1 \\ 2 & x-1 & 2 \\ 3 & x-2 & 3 \end{vmatrix} = 0$$

$$(C_3 \to C_3 - C_2)$$

$\therefore \quad f(100) = 0.$

27. Put $\lambda = 0$ on both sides. Then

$$\begin{vmatrix} 0 & -1 & 3 \\ 1 & 0 & -4 \\ -3 & 4 & 0 \end{vmatrix} = t.$$

28. $\Delta = (1+x)^2\begin{vmatrix} 1 & x & x^2 \\ 1 & 1+x & x^2 \\ 1 & x & 1+x \end{vmatrix}.$

Differentiating both sides of the given equality w.r.t. x, we get

$$2(1+x)\begin{vmatrix} 1 & x & x^2 \\ 1 & 1+x & x^2 \\ 1 & x & 1+x \end{vmatrix}$$

$$+ (1+x)^2\left\{\begin{vmatrix} 1 & 1 & x^2 \\ 1 & 1 & x^2 \\ 1 & 1 & 1+x \end{vmatrix} + \begin{vmatrix} 1 & x & 2x \\ 1 & 1+x & 2x \\ 1 & x & 1 \end{vmatrix}\right\}$$

$$= 5ax^4 + 4bx^3 + 3cx^2 + 2dx + \lambda.$$

Putting $x = 0$,

$$2\begin{vmatrix} 1 & 0 & 0 \\ 1 & 1 & 0 \\ 1 & 0 & 1 \end{vmatrix} + \begin{vmatrix} 1 & 0 & 0 \\ 1 & 1 & 0 \\ 1 & 0 & 1 \end{vmatrix} = \lambda$$

$\therefore \quad 2 + 1 = \lambda.$

29. As the determinant is of the third degree in a, b, c, we get

$$\begin{vmatrix} -2a & a+b & a+c \\ b+a & -2b & b+c \\ c+a & c+b & -2c \end{vmatrix} = \lambda(b+c)(c+a)(a+b),$$

where λ is independent of a, b, c.

Putting $a = 0, b = 1, c = 2,$

$$\begin{vmatrix} 0 & 1 & 2 \\ 1 & -2 & 3 \\ 2 & 3 & -4 \end{vmatrix} = \lambda \cdot 3 \cdot 2 \cdot 1. \quad \text{Now find } \lambda.$$

30. The constant term = value of the determinant when $x = 0$.

31. $C_3 \to C_3 + C_2 - C_1$ gives $\Delta(x) = \begin{vmatrix} 1 & \cos x & 0 \\ 1+\sin x & \cos x & 0 \\ \sin x & \sin x & 1 \end{vmatrix}$

$$= \cos x - \cos x(1+\sin x) = -\sin x \cdot \cos x$$

$$\therefore \quad \int_0^{\pi/2} \Delta(x)\,dx = -\frac{1}{2}\int_0^{\pi/2} \sin 2x\,dx$$

$$= -\frac{1}{2}\left[-\frac{\cos 2x}{2}\right]_0^{\pi/2} = \frac{1}{4}(\cos\pi - \cos 0) = -\frac{1}{2}.$$

32. $x^4 - 1 = 0$ has roots $\alpha, \beta, \gamma, \delta$. Therefore, $\alpha + \beta + \gamma + \delta = 0$.

Use $R_1 \to R_1 + R_2 + R_3 + R_4$.

33. Use $R_1 \to R_1 - R_2$, $R_2 \to R_2 - R_3$, $R_3 \to R_3 - R_4$. Then

$$(x-\lambda)(x-\mu)(x-\nu) = 0.$$

34. The elements in the leading diagonal are 1, 2, 3, 4, 5. On one side of the leading diagonal all the elements are zero.

$\therefore$ the value of the determinant = the product of the elements in the leading diagonal.

35. Check $\begin{vmatrix} 0 & c & b \\ c & 0 & a \\ b & a & 0 \end{vmatrix} \times \begin{vmatrix} 0 & c & b \\ c & 0 & a \\ b & a & 0 \end{vmatrix}$ = the given determinant.

37. $a(x+y+z) = (1+a)x$, $b(x+y+z) = (1+b)y$, $c(x+y+z) = (1+c)z$.

In case of nontrivial solution, $x + y + z \neq 0$.

$$\therefore \quad \frac{a}{1+a} = \frac{x}{x+y+z}, \text{ etc.}$$

Adding, $\dfrac{a}{1+a} + \dfrac{b}{1+b} + \dfrac{c}{1+c} = 1.$

$$\therefore \quad \left(1 - \frac{1}{1+a}\right) + \left(1 - \frac{1}{1+b}\right) + \left(1 - \frac{1}{1+c}\right) = 1.$$

39. Each of the first three options contains $m = 3$.

When $m = 3$, the last two equations become $x + 2y + 3z = 10$ and $x + 2y + 3z = n$.

Obviously, when $n = 10$ these equations become the same. So, we are left with only two independent equations to find the values of the three unknowns.

Consequently, there will be infinite solutions.

40. Equations are solvable when they are consistent.

41. For existence of a solution of the first system,

$$\begin{vmatrix} a & b & c \\ b & c & a \\ c & a & b \end{vmatrix} = 0.$$

The second system will have a nontrivial solution if we can prove that

$$\begin{vmatrix} b+c & c+a & a+b \\ c+a & a+b & b+c \\ a+b & b+c & c+a \end{vmatrix} = 0.$$

Establish $\begin{vmatrix} b+c & c+a & a+b \\ c+a & a+b & b+c \\ a+b & b+c & c+a \end{vmatrix} = 2\begin{vmatrix} a & b & c \\ b & c & a \\ c & a & b \end{vmatrix} = 0.$

Remember that the existence of one nontrivial solution implies existence of infinite number of nontrivial solutions.

42. The number of third-order determinants = the number of arrangements of nine different numbers in nine places = 9 !.

Corresponding to each determinant made, there is a determinant obtained by interchanging two consecutive rows (or columns). So, the sum of this pair will be 0.

$\therefore$ the sum of all the determinants

$$= 0 + 0 + 0 + \ldots \text{ to } \frac{9\,!}{2} \text{ times} = 0.$$

43. Using $C_2 \to C_2 - 2 \times C_3$, $\Delta = \begin{vmatrix} x^2 & y^2+z^2 & yz \\ y^2 & z^2+x^2 & zx \\ z^2 & x^2+y^2 & xy \end{vmatrix}$

$$= (x^2+y^2+z^2)\begin{vmatrix} x^2 & 1 & yz \\ y^2 & 1 & zx \\ z^2 & 1 & xy \end{vmatrix}.$$

Now use $R_1 \to R_1 - R_2$, $R_2 \to R_2 - R_3$.

44. Using $R_1 \to R_1 + R_2 + R_3$, $\Delta = (1+x+x^2)\begin{vmatrix} 1 & 1 & 1 \\ x^2 & 1 & x \\ x & x^2 & 1 \end{vmatrix}$

$$\therefore \quad \Delta = (1+x+x^2)\begin{vmatrix} 0 & 0 & 1 \\ x^2-x & 1-x & x \\ x-1 & x^2-1 & 1 \end{vmatrix}$$

$$= (1+x+x^2)(x-1)^2\begin{vmatrix} 0 & 0 & 1 \\ x & -1 & x \\ 1 & x+1 & 1 \end{vmatrix}$$

$$= (1+x+x^2)^2(x-1)^2$$

$\therefore \quad \Delta = 0$ has roots $\omega, \omega, \omega^2, \omega^2, 1, 1$.

45. $f(n) = \begin{vmatrix} n & n+1 & n+2 \\ n! & (n+1)! & (n+2)! \\ 1 & 1 & 1 \end{vmatrix} = \begin{vmatrix} n & 1 & 1 \\ n! & n \cdot n! & (n+1)\cdot(n+1)! \\ 1 & \cdot 0 & 0 \end{vmatrix}$

(using $C_3 \to C_3 - C_2$, $C_2 \to C_2 - C_1$)

$$= (n+1)\cdot(n+1)! - n \cdot n! = n! \cdot \{(n+1)^2 - n\}.$$

47. Using $C_3 \to C_3 - C_1$, $\Delta = \begin{vmatrix} 1 & \alpha & \alpha^2 - 1 \\ \cos(p-d)a & \cos pa & 0 \\ \sin(p-d)a & \sin pa & 0 \end{vmatrix}$

$$= (\alpha^2 - 1)\sin\{pa - (p-d)a\} = (\alpha^2 - 1)\sin da.$$

48. Using $R_3 \to R_3 - R_2$, $R_2 \to R_2 - R_1$, we get

$\Delta = \begin{vmatrix} x+a & x+b & x+a-c \\ k & k & 2k-1 \\ k & k & 1+2k \end{vmatrix}$, where k = common difference.

$$\therefore \quad \Delta = \begin{vmatrix} x+a & k & -c \\ k & 0 & k-1 \\ k & 0 & k+1 \end{vmatrix} = -k\{k(k+1) - k(k-1)\} = -2k^2.$$

$$\therefore \quad \int_0^2 \Delta(x)\,dx = \int_0^2 -2k^2 dx = -2k^2[x]_0^2 = -4k^2 = -16$$

$\therefore \quad k = 2, -2.$

49. $z = e^{iA} \cdot e^{iB} \cdot e^{iC}\begin{vmatrix} e^{iA} & e^{-i(C+A)} & e^{-i(B+A)} \\ e^{-i(C+B)} & e^{iB} & e^{-i(A+B)} \\ e^{-i(B+C)} & e^{-i(A+C)} & e^{iC} \end{vmatrix}$

$$= -1 \cdot \begin{vmatrix} e^{iA} & -e^{iB} & -e^{iC} \\ -e^{iA} & e^{iB} & -e^{iC} \\ -e^{iA} & -e^{iB} & e^{iC} \end{vmatrix}$$

because $e^{i(A+B+C)} = e^{i\pi} = \cos\pi + i\sin\pi = -1$, etc.

$$= -\begin{vmatrix} 0 & -2e^{iB} & 0 \\ -2e^{iA} & 0 & 0 \\ -e^{iA} & -e^{iB} & e^{iC} \end{vmatrix} \quad \text{(using } R_1 \to R_1 + R_3, R_2 \to R_2 + R_3\text{)}$$

$$= -2e^{iB}\{-2e^{i(A+C)}\} = 4e^{i(A+B+C)} = 4e^{i\pi} = -4.$$

52. Here $a - xb + xc = 0$

$$-ya - b + yc = 0$$

$$za - zb - c = 0$$

Eliminating a, b, c, we get $\begin{vmatrix} 1 & -x & x \\ -y & -1 & y \\ z & -z & -1 \end{vmatrix} = 0.$

Determinants in options (b) and (c) are equal to this determinant.

53. For existence of nontrivial solutions,

$$\begin{vmatrix} 6 & 5 & \lambda \\ 3 & -1 & 4 \\ 1 & 2 & -3 \end{vmatrix} = 0 \Rightarrow \lambda = -5.$$

If $\lambda \neq -5$, $x = 0, y = 0, z = 0$ is the only solution.

□

6

Binomial Theorem for Positive Integral Index

• *Type 1* •

Choose the most appropriate option (a, b, c or d).

1. The number of terms in the expansion of $(1+3x+3x^2+x^3)^6$ is

   (a) 18 (b) 9

   (c) 19 (d) 24

2. The number of distinct terms in the expansion of $(x+y-z)^{16}$ is

   (a) 136 (b) 153

   (c) 16 (d) 17

3. The number of irrational terms in the expansion of $(\sqrt[8]{5}+\sqrt[6]{2})^{100}$ is

   (a) 97 (b) 98

   (c) 96 (d) 99

4. The number of terms whose values depend on x in the expansion of $\left(x^2-2+\frac{1}{x^2}\right)^n$ is

   (a) $2n+1$ (b) $2n$

   (c) n (d) none of these

5. The number of real negative terms in the binomial expansion of $(1+ix)^{4n-2}, n \in N, x>0$, is

   (a) n (b) $n+1$

   (c) $n-1$ (d) $2n$

6. In the expansion of $(x+\sqrt{x^2-1})^6+(x-\sqrt{x^2-1})^6$, the number of terms is

(a) 7 (b) 14
(c) 6 (d) 4

7. The number of terms in the expansion of $\left(x^2+1+\frac{1}{x^2}\right)^n, n \in N$, is

(a) $2n$ (b) $3n$
(c) $2n+1$ (d) $3n+1$

8. The number of rational terms in the expansion of $(1+\sqrt{2}+\sqrt[3]{3})^6$ is

(a) 6 (b) 7
(c) 5 (d) 8

9. The number of terms with integral coefficients in the expansion of $(7^{1/3}+5^{1/2}\cdot x)^{600}$ is

(a) 100 (b) 50
(c) 101 (d) none of these

10. The sum of the rational terms in the expansion of $(\sqrt{2}+\sqrt[5]{3})^{10}$ is

(a) 32 (b) 9
(c) 41 (d) none of these

11. The last term in the binomial expansion of $\left(\sqrt[3]{2}-\frac{1}{\sqrt{2}}\right)^n$ is $\left(\frac{1}{3\cdot\sqrt[3]{9}}\right)^{\log_3 8}$. Then the 5th term from the beginning is

(a) ${}^{10}C_6$ (b) $2\cdot{}^{10}C_4$
(c) $\frac{1}{2}\cdot{}^{10}C_4$ (d) none of these

12. If the 4th term in the expansion of $(px+x^{-1})^m$ is 2.5 for all $x \in R$ then

(a) $p=\frac{5}{2}, m=3$ (b) $p=\frac{1}{2}, m=6$
(c) $p=-\frac{1}{2}, m=6$ (d) none of these

13. In the expansion of $(1+ax)^n, n \in N$, the coefficient of x and x^2 are 8 and 24 respectively. Then

(a) $a=2, n=4$ (b) $a=4, n=2$
(c) $a=2, n=6$ (d) $a=-2, n=4$

14. In the expansion of $\left(x^3 - \frac{1}{x^2}\right)^n$, $n \in N$, if the sum of the coefficients of x^5 and x^{10} is 0 then n is

(a) 25 (b) 20

(c) 15 (d) none of these

15. The coefficient of x^{20} in the expansion of

$$(1 + x^2)^{40} \cdot \left(x^2 + 2 + \frac{1}{x^2}\right)^{-5} \text{ is}$$

(a) ${}^{30}C_{10}$ (b) ${}^{30}C_{25}$

(c) 1 (d) none of these

16. The coefficient of $a^8 b^{10}$ in the expansion of $(a + b)^{18}$ is

(a) ${}^{18}C_8$ (b) ${}^{18}P_{10}$

(c) 2^{18} (d) none of these

17. If the coefficients of the $(m + 1)$th term and the $(m + 3)$th term in the expansion of $(1 + x)^{20}$ are equal then the value of m is

(a) 10 (b) 8

(c) 9 (d) none of these

18. The coefficient of x^3 in the expansion of $(1 - x + x^2)^5$ is

(a) 10 (b) –20

(c) –50 (d) –30

19. If the coefficients of the 2nd, 3rd and 4th terms in the expansion of $(1 + x)^n$, $n \in N$, are in AP then n is

(a) 7 (b) 14

(c) 2 (d) none of these

20. The coefficient of x^6 in $\{(1 + x)^6 + (1 + x)^7 + \ldots + (1 + x)^{15}\}$ is

(a) ${}^{16}C_9$ (b) ${}^{16}C_5 - {}^6C_5$

(c) ${}^{16}C_6 - 1$ (d) none of these

21. The coefficient of $x^3 y^4 z$ in the expansion of $(1 + x + y - z)^9$ is

(a) $2 \cdot {}^9C_7 \cdot {}^7C_4$ (b) $-2 \cdot {}^9C_2 \cdot {}^7C_3$

(c) ${}^9C_7 \cdot {}^7C_4$ (d) none of these

22. The coefficient of x^{13} in the expansion of $(1-x)^5(1+x+x^2+x^3)^4$ is

(a) 4 (b) –4

(c) 0 (d) none of these

23. The coefficient of $x^6 \cdot y^{-2}$ in the expansion of $\left(\frac{x^2}{y}-\frac{y}{x}\right)^{12}$ is

(a) ${}^{12}C_6$ (b) $-{}^{12}C_5$

(c) 0 (d) none of these

24. The greatest value of the term independent of x in the expansion of $(x\sin\alpha + x^{-1}\cos\alpha)^{10}, \alpha \in R$, is

(a) 2^5 (b) $\frac{10!}{(5!)^2}$

(c) $\frac{1}{2^5}\cdot\frac{10!}{(5!)^2}$ (d) none of these

25. In the expansion of $\left(x^3-\frac{1}{x^2}\right)^{15}$, the constant term is

(a) ${}^{15}C_6$ (b) 0

(c) $-{}^{15}C_6$ (d) 1

26. The constant term in the expansion of $(1+x)^{10}\cdot\left(1+\frac{1}{x}\right)^{12}$ is

(a) ${}^{22}C_{10}$ (b) 0

(c) ${}^{22}C_{11}$ (d) none of these

27. The term independent of x in the expansion of $(1-x)^2\cdot\left(x+\frac{1}{x}\right)^{10}$ is

(a) ${}^{11}C_5$ (b) ${}^{10}C_5$

(c) ${}^{10}C_4$ (d) none of these

28. The middle term in the expansion of $\left(\frac{2x}{3}-\frac{3}{2x^2}\right)^{2n}$ is

(a) ${}^{2n}C_n$ (b) $(-1)^n\frac{(2n)!}{(n!)^2}\cdot x^{-n}$

(c) ${}^{2n}C_n\cdot\frac{1}{x^n}$ (d) none of these

29. The middle term in the expansion of $\left(1-\frac{1}{x}\right)^n \cdot (1-x)^n$ is

(a) ${}^{2n}C_n$ (b) $-{}^{2n}C_n$

(c) $-{}^{2n}C_{n-1}$ (d) none of these

30. If the rth term is the middle term in the expansion of $\left(x^2-\frac{1}{2x}\right)^{20}$ then the $(r+3)$th term is

(a) ${}^{20}C_{14} \cdot \frac{1}{2^{14}} \cdot x$ (b) ${}^{20}C_{12} \cdot \frac{1}{2^{12}} \cdot x^2$

(c) $-\frac{1}{2^{13}} \cdot {}^{20}C_7 \cdot x$ (d) none of these

31. Let $n \in N$ and $n < (\sqrt{2}+1)^6$. Then the greatest value of n is

(a) 199 (b) 198

(c) 197 (d) 196

32. If the coefficient of the 5th term be the numerically greatest coefficient in the expansion of $(1-x)^n$ then the positive integral value of n is

(a) 9 (b) 8

(c) 7 (d) 10

33. The greatest coefficient in the expansion of $(1+x)^{2n}$ is

(a) $\frac{1 \cdot 3 \cdot 5 \cdot \ldots \cdot (2n-1)}{n!} \cdot 2^n$ (b) ${}^{2n}C_{n-1}$

(c) ${}^{2n}C_{n+1}$ (d) none of these

34. Let n be an odd natural number greater than 1. Then the number of zeros at the end of the sum $99^n + 1$ is

(a) 3 (b) 4

(c) 2 (d) none of these

35. Let $f(n) = 10^n + 3 \cdot 4^{n+2} + 5, n \in N$. The greatest value of the integer which divides $f(n)$ for all n is

(a) 27 (b) 9

(c) 3 (d) none of these

36. 2^{60} when divided by 7 leaves the remainder

(a) 1 (b) 6

(c) 5 (d) 2

37. If $\{x\}$ denotes the fractional part of x then $\left\{\frac{3^{2n}}{8}\right\}$, $n \in N$, is

(a) $\frac{3}{8}$ (b) $\frac{7}{8}$

(c) $\frac{1}{8}$ (d) none of these

38. The sum of the coefficients in the binomial expansion of $\left(\frac{1}{x}+2x\right)^n$ is equal to 6561. The constant term in the expansion is

(a) ${}^8C_4$ (b) $16 \cdot {}^8C_4$

(c) ${}^6C_4 \cdot 2^4$ (d) none of these

39. The sum of the numerical coefficients in the expansion of $\left(1+\frac{x}{3}+\frac{2y}{3}\right)^{12}$ is

(a) 1 (b) 2

(c) 2^{12} (d) none of these

40. The sum of the last ten coefficients in the expansion of $(1+x)^{19}$ when expanded in ascending powers of x is

(a) 2^{18} (b) 2^{19}

(c) $2^{18} - {}^{19}C_{10}$ (d) none of these

41. The sum of the coefficients of x^{2r}, $r = 1, 2, 3, \ldots$, in the expansion of $(1+x)^n$ is

(a) 2^n (b) $2^{n-1} - 1$

(c) $2^n - 1$ (d) $2^{n-1} + 1$

42. The sum of the coefficients in the polynomial expansion of $(1+x-3x^2)^{2163}$ is

(a) 1 (b) –1

(c) 0 (d) none of these

43. The sum of the coefficients of all the integral powers of x in the expansion of $(1+2\sqrt{x})^{40}$ is

(a) $3^{40}+1$ (b) $3^{40}-1$

(c) $\frac{1}{2}(3^{40}-1)$ (d) $\frac{1}{2}(3^{40}+1)$

44. If $(1+x-2x^2)^8 = a_0 + a_1x + a_2x^2 + \dots + a_{16}x^{16}$ then the sum

$$a_1 + a_3 + a_5 + \dots + a_{15}$$

is equal to

(a) -2^7 (b) 2^7

(c) 2^8 (d) none of these

45. The sum $^{20}C_0 + {}^{20}C_1 + {}^{20}C_2 + \dots + {}^{20}C_{10}$ is equal to

(a) $2^{20} + \dfrac{20\,!}{(10\,!)^2}$ (b) $2^{19} - \dfrac{1}{2} \cdot \dfrac{20\,!}{(10\,!)^2}$

(c) $2^{19} + {}^{20}C_{10}$ (d) none of these

46. The sum $^{10}C_3 + {}^{11}C_3 + {}^{12}C_3 + \dots + {}^{20}C_3$ is equal to

(a) $^{21}C_4$ (b) $^{21}C_4 + {}^{10}C_4$

(c) $^{21}C_{17} - {}^{10}C_6$ (d) none of these

47. If $(1+x)^{10} = a_0 + a_1x + a_2x^2 + \dots + a_{10}x^{10}$ then

$(a_0 - a_2 + a_4 - a_6 + a_8 - a_{10})^2 + (a_1 - a_3 + a_5 - a_7 + a_9)^2$ is equal to

(a) 3^{10} (b) 2^{10}

(c) 2^9 (d) none of these

48. The sum $\dfrac{1}{2}\,{}^{10}C_0 - {}^{10}C_1 + 2 \cdot {}^{10}C_2 - 2^2 \cdot {}^{10}C_3 + \dots + 2^9 \cdot {}^{10}C_{10}$ is equal to

(a) $\dfrac{1}{2}$ (b) 0

(c) $\dfrac{1}{2} \cdot 3^{10}$ (d) none of these

49. $1 \cdot {}^nC_1 + 2 \cdot {}^nC_2 + 3 \cdot {}^nC_3 + \dots + n \cdot {}^nC_n$ is equal to

(a) $\dfrac{n(n+1)}{4} \cdot 2^n$ (b) $2^{n+1} - 3$

(c) $n \cdot 2^{n-1}$ (d) none of these

50. If $a_n = \sum_{r=0}^{n} \dfrac{1}{^nC_r}$ then $\sum_{r=0}^{n} \dfrac{r}{^nC_r}$ equals

(a) $(n-1)a_n$ (b) na_n

(c) $\frac{1}{2} na_n$ (d) none of these

51. The sum of the series $\sum_{r=1}^{n} (-1)^{r-1} \cdot {}^nC_r(a-r)$ is equal to

(a) $n \cdot 2^{n-1} + a$ (b) 0

(c) a (d) none of these

52. Let $(1+x)^n = \sum_{r=0}^{n} a_r x^r$. Then $\left(1+\frac{a_1}{a_0}\right)\left(1+\frac{a_2}{a_1}\right)\ldots\left(1+\frac{a_n}{a_{n-1}}\right)$ is equal to

(a) $\frac{(n+1)^{n+1}}{n!}$ (b) $\frac{(n+1)^n}{n!}$

(c) $\frac{n^{n-1}}{(n-1)!}$ (d) $\frac{(n+1)^{n-1}}{(n-1)!}$

53. The value of $\sum_{r=1}^{10} r \cdot \frac{{}^nC_r}{{}^nC_{r-1}}$ is equal to

(a) $5(2n-9)$ (b) $10n$

(c) $9(n-4)$ (d) none of these

54. The sum $\sum_{r=1}^{n} r \cdot {}^{2n}C_r$ is equal to

(a) $n \cdot 2^{2n-1}$ (b) 2^{2n-1}

(c) $2^{n-1}+1$ (d) none of these

55. The sum $1 \cdot {}^{20}C_1 - 2 \cdot {}^{20}C_2 + 3 \cdot {}^{20}C_3 - \ldots - 20 \cdot {}^{20}C_{20}$ is equal to

(a) 2^{19} (b) 0

(c) $2^{20}-1$ (d) none of these

• Type 2 •

Choose the correct options. One or more options may be correct.

56. Let $f(x) = (\sqrt{x^2+1} + \sqrt{x^2-1})^6 + \left(\dfrac{2}{\sqrt{x^2+1}+\sqrt{x^2-1}}\right)^6$. Then

(a) $f(x)$ is a polynomial of the sixth degree in x

(b) $f(x)$ has exactly two terms

(c) $f(x)$ is not a polynomial in x

(d) coefficient of x^6 is 64

57. The coefficient of $a^8b^6c^4$ in the expansion of $(a+b+c)^{18}$ is

(a) ${}^{18}C_{14} \cdot {}^{14}C_8$ (b) ${}^{18}C_{10} \cdot {}^{10}C_6$

(c) ${}^{18}C_6 \cdot {}^{12}C_8$ (d) ${}^{18}C_4 \cdot {}^{14}C_6$

58. The term independent of x in the expansion of $(1+x)^n \cdot \left(1-\frac{1}{x}\right)^n$ is

(a) 0, if n is odd (b) $(-1)^{\frac{n-1}{2}} \cdot {}^nC_{\frac{n-1}{2}}$, if n is odd

(c) $(-1)^{n/2} \cdot {}^nC_{n/2}$, if n is even (d) none of these

59. The coefficient of the $(r+1)$th term of $\left(x+\frac{1}{x}\right)^{20}$ when expanded in the descending powers of x is equal to the coefficient of the 6th term of $\left(x^2+2+\frac{1}{x^2}\right)^{10}$ when expanded in ascending powers of x. The value of r is

(a) 5 (b) 6

(c) 14 (d) 15

60. If $(1+x)^{2n} = a_0 + a_1x + a_2x^2 + \ldots + a_{2n}x^{2n}$ then

(a) $a_0 + a_2 + a_4 + \ldots = \frac{1}{2}(a_0 + a_1 + a_2 + a_3 + \ldots)$

(b) $a_{n+1} < a_n$

(c) $a_{n-3} = a_{n+3}$

(d) none of these

61. In the expansion of $\left(\sqrt[3]{4}+\frac{1}{\sqrt[4]{6}}\right)^{20}$,

(a) the number of rational terms = 4

(b) the number of irrational terms = 18

(c) the middle term is irrational

(d) the number of irrational terms = 17

62. Let $n \in N$. If $(1+x)^n = a_0 + a_1x + a_2x^2 + \dots + a_nx^n$, and $a_{n-3}, a_{n-2}, a_{n-1}$ are in AP then

(a) a_1, a_2, a_3 are in AP

(b) a_1, a_2, a_3 are in HP

(c) $n = 7$

(d) $n = 14$

63. Let $R = (8+3\sqrt{7})^{20}$ and $[R]$ = the greatest integer less than or equal to R. Then

(a) $[R]$ is even

(b) $[R]$ is odd

(c) $R - [R] = 1 - \frac{1}{(8+3\sqrt{7})^{20}}$

(d) none of these

64. $\frac{1}{1!\cdot(n-1)!}+\frac{1}{3!\cdot(n-3)!}+\frac{1}{5!\cdot(n-5)!}+\dots$ is equal to

(a) $\frac{2^{n-1}}{n!}$ for even values of n only

(b) $\frac{2^{n-1}+1}{n!}-1$ for odd values of n only

(c) $\frac{2^{n-1}}{n!}$ for all $n \in N$

(d) none of these

65. In the expansion of $(x+y+z)^{25}$

(a) every term is of the form ${}^{25}C_r \cdot {}^{r}C_k \cdot x^{25-r} \cdot y^{r-k} \cdot z^k$

(b) the coefficient of $x^8y^9z^9$ is 0

(c) the number of terms is 325

(d) none of these

Answers

1. c	**2.** b	**3.** a	**4.** b	**5.** a
6. d	**7.** c	**8.** b	**9.** c	**10.** c
11. a	**12.** b	**13.** a	**14.** c	**15.** b
16. a	**17.** c	**18.** d	**19.** a	**20.** a
21. b	**22.** a	**23.** c	**24.** c	**25.** c
26. a	**27.** a	**28.** b	**29.** a	**30.** c
31. c	**32.** b	**33.** a	**34.** c	**35.** b
36. a	**37.** c	**38.** b	**39.** c	**40.** a
41. b	**42.** b	**43.** d	**44.** a	**45.** d
46. c	**47.** b	**48.** a	**49.** c	**50.** c
51. c	**52.** b	**53.** a	**54.** a	**55.** b
56. a, b, d	**57.** a, b, c, d	**58.** a, c	**59.** a, d	**60.** a, b, c
61. b, c	**62.** a, c	**63.** b, c	**64.** c	**65.** a, b

Hints and Solutions to Selected Questions

2. $(x+y-z)^{16} = {}^{16}C_0 x^{16} + {}^{16}C_1 x^{15}(y-z) + \dots$

$$+ {}^{16}C_r x^{16-r}(y-z)^r + \dots + {}^{16}C_{16}(y-z)^{16}.$$

Clearly, all the terms are distinct.

$\therefore$ the number of distinct terms

$$= 1+2+3+\dots+17 = \frac{17\times 18}{2}.$$

3. $t_{r+1} = {}^{100}C_r(\sqrt[8]{5})^{100-r}\cdot(\sqrt[6]{2})^r.$

As 2 and 5 are coprime, t_{r+1} will be rational if $100-r$ is a multiple of 8 and r is a multiple of 6. Also $0 \le r \le 100$.

$\therefore \quad r = 0, 6, 12, \dots, 96$

$\therefore \quad 100 - r = 4, 10, 16, \dots, 100$ ... (1)

But $100 - r$ is to be a multiple of 8.

So, $100 - r = 0, 8, 16, 24, \dots, 96$ ... (2)

The common terms in (1) and (2) are 16, 40, 64 and 88.

$\therefore \quad r = 84, 60, 36, 12$ give rational terms.

$\therefore \quad$ the number of irrational terms $= 101 - 4 = 97$.

4. $\left(x^2 - 2 + \frac{1}{x^2}\right)^n = \left(x - \frac{1}{x}\right)^{2n}$

$$\therefore \quad t_{r+1} = {}^{2n}C_r x^{2n-r}\cdot\left(-\frac{1}{x}\right)^r = (-1)^r\cdot{}^{2n}C_r\cdot x^{2n-2r}.$$

If $2n - 2r = 0$, the value of the term will not depend on x. So, $r = n$.

$\therefore$ the required number of terms $= (2n+1) - 1 = 2n$.

5. $t_{r+1} = {}^{4n-2}C_r(ix)^r$. The term is real negative if $r = 2, 6, 10, \dots$.

But $0 \le r \le 4n-2$ and $4n - 2 = 2 + (n-1)4$.

$\therefore$ the required number of terms $= n$.

6. On expansion and simplification,

$$\text{expression} = 2\{{}^6C_0x^6 + {}^6C_2x^4(x^2-1) + {}^6C_4x^2(x^2-1)^2 + {}^6C_6(x^2-1)^3\}$$

$$= 2\{({}^6C_0 + {}^6C_2 + {}^6C_4 + {}^6C_6)x^6 + (-{}^6C_2 - {}^6C_4\times 2 - {}^6C_6\times 3)x^4 + ({}^6C_4 + {}^6C_6\times 3)x^2 - {}^6C_6\}.$$

7. $\left\{1+\left(x^2+\frac{1}{x^2}\right)\right\}^n = {}^nC_0 + {}^nC_1\left(x^2+\frac{1}{x^2}\right) + {}^nC_2\left(x^2+\frac{1}{x^2}\right)^2 + \dots + {}^nC_n\left(x^2+\frac{1}{x^2}\right)^n$

Here, all the terms are positive and will contain powers

$(x^2)^0, (x^2)^1, (x^2)^2, \dots, (x^2)^n, (x^2)^{-n}, (x^2)^{-(n-1)}, \dots, (x^2)^{-1}$ only.

$\therefore$ the number of terms will be $2n+1$.

8. $(1+\sqrt{2}+\sqrt[3]{3})^6 = {}^6C_0 + {}^6C_1(\sqrt{2}+\sqrt[3]{3}) + {}^6C_2(\sqrt{2}+\sqrt[3]{3})^2 + \dots + {}^6C_6(\sqrt{2}+\sqrt[3]{3})^6$

In $(\sqrt{2}+\sqrt[3]{3})^6$, $t_{r+1} = {}^6C_r\cdot(\sqrt{2})^{6-r}\cdot(\sqrt[3]{3})^r$.

It is rational if $r = 0, 6$. Similarly in other cases.

$\therefore$ the number of rational terms $= 1+0+1+1+1+1+2 = 7$.

9. $t_{r+1} = {}^{600}C_r\cdot 7^{\frac{600-r}{3}}\cdot 5^{\frac{r}{2}}x^r$.

Here, $0 \le r \le 600$ and $\frac{r}{2}$, $200-\frac{r}{3}$ are integers.

$\therefore$ r should be a multiple of 6

$\therefore$ $r = 0, 6, 12, \dots, 600$.

10. $t_{r+1} = {}^{10}C_r\cdot 2^{\frac{10-r}{2}}\cdot 3^{\frac{r}{5}}$. This is rational if $\frac{r}{5}$ and $5-\frac{r}{2}$ are integers.

Also $0 \le r \le 10$.

Now, r must be a multiple of 10(LCM of 2 and 5). So $r = 0, 10$.

$\therefore$ the rational terms are ${}^{10}C_0\cdot 2^5\cdot 3^0$, ${}^{10}C_{10}\cdot 2^0\cdot 3^2$.

11. The last term $= {}^nC_n\cdot\left(-\frac{1}{\sqrt{2}}\right)^n = \left(\frac{1}{3\cdot\sqrt[3]{9}}\right)^{\log_3 8}$ (from the question)

$\therefore$ $(-1)^n\cdot\left(\frac{1}{2}\right)^{n/2} = \left(\frac{1}{3^{5/3}}\right)^{\log_3 8}$

$$= 3^{-5/3\cdot 3\log_3 2} = 3^{\log_3 2^{-5}} = 2^{-5} = \left(\frac{1}{2}\right)^5.$$

$\therefore$ $n = 10$.

So, $t_5 = {}^{10}C_4 \cdot (2^{1/3})^6 \cdot \left(-\frac{1}{\sqrt{2}}\right)^4 = {}^{10}C_4 = {}^{10}C_{10-4}$.

15. Expression $= (1+x^2)^{40} \cdot \left(x+\frac{1}{x}\right)^{-10} = (1+x^2)^{30} \cdot x^{10}$.

The coefficient of x^{20} in $x^{10}(1+x^2)^{30}$

$= \text{the coefficient of } x^{10} \text{ in } (1+x^2)^{30}$

$= {}^{30}C_5 = {}^{30}C_{30-5}$.

18. $(1-x+x^2)^5 = \{1+x(x-1)\}^5$

$$= {}^5C_0 + {}^5C_1 x(x-1) + {}^5C_2 x^2(x-1)^2 + {}^5C_3 x^3(x-1)^3 + \ldots$$

$\therefore$ the coefficient of $x^3 = -2 \cdot {}^5C_2 - {}^5C_3$.

20. Expression $= \dfrac{(1+x)^6\{1-(1+x)^{10}\}}{1-(1+x)} = \dfrac{(1+x)^{16}-(1+x)^6}{x}$

$\therefore$ the required coefficient

$= \text{the coefficient of } x^7 \text{ in } \{(1+x)^{16} - (1+x)^6\}$

$= {}^{16}C_7 = {}^{16}C_{16-7}$.

21. $(1+x+y-z)^9 = \{(1-z)+(x+y)\}^9$

$$= {}^9C_0(1-z)^9 + {}^9C_1(1-z)^8(x+y) + {}^9C_2(1-z)^7(x+y)^2 + \ldots + {}^9C_7(1-z)^2 \cdot (x+y)^7 + \ldots.$$

x^3y^4 appears in $(x+y)^7$ only.

$\therefore$ the required coefficient $= {}^9C_7(-2) \cdot {}^7C_4$.

22. Expression $= (1-x)^5 \cdot (1+x)^4(1+x^2)^4 = (1-x)(1-x^2)^4(1+x^2)^4$

$= (1-x)(1-x^4)^4$.

$\therefore$ the coefficient of $x^{13} = -{}^4C_3(-1)^3 = 4$.

24. $t_{r+1} = {}^{10}C_r(x\sin\alpha)^{10-r} \cdot \left(\dfrac{\cos\alpha}{x}\right)^r$.

It is independent of x if $r = 5$.

$\therefore$ the term independent of $x = {}^{10}C_5 \cdot \sin^5\alpha \cdot \cos^5\alpha$

$$= {}^{10}C_5 \cdot \frac{1}{2^5}(\sin 2\alpha)^5 \le {}^{10}C_5 \cdot \frac{1}{2^5}.$$

27. Expression $= \dfrac{1}{x^{10}}(1-2x+x^2)(1+x^2)^{10}$

$$= \frac{1}{x^{10}}(1+x^2)^{10} - \frac{2}{x^9}(1+x^2)^{10} + \frac{1}{x^8}(1+x^2)^{10}.$$

$\therefore$ the term independent of $x = {}^{10}C_5 - 2\times 0 + {}^{10}C_4$.

30. As the index = 20, the 11th term is the middle term. So, we have to find the 14th term.

31. $(\sqrt{2}+1)^6 = P + g = {}^6C_0(\sqrt{2})^6 + {}^6C_1(\sqrt{2})^5 + \dots + {}^6C_6$, where P = integer, $0 < g < 1$.

$(\sqrt{2}-1)^6 = f = {}^6C_0(\sqrt{2})^6 - {}^6C_1(\sqrt{2})^5 + \dots, 0 < f < 1$

$$\therefore \quad P + f + g = 2\{{}^6C_0 \cdot 2^3 + {}^6C_2 \cdot 2^2 + {}^6C_4 \cdot 2 + {}^6C_6\}$$
$$= 2(8 + 15\times 4 + 15\times 2 + 1) = 198$$

$\therefore \quad 197 < P + g < 198$

$\therefore \quad 197 < (\sqrt{2}+1)^6 < 198$

$\therefore \quad n < (\sqrt{2}+1)^6 \Rightarrow$ the greatest value of the natural number $n = 197$.

32. In $(1-x)^n$ the numerically greatest coefficient is the middle coefficient. From the question the middle term is the 5th term.

$\therefore$ there will be 9 terms. So, $n = 8$.

33. The greatest coefficient

= the coefficient of the middle term

$= {}^{2n}C_n$.

Now simplify.

34. $1 + 99^n = 1 + (100-1)^n = 1 + \{{}^nC_0 100^n - {}^nC_1 \cdot 100^{n-1} + \dots - {}^nC_n\}$

because n is odd

$$= 100\{{}^nC_0 \cdot 100^{n-1} - {}^nC_1 \cdot 100^{n-2} + \dots - {}^nC_{n-2} \cdot 100 + {}^nC_{n-1}\}$$

= 100 × integer whose units place is different from 0

(n having odd digit in units place).

35. Expression $= 10^n - 1 + 3\cdot 4^{n+2} + 6 = (10^n - 1) + 6(2^{2n+3}+1)$

$$= (10-1)\{10^{n-1} + 10^{n-2} + \dots + 10 + 1\}$$
$$+ 6(2+1)\{2^{2n+2} - 2^{2n+1} + \dots - 2 + 1\}.$$

$10^{n-1} + 10^{n-2} + \ldots + 10 + 1$ is an odd number which is not divisible by any odd number for all n. $2^{2n+2} - 2^{2n+1} + \ldots - 2 + 1$ is also an odd number which is not divisible by any odd number for all n.

$\therefore$ the expression is divisible by 9.

36. $2^{60} = (1+7)^{20} = {}^{20}C_0 + {}^{20}C_1 \cdot 7 + {}^{20}C_2 \cdot 7^2 + \ldots + {}^{20}C_{20} \cdot 7^{20}$

$\therefore$ the remainder $= {}^{20}C_0 = 1$.

37. $3^{2n} = (1+8)^n = {}^nC_0 + {}^nC_1 \cdot 8 + {}^nC_2 \cdot 8^2 + \ldots + {}^nC_n 8^n$

$\therefore \quad \frac{3^{2n}}{8} = \frac{1}{8} + ({}^nC_1 + {}^nC_2 \cdot 8 + \ldots + {}^nC_n \cdot 8^7) = \frac{1}{8} + \text{integer}.$

38. The sum of all the coefficients in an expansion is obtained by putting $x = 1$ in the expression.

$\therefore \quad \left(\frac{1}{1} + 2 \cdot 1\right)^n = 6\,561$

$\therefore \quad 3^n = 3^8 \quad \therefore \; n = 8.$

In $\left(\frac{1}{x} + 2x\right)^8$, $t_{r+1} = {}^8C_r \cdot \left(\frac{1}{x}\right)^{8-r} \cdot (2x)^r$.

This is a constant if $2r - 8 = 0$, i.e., $r = 4$.

$\therefore$ the constant term $= t_5 = {}^8C_4 \cdot 2^4$.

39. The sum of all the numerical coefficients in the expansion is obtained by putting $x = 1, y = 1$.

40. The required sum $= {}^{19}C_{10} + {}^{19}C_{11} + \ldots + {}^{19}C_{19}$

$= {}^{19}C_0 + {}^{19}C_1 + {}^{19}C_2 + \ldots + {}^{19}C_9 \quad (\because \; {}^nC_r = {}^nC_{n-r})$

Adding, $2 \times (\text{required sum}) = {}^{19}C_0 + {}^{19}C_1 + \ldots + {}^{19}C_{19} = 2^{19}$.

42. See the hint to Q. No. 38.

43. The coefficients of the integral powers of x are ${}^{40}C_0$, ${}^{40}C_2 \cdot 2^2$, ${}^{40}C_4 \cdot 2^4, \ldots, {}^{40}C_{40} \cdot 2^{40}$.

$(1+2)^{40} = {}^{40}C_0 + {}^{40}C_1 \cdot 2 + {}^{40}C_2 \cdot 2^2 + \ldots + {}^{40}C_{40} \cdot 2^{40}$.

$(1-2)^{40} = {}^{40}C_0 - {}^{40}C_1 \cdot 2 + {}^{40}C_2 \cdot 2^2 - \ldots + {}^{40}C_{40} \cdot 2^{40}$.

Adding, $3^{40} + 1 = 2 \times (\text{required sum})$.

44. Sum $= \frac{1}{2}\{(a_0 + a_1 + a_2 + \ldots + a_{16}) - (a_0 - a_1 + a_2 - \ldots + a_{16})\}$

$$= \frac{1}{2}\{(1+1-2)^8 - (1-1-2)^8\} = \frac{1}{2}(-2^8) = -2^7.$$

45. Let $S = {}^{20}C_0 + {}^{20}C_1 + {}^{20}C_2 + \ldots + {}^{20}C_{10}$

$$= {}^{20}C_{20} + {}^{20}C_{19} + {}^{20}C_{18} + \ldots + {}^{20}C_{10} \qquad (\because\ {}^nC_r = {}^nC_{n-r})$$

Adding, $2S = ({}^{20}C_0 + {}^{20}C_1 + \ldots + {}^{20}C_{20}) + {}^{20}C_{10} = 2^{20} + {}^{20}C_{10}$.

46. Expression = coefficient of x^3 in

$$\{(1+x)^{10} + (1+x)^{11} + (1+x)^{12} + \ldots + (1+x)^{20}\}$$

$$= \text{coefficient of } x^3 \text{ in } \frac{(1+x)^{10}\{1-(1+x)^{11}\}}{1-(1+x)}$$

$$= \text{coefficient of } x^4 \text{ in } \{(1+x)^{21} - (1+x)^{10}\}$$

$$= {}^{21}C_4 - {}^{10}C_4.$$

47. Putting $x = i, -i$ and multiplying both the results, we get the value of the required expression.

48. Sum $= \frac{1}{2}\{{}^{10}C_0 - {}^{10}C_1 \cdot 2 + {}^{10}C_2 \cdot 2^2 - {}^{10}C_3 \cdot 2^3 + \ldots + {}^{10}C_{10} \cdot 2^{10}\}$

$$= \frac{1}{2}(1-2)^{10} = \frac{1}{2}.$$

49. $r \cdot {}^nC_r = n \cdot {}^{n-1}C_{r-1}$.

$\therefore$ the sum $= n\{{}^{n-1}C_0 + {}^{n-1}C_1 + {}^{n-1}C_2 + \ldots + {}^{n-1}C_{n-1}\}$

$$= n \cdot 2^{n-1}.$$

50. Take $n = 2m$. Then

$$a_n = \frac{1}{{}^{2m}C_0} + \frac{1}{{}^{2m}C_1} + \ldots + \frac{1}{{}^{2m}C_{2m}}$$

$$= 2\left(\frac{1}{{}^{2m}C_0} + \frac{1}{{}^{2m}C_1} + \ldots + \frac{1}{{}^{2m}C_{m-1}}\right) + \frac{1}{{}^{2m}C_m} \qquad \ldots (1)$$

$$\sum_{r=0}^{n} \frac{r}{{}^nC_r} = \sum_{r=0}^{2m} \frac{r}{{}^{2m}C_r} = \frac{1}{{}^{2m}C_1} + \frac{2}{{}^{2m}C_2} + \ldots + \frac{2m}{{}^{2m}C_{2m}}$$

$$= \left(\frac{1}{^{2m}C_1} + \frac{2m-1}{^{2m}C_{2m-1}}\right) + \left(\frac{2}{^{2m}C_2} + \frac{2m-2}{^{2m}C_{2m-2}}\right) + \dots$$

$$+ \left(\frac{m-1}{^{2m}C_{m-1}} + \frac{m+1}{^{2m}C_{m+1}}\right) + \frac{m}{^{2m}C_m} + \frac{2m}{^{2m}C_{2m}}$$

$$= 2m\left\{\frac{1}{^{2m}C_1} + \frac{1}{^{2m}C_2} + \dots + \frac{1}{^{2m}C_{m-1}}\right\} + \frac{m}{^{2m}C_m} + 2m$$

$$= 2m\left\{\frac{1}{^{2m}C_0} + \frac{1}{^{2m}C_1} + \frac{1}{^{2m}C_2} + \dots + \frac{1}{^{2m}C_{m-1}}\right\} + \frac{m}{^{2m}C_m}$$

$$= m\left(a_n - \frac{1}{^{2m}C_m}\right) + \frac{m}{^{2m}C_m} = ma_n = \frac{n}{2}a_n.$$

Similarly for $n = 2m + 1$.

51. Sum $= a\sum_{r=1}^{n}(-1)^{r-1}\cdot {^nC_r} - \sum_{r=1}^{n}(-1)^{r-1}\cdot r\cdot {^nC_r}$

$$= a\{^nC_1 - {^nC_2} + {^nC_3} - \dots\} - n\{^{n-1}C_0 - {^{n-1}C_1} + \dots\}$$

(because $r\cdot {^nC_r} = n\cdot {^{n-1}C_{r-1}}$)

$$= a\cdot {^nC_0} \qquad (\because\ {^nC_0} - {^nC_1} + {^nC_2} - \dots = 0).$$

52. Expression $= \frac{^nC_0 + {^nC_1}}{^nC_0}\cdot\frac{^nC_1 + {^nC_2}}{^nC_1}\dots\frac{^nC_{n-1} + {^nC_n}}{^nC_{n-1}}$

$$= \frac{^{n+1}C_1}{^nC_0}\cdot\frac{^{n+1}C_2}{^nC_1}\dots\frac{^{n+1}C_n}{^nC_{n-1}}$$

$$= \frac{1\cdot {^{n+1}C_1}}{^nC_0}\cdot\frac{2\cdot {^{n+1}C_2}}{^nC_1}\dots\frac{n\cdot {^{n+1}C_n}}{^nC_{n-1}}\cdot\frac{1}{n!}$$

$$= \frac{(n+1)\cdot {^nC_0}}{^nC_0}\cdot\frac{(n+1)\cdot {^nC_1}}{^nC_1}\dots\frac{(n+1)\cdot {^nC_{n-1}}}{^nC_{n-1}}\cdot\frac{1}{n!}.$$

53. $\frac{r\cdot {^nC_r}}{^nC_{r-1}} = \frac{n\cdot {^{n-1}C_{r-1}}}{^nC_{r-1}} = n\cdot\frac{(n-1)!}{(r-1)!\,(n-r)!}\times\frac{(r-1)!\,(n-r+1)!}{n!} = n - r + 1.$

$\therefore$ sum $= n + (n-1) + \dots + (n-9).$

54. $r\cdot {^{2n}C_r} = 2n\cdot {^{2n-1}C_{r-1}}$

$\therefore$ sum $= 2n\{^{2n-1}C_0 + {}^{2n-1}C_1 + \ldots + {}^{2n-1}C_{n-1}\}$

$= 2n \cdot 2^{(2n-1)-1}$. (See the hint to Q. No. 40.)

55. Using $r \cdot {}^nC_r = n \cdot {}^{n-1}C_{r-1}$,

sum $= 20\{^{19}C_0 - {}^{19}C_1 + {}^{19}C_2 - \ldots - {}^{19}C_{19}\} = 20 \times 0 = 0.$

56. $f(x) = (\sqrt{x^2+1} + \sqrt{x^2-1})^6 + (\sqrt{x^2+1} - \sqrt{x^2-1})^6$.

After expansion and simplification, $f(x) = 64x^6 - 48x^2$.

57. Expression $= \{a + (b+c)\}^{18} = {}^{18}C_0 a^{18} + {}^{18}C_1 a^{17}(b+c)$

$+ \ldots + {}^{18}C_{10} a^8 (b+c)^{10} + \ldots$

$\therefore$ the coefficient of $a^8 b^6 c^4 = {}^{18}C_{10} \cdot {}^{10}C_6 = \dfrac{18!}{8!6!4!}$ = all options.

58. Expression $= (-1)^n \cdot \dfrac{(1-x^2)^n}{x^n}$

$\therefore$ the term independent of x

= the coefficient of x^n in $(-1)^n \cdot (1-x^2)^n$

= the coefficient of x^n in

$(-1)^n\{^nC_0 + {}^nC_1(-x^2) + {}^nC_2(-x^2)^2 + \ldots + {}^nC_n(-x^2)^n\}$,

and the expansion contains only even powers of x.

60. $a_0 + a_1 + a_2 + \ldots = 2^{2n}$ and $a_0 + a_2 + a_4 + \ldots = 2^{2n-1}$.

$a_n = {}^{2n}C_n$ = the greatest coefficient, being the middle coefficient.

$a_{n-3} = {}^{2n}C_{n-3} = {}^{2n}C_{2n-(n-3)} = {}^{2n}C_{n+3} = a_{n+3}$

61. $t_{r+1} = {}^{20}C_r \cdot 4^{\frac{20-r}{3}} \cdot 6^{-\frac{r}{4}} = {}^{20}C_r \cdot 2^{\frac{40-2r}{3}} \cdot 2^{-\frac{r}{4}} \cdot 3^{-\frac{r}{4}}$

$= {}^{20}C_r \cdot 2^{\frac{160-11r}{12}} \cdot 3^{-\frac{r}{4}}$.

For rational terms, $\dfrac{r}{4}$ and $\dfrac{160-11r}{12}$ must be integers and $0 \le r \le 20$.

So, $\dfrac{r}{4}$ = integer $\Rightarrow r = 0, 4, 8, 12, 16, 20.$

Clearly, for $r = 8$, 16 and 20, $\frac{160 - 11r}{12}$ is also an integer.

$\therefore$ only three terms are rational. So 21 – 3, i.e., 18 terms are irrational.

62. In an expansion, binomial coefficients equidistant from the beginning and the end are equal.

$\therefore\ a_{n-1} = a_1, a_{n-2} = a_2, a_{n-3} = a_3.$

63. $R = [R] + g = (8 + 3\sqrt{7})^{20} = {}^{20}C_0 8^{20} + {}^{20}C_1 8^{19}(3\sqrt{7}) + \dots,$ where $0 < g < 1$.

$f = (8 - 3\sqrt{7})^{20} = {}^{20}C_0 \cdot 8^{20} - {}^{20}C_1 8^{19}(3\sqrt{7}) + \dots,$ where $0 < f < 1$.

$\therefore\quad [R] + g + f = 2[{}^{20}C_0 \cdot 8^{20} + {}^{20}C_2\, 8^{18}(3\sqrt{7})^2 + \dots] =$ an even integer.

$\therefore\quad g + f =$ an integer $= 1$ as $0 < g < 1, 0 < f < 1$.

$\therefore\quad [R] =$ an even integer $-1 =$ an odd integer.

Also, $R - [R] = g = 1 - f = 1 - (8 - 3\sqrt{7})^{20} = 1 - \frac{1}{(8 + 3\sqrt{7})^{20}}$.

64. Expression $= \frac{1}{n!}\{{}^nC_1 + {}^nC_3 + {}^nC_5 + \dots\} = \frac{1}{n!} \cdot 2^{n-1}$ for all $n \in N$.

65. $(x + y + z)^{25} = \{x + (y + z)\}^{25} = {}^{25}C_0 x^{25} + {}^{25}C_1 x^{24}(y + z)$

$$+ \dots + {}^{25}C_r x^{25-r}(y + z)^r + \dots$$

$$= \dots + {}^{25}C_r\, x^{25-r} \cdot (\dots + {}^rC_k\, y^{r-k} z^k + \dots) + \dots.$$

$8 + 9 + 9 \neq 25$. So, there is no term like $x^8 y^9 z^9$.

The number of terms $= 1 + 2 + 3 + \dots + 26 = \frac{26 \times 27}{2}$.

□

7

Matrices

• Type 1 •

Choose the most appropriate option (a, b, c or d).

1. If $A = \begin{bmatrix} 1 & -2 & 4 \\ 2 & 3 & 2 \\ 3 & 1 & 5 \end{bmatrix}$ and $B = \begin{bmatrix} 0 & -2 & 4 \\ 1 & 3 & 2 \\ -1 & 1 & 5 \end{bmatrix}$ then $A + B$ is

(a) $\begin{bmatrix} 1 & -2 & 4 \\ 3 & 3 & 2 \\ 2 & 1 & 5 \end{bmatrix}$ (b) $\begin{bmatrix} 1 & -2 & 8 \\ 3 & 3 & 4 \\ 2 & 1 & 10 \end{bmatrix}$

(c) $\begin{bmatrix} 1 & -4 & 8 \\ 3 & 6 & 4 \\ 2 & 2 & 10 \end{bmatrix}$ (d) none of these

2. If $A^2 = 8A + kI$ where $A = \begin{bmatrix} 1 & 0 \\ -1 & 7 \end{bmatrix}$ then k is

(a) 7 (b) –7 (c) 1 (d) –1

3. The matrix $\begin{bmatrix} \lambda & 7 & -2 \\ 4 & 1 & 3 \\ 2 & -1 & 2 \end{bmatrix}$ is a singular matrix if λ is

(a) $\frac{2}{5}$ (b) $\frac{5}{2}$ (c) –5 (d) none of these

4. If the matrix $A = \begin{bmatrix} a & b \\ c & d \end{bmatrix}$ then A^2 is

(a) $\begin{bmatrix} a^2 & b^2 \\ c^2 & d^2 \end{bmatrix}$ (b) $\begin{bmatrix} a^2 + bc & ab + bd \\ ac + dc & bc + d^2 \end{bmatrix}$

(c) nonexistent (d) none of these

5. If $A=\begin{bmatrix} \alpha & 0 \\ 1 & 1 \end{bmatrix}$ and $B=\begin{bmatrix} 1 & 0 \\ 5 & 1 \end{bmatrix}$ such that $A^2=B$ then α is

(a) 1 (b) –1

(c) 4 (d) none of these

6. If $\begin{bmatrix} 2 & -3 \\ 1 & \lambda \end{bmatrix}\times\begin{bmatrix} 1 & 5 & \mu \\ 0 & 2 & -3 \end{bmatrix}=\begin{bmatrix} 2 & 4 & 1 \\ 1 & -1 & 13 \end{bmatrix}$ then

(a) $\lambda=3, \mu=4$

(b) $\lambda=4, \mu=-3$

(c) no real values of λ, μ are possible

(d) none of these

7. If $AB=0$ where $A=\begin{bmatrix} \cos^2\theta & \cos\theta\sin\theta \\ \cos\theta\sin\theta & \sin^2\theta \end{bmatrix}$

and $B=\begin{bmatrix} \cos^2\phi & \cos\phi\sin\phi \\ \cos\phi\sin\phi & \sin^2\phi \end{bmatrix}$ then $|\theta-\phi|$ is equal to

(a) 0 (b) $\frac{\pi}{2}$

(c) $\frac{\pi}{4}$ (d) π

8. If $A=\begin{bmatrix} 0 & -4 & 1 \\ 2 & \lambda & -3 \\ 1 & 2 & -1 \end{bmatrix}$ then A^{-1} exists (i.e., A is invertible) if

(a) $\lambda\neq 4$ (b) $\lambda\neq 8$

(c) $\lambda=4$ (d) none of these

9. The reciprocal matrix of $\begin{bmatrix} 1 & 0 & 2 \\ 0 & 1 & -1 \\ 1 & 2 & 1 \end{bmatrix}$ is

(a) $\begin{bmatrix} -3 & -4 & 2 \\ -1 & 1 & -1 \\ 1 & 2 & -1 \end{bmatrix}$ (b) $\begin{bmatrix} 3 & 4 & -2 \\ 1 & -1 & 1 \\ -1 & -2 & 1 \end{bmatrix}$

(c) $\begin{bmatrix} -3 & -1 & 1 \\ -4 & 1 & 2 \\ 2 & -1 & -1 \end{bmatrix}$ (d) none of these

10. If $A=\begin{bmatrix}1 & -1 & 1\\ 1 & 2 & 0\\ 1 & 3 & 0\end{bmatrix}$ then the value of $|\text{adj } A|$ is equal to

(a) 5 (b) 0

(c) 1 (d) none of these

11. If $A=\begin{bmatrix}\sin\alpha & -\cos\alpha & 0\\ \cos\alpha & \sin\alpha & 0\\ 0 & 0 & 1\end{bmatrix}$ then A^{-1} is equal to

(a) A^T (b) A

(c) adj A (d) none of these

12. If $A=\begin{bmatrix}4 & -1 & -4\\ 3 & 0 & -4\\ 3 & -1 & -3\end{bmatrix}$ then A^2 is equal to

(a) A (b) I

(c) A^T (d) none of these

13. If $f(x)=\begin{bmatrix}\cos x & -\sin x & 0\\ \sin x & \cos x & 0\\ 0 & 0 & 1\end{bmatrix}$ then $f(x+y)$ is equal to

(a) $f(x)+f(y)$ (b) $f(x)-f(y)$

(c) $f(x)\cdot f(y)$ (d) none of these

14. If $A=\begin{bmatrix}1 & \omega & \omega^2\\ \omega & \omega^2 & 1\\ \omega^2 & 1 & \omega\end{bmatrix}$, $B=\begin{bmatrix}\omega & \omega^2 & 1\\ \omega^2 & 1 & \omega\\ \omega & \omega^2 & 1\end{bmatrix}$ and $C=\begin{bmatrix}1\\ \omega\\ \omega^2\end{bmatrix}$ where ω is the complex cube root of 1 then $(A+B)C$ is equal to

(a) $\begin{bmatrix}0\\0\\0\end{bmatrix}$ (b) $\begin{bmatrix}1 & 0 & 0\\ 0 & 1 & 0\\ 0 & 0 & 1\end{bmatrix}$

(c) $\begin{bmatrix}1\\0\\1\end{bmatrix}$ (d) $\begin{bmatrix}1\\1\\1\end{bmatrix}$

15. If $A=\begin{bmatrix}0 & c & -b\\ -c & 0 & a\\ b & -a & 0\end{bmatrix}$ and $B=\begin{bmatrix}a^2 & ab & ac\\ ba & b^2 & bc\\ ca & cb & c^2\end{bmatrix}$ then AB is equal to

(a) 0 (b) I

(c) $2I$ (d) none of these

16. If A be a matrix such that $A \times \begin{bmatrix} 1 & -2 \\ 1 & 4 \end{bmatrix} = \begin{bmatrix} 6 & 0 \\ 0 & 6 \end{bmatrix}$ then A is

(a) $\begin{bmatrix} 2 & 4 \\ 1 & -1 \end{bmatrix}$ (b) $\begin{bmatrix} -1 & 1 \\ 4 & 2 \end{bmatrix}$

(c) $\begin{bmatrix} 4 & 2 \\ -1 & 1 \end{bmatrix}$ (d) none of these

17. The rank of the matrix $\begin{bmatrix} -5 & 3 & 2 \\ 3 & 2 & -5 \\ 4 & -1 & -3 \end{bmatrix}$ is

(a) 3 (b) 2

(c) 1 (d) none of these

18. The rank of the matrix $\begin{bmatrix} 1 & 2 & 3 \\ \lambda & 2 & 4 \\ 2 & -3 & 1 \end{bmatrix}$ is 3 if

(a) $\lambda \neq \frac{18}{11}$ (b) $\lambda = \frac{18}{11}$

(c) $\lambda = -\frac{18}{11}$ (d) none of these

19. The rank of the matrix $\begin{bmatrix} 4 & 1 & 0 & 0 \\ 3 & 0 & 1 & 0 \\ 5 & 0 & 0 & 1 \end{bmatrix}$ is

(a) 4 (b) 3 (c) 2 (d) none of these

20. The system of equations

$$x + y + z = 2$$
$$2x - y + 3z = 5$$
$$x - 2y - z + 1 = 0$$

written in matrix form is

(a) $\begin{bmatrix} x \\ y \\ z \end{bmatrix} \begin{bmatrix} 1 & 1 & 1 \\ 2 & -1 & 3 \\ 1 & -2 & -1 \end{bmatrix} = \begin{bmatrix} 2 \\ 5 \\ -1 \end{bmatrix}$ (b) $\begin{bmatrix} 1 & 1 & 1 \\ 2 & -1 & 3 \\ 1 & -2 & -1 \end{bmatrix} \begin{bmatrix} x \\ y \\ z \end{bmatrix} = \begin{bmatrix} -2 \\ -5 \\ 1 \end{bmatrix}$

(c) $\begin{bmatrix} 1 & 1 & 1 \\ 2 & -1 & 3 \\ 1 & -2 & -1 \end{bmatrix} \begin{bmatrix} x \\ y \\ z \end{bmatrix} = \begin{bmatrix} 2 \\ 5 \\ -1 \end{bmatrix}$ (d) none of these

21. If $[1 \;\; x \;\; 1] \begin{bmatrix} 1 & 3 & 2 \\ 2 & 5 & 1 \\ 15 & 3 & 2 \end{bmatrix} \begin{bmatrix} 1 \\ 2 \\ x \end{bmatrix} = 0$ then x is

(a) 2 (b) −2 (c) 14 (d) none of these

22. If $\begin{bmatrix} x+y & y \\ 2x & x-y \end{bmatrix}\begin{bmatrix} 2 \\ -1 \end{bmatrix}=\begin{bmatrix} 3 \\ 2 \end{bmatrix}$ then $x \cdot y$ is equal to

(a) –5 (b) 5

(c) 4 (d) 6

• *Type 2* •

Choose the correct options. One or more options may be correct.

23. $\begin{bmatrix} 1 & -2 & 3 \\ 2 & -1 & 4 \\ 3 & 4 & 1 \end{bmatrix}$ is a

(a) rectangular matrix (b) singular matrix

(c) square matrix (d) nonsingular matrix

24. If $A=\begin{bmatrix} 3 & 1 \\ -1 & 2 \\ 0 & 6 \end{bmatrix}$ and $B=\begin{bmatrix} 5 & 4 & 6 \\ 4 & 1 & 2 \\ -5 & -1 & 1 \end{bmatrix}$ then

(a) $A+B$ exists (b) AB exists

(c) BA exists (d) none of these

25. If $A=\begin{bmatrix} 1 & 1 & 1 \\ 1 & 1 & 1 \\ 1 & 1 & 1 \end{bmatrix}$ then

(a) $A^3=9A$ (b) $A^3=27A$

(c) $A+A=A^2$ (d) A^{-1} does not exist

Answers

1. c	**2.** b	**3.** a	**4.** b	**5.** d
6. d	**7.** b	**8.** b	**9.** a	**10.** c
11. c	**12.** b	**13.** c	**14.** a	**15.** a
16. c	**17.** b	**18.** a	**19.** b	**20.** c
21. b	**22.** a	**23.** c, d	**24.** c	**25.** a, c, d

Hints and Solutions to Selected Questions

2. $\begin{bmatrix} 1 & 0 \\ -1 & 7 \end{bmatrix}\begin{bmatrix} 1 & 0 \\ -1 & 7 \end{bmatrix} = 8\begin{bmatrix} 1 & 0 \\ -1 & 7 \end{bmatrix} + k\begin{bmatrix} 1 & 0 \\ 0 & 1 \end{bmatrix}$

or $\begin{bmatrix} 1+0 & 0+0 \\ -1-7 & 0+49 \end{bmatrix} = \begin{bmatrix} 8 & 0 \\ -8 & 56 \end{bmatrix} + \begin{bmatrix} k & 0 \\ 0 & k \end{bmatrix} = \begin{bmatrix} 8+k & 0 \\ -8 & 56+k \end{bmatrix}$

$\Rightarrow \quad 1 = 8 + k,\ 49 = 56 + k \Rightarrow k = -7.$

5. $\begin{bmatrix} \alpha & 0 \\ 1 & 1 \end{bmatrix}\begin{bmatrix} \alpha & 0 \\ 1 & 1 \end{bmatrix} = \begin{bmatrix} 1 & 0 \\ 5 & 1 \end{bmatrix}$

or $\begin{bmatrix} \alpha^2 & 0 \\ \alpha+1 & 1 \end{bmatrix} = \begin{bmatrix} 1 & 0 \\ 5 & 1 \end{bmatrix} \Rightarrow \alpha^2 = 1, \alpha + 1 = 5$

$\Rightarrow \alpha = 1, -1$ satisfying $\alpha = 4$ (absurd).

6. Here $\begin{bmatrix} 2 & 4 & 2\mu+9 \\ 1 & 5+2\lambda & \mu-3\lambda \end{bmatrix} = \begin{bmatrix} 2 & 4 & 1 \\ 1 & -1 & 13 \end{bmatrix}$

$\Rightarrow \quad 2\mu + 9 = 1, 5 + 2\lambda = -1, \mu - 3\lambda = 13$

$\therefore \quad \mu = -4, \lambda = -3$, which does not satisfy $\mu - 3\lambda = 13$.

7. $AB = \begin{bmatrix} \cos^2\theta & \cos\theta \cdot \sin\theta \\ \cos\theta\sin\theta & \sin^2\theta \end{bmatrix}\begin{bmatrix} \cos^2\phi & \cos\phi\sin\phi \\ \cos\phi\sin\phi & \sin^2\phi \end{bmatrix}$

$$= \begin{bmatrix} \cos^2\theta\cos^2\phi + \cos\theta\sin\theta\cos\phi\sin\phi & \cos^2\theta\cos\phi\sin\phi + \cos\theta\sin\theta\sin^2\phi \\ \cos\theta\sin\theta\cos^2\phi + \cos\phi\sin\phi \cdot \sin^2\theta & \cos\theta\sin\theta\cos\phi\sin\phi + \sin^2\theta\sin^2\phi \end{bmatrix}$$

$$= \begin{bmatrix} \cos\theta\cos\phi\cos(\theta-\phi) & \cos\theta\sin\phi\cos(\theta-\phi) \\ \sin\theta\cos\phi\cos(\theta-\phi) & \sin\theta\sin\phi\cos(\theta-\phi) \end{bmatrix} = \begin{bmatrix} 0 & 0 \\ 0 & 0 \end{bmatrix}$$

$\Rightarrow \quad \cos(\theta - \phi) = 0 \Rightarrow |\theta - \phi| = \text{odd multiple of } \frac{\pi}{2}.$

8. A is invertible if $|A| \neq 0$, i.e., $\begin{vmatrix} 0 & -4 & 1 \\ 2 & \lambda & -3 \\ 1 & 2 & -1 \end{vmatrix} \neq 0.$

11. Here $|A| = \begin{vmatrix} \sin\alpha & -\cos\alpha & 0 \\ \cos\alpha & \sin\alpha & 0 \\ 0 & 0 & 1 \end{vmatrix} = 1$

$\therefore$ by definition, $A^{-1} = \dfrac{1}{|A|}$ adj A. We get $A^{-1} =$ adj A.

13. $f(x+y) = \begin{bmatrix} \cos(x+y) & -\sin(x+y) & 0 \\ \sin(x+y) & \cos(x+y) & 0 \\ 0 & 0 & 1 \end{bmatrix}$

$$= \begin{bmatrix} \cos x & -\sin x & 0 \\ \sin x & \cos x & 0 \\ 0 & 0 & 1 \end{bmatrix} \begin{bmatrix} \cos y & -\sin y & 0 \\ \sin y & \cos y & 0 \\ 0 & 0 & 1 \end{bmatrix} = f(x) \cdot f(y).$$

14. $A + B = \begin{bmatrix} 1+\omega & \omega+\omega^2 & \omega^2+1 \\ \omega+\omega^2 & \omega^2+1 & 1+\omega \\ \omega^2+\omega & 1+\omega^2 & \omega+1 \end{bmatrix} = \begin{bmatrix} -\omega^2 & -1 & -\omega \\ -1 & -\omega & -\omega^2 \\ -1 & -\omega & -\omega^2 \end{bmatrix}$

$$\therefore \quad (A+B)C = \begin{bmatrix} -\omega^2 & -1 & -\omega \\ -1 & -\omega & -\omega^2 \\ -1 & -\omega & -\omega^2 \end{bmatrix} \begin{bmatrix} 1 \\ \omega \\ \omega^2 \end{bmatrix} = \begin{bmatrix} -\omega^2 - \omega - \omega^3 \\ -1 - \omega^2 - \omega^4 \\ -1 - \omega^2 - \omega^4 \end{bmatrix} = \begin{bmatrix} 0 \\ 0 \\ 0 \end{bmatrix},$$

using $\omega^3 = 1, 1 + \omega + \omega^2 = 0$.

16. Let $A = \begin{bmatrix} a & b \\ c & d \end{bmatrix}$. Then $\begin{bmatrix} a & b \\ c & d \end{bmatrix} \begin{bmatrix} 1 & -2 \\ 1 & 4 \end{bmatrix} = \begin{bmatrix} 6 & 0 \\ 0 & 6 \end{bmatrix}$

$$\Rightarrow \quad \begin{bmatrix} a+b & -2a+4b \\ c+d & -2c+4d \end{bmatrix} = \begin{bmatrix} 6 & 0 \\ 0 & 6 \end{bmatrix}$$

$$\Rightarrow \quad a+b = 6, -2a+4b = 0 \quad \Rightarrow \quad a = 4, b = 2$$

$$c+d = 0, \ -2c+4d = 6 \quad \Rightarrow \quad c = -1, d = 1.$$

17. $\begin{vmatrix} -5 & 3 & 2 \\ 3 & 2 & -5 \\ 4 & -1 & -3 \end{vmatrix} = 0$ and $\begin{vmatrix} -5 & 3 \\ 3 & 2 \end{vmatrix} \neq 0$. So, the only possible third-order minor is 0, while one second-order minor is nonzero. So, the rank is 2.

18. For the rank to be 3, $\begin{vmatrix} 1 & 2 & 3 \\ \lambda & 2 & 4 \\ 2 & -3 & 1 \end{vmatrix} \neq 0$.

19. $\begin{vmatrix} 1 & 0 & 0 \\ 0 & 1 & 0 \\ 0 & 0 & 1 \end{vmatrix} = 1 \neq 0$. So, the rank $= 3$.

□

8

Some Additional Topics

• *Type 1* •

Choose the most appropriate option (a, b, c or d).

INFINITE SERIES

1. $(1-x)^{3/2}$ can be expanded in ascending powers of x if

(a) $-1<x<1$ (b) $x<-1$

(c) $x>1$ (d) none of these

2. $(3+x)^{p/q}$ can be expanded in ascending powers of x if

(a) $-1<x<1$ (b) $x>3$

(c) $-3<x<3$ (d) $x<-3$

3. If x is positive, the first negative term in the expansion of $(1+x)^{27/6}$ is the

(a) 5th term (b) 7th term

(c) 6th term (d) 8th term

4. In the expansion of $(1-x)^{-3}$, $|x|<1$, the coefficient of x^7 is

(a) 36 (b) ${}^8C_7$

(c) 45 (d) none of these

5. In the expansion of $(2+x)^{-5/2}$, the coefficient of x^4, if it exists, is

(a) $\dfrac{5\cdot 7\cdot 9\cdot 11}{2^{5/2}\cdot 4\,!}\left(\dfrac{1}{2}\right)^4$ (b) $\dfrac{5\cdot 7\cdot 9\cdot 11}{4\,!}\left(\dfrac{1}{2}\right)^4$

(c) $\dfrac{5 \cdot 7 \cdot 9 \cdot 11}{(\sqrt{2})^{21}}$ (d) none of these

6. The coefficient of x^5 in the expansion of $\dfrac{1+x^2}{1+x}$, $|x| < 1$, is

(a) -1 (b) 2

(c) 0 (d) -2

7. If $\dfrac{x^2+x}{1-x} = a_1 x + a_2 x^2 + \ldots$ to ∞, $|x| < 1$, then

(a) $a_1 + a_2 = 4$ (b) $a_1 - a_2 = 3$

(c) $a_p = a_q$ (d) none of these

8. The coefficient of x^n in the expansion of e^{2x+3} is

(a) $\dfrac{2^n}{n!}$ (b) $\dfrac{e^3 \cdot 2^n}{n!}$

(c) $\dfrac{e^2 \cdot 3^n}{n!}$ (d) none of these

9. The coefficient of x^{10} in the expansion of 10^x in ascending powers of x is

(a) $\dfrac{(\log_e 10)^{10}}{10!}$ (b) $\dfrac{1}{10!}$

(c) $\dfrac{(\log_{10} e)^{10}}{10!}$ (d) none of these

10. In the expansion of $\dfrac{e^x - 1 - x}{x^2}$ in ascending powers of x, the fourth term is

(a) $\dfrac{1}{5!}x^3$ (b) $\dfrac{1}{4!}x^4$

(c) $\dfrac{1}{3!}x^3$ (d) none of these

11. The constant term in the expansion of $\dfrac{3^x - 2^x}{x^2}$ is

(a) $\log_e 3$ (b) $\log_e 6 \cdot \log_e \dfrac{3}{2}$

(c) $\frac{1}{2}\log_e 6 \cdot \log_e \frac{3}{2}$ (d) none of these

12. If $|x| < 1$, the coefficient of x^3 in the expansion of $\frac{1}{e^x \cdot (1+x)}$ is

(a) $\frac{17}{6}$ (b) $-\frac{17}{6}$

(c) $-\frac{11}{6}$ (d) none of these

13. The constant term in the expansion of $\frac{x + \log_e (1-x)}{x^3}$ is

(a) $-\frac{1}{3}$ (b) 0

(c) $-\frac{1}{2}$ (d) $\frac{1}{3}$

14. In the expansion of $\log_{10} (1-x)$, $|x| < 1$, the coefficient of x^n is

(a) $-\frac{1}{n}$ (b) $-\frac{1}{n}\log_{10} e$

(c) $\frac{1}{n}$ (d) none of these

15. If $|x| < 1$, the coefficient of x^2 in the expansion of $\frac{\log_e (1+x)}{(1-x)^2}$ is

(a) $\frac{3}{2}$ (b) ${}^2C_1$

(c) $-\frac{1}{2}$ (d) none of these

16. The sum of the series ${}^4C_0 + {}^5C_1 x + {}^6C_2 x^2 + {}^7C_3 x^3 + \ldots$ to ∞ is

(a) $(1-x)^{-4}$ (b) $\frac{1}{(1-x)^5}$

(c) $(1+x)^{-5}$ (d) none of these

17. The sum of the series ${}^2C_0 - {}^3C_1 x^2 + {}^4C_2 x^4 - {}^5C_3 x^6 + \ldots$ to ∞ is

(a) $\frac{1}{(1+x^2)^3}$ (b) $(1-x^2)^{-3}$

(c) $\dfrac{1}{(1+x)^3}$ (d) none of these

18. The sum of the series $1+\dfrac{1}{1!}\cdot\dfrac{1}{4}+\dfrac{1\cdot 3}{2!}\cdot\left(\dfrac{1}{4}\right)^2+\dfrac{1\cdot 3\cdot 5}{3!}\cdot\left(\dfrac{1}{4}\right)^3+\ldots$ to ∞ is

(a) $\sqrt{2}$ (b) 2

(c) $\dfrac{1}{\sqrt{2}}$ (d) none of these

19. The sum of the series ${}^3C_0-{}^4C_1\cdot\dfrac{1}{2}+{}^5C_2\cdot\left(\dfrac{1}{2}\right)^2-{}^6C_3\cdot\left(\dfrac{1}{2}\right)^3+\ldots$ to ∞ is

(a) 16 (b) 8

(c) $\dfrac{16}{81}$ (d) none of these

20. $1+\dfrac{1}{3!}+\dfrac{1}{5!}+\dfrac{1}{7!}+\ldots$ to ∞ is equal to

(a) $\dfrac{1}{2}(e-e^{-1})$ (b) $\dfrac{1}{2}(e+e^{-1})$

(c) $\dfrac{1}{2}e$ (d) none of these

21. $\dfrac{1}{0!}+\dfrac{2}{1!}+\dfrac{3}{2!}+\dfrac{4}{3!}+\ldots$ to ∞ is equal to

(a) $4e$ (b) $3e$

(c) $2e$ (d) none of these

22. $\dfrac{1}{3!}+\dfrac{2}{5!}+\dfrac{3}{7!}+\ldots$ to ∞ is equal to

(a) $\dfrac{1}{2}e$ (b) $\dfrac{1}{2e}$

(c) $\dfrac{3}{2e}$ (d) none of these

23. $(a^2-b^2)+\dfrac{a^4-b^4}{2!}+\dfrac{a^6-b^6}{3!}+\ldots$ to ∞ is equal to

(a) $e^{a^2}-e^{b^2}$ (b) $e^{b^2}-e^{a^2}$

(c) $\dfrac{e^{a^2}}{e^{b^2}}$ (d) none of these

24. $\frac{a-b}{a}+\frac{1}{2}\left(\frac{a-b}{a}\right)^2+\frac{1}{3}\left(\frac{a-b}{a}\right)^3+\ldots$ to ∞ is equal to

(a) $\log a+\log b$ (b) $\log\frac{b}{a}$

(c) $\log a-\log b$ (d) none of these

25. $\sum_{n=1}^{\infty}\frac{1}{2n-1}\cdot x^{2n}$ is equal to

(a) $\frac{x}{2}\log\left(\frac{1+x}{1-x}\right)^x$ (b) $\frac{1}{2}\log\left(\frac{1+x^2}{1-x^2}\right)$

(c) $x\log\frac{1+x}{1-x}$ (d) none of these

26. $\sum_{n=1}^{\infty}\frac{1}{2n(2n+1)}$ is equal to

(a) $\log_e\left(\frac{2}{e}\right)$ (b) $1-\log_e 2$

(c) $1-\log_e\frac{1}{2e}$ (d) none of these

27. (i) If $x-\frac{x^2}{2}+\frac{x^3}{3}-\frac{x^4}{4}+\ldots$ to $\infty=y$ then $y+\frac{y^2}{2!}+\frac{y^3}{3!}+\ldots$ to ∞ is equal to

(a) $-x$ (b) x

(c) $x+1$ (d) none of these

(ii) The sum of the series $\frac{1}{1\cdot 2}-\frac{1}{2\cdot 3}+\frac{1}{3\cdot 4}-\ldots$ up to ∞ is equal to

(a) $\log_e 2-1$ (b) $\log_e 2$

(c) $\log_e\frac{4}{e}$ (d) $2\log_e 2$

PARTIAL FRACTIONS

28. If $\frac{3x+5}{2x^2+x-6}=\frac{A}{2x-3}+\frac{B}{x+2}$ identically then

(a) $A=\frac{1}{7}, B=\frac{19}{7}$ (b) $A=\frac{-19}{7}, B=\frac{1}{7}$

(c) $A = \frac{19}{7}, B = \frac{1}{7}$ (d) none of these

29. If $f(x)$ is a function of x such that $\frac{1}{(1+x)(1+x^2)} = \frac{A}{1+x} + \frac{f(x)}{1+x^2}$ for all $x \in R$ then $f(x)$ is

(a) $\frac{1-x}{2}$ (b) $\frac{x+1}{2}$

(c) $1-x$ (d) none of these

30. If $|x| < 1$, the coefficient of x^n in the expansion of $\frac{1}{x^2 - 3x + 2}$ is

(a) $1 - \frac{1}{2^n}$ (b) $1 - \frac{1}{2^{n+1}}$

(c) $1 - 2^n$ (d) none of these

31. If $|x| < \frac{1}{2}$, the coefficient of x^4 in the expansion of $\frac{1}{(1+2x)(1-x^2)}$ is

(a) 1 (b) 22

(c) 21 (d) none of these

32. The sum to infinite terms of the series

$$\frac{1}{(1+a)(2+a)} + \frac{1}{(2+a)(3+a)} + \frac{1}{(3+a)(4+a)} + \dots,$$

where a is a constant, is

(a) $\frac{1}{1+a}$ (b) $\frac{2}{1+a}$

(c) ∞ (d) none of these

SURDS

33. If $x = 1 + \sqrt{2} - \sqrt{3}$ then the reciprocal of x is

(a) $\frac{1}{2} - \frac{1}{4}(\sqrt{6} + \sqrt{2})$ (b) $\sqrt{3} - \sqrt{2} - 1$

(c) $\frac{1}{2} + \frac{1}{4}(\sqrt{6} + \sqrt{2})$ (d) none of these

34. The value of $\frac{1}{\sqrt{40} + \sqrt{20} + \sqrt{10} - \sqrt{80}}$ is equal to

(a) $\frac{1}{70}(3\sqrt{10}+2\sqrt{5})$ (b) $\frac{3\sqrt{10}-2\sqrt{5}}{70}$

(c) $\frac{3\sqrt{10}+2\sqrt{5}}{50}$ (d) none of these

35. The square root of $2x+2\sqrt{x^2-1}$ is

(a) $\sqrt{x+1}-\sqrt{x-1}$ (b) $\sqrt{x+1}+\sqrt{x-1}$

(c) $\sqrt{x-1}-\sqrt{x+1}$ (d) none of these

36. If $\frac{4+3\sqrt{3}}{\sqrt{7+4\sqrt{3}}}=x+\sqrt{y}, x\in Z, y\in Z$ (where Z = the set of integers) then

(a) $x=-1, y=12$ (b) $x=1, y=-12$

(c) $x=1, y=12$ (d) none of these

37. If $x=\frac{\sqrt{3}-\sqrt{2}}{\sqrt{3}+\sqrt{2}}$, $y=\frac{\sqrt{3}+\sqrt{2}}{\sqrt{3}-\sqrt{2}}$ then the value of x^2+xy+y^2 is

(a) 5 (b) 99

(c) 98 (d) none of these

38. If $x=\frac{\sqrt{3}}{2}$ then $\frac{1+x}{1+\sqrt{1+x}}+\frac{1-x}{1-\sqrt{1-x}}$ is equal to

(a) 1 (b) $\frac{\sqrt{3}}{2}$

(c) $\frac{1}{\sqrt{3}}$ (d) none of these

VARIATION

39. If x varies directly as y, and $x=2$ when y is 3 then $x=3$ when y is

(a) 2 (b) $\frac{2}{3}$ (c) $\frac{9}{2}$ (d) none of these

40. If x varies inversely as y, and $x=5$ when $y=2$ then for $y=5$, x is

(a) $\frac{2}{5}$ (b) 2 (c) $\frac{5}{2}$ (d) none of these

41. If $x\propto yz$ and $y\propto xz$ then

(a) $z\propto xy$ (b) z is a constant

(c) xyz is a constant (d) none of these

42. A varies as B and C jointly, and $A = 2$ when $B = \frac{3}{5}$, $C = \frac{10}{27}$. The value of A, when $B = 2$, $C = \frac{5}{3}$, is

(a) 30 (b) 10

(c) $\frac{36}{5}$ (d) none of these

43. Let $y \propto p + q$ where p varies directly as x and q varies inversely as x^2. If $y = 19$ when $x = 2$ or 3 then y in terms of x is

(a) $36x + \frac{5}{x^2}$ (b) $\frac{5}{x} + 36x^2$

(c) $5x + \frac{36}{x^2}$ (d) none of these

• Type 2 •

Choose the correct options. One or more options may be correct.

INFINITE SERIES

44. In the expansion of $(1 + x)^{-3}$

(a) the third term is $^4C_2 x^2$, if $|x| < 1$

(b) the third term is $^4C_2 \left(\frac{1}{x}\right)^2$, if $x > 1$

(c) the value of the third term is $-\frac{3}{2}$, when $x = \frac{1}{2}$

(d) the value of the third term is $-\frac{2}{81}$, when $x = -3$

45. If $\frac{3ax}{4} + \sqrt{4 + ax} - \frac{2}{\sqrt{1 - ax}} = -x^2 + bx^3 + \ldots$ to ∞ then

(a) $a = \frac{8}{7}$, $b = -\frac{319}{343}$ (b) $a = -\frac{8}{7}$, $b = \frac{319}{343}$

(c) $a = \frac{8}{7}$, $b = \frac{319}{343}$ (d) $a = -\frac{8}{7}$, $b = -\frac{319}{343}$

46. If $e^{ax} + e^{-bx} = p_0 + p_1 x + p_2 x^2 + \dots$ to ∞ then

(a) $p_1 = a + b, p_2 = \dfrac{a^2 - b^2}{2}$ (b) $p_0 = 2, p_3 = \dfrac{a^3 - b^3}{6}$

(c) $p_1 = a - b, p_2 = \dfrac{a^2 + b^2}{2}$ (d) $p_0 = 2, p_3 = \dfrac{a^3 + b^3}{6}$

47. If $\log_e \dfrac{1+x}{1-x} = a_0 + a_1 x + a_2 x^2 + \dots$ to ∞ then

(a) $a_1 = 2, a_2 = 0$ (b) $a_1 = 0, a_2 = 2$

(c) $a_0 + a_1 + a_2 = 2$ (d) a_1, a_3, a_5 are in HP

48. The coefficient of x^n, $n \in N$, in the expansion of $\dfrac{1+x}{1+x^2}$ is

(a) $(-1)^{n/2}$, when n is an even integer
(b) 0, when n is an odd integer
(c) 1, when n is an even integer
(d) $(-1)^{(n-1)/2}$, when n is an odd integer

SURDS

49. Let $a = \sqrt{5} - 2$, $b = \sqrt{7} - \sqrt{6}$ and $c = \sqrt{13} - \sqrt{12}$. Then

(a) $a > b$ (b) $b > c$

(c) $c > b$ (d) $a > c$

VARIATION

50. Let $x + y \propto z + \dfrac{1}{z}$, $x - y \propto z - \dfrac{1}{z}$ and $z = 2$, when $y = 1, x = 3$. Then

(a) $x = \dfrac{2}{15}\left(11z + \dfrac{1}{z}\right)$ (b) $x = \dfrac{22}{15} z - \dfrac{2}{15} \cdot \dfrac{1}{z}$

(c) $y = \dfrac{2}{15} z - \dfrac{22}{15} \cdot \dfrac{1}{z}$ (d) $y = \dfrac{2}{15}\left(z + \dfrac{11}{z}\right)$

51. If $2x + 3y \propto \sqrt{xy}$ then

(a) $x^2 + y^2 \propto xy$ (b) $x^3 + y^3 \propto (xy)^{3/2}$

(c) $x^4 + y^4 \propto x^2 y^2$ (d) none of these

Answers

1. a	**2.** c	**3.** b	**4.** a	**5.** c
6. d	**7.** c	**8.** b	**9.** a	**10.** a
11. c	**12.** b	**13.** a	**14.** b	**15.** a
16. b	**17.** a	**18.** a	**19.** c	**20.** a
21. c	**22.** b	**23.** a	**24.** c	**25.** a
26. b	**27.** (i) b (ii) a	**28.** c	**29.** a	**30.** b
31. c	**32.** a	**33.** c	**34.** a	**35.** b
36. a	**37.** b	**38.** a	**39.** c	**40.** b
41. b	**42.** a	**43.** c	**44.** a, d	**45.** a, b
46. b, c	**47.** a, c, d	**48.** a, d	**49.** a, b, d	**50.** a, d
51. a, b, c				

Hints and Solutions to Selected Questions

2. $(3+x)^{p/q} = 3^{p/q}\left(1+\frac{x}{3}\right)^{p/q}$. It can be expanded if $\left|\frac{x}{3}\right| < 1$.

6. $\frac{1+x^2}{1+x} = (1+x^2)(1+x)^{-1}$

$$= (1+x^2)(1-x+x^2-x^3+x^4-x^5+\ldots \text{ to } \infty).$$

7. $\frac{x^2+x}{1-x} = (x^2+x)(1-x)^{-1} = (x^2+x)(1+x+x^2+x^3+\ldots+x^r+\ldots).$

The coefficient of $x^r = 1+1 = 2.$

$\therefore \quad a_r = 2$ for all $r = 1, 2, 3, \ldots$.

8. $e^{2x+3} = e^3 \cdot e^{2x} = e^3\left\{1+\frac{2x}{1!}+\frac{(2x)^2}{2!}+\ldots\right\}.$

9. $10^x = e^{\log_e 10^x} = e^{x\log_e 10} = 1+\frac{x\log_e 10}{1!}+\frac{(x\log_e 10)^2}{2!}+\ldots$.

$\therefore \quad$ the coefficient of $x^{10} = \frac{(\log_e 10)^{10}}{10!}$.

12. $\frac{1}{e^x \cdot (1+x)} = e^{-x} \cdot (1+x)^{-1}$

$$= \left\{1-\frac{x}{1!}+\frac{x^2}{2!}-\frac{x^3}{3!}+\ldots\right\}\{1-x+x^2-x^3+\ldots\}.$$

$\therefore$ the coefficient of $x^3 = -1-\frac{1}{1!}-\frac{1}{2!}-\frac{1}{3!}$.

14. $\log_{10}(1-x) = \log_{10} e \cdot \log_e(1-x) = \log_{10} e \cdot \left\{-x-\frac{1}{2}x^2-\frac{1}{3}x^3-\ldots\right\}$.

15. $\frac{\log_e (1+x)}{(1-x)^2} = \log_e(1+x) \cdot (1-x)^{-2}$

$$= \left\{x-\frac{1}{2}x^2+\frac{1}{3}x^2-\ldots\right\}\left\{{}^1C_0+{}^2C_1x+{}^3C_2x^2+\ldots\right\}.$$

$\therefore$ the coefficient of $x^2 = {}^2C_1 - \frac{1}{2} \cdot {}^1C_0$.

16. $(1-x)^{-n} = {}^{n-1}C_0 + {}^nC_1 x + {}^{n+1}C_2 x^2 + \ldots$; here $n = 5$.

17. $(1+x)^{-n} = {}^{n-1}C_0 - {}^nC_1 x + {}^{n+1}C_2 x^2 - \ldots$; here $n = 3$ and x is x^2.

18. $(1-x)^{-p/q} = 1 + \frac{p}{1\,!} \cdot \frac{x}{q} + \frac{p(p+q)}{2\,!}\left(\frac{x}{q}\right)^2 + \ldots$; here $p = 1, q = 2, \frac{x}{q} = \frac{1}{4}$.

21. $t_n = \frac{n}{(n-1)\,!} = \frac{(n-1)+1}{(n-1)\,!} = \frac{1}{(n-2)\,!} + \frac{1}{(n-1)\,!}$.

$$\therefore \text{ sum} = \sum_{n=1}^{\infty} \frac{1}{(n-2)\,!} + \sum_{n=1}^{\infty} \frac{1}{(n-1)\,!} = e + e.$$

22. $t_n = \frac{n}{(2n+1)\,!} = \frac{1}{2} \cdot \frac{(2n+1)-1}{(2n+1)\,!} = \frac{1}{2}\left\{\frac{1}{2n\,!} - \frac{1}{(2n+1)\,!}\right\}$.

$$\therefore \text{ sum} = \frac{1}{2}\left(\frac{1}{2\,!} - \frac{1}{3\,!}\right) + \frac{1}{2}\left(\frac{1}{4\,!} - \frac{1}{5\,!}\right) + \ldots$$

$$= \frac{1}{2}\left\{1 - \frac{1}{1\,!} + \frac{1}{2\,!} - \frac{1}{3\,!} + \frac{1}{4\,!} - \frac{1}{5\,!} + \ldots\right\} = \frac{1}{2} \cdot e^{-1}.$$

23. Series $= \left(a^2 + \frac{a^4}{2\,!} + \frac{a^6}{3\,!} + \ldots\right) - \left(b^2 + \frac{b^4}{2\,!} + \frac{b^6}{3\,!} + \ldots\right)$

$$= \left(e^{a^2} - 1\right) - \left(e^{b^2} - 1\right).$$

24. Expression $= -\log\left(1 - \frac{a-b}{a}\right) = -\log\frac{b}{a} = -(\log b - \log a)$.

25. Expression $= x^2 + \frac{1}{3}x^4 + \frac{1}{5}x^6 + \ldots = x\left(x + \frac{1}{3}x^3 + \frac{1}{5}x^5 + \ldots\right)$

$$= x \cdot \frac{1}{2}\{\log(1+x) - \log(1-x)\} = \frac{x}{2}\log\frac{1+x}{1-x}.$$

26. $\frac{1}{2n(2n+1)} = \frac{1}{2n} - \frac{1}{2n+1}$.

$$\therefore \text{ sum} = \frac{1}{2} - \frac{1}{3} + \frac{1}{4} - \frac{1}{5} + \ldots = 1 - \left(1 - \frac{1}{2} + \frac{1}{3} - \frac{1}{4} + \ldots\right) = 1 - \log_e 2.$$

27. $\log_e(1+x) = y$. $\quad \therefore\ 1 + x = e^y = 1 + y + \frac{y^2}{2\,!} + \frac{y^3}{3\,!} + \ldots$.

30. $\dfrac{1}{x^2-3x+2}=\dfrac{1}{(x-1)(x-2)}=\dfrac{1}{x-2}-\dfrac{1}{x-1}=(1-x)^{-1}-\dfrac{1}{2}\left(1-\dfrac{x}{2}\right)^{-1}.$

$\therefore$ the coefficient of $x^n=1-\dfrac{1}{2}\cdot\left(\dfrac{1}{2}\right)^n.$

31. Fraction $=\dfrac{1}{(1+x)(1-x)(1+2x)}=\dfrac{A}{1+x}+\dfrac{B}{1-x}+\dfrac{C}{1+2x}$, where

$$A=\frac{1}{\{1-(-1)\}\{1+2(-1)\}}=-\frac{1}{2};\ B=\frac{1}{(1+1)(1+2\cdot 1)}=\frac{1}{6};$$

$$C=\frac{1}{\left\{1+\left(-\frac{1}{2}\right)\right\}\left\{1-\left(-\frac{1}{2}\right)\right\}}=\frac{4}{3}.$$

$\therefore$ fraction $=-\dfrac{1}{2}(1+x)^{-1}+\dfrac{1}{6}(1-x)^{-1}+\dfrac{4}{3}(1+2x)^{-1}$

$$=-\frac{1}{2}\{1-x+x^2-x^3+x^4-\ldots\}+\frac{1}{6}\{1+x+x^2+x^3+x^4+\ldots\}$$
$$+\frac{4}{3}\{1-2x+(2x)^2-(2x)^3+(2x)^4-\ldots\}.$$

$\therefore$ the coefficient of $x^4=-\dfrac{1}{2}+\dfrac{1}{6}+\dfrac{4}{3}\times 2^4.$

32. $S_n=\left(\dfrac{1}{1+a}-\dfrac{1}{2+a}\right)+\left(\dfrac{1}{2+a}-\dfrac{1}{3+a}\right)+\ldots+\left(\dfrac{1}{n+a}-\dfrac{1}{n+1+a}\right)$

$$=\frac{1}{1+a}-\frac{1}{n+1+a}.$$

$\therefore$ sum $=\lim\limits_{n\to\infty} S_n=\dfrac{1}{1+a}.$

33. The reciprocal of $x=\dfrac{1}{1+\sqrt{2}-\sqrt{3}}=\dfrac{1+\sqrt{2}+\sqrt{3}}{(1+\sqrt{2})^2-(\sqrt{3})^2}$

$$=\frac{1+\sqrt{2}+\sqrt{3}}{2\sqrt{2}}=\frac{1}{2}+\frac{1}{4}(\sqrt{2}+\sqrt{6}).$$

34. Value $=\dfrac{1}{2\sqrt{10}+2\sqrt{5}+\sqrt{10}-4\sqrt{5}}=\dfrac{1}{3\sqrt{10}-2\sqrt{5}}=\dfrac{3\sqrt{10}+2\sqrt{5}}{90-20}.$

35. $2x+2\sqrt{x^2-1}=(x+1)+2\sqrt{x+1}\,\sqrt{x-1}+(x-1)$

$$=(\sqrt{x+1}+\sqrt{x-1})^2.$$

36. $7 + 4\sqrt{3} = 2^2 + (\sqrt{3})^2 + 2 \cdot 2 \cdot \sqrt{3} = (2 + \sqrt{3})^2.$

$\therefore$ we have $\dfrac{4 + 3\sqrt{3}}{2 + \sqrt{3}} = x + \sqrt{y}$ or $\dfrac{(4 + 3\sqrt{3})(2 - \sqrt{3})}{4 - 3} = x + \sqrt{y}$

or $-1 + 2\sqrt{3} = x + \sqrt{y} \Rightarrow x = -1, y = 12.$

37. $xy = 1, x = (\sqrt{3} - \sqrt{2})^2 = 5 - 2\sqrt{6}, y = (\sqrt{3} + \sqrt{2})^2 = 5 + 2\sqrt{6}.$

38. As $x = \dfrac{\sqrt{3}}{2}$, $1 + x = 1 + \dfrac{\sqrt{3}}{2} = \dfrac{2 + \sqrt{3}}{2} = \dfrac{4 + 2\sqrt{3}}{4} = \left(\dfrac{1 + \sqrt{3}}{2}\right)^2.$

$1 - x = 1 - \dfrac{\sqrt{3}}{2} = \left(\dfrac{\sqrt{3} - 1}{2}\right)^2.$

$$\therefore \text{ expression} = \frac{\frac{2 + \sqrt{3}}{2}}{1 + \frac{1 + \sqrt{3}}{2}} + \frac{\frac{2 - \sqrt{3}}{2}}{1 - \frac{\sqrt{3} - 1}{2}} = \frac{2 + \sqrt{3}}{3 + \sqrt{3}} + \frac{2 - \sqrt{3}}{3 - \sqrt{3}}$$

$$= \frac{(2 + \sqrt{3})(3 - \sqrt{3})}{6} + \frac{(2 - \sqrt{3})(3 + \sqrt{3})}{6}$$

$$= \frac{1}{6}(3 + \sqrt{3} + 3 - \sqrt{3}) = 1.$$

41. $x = yzk, y = xzk' \Rightarrow xy = xyz^2kk' \Rightarrow z^2 = \dfrac{1}{kk'} \Rightarrow z = \text{constant}.$

43. $y = k_1(p + q), p = k_2x, qx^2 = k_3.$

$\therefore \quad y = k_1\left\{k_2x + \dfrac{k_3}{x^2}\right\} = \lambda x + \dfrac{\mu}{x^2}.$

When $y = 19, x = 2$ we get,

$$19 = 2\lambda + \frac{\mu}{4}. \qquad \ldots (1)$$

When $y = 19, x = 3$ we get,

$$19 = 3\lambda + \frac{\mu}{9}. \qquad \ldots (2)$$

(1), (2) $\Rightarrow 2\lambda + \dfrac{\mu}{4} = 3\lambda + \dfrac{\mu}{9}$ or $\lambda = \left(\dfrac{1}{4} - \dfrac{1}{9}\right)\mu = \dfrac{5}{36}\mu.$

Putting in (1), $19 = \dfrac{5}{18}\mu + \dfrac{\mu}{4} = \dfrac{19}{36}\mu.$ $\quad \therefore \mu = 36$ and $\lambda = 5.$

44. When $|x|<1$, $(1+x)^{-3}={}^2C_0-{}^3C_1x+{}^4C_2x^2-{}^5C_3x^3+\ldots$.

When $|x|>1$, $(1+x)^{-3}=\frac{1}{x^3}\left(1+\frac{1}{x}\right)^{-3}=\frac{1}{x^3}\left\{{}^2C_0-{}^3C_1\cdot\frac{1}{x}+{}^4C_2\cdot\frac{1}{x^2}-\ldots\right\}$.

45. $\frac{3ax}{4}+\sqrt{4+ax}-\frac{2}{\sqrt{1-ax}}=\frac{3ax}{4}+2\left(1+\frac{ax}{4}\right)^{1/2}-2(1-ax)^{-1/2}$

$$=\frac{3ax}{4}+2\left\{1+\frac{1}{2}\cdot\frac{ax}{4}+\frac{\frac{1}{2}\left(\frac{1}{2}-1\right)}{2\,!}\cdot\left(\frac{ax}{4}\right)^2+\ldots\right\}$$

$$-2\left\{1+\left(-\frac{1}{2}\right)(-ax)+\frac{-\frac{1}{2}\left(-\frac{1}{2}-1\right)}{2\,!}(-ax)^2+\ldots\right\}.$$

$\therefore$ the coefficient of $x=\frac{3a}{4}+\frac{a}{4}-a=0.$

The coefficient of $x^2=-\frac{1}{4}\cdot\frac{a^2}{16}-\frac{3}{4}a^2=-1$ (from the question).

$\therefore$ $a^2\left(\frac{1}{64}+\frac{3}{4}\right)=1$ or $a=\pm\frac{8}{7}$.

The coefficient of x^3

$$=2\cdot\frac{\frac{1}{2}\left(\frac{1}{2}-1\right)\left(\frac{1}{2}-2\right)}{3\,!}\left(\frac{a}{4}\right)^3-2\cdot\frac{-\frac{1}{2}\left(-\frac{1}{2}-1\right)\left(-\frac{1}{2}-2\right)}{3\,!}(-a)^3$$

$$=\left(\frac{a}{8}\right)^3-\frac{5}{8}a^3=-\frac{319}{8^3}a^3=b \quad \text{(from the question).}$$

47. $\log\frac{1+x}{1-x}=\log(1+x)-\log(1-x)=2\left\{x+\frac{1}{3}x^3+\frac{1}{5}x^5+\ldots\right\}$.

$\therefore$ $a_0=0, a_1=2, a_2=0, a_3=\frac{2}{3}, a_4=0, a_5=\frac{2}{5}$.

48. $\frac{1+x}{1+x^2}=(1+x)(1+x^2)^{-1}$

$$=(1+x)\{1-x^2+(x^2)^2-(x^2)^3+\ldots+(-1)^r(x^2)^r+\ldots\}.$$

If n is even, say $2m$, then

the coefficient of x^n = the coefficient of $x^{2m}=(-1)^m+0=(-1)^{n/2}$.

If n is odd, say $2m+1$, then

$$\text{the coefficient of } x^n = \text{the coefficient of } x^{2m+1}$$

$$= 0 + (-1)^m = (-1)^{\frac{n-1}{2}}.$$

49. Clearly, $\sqrt{13} + \sqrt{12} > \sqrt{7} + \sqrt{6} > 0.$

$\therefore \quad \dfrac{1}{\sqrt{7}+\sqrt{6}} > \dfrac{1}{\sqrt{13}+\sqrt{12}}$ or $\sqrt{7} - \sqrt{6} > \sqrt{13} - \sqrt{12}$

(rationalising both sides).

Similarly, $\sqrt{5} - \sqrt{4} > \sqrt{7} - \sqrt{6}$ and $\sqrt{5} - \sqrt{4} > \sqrt{13} - \sqrt{12}$.

50. $x + y = k_1\left(z + \dfrac{1}{z}\right), x - y = k_2\left(z - \dfrac{1}{z}\right)\cdot$ ... (1)

As $z = 2, y = 1, x = 3$, we get $3 + 1 = k_1\left(2 + \dfrac{1}{2}\right), 3 - 1 = k_2\left(2 - \dfrac{1}{2}\right)\cdot$

$\therefore \quad k_1 = \dfrac{8}{5}, k_2 = \dfrac{4}{3}\cdot$ ... (2)

Using (1) and (2), get x, y in terms of z.

51. $2x + 3y = k\sqrt{xy}$ or $4x^2 + 9y^2 + 12xy = k^2xy$

or $\quad 4\left(\dfrac{x}{y}\right)^2 + (12 - k^2)\dfrac{x}{y} + 9 = 0 \Rightarrow \dfrac{x}{y} = \text{constant} = c.$

$\therefore \quad x^2 + y^2 = x \cdot cy + y \cdot \dfrac{x}{c} = \left(c + \dfrac{1}{c}\right)xy$

$$x^3 + y^3 = x^{3/2} \cdot (cy)^{3/2} + y^{3/2} \cdot \left(\frac{x}{c}\right)^{3/2} = \left(c^{3/2} + \frac{1}{c^{3/2}}\right)(xy)^{3/2}$$

$$x^4 + y^4 = x^2 \cdot (cy)^2 + y^2 \cdot \left(\frac{x}{c}\right)^2 = \left(c^2 + \frac{1}{c^2}\right)x^2y^2.$$

□

Part 2

Trigonometry

1

Trigonometrical Functions and Identities

• Type 1 •

Choose the most appropriate option (a, b, c or d).

1. If $\tan\theta = a - \frac{1}{4a}$ then $\sec\theta - \tan\theta$ is equal to

(a) $-2a, \frac{1}{2a}$ (b) $-\frac{1}{2a}, 2a$

(c) $2a$ (d) $\frac{1}{2a}, 2a$

2. $\sec^2\theta = \frac{4xy}{(x+y)^2}$, where $x \in R, y \in R$, is true if and only if

(a) $x + y \neq 0$ (b) $x = y, x \neq 0$

(c) $x = y$ (d) $x \neq 0, y \neq 0$

3. $\sin^2\theta = \frac{(x+y)^2}{4xy}$, where $x \in R, y \in R$, gives real θ if and only if

(a) $x + y = 0$ (b) $x = y$

(c) $|x| = |y| \neq 0$ (d) none of these

4. $\operatorname{cosec}\theta = \frac{x^2 - y^2}{x^2 + y^2}$, where $x \in R, y \in R$, gives real θ if and only if

(a) $x = y \neq 0$ (b) $|x| = |y| \neq 0$

(c) $x + y = 0, x \neq 0$ (d) none of these

5. If $\sin\theta + \operatorname{cosec}\theta = 2$ then the value of $\sin^8\theta + \operatorname{cosec}^8\theta$ is equal to

(a) 2 (b) 2^8

(c) 2^4 (d) none of these

6. If $x = r\sin\theta \cdot \cos\phi$, $y = r\sin\theta \cdot \sin\phi$ and $z = r\cos\theta$ then the value of $x^2 + y^2 + z^2$ is independent of

(a) θ, ϕ (b) r, θ

(c) r, ϕ (d) r

7. Let $p = a\cos\theta - b\sin\theta$. Then for all real θ

(a) $p > \sqrt{a^2 + b^2}$ (b) $p < -\sqrt{a^2 + b^2}$

(c) $-\sqrt{a^2 + b^2} \le p \le \sqrt{a^2 + b^2}$ (d) none of these

8. If $0° < \theta < 180°$ then

$$\sqrt{2 + \sqrt{2 + \sqrt{2 + \ldots + \sqrt{2(1 + \cos\theta)}}}},$$

there being n number of 2's, is equal to

(a) $2\cos\dfrac{\theta}{2^n}$ (b) $2\cos\dfrac{\theta}{2^{n-1}}$

(c) $2\cos\dfrac{\theta}{2^{n+1}}$ (d) none of these

9. The value of $\tan\dfrac{\pi}{16} + 2\tan\dfrac{\pi}{8} + 4$ is equal to

(a) $\cot\dfrac{\pi}{8}$ (b) $\cot\dfrac{\pi}{16}$

(c) $\cot\dfrac{\pi}{16} - 4$ (d) none of these

10. The value of $\sin 78° - \sin 66° - \sin 42° + \sin 6°$ is

(a) $\dfrac{1}{2}$ (b) $-\dfrac{1}{2}$

(c) -1 (d) none of these

11. The value of $\sqrt{3}\operatorname{cosec} 20° - \sec 20°$ is equal to

(a) 2 (b) 4

(c) $2 \cdot \dfrac{\sin 20°}{\sin 40°}$ (d) $4 \cdot \dfrac{\sin 20°}{\sin 40°}$

12. The maximum value of $1 + \sin\left(\dfrac{\pi}{4} + \theta\right) + 2\cos\left(\dfrac{\pi}{4} - \theta\right)$ for real values of θ is

(a) 3 (b) 5
(c) 4 (d) none of these

13. The minimum value of $\cos 2\theta + \cos \theta$ for real values of θ is

(a) $-\dfrac{9}{8}$ (b) 0
(c) -2 (d) none of these

14. The value of $\operatorname{cosec} 10° - \sqrt{3} \sec 10°$ is equal to

(a) $\dfrac{1}{2}$ (b) 2
(c) 4 (d) 8

15. The least value of $\cos^2\theta - 6\sin\theta \cdot \cos\theta + 3\sin^2\theta + 2$ is

(a) $4 + \sqrt{10}$ (b) $4 - \sqrt{10}$
(c) 0 (d) none of these

16. If $\cos^4\theta \cdot \sec^2\alpha, \dfrac{1}{2}$ and $\sin^4\theta \cdot \operatorname{cosec}^2\alpha$ are in AP then

$$\cos^8\theta \cdot \sec^6\alpha,\ \frac{1}{2} \text{ and } \sin^8\theta \cdot \operatorname{cosec}^6\alpha \text{ are in}$$

(a) AP (b) GP
(c) HP (d) none of these

17. If $\tan\dfrac{\pi}{9}, x$ and $\tan\dfrac{5\pi}{18}$ are in AP and $\tan\dfrac{\pi}{9}, y$ and $\tan\dfrac{7\pi}{18}$ are also in AP then

(a) $2x = y$ (b) $x > y$
(c) $x = y$ (d) none of these

18. If $\cos(x - y)$, $\cos x$ and $\cos(x + y)$ are in HP then $\left|\cos x \cdot \sec\dfrac{y}{2}\right|$ equals

(a) 1 (b) 2
(c) $\sqrt{2}$ (d) none of these

19. If $2\sin\alpha\cdot\cos\beta\cdot\sin\gamma = \sin\beta\cdot\sin(\alpha+\gamma)$ then $\tan\alpha$, $\tan\beta$ and $\tan\gamma$ are in

(a) AP
(b) GP
(c) HP
(d) none of these

20. If $\tan\alpha = \sqrt{a}$, where a is a rational number which is not a perfect square, then which of the following is a rational number?

(a) $\sin 2\alpha$
(b) $\tan 2\alpha$
(c) $\cos 2\alpha$
(d) none of these

21. Let $f(\theta) = \dfrac{\cot\theta}{1+\cot\theta}$ and $\alpha+\beta = \dfrac{5\pi}{4}\cdot$ Then the value of $f(\alpha)\cdot f(\beta)$ is

(a) 2
(b) $-\dfrac{1}{2}$
(c) $\dfrac{1}{2}$
(d) none of these

22. If $\tan\dfrac{\alpha}{2}$ and $\tan\dfrac{\beta}{2}$ are the roots of the equation $8x^2 - 26x + 15 = 0$ then $\cos(\alpha+\beta)$ is equal to

(a) $-\dfrac{627}{725}$
(b) $\dfrac{627}{725}$
(c) -1
(d) none of these

23. If $\sin\alpha + \sin\beta = a$ and $\cos\alpha - \cos\beta = b$ then $\tan\dfrac{\alpha-\beta}{2}$ is equal to

(a) $-\dfrac{a}{b}$
(b) $-\dfrac{b}{a}$
(c) $\sqrt{a^2+b^2}$
(d) none of these

24. If $0 < \beta < \alpha < \dfrac{\pi}{4}$, $\cos(\alpha+\beta) = \dfrac{3}{5}$ and $\cos(\alpha-\beta) = \dfrac{4}{5}$ then $\sin 2\alpha$ is equal to

(a) 1
(b) 0
(c) 2
(d) none of these

25. If $\cos\alpha = \dfrac{1}{2}\left(x+\dfrac{1}{x}\right)$, $\cos\beta = \dfrac{1}{2}\left(y+\dfrac{1}{y}\right)$ then $\cos(\alpha-\beta)$ is equal to

(a) $\dfrac{x}{y}+\dfrac{y}{x}$
(b) $xy+\dfrac{1}{xy}$
(c) $\dfrac{1}{2}\left(\dfrac{x}{y}+\dfrac{y}{x}\right)$
(d) none of these

26. If $\frac{2\sin\alpha}{1+\sin\alpha+\cos\alpha}=\lambda$ then $\frac{1+\sin\alpha-\cos\alpha}{1+\sin\alpha}$ is equal to

(a) $\frac{1}{\lambda}$ (b) λ

(c) $1-\lambda$ (d) $1+\lambda$

27. If $|\tan A|<1$, and $|A|$ is acute then $\frac{\sqrt{1+\sin 2A}+\sqrt{1-\sin 2A}}{\sqrt{1+\sin 2A}-\sqrt{1-\sin 2A}}$ is equal to

(a) $\tan A$ (b) $-\tan A$

(c) $\cot A$ (d) $-\cot A$

28. $\tan\theta\cdot\tan\left(\frac{\pi}{3}+\theta\right)\cdot\tan\left(\frac{\pi}{3}-\theta\right)$ is equal to

(a) $\tan 2\theta$ (b) $\tan 3\theta$

(c) $\tan^3\theta$ (d) none of these

29. The set of all possible values of α in $[-\pi,\pi]$ such that $\sqrt{\frac{1-\sin\alpha}{1+\sin\alpha}}$ is equal to $\sec\alpha-\tan\alpha$ is

(a) $\left[0,\frac{\pi}{2}\right)$ (b) $\left[0,\frac{\pi}{2}\right)\cup\left(\frac{\pi}{2},\pi\right)$

(c) $[-\pi,0]$ (d) $\left(-\frac{\pi}{2},\frac{\pi}{2}\right)$

30. For all real values of θ, $\cot\theta-2\cot 2\theta$ is equal to

(a) $\tan 2\theta$ (b) $\tan\theta$

(c) $-\cot 3\theta$ (d) none of these

31. Let $a=\cos A+\cos B-\cos(A+B)$ and $b=4\sin\frac{A}{2}\sin\frac{B}{2}\cos\frac{A+B}{2}$. Then $a-b$ is equal to

(a) 1 (b) 0

(c) -1 (d) none of these

32. If $\tan\theta+\tan\left(\theta+\frac{\pi}{3}\right)+\tan\left(\theta-\frac{\pi}{3}\right)=k\tan 3\theta$ then k is equal to

(a) 1 (b) 3

(c) $\frac{1}{3}$ (d) none of these

33. If $a\sec\alpha - c\tan\alpha = d$ and $b\sec\alpha + d\tan\alpha = c$ then

(a) $a^2 + c^2 = b^2 + d^2$ (b) $a^2 + d^2 = b^2 + c^2$

(c) $a^2 + b^2 = c^2 + d^2$ (d) $ab = cd$

34. If $\cos 20° - \sin 20° = p$ then $\cos 40°$ is equal to

(a) $-p\sqrt{2-p^2}$ (b) $p\sqrt{2-p^2}$

(c) $p + \sqrt{2-p^2}$ (d) none of these

35. If $3\sin\theta + 4\cos\theta = 5$ then the value of $4\sin\theta - 3\cos\theta$ is

(a) 0 (b) 5

(c) 1 (d) none of these

36. If $\cos 2x + 2\cos x = 1$ then $\sin^2 x(2 - \cos^2 x)$ is equal to

(a) 1 (b) -1

(c) $-\sqrt{5}$ (d) $\sqrt{5}$

37. If $0 < \phi < \frac{\pi}{2}$, $x = \sum_{n=0}^{\infty} \cos^{2n}\phi$, $y = \sum_{n=0}^{\infty} \sin^{2n}\phi$ and $z = \sum_{n=0}^{\infty} \cos^{2n}\phi \cdot \sin^{2n}\phi$ then

(a) $xyz = xz + y$ (b) $xyz = xy + z$

(c) $xyz = x + y + z$ (d) $xyz = yz + x$

38. Let n be an odd integer. If $\sin n\theta = \sum_{r=0}^{n} b_r \sin^r\theta$ for all real θ then

(a) $b_0 = 1, b_1 = 3$ (b) $b_0 = 0, b_1 = n$

(c) $b_0 = -1, b_1 = n$ (d) $b_0 = 0, b_1 = n^2 - 3n - 3$

39. If $\cos 5\theta = a\cos^5\theta + b\cos^3\theta + c\cos\theta$ then c is equal to

(a) -5 (b) 1

(c) 5 (d) none of these

40. If $\sin^3 x \cdot \sin 3x = \sum_{m=0}^{n} c_m \cdot \cos mx$ is an identity in x, where c_m's are constants then the value of n is

(a) 4 (b) 6

(c) 9 (d) none of these

41. The value of $\sin\frac{\pi}{14} \cdot \sin\frac{3\pi}{14} \cdot \sin\frac{5\pi}{14} \cdot \sin\frac{7\pi}{14} \cdot \sin\frac{9\pi}{14} \cdot \sin\frac{11\pi}{14} \cdot \sin\frac{13\pi}{14}$ is equal to

(a) 1 (b) $\frac{1}{16}$

(c) $\frac{1}{64}$ (d) none of these

42. The numerical value of $\sin\frac{\pi}{18} \cdot \sin\frac{5\pi}{18} \cdot \sin\frac{7\pi}{18}$ is equal to

(a) 1 (b) $\frac{1}{8}$

(c) $\frac{1}{4}$ (d) $\frac{1}{2}$

43. The value of $\tan 63° - \cot 63°$ is equal to

(a) $\frac{2}{\sqrt{5}+1} \cdot \sqrt{10-2\sqrt{5}}$ (b) $\frac{2}{\sqrt{5}+1} \cdot \sqrt{10+2\sqrt{5}}$

(c) $\frac{\sqrt{5}-1}{4} \cdot \sqrt{10-2\sqrt{5}}$ (d) none of these

44. The value of $\cos 9° - \sin 9°$ is

(a) $-\frac{\sqrt{5-\sqrt{5}}}{2}$ (b) $\frac{5+\sqrt{5}}{4}$

(c) $\frac{1}{2}\sqrt{5-\sqrt{5}}$ (d) none of these

45. The value of $2\tan\frac{\pi}{10} + 3\sec\frac{\pi}{10} - 4\cos\frac{\pi}{10}$ is

(a) 0 (b) $\sqrt{5}$

(c) 1 (d) none of these

46. The value of $\tan 20° + 2\tan 50° - \tan 70°$ is

(a) 1 (b) 0

(c) tan 50° (d) none of these

47. If α, β, γ and δ be four angles of a cyclic quadrilateral then the value of $\cos\alpha + \cos\beta + \cos\gamma + \cos\delta$ is

(a) 1 (b) 0

(c) –1 (d) none of these

48. If $4n\alpha = \pi$ then $\cot\alpha \cdot \cot 2\alpha \cdot \cot 3\alpha \cdot \ldots \cot(2n-1)\alpha$ is equal to

(a) 1 (b) –1

(c) ∞ (d) none of these

49. The value of $\cos 12° \cdot \cos 24° \cdot \cos 36° \cdot \cos 48° \cdot \cos 72° \cdot \cos 84°$ is

(a) $\frac{1}{64}$ (b) $\frac{1}{32}$

(c) $\frac{1}{16}$ (d) $\frac{1}{128}$

50. The value of $\cos\frac{\pi}{11} + \cos\frac{3\pi}{11} + \cos\frac{5\pi}{11} + \cos\frac{7\pi}{11} + \cos\frac{9\pi}{11}$ is

(a) 0 (b) 1

(c) $\frac{1}{2}$ (d) none of these

51. $\sum_{r=1}^{n-1} \cos^2\frac{r\pi}{n}$ is equal to

(a) $\frac{n}{2}$ (b) $\frac{n-1}{2}$

(c) $\frac{n}{2}-1$ (d) none of these

52. The value of $\sin\frac{\pi}{n} + \sin\frac{3\pi}{n} + \sin\frac{5\pi}{n} + \ldots$ to n terms is equal to

(a) 1 (b) 0

(c) $\frac{n}{2}$ (d) none of these

53. The sum of the real roots of $\cos^6 x + \sin^4 x = 1$ in the interval $-\pi \le x \le \pi$ is equal to

(a) 0
(b) π
(c) $-\pi$
(d) none of these

54. The number of real solutions of the equation $\cos^7 x + \sin^4 x = 1$ in the interval $[-\pi, \pi]$ is

(a) 2
(b) 3
(c) 5
(d) none of these

55. If the solutions for θ from the equation $\sin^2\theta - 2\sin\theta + \lambda = 0$ lie in $\bigcup_{n \in Z}\left(2n\pi - \frac{\pi}{6}, \overline{2n+1}\,\pi + \frac{\pi}{6}\right)$ then the set of possible values of λ is

(a) $\left(-\frac{5}{4}, 1\right]$
(b) $(-\infty, 1]$
(c) $\left(-\frac{5}{4}, +\infty\right]$
(d) $\{1\}$

56. If ABCD is a convex quadrilateral such that $4\sec A + 5 = 0$ then the quadratic equation whose roots are $\tan A$ and $\operatorname{cosec} A$ is

(a) $12x^2 - 29x + 15 = 0$
(b) $12x^2 - 11x - 15 = 0$
(c) $12x^2 + 11x - 15 = 0$
(d) none of these

57. If ABCD is a cyclic quadrilateral such that $12\tan A - 5 = 0$ and $5\cos B + 3 = 0$ then the quadratic equation whose roots are $\cos C$, $\tan D$ is

(a) $39x^2 - 16x - 48 = 0$
(b) $39x^2 + 88x + 48 = 0$
(c) $39x^2 - 88x + 48 = 0$
(d) none of these

58. The number of real solutions of the equation $\sin(e^x) = 2^x + 2^{-x}$ is

(a) 1
(b) 0
(c) 2
(d) infinite

59. The equation $(\cos p - 1)x^2 + (\cos p)x + \sin p = 0$ in x has real roots. Then the set of values of p is

(a) $[0, 2\pi]$
(b) $[-\pi, 0]$
(c) $\left[-\frac{\pi}{2}, \frac{\pi}{2}\right]$
(d) $[0, \pi]$

60. If $e^{\sin x} - e^{-\sin x} - 4 = 0$ then the number of real values of x is

(a) 0
(b) 2
(c) 1
(d) infinite

61. If $\sin\alpha = p$, where $|p| \le 1$ then the quadratic equation whose roots are $\tan\frac{\alpha}{2}$ and $\cot\frac{\alpha}{2}$ is

(a) $px^2 + 2x + p = 0$
(b) $px^2 - x + p = 0$
(c) $px^2 - 2x + p = 0$
(d) none of these

62. If sec α and cosec α are the roots of $x^2 - px + q = 0$ then

(a) $p^2 = q(q-2)$
(b) $p^2 = q(q+2)$
(c) $p^2 + q^2 = 2q$
(d) none of these

63. The number of values of x in the interval $[0, 5\pi]$ satisfying the equation $3\sin^2 x - 7\sin x + 2 = 0$ is

(a) 0
(b) 5
(c) 6
(d) 10

64. If $x = \alpha, \beta$ satisfies both the equations $\cos^2 x + a\cos x + b = 0$ and $\sin^2 x + p\sin x + q = 0$ then the relation between a, b, p and q is

(a) $1 + b + a^2 = p^2 - q - 1$
(b) $a^2 + b^2 = p^2 + q^2$
(c) $2(b+q) = a^2 + p^2 - 2$
(d) none of these

65. If $0 \le a \le 3, 0 \le b \le 3$ and the equation $x^2 + 4 + 3\cos(ax+b) = 2x$ has at least one solution then the value of $a+b$ is

(a) 0
(b) $\frac{\pi}{2}$
(c) π
(d) none of these

66. The equation $\cos\theta = x + \frac{p}{x}$ for all $x \in R$ has a real solution for θ. Then

(a) $p = \frac{1}{2}$
(b) $p \le \frac{1}{4}$
(c) $p \ge \frac{1}{4}$
(d) none of these

67. If $f(x) = \frac{\sin 3x}{\sin x}$, where $x \ne n\pi$, then the range of values of $f(x)$ for real values of x is

(a) $[-1, 3]$
(b) $(-\infty, -1]$
(c) $(3, +\infty)$
(d) $[-1, 3)$

68. The set of values of $k \in R$ such that the equation $\cos 2\theta + \cos\theta + k = 0$ admits of a solution for θ is

(a) $\left[0, \frac{9}{8}\right]$ (b) $[0, +\infty)$

(c) $[-2, 0]$ (d) none of these

69. The set of values of $\lambda \in R$ such that $\tan^2\theta + \sec\theta = \lambda$ holds for some θ is

(a) $(-\infty, 1]$ (b) $(-\infty, -1]$

(c) ϕ (d) $[-1, +\infty)$

70. If $\tan A + \tan B + \tan C = \tan A \cdot \tan B \cdot \tan C$ then

(a) A, B, C must be angles of a triangle

(b) the sum of any two of A, B, C is equal to the third

(c) $A + B + C$ must be an integral multiple of π

(d) none of these

• *Type 2* •

Choose the correct options. One or more options may be correct.

71. If $x = \sin(\alpha - \beta) \cdot \sin(\gamma - \delta)$, $y = \sin(\beta - \gamma) \cdot \sin(\alpha - \delta)$ and $z = \sin(\gamma - \alpha) \cdot \sin(\beta - \delta)$ then

(a) $x + y + z = 0$ (b) $x + y - z = 0$

(c) $y + z - x = 0$ (d) $x^3 + y^3 + z^3 = 3xyz$

72. $\sin\frac{15\pi}{32} \cdot \sin\frac{7\pi}{16} \cdot \sin\frac{3\pi}{8}$ is equal to

(a) $\dfrac{1}{8\sqrt{2}\cos\frac{15\pi}{32}}$ (b) $\dfrac{1}{8\sin\frac{\pi}{32}}$

(c) $\frac{1}{4\sqrt{2}}\operatorname{cosec}\frac{\pi}{16}$ (d) $\frac{1}{8\sqrt{2}}\operatorname{cosec}\frac{\pi}{32}$

73. Which of the following is a rational number?

(a) $\sin 15°$ (b) $\cos 15°$

(c) $\sin 15° \cdot \cos 15°$ (d) $\sin 15° \cdot \cos 75°$

74. If $a = \dfrac{1}{5\cos x + 12\sin x}$ then for all real x

(a) the least positive value of a is $\dfrac{1}{13}$

(b) the greatest negative value of a is $-\dfrac{1}{13}$

(c) $a \le \dfrac{1}{13}$

(d) $-\dfrac{1}{13} \le a \le \dfrac{1}{13}$

75. Let $y = \sin^2 x + \cos^4 x$. Then for all real x

(a) the maximum value of y is 2 (b) the minimum value of y is $\dfrac{3}{4}$

(c) $y \le 1$ (d) $y \ge \dfrac{1}{4}$

76. Let $y = \sin x \cdot \sin(60° + x) \cdot \sin(60° - x)$. Then for all real x

(a) the minimum value of y is $-\dfrac{1}{4}$ (b) the maximum value of y is 1

(c) $y \le \dfrac{1}{4}$ (d) $y \ge -1$

77. Let $f_n(\theta) = \tan\dfrac{\theta}{2} \cdot (1 + \sec\theta)(1 + \sec 2\theta)(1 + \sec 4\theta) \ldots (1 + \sec 2^n\theta)$. Then

(a) $f_2\left(\dfrac{\pi}{16}\right) = 1$ (b) $f_3\left(\dfrac{\pi}{32}\right) = 1$

(c) $f_4\left(\dfrac{\pi}{64}\right) = 1$ (d) $f_5\left(\dfrac{\pi}{128}\right) = 1$

78. Let $0 \le \theta \le \dfrac{\pi}{2}$ and $x = X\cos\theta + Y\sin\theta$, $y = X\sin\theta - Y\cos\theta$ such that $x^2 + 4xy + y^2 = aX^2 + bY^2$, where a, b are constants. Then

(a) $a = -1, b = 3$ (b) $\theta = \dfrac{\pi}{4}$

(c) $a = 3, b = -1$ (d) $\theta = \dfrac{\pi}{3}$

79. If $7\cos x - 24\sin x = \lambda\cos(x + \alpha)$, $0 < \alpha < \dfrac{\pi}{2}$, be true for all $x \in R$ then

(a) $\lambda = 25$

(b) $\alpha = \sin^{-1}\dfrac{24}{25}$

(c) $\lambda = -25$

(d) $\alpha = \cos^{-1}\dfrac{7}{25}$

80. If $A + B = \dfrac{\pi}{3}$ and $\cos A + \cos B = 1$ then

(a) $\cos(A - B) = \dfrac{1}{3}$

(b) $|\cos A - \cos B| = \sqrt{\dfrac{2}{3}}$

(c) $\cos(A - B) = -\dfrac{1}{3}$

(d) $|\cos A - \cos B| = \dfrac{1}{2\sqrt{3}}$

81. If $\tan\theta = a \neq 0$, $\tan 2\theta = b \neq 0$ and $\tan\theta + \tan 2\theta = \tan 3\theta$ then

(a) $a = b$

(b) $ab = 1$

(c) $a + b = 0$

(d) $b = 2a$

82. $\cos^3 x \cdot \sin 2x = \sum_{m=1}^{n} a_m \sin mx$ is an identity in x. Then

(a) $a_3 = \dfrac{3}{8}, a_2 = 0$

(b) $n = 6, a_1 = \dfrac{1}{2}$

(c) $n = 5, a_1 = \dfrac{1}{4}$

(d) $\Sigma a_m = \dfrac{3}{4}$

83. If $1 + \cos(x - y) = 0$ then

(a) $\cos x - \cos y = 0$

(b) $\cos x + \cos y = 0$

(c) $\sin x + \sin y = 0$

(d) $\cos x + \sin y = 1$

84. If $A \geq 0, B \geq 0, A + B = \dfrac{\pi}{3}$ and $y = \tan A \cdot \tan B$ then

(a) the maximum value of y is 3

(b) the minimum value of y is $\dfrac{1}{3}$

(c) the maximum value of y is $\dfrac{1}{3}$

(d) the minimum value of y is 0

85. $\cos\dfrac{2\pi}{7} + \cos\dfrac{4\pi}{7} + \cos\dfrac{6\pi}{7}$ is equal to

(a) an integer

(b) a positive rational number

(c) a negative rational number

(d) an irrational number

Answers

1. a	**2.** b	**3.** c	**4.** d	**5.** a
6. a	**7.** c	**8.** a	**9.** b	**10.** b
11. b	**12.** c	**13.** a	**14.** c	**15.** b
16. a	**17.** a	**18.** c	**19.** c	**20.** c
21. c	**22.** a	**23.** b	**24.** a	**25.** c
26. b	**27.** c	**28.** b	**29.** d	**30.** b
31. a	**32.** b	**33.** c	**34.** b	**35.** a
36. a	**37.** b	**38.** b	**39.** c	**40.** b
41. c	**42.** b	**43.** a	**44.** c	**45.** a
46. b	**47.** b	**48.** a	**49.** a	**50.** c
51. c	**52.** b	**53.** a	**54.** b	**55.** a
56. b	**57.** a	**58.** b	**59.** d	**60.** a
61. c	**62.** b	**63.** c	**64.** c	**65.** c
66. b	**67.** d	**68.** a	**69.** d	**70.** c
71. a, d	**72.** a, d	**73.** c	**74.** a, b	**75.** b, c
76. a, c	**77.** a, b, c, d	**78.** b, c	**79.** a, b, d	**80.** b, c
81. c	**82.** a, c, d	**83.** b, c	**84.** c, d	**85.** c

Hints and Solutions to Selected Questions

1. $\sec\theta - \tan\theta = \lambda \quad \Rightarrow \sec\theta + \tan\theta = \frac{1}{\lambda}$.

$\therefore$ subtracting, $2\tan\theta = \frac{1}{\lambda} - \lambda$ or $2\left(a - \frac{1}{4a}\right) = \frac{1}{\lambda} - \lambda$

or $2a - \frac{1}{2a} = \frac{1}{\lambda} - \lambda \quad \Rightarrow \lambda = \frac{1}{2a}, -2a.$

2. $(x+y)^2 - 4xy = (x-y)^2 \geq 0$

$\therefore$ $(x+y)^2 \geq 4xy$ and equality holds when $x = y$.

But $\sec^2\theta \geq 1$. Hence, $\sec^2\theta = \frac{4xy}{(x+y)^2}$ can hold only if $(x+y)^2 = 4xy$.

3. See the hint to Q. No. 2.

As $0 \leq \sin^2\theta \leq 1$, $\sin^2\theta = \frac{(x+y)^2}{4xy}$ can hold if $x = y$ or $x + y = 0$.

5. $\sin\theta + \operatorname{cosec}\theta = 2 \quad \Rightarrow \sin^2\theta - 2\sin\theta + 1 = 0 \quad \Rightarrow \sin\theta = 1.$

7. $p = \sqrt{a^2+b^2}\left\{\frac{a}{\sqrt{a^2+b^2}}\cos\theta - \frac{b}{\sqrt{a^2+b^2}}\sin\theta\right\}$

$= \sqrt{a^2+b^2}\cos(\theta+\alpha)$, where $\cos\alpha = \frac{a}{\sqrt{a^2+b^2}}$.

But $-1 \leq \cos(\theta+\alpha) \leq 1$.

8. $\sqrt{2+\sqrt{2+\sqrt{2+\ldots+\sqrt{2(1+\cos\theta)}}}}$, there being n number of 2's

$= \sqrt{2+\sqrt{2+\sqrt{2+\ldots+\sqrt{2(1+\cos\theta/2)}}}}$,

there being $(n-1)$ number of 2's

$= \sqrt{2+\sqrt{2+\sqrt{2+\ldots+\sqrt{2(1+\cos\theta/2^2)}}}}$,

there being $(n-2)$ number of 2's

$$= \ldots = \sqrt{2\left(1+\cos\frac{\theta}{2^{n-1}}\right)} = 2\cos\frac{\theta}{2^n}.$$

9. We know that $\tan\theta = \cot\theta - 2\cot 2\theta$.

$\therefore$ the expression

$$= \left(\cot\frac{\pi}{16} - 2\cot\frac{\pi}{8}\right) + 2\left(\cot\frac{\pi}{8} - 2\cot\frac{\pi}{4}\right) + 4 = \cot\frac{\pi}{16}.$$

10. The expression $= (\sin 78° - \sin 42°) - (\sin 66° - \sin 6°)$

$$= 2\cdot\cos 60°\cdot\sin 18° - 2\cdot\cos 36°\cdot\sin 30°$$

$= \sin 18° - \cos 36°$. Put values and simplify.

11. Value $= \dfrac{\sqrt{3}\cos 20° - \sin 20°}{\sin 20°\cdot\cos 20°} = \dfrac{4\left(\dfrac{\sqrt{3}}{2}\cos 20° - \dfrac{1}{2}\sin 20°\right)}{\sin 40°}$

$$= \frac{4\sin(60° - 20°)}{\sin 40°} = 4.$$

12. Value $= 1 + \dfrac{1}{\sqrt{2}}(\cos\theta + \sin\theta) + \sqrt{2}(\cos\theta + \sin\theta)$.

$$= 1 + \left(\frac{1}{\sqrt{2}} + \sqrt{2}\right)\cdot(\cos\theta + \sin\theta)$$

$$= 1 + \left(\frac{1}{\sqrt{2}} + \sqrt{2}\right)\cdot\sqrt{2}\cos\left(\theta - \frac{\pi}{4}\right).$$

$\therefore$ the maximum value $= 1 + \left(\dfrac{1}{\sqrt{2}} + \sqrt{2}\right)\sqrt{2} = 4.$

13. Value $= 2\cos^2\theta - 1 + \cos\theta = -1 + 2\left(\cos^2\theta + \dfrac{1}{2}\cos\theta + \dfrac{1}{16}\right) - \dfrac{1}{8}$

$$= -\frac{9}{8} + 2\left(\cos\theta + \frac{1}{4}\right)^2 \geq -\frac{9}{8}.$$

So, the minimum value $= -\dfrac{9}{8}$.

14. See the hint to Q. No. 11.

15. The expression $= 2\sin^2\theta - 3\sin 2\theta + 3 = 4 - (\cos 2\theta + 3\sin 2\theta)$

$$= 4 - \sqrt{1^2 + 3^2}\left(\frac{1}{\sqrt{10}}\cos 2\theta + \frac{3}{\sqrt{10}}\sin 2\theta\right)$$

$$= 4 - \sqrt{10}\cos(2\theta - \alpha), \text{ where } \cos\alpha = \frac{1}{\sqrt{10}}.$$

$\therefore$ the least value $= 4 - \sqrt{10} \times$ the greatest value of $\cos(2\theta - \alpha)$

$$= 4 - \sqrt{10} \times 1 = 4 - \sqrt{10}.$$

16. Here $\cos^4\theta \cdot \sec^2\alpha + \sin^4\theta \cdot \operatorname{cosec}^2\alpha = 1$

or $\cos^4\theta \cdot \sin^2\alpha + \sin^4\theta \cdot \cos^2\alpha = \sin^2\alpha \cdot \cos^2\alpha$

or $(1 - \sin^2\theta)\cos^2\theta \cdot \sin^2\alpha + \sin^4\theta \cdot (1 - \sin^2\alpha) = \sin^2\alpha(1 - \sin^2\alpha)$

or $\cos^2\theta \sin^2\alpha + \sin^4\theta - \sin^2\theta \sin^2\alpha(\cos^2\theta + \sin^2\theta) = \sin^2\alpha - \sin^4\alpha$

or $\sin^4\theta + \sin^4\alpha - 2\sin^2\theta \cdot \sin^2\alpha = 0$

$\Rightarrow$ $\sin^2\theta = \sin^2\alpha$ and so $\cos^2\theta = \cos^2\alpha$.

$\therefore$ $\cos^8\theta \cdot \sec^6\alpha + \sin^8\theta \cdot \operatorname{cosec}^6\alpha = \cos^2\theta + \sin^2\theta = 1.$

17. Here, $2x = \tan 20° + \tan 50° = \dfrac{\sin 70°}{\cos 20° \cdot \cos 50°} = \sec 50° = \operatorname{cosec} 40°$

and $2y = \tan 20° + \tan 70° = \dfrac{\sin 90°}{\cos 20° \cdot \cos 70°}$

$$= \frac{1}{\cos 20° \cdot \sin 20°} = \frac{2}{\sin 40°}.$$

18. Here, $\cos x = \dfrac{2\cos(x-y)\cos(x+y)}{\cos(x-y) + \cos(x+y)} = \dfrac{\cos^2 x - \sin^2 y}{\cos x \cdot \cos y}$

or $\cos^2 x(1 - \cos y) = \sin^2 y$ or $\cos^2 x = 2\cos^2 \dfrac{y}{2}$

or $\left(\cos x \cdot \sec\dfrac{y}{2}\right)^2 = 2.$

19. $2\sin\alpha \cdot \cos\beta \cdot \sin\gamma = \sin\alpha \cdot \sin\beta \cdot \cos\gamma + \sin\beta \cdot \sin\gamma \cdot \cos\alpha.$

Dividing by $\sin\alpha \cdot \sin\beta \cdot \sin\gamma$, $2\cot\beta = \cot\gamma + \cot\alpha.$

22. Use $\cos(\alpha+\beta) = \dfrac{1 - \tan^2\dfrac{\alpha+\beta}{2}}{1 + \tan^2\dfrac{\alpha+\beta}{2}}$, $\tan\dfrac{\alpha+\beta}{2} = \dfrac{\tan\dfrac{\alpha}{2} + \tan\dfrac{\beta}{2}}{1 - \tan\dfrac{\alpha}{2} \cdot \tan\dfrac{\beta}{2}}$

and $\tan\dfrac{\alpha}{2} + \tan\dfrac{\beta}{2} = \dfrac{13}{4}$, $\tan\dfrac{\alpha}{2}\tan\dfrac{\beta}{2} = \dfrac{15}{8}.$

23. Here $2\sin\frac{\alpha+\beta}{2}\cdot\cos\frac{\alpha-\beta}{2}=a,\ 2\sin\frac{\alpha+\beta}{2}\cdot\sin\frac{\beta-\alpha}{2}=b.$

Now, divide and get the value.

24. Use $\sin 2\alpha=\sin\{(\alpha+\beta)+(\alpha-\beta)\}$

$$=\sin(\alpha+\beta)\cdot\cos(\alpha-\beta)+\cos(\alpha+\beta)\cdot\sin(\alpha-\beta).$$

25. $\cos\alpha=\frac{1}{2}\left(x+\frac{1}{x}\right)\ \Rightarrow\ x=\cos\alpha\pm i\sin\alpha.$

Similarly, $y=\cos\beta\pm i\sin\beta.$

$$\therefore\quad \frac{x}{y}=\cos(\alpha-\beta)\pm i\sin(\alpha-\beta),$$

$$xy=\cos(\alpha+\beta)\pm i\sin(\alpha+\beta)$$

and $\frac{y}{x}=\cos(\alpha-\beta)\mp i\sin(\alpha-\beta).$

26. $$\frac{1+\sin\alpha-\cos\alpha}{1+\sin\alpha}=\frac{(1+\sin\alpha)^2-\cos^2\alpha}{(1+\sin\alpha)(1+\sin\alpha+\cos\alpha)}$$

$$=\frac{2\sin\alpha+2\sin^2\alpha}{(1+\sin\alpha)(1+\sin\alpha+\cos\alpha)}$$

$$=\frac{2\sin\alpha}{1+\sin\alpha+\cos\alpha}=\lambda.$$

27. The expression $=\frac{|\cos A+\sin A|+|\cos A-\sin A|}{|\cos A+\sin A|-|\cos A-\sin A|}$

$$=\frac{\cos A+\sin A+\cos A-\sin A}{\cos A+\sin A-(\cos A-\sin A)},$$

because $-\frac{\pi}{4}<A<\frac{\pi}{4}$ and in this interval $\cos A>\sin A$.

28. The expression $=\tan\theta\cdot\frac{\sqrt{3}+\tan\theta}{1-\sqrt{3}\tan\theta}\cdot\frac{\sqrt{3}-\tan\theta}{1+\sqrt{3}\tan\theta}$

$$=\frac{3\tan\theta-\tan^3\theta}{1-3\tan^2\theta}=\tan 3\theta.$$

29. Clearly $\alpha\neq\pm\frac{\pi}{2}\cdot$

$\sec\alpha-\tan\alpha=\frac{1-\sin\alpha}{\cos\alpha}$ and

$$\sqrt{\frac{1-\sin\alpha}{1+\sin\alpha}}=\sqrt{\frac{(1-\sin\alpha)^2}{\cos^2\alpha}}=\left|\frac{1-\sin\alpha}{\cos\alpha}\right|=\frac{1-\sin\alpha}{|\cos\alpha|}.$$

Hence, these will be equal if $\cos\alpha>0$, i.e., $-\frac{\pi}{2}<\alpha<\frac{\pi}{2}$.

31. $a=2\cos\frac{A+B}{2}\cdot\cos\frac{A-B}{2}-2\cos^2\frac{A+B}{2}+1$

$=2\cos\frac{A+B}{2}\left\{\cos\frac{A-B}{2}-\cos\frac{A+B}{2}\right\}+1$

$=2\cos\frac{A+B}{2}\cdot 2\sin\frac{A}{2}\cdot\sin\frac{B}{2}+1=b+1.$

32. $\text{LHS}=\tan\theta+\frac{\tan\theta+\sqrt{3}}{1-\sqrt{3}\tan\theta}+\frac{\tan\theta-\sqrt{3}}{1+\sqrt{3}\tan\theta}$

$$=\frac{\left\{\begin{array}{l}\tan\theta-3\tan^3\theta+(\tan\theta+\sqrt{3})(1+\sqrt{3}\tan\theta)\\ \qquad\qquad+(\tan\theta-\sqrt{3})(1-\sqrt{3}\tan\theta)\end{array}\right\}}{1-3\tan^2\theta}$$

$=\frac{3(3\tan\theta-\tan^3\theta)}{1-3\tan^2\theta}=3\tan 3\theta;\quad\therefore\ k=3.$

33. $a=d\cos\alpha+c\sin\alpha,\ b=c\cos\alpha-d\sin\alpha\ \Rightarrow\ a^2+b^2=c^2+d^2.$

34. $(\cos 20°+\sin 20°)^2+(\cos 20°-\sin 20°)^2=2$

$\therefore\quad \cos 20°+\sin 20°=\sqrt{2-p^2}>0$

$\therefore\quad \cos 40°=(\cos 20°-\sin 20°)(\cos 20°+\sin 20°)=p\sqrt{2-p^2}.$

35. $(3\sin\theta+4\cos\theta)^2+(4\sin\theta-3\cos\theta)^2=16+9=25$

$\therefore\quad 25+(4\sin\theta-3\cos\theta)^2=25$

$\therefore\quad 4\sin\theta-3\cos\theta=0.$

36. Here $\cos^2x+\cos x-1=0$ or $\cos x=\frac{-1+\sqrt{5}}{2}$

$$\left\{\because\ \cos x\neq\frac{-1-\sqrt{5}}{2}<-1\right\}$$

$\therefore\quad \cos^2x=\left(\frac{\sqrt{5}-1}{2}\right)^2=\frac{6-2\sqrt{5}}{4}=\frac{3-\sqrt{5}}{2}$

$\therefore\quad \text{value}=\left(1-\frac{3-\sqrt{5}}{2}\right)\left(2-\frac{3-\sqrt{5}}{2}\right)=\frac{\sqrt{5}-1}{2}\cdot\frac{\sqrt{5}+1}{2}=1.$

37. $x = 1 + \cos^2\phi + \cos^4\phi + \dots \text{ to } \infty = \dfrac{1}{1-\cos^2\phi} = \dfrac{1}{\sin^2\phi}$.

Similarly, $y = \dfrac{1}{1-\sin^2\phi} = \dfrac{1}{\cos^2\phi}$, $z = \dfrac{1}{1-\cos^2\phi \cdot \sin^2\phi}$

$$\therefore \quad xyz = \frac{1}{\sin^2\phi \cdot \cos^2\phi(1-\sin^2\phi\cos^2\phi)}$$

$$= \frac{(1-\sin^2\phi \cdot \cos^2\phi) + \sin^2\phi \cdot \cos^2\phi}{\sin^2\phi \cdot \cos^2\phi(1-\sin^2\phi\cos^2\phi)}$$

$$= \frac{1}{\sin^2\phi\cos^2\phi} + \frac{1}{1-\sin^2\phi\cos^2\phi} = xy + z.$$

38. $\sin n\theta = b_0 + b_1\sin\theta + b_2\sin^2\theta + \dots$.

This is possible when n is an odd integer.

Put $\theta = 0$ to get b_0. After differentiating w.r.t. θ, put $\theta = 0$ to get b_1.

39. Differentiate w.r.t. θ and put $\theta = \dfrac{\pi}{2}$.

40. $\sin^3 x \cdot \sin 3x = \dfrac{3\sin x - \sin 3x}{4}\sin 3x$

$$= \frac{3}{8}(2\sin x \cdot \sin 3x) - \frac{1}{4}\sin^2 3x$$

$$= \frac{3}{8}(\cos 2x - \cos 4x) - \frac{1}{8}(1-\cos 6x)$$

$$= -\frac{1}{8} + \frac{3}{8}\cos 2x - \frac{3}{8}\cos 4x + \frac{1}{8}\cos 6x.$$

41. The expression $= \left(\sin\dfrac{\pi}{14} \cdot \sin\dfrac{3\pi}{14}\sin\dfrac{5\pi}{14}\right)^2 \cdot \sin\dfrac{7\pi}{14}$.

Clearly, $\sin\dfrac{\pi}{14} = \cos\left(\dfrac{\pi}{2} - \dfrac{\pi}{14}\right) = \cos\dfrac{3\pi}{7} = -\cos\left(x - \dfrac{3\pi}{7}\right) = -\cos\dfrac{4\pi}{7}$.

$$\sin\frac{3\pi}{14} = \cos\frac{2\pi}{7}, \ \sin\frac{5\pi}{7} = \cos\frac{\pi}{7}.$$

$$\therefore \quad \text{value} = \left(\cos\frac{\pi}{7}\cos\frac{2\pi}{7} \cdot \frac{4\pi}{7}\right)^2 \cdot 1$$

$$= \left\{\frac{2\sin\frac{\pi}{7} \cdot \cos\frac{\pi}{7} \cdot \cos\frac{2\pi}{7} \cdot \cos\frac{4\pi}{7}}{2\sin\frac{\pi}{7}}\right\}^2 = \dots = \left(\frac{\sin\frac{8\pi}{7}}{8\sin\frac{\pi}{7}}\right)^2 = \frac{1}{8^2}.$$

42. Value $= \sin 10° \cdot \sin 50° \cdot \sin 70° = \cos 20° \cdot \cos 40° \cdot \cos 80°$

(See the hint to Q. No. 41.)

43. Value $= -\dfrac{\cos 126°}{\frac{1}{2}\sin 126°} = -2\cot 126° = 2\tan 36° = \dfrac{4 \cdot \sin 18° \cdot \cos 18°}{\cos 36°}$

$$= \frac{4 \cdot \dfrac{\sqrt{5}-1}{4} \cdot \sqrt{1-\left(\dfrac{\sqrt{5}-1}{4}\right)^2}}{\dfrac{\sqrt{5}+1}{4}} = \frac{\sqrt{5}-1}{\sqrt{5}+1}\sqrt{10+2\sqrt{5}}$$

$$= \frac{\sqrt{(\sqrt{5}-1)^2(10+2\sqrt{5})}}{\sqrt{5}+1}.$$

44. $(\cos 9° - \sin 9°)^2 = 1 - \sin 18°$

$$= 1 - \frac{\sqrt{5}-1}{4} = \frac{5-\sqrt{5}}{4}; \text{ also } \cos 9° > \sin 9°.$$

45. Value $= \dfrac{2\sin 18° + 3 - 4\cos^2 18°}{\cos 18°} = \dfrac{2\sin 18° + 3 - 2(1+\cos 36°)}{\cos 18°}$

$$= \frac{2 \cdot \dfrac{\sqrt{5}-1}{4} + 3 - 2 - 2 \cdot \dfrac{\sqrt{5}+1}{4}}{\cos 18°} = 0.$$

46. Value $= 2\tan 50° - (\tan 70° - \tan 20°)$

$$= 2\tan 50° - \frac{\sin 50°}{\cos 70° \cdot \cos 20°}$$

$$= 2\tan 50° - \frac{2\sin 50°}{2\sin 20° \cdot \cos 20°}$$

$$= 2\tan 50° - 2 \cdot \frac{\sin 50°}{\sin 40°} = 2\tan 50° - 2 \cdot \frac{\sin 50°}{\cos 50°} = 0.$$

47. Use the fact that opposite angles are supplementary.

48. Here $2n\alpha = \dfrac{\pi}{2} \cdot$ So,

$$\cot\alpha \cdot \cot(2n-1)\alpha = \cot\alpha \cdot \cot\left(\frac{\pi}{2} - \alpha\right) = \cot\alpha \cdot \tan\alpha = 1.$$

49. Value $= \cos 12° \cdot \cos 24° \cdot \cos 48° \cdot \cos(180° - 96°) \cdot \cos 36° \cdot \cos 72°$

$$= -(\cos 12° \cdot \cos 24° \cdot \cos 48° \cdot \cos 96°)(\cos 36° \cdot \cos 72°)$$

$$= -\left(\frac{2\sin 12^\circ \cdot \cos 12^\circ \cdot \cos 24^\circ \cdot \cos 48^\circ \cdot \cos 96^\circ}{2\sin 12^\circ}\right)$$

$$\times \left(\frac{2\sin 36^\circ \cdot \cos 36^\circ \cdot \cos 72^\circ}{2\sin 36^\circ}\right)$$

$$= -\frac{\sin 192^\circ}{2^4 \cdot \sin 12^\circ} \cdot \frac{\sin 144^\circ}{2^2 \sin 36^\circ} = \frac{1}{2^4} \cdot \frac{1}{2^2}.$$

50. Use $\cos\alpha + \cos(\alpha+\beta) + \cos(\alpha+2\beta) + \ldots + \cos(\alpha + \overline{n-1}\,\beta)$

$$= \frac{\sin\frac{n\beta}{2}}{\sin\frac{\beta}{2}} \cos\frac{2\alpha + (n-1)\beta}{2}.$$

Here $\alpha = \frac{\pi}{11}, \beta = \frac{2\pi}{11}, n = 5.$

51. Sum $= \frac{1}{2}\sum_{r=1}^{n-1}\left(1 + \cos\frac{2r\pi}{n}\right)$

$$= \frac{1}{2}(n-1) + \frac{1}{2}\left\{\cos\frac{2\pi}{n} + \cos\frac{4\pi}{n} + \cos\frac{6\pi}{n} + \ldots + \cos\frac{(2n-2)\pi}{n}\right\}$$

Now, see the hint to Q. No. 50.

52. Use $\sin\alpha + \sin(\alpha+\beta) + \sin(\alpha+2\beta) + \ldots + \sin(\alpha + \overline{n-1}\,\beta)$

$$= \frac{\sin\frac{n\beta}{2}}{\sin\frac{\beta}{2}} \sin\frac{2\alpha + (n-1)\beta}{2}.$$

Here $\alpha = \frac{\pi}{n}, \beta = \frac{2\pi}{n}.$

53. $\cos^6 x = 1 - \sin^4 x = \cos^2 x(1 + \sin^2 x)$

$\Rightarrow \quad \cos^2 x = 0 \quad$ or $\quad \cos^4 x = 1 + \sin^2 x$

$1 + \sin^2 x = \cos^4 x \quad \Rightarrow \sin^2 x = -(1 - \cos^4 x) = -\sin^2 x(1 + \cos^2 x)$

$\Rightarrow \quad \sin^2 x = 0 \quad$ or $\quad 2 + \cos^2 x = 0$ (absurd).

$\therefore \quad \cos x = 0$ or $\sin x = 0 \quad \Rightarrow x = \frac{\pi}{2}, -\frac{\pi}{2}, 0, \pi, -\pi.$

54. As $\cos^2 x + \sin^2 x = 1$, $\sin^2 x \geq \sin^4 x$ (equality holding when $\sin x = 0$) and $\cos^2 x \geq \cos^7 x$ (equality holding when $\cos x = 0$)

we find that $\cos^7 x + \sin^4 x = 1$ can hold only when

$$\cos x = 0 \quad \text{or} \quad \sin x = 0\ (x \neq \pi, -\pi).$$

So, $\quad x = -\frac{\pi}{2}, \frac{\pi}{2}, 0.$

55. $\sin\theta = \frac{2 \pm \sqrt{4 - 4\lambda}}{2} = 1 \pm \sqrt{1-\lambda}$. For real values, $1 - \lambda \geq 0$, i.e., $\lambda \leq 1$.

As $-1 \leq \sin\theta \leq 1$, $\sin\theta = 1 - \sqrt{1-\lambda}$.

From the question, $\sin\theta > -\frac{1}{2}$. Thus $-\frac{1}{2} < 1 - \sqrt{1-\lambda} \leq 1$

or $\quad -\frac{3}{2} < -\sqrt{1-\lambda} \leq 0 \quad \Rightarrow \sqrt{1-\lambda} < \frac{3}{2} \quad \Rightarrow 1 - \lambda < \frac{9}{4}$

$\Rightarrow \quad \lambda > -\frac{5}{4}$.

56. $\sec A = -\frac{5}{4}$. So, $\frac{\pi}{2} < A < \pi$. Hence, $\tan A = -\frac{3}{4}$ and $\operatorname{cosec} A = \frac{5}{3}$.

$\therefore\quad$ the required equation is $x^2 - \left(-\frac{3}{4} + \frac{5}{3}\right)x + \left(-\frac{3}{4}\right) \times \frac{5}{3} = 0.$

57. In a convex quadrilateral no angle is greater than 180°.

Here $\tan A = \frac{5}{12}$. So $0 < A < \frac{\pi}{2}$ and $\frac{\pi}{2} < C < \pi \qquad (\because A + C = 180°)$

$\therefore \quad \tan(\pi - C) = \frac{5}{12}$, i.e., $\tan C = -\frac{5}{12} \qquad \therefore \cos C = -\frac{12}{13}$.

Also, $\cos B = -\frac{3}{5}$. So $\frac{\pi}{2} < B < \pi$ and $0 < D < \frac{\pi}{2} \qquad (\because B + D = 180°)$

$\therefore \quad \cos(\pi - D) = -\frac{3}{5}$, i.e., $\cos D = \frac{3}{5} \qquad \therefore \tan D = \frac{4}{3}$.

$\therefore \quad$ the required equation is $x^2 - \left(-\frac{12}{13} + \frac{4}{3}\right)x + \left(-\frac{12}{13}\right)\cdot\frac{4}{3} = 0.$

58. $2^x + 2^{-x} \geq 2$ (using AM $\geq$ GM). But $\sin(e^x) \leq 1$. So, there is no solution.

59. $D \geq 0 \Rightarrow \cos^2 p - 4\sin p \cdot (\cos p - 1) \geq 0$

or $\quad \cos^2 p + 4\sin p(1 - \cos p) \geq 0 \qquad \ldots (1)$

As $\cos^2 p \geq 0$ and $1 - \cos p \geq 0$, (1) will hold for all $\sin p \geq 0$. So, $p \in [0, \pi]$.

60. $(e^{\sin x})^2 - 4e^{\sin x} - 1 = 0$

$\therefore \quad e^{\sin x} = \frac{4 \pm \sqrt{16+4}}{2} = 2 \pm \sqrt{5},$

which is absurd because $0 < e^{\sin x} \le e = 2.72$ (nearly).

61. $\tan\frac{\alpha}{2} + \cot\frac{\alpha}{2} = \frac{1}{\cos\frac{\alpha}{2}\cdot\sin\frac{\alpha}{2}} = \frac{2}{\sin\alpha} = \frac{2}{p}$ and $\tan\frac{\alpha}{2}\cdot\cot\frac{\alpha}{2} = 1.$

$\therefore$ the required equation is $x^2 - \frac{2}{p}x + 1 = 0.$

62. $\sec\alpha + \operatorname{cosec}\alpha = p$ and $\sec\alpha\cdot\operatorname{cosec}\alpha = q.$

$\therefore$ $\sin\alpha + \cos\alpha = p\sin\alpha\cdot\cos\alpha$ and $\sin\alpha\cdot\cos\alpha = \frac{1}{q}\cdot$

$\therefore$ $\sin\alpha + \cos\alpha = \frac{p}{q}$

$\therefore$ $\frac{p^2}{q^2} = 1 + 2\sin\alpha\cdot\cos\alpha = 1 + \frac{2}{q}\cdot$

63. $\sin x = \frac{7 \pm \sqrt{49 - 24}}{6} = \frac{7 \pm 5}{6} = \frac{1}{3}$, 2 (absurd).

$\therefore$ $\sin x = \frac{1}{3} = \sin\alpha,$ where $0 < \alpha < \frac{\pi}{2}\cdot$

$\therefore$ $x = \alpha, \pi - \alpha, 2\pi + \alpha, 3\pi - \alpha, 4\pi + \alpha, 5\pi - \alpha.$

64. $\cos\alpha + \cos\beta = -a, \cos\alpha\cdot\cos\beta = b$ and $\sin\alpha + \sin\beta = -p, \sin\alpha\cdot\sin\beta = q.$

$\therefore$ $a^2 + p^2 = 2 + 2\cos(\alpha - \beta)$ and $b + q = \cos(\alpha - \beta).$

$\therefore$ $a^2 + p^2 = 2 + 2(b + q).$

65. $x^2 - 2x + 4 = -3\cos(ax + b) \Rightarrow (x - 1)^2 + 3 = -3\cos(ax + b).$

As $-1 \le \cos(ax + b) \le 1$ and $(x - 1)^2 \ge 0$, the above is possible only if $\cos(ax + b) = -1$ when $x = 1$. So, $a + b = \pi, 3\pi, 5\pi$, etc., and $3\pi > 6$.

66. $x^2 - \cos\theta\cdot x + p = 0.$ $\therefore$ $x = \frac{\cos\theta \pm \sqrt{\cos^2\theta - 4p}}{2}.$

For real x, $\cos^2\theta - 4p \ge 0.$ $\therefore$ $4p \le \cos^2\theta \le 1.$

67. $3 - 4\sin^2 x = y$

$\therefore$ $\sin^2 x = \frac{3 - y}{4}\cdot$ But $0 < \sin^2 x \le 1$ $(\because \sin x = 0 \Rightarrow x = n\pi).$

$\therefore$ $0 < \frac{3 - y}{4} \le 1$ or $0 < 3 - y \le 4.$

68. $2\cos^2\theta + \cos\theta + (k-1) = 0$

$$\therefore \quad \cos\theta = \frac{-1 \pm \sqrt{1-8(k-1)}}{4} = \frac{-1 \pm \sqrt{9-8k}}{4}.$$

For real $\cos\theta$, $9 - 8k \geq 0$, i.e., $k \leq \frac{9}{8}$.

Also, $-1 \leq \frac{-1 \pm \sqrt{9-8k}}{4} \leq 1$ or $-4 \leq -1 \pm \sqrt{9-8k} \leq 4$

or $\quad -3 \leq \pm\sqrt{9-8k} \leq 5$

$\therefore \quad -3 \leq -\sqrt{9-8k}$ and $\sqrt{9-8k} \leq 5$

$\Rightarrow \quad 9 - 8k \leq 9$ and $9 - 8k \leq 25 \quad \therefore\ k \geq 0$ and $k \geq -2 \quad \Rightarrow\ k \geq 0$.

69. $\sec^2\theta + \sec\theta - (\lambda + 1) = 0$

$$\therefore \quad \sec\theta = \frac{-1 \pm \sqrt{1+4(\lambda+1)}}{2} = \frac{-1 \pm \sqrt{4\lambda+5}}{2}.$$

For real $\sec\theta$, $4\lambda + 5 \geq 0$, i.e., $\lambda \geq -\frac{5}{4}$.

Also $\sec\theta \geq 1$ or $\sec\theta \leq -1$.

$$\therefore \quad \frac{-1 \pm \sqrt{4\lambda+5}}{2} \geq 1 \quad \text{or} \quad \frac{-1 \pm \sqrt{4\lambda+5}}{2} \leq -1$$

$\Rightarrow \quad -1 + \sqrt{4\lambda+5} \geq 2$ or $-1 - \sqrt{4\lambda+5} \leq -2$

$\Rightarrow \quad 4\lambda + 5 \geq 9$ or $4\lambda + 5 \geq 1 \quad \Rightarrow \lambda \geq 1$ or $\lambda \geq -1$

$\therefore \quad \lambda \geq -\frac{5}{4}$ and $\lambda \geq 1$ or $\lambda \geq -\frac{5}{4}$ and $\lambda \geq -1$

$\therefore \quad \lambda \geq 1$ or $\lambda \geq -1 \quad \therefore\ \lambda \in [-1, +\infty)$.

70. $\tan(A + B + C) = 0$ (from the question) $\quad \therefore\ A + B + C = n\pi$.

71. $2x = \cos(\alpha - \beta - \gamma + \delta) - \cos(\alpha - \beta + \gamma - \delta)$
$2y = \cos(\beta - \gamma - \alpha + \delta) - \cos(\beta - \gamma + \alpha - \delta)$ and similarly for $2x$.
Adding, $2x + 2y + 2z = 0$.

72. The expression $= \cos\left(\frac{\pi}{2} - \frac{15\pi}{32}\right) \cdot \cos\left(\frac{\pi}{2} - \frac{7\pi}{16}\right) \cdot \cos\left(\frac{\pi}{2} - \frac{3\pi}{8}\right)$

$$= \cos\frac{\pi}{32}\cos\frac{\pi}{16} \cdot \cos\frac{\pi}{8}$$

$$= \frac{\left(2\sin\frac{\pi}{32} \cdot \cos\frac{\pi}{32}\right)\cos\frac{\pi}{16} \cdot \cos\frac{\pi}{8}}{2\sin\frac{\pi}{32}} \text{ and so on.}$$

74. We know that $-\sqrt{5^2+12^2} \le 5\cos x + 12\sin x \le \sqrt{5^2+12^2}$

$\therefore \quad -13 \le 5\cos x + 12\sin x < 0 \;\Rightarrow\; -13 \le \frac{1}{a} < 0$ and

$0 < 5\cos x + 12\sin x \le 13 \quad \Rightarrow \; 0 < \frac{1}{a} \le 13$

$\Rightarrow \quad \frac{a}{-13}(-13) \le \frac{a}{-13}\cdot\frac{1}{a} \left(\because \; \frac{a}{-13} \text{ is positive}\right) \quad \therefore \; a \le -\frac{1}{13}$

and $a \ge \frac{1}{13}$ ($\because$ a is positive).

75. $y = \cos^4 x - \cos^2 x + 1 = \left(\cos^2 x - \frac{1}{2}\right)^2 + \frac{3}{4}$.

$\therefore \quad y_{\min} = \frac{3}{4}$ and y is maximum when $\left(\cos^2 x - \frac{1}{2}\right)^2$ is the maximum

$\therefore \quad y_{\max} = \frac{1}{4} + \frac{3}{4} = 1.$

76. $y = \sin x \cdot (\sin^2 60° - \sin^2 x) = \sin x \left(\frac{3}{4} - \sin^2 x\right) = \frac{1}{4}\sin 3x.$

77. $1 + \sec\theta = \dfrac{1+\cos\theta}{\cos\theta} = \dfrac{2\cos^2\frac{\theta}{2}}{\cos\theta}$ · Similarly for others.

$$\therefore \quad f_n(\theta) = \tan\frac{\theta}{2}\cdot\frac{2\cos^2\frac{\theta}{2}}{\cos\theta}\cdot\frac{2\cos^2\theta}{\cos 2\theta}\cdot\ldots\cdot\frac{2\cos^2 2^{n-1}\theta}{\cos 2^n\theta}$$

$$= \tan\frac{\theta}{2}\cdot 2^{n+1}\cdot(\cos\theta\cdot\cos 2\theta\cdot\ldots\cdot\cos 2^{n-1}\theta)\cdot\frac{\cos^2\frac{\theta}{2}}{\cos 2^n\theta}$$

$$= \frac{\sin\theta}{\cos 2^n\theta}\cdot 2^n\cdot\left\{\frac{2\sin\theta\cdot\cos\theta\cdot\cos 2\theta\cdot\ldots\cdot\cos 2^{n-1}\theta}{2\sin\theta}\right\} = \ldots$$

$$= \frac{\sin\theta}{\cos 2^n\theta}\cdot 2^n\cdot\frac{\sin 2^n\theta}{2^n\sin\theta} = \tan 2^n\theta.$$

$\therefore \quad f_2\left(\frac{\pi}{16}\right) = \tan\left(4\cdot\frac{\pi}{16}\right) = 1, f_3\left(\frac{\pi}{32}\right) = \tan\left(8\cdot\frac{\pi}{32}\right) = 1$ and so on.

78. $x^2 + y^2 = X^2 + Y^2, xy = (X^2 - Y^2)\sin\theta\cdot\cos\theta - XY(\cos^2\theta - \sin^2\theta)$

$x^2 + 4xy + y^2 = X^2 + Y^2 + 2(X^2 - Y^2)\sin 2\theta - 4XY\cos 2\theta$

$$= (1 + 2\sin 2\theta)X^2 + (1 - 2\sin 2\theta)Y^2 - 4\cos 2\theta\cdot XY$$

From the question $a = 1 + 2\sin 2\theta$, $b = 1 - 2\sin 2\theta$, $\cos 2\theta = 0$.

$\cos 2\theta = 0 \quad \Rightarrow \quad \theta = \frac{\pi}{4}$. Then $a = 1 + 2\sin\frac{\pi}{2}$, $b = 1 - 2\sin\frac{\pi}{2}$.

79. $7\cos x - 24\sin x = 25 \cdot \left(\frac{7}{25}\cos x - \frac{24}{25}\sin x\right) = 25\cos(x + \alpha)$, where

$\cos\alpha = \frac{7}{25}$.

80. $2\cos\frac{A+B}{2} \cdot \cos\frac{A-B}{2} = 1 \quad$ or $\quad 2 \cdot \frac{\sqrt{3}}{2} \cdot \frac{A-B}{2} = 1$

$\therefore \quad \cos\frac{A-B}{2} = \frac{1}{\sqrt{3}}$

$\therefore \quad \cos(A-B) = 2\cos^2\frac{A-B}{2} - 1 = \frac{2}{3} - 1 = -\frac{1}{3}$.

$$|\cos A - \cos B| = \left|2\sin\frac{A+B}{2} \cdot \sin\frac{A-B}{2}\right|$$

$$= 2 \cdot \frac{1}{2} \cdot \sqrt{1 - \frac{1}{3}} = \sqrt{\frac{2}{3}}.$$

81. $\because \quad \tan 3\theta - \tan 2\theta - \tan\theta = \tan 3\theta \cdot \tan 2\theta \cdot \tan\theta$

$\therefore \quad 0 = ab\tan 3\theta \quad \Rightarrow \tan 3\theta = 0$

$\Rightarrow \tan\theta + \tan 2\theta = 0 \quad \Rightarrow a + b = 0.$

82. $\cos^3 x \cdot \sin 2x = \frac{\cos 3x + 3\cos x}{4}\sin 2x$

$$= \frac{1}{8}(\sin 5x - \sin x) + \frac{3}{8}(\sin 3x + \sin x)$$

$$= \frac{1}{4}\sin x + \frac{3}{8}\sin 3x + \frac{1}{8}\sin 5x.$$

$\therefore \quad n = 5, a_1 = \frac{1}{4}, a_2 = 0, a_3 = \frac{3}{8}, a_4 = 0, a_5 = \frac{1}{8}$.

84. $y = \tan A \cdot \tan\left(\frac{\pi}{3} - A\right) \quad \Rightarrow \tan^2 A + \sqrt{3}(y-1)\tan A + y = 0.$

As $\tan A$ is real, $D \geq 0 \quad \Rightarrow 3(y-1)^2 - 4y \geq 0$

$\therefore \quad 3y^2 - 10y + 3 \geq 0 \quad \Rightarrow y \leq \frac{1}{3}$ or $y \geq 3$.

But each of $\tan A$, $\tan B$ is less than $\sqrt{3}$ or one is 0 and the other is $\sqrt{3}$.

$\therefore \quad y \geq 3$ is not possible. So $y \leq \frac{1}{3} \cdot$ Also, $\tan A \cdot \tan B \geq 0$.

85. $$\text{Sum} = \frac{\sin 3 \cdot \frac{\pi}{7}}{\sin \frac{\pi}{7}} \cdot \cos \frac{4\pi}{7}$$ (See the hint to Q. No. 52)

$$= \frac{\sin \pi - \sin \frac{\pi}{7}}{2\sin \frac{\pi}{7}} = -\frac{1}{2} \cdot$$

□

2

Solutions of Equations and Inequations

• Type 1 •

Choose the most appropriate option (a, b, c or d).

1. The number of distinct solutions of $\sin 5\theta \cdot \cos 3\theta = \sin 9\theta \cdot \cos 7\theta$ in $[0, \pi/2]$ is

(a) 4 (b) 5

(c) 8 (d) 9

2. The number of solutions of $\sin^2\theta + 3\cos\theta = 3$ in $[-\pi, \pi]$ is

(a) 4 (b) 2

(c) 0 (d) none of these

3. The number of distinct solutions of $\sec\theta + \tan\theta = \sqrt{3}, 0 \le \theta \le 3\pi$, is

(a) 3 (b) 5

(c) 4 (d) none of these

4. The number of solutions of the equation $5\sec\theta - 13 = 12\tan\theta$ in $[0, 2\pi]$ is

(a) 2 (b) 1

(c) 4 (d) 0

5. The equation $p\cos x - q\sin x = r$ admits of a solution for x only if

(a) $r < \max\{p, q\}$ (b) $-\sqrt{p^2 + q^2} < r < \sqrt{p^2 + q^2}$

(c) $r^2 = p^2 + q^2$ (d) none of these

6. The equation $k\cos x - 3\sin x = k + 1$ is solvable only if k belongs to the interval

   (a) $[4, +\infty)$ (b) $[-4, 4]$
   (c) $(-\infty, 4]$ (d) none of these

7. The number of solutions of $\cos\theta + \sqrt{3}\sin\theta = 5, 0 \le \theta \le 5\pi$, is

   (a) 4 (b) 0
   (c) 5 (d) none of these

8. If the equation $a_1 + a_2\cos 2x + a_3\sin^2 x = 1$ is satisfied by every real value of x then the number of possible values of the triplet (a_1, a_2, a_3) is

   (a) 0 (b) 1
   (c) 3 (d) infinite

9. The number of solutions of the equation $\tan x + \sec x = 2\cos x$ lying in the interval $[0, 2\pi]$ is

   (a) 0 (b) 1
   (c) 2 (d) 3

10. The number of values of x in $[0, 5\pi]$ satisfying the equation $3\cos 2x - 10\cos x + 7 = 0$ is

    (a) 5 (b) 6
    (c) 8 (d) 10

11. The smallest positive integral value of p for which the equation $\cos(p\sin x) = \sin(p\cos x)$ in x has a solution in $[0, 2\pi]$ is

    (a) 2 (b) 1
    (c) 3 (d) none of these

12. If $\sin^2\theta - 2\sin\theta - 1 = 0$ is to be satisfied for exactly 4 distinct values of $\theta \in [0, n\pi], n \in N$, then the least value of n is

    (a) 2 (b) 6
    (c) 4 (d) 1

13. If $2\tan^2 x - 5\sec x$ is equal to 1 for exactly 7 distinct values of $x \in \left[0, \frac{n\pi}{2}\right], n \in N$, then the greatest value of n is

    (a) 6 (b) 12
    (c) 13 (d) 15

14. The number of values of $x \in [0, 2\pi]$ that satisfy $\cot x - \operatorname{cosec} x = 2\sin x$ is

(a) 3 (b) 2

(c) 1 (d) 0

15. If $\sin \alpha$, 1 and $\cos 2\alpha$ are in GP then α is equal to

(a) $n\pi + (-1)^n \frac{\pi}{2}, n \in Z$ (b) $n\pi + (-1)^{n-1} \cdot \frac{\pi}{2}, n \in Z$

(c) $2n\pi, n \in Z$ (d) none of these

16. If $\frac{1}{6} \sin \theta$, $\cos \theta$ and $\tan \theta$ are in GP then the general value of θ is

(a) $2n\pi \pm \frac{\pi}{3}, n \in Z$ (b) $2n\pi \pm \frac{\pi}{6}, n \in Z$

(c) $n\pi + (-1)^n \frac{\pi}{3}, n \in Z$ (d) $n\pi + \frac{\pi}{3}, n \in Z$

17. The general values of x for which $\cos 2x$, $\frac{1}{2}$ and $\sin 2x$ are in AP are given by

(a) $n\pi, n\pi + \frac{\pi}{2}$ (b) $n\pi, n\pi + \frac{\pi}{4}$

(c) $n\pi + \frac{\pi}{4}$ (d) $n\pi$

18. The most general values of θ satisfying $\tan \theta + \tan \left(\frac{3\pi}{4} + \theta\right) = 2$ are

(a) $n\pi \pm \frac{\pi}{3}, n \in Z$ (b) $2n\pi + \frac{\pi}{3}, n \in Z$

(c) $2n\pi \pm \frac{\pi}{3}, n \in Z$ (d) $n\pi + (-1)^n \frac{\pi}{3}, n \in Z$

19. The most general solutions of the equation $\sec x - 1 = (\sqrt{2} - 1)\tan x$ are given by

(a) $n\pi + \frac{\pi}{8}$ (b) $2n\pi, 2n\pi + \frac{\pi}{4}$

(c) $2n\pi$ (d) none of these

20. The most general solutions of the equation $\sec^2 x = \sqrt{2}(1 - \tan^2 x)$ are given by

(a) $n\pi + \frac{\pi}{8}$ (b) $n\pi \pm \frac{\pi}{4}$

(c) $n\pi \pm \frac{\pi}{8}$ (d) none of these

21. The most general values of x for which

$$\sin x + \cos x = \min_{a \in R} \{1, a^2 - 4a + 6\}$$

are given by

(a) $2n\pi$ (b) $2n\pi + \frac{\pi}{2}$

(c) $n\pi + (-1)^n \cdot \frac{\pi}{4} - \frac{\pi}{4}$ (d) none of these

22. If $\max_{\theta \in R} \{5\sin\theta + 3\sin(\theta - \alpha)\} = 7$ then the set of possible values of α is

(a) $\left\{x \mid x = 2n\pi \pm \frac{\pi}{3}; n \in Z\right\}$ (b) $\left\{x \mid x = 2n\pi \pm \frac{2\pi}{3}; n \in Z\right\}$

(c) $\left[\frac{\pi}{3}, \frac{2\pi}{3}\right]$ (d) none of these

23. Let α, β be any two positive values of x for which $2\cos x$, $|\cos x|$ and $1 - 3\cos^2 x$ are in GP. The minimum value of $|\alpha - \beta|$ is

(a) $\frac{\pi}{3}$ (b) $\frac{\pi}{4}$

(c) $\frac{\pi}{2}$ (d) none of these

24. The sum of all the solutions of the equation

$$\cos x \cdot \cos\left(\frac{\pi}{3} + x\right) \cdot \cos\left(\frac{\pi}{3} - x\right) = \frac{1}{4}, x \in [0, 6\pi]$$

is

(a) 15π (b) 30π

(c) $\frac{110\pi}{3}$ (d) none of these

25. The values of $x \in [-2\pi, 2\pi]$ such that $\frac{\sin x + i\cos x}{1 + i}$, $i = \sqrt{-1}$, is purely imaginary, are given by

(a) $n\pi - \frac{\pi}{4}$ (b) $n\pi + \frac{\pi}{4}$

(c) $n\pi$ (d) none of these

26. If $x \in \left[-\frac{5\pi}{2}, \frac{5\pi}{2}\right]$, the greatest positive solution of $1 + \sin^4 x = \cos^2 3x$ is

(a) π (b) 2π

(c) $\frac{5\pi}{2}$ (d) none of these

27. The number of solutions of $|\cos x| = \sin x, 0 \le x \le 4\pi$, is

(a) 8 (b) 4

(c) 2 (d) none of these

28. The number of solutions of $\cos x = |1 + \sin x|, 0 \le x \le 3\pi$, is

(a) 3 (b) 2

(c) 4 (d) none of these

29. Let $[x]$ – the greatest integer less than or equal to x and let $f(x) = \sin x + \cos x$. Then the most general solutions of $f(x) = \left[f\left(\frac{\pi}{10}\right)\right]$ are

(a) $2n\pi + \frac{\pi}{2}, n \in Z$ (b) $n\pi, n \in Z$

(c) $2n\pi, n \in Z$ (d) none of these

30. The most general solutions of $2^{1 + |\cos x| + \cos^2 x + |\cos x|^3 + \ldots \text{ to } \infty} = 4$ are given by

(a) $n\pi \pm \frac{\pi}{3}, n \in Z$ (b) $2n\pi \pm \frac{\pi}{3}, n \in Z$

(c) $2n\pi \pm \frac{2\pi}{3}, n \in Z$ (d) none of these

31. If $x \ne \frac{n\pi}{2}$ and $(\cos x)^{\sin^2 x - 3\sin x + 2} = 1$ then all solutions of x are given by

(a) $2n\pi + \frac{\pi}{2}$ (b) $(2n + 1)\pi - \frac{\pi}{2}$

(c) $n\pi + (-1)^n \frac{\pi}{2}$ (d) none of these

32. The solution of the equation $(\sin x + \cos x)^{1 + \sin 2x} = 2, -\pi \le x \le \pi$, is

(a) $\frac{\pi}{2}$ (b) π

(c) $\frac{\pi}{4}$ (d) none of these

33. The most general solutions of $2^{\sin x} + 2^{\cos x} = 2^{1 - 1/\sqrt{2}}$ are

(a) $n\pi - \frac{\pi}{4}$ (b) $n\pi + \frac{\pi}{4}$

(c) $n\pi + (-1)^n \frac{\pi}{4}$ (d) $2n\pi \pm \frac{\pi}{4}$

34. The number of solutions of $16^{\sin^2 x} + 16^{\cos^2 x} = 10, 0 \le x \le 2\pi$, is

(a) 8 (b) 6

(c) 4 (d) 2

35. If $3\sin^2\theta + 2\sin^2\phi = 1$ and $3\sin 2\theta = 2\sin 2\phi$, $0 < \theta < \frac{\pi}{2}$ and $0 < \phi < \frac{\pi}{2}$, then the value of $\theta + 2\phi$ is

(a) $\frac{\pi}{2}$ (b) $\frac{\pi}{4}$

(c) 0 (d) none of these

36. The most general values of θ satisfying the equation $(1 + 2\sin\theta)^2 + (\sqrt{3}\tan\theta - 1)^2 = 0$ are given by

(a) $n\pi \pm \frac{\pi}{6}$ (b) $n\pi + (-1)^n \frac{7\pi}{6}$

(c) $2n\pi + \frac{7\pi}{6}$ (d) $2n\pi + \frac{11\pi}{6}$

37. If $r > 0, -\pi \le \theta \le \pi$ and r, θ satisfy $r\sin\theta = 3$ and $r = 4(1 + \sin\theta)$ then the number of possible solutions of the pair (r, θ) is

(a) 2 (b) 4

(c) 0 (d) infinite

38. The number of solutions of the equation $x^3 + x^2 + 4x + 2\sin x = 0$ in $0 \le x \le 2\pi$ is

(a) zero (b) one

(c) two (d) four

39. The number of real solutions of $\sin e^x \cdot \cos e^x = 2^{x-2} + 2^{-x-2}$ is

(a) zero (b) one

(c) two (d) infinite

40. The least positive nonintegral solution of $\sin \pi(x^2 + x) - \sin \pi x^2 = 0$ is

(a) rational

(b) irrational of the form $\sqrt{p}$

(c) irrational of the form $\frac{\sqrt{p}-1}{4}$, where p is an odd integer

(d) irrational of the form $\frac{\sqrt{p}+1}{4}$, where p is an even integer

41. If the equation $2\cos x + \cos 2\lambda x = 3$ has only one solution then λ is

(a) 1 (b) a rational number

(c) an irrational number (d) none of these

42. If $0 \le x \le 2\pi$, $0 \le y \le 2\pi$ and $\sin x + \sin y = 2$ then the value of $x + y$ is

(a) π (b) $\frac{\pi}{2}$

(c) 3π (d) none of these

43. If $-\pi \le x \le \pi$, $-\pi \le y \le \pi$ and $\cos x + \cos y = 2$ then the value $\cos(x - y)$ is

(a) -1 (b) 0

(c) 1 (d) none of these

44. If $0 \le x \le 3\pi$, $0 \le y \le 3\pi$ and $\cos x \cdot \sin y = 1$ then the possible number of values of the ordered pair (x, y) is

(a) 6 (b) 12

(c) 8 (d) 15

45. If $\theta \in [0, 5\pi]$ and $r \in R$ such that $2\sin\theta = r^4 - 2r^2 + 3$ then the maximum number of values of the pair (r, θ) is

(a) 8 (b) 10

(c) 6 (d) none of these

46. The number of values of x for which $\sin 2x + \cos 4x = 2$ is

(a) 0 (b) 1

(c) 2 (d) infinite

47. If $2\sin x + 1 \ge 0$ and $x \in [0, 2\pi]$ then the solution set for x is

(a) $\left[0, \frac{7\pi}{6}\right]$ (b) $\left[0, \frac{7\pi}{6}\right] \cup \left[\frac{11\pi}{6}, 2\pi\right]$

(c) $\left[\frac{11\pi}{6}, 2\pi\right]$ (d) none of these

48. If $2\cos x < \sqrt{3}$ and $x \in [-\pi, \pi]$ then the solution set for x is

(a) $\left[-\pi, -\frac{\pi}{6}\right) \cup \left(\frac{\pi}{6}, \pi\right]$ (b) $\left(-\frac{\pi}{6}, \frac{\pi}{6}\right)$

(c) $\left[-\pi, -\frac{\pi}{6}\right] \cup \left[\frac{\pi}{6}, \pi\right]$ (d) none of these

49. If $\cos x - \sin x \geq 1$ and $0 \leq x \leq 2\pi$ then the solution set for x is

(a) $\left[0, \frac{\pi}{4}\right] \cup \left[\frac{7\pi}{4}, 2\pi\right]$ (b) $\left[\frac{3\pi}{2}, \frac{7\pi}{4}\right] \cup \{0\}$

(c) $\left[\frac{3\pi}{2}, 2\pi\right] \cup \{0\}$ (d) none of these

50. If $|\tan x| \leq 1$ and $x \in [-\pi, \pi]$ then the solution set for x is

(a) $\left[-\pi, -\frac{3\pi}{4}\right] \cup \left[-\frac{\pi}{4}, \frac{\pi}{4}\right] \cup \left[\frac{3\pi}{4}, \pi\right]$

(b) $\left[-\frac{\pi}{4}, \frac{\pi}{4}\right] \cup \left[\frac{3\pi}{4}, \pi\right]$

(c) $\left[-\frac{\pi}{4}, \frac{\pi}{4}\right]$

(d) none of these

51. If $4\sin^2 x - 8\sin x + 3 \leq 0$, $0 \leq x \leq 2\pi$, then the solution set for x is

(a) $\left[0, \frac{\pi}{6}\right]$ (b) $\left[0, \frac{5\pi}{6}\right]$

(c) $\left[\frac{5\pi}{6}, 2\pi\right]$ (d) $\left[\frac{\pi}{6}, \frac{5\pi}{6}\right]$

52. The set of values of x for which $\sin x \cdot \cos^3 x > \cos x \cdot \sin^3 x$, $0 \leq x \leq 2\pi$, is

(a) $(0, \pi)$ (b) $\left(0, \frac{\pi}{4}\right)$

(c) $\left(\frac{\pi}{4}, \pi\right)$ (d) none of these

53. The number of values of $x \in [0, 4\pi]$ satisfying

$$|\sqrt{3}\cos x - \sin x| \geq 2$$

is

(a) 2 (b) 0

(c) 4 (d) 8

• Type 2 •

Choose the correct options. One or more options may be correct.

54. If $\alpha \in [-2\pi, 2\pi]$ and $\cos\frac{\alpha}{2} + \sin\frac{\alpha}{2} = \sqrt{2}(\cos 36° - \sin 18°)$ then a value of α is

(a) $\frac{7\pi}{6}$ (b) $\frac{\pi}{6}$

(c) $-\frac{5\pi}{6}$ (d) $-\frac{\pi}{6}$

55. If $\cos x = \sqrt{1 - \sin 2x}$, $0 \le x \le \pi$, then a value of x is

(a) π (b) 0

(c) $\tan^{-1}2$ (d) none of these

56. If $\sin 3\theta = \cos 2\theta$ then θ is equal to

(a) $(4n+1)\frac{\pi}{2}, (4n+1)\frac{\pi}{10}$; when n is an even integer only

(b) $(4n+1)\frac{\pi}{2}, (4n+1)\frac{\pi}{10}$; when n is an any integer

(c) $(4n+1)\frac{\pi}{2}, (4n+1)\frac{\pi}{10}$; when n is an odd integer only

(d) none of these

57. $\sin\theta + \sqrt{3}\cos\theta = 6x - x^2 - 11$, $0 \le \theta \le 4\pi$, $x \in R$, holds for

(a) no value of x and θ

(b) one value of x and two values of θ

(c) two values of x and two values of θ

(d) two pairs of values of (x, θ)

58. If $0 \le x \le 2\pi$ and $|\cos x| \le \sin x$ then

(a) the set of value of x is $\left[\frac{\pi}{4}, \frac{\pi}{2}\right]$

(b) the number of solutions that are integral multiples of $\frac{\pi}{4}$ is three

(c) the sum of the largest and the smallest solution is $\frac{3\pi}{4}$

(d) $x \in \left[\frac{\pi}{4}, \frac{\pi}{2}\right] \cup \left(\frac{\pi}{2}, \frac{3\pi}{4}\right]$

59. $3^{\sin 2x + 2\cos^2 x} + 3^{1 - \sin 2x + 2\sin^2 x} = 28$ is satisfied by

(a) those values of x for which $\tan x = -1$

(b) those values of x for which $\tan x = -\frac{1}{2}$

(c) those values of x for which $\cos x = 0$

(d) those values of x for which $\tan x = 1$

60. Let $[x]$ = the greatest integer less than or equal to x. The equation $\sin x = [1 + \sin x] + [1 - \cos x]$ has

(a) no solution in $\left[-\frac{\pi}{2}, \frac{\pi}{2}\right]$

(b) no solution in $\left[\frac{\pi}{2}, \pi\right]$

(c) no solution in $\left[\pi, \frac{3\pi}{2}\right]$

(d) no solution for $x \in R$

61. If $\sin\theta = a$ for exactly one value of $\theta \in \left[0, \frac{7\pi}{3}\right]$ then the value of a is

(a) $\frac{\sqrt{3}}{2}$

(b) 1

(c) 0

(d) -1

Answers

1. d	**2.** d	**3.** d	**4.** a	**5.** d
6. c	**7.** b	**8.** d	**9.** c	**10.** c
11. a	**12.** c	**13.** d	**14.** d	**15.** b
16. a	**17.** b	**18.** a	**19.** b	**20.** c
21. c	**22.** a	**23.** d	**24.** b	**25.** a
26. b	**27.** b	**28.** a	**29.** d	**30.** a
31. d	**32.** c	**33.** b	**34.** a	**35.** a
36. c	**37.** a	**38.** b	**39.** a	**40.** c
41. c	**42.** a	**43.** c	**44.** a	**45.** c
46. a	**47.** b	**48.** a	**49.** c	**50.** a
51. d	**52.** b	**53.** c	**54.** a, d	**55.** b, c
56. b	**57.** b, d	**58.** b, d	**59.** a, c	**60.** a, b, c, d
61. b, d				

Hints and Solutions to Selected Questions

1. $\sin 8\theta + \sin 2\theta = \sin 16\theta + \sin 2\theta$ or $\sin 16\theta = \sin 8\theta$

$\therefore \quad 16\theta = n\pi + (-1)^n 8\theta \quad \Rightarrow 8\theta = 2m\pi$, when n is even

$24\theta = (2m+1)\pi$, when n is odd

$\therefore \quad \theta = \frac{m\pi}{4}, \frac{(2m+1)\pi}{24}$, when $m \in Z$

$= 0, \frac{\pi}{4}, \frac{\pi}{2}$ and $\frac{\pi}{24}, \frac{\pi}{8}, \frac{5\pi}{24}, \frac{7\pi}{24}, \frac{3\pi}{8}, \frac{11\pi}{24}$.

3. $1 + \sin\theta = \sqrt{3}\cos\theta$ or $\sqrt{3}\cos\theta - \sin\theta = 1$

or $\cos\left(\theta + \frac{\pi}{6}\right) = \frac{1}{2}$ or $\theta + \frac{\pi}{6} = 2n\pi \pm \frac{\pi}{3}$ or $\theta = 2n\pi + \frac{\pi}{6}, 2n\pi - \frac{\pi}{2}$.

$\therefore$ when $\theta = 2n\pi + \frac{\pi}{6}$ there are solutions for $n = 0, 1$ and

when $\theta = 2n\pi - \frac{\pi}{2}$, for $n = 1$. But this value does not satisfy the given equation.

4. $13\cos\theta + 12\sin\theta = 5$

or $\frac{13}{\sqrt{13^2+12^2}}\cos\theta + \frac{12}{\sqrt{13^2+12^2}}\sin\theta = \frac{5}{\sqrt{13^2+12^2}}$

or $\cos(\theta - \alpha) = \frac{5}{\sqrt{313}}$, where $\cos\alpha = \frac{13}{\sqrt{313}}$

$\therefore \quad \theta = 2n\pi \pm \cos^{-1}\frac{5}{\sqrt{313}} + \alpha = 2n\pi \pm \cos^{-1}\frac{5}{\sqrt{313}} + \cos^{-1}\frac{13}{\sqrt{313}}$.

As $\cos^{-1}\frac{5}{\sqrt{313}} > \cos^{-1}\frac{13}{\sqrt{313}}$, we get $\theta \in [0, 2\pi]$, when $n = 0$ (one value, taking positive sign) and when $n = 1$ (one value, taking negative sign).

5. The condition for solutions is $-\sqrt{p^2+q^2} \le r \le \sqrt{p^2+q^2}$.

6. The condition is $\left|\frac{k+1}{\sqrt{k^2+9}}\right| \le 1$, i.e., $(k+1)^2 \le k^2 + 9$. $\quad \therefore \; k \le 4$.

7. $\left|\frac{5}{\sqrt{1^2+(\sqrt{3})^2}}\right| \le 1$ is not true. So, there is no solution.

8. $a_1 + a_2(2\cos^2 x - 1) + a_3(1 - \cos^2 x) = 1$

or $(2a_2 - a_3)\cos^2 x + (a_1 - a_2 + a_3 - 1) = 0.$

This can hold for all x if $2a_2 - a_3 = 0$ and $a_1 - a_2 + a_3 - 1 = 0$.

As there are two equations in three unknowns, the number of solutions is infinite.

9. Here, $2\sin^2 x + \sin x - 1 = 0 \Rightarrow \sin x = \frac{1}{2}, -1$. But $\sin x = -1 \Rightarrow x = \frac{3\pi}{2}$ which does not satisfy the equation.

10. Here, $6\cos^2 x - 10\cos x + 4 = 0$ or $(3\cos x - 2)(\cos x - 1) = 0$

$\therefore \quad \cos x = \frac{2}{3}, 1 \quad \Rightarrow x = 2n\pi \pm \cos^{-1}\frac{2}{3}, 2n\pi.$

From $x - 2n\pi$ we get one solution each for $n = 0, 1, 2$.

From $x = 2n\pi \pm \cos^{-1}\frac{2}{3}$ we get the following solutions:

when $n = 0$, one solution; when $n = 1$, two solutions; when $n = 2$, two solutions.

11. Here, $p\sin x = \frac{\pi}{2} \pm p\cos x$ or $\sin x \mp \cos x = \frac{\pi}{2p}$

or $\sin\left(x \mp \frac{\pi}{4}\right) = \frac{\pi}{2\sqrt{2}p} \quad \Rightarrow \left|\frac{\pi}{2\sqrt{2}p}\right| \le 1$

$\therefore$ for positive p, $p \ge \frac{\pi}{2\sqrt{2}}$. But $1 < \frac{\pi}{2\sqrt{2}} < 2$.

12. Here, $\sin x = 1 \pm \sqrt{2}$; but $|\sin x| \le 1$. So, $\sin x = 1 - \sqrt{2}$, which gives two values of θ in each of $[0, 2\pi]$, $(2\pi, 4\pi]$, $(4\pi, 6\pi]$, etc. Hence, the least value of $n = 4$.

13. Here, $(2\sec x + 1)(\sec x - 3) = 0$; but $|\sec x| \ge 1$. So, $\sec x = 3$, which gives two values of θ in each of $[0, 2\pi]$, $(2\pi, 4\pi]$, $(4\pi, 6\pi]$, and one value in $(6\pi, 6\pi + 3\pi/2]$.

14. Here, $(2\cos x + 3)(\cos x - 1) = 0$; but $|\cos x| \le 1$.

So, $\cos x = 1 \quad \Rightarrow x = 2n\pi \quad \therefore x = 0, 2\pi.$

None of these values satisfies the given equation.

16. $\cos^2\theta = \frac{1}{6}\sin\theta \cdot \tan\theta$ or $6\cos^3\theta + \cos^2\theta - 1 = 0.$

As $\cos\theta = \frac{1}{2}$ satisfies the equation (by trial),

$$(2\cos\theta - 1)(3\cos^2\theta + 2\cos\theta + 1) = 0$$

$\Rightarrow \quad \cos\theta = \frac{1}{2}$ (other values of $\cos\theta$ are imaginary).

So, $\quad \theta = 2n\pi \pm \frac{\pi}{3}, n \in Z$.

19. $1 - \cos x = (\sqrt{2} - 1)\sin x \quad \Rightarrow \sin\frac{x}{2}\left\{\sin\frac{x}{2} - (\sqrt{2}-1)\cos\frac{x}{2}\right\} = 0$

$\therefore \quad \sin\frac{x}{2} = 0 \quad$ or $\quad \tan\frac{x}{2} = \sqrt{2} - 1 = \tan\frac{45°}{2}$

$\Rightarrow \quad \frac{x}{2} = n\pi \quad$ or $\quad \frac{x}{2} = n\pi + \frac{\pi}{8} \quad \therefore \; x = 2n\pi, 2n\pi + \frac{\pi}{4}$.

20. $\tan^2 x = \frac{\sqrt{2}-1}{\sqrt{2}+1} = (\sqrt{2}-1)^2 = \tan^2\frac{\pi}{8} \quad \Rightarrow \tan x = \tan\left(\pm\frac{\pi}{8}\right)$.

21. $a^2 - 4a + 6 = (a-2)^2 + 2 > 1$ for all a. So, $\min\limits_{a \in R}\{1, a^2 - 4a + 6\} = 1$.

$\therefore \quad$ the equation is $\sin x + \cos x = 1$.

22. $5\sin\theta + 3(\sin\theta \cdot \cos\alpha - \cos\theta \cdot \sin\alpha) = (5 + 3\cos\alpha)\sin\theta - 3\sin\alpha \cdot \cos\theta$.

$\therefore \quad \max\limits_{\theta \in R}\{5\sin\theta + 3\sin(\theta - \alpha)\} = \sqrt{(5 + 3\cos\alpha)^2 + 9\sin^2\alpha}$

$$= \sqrt{34 + 30\cos\alpha}.$$

$\therefore \quad$ the given equation is $34 + 30\cos\alpha = 49$.

23. $\cos^2 x = 2\cos x \cdot (1 - 3\cos^2 x) \quad \Rightarrow \cos x\{6\cos^2 x + \cos x - 2\} = 0$

$\therefore \; \cos x = 0, \frac{1}{2}, -\frac{2}{3}$. The two smallest positive values of x are $\frac{\pi}{3}$ and $\frac{\pi}{2}$.

24. Here, $\cos x\left(\frac{1}{4}\cos^2 x - \frac{3}{4}\sin^2 x\right) = \frac{1}{4} \quad$ or $\quad \frac{\cos x}{4}(4\cos^2 x - 3) = \frac{1}{4}$

or $\quad \cos 3x = 1 \quad \Rightarrow 3x = 2n\pi \quad \Rightarrow x = \frac{2n\pi}{3}$, where

$n = 0, 1, 2, 3, 4, 5, 6, 7, 8, 9$.

$\therefore \quad$ the required sum $= \frac{2\pi}{3}\sum\limits_{n=0}^{9} n$.

25. $\dfrac{\sin x + i\cos x}{1+i} = \dfrac{(1-i)(\sin x + i\cos x)}{(1+i)(1-i)}$

$= \dfrac{\sin x + \cos x + i(\cos x - \sin x)}{2}$, which is purely imaginary

$\Rightarrow \quad \sin x + \cos x = 0 \quad \Rightarrow \tan x = -1$

$\therefore \quad x = n\pi - \dfrac{\pi}{4}.$

26. $\sin^2 3x + \sin^4 x = 0 \quad$ or $\quad \sin^2 x\{(3 - 4\sin^2 x)^2 + \sin^2 x\} = 0.$

$\therefore \quad \sin x = 0 \quad \Rightarrow x = n\pi.$

27. If $\cos x \geq 0$, i.e., $x \in \left[0, \dfrac{\pi}{2}\right] \cup \left[\dfrac{3\pi}{2}, \dfrac{5\pi}{2}\right] \cup \left[\dfrac{7\pi}{2}, 4\pi\right]$ then $\cos x = \sin x$.

$\therefore \quad \tan x = 1 \quad$ or $\quad x = n\pi + \dfrac{\pi}{4} = \dfrac{\pi}{4}, \dfrac{5\pi}{4}, \dfrac{9\pi}{4}, \dfrac{13\pi}{4}.$

$\therefore$ if $\cos x \geq 0$, the possible values of x are $\dfrac{\pi}{4}, \dfrac{9\pi}{4}.$

If $\cos x < 0$, i.e., $x \in \left(\dfrac{\pi}{2}, \dfrac{3\pi}{2}\right) \cup \left(\dfrac{5\pi}{2}, \dfrac{7\pi}{2}\right)$ then $-\cos x = \sin x$.

$\therefore \quad \tan x = -1 \quad$ or $\quad x = n\pi - \dfrac{\pi}{4} = \dfrac{3\pi}{4}, \dfrac{7\pi}{4}, \dfrac{11\pi}{4}, \dfrac{15\pi}{4}.$

$\therefore$ if $\cos x < 0$ the possible values of x are $\dfrac{3\pi}{4}, \dfrac{11\pi}{4}.$

28. Clearly, $1 + \sin x \geq 0$. So, the equation is $\cos x - \sin x = 1$.

29. $f\left(\dfrac{\pi}{10}\right) = \sin 18° + \cos 18° = \sqrt{2}\sin(45° + 18°) = \sqrt{2}\sin 63°.$

As $\sin 63° > \sin 45° = \dfrac{1}{\sqrt{2}}$ and $\sin 63° < 1$, we get $1 < f\left(\dfrac{\pi}{10}\right) < \sqrt{2}.$

$\therefore \quad \left[f\left(\dfrac{\pi}{10}\right)\right] = 1$. So, the equation is $\sin x + \cos x = 1$.

$\therefore \quad \cos\left(x - \dfrac{\pi}{4}\right) = \dfrac{1}{\sqrt{2}} \quad$ or $\quad x - \dfrac{\pi}{4} = 2n\pi \pm \dfrac{\pi}{4}.$

30. $2^{\frac{1}{1-|\cos x|}} = 2^2 \quad \Rightarrow \dfrac{1}{1 - |\cos x|} = 2 \quad \Rightarrow |\cos x| = \dfrac{1}{2}$

$\Rightarrow \quad \cos x = \pm\dfrac{1}{2}.$

$$\therefore \quad x = 2n\pi \pm \frac{\pi}{3},\ 2n\pi \pm \left(\pi - \frac{\pi}{3}\right) = 2n\pi \pm \frac{\pi}{3},\ (2n \pm 1)\pi \mp \frac{\pi}{3} = n\pi \pm \frac{\pi}{3}.$$

31. As $x \neq \frac{n\pi}{2}$, we get $\cos x \neq 0, 1, -1$. So, $\sin^2 x - 3\sin x + 2 = 0$.

$\therefore$ $(\sin x - 2)(\sin x - 1) = 0 \Rightarrow \sin x = 1$. But this does not satisfy the equation because $0^0 = 1$ is not true.

32. The maximum values of $\sin x + \cos x$ and $1 + \sin 2x$ are $\sqrt{2}$ and 2 respectively. Also, $(\sqrt{2})^2 = 2$.

$\therefore$ the equation can hold only when $\sin x + \cos x = \sqrt{2}$ and $1 + \sin 2x = 2$.

Now, $\sin x + \cos x = \sqrt{2} \quad \Rightarrow \cos\left(x - \frac{\pi}{4}\right) = 1 \quad \Rightarrow x = 2n\pi + \frac{\pi}{4}.$

$1 + \sin 2x = 2 \quad \Rightarrow \sin 2x = 1 \quad \Rightarrow x = \frac{n\pi}{2} + (-1)^n \cdot \frac{\pi}{4}.$

The value in $[-\pi, \pi]$ satisfying both the equations is $\frac{\pi}{4}$ (when $n = 0$).

33. $\frac{2^{\sin x} + 2^{\cos x}}{2} \geq \sqrt{2^{\sin x} \cdot 2^{\cos x}}$

(using AM $\geq$ GM where the equality holds when $2^{\sin x} = 2^{\cos x}$)

$\therefore \quad 2^{\sin x} + 2^{\cos x} \geq 2 \cdot \sqrt{2^{\sin x + \cos x}}.$

But the minimum value of $\sin x + \cos x$ is $-\sqrt{2}$.

$\therefore \quad 2^{\sin x} + 2^{\cos x} \geq 2 \cdot \sqrt{2^{-\sqrt{2}}} = 2^{1 - 1/\sqrt{2}}.$

The equation holds when the equality holds, i.e., $2^{\sin x} = 2^{\cos x}$.

$\therefore \quad \sin x = \cos x$ or $\tan x = 1$

34. $16^{\sin^2 x} + \frac{16}{16^{\sin^2 x}} = 10$ or $y^2 - 10y + 16 = 0$, where $y = 16^{\sin^2 x}$.

$\therefore \quad y = 2, 8 \quad \Rightarrow 16^{\sin^2 x} = 2^1, 2^3 \quad \Rightarrow 4\sin^2 x = 1, 3$

$\therefore \quad \sin x = \frac{1}{2}, -\frac{1}{2}, \frac{\sqrt{3}}{2}, -\frac{\sqrt{3}}{2}.$

Each value will give two values of x in $[0, 2\pi]$.

35. Here, $3\sin^2\theta = \cos 2\phi$ and $3\sin\theta \cdot \cos\theta = \sin 2\phi$. Squaring and adding, $9\sin^2\theta(\sin^2\theta + \cos^2\theta) = 1$, i.e., $\sin\theta = \frac{1}{3}$ and $\cos\theta = \frac{2\sqrt{2}}{3}$.

$\therefore \quad \cos 2\phi = 3 \cdot \frac{1}{9} = \frac{1}{3}$ and $\sin 2\phi = \frac{2\sqrt{2}}{3}$.

$\therefore \quad \cos(\theta + 2\phi) = \cos\theta \cdot \cos 2\phi - \sin\theta \cdot \sin 2\phi$

$$= \frac{2\sqrt{2}}{3} \cdot \frac{1}{3} - \frac{1}{3} \cdot \frac{2\sqrt{2}}{3} = 0 \text{ and } \theta + 2\phi < \frac{3\pi}{2}.$$

36. Here, $\sin\theta = -\frac{1}{2}$ and $\tan\theta = \frac{1}{\sqrt{3}}$.

$\sin\theta = -\frac{1}{2} \Rightarrow \theta = m\pi + (-1)^m \cdot \left(-\frac{\pi}{6}\right)$ and $\tan\theta = \frac{1}{\sqrt{3}} \Rightarrow \theta = m\pi + \frac{\pi}{6}$.

For common values, m must be odd, i.e., $m = 2n + 1$.

37. Eliminating θ, $r = 4\left(1 + \frac{3}{r}\right) \Rightarrow (r-6)(r+2) = 0 \Rightarrow r = 6$.

$\therefore \quad \sin\theta = \frac{1}{2} \Rightarrow \theta = \frac{\pi}{6}, \frac{5\pi}{6}$. So, $(r, \theta) = \left(6, \frac{\pi}{6}\right), \left(6, \frac{5\pi}{6}\right)$.

38. Here, $x^3 + (x+2)^2 + 2\sin x = 4$. Clearly, $x = 0$ satisfies the equation.

If $0 < x \le \pi$, $x^3 + (x+2)^2 + 2\sin x > 4$.

If $\pi < x \le 2\pi$, $x^3 + (x+2)^2 + 2\sin x > 27 + 25 - 2$.

So, $x = 0$ is the only solution.

39. Here, $\frac{1}{2}\sin(2e^x) = \frac{1}{4}(2^x + 2^{-x}) \ge \frac{1}{4} \cdot 2 = \frac{1}{2}$; so, $\sin(2e^x) \ge 1$.

But $\sin(2e^x) \not> 1$. Therefore, $\sin(2e^x) = 1$.

But equality can hold when $2^x = 2^{-x} = 1$, i.e., $x = 0$. Then $\sin(2 \cdot e^0) = 1$, which is not true.

40. $\sin\pi(x^2 + x) = \sin\pi x^2 \quad \Rightarrow \pi(x^2 + x) = n\pi + (-1)^n \pi x^2$

$\therefore \quad x^2 + x = 2m + x^2 \quad \Rightarrow x = 2m \in Z$

or $\quad x^2 + x = p' - x^2$, where p' is an odd integer $\Rightarrow 2x^2 + x - p' = 0$.

For nonintegral solution, $x = \frac{-1 \pm \sqrt{1+8p'}}{4} = \frac{-1 + \sqrt{1+8p'}}{4}$

(taking the positive value)

$$= \frac{\sqrt{p} - 1}{4}, \text{ where } p \text{ is an odd integer.}$$

41. As $\quad \max \cos\theta = 1$, $2\cos x + \cos 2\lambda x = 3$ is possible only when

$\cos x = 1$ and $\cos 2\lambda x = 1$, i.e., $\cos x = 1$ and $\sin\lambda x = 0$.

Clearly, if λ is rational, say $\frac{p}{q}$, then $x = 2q\pi, q \in Z$, satisfies both the equations. Therefore, for exactly one solution, $x = 0$, λ should be irrational.

42. max $\sin\theta = 1$. So, the equation can have a solution only when

$\sin x = 1, \sin y = 1 \;\Rightarrow\; x = \frac{\pi}{2}, y = \frac{\pi}{2}$

$\therefore\; x + y = \pi.$

43. max $\cos\theta = 1$. So, the equation can have solution only when

$\cos x = 1, \cos y = 1$

$\Rightarrow\; x = 0, y = 0 \;\Rightarrow\; \cos(x - y) = \cos 0° = 1.$

44. max $\cos\theta = 1$. So, $\cos x \cdot \sin y = 1$

$\Rightarrow\; \cos x = 1, \sin y = 1$ or $\cos x = -1, \sin y = -1.$

$\cos x = 1, \sin y = 1 \;\Rightarrow\; x = 0, 2\pi$ and $y = \frac{\pi}{2}, \frac{5\pi}{2}$ (from the question).

$\cos x = -1, \sin y = -1 \;\Rightarrow\; x = \pi, 3\pi$ and $y = \frac{3\pi}{2}$ (from the question).

$\therefore$ the required number of ordered pair $= 2 \times 2 + 2 \times 1 = 6.$

45. $2\sin\theta = (r^2 - 1)^2 + 2 \geq 2$. But max $\sin\theta = 1$.

$\therefore$ the equation implies $\sin\theta = 1, r^2 = 1.$

$\sin\theta = 1 \;\Rightarrow\; \theta = \frac{\pi}{2}, \frac{5\pi}{2}, \frac{9\pi}{2}$ (from the question).

$r^2 = 1 \;\Rightarrow\; r = 1, -1.$

$\therefore$ the number of values of the pair $(r, \theta) = 2 \times 3 = 6.$

46. $\sin 2x + \cos 4x = 2 \;\Rightarrow\; \sin 2x = 1, \cos 4x = 1$

$\therefore\; 1 - 2\sin^2 2x = 1$ or $1 - 2 = 1$ (absurd).

47. $2\sin x + 1 \geq 0 \Rightarrow \sin x \geq -\frac{1}{2}\cdot$ The value scheme for this is shown below.

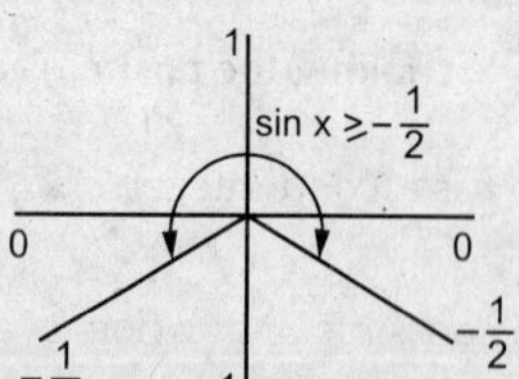

From the figure,

$\frac{11\pi}{6} \leq x \leq 2\pi$ or $0 \leq x \leq \frac{7\pi}{6}\cdot$

$\therefore\; x \in \left[0, \frac{7\pi}{6}\right] \cup \left[\frac{11\pi}{6}, 2\pi\right].$

48. Here, $\cos x < \frac{\sqrt{3}}{2}\cdot$ The value scheme for this is shown below.

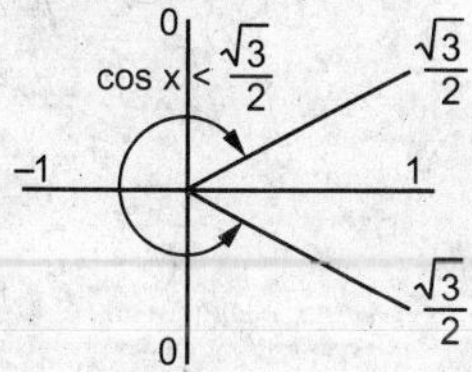

From the figure, $-\pi \le x < -\frac{\pi}{6}$ or $\frac{\pi}{6} < x \le \pi$.

$\therefore \quad x \in \left[-\pi, -\frac{\pi}{6}\right) \cup \left(\frac{\pi}{6}, \pi\right]$.

49. $\cos\left(x + \frac{\pi}{4}\right) \ge \frac{1}{\sqrt{2}}\cdot$ The value scheme for this is shown below.

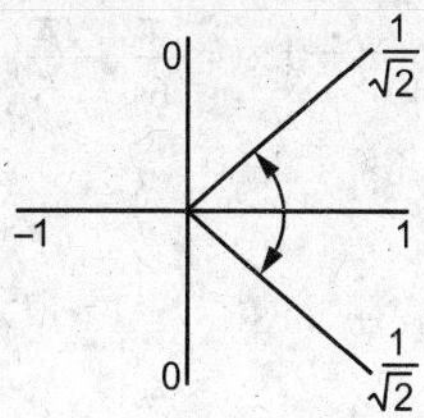

From the figure, $-\frac{\pi}{4} \le x + \frac{\pi}{4} \le \frac{\pi}{4}$ and

in general, $2n\pi - \frac{\pi}{4} \le x + \frac{\pi}{4} \le 2n\pi + \frac{\pi}{4}$

$\therefore \quad 2n\pi - \frac{\pi}{2} \le x \le 2n\pi.$

For $n = 0, -\frac{\pi}{2} \le x \le 0$; for $n = 1, \frac{3\pi}{2} \le x \le 2\pi$.

50. $-1 \le \tan\theta \le 1$. The value scheme for this is shown below.

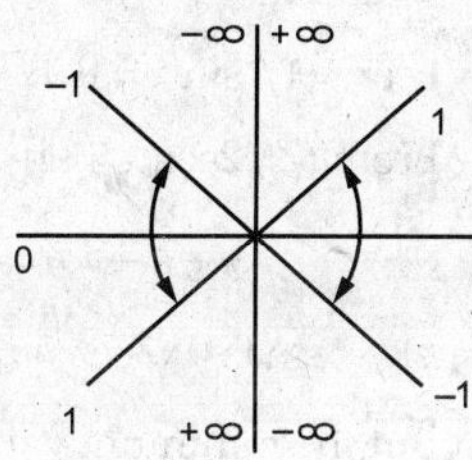

From the figure, $-\frac{\pi}{4} \le x \le \frac{\pi}{4}$

or $\quad -\pi \le x \le -\frac{3\pi}{4} \quad$ or $\quad \frac{3\pi}{4} \le x \le \pi$.

$\therefore \quad x \in \left[-\pi, -\frac{3\pi}{4}\right] \cup \left[-\frac{\pi}{4}, \frac{\pi}{4}\right] \cup \left[\frac{3\pi}{4}, \pi\right]$.

51. Here, $(2\sin x - 1)(2\sin x - 3) \le 0$. But $2\sin x - 3$ is always negative.

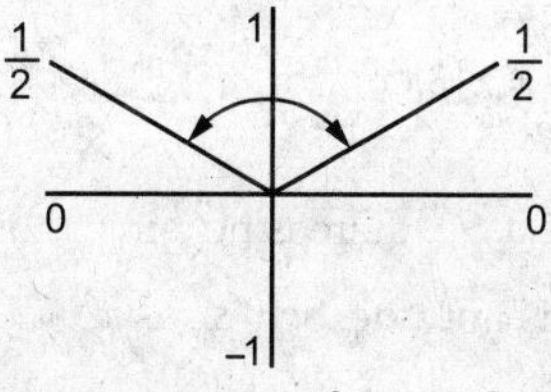

$\therefore \quad 2\sin x - 1 \ge 0$, i.e., $\sin x \ge \frac{1}{2}\cdot$

$\therefore$ from the figure, $\frac{\pi}{6} \le x \le \frac{5\pi}{6}\cdot$

52. $\sin x \cdot \cos x\,(\cos^2 x - \sin^2 x) > 0 \quad$ or $\quad \frac{1}{2}\sin 2x \cdot \cos 2x > 0$

or $\quad \sin 4x > 0$

$\therefore \quad 0 < 4x < \pi.$

53. We have $|\sqrt{3}\cos x - \sin x| \le \sqrt{3+1} = 2$. Thus, we must have

$$|\sqrt{3}\cos x - \sin x| = 2$$

$\therefore \quad \left|\cos\left(x+\frac{\pi}{6}\right)\right| = 1 \quad \therefore \cos\left(x+\frac{\pi}{6}\right) = 1, -1$

$\therefore \quad x+\frac{\pi}{6} = 0, 2\pi, 4\pi, 6\pi, \ldots, \pi, 3\pi, 5\pi, \ldots$

$\therefore \quad x = \frac{11\pi}{6}, \frac{23\pi}{6}, \frac{5\pi}{6}, \frac{17\pi}{6}.$

54. Here, $\cos\left(\frac{\alpha}{2}-\frac{\pi}{4}\right) = \frac{\sqrt{5}+1}{4} - \frac{\sqrt{5}-1}{4} = \frac{1}{2} \quad \Rightarrow \frac{\alpha}{2}-\frac{\pi}{4} = 2n\pi \pm \frac{\pi}{3}$

$\therefore \quad \alpha = 4n\pi + \frac{\pi}{2} \pm \frac{2\pi}{3} = 4n\pi + \frac{7\pi}{6}, 4n\pi - \frac{\pi}{6}.$ For $n = 0$, $\alpha = \frac{7\pi}{6}, -\frac{\pi}{6}.$

55. Here, $\cos x = |\cos x - \sin x|$.

In $\left[0, \frac{\pi}{4}\right]$, $\cos x \geq \sin x$ and in $\left(\frac{\pi}{4}, \pi\right]$, $\cos x < \sin x$.

$\therefore \quad \cos x = \cos x - \sin x, 0 \leq x \leq \frac{\pi}{4}$ ... (1)

and $\cos x = \sin x - \cos x, \frac{\pi}{4} < x \leq \pi$ ... (2)

From (1), $\sin x = 0 \quad \therefore x = 0.$

From (2), $2\cos x = \sin x \quad \therefore \tan x = 2 \quad \therefore x = \tan^{-1}2.$

57. $\sin\theta + \sqrt{3}\cos\theta = -2 - (x^2 - 6x + 9) = -2 - (x-3)^2$

$\therefore \quad \sin\theta + \sqrt{3}\cos\theta \leq -2.$

But the minimum value of $\sin\theta + \sqrt{3}\cos\theta = -\sqrt{1+3} = -2.$

$\therefore$ we have, $\sin\theta + \sqrt{3}\cos\theta = -2$ and then $x = 3$.

$\therefore \quad x = 3$ and $\cos\left(\theta - \frac{\pi}{6}\right) = -1$, i.e., $\theta - \frac{\pi}{6} = \pi, 3\pi.$

58. $|\cos x| \geq 0$. So $|\cos x| \leq \sin x \Rightarrow \sin x \geq 0$. So, there is no x in $(\pi, 2\pi)$.

Now, if $x = 2\pi$, $|\cos 2\pi| \leq \sin 2\pi$, which is not true. So $0 \leq x \leq \pi$.

If $0 \leq x \leq \frac{\pi}{2}$ then $\quad |\cos x| \leq \sin x \quad \Rightarrow \cos x \leq \sin x \quad \Rightarrow \tan x \geq 1.$

$\therefore \quad \frac{\pi}{4} \leq x \leq \frac{\pi}{2}.$

If $\frac{\pi}{2} < x \leq \pi$ then $|\cos x| \leq \sin x$

$\Rightarrow \quad -\cos x \leq \sin x \quad \Rightarrow \tan x \leq -1 \qquad (\because \cos x < 0).$

$\therefore \quad \frac{\pi}{2} < x \le \frac{3\pi}{4}$.

$\therefore \quad x \in \left[\frac{\pi}{4}, \frac{\pi}{2}\right] \cup \left[\frac{\pi}{2}, \frac{3\pi}{4}\right]$.

59. $3^{\sin 2x + \cos 2x + 1} + 3^{2 - \sin 2x - \cos 2x} = 28.$

Taking $3^{\sin 2x + \cos 2x + 1} = y$,

$$y + \frac{27}{y} = 28 \quad \text{or} \quad y^2 - 28y + 27 = 0 \quad \text{or} \quad y = 1, 27 = 3^0, 3^3$$

$\therefore \quad \sin 2x + \cos 2x + 1 = 0, 3.$

But the maximum value of $\sin 2x + \cos 2x = \sqrt{2}$.

$\therefore \quad$ we get only $\sin 2x + \cos 2x + 1 = 0$

$$\Rightarrow \quad \frac{2\tan x}{1 + \tan^2 x} + \frac{1 - \tan^2 x}{1 + \tan^2 x} + 1 = 0 \quad \Rightarrow \tan x = -1.$$

Also, $\sin 2x + \cos 2x + 1 = 0$

$\Rightarrow \quad 2\sin x \cos x + \cos^2 x = 0$

$\Rightarrow \quad \cos x = 0 \quad \text{or} \quad 2\sin x + \cos x = 0.$

$\cos x = 0$ satisfies the equation, but $\tan x = -\frac{1}{2}$ does not.

60. At $x = -\frac{\pi}{2}, \frac{3\pi}{2}$; $\quad [1 + \sin x] = 0, [1 - \cos x] = 1$

$\therefore \quad \sin x = 0 + 1 \quad \Rightarrow -1 = 1$ (absurd).

At $x = 0$, $\quad [1 + \sin x] = 1, [1 - \cos x] = 0$

$\therefore \quad \sin x = 1 + 0 \quad \Rightarrow 0 = 1$ (absurd).

At $x = \frac{\pi}{2}$, $\quad [1 + \sin x] = 2, [1 - \cos x] = 1$

$\therefore \quad \sin x = 2 + 1 = 3$ (absurd).

At $x = \pi$, $\quad [1 + \sin x] = 1, [1 - \cos x] = 2$

$\therefore \quad \sin x = 1 + 2 = 3$ (absurd).

In $\left(-\frac{\pi}{2}, 0\right)$, $\quad [1 + \sin x] = 0, [1 - \cos x] = 0$

$\therefore \quad \sin x = 0 + 0 = 0$ (absurd).

In $\left(0, \frac{\pi}{2}\right)$, $\quad [1 + \sin x] = 1, [1 - \cos x] = 0$

$\therefore \quad \sin x = 1 + 0 = 1$ (absurd).

In $\left(\frac{\pi}{2}, \pi\right)$, $[1 + \sin x] = 1, [1 - \cos x] = 1$

$\therefore \quad \sin x = 1 + 1 = 2$ (absurd).

In $\left(\pi, \frac{3\pi}{2}\right)$, $[1 + \sin x] = 0, [1 - \cos x] = 1$

$\therefore \quad \sin x = 0 + 1 = 1$ (absurd).

61. Clearly $-1 \le a \le 1$. For any value of a other than 1, –1 we know $\sin\theta$ has two values (in quadrants, I, II or III, IV).

□

3

Inverse Circular Functions

• *Type 1* •

Choose the most appropriate option (a, b, c or d).

1. If α satisfies the inequation $x^2 - x - 2 > 0$ then a value exists for

(a) $\sin^{-1}\alpha$ (b) $\sec^{-1}\alpha$

(c) $\cos^{-1}\alpha$ (d) none of these

2. If α, β are roots of the equation $6x^2 + 11x + 3 = 0$ then

(a) both $\cos^{-1}\alpha$ and $\cos^{-1}\beta$ are real

(b) both $\operatorname{cosec}^{-1}\alpha$ and $\operatorname{cosec}^{-1}\beta$ are real

(c) both $\cot^{-1}\alpha$ and $\cot^{-1}\beta$ are real

(d) none of these

3. Let $f(x) = \sec^{-1}x + \tan^{-1}x$. Then $f(x)$ is real for

(a) $x \in [-1, 1]$ (b) $x \in R$

(c) $x \in (-\infty, -1] \cup [1, +\infty)$ (d) none of these

4. If $\cos^{-1}x - \sin^{-1}x = 0$ then x is equal to

(a) $\pm\frac{1}{\sqrt{2}}$ (b) 1

(c) $\sqrt{2}$ (d) $\frac{1}{\sqrt{2}}$

5. $\operatorname{cosec}^{-1}(\cos x)$ is real if

(a) $x \in [-1, 1]$ (b) $x \in R$

(c) x is an odd multiple of $\frac{\pi}{2}$ (d) x is a multiple of π

6. The principal value of $\sin^{-1}\left\{\sin\frac{5\pi}{6}\right\}$ is

(a) $\frac{\pi}{6}$ (b) $\frac{5\pi}{6}$

(c) $\frac{7\pi}{6}$ (d) none of these

7. The principal value of $\cos^{-1}\left(-\sin\frac{7\pi}{6}\right)$ is

(a) $\frac{5\pi}{3}$ (b) $\frac{7\pi}{6}$

(c) $\frac{\pi}{3}$ (d) none of these

8. The principal value of $\sin^{-1}\left\{\cos\left(\sin^{-1}\frac{\sqrt{3}}{2}\right)\right\}$ is

(a) $\frac{\pi}{6}$ (b) $\frac{\pi}{3}$

(c) $-\frac{\pi}{3}$ (d) none of these

9. The principal value of $\cos^{-1}\left\{\frac{1}{\sqrt{2}}\left(\cos\frac{9\pi}{10} - \sin\frac{9\pi}{10}\right)\right\}$ is

(a) $\frac{3\pi}{20}$ (b) $\frac{7\pi}{20}$

(c) $\frac{7\pi}{10}$ (d) none of these

10. The value of $\cos\left\{\tan^{-1}\left(\tan\frac{15\pi}{4}\right)\right\}$ is

(a) $\frac{1}{\sqrt{2}}$ (b) $-\frac{1}{\sqrt{2}}$

(c) 1 (d) none of these

11. $-\frac{2\pi}{5}$ is the principal value of

(a) $\cos^{-1}\left(\cos\frac{7\pi}{5}\right)$
(b) $\sin^{-1}\left(\sin\frac{7\pi}{5}\right)$
(c) $\sec^{-1}\left(\sec\frac{7\pi}{5}\right)$
(d) none of these

12. If $\cos^{-1}\lambda + \cos^{-1}\mu + \cos^{-1}\nu = 3\pi$ then $\lambda\mu + \mu\nu + \nu\lambda$ is equal to

(a) -3
(b) 0
(c) 3
(d) -1

13. If $\sum_{i=1}^{10} \cos^{-1}x_i = 0$ then $\sum_{i=1}^{10} x_i$ is

(a) 0
(b) 10
(c) 5
(d) none of these

14. If $\sum_{i=1}^{2n} \sin^{-1}x_i = n\pi$ then $\sum_{i=1}^{2n} x_i$ is equal to

(a) n
(b) $2n$
(c) $\frac{n(n+1)}{2}$
(d) none of these

15. The value of $\cos^{-1}\left(-\frac{1}{2}\right) + \sin^{-1}\left(-\frac{\sqrt{3}}{2}\right)$ is

(a) $\frac{\pi}{3}$
(b) 0
(c) $\frac{2\pi}{3}$
(d) none of these

16. The value of $\tan\left\{2\tan^{-1}\frac{1}{5} - \frac{\pi}{4}\right\}$ is

(a) 0
(b) 1
(c) $\frac{7}{17}$
(d) none of these

17. The formula $\cos^{-1}\frac{1-x^2}{1+x^2} = 2\tan^{-1}x$ holds only for

(a) $x \in R$
(b) $|x| \leq 1$
(c) $x \in (-1, 1]$
(d) $x \in [1, +\infty)$

18. $\tan^{-1} a + \tan^{-1} b$, where $a > 0, b > 0, ab > 1$, is equal to

(a) $\tan^{-1} \frac{a+b}{1-ab}$ (b) $\tan^{-1} \frac{a+b}{1-ab} - \pi$

(c) $\pi + \tan^{-1} \frac{a+b}{1-ab}$ (d) none of these

19. The set of values of x for which $\tan^{-1} \frac{x}{\sqrt{1-x^2}} = \sin^{-1} x$ holds is

(a) R (b) $[-1, 1]$

(c) $[0, 1]$ (d) $[-1, 0]$

20. $\cos^{-1} \left\{ \frac{1}{2} x^2 + \sqrt{1-x^2} \cdot \sqrt{1 - \frac{x^2}{4}} \right\} = \cos^{-1} \frac{x}{2} - \cos^{-1} x$ holds for

(a) $|x| \le 1$ (b) $x \in R$

(c) $0 \le x \le 1$ (d) $-1 \le x \le 0$

21. If $f(x) = \sin^{-1} \left\{ \frac{\sqrt{3}}{2} x - \frac{1}{2} \sqrt{1-x^2} \right\}, -\frac{1}{2} \le x \le 1$, then $f(x)$ is equal to

(a) $\sin^{-1} \frac{1}{2} - \sin^{-1} x$ (b) $\sin^{-1} x - \frac{\pi}{6}$

(c) $\sin^{-1} x + \frac{\pi}{6}$ (d) none of these

22. The formula $2\sin^{-1} x = \sin^{-1}(2x\sqrt{1-x^2})$ holds for

(a) $x \in [0, 1]$ (b) $x \in \left[-\frac{1}{\sqrt{2}}, \frac{1}{\sqrt{2}} \right]$

(c) $x \in (-1, 0)$ (d) $x \in \left[-\frac{\sqrt{3}}{2}, \frac{\sqrt{3}}{2} \right]$

23. $\cos^{-1}(\cos x) = x$ is satisfied by

(a) $x \in R$ (b) $x \in [0, \pi]$

(c) $x \in [-1, 1]$ (d) none of these

24. If $2\tan^{-1} x + \sin^{-1} \frac{2x}{1+x^2}$ is independent of x then

(a) $x \in [1, +\infty)$ (b) $x \in [-1, 1]$

(c) $x \in (-\infty, -1]$ (d) none of these

25. If $\tan^{-1}2$, $\tan^{-1}3$ are two angles of a triangle then the third angle is

(a) $\frac{\pi}{4}$ (b) $\frac{3\pi}{4}$

(c) $\frac{\pi}{2}$ (d) none of these

26. $\sum_{n=1}^{3} \tan^{-1}\frac{1}{n}$ is equal to

(a) 0 (b) π

(c) $\frac{\pi}{2}$ (d) none of these

27. The value of $\tan\left\{\cos^{-1}\left(\frac{4}{5}\right)+\sin^{-1}\left(\frac{2}{\sqrt{13}}\right)\right\}$ is

(a) $\frac{7}{16}$ (b) $\frac{17}{6}$

(c) $\frac{6}{17}$ (d) none of these

28. If $\sin^{-1}x-\cos^{-1}x=\frac{\pi}{6}$ then x is

(a) $\frac{1}{2}$ (b) $\frac{\sqrt{3}}{2}$

(c) $-\frac{1}{2}$ (d) none of these

29. The value of $2\tan^{-1}\frac{\sqrt{1+x^2}-1}{x}$ is equal to

(a) $\cot^{-1}x$ (b) $\sec^{-1}x$

(c) $\tan^{-1}x$ (d) none of these

30. The value of $2\tan^{-1}\frac{1}{3}+\tan^{-1}\frac{1}{7}$ is

(a) $\frac{\pi}{4}$ (b) $\frac{\pi}{2}$

(c) π (d) none of these

31. The value of $\cot^{-1}3 + \text{cosec}^{-1}\sqrt{5}$ is

(a) $\frac{\pi}{3}$ (b) $\frac{\pi}{2}$

(c) $\frac{\pi}{4}$ (d) none of these

32. $\sin \cot^{-1} \tan \cos^{-1} x$ is equal to

(a) x (b) $\sqrt{1-x^2}$

(c) $\frac{1}{x}$ (d) none of these

33. The value of $\tan^{-1}\frac{1}{3} + \tan^{-1}\frac{1}{5} + \tan^{-1}\frac{1}{7} + \tan^{-1}\frac{1}{8}$ is

(a) π (b) $\frac{\pi}{4}$

(c) $\frac{3\pi}{4}$ (d) none of these

34. The value of $\tan^2(\sec^{-1}2) + \cot^2(\text{cosec}^{-1}3)$ is

(a) 13 (b) 15

(c) 11 (d) none of these

35. $\tan\left(\frac{\pi}{4} + \frac{1}{2}\cos^{-1}x\right) + \tan\left(\frac{\pi}{4} - \frac{1}{2}\cos^{-1}x\right)$, $x \neq 0$, is equal to

(a) x (b) $2x$

(c) $\frac{2}{x}$ (d) none of these

36. The number of real solutions of the equation

$$\sqrt{1+\cos 2x} = \sqrt{2}\sin^{-1}(\sin x), -\pi \leq x \leq \pi,$$

is

(a) 0 (b) 1

(c) 2 (d) infinite

37. The number of real solutions of $\tan^{-1}\sqrt{x(x+1)} + \sin^{-1}\sqrt{x^2+x+1} = \frac{\pi}{2}$ is

(a) zero (b) one

(c) two (d) infinite

38. The number of positive integral solutions of the equation

$$\tan^{-1}x + \cos^{-1}\frac{y}{\sqrt{1+y^2}} = \sin^{-1}\frac{3}{\sqrt{10}}$$

is

(a) one (b) two
(c) zero (d) none of these

39. Considering principal values, the number of solutions of

$$\tan^{-1}2x + \tan^{-1}3x = \frac{\pi}{4}$$

is

(a) two (b) three
(c) one (d) none of these

40. The number of real solutions of (x, y), where $|y| = \sin x$, $y = \cos^{-1}(\cos x)$, $-2\pi \le x \le 2\pi$, is

(a) 2 (b) 1
(c) 3 (d) 4

41. If $\cos^{-1}x > \sin^{-1}x$ then

(a) $x < 0$ (b) $-1 < x < 0$
(c) $0 \le x < \frac{1}{\sqrt{2}}$ (d) $-1 \le x < \frac{1}{\sqrt{2}}$

42. If $\cot^{-1}\frac{n}{\pi} > \frac{\pi}{6}$, $n \in N$, then the maximum value of n is

(a) 1 (b) 5
(c) 9 (d) none of these

43. The set of values of k for which $x^2 - kx + \sin^{-1}(\sin 4) > 0$ for all real x is

(a) ϕ (b) $(-2, 2)$
(c) R (d) none of these

• Type 2 •

Choose the correct options. One or more options may be correct.

44. Let $f(x) = \sin^{-1}x + \cos^{-1}x$. Then $\frac{\pi}{2}$ is equal to

(a) $f\left(-\frac{1}{2}\right)$ (b) $f(k^2-2k+3), k \in R$

(c) $f\left(\frac{1}{1+k^2}\right), k \in R$ (d) $f(-2)$

45. At $x=\frac{3}{2}$, the value is real for

(a) $\tan^{-1}x$ (b) $\text{cosec}^{-1}x$

(c) $\cos^{-1}2x$ (d) none of these

46. If $\frac{1}{2} < |x| < 1$ then which of the following are real?

(a) $\sin^{-1}x$ (b) $\tan^{-1}x$

(c) $\sec^{-1}x$ (d) $\cos^{-1}x$

47. Let $\tan^{-1}\left(\tan\frac{5\pi}{4}\right)=\alpha$, $\tan^{-1}\left(-\tan\frac{2\pi}{3}\right)=\beta$. Then

(a) $\alpha > \beta$ (b) $4\alpha - 3\beta = 0$

(c) $\alpha+\beta=\frac{7\pi}{12}$ (d) none of these

48. Let $f(x)=e^{\cos^{-1}\sin(x+\pi/3)}$. Then

(a) $f\left(\frac{8\pi}{9}\right)=e^{5\pi/18}$ (b) $f\left(\frac{8\pi}{9}\right)=e^{13\pi/18}$

(c) $f\left(-\frac{7\pi}{4}\right)=e^{\pi/12}$ (d) $f\left(-\frac{7\pi}{4}\right)=e^{11\pi/12}$

49. If $f(x)=\cos^{-1}x+\cos^{-1}\left\{\frac{x}{2}+\frac{1}{2}\sqrt{3-3x^2}\right\}$ then

(a) $f\left(\frac{2}{3}\right)=\frac{\pi}{3}$ (b) $f\left(\frac{2}{3}\right)=2\cos^{-1}\frac{2}{3}-\frac{\pi}{3}$

(c) $f\left(\frac{1}{3}\right)=\frac{\pi}{3}$ (d) $f\left(\frac{1}{3}\right)=2\cos^{-1}\frac{1}{3}-\frac{\pi}{3}$

50. The value of $\tan\left\{\frac{1}{2}\sin^{-1}\frac{2x}{1+x^2}+\frac{1}{2}\cos^{-1}\frac{1-x^2}{1+x^2}\right\}$ is

(a) $\frac{2x}{1-x^2}$ if $0 \le x \le 1$ (b) $\frac{2x}{1-x^2}$ if $x<1$

(c) not finite if $x>1$ (d) none of these

51. α, β and γ are three angles given by

$\alpha = 2\tan^{-1}(\sqrt{2}-1)$, $\beta = 3\sin^{-1}\frac{1}{\sqrt{2}} + \sin^{-1}\left(-\frac{1}{2}\right)$ and $\gamma = \cos^{-1}\frac{1}{3}$. Then

(a) $\alpha > \beta$ (b) $\beta > \gamma$

(c) $\alpha < \gamma$ (d) none of these

52. If $0 < x < 1$ then $\tan^{-1}\frac{\sqrt{1-x^2}}{1+x}$ is equal to

(a) $\frac{1}{2}\cos^{-1}x$ (b) $\cos^{-1}\sqrt{\frac{1+x}{2}}$

(c) $\sin^{-1}\sqrt{\frac{1-x}{2}}$ (d) none of these

53. One of the values of x satisfying $\tan(\sec^{-1}x) = \sin\cos^{-1}\frac{1}{\sqrt{5}}$ is

(a) $\frac{\sqrt{5}}{3}$ (b) $\frac{3}{\sqrt{5}}$

(c) $-\frac{\sqrt{5}}{3}$ (d) $-\frac{3}{\sqrt{5}}$

54. If $\sin^{-1}x + \sin^{-1}y = \frac{2\pi}{3}$, $\cos^{-1}x - \cos^{-1}y = \frac{\pi}{3}$ then the number of values of (x, y) is

(a) two (b) four

(c) zero (d) none of these

55. The solution set of the equation $\cos^{-1}x - \sin^{-1}x = \sin^{-1}(1-x)$ is

(a) $[-1, 1]$ (b) $\left[0, \frac{1}{2}\right]$

(c) $[-1, 0]$ (d) none of these

Answers

1. b	**2.** c	**3.** c	**4.** d	**5.** d
6. a	**7.** c	**8.** a	**9.** d	**10.** a
11. b	**12.** c	**13.** b	**14.** b	**15.** a
16. d	**17.** d	**18.** c	**19.** b	**20.** c
21. b	**22.** b	**23.** b	**24.** a	**25.** a
26. c	**27.** b	**28.** b	**29.** c	**30.** a
31. c	**32.** a	**33.** b	**34.** c	**35.** c
36. c	**37.** c	**38.** b	**39.** a	**40.** c
41. d	**42.** b	**43.** a	**44.** a, c	**45.** a, b
46. a, b, d	**47.** b, c	**48.** b, c	**49.** a, d	**50.** a, c
51. b, c	**52.** a, b, c	**53.** b, d	**54.** d	**55.** d

Hints and Solutions to Selected Questions

1. $x^2 - x - 2 > 0 \quad \Rightarrow (x-2)(x+1) > 0$

$\Rightarrow \quad x < -1$ or $x > 2$, using sign scheme.

We know that $\sin^{-1} x$ is defined for $|x| \le 1$, $\cos^{-1} x$ is defined for $|x| \le 1$ and $\sec^{-1} x$ is defined for $|x| \ge 1$, i.e., $x \le -1$ or $x \ge 1$.

2. $6x^2 + 11x + 3 = 0 \quad \Rightarrow x = -\frac{1}{3}, -\frac{3}{2}$ and $-1 < -\frac{1}{3} < 1, -\frac{3}{2} < -1$.

$\therefore \quad \cos^{-1}\left(-\frac{1}{3}\right)$ exists but $\cos^{-1}\left(-\frac{3}{2}\right)$ does not;

$\operatorname{cosec}^{-1}\left(-\frac{3}{2}\right)$ exists but $\operatorname{cosec}^{-1}\left(-\frac{1}{3}\right)$ does not;

$\cot^{-1}\left(-\frac{1}{3}\right)$ and $\cot^{-1}\left(-\frac{3}{2}\right)$ exist.

3. $\sec^{-1} x$ is real when $x \le -1$ or $x \ge 1$. $\tan^{-1} x$ exists for all $x \in R$.

5. $\operatorname{cosec}^{-1}(\cos x)$ exists if $\cos x \le -1$ or $\cos x \ge 1$. But $-1 \le \cos x \le 1$.

$\therefore \quad \operatorname{cosec}^{-1}(\cos x)$ exists if $\cos x = 1$ or $-1 \quad \Rightarrow x = n\pi, n \in Z$.

6. $\sin \frac{5\pi}{6}$ is to be written as $\sin \alpha$, where $-\frac{\pi}{2} \le \alpha \le \frac{\pi}{2}$.

$$\therefore \quad \sin^{-1}\left(\sin \frac{5\pi}{6}\right) = \sin^{-1}\left\{\sin\left(\pi - \frac{\pi}{6}\right)\right\} = \sin^{-1}\left(\sin \frac{\pi}{6}\right) = \frac{\pi}{6}.$$

7.
$$\cos^{-1}\left(-\sin \frac{7\pi}{6}\right) = \cos^{-1}\left\{\cos\left(\frac{\pi}{2} + \frac{7\pi}{6}\right)\right\}$$
$$= \cos^{-1}\left(\cos \frac{5\pi}{3}\right) = \cos^{-1}\left\{\cos\left(2\pi - \frac{5\pi}{3}\right)\right\}$$
$$= \cos^{-1}\left(\cos \frac{\pi}{3}\right) = \frac{\pi}{3}.$$

Remember, $\cos^{-1}(\cos x) = x$ if $0 \le x \le \pi$.

9. $\frac{1}{\sqrt{2}}\left(\cos\frac{9\pi}{10}-\sin\frac{9\pi}{10}\right)=\cos\left(\frac{\pi}{4}+\frac{9\pi}{10}\right)\cdot$

$$=\cos\frac{23\pi}{20}=\cos\left(2\pi-\frac{23\pi}{20}\right)=\cos\frac{17\pi}{20}$$

$\therefore$ the value $=\cos^{-1}\left(\cos\frac{17\pi}{20}\right)=\frac{17\pi}{20}$

10. $\tan^{-1}(\tan x)=x$ if $-\frac{\pi}{2}<x<\frac{\pi}{2}\cdot$

$\tan\frac{15\pi}{4}=\tan\left(4\pi-\frac{\pi}{4}\right)=\tan\left(-\frac{\pi}{4}\right)\cdot$

$\therefore$ the value $=\cos\left\{\tan^{-1}\left(\tan\frac{-\pi}{4}\right)\right\}=\cos\left(-\frac{\pi}{4}\right)=\frac{1}{\sqrt{2}}\cdot$

11. $\cos^{-1}\left(\cos\frac{7\pi}{5}\right)=\cos^{-1}\left\{\cos\left(2\pi-\frac{3\pi}{5}\right)\right\}=\cos^{-1}\cos\left(\frac{3\pi}{5}\right)=\frac{3\pi}{5}\cdot$

$\sin^{-1}\left(\sin\frac{7\pi}{5}\right)=\sin^{-1}\left\{\sin\left(\pi+\frac{2\pi}{5}\right)\right\}$

$$=\sin^{-1}\left\{-\sin\frac{2\pi}{5}\right\}=\sin^{-1}\left\{\sin\left(-\frac{2\pi}{5}\right)\right\}=-\frac{2\pi}{5}\cdot$$

$\sec^{-1}\left(\sec\frac{7\pi}{5}\right)=\sec^{-1}\left\{\sec\left(2\pi-\frac{3\pi}{5}\right)\right\}=\sec^{-1}\left(\sec\frac{3\pi}{5}\right)=\frac{3\pi}{5}\cdot$

12. $0\le\cos^{-1}x\le\pi$. Hence, from the question,

$$\cos^{-1}\lambda=\pi,\ \cos^{-1}\mu=\pi,\ \cos^{-1}\nu=\pi$$

$\therefore\quad \lambda=\mu=\nu=-1.$

13. $0\le\cos^{-1}x\le\pi$. Hence, from the question, $\cos^{-1}x_i=0$ for all i.

$\therefore\quad x_i=1$ for all i.

14. $-\frac{\pi}{2}\le\sin^{-1}x\le\frac{\pi}{2}\cdot$ Hence, from the question, $\sin^{-1}x_i=\frac{\pi}{2}$ for all i.

$\therefore\quad x_i=1$ for all i.

16. $2\tan^{-1}\frac{1}{5}=\tan^{-1}\frac{2\cdot\frac{1}{5}}{1-\left(\frac{1}{5}\right)^2}=\tan^{-1}\frac{5}{12}$

$$\therefore \quad \text{the value} = \frac{\tan\left(\tan^{-1}\frac{5}{12}\right) - \tan\frac{\pi}{4}}{1 + \tan\left(\tan^{-1}\frac{5}{12}\right)\cdot\tan\frac{\pi}{4}} = \frac{\frac{5}{12} - 1}{1 + \frac{5}{12}\cdot 1}.$$

17. If $x = -1$, LHS $= \frac{\pi}{2}$, RHS $= 2 \times \left(-\frac{\pi}{2}\right)$. So, the formula does not hold.

If $x < -1$, the angle on the LHS is in the second quadrant while the angle on the RHS is $2 \times$ (angle in the fourth quadrant), which cannot be equal.

If $x > 1$, the angle on the LHS is in the second quadrant while the angle on the RHS is $2 \times$ (angle in the first quadrant) and these two may be equal.

If $-1 < x < 0$, the angle on the LHS is positive and that on the RHS is negative and the two cannot be equal.

20. Clearly, $0 \le \text{LHS} \le \pi/2$ because $\cos^{-1}x$ is in the first quadrant when x is positive.

$\therefore \quad$ RHS ≥ 0. So, $\cos^{-1}\frac{x}{2} \ge \cos^{-1}x$.

Also $\left|\frac{x}{2}\right| \le 1$, $|x| \le 1$. This means $|x| \le 1$.

In $0 \le x \le 1$, clearly $\cos^{-1}\frac{x}{2} \ge \cos^{-1}x$ because $\cos^{-1}\alpha$ increases as α decreases.

21. Let $x = \sin\theta$. Then $f(x) = \sin^{-1}\left\{\sin\left(\theta - \frac{\pi}{6}\right)\right\}$.

$$-\frac{1}{2} \le x \le 1 \quad \Rightarrow -\frac{1}{2} \le \sin\theta \le 1 \quad \Rightarrow -\frac{\pi}{6} \le \theta \le \frac{\pi}{2}.$$

So, $\theta - \frac{\pi}{6}$ is in the fourth or the first quadrant. Hence, $f(x) = \theta - \frac{\pi}{6}$.

22. Let $x = \sin\theta$. Then $\sin^{-1}(2x\sqrt{1-x^2}) = \sin^{-1}(\sin 2\theta) = 2\theta = 2\sin^{-1}x$

$$\text{if } -\frac{\pi}{2} \le 2\theta \le \frac{\pi}{2}, \text{ i.e., } \sin\left(-\frac{\pi}{4}\right) \le \sin\theta \le \sin\left(\frac{\pi}{4}\right), \text{ i.e., } -\frac{1}{\sqrt{2}} \le x \le \frac{1}{\sqrt{2}}.$$

24. Let $x = \tan\theta$. Then $\sin^{-1}\frac{2x}{1+x^2} = \sin^{-1}\frac{2\tan\theta}{1+\tan^2\theta} = \sin^{-1}(\sin 2\theta)$.

$$\therefore \quad 2\tan^{-1}x + \sin^{-1}\frac{2x}{1+x^2} = 2\theta + \sin^{-1}(\sin 2\theta).$$

If $-\frac{\pi}{2} \le 2\theta \le \frac{\pi}{2}$, $2\tan^{-1}x + \sin^{-1}\frac{2x}{1+x^2} = 2\theta + 2\theta = 4\tan^{-1}x$

$\neq$ independent of x.

If $-\frac{\pi}{2} \le \pi - 2\theta \le \frac{\pi}{2}$, $2\tan^{-1}x + \sin^{-1}\frac{2x}{1+x^2}$

$$= 2\theta + \sin^{-1}\{\sin(\pi - 2\theta)\} = 2\theta + \pi - 2\theta$$

$$= \pi = \text{independent of } x.$$

$\therefore \quad \theta \notin \left[-\frac{\pi}{4}, \frac{\pi}{4}\right]$ but $\theta \in \left[\frac{\pi}{4}, \frac{3\pi}{4}\right]$ and from the principal value of $\tan^{1}x$,

$\theta \in \left(-\frac{\pi}{2}, \frac{\pi}{2}\right)$. Hence, $\theta \in \left(\frac{\pi}{4}, \frac{\pi}{2}\right)$.

$\therefore \quad \theta \in \left(\frac{\pi}{4}, \frac{\pi}{2}\right) \quad \Rightarrow \quad \tan^{-1}x + \sin^{-1}\frac{2x}{1+x^2} = \pi.$

Also at $\theta = \frac{\pi}{4}$, $2\tan^{-1}x + \sin^{-1}\frac{2x}{1+x^2} = 2\cdot\frac{\pi}{4} + \sin^{-1}1 = \frac{\pi}{2} + \frac{\pi}{2} = \pi.$

$\therefore \quad$ the given function $= \pi =$ constant if $\theta \in \left[\frac{\pi}{4}, \frac{\pi}{2}\right)$, i.e., $x \in [1, +\infty)$.

28. $\left(\frac{\pi}{2} - \cos^{-1}x\right) - \cos^{-1}x = \frac{\pi}{6} \quad$ or $\quad 2\cos^{-1}x = \frac{\pi}{2} - \frac{\pi}{6} = \frac{\pi}{3}$

or $\quad \cos^{-1}x = \frac{\pi}{6} \quad \therefore \; x = \frac{\sqrt{3}}{2}$.

29. Put $x = \tan\theta$.

31. Value $= \tan^{-1}\frac{1}{3} + \tan^{-1}\frac{1}{2} = \tan^{-1}\frac{\frac{1}{3}+\frac{1}{2}}{1-\frac{1}{3}\cdot\frac{1}{2}} = \tan^{-1}1 = \frac{\pi}{4}$.

34. Value $= \sec^2(\sec^{-1}2) - 1 + \text{cosec}^2(\text{cosec}^{-1}3) - 1 = 2^2 - 1 + 3^2 - 1 = 11.$

35. Let $\cos^{-1}x = \theta$. Then the expression

$$= \frac{1+\tan\theta/2}{1-\tan\theta/2} + \frac{1-\tan\theta/2}{1+\tan\theta/2} = \frac{2(1+\tan^2\theta/2)}{1-\tan^2\theta/2} = \frac{2}{\cos\theta} = \frac{2}{x}.$$

36. Here $|\cos x| = \sin^{-1}(\sin x)$.

If $-\dfrac{\pi}{2} \le x \le \dfrac{\pi}{2}$ then $\cos x = x$.

In the case there is one solution, obtained graphically.

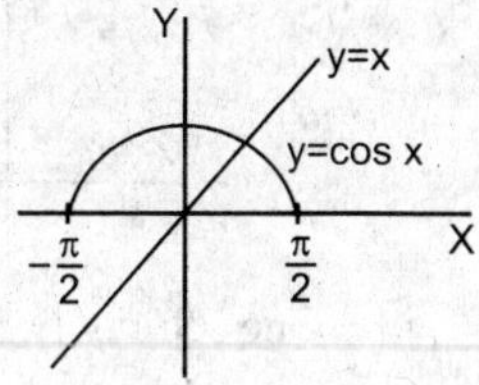

If $\dfrac{\pi}{2} < x \le \pi$ then $-\cos x = \sin^{-1}\{\sin(\pi - x)\} = \pi - x$.

$\therefore \quad \cos x = x - \pi$.

In this case there is one solution, obtained graphically.

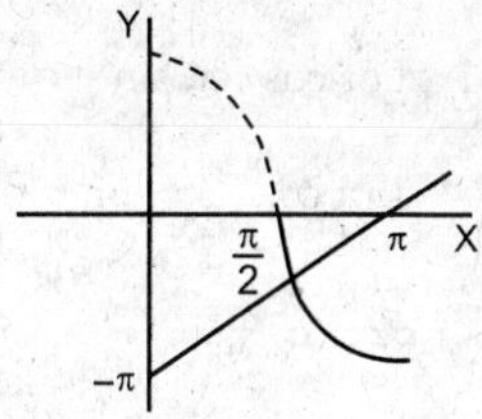

If $-\pi \le x < -\dfrac{\pi}{2}$ then $-\cos x = \sin^{-1}\{\sin(-\pi - x)\}$

$= -x - \pi$, i.e., $\cos x = x + \pi$.

This gives no solution as can be seen from their graphs.

37. Clearly, $x(x+1) \ge 0$ and $x^2 + x + 1 \le 1$. Together they imply $x(x+1) = 0$.

$\therefore \quad x = 0, -1$. When $x = 0$, LHS $= \tan^{-1}0 + \sin^{-1}1 = \dfrac{\pi}{2}$.

When $x = -1$, LHS $= \tan^{-1}0 + \sin^{-1}\sqrt{1-1+1} = 0 + \sin^{-1}1 = \dfrac{\pi}{2}$.

38. $\tan^{-1}x + \tan^{-1}\dfrac{1}{y} = \tan^{-1}3 \quad$ or $\quad \tan^{-1}\dfrac{1}{y} = \tan^{-1}3 - \tan^{-1}x$

or $\quad \tan^{-1}\dfrac{1}{y} = \tan^{-1}\dfrac{3-x}{1+3x} \quad \Rightarrow y = \dfrac{1+3x}{3-x}$.

As x, y are positive integers, $x = 1, 2$ and corresponding $y = 2, 7$.

$\therefore \quad$ solutions are $(x, y) = (1, 2), (2, 7)$.

40. In $[0, \pi]$, $|y| = \sin x$, $y = \cos^{-1}(\cos x) = x$.
In $[\pi, 2\pi]$, $|y| = \sin x$, $y = \cos^{-1}\{\cos(2\pi - x)\} = 2\pi - x$.
In $[-\pi, 0]$, $|y| = \sin x$, $y = \cos^{-1}\{\cos(-x)\} = -x$.
In $[-2\pi, -\pi]$, $|y| = \sin x$, $y = \cos^{-1}\{\cos(2\pi + x)\} = 2\pi + x$.
Plotting the graphs, we have

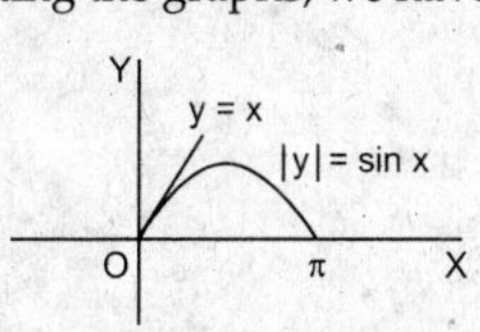

One solution: $x = 0, y = 0$

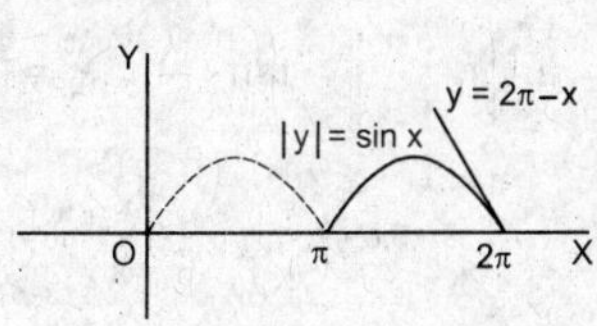

One solution: $x = 2\pi, y = 0$

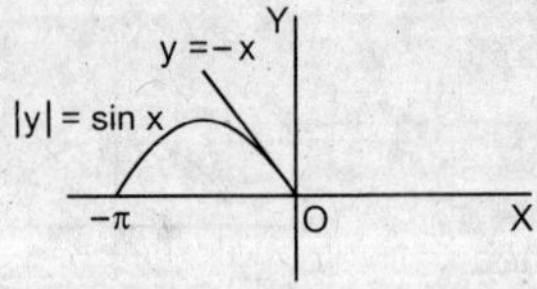

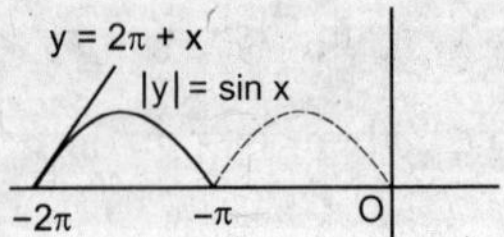

One solution: $x = 0, y = 0$ One solution: $x = -2\pi, y = 0$

$\therefore$ there are 3 solutions (0 , 0), (2π, 0), (–2π, 0).

41. For $\cos^{-1}x$, $\sin^{-1}x$ to be real, $-1 \le x \le 1$.

But $\cos^{-1}x > \sin^{-1}x \quad \Rightarrow 2\cos^{-1}x > \dfrac{\pi}{2} \quad \Rightarrow \cos^{-1} x > \dfrac{\pi}{4}$

$\therefore \quad \cos(\cos^{-1}x) < \cos\dfrac{\pi}{4} \quad$ or $\quad x < \dfrac{1}{\sqrt{2}}$.

The common values are $-1 \le x < \dfrac{1}{\sqrt{2}}$.

42. $\cot^{-1}\dfrac{n}{\pi} > \dfrac{\pi}{6} \quad \Rightarrow \cot\left(\cot^{-1}\dfrac{n}{\pi}\right) < \cot\dfrac{\pi}{6}$

$\Rightarrow \quad \dfrac{n}{\pi} < \sqrt{3} \quad \Rightarrow n < \sqrt{3}\pi = 5.5$ (nearly).

So, the maximum value of n is 5.

43. $\sin^{-1}(\sin 4) = \sin^{-1}\{\sin(\pi - 4)\} = \pi - 4 \quad \left(\because \; -\dfrac{\pi}{2} < \pi - 4 < \dfrac{\pi}{2}\right)$

$\therefore$ we have $x^2 - kx + \pi - 4 > 0$ for all $x \in R$.

$\therefore$ $D < 0$, i.e., $k^2 - 4(\pi - 4) < 0$

or $k^2 + 4(4 - \pi) < 0$, which is not true for any real k.

44. $\sin^{1}x + \cos^{-1}x = \dfrac{\pi}{2}$ if $-1 \le x \le 1$. Use this to solve the problem.

47. $\tan^{-1}\left\{\tan\left(\pi + \dfrac{\pi}{4}\right)\right\} = \tan^{-1}\tan\dfrac{\pi}{4} = \dfrac{\pi}{4} = \alpha.$

$\tan^{-1}\left\{-\tan\left(\pi - \dfrac{\pi}{3}\right)\right\} = \tan^{-1}\tan\dfrac{\pi}{3} = \dfrac{\pi}{3} = \beta.$

48. $f\left(\dfrac{8\pi}{9}\right) = e^{\cos^{-1}\sin(8\pi/9 + \pi/3)} = e^{\cos^{-1}\sin 11\pi/9}$

$= e^{\cos^{-1}\cos 13\pi/18} = e^{13\pi/18}.$

$$f\left(-\frac{7\pi}{4}\right) = e^{\cos^{-1}\sin(-7\pi/4 + \pi/3)} = e^{\cos^{-1}\sin(-17\pi/12)}$$

$$= e^{\cos^{-1}\cos \pi/12} = e^{\pi/12}.$$

49. $f(x) = \cos^{-1}x + \cos^{-1}\left\{\frac{1}{2}\cdot x + \frac{\sqrt{3}}{2}\cdot\sqrt{1-x^2}\right\}$

$= \cos^{-1}x \pm \left(\cos^{-1}\frac{1}{2} - \cos^{-1}x\right)$, according as $\cos^{-1}\frac{1}{2} >$ or $< \cos^{-1}x$

$= \cos^{-1}\frac{1}{2}$ if $\cos^{-1}\frac{1}{2} > \cos^{-1}x$, which holds for $x = \frac{2}{3}$

$= 2\cos^{-1}x - \cos^{-1}\frac{1}{2}$ if $\cos^{-1}\frac{1}{2} < \cos^{-1}x$, which holds for $x = \frac{1}{3}\cdot$

50. Let $x = \tan\theta$. Then the expression $= \tan\left\{\frac{1}{2}\sin^{-1}\sin 2\theta + \frac{1}{2}\cos^{-1}\cos 2\theta\right\}$.

If $0 \le 2\theta \le \frac{\pi}{2}$, the expression $= \tan\left\{\frac{1}{2}\cdot 2\theta + \frac{1}{2}\cdot 2\theta\right\} = \tan 2\theta = \frac{2x}{1+x^2}\cdot$

But $0 \le 2\theta \le \frac{\pi}{2} \Rightarrow 0 \le \theta \le \frac{\pi}{4} \Rightarrow 0 \le \tan\theta \le 1$.

If $\frac{\pi}{2} < 2\theta \le \pi$, i.e., $\tan\theta > 1$ then the expression

$$= \tan\left\{\frac{1}{2}\sin^{-1}\sin(\pi - 2\theta) + \frac{1}{2}\cos^{-1}\cos 2\theta\right\}$$

$$= \tan\left\{\frac{\pi - 2\theta}{2} + \frac{1}{2}\cdot 2\theta\right\} = \tan\frac{\pi}{2}\cdot$$

51. $\alpha = 2\tan^{-1}(\sqrt{2} - 1) = 2\tan^{-1}\tan\frac{\pi}{8} = 2\times\frac{\pi}{8} = \frac{\pi}{4} = \cos^{-1}\frac{1}{\sqrt{2}}\cdot$

$\beta = 3\cdot\frac{\pi}{4} - \frac{\pi}{6} = \frac{7\pi}{12}\cdot$

$\therefore \quad \beta > \alpha$. Also $\frac{1}{3} < \frac{1}{\sqrt{2}} \Rightarrow \cos^{-1}\frac{1}{3} > \cos^{-1}\frac{1}{\sqrt{2}} = \frac{\pi}{4}\cdot$

So $\gamma > \alpha$.

Again $\cos^{-1}\frac{1}{3}$ belongs to the first quadrant and β is in the second quadrant.

$\therefore \quad \beta > \gamma$.

53. $\sqrt{\sec^2(\sec^{-1}x)-1} = \sin\sin^{-1}\frac{2}{\sqrt{5}} \Rightarrow \sqrt{x^2-1} = \frac{2}{\sqrt{5}} \Rightarrow x = \pm\frac{3}{\sqrt{5}}$.

54. Adding, $\frac{\pi}{2} + \sin^{-1}y - \cos^{-1}y = \frac{\pi}{3}$

or $\pi - 2\cos^{-1}y = \frac{\pi}{3}$ or $\cos^{-1}y = \frac{\pi}{3} \Rightarrow y = \frac{1}{2}$

$\therefore$ $\sin^{-1}x + \sin^{-1}\frac{1}{2} = \frac{2\pi}{3}$ or $\sin^{-1}x = \frac{2\pi}{3} - \frac{\pi}{3} = \frac{\pi}{3} \Rightarrow x = \frac{\sqrt{3}}{2}$.

So, there is one solution.

55. $\sin^{-1}x, \cos^{-1}x$ exist for $-1 \le x \le 1$. But for $\sin^{-1}(1-x)$ we must have $-1 \le 1-x \le 1$, i.e., $0 \le x \le 2$. So the equation may hold for $0 \le x \le 1$. Therefore, the options *a*, *b* and *c* are incorrect.

□

4

Logarithms

• Type 1 •

Choose the most appropriate option (a, b, c or d).

1. If $\log_4 5 = a$ and $\log_5 6 = b$ then $\log_3 2$ is equal to

(a) $\frac{1}{2a+1}$ (b) $\frac{1}{2b+1}$

(c) $2ab+1$ (d) $\frac{1}{2ab-1}$

2. If $\log_{ax} x, \log_{bx} x, \log_{cx} x$ are in HP, where a, b, c, x belong to $(1, +\infty)$, then a, b, c are in

(a) AP (b) GP

(c) HP (d) none of these

3. If $\log_5 a \cdot \log_a x = 2$ then x is equal to

(a) 125 (b) a^2

(c) 25 (d) none of these

4. The value of $3\log\frac{81}{80} + 5\log\frac{25}{24} + 7\log\frac{16}{15}$ is

(a) $\log 2$ (b) $\log 3$

(c) 1 (d) 0

5. Let $f(x) = \sqrt{\log_{10} x^2}$. The set of all values of x for which $f(x)$ is real, is

(a) $[-1, 1]$ (b) $[1, +\infty)$

(c) $(-\infty, -1]$ (d) $(-\infty, -1] \cup [1, +\infty)$

6. The set of real values of x for which $2^{\log_{\sqrt{2}}(x-1)} > x+5$ is

(a) $(-\infty, -1) \cup (4, +\infty)$ (b) $(4, +\infty)$

(c) $(-1, 4)$ (d) none of these

7. If $a_1, a_2, a_3, \ldots$ are positive numbers in GP then $\log a_n, \log a_{n+1}, \log a_{n+2}$ are in

(a) AP (b) GP

(c) HP (d) none of these

8. If $x = \log_a(bc)$, $y = \log_b(ca)$ and $z = \log_c(ab)$ then which of the following is equal to 1?

(a) $x+y+z$ (b) $(1+x)^{-1} + (1+y)^{-1} + (1+z)^{-1}$

(c) xyz (d) none of these

9. If $x_n > x_{n-1} > \ldots > x_2 > x_1 > 1$ then the value of

$$\log_{x_1} \log_{x_2} \log_{x_3} \ldots \log_{x_n} x_n^{x_{n-1}^{\cdot^{\cdot^{x_1}}}}$$

is equal to

(a) 0 (b) 1

(c) 2 (d) none of these

10. If $\log x : \log y : \log z = (y-z) : (z-x) : (x-y)$ then

(a) $x^y \cdot y^z \cdot z^x = 1$ (b) $x^x y^y z^z = 1$

(c) $\sqrt[x]{x} \cdot \sqrt[y]{y} \cdot \sqrt[z]{z} = 1$ (d) none of these

11. $x^{\log_x a \times \log_a y \times \log_y z}$ is equal to

(a) x (b) y

(c) z (d) none of these

12. The number of zeros coming immediately after the decimal point in the value of $(0.2)^{25}$ is (given $\log_{10} 2 = 0.30103$)

(a) 16 (b) 17

(c) 18 (d) none of these

13. If $[x]$ = the greatest integer less than or equal to x then $[\log_{10} 6730.4]$ has the value

(a) 6 (b) 4

(c) 5 (d) none of these

14. The number of solutions of $\log_2 (x+5) = 6 - x$ is

(a) 2 (b) 0

(c) 3 (d) none of these

15. The number of real values of the parameter k for which $(\log_{16} x)^2 - \log_{16} x + \log_{16} k = 0$ with real coefficients will have exactly one solution is

(a) 2 (b) 1

(c) 4 (d) none of these

16. If $\log_{0.5} \sin x = 1 - \log_{0.5} \cos x$ then the number of solutions of $x \in [-2\pi, 2\pi]$ is

(a) 3 (b) 2

(c) 1 (d) 4

17. If $\log_{\cos x} \tan x + \log_{\sin x} \cot x = 0$ then the most general solutions of x are

(a) $n\pi + \frac{\pi}{4}, n \in Z$ (b) $2n\pi + \frac{\pi}{4}, n \in Z$

(c) $2n\pi - \frac{3\pi}{4}, n \in Z$ (d) none of these

18. The number of values of $x \in [0, n\pi], n \in Z$, that satisfy $\log_{|\sin x|} (1 + \cos x) = 2$, is

(a) 0 (b) n

(c) $2n$ (d) none of these

19. The value of $\sum_{r=1}^{89} \log_{10} \tan \frac{\pi r}{180}$ is equal to

(a) 10 (b) 1

(c) 0 (d) none of these

20. The solution set of $\log_2 |4 - 5x| > 2$ is

(a) $\left(\frac{8}{5}, +\infty\right)$ (b) $\left(\frac{4}{5}, \frac{8}{5}\right)$

(c) $(-\infty, 0) \cup \left(\frac{8}{5}, +\infty\right)$ (d) none of these

21. The set of real values of x for which $\log_{0.2} \frac{x+2}{x} \le 1$ is

(a) $\left(-\infty, -\frac{5}{2}\right] \cup (0, +\infty)$ (b) $\left[\frac{5}{2}, +\infty\right)$

(c) $(-\infty, -2) \cup (0, +\infty)$ (d) none of these

22. The set of real values of x satisfying $\log_{1/2}(x^2 - 6x + 12) \geq -2$ is

(a) $(-\infty, 2]$ (b) $[2, 4]$

(c) $[4, +\infty)$ (d) none of these

23. If $\log_{0.04}(x-1) \geq \log_{0.2}(x-1)$ then x belongs to the interval

(a) $(1, 2]$ (b) $(-\infty, 2]$

(c) $[2, +\infty)$ (d) none of these

24. If $\log_{1/\sqrt{2}} \sin x > 0$, $x \in [0, 4\pi]$, then the number of values of x which are integral multiples of $\frac{\pi}{4}$, is

(a) 6 (b) 12

(c) 3 (d) none of these

25. If $\log_{\cos x} \sin x \geq 2$ and $0 \leq x \leq 3\pi$ then $\sin x$ lies in the interval

(a) $\left[\frac{\sqrt{5}-1}{2}, 1\right]$ (b) $\left(0, \frac{\sqrt{5}-1}{2}\right]$

(c) $\left[0, \frac{1}{2}\right]$ (d) none of these

26. If $\log_{\sqrt{3}}(\sin x + 2\sqrt{2}\cos x) \geq 2$, $-2\pi \leq x \leq 2\pi$, then the number of solutions of x is

(a) 0 (b) infinite

(c) 3 (d) none of these

• Type 2 •

Choose the correct options. One or more options may be correct.

27. If $\log_k x \cdot \log_5 k = \log_x 5$, $k \neq 1$, $k > 0$, then x is equal to

(a) k (b) $\frac{1}{5}$

(c) 5 (d) none of these

28. If $x^2 + 4y^2 = 12xy$, $x \in [1, 4]$, $y \in [1, 4]$, then

(a) the greatest value of $\log_2 (x + 2y)$ is 4

(b) the least value of $\log_2 (x + 2y)$ is 3

(c) the range of values of $\log_2 (x + 2y)$ is [2, 4]

(d) the number of integral values of (x, y) is 2 such that $\log_2 (x + 2y)$ is equal to 3

29. If $\frac{1}{2} \leq \log_{0.1} x \leq 2$ then

(a) the maximum value of x is $\frac{1}{\sqrt{10}}$

(b) x lies between $\frac{1}{100}$ and $\frac{1}{\sqrt{10}}$

(c) x does not lie between $\frac{1}{100}$ and $\frac{1}{\sqrt{10}}$

(d) the minimum value of x is $\frac{1}{100}$

30. If $x^{3/4(\log_3 x)^2 + \log_3 x - 5/4} = \sqrt{3}$ then x has

(a) one positive integral value (b) one irrational value

(c) two positive rational values (d) none of these

Answers

1. d	**2.** b	**3.** c	**4.** a	**5.** d
6. a	**7.** a	**8.** b	**9.** b	**10.** b
11. c	**12.** b	**13.** d	**14.** d	**15.** b
16. b	**17.** b	**18.** a	**19.** c	**20.** c
21. a	**22.** b	**23.** c	**24.** a	**25.** b
26. d	**27.** b, c	**28.** a, c, d	**29.** a, b, d	**30.** a, b, c

Hints and Solutions to Selected Questions

1. Here, $5 = 4^a$ and $6 = 5^b$.

Let $\log_3 2 = x$. Then $2 = 3^x$.

Now, $6 = 5^b = (4^a)^b = 4^{ab}$ or $3 = 2^{2ab-1}$

$\therefore \quad 2 = (2^{2ab-1})^x = 2^{x(2ab-1)} \quad \Rightarrow x(2ab-1) = 1.$

2. Clearly, $\log_x(ax), \log_x(bx), \log_x(cx)$ are in AP

$\Rightarrow \quad 1 + \log_x a, 1 + \log_x b, 1 + \log_x c$ are in AP

$\Rightarrow \quad \log_x a, \log_x b, \log_x c$ are in AP

$$\Rightarrow \frac{\log a}{\log x} + \frac{\log c}{\log x} = 2 \cdot \frac{\log b}{\log x} \quad \Rightarrow \log a + \log c = 2\log b \quad \Rightarrow ac = b^2.$$

5. $\log_{10} x^2 \geq 0 \quad \Rightarrow \log_{10} x^2 \geq \log_{10} 1$

$\Rightarrow \quad x^2 \geq 1 \quad \Rightarrow x \geq 1$ or $x \leq -1$.

6. $\{(\sqrt{2})^2\}^{\log_{\sqrt{2}}(x-1)} > x + 5$ or $(\sqrt{2})^{\log_{\sqrt{2}}(x-1)^2} > x + 5$

or $(x-1)^2 > x + 5$ or $x^2 - 3x - 4 > 0$ or $(x-4)(x+1) > 0$.

9. $$\text{Value} = \log_{x_1} \log_{x_2} \log_{x_3} \ldots \log_{x_{n-1}} \left(x_{n-1}^{x_{n-2}^{\cdot^{\cdot^{x_1}}}} \log_{x_n} x_n\right)$$

$$= \log_{x_1} \log_{x_2} \log_{x_3} \ldots \log_{x_{n-1}} x_{n-1}^{x_{n-2}^{\cdot^{\cdot^{x_1}}}}$$

$$= \ldots = \log_{x_1} x_1 = 1.$$

13. $\log_{10} 6730.4 = 3 \cdot \times\times\times\times \quad \Rightarrow \quad [\log_{10} 6730.4] = 3.$

14. Here $x + 5 = 2^{6-x}$. Its graph is as follows.

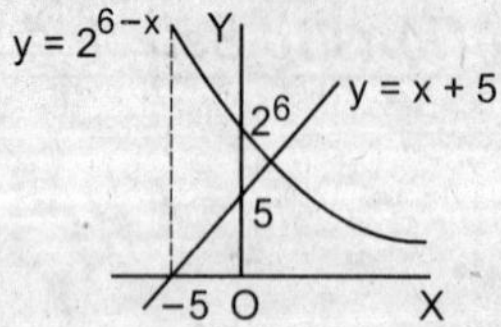

Clearly, there cannot be more than one solution and by trial, the solution is $x = 3$.

15. $\log_{16} x = \dfrac{1 \pm \sqrt{1 - 4\log_{16} k}}{2}$. For exactly one solution, $4\log_{16} k = 1$

$\therefore \quad k^4 = 16 \quad \therefore \; k = 2, -2, 2i, -2i$. But k is positive and real.

16. $\log_{0.5}(\sin x \cdot \cos x) = 1 \quad \Rightarrow \sin x \cdot \cos x = \dfrac{1}{2} \quad \Rightarrow \sin 2x = 1.$

$$\therefore \quad 2x = \frac{\pi}{2}, \pm 2\pi + \frac{\pi}{2}, \pm 4\pi + \frac{\pi}{2}, \ldots.$$

$$\therefore \quad x = \frac{\pi}{4}, \pm \pi + \frac{\pi}{4}, \pm 2\pi + \frac{\pi}{4}, \ldots$$

But $\log_{0.5} \sin x$, $\log_{0.5} \cos x$ have to be real at the same time. So angles in the first quadrant only can be considered.

$$\therefore \quad x = \frac{\pi}{4}, -2\pi + \frac{\pi}{4}.$$

17. $\log_{\cos x} \sin x - 1 + \log_{\sin x} \cos x - 1 = 0$

or $\quad y + \dfrac{1}{y} = 2$, where $y = \log_{\cos x} \sin x$.

$\therefore \quad (y-1)^2 = 0 \quad$ or $\log_{\cos x} \sin x = 1 \quad$ or $\sin x = \cos x$.

Also $\sin x$, $\cos x$, $\tan x$, $\cot x$ must be positive and $\sin x \neq 1$, $\cos x \neq 1$.

18. $1 + \cos x = |\sin x|^2 = \sin^2 x \quad$ or $\quad \cos x(1 + \cos x) = 0$.

But $1 + \cos x \neq 0$ for the value of the logarithm to be real.

$\cos x = 0 \quad \Rightarrow \quad \sin x = 1$ (not possible, because the base of the logarithm cannot be 1).

19. Value $= \log_{10}\{\tan 1° \cdot \tan 2° \cdot \tan 3° \ldots \tan 89°\}$

$$= \log_{10}\{(\tan 1° \cdot \tan 89°) \ldots (\tan 44° \cdot \tan 46°) \cdot \tan 45°\} = \log_{10} 1 = 0.$$

20. $|4-5x|>2^2=4 \quad \Rightarrow \left|\frac{5x}{4}-1\right|>1 \quad \Rightarrow \frac{5x}{4}-1>1$ or $\frac{5x}{4}-1<-1$

$\therefore \quad x>\frac{8}{5}$ or $x<0$. So, the solution set $=(-\infty, 0) \cup \left(\frac{8}{5}, +\infty\right)$.

21. $\frac{x+2}{x} \ge (0.2)^1 \quad$ or $\quad \frac{x+2}{x} \ge \frac{1}{5}$.

Multiplying by $5x^2$ (positive),

$5x(x+2) \ge x^2 \quad$ or $\quad 4x^2+10x \ge 0 \quad \therefore \; x \ge 0$ or $x \le -\frac{5}{2}$.

Also $\frac{x+2}{x}>0 \quad \Rightarrow x(x+2)>0 \quad \therefore \; x<-2$ or $x>0$

$\therefore \quad$ the solution set is $\left(-\infty, -\frac{5}{2}\right] \cup (0, +\infty)$.

22. $x^2-6x+12 \le \left(\frac{1}{2}\right)^{-2} \quad$ or $\quad x^2-6x+8 \le 0 \quad$ or $\quad (x-2)(x-4) \le 0$.

23. $\frac{\log_{10}(x-1)}{\log_{10}(0.04)} \ge \frac{\log_{10}(x-1)}{\log_{10}(0.2)} \quad \Rightarrow \log_{10}(x-1) \cdot \left\{\frac{1}{2\log_{10}(0.2)} - \frac{1}{\log_{10}(0.2)}\right\} \ge 0$

or $\quad \log_{10}(x-1) \cdot \frac{-1}{2\log_{10}(0.2)} \ge 0$

or $\quad \log_{10}(x-1) \ge 0$, because $\log_{10}(0.2)<0$

$\therefore \quad x-1 \ge 10^0 \quad \therefore \; x \ge 2$.

Also $x-1>0$, i.e., $x>1$.

24. $\sin x < \left(\frac{1}{\sqrt{2}}\right)^0 = 1$ and $\sin x > 0$

$\therefore \quad \sin x$ has all values in $(0, \pi)$, $(2\pi, 3\pi)$, $(4\pi, 5\pi)$, …

$\therefore \quad$ required $x = \frac{\pi}{4}, 2\times\frac{\pi}{4}, 3\times\frac{\pi}{4}, 9\times\frac{\pi}{4}, 10\times\frac{\pi}{4}, 11\times\frac{\pi}{4}$.

25. $\sin x \le \cos^2 x$, because $\cos x$ must be a positive proper fraction.

or $\quad \sin^2 x+\sin x-1 \le 0 \quad$ or $\quad \left(\sin x+\frac{1}{2}\right)^2-\frac{5}{4} \le 0$.

From the definition of logarithm, $\sin x>0$, $\cos x>0$, $\cos x \ne 1$

$\therefore \quad \sin x + \frac{1}{2} \le \frac{\sqrt{5}}{2} \qquad \therefore \; 0 < \sin x \le \frac{\sqrt{5}-1}{2}.$

26. $\sin x + 2\sqrt{2}\cos x \ge (\sqrt{3})^2$ or $\sin x + 2\sqrt{2}\cos x \ge 3$

or $\sin\left(x + \cos^{-1}\frac{1}{3}\right) \ge 1 \quad \Rightarrow \sin\left(x + \cos^{-1}\frac{1}{3}\right) = 1.$

$\therefore \quad x = n\pi + (-1)^n \frac{\pi}{2} - \cos^{-1}\frac{1}{3}.$

For solutions in $[-2\pi, 2\pi]$, $n = 0, 1, -1, -2$.

27. $\log_5 x = \log_x 5 \quad \Rightarrow (\log_5 x)^2 = 1 \quad \Rightarrow \log_5 x = \pm 1 \quad \Rightarrow x = 5, 5^{-1}.$

28. $(x+2y)^2 = 16xy$ or $2\log_2(x+2y) = 4 + \log_2 x + \log_2 y$

$\therefore \quad \log_2(x+2y) = 2 + \frac{1}{2}(\log_2 x + \log_2 y).$

But $1 \le x \le 4$, $1 \le y \le 4$.

$\therefore \quad \max \log_2(x+2y) = 2 + \frac{1}{2}(\log_2 4 + \log_2 4).$

$\min \log_2(x+2y) = 2 + \frac{1}{2}(\log_2 1 + \log_2 1)$

Also $\log_2(x+2y) = 3 \quad \Rightarrow x + 2y = 8.$

This is satisfied by $x = 2, y = 3$; $x = 4, y = 2$.

29. $\frac{1}{2} \le \log_{1/10} x \le 2 \quad \Rightarrow \left(\frac{1}{10}\right)^{1/2} \ge x, \left(\frac{1}{10}\right)^2 \le x.$

So, $\frac{1}{100} \le x \le \frac{1}{\sqrt{10}}.$

30. Taking logarithm, $\left\{\frac{3}{4}(\log_3 x)^2 + \log_3 x - \frac{5}{4}\right\}\log_3 x = \log_3 \sqrt{3}$

or $\frac{3}{4}y^3 + y^2 - \frac{5}{4}y = \frac{1}{2}$, (writing y for $\log_3 x$)

or $3y^3 + 4y^2 - 5y - 2 = 0$ or $(y-1)(3y^2 + 7y + 2) = 0$

or $(y-1)(3y+1)(y+2) = 0$

$\therefore \quad \log_3 x = 1, -\frac{1}{3}, -2 \quad \Rightarrow x = 3, 3^{-1/3}, 3^{-2}.$

□

5

Properties and Solution of Triangles

• Type 1 •

Choose the most appropriate option (a, b, c or d).

1. In a ΔABC, $\dfrac{c+b}{c-b} \cdot \tan\dfrac{A}{2}$ is equal to

(a) $\tan\left(\dfrac{A}{2}+B\right)$ (b) $\cot\left(\dfrac{A}{2}+B\right)$

(c) $\tan\left(A+\dfrac{B}{2}\right)$ (d) none of these

2. If the area of a ΔABC be λ then $a^2 \sin 2B + b^2 \sin 2A$ is equal to

(a) 2λ (b) λ

(c) 4λ (d) none of these

3. If k be the perimeter of the ΔABC then $b\cos^2\dfrac{C}{2} + c\cos^2\dfrac{B}{2}$ is equal to

(a) k (b) $2k$

(c) $\dfrac{k}{2}$ (d) none of these

4. If R denotes circumradius then in a ΔABC, $\dfrac{b^2-c^2}{2aR}$ is equal to

(a) $\cos(B-C)$ (b) $\sin(B-C)$

(c) $\cos B - \cos C$ (d) none of these

5. In a ΔABC, $\cot \dfrac{A-B}{2} \cdot \tan \dfrac{A+B}{2}$ is equal to

(a) $\dfrac{a+b}{a-b}$ (b) $\dfrac{a-b}{a+b}$

(c) $\dfrac{a(a-b)}{b(a+b)}$ (d) none of these

6. In a ΔABC, $(c+a+b)(a+b-c)=ab$. The measure of $\angle C$ is

(a) $\dfrac{\pi}{3}$ (b) $\dfrac{\pi}{6}$

(c) $\dfrac{2\pi}{3}$ (d) none of these

7. In a ΔABC, $A:B:C=3:5:4$. Then $a+b+c\sqrt{2}$ is equal to

(a) $2b$ (b) $2c$

(c) $3b$ (d) $3a$

8. The equation $ax^2+bx+c=0$, where a, b, c are the sides of a ΔABC, and the equation $x^2+\sqrt{2}x+1=0$ have a common root. The measure of $\angle C$ is

(a) 90° (b) 45°

(c) 60° (d) none of these

9. If in a ΔABC, $a^2\cos^2 A=b^2+c^2$ then

(a) $A<\dfrac{\pi}{4}$ (b) $\dfrac{\pi}{4}<A<\dfrac{\pi}{2}$

(c) $A>\dfrac{\pi}{2}$ (d) $A=\dfrac{\pi}{2}$

10. In a ΔABC, $\tan\dfrac{A}{2}$ and $\tan\dfrac{B}{2}$ satisfy $6x^2-5x+1=0$. Then

(a) $a^2+b^2>c^2$ (b) $a^2-b^2=c^2$

(c) $a^2+b^2=c^2$ (d) none of these

11. Two sides of a triangle are given by the roots of the equation $x^2-2\sqrt{3}x+2=0$. The angle between the sides is $\pi/3$ · The perimeter of the triangle is

(a) $6+\sqrt{3}$ (b) $2\sqrt{3}+\sqrt{6}$

(c) $2\sqrt{3}+\sqrt{10}$ (d) none of these

12. The sides of a $\triangle ABC$ are $AB = \sqrt{13}$ cm, $BC = 4\sqrt{3}$ cm and $CA = 7$ cm. Then $\sin\theta$, where θ is the smallest angle of the triangle, is equal to

(a) $\dfrac{\sqrt{3}}{2}$ (b) $\dfrac{1}{2}$

(c) $\dfrac{\sqrt{3}-1}{2\sqrt{2}}$ (d) none of these

13. Two sides of a triangle are $2\sqrt{2}$ cm and $2\sqrt{3}$ cm and the angle opposite to the shorter side of the two is $\pi/4$. The largest possible length of the third side is

(a) $\sqrt{2}(\sqrt{3}+1)$ cm (b) $(6+\sqrt{2})$ cm

(c) $(\sqrt{6}-\sqrt{2})$ cm (d) none of these

14. In a $\triangle ABC$, $a = 8$, $b = 10$ and $c = 12$. Then C is equal to

(a) $\dfrac{A}{2}$ (b) $2A$

(c) $3A$ (d) none of these

15. In a $\triangle ABC$, if $\tan\dfrac{A}{2} = \dfrac{5}{6}$ and $\tan\dfrac{B}{2} = \dfrac{20}{37}$ then

(a) $2a = b + c$ (b) $a > b > c$

(c) $2c = a + b$ (d) none of these

16. In a $\triangle ABC$, $a = 2b$ and $|A - B| = \dfrac{\pi}{3}$. The measure of $\angle C$ is

(a) $\dfrac{\pi}{4}$ (b) $\dfrac{\pi}{3}$

(c) $\dfrac{\pi}{6}$ (d) none of these

17. In a $\triangle ABC$, the tangent of half the difference of two angles is one-third the tangent of half the sum of the two angles. The ratio of the sides opposite the angles is

(a) $2:3$ (b) $1:3$

(c) $1:2$ (d) $3:4$

18. In a $\triangle ABC$, $A = \dfrac{2\pi}{3}$, $b - c = 3\sqrt{3}$ cm and ar $(\triangle ABC) = \dfrac{9\sqrt{3}}{2}$ cm^2. Then a is

(a) $6\sqrt{3}$ cm (b) 9 cm
(c) 18 cm (d) none of these

19. The area of a ΔABC is $a^2 - (b-c)^2$. Then $\tan A$ is equal to

(a) $\frac{4}{3}$ (b) $\frac{3}{4}$
(c) $\frac{8}{15}$ (d) none of these

20. In a ΔABC, $B = 90°$, $AC = h$ and the length of the perpendicular from B to AC is p such that $h = 4p$. If $AB < BC$ then $\angle C$ has the measure

(a) $\frac{5\pi}{12}$ (b) $\frac{\pi}{6}$
(c) $\frac{\pi}{12}$ (d) none of these

21. In a ΔABC, $a = 5$, $b = 4$ and $\tan \frac{C}{2} = \sqrt{\frac{7}{9}}$ · The side c is

(a) 6 (b) 3
(c) 2 (d) none of these

22. If in a ΔABC, $AC = 12$, $BC = 13$ and $AB = 5$, then the distance of A from BC is

(a) $\frac{25}{13}$ (b) $\frac{60}{13}$
(c) $\frac{65}{12}$ (d) none of these

23. In a ΔABC, $\cos A = \frac{3}{5}$ and $\cos B = \frac{5}{13}$ · The value of $\cos C$ can be

(a) $\frac{7}{13}$ (b) $\frac{12}{13}$
(c) $\frac{33}{65}$ (d) none of these

24. In a ΔABC, $B = \frac{\pi}{8}$ and $C = \frac{5\pi}{8}$ · The altitude from A to the side BC is

(a) $\frac{a}{2}$ (b) $2a$

(c) $\frac{1}{2}(b+c)$ (d) none of these

25. Two angles of a triangle are $\frac{\pi}{6}$ and $\frac{\pi}{4}$, and the length of the included side is $(\sqrt{3}+1)$ cm. The area of the triangle is

(a) $\frac{\sqrt{3}-1}{2}\text{ cm}^2$ (b) $\frac{\sqrt{3}}{2}\text{ cm}^2$

(c) $\frac{\sqrt{3}+1}{2}\text{ cm}^2$ (d) none of these

26. If in a ΔABC, $\sin^3 A+\sin^3 B+\sin^3 C=3\sin A\cdot\sin B\cdot\sin C$, then the value of the determinant

$$\begin{vmatrix} a & b & c \\ b & c & a \\ c & a & b \end{vmatrix} \text{ is}$$

(a) 0 (b) $(a+b+c)^3$

(c) $(a+b+c)(ab+bc+ca)$ (d) none of these

27. In a ΔABC, $A=90°$. Then $\tan^{-1}\frac{b}{a+c}+\tan^{-1}\frac{c}{a+b}$ is equal to

(a) $\frac{\pi}{4}$ (b) $\frac{\pi}{2}$

(c) $\tan^{-1}\frac{a}{b+c}$ (d) none of these

28. If in a ΔABC, the values of $\cot A$, $\cot B$, $\cot C$ are in AP, then

(a) a, b, c are in AP (b) a^2, b^2, c^2 are in AP

(c) $\cos A$, $\cos B$, $\cos C$ are in AP (d) none of these

29. If in a ΔABC, $\frac{a}{\cos A}=\frac{b}{\cos B}$, then

(a) $2\sin A\sin B\sin C-1$ (b) $\sin^2 A+\sin^2 B=\sin^2 C$

(c) $2\sin A\cos B=\sin C$ (d) none of these

30. If the sides of a triangle are in GP and the largest angle is twice the smallest angle then the common ratio, which is greater than 1, lies in the interval

(a) $(1, \sqrt{3})$ (b) $(1, \sqrt[4]{3})$

(c) $\left(1, \frac{\sqrt{5}+1}{2}\right)$ (d) none of these

31. In a ΔABC, the sides a, b and c are such that they are the roots of $x^3 - 11x^2 + 38x - 40 = 0$. Then $\frac{\cos A}{a} + \frac{\cos B}{b} + \frac{\cos C}{c}$ is equal to

(a) $\frac{3}{4}$ (b) 1

(c) $\frac{9}{16}$ (d) none of these

32. If $\cos A + \cos B + 2\cos C = 2$ then the sides of the ΔABC are in

(a) AP (b) GP

(c) HP (d) none of these

33. If in the ΔABC, the incentre is the middle point of the median AD then $\cos A$ has the value

(a) $\frac{7}{8}$ (b) $\frac{1}{4}$

(c) $\frac{1}{3}$ (d) $\frac{1}{\sqrt{2}}$

34. If in a ΔABC, $3a = b + c$ then $\tan \frac{B}{2} \cdot \tan \frac{C}{2}$ is equal to

(a) $\tan \frac{A}{2}$ (b) 1

(c) 2 (d) none of these

35. If in a ΔABC, $2\cos A \sin C = \sin B$ then the triangle is

(a) equilateral (b) isosceles

(c) right angled (d) none of these

36. In a ΔABC, a = 1 and the perimeter is six times the AM of the sines of the angles. The measure of $\angle A$ is

(a) $\frac{\pi}{3}$ (b) $\frac{\pi}{2}$

(c) $\frac{\pi}{6}$ (d) $\frac{\pi}{4}$

37. In a ΔABC, $\angle A > \angle B$. If $\sin A$ and $\sin B$ satisfy the equation $3\sin x - 4\sin^3 x - k = 0, 0 < k < 1$, then $\angle C$ is

(a) $\frac{\pi}{3}$ (b) $\frac{\pi}{2}$

(c) $\frac{2\pi}{3}$ (d) $\frac{5\pi}{6}$

38. In a ΔABC the sides a, b and c are in AP. Then $\left(\tan\frac{A}{2}+\tan\frac{C}{2}\right):\cot\frac{B}{2}$ is equal to

(a) $3:2$ (b) $1:2$

(c) $3:4$ (d) none of these

39. If the sides of a triangle are proportional to the cosines of the opposite angles then the triangle is

(a) right angled (b) equilateral

(c) obtuse angled (d) none of these

40. If in a ΔABC, $c\cos^2\frac{A}{2}+a\cos^2\frac{C}{2}=\frac{3b}{2}$, then a, b, c are in

(a) GP (b) HP

(c) AP (d) none of these

41. The sides of a triangle are in AP and its area is $\frac{3}{5}\times$ (area of an equilateral triangle of the same perimeter). Then the ratio of the sides is

(a) $1:2:3$ (b) $3:5:7$

(c) $1:3:5$ (d) none of these

42. If BD, BE and CF are the medians of a ΔABC then

$$(AD^2+BE^2+CF^2):(BC^2+CA^2+AB^2)$$

is equal to

(a) $4:3$ (b) $3:2$

(c) $3:4$ (d) $2:3$

43. $\sin A$, $\sin B$ and $\sin C$ are in AP for the ΔABC Then

(a) the altitudes are in AP (b) the altitudes are in HP

(c) the medians are in GP (d) the medians are in AP

44. AD is a median of the ΔABC. If AE and AF are medians of the triangles ABD and ADC respectively, and $AD=m_1$, $AE=m_2$, $AF=m_3$, then $\frac{a^2}{8}$ is equal to

(a) $m_2^2+m_3^2-2m_1^2$ (b) $m_1^2+m_2^2-2m_3^2$

(c) $m_2^2 + m_3^2 - m_1^2$ (d) none of these

45. The ratio of the distances of the orthocentre of an acute-angled $\triangle ABC$ from the sides BC, AC and AB is

(a) $\cos A : \cos B : \cos C$ (b) $\sin A : \sin B : \sin C$

(c) $\sec A : \sec B : \sec C$ (d) none of these

46. In a $\triangle ABC$, I is the incentre. The ratio $IA : IB : IC$ is equal to

(a) $\operatorname{cosec} \frac{A}{2} : \operatorname{cosec} \frac{B}{2} : \operatorname{cosec} \frac{C}{2}$ (b) $\sin \frac{A}{2} : \sin \frac{B}{2} : \sin \frac{C}{2}$

(c) $\sec \frac{A}{2} : \sec \frac{B}{2} : \sec \frac{C}{2}$ (d) none of these

47. In the figure, ABC is a triangle in which $C = 90°$ and $AB = 5$ cm.

D is a point on AB such that $AD = 3$ cm and $\angle ACD = 60°$. Then the length of AC is

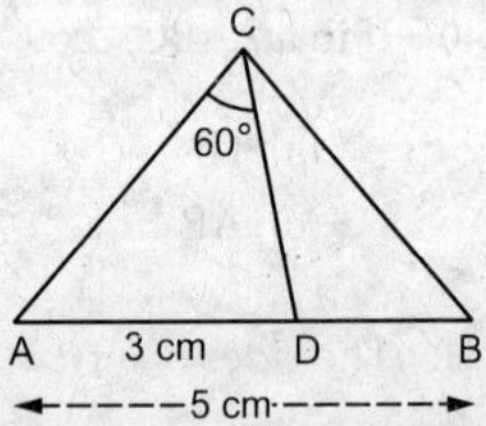

(a) $5\sqrt{\frac{3}{7}}$ cm (b) $\sqrt{\frac{7}{3}}$ cm

(c) $\frac{3}{\sqrt{7}}$ cm (d) none of these

48. If α, β, γ are the altitudes of a $\triangle ABC$ and $2s$ denotes its perimeter then $\alpha^{-1} + \beta^{-1} + \gamma^{-1}$ is equal to

(a) $\frac{\Delta}{s}$ (b) $\frac{s}{\Delta}$

(c) $s \cdot \Delta$ (d) none of these

49. In a $\triangle ABC$, the sides are in the ratio 4 : 5 : 6. The ratio of the circumradius and the inradius is

(a) 8 : 7 (b) 3 : 2

(c) 7 : 3 (d) 16 : 7

50. In a $\triangle ABC$, R = circumradius and r = inradius. The value of

$$\frac{a\cos A + b\cos B + c\cos C}{a + b + c}$$

is equal to

(a) $\frac{R}{r}$ (b) $\frac{R}{2r}$

(c) $\frac{r}{R}$ (d) $\frac{2r}{R}$

51. In a $\triangle ABC$, $2s$ = perimeter and R = circumradius. Then s/R is equal to

(a) $\sin A + \sin B + \sin C$ (b) $\cos A + \cos B + \cos C$

(c) $\sin\frac{A}{2} + \sin\frac{B}{2} + \sin\frac{C}{2}$ (d) none of these

52. The ratio of the circumradius and inradius of an equilateral triangle is

(a) $3 : 1$ (b) $1 : 1$

(c) $2 : \sqrt{3}$ (d) $2 : 1$

53. If in a $\triangle ABC, a^2 + b^2 + c^2 = 8R^2$, where R = circumradius, then the triangle is

(a) equilateral (b) isosceles

(c) right angled (d) none of these

54. In an equilateral triangle, (circumradius) : (inradius) : (exradius) is equal to

(a) $1 : 1 : 1$ (b) $1 : 2 : 3$

(c) $2 : 1 : 3$ (d) $3 : 2 : 4$

55. The diameter of the circumcircle of a triangle with sides 5 cm, 6 cm and 7 cm is

(a) $\frac{3\sqrt{6}}{2}$ cm (b) $2\sqrt{6}$ cm

(c) $\frac{35}{48}$ cm (d) none of these

56. If in a triangle, R and r are the circumradius and inradius respectively then the HM of the exradii of the triangle is

(a) $3r$ (b) $2R$

(c) $R + r$ (d) none of these

57. The angles of a right-angled triangle are in AP. The ratio of the inradius and the perimeter is

(a) $(2 - \sqrt{3}) : 2\sqrt{3}$ (b) $1 : 8\sqrt{3}(2 + \sqrt{3})$

(c) $(2 + \sqrt{3}) : 4\sqrt{3}$ (d) none of these

58. A $\triangle ABC$ is right angled at B. Then the diameter of the incircle of the triangle is

(a) $2(c+a-b)$ (b) $c+a-2b$

(c) $c+a-b$ (d) none of these

59. If the exradii of a triangle are in HP then the corresponding sides are in

(a) AP (b) GP

(c) HP (d) none of these

60. In a ΔABC, the inradius and three exradii are r, r_1, r_2 and r_3 respectively. In usual notations the value of $r \cdot r_1 \cdot r_2 \cdot r_3$ is equal to

(a) 2Δ (b) Δ^2

(c) $\frac{abc}{4R}$ (d) none of these

61. In a ΔABC, the perimeter $= 2s$ and the exradii are r_1, r_2 and r_3. Then $r_1r_2 + r_2r_3 + r_3r_1$ is equal to

(a) s^2 (b) $2s^2$

(c) $3s^2$ (d) $4s^2$

62. In a ΔABC, $\tan A \cdot \tan B \cdot \tan C = 9$. For such triangles, if

$$\tan^2 A + \tan^2 B + \tan^2 C = k$$

then

(a) $9 \cdot \sqrt[3]{3} < k < 27$ (b) $k \nleq 27$

(c) $k < 9 \cdot \sqrt[3]{3}$ (d) $k \nless 27$

63. If the ΔABC is acute angled at C then

(a) $\cos 2A + \cos 2B - \cos 2C < 1$ (b) $\cos 2A + \cos 2B + \cos 2C > 1$

(c) $\cos^2 A + \cos^2 B + \cos^2 C > 1$ (d) none of these

64. If for a ΔABC, $\cot A \cdot \cot B \cdot \cot C > 0$ then the triangle is

(a) right angled (b) acute angled

(c) obtuse angled (d) all these options are possible

65. In a ΔABC, $\cos B \cdot \cos C + \sin B \cdot \sin C \cdot \sin^2 A = 1$. Then the triangle is

(a) right-angled isosceles

(b) isosceles whose equal angles are greater than $\pi/4$

(c) equilateral

(d) none of these

66. In a ΔABC, the angles A and B are two values of θ satisfying $\sqrt{3}\cos\theta + \sin\theta = k$, $|k| < 2$. The triangle

(a) is acute angled (b) is right angled

(c) is obtuse angled (d) has one angle $= \pi/3$

67. If in an obtuse-angled triangle the obtuse angle is $3\pi/4$ and the other two angles are equal to two values of θ satisfying $a\tan\theta + b\sec\theta = c$, where $|b| \le \sqrt{a^2 + c^2}$, then $a^2 - c^2$ is equal to

(a) ac (b) $2ac$

(c) a/c (d) none of these

68. Let A_0, A_1, A_2, A_3, A_4 and A_5 be the consecutive vertices of a regular hexagon inscribed in a unit circle. The product of the lengths of A_0A_1, A_0A_2 and A_0A_4 is

(a) $\dfrac{3}{4}$ (b) $3\sqrt{3}$

(c) 3 (d) $\dfrac{3\sqrt{3}}{2}$

69. The area of a circle is A_1 and the area of a regular pentagon inscribed in the circle is A_2. Then $A_1 : A_2$ is

(a) $\dfrac{\pi}{5}\cos\dfrac{\pi}{10}$ (b) $\dfrac{2\pi}{5}\sec\dfrac{\pi}{10}$

(c) $\dfrac{2\pi}{5}\operatorname{cosec}\dfrac{\pi}{10}$ (d) none of these

70. The area of a cyclic quadrilateral ABCD is $(3\sqrt{3})/4$. The radius of the circle circumscribing it is 1. If $AB = 1$, $BD = \sqrt{3}$ then $BC \cdot CD$ is equal to

(a) 2 (b) $3 - \dfrac{1}{\sqrt{3}}$

(c) $3\sqrt{3} + 1$ (d) none of these

• *Type 2* •

Choose the correct options. One or more options may be correct.

71. If in a ΔABC, $a = 6$, $b = 3$ and $\cos(A - B) = \frac{4}{5}$ then

(a) $C = \frac{\pi}{4}$ (b) $A = \sin^{-1}\frac{2}{\sqrt{5}}$

(c) ar $(\Delta ABC) = 9$ (d) none of these

72. The number of possible triangles ABC in which $BC = \sqrt{11}$ cm, $CA = \sqrt{13}$ cm and $A = 60°$ is

(a) 0 (b) 1

(c) 2 (d) none of these

73. In a ΔABC, $A = \frac{\pi}{3}$ and $b : c = 2 : 3$. If $\tan\alpha = \frac{\sqrt{3}}{5}$, $0 < \alpha = \frac{\pi}{2}$, then

(a) $B = 60° + \alpha$ (b) $C = 60° + \alpha$

(c) $B = 60° - \alpha$ (d) $C = 60° - \alpha$

74. In a triangle the cosines of two angles are inversely proportional to the sides opposite the angles. The triangle is

(a) isosceles (b) equilateral

(c) right angled (d) none of these

75. In a ΔABC, the line segments AD, BE and CF are three altitudes. If R is the circumradius of the ΔABC, a side of the ΔDEF will be

(a) $R\sin 2A$ (b) $c\cos B$

(c) $a\sin A$ (d) $b\cos B$

76. In a ΔABC, $\tan A$ and $\tan B$ are the roots of the equation $ab(x^2 + 1) = c^2x$, where a, b and c are the sides of the triangle. Then

(a) $\tan(A - B) = \frac{a^2 - b^2}{2ab}$ (b) $\cot C = 0$

(c) $\sin^2 A + \sin^2 B = 1$ (d) none of these

77. The distances of the circumcentre of the acute-angled ΔABC from the sides BC, CA and AB are in the ratio

(a) $a\sin A : b\sin B : c\sin C$ (b) $\cos A : \cos B : \cos C$

(c) $a\cot A : b\cot B : c\cot C$ (d) none of these

78. In any $\triangle ABC$, $\Sigma a^3 \sin(B - C)$ is equal to

(a) $\Sigma a(\sin B - \sin C)$ (b) $\Sigma a^2(\cos^2 B - \cos^2 C)$

(c) 0 (d) none of these

79. In a $\triangle ABC$, $2\cos\dfrac{A-C}{2} = \dfrac{a+c}{\sqrt{a^2+c^2-ac}}$. Then

(a) $B = \dfrac{\pi}{3}$ (b) $B = C$

(c) A, B, C are in AP (d) $B + C = A$

80. Let $L\sin\theta = 10 + \log\sin\theta$. The number of triangles ABC such that $\log b + 10 = \log c + L\sin B$ is

(a) one (b) two

(c) infinite (d) none of these

81. In any triangle ABC, $\sin\dfrac{A}{2}$ is

(a) less than $\dfrac{b+c}{a}$ (b) less than or equal to $\dfrac{a}{b+c}$

(c) greater than $\dfrac{2a}{a+b+c}$ (d) none of these

82. In a $\triangle ABC$, $\tan C < 0$. Then

(a) $\tan A \cdot \tan B < 1$ (b) $\tan A \cdot \tan B > 1$

(c) $\tan A + \tan B + \tan C < 0$ (d) $\tan A + \tan B + \tan C > 0$

83. In a $\triangle ABC$, $\tan A$ and $\tan B$ satisfy the inequation $\sqrt{3}x^2 - 4x + \sqrt{3} < 0$. Then

(a) $a^2 + b^2 + ab > c^2$ (b) $a^2 + b^2 - ab < c^2$

(c) $a^2 + b^2 > c^2$ (d) none of these

84. In a $\triangle ABC$, $\cos A + \cos B + \cos C > 1$ only if the triangle is

(a) acute angled

(b) obtuse angled

(c) right angled

(d) the nature of the triangle cannot be determined

Answers

1. a	**2.** c	**3.** c	**4.** b	**5.** a
6. c	**7.** c	**8.** b	**9.** c	**10.** c
11. b	**12.** b	**13.** a	**14.** b	**15.** b
16. b	**17.** c	**18.** b	**19.** c	**20.** c
21. a	**22.** b	**23.** c	**24.** a	**25.** c
26. a	**27.** a	**28.** b	**29.** c	**30.** b
31. c	**32.** a	**33.** b	**34.** d	**35.** b
36. c	**37.** c	**38.** d	**39.** b	**40.** c
41. b	**42.** c	**43.** b	**44.** a	**45.** c
46. a	**47.** a	**48.** b	**49.** d	**50.** c
51. a	**52.** d	**53.** c	**54.** c	**55.** d
56. a	**57.** a	**58.** c	**59.** a	**60.** b
61. a	**62.** b	**63.** a	**64.** b	**65.** a
66. c	**67.** b	**68.** c	**69.** b	**70.** a
71. b, c	**72.** c	**73.** b, c	**74.** a, c	**75.** a, d
76. a, b, c	**77.** b, c	**78.** a, b, c	**79.** a, c	**80.** a
81. b	**82.** a, c	**83.** a, b	**84.** d	

Hints and Solutions to Selected Questions

1. $$\frac{c+b}{c-b}\tan\frac{A}{2}=\frac{\sin C+\sin B}{\sin C-\sin B}\tan\frac{A}{2}=\frac{2\sin\frac{B+C}{2}\cdot\cos\frac{C-B}{2}}{2\cos\frac{B+C}{2}\cdot\sin\frac{C-B}{2}}\cdot\tan\frac{A}{2}$$

$$=\tan\frac{B+C}{2}\cdot\cot\frac{C-B}{2}\cdot\tan\frac{A}{2}=\cot\frac{C-B}{2}=\cot\frac{\pi-A-B-B}{2}$$

$$=\cot\left(\frac{\pi}{2}-\frac{A}{2}+B\right)=\tan\left(\frac{A}{2}+B\right).$$

2. $$a^2\sin 2B+b^2\sin 2A=2a^2\sin B\cdot\cos B+2b^2\sin A\cos A$$

$$=\frac{a^2b}{R}\cos B+\frac{b^2a}{R}\cos A$$

$$=\frac{ab}{R}(a\cos B+b\cos A)=\frac{abc}{R}=2bc\sin A$$

$$=4\left(\frac{1}{2}bc\sin A\right)=4\lambda.$$

3. $$b\cos^2\frac{C}{2}+c\cos^2\frac{B}{2}=\frac{b}{2}(1+\cos C)+\frac{c}{2}(1+\cos B)$$

$$=\frac{b}{2}+\frac{c}{2}+\frac{1}{2}(b\cos C+c\cos B)=\frac{a+b+c}{2}=\frac{k}{2}.$$

4. $$\frac{b^2-c^2}{2aR}=\frac{4R^2(\sin^2 B-\sin^2 C)}{4R^2\sin A}=\frac{\sin(B+C)\sin(B-C)}{\sin A}=\sin(B-C).$$

6. $2s(2s-2c)=ab$ or $\dfrac{s(s-c)}{ab}=\dfrac{1}{4}$

or $\cos^2\dfrac{C}{2}=\dfrac{1}{4}$ or $\cos\dfrac{C}{2}=\dfrac{1}{2}$ $\left(\because \dfrac{C}{2}\text{ must be acute}\right)$

7. Clearly, $A=45°, B=75°, C=60°$. So $\dfrac{a}{\sin 45°}=\dfrac{b}{\sin 75°}=\dfrac{c}{\sin 60°}=2R.$

$$\therefore\quad a=\sqrt{2}R,\ b=\frac{\sqrt{3}+1}{\sqrt{2}}R,\ c=\sqrt{3}R.$$

Now, $a+b+c\sqrt{2}=\left(\sqrt{2}+\dfrac{\sqrt{3}+1}{\sqrt{2}}+\sqrt{6}\right)R=3b.$

8. Clearly, the roots of $x^2+\sqrt{2}x+1=0$ are nonreal complex. So, one root common implies both roots are common.

So, $\dfrac{a}{1}=\dfrac{b}{\sqrt{2}}=\dfrac{c}{1}=k.$

$$\therefore \quad \cos C=\frac{a^2+b^2-c^2}{2ab}=\frac{k^2+2k^2-k^2}{2\cdot k\cdot\sqrt{2}k}=\frac{1}{\sqrt{2}}.$$

9. $\cos A=\dfrac{b^2+c^2-a^2}{2bc}=\dfrac{a^2\cos^2 A-a^2}{2bc}=\dfrac{a^2(\cos^2 A-1)}{2bc}<0.$

So, $A>\dfrac{\pi}{2}\cdot$

10. $\tan\dfrac{A}{2}+\tan\dfrac{B}{2}=\dfrac{5}{6}$, $\tan\dfrac{A}{2}\cdot\tan\dfrac{B}{2}=\dfrac{1}{6}\Rightarrow\tan\left(\dfrac{A+B}{2}\right)=1$

$$\Rightarrow\quad A+B=2\times\frac{\pi}{4}\cdot$$

11. Here $a+b=2\sqrt{3}$, $ab=2$ and $C=\dfrac{\pi}{3}\cdot$

$$\cos C=\frac{a^2+b^2-c^2}{2ab}\quad\Rightarrow\frac{1}{2}=\frac{a^2+b^2-c^2}{2ab}\quad\Rightarrow a^2+b^2-c^2=ab$$

or $\quad (a+b)^2-2ab-c^2=ab\quad$ or $\;12-4-c^2=2\quad$ or $\;c=\sqrt{6}.$

$\therefore\quad$ the perimeter $=a+b+c=2\sqrt{3}+\sqrt{6}.$

13. $\cos\dfrac{\pi}{4}=\dfrac{(2\sqrt{3})^2+x^2-(2\sqrt{2})^2}{2\cdot 2\sqrt{3}\cdot x}\;$ or $\;\dfrac{4\sqrt{3}}{\sqrt{2}}x=x^2+4\;$ or $\;x^2-\dfrac{4\sqrt{3}}{\sqrt{2}}x+4=0.$

$\therefore\quad x=\sqrt{6}\pm\sqrt{2}.$

14. $\cos C=\dfrac{8^2+10^2-12^2}{2\cdot 8\cdot 10}=\dfrac{1}{8}\cdot$

$$\cos A=\frac{10^2+12^2-8^2}{2\cdot 10\cdot 12}=\frac{3}{4}\cdot$$

$$\cos 2A=2\cos^2 A-1=2\times\frac{9}{16}-1=\frac{1}{8}\cdot$$

So, $\cos C=\cos 2A\quad\Rightarrow\quad C=2A.$

15. Clearly, $\tan\frac{A}{2} > \tan\frac{B}{2}$; $\therefore\ \frac{A}{2} > \frac{B}{2}$ or $A > B$.

$$\tan\frac{A+B}{2} = \frac{\frac{5}{6}+\frac{20}{37}}{1-\frac{5}{6}\cdot\frac{20}{37}} = \frac{305}{122}$$

or $\tan\frac{C}{2} = \frac{122}{305}$.

As $\frac{20}{37} > \frac{122}{305}$, we get

$\tan\frac{B}{2} > \tan\frac{C}{2} \quad \Rightarrow \quad \frac{B}{2} > \frac{C}{2}$ i.e., $B > C$.

$\therefore \quad A > B > C \qquad \therefore \quad a > b > c$.

$\therefore \quad 2a > b + c$ and $a + b > 2c$.

16. Clearly, $A > B$ $(\because a > b)$.

Now $\tan\frac{A-B}{2} = \frac{a-b}{a+b}\cot\frac{C}{2} \quad \Rightarrow \tan 30° = \frac{1}{3}\cot\frac{C}{2}$

$\therefore \quad \sqrt{3} = \cot\frac{C}{2}$ or $\frac{C}{2} = \frac{\pi}{6}$.

17. $\tan\frac{A-B}{2} = \frac{1}{3}\tan\frac{A+B}{2} = \frac{1}{3}\cot\frac{C}{2}$

$\Rightarrow \quad \frac{a-b}{a+b} = \frac{1}{3}$.

18. $\frac{1}{2}bc\sin\frac{2\pi}{3} = \frac{9\sqrt{3}}{2}$ or $\frac{1}{2}\cdot\frac{\sqrt{3}}{2}\cdot bc = \frac{9\sqrt{3}}{2} \quad \Rightarrow bc = 18$.

Also, $\cos\frac{2\pi}{3} = \frac{b^2+c^2-a^2}{2bc} \quad \Rightarrow -\frac{1}{2} = \frac{(b-c)^2+2bc-a^2}{2bc}$

or $\quad (b-c)^2 + 3bc - a^2 = 0$ or $27 + 54 = a^2$.

19. Here $\Delta = a^2 - (b-c)^2$

$\Rightarrow \quad \Delta = (2s-2c)(2s-2b) = 4(s-b)(s-c)$

$\therefore \quad \frac{1}{4} = \frac{(s-b)(s-c)}{\Delta} = \tan\frac{A}{2}$

$$\Rightarrow \quad \tan A = \frac{2\tan\frac{A}{2}}{1-\tan^2\frac{A}{2}} = \frac{2\times\frac{1}{4}}{1-\left(\frac{1}{4}\right)^2}.$$

20. Here, $CD = p\cot C$ and $AD = p\cot A = p\tan C$

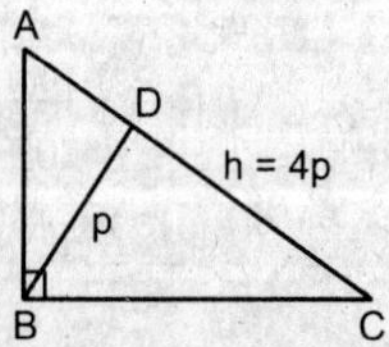

Adding, $4p = p\cot C + p\tan C$

$$\Rightarrow \quad \tan C + \cot C = 4$$

or $\dfrac{1}{\sin C\cdot\cos C} = 4$ or $\sin 2C = \dfrac{1}{2}$

$$\Rightarrow \quad C = 15°.$$

21. $\tan^2\dfrac{C}{2} = \dfrac{(s-a)(s-b)}{s(s-c)} = \dfrac{(9+c-10)(9+c-8)}{(9+c)(9-c)} = \dfrac{c^2-1}{81-c^2}$

$$\therefore \quad \frac{7}{9} = \frac{c^2-1}{81-c^2} \quad \Rightarrow c = 6.$$

22. $\dfrac{1}{2}\cdot p\cdot a = \text{area} = \sqrt{s(s-a)(s-b)(s-c)}.$

23. $\tan A = \dfrac{4}{3}$ and $\tan B = \dfrac{12}{5}$. Clearly, $\tan C$ should be such that

$$\tan A + \tan B + \tan C = \tan A \tan B \tan C$$

$\therefore \quad \dfrac{4}{3}+\dfrac{12}{5}+\tan C = \dfrac{4}{3}\cdot\dfrac{12}{5}\cdot\tan C$ or $\dfrac{56}{15}+\tan C = \dfrac{16}{5}\tan C$

or $\quad \tan C = \dfrac{56}{33}$

$\therefore \quad \cos C = \dfrac{33}{65}.$

24. $\dfrac{1}{2}pa = \text{area} = \dfrac{1}{2}bc\cdot\sin\dfrac{\pi}{4}; \quad \therefore\ p = \dfrac{bc}{a\sqrt{2}}.$

Also $\dfrac{a}{\sin\frac{-\pi}{4}} = \dfrac{b}{\sin\frac{\pi}{8}} = \dfrac{c}{\sin\frac{5\pi}{8}} \quad \Rightarrow b = \sqrt{2}a\sin\dfrac{\pi}{8},\ c = \sqrt{2}a\sin\dfrac{5\pi}{8}$

$$\therefore \quad p = \frac{2a^2\cdot\sin\frac{\pi}{8}\cdot\sin\frac{5\pi}{8}}{a\sqrt{2}} = \frac{a}{\sqrt{2}}\left(\cos\frac{4\pi}{8}-\cos\frac{6\pi}{8}\right).$$

25. 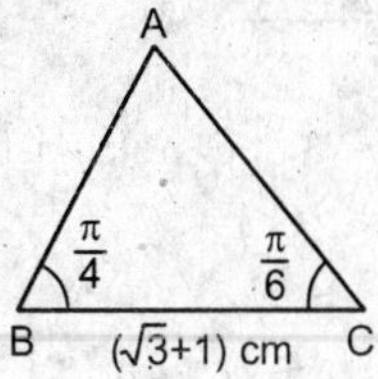

$$A = \pi - \frac{\pi}{4} - \frac{\pi}{6} = 105°.$$

$$\text{So, } \frac{\sqrt{3}+1}{\sin 105°} = \frac{c}{\sin\frac{\pi}{6}} \Rightarrow c = \frac{\frac{\sqrt{3}+1}{2}}{\frac{\sqrt{3}+1}{2\sqrt{2}}} = \sqrt{2}$$

$$\therefore \quad \text{area} = \frac{1}{2} ac \sin B = \frac{1}{2} \cdot (\sqrt{3}+1)\sqrt{2} \cdot \sin\frac{\pi}{4}.$$

26. $$\begin{vmatrix} a & b & c \\ b & c & a \\ c & a & b \end{vmatrix} = (a+b+c) \begin{vmatrix} 1 & 1 & 1 \\ b & c & a \\ c & a & b \end{vmatrix}$$

$$= (a+b+c)(bc+ca+ab-a^2-b^2-c^2)$$

$$= -(a^3+b^3+c^3-3abc)$$

$$= -8R^3(\sin^3 A + \sin^3 B + \sin^3 C - 3\sin A \sin B \sin C)$$

$= 0$, from the question.

28. $\cot A, \cot B, \cot C$ are in AP

$$\Rightarrow \quad \frac{b^2+c^2-a^2}{2bc \cdot \frac{a}{2R}}, \frac{c^2+a^2-b^2}{2ca \cdot \frac{b}{2R}}, \frac{a^2+b^2-c^2}{2ab \cdot \frac{c}{2R}} \text{ are in AP}$$

$\Rightarrow \quad b^2+c^2-a^2, c^2+a^2-b^2, a^2+b^2-c^2$ are in AP

$\Rightarrow \quad -2a^2, -2b^2, -2c^2$ are in AP (subtracting $a^2+b^2+c^2$ from each).

29. $\dfrac{2R\sin A}{\cos A} = \dfrac{2R\sin B}{\cos B}$ or $\sin A \cos B = \cos A \sin B$

or $\quad 2\sin A \cos B = \cos A \sin B + \sin A \cos B = \sin(A+B) = \sin C.$

30. $b = ar, c = ar^2$, where $r > 1$.

From the question, $C = 2A$. So, $B = \pi - A - C = \pi - 3A$.

$$\therefore \quad \frac{a}{\sin A} = \frac{b}{\sin B} = \frac{c}{\sin C} \Rightarrow \frac{1}{\sin A} = \frac{r}{\sin 3A} = \frac{r^2}{\sin 2A}$$

$$\therefore \quad r^2 = 2\cos A \text{ and } r = \frac{\sin 3A}{\sin A} = 3 - 4\sin^2 A = 4\cos^2 A - 1$$

$$\therefore \quad r = r^4 - 1 \quad \text{or} \quad r^4 = 1 + r.$$

Among $\sqrt{3}, \sqrt[4]{3}, \dfrac{\sqrt{5}+1}{2}$ we find $\sqrt[4]{3}$ is the smallest because

$$(\sqrt{3})^4 = 9, (\sqrt[4]{3})^4 = 3, \left(\frac{\sqrt{5}+1}{2}\right)^4 = \left(\frac{3+\sqrt{5}}{2}\right)^2 = \frac{14+6\sqrt{5}}{4} > 3.$$

Now, putting $r = \sqrt[4]{3}$, we have $3 > 1 + \sqrt[4]{3}$, implying $\sqrt[4]{3}$ does not satisfy $r^4 = 1 + r$. So, $r < \sqrt[4]{3}$.

31. Expression $= \dfrac{b^2+c^2-a^2}{2abc} + \dfrac{c^2+a^2-b^2}{2abc} + \dfrac{a^2+b^2-c^2}{2abc} = \dfrac{a^2+b^2+c^2}{2abc}$

$$= \frac{(a+b+c)^2 - 2(ab+bc+ca)}{2abc} = \frac{11^2 - 2\cdot 38}{2\cdot 40} = \frac{9}{16}.$$

32. $2\cos\dfrac{A+B}{2}\cdot\cos\dfrac{A-B}{2} = 2\sin^2\dfrac{C}{2}$

$\Rightarrow \quad \cos\dfrac{A-B}{2} = 2\cos\dfrac{A+B}{2} \quad \left(\because \cos\dfrac{A+B}{2} \neq 0\right)$

or $\quad \cos\dfrac{A}{2}\cos\dfrac{B}{2} + \sin\dfrac{A}{2}\cdot\sin\dfrac{B}{2} = 2\left(\cos\dfrac{A}{2}\cos\dfrac{B}{2} - \sin\dfrac{A}{2}\cdot\sin\dfrac{B}{2}\right)$

or $\quad 3\sin\dfrac{A}{2}\cdot\sin\dfrac{B}{2} = \cos\dfrac{A}{2}\cos\dfrac{B}{2}$

$\Rightarrow \quad \tan\dfrac{A}{2}\cdot\tan\dfrac{B}{2} = \dfrac{1}{3}$

or $\quad \sqrt{\dfrac{(s-b)(s-c)}{s(s-a)}}\cdot\sqrt{\dfrac{(s-c)(s-a)}{s(s-b)}} = \dfrac{1}{3} \quad \Rightarrow \quad \dfrac{s-c}{s} = \dfrac{1}{3}$

$\Rightarrow \quad \dfrac{a+b-c}{a+b+c} = \dfrac{1}{3} \quad \Rightarrow 3a+3b-3c = a+b+c$

$\Rightarrow \quad a+b = 2c \quad \Rightarrow a, c, b$ are in AP.

33.

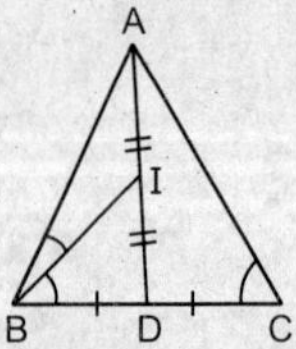

Here, $\dfrac{BD}{AB} = \dfrac{DI}{AI} \quad \Rightarrow \quad \dfrac{a/2}{c} = \dfrac{1}{1} \quad \therefore \; a = 2c.$

Similarly, $b = 2c$.

$\therefore \quad \cos A = \dfrac{b^2+c^2-a^2}{2bc} = \dfrac{4c^2+c^2-4c^2}{2\cdot 2c\cdot c}$

$$= \frac{1}{4}.$$

34. $3\sin A = \sin B + \sin C \quad \Rightarrow \quad 6\sin\dfrac{A}{2}\cdot\cos\dfrac{A}{2} = 2\sin\dfrac{B+C}{2}\cdot\cos\dfrac{B-C}{2}$

$$\Rightarrow \quad 3\cos\frac{B+C}{2} = \cos\frac{B-C}{2}$$

$$\Rightarrow \quad 3\left(\cos\frac{B}{2}\cdot\cos\frac{C}{2} - \sin\frac{B}{2}\cdot\sin\frac{C}{2}\right) = \cos\frac{B}{2}\cdot\cos\frac{C}{2} + \sin\frac{B}{2}\sin\frac{C}{2}$$

$$\Rightarrow \quad 2\cos\frac{B}{2}\cdot\cos\frac{C}{2} = 4\sin\frac{B}{2}\cdot\sin\frac{C}{2} \quad \Rightarrow \tan\frac{B}{2}\cdot\tan\frac{C}{2} = \frac{1}{2}\cdot$$

35. $2\cos A\cdot\sin C = \sin(C+A) = \sin C\cdot\cos A + \cos C\cdot\sin A$

$$\Rightarrow \quad \sin(C-A) = 0 \quad \Rightarrow C = A.$$

36. $1+b+c = 6\cdot\dfrac{\sin A+\sin B+\sin C}{3} = 2\left(\dfrac{1}{2R}+\dfrac{b}{2R}+\dfrac{c}{2R}\right) = \dfrac{1}{R}(1+b+c)$

$\therefore \quad R = 1 \quad (\because \quad 1+b+c \neq 0).$

So, $\dfrac{1}{\sin A} = 2R = 2 \quad \Rightarrow A = \dfrac{\pi}{6}\cdot$

37. From the question, $\sin 3A = k$, $\sin 3B = k$

$$\Rightarrow \quad \sin 3A = \sin 3B \quad \Rightarrow 3A+3B = \pi \quad \Rightarrow 3(\pi - C) = \pi.$$

38. $2b = a+c \quad \Rightarrow 2\sin B = \sin A + \sin C \quad \Rightarrow 2\sin\dfrac{B}{2} = \cos\dfrac{A-C}{2}$

$$\Rightarrow \quad 2\cos\frac{A+C}{2} = \cos\frac{A-C}{2}$$

$$\Rightarrow \quad \cos\frac{A}{2}\cdot\cos\frac{C}{2} = 3\sin\frac{A}{2}\cdot\sin\frac{C}{2} \quad \Rightarrow \tan\frac{A}{2}\cdot\tan\frac{C}{2} = \frac{1}{3}$$

$$\therefore \quad \frac{\tan\frac{A}{2}+\tan\frac{C}{2}}{\cot\frac{B}{2}} = \frac{\cos\frac{B}{2}}{\cos\frac{A}{2}\cdot\cos\frac{C}{2}}\cdot\frac{\sin\frac{B}{2}}{\cos\frac{B}{2}} = \frac{\cos\frac{A+C}{2}}{\cos\frac{A}{2}\cdot\cos\frac{C}{2}}$$

$$= 1 - \tan\frac{A}{2}\cdot\tan\frac{C}{2} = 1 - \frac{1}{3} = \frac{2}{3}\cdot$$

40. $c(1+\cos A) + a(1+\cos C) = 3b$

$$\Rightarrow \quad a+c+(c\cos A + a\cos C) = 3b \quad \Rightarrow a+c+b = 3b.$$

41. Here $2b = a+c$ and $\Delta = \dfrac{3}{5}\times\dfrac{\sqrt{3}}{4}\cdot\left(\dfrac{a+b+c}{3}\right)^2 \quad \therefore \quad \Delta = \dfrac{3\sqrt{3}}{20}\cdot b^2.$

$$\therefore \quad \sqrt{s(s-a)(s-b)(s-c)} = \frac{3\sqrt{3}}{20}b^2$$

or $$\sqrt{\frac{1}{16}(a+b+c)(b+c-a)(c+a-b)(a+b-c)} = \frac{3\sqrt{3}}{20}b^2$$

or $$\sqrt{3b\cdot(b+c+c-2b)\,b(2b-c+b-c)} = \frac{3\sqrt{3}}{5}b^2$$

or $$\sqrt{(2c-b)(3b-2c)} = \frac{3}{5}b \quad \text{or} \quad 8bc-4c^2-3b^2 = \frac{9}{25}b^2$$

or $$\frac{84}{25}b^2 - 8bc + 4c^2 = 0 \quad \text{or} \quad \frac{b}{c} = \frac{8 \pm \sqrt{64-16\cdot\frac{84}{25}}}{2\cdot\frac{84}{25}} = \frac{5}{7}, \frac{5}{3}.$$

Also $2b = a + c \quad \Rightarrow 2\frac{b}{c} = \frac{a}{c} + 1$

$$\Rightarrow \quad \frac{a}{c} = \frac{2b}{c} - 1 = \frac{3}{7}, \frac{7}{3}$$

42. We have, $AB^2 + AC^2 = 2(AD^2 + BD^2)$

$$\Rightarrow \quad \frac{c^2+b^2}{2} - \frac{a^2}{4} = AD^2, \text{etc.}$$

Adding, $a^2+b^2+c^2 - \dfrac{a^2+b^2+c^2}{4} = AD^2 + BE^2 + CF^2$

$$\Rightarrow \quad (AD^2 + BE^2 + CF^2) : (a^2+b^2+c^2) = 3:4.$$

43. $\sin A, \sin B, \sin C$ are in AP

$\Rightarrow \quad \dfrac{a}{2R}, \dfrac{b}{2R}, \dfrac{c}{2R}$ are in AP $\quad \Rightarrow a, b, c$ are in AP

$\Rightarrow \quad \dfrac{2\Delta}{h_1}, \dfrac{2\Delta}{h_2}, \dfrac{2\Delta}{h_3}$ are in AP $\quad \Rightarrow \quad \dfrac{1}{h_1}, \dfrac{1}{h_2}, \dfrac{1}{h_3}$ are in AP.

Now, medians are $\sqrt{\dfrac{c^2+b^2}{2} - \dfrac{a^2}{4}}, \sqrt{\dfrac{a^2+c^2}{2} - \dfrac{b^2}{4}}, \sqrt{\dfrac{a^2+b^2}{2} - \dfrac{c^2}{4}}$

(see the hint to Q. No. 42).

These are neither in AP nor GP.

44. In ΔABC, $AD^2 = m_1^2 = \dfrac{c^2+b^2}{2} - \dfrac{a^2}{4}$ (see the hint of Q. No. 42).

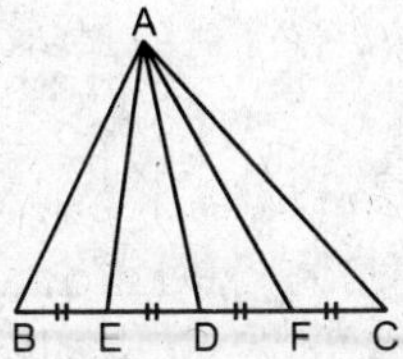

In $\triangle ABD$, $AE^2 = m_2^2 = \dfrac{c^2 + AD^2}{2} - \dfrac{\left(\frac{a}{2}\right)^2}{4}$

In $\triangle ADC$, $AF^2 = m_3^2 = \dfrac{AD^2 + b^2}{2} - \dfrac{\left(\frac{a}{2}\right)^2}{4}$

$$\therefore \quad m_2^2 + m_3^2 = AD^2 + \frac{b^2 + c^2}{2} - \frac{a^2}{8} = m_1^2 + m_1^2 + \frac{a^2}{4} - \frac{a^2}{8} = 2m_1^2 + \frac{a^2}{8}.$$

45.

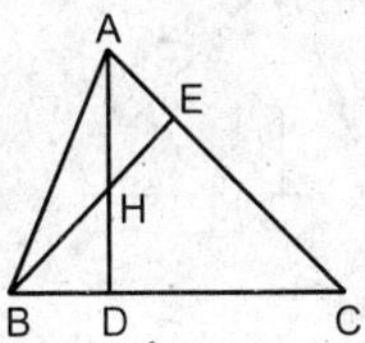

$$HD = BD \cdot \tan \angle EBC = c\cos B \cdot \tan(90° - C)$$

$$= \frac{c\cos B \cdot \cos C}{\sin C} = 2R\cos B\cos C$$

$$= \frac{2R\cos A\cos B\cos C}{\cos A}.$$

Similarly for others. So, the ratio of the distances of the orthocentre from the sides $= \dfrac{1}{\cos A} : \dfrac{1}{\cos B} : \dfrac{1}{\cos C}$.

46. Here $BD : DC = c : b$.

But $BD + DC = a$; $\quad \therefore \quad BD = \dfrac{c}{b+c} \cdot a$.

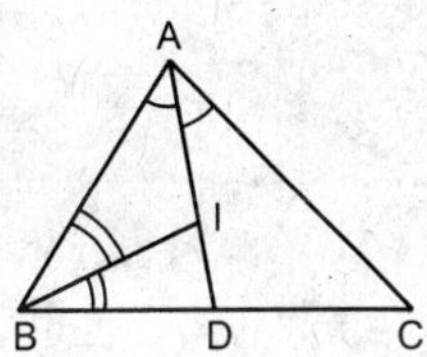

In $\triangle ABD$, $\dfrac{BD}{\sin \frac{A}{2}} = \dfrac{AD}{\sin B}$

$$\therefore \quad AD = \frac{ca}{b+c} \cdot \frac{\sin B}{\sin \frac{A}{2}} = \frac{2\Delta}{b+c} \operatorname{cosec} \frac{A}{2}.$$

Also, $\dfrac{AI}{ID} = \dfrac{AB}{BD} = \dfrac{c}{ca/(b+c)} = \dfrac{b+c}{a} \quad \Rightarrow \quad AI = \dfrac{b+c}{a+b+c} \cdot AD = \dfrac{\Delta}{s} \operatorname{cosec} \dfrac{A}{2}$.

Similarly for BI and CI.

47. Using $(m+n)\cot\theta = n\cot\beta - m\cot\alpha$, we get

$$(3+2)\cot \angle CDA = 2\cot 30° - 3\cot 60° \quad \Rightarrow \quad \cot \angle CDA = \frac{\sqrt{3}}{5}.$$

Now use sine rule in $\triangle CDA$.

48. $\dfrac{1}{2}\alpha \cdot a = \Delta$; $\quad \therefore \quad \dfrac{1}{\alpha} = \dfrac{a}{2\Delta}$. Similarly for others.

$$\therefore \quad \alpha^{-1}+\beta^{-1}+\gamma^{-1}=\frac{1}{2\Delta}(a+b+c)=\frac{s}{\Delta}.$$

49. Here $a=4k, b=5k, c=6k$. $\quad \therefore \; s=\frac{15k}{2}$.

$$\therefore \quad \Delta=\sqrt{\frac{15k}{2}\left(\frac{15k}{2}-4k\right)\left(\frac{15k}{2}-5k\right)\left(\frac{15k}{2}-6k\right)}=\frac{15\sqrt{7}}{4}k^2.$$

But $R=\frac{abc}{4\Delta}=\frac{4k\cdot 5k\cdot 6k}{15\sqrt{7}k^2}=\frac{8}{\sqrt{7}}k$

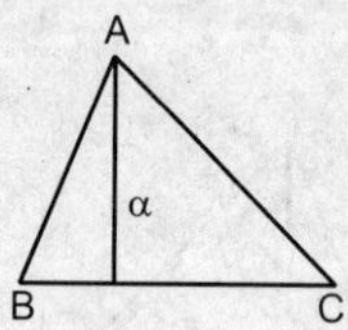

and $r=\frac{\Delta}{s}=\frac{15\sqrt{7}}{4}k^2\cdot\frac{2}{15k}=\frac{\sqrt{7}}{2}k.$

$$\therefore \quad \frac{R}{r}=\frac{8k/\sqrt{7}}{\sqrt{7}k/2}=\frac{16}{7}.$$

50. $$\frac{a\cos A+b\cos B+c\cos C}{a+b+c}=\frac{2R\sin A\cos A+2R\sin B\cos B+2r\sin C\cos C}{2s}$$

$$=\frac{R}{2s}\cdot(\sin 2A+\sin 2B+\sin 2C)$$

$$=\frac{R}{2s}\cdot 4\sin A\sin B\sin C$$

$$=\frac{2R}{s}\cdot\frac{abc}{8R^3}=\frac{abc}{4sR^2}.$$

But $R=\frac{abc}{4\Delta}$, $r=\frac{\Delta}{s}$. So, the value $=\frac{4\Delta R}{4\cdot\frac{\Delta}{r}\cdot R^2}=\frac{r}{R}$.

51. $\frac{s}{R}=\frac{(a+b+c)}{2R}=\frac{a}{2R}+\frac{b}{2R}+\frac{c}{2R}=\sin A+\sin B+\sin C.$

52. See the hint to Q. No. 50.

$\frac{r}{R}=\frac{a\cos A+b\cos B+c\cos C}{a+b+c}=\frac{3k\cos 60°}{3k}=\frac{1}{2}$, where side $=k$, angle $=60°$ in an equilateral triangle.

53. $a^2+b^2+c^2=8R^2 \quad \Rightarrow \sin^2 A+\sin^2 B+\sin^2 C=2$

$\Rightarrow \quad \cos 2A+\cos 2B+\cos 2C+1=0$

$\therefore \quad -4\cos A\cdot\cos B\cdot\cos C=0 \quad \Rightarrow A$ or B or $C=\frac{\pi}{2}$.

54. See the hint to Q. No. 52.

The exradius $= 4R \sin \cdot \dfrac{A}{2} \cdot \cos \dfrac{B}{2} \cdot \cos \dfrac{C}{2} = 4R \cdot \dfrac{1}{2} \cdot \dfrac{\sqrt{3}}{2} \cdot \dfrac{\sqrt{3}}{2} = \dfrac{3}{2} R.$

56. $\Sigma \dfrac{1}{r_1} = \Sigma \dfrac{s-a}{\Delta} = \dfrac{3s-(a+b+c)}{\Delta} = \dfrac{s}{\Delta}.$

$\therefore$ HM of exradii $= \dfrac{1}{\dfrac{\Sigma 1/r_1}{3}} = \dfrac{3}{s/\Delta} = \dfrac{3\Delta}{s} = 3r.$

57. $(\alpha - \beta) + \alpha + (\alpha + \beta) = 180°$ and $\alpha + \beta = 90°$.

$\therefore$ the angles are 30°, 60°, 90°.

$\therefore \quad \dfrac{a}{\sin 30°} = \dfrac{b}{\sin 60°} = \dfrac{c}{\sin 90°} = 2R$ or $a = R, b = \sqrt{3}R, c = 2R.$

$$r = \frac{\Delta}{s} = \frac{\frac{1}{2}R \cdot \sqrt{3}R}{\frac{1}{2}(R + \sqrt{3}R + 2R)} = \frac{\sqrt{3}R}{3+\sqrt{3}}$$

and $2s = R + \sqrt{3}R + 2R = (3+\sqrt{3})R.$

$$\therefore \quad \frac{r}{2s} = \frac{\sqrt{3}R}{3+\sqrt{3}} \times \frac{1}{(3+\sqrt{3})R} = \frac{1}{(3+\sqrt{3})(\sqrt{3}+1)}$$

$$= \frac{1}{6+4\sqrt{3}} = \frac{1}{2\sqrt{3}(2+\sqrt{3})} = \frac{2-\sqrt{3}}{2\sqrt{3}}.$$

58. $r = \dfrac{\Delta}{s} = \dfrac{(1/2) \cdot ac}{(1/2) \cdot (a+b+c)} = \dfrac{ac}{a+b+c} = \dfrac{ac(c+a-b)}{(c+a)^2 - b^2} = \dfrac{ac(c+a-b)}{c^2 + 2ca + a^2 - b^2}$

$= \dfrac{ac(c+a-b)}{2ca + b^2 - b^2} = \dfrac{c+a-b}{2} \quad (\because \; a^2 + c^2 = b^2).$

59. $\dfrac{1}{r_1}, \dfrac{1}{r_2}, \dfrac{1}{r_3}$ are in AP $\Rightarrow \dfrac{s-a}{\Delta}, \dfrac{s-b}{\Delta}, \dfrac{s-c}{\Delta}$ are in AP

$\Rightarrow \quad s-a, s-b, s-c$ are in AP $\quad \Rightarrow -a, -b, -c$ are in AP.

60. $r \cdot r_1 \cdot r_2 \cdot r_3 = \dfrac{\Delta}{s} \cdot \dfrac{\Delta}{s-a} \cdot \dfrac{\Delta}{s-b} \cdot \dfrac{\Delta}{s-c} = \dfrac{\Delta^4}{\Delta^2} = \Delta^2.$

61. See the hint to Q. No. 60. Here $r_1 r_2 r_3 = \dfrac{\Delta^2}{r}.$

See the hint to Q. No. 56. Here $\frac{1}{r_1}+\frac{1}{r_2}+\frac{1}{r_3}=\frac{s}{\Delta}$.

$$\therefore \quad r_1r_2+r_2r_3+r_3r_1=r_1r_2r_3\left(\frac{1}{r_1}+\frac{1}{r_2}+\frac{1}{r_3}\right)$$

$$=\frac{\Delta^2}{r}\cdot\frac{s}{\Delta}=\frac{\Delta s}{r}=\frac{\Delta s}{\Delta/s}=s^2.$$

62. $3(\tan^2 A+\tan^2 B+\tan^2 C)-(\tan A+\tan B+\tan C)^2$

$=(\tan A-\tan B)^2+(\tan B-\tan C)^2+(\tan C-\tan A)^2>0$

for here $\tan A=\tan B=\tan C=\sqrt{3}$ is not true

or $\quad 3k-(\tan A\cdot\tan B\cdot\tan C)^2>0$

because in the ΔABC, $\tan A+\tan B+\tan C=\tan A\cdot\tan B\cdot\tan C$

or $\quad 3k-81>0 \quad$ or $\quad k>27$.

63. In ΔABC, $\cos 2A+\cos 2B-\cos 2C=1-4\sin A\sin B\cos C<1$ for all A, B.

$\cos 2A+\cos 2B+\cos 2C=-1-4\cos A\cos B\cos C<-1$ for acute A, B.

$\cos^2 A+\cos^2 B+\cos^2 C=1-2\cos A\cos B\cos C<1$ for acute A, B.

64. $\cot A\cdot\cot B\cdot\cot C>0 \Rightarrow \cot A>0, \quad \cot B>0, \quad \cot C>0$
because two or more of $\cot A$, $\cot B$, $\cot C$ cannot be negative at the same time in a triangle.

65. $\cos B\cdot\cos C+\sin B\cdot\sin C\geq 1$

because $\sin B\cdot\sin C\cdot\sin^2 A$ is positive and $\sin^2 A\leq 1$

or $\cos(B-C)\geq 1$. But $\cos(B-C)\ngtr 1$. So, $\cos(B-C)=1$. Therefore, $B=C$ and then $\sin A=1$.

$$\therefore \quad A=\frac{\pi}{2}, B=C=\frac{\pi}{4}.$$

66. $\sqrt{3}\cos A+\sin A=\sqrt{3}\cos B+\sin B$

or $\quad \dfrac{\sin A-\sin B}{\cos B-\cos A}=\sqrt{3}$

$$\therefore \quad \sqrt{3}=\frac{2\cos\frac{A+B}{2}\cdot\sin\frac{A-B}{2}}{2\sin\frac{A+B}{2}\cdot\sin\frac{A-B}{2}}=\cot\frac{A+B}{2}$$

or $\quad \dfrac{A+B}{2}=\dfrac{\pi}{6} \quad$ or $\quad A+B=\dfrac{\pi}{3}$

$\therefore \quad C = \pi - \frac{\pi}{3}$.

67. $c\cos\theta - a\sin\theta = b$. Therefore, $c\cos\alpha - a\sin\alpha = c\cos\beta - a\sin\beta$, where α, β are the other two angles of the triangle.

$\therefore \quad c(\cos\alpha - \cos\beta) = a(\sin\alpha - \sin\beta)$

or $$\frac{c}{a} = \frac{2\cos\frac{\alpha+\beta}{2}\sin\frac{\alpha-\beta}{2}}{2\sin\frac{\alpha+\beta}{2}\sin\frac{\beta-\alpha}{2}} = -\cot\frac{\alpha+\beta}{2}$$

$$\therefore \quad \tan\frac{\alpha+\beta}{2} = \frac{-a}{c}$$

$$\therefore \quad \tan(\alpha+\beta) = \frac{2\times(-a/c)}{1-a^2/c^2} = \frac{2ac}{a^2 \quad c^2}$$

$$\therefore \quad \tan\left(\pi - \frac{3\pi}{4}\right) = \frac{2ac}{a^2-c^2} \quad \text{or} \quad a^2 - c^2 = 2ac.$$

68. $A_0A_1 = 2\times1\times\cos 60° = 1 = A_1A_2$

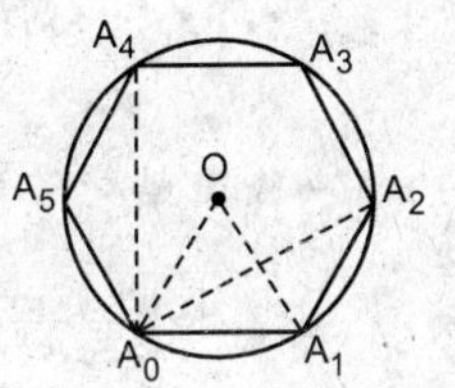

$$\cos 120° = \frac{A_0A_1^2 + A_1A_2^2 - A_0A_2^2}{2\cdot A_0A_1\cdot A_1A_2} = \frac{1+1-A_0A_2^2}{2\cdot1\cdot1}$$

$\therefore \quad A_0A_2 = \sqrt{3} = A_0A_4$

$\therefore \quad A_0A_1\times A_0A_2\times A_0A_4 = 1\times\sqrt{3}\times\sqrt{3} = 3.$

69. In the ΔOAB, $OA = OB = r$ and $\angle AOB = \frac{360°}{5} = 72°$.

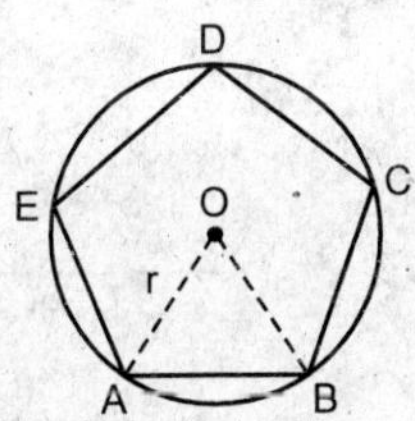

$$\therefore \quad \text{ar}(\Delta AOB) = \frac{1}{2}\cdot r\cdot r\cdot\sin 72° = \frac{1}{2}r^2\cos 18°$$

$$\therefore \quad \text{ar (pentagon)} = \frac{5}{2}r^2\cos 18°$$

$$\therefore \quad A_1 : A_2 = \frac{2\pi r^2}{5r^2\cos 18°} = \frac{2\pi}{5}\sec\frac{\pi}{10}.$$

70. Clearly, $\angle BOD = 2C \quad \therefore \quad \cos 2C = \frac{1^2+1^2-(\sqrt{3})^2}{2\cdot1\cdot1}$

or $\quad \cos 2C = -\frac{1}{2} \quad \therefore \quad C = 60°$

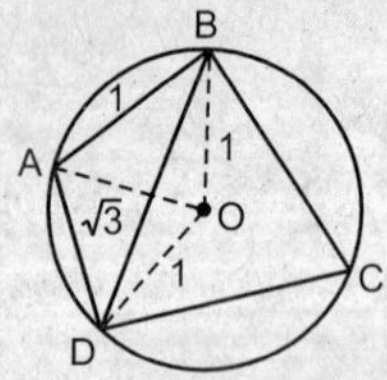

$\therefore \quad A = 180° - C = 120°.$

So, $\cos 120° = \dfrac{AD^2 + 1^2 - (\sqrt{3})^2}{2 \cdot AD \cdot 1}$

$\therefore \quad -\dfrac{1}{2} = \dfrac{AD^2 - 2}{2AD} \quad$ or $\quad AD^2 + AD - 2 = 0$

or $\quad AD = 1.$

Now, ar (ABCD) = ar (△ADB) + ar (△BCD)

or $\quad \dfrac{3\sqrt{3}}{4} = \dfrac{1}{2} \cdot 1 \cdot 1 \cdot \sin 120° + \dfrac{1}{2} \cdot BC \cdot CD \cdot \sin 60° \quad$ or $\quad BC \cdot CD = 2.$

71. $\tan \dfrac{A-B}{2} = \dfrac{a-b}{a+b} \cot \dfrac{C}{2} = \dfrac{1}{3} \cot \dfrac{C}{2}$

$$\therefore \quad \frac{4}{5} = \cos(A-B) = \frac{1 - \tan^2 \dfrac{A-B}{2}}{1 + \tan^2 \dfrac{A-B}{2}} = \frac{1 - \dfrac{1}{9}\cot^2 \dfrac{C}{2}}{1 + \dfrac{1}{9}\cot^2 \dfrac{C}{2}}$$

$\therefore \quad \tan^2 \dfrac{C}{2} = 1 \quad \Rightarrow C = \dfrac{\pi}{2}.$

$\therefore \quad \text{ar}(\triangle ABC) = \dfrac{1}{2} ab = \dfrac{1}{2} \cdot 6 \cdot 3 = 9.$

Also $\sin A = \dfrac{6}{\sqrt{3^2 + 6^2}}.$

72. Using $\cos A = \dfrac{b^2 + c^2 - a^2}{2bc}$, $\quad \dfrac{1}{2} = \dfrac{13 + c^2 - 11}{2 \cdot \sqrt{13} \cdot c}$

So, $c^2 - \sqrt{13} \cdot c + 2 = 0$. This gives two values of c.

73. $\tan \dfrac{C-B}{2} = \dfrac{c-b}{c+b} \cot \dfrac{A}{2}$

$\Rightarrow \quad \tan \dfrac{C-B}{2} = \dfrac{1}{5} \cot 30° = \dfrac{\sqrt{3}}{5} = \tan \alpha$

$\therefore \quad C - B = 2\alpha$ and $C + B = 180° - 60° = 120°.$

74. Here, $a\cos A = b\cos B$

$\Rightarrow \quad \sin 2A = \sin 2B$

$\Rightarrow \quad 2A = 2B \quad$ or $\quad 2A + 2B = 180°.$

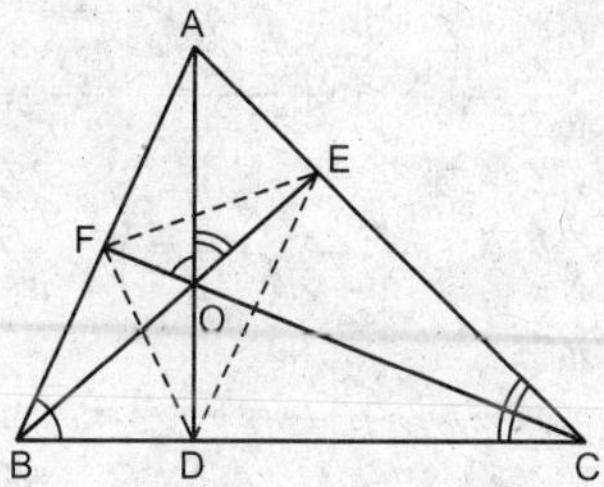

75. From geometry, $\angle AOF = B, \angle AOE = C$.

Also $OF = b\cos A \cdot \tan(90° - B)$

$= b\cos A \cdot \cot B = 2R\cos A\cos B$.

Similarly, $OE = 2R\cos A\cos C$.

In ΔOEF, $\cos(B+C) = \dfrac{OE^2 + OF^2 - EF^2}{2 \cdot OE \cdot OF}$

$$\Rightarrow \quad -\cos A = \frac{4R^2\cos^2 A(\cos^2 B + \cos^2 C) - EF^2}{8R^2\cos^2 A\cos B\cos C}$$

$$\therefore \quad EF^2 = 4R^2\cos^2 A[\cos^2 B + \cos^2 C + 2\cos A\cos B\cos C]$$

$= 4R^2\cos^2 A \cdot \sin^2 A$ because in a ΔABC.

$$\cos^2 B + \cos^2 C + \cos^2 A = 1 - 2\cos A\cos B\cos C$$

$$\therefore \quad EF = R\sin 2A = \frac{a}{2\sin A}\sin 2A = a\cos A.$$

So, $DF = b\cos B$.

76. $\tan A, \tan B$ are the roots of $abx^2 - c^2x + ab = 0$.

So, $\tan A + \tan B = \dfrac{c^2}{ab}$, $\tan A \cdot \tan B = 1$.

$\tan A \cdot \tan B = 1 \quad \Rightarrow \quad A, B$ are both acute angles and $\cos(A+B) = 0$.

$\therefore \quad A + B = \dfrac{\pi}{2}$, i.e., $\cot C = \cot\dfrac{\pi}{2} = 0$. So, ΔABC is right-angled at C.

Also $\sin A = \cos B$, i.e., $\sin^2 A + \sin^2 B = 1$.

Also $\tan(A - B) = \dfrac{\tan A - \tan B}{1 + \tan A\tan B} = \dfrac{\frac{a}{b} - \frac{b}{a}}{1+1} = \dfrac{a^2 - b^2}{2ab}$.

77.

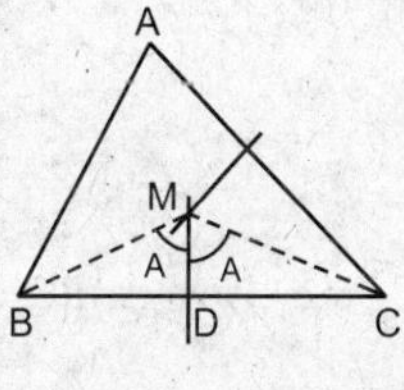

$MD = MB\cos A = R\cos A$, etc.

$\therefore$ the required ratio

$= R\cos A : R\cos B : R\cos C$

$= \dfrac{a}{2}\cot A : \dfrac{b}{2}\cot B : \dfrac{c}{2}\cot C$.

79. $\cos\dfrac{A - C}{2} = \dfrac{a + c}{2\sqrt{a^2 + c^2 - ac}}$

$$\Rightarrow \quad \tan\frac{A-C}{2} = \frac{\sqrt{4(a^2+c^2-ac)-(a+c)^2}}{a+c} = \sqrt{3}\cdot\frac{a-c}{a+c}$$

$$\text{or} \quad \frac{a-c}{a+c}\cot\frac{B}{2} = \sqrt{3}\cdot\frac{a-c}{a+c} \quad \Rightarrow \cot\frac{B}{2} = \sqrt{3} \quad \Rightarrow B = \frac{\pi}{3}\cdot$$

$$\therefore \quad A + C = \pi - B = \pi - \frac{\pi}{3} = \frac{2\pi}{3} = 2B.$$

80. $\log b + 10 = \log c + 10 + \log \sin B$

$$\Rightarrow \quad b = c\sin B \quad \Rightarrow b = b\sin C \quad \Rightarrow \sin C = 1.$$

81. $\dfrac{a}{b+c} = \dfrac{\sin A}{\sin B + \sin C}$

$$= \frac{2\sin\frac{A}{2}\cos\frac{A}{2}}{2\sin\frac{B+C}{2}\cdot\cos\frac{B-C}{2}} = \frac{\sin\frac{A}{2}}{\cos\frac{B-C}{2}}$$

$$\therefore \quad \sin\frac{A}{2} = \frac{a}{b+c}\cos\frac{B-C}{2} \le \frac{a}{b+c}\cdot$$

$$\frac{2a}{a+b+c} = \frac{2\sin A}{\sin A + \sin B + \sin C} = \frac{2\cdot 2\sin\frac{A}{2}\cos\frac{A}{2}}{4\cos\frac{A}{2}\cdot\cos\frac{B}{2}\cdot\cos\frac{C}{2}}$$

$$= \frac{\sin\frac{A}{2}}{\cos\frac{B}{2}\cos\frac{C}{2}} \ge \sin\frac{A}{2}\cdot$$

82. Here, $\tan C < 0 \Rightarrow C$ is obtuse.

$$\Rightarrow \quad A, B \text{ are acute and } A + B < \frac{\pi}{2}\cdot$$

$$\tan(A+B) = \frac{\tan A + \tan B}{1 - \tan A \tan B} > 0$$

$$\Rightarrow \quad 1 - \tan A\cdot\tan B > 0 \quad \Rightarrow \tan A \tan B < 1.$$

$\tan A + \tan B + \tan C = (+)(+)(-) < 0 = \tan A\cdot\tan B\cdot\tan C$ in a ΔABC.

83. $\sqrt{3}x^2 - 4x + \sqrt{3} < 0$

$$\Rightarrow \quad (x-\sqrt{3})(\sqrt{3}x - 1) < 0 \quad \Rightarrow \frac{1}{\sqrt{3}} < x < \sqrt{3}.$$

$\therefore \quad \frac{1}{\sqrt{3}} < \tan A < \sqrt{3}, \frac{1}{\sqrt{3}} < \tan B < \sqrt{3} \Rightarrow 30° < A < 60°, 30° < B < 60°.$

$\therefore \quad 60° < C < 120° \Rightarrow -\frac{1}{2} < \cos C < \frac{1}{2} \quad \Rightarrow -\frac{1}{2} < \frac{a^2 + b^2 - c^2}{2ab} < \frac{1}{2}$

$\therefore \quad -ab < a^2 + b^2 - c^2 < ab.$

84. $\cos A + \cos B + \cos C > 1 \quad \Rightarrow 1 + 4\sin\frac{A}{2} \cdot \sin\frac{B}{2} \cdot \sin\frac{C}{2} > 1$

$\therefore \quad \sin\frac{A}{2} \cdot \sin\frac{B}{2} \cdot \sin\frac{C}{2} > 0 \quad \Rightarrow \frac{A}{2}, \frac{B}{2}, \frac{C}{2}$ are acute.

□

6

Heights and Distances

(excluded from IIT entrance examination)

• *Type 1* •

Choose the most appropriate option (a, b, c or d).

1. A 6-ft-tall man finds that the angle of elevation of the top of a 24-ft-high pillar and the angle of depression of its base are complementary angles. The distance of the man from the pillar is

(a) $2\sqrt{3}$ ft (b) $8\sqrt{3}$ ft

(c) $6\sqrt{3}$ ft (d) none of these

2. A rocket of height h metres is fired vertically upwards. Its velocity at time t seconds is $(2t + 3)$ metres/second. If the angle of elevation of the top of the rocket from a point on the ground after 1 second of firing is $\pi/6$ and after 3 seconds it is $\pi/3$ then the distance of the point from the rocket is

(a) $14\sqrt{3}$ metres

(b) $7\sqrt{3}$ metres

(c) $2\sqrt{3}$ metres

(d) cannot be found without the value of h

3. Three vertical poles of heights h_1, h_2 and h_3 at the vertices A, B and C of a ΔABC subtend angles α, β and γ respectively at the circumcentre of the triangle. If $\cot\alpha$, $\cot\beta$ and $\cot\gamma$ are in AP then h_1, h_2, h_3 are in

(a) AP (b) GP (c) HP (d) none of these

4. The angle of elevation of the top of a hill from each of the vertices A, B, C of a horizontal triangle is α. The height of the hill is

(a) $b\tan\alpha \cdot \operatorname{cosec} B$
(b) $\frac{1}{2}a\tan\alpha \cdot \operatorname{cosec} A$
(c) $\frac{1}{2}c\tan\alpha \cdot \operatorname{cosec} C$
(d) none of these

5. A piece of paper in the shape of a sector of a circle of radius 10 cm and of angle 216° just covers the lateral surface of a right circular cone of vertical angle 2θ. Then sin θ is

(a) $\frac{3}{5}$
(b) $\frac{4}{5}$
(c) $\frac{3}{4}$
(d) none of these

6. The angle of elevation of the top of a vertical pole when observed from each vertex of a regular hexagon is π/3. If the area of the circle circumscribing the hexagon be A metre^2 then the area of the hexagon is

(a) $\frac{3\sqrt{3}}{8}A$ metre^2
(b) $\frac{\sqrt{3}}{\pi}A$ metre^2
(c) $\frac{3\sqrt{3}}{4\pi}A$ metre^2
(d) $\frac{3\sqrt{3}}{2\pi}A$ metre^2

7. A vertical pole PO is standing at the centre O of a square ABCD. If AC subtends an angle 90° at the top, P, of the pole then the angle subtended by a side of the square at P is

(a) 45°
(b) 30°
(c) 60°
(d) none of these

8. A vertical lamp-post of height 9 metres stands at the corner of a rectangular field. The angle of elevation of its top from the farthest corner is 30°, while from another corner it is 45°. The area of the field is

(a) $81\sqrt{2}$ metre^2
(b) $9\sqrt{2}$ metre^2
(c) $81\sqrt{3}$ metre^2
(d) $9\sqrt{3}$ metre^2

9. A vertical lamp-post, 6 m high, stands at a distance of 2 m from a wall, 4 m high. A 1.5-m-tall man starts to walk away from the wall on the other side of the wall, in line with the lamp-post. The maximum distance to which the man can walk remaining in the shadow is

(a) $\frac{5}{2}$ m
(b) $\frac{3}{2}$ m
(c) 4 m
(d) none of these

10. A circular ring of radius 3 cm is suspended horizontally from a point 4 cm vertically above the centre by 4 strings attached at equal intervals to its circumference. If the angle between two consecutive strings be θ then cos θ is

(a) $\frac{4}{5}$ (b) $\frac{4}{25}$

(c) $\frac{16}{25}$ (d) none of these

• *Type 2* •

Choose the correct options. One or more options may be correct.

11. A man standing between two vertical posts finds that the angle subtended at his eyes by the tops of the posts is a right angle. If the heights of the two posts are two times and four times the height of the man, and the distance between them is equal to the length of the longer post, then the ratio of the distances of the man from the shorter and the longer post is

(a) 3 : 1 (b) 2 : 3

(c) 3 : 2 (d) 1 : 3

12. A flagstaff stands vertically on a pillar, the height of the flagstaff being double the height of the pillar. A man on the ground at a distance finds that both the pillar and the flagstaff subtend equal angles at his eyes. The ratio of the height of the pillar and the distance of the man from the pillar, is

(a) $\sqrt{3} : 1$ (b) 1 : 3

(c) $1 : \sqrt{3}$ (d) $\sqrt{3} : 2$

Answers

1. c	**2.** b	**3.** c	**4.** b	**5.** a
6. d	**7.** c	**8.** a	**9.** a	**10.** c
11. a, d	**12.** c			

Hints and Solutions to Selected Questions

1. Here,

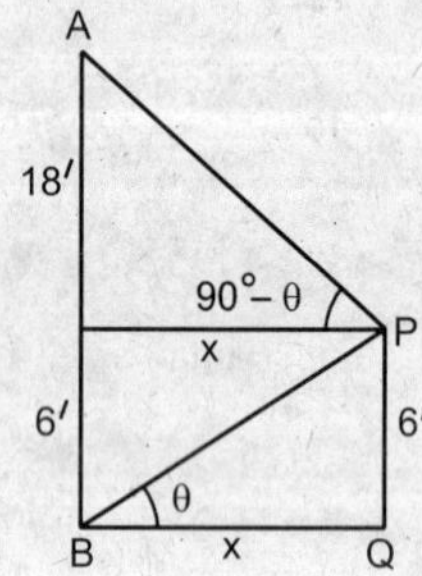

$$\tan\theta = \frac{6}{x},\ \tan(90° - \theta) = \frac{18}{x}$$

$\therefore \quad \frac{6}{x} \cdot \frac{18}{x} = 1$

or $\quad x^2 = 6 \times 18$

or $\quad x = 6\sqrt{3}.$

2.

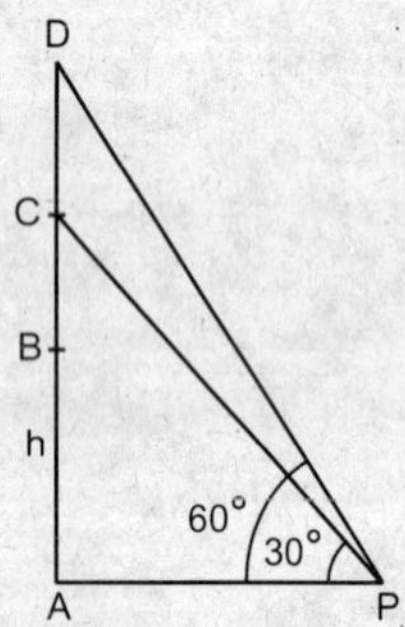

$$\frac{dx}{dt} = \frac{18}{x} = 2t + 3 \quad \Rightarrow x = t^2 + 3t + c$$

$\therefore \quad BC = 1^2 + 3.1 + c = 4 + c$

$BD = 3^2 + 3.3 + c = 18 + c$

$\therefore \quad \frac{h + 4 + c}{AP} = \frac{18}{x} = \tan 30°$

$\frac{h + 18 + c}{AP} = \tan 60°$

$\therefore \quad \frac{14}{AP} = \tan 60° - \tan 30°$

$= \sqrt{3} - \frac{1}{\sqrt{3}} = \frac{2}{\sqrt{3}}$

$\therefore \quad AP = 7\sqrt{3}.$

3. $\frac{R}{h_1} = \cot\alpha$, etc., where R is the circumradius.

$\cot\alpha, \cot\beta, \cot\gamma$ are in AP $\quad \Rightarrow \quad \frac{1}{h_1}, \frac{1}{h_2}, \frac{1}{h_3}$ are in AP.

4. The distance of the foot from each vertex $= h\cot\alpha$.

$\therefore$ the foot is at the circumcentre of the triangle. So, $R = h\cot\alpha$.

$\therefore \quad h = R\tan\alpha = \frac{a}{2\sin A}\tan\alpha.$

5. The area of the sector $= \frac{\pi \cdot 10^2}{360} \times 216 \text{ cm}^2$

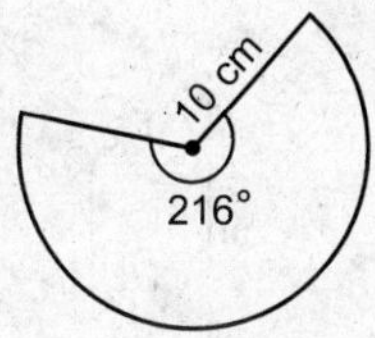

$= \text{lateral surface area of the cone.}$

The circumference of the base circle of the cone

$$= \frac{2\pi \times 10}{360} \times 216 \text{ cm}$$

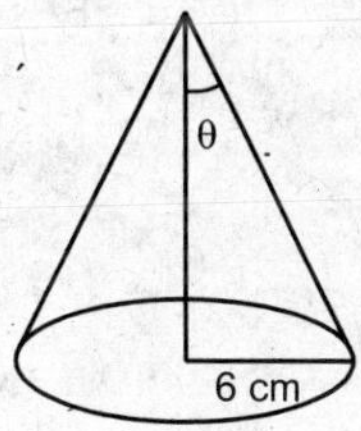

$$\therefore \quad 2\pi r = \frac{2\pi \times 10}{360} \times 216 \quad \text{or} \quad r = 6 \text{ cm.}$$

The lateral height of the cone $= 6\operatorname{cosec}\theta$ cm.

$$\therefore \quad \text{lateral surface area} = \pi r l = \pi \cdot 6 \cdot 6\operatorname{cosec}\theta \text{ cm}^2$$

$$\therefore \quad \pi \cdot 36\operatorname{cosec}\theta = \frac{\pi \cdot 10^2 \cdot 216}{360} \quad \Rightarrow \sin\theta = \frac{3}{5}.$$

6. If the height of the pole is h metres then $\frac{h}{\text{circumradius}} = \tan 60°$.

$$\therefore \quad \text{circumradius} = \frac{h}{\sqrt{3}} \text{ metres} \quad \therefore \quad A = \pi\left(\frac{h}{\sqrt{3}}\right)^2 \text{ metre}^2.$$

The area of the hexagon $= 6 \times \frac{\sqrt{3}}{4} \cdot \left(\frac{h}{\sqrt{3}}\right)^2$, i.e., $\frac{3\sqrt{3}}{2} \cdot \frac{A}{\pi}$ metre^2.

7.

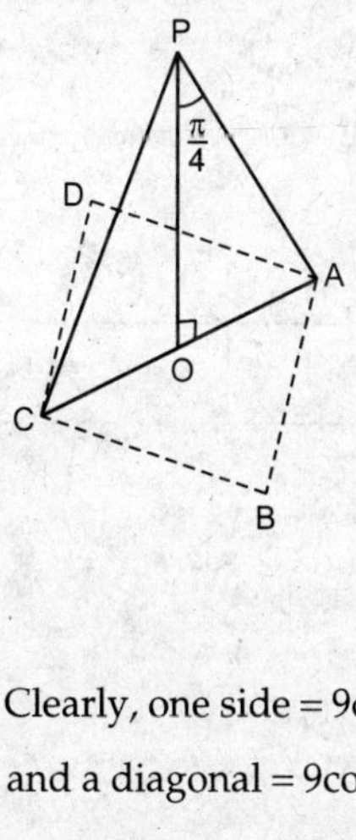

Clearly, $\angle OPA = \angle OPC = \frac{\pi}{4}$. Let $OP = h$.

So, $\frac{OA}{OP} = \tan 45° = 1 \quad \therefore \quad OA = h$ and $AC = 2h$.

Also $\frac{OP}{PA} = \cos\frac{\pi}{4} \quad \Rightarrow PA = \sqrt{2}h$.

Also $AC = \sqrt{2}AB \quad \Rightarrow 2h = \sqrt{2}AB$, i.e., $AB = \sqrt{2}h$

$\therefore \quad PA = AB = PB$ (similarly).

So, ΔBPA is equilateral.

Hence, the required angle $= 60°$.

8. Clearly, one side $= 9\cot 45° = 9$ m

and a diagonal $= 9\cot 30° = 9\sqrt{3}$ m.

$\therefore$ the other side $= \sqrt{(9\sqrt{3})^2 - 9^2}$, i.e., $9\sqrt{2}$ m.

So, area $= 9 \times 9\sqrt{2}$ m^2.

9.

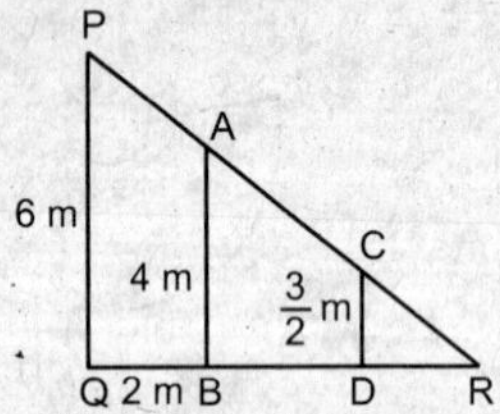

BD is the required distance. From similar triangles,

$$\frac{6}{4}=\frac{QR}{BR},\ \frac{4}{3/2}=\frac{BR}{DR}.$$

$$\frac{6}{4}=\frac{QR}{BR} \Rightarrow \frac{3}{2}=\frac{2+BR}{BR}$$

$$\Rightarrow BR=4.$$

$$\frac{4}{3/2}=\frac{BR}{DR} \Rightarrow \frac{8}{3}=\frac{4}{DR}$$

$$\Rightarrow DR=\frac{3}{2}.$$

$$\therefore \quad BD=BR-DR=4-\frac{3}{2}=\frac{5}{2}.$$

10. Clearly,

$$AC=\sqrt{4^2+3^2}=5=AD$$

and $\angle CBD=90°$.

So, $CD=\sqrt{3^2+3^2}=3\sqrt{2}$.

$$\therefore \quad \cos\theta=\frac{5^2+5^2-(3\sqrt{2})^2}{2\cdot 5\cdot 5}$$

$$=\frac{16}{25}.$$

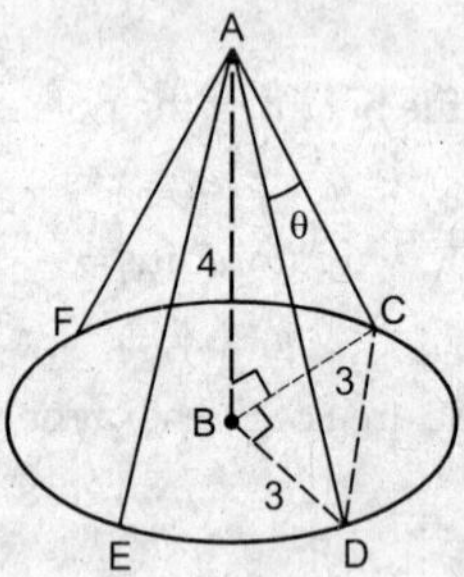

11. $\frac{h}{x}=\tan\theta,\ \frac{3h}{y}=\tan(90°-\theta)=\cot\theta$

$$\therefore \quad \frac{3h^2}{xy}=1.$$

So, $xy=3h^2$.

Also $x+y=4h$. Solving these,

$$x=3h,\ y=h \text{ or } x=h,\ y=3h.$$

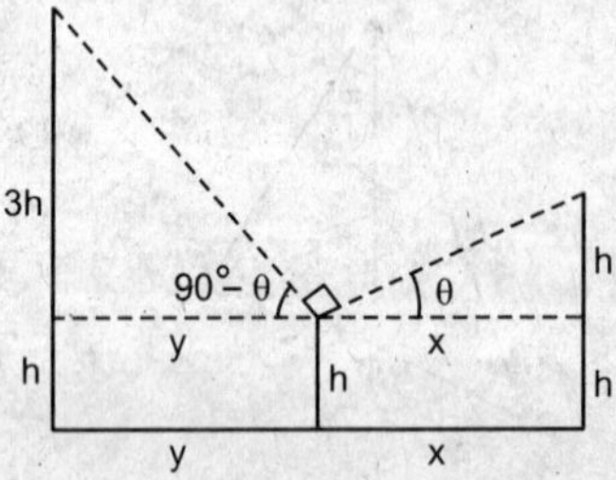

12.

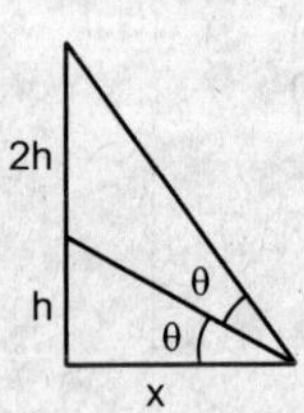

Let $\frac{h}{x}=a$. Then $\tan\theta=a$, $\tan 2\theta=3a$.

$$\therefore \quad \frac{2a}{1-a^2}=3a$$

$$\Rightarrow \quad a=\frac{1}{\sqrt{3}}.$$

□

Part 3

Coordinate Geometry

1

Coordinates and Straight Lines

• Type 1 •

Choose the most appropriate option (a, b, c or d).

1. If two vertices of an equilateral triangle have integral coordinates then the third vertex will have

(a) integral coordinates

(b) coordinates which are rational

(c) at least one coordinate irrational

(d) coordinates which are irrational

2. If the line segment joining (2, 3) and (–1, 2) is divided internally in the ratio 3 : 4 by the line $x + 2y = k$ then k is

(a) $\frac{41}{7}$ (b) $\frac{5}{7}$

(c) $\frac{36}{7}$ (d) $\frac{31}{7}$

3. The polar coordinates of the vertices of a triangle are (0, 0), $(3, \pi/2)$ and $(3, \pi/6)$. Then the triangle is

(a) right angled (b) isosceles

(c) equilateral (d) none of these

4. The points $(a, b + c)$, $(b, c + a)$, $(c, a + b)$ are

(a) vertices of an equilateral triangle

(b) collinear

(c) concyclic

(d) none of these

5. The incentre of the triangle formed by the axes and the line $\frac{x}{a}+\frac{y}{b}=1$ is

(a) $\left(\frac{a}{2}, \frac{b}{2}\right)$

(b) $\left(\frac{ab}{a+b+\sqrt{ab}}, \frac{ab}{a+b+\sqrt{ab}}\right)$

(c) $\left(\frac{a}{3}, \frac{b}{3}\right)$

(d) $\left(\frac{ab}{a+b+\sqrt{a^2+b^2}}, \frac{ab}{a+b+\sqrt{a^2+b^2}}\right)$

6. In the ΔABC, the coordinates of B are (0, 0), $AB=2$, $\angle ABC=\frac{\pi}{3}$ and the middle point of BC has the coordinates (2, 0). The centroid of the triangle is

(a) $\left(\frac{1}{2}, \frac{\sqrt{3}}{2}\right)$ (b) $\left(\frac{5}{3}, \frac{1}{\sqrt{3}}\right)$

(c) $\left(\frac{4+\sqrt{3}}{3}, \frac{1}{3}\right)$ (d) none of these

7. The coordinates of three consecutive vertices of a parallelogram are (1, 3), (–1, 2) and (2, 5). The coordinates of the fourth vertex are

(a) (6, 4) (b) (4, 6)

(c) (–2, 0) (d) none of these

8. The area of the pentagon whose vertices are (4, 1), (3, 6), (–5, 1), (–3, –3) and (–3, 0) is

(a) 30 unit^2 (b) 60 unit^2

(c) 120 unit^2 (d) none of these

9. A point moves in the x-y plane such that the sum of its distances from two mutually perpendicular lines is always equal to 3. The area enclosed by the locus of the point is

(a) 18 unit^2 (b) $\frac{9}{2}$ unit^2

(c) 9 unit^2 (d) none of these

10. Let $A=(1, 2)$, $B=(3, 4)$ and let $C=(x, y)$ be a point such that $(x-1)(x-3)+(y-2)(y-4)=0$. If ar $(\Delta ABC)=1$ then maximum number of positions of C in the x-y plane is

(a) 2 (b) 4
(c) 8 (d) none of these

11. The points (α, β), (γ, δ), (α, δ) and (γ, β) taken in order, where $\alpha, \beta, \gamma, \delta$ are different real numbers, are

(a) collinear (b) vertices of a square
(c) vertices of a rhombus (d) concyclic

12. The diagonals of a parallelogram PQRS are along the lines $x + 3y = 4$ and $6x - 2y = 7$. Then PQRS must be a

(a) rectangle (b) square
(c) cyclic quadrilateral (d) rhombus

13. The coordinates of the four vertices of a quadrilateral are $(-2, 4)$, $(-1, 2)$, $(1, 2)$ and $(2, 4)$ taken in order. The equation of the line passing through the vertex $(-1, 2)$ and dividing the quadrilateral in two equal areas is

(a) $x + 1 = 0$ (b) $x + y = 1$
(c) $x - y + 3 = 0$ (d) none of these

14. The equation of the straight line which passes through the point $(-4, 3)$ such that the portion of the line between the axes is divided internally by the point in the ratio $5 : 3$ is

(a) $9x - 20y + 96 = 0$ (b) $9x + 20y = 24$
(c) $20x + 9y + 53 = 0$ (d) none of these

15. The equation of the straight line which bisects the intercepts made by the axes on the lines $x + y = 2$ and $2x + 3y = 6$ is

(a) $2x = 3$ (b) $y = 1$
(c) $2y = 3$ (d) $x = 1$

16. The equation of a straight line passing through the point $(-2, 3)$ and making intercepts of equal length on the axes is

(a) $2x + y + 1 = 0$ (b) $x - y = 5$
(c) $x - y + 5 = 0$ (d) none of these

17. The foot of the perpendicular on the line $3x + y = \lambda$ drawn from the origin is C. If the line cuts the x-axis and y-axis at A and B respectively then BC : CA is

(a) $1 : 3$ (b) $3 : 1$
(c) $1 : 9$ (d) $9 : 1$

18. The distance of the line $2x - 3y = 4$ from the point (1, 1) in the direction of the line $x + y = 1$ is

(a) $\sqrt{2}$
(b) $5\sqrt{2}$
(c) $\frac{1}{\sqrt{2}}$
(d) none of these

19. The four sides of a quadrilateral are given by the equation $xy(x-2)(y-3) = 0$. The equation of the line parallel to $x - 4y = 0$ that divides the quadrilateral in two equal areas is

(a) $x - 4y + 5 = 0$
(b) $x - 4y - 5 = 0$
(c) $4y = x + 1$
(d) $4y + 1 = x$

20. The coordinates of two consecutive vertices A and B of a regular hexagon ABCDEF are (1, 0) and (2, 0) respectively. The equation of the diagonal CE is

(a) $\sqrt{3}x + y = 4$
(b) $x + \sqrt{3}y + 4 = 0$
(c) $x + \sqrt{3}y = 4$
(d) none of these

21. ABC is an isosceles triangle in which A is (–1, 0), $\angle A = 2\pi/3$, $AB = AC$ and AB is along the x-axis. If $BC = 4\sqrt{3}$ then the equation of the line BC is

(a) $x + \sqrt{3}y = 3$
(b) $\sqrt{3}x + y = 3$
(c) $x + y = \sqrt{3}$
(d) none of these

22. The graph of the function $\cos x \cdot \cos(x+2) - \cos^2(x+1)$ is a

(a) straight line passing through the point $(0, -\sin^2 1)$ with slope 2
(b) straight line passing through the origin
(c) parabola with vertex $(1, -\sin^2 1)$
(d) straight line passing through the point $(\pi/2, -\sin^2 1)$ and parallel to the x-axis

23. If the points (–2, 0), $(-1, 1/\sqrt{3})$ and $(\cos\theta, \sin\theta)$ are collinear then the number of values of $\theta \in [0, 2\pi]$ is

(a) 0
(b) 1
(c) 2
(d) infinite

24. The limiting position of the point of intersection of the lines $3x + 4y = 1$ and $(1+c)x + 3c^2y = 2$ as c tends to 1 is

(a) (–5, 4)
(b) (5, –4)
(c) (4, –5)
(d) none of these

25. The coordinates of the point on the x-axis which is equidistant from the points (–3, 4) and (2, 5) are

(a) (20, 0) (b) (–23 , 0)

(c) $\left(\frac{4}{5}, 0\right)$ (d) none of these

26. The distance between the lines $3x + 4y = 9$ and $6x + 8y + 15 = 0$ is

(a) $\frac{3}{10}$ (b) $\frac{33}{10}$

(c) $\frac{33}{5}$ (d) none of these

27. If a vertex of an equilateral triangle is the origin and the side opposite to it has the equation $x + y = 1$ then the orthocentre of the triangle is

(a) $\left(\frac{1}{3}, \frac{1}{3}\right)$ (b) $\left(\frac{\sqrt{2}}{3}, \frac{\sqrt{2}}{3}\right)$

(c) $\left(\frac{2}{3}, \frac{2}{3}\right)$ (d) none of these

28. The equations of the three sides of a triangle are $x = 2, y + 1 = 0$ and $x + 2y = 4$. The coordinates of the circumcentre of the triangle are

(a) (4, 0) (b) (2, –1)

(c) (0, 4) (d) none of these

29. L is a variable line such that the algebraic sum of the distances of the points (1, 1), (2, 0) and (0, 2) from the line is equal to zero. The line L will always pass through

(a) (1, 1) (b) (2, 1)

(c) (1, 2) (d) none of these

30. ABC is an equilateral triangle such that the vertices B and C lie on two parallel lines at a distance 6. If A lies between the parallel lines at a distance 4 from one of them then the length of a side of the equilateral triangle is

(a) 8 (b) $\sqrt{\frac{88}{3}}$

(c) $\frac{4\sqrt{7}}{\sqrt{3}}$ (d) none of these

31. If p and p' are the perpendiculars from the origin upon the lines $x\sec\theta + y\operatorname{cosec}\theta = a$ and $x\cos\theta - y\sin\theta = a\cos 2\theta$ respectively then

(a) $4p^2+p'^2=a^2$ (b) $p^2+4p'^2=a^2$

(c) $p^2+p'^2=a^2$ (d) none of these

32. Let the perpendiculars from any point on the line $2x+11y=5$ upon the lines $24x+7y=20$ and $4x-3y=2$ have the lengths p and p' respectively. Then

(a) $2p=p'$ (b) $p=p'$

(c) $p=2p'$ (d) none of these

33. If $P(1+t/\sqrt{2}, 2+t/\sqrt{2})$ be any point on a line then the range of values of t for which the point P lies between the parallel lines $x+2y=1$ and $2x+4y=15$ is

(a) $-\frac{4\sqrt{2}}{5}<t<\frac{5\sqrt{2}}{6}$ (b) $0<t<\frac{5\sqrt{2}}{6}$

(c) $-\frac{4\sqrt{2}}{5}<t<0$ (d) none of these

34. There are two parallel lines, one of which has the equation $3x+4y=2$. If the lines cut an intercept of length 5 on the line $x+y=1$ then the equation of the other line is

(a) $3x+4y=\frac{\sqrt{6}-2}{2}$ (b) $3x+4y=\frac{\sqrt{6}+2}{2}$

(c) $3x+4y=7$ (d) none of these

35. If the intercept made on the line $y=mx$ by lines $y=2$ and $y=6$ is less than 5 then the range of values of m is

(a) $\left(-\infty,-\frac{4}{3}\right)\cup\left(\frac{4}{3},+\infty\right)$ (b) $\left(-\frac{4}{3},\frac{4}{3}\right)$

(c) $\left(-\frac{3}{4},\frac{3}{4}\right)$ (d) none of these

36. If a, b, c are any three terms of an AP then the line $ax+by+c=0$

(a) has a fixed direction

(b) always passes through a fixed point

(c) always cuts intercepts on the axes such that their sum is zero

(d) forms a triangle with the axes whose area is constant

37. If a, c, b are in GP then the line $ax+by+c=0$

(a) has a fixed direction

(b) always passes through a fixed point

(c) forms a triangle with the axes whose area is constant
(d) always cuts intercepts on the axes such that their sum is zero

38. The number of real values of k for which the lines $x - 2y + 3 = 0$, $kx + 3y + 1 = 0$ and $4x - ky + 2 = 0$ are concurrent is

(a) 0 (b) 1
(c) 2 (d) infinite

39. A family of lines is given by $(1 + 2\lambda)x + (1 - \lambda)y + \lambda = 0$, λ being the parameter. The line belonging to this family at the maximum distance from the point (1, 4) is

(a) $4x - y + 1 = 0$ (b) $33x + 12y + 7 = 0$
(c) $12x + 33y = 7$ (d) none of these

40. The members of the family of lines $(\lambda + \mu)x + (2\lambda + \mu)y = \lambda + 2\mu$, where $\lambda \neq 0, \mu \neq 0$, pass through the point

(a) $(3, -1)$ (b) $(-3, 1)$
(c) $(1, 1)$ (d) none of these

41. The equations of the sides AB, BC and CA of the ΔABC are $y - x = 2$, $x + 2y = 1$ and $3x + y + 5 = 0$ respectively. The equation of the altitude through B is

(a) $x - 3y + 1 = 0$ (b) $x - 3y + 4 = 0$
(c) $3x - y + 2 = 0$ (d) none of these

42. The range of values of the ordinate of a point moving on the line $x = 1$, and always remaining in the interior of the triangle formed by the lines $y = x$, the x-axis and $x + y = 4$, is

(a) $(0, 1)$ (b) $[0, 1]$
(c) $[0, 4]$ (d) none of these

43. If the point (a, a) falls between the lines $|x + y| = 2$ then

(a) $|a| = 2$ (b) $|a| = 1$
(c) $|a| < 1$ (d) $|a| < \frac{1}{2}$

44. If A$(\sin\alpha, 1/\sqrt{2})$ and B$(1/\sqrt{2}, \cos\alpha)$, $-\pi \leq \alpha \leq \pi$, are two points on the same side of the line $x - y = 0$ then α belongs to the interval

(a) $\left(-\frac{\pi}{4}, \frac{\pi}{4}\right) \cup \left(\frac{\pi}{4}, \frac{3\pi}{4}\right)$ (b) $\left(-\frac{\pi}{4}, \frac{\pi}{4}\right)$

(c) $\left(\frac{\pi}{4}, \frac{3\pi}{4}\right)$ (d) none of these

45. The straight lines $L_1 \equiv 4x - 3y + 2 = 0$, $L_2 \equiv 3x + 4y - 4 = 0$ and $L_3 \equiv x - 7y + 6 = 0$

(a) form a right-angled triangle

(b) form a right-angled isosceles triangle

(c) are concurrent

(d) none of these

46. The equation of the bisector of the acute angle between the lines $2x - y + 4 = 0$ and $x - 2y = 1$ is

(a) $x + y + 5 = 0$ (b) $x - y + 1 = 0$

(c) $x - y = 5$ (d) none of these

47. The equation of the bisector of that angle between the lines $x + y = 3$ and $2x - y = 2$ which contains the point (1, 1) is

(a) $(\sqrt{5} - 2\sqrt{2})x + (\sqrt{5} + \sqrt{2})y = 3\sqrt{5} - 2\sqrt{2}$

(b) $(\sqrt{5} + 2\sqrt{2})x + (\sqrt{5} - \sqrt{2})y = 3\sqrt{5} + 2\sqrt{2}$

(c) $3x = 10$

(d) none of these

48. Two lines $2x - 3y = 1$ and $x + 2y + 3 = 0$ divide the x-y plane in four compartments which are named as shown in the figure. Consider the locations of the points (2, –1), (3, 2) and (–1, –2). We get

(a) $(2, -1) \in$ IV

(b) $(3, 2) \in$ III

(c) $(-1, -2) \in$ II

(d) none of these

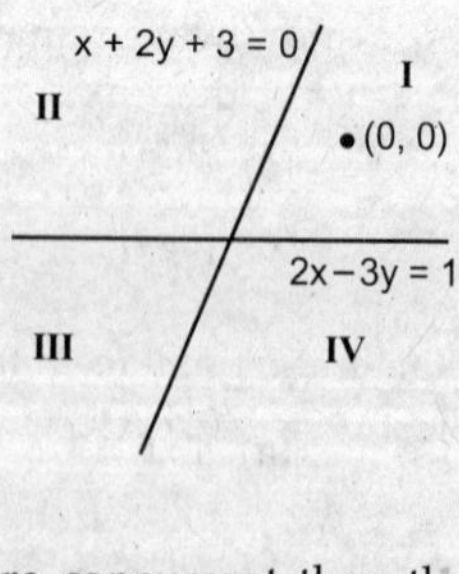

49. If the lines $y - x = 5$, $3x + 4y = 1$ and $y = mx + 3$ are concurrent then the value of m is

(a) $\frac{19}{5}$ (b) 1

(c) $\frac{5}{19}$ (d) none of these

50. If the point $(\cos\theta, \sin\theta)$ does not fall in that angle between the lines $y = |x-1|$ in which the origin lies then θ belongs to

(a) $\left(\frac{\pi}{2}, \frac{3\pi}{2}\right)$ (b) $\left(-\frac{\pi}{2}, \frac{\pi}{2}\right)$

(c) $(0, \pi)$ (d) none of these

51. The points $(-1, 1)$ and $(1, -1)$ are symmetrical about the line

(a) $y + x = 0$ (b) $y = x$

(c) $x + y = 1$ (d) none of these

52. The equation of the line segment AB is $y = x$. If A and B lie on the same side of the line mirror $2x - y = 1$, the image of AB has the equation

(a) $x + y = 2$ (b) $8x + y = 9$

(c) $7x - y = 6$ (d) none of these

53. Let $P = (1, 1)$ and $Q = (3, 2)$. The point R on the x-axis such that $PR + RQ$ is the minimum is

(a) $\left(\frac{5}{3}, 0\right)$ (b) $\left(\frac{1}{3}, 0\right)$

(c) $(3, 0)$ (d) none of these

54. If a ray travelling along the line $x = 1$ gets reflected from the line $x + y = 1$ then the equation of the line along which the reflected ray travels is

(a) $y = 0$ (b) $x - y = 1$

(c) $x = 0$ (d) none of these

55. The point P(2, 1) is shifted by $3\sqrt{2}$ parallel to the line $x + y = 1$, in the direction of increasing ordinate, to reach Q. The image of Q by the line $x + y = 1$ is

(a) $(5, -2)$ (b) $(-1, 4)$

(c) $(3, -4)$ (d) $(-3, 2)$

56. Let $A = (1, 0)$ and $B = (2, 1)$. The line AB turns about A through an angle $\pi/6$ in the clockwise sense, and the new position of B is B′. Then B′ has the coordinates

(a) $\left(\frac{3+\sqrt{3}}{2}, \frac{\sqrt{3}-1}{2}\right)$ (b) $\left(\frac{3-\sqrt{3}}{2}, \frac{\sqrt{3}+1}{2}\right)$

(c) $\left(\frac{1-\sqrt{3}}{2}, \frac{1+\sqrt{3}}{2}\right)$ (d) none of these

57. A line has intercepts a, b on the coordinate axes. If the axes are rotated about the origin through an angle α then the line has intercepts p, q on the new position of the axes respectively. Then

(a) $\frac{1}{p^2}+\frac{1}{q^2}=\frac{1}{a^2}+\frac{1}{b^2}$ (b) $\frac{1}{p^2}-\frac{1}{q^2}=\frac{1}{a^2}-\frac{1}{b^2}$

(c) $\frac{1}{p^2}+\frac{1}{a^2}=\frac{1}{q^2}+\frac{1}{b^2}$ (d) none of these

58. Two points A and B move on the x-axis and the y-axis respectively such that the distance between the two points is always the same. The locus of the middle point of AB is

(a) a straight line (b) a pair of straight lines

(c) a circle (d) none of these

59. Three vertices of a quadrilateral in order are (6, 1), (7, 2) and (–1, 0). If the area of the quadrilateral is 4 unit ^{2} then the locus of the fourth vertex has the equation

(a) $x-7y=1$ (b) $x-7y+15=0$

(c) $(x-7y)^2+14(x-7y)-15=0$ (d) none of these

60. A variable line through the point (a, b) cuts the axes of reference at A and B respectively. The lines through A and B parallel to the y-axis and the x-axis respectively meet at P. Then the locus of P has the equation

(a) $\frac{x}{a}+\frac{y}{b}=1$ (b) $\frac{x}{b}+\frac{y}{a}=1$

(c) $\frac{a}{x}+\frac{b}{y}=1$ (d) $\frac{b}{x}+\frac{a}{y}=1$

• Type 2 •

Choose the correct options. One or more options may be correct.

61. If the coordinates of the vertices of a triangle are rational numbers then which of the following points of the triangle will always have rational coordinates?

(a) Centroid (b) Incentre

(c) Circumcentre (d) Orthocentre

62. Two consecutive vertices of a rectangle of area 10 unit ^{2} are (1, 3) and (–2, –1). Other two vertices are

(a) $\left(-\frac{3}{5}, \frac{21}{5}\right), \left(-\frac{18}{5}, \frac{1}{5}\right)$ (b) $\left(-\frac{3}{5}, \frac{21}{5}\right), \left(-\frac{2}{5}, -\frac{11}{5}\right)$

(c) $\left(-\frac{2}{5}, -\frac{11}{5}\right), \left(\frac{13}{5}, \frac{9}{5}\right)$ (d) $\left(\frac{13}{5}, \frac{9}{5}\right), \left(-\frac{18}{5}, \frac{1}{5}\right)$

63. The ends of a diagonal of a square are (2, –3) and (–1, 1). Another vertex of the square can be

(a) $\left(-\frac{3}{2}, -\frac{5}{2}\right)$ (b) $\left(\frac{5}{2}, \frac{1}{2}\right)$

(c) $\left(\frac{1}{2}, \frac{5}{2}\right)$ (d) none of these

64. If each of the vertices of a triangle has integral coordinates then the triangle may be

(a) right angled (b) equilateral

(c) isosceles (d) none of these

65. If (–1, 2), (2, –1) and (3, 1) are any three vertices of a parallelogram then the fourth vertex (a, b) will be such that

(a) $a = 2, b = 0$ (b) $a = -2, b = 0$

(c) $a = -2, b = 6$ (d) $a = 6, b = -2$

66. If (α, β) be an end of a diagonal of a square and the other diagonal has the equation $x - y = \alpha$ then another vertex of the square can be

(a) $(\alpha - \beta, \alpha)$ (b) $(\alpha, 0)$

(c) $(0, -\alpha)$ (d) $(\alpha + \beta, \beta)$

67. A point on the line $y = x$ whose perpendicular distance from the line $\frac{x}{4} + \frac{y}{3} = 1$ is 4 has the coordinates

(a) $\left(-\frac{8}{7}, -\frac{8}{7}\right)$ (b) $\left(\frac{32}{7}, \frac{32}{7}\right)$

(c) $\left(\frac{3}{2}, \frac{3}{2}\right)$ (d) none of these

68. The parametric equation of a line is given by

$$x = -2 + \frac{r}{\sqrt{10}} \text{ and } y = 1 + 3\frac{r}{\sqrt{10}}.$$

Then, for the line

(a) intercept on the x-axis $= \frac{7}{3}$ (b) intercept on the y-axis $= -7$

(c) slope of the line = $\tan^{-1}\frac{1}{3}$

(d) slope of the line = $\tan^{-1}3$

69. One side of a square of length a is inclined to the x-axis at an angle α with one of the vertices of the square at the origin. The equation of a diagonal of the square is

(a) $y(\cos\alpha - \sin\alpha) = x(\cos\alpha + \sin\alpha)$

(b) $y(\cos\alpha + \sin\alpha) = x(\cos\alpha - \sin\alpha)$

(c) $y(\sin\alpha + \cos\alpha) - x(\sin\alpha - \cos\alpha) = a$

(d) $y(\sin\alpha + \cos\alpha) + x(\sin\alpha - \cos\alpha) = a$

70. If the equations of the hypotenuse and a side of a right-angled isosceles triangle be $x + my = 1$ and $x = k$ respectively then

(a) $m = 1$

(b) $m = k$

(c) $m = -1$

(d) $m + k = 0$

71. The centroid and a vertex of an equilateral triangle are (1, 1) and (1, 2) respectively. Another vertex of the triangle can be

(a) $\left(\frac{2-\sqrt{3}}{2}, \frac{1}{2}\right)$

(b) $\left(\frac{2+3\sqrt{3}}{2}, \frac{1}{2}\right)$

(c) $\left(\frac{2+\sqrt{3}}{2}, \frac{1}{2}\right)$

(d) none of these

72. If one vertex of an equilateral triangle of side 2 is the origin and another vertex lies on the line $x = \sqrt{3}y$ then the third vertex can be

(a) $(0, 2)$

(b) $(\sqrt{3}, -1)$

(c) $(0, -2)$

(d) $(\sqrt{3}, 1)$

73. A line passing through the point (2, 2) and the axes enclose an area λ. The intercepts on the axes made by the line are given by the two roots of

(a) $x^2 - 2|\lambda|x + |\lambda| = 0$

(b) $x^2 + |\lambda|x + 2|\lambda| = 0$

(c) $x^2 - |\lambda|x + 2|\lambda| = 0$

(d) none of these

74. A line passing through the origin and making an angle $\pi/4$ with the line $y - 3x = 5$ has the equation

(a) $x + 2y = 0$

(b) $2x = y$

(c) $x = 2y$

(d) $y + 2x = 0$

75. The coordinates of a point on the line $x + y = 3$ such that the point is at equal distances from the lines $|x| = |y|$ are

(a) $(3, 0)$ (b) $(0, 3)$

(c) $(-3, 0)$ (d) $(0, -3)$

76. A line perpendicular to the line $3x - 2y = 5$ cuts off an intercept 3 on the positive side of the x-axis. Then

(a) the slope of the line is $\frac{2}{3}$

(b) the intercept on the y-axis is 2

(c) the area of the triangle formed by the line with the axes is 3 unit ^{2}

(d) none of these

77. One diagonal of a square is the portion of the line $\sqrt{3}x + y = 2\sqrt{3}$ intercepted by the axes. Then an extremity of the other diagonal is

(a) $(1 + \sqrt{3}, \sqrt{3} - 1)$ (b) $(1 + \sqrt{3}, \sqrt{3} + 1)$

(c) $(1 - \sqrt{3}, \sqrt{3} - 1)$ (d) $(1 - \sqrt{3}, \sqrt{3} + 1)$

78. If $bx + cy = a$, where a, b, c are of the same sign, be a line such that the area enclosed by the line and the axes of reference is $\frac{1}{8}$ unit ^{2} then

(a) b, a, c are in GP (b) $b, 2a, c$ are in GP

(c) $b, \frac{a}{2}, c$ are in AP (d) $b, -2a, c$ are in GP

79. The sides of a triangle are $x + y = 1, 7y = x$ and $\sqrt{3}y + x = 0$. Then the following is an interior point of the triangle.

(a) Circumcentre (b) Centroid

(c) Incentre (d) Orthocentre

80. If (x, y) be a variable point on the line $y = 2x$ lying between the lines $2(x + 1) + y = 0$ and $x + 3(y - 1) = 0$ then

(a) $x \in \left(-\frac{1}{2}, \frac{6}{7}\right)$ (b) $x \in \left(-\frac{1}{2}, \frac{3}{7}\right)$

(c) $y \in \left(-1, \frac{3}{7}\right)$ (d) $y \in \left(-1, \frac{6}{7}\right)$

81. If the equations of the three sides of a triangle are $x + y = 1, 3x + 5y = 2$ and $x - y = 0$ then the orthocentre of the triangle lies on the line

(a) $5x - 3y = 2$ (b) $3x - 5y + 1 = 0$

(c) $2x - 3y = 1$ (d) $5x - 3y = 1$

82. A ray travelling along the line $3x - 4y = 5$ after being reflected from a line l travels along the line $5x + 12y = 13$. Then the equation of the line l is

(a) $x + 8y = 0$
(b) $x = 8y$
(c) $32x + 4y = 65$
(d) $32x - 4y + 65 = 0$

83. A ray of light travelling along the line $x + y = 1$ is incident on the x-axis and after refraction it enters the other side of the x-axis by turning $\pi/6$ away from the x-axis. The equation of the line along which the refracted ray travels is

(a) $x + (2 - \sqrt{3})y = 1$
(b) $(2 - \sqrt{3})x + y = 1$
(c) $y + (2 + \sqrt{3})x = 2 + \sqrt{3}$
(d) none of these

Answers

1. c	**2.** a	**3.** c	**4.** b	**5.** d
6. b	**7.** b	**8.** a	**9.** a	**10.** b
11. d	**12.** d	**13.** c	**14.** a	**15.** b
16. c	**17.** d	**18.** a	**19.** a	**20.** c
21. a	**22.** d	**23.** b	**24.** a	**25.** d
26. b	**27.** a	**28.** a	**29.** a	**30.** c
31. a	**32.** b	**33.** a	**34.** d	**35.** a
36. b	**37.** c	**38.** a	**39.** c	**40.** a
41. b	**42.** a	**43.** c	**44.** a	**45.** c
46. b	**47.** a	**48.** a	**49.** c	**50.** b
51. b	**52.** c	**53.** a	**54.** a	**55.** d
56. a	**57.** a	**58.** c	**59.** c	**60.** c
61. a, c, d	**62.** a, c	**63.** a, b	**64.** a, c, d	**65.** b, d
66. b, d	**67.** a, b	**68.** d	**69.** a, c	**70.** a, c
71. a, c	**72.** a, b	**73.** c	**74.** c, d	**75.** a, b
76. b, c	**77.** b, c	**78.** b, d	**79.** b, c	**80.** b, d
81. b, d	**82.** b, c	**83.** a, c		

Hints and Solutions to Selected Questions

1. If none of the coordinates are irrational then

$$\Delta = \frac{1}{2}\begin{vmatrix} x_1 & y_1 & 1 \\ x_2 & y_2 & 1 \\ x_3 & y_3 & 1 \end{vmatrix} = \frac{1}{2} \times \text{rational} = \text{rational}$$

But the area of an equilateral triangle

$$= \frac{\sqrt{3}}{4} \times (\text{side})^2 = \frac{\sqrt{3}}{4} \times \text{rational} = \text{irrational}.$$

So, both the coordinates of the third vertex cannot be rational.

3. The relations between polar coordinates and cartesian coordinates are $x = r\cos\theta$, $y = r\sin\theta$. So, the cartesian coordinates of the points are (0, 0), (0, 3) and $\left(\frac{3\sqrt{3}}{2}, \frac{3}{2}\right)$ respectively. Hence the sides are equal.

8. As the points are in order, the area

$$= \left|\frac{1}{2}\left\{\begin{vmatrix} 4 & 1 \\ 3 & 6 \end{vmatrix} + \begin{vmatrix} 3 & 6 \\ -5 & 1 \end{vmatrix} + \begin{vmatrix} -5 & 1 \\ -3 & -3 \end{vmatrix} + \begin{vmatrix} -3 & -3 \\ -3 & 0 \end{vmatrix} + \begin{vmatrix} -3 & 0 \\ 4 & 1 \end{vmatrix}\right\}\right| = 30 \text{ unit}^2.$$

9. With the mutually perpendicular lines as axes, we get $|x| + |y| = 1$. So, the locus consists of four lines, each enclosing the same area $\frac{1}{2} \cdot 3 \cdot 3$ with the axes.

10. $(x-1)(x-3) + (y-2)(y-4) = 0$

$\Rightarrow$ AC $\perp$ BC $\Rightarrow$ $\angle ACB = 90°$.

$\therefore$ C is on the circle whose diameter is AB.

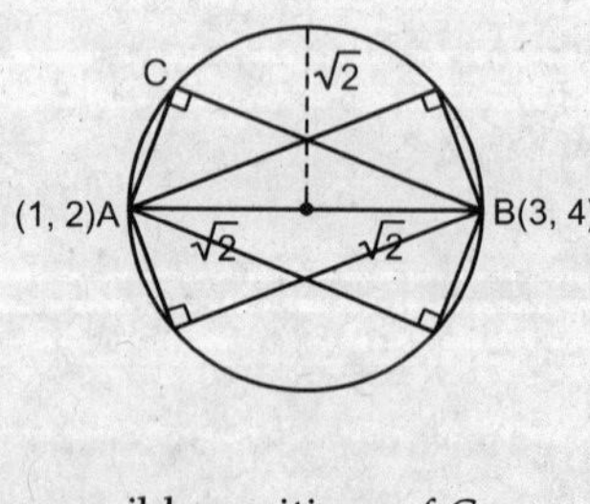

$AB = 2\sqrt{2}$.

As ar (ΔABC) = 1, $\frac{1}{2} \cdot 2\sqrt{2} \cdot (\text{altitude}) = 1$

$\therefore$ altitude $= \frac{1}{\sqrt{2}} <$ radius. So, there are four possible positions of C.

Note If ar(ΔABC) = 2, altitude = radius $\Rightarrow$ two possible positions of C.

11. Let the points be A, B, C and D. Clearly,

'm' of AD = 0 = 'm' of BC

$\therefore$ AD || BC || x-axis. Also $AD^2 = BC^2$.

'm' of AC = ∞ = 'm' of BD

$\therefore$ AC || BD || y-axis. Also, $AC^2 = BD^2$.

12. The diagonals are perpendicular to each other.

13. The points A, B,C, D form a trapezium whose line of symmetry is the y-axis.

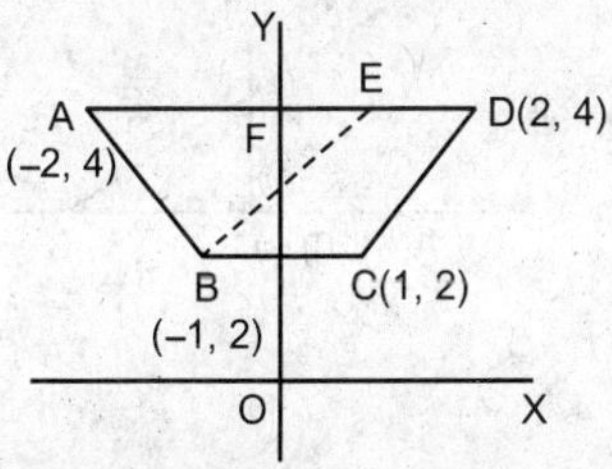

Let BE be the required line where $ED = \lambda$.

$$\text{ar (BCDE)} = \frac{1}{2}\text{ ar (ABCD)}$$

$$\Rightarrow \frac{1}{2}\cdot(2+\lambda)\cdot 2 = \frac{1}{2}\cdot\frac{1}{2}(4+2)\cdot 2$$

$$\Rightarrow \lambda = 1$$

$\therefore$ the point $E = (1, 4)$.

So, the required line BE is $y - 2 = \dfrac{4-2}{1+1}(x+1)$.

17. $\tan(180° - \theta)$ = slope of AB = -3

$\therefore$ $\tan\theta = 3$.

$\therefore$ $\dfrac{OC}{AC} = \tan\theta, \dfrac{OC}{BC} = \cot\theta$

$\Rightarrow$ $\dfrac{BC}{AC} = \dfrac{\tan\theta}{\cot\theta}$

$= \tan^2\theta = 9.$

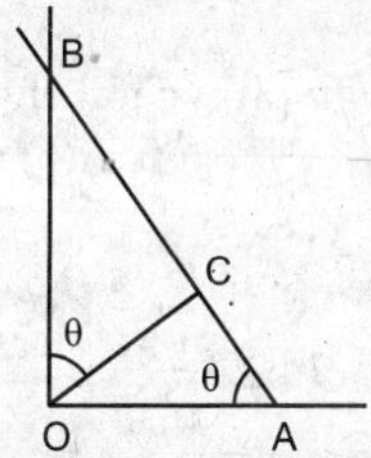

18. Any point on the line through $(1, 1)$ and parallel to $x + y = 1$ is $P(1 + r\cos\theta, 1 + r\sin\theta)$ where $\tan\theta = -1$. Use the fact that P is on the line $2x - 3y = 4$ and find $|r|$.

20.

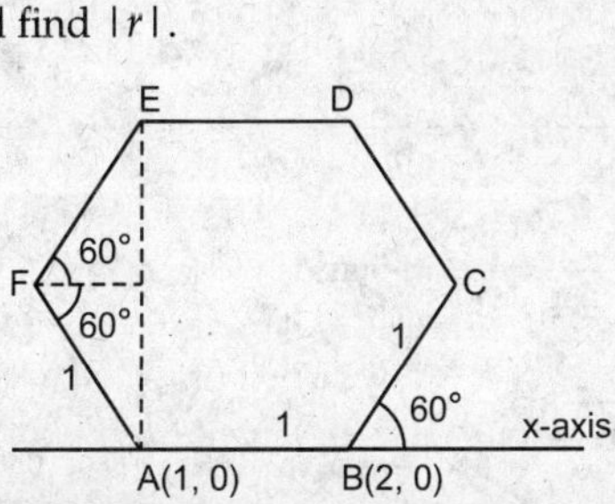

$$C = (2 + 1 \cdot \cos 60°, 1 \cdot \sin 60°) = \left(\frac{5}{2}, \frac{\sqrt{3}}{2}\right)$$

$$E = (1, 1 \cdot \sin 60° + 1 \cdot \sin 60°) = (1, \sqrt{3})$$

$\therefore$ the equation of CE is $y - \sqrt{3} = \dfrac{\sqrt{3} - \frac{\sqrt{3}}{2}}{1 - \frac{5}{2}} \cdot (x - 1)$.

21. $(1 + OB)\cos 30° = 2\sqrt{3}$

$\Rightarrow \quad 1 + OB = 2\sqrt{3} \times \dfrac{2}{\sqrt{3}} = 4$

$\therefore \quad OB = 3$

$\therefore \quad B = (3, 0)$ and

'm' of BC $= \tan 150° = -\dfrac{1}{\sqrt{3}}$

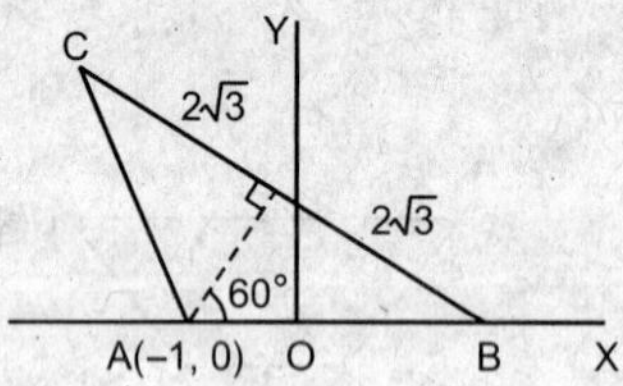

$\therefore$ the equation of BC is $y - 0 = -\dfrac{1}{\sqrt{3}}(x - 3)$.

22. The equation of the graph is $y = \cos x \cdot \cos(x + 2) - \cos^2(x + 1)$

or $\quad y = \dfrac{1}{2}\{\cos 2 + \cos 2(x + 1)\} - \dfrac{1}{2}\{1 + \cos 2(x + 1)\}$

$$= \frac{1}{2}(\cos 2 - 1) = -\frac{1}{2} \cdot 2\sin^2 1 = -\sin^2 1.$$

The graph is parallel to the x-axis, and $\left(\dfrac{\pi}{2}, -\sin^2 1\right)$ satisfies it.

23. $\dfrac{1}{2}\begin{vmatrix} -2 & 0 & 1 \\ -1 & \dfrac{1}{\sqrt{3}} & 1 \\ \cos\theta & \sin\theta & 1 \end{vmatrix} = 0$

$\Rightarrow \quad \sqrt{3}\sin\theta - \cos\theta = 2 \quad \Rightarrow \sin\left(\theta - \dfrac{\pi}{6}\right) = 1 \quad \Rightarrow \theta - \dfrac{\pi}{6} = \dfrac{\pi}{2}$.

24. Solving, $x = \dfrac{3c^2 - 8}{9c^2 - 4c - 4}, y = \dfrac{5 - c}{9c^2 - 4c - 4}$.

Now, $\lim\limits_{c \to 1} \dfrac{3c^2 - 8}{9c^2 - 4c - 4} = \dfrac{-5}{1} = -5$ and

$\lim\limits_{c \to 1} \dfrac{5 - c}{9c^2 - 4c - 4} = \dfrac{4}{1} = 4.$

27. In an equilateral triangle the orthocentre and the centroid are the same. OPQ is the equilateral triangle in which $OC \perp PQ$.

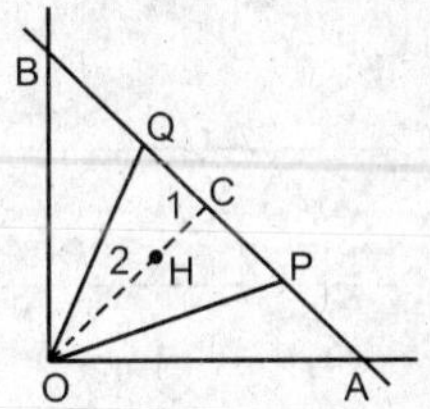

Clearly, the point H which divides OC internally in the ratio 2 : 1 is the orthocentre.

Clearly, $OC = \frac{1}{\sqrt{2}}$. So, $OH = \frac{2}{3} \times \frac{1}{\sqrt{2}}$.

$\therefore \quad H = \left(\frac{2}{3\sqrt{2}} \cos 45°, \frac{2}{3\sqrt{2}} \sin 45°\right)$.

28. One side of the triangle is parallel to the y-axis and another side is parallel to the x-axis. So, the triangle is a right-angled triangle. Hence, the middle point of the hypotenuse is the circumcentre. Solving $x = 2, x + 2y = 4$ we get one end of the hypotenuse, and solving $y + 1 = 0, x + 2y = 4$ we get the other end. Their coordinates are (2, 1) and (6, –1).

$\therefore \quad \text{circumcentre} = \left(\frac{2+6}{2}, \frac{1-1}{2}\right)$.

29. Let the line be $y = mx + c$ or $mx - y + c = 0$.

The algebraic sum of the distances $= \frac{m-1+c}{\sqrt{1+m^2}} + \frac{2m+c}{\sqrt{1+m^2}} + \frac{-2+c}{\sqrt{1+m^2}} = 0$

$\Rightarrow \quad 3m + 3c = 3 \quad \text{or} \quad 1 = m + c.$

So, (1, 1) satisfies $y = mx + c$ for all m, c.

30. From the choice of the axes,

$A = (0, 0), B = (2\cot\theta, 2), C = (4\cot\overline{60° - \theta}, -4)$.

$\therefore$ (side of the equilateral triangle)^2

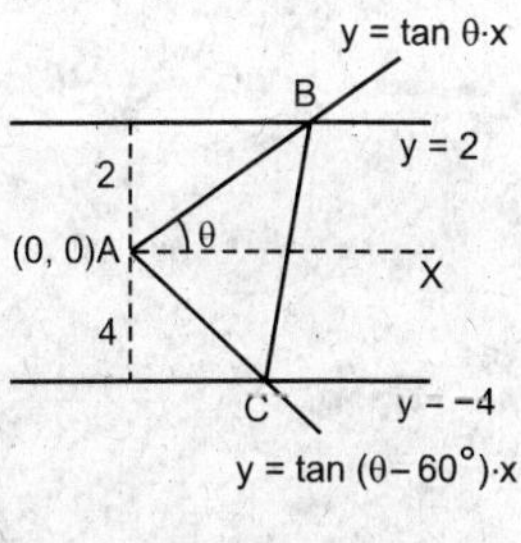

$= 4\cot^2\theta + 4$

$= 16\cot^2(60° - \theta) + 16$

$\Rightarrow \quad 4\operatorname{cosec}^2\theta = 16\operatorname{cosec}^2(60° - \theta)$

$\therefore \quad \operatorname{cosec}\theta = 2\operatorname{cosec}(60° - \theta)$

or $\quad 2\sin\theta = \sin(60° - \theta)$

$\therefore \quad 2\sin\theta = \frac{\sqrt{3}}{2}\cos\theta - \frac{1}{2}\sin\theta$

or $\quad 5\sin\theta = \sqrt{3}\cos\theta \quad \therefore \tan\theta = \frac{\sqrt{3}}{5}$

$\therefore \quad$ the required length $= 2\operatorname{cosec}\theta = 2 \times \frac{\sqrt{25+3}}{\sqrt{3}} = \frac{2\sqrt{28}}{\sqrt{3}}$.

32. $2x + 11y = 5$ is one of the bisectors of the angles between the given lines.

33. Let P be on $x + 2y = 1$

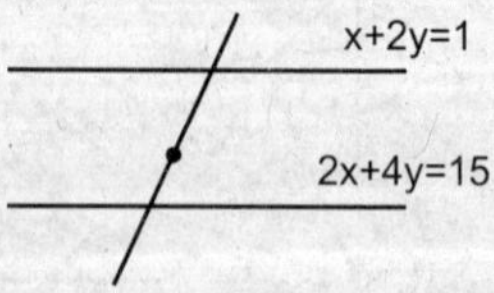

$$\Rightarrow \quad 1 + \frac{t}{\sqrt{2}} + 2\left(2 + \frac{t}{\sqrt{2}}\right) = 1 \quad \text{or} \quad t = -\frac{4\sqrt{2}}{3}.$$

Let P be on $2x + 4y = 15$

$$\Rightarrow \quad 2\left(1 + \frac{t}{\sqrt{2}}\right) + 4\left(2 + \frac{t}{\sqrt{2}}\right) = 15 \text{ or } t = \frac{5\sqrt{2}}{6}.$$

35. The distance between $\left(\frac{2}{m}, 2\right)$ and $\left(\frac{6}{m}, 6\right)$ is less than 5

$$\Rightarrow \quad \left(\frac{2}{m} - \frac{6}{m}\right)^2 + (2-6)^2 < 25$$

$$\Rightarrow \quad \frac{16}{m^2} < 9 \quad \Rightarrow m^2 > \frac{16}{9} \quad \Rightarrow m > \frac{4}{3} \text{ or } m < \frac{-4}{3}.$$

36. If a, b, c are the pth, qth and rth terms of an AP whose first term = λ and the common difference = μ then the line is

$$\{\lambda + (p-1)\mu\}x + \{\lambda + (q-1)\mu\}y + \lambda + (r-1)\mu = 0$$

or
$$\lambda\{x + y + 1\} + \mu\{(p-1)x + (q-1)y + r - 1\} = 0,$$

which is of the form $L_1 + kL_2 = 0$.

37. The area of the triangle $= \frac{1}{2} \times (x\text{-intercept}) \times (y\text{-intercept})$

$$= \frac{1}{2} \times \left(-\frac{c}{a}\right) \times \left(-\frac{c}{b}\right)$$

$$= \frac{1}{2} \cdot \frac{c^2}{ab} = \frac{1}{2} \quad (\because \ a, c, b \text{ are in GP}).$$

39. Here $x + y + \lambda(2x - y + 1) = 0$

$\Rightarrow$ each line passes through the point of intersection of the lines $x + y = 0$ and $2x - y + 1 = 0$, which is $\left(-\frac{1}{3}, \frac{1}{3}\right)$.

The required line passes through $\left(-\frac{1}{3}, \frac{1}{3}\right)$ and is perpendicular to the line joining (1, 4) and $\left(-\frac{1}{3}, \frac{1}{3}\right)$.

$\therefore$ 'm' of the required line $= \dfrac{-1}{\dfrac{4-1/3}{1+1/3}} = -\dfrac{4}{11}$.

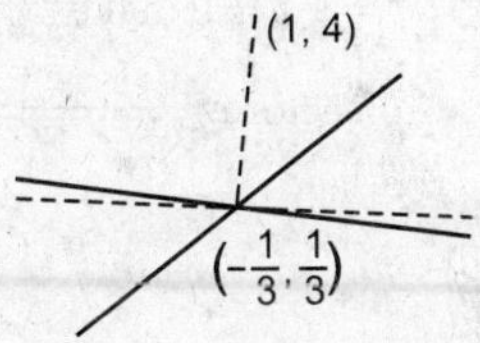

$\therefore$ the required line is $y - \dfrac{1}{3} = -\dfrac{4}{11}\left(x + \dfrac{1}{3}\right)$

or $12x + 33y = 7$.

42.

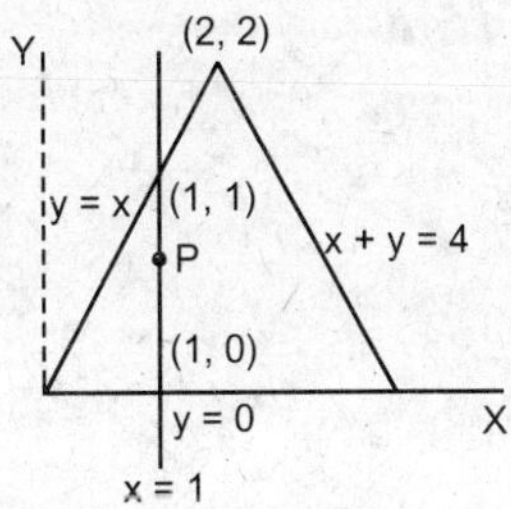

From the figure, y-coordinates of P vary from 0 to 1. So, to be an interior point,

$$0 < \text{ordinate of P} < 1.$$

43. From the figure, $-1 < a < 1$.

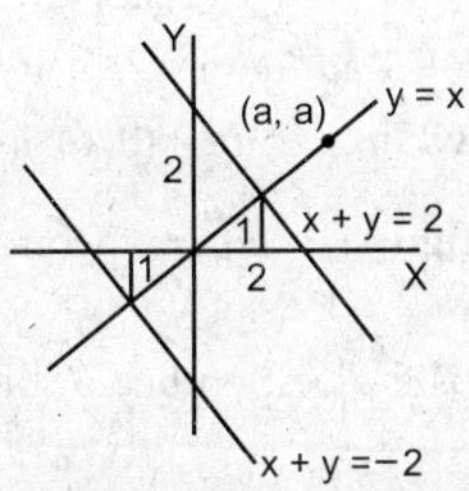

44. $\sin\alpha - \dfrac{1}{\sqrt{2}} > 0$ and $\dfrac{1}{\sqrt{2}} - \cos\alpha > 0$ ... (1)

or $\sin\alpha - \dfrac{1}{\sqrt{2}} < 0$ and $\dfrac{1}{\sqrt{2}} - \cos\alpha < 0$ ... (2)

(1) $\Rightarrow \sin\alpha > \dfrac{1}{\sqrt{2}}$ and $\cos\alpha < \dfrac{1}{\sqrt{2}} \Rightarrow \dfrac{\pi}{4} < \alpha < \dfrac{3\pi}{4}$.

(2) $\Rightarrow \sin\alpha < \dfrac{1}{\sqrt{2}}$ and $\cos\alpha > \dfrac{1}{\sqrt{2}} \Rightarrow -\dfrac{\pi}{4} < \alpha < \dfrac{\pi}{4}$.

45. $1 \cdot L_1 - 1 \cdot L_2 - 1 \cdot L_3 \equiv 0$

$\Rightarrow$ the straight lines are concurrent.

46.

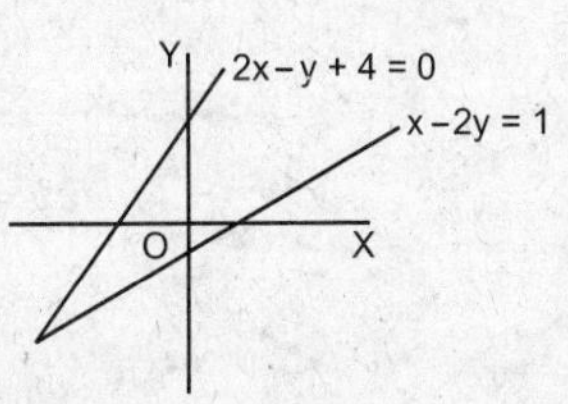

Clearly from the figure, the origin is contained in the acute angle. Writing the equations of the lines as $2x - y + 4 = 0$ and $-x + 2y + 1 = 0$, the required bisector is

$$\frac{2x - y + 4}{\sqrt{5}} = \frac{-x + 2y + 1}{\sqrt{5}}.$$

47. $x = 1, y = 1$ make both $x + y - 3$ and $2x - y - 2$ negative. So the required bisector is $\frac{x+y-3}{\sqrt{2}} = \frac{2x-y-2}{\sqrt{5}}$.

50.

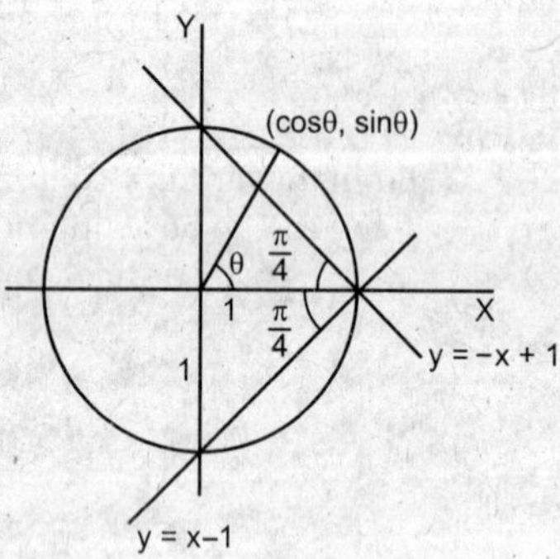

The lines are $y = x - 1$ and $y = -x + 1$ and $(\cos\theta, \sin\theta)$ is any point on the circle $x^2 + y^2 = 1$ where centre = (0, 0) and radius = 1.

Clearly from the figure, θ can vary from $-\frac{\pi}{2}$ to $\frac{\pi}{2}$.

51. The required line is the perpendicular bisector of the line segment joining the given points.

52. The image of (0, 0) on $y - x = 0$ by the line $2x - y = 1$ is $\left(\frac{4}{5}, -\frac{2}{5}\right)$.

The point of intersection of $y - x = 0$ and $2x - y = 1$ is (1, 1).

$\therefore$ the required line is the line joining the points (1, 1) and $\left(\frac{4}{5}, -\frac{2}{5}\right)$.

53. The point R should be such that PR is reflected along RQ from the line $y = 0$.

The equation of P′Q (where P′ is the image of P) is $3x - 2y = 5$.

R is the point of intersection of $y = 0$ and $3x - 2y = 5$.

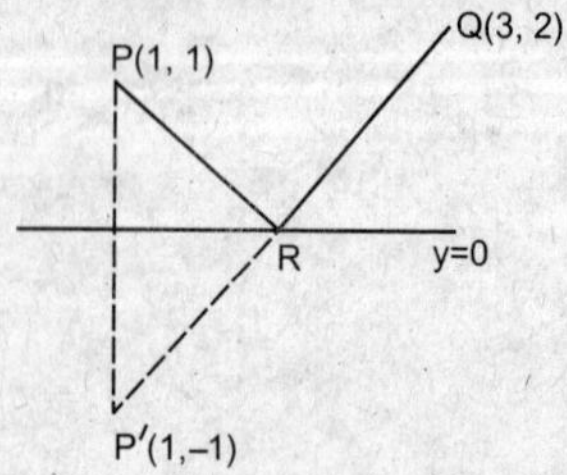

54. 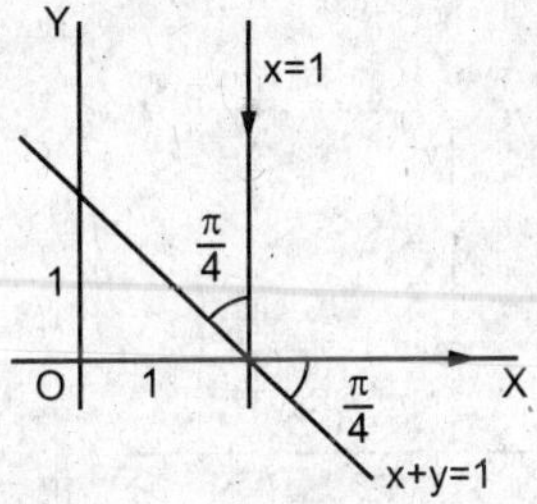

Clearly from the figure, the reflected ray moves along the x-axis.

55. $Q = (2 \pm 3\sqrt{2}\cos\theta, 1 \pm 3\sqrt{2}\sin\theta)$, where $\tan\theta = -1$

$$= \left(2 \pm 3\sqrt{2} \times \frac{-1}{\sqrt{2}}, 1 \pm 3\sqrt{2} \times \frac{1}{\sqrt{2}}\right) = (2 \mp 3, 1 \pm 3).$$

From the question, the y-coordinate of Q should be more than 1 (= y-coordinate of P).

56. $B' = \left(1 + AB \cdot \cos\left(\theta - \frac{\pi}{6}\right), 0 + AB \cdot \sin\left(\theta - \frac{\pi}{6}\right)\right)$,

where $\tan\theta$ = slope of AB $= \frac{1-0}{2-1} = 1$ and $AB = \sqrt{(2-1)^2 + (1-0)^2} = \sqrt{2}$.

$\therefore \quad B' = (1 + \sqrt{2}\cos 15°, 0 + \sqrt{2} \cdot \sin 15°)$.

57. The length of the perpendicular from the origin to the line in both the cases is the same. So

$$\frac{\frac{0}{a} + \frac{0}{b} - 1}{\sqrt{\frac{1}{a^2} + \frac{1}{b^2}}} = \frac{\frac{0}{p} + \frac{0}{q} - 1}{\sqrt{\frac{1}{p^2} + \frac{1}{q^2}}}$$

$$\Rightarrow \quad \frac{1}{a^2} + \frac{1}{b^2} = \frac{1}{p^2} + \frac{1}{q^2}.$$

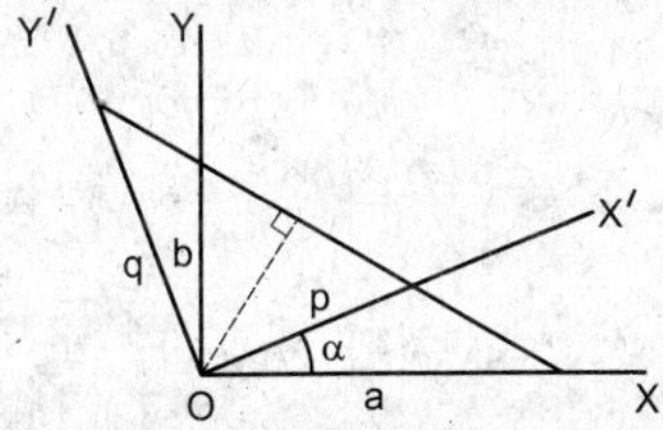

58. Let A = $(a, 0)$, B = $(0, b)$. Then $\sqrt{a^2 + b^2} = AB = k$ (constant).

If the middle point of AB be P = (x_1, y_1) then $x_1 = \frac{a}{2}, y_1 = \frac{b}{2}$.

$\therefore (2x_1)^2 + (2y_1)^2 = k^2$, i.e., the equation of the locus of P is $x^2 + y^2 = \left(\frac{k}{2}\right)^2$.

60. Clearly,

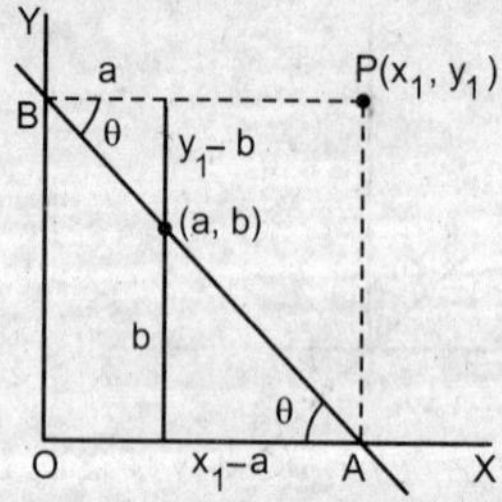

$$\frac{b}{x_1 - a} = \tan\theta = \frac{y_1 - b}{a}$$

$$\therefore \quad ab = x_1y_1 - bx_1 - ay_1 + ab$$

$$\Rightarrow \quad \frac{a}{x_1} + \frac{b}{y_1} = 1.$$

61. If $A = (x_1, y_1), B = (x_2, y_2), C = (x_3, y_3)$, where x_1, y_1, etc., are rational numbers then $\Sigma x_1, \Sigma y_1$ are also rational. So, the coordinates of the centroid $\left(\frac{\Sigma x_1}{3}, \frac{\Sigma y_1}{3}\right)$ will be rational.

As $AB = c = \sqrt{(x_1 - x_2)^2 + (y_1 - y_2)^2}$, c may or may not be rational and it may be an irrational number of the form $\sqrt{p}$. Hence, the coordinates of the incentre $\left(\frac{\Sigma ax_1}{\Sigma a}, \frac{\Sigma ay_1}{\Sigma a}\right)$ may or may not be rational.

If (α, β) be the circumcentre or orthocentre, α and β are found by solving two linear equations in α, β with rational coefficients. So α, β must be rational numbers.

62. From the figure,

P or $P' = (1 \pm 2 \cdot \cos\theta, 3 \pm 2 \cdot \sin\theta)$, where $\tan\theta = -\frac{3}{4}$

$$= \left(1 \pm 2 \cdot \frac{-4}{5}, 3 \pm 2 \cdot \frac{3}{5}\right).$$

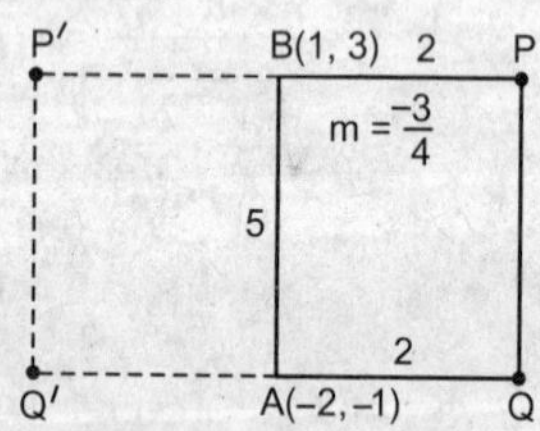

$$Q \text{ or } Q' = \left(-2 \pm 2 \cdot \frac{-4}{5}, -1 \pm 2 \cdot \frac{3}{5}\right).$$

$\therefore$ the other two vertices are

$$\left(1 + 2 \cdot \frac{-4}{5}, 3 + 2 \cdot \frac{3}{5}\right), \left(-2 + 2 \cdot \frac{-4}{5}, -1 + 2 \cdot \frac{3}{5}\right)$$

or $$\left(1 - 2 \cdot \frac{-4}{5}, 3 - 2 \cdot \frac{3}{5}\right), \left(-2 - 2 \cdot \frac{-4}{5}, -1 - 2 \cdot \frac{3}{5}\right).$$

63. 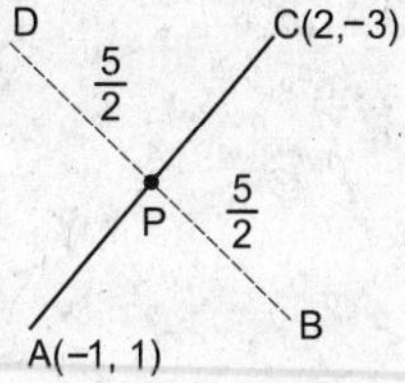

$AC = \sqrt{3^2 + 4^2} = 5$

The midpoint P of AC $= \left(\frac{1}{2}, -1\right)$

'm' of AC $= \frac{4}{-3}$ $\quad\therefore$ 'm' of BD $= \frac{3}{4} = \tan\theta$

$\therefore$ B or D $= \left(\frac{1}{2} \pm \frac{5}{2}\cos\theta, -1 \pm \frac{5}{2}\sin\theta\right)$.

64. See the hint to Q. No. 61.

67. Let the point be (t, t).

So, $\left|\dfrac{\frac{t}{4} + \frac{t}{3} - 1}{\sqrt{\frac{1}{4^2} + \frac{1}{3^2}}}\right| = 4 \quad \therefore \quad t\left(\frac{1}{4} + \frac{1}{3}\right) = 1 \pm 4 \cdot \sqrt{\frac{1}{4^2} + \frac{1}{3^2}}$.

68. The cartesian equation is $x + 2 = \frac{y-1}{3}$, i.e., $3x - y + 7 = 0$.

69.

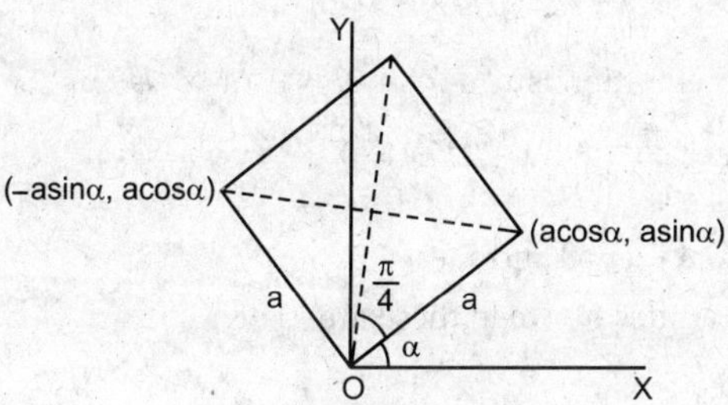

The diagonals are $y = \tan\left(\frac{\pi}{4} + \alpha\right) \cdot x$ and

$$y - a\sin\alpha = \frac{a(\sin\alpha - \cos\alpha)}{a(\cos\alpha + \sin\alpha)}(x - a\cos\alpha).$$

71. $AG = 1$ and the equation of AD is $x = 1$.

So, D $= \left(1, \frac{1}{2}\right)$, because $\frac{AG}{GD} = \frac{2}{1}$

$AD = 1 + \frac{1}{2} = \frac{3}{2}$

$\therefore \quad \tan 60° = \frac{3/2}{BD} \quad \therefore \quad BD = \frac{\sqrt{3}}{2} = DC$

$\therefore \quad$ B $= \left(1 - \frac{\sqrt{3}}{2}, \frac{1}{2}\right)$ and C $= \left(1 + \frac{\sqrt{3}}{2}, \frac{1}{2}\right)$.

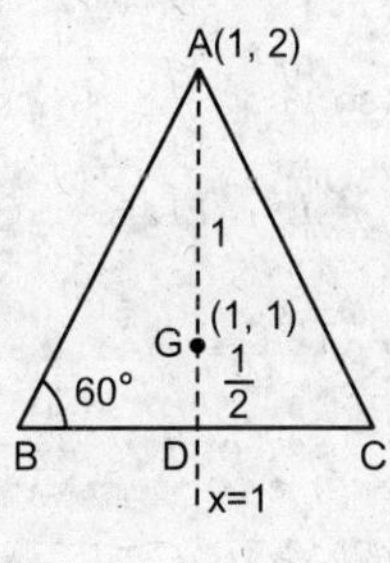

72. Clearly Q, R are the two possible positions of the third vertex.

P = (2cos 30°, 2sin 30°), Q = (0, 2), R = (2cos 30°, –2sin 30°).

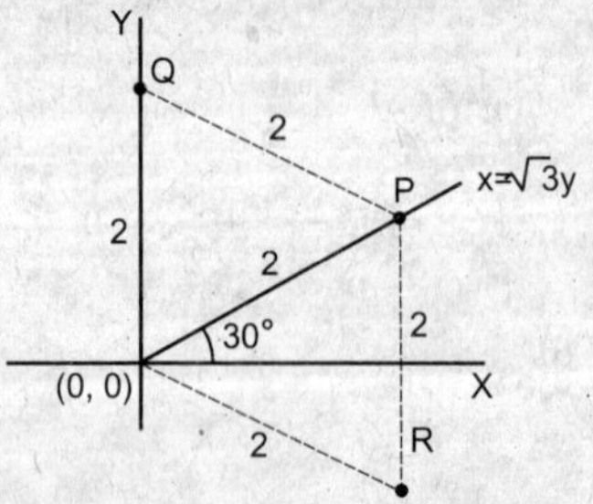

73. If a, b are intercepts then the equation of the line is $\frac{x}{a}+\frac{y}{b}=1$. It passes through (2, 2).

$\therefore \quad \frac{2}{a}+\frac{2}{b}=1.$ Also $\frac{1}{2}ab=\pm\lambda$ or $ab=\pm 2\lambda$

$\therefore \quad \frac{a+b}{ab}=\frac{1}{2}$ or $\frac{a+b}{\pm 2\lambda}=\frac{1}{2}$ or $a+b=\pm\lambda$

$\therefore$ the required equation is $x^2-\lambda x+2\lambda=0$ or $x^2+\lambda x-2\lambda=0$

Combining these, $x^2-|\lambda|x+2|\lambda|=0$.

75. $|x|=|y| \quad \Rightarrow \quad x-y=0$ and $x+y=0$.

Bisectors of the angles between these are

$$\frac{x-y}{\sqrt{2}}=\pm\frac{x+y}{\sqrt{2}} \quad \Rightarrow \quad y=0 \text{ and } x=0.$$

By solving $y=0, x+y=3$ we get one of the required points while by solving $x=0, x+y=3$ we get the other point.

77. Clearly, $OA=2, OC=2\sqrt{3}$

$\therefore \quad E=(1, \sqrt{3})$.

Also, $AC=\sqrt{2^2+(2\sqrt{3})^2}=4$

$\therefore \quad EB=ED=\frac{1}{2}AC=2.$

'm' of $AC=-\sqrt{3}$

$\therefore \quad$ 'm' of $BD=\frac{1}{\sqrt{3}}=\tan 30°.$

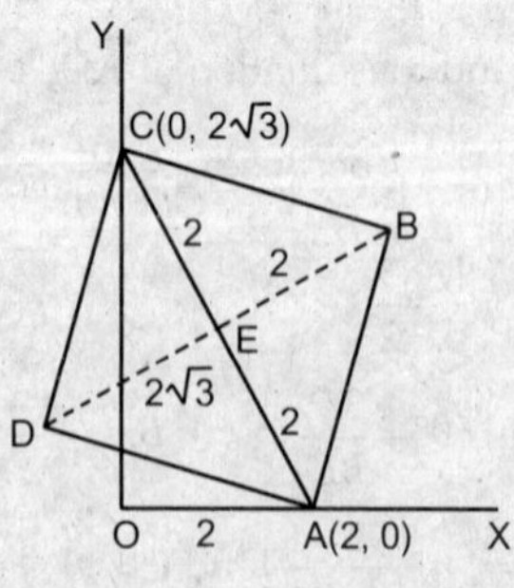

$\therefore \quad$ B or D = $(1\pm 2\cdot\cos 30°, \sqrt{3}\pm 2\cdot\sin 30°)$.

79. In any triangle the centroid and the incentre lie within the triangle. If the triangle is obtuse angled then the orthocentre and the circumcentre fall outside the triangle. Here, the angle between $7y = x$ and $\sqrt{3}y + x = 0$ is obtuse because the first line is inclined at an angle less than 45° with the x-axis, while the second line makes 150° with the x-axis.

80.

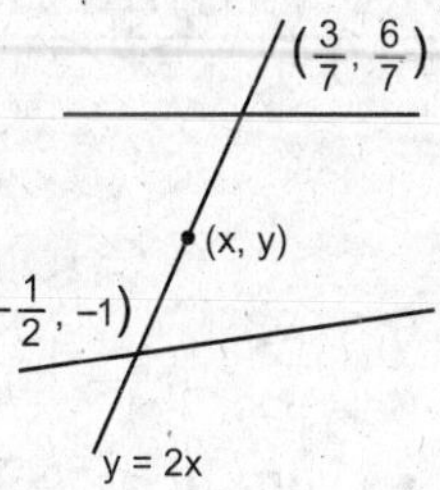

Solving $y = 2x$, $2(x + 1) + y = 0$, we get, $x = -\frac{1}{2}$, $y = -1$.

Solving $y = 2x$, $x + 3(y - 1) = 0$, we get, $x = \frac{3}{7}$, $y = \frac{6}{7}$.

81. As $x + y = 1$, $x - y = 0$ represent perpendicular lines, the triangle is right angled and the orthocentre is the point of intersection of these lines.

82.

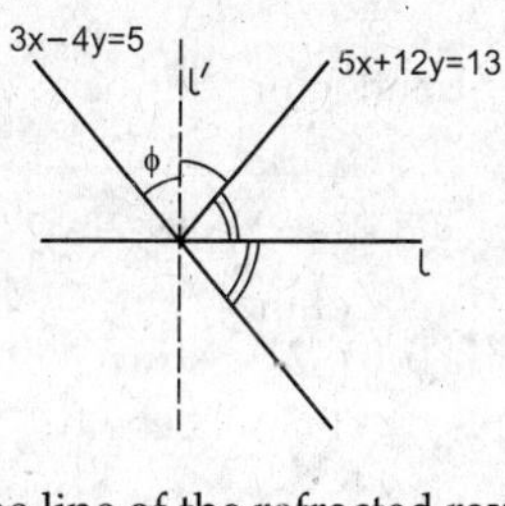

Clearly, the line l can be any one of the bisectors of the angles between the lines $3x - 4y = 5$ and $5x + 12y = 13$.

83. The line of the refracted ray passes through the point (1, 0) and its slope is tan 105°.

$\therefore$ the equation of the line of the refracted ray is

$$y - 0 = \tan 105° \cdot (x - 1).$$

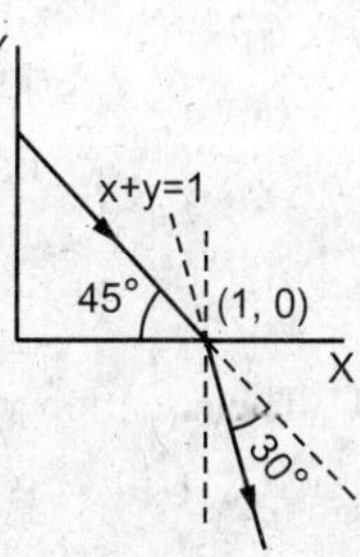

□

2

Pair of Straight Lines

• *Type 1* •

Choose the most appropriate option (a, b, c or d).

1. The equation $4x^2 + mxy - 3y^2 = 0$ represents a pair of real and distinct lines if

(a) $m \in R$ (b) $m \in (3, 4)$

(c) $m \in (-3, 4)$ (d) $m > 4$

2. If $2x^2 + 3xy + my^2 = 0$ represents two real and mutually perpendicular lines then m is

(a) any negative real number (b) any positive real number

(c) -2 (d) none of these

3. The equation $kx^2 + 4xy + 5y^2 = 0$ represents two lines inclined at an angle π if k is

(a) $\frac{5}{4}$ (b) $\frac{4}{5}$

(c) $-\frac{4}{5}$ (d) none of these

4. The equation $x^3 + y^3 = 0$ represents

(a) three real straight lines

(b) three points

(c) the combined equation of a straight line and a circle

(d) none of these

5. The angle between the pair of lines $y^2 - 2xy\operatorname{cosec}\theta + x^2 = 0, 0 \le \theta \le \frac{\pi}{2}$, is

(a) $\frac{\pi}{2}$ (b) θ

(c) $\frac{\pi}{2} - \theta$ (d) none of these

6. If the slope of one line is double the slope of another line and the combined equation of the pair of lines is $\frac{x^2}{a} + \frac{2xy}{h} + \frac{y^2}{b} = 0$ then $ab : h^2$ is equal to

(a) $9:8$ (b) $3:2$

(c) $8:3$ (d) none of these

7. The triangle formed by the lines whose combined equation is $(y^2 - 4xy - x^2)(x + y - 1) = 0$ is

(a) equilateral (b) right angled

(c) isosceles (d) obtuse angled

8. The three lines whose combined equation is $y^3 - 4x^2y = 0$ form a triangle which is

(a) isosceles (b) equilateral

(c) right angled (d) none of these

9. The combined equation of the lines l_1, l_2 is $2x^2 + 6xy + y^2 = 0$ and that of the lines m_1, m_2 is $4x^2 + 18xy + y^2 = 0$. If the angle between l_1 and m_2 be α then the angle between l_2 and m_1 will be

(a) $\frac{\pi}{2} - \alpha$ (b) 2α

(c) $\frac{\pi}{4} + \alpha$ (d) α

10. The lines represented by $x^2 + 2\lambda xy + 2y^2 = 0$ and the lines represented $(1 + \lambda)x^2 - 8xy + y^2 = 0$ are equally inclined then λ is

(a) any real number (b) greater than 2

(c) ± 2 (d) less than -2

11. The area of the triangle formed by two rays whose combined equation is $y = |x|$ and the line $x + 2y = 2$ is

(a) $\frac{8}{3}$ unit^2 (b) $\frac{4}{3}$ unit^2

(c) 4 unit^2 (d) $\frac{16}{3}$ unit^2

12. The centroid of the triangle whose three sides are given by the combined equation $(x^2+7xy+2y^2)(y-1)=0$ is

(a) $\left(\frac{2}{3}, 0\right)$ (b) $\left(\frac{7}{3}, \frac{2}{3}\right)$

(c) $\left(-\frac{7}{3}, \frac{2}{3}\right)$ (d) none of these

13. The orthocentre of the triangle formed by the pair of lines $2x^2-xy-y^2+x+2y-1=0$ and the line $x+y+1=0$ is

(a) $(-1, 0)$ (b) $(0, 1)$

(c) $(-1, 1)$ (d) none of these

14. The angle between the pair of lines whose equation is $4x^2+10xy+my^2+5x+10y=0$ is

(a) $\tan^{-1}\frac{3}{8}$ (b) $\tan^{-1}\frac{3}{4}$

(c) $\tan^{-1}\frac{2\sqrt{25-4m}}{m+4}$, $m \in R$ (d) none of these

15. The combined equation of the pair of lines through the point (1, 0) and parallel to the lines represented by $2x^2-xy-y^2=0$ is

(a) $2x^2-xy-2y^2+4x-y=6$ (b) $2x^2-xy-y^2-4x-y+2=0$

(c) $2x^2-xy-y^2-4x+y+2=0$ (d) none of these

16. The equation $x^2+(\lambda+\mu)xy+\lambda\mu y^2+x+\mu y=0$ represents two parallel straight lines if

(a) $\lambda+\mu=0$ (b) $\lambda=4\mu$

(c) $\lambda=\mu$ (d) none of these

17. The product of perpendiculars drawn from the point (1, 2) to the pair of lines $x^2+4xy+y^2=0$ is

(a) $\frac{9}{4}$ (b) $\frac{3}{4}$

(c) $\frac{9}{16}$ (d) none of these

18. If pair of lines represented by $ax^2 + 2hxy + by^2 = 0, b \neq 0$, are such that the sum of the slopes of the lines is three times the product of their slopes then

(a) $3b + 2h = 0$ (b) $3a + 2h = 0$

(c) $3h + 2a = 0$ (d) none of these

19. The pair of lines $\sqrt{3}x^2 - 4xy + \sqrt{3}y^2 = 0$ are rotated about the origin by $\frac{\pi}{6}$ in the anticlockwise sense. The equation of the pair in the new position is

(a) $\sqrt{3}x^2 - xy = 0$ (b) $x^2 - \sqrt{3}xy = 0$

(c) $xy - \sqrt{3}y^2 = 0$ (d) none of these

20. The equation of the image of the pair of rays $y = |x|$ by the line $x = 1$ is

(a) $|y| = x + 2$ (b) $|y| + 2 = x$

(c) $y = |x - 2|$ (d) none of these

21. Two lines represented by the equation $x^2 - y^2 - 2x + 1 = 0$ are rotated about the point (1, 0), the line making the bigger angle with the positive direction of the x-axis being turned by 45° in the clockwise sense and the other line being turned by 15° in the anticlockwise sense. The combined equation of the pair of lines in their new positions is

(a) $\sqrt{3}x^2 - xy + 2\sqrt{3}x - y + \sqrt{3} = 0$ (b) $\sqrt{3}x^2 - xy - 2\sqrt{3}x + y + \sqrt{3} = 0$

(c) $\sqrt{3}x^2 - xy - 2\sqrt{3}x + \sqrt{3} = 0$ (d) none of these

• Type 2 •

Choose the correct options. One or more options may be correct.

22. The equation $x^3 + x^2y - xy^2 = y^3$ represents

(a) three real straight lines

(b) lines in which two of them are perpendicular to each other

(c) lines in which two of them are coincident

(d) none of these

23. The combined equation of two sides of an equilateral triangle is $x^2 - 3y^2 - 2x + 1 = 0$. If the length of a side of the triangle is 4 then the equation of the third side is

(a) $x=2\sqrt{3}+1$ (b) $y=2\sqrt{3}+1$

(c) $x+2\sqrt{3}=1$ (d) $x=2\sqrt{3}$

24. Two pairs of straight lines have the equations $y^2+xy-12x^2=0$ and $ax^2+2hxy+by^2=0$. One line will be common among them if

(a) $a=-3(2h+3b)$ (b) $a=8(h-2b)$

(c) $a=2(b+h)$ (d) $a=-3(b+h)$

25. If one of the lines of $my^2+(1-m^2)xy-mx^2=0$ is a bisector of the angle between the lines $xy=0$ then m is

(a) 1 (b) 2

(c) $-\frac{1}{2}$ (d) -1

26. The straight lines represented by $x^2+mxy-2y^2+3y-1=0$ meet at

(a) $\left(-\frac{1}{3},\frac{2}{3}\right)$ (b) $\left(-\frac{1}{3},-\frac{2}{3}\right)$

(c) $\left(\frac{1}{3},\frac{2}{3}\right)$ (d) none of these

27. If the chord $y=mx+1$ of the circle $x^2+y^2=1$ subtends an angle of measure 45° at the major segment of the circle then the value of m is

(a) 2 (b) 1

(c) -1 (d) none of these

28. The equation $2x^2-3xy-py^2+x+qy-1=0$ represents two mutually perpendicular lines if

(a) $p=3, q=2$ (b) $p=2, q=3$

(c) $p=-2, q=3$ (d) $p=2, q=-\frac{9}{2}$

29. The diagonals of a square are along the pair of lines whose equation is $2x^2-3xy-2y^2=0$. If (2, 1) is a vertex of the square then another vertex consecutive to it can be

(a) (1, –2) (b) (1, 4)

(c) (–1, 2) (d) (–1, –4)

30. There is a pair of points, one on each of the lines whose combined equation is $(4x-3y+5)(6x+8y+5)=0$. If they are such that the distance of the point on one line is 2 units from the other line then the points are

(a) $\left(\frac{1}{10}, \frac{9}{5}\right)$ and $\left(\frac{1}{2}, -1\right)$ (b) $\left(\frac{1}{2}, -1\right)$ and $\left(-\frac{23}{10}, -\frac{7}{3}\right)$

(c) $\left(\frac{1}{10}, \frac{9}{5}\right)$ and $\left(-\frac{23}{10}, -\frac{7}{5}\right)$ (d) none of these

31. The pairs of straight lines $ax^2 + 2hxy - ay^2 = 0$ and $hx^2 - 2axy - hy^2 = 0$ are such that

(a) one pair bisects the angles between the other pair

(b) the lines of one pair are equally inclined to the lines of the other pair

(c) the lines of one pair are perpendicular to the lines of the other pair

(d) none of these

32. If the pair of lines $ax^2 + 2hxy + by^2 + 2gx + 2fy + c = 0$ intersect on the y-axis then

(a) $2fgh = bg^2 + ch^2$ (b) $bg^2 = ch^2$

(c) $abc = 2fgh$ (d) none of these

33. The combined equation of three sides of a triangle is $(x^2 - y^2)(2x + 3y - 6) = 0$. If $(-2, a)$ is an interior point and $(b, 1)$ is an exterior point of the triangle then

(a) $2 < a < \frac{10}{3}$ (b) $-2 < a < \frac{10}{3}$

(c) $-1 < b < \frac{9}{2}$ (d) $-1 < b < 1$

Answers

1. a	**2.** c	**3.** b	**4.** d	**5.** c
6. a	**7.** b	**8.** d	**9.** d	**10.** c
11. b	**12.** c	**13.** a	**14.** b	**15.** c
16. c	**17.** a	**18.** b	**19.** a	**20.** c
21. b	**22.** a, b, c	**23.** a, c	**24.** a, b	**25.** a, d
26. a, c	**27.** b, c	**28.** b, d	**29.** a, c	**30.** a, b
31. a, b	**32.** a, b	**33.** a, d		

Hints and Solutions to Selected Questions

3. Here the lines are coincident. So, $5\left(\frac{y}{x}\right)^2 + 4\left(\frac{y}{x}\right) + k = 0$ will have equal roots.

$\therefore \quad 4^2 - 4 \cdot 5 \cdot k = 0 \quad \Rightarrow k = \frac{4}{5}$.

4. Here $(x+y)(x^2 - xy + y^2) = 0$. Clearly, $x + y = 0$ is a real straight line. But $x^2 - xy + y^2 = 0$ represents two imaginary straight lines because $D = (-1)^2 - 4 \cdot 1 \cdot 1 < 0$.

6. If m, $2m$ are slopes then

$$m + 2m = -\frac{2/h}{1/b} = -\frac{2b}{h} \quad \text{and} \quad m \cdot 2m = \frac{1/a}{1/b} = \frac{b}{a}.$$

Eliminating m, $2\left(-\frac{2b}{3h}\right)^2 = \frac{b}{a} \quad \Rightarrow \frac{ab}{h^2} = \frac{9}{8}$.

7. The pair $y^2 - 4xy - x^2 = 0$ contains lines which are at right angles because $a + b = 0$.

8. Here $y(y^2 - 4x^2) = 0$ gives the lines $y = 0$, $y = 2x$ and $y = -2x$, which are concurrent at (0, 0).

9. The combined equation of bisectors of angles between the lines of the first pair is $\frac{x^2 - y^2}{2-1} = \frac{xy}{3}$,

and that of the other pair is $\frac{x^2 - y^2}{4-1} = \frac{xy}{9}$.

As these equations are the same, the two pairs are equally inclined to each other.

10. Use the fact that the two pairs have the same bisectors of angles.

11. The lines are $y = x$, $y = -x$ and $x + 2y = 2$ as shown in the figure.

Solving $y = x$, $x + 2y = 2$, the point $\text{A} = \left(\frac{2}{3}, \frac{2}{3}\right)$

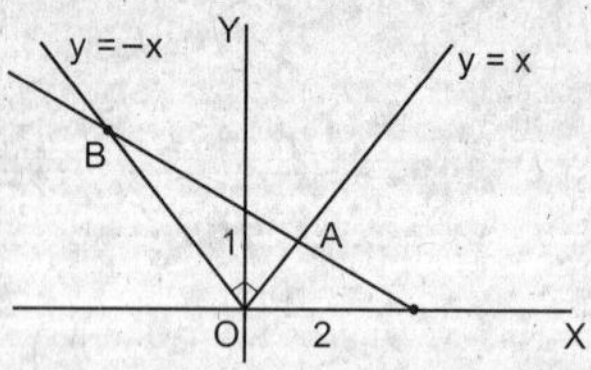

$\therefore \quad OA = \sqrt{\frac{4}{9} + \frac{4}{9}} = \frac{2\sqrt{2}}{3}.$

Solving $y = -x$, $x + 2y = 2$, the point B = (−2, 2)

$\therefore \quad OB = \sqrt{4 + 4} = 2\sqrt{2}.$

$\therefore \quad \text{ar}\ (\Delta OAB) = \frac{1}{2} \cdot OA \cdot OB = \frac{1}{2} \cdot \frac{2\sqrt{2}}{3} \cdot 2\sqrt{2} = \frac{4}{3}.$

12. The sides are $y = 1$ and the pair $x^2 + 7xy + 2y^2 = 0$. Clearly, one vertex is (0, 0) and the y-coordinate of each of the other two vertices is 1. Putting $y = 1$ in the second equation, we get $x^2 + 7x + 2 = 0$.

If x_1, x_2 are the roots then $x_1 + x_2 = -7$.

$\therefore$ the centroid $= \left(\frac{0 + x_1 + x_2}{3}, \frac{0 + 1 + 1}{3}\right) = \left(-\frac{7}{3}, \frac{2}{3}\right).$

13. $2x^2 - xy - y^2 + x + 2y - 1 = 0 \quad \Rightarrow (2x + y - 1)(x - y + 1) = 0.$

So, the sides $x - y + 1 = 0$ and $x + y + 1 = 0$ are perpendicular to each other.

$\therefore$ the orthocentre is the intersection of the perpendicular sides.

14. Being a pair of lines, $abc + 2fgh - af^2 - bg^2 - ch^2 = 0$. This gives $m = 4$.

15. The given pair is $(2x + y)(x - y) = 0$.

So, the required pair is $(2x + y + k)(x - y + k') = 0$,

where $2x + y + k = 0$ and $x - y + k' = 0$ pass through (1, 0).

$\therefore \quad k = -2, k' = -1$

$\therefore \quad$ the required pair is $(2x + y - 2)(x - y - 1) = 0$.

19. The lines are $(\sqrt{3}x - y)(x - \sqrt{3}y) = 0$.

Their separate equations are $y = \tan 60° \cdot x$, $y = \tan 30° \cdot x$.

After rotation, the separate equations are $y = \tan 90° \cdot x$, $y = \tan 60° \cdot x$.

$\therefore$ the combined equation in the new position is $x(\sqrt{3}x - y) = 0$.

20.

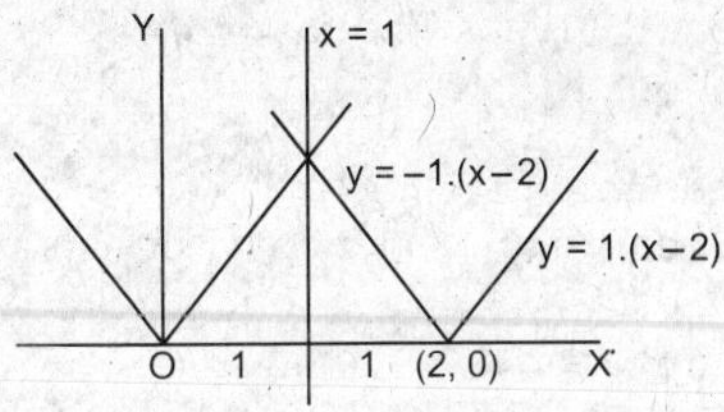

Clearly from the figure, the image is the combined rays given by

$y = 1 \cdot (x-2)$ and $y = -1 \cdot (x-2)$, i.e, $y = |x-2|$.

21.

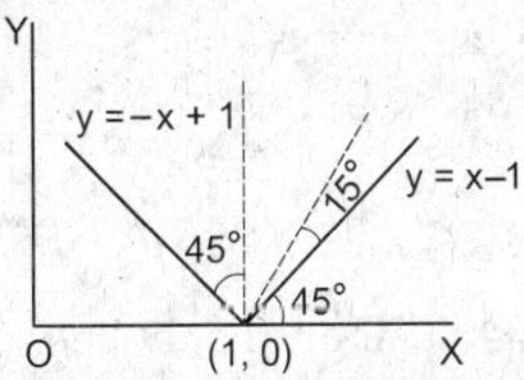

The combined equation is $(x+y-1)(x-y-1)=0$. After rotation, the lines are $x=1$ and $y=\sqrt{3}(x-1)$. So, the combined equation after rotation is $(x-1)(\sqrt{3}x-y-\sqrt{3})=0$.

22. The equation is $x^2(x+y)-y^2(x+y)=0$

or $$(x+y)^2(x-y)=0.$$

It represents the lines $x+y=0, x+y=0, x-y=0$.

23.

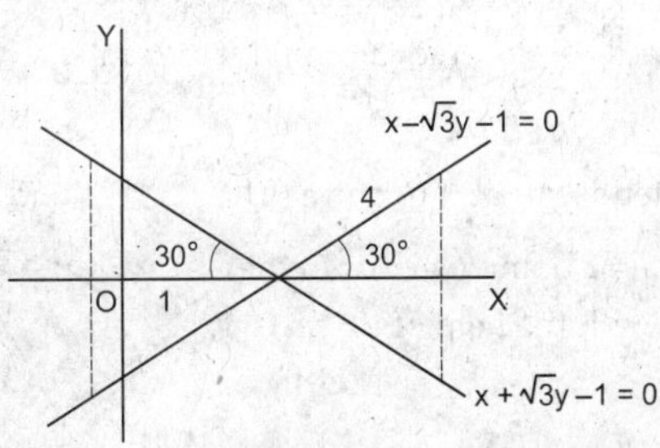

Here, $(x-1)^2 - 3y^2 = 0$

$\Rightarrow \quad (x+\sqrt{3}y-1)(x-\sqrt{3}y-1)=0.$

Now, $4\cos 30° = 4 \times \dfrac{\sqrt{3}}{2} = 2\sqrt{3}.$

$\therefore$ from the figure, the equation of the third side is

$$x = 1 + 2\sqrt{3} \quad \text{or} \quad x = -(2\sqrt{3}-1).$$

24. $y^2 + xy - 12x^2 = 0 \quad \Rightarrow (y + 4x)(y - 3x) = 0$

$\therefore \quad \frac{y}{x} = 3, -4.$

The two pairs will have a line common if 3 or –4 is a root of

$$b\left(\frac{y}{x}\right)^2 + 2h\left(\frac{y}{x}\right) + a = 0.$$

$\therefore \quad 9b + 6h + a = 0 \quad$ or $\quad 16b - 8h + a = 0.$

25. Here $my(y - mx) + x(y - mx) = 0$, i.e., $(y - mx)(my + x) = 0$.

So, the lines are $y = mx$ or $y = \frac{-1}{m}x$.

Bisectors between the lines $xy = 0$ are $y = x$ or $y = -x$.

$\therefore \; m = 1$ or -1

26. The equation represents a pair of straight lines

$$\Rightarrow \quad 1 \cdot (-2)(-1) + 2\left(\frac{3}{2}\right) \cdot 0 \cdot \frac{m}{2} - 1 \cdot \left(\frac{3}{2}\right)^2 - (-2) \cdot 0^2 - (-1) \cdot \left(\frac{m}{2}\right)^2 = 0$$

$\Rightarrow \quad m = 1, -1.$

The points of intersection of the pair of lines are obtained by solving

$$\frac{\partial S}{\partial x} \equiv 2x + my = 0 \text{ and } \frac{\partial S}{\partial y} \equiv mx - 4y + 3 = 0$$

When $m = 1$, the required point is the intersection of $2x + y = 0$, $x - 4y + 3 = 0$.

When $m = -1$, the required point is the intersection of $2x - y = 0$, $-x - 4y + 3 = 0$.

27. The chord subtends 90° at the centre (0, 0).

Making $x^2 + y^2 = 1$ homogeneous in the second degree with the help of $y = mx + 1$, we get

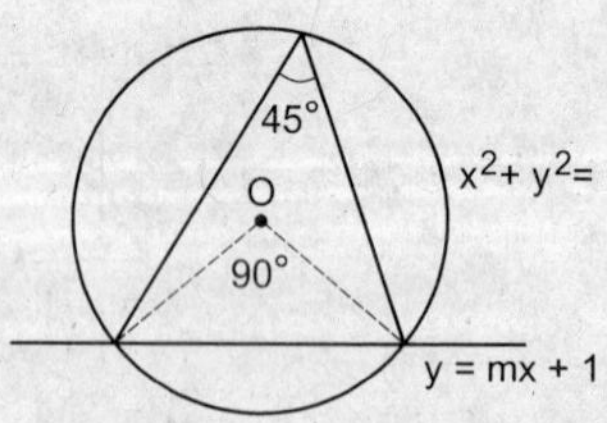

$$x^2 + y^2 = (y - mx)^2$$

or $\quad (1 - m^2)x^2 + 2mxy = 0.$

The angle between these is 90° if

$$1 - m^2 + 0 = 0, \text{ i.e., } m = \pm 1.$$

28. For the lines to be perpendicular,

$$2 + (-p) = 0, \text{ i.e., } p = 2.$$

To represent two lines,

$$2\cdot(-p)\cdot(-1)+2\cdot\frac{q}{2}\cdot\frac{1}{2}\cdot\left(-\frac{3}{2}\right)-2\cdot\left(\frac{q}{2}\right)^2-(-p)\cdot\left(\frac{1}{2}\right)^2-(-1)\cdot\left(-\frac{3}{2}\right)^2=0$$

or $\quad 4-\frac{3}{4}q-\frac{q^2}{2}+\frac{1}{2}+\frac{9}{4}=0$

or $\quad 2q^2+3q-27=0.$

29.

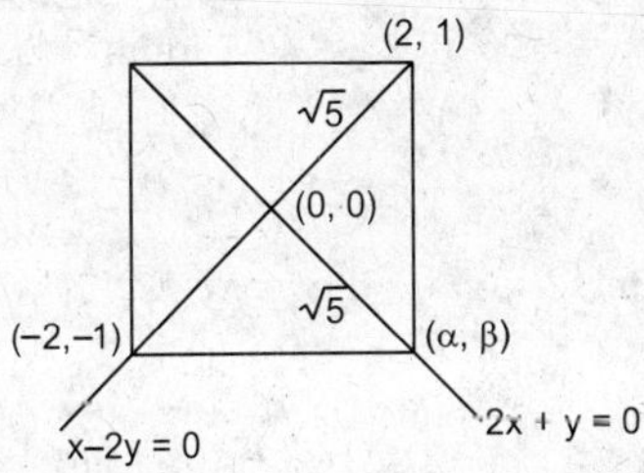

The diagonals are $(2x+y)(x-2y)=0$, i.e., $2x+y=0$ and $x-2y=0$.

If (α, β) is a consecutive vertex then $\alpha^2+\beta^2=5$ and $2\alpha+\beta=0$.

Solving, we get $\alpha=1,-1$ and $\beta=-2, 2$.

30.

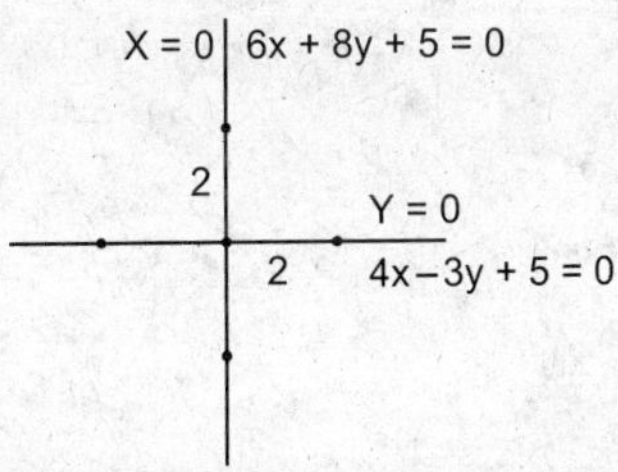

The lines are $4x-3y+5=0$ and $6x+8y+5=0$.

They are mutually perpendicular.

Let $X=\frac{6x+8y+5}{\sqrt{6^2+8^2}}$ and $Y=\frac{4x-3y+5}{\sqrt{4^2+(-3)^2}}$

$\therefore$ in X, Y coordinates the pair of points are

(i) (2, 0), (0, 2) (ii) (−2, 0), (0, 2) (iii) (2, 0), (0, −2) (iv) (−2, 0), (0, −2)

Now, $X=2, Y=0 \quad\Rightarrow\quad 6x+8y+5=20,\ 4x-3y+5=0$

$$\Rightarrow\quad x=\frac{1}{10},\ y=\frac{9}{5}.$$

$$X = 0, Y = 2 \Rightarrow 6x + 8y + 5 = 0, 4x - 3y + 5 = 10$$
$$\Rightarrow x = \frac{1}{2}, y = -1.$$

$$X = -2, Y = 0 \Rightarrow 6x + 8y + 5 = -20, 4x - 3y + 5 = 0$$
$$\Rightarrow x = -\frac{23}{10}, y = -\frac{7}{3}.$$

$$X = 0, Y = -2 \Rightarrow 6x + 8y + 5 = 0, 4x - 3y + 5 = -10$$
$$\Rightarrow x = -\frac{27}{10}, y = \frac{7}{5}.$$ Similarly, in other cases.

32. Here $abc + 2fgh - af^2 - bg^2 - ch^2 = 0$. ... (1)

Let the point of intersection on the y-axis be $(0, \lambda)$.

The point of intersection of the pair of lines is obtained by solving

$$\frac{\partial S}{\partial x} \equiv 2ax + 2hy + 2g = 0 \text{ and } \frac{\partial S}{\partial y} \equiv 2hx + 2by + 2f = 0.$$

$\therefore$ $h\lambda + g = 0$ and $b\lambda + f = 0$.

$\therefore$ $fh = bg$ ... (2)

The equation of the y-axis is $x = 0$. Solving this and the equation of the pair of lines, we get $by^2 + 2fy + c = 0$ which must have equal roots.

$\therefore$ $f^2 = bc$ ... (3)

(2) $\Rightarrow fgh = bg^2$

$\therefore$ (1) $\Rightarrow abc + fgh - af^2 - ch^2 = 0 \quad \Rightarrow fgh = ch^2$, using (3)

Adding, $2fgh = bg^2 + ch^2$

Also, $fgh = bg^2 = ch^2$.

33.

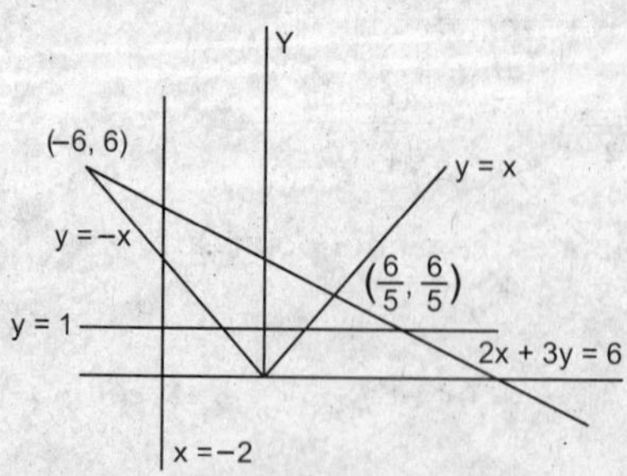

The separate equations of the sides are $x - y = 0, x + y = 0$ and $2x + 3y - 6 = 0$.

The point $(-2, a)$ moves on the line $x = -2$ and $(b, 1)$ moves on the line $y = 1$. From the figure, the y-coordinates of points of intersection of $x = -2$ with $y = -x$ and $2x + 3y = 6$ give the range of values of a.

The x-coordinates of the points of intersection of $y = 1$ with $y = -x$ and $y = x$ give the range of values of b.

□

3

Circle

• Type 1 •

Choose the most appropriate option (a, b, c or d).

1. If the equation of a circle is $ax^2 + (2a - 3)y^2 - 4x - 1 = 0$ then its centre is

(a) $(2, 0)$ (b) $(2/3, 0)$

(c) $(-2/3, 0)$ (d) none of these

2. If $2x^2 + \lambda xy + 2y^2 + (\lambda - 4)x + 6y - 5 = 0$ is the equation of a circle then its radius is

(a) $3\sqrt{2}$ (b) $2\sqrt{3}$

(c) $2\sqrt{2}$ (d) none of these

3. The equation $x^2 + y^2 - 2x + 4y + 5 = 0$ represents

(a) a point (b) a pair of straight lines

(c) a circle of nonzero radius (d) none of these

4. Three sides of a triangle have the equations $L_r \equiv y - m_r x - c_r = 0$; $r = 1, 2, 3$. Then $\lambda L_2 L_3 + \mu L_3 L_1 + \nu L_1 L_2 = 0$, where $\lambda \neq 0, \mu \neq 0, \nu \neq 0$, is the equation of the circumcircle of the triangle if

(a) $\lambda(m_2 + m_3) + \mu(m_3 + m_1) + \nu(m_1 + m_2) = 0$

(b) $\lambda(m_2 m_3 - 1) + \mu(m_3 m_1 - 1) + \nu(m_1 m_2 - 1) = 0$

(c) both (a) and (b) hold together

(d) none of these

5. The number of integral values of λ for which

$$x^2 + y^2 + \lambda x + (1 - \lambda)y + 5 = 0$$

is the equation of a circle whose radius cannot exceed 5, is

(a) 14 (b) 18
(c) 16 (d) none of these

6. If $2(x^2+y^2)+4\lambda x+\lambda^2=0$ represents a circle of meaningful radius then the range of real values of λ is

(a) R (b) $(0, +\infty)$
(c) $(-\infty, 0)$ (d) none of these

7. If a circle passes through the points of intersection of the lines $2x-y+1=0$ and $x+\lambda y-3=0$ with the axes of reference then the value of λ is

(a) 1/2 (b) 2
(c) 1 (d) –2

8. The equation of the circle passing through the point (1, 1) and having two diameters along the pair of lines $x^2-y^2-2x+4y-3=0$ is

(a) $x^2+y^2-2x-4y+4=0$ (b) $x^2+y^2+2x+4y-4=0$
(c) $x^2+y^2-2x+4y+4=0$ (d) none of these

9. Two vertices of an equilateral triangle are (–1, 0) and (1, 0), and its third vertex lies above the x-axis. The equation of the circumcircle of the triangle is

(a) $x^2+y^2=1$ (b) $\sqrt{3}(x^2+y^2)+2y-\sqrt{3}=0$
(c) $\sqrt{3}(x^2+y^2)-2y-\sqrt{3}=0$ (d) none of these

10. A triangle is formed by the lines whose combined equation is given by $(x+y-4)(xy-2x-y+2)=0$. The equation of its circumcircle is

(a) $x^2+y^2-5x-3y+8=0$ (b) $x^2+y^2-3x-5y+8=0$
(c) $x^2+y^2-3x-5y-8=0$ (d) none of these

11. If the centroid of an equilateral triangle is (1, 1) and its one vertex is (–1, 2) then the equation of its circumcircle is

(a) $x^2+y^2-2x-2y-3=0$ (b) $x^2+y^2+2x-2y-3=0$
(c) $x^2+y^2+2x+2y-3=0$ (d) none of these

12. The equation of the circle whose one diameter is PQ, where the ordinates of P, Q are the roots of the equation $x^2+2x-3=0$ and the abscissae are the roots of the equation $y^2+4y-12=0$, is

(a) $x^2+y^2+2x+4y-15=0$ (b) $x^2+y^2-4x-2y-15=0$

(c) $x^2+y^2+4x+2y-15=0$ (d) none of these

13. The maximum number of points with rational coordinates on a circle whose centre is $(\sqrt{3}, 0)$ is

(a) one (b) two

(c) four (d) infinite

14. A circle touches the y-axis at (0, 2) and has an intercept of 4 units on the positive side of the x-axis. Then the equation of the circle is

(a) $x^2+y^2-4(\sqrt{2}x+y)+4=0$ (b) $x^2+y^2-4(x+\sqrt{2}y)+4=0$

(c) $x^2+y^2-2(\sqrt{2}x+y)+4=0$ (d) none of these

15. C_1 is a circle of radius 1 touching the x-axis and the y-axis. C_2 is another circle of radius > 1 and touching the axes as well as the circle C_1. Then the radius of C_2 is

(a) $3-2\sqrt{2}$ (b) $3+2\sqrt{2}$

(c) $3+2\sqrt{3}$ (d) none of these

16. The intercept on the line $y=x$ by the circle $x^2+y^2-2x=0$ is AB. The equation of the circle with AB as a diameter is

(a) $x^2+y^2+x+y=0$ (b) $x^2+y^2=x+y$

(c) $x^2+y^2-3x+y=0$ (d) none of these

17. Two circles, each of radius 5, have a common tangent at (1, 1) whose equation is $3x+4y-7=0$. Then their centres are

(a) (4, –5), (–2, 3) (b) (4, –3), (–2, 5)

(c) (4, 5), (–2, –3) (d) none of these

18. The equation of the circumcircle of the regular hexagon whose two consecutive vertices have the coordinates (–1, 0) and (1, 0) and which lies wholly above the x-axis, is

(a) $x^2+y^2-2\sqrt{3}y-1=0$ (b) $x^2+y^2-\sqrt{3}y-1=0$

(c) $x^2+y^2-2\sqrt{3}x-1=0$ (d) none of these

19. The equation of the incircle of the triangle formed by the axes and the line $4x+3y=6$ is

(a) $x^2+y^2-6x-6y+9=0$ (b) $4(x^2+y^2-x-y)+1=0$

(c) $4(x^2+y^2+x+y)+1=0$ (d) none of these

20. If p and q be the longest distance and the shortest distance respectively of the point $(-7, 2)$ from any point (α, β) on the curve whose equation is $x^2 + y^2 - 10x - 14y - 51 = 0$ then GM of p and q is equal to

(a) $2\sqrt{11}$ (b) $5\sqrt{5}$

(c) 13 (d) none of these

21. The equation of the circumcircle of an equilateral triangle is $x^2 + y^2 + 2gx + 2fy + c = 0$ and one vertex of the triangle is (1, 1). The equation of incircle of the triangle is

(a) $4(x^2 + y^2) = g^2 + f^2$

(b) $4(x^2 + y^2) + 8gx + 8fy = (1 - g)(1 + 3g) + (1 - f)(1 + 3f)$

(c) $4(x^2 + y^2) + 8gx + 8fy = g^2 + f^2$

(d) none of these

22. Let $f(x, y) = 0$ be the equation of a circle. If $f(0, \lambda) = 0$ has equal roots $\lambda = 2, 2$, and $f(\lambda, 0) = 0$ has roots $\lambda = \frac{4}{5}, 5$ then the centre of the circle is

(a) (2, 29/10) (b) (29/10, 2)

(c) (– 2, 29/10) (d) none of these

23. For each $k \in N$, let C_k denote the circle whose equation is $x^2 + y^2 = k^2$. On the circle C_k, a particle moves k units in the anticlockwise direction. After completing its motion on C_k, the particle moves to C_{k+1} in the radial direction. The motion of the particle continues in this manner. The particle starts at (1, 0). If the particle crosses the positive direction of the x-axis for the first time on the circle C_n then n is

(a) 7 (b) 6

(c) 2 (d) none of these

24. Two distinct chords drawn from the point (p, q) on the circle $x^2 + y^2 = px + qy$, where $pq \neq 0$, are bisected by the x-axis. Then

(a) $|p| = |q|$ (b) $p^2 = 8q^2$

(c) $p^2 < 8q^2$ (d) $p^2 > 8q^2$

25. The length of the chord of the circle $x^2 + y^2 + 4x - 7y + 12 = 0$ along the y-axis is

(a) 1 (b) 2

(c) 1/2 (d) none of these

26. If the line $y-1=m(x-1)$ cuts the circle $x^2+y^2=4$ at two real points then the number of possible values of m is

(a) 1 (b) 2

(c) infinite (d) none of these

27. The length of the chord of the circle $x^2+y^2=9$ passing through (3, 0) and perpendicular to the line $y+x=0$ is

(a) $3/\sqrt{2}$ (b) $3\sqrt{2}$

(c) $2\sqrt{3}$ (d) none of these

28. The number of points on the circle $2x^2+2y^2-3x=0$ which are at a distance 2 from the point (–2, 1) is

(a) 2 (b) 0

(c) 1 (d) none of these

29. The equation of the diameter of the circle $3(x^2+y^2)-2x+6y-9=0$ which is perpendicular to the line $2x+3y=12$ is

(a) $3x-2y=3$ (b) $3x-2y+1=0$

(c) $3x-2y=9$ (d) none of these

30. The equation of a circle C is $x^2+y^2-6x-8y-11=0$. The number of real points at which the circle drawn with the points (1, 8) and (0, 0) at the ends of a diameter cuts the circle C is

(a) 0 (b) 1

(c) 2 (d) none of these

31. The equation of the circle of radius $2\sqrt{2}$ whose centre lies on the line $x-y=0$ and which touches the line $x+y=4$, and whose centre's coordinates satisfy the inequality $x+y>4$ is

(a) $x^2+y^2-8x-8y+24=0$ (b) $x^2+y^2=8$

(c) $x^2+y^2-8x+8y=24$ (d) none of these

32. The equation of the chord of the circle $x^2+y^2=25$ of length 8 that passes through the point $(2\sqrt{3}, 2)$ and makes an acute angle with the positive direction of the x-axis is

(a) $(4\sqrt{3}-3\sqrt{7})x+3y=18-6\sqrt{21}$

(b) $(4\sqrt{3}+3\sqrt{7})x-3y=18+6\sqrt{21}$

(c) $(4\sqrt{3}+3\sqrt{7})x-3y+18+6\sqrt{21}=0$

(d) none of these

33. If (a, b) is a point on the chord AB of the circle, where the ends of the chord are A = (2, –3) and B = (3, 2), then

(a) $a \in [-3, 2], b \in [2, 3]$ (b) $a \in [2, 3], b \in [-3, 2]$

(c) $a \in [-2, 2], b \in [-3, 3]$ (d) none of these

34. The number of points with integral coordinates that are interior to the circle $x^2 + y^2 = 16$ is

(a) 43 (b) 49

(c) 45 (d) 51

35. The range of values of a for which the point $(a, 4)$ is outside the circles $x^2 + y^2 + 10x = 0$ and $x^2 + y^2 - 12x + 20 = 0$ is

(a) $(-\infty, -8) \cup (-2, 6) \cup (6, +\infty)$ (b) $(-8, -2)$

(c) $(-\infty, -8) \cup (-2, +\infty)$ (d) none of these

36. A region in the x-y plane is bounded by the curve $y = \sqrt{25 - x^2}$ and the line $y = 0$. If the point $(a, a + 1)$ lies in the interior of the region then

(a) $a \in (-4, 3)$ (b) $a \in (-\infty, -1) \cup (3, +\infty)$

(c) $a \in (-1, 3)$ (d) none of these

37. If (2, 4) is a point interior to the circle $x^2 + y^2 - 6x - 10y + \lambda = 0$ and the circle does not cut the axes at any point then λ belongs to the interval

(a) (25 , 32) (b) (9, 32)

(c) $(32, +\infty)$ (d) none of these

38. The range of values of $\theta \in [0, 2\pi]$ for which $(1 + \cos\theta, \sin\theta)$ is an interior point of the circle $x^2 + y^2 = 1$ is

(a) $(\pi/6, 5\pi/6)$ (b) $(2\pi/3, 5\pi/3)$

(c) $(\pi/6, 7\pi/6)$ (d) $(2\pi/3, 4\pi/3)$

39. The range of values of r for which the point $\left(-5 + \frac{r}{\sqrt{2}}, -3 + \frac{r}{\sqrt{2}}\right)$ is an interior point of the major segment of the circle $x^2 + y^2 = 16$, cut off by the line $x + y = 2$, is

(a) $(-\infty, 5\sqrt{2})$ (b) $(4\sqrt{2} - \sqrt{14}, 5\sqrt{2})$

(c) $(4\sqrt{2} - \sqrt{14}, 4\sqrt{2} + \sqrt{14})$ (d) none of these

40. There are two circles whose equations are $x^2 + y^2 = 9$ and $x^2 + y^2 - 8x - 6y + n^2 = 0$, $n \in Z$. If the two circles have exactly two common tangents then the number of possible values of n is

(a) 2 (b) 8
(c) 9 (d) none of these

41. The number of common tangents to the circles $x^2+y^2=4$ and $x^2+y^2-6x-8y=24$ is

(a) 0 (b) 1
(c) 3 (d) 4

42. The number of common tangents to the circles $x^2+y^2+2x+8y-23=0$ and $x^2+y^2-4x-10y+19=0$ is

(a) 1 (b) 2
(c) 3 (d) 4

43. If the circles $x^2+y^2+2ax+c=0$ and $x^2+y^2+2by+c=0$ touch each other then

(a) $a^{-2}+b^{-2}=c^{-1}$ (b) $a^{-2}+b^{-2}=c^{-2}$
(c) $a+b=2c$ (d) $\frac{1}{a}+\frac{1}{b}=\frac{2}{c}$

44. The number of common tangents to the circles one of which passes through the origin and cuts off intercepts 2 from each of the axes, and the other circle has the line segment joining the origin and the point (1, 1) as a diameter, is

(a) 0 (b) 1
(c) 3 (d) 2

45. The range of values of λ for which the circles $x^2+y^2=4$ and $x^2+y^2-4\lambda x+9=0$ have two common tangents, is

(a) $\lambda \in \left[-\frac{13}{8}, \frac{13}{8}\right]$ (b) $\lambda > \frac{13}{8}$ or $\lambda < -\frac{13}{8}$
(c) $1 < \lambda < \frac{13}{8}$ (d) none of these

46. The number of common tangents to the circles $x^2+y^2-6x-14y+48=0$ and $x^2+y^2-6x=0$ is

(a) 1 (b) 2
(c) 0 (d) 4

47. Two circles have the equations $x^2+y^2-4x-6y-8=0$ and $x^2+y^2-2x-3=0$. Then

(a) they cut each other
(b) they touch each other

(c) one circle lies inside the other

(d) one circle lies wholly outside the other

48. The equations of two circles are $x^2 + y^2 - 26y + 25 = 0$ and $x^2 + y^2 = 25$. Then

(a) they touch each other

(b) they cut each other orthogonally

(c) one circle is inside the other circle

(d) none of these

49. A tangent is drawn to the circle $2(x^2 + y^2) - 3x + 4y = 0$ and it touches the circle at point A. The tangent passes through the point P(2, 1). Then *PA* is equal to

(a) 4 (b) 2

(c) $2\sqrt{2}$ (d) none of these

50. If the points A(1, 4) and B are symmetrical about the tangent to the circle $x^2 + y^2 - x + y = 0$ at the origin then coordinates of B are

(a) (1, 2) (b) $(\sqrt{2}, 1)$

(c) (4, 1) (d) none of these

51. The range of values of m for which the line $y = mx + 2$ cuts the circle $x^2 + y^2 = 1$ at distinct or coincident points is

(a) $(-\infty, -\sqrt{3}] \cup [\sqrt{3}, +\infty)$ (b) $[-\sqrt{3}, \sqrt{3}]$

(c) $[\sqrt{3}, +\infty)$ (d) none of these

52. The line $y = mx + \sqrt{4 + 4m^2}$, $m \in R$, is a tangent to the circle

(a) $x^2 + y^2 = 2$ (b) $x^2 + y^2 = 4$

(c) $x^2 + y^2 = 1$ (d) none of these

53. The equation of any tangent to the circle $x^2 + y^2 - 2x + 4y - 4 = 0$ is

(a) $y = m(x-1) + 3\sqrt{1 + m^2} - 2$ (b) $y = mx + 3\sqrt{1 + m^2}$

(c) $y = mx + 3\sqrt{1 + m^2} - 2$ (d) none of these

54. Two tangents to the circle $x^2 + y^2 = 4$ at the points A and B meet at P(–4, 0). The area of the quadrilateral PAOB, where O is the origin, is

(a) 4 (b) $6\sqrt{2}$

(c) $4\sqrt{3}$ (d) none of these

55. The angle between the pair of tangents from the point (1, 1/2) to the circle $x^2 + y^2 + 4x + 2y - 4 = 0$ is

(a) $\cos^{-1}\frac{4}{5}$
(b) $\sin^{-1}\frac{4}{5}$
(c) $\sin^{-1}\frac{3}{5}$
(d) none of these

56. The chords of contact of the pair of tangents to the circle $x^2 + y^2 = 1$ drawn from any point on the line $2x + y = 4$ pass through the point

(a) (1/2, 1/4)
(b) (1/4, 1/2)
(c) (1, 1/2)
(d) (1/2, 1)

57. A foot of the normal from the point (4, 3) to a circle is (2, 1), and a diameter of the circle has the equation $2x - y = 2$. Then the equation of the circle is

(a) $x^2 + y^2 + 2x - 1 = 0$
(b) $x^2 + y^2 - 2x - 1 = 0$
(c) $x^2 + y^2 - 2y - 1 = 0$
(d) none of these

58. The line $\lambda x + \mu y = 1$ is a normal to the circle $2x^2 + 2y^2 - 5x + 6y - 1 = 0$ if

(a) $5\lambda - 6\mu = 2$
(b) $4 + 5\mu = 6\lambda$
(c) $4 + 6\mu = 5\lambda$
(d) none of these

59. The number of feet of normals from the point (7, –4) to the circle $x^2 + y^2 = 5$ is

(a) 1
(b) 2
(c) 3
(d) 4

60. The equation of a chord of the circle $x^2 + y^2 - 4x = 0$ which is bisected at the point (1, 1) is

(a) $x + y = 2$
(b) $3x - y = 2$
(c) $x - 2y + 1 = 0$
(d) $x - y = 0$

61. Lines are drawn through the point P(–2, –3) to meet the circle $x^2 + y^2 - 2x - 10y + 1 = 0$. The length of the line segment PA, A being the point on the circle where the line meets the circle at coincident points, is

(a) 16
(b) $4\sqrt{3}$
(c) 48
(d) none of these

62. The equations of two circles are $x^2 + y^2 + 2\lambda x + 5 = 0$ and $x^2 + y^2 + 2\lambda y + 5 = 0$. P is any point on the line $x - y = 0$. If PA and PB are the lengths of the tangents from P to the two circles and $PA = 3$ then PB is equal to

(a) 1.5 (b) 6

(c) 3 (d) none of these

63. The common chord of the circle $x^2 + y^2 + 6x + 8y - 7 = 0$ and a circle passing through the origin, and touching the line $y = x$, always passes through the point

(a) $(-1/2, 1/2)$ (b) $(1, 1)$

(c) $(1/2, 1/2)$ (d) none of these

64. A tangent to the circle $x^2 + y^2 = 1$ through the point (0, 5) cuts the circle $x^2 + y^2 = 4$ at at A and B. The tangents to the circle $x^2 + y^2 = 4$ at A and B meet at C. The coordinates of C are

(a) $\left(\frac{8}{5}\sqrt{6}, \frac{4}{5}\right)$ (b) $\left(\frac{8}{5}\sqrt{6}, -\frac{4}{5}\right)$

(c) $\left(-\frac{8}{5}\sqrt{6}, -\frac{4}{5}\right)$ (d) none of these

65. If the common chord of the circles $x^2 + (y - \lambda)^2 = 16$ and $x^2 + y^2 = 16$ subtend a right angle at the origin then λ is equal to

(a) 4 (b) $4\sqrt{2}$

(c) $\pm 4\sqrt{2}$ (d) 8

66. The equation of the smallest circle passing through the intersection of the line $x + y = 1$ and the circle $x^2 + y^2 = 9$ is

(a) $x^2 + y^2 + x + y - 8 = 0$ (b) $x^2 + y^2 - x - y - 8 = 0$

(c) $x^2 + y^2 - x + y - 8 = 0$ (d) none of these

67. The equation of a circle is $x^2 + y^2 = 4$. The centre of the smallest circle touching this circle and the line $x + y = 5\sqrt{2}$ has the coordinates

(a) $\left(\frac{7}{2\sqrt{2}}, \frac{7}{2\sqrt{2}}\right)$ (b) $\left(\frac{3}{2}, \frac{3}{2}\right)$

(c) $\left(-\frac{7}{2\sqrt{2}}, -\frac{7}{2\sqrt{2}}\right)$ (d) none of these

68. The members of a family of circles are given by the equation $2(x^2 + y^2) + \lambda x - (1 + \lambda^2)y - 10 = 0$. The number of circles belonging to the family that are cut orthogonally by the fixed circle $x^2 + y^2 + 4x + 6y + 3 = 0$ is

(a) 2 (b) 1

(c) 0 (d) none of these

69. The equation of the circle with the chord $y = 2x$ of the circle $x^2 + y^2 - 10x = 0$ as its diameter is

(a) $x^2 + y^2 - 2x - 4y - 5 = 0$ (b) $x^2 + y^2 = 2x + 4y$

(c) $x^2 + y^2 = 4x + 2y$ (d) none of these

70. A circle of radius 2 touches the coordinate axes in the first quadrant. If the circle makes a complete rotation on the x-axis along the positive direction of the x-axis then the equation of the circle in the new position is

(a) $x^2 + y^2 - 4(x + y) - 8\pi x + (2 + 4\pi)^2 = 0$

(b) $x^2 + y^2 - 4x - 4y + (2 + 4\pi)^2 = 0$

(c) $x^2 + y^2 - 8\pi x - 4y + (2 + 4\pi)^2 = 0$

(d) none of these

71. A ray of light incident at the point $(-2, -1)$ gets reflected from the tangent at $(0, -1)$ to the circle $x^2 + y^2 = 1$. The reflected ray touches the circle. The equation of the line along which the incident ray moved is

(a) $4x - 3y + 11 = 0$ (b) $4x + 3y + 11 = 0$

(c) $3x + 4y + 11 = 0$ (d) none of these

72. The locus of the centres of the circles for which one end of a diameter is $(1, 1)$ while the other end is on the line $x + y = 3$ is

(a) $x + y = 1$ (b) $2(x - y) = 5$

(c) $2x + 2y = 5$ (d) none of these

73. The angle between a pair of tangents drawn from a point P to the curve $x^2 + y^2 + 4x - 6y + 9\sin^2\alpha + 13\cos^2\alpha = 0$ is 2α. The locus of P is

(a) $x^2 + y^2 + 4x - 6y + 4 = 0$ (b) $x^2 + y^2 + 4x - 6y - 9 = 0$

(c) $x^2 + y^2 + 4x - 6y - 4 = 0$ (d) $x^2 + y^2 + 4x - 6y + 9 = 0$

74. The point P moves in the plane of a regular hexagon such that the sum of the squares of its distances from the vertices of the hexagon is $6a^2$. If the radius of the circumcircle of the hexagon is $r(< a)$ then the locus of P is

(a) a pair of straight lines

(b) an ellipse

(c) a circle of radius $\sqrt{a^2 - r^2}$

(d) an ellipse of major axis a and minor axis r

75. The locus of the middle points of chords of length 4 of the circle $x^2 + y^2 = 16$ is

(a) a straight line (b) a circle of radius 2

(c) a circle of radius $2\sqrt{3}$ (d) an ellipse

76. The equation of the locus of the middle point of a chord of the circle $x^2 + y^2 = 2(x + y)$ such that the pair of lines joining the origin to the point of intersection of the chord and the circle are equally inclined to the x-axis is

(a) $x + y = 2$ (b) $x - y = 2$

(c) $2x - y = 1$ (d) none of these

77. The locus of the centres of circles passing through the origin and intersecting the fixed circle $x^2 + y^2 - 5x + 3y - 1 = 0$ orthogonally is

(a) a straight line of the slope $\frac{3}{5}$ (b) a circle

(c) a pair of straight lines (d) none of these

78. The equation of a circle C_1 is $x^2 + y^2 - 4x - 2y - 11 = 0$. A circle C_2 of radius 1 rolls on the outside of the circle C_1. The locus of the centre of C_2 has the equation

(a) $x^2 + y^2 - 4x - 2y - 20 = 0$ (b) $x^2 + y^2 + 4x + 2y - 20 = 0$

(c) $x^2 + y^2 - 3x - y - 11 = 0$ (d) none of these

79. Circles are drawn through the point (3, 0) to cut an intercept of length 6 units on the negative direction of the x-axis. The equation of the locus of their centres is

(a) the x-axis (b) $x - y = 0$

(c) the y-axis (d) $x + y = 0$

80. The locus of the centre of a circle touching the lines $x + 2y = 0$ and $x - 2y = 0$ is

(a) $xy = 0$ (b) $x = 0$

(c) $y = 0$ (d) none of these

81. The locus of a point from which the lengths of the tangents to the circles $x^2 + y^2 = 4$ and $2(x^2 + y^2) - 10x + 3y - 2 = 0$ are equal is

(a) a straight line inclined at $\pi/4$ with the line joining the centres of the circles

(b) a circle

(c) an ellipse

(d) a straight line perpendicular to the line joining the centres of the circles

82. The locus of the centres of the circles passing through the intersection of the circles $x^2 + y^2 = 1$ and $x^2 + y^2 - 2x + y = 0$ is

(a) a line whose equation is $x + 2y = 0$

(b) a line whose equation is $2x - y = 1$

(c) a circle

(d) a pair of lines

• *Type 2* •

Choose the correct options. One or more options may be correct.

83. The equation of a circle is $x^2 + y^2 = 4$. A regular hexagon is inscribed in the circle whose one vertex is (2, 0). Then a consecutive vertex has the coordinates

(a) $(\sqrt{3}, 1)$ (b) $(1, -\sqrt{3})$

(c) $(\sqrt{3}, -1)$ (d) $(1, \sqrt{3})$

84. $P(\sqrt{2}, \sqrt{2})$ is a point on the circle $x^2 + y^2 = 4$ and Q is another point on the circle such that arc $PQ = \frac{1}{4} \times$ circumference. The coordinates of Q are

(a) $(-\sqrt{2}, -\sqrt{2})$ (b) $(\sqrt{2}, -\sqrt{2})$

(c) $(-\sqrt{2}, \sqrt{2})$ (d) none of these

85. Circles $x^2 + y^2 = 1$ and $x^2 + y^2 - 8x + 11 = 0$ cut off equal intercepts on a line through the point $\left(-2, \frac{1}{2}\right)$. The slope of the line is

(a) $\frac{-1 + \sqrt{29}}{14}$ (b) $\frac{1 + \sqrt{7}}{4}$

(c) $\frac{-1 - \sqrt{29}}{14}$ (d) none of these

86. Let L_1 be a straight line passing through the origin and L_2 be the straight line $x + y = 1$. If the intercepts made by the circle $x^2 + y^2 - x + 3y = 0$ on L_1 and L_2 are equal then which of the following equations can represent L_1?

(a) $x+y=0$ (b) $x-y=0$
(c) $x+7y=0$ (d) $x-7y=0$

87. The parametric equation of a circle is given by $x=3\cos\phi+2,\ y=3\sin\phi$. Then its

(a) centre = $(-2, 0)$ (b) radius = 3
(c) centre = $(2, 0)$ (d) radius = 1

88. If A and B are two points on the circle $x^2+y^2-4x+6y-3=0$ which are farthest and nearest respectively from the point $(7, 2)$ then

(a) $A=(2-2\sqrt{2}, -3-2\sqrt{2})$ (b) $B=(2+2\sqrt{2}, -3+2\sqrt{2})$
(c) $A=(2+2\sqrt{2}, -3+2\sqrt{2})$ (d) $B=(2-2\sqrt{2}, -3+2\sqrt{2})$

89. A point $P(\sqrt{3}, 1)$ moves on the circle $x^2+y^2=4$ and after covering a quarter of the circle leaves it tangentially. The equation of a line along which the point moves after leaving the circle is

(a) $y=\sqrt{3}x+4$ (b) $\sqrt{3}y=x+4$
(c) $\sqrt{3}y=x-4$ (d) $y=\sqrt{3}x-4$

90. The equation of a circle of radius 1 touching the circles $x^2+y^2-2|x|=0$ is

(a) $x^2+y^2+2\sqrt{3}x-2=0$ (b) $x^2+y^2-2\sqrt{3}y+2=0$
(c) $x^2+y^2+2\sqrt{3}y+2=0$ (d) $x^2+y^2+2\sqrt{3}x+2=0$

91. The line $4y-3x+\lambda=0$ touches the circle $x^2+y^2-4x-8y-5=0$. The value of λ is

(a) 29 (b) 10
(c) –35 (d) none of these

92. A circle which touches the axes, and whose centre is at distance $2\sqrt{2}$ from the origin, has the equation

(a) $x^2+y^2-4x+4y+4=0$ (b) $x^2+y^2+4x-4y+4=0$
(c) $x^2+y^2+4x+4y+4=0$ (d) none of these

93. Let the equation of a circle be $x^2+y^2=a^2$. If $h^2+k^2-a^2<0$ then the line $hx+ky=a^2$ is the

(a) polar line of the point (h, k) with respect to the circle

(b) real chord of contact of the tangents from (h, k) to the circle

(c) equation of a tangent to the circle from the point (h, k)
(d) none of these

94. For the equation $x^2 + y^2 + 2\lambda x + 4 = 0$ which of the following can be true?
(a) It represents a real circle for all $\lambda \in R$.
(b) It represents a real circle for $|\lambda| > 2$.
(c) The radical axis of any two circles of the family is the y-axis.
(d) The radical axis of any two circles of the family is the x-axis.

95. The point of contact of a tangent from the point (1, 2) to the circle $x^2 + y^2 = 1$ has the coordinates
(a) $\left(\frac{1-2\sqrt{19}}{5}, \frac{2+\sqrt{19}}{5}\right)$
(b) $\left(\frac{1-2\sqrt{19}}{5}, \frac{2-\sqrt{19}}{5}\right)$
(c) $\left(\frac{1+2\sqrt{19}}{5}, \frac{2+\sqrt{19}}{5}\right)$
(d) $\left(\frac{1+2\sqrt{19}}{5}, \frac{2-\sqrt{19}}{5}\right)$

96. The equation of a circle C_1 is $x^2 + y^2 = 4$. The locus of the intersection of orthogonal tangents to the circle is the curve C_2 and the locus of the intersection of perpendicular tangents to the curve C_2 is the curve C_3. Then
(a) C_3 is a circle
(b) the area enclosed by the curve C_3 is 8π
(c) C_2 and C_3 are circles with the same centre
(d) none of these

97. A line parallel to the line $x - 3y = 2$ touches the circle $x^2 + y^2 - 4x + 2y - 5 = 0$ at the point
(a) (1, –4)
(b) (1, 2)
(c) (3, –4)
(d) (3, 2)

98. A point on the line $x = 3$ from which the tangents drawn to the circle $x^2 + y^2 = 8$ are at right angles is
(a) $(3, -\sqrt{7})$
(b) $(3, \sqrt{23})$
(c) $(3, \sqrt{7})$
(d) $(3, -\sqrt{23})$

99. Tangents drawn from (2, 0) to the circle $x^2 + y^2 = 1$ touch the circle at A and B. Then
(a) $A = \left(\frac{1}{2}, \frac{\sqrt{3}}{2}\right), B = \left(-\frac{1}{2}, -\frac{\sqrt{3}}{2}\right)$
(b) $A = \left(-\frac{1}{2}, \frac{-\sqrt{3}}{2}\right), B = \left(\frac{1}{2}, \frac{\sqrt{3}}{2}\right)$

(c) $A = \left(\frac{1}{2}, \frac{\sqrt{3}}{2}\right), B = \left(\frac{1}{2}, -\frac{\sqrt{3}}{2}\right)$ (d) $A = \left(\frac{1}{2}, -\frac{\sqrt{3}}{2}\right), B = \left(\frac{1}{2}, \frac{\sqrt{3}}{2}\right)$

100. The equations of four circles are $(x \pm a)^2 + (y \pm a)^2 = a^2$. The radius of a circle touching all the four circles is

(a) $(\sqrt{2} - 1)a$ (b) $2\sqrt{2}a$

(c) $(\sqrt{2} + 1)a$ (d) $(2 + \sqrt{2})a$

101. Let a line through the point P(5, 10) cut the line l whose equation is $x + 2y = 5$, at Q and the circle C whose equation is $x^2 + y^2 = 25$, at A and B. Then

(a) P is the pole of the line l with respect to the circle C

(b) l is the pllar of the point P with respect to the circle C

(c) PA, PQ, PB are in AP

(d) PQ is the HM of PA and PB

102. The equation of a circle is $x^2 + y^2 - 4x + 2y - 4 = 0$. With respect to the circle

(a) the pole of the line $x - 2y + 5 = 0$ is (1, 1)

(b) the chord of contact of real tangents from (1, 1) is the line $x - 2y + 5 = 0$

(c) the polar of the point (1, 1) is $x - 2y + 5 = 0$

(d) none of these

Answers

1. b	**2.** d	**3.** a	**4.** c	**5.** c
6. a	**7.** d	**8.** a	**9.** c	**10.** b
11. a	**12.** c	**13.** b	**14.** a	**15.** b
16. b	**17.** c	**18.** a	**19.** b	**20.** a
21. b	**22.** b	**23.** a	**24.** d	**25.** a
26. c	**27.** b	**28.** b	**29.** a	**30.** c
31. a	**32.** b	**33.** b	**34.** c	**35.** a
36. c	**37.** a	**38.** d	**39.** b	**40.** c
41. b	**42.** c	**43.** a	**44.** b	**45.** b
46. d	**47.** a	**48.** b	**49.** b	**50.** c
51. a	**52.** b	**53.** a	**54.** c	**55.** b
56. a	**57.** b	**58.** c	**59.** b	**60.** d
61. b	**62.** c	**63.** c	**64.** a	**65.** c
66. b	**67.** a	**68.** a	**69.** b	**70.** a
71. b	**72.** c	**73.** d	**74.** c	**75.** c
76. a	**77.** d	**78.** a	**79.** c	**80.** a
81. d	**82.** a	**83.** b, d	**84.** b, c	**85.** a, c
86. b, d	**87.** b, c	**88.** a, b	**89.** b, c	**90.** b, c
91. c	**92.** a, b, c	**93.** a	**94.** b, c	**95.** a, d
96. a, c	**97.** b, c	**98.** a, c	**99.** c, d	**100.** a, c
101. a, b, d	**102.** a, c			

Hints and Solutions to Selected Questions

3. Here, $(x-1)^2+(y+2)^2=0$

$\Rightarrow \quad x=1, y=-2$ which give a point.

4. $\lambda L_2L_3+\mu L_3L_1+\nu L_1L_2=0$ is a curve passing through the vertices, which are obtained by solving any two of $L_1=0, L_2=0, L_3=0$ together.

It will be the circumcircle if the coefficient of x^2 = the coefficient of y^2 and the coefficient of $xy=0$.

5. Here, radius $=\sqrt{\left(\frac{\lambda}{2}\right)^2+\left(\frac{1-\lambda}{2}\right)^2-5}\leq 5$

$\Rightarrow \quad 2\lambda^2-2\lambda-119\leq 0$

$\therefore \quad \frac{1-\sqrt{239}}{2}\leq\lambda\leq\frac{1+\sqrt{239}}{2}$

$\Rightarrow \quad -7.2\leq\lambda\leq 8.2$ (nearly) $\quad\therefore\ \lambda=-7,-6,\ldots,8.$

7. The points of intersection of $2x-y+1=0$ with the axes are $(-1/2, 0)$ and $(0, 1)$.

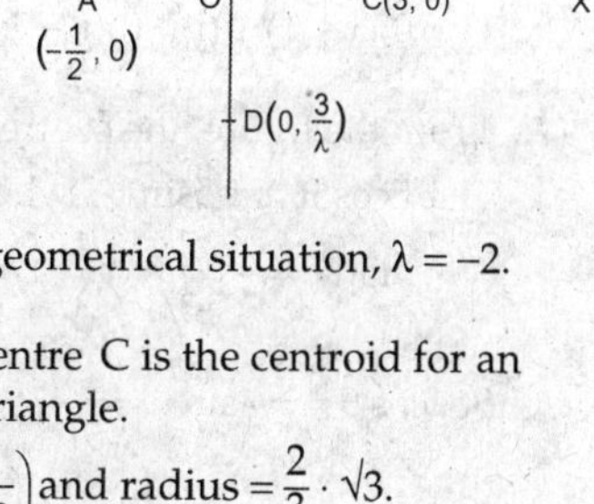

The points of intersection of

$x+\lambda y-3=0$

and the axes are $(3, 0)$, $\left(0, \frac{3}{\lambda}\right)$.

The points are concyclic if

$OA\cdot OC=OB\cdot OD.$

$\Rightarrow \quad \frac{1}{2}\cdot 3=1\cdot\left|\frac{3}{\lambda}\right| \quad$ or $\quad |\lambda|=2.$

As λ must be negative according to the geometrical situation, $\lambda=-2$.

9.

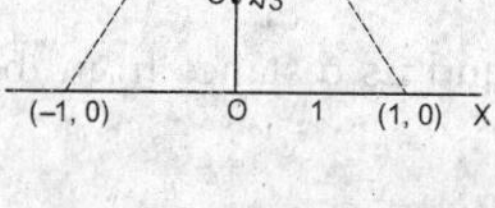

The circumcentre C is the centroid for an equilateral triangle.

$\therefore\ C=\left(0, \frac{1}{\sqrt{3}}\right)$ and radius $=\frac{2}{3}\cdot\sqrt{3}$.

So, the circumcircle is

$$(x-0)^2+\left(y-\frac{1}{\sqrt{3}}\right)^2=\left(\frac{2}{3}\cdot\sqrt{3}\right)^2.$$

10. The sides are $x+y-4=0,\ x-1=0, y-2=0$. So, the triangle is right angled at (1, 2). The hypotenuse is $x+y-4=0$ whose ends are (1, 3) and (2, 2). The circumcentre $=\left(\frac{1+2}{2}, \frac{3+2}{2}\right)$ and circumradius $=\frac{1}{2}\sqrt{(1-2)^2+(3-2)^2}=\frac{1}{\sqrt{2}}$.

13. There cannot be 3 points on the circle with rational coordinates for then the centre of the circle, being the circumcentre of a triangle whose vertices have rational coordinates, must have rational coordinates ($\because$ the coordinates will be obtained by solving two linear equations in x, y having rational coefficients). But the point $(\sqrt{3}, 0)$ does not have rational coordinates. Also the equation of the circle is

$$(x-\sqrt{3})^2+y^2=r^2 \quad \Rightarrow \quad x=\sqrt{3}\pm\sqrt{r^2-y^2}.$$

For suitable r, x, where x is rational, y may have two rational values. For example, $r=2, x=0, y=1, -1$ satisfy $x=\sqrt{3}\pm\sqrt{r^2-y^2}$.

So we get two points (0 , 1), (0, –1) which have rational coordinates.

15.

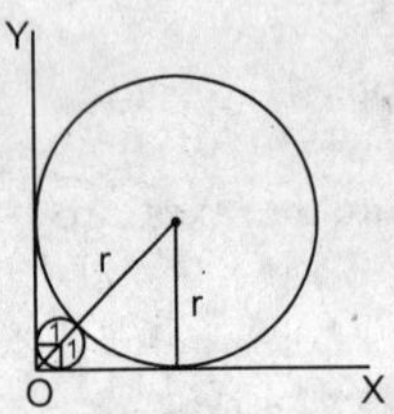

Clearly from the figure,

$$\frac{r}{\sqrt{2}+1+r}=\sin 45°$$

$$\Rightarrow \quad r=3+2\sqrt{2}.$$

16. Solving $y=x, x^2+y^2-2x=0$, we get $(x, y)=(0, 0), (1, 1)$. So, the ends of the diameter are (0, 0), (1, 1). Hence, the equation of the required circle is $(x-0)(x-1)+(y-0)(y-1)=0$.

17. From the figure, the centres are $(1\pm 5\cos\theta, 1\pm 5\sin\theta)$, where

$$\tan\theta=-\frac{1}{-3/4}=\frac{4}{3}.$$

So, $\cos\theta=\frac{3}{5}, \sin\theta=\frac{4}{5}$.

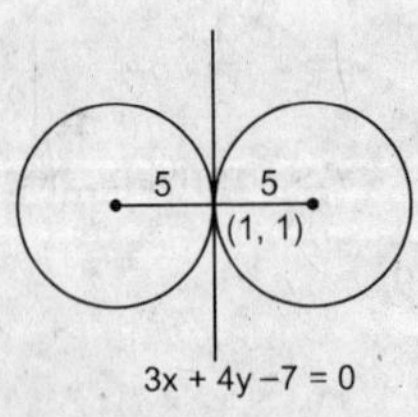

18. Draw the figure. Clearly, centre $=(0, \sqrt{3})$ and radius $=2$.

19. If the inradius $=r$ then the centre $=(r, r)$ and its distance from the line $4x+3y=6$ is r.

So, $\left|\frac{4r+3r-6}{\sqrt{4^2+3^2}}\right| = r$ or $|7r-6| = 5r$

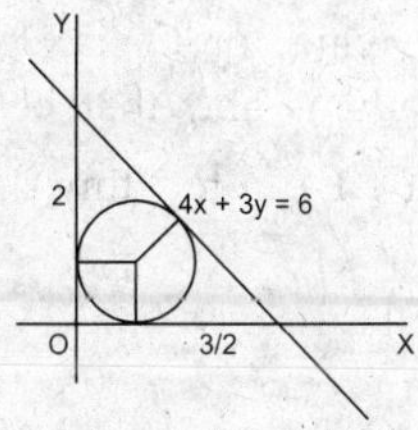

$\therefore \quad 7r - 6 = \pm 5r \quad \Rightarrow r = 3, \frac{1}{2}.$

Clearly from the figure, $r \neq 3$.

$\therefore$ the equation of the incircle is

$$\left(x-\frac{1}{2}\right)^2 + \left(y-\frac{1}{2}\right)^2 = \left(\frac{1}{2}\right)^2.$$

20.

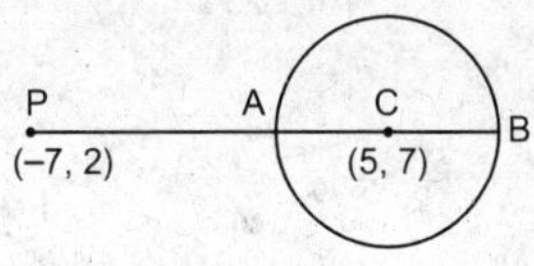

The centre C of the circle = (5, 7) and the radius $= \sqrt{5^2+7^2+51} = 5\sqrt{5}$

$PC = \sqrt{12^2+5^2} = 13$

$\therefore \; PA = 13 - 5\sqrt{5}, PB = 13 + 5\sqrt{5}.$

21. In an equilateral triangle, the circumcentre and the incentre are the same point.

$\therefore$ incentre $= (-g, -f)$.

Also $1^2 + 1^2 + 2g + 2f + c = 0$ (given), i.e., $c = -2(g+f+1)$.

Also, in an equilateral triangle, circumradius $= 2 \times$ inradius.

$\therefore$ inradius $= \frac{1}{2} \times \sqrt{g^2+f^2-c}$.

$\therefore$ the equation of the incircle is

$$(x+g)^2 + (y+f)^2 = \frac{1}{4}(g^2+f^2-c) = \frac{1}{4}(g^2+f^2) + \frac{1}{4} \cdot 2(g+f+1).$$

Simplify.

22. $f(x, y) \equiv x^2 + y^2 + 2gx + 2fy + c = 0.$

Now, $f(0, \lambda) \equiv \lambda^2 + 2f\lambda + c = 0$, and its roots are 2, 2.

$\therefore \quad 2 + 2 = -2f, 2 \times 2 = c$, i.e., $f = -2, c = 4$.

$f(\lambda, 0) \equiv \lambda^2 + 2g\lambda + c = 0$, and its roots are $\frac{4}{5}, 5$.

$\therefore \quad \frac{4}{5} + 5 = -2g, \frac{4}{5} \times 5 = c$. Thus $g = -\frac{29}{10}, f = -2$.

23. The angle described anticlockwise before leaving for C_n is $\{1+1+\ldots$ to $(n-1)$ times$\}$ radians and that before leaving for C_{n+1} is $\{1+1+\ldots$ to n times$\}$ radians

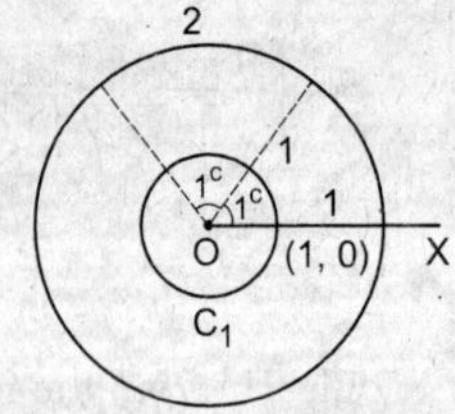

$\therefore \quad (n-1) < 2\pi < n \quad \Rightarrow \quad n-1 < \frac{44}{7} < n.$

24. Here, the centre $= \left(\frac{p}{2}, \frac{q}{2}\right)$.

So (p, q) and $(0, 0)$ are the ends of a diameter. As the two chords are bisected by the line $y = 0$, the chords will cut the circle at the points $(x_1, -q)$ and $(x_2, -q)$ where x_1, x_2 are real.

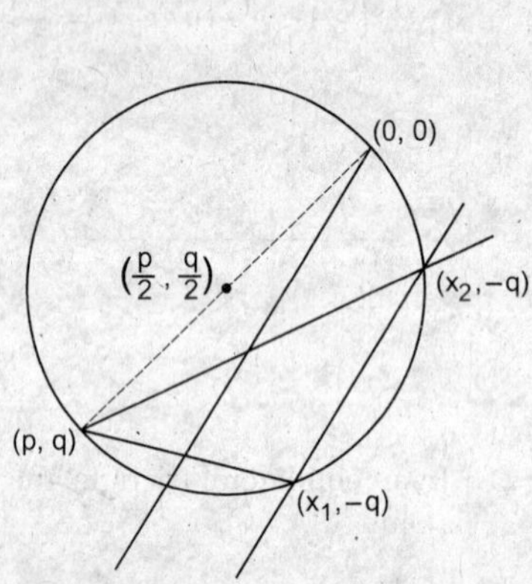

Clearly, the line joining these points is $y = -q$. Solving $y = -q$ and $x^2 + y^2 = px + qy$, we get $x^2 - px + 2q^2 = 0$, whose roots are x_1, x_2.

For real distinct roots, $D > 0$, i.e., $p^2 - 8q^2 > 0$.

26. The line passes through the interior point (1, 1). So, m can have any real value.

28. Here, centre $= \left(\frac{3}{4}, 0\right)$ and radius $= \frac{3}{4}$.

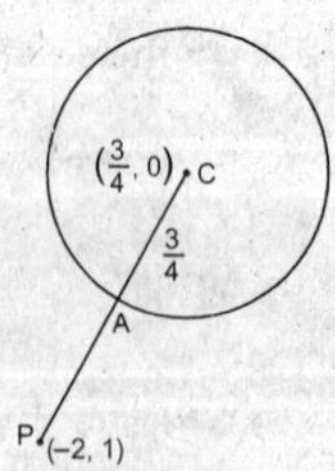

$$PC = \sqrt{\left(\frac{3}{4}+2\right)^2 + (0-1)^2} = \sqrt{\frac{121}{16}+1} = \frac{\sqrt{137}}{4}$$

$$\therefore \quad PA = \frac{\sqrt{137}}{4} - \frac{3}{4} = \frac{\sqrt{137}-3}{4} > 2.$$

31. The centre $= (t, t)$.

The radius $= 2\sqrt{2} = \left|\frac{t+t-4}{\sqrt{2}}\right|$. So, $|2t-4| = 4$.

$\therefore \quad 2t - 4 = \pm 4$, i.e., $t = 4, 0$. So, the centre $= (4, 4), (0, 0)$.

$\therefore \quad$ satisfying $x + y > 4$, the centre $= (4, 4)$.

33. (a, b) should be an internal or end point of the line segment joining A, B. So, the x-coordinate a will vary from 2 to 3, i.e., from the x-coordinate of A to the x-coordinate of B. Similarly, for y-coordinate.

34. The number of points is equal to the number of integral solutions (x, y) such that $x^2 + y^2 < 16$. So x, y are integers such that $-3 \le x \le 3, -3 \le y \le 3$ satisfying the inequation $x^2 + y^2 < 16$. The number of selections of values of x is 7, namely $-3, -2, -1, 0, 1, 2, 3$. The same is true for y. So the number of ordered pairs (x, y) is 7×7. But (3, 3), (3, –3), (–3, 3), (–3, –3) are rejected because they do not satisfy the inequation $x^2 + y^2 < 16$. So, the number of points is 45.

35. $a^2 + 4^2 + 10a > 0, a^2 + 4^2 - 12a + 20 > 0$, i.e., the inequations in a are $a^2 + 10a + 16 > 0$ and $a^2 - 12a + 36 > 0$. Solve these and take the common values.

36. $y = \sqrt{25 - x^2}, y = 0$ bound the semicircle above the x-axis.

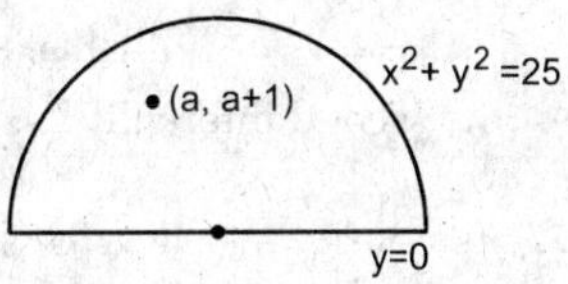

$\therefore \quad a + 1 > 0$

and $a^2 + (a+1)^2 - 25 < 0$.

Solve these and take the common values.

37. $2^2 + 4^2 - 6 \times 2 - 10 \times 4 + \lambda < 0 \quad \Rightarrow \lambda - 32 < 0 \quad$ or $\lambda < 32$.

Solving $y = 0, x^2 + y^2 - 6x - 10y + \lambda = 0$, we get $x^2 - 6x + \lambda = 0$, which must have imaginary roots, i.e., $36 - 4\lambda < 0$, i.e., $\lambda > 9$.

Solving $x = 0, x^2 + y^2 - 6x - 10y + \lambda = 0$, we get $y^2 - 10y + \lambda = 0$, which must have imaginary roots, i.e., $100 - 4\lambda < 0$, i.e., $\lambda > 25$. So $25 < \lambda < 32$.

38. $(1 + \cos\theta)^2 + \sin^2\theta - 1 < 0$ or $1 + 2\cos\theta < 0$ or $\cos\theta < -1/2$.

39. The given point is an interior point $\Rightarrow \left(-5 + \frac{r}{\sqrt{2}}\right)^2 + \left(-3 + \frac{r}{\sqrt{2}}\right)^2 - 16 < 0$

$\Rightarrow \quad r^2 - 8\sqrt{2}r + 18 < 0 \quad \Rightarrow 4\sqrt{2} - \sqrt{14} < r < 4\sqrt{2} + \sqrt{14}$.

The point is on the major segment $\Rightarrow$ the centre and the point are on the same side of the line $x + y = 2$.

$\therefore \quad -5 + \frac{r}{\sqrt{2}} - 3 + \frac{r}{\sqrt{2}} - 2 < 0$

$\Rightarrow \quad r < 5\sqrt{2}$. So, $4\sqrt{2} - \sqrt{14} < r < 5\sqrt{2}$.

40. For $x^2+y^2=9$, the centre = (0, 0) and the radius = 3.

For $x^2+y^2-8x-6y+n^2=0$,

the centre = (4, 3) and the radius $=\sqrt{4^2+3^2-n^2}$

$\therefore \quad 4^2+3^2-n^2>0 \quad$ or $\quad n^2<5^2 \quad$ or $\quad -5<n<5$.

Circles should cut to have exactly two common tangents. So, $r_1+r_2>d$.

$\therefore \quad 3+\sqrt{25-n^2}>\sqrt{4^2+3^2} \quad$ or $\quad \sqrt{25-n^2}>2 \quad$ or $\quad 25-n^2>4$

$\therefore \quad n^2<21 \quad$ or $\quad -\sqrt{21}<n<\sqrt{21}$.

Therefore, common values of n should satisfy $-\sqrt{21}<n<\sqrt{21}$. But $n \in Z$. So, $n=-4,-3,\ldots,4$.

41. For the first circle, centre = (0, 0), radius $r_1=2$.

For the second circle, centre = (3, 4), radius $r_2=\sqrt{3^2+4^2+24}=7$.

$\therefore \ |r_1-r_2|=d$ (= distance between centres). So the first circle touches the second internally.

42. Establish $r_1+r_2=d$. This would mean they touch externally. Hence there will be 3 tangents.

43. Subtracting the equations, we get $ax=by$. Putting this in the first equation, $x^2+\left(\frac{a}{b}x\right)^2+2ax+c=0$, which should have equal roots. This gives $\frac{1}{a^2}+\frac{1}{b^2}=\frac{1}{c}$.

44. Circles are $(x-1)^2+(y-1)^2=2$ and $x(x-1)+y(y-1)=0$. Here $|r_1-r_2|=d$.

45. Solving the equations, $x^2+\left(\frac{13}{4\lambda}\right)^2=4$ or $x=\sqrt{4-\left(\frac{13}{4\lambda}\right)^2}$.

It should have two real and distinct values. So, $4-\left(\frac{13}{4\lambda}\right)^2>0$.

46. For the first circle, centre = (3, 7), radius $r_1=\sqrt{3^2+7^2-48}=\sqrt{10}$.

For the second circle, centre = (3, 0), radius $r_2=3$.

So, $r_1+r_2<d$ (= the distance between centres).

$\therefore$ circles do not cut and hence the number of common tangents = 4.

47. Here $r_1 + r_2 > d$. So, the circles cut each other.

48. $2gg' + 2ff' = 0 + 0 = 25 - 25 = c + c'$

$\therefore$ the circles cut orthogonally.

50. The tangent at the origin is $y - x = 0$. B is the image of A(1, 4) by this line.

51. The length of the perpendicular from the centre (0, 0) to the line $= \frac{2}{\sqrt{1+m^2}}$.

The radius of the circle = 1. For the line to cut the circle at distinct or coincident points, $\frac{2}{\sqrt{1+m^2}} \le 1$ or $1 + m^2 \ge 4$ or $m^2 \ge 3$.

52. $y = mx + \sqrt{a^2 + a^2m^2}$ touches the circle $x^2 + y^2 = a^2$. Here $a = 2$.

53. The circle is $(x-1)^2 + (y+2)^2 = 3^2$. As any tangent to $x^2 + y^2 = 3^2$ is given by $y = mx + 3\sqrt{1+m^2}$, any tangent to the given circle will be $y + 2 = m(x-1) + 3\sqrt{1+m^2}$.

54. Clearly, $\sin\theta = \frac{2}{4} = \frac{1}{2}$ $\therefore$ $\theta = 30°$.

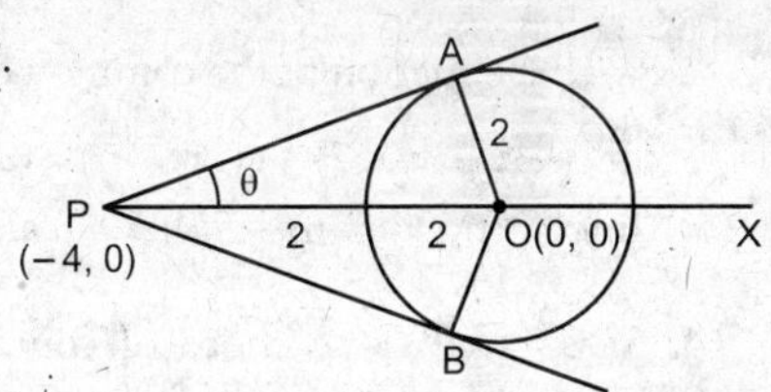

So, ar ($\triangle$POA) $= \frac{1}{2} \cdot 2 \cdot 4 \cdot \sin 60°$

$\therefore$ ar (quad. PAOB)

$= 2 \cdot \frac{1}{2} \cdot 2 \cdot 4 \sin 60° = 8 \cdot \frac{\sqrt{3}}{2} = 4\sqrt{3}$.

55. The equation of the pair of tangents is $S \cdot S_1 = T^2$, i.e.,

$$(x^2 + y^2 + 4x + 2y - 4)\left\{1^2 + \left(\frac{1}{2}\right)^2 + 4 \cdot 1 + 2 \cdot \frac{1}{2} - 4\right\}$$

$$= \left\{x \cdot 1 + y \cdot \frac{1}{2} + 2(x+1) + \left(y + \frac{1}{2}\right) - 4\right\}^2$$

or $\frac{9}{4}(x^2 + y^2 + 4x + 2y - 4) = \frac{9}{4}(2x + y - 1)^2$

or $3x^2 + 4xy - 8x - 4y + 5 = 0$

$\therefore$ the required angle $= \tan^{-1}\left|\dfrac{2\sqrt{(2)^2 - 3\times 0}}{3+0}\right| = \tan^{-1}\dfrac{4}{3}$.

56. Any point on $2x + y = 4$ is $(t, 4 - 2t)$. The equation of the chord of contact of the point is $x \cdot t + y \cdot (4 - 2t) = 1$ or $4y - 1 + t(x - 2y) = 0$.

This line passes through the intersection of $4y - 1 = 0$ and $x - 2y = 0$.

57. The line joining (4, 3) and (2, 1) is also along a diameter. So, the centre is the intersection of the diameters $2x - y = 2$ and $y - 3 = \dfrac{3-1}{4-2} = (x - 4)$.

Solving these, the centre = (1, 0).

Also, the radius = the distance between (1, 0) and (2, 1) = $\sqrt{2}$.

61. PA = the length of the tangent from P to the circle

$= \sqrt{(-2)^2 + (-3)^2 - 2(-2) - 10(-3) + 1}$.

62. The radical axis of the circles is $x - y = 0$. So, $PA = PB = 3$.

63. Let the second circle be $x^2 + y^2 + 2gx + 2fy = 0$. The common chord has the equation $2(g - 3)x + 2(f - 4)y + 7 = 0$.

But $y = x$ touches the circle.

Hence, $x^2 + x^2 + 2gx + 2fx = 0$ has equal roots, i.e., $f + g = 0$.

$\therefore$ the equation of the common chord is $2(g - 3)x + 2(-g - 4)y + 7 = 0$

or $(-6x - 8y + 7) + g(2x - 2y) = 0$, which passes through the point of intersection of $-6x - 8y + 7 = 0$ and $2x - 2y = 0$.

64. If C = (α, β), the chord of contact of the tangents from it to the circle $x^2 + y^2 = 4$, i.e., $\alpha x + \beta y = 4$ passes through (0, 5). So, $\beta = \dfrac{4}{5}$.

Also $\alpha x + \beta y = 4$ touches the circle $x^2 + y^2 = 1$.

Hence, $\dfrac{4}{\sqrt{\alpha^2 + \beta^2}} = 1$.

$\therefore \quad 16 = \alpha^2 + \left(\dfrac{4}{5}\right)^2$ or $\alpha^2 = 16 - \dfrac{16}{25} = \dfrac{384}{25}$ $\quad\therefore\ \alpha = \pm\dfrac{8\sqrt{6}}{5}$.

65. The common chord is

$$x^2 + (y - \lambda)^2 - 16 - \{x^2 + y^2 - 16\} = 0, \text{ i.e., } y = \frac{\lambda}{2} \quad (\because\ \lambda \neq 0).$$

The pair of straight lines joining the origin to the points of intersection of $y=\frac{\lambda}{2}$ and $x^2+y^2=16$ is

$$x^2+y^2=16\cdot\left(\frac{2y}{\lambda}\right)^2, \text{ i.e., } \lambda^2x^2+(\lambda^2-64)y^2=0.$$

These lines are at right angles if $\lambda^2+\lambda^2-64=0$, i.e., $\lambda=\pm 4\sqrt{2}$.

66. Any circle passing through the points of intersection of the given line and circle has the equation $x^2+y^2-9+\lambda(x+y-1)=0$. Its centre $=\left(-\frac{\lambda}{2},-\frac{\lambda}{2}\right)$.

The circle is the smallest if $\left(-\frac{\lambda}{2},-\frac{\lambda}{2}\right)$ is on the chord $x+y=1$.

$$\therefore \quad -\frac{\lambda}{2}-\frac{\lambda}{2}=1 \quad \Rightarrow \lambda=-1.$$

Putting this value for λ, the equation of the smallest circle is $x^2+y^2-9-(x+y-1)=0$.

67. Here, OB = radius = 2.

The distance of (0, 0) from $x+y=5\sqrt{2}$ is 5.

$\therefore$ the radius of the smallest circle $=\frac{5-2}{2}=\frac{3}{2}$

and $OC=2+\frac{3}{2}=\frac{7}{2}$.

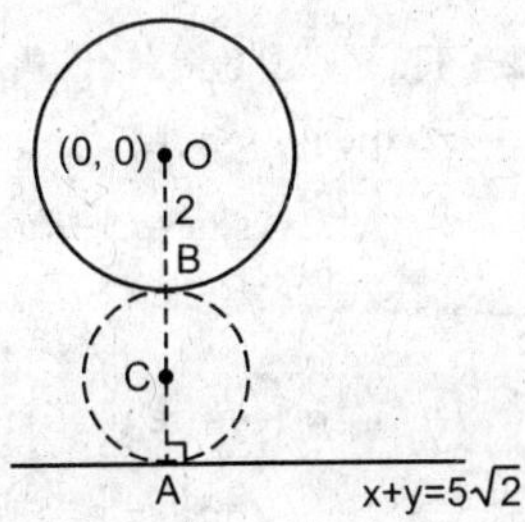

The slope of OA = 1 = tan θ

$$\therefore \quad \cos\theta=\frac{1}{\sqrt{2}}, \sin\theta=\frac{1}{\sqrt{2}}$$

$$\therefore \quad C=(0+OC\cdot\cos\theta, 0+OC\cdot\sin\theta)$$

$$=\left(\frac{7}{2\sqrt{2}},\frac{7}{2\sqrt{2}}\right).$$

68. $x^2+y^2+\frac{\lambda}{2}x-\frac{1+\lambda^2}{2}y-5=0$

and $x^2+y^2+4x+6y+3=0$ cut orthogonally if

$$2\cdot\frac{\lambda}{4}\cdot 2+2\cdot\left(-\frac{1+\lambda^2}{4}\right)\cdot 3=-5+3 \quad \Rightarrow \text{two real values of } \lambda.$$

69. See the hint to Q. No. 65.

70. In the beginning, coordinates of the centre = (2, 2), radius = 2. After one complete revolution on the x-axis in the positive direction, the x-coordinate increases by $2\pi r$, i.e., 4π, whereas the y-coordinate remains the same.

71.

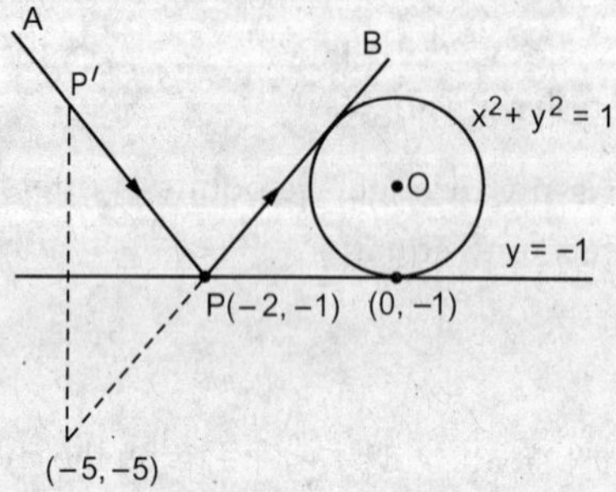

Any line through P(–2, –1) is $y+1=m(x+2)$.

It touches the circle if $\left|\dfrac{2m-1}{\sqrt{1+m^2}}\right|=1$, i.e., $m=0, \dfrac{4}{3}$.

$\therefore$ the equation of PB is $y+1=\dfrac{4}{3}(x+2)$, i.e., $4x-3y+5=0$.

A point on PB is (–5, –5). Its image by the line $y=-1$ is P′(–5, 3). So, the equation of the incident ray P′P is

$$y-3=\frac{3+1}{-5+2}(x+5) \quad \text{or} \quad 4x+3y+11=0.$$

72. The other end is $(t, 3-t)$.

So the equation of the variable circle is

$$(x-1)(x-t)+(y-1)(y-3+t)=0$$

or $\quad x^2+y^2-(1+t)x-(4-t)y+3=0.$

$\therefore$ the centre (α, β) is given by $\alpha=\dfrac{1+t}{2}, \beta=\dfrac{4-t}{2}$

$\Rightarrow \quad 2\alpha+2\beta=5.$

73. Here, the centre = (–2, 3) and

the radius $=\sqrt{(-2)^2+3^2-9\sin^2\alpha-13\cos^2\alpha}$

$=2\sin\alpha.$

$\therefore \quad \sin\alpha=\dfrac{2\sin\alpha}{\sqrt{(x_1+2)^2+(y_1-3)^2}}$

$\Rightarrow \quad (x_1+2)^2+(y_1-3)^2=4.$

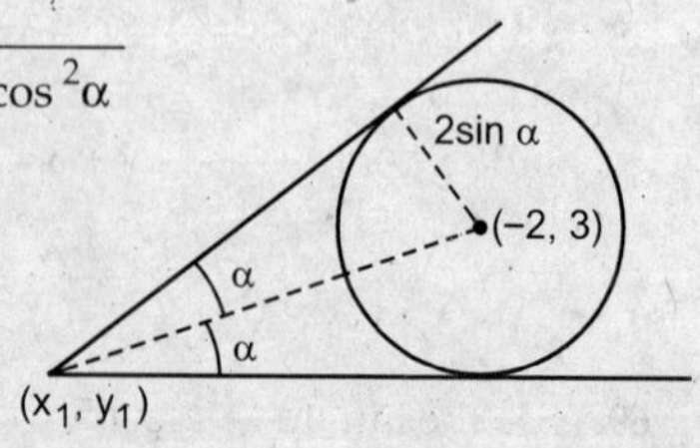

74. The vertices are $A(r, 0)$, $B(r\cos 60°, r\sin 60°)$, $C(-r\cos 60°, r\sin 60°)$, $D(-r, 0)$, $E(-r\cos 60°, -r\sin 60°)$, $F(r\cos 60°, -r\sin 60°)$.

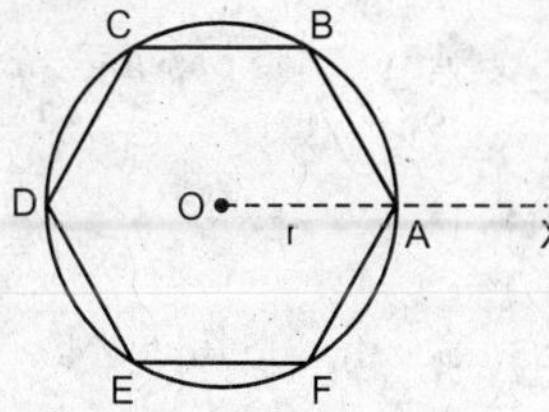

If $P = (x, y)$ then $\Sigma PA^2 = 6a^2$ implies

$$\{(x-r)^2 + y^2\} + \left\{\left(x - \frac{r}{2}\right)^2 + \left(y - \frac{r\sqrt{3}}{2}\right)^2\right\}$$

$$+ \ldots + \left\{\left(x - \frac{r}{2}\right)^2 + \left(x + \frac{r\sqrt{3}}{2}\right)^2\right\} = 6a^2$$

or $2(x^2 + y^2 + r^2) + 2\left(x^2 + y^2 + \frac{r^2}{4} + \frac{3r^2}{4}\right) + 2\left(x^2 + y^2 + \frac{r^2}{4} + \frac{3r^2}{4}\right) = 6a^2$

or $x^2 + y^2 + r^2 = a^2$ or $x^2 + y^2 = (\sqrt{a^2 - r^2})^2$.

76. Solving $y = mx$ and $x^2 + y^2 - 2x - 2y = 0$, we get

$$x^2 + m^2x^2 - 2x - 2mx = 0$$

$$\Rightarrow \quad x = 0, \frac{2(1+m)}{1+m^2}.$$

Similarly, solving $y = -mx$ and the equation of the circle, we get

$$x = 0, \frac{2(1-m)}{1+m^2}$$

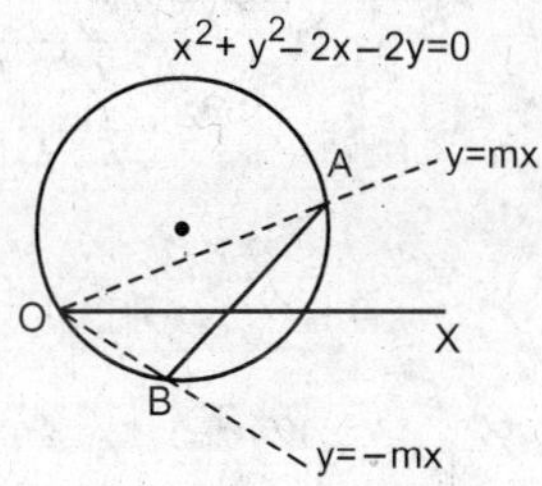

$$\therefore \quad A = \left(\frac{2(1+m)}{1+m^2}, \frac{2m(1+m)}{1+m^2}\right)$$

and $B = \left(\frac{2(1-m)}{1+m^2}, \frac{-2m(1-m)}{1+m^2}\right)$

Let the middle point of AB be (α, β). Then

$$\alpha = \frac{1}{2}\left\{\frac{2(1+m)}{1+m^2} + \frac{2(1-m)}{1+m^2}\right\} \text{ and } \beta = \frac{1}{2}\left\{\frac{2m(1+m)}{1+m^2} + \frac{-2m(1-m)}{1+m^2}\right\}$$

$\therefore \quad \alpha = \frac{2}{1+m^2}, \beta = \frac{2m^2}{1+m^2}$. Eliminating m from these, $\alpha + \beta = 2$.

77. Let the circle be $x^2 + y^2 + 2gx + 2fy = 0$. It cuts the other circle orthogonally if $2g(-5/2) + 2f \cdot 3/2 = 0 + (-1)$, i.e., $5(-g) - 3(-f) + 1 = 0$. Hence, the locus of the centre $(-g, -f)$ is $5x - 3y + 1 = 0$.

78. The centre of $C_1 = (2, 1)$ and the radius $= \sqrt{2^2 + 1^2 + 11} = 4$.

If (α, β) be the centre of C_2, $\sqrt{(\alpha - 2)^2 + (\beta - 1)^2} = 4 + 1$.

81. The locus is the radical axis which is perpendicular to the line joining the centres of the circles.

82. Any circle passing through the points of intersection of the given circles is $x^2+y^2-2x+y+\lambda(x^2+y^2-1)=0$. If (α, β) be its centre then

$$\alpha=\frac{1}{1+\lambda}, \beta=\frac{-1/2}{1+\lambda} \quad \Rightarrow \alpha+2\beta=0.$$

83. See the hint to Q. No. 74.

85. Let the line be $y-\frac{1}{2}=m(x+2)$.

Its distance from the centre $(0, 0)=\dfrac{2m+\frac{1}{2}}{\sqrt{1+m^2}}$.

So, the length of the intercept $=2\times\sqrt{1^2-\left(\dfrac{2m+\frac{1}{2}}{\sqrt{1+m^2}}\right)^2}$ ... (1)

The distance of the line from the other centre $(4, 0)=\dfrac{4m+2m+\frac{1}{2}}{\sqrt{1+m^2}}$

$\therefore$ the length of the intercept in this case

$$=2\times\sqrt{(\sqrt{5})^2-\left(\frac{6m+\frac{1}{2}}{\sqrt{1+m^2}}\right)^2} \quad \text{... (2)}$$

(1) and (2) are equal. Hence, $1-\dfrac{(4m+1)^2}{4(1+m^2)}=5-\dfrac{(12m+1)^2}{4(1+m^2)}$

$$\Rightarrow \quad \frac{(12m+1)^2-(4m+1)^2}{4(1+m^2)}=4 \quad \Rightarrow 7m^2+m-1=0.$$

86. Let L_1 be $y=mx$. Now, see the hint to Q. No. 85.

88. See the hint to Q. No. 20.

89.

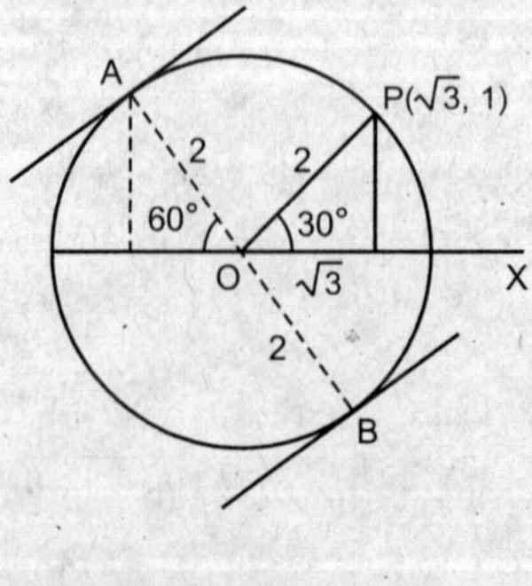

Clearly, A = $(-2\cos 60°, 2\sin 60°)$ and B = $(2\cos 60°, -2\sin 60°)$.

The tangent at A is $x(-2\cos 60°) + y(2\sin 60°) = 4$ and the tangent at B is $x(2\cos 60°) + y(-2\sin 60°) = 4$.

90.

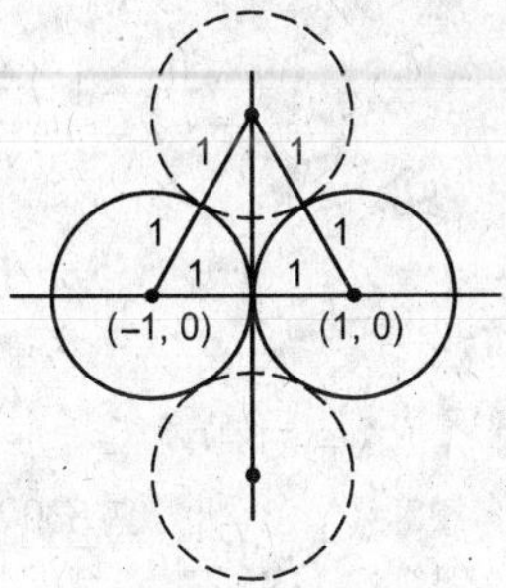

The given circles are $x^2 + y^2 - 2x = 0, x > 0$, and $x^2 + y^2 + 2x = 0, x < 0$.

From the figure, the centres of the required circles will be $(0, \sqrt{3})$ and $(0, -\sqrt{3})$.

$\therefore$ the equations of the circles are $(x-0)^2 + (y \mp \sqrt{3})^2 = 1^2$.

93. (h, k) is a point in the interior of the circle $x^2 + y^2 = a^2$. So, $hx + ky = a^2$ neither can be a real tangent nor a chord of contact of tangents.

95. If (x_1, y_1) is the point of contact then $xx_1 + yy_1 = 1$ which passes through $(1, 2)$.

So, $x_1 + 2y_1 = 1$.

Also, $x_1^2 + y_1^2 = 4$.

Solving these equations in x_1, y_1 we get $x_1 = \dfrac{1 \mp 2\sqrt{19}}{5}$, $y_1 = \dfrac{2 \pm \sqrt{19}}{5}$.

96. C_2 is the director circle of C_1. So, the equation of C_2 is $x^2 + y^2 = (\sqrt{2} \cdot 2)^2$. Again C_3 is the director circle of C_2. Hence, the equation of C_3 is $x^2 + y^2 = (\sqrt{2} \cdot \sqrt{2} \cdot 2)^2$.

97. Let $x - 3y + \lambda = 0$ touch the circle at (x_1, y_1).

Then $xx_1 + yy_1 - 2(x + x_1) + y + y_1 - 5 = 0$ and $x - 3y + \lambda = 0$ are identical.

Hence, comparing these

$$\frac{x_1 - 2}{1} = \frac{y_1 + 1}{-3} = \frac{-2x_1 + y_1 - 5}{\lambda}$$

$$= \frac{2(x_1-2)-1(y_1+1)+(-2x_1+y_1-5)}{2\times 1-1\times(-3)+\lambda}$$

$$\therefore \quad \frac{x_1-2}{1} = = \frac{y_1+1}{-3} = \frac{-10}{5+\lambda}$$

$$\Rightarrow \quad x_1 = \frac{-10}{5+\lambda}+2 = \frac{2\lambda}{5+\lambda},\ y_1 = \frac{30}{5+\lambda}-1 = \frac{25-\lambda}{5+\lambda}.$$

(x_1, y_1) is on the circle. So,

$$\left(\frac{2\lambda}{5+\lambda}\right)^2 + \left(\frac{25-\lambda}{5+\lambda}\right)^2 - 4\cdot\frac{2\lambda}{5+\lambda} + 2\cdot\frac{25-\lambda}{5+\lambda} - 5 = 0.$$

Onsimplification, $\lambda^2+10\lambda-75=0 \quad \Rightarrow \lambda = 5, -15.$

$$\therefore \quad x_1 = \frac{10}{10} = 1,\ y_1 = \frac{20}{10} = 2 \quad \text{or} \quad x_1 = \frac{-30}{-10} = 3,\ y_1 = \frac{25+15}{5-15} = -4.$$

So, $(x_1, y_1) = (1, 2), (3, -4)$.

98. The point on $x = 3$ should also be on the director circle
$$x^2 + y^2 = (\sqrt{2}\cdot\sqrt{8})^2 = 16.$$

Solving $x = 3$, $x^2 + y^2 = 16$, we get $x = 3$, $y = \pm\sqrt{7}$.

99. AB is the chord of contact of the tangents from (2, 0). So, the equation of AB is $2x = 1$.

Solving $2x = 1, x^2 + y^2 = 1$, we get A, B.

100.

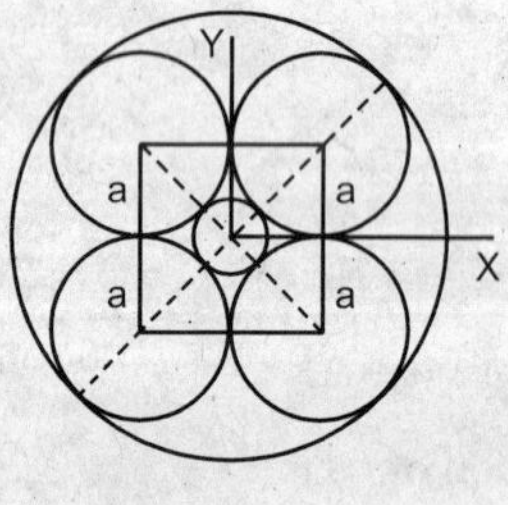

The four circles are as shown in the figure. The smallest circle touching all of them has the radius $= \sqrt{2}a - a$ and the greatest circle touching all of them has the radius $= \sqrt{2}a + a$.

101. The polar of the point P(5,10) with respect to the circle $x^2 + y^2 = 25$ is $x + 2y = 5$ which is the given line l.

Also, from the property of polar line of a point we have PA, PQ, PB are in HP.

102. The point (1, 1) is an interior point of the circle. So the chord of contact is not real.

□

4

Parabola

• Type 1 •

Choose the most appropriate option (a, b, c or d).

1. Given the two ends of the latus rectum, the maximum number of parabolas that can be drawn is

(a) 1 (b) 2

(c) 0 (d) infinite

2. If the focus of a parabola is $(-2, 1)$ and the directrix has the equation $x + y = 3$ then the vertex is

(a) $(0, 3)$ (b) $(-1, 1/2)$

(c) $(-1, 2)$ (d) $(2, -1)$

3. If the vertex and the focus of a parabola are $(-1, 1)$ and $(2, 3)$ respectively then the equation of the directrix is

(a) $3x + 2y + 14 = 0$ (b) $3x + 2y - 25 = 0$

(c) $2x - 3y + 10 = 0$ (d) none of these

4. The vertex of a parabola is $(a, 0)$ and the directrix is $x + y = 3a$. The equation of the parabola is

(a) $x^2 + 2xy + y^2 + 6ax + 10ay + 7a^2 = 0$

(b) $x^2 - 2xy + y^2 + 6ax + 10ay + 2u^2 = 0$

(c) $x^2 - 2xy + y^2 - 6ax + 10ay = 2a^2$

(d) none of these

5. If the vertex $= (2, 0)$ and the extremities of the latus rectum are $(3, 2)$ and $(3, -2)$ then the equation of the parabola is

(a) $y^2 = 2x - 4$ (b) $x^2 = 4y - 8$

(c) $y^2 = 4x - 8$ (d) none of these

6. Any point on the parabola whose focus is (0, 1) and the directrix is $x + 2 = 0$ is given by

(a) $(t^2 + 1, 2t - 1)$ (b) $(t^2 + 1, 2t + 1)$

(c) $(t^2, 2t)$ (d) $(t^2 - 1, 2t + 1)$

7. The equation of the parabola whose vertex and focus are on the positive side of the x-axis at distances a and b respectively from the origin is

(a) $y^2 = 4(b - a)(x - a)$ (b) $y^2 = 4(a - b)(x - b)$

(c) $x^2 = 4(b - a)(y - a)$ (d) none of these

8. The equation $x^2 + 4xy + 4y^2 - 3x - 6y - 4 = 0$ represents a

(a) circle (b) parabola

(c) a pair of lines (d) none of these

9. The equation $\lambda x^2 + 4xy + y^2 + \lambda x + 3y + 2 = 0$ represents a parabola if λ is

(a) –4 (b) 4

(c) 0 (d) none of these

10. The focus of the parabola $y^2 - x - 2y + 2 = 0$ is

(a) $\left(\frac{5}{4}, 1\right)$ (b) $\left(\frac{1}{4}, 0\right)$

(c) (1, 1) (d) none of these

11. The vertex of the parabola $(y - a)^2 = 4a(x + a)$ is

(a) $(-a, a)$ (b) $(a, -a)$

(c) $(-2a, 2a)$ (d) $\left(-\frac{a}{2}, \frac{a}{2}\right)$

12. The equation of the axis of the parabola $9y^2 - 16x - 12y - 57 = 0$ is

(a) $2x = 3$ (b) $y = 3$

(c) $3y = 2$ (d) $x + 3y = 3$

13. The length of the latus rectum of the parabola

$$169\{(x - 1)^2 + (y - 3)^2\} = (5x - 12y + 17)^2 \text{ is}$$

(a) $\dfrac{14}{13}$ (b) $\dfrac{28}{13}$

(c) $\dfrac{12}{13}$ (d) none of these

14. The length of the latus rectum of the parabola $x = ay^2 + by + c$ is

(a) $\dfrac{a}{4}$ (b) $\dfrac{a}{3}$

(c) $\dfrac{1}{a}$ (d) $\dfrac{1}{4a}$

15. The parametric equation of a parabola is $x = t^2 + 1,\ y = 2t + 1$. The Cartesian equation of its directrix is

(a) $x = 0$ (b) $x + 1 = 0$

(c) $y = 0$ (d) none of these

16. If $(2, -8)$ is at an end of a focal chord of the parabola $y^2 = 32x$ then the other end of the chord is

(a) $(32, 32)$ (b) $(32, -32)$

(c) $(-2, 8)$ (d) none of these

17. A line L passing through the focus of the parabola $y^2 = 4(x-1)$ intersects the parabola in two distinct points. If 'm' be the slope of the line L then

(a) $-1 < m < 1$ (b) $m < -1$ or $m > 1$

(c) $m \in R$ (d) none of these

18. The HM of the segments of a focal chord of the parabola $y^2 = 4ax$ is

(a) $4a$ (b) $2a$

(c) a (d) a^2

19. The length of a focal chord of the parabola $y^2 = 4ax$ at a distance b from the vertex is c. Then

(a) $2a^2 = bc$ (b) $a^3 = b^2c$

(c) $ac = b^2$ (d) $b^2c = 4a^3$

20. The parabola $y^2 = kx$ makes an intercept of length 4 on the line $x - 2y = 1$. Then k is

(a) $\frac{\sqrt{105}-5}{10}$ (b) $\frac{5-\sqrt{105}}{10}$

(c) $\frac{5+\sqrt{105}}{10}$ (d) none of these

21. A double ordinate of the parabola $y^2 = 8px$ is of length $16p$. The angle subtended by it at the vertex of the parabola is

(a) $\frac{\pi}{4}$ (b) $\frac{\pi}{2}$

(c) $\frac{\pi}{3}$ (d) none of these

22. The chord AB of the parabola $y^2 = 4ax$ cuts the axis of the parabola at C. If $A = (at_1^2, 2at_1)$, $B = (at_2^2, 2at_2)$ and $AC : AB = 1 : 3$ then

(a) $t_2 = 2t_1$ (b) $t_2 + 2t_1 = 0$

(c) $t_1 + 2t_2 = 0$ (d) none of these

23. AB is a chord of the parabola $y^2 = 4ax$. If its equation is $y = mx + c$ and it subtends a right angle at the vertex of the parabola then

(a) $c = 4am$ (b) $a = 4mc$

(c) $c = -4am$ (d) $a + 4mc = 0$

24. If 't_1' and 't_2' are the ends of a focal chord of the parabola $y^2 = 2x$ then

(a) $t_1^2 + t_2^2 = 2$ (b) $t_1 + t_2 = 1$

(c) $t_1 t_2 = -1$ (d) none of these

25. A ray of light moving parallel to the x-axis gets reflected from a parabolic mirror whose equation is $(y-2)^2 = 4(x+1)$. After reflection, the ray must pass through the point

(a) (0, 2) (b) (2, 0)

(c) (0, –2) (d) (–1, 2)

26. The equation of a parabola is $y^2 = 4x$. P(1, 3) and Q(1, 1) are two points in the x-y plane. Then, for the parabola

(a) P and Q are exterior points

(b) P is an interior point while Q is an exterior point

(c) P and Q are interior points

(d) P is an exterior point while Q is an interior point

27. The point $(a, 2a)$ is an interior point of the region bounded by the parabola $y^2 = 16x$ and the double ordinate through the focus. Then a belongs to the open interval

(a) $a < 4$ (b) $0 < a < 4$

(c) $0 < a < 2$ (d) $a > 4$

28. The ends of a line segment are P(1, 3) and Q(1, 1). R is a point on the line segment PQ such that $PR : QR = 1 : \lambda$. If R is an interior point of the parabola $y^2 = 4x$ then

(a) $\lambda \in (0, 1)$ (b) $\lambda \in \left(-\frac{3}{5}, 1\right)$

(c) $\lambda \in \left(\frac{1}{2}, \frac{3}{5}\right)$ (d) none of these

29. The range of values of λ for which the point $(\lambda, -1)$ is exterior to both the parabolas $y^2 = |x|$ is

(a) (0, 1) (b) (−1, 1)

(c) (−1, 0) (d) none of these

30. The number of points with integral coordinates that lie in the interior of the region common to the circle $x^2 + y^2 = 16$ and the parabola $y^2 = 4x$ is

(a) 8 (b) 10

(c) 16 (d) none of these

31. The number of distinct real tangents that can be drawn from (0, −2) to the parabola $y^2 = 4x$ is

(a) one (b) two

(c) zero (d) none of these

32. The tangents to the parabola $y^2 = 4x$ at the points (1, 2) and (4, 4) meet on the line

(a) $x = 3$ (b) $x + y = 4$

(c) $y = 3$ (d) none of these

33. The point of intersection of the tangents to the parabola $y^2 = 4x$ at the points, where the parameter 't' has the value 1 and 2, is

(a) (3, 8) (b) (1, 5)

(c) (2, 3) (d) (4, 6)

34. The triangle formed by the tangents to a parabola $y^2 = 4ax$ at the ends of the latus rectum and the double ordinate through the focus is

(a) equilateral
(b) isosceles
(c) right-angled isosceles
(d) dependent on the value of a for its classification

35. If two tangents drawn from the point (α, β) to the parabola $y^2 = 4x$ be such that the slope of one tangent is double of the other then

(a) $\beta = \frac{2}{9}\alpha^2$ (b) $\alpha = \frac{2}{9}\beta^2$

(c) $2\alpha = 9\beta^2$ (d) none of these

36. The tangents from the origin to the parabola $y^2 + 4 = 4x$ are inclined at

(a) $\frac{\pi}{6}$ (b) $\frac{\pi}{4}$

(c) $\frac{\pi}{3}$ (d) $\frac{\pi}{2}$

37. If $y + b = m_1(x + a)$ and $y + b = m_2(x + a)$ are two tangents to the parabola $y^2 = 4ax$ then

(a) $m_1 + m_2 = 0$ (b) $m_1 m_2 = 1$

(c) $m_1 m_2 = -1$ (d) none of these

38. The tangents to a parabola at the vertex V and any point P meet at Q. If S be the focus then SP, SQ, SV are in

(a) AP (b) GP

(c) HP (d) none of these

39. The equation of the common tangent to the equal parabolas $y^2 = 4ax$ and $x^2 = 4ay$ is

(a) $x + y + a = 0$ (b) $x + y = a$

(c) $x - y = a$ (d) none of these

40. 't_1' and 't_2' are two points on the parabola $y^2 = 4x$. If the chord joining them is a normal to the parabola at 't_1' then

(a) $t_1 + t_2 = 0$ (b) $t_1(t_1 + t_2) = 1$

(c) $t_1(t_1 + t_2) + 2 = 0$ (d) $t_1 t_2 + 1 = 0$

41. The normal to the curve $x = at^2, y = 2at$ at the point $P(t)$ meets the curve again at $Q(t')$. Then t' is

(a) $t+\frac{1}{t}$ (b) $-t-\frac{2}{t}$

(c) $t+\frac{2}{t}$ (d) $t-\frac{1}{t}$

42. The set of points on the axis of the parabola $y^2 = 4x + 8$ from which the 3 normals to the parabola are all real and different is

(a) $\{(k, 0) \mid k \le -2\}$ (b) $\{(k, 0) \mid k > -2\}$

(c) $\{(0, k) \mid k > -2\}$ (d) none of these

43. The number of distinct normals that can be drawn from $(-2, 1)$ to the parabola $y^2 - 4x - 2y - 3 = 0$ is

(a) 1 (b) 2

(c) 3 (d) 0

44. If the line $y = x + k$ is a normal to the parabola $y^2 = 4x$ then k can have the value

(a) $2\sqrt{2}$ (b) 4

(c) -3 (d) 3

45. The arithmetic mean of the ordinates of the feet of the normals from $(3, 5)$ to the parabola $y^2 = 8x$ is

(a) 4 (b) 0

(c) 8 (d) none of these

46. The area of the triangle formed by the tangent and the normal to the parabola $y^2 = 4ax$, both drawn at the same end of the latus rectum, and the axis of the parabola is

(a) $2\sqrt{2}a^2$ (b) $2a^2$

(c) $4a^2$ (d) none of these

47. If two of the three feet of normals drawn from a point to the parabola $y^2 = 4x$ be $(1, 2)$ and $(1, -2)$ then the third foot is

(a) $(2, 2\sqrt{2})$ (b) $(2, -2\sqrt{2})$

(c) $(0, 0)$ (d) none of these

48. Let P, Q, R be three points on a parabola, normals at which are concurrent. The centroid of the ΔPQR must lie on

(a) a line parallel to the directrix

(b) the axis of the parabola

(c) a line of slope 1 passing through the vertex

(d) none of these

49. The vertex of the parabola $y^2 = 8x$ is at the centre of a circle and the parabola cuts the circle at the ends of its latus rectum. Then the equation of the circle is

(a) $x^2 + y^2 = 4$ (b) $x^2 + y^2 = 20$

(c) $x^2 + y^2 = 80$ (d) none of these

50. The length of the common chord of the parabola $2y^2 = 3(x+1)$ and the circle $x^2 + y^2 + 2x = 0$ is

(a) $\sqrt{3}$ (b) $2\sqrt{3}$

(c) $\frac{\sqrt{3}}{2}$ (d) none of these

51. The circle $x^2 + y^2 + 2\lambda x = 0, \lambda \in R,$ touches the parabola $y^2 = 4x$ externally. Then

(a) $\lambda > 0$ (b) $\lambda < 0$

(c) $\lambda > 1$ (d) none of these

52. The locus of the middle points of chords of a parabola which subtend a right angle at the vertex of the parabola is

(a) a circle (b) an ellipse

(c) a parabola (d) none of these

53. The locus of a point from which tangents to a parabola are at right angles is a

(a) straight line (b) pair of straight lines

(c) circle (d) parabola

54. P is a point. Two tangents are drawn from it to the parabola $y^2 = 4x$ such that the slope of one tangent is three times the slope of the other. The locus of P is

(a) a straight line (b) a circle

(c) a parabola (d) an ellipse

55. The locus of the middle points of parallel chords of a parabola $x^2 = 4ay$ is a

(a) straight line parallel to the x-axis

(b) straight line parallel to the y-axis

(c) circle

(d) straight line parallel to a bisector of the angles between the axes

56. The locus of the middle points of chords of the parabola $y^2 = 8x$ drawn through the vertex is a parabola whose

(a) focus is (2, 0)
(b) latus rectum = 8
(c) focus is (0, 2)
(d) latus rectum = 4

57. The locus of the points of trisection of the double ordinates of a parabola is a

(a) pair of lines
(b) circle
(c) parabola
(d) straight line

• *Type 2* •

Choose the correct options. One or more options may be correct.

58. The parabola $x^2 + 2x - 4y = 0$ has

(a) vertex = $(-1, -1)$
(b) latus rectum = 4
(c) focus = $\left(-1, \frac{3}{4}\right)$
(d) focus = $\left(0, -\frac{1}{4}\right)$

59. The equation of a parabola is $25\{(x-2)^2 + (y+5)^2\} = (3x + 4y - 1)^2$. For this parabola

(a) vertex = $(2, -5)$

(b) focus = $(2, -5)$

(c) directrix has the equation $3x + 4y - 1 = 0$

(d) axis has the equation $3x + 4y - 1 = 0$

60. Let PQ be a chord of the parabola $y^2 = 4x$. A circle drawn with PQ as a diameter passes through the vertex V of the parabola. If ar $(\Delta PVQ) = 20$ unit^2 then the coordinates of P are

(a) (16, 8)
(b) (16, –8)
(c) (–16, 8)
(d) (–16, –8)

61. The equation of a tangent to the parabola $y^2 = 9x$ from the point (4, 10) is

(a) $x - 4y + 36 = 0$
(b) $81x - 8y - 162 = 0$

(c) $9x - 4y + 4 = 0$ (d) $x - 4y - 36 = 0$

62. If the tangents drawn from the point (0, 2) to the parabola $y^2 = 4ax$ are inclined at an angle $\frac{3\pi}{4}$ then the value of a is

(a) 2 (b) –2

(c) 1 (d) none of these

63. If the tangents to the parabola $y^2 = 4ax$ at (x_1, y_1), (x_2, y_2) cut at (x_3, y_3) then

(a) x_1, x_3, x_2 are in AP (b) x_1, x_3, x_2 are in GP

(c) y_1, y_3, y_2 are in AP (d) y_1, y_3, y_2 are in GP

64. The equation of a locus is $y^2 + 2ax + 2by + c = 0$. Then

(a) it is an ellipse (b) it is a parabola

(c) its latus rectum $= a$ (d) its latus rectum $= 2a$

65. A tangent to the parabola $y^2 = 4ax$ is inclined at $\frac{\pi}{3}$ with the axis of the parabola. The point of contact is

(a) $\left(\frac{a}{3}, -\frac{2a}{\sqrt{3}}\right)$ (b) $(3a, -2\sqrt{3}a)$

(c) $(3a, 2\sqrt{3}a)$ (d) $\left(\frac{a}{3}, \frac{2a}{\sqrt{3}}\right)$

66. A chord PP′ of a parabola cuts the axis of the parabola at O. The feet of the perpendiculars from P and P′ on the axis are M and M′ respectively. If V is the vertex then VM, VO, VM′ are in

(a) AP (b) GP

(c) HP (d) none of these

67. Let the equations of a circle and a parabola be $x^2 + y^2 - 4x - 6 = 0$ and $y^2 = 9x$ respectively. Then

(a) (1, –1) is a point on the common chord of contact

(b) the equation of the common chord is $y + 1 = 0$

(c) the length of the common chord is 6

(d) none of these

68. The equation of a common tangent to the parabola $y^2 = 2x$ and the circle $x^2 + y^2 + 4x = 0$ is

(a) $2\sqrt{6}x + y = 12$ (b) $x + 2\sqrt{6}y + 12 = 0$

(c) $x - 2\sqrt{6}y + 12 = 0$ (d) $2\sqrt{6}x - y = 12$

69. Let there be two parabolas with the same axis, focus of each being exterior to the other and the latus recta being $4a$ and $4b$. The locus of the middle points of the intercepts between the parabolas made on the lines parallel to the common axis is a

(a) straight line if $a = b$ (b) parabola if $a \neq b$

(c) parabola for all a, b (d) none of these

70. P is a point which moves in the x-y plane such that the point P is nearer to the centre of a square than any of the sides. The four vertices of the square are $(\pm a, \pm a)$. The region in which P will move is bounded by parts of parabolas of which one has the equation

(a) $y^2 = a^2 + 2ax$ (b) $x^2 = a^2 + 2ay$

(c) $y^2 + 2ax = a^2$ (d) none of these

Answers

1. b	**2.** c	**3.** a	**4.** b	**5.** c
6. d	**7.** a	**8.** c	**9.** b	**10.** a
11. a	**12.** c	**13.** b	**14.** c	**15.** a
16. a	**17.** d	**18.** b	**19.** d	**20.** a
21. b	**22.** b	**23.** c	**24.** c	**25.** a
26. d	**27.** b	**28.** a	**29.** b	**30.** a
31. b	**32.** c	**33.** c	**34.** c	**35.** b
36. d	**37.** c	**38.** b	**39.** a	**40.** c
41. b	**42.** d	**43.** a	**44.** c	**45.** b
46. c	**47.** c	**48.** b	**49.** b	**50.** a
51. a	**52.** c	**53.** a	**54.** c	**55.** b
56. d	**57.** c	**58.** b, c	**59.** b, c	**60.** a, b
61. a, c	**62.** a, b	**63.** b, c	**64.** b, d	**65.** a, d
66. b	**67.** a, c	**68.** b, c	**69.** a, b	**70.** a, b, c

Hints and Solutions to Selected Questions

1. L, L′ are the ends of the latus rectum. S bisects LL'. VSV′ is the perpendicular bisector of LL′, where $VS = \frac{1}{4} LL' = V'S$. Clearly, two parabolas are possible.

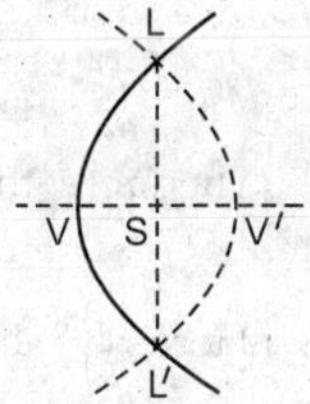

2. The vertex is the middle point of the perpendicular dropped from the focus to the directrix.

3. See the hint to Q. No. 2.

4. See to hint to Q. No. 2.

5. The focus $= \left(\frac{3+3}{2}, \frac{2-2}{2}\right) = (3, 0)$. The vertex $= (2, 0)$.

As $MV = VS$, $M = (1, 0)$. Clearly, the directrix is perpendicular to VS whose equation is $y = 0$. So, the directrix is $x = k$ which passes through M(1, 0). Therefore, we get $x = 1$.

$\therefore$ the equation of the parabola is

$$(x-3)^2 + (y-0)^2 = \left(\frac{x-1}{\sqrt{1}}\right)^2.$$

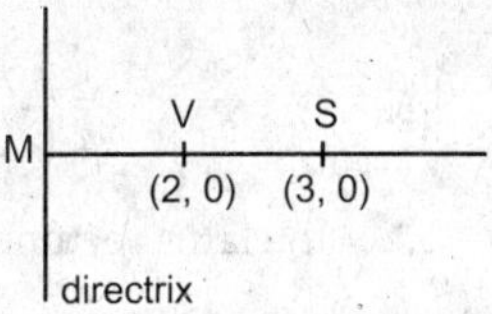

6. The equation is $(x-0)^2 + (y-1)^2 = \left(\frac{x+2}{\sqrt{1}}\right)^2$

$$\text{or } (y-1)^2 = 4(x+1).$$

Clearly, $x = t^2 - 1$ and $y = 2t + 1$ satisfy it for all t.

7. Here, the vertex $= (a, 0)$, the focus $= (b, 0)$. Now, see the hint to Q. No. 5.

8. $\Delta = abc + 2fgh - af^2 - bg^2 - ch^2 = 0$ though $h^2 = ab$ is satisfied. So the equation represents a pair of lines.

9. $\Delta \neq 0$ for all real λ and $h^2 = ab \quad \Rightarrow \; 2^2 = \lambda \cdot 1$.

10. $y^2 - 2y + 1 = x - 1$ or $(y-1)^2 = x - 1.$

Putting $x - 1 = X, y - 1 = Y$ the equation becomes $Y^2 = X$, i.e., $Y^2 = 4 \cdot \frac{1}{4} \cdot X$. So, the focus $= \left(\frac{1}{4}, 0\right)_{X,Y} = \left(\frac{5}{4}, 1\right).$

12. $9y^2 - 12y + 4 = 16x + 61$ or $(3y-2)^2 = 4 \cdot 4\left(x + \frac{61}{16}\right).$

Therefore, the equation of the axis is $3y - 2 = 0$.

13. Here, $(x-1)^2 + (y-3)^2 = \left\{\frac{5x - 12y + 17}{\sqrt{5^2 + (-12)^2}}\right\}^2$

$\therefore$ the focus $= (1, 3)$ and the directrix is $5x - 12y + 17 = 0$.

The distance of the focus from the directrix $= \left|\frac{5 \times 1 - 12 \times 3 + 17}{\sqrt{5^2 + (-12)^2}}\right| = \frac{14}{13}$

$\therefore$ latus rectum $= 2 \times \frac{14}{13} = \frac{28}{13}.$

14. $x = a\left(y^2 + \frac{b}{a}y + \frac{b^2}{4a^2}\right) - \frac{b^2}{4a} + c$

or $a\left(y + \frac{b}{2a}\right)^2 = x + \frac{b^2}{4a} - c$

or $\left(y + \frac{b}{2a}\right)^2 = \frac{1}{a}\left\{x + \frac{b^2 - 4ac}{4a}\right\}$

$\therefore$ the latus rectum $= \frac{1}{a}.$

15. Eliminating t, $x = \left(\frac{y-1}{2}\right)^2 + 1$ or $(y-1)^2 = 4(x-1).$

Putting $y - 1 = Y, x - 1 = X$ the equation becomes $Y^2 = 4X$.

So, the equation of the directrix is $X + 1 = 0 \Rightarrow x = 0.$

17. Let $y = Y, x - 1 = X$. Then the equation is $Y^2 = 4X$.

So, the focus $= (1, 0)_{X,Y} = (2, 0).$

Any line through the focus is $y = m(x-2)$. Solving this with $y^2 = 4(x-1)$,

$$m^2(x-2)^2 = 4(x-1) \quad \text{or} \quad m^2x^2 - 4(m^2+1)x + 4(m^2+1) = 0.$$

If $m \neq 0$, $D = 16(m^2+1)^2 - 16m^2(m^2+1) = 16(m^2+1) > 0$ for all m.

But, if $m = 0$ then x does not have two real distinct values.

So $m \in R$ except $m = 0$.

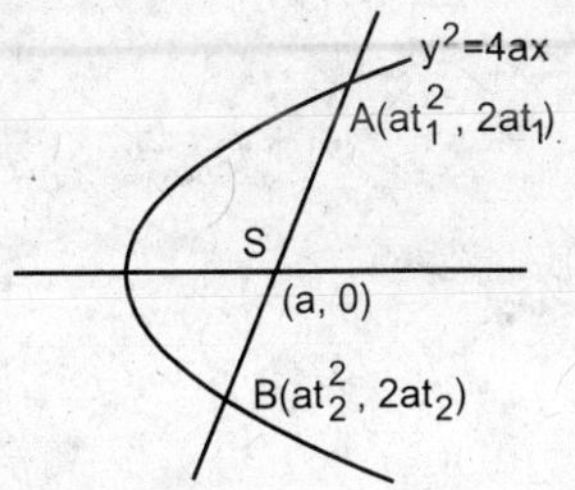

18. $SA^2 = (a - at_1^2)^2 + (0 - 2at_1)^2$

$= a^2(1 - t_1^2)^2 + 4a^2t_1^2 = a^2(1 + t_1^2)^2.$

Similarly, $SB^2 = a^2(1 + t_2^2)^2$.

But A, S, B are collinear.

Hence, $\dfrac{2at_1 - 0}{at_1^2 - a} = \dfrac{2at_2 - 0}{at_2^2 - a}$

or $\dfrac{2t_1}{t_1^2 - 1} = \dfrac{2t_2}{t_2^2 - 1}$ or $t_1t_2^2 - t_1 = t_1^2t_2 - t_2$

or $t_1t_2(t_2 - t_1) + (t_2 - t_1) = 0.$

But $t_1 \neq t_2$. So, $t_1t_2 + 1 = 0$.

HM of $SA, SB = \dfrac{2SA \cdot SB}{SA + SB} = \dfrac{2 \cdot a(1 + t_1^2) \cdot a(1 + t_2^2)}{a(1 + t_1^2) + a(1 + t_2^2)}$

$$= \frac{2a^2\{1 + (t_1^2 + t_2^2) + t_1^2t_2^2\}}{a\{2 + (t_1^2 + t_2^2)\}} = \frac{2a^2\{1 + (t_1^2 + t_2^2) + 1\}}{a\{2 + (t_1^2 + t_2^2)\}} = 2a.$$

19. Let the ends of the focal chord be $(at_1^2, 2at_1)$ and $(at_2^2, 2at_2)$. Then, as in Q. No. 18., $t_1t_2 + 1 = 0$. Then $c^2 = \{a(t_1^2 - t_2^2)\}^2 + \{2a(t_1 - t_2\}^2$.

The equation of the focal chord is $y - 2at_1 = \dfrac{2at_1 - 2at_2}{at_1^2 - at_2^2}(x - at_1^2)$

$$\text{or} \quad y - 2at_1 = \frac{2}{t_1 + t_2}(x - at_1^2).$$

So, $b = \dfrac{\dfrac{2}{t_1 + t_2}(-at_1^2) + 2at_1}{\sqrt{1 + \dfrac{4}{(t_1 + t_2)^2}}} = \dfrac{2at_1t_2}{\sqrt{4 + (t_1 + t_2)^2}}$

$\therefore \quad b = \dfrac{-2a}{\sqrt{2 + t_1^2 + t_2^2}}.$

Also, $c^2 = a^2(t_1 - t_2)^2\{(t_1 + t_2)^2 + 4\} = a^2(t_1^2 + t_2^2 + 2)(t_1^2 + t_2^2 + 2)$

$\therefore \quad c = a(2 + t_1^2 + t_2^2).$

Hence, $b^2 = \dfrac{4a^2}{2 + t_1^2 + t_2^2} = \dfrac{4a^2}{c/a} = \dfrac{4a^3}{c}$.

20. Solving $x - 2y = 1, y^2 = kx$, we get

$$y^2 = k(1 + 2y) \quad \text{or} \quad y^2 - 2ky - k = 0$$

$$\therefore \quad y_1 + y_2 = 2k, \; y_1 \cdot y_2 = -k$$

$$\therefore \quad 16 = (x_1 - x_2)^2 + (y_1 - y_2)^2 = \left(\frac{y_1^2}{k} - \frac{y_2^2}{k}\right)^2 + (y_1 - y_2)^2$$

$$= (y_1 - y_2)^2 \cdot \left\{\frac{(y_1 + y_2)^2}{k^2} + 1\right\}$$

$$= \{(y_1 + y_2)^2 - 4y_1y_2\}\left\{\frac{4k^2}{k^2} + 1\right\}$$

$$= 5\{4k^2 + 4k\} = 20k^2 + 20k; \quad \therefore \quad 5k^2 + 5k - 4 = 0.$$

21.

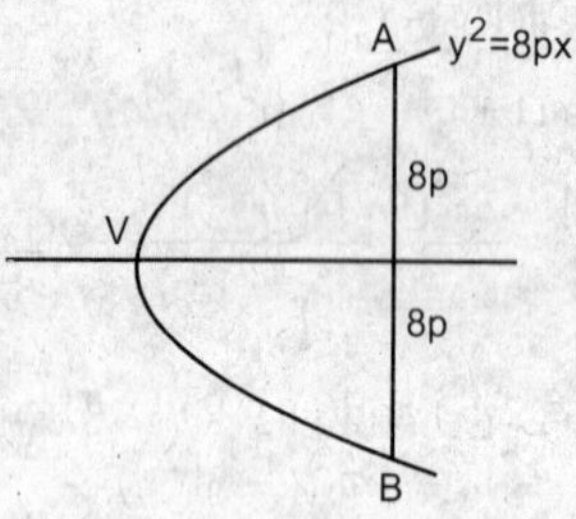

Clearly, $A = (8p, 8p)$ and $B = (8p, -8p)$.

Also, $V = (0, 0)$.

'm' of AV = 1; 'm' of BV = –1.

Therefore, AV ⊥ BV.

22. Here, $C = \left(\dfrac{2at_1^2 + at_2^2}{3}, \dfrac{4at_1 + 2at_2}{3}\right)$.

It lies on $y = 0$.

$$\therefore \quad \frac{4at_1 + 2at_2}{3} = 0.$$

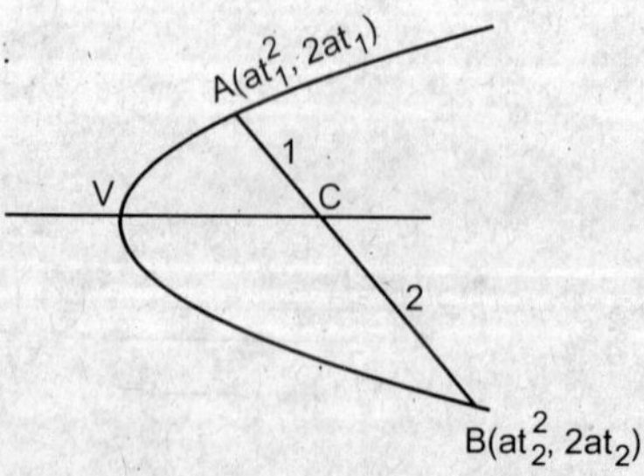

23. The equation of the pair of lines VA and VB, where V is the vertex = (0, 0), is

$$y^2 = 4ax \cdot \frac{y - mx}{c} \quad \text{or} \quad cy^2 - 4axy + 4amx^2 = 0.$$

The pair is at right angles if $c + 4am = 0$.

25. The equation of the axis of the parabola is $y - 2 = 0$, which is parallel to the x-axis. So, a ray parallel to the x-axis is parallel to the axis of the parabola. We know that any ray parallel to the axis of a parabola passes through the focus after reflection. Here (0, 2) is the focus.

26. Here, $S \equiv y^2 - 4x = 0$.

$S(1, 3) = 3^2 - 4 \cdot 1 > 0 \quad \Rightarrow$ P(1, 3) is an exterior point.

$S(1, 1) = 1^2 - 4 \cdot 1 < 0 \quad \Rightarrow$ Q(1, 1) is an interior point.

27. $(a, 2a)$ is an interior point of $y^2 - 16x = 0$ if $(2a)^2 - 16a < 0$, i.e., $a^2 - 4a < 0$.

V(0, 0) and $(a, 2a)$ are on the same side of $x - 4 = 0$. So, $a - 4 < 0$, i.e., $a < 4$.

Now, $a^2 - 4a < 0 \Rightarrow 0 < a < 4$.

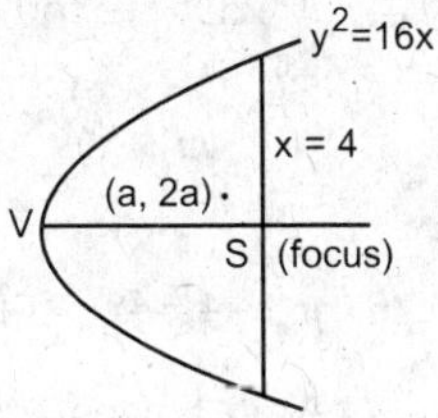

28. $R = \left(1, \dfrac{1+3\lambda}{1+\lambda}\right)$. It is an interior point of $y^2 - 4x = 0$ if $\left(\dfrac{1+3\lambda}{1+\lambda}\right)^2 - 4 < 0$.

Therefore, $-\dfrac{3}{5} < \lambda < 1$. But $\lambda > 0$.

29. The parabolas are $y^2 - x = 0$ and $y^2 + x = 0$. The point $(\lambda, -1)$ is an exterior point if

$1 - \lambda > 0$ and $1 + \lambda > 0 \quad \Rightarrow \lambda < 1$ and $\lambda > -1 \quad \Rightarrow -1 < \lambda < 1$.

30. (λ, μ) is interior to both the curves if $\lambda^2 + \mu^2 - 16 < 0$ and $\mu^2 - 4\lambda < 0$.

Now, $\mu^2 - 4\lambda < 0 \quad \Rightarrow \lambda > \left(\dfrac{\mu}{2}\right)^2$.

Hence, if $\mu = 0, \lambda = 1, 2, 3, \ldots$; if $\mu = 1, \lambda = 1, 2, 3, \ldots$; if $\mu = 2, \lambda = 2, 3, \ldots$; if $\mu = 3, \lambda = 3, 4, \ldots$.

Also $\lambda^2 + m^2 - 16 < 0 \quad \Rightarrow \lambda^2 < 16 - \mu^2$.

Hence, if $\mu = 0$, $\lambda = 1, 2, 3$; if $\mu = 1$, $\lambda = 1, 2, 3$; if $\mu = 2$, $\lambda = 2, 3$; if $\mu = 3$, λ has no integral value.

$\therefore$ (1, 0), (2, 0), (3, 0), (1, 1), (2, 1), (3, 1), (2, 2), (3, 2) are the possible points.

32. The tangents to the parabola $y^2 = 4ax$ at the points $(at_1^2, 2at_1)$ and $(at_2^2, 2at_2)$ meet at $\left(at_1t_2, a(t_1 + t_2)\right)$. Here $a = 1, t_1 = 1, t_2 = 2$. So they meet at (2, 3), which is on the line $y = 3$.

33. See the hint to Q. No. 32.

34. The tangents to the parabola $y^2 = 4ax$ at the points $(a, 2a), (a, -2a)$ are $y = x + a$ and $y = -x - a$. The third side of the triangle is $x = a$. Clearly, these lines form a right-angled triangle whose two sides are equal.

35. Any tangent to the parabola $y^2 = 4x$ is $y = mx + \frac{1}{m}\cdot$ It passes through (α, β) if

$$\beta = m\alpha + \frac{1}{m} \quad \text{or} \quad \alpha m^2 - \beta m + 1 = 0.$$

It will have roots $m_1, 2m_1$ if

$$m_1 + 2m_1 = \frac{\beta}{\alpha},\ m_1 \cdot 2m_1 = \frac{1}{\alpha} \quad \Rightarrow 2\cdot\left(\frac{\beta}{3\alpha}\right)^2 = \frac{1}{\alpha}\cdot$$

36. $y^2 + 4 = 4x \quad \text{or} \quad y^2 = 4(x-1).$

Any tangent to it is $y = m(x-1) + \frac{1}{m}\cdot$ It is through the origin if $0 = -m + \frac{1}{m} \quad \Rightarrow m = 1, -1.$

So, the two tangents are inclined at $\frac{\pi}{2}\cdot$

37. Here the lines are two tangents to the parabola $y^2 = 4ax$ passing through the point $(-a, -b)$, their slopes being m_1 and m_2. But $(-a, -b)$ is on the directrix $x + a = 0$. So, the tangents are at right angles.

38. Let the parabola be $y^2 = 4ax$. Q is the intersection of the lines $x = 0$ and $ty = x + at^2$, where $P = (at^2, 2at)$. Solving these, $Q = (0, at)$. Also $S = (a, 0)$.

$$\therefore \quad SP^2 = \{a(t^2-1)\}^2 + 4a^2t^2 = a^2(t^2+1)^2$$

$$SQ^2 = a^2 + a^2t^2 = a^2(t^2+1) \quad \text{and} \quad SV = a$$

$$\therefore \quad SQ^2 = SP \cdot SV.$$

39. Any tangent to $y^2 = 4ax$ is $y = mx + \frac{a}{m}\cdot$ It touches $x^2 = 4ay$ if $x^2 = 4a\left(mx + \frac{a}{m}\right)$, i.e., $x^2 - 4amx - \frac{4a^2}{m} = 0$ has equal roots.

So, $m^3 + 1 = 0$, i.e., $m = -1$. Hence, the common tangent is $y = -x - a$.

40. The equation of the chord is $y - 2t_1 = \frac{2}{t_1 + t_2}(x - t_1^2)$.

The equation of the tangent at 't_1' is $y \cdot 2t_1 = 2(x + t_1^2)$. Therefore, the slope of the normal at 't_1' is $-t_1$. So, the chord will be the normal at 't_1' if

$$\frac{2}{t_1 + t_2} = -t_1.$$

41. See the hint to Q. No. 40.

42. The equation of the normal at $(-2 + t^2, 2t)$ is

$$y - 2t = \frac{-1}{\left(\frac{dy}{dx}\right)_{(-2+t^2,\,2t)}} \cdot (x + 2 - t^2) \quad \text{or} \quad y - 2t = -\frac{2t}{2} \cdot (x + 2 - t^2)$$

or $\quad tx + y = 2t - 2t + t^3 \quad$ or $\quad tx + y = t^3$.

It passes through $(k, 0)$ if $tk = t^3$.

or $\quad t(t^2 - k) = 0.$

It has three real values of t if $k > 0$. So, the set of points $(k, 0)$ is such that $k > 0$.

43. Here, $y^2 - 2y + 1 = 4(x + 1) \quad$ or $\quad (y - 1)^2 = 4(x + 1)$.

So, the axis is $y - 1 = 0$. Also $(-2, 1)$ lies on the axis, and it is exterior to the parabola because $1^2 - 4(-2) - 2(1) - 3 > 0$. Hence, only one normal is possible.

44. Any normal to $y^2 = 4x$ at $(t^2, 2t)$ is $y + tx = 2t + t^3$. Comparing this with $y = x + k$, we get

$$\frac{t}{-1} = \frac{2t + t^3}{k} \quad \text{or} \quad 2 + t^2 = -k \quad \therefore\ k \text{ is negative.}$$

45. The sum of the ordinates of the feet of three normals $= y_1 + y_2 + y_3 = 0$.

46. An end of the latus rectum $= (a, 2a)$. The equation of the tangent at $(a, 2a)$ is $y \cdot 2a = 2a(x + a)$, i.e., $y = x + a$. The normal at $(a, 2a)$ is $y + x = 2a + a$, i.e., $y + x = 3a$. Solving $y = 0$ and $y = x + a$, we get $x = -a, y = 0$. Solving $y = 0, y + x = 3a$, we get $x = 3a, y = 0$. The area of the triangle with vertices $(a, 2a), (-a, 0), (3a, 0)$ is $4a^2$.

47. The sum of the ordinates of the feet $= y_1 + y_2 + y_3 = 0$

$$\therefore\ 2 + (-2) + y_3 = 0 \quad \therefore\ y_3 = 0.$$

48. y-coordinate of the centroid $= \dfrac{y_1 + y_2 + y_3}{3} = \dfrac{0}{3} = 0.$

49. Vertex $= (0, 0)$. The ends of latus rectum are $(2, 4), (2, -4)$.

$\therefore$ centre $= (0, 0)$ radius $= \sqrt{2^2 + 4^2} = \sqrt{20}$.

50. Solving the equations, $x^2 + \frac{3(x+1)}{2} + 2x = 0$ or $x^2 + \frac{7x}{2} + \frac{3}{2} = 0$

or $2x^2 + 7x + 3 = 0$ or $(2x+1)(x+3) = 0$ or $x = -\frac{1}{2}, -3$.

But $x = -3$ makes y imaginary because $2y^2 = 3(x+1)$.

So, $x = -\frac{1}{2} \Rightarrow y = \pm\frac{\sqrt{3}}{2}$.

$\therefore$ the length of the chord = the distance between $\left(-\frac{1}{2}, \frac{\sqrt{3}}{2}\right)$ and $\left(-\frac{1}{2}, -\frac{\sqrt{3}}{2}\right)$.

51.

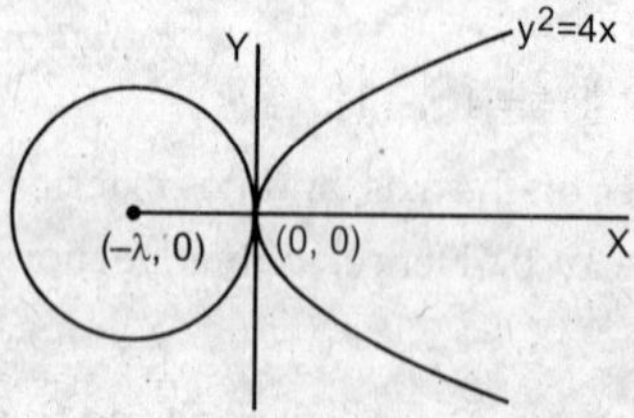

The graph shows $\lambda > 0$.

52. Here, $\alpha = \frac{at_1^2 + at_2^2}{2}$, $\beta = \frac{2at_1 + 2at_2}{2} = a(t_1 + t_2)$.

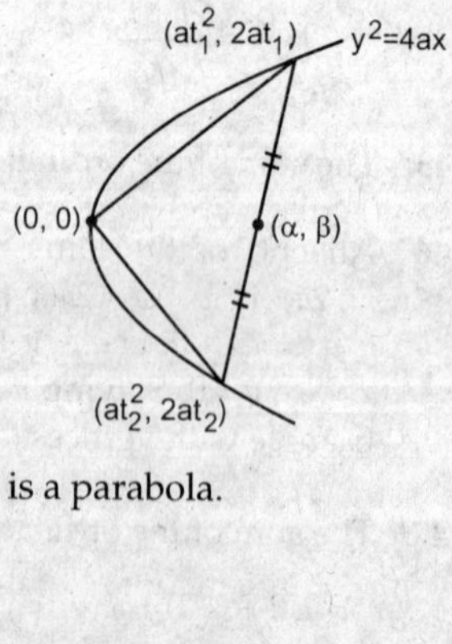

Also, $\frac{2at_1 - 0}{at_1^2 - 0} \times \frac{2at_2 - 0}{at_2^2 - 0} = -1 \Rightarrow t_1 t_2 = -4$.

Now, $\frac{2\alpha}{a} = t_1^2 + t_2^2 = (t_1 + t_2)^2 - 2t_1 t_2$

$= \left(\frac{\beta}{a}\right)^2 - 2 \cdot (-4)$.

$\therefore$ the locus of (α, β) is $\frac{2x}{a} = \frac{y^2}{a^2} + 8$, which is a parabola.

53. The locus is the directrix, which is a line.

54. Let $P = (\alpha, \beta)$. Any tangent to the parabola is $y = mx + \frac{a}{m}$. It passes through (α, β). So,

$$\beta = m\alpha + \frac{1}{m} \quad (\because \text{ here } a = 1); \qquad \therefore \quad m^2\alpha - \beta m + 1 = 0.$$

Its roots are $m_1, 3m_1$. So, $m_1 + 3m_1 = \frac{\beta}{\alpha}$, $m_1 \cdot 3m_1 = \frac{1}{\alpha}$.

$$\therefore \quad 3 \cdot \left(\frac{\beta}{4\alpha}\right)^2 = \frac{1}{\alpha} \quad \text{or} \quad 3\beta^2 = 16\alpha.$$

Thus, the locus is $3y^2 = 16x$, which is a parabola.

55. The locus of the middle points of the parallel chords is a line parallel to the axis of the parabola.

56. If the middle point of a chord is (α, β) then $\alpha = \frac{2t^2 + 0}{2}$, $\beta = \frac{4t + 0}{2}$.

Eliminating t, $\alpha = \left(\frac{\beta}{2}\right)^2$. So, the locus is $y^2 = 4x$.

60. 'm' of PV $= \frac{2t - 0}{t^2 - 0} = \frac{2}{t}$

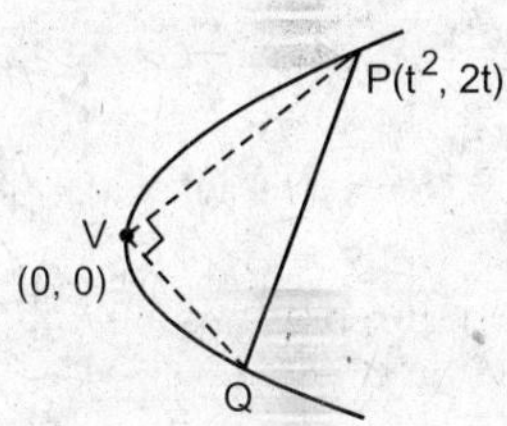

$\therefore$ the equation of QV is $y = -\frac{t}{2}x$.

Solving it with $y^2 = 4x$, $\text{Q} = \left(\frac{16}{t^2}, \frac{-8}{t}\right)$.

Now, ar $(\triangle \text{PVQ}) = \frac{1}{2} PV \cdot VQ = 20$ (given)

$$\therefore \quad PV^2 \cdot VQ^2 = 40^2$$

$$\text{or} \quad \{(t^2)^2 + (2t)^2\} \left\{\left(\frac{16}{t^2}\right)^2 + \left(\frac{-8}{t}\right)^2\right\} = 40^2$$

$$\text{or} \quad t^2(4 + t^2)\frac{1}{t^2}\left(\frac{256}{t^2} + 64\right) = 40^2$$

$$\text{or} \quad \frac{256 \times 4}{t^2} + 256 + 256 + 64t^2 = 40^2$$

$$\text{or} \quad (t^2 - 16)(t^2 - 1) = 0.$$

$$\therefore \quad t = \pm 4, \pm 1.$$

61. Any tangent to the parabola is $y = mx + \frac{9/4}{m}$. It passes through (4, 10).

$\therefore \quad 10 = 4m + \frac{9}{4m} \quad \Rightarrow m = \frac{1}{4}, \frac{9}{4}.$

62.

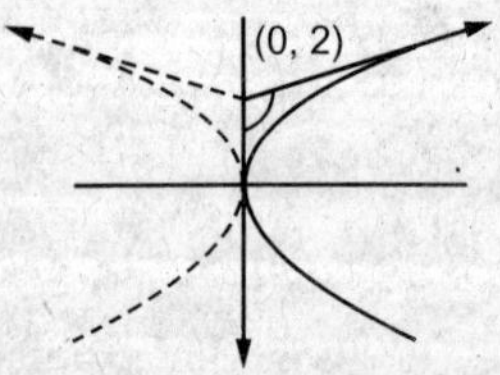

Any tangent is $y = mx + \frac{a}{m}\cdot$ It passes through (0, 2) if $2 = \frac{a}{m} + 0 \cdot m.$

$\therefore \quad m = \infty, \frac{a}{2}.$

Now, $\frac{3\pi}{4} = \frac{\pi}{2} \pm \tan^{-1}\frac{a}{2}$

$\therefore \quad a = 2, -2.$

63. Let $(x_1, y_1) = (at_1^2, 2at_1)$, $(x_2, y_2) = (at_2^2, 2at_2)$. Then
$(x_3, y_3) = \left(at_1t_2, a(t_1 + t_2)\right)$

$\therefore \quad x_1 \cdot x_2 = at_1^2 \cdot at_2^2 = (at_1t_2)^2 = x_3^2$ and $y_3 = a(t_1 + t_2) = \frac{1}{2}(y_1 + y_2),$

64. Here, $y^2 + 2by + b^2 = b^2 - c - 2ax$

or $\quad (y + b)^2 = -2a \cdot \left(x - \frac{b^2 - c}{2a}\right).$

It represents a parabola and its latus rectum $= 2a$.

65. If the point of contact is $(at^2, 2at)$ then the tangent is $y \cdot 2at = 2a(x + at^2)$.

$\therefore \quad 'm' = \frac{1}{t}\cdot$ From the question, $\frac{1}{t} = \tan\left(\pm\frac{\pi}{3}\right) = \pm\sqrt{3}.$

66.

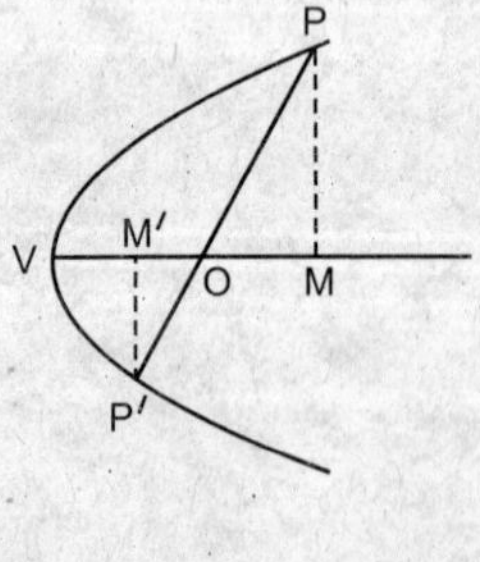

Let the parabola be $y^2 = 4ax$ and $P = (at_1^2, 2at_1)$, $P' = \left(at_2^2, 2at_2\right)$

$\therefore \quad VM = at_1^2$, $VM' = at_2^2$ and $VO = k$, where

$$\begin{vmatrix} a_1^2 & 2at_1 & 1 \\ at_2^2 & 2at_2 & 1 \\ k & 0 & 1 \end{vmatrix} = 0$$

($\because$ P, P′, O are collinear)

$\Rightarrow \quad k + at_1t_2 = 0$

$\therefore \quad VM \cdot VM' = (at_1t_2)^2 = k^2 = OV^2.$

67. Solving the equations, $x^2 + 5x - 6 = 0$, i.e., $x = 1, -6$.

But x cannot be negative. So $x = 1$. Then $y = 3, -3$. So, the common chord is the line segment joining (1, 3) and (1, −3).

68. Any tangent to $y^2 = 2x$ is $y = mx + \frac{1/2}{m}$ ... (1)

For the circle, centre = (−2, 0) and radius = 2.

$\therefore$ (1) is a common tangent if $\left|\dfrac{-2m + \frac{1}{2m}}{\sqrt{1+m^2}}\right| = 2$

or $\left(-2m + \dfrac{1}{2m}\right)^2 = 4(1+m^2)$ or $4m^2 + \dfrac{1}{4m^2} - 2 = 4 + 4m^2$

or $m^2 = \dfrac{1}{24}$, $\therefore$ $m = \pm\dfrac{1}{2\sqrt{6}}$.

So, the common tangents are $y = \pm\dfrac{1}{2\sqrt{6}}x \pm \sqrt{6}$.

Note There are 3 common tangents, the third one being $x = 0$, which corresponds to $m = \infty$.

69. With choice of axes as shown in the figure, the equations of the parabolas can be taken as $y^2 = 4a(x-k)$ and $y^2 = -4b(x+k)$.

A line parallel to the common axis is $y = h$.

Then $A = \left(\dfrac{h^2}{4a} + k, h\right)$ and

$B = \left(-k - \dfrac{h^2}{4b}, h\right)$.

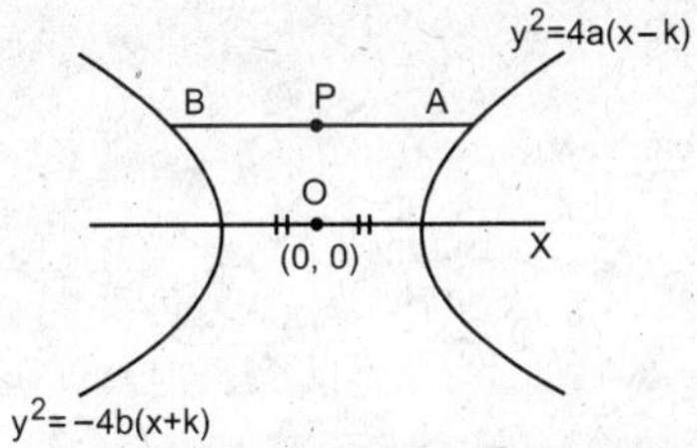

$\therefore$ if $P = (\alpha, \beta)$ then

$$\alpha = \frac{1}{2}\left(\frac{h^2}{4a} + k - k - \frac{h^2}{4b}\right), \beta = h$$

$$\therefore \quad 2\alpha = \frac{h^2}{4}\left(\frac{1}{a} - \frac{1}{b}\right)$$

$\therefore$ the locus of P is $2x = \dfrac{y^2}{4}\left(\dfrac{b-a}{ab}\right)$.

70. If $P = (x, y)$ then

$$\sqrt{x^2+y^2} < |a-x|, \sqrt{x^2+y^2} < |a+x|,$$

$$\sqrt{x^2+y^2} < |a-y|, \sqrt{x^2+y^2} < |a+y|.$$

$\therefore$ the region is bounded by the curves

$$x^2+y^2=(a-x)^2,\ x^2+y^2=(a+x)^2,$$
$$x^2+y^2=(a-y)^2,\ x^2+y^2=(a+y)^2.$$

□

5

Ellipse and Hyperbola

• Type 1 •

Choose the most appropriate option (a, b, c or d).

1. The equation $2x^2 - 3xy + 5y^2 + 6x - 3y + 5 = 0$ represents

(a) a parabola (b) an ellipse

(c) a hyperbola (d) a pair of straight lines

2. The set of real values of k for which the equation

$$(k+1)x^2 + 2(k-1)xy + y^2 - x + 2y + 3 = 0$$

represents an ellipse is

(a) $(0, 3)$ (b) $(-\infty, 0)$

(c) $(3, +\infty)$ (d) $(-\infty, \infty)$

3. The centre of the conic section $14x^2 - 4xy + 11y^2 - 44x - 58y + 71 = 0$ is

(a) $(2, 3)$ (b) $(2, -3)$

(c) $(-2, 3)$ (d) $(-2, -3)$

4. The eccentricity of the ellipse $\frac{x^2}{4} + \frac{y^2}{9} = 1$ is

(a) $\frac{\sqrt{5}}{2}$ (b) $\frac{2}{3}$ (c) $\frac{\sqrt{5}}{3}$ (d) $\frac{4}{9}$

5. The eccentricity of the hyperbola $x^2 - 4y^2 = 16$ is

(a) 2 (b) $\frac{\sqrt{5}}{2}$ (c) 4 (d) $\frac{\sqrt{3}}{2}$

6. The eccentricity of the conic section $4(x^2 - y^2) = 1$ is

(a) $\sqrt{2}$ (b) 2

(c) 4 (d) $\frac{1}{4}$

7. The latus rectum of the conic section $\frac{x^2}{a^2} + \frac{y^2}{b^2} = 1$ whose eccentricity $= e$, is

(a) $\frac{2a^2}{b}$ (b) $\frac{2b}{a^2}$

(c) $2a(1 - e^2)$ (d) $2b(1 - e^2)$

8. The ellipse $\frac{x^2}{a^2} + \frac{y^2}{b^2} = 1$ passes through the point (–3, 1) and has the eccentricity $\sqrt{\frac{2}{5}}$. Then the major axis of the ellipse has the length

(a) $4 \cdot \sqrt{\frac{2}{5}}$ (b) $8 \cdot \sqrt{\frac{2}{3}}$

(c) $4 \cdot \sqrt{\frac{2}{3}}$ (d) $8 \cdot \sqrt{\frac{2}{5}}$

9. The hyperbola $\frac{x^2}{a^2} - \frac{y^2}{b^2} = 1$ passes through the point (2, 3) and has the eccentricity 2. Then the transverse axis of the hyperbola has the length

(a) 1 (b) 3

(c) 2 (d) 4

10. In the ellipse $x^2 + 3y^2 = 9$ the distance between the foci is

(a) $\sqrt{6}$ (b) 3

(c) $\frac{2}{3}\sqrt{6}$ (d) $2\sqrt{6}$

11. The minor axis of the ellipse $9x^2 + 5y^2 = 30y$ is

(a) 6 (b) $2\sqrt{5}$

(c) $\sqrt{6}$ (d) $\sqrt{5}$

12. The foci of the ellipse $25x^2 + 36y^2 = 225$ are

(a) $\left(\pm\frac{1}{2}\sqrt{11}, 0\right)$ (b) $\left(\pm\frac{5}{2}, 0\right)$

(c) $\left(0, \pm\frac{1}{2}\sqrt{11}\right)$ (d) $\left(0, \pm\frac{5}{2}\right)$

13. If the eccentricity of the hyperbola $\frac{x^2}{a^2}-\frac{y^2}{b^2}=1$ is e then the eccentricity of the hyperbola $\frac{y^2}{b^2}-\frac{x^2}{a^2}=1$ is

(a) e (b) $\frac{e}{\sqrt{e^2-1}}$

(c) $e\sqrt{e^2-1}$ (d) e^2-e

14. If in an ellipse the minor axis = the distance between the foci and its latus rectum = 10 then the equation of the ellipse in the standard form is

(a) $\frac{x^2}{(10)^2}+\frac{y^2}{(5\sqrt{2})^2}=1$ (b) $\frac{x^2}{(5\sqrt{2})^2}+\frac{y^2}{(10)^2}=1$

(c) $\frac{x^2}{25}+\frac{y^2}{(5/\sqrt{2})^2}=1$ (d) none of these

15. If in a hyperbola the eccentricity is $\sqrt{3}$, and the distance between the foci is 9 then the equation of the hyperbola in the standard form is

(a) $\frac{x^2}{\left(\frac{\sqrt{3}}{2}\right)^2}-\frac{y^2}{\left(\sqrt{\frac{3}{2}}\right)^2}=1$ (b) $\frac{x^2}{\left(\frac{3\sqrt{3}}{2}\right)^2}-\frac{y^2}{\left(\frac{3\sqrt{3}}{\sqrt{2}}\right)^2}=1$

(c) $\frac{x^2}{\left(\frac{3\sqrt{3}}{\sqrt{2}}\right)^2}-\frac{y^2}{\left(\frac{3\sqrt{2}}{2}\right)^2}=1$ (d) none of these

16. If in an ellipse, a focus is (6, 7), the corresponding directrix is $x+y+2=0$ and the eccentricity $=\frac{1}{2}$ then the equation of the ellipse is

(a) $7x^2+2xy+7y^2-44x-108y+684=0$

(b) $7x^2-2xy+7y^2-52x-116y+676=0$

(c) $9x^2-2xy+9y^2-44x-108y+684=0$

(d) none of these

17. If for a rectangular hyperbola a focus is (1, 2) and the corresponding directrix is $x+y=1$ then the equation of the rectangular hyperbola is

(a) $x^2-y^2=2$ (b) $xy-y+2=0$

(c) $xy+y-2=0$ (d) none of these

18. If two foci of an ellipse be (–2, 0) and (2, 0) and its eccentricity is $\frac{2}{3}$ then the ellipse has the equation

(a) $5x^2+9y^2=45$ (b) $9x^2+5y^2=45$

(c) $5x^2+9y^2=90$ (d) $9x^2+5y^2=90$

19. If for a conic section a focus is (–1, 1), eccentricity = 3 and the equation of the corresponding directrix is $x-y+3=0$ then the equation of the conic section is

(a) $7x^2-18xy+7y^2+50x-50y+77=0$

(b) $7x^2+18xy+7y^2=1$

(c) $7x^2+18xy+7y^2-50x+50y+77=0$

(d) none of these

20. An ellipse having foci at (3, 1) and (1, 1) passes through the point (1, 3). Its eccentricity is

(a) $\sqrt{2}-1$ (b) $\sqrt{3}-1$

(c) $\frac{1}{2}(\sqrt{2}-1)$ (d) $\frac{1}{2}(\sqrt{3}-1)$

21. A point on the ellipse $\frac{x^2}{6}+\frac{y^2}{2}=1$ at a distance 2 from the centre of the ellipse has the eccentric angle

(a) $\frac{\pi}{4}$ (b) $\frac{\pi}{3}$

(c) $\frac{\pi}{6}$ (d) $\frac{\pi}{2}$

22. A point P on the ellipse $\frac{x^2}{25}+\frac{y^2}{9}=1$ has the eccentric angle $\frac{\pi}{8}$. The sum of the distances of P from the two foci is

(a) 5 (b) 6

(c) 10 (d) 3

23. If any point on a hyperbola has the coordinates (5tan ϕ, 4sec ϕ) then the eccentricity of the hyperbola is

(a) $\frac{5}{4}$ (b) $\frac{\sqrt{41}}{5}$

(c) $\frac{25}{16}$ (d) $\frac{\sqrt{41}}{4}$

24. The slope of the diameter of the ellipse $\frac{x^2}{a^2}+\frac{y^2}{b^2}=1$, whose length is the GM of the major and minor axes, is

(a) $\sqrt{\frac{a}{b}}$ (b) $\sqrt{ab}$

(c) $\sqrt{\frac{b}{a}}$ (d) $\frac{a}{b}$

25. PP′ is a diameter of the ellipse $b^2x^2+a^2y^2=a^2b^2$ such that PP'^2 is the AM of the squares of the major and minor axes. Then the slope of PP′ is

(a) $\frac{b}{a}$ (b) $\frac{a}{b}$

(c) $\frac{\pi}{4}$ (d) $\frac{\pi}{3}$

26. P is a variable point on the ellipse $\frac{x^2}{a^2}+\frac{y^2}{b^2}=2$ whose foci are F_1 and F_2. The maximum area (in unit^2) of the ΔPFF′ is

(a) $2b\sqrt{a^2-b^2}$ (b) $\sqrt{2}b\sqrt{a^2-b^2}$

(c) $b\sqrt{a^2-b^2}$ (d) $2a\sqrt{a^2-b^2}$

27. Which of the following points is an exterior point of the ellipse $16x^2+9y^2-16x-32=0$?

(a) $\left(\frac{1}{2}, 2\right)$ (b) $\left(\frac{1}{4}, 1\right)$

(c) $(3, -2)$ (d) none of these

28. For the hyperbola $\frac{x^2}{\cos^2\alpha}-\frac{y^2}{\sin^2\alpha}=1$, which of the following remains constant when α varies?

(a) abscissae of vertices (b) abscissae of foci

(c) eccentricity (d) directrix

29. The foci of the ellipse $\frac{x^2}{16}+\frac{y^2}{b^2}=1$ and the hyperbola $\frac{x^2}{144}-\frac{y^2}{81}=\frac{1}{25}$ coincide. Then the value of b^2 is

(a) 5 (b) 7 (c) 9 (d) 1

30. The equation of the tangent to the ellipse $4x^2+3y^2=12$ at the point whose eccentric angle is $\frac{\pi}{4}$ is

(a) $\sqrt{3}x+2y=2\sqrt{6}$
(b) $2x+\sqrt{3}y=2\sqrt{6}$
(c) $2x-\sqrt{3}y=2\sqrt{6}$
(d) none of these

31. The number of values of m for which the line $y=mx+\sqrt{m^2-4}$ touches the hyperbola $4(x^2-1)=y^2$ is

(a) two
(b) zero
(c) one
(d) infinite

32. The value of c for which the line $y=3x+c$ touches the ellipse $16x^2+y^2=16$ is

(a) 5
(b) 1
(c) 4
(d) 3

33. The number of values of $\phi \in [0, 2\pi]$ for which the line $2x\cos\phi+3y\sin\phi=6$ touches the ellipse $4x^2+9y^2=36$ is

(a) four
(b) two
(c) one
(d) infinite

34. The line $3x+5y=k$ is a tangent to the ellipse $16x^2+25y^2=400$ if k is

(a) $\pm\sqrt{5}$
(b) $\pm\sqrt{15}$
(c) ± 25
(d) $\pm\sqrt{10}$

35. The line $px+qy=r$ touches the hyperbola $b^2x^2-a^2y^2=a^2b^2$ if

(a) $a^2p^2+b^2q^2=r^2$
(b) $a^2p^2-b^2q^2=r^2$
(c) $a^2q^2+b^2p^2=r^2$
(d) $a^2q^2-b^2p^2=r^2$

36. The equation of the tangent to the ellipse $\frac{x^2}{25}+\frac{y^2}{16}=1$, which is parallel to the line $y=3x$, is

(a) $y=3x+\sqrt{241}$
(b) $y=3x+13$
(c) $y=3x+\sqrt{209}$
(d) none of these

37. The equation of a tangent to the hyperbola $x^2-2y^2=18$, which is perpendicular to the line $x-y=0$, is

(a) $x+y=3$
(b) $x+y+2=0$
(c) $x+y=3\sqrt{2}$
(d) $x+y+3\sqrt{2}=0$

38. If the tangents from the point $(\lambda, 3)$ to the ellipse $\frac{x^2}{9}+\frac{y^2}{4}=1$ are at right angles then λ is

(a) ± 1 (b) ± 3

(c) ± 2 (d) none of these

39. A point on the ellipse $x^2+3y^2=9$, where the tangent is parallel to the line $y-x=0$, is

(a) $(\sqrt{3}, \sqrt{2})$ (b) $\left(-\frac{3\sqrt{3}}{2}, -\frac{\sqrt{3}}{2}\right)$

(c) $\left(-\frac{3\sqrt{3}}{2}, \frac{\sqrt{3}}{2}\right)$ (d) $(-\sqrt{3}, \sqrt{2})$

40. The ordinate of the point of contact of a tangent is 2. Then the equation of the tangent to $x^2+4y^2=25$ is

(a) $3x+8y=25$ (b) $8x+3y=25$

(c) $3x-8y=25$ (d) none of these

41. The tangent to the ellipse $16x^2+9y^2=144$, making equal intercepts on both the axes, is

(a) $y=x+3$ (b) $y=x-2$

(c) $x+y=5$ (d) $y=-x+4$

42. If the tangent to the ellipse $x^2+4y^2=16$ at the point 'ϕ' is a normal to the circle $x^2+y^2-8x-4y=0$ then ϕ is equal to

(a) $\frac{\pi}{2}$ (b) $\frac{\pi}{4}$

(c) $\frac{\pi}{3}$ (d) $-\frac{\pi}{4}$

43. The area of the quadrilateral formed by tangents at the end points of latus recta of the ellipse $\frac{x^2}{9}+\frac{y^2}{5}=1$ is

(a) $\frac{27}{4}$ unit^2 (b) 9 unit^2

(c) $\frac{27}{2}$ unit^2 (d) 27 unit^2

44. The tangent at $(3\sqrt{3}\cos\theta, \sin\theta)$ is drawn to the ellipse $\frac{x^2}{27}+y^2=1$. Then the value of θ such that the sum of intercepts on axes made by the tangent is minimum is

(a) $\frac{\pi}{3}$ (b) $\frac{\pi}{6}$

(c) $\frac{\pi}{8}$ (d) $\frac{\pi}{4}$

45. The number of normals that can be drawn to the curve $4x^2+9y^2=36$ from an external point, in general, is

(a) 1 (b) 3

(c) 4 (d) infinite

46. The number of distinct normal lines from the exterior point $(0, c)$, $c > b$, to the ellipse $\frac{x^2}{a^2}+\frac{y^2}{b^2}=1$ is

(a) 3 (b) 4

(c) 2 (d) 1

47. The equation of the normal to the ellipse $x^2+4y^2=16$ at the end of the latus rectum in the first quadrant is

(a) $2x+\sqrt{3}(y+3)=0$ (b) $2x=\sqrt{3}(y+3)$

(c) $\sqrt{3}x=2(y+3)$ (d) none of these

48. If the tangent and the normal to $x^2-y^2=4$ at a point cut off intercepts a_1, a_2 on the x-axis respectively and b_1, b_2 on the y-axis respectively then the value of $a_1a_2+b_1b_2$ is

(a) 1 (b) –1

(c) 0 (d) 4

49. The normal to the rectangular hyperbola $xy=c^2$ at the point 't_1' meets the curve again at the point 't_2'. The value of $t_1^3 \cdot t_2$ is

(a) 1 (b) c

(c) $-c$ (d) –1

50. If P and Q are the ends of a pair of conjugate diameters and C is the centre of the ellipse $4x^2+9y^2=36$ then the area of the ΔCPQ is

(a) 6 unit^2 (b) 3 unit^2 (c) 2 unit^2 (d) 12 unit^2

51. If $y = x$ and $3y + 2x = 0$ are the equations of a pair of conjugate diameters of the ellipse $\frac{x^2}{a^2} + \frac{y^2}{b^2} = 1$ then its eccentricity is

(a) $\frac{1}{2}$ (b) $\frac{1}{3}$

(c) $\frac{1}{\sqrt{3}}$ (d) $\frac{\sqrt{3}}{2}$

52. The locus of a point $P(\alpha, \beta)$ moving under the condition that the line $y = \alpha x + \beta$ is a tangent to the hyperbola $\frac{x^2}{a^2} - \frac{y^2}{b^2} = 1$ is

(a) a circle (b) an ellipse

(c) a hyperbola (d) a parabola

• Type 2 •

Choose the correct options. One or more options may be correct.

53. A focus of the hyperbola $25x^2 - 36y^2 = 225$ is

(a) $(\sqrt{61}, 0)$ (b) $\left(\frac{1}{2}\sqrt{61}, 0\right)$

(c) $(-\sqrt{61}, 0)$ (d) $\left(-\frac{1}{2}\sqrt{61}, 0\right)$

54. The point P on the ellipse $4x^2 + 9x^2 = 36$ is such that the area of the $\Delta PF_1F_2 = \sqrt{10}$ where F_1, F_2 are foci. Then P has the coordinates

(a) $\left(\frac{3}{\sqrt{2}}, \sqrt{2}\right)$ (b) $\left(\frac{3}{2}, 2\right)$

(c) $\left(-\frac{3}{2}, -2\right)$ (d) $\left(-\frac{3}{\sqrt{2}}, -\sqrt{2}\right)$

55. A point on the ellipse $x^2 = 3y^2 = 37$, where the normal is parallel to the line $6x - 5y = 2$, is

(a) $(5, -2)$ (b) $(5, 2)$

(c) $(-5, 2)$ (d) $(-5, -2)$

Answers

1. b	**2.** a	**3.** a	**4.** c	**5.** b
6. a	**7.** c	**8.** b	**9.** c	**10.** d
11. b	**12.** a	**13.** b	**14.** a	**15.** b
16. b	**17.** c	**18.** a	**19.** a	**20.** a
21. a	**22.** c	**23.** d	**24.** c	**25.** a
26. a	**27.** c	**28.** b	**29.** b	**30.** b
31. d	**32.** a	**33.** d	**34.** c	**35.** b
36. a	**37.** a	**38.** c	**39.** c	**40.** a
41. c	**42.** a	**43.** d	**44.** b	**45.** c
46. d	**47.** b	**48.** c	**49.** d	**50.** b
51. c	**52.** c	**53.** b, d	**54.** a, d	**55.** b, d

Hints and Solutions to Selected Questions

2. For ellipse, $ab-h^2>0 \Rightarrow (k+1)\cdot 1-(k-1)^2>0$ or $k(k-3)<0$.

3. For centre, $\frac{\partial S}{\partial x}=0,\ \frac{\partial S}{\partial y}=0 \Rightarrow 7x-y-11=0,\ 2x-11y+29=0$. Solve.

7. Latus rectum $=\frac{2b^2}{a}=\frac{2}{a}\cdot a^2(1-e^2)$.

9. $\frac{4}{a^2}-\frac{9}{b^2}=1$ and $b^2=a^2(e^2-1)=a^2\cdot 3$. So, $\frac{4}{a^2}-\frac{9}{3a^2}=1 \Rightarrow a^2=1$.

Transverse axis $=2a=2$.

10. $\frac{x^2}{9}+\frac{y^2}{3}=1 \Rightarrow a^2=9, b^2=3$.

So, $b^2=a^2(1-e^2) \Rightarrow 3=9(1-e^2) \Rightarrow e=\sqrt{\frac{2}{3}}$.

$\therefore$ distance between foci $=2ae=2\cdot 3\sqrt{\frac{2}{3}}$.

14. $2b=2ae,\ \frac{2b^2}{a}=10,\ b^2=a^2(1-e^2) \Rightarrow e=\frac{1}{\sqrt{2}},\ a=10,\ b^2=50$.

15. For rectangular hyperbola, $e=\sqrt{2}$. So, $(x-1)^2+(y-2)^2=\left(\sqrt{2}\cdot\frac{x+y-1}{\sqrt{2}}\right)^2$.

18. Here, the situation is standard. The distance between foci $=2ae=4$ and $e=\frac{2}{3}$.

$\Rightarrow a=3$. So, $b^2=a^2(1-e^2)=9\cdot\left(1-\frac{4}{9}\right)=5$.

20. The distance between foci $=2ae=\sqrt{(3-1)^2+(1-1)^2}=2$. $\therefore$ $ae=1$.

The sum of the distances of foci from $(1,3)$ on the ellipse $=2a$

$\Rightarrow \sqrt{(3-1)^2+(1-3)^2}+\sqrt{(1-1)^2+(1-3)^2}=2a \Rightarrow 2a=2\sqrt{2}+2$

$\therefore \quad e=\frac{1}{a}=\frac{2}{2\sqrt{2}+2}=\frac{1}{\sqrt{2}+1}=\sqrt{2}-1$.

21. Any point on the ellipse $=(\sqrt{6}\cos\phi, \sqrt{2}\sin\phi)$. So,

$(\sqrt{6}\cos\phi-0)^2+(\sqrt{2}\sin\phi-0)^2=4 \Rightarrow \cos\phi=\pm\frac{1}{\sqrt{2}}$.

22. The sum of distances of P from the foci $= 2a = 2 \times 5 = 10$.

23. $x = 5\tan\phi, y = 4\sec\phi \Rightarrow \frac{y^2}{16} - \frac{x^2}{25} = 1 \Rightarrow a^2 = 16, b^2 = 25.$

Also, $b^2 = a^2(e^2 - 1)$.

24. The length of the diameter $= \sqrt{2a \cdot 2b} = 2\sqrt{ab}$.

Let $P(a\cos\phi, b\sin\phi)$ be one end of the diameter.

Then $2\{(a\cos\phi - 0)^2 + (b\sin\phi - 0)^2\} = 2\sqrt{ab}$

or $a^2\cos^2\phi + b^2\sin^2\phi = ab$. The slope of the diameter $= \frac{b\sin\phi - 0}{a\cos\phi - 0} = \frac{b}{a}\tan\phi$.

Now, $a^2 + b^2\tan^2\phi = ab\sec^2\phi = ab(1 + \tan^2\phi)$.

$\therefore\ \tan^2\phi = \frac{ab - a^2}{b^2 - ab} = \frac{a(b-a)}{b(b-a)} = \frac{a}{b} \Rightarrow \frac{b^2}{a^2}\tan^2\phi = \frac{b}{a}. \quad \therefore\ (\text{slope})^2 = \frac{b}{a}.$

25. See Q. No. 24.

26. Let $P = (\sqrt{2}a\cos\phi, \sqrt{2}b\sin\phi)$. F_1 and $F_2 = (\pm\sqrt{2}ae, 0)$.

$$\text{The area of } \Delta PFF' = \left| \frac{1}{2} \begin{vmatrix} \sqrt{2}a\cos\phi & \sqrt{2}b\sin\phi & 1 \\ \sqrt{2}ae & 0 & 1 \\ -\sqrt{2}ae & 0 & 1 \end{vmatrix} \right| = \frac{1}{2} \cdot \sqrt{2}b\sin\phi \cdot \sqrt{2}ae$$

$$= 2abe\sin\phi.$$

$\therefore$ maximum area $= 2abc = 2ab \cdot \frac{\sqrt{a^2 - b^2}}{a}$.

27. (α, β) is an exterior point if $16\alpha^2 + 9\beta^2 - 16\alpha - 32 > 0$.

28. Here, $a = \cos\alpha, b = \sin\alpha$.

$b^2 = a^2(e^2 - 1) \Rightarrow \sin^2\alpha = \cos^2\alpha(e^2 - 1) \Rightarrow e^2 - 1 = \tan^2\alpha \Rightarrow e = \sec\alpha.$

So, $ae = 1$.

$\therefore$ abscissae of foci $= \pm ae = \pm 1$.

29. For the ellipse, $a^2 = 16;\ b^2 = a^2(1 - e^2)$

$\Rightarrow e = \frac{\sqrt{16 - b^2}}{4} \Rightarrow ae = \sqrt{16 - b^2}.$

For the hyperbola, $a^2 = \frac{144}{25},\ b^2 = \frac{81}{25};\ b^2 = a^2(e^2 - 1) \Rightarrow e = \frac{5}{4} \Rightarrow ae = 3.$

$\therefore\ \sqrt{16 - b^2} = 3.$

31. Here, $\frac{x^2}{1}-\frac{y^2}{4}=1$. Any tangent to the hyperbola is $y=mx\pm\sqrt{a^2m^2-b^2}$,

i.e., $y=mx\pm\sqrt{m^2-4}$ in which m is any real number.

32. $c=\sqrt{a^2m^2+b^2}$ where $a=1, b=4, m=3$.

38. The equation of the pair of tangents is $SS_1=T^2$

or $\left(\frac{x^2}{9}+\frac{y^2}{4}-1\right)\left(\frac{\lambda^2}{9}+\frac{9}{4}-1\right)=\left(\frac{\lambda x}{9}+\frac{3y}{4}-1\right)^2$.

For right angle, $a+b=0 \Rightarrow \left\{\frac{1}{9}\left(\frac{\lambda^2}{9}+\frac{5}{4}\right)-\frac{\lambda^2}{81}\right\}+\left\{\frac{1}{4}\left(\frac{\lambda^2}{9}+\frac{5}{4}\right)-\frac{9}{16}\right\}=0.$

39. The point is (x_1, y_1) if $xx_1+3yy_1=9$ has the slope 1, i.e., $-\frac{x_1}{3y_1}=1$ and $x_1^2+3y_1^2=9$. Solve the two equations.

41. Tangent at (x_1, y_1) is $16x_1x+9y_1y=144$. Intercepts on axes are equal

$\Rightarrow \frac{144}{16x_1}=\frac{144}{9y_1}$ or $9y_1=16x_1$. Also, $16x_1^2+9y_1^2=144$.

Solving these, $x_1=\frac{9}{5}, -\frac{9}{5}$ and $y_1=\frac{16}{5}, -\frac{16}{5}$.

So, the tangent is $16\times\frac{9}{5}x+9\times\frac{16}{5}y=144$ or $16\times\frac{-9}{5}x+9\times\frac{-16}{5}y=144$

$\Rightarrow x+y=5$ or $x+y+5=0$.

42. The tangent at $(4\cos\phi, 2\sin\phi)$ is $4\cos\phi\cdot x+4\cdot 2\sin\phi\, y=16$.
Being normal, it passes through the centre (4, 2).

So, $16\cos\phi+16\sin\phi=16$ or $\cos\phi+\sin\phi=1$. It is satisfied by $\phi=\frac{\pi}{2}$.

43. Here $a^2=9, b^2=5$. So, latus rectum $=\frac{2b^2}{a}=\frac{2\times 5}{3}$.

$\therefore$ $P=\left(\alpha, \frac{5}{3}\right)$. It is on the ellipse. So,

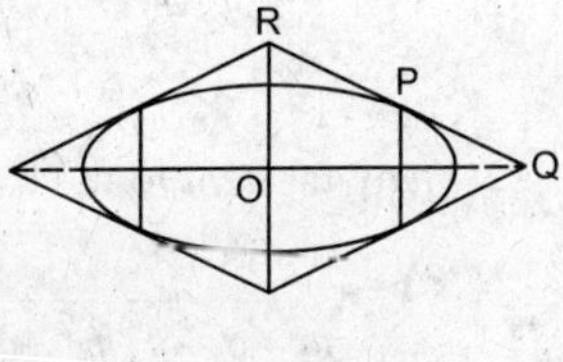

$\frac{\alpha^2}{9}+\frac{\left(\frac{5}{3}\right)^2}{5}=1$ or $\alpha=\pm 2$. So, $P=\left(2, \frac{5}{3}\right)$.

The tangent at P is $\frac{2x}{9}+\frac{\frac{5}{3}y}{5}=1$ or $\frac{x}{\frac{9}{2}}+\frac{y}{3}=1$.

$\therefore$ area $=4\times$ ar $(\Delta ROQ)=4\times\left(\frac{1}{2}\times\frac{9}{2}\times 3\right)=27$.

44. The tangent is $\frac{3\sqrt{3}\cos\phi \cdot x}{27} + \frac{\sin\phi \cdot y}{1} = 1$ or $\frac{x}{3\sqrt{3}\sec\theta} + \frac{y}{\operatorname{cosec}\theta} = 1.$

$\therefore$ z = sum of intercepts on axes = $3\sqrt{3}\sec\theta + \operatorname{cosec}\theta.$
For maximum or minimum,

$$\frac{dz}{d\theta} = 0 \Rightarrow \cot^3\theta = 3\sqrt{3} \Rightarrow \cot\theta = \sqrt{3} \Rightarrow \theta = \frac{\pi}{6}.$$

48. The tangent at $(2\sec\phi, 2\tan\phi)$ is $x\sec\phi - y\tan\phi = 2.$

$\therefore$ $a_1 = 2\cos\phi, b_1 = -2\cot\phi.$

The normal at $(2\sec\phi, 2\tan\phi)$ is $x\cos\phi + y\cot\phi = 8.$

$\therefore$ $a_2 = 8\sec\phi, b_2 = 8\tan\phi.$

$\therefore$ $a_1a_2 + b_1b_2 = 16\cos\phi\sec\phi + (-2\cot\phi)(8\tan\phi) = 16 - 16 = 0.$

50. The ellipse is $\frac{x^2}{3^2} + \frac{y^2}{2^2} = 1$. So, $P = (3\cos\phi, 2\sin\phi)$ and

$$Q = \left(3\cos\overline{\phi + \frac{\pi}{2}},\ 2\sin\overline{\phi + \frac{\pi}{2}}\right).$$

$$\therefore \text{ area of the } \Delta CPQ = \frac{1}{2}\begin{vmatrix} 0 & 0 & 1 \\ 3\cos\phi & 2\sin\phi & 1 \\ -3\sin\phi & 2\cos\phi & 1 \end{vmatrix}$$

$$= \frac{1}{2}(6\cos^2\phi + 6\sin^2\phi) = \frac{1}{2}\cdot 6 = 3.$$

51. We know that two diameters $y = m_1x, y = m_2x$ are conjugate diameters of $\frac{x^2}{a^2} + \frac{y^2}{b^2} = 1$ if $m_1m_2 = -\frac{b^2}{a^2}.$

Hence, we have $1\cdot\left(-\frac{2}{3}\right) = -\frac{b^2}{a^2}$ or $b^2 = \frac{2}{3}a^2.$

$\therefore$ $b^2 = a^2(1-e^2) \Rightarrow \frac{2}{3}a^2 = a^2(1-e^2) \Rightarrow \frac{2}{3} = 1 - e^2.$

52. Any tangent to the hyperbola is $y = mx + \sqrt{a^2m^2 - b^2}$. This is the same as $y = \alpha x + \beta.$

$\therefore$ $m = \alpha, \sqrt{a^2m^2 - b^2} = \beta \Rightarrow a^2\alpha^2 - \beta^2 = b^2$

$\therefore$ the locus is $a^2x^2 - y^2 = b^2.$

□

Part 4

Calculus

1

Functions

• Type 1 •

Choose the most appropriate option (a, b, c or d).

1. If $f(x) = x^2 + \lambda x + \mu$ be an integral function of the integral variable x then

(a) λ is an integer and μ is a rational fraction

(b) λ and μ are integers

(c) μ is an integer and λ is a rational fraction

(d) λ and μ are rational fractions

2. Let $f(x) = ax^2 + bx + c$, where a, b, c are rational, and $f: Z \to Z$, where Z is the set of integers. Then $a + b$ is

(a) a negative integer (b) an integer

(c) nonintegral rational number (d) none of these

3. If $f(x) = \cos[\pi]x + \cos[\pi x]$, where $[y]$ is the greatest integer function of y then $f\left(\frac{\pi}{2}\right)$ is equal to

(a) cos 3 (b) 0

(c) cos 4 (d) none of these

4. Let $f(x) = \sin(\tan^{-1} x)$. Then $[f(-\sqrt{3})]$, where $[\cdot]$ denotes the greatest integer function, is

(a) $-\frac{\sqrt{3}}{2}$ (b) 0

(c) -1 (d) none of these

5. If $f(x)=\dfrac{x-1}{x+1}$ then $f(ax)$ in term of $f(x)$ is equal to

(a) $\dfrac{f(x)+a}{1+af(x)}$

(b) $\dfrac{(a-1)f(x)+a+1}{(a+1)f(x)+a-1}$

(c) $\dfrac{(a+1)f(x)+a-1}{(a-1)f(x)+a+1}$

(d) none of these

6. Let $f(1)=1$ and $f(n)=2\sum_{r=1}^{n-1} f(r)$. Then $\sum_{n=1}^{m} f(n)$ is equal to

(a) 3^m-1

(b) 3^m

(c) 3^{m-1}

(d) none of these

7. If $f(x+1)+f(x-1)=2f(x)$ and $f(0)=0$ then $f(n), n \in N$, is

(a) $nf(1)$

(b) $\{f(1)\}^n$

(c) 0

(d) none of these

8. If $af(x+1)+bf\left(\dfrac{1}{x+1}\right)=x, x\neq -1, a\neq b$ then $f(2)$ is equal to

(a) $\dfrac{2a+b}{2(a^2-b^2)}$

(b) $\dfrac{a}{a^2-b^2}$

(c) $\dfrac{a+2b}{a^2-b^2}$

(d) none of these

9. Let f be a function satisfying $f(x+y)=f(x)+f(y)$ for all $x, y \in R$. If $f(1)=k$ then $f(n), n \in N$, is equal to

(a) k^n

(b) nk

(c) n^k

(d) none of these

10. Let f be a function satisfying $f(x+y)=f(x)\cdot f(y)$ for all $x, y \in R$. If $f(1)=3$ then $\sum_{r=1}^{n} f(r)$ is equal to

(a) $\dfrac{3}{2}(3^n-1)$

(b) $\dfrac{3}{2}n(n+1)$

(c) $3^{n+1}-3$

(d) none of these

11. If $f(x+y) = f(x) + f(y) - xy - 1$ for all x, y, and $f(1) = 1$ then the number of solutions of $f(n) = n, n \in N$, is

(a) one (b) two

(c) four (d) none of these

12. Let $f(x) = 1 + |x|, x < -1$

$[x], x \geq -1$, where $[\cdot]$ denotes the greatest integer function.

Then $f\{f(-2.3)\}$ is equal to

(a) 4 (b) 2

(c) –3 (d) 3

13. The domain of the function

$$y = \underbrace{\log_{10} \log_{10} \log_{10} \ldots \log_{10}}_{n \text{ times}} x \quad \text{is}$$

(a) $[10^n, +\infty)$ (b) $(10^{n-1}, +\infty)$

(c) $(10^{n-2}, +\infty)$ (d) none of these

14. The largest set of real values of x for which

$$f(x) = \sqrt{(x+2)(5-x)} - \frac{1}{\sqrt{x^2-4}}$$

is a real function is

(a) $[1, 2) \cup (2, 5]$ (b) $(2, 5]$

(c) $[3, 4]$ (d) none of these

15. Let $f(x) = (x^{12} - x^9 + x^4 - x + 1)^{-1/2}$. The domain of the function is

(a) $(1, +\infty)$ (b) $(-\infty, -1)$

(c) $(-1, 1)$ (d) $(-\infty, +\infty)$

16. The domain of the function $f(x) = \sqrt{x - \sqrt{1 - x^2}}$ is

(a) $\left[-1, -\frac{1}{\sqrt{2}}\right] \cup \left[\frac{1}{\sqrt{2}}, 1\right]$ (b) $[-1, 1]$

(c) $\left(-\infty, -\frac{1}{2}\right] \cup \left[\frac{1}{\sqrt{2}}, +\infty\right)$ (d) $\left[\frac{1}{\sqrt{2}}, 1\right]$

17. The domain of the function $f(x) = \sqrt{1 - \sqrt{1 - \sqrt{1 - x^2}}}$ is

(a) $\{x \mid x < 1\}$ (b) $\{x \mid x > -1\}$

(c) $[0, 1]$ (d) $[-1, 1]$

18. The domain of the function $f(x) = \log_{10} \log_{10} (1 + x^3)$ is

(a) $(-1, +\infty)$ (b) $(0, +\infty)$

(c) $[0, +\infty)$ (d) $(-1, 0)$

19. The domain of the function $f(x) = \sqrt{x^2 - [x]^2}$, where $[x]$ = the greatest integer less than or equal to x, is

(a) R (b) $[0, +\infty)$

(c) $(-\infty, 0]$ (d) none of these

20. The domain of $f(x) = \dfrac{1}{\sqrt{|\cos x| + \cos x}}$ is

(a) $[-2n\pi, 2n\pi]$ (b) $(2n\pi, \overline{2n+1}\,\pi)$

(c) $\left(\dfrac{(4n+1)\pi}{2}, \dfrac{(4n+3)\pi}{2}\right)$ (d) $\left(\dfrac{(4n-1)\pi}{2}, \dfrac{(4n+1)\pi}{2}\right)$

21. The domain of $f(x) = \sqrt{\log_{x^2 - 1}(x)}$ is

(a) $(\sqrt{2}, +\infty)$ (b) $(0, +\infty)$

(c) $(1, +\infty)$ (d) none of these

22. The domain of the function $f(x) = {}^{16-x}C_{2x-1} + {}^{20-3x}P_{4x-5}$, where the symbols have their usual meanings, is the set

(a) $\{1, 2, 3, 4, 5\}$ (b) $\{2, 3, 4\}$

(c) $\{2, 3\}$ (d) none of these

23. The domain of $f(x) = \sin^{-1}\left(\dfrac{1+x^2}{2x}\right) + \sqrt{1-x^2}$ is

(a) $\{1\}$ (b) $(-1, 1)$

(c) $\{1, -1\}$ (d) none of these

24. The domain of the function $f(x) = \sqrt{\sec^{-1}\left\{\dfrac{1-|x|}{2}\right\}}$ is

(a) $(-\infty, -3] \cup [3, +\infty)$ (b) $[3, +\infty)$

(c) ϕ (d) R

25. The function $f(x) = \sqrt{e^{\cos^{-1}(\log_4 x^2)}}$ is real valued. It is defined if

(a) $x \in \left[\frac{1}{2}, 2\right]$.

(b) $x \in \left[-2, -\frac{1}{2}\right] \cup \left[\frac{1}{2}, 2\right]$

(c) $x \in \left[-2, -\frac{1}{2}\right]$

(d) none of these

26. The domain of the real-valued function $f(x) = \log_e |\log_e x|$ is

(a) $(1, +\infty)$

(b) $(0, +\infty)$

(c) $(e, +\infty)$

(d) none of these

27. If $[\cdot]$ denotes the greatest integer function then the domain of the real-valued function $\log_{[x+½]} |x^2 - x - 2|$ is

(a) $\left[\frac{3}{2}, +\infty\right)$

(b) $\left[\frac{3}{2}, 2\right) \cup (2, +\infty)$

(c) $\left(\frac{1}{2}, 2\right) \cup (2, +\infty)$

(d) none of these

28. The domain of the function $f(x) = \log_e (x - [x])$, where $[\cdot]$ denotes the greatest integer function, is

(a) R

(b) $R - Z$

(c) $(0, +\infty)$

(d) none of these

29. The domain of the function $f(x) = \sin^{-1}(x + [x])$, where $[\cdot]$ denotes the greatest integer function, is

(a) $[0, 1)$

(b) $[-1, 1]$

(c) $(-1, 0)$

(d) none of these

30. Let $f(x) = \log_{x^2} 25$ and $g(x) = \log_x 5$ then $f(x) = g(x)$ holds for x belonging to

(a) R

(b) $(0, 1) \cup (1, +\infty)$

(c) ϕ

(d) none of these

31. Let $f(x) = \sin^2 \frac{x}{2} + \cos^2 \frac{x}{2}$ and $g(x) = \sec^2 x - \tan^2 x$. The two functions are equal over the set

(a) ϕ

(b) R

(c) $R - \left\{x \mid x = (2n+1)\frac{\pi}{2}, n \in Z\right\}$

(d) none of these

32. The range of the function $f(x) = x^2 + \frac{1}{x^2+1}$ is

(a) $[1, +\infty)$
(b) $[2, +\infty)$
(c) $\left[\frac{3}{2}, +\infty\right)$
(d) none of these

33. Let $f(x) = \cos^{-1}\left(\frac{x^2}{1+x^2}\right)$. The range of f is

(a) $\left[0, \frac{\pi}{2}\right]$
(b) $\left[-\frac{\pi}{2}, \frac{\pi}{2}\right]$
(c) $\left[-\frac{\pi}{2}, 0\right]$
(d) none of these

34. The range of the real-valued function $f(x) = \sqrt{9-x^2}$ is

(a) $[0, 3]$
(b) $[-3, 3]$
(c) $[-3, 0]$
(d) none of these

35. The range of the function $f(x) = |x-1| + |x-2|, -1 \le x \le 3$, is

(a) $[1, 3]$
(b) $[1, 5]$
(c) $[3, 5]$
(d) none of these

36. The range of the function $y = \log_3(5 + 4x - x^2)$ is

(a) $(0, 2]$
(b) $(-\infty, 2]$
(c) $(0, 9]$
(d) none of these

37. Let $f: \{x, y, z\} \to \{a, b, c\}$ be a one-one function and only one of the conditions (i) $f(x) \ne b$, (ii) $f(y) = b$, (iii) $f(z) \ne a$ is true then the function f is given by the set

(a) $\{(x, a), (y, b), (z, c)\}$
(b) $\{(x, a), (y, c), (z, b)\}$
(c) $\{(x, b), (y, a), (z, c)\}$
(d) $\{(x, c), (y, b), (z, a)\}$

38. Let $f: R \to R$ be a function such that $f(x) = x^3 - 6x^2 + 11x - 6$. Then

(a) f is one-one and into
(b) f is many-one and into
(c) f is one-one and onto
(d) f is many-one and onto

39. Let $f: R \to R$ be a function such that $f(x) = x^3 + x^2 + 3x + \sin x$. Then

(a) f is one-one and into
(b) f is one-one and onto
(c) f is many-one and into
(d) f is many-one and onto

40. The function $f : R \to R$ defined by $f(x) = 6^x + 6^{|x|}$ is

(a) one-one and onto
(b) many-one and onto
(c) one-one and into
(d) many-one and into

41. If the real-valued function $f(x) = px + \sin x$ is a bijective function then the set of possible values of $p \in R$ is

(a) $R - \{0\}$
(b) R
(c) $(0, +\infty)$
(d) none of these

42. Let $f(x) = 2x + |\cos x|$. Then f is

(a) one-one and into
(b) one-one and onto
(c) many-one and into
(d) many-one and onto

43. Let f be a function from R to R given by $f(x) = \dfrac{x^2 - 4}{x^2 + 1}$. Then $f(x)$ is

(a) one-one and into
(b) one-one and onto
(c) many-one and into
(d) many-one and onto

44. Let $f : R \to A = \left\{ y \mid 0 \le y < \dfrac{\pi}{2} \right\}$ be a function such that $f(x) = \tan^{-1}(x^2 + x + k)$, where k is a constant. The value of k for which f is an onto function, is

(a) 1
(b) 0
(c) $\dfrac{1}{4}$
(d) none of these

45. $f(x) = x + \sqrt{x^2}$ is a function from $R \to R$. Then $f(x)$ is

(a) injective
(b) surjective
(c) bijective
(d) none of these

46. Which of the following is an even function?

Here $[\cdot]$ denotes the greatest integer function and f is any function.

(a) $[x] - x$
(b) $f(x) - f(-x)$
(c) $e^{3-2x} \cdot \tan^2 x$
(d) $f(x) + f(-x)$

47. Let $f(x) = |x - 2| + |x - 3| + |x - 4|$ and $g(x) = f(x + 1)$. Then

(a) $g(x)$ is an even function
(b) $g(x)$ is an odd function
(c) $g(x)$ is neither even nor odd
(d) $g(x)$ is periodic

48. $f(x) = \log_{10}(x + \sqrt{x^2 + 1})$ is

(a) an odd function
(b) a periodic function
(c) an even function
(d) none of these

49. A function whose graph is symmetrical about the y-axis is given by

(a) $f(x) = \log_e(x + \sqrt{x^2 + 1})$

(b) $f(x + y) = f(x) + f(y)$ for all $x, y \in R$

(c) $f(x) = \cos x + \sin x$

(d) none of these

50. A function whose graph is symmetrical about the origin is given by

(a) $f(x) = e^x + e^{-x}$
(b) $f(x) = \log_e x$
(c) $f(x + y) = f(x) + f(y)$
(d) none of these

51. Let $f(x) = 4, x < -1$

$-4x, -1 \le x \le 0.$

If $f(x)$ is an even function in R then the definition of $f(x)$ in $(0, +\infty)$ is

(a) $f(x) = 4x, 0 < x \le 1$
$4, x > 1$

(b) $f(x) = 4x, 0 < x \le 1$
$-4, x > 1$

(c) $f(x) = 4, 0 < x \le 1$
$4x, x > 1$

(d) none of these

52. If $f(x) = x^2 \sin \frac{\pi x}{2}$, $|x| < 1$

$x|x|$, $|x| \ge 1$ then $f(x)$ is

(a) an even function
(b) an odd function
(c) a periodic function
(d) none of these

53. The period of the function $f(x) = \left|\sin \frac{x}{2}\right| + |\cos x|$ is

(a) 2π
(b) π
(c) 4π
(d) none of these

54. If $f(x)$ is a periodic function of the period k then $f(kx + a)$, where a is a constant, is a periodic function of the period

(a) k
(b) 1
(c) $\frac{k}{a}$
(d) none of these

55. The period of the function $f(x) = 4\cos(2x + 3)$ is

(a) 2π (b) $\frac{\pi}{2}$

(c) π (d) none of these

56. The period of the function $f(x) = 3\sin\frac{\pi x}{3} + 4\cos\frac{\pi x}{4}$ is

(a) 6 (b) 24

(c) 8 (d) 2π

57. Let $f(x) = \cos\sqrt{p}x$, where $p = [a]$ = the greatest integer less than or equal to a. If the period of $f(x)$ is π then

(a) $a \in [4, 5]$ (b) $a = 4, 5$

(c) $a \in [4, 5)$ (d) none of these

58. Let $f(x) = \cos 3x + \sin\sqrt{3}x$. Then $f(x)$ is

(a) a periodic function of period 2π

(b) a periodic function of period $\sqrt{3}\pi$

(c) not a periodic function

(d) none of these

59. The function $f(x) = \sin\frac{\pi x}{n!} - \cos\frac{\pi x}{(n+1)!}$ is

(a) not periodic (b) periodic, with period $2\,(n\,!)$

(c) periodic, with period $(n+1)$ (d) none of these

60. The function $f(x) = x - [x] + \cos x$, where $[x]$ = the greatest integer less than or equal to x, is a

(a) periodic function of indeterminate period

(b) periodic function of period 2π

(c) nonperiodic function

(d) periodic function of period 1

61. Let $f(x) = nx + n - [nx + n] + \tan\frac{\pi x}{2}$, where $[x]$ is the greatest integer $\leq x$ and $n \in N$. It is

(a) a periodic function of period 1

(b) a periodic function of period 4

(c) not periodic
(d) a periodic function of period 2

62. Let $f(x) = x(2 - x)$, $0 \le x \le 2$. If the definition of f is extended over the set $R - [0, 2]$ by $f(x + 2) = f(x)$ then f is a

(a) periodic function of period 1 (b) nonperiodic function

(c) periodic function of period 2 (d) periodic function of period $\frac{1}{2}$

63. If $f(x) = \sin^2 x + \sin^2\left(x + \frac{\pi}{3}\right) + \cos x \cdot \cos\left(x + \frac{\pi}{3}\right)$ and $g\left(\frac{5}{4}\right) = 1$ then $(g \circ f)(x)$ is

(a) a polynomial of the first degree in $\sin x$, $\cos x$
(b) a constant function
(c) a polynomial of the second degree in $\sin x$, $\cos x$
(d) none of these

64. If $f(x) = x^n$, $n \in N$ and $(g \circ f)(x) = ng(x)$ then $g(x)$ can be

(a) $n\,|x|$ (b) $3 \cdot \sqrt[3]{x}$

(c) e^x (d) $\log |x|$

65. If $g\{f(x)\} = |\sin x|$ and $f\{g(x)\} = (\sin \sqrt{x})^2$ then

(a) $f(x) = \sin^2 x$, $g(x) = \sqrt{x}$ (b) $f(x) = \sin x$, $g(x) = |x|$'

(c) $f(x) = x^2$, $g(x) = \sin \sqrt{x}$ (d) f and g cannot be determined

66. If $f(x) = \frac{1}{1 - x}$, $x \ne 0, 1$, then the graph of the function $y = f\{f(f(x))\}$, $x > 1$, is

(a) a circle (b) an ellipse

(c) a straight line (d) a pair of straight lines

67. If $f(x)$ is a polynomial function of the second degree such that $f(-3) = 6$, $f(0) = 6$ and $f(2) = 11$ then the graph of the function $f(x)$ cuts the ordinate $x = 1$ at the point

(a) $(1, 8)$ (b) $(1, 4)$

(c) $(1, -2)$ (d) none of these

68. Let $f(x)$ be a function whose domain is $[-5, 7]$. Let $g(x) = |2x + 5|$. Then the domain of $(f \circ g)(x)$ is

(a) $[-5, 1]$ (b) $[-4, 0]$

(c) $[-6, 1]$ (d) none of these

69. Let $f : (-\infty, 1] \to (-\infty, 1]$ such that $f(x) = x(2 - x)$. Then $f^{-1}(x)$ is

(a) $1 + \sqrt{1 - x}$ (b) $1 - \sqrt{1 - x}$

(c) $\sqrt{1 - x}$ (d) none of these

70. If $f(x) = 3x - 5$ then $f^{-1}(x)$

(a) is given by $\dfrac{1}{3x - 5}$

(b) is given by $\dfrac{x + 5}{3}$

(c) does not exist because f is not one-one

(d) does not exist because f is not onto

71. If the function $f : [1, +\infty) \to [1, +\infty)$ is defined by $f(x) = 2^{x(x-1)}$ then $f^{-1}(x)$ is

(a) $\left(\dfrac{1}{2}\right)^{x(x-1)}$ (b) $\dfrac{1}{2}(1 + \sqrt{1 + 4\log_2 x})$

(c) $\dfrac{1}{2}(1 - \sqrt{1 + 4\log_2 x})$ (d) not defined

72. If the function $f : R \to R$ be such that $f(x) = x - [x]$, where $[y]$ denotes the greatest integer less than or equal to y, then $f^{-1}(x)$ is

(a) $\dfrac{1}{x - [x]}$ (b) $[x] - x$

(c) not defined (d) none of these

73. The inverse function of the function $f(x) = \dfrac{e^x - e^{-x}}{e^x + e^{-x}}$ is

(a) $\dfrac{1}{2}\log\dfrac{1 + x}{1 - x}$ (b) $\dfrac{1}{2}\log\dfrac{2 + x}{2 - x}$

(c) $\dfrac{1}{2}\log\dfrac{1 - x}{1 + x}$ (d) none of these

74. The graph of a real-valued function $f(x)$ is the following.

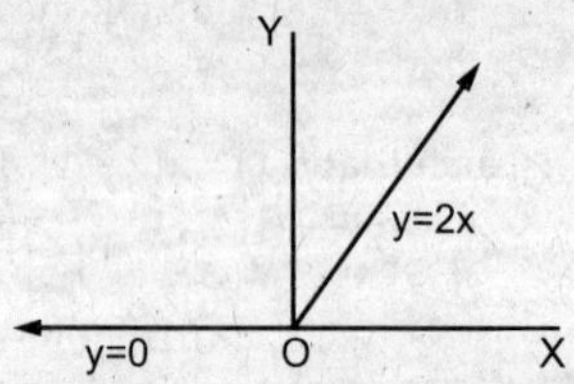

The function is

(a) $f(x) = x - |x|$ (b) $f(x) = x + |x|$

(c) $f(x) = 2x$ (d) none of these

75. If $f(x+y, x-y) = xy$ then the arithmetic mean of $f(x, y)$ and $f(y, x)$ is

(a) x (b) y

(c) 0 (d) none of these

76. The graph of the function $y = f(x)$ is symmetrical about the line $x = 2$. Then

(a) $f(x+2) = f(x-2)$ (b) $f(2+x) = f(2-x)$

(c) $f(x) = f(-x)$ (d) none of these

• Type 2 •

Choose the correct options. One or more options may be correct.

77. Let $f(x) = x^2, 0 < x < 2$

$2x - 3, 2 \le x < 3$

$x + 2, x \ge 3$. Then

(a) $f\left\{f\left(f\left(\frac{3}{2}\right)\right)\right\} = f\left(\frac{3}{2}\right)$ (b) $1 + f\left\{f\left(f\left(\frac{5}{2}\right)\right)\right\} = f\left(\frac{5}{2}\right)$

(c) $f\{f(1)\} = f(1) = 1$ (d) none of these

78. If $f(x) = \cos^2 x + \cos^2\left(\frac{\pi}{3} + x\right) - \cos x \cdot \cos\left(\frac{\pi}{3} + x\right)$ then

(a) $f(x)$ is an even function (b) $f\left(\frac{\pi}{8}\right) = f\left(\frac{\pi}{4}\right)$

(c) $f(x)$ is a constant function (d) $f(x)$ is not a periodic function

79. If one of the roots of $x^2 + f(a) \cdot x + a = 0$ is equal to the third power of the other for all real a then

(a) the domain of the real-valued function f is the set of non-negative real numbers

(b) $f(x) = -x^{1/4}(1 + x^{1/2})$

(c) $f(x) = x^{1/4} + x^{3/4}$

(d) none of these

80. If f is an even function defined on the interval $(-5, 5)$ then a value of x satisfying the equation $f(x) = f\left(\frac{x+1}{x+2}\right)$ is

(a) $\frac{-1+\sqrt{5}}{2}$ (b) $\frac{-3+\sqrt{5}}{2}$

(c) $\frac{1-\sqrt{5}}{2}$ (d) $\frac{-3-\sqrt{5}}{2}$

81. Let $f(x) = [x]$ = the greatest integer less than or equal to x and $g(x) = x - [x]$. Then for any two real numbers x and y

(a) $f(x + y) = f(x) + f(y)$ (b) $g(x + y) = g(x) + g(y)$

(c) $f(x + y) = f(x) + f\{y + g(x)\}$ (d) none of these

82. Let $x \in N$ and let x be a perfect square. Let $f(x)$ = the quotient when x is divided by 5 and $g(x)$ = the remainder when x is divided by 5. Then $\sqrt{x} = f(x) + g(x)$ holds for x equal to

(a) 0 (b) 16

(c) 25 (d) none of these

83. If $f(x) = 27x^3 + \frac{1}{x^3}$ and α, β are the roots of $3x + \frac{1}{x} = 2$ then

(a) $f(\alpha) = f(\beta)$ (b) $f(\alpha) = 10$

(c) $f(\beta) = -10$ (d) none of these

84. If $f(x) = \sin^{-1}(\sin x)$ then

(a) $f(x) = \pi - x, 0 \le x \le \frac{\pi}{2}$ (b) $f(x) = \pi - x, \frac{\pi}{2} \le x \le \pi$

(c) $f(x) = x, 0 \le x \le \pi$ (d) $f(x) = -x, -\frac{\pi}{2} \le x \le 0$

85. If $e^x + e^{f(x)} = e$ then for $f(x)$

(a) domain $= (-\infty, 1)$ (b) range $= (-\infty, 1)$

(c) domain $= (-\infty, 0]$ (d) range $= (-\infty, 1]$

86. If $f(x)$ is an odd function then

(a) $\dfrac{f(-x)+f(x)}{2}$ is an even function

(b) $[\,|f(x)|+1]$ is even, where $[x]$ = the greatest integer $\le x$

(c) $\dfrac{f(x)-f(-x)}{2}$ is neither even nor odd

(d) none of these

87. Let $f(x) = \sec^{-1}[1+\cos^2 x]$ where $[\cdot]$ denotes the greatest integer function. Then

(a) the domain of f is R (b) the domain of f is $[1, 2]$

(c) the range of f is $[1, 2]$ (d) the range of f is $\{\sec^{-1}1, \sec^{-1}2\}$

88. If $f(x)$ and $g(x)$ are two functions of x such that $f(x)+g(x)=e^x$ and $f(x)-g(x)=e^{-x}$ then

(a) $f(x)$ is an odd function (b) $g(x)$ is an odd function

(c) $f(x)$ is an even function (d) $g(x)$ is an even function

89. Let $f(x) = 4\cos\sqrt{x^2 - \dfrac{\pi^2}{9}}$ · Then

(a) the domain of f is $\left[\dfrac{\pi}{3}, +\infty\right)$

(b) the range of f is $[-1, 1]$

(c) the domain of f is $\left(-\infty, -\dfrac{\pi}{3}\right] \cup \left[\dfrac{\pi}{3}, +\infty\right)$

(d) the range of f is $[-4, 4]$

90. Let $f(x+y) = f(x) + f(y)$ for all $x, y \in R$. Then

(a) $f(x)$ is an even function (b) $f(x)$ is an odd function

(c) $f(0) = 0$ (d) $f(n) = nf(1), n \in N$

91. Let $f(x) = [x]^2 + [x+1] - 3$, where $[x]$ = the greatest integer $\leq x$. Then

(a) $f(x)$ is a many-one and into function

(b) $f(x) = 0$ for infinite number of values of x

(c) $f(x) = 0$ for only two real values

(d) none of these

92. Let f and g be functions from the interval $[0, \infty)$ to the interval $[0, \infty)$, f being an increasing function and g being a decreasing function. If $f\{g(0)\} = 0$ then

(a) $f\{g(x)\} \geq f\{g(0)\}$

(b) $g\{f(x)\} \leq g\{f(0)\}$

(c) $f\{g(2)\} = 0$

(d) none of these

Answers

1. b	**2.** b	**3.** c	**4.** c	**5.** c
6. c	**7.** a	**8.** a	**9.** b	**10.** a
11. a	**12.** d	**13.** d	**14.** b	**15.** d
16. d	**17.** d	**18.** b	**19.** d	**20.** d
21. a	**22.** c	**23.** c	**24.** a	**25.** b
26. d	**27.** b	**28.** b	**29.** a	**30.** b
31. c	**32.** a	**33.** d	**34.** a	**35.** b
36. b	**37.** c	**38.** d	**39.** b	**40.** c
41. a	**42.** b	**43.** c	**44.** c	**45.** d
46. d	**47.** c	**48.** a	**49.** d	**50.** c
51. a	**52.** b	**53.** a	**54.** b	**55.** c
56. b	**57.** c	**58.** c	**59.** d	**60.** c
61. d	**62.** c	**63.** b	**64.** d	**65.** a
66. c	**67.** a	**68.** c	**69.** b	**70.** b
71. b	**72.** c	**73.** a	**74.** b	**75.** c
76. b	**77.** a, b, c	**78.** a, b, c	**79.** a, b	**80.** a, b, c, d
81. c	**82.** b, c	**83.** a, c	**84.** b	**85.** a, b
86. a, b	**87.** a, d	**88.** b, c	**89.** c, d	**90.** b, c, d
91. a, b	**92.** b, c			

Hints and Solutions to Selected Questions

1. $x^2 + \lambda x + \mu =$ integer for all integer x.

When $x = 0, x^2 + \lambda x + \mu = 0 =$ integer. So, $x(x + \lambda)$ is an integer for all integral values of x.

Therefore, λ is an integer.

2. $f(0) = c =$ integer and $f(1) = a + b + c =$ integer. Therefore, $a + b =$ integer.

3. $f\left(\frac{\pi}{2}\right) = \cos[\pi]\cdot\frac{\pi}{2} + \cos\left[\pi\cdot\frac{\pi}{2}\right] = \cos\frac{3\pi}{2} + \cos 4 = \cos 4.$

6. $f(n) = 2\{f(1) + f(2) + \ldots + f(n-1)\}.$

$\therefore\quad f(n+1) = 2\{f(1) + f(2) + \ldots + f(n)\} \Rightarrow f(n+1) = 3f(n)$ for $n \geq 2$.

Also $f(2) = 2f(1) = 2 \quad \therefore\; f(3) = 3f(2) = 2\cdot 3$, etc.

$$\therefore\quad \sum_{n=1}^{m} f(n) = f(1) + f(2) + \ldots + f(m)$$

$$= 1 + 2 + 2\cdot 3 + 2\cdot 3^2 + \ldots + 2\cdot 3^{m-2}$$

$$= 1 + 2(1 + 3 + 3^2 + \ldots + 3^{m-2}).$$

7. Putting $x = 1, f(2) + f(0) = 2f(1)$

$\Rightarrow\quad f(2) = 2f(1).$

Putting $x = 2, f(3) + f(1) = 2f(2)$

$\Rightarrow\quad f(3) = 2\times 2f(1) - f(1) = 3f(1)$, etc.

$\therefore\quad f(n) = nf(1)$, for $n = 1, 2, \ldots, n$.

$f(n+1) + f(n-1) = 2f(n)$

$\Rightarrow\quad f(n+1) + (n-1)f(1) = 2nf(1) \quad \Rightarrow f(n+1) = (n+1)f(1).$

8. $af(x+1) + bf\left(\frac{1}{x+1}\right) = (x+1) - 1 \qquad \ldots (1)$

$\therefore\quad af\left(\frac{1}{x+1}\right) + bf(x+1) = \frac{1}{x+1} - 1 \qquad \ldots (2)$

$$(1)\times a-(2)\times b \;\Rightarrow\; (a^2-b^2)f(x+1)=a(x+1)-a-\frac{b}{x+1}+b.$$

Putting $x=1$, $(a^2-b^2)f(2)=2a-a-\frac{b}{2}+b=a+\frac{b}{2}=\frac{2a+b}{2}$.

9. Putting $x=1, y=1$ we get, $f(2)=f(1)+f(1)=2f(1)=2k$.

Putting $x=2, y=1$ we get, $f(3)=f(2)+f(1)=3f(1)=3k$, etc.

10. Putting $x=1, y=1$ we get, $f(2)=\{f(1)\}^2=3^2$.

Putting $x=2, y=1$ we get, $f(3)==f(2)\cdot f(1)=\{f(1)\}^3=3^3$, etc.

$$\therefore \quad \sum_{r=1}^{n} f(r)=\sum_{r=1}^{n} 3^r=\frac{3(3^n-1)}{3-1}.$$

11. Putting $y=1, f(x+1)=f(x)+f(1)-x-1=f(x)-x$.

$\therefore \quad f(n+1)=f(n)-n<f(n)$.

So, $f(n)<f(n-1)<f(n-2)<\dots<f(1)=1$.

$\therefore \quad f(n)=n$ holds for $n=1$ only.

12. $f(-2.3)=1+|-2.3|=3.3$

$\therefore \quad f\{f(-2.3)\}=f(3.3)=[3.3]=3.$

13. $\underbrace{\log_{10}\log_{10}\dots\log_{10}}_{(n-1)\text{ times}} x>0$

or $\underbrace{\log_{10}\log_{10}\dots\log_{10}}_{(n-2)\text{ times}} x>10^0=1$

or $\underbrace{\log_{10}\log_{10}\dots\log_{10}}_{(n-3)\text{ times}} x>10^1$

or $\underbrace{\log_{10}\log_{10}\dots\log_{10}}_{(n-4)\text{ times}} x>10^{10}$, etc.

14. The domain of $\sqrt{(x+2)(5-x)}$ is the set of values of x for which $(x+2)(5-x)\geq 0$, i.e., $x\in[-2,5]$.

The domain of $\frac{1}{\sqrt{x^2-4}}$ is the set of values of x for which $x^2-4>0$, i.e., $x\in(-\infty,-2)\cup(2,+\infty)$. The common values are in $(2,5]$.

15. $f(x)$ is real for those values of x for which $x^{12}-x^9+x^4-x+1>0$

or $\quad x^4(x^8+1)-x(x^8+1)+1>0$

or $\quad (x^8+1)\,x(x^3-1)+1>0.$

If $x \geq 1$ or $x \leq -1$ then the above expression is obviously positive.

If $-1 < x \leq 0$, the above holds.

If $0 < x < 1$, $x^{12} - x(x^8+1) + (x^4+1) > 0$ because $x^4+1 > x^8+1$ and so $x^4+1 > x(x^8+1)$.

16. Here, $x - \sqrt{1-x^2} \geq 0 \quad \therefore \quad x \geq \sqrt{1-x^2} > 0$

$\therefore \quad x^2 \geq 1 - x^2 \quad$ or $\quad x^2 \geq \frac{1}{2}. \qquad$ Also $1 - x^2 \geq 0 \quad$ or $\quad x^2 \leq 1.$

$x^2 \geq \frac{1}{2} \;\Rightarrow\; x < -\frac{1}{\sqrt{2}}$ or $x \geq \frac{1}{\sqrt{2}}. \qquad x^2 \leq 1 \;\Rightarrow\; -1 \leq x \leq 1.$

Thus $x > 0$, $x^2 \geq \frac{1}{2}$ and $x^2 \leq 1 \;\Rightarrow\; x \in \left[\frac{1}{\sqrt{2}}, 1\right].$

17. Clearly, $1 - x^2 \geq 0$, $1 - \sqrt{1-x^2} \geq 0$, $1 - \sqrt{1 - \sqrt{1-x^2}} \geq 0.$

$1 - x^2 \geq 0 \Rightarrow x^2 \leq 1 \Rightarrow -1 \leq x \leq 1$. For these values, the other two hold.

18. $\log_{10}(1+x^3) > 0$

or $\quad 1 + x^3 > 10^0 = 1 \quad$ or $\quad x^3 > 0 \quad \Rightarrow x \in (0, +\infty).$

19. $x^2 - [x]^2 \geq 0 \quad \Rightarrow x^2 \geq [x]^2.$

This is true for all positive x and all negative integer x.

20. $|\cos x| + \cos x > 0$

$\Rightarrow \quad \cos x > 0 \quad \Rightarrow 2n\pi - \frac{\pi}{2} < x < 2n\pi + \frac{\pi}{2}.$

21. $x^2 - 1 > 0$, $x^2 - 1 \neq 1$, $x > 0$ and $\log_{x^2-1}(x) \geq 0.$

The first three imply $x > 1$ but $x \neq \sqrt{2}$.

$\log_{x^2-1}(x) \geq 0 \quad \Rightarrow x \geq 1$ if $x^2 - 1 > 1$

$\qquad\qquad x \leq 1$ if $x^2 - 1 < 1.$

Now, $x^2 - 1 > 1$ or $x > \sqrt{2}$ $(\because x > 1)$ then $x \geq 1$ which is true for all $x > \sqrt{2}$.

$x^2 - 1 < 1$ or $1 < x < \sqrt{2}$ $(\because x > 1)$ then $x \leq 1$ which is not true for any x in $(1, \sqrt{2})$.

This gives no value of x. Thus $x > \sqrt{2}$.

22. $16-x>0, 2x-1\geq 0, 16-x\geq 2x-1, 20-3x>0, 4x-5\geq 0, 20-3x\geq 4x-5$

$\Rightarrow \quad x<16, x\geq \frac{1}{2}, x\leq \frac{17}{3}, x<\frac{20}{3}, x\geq \frac{5}{4}, x\leq \frac{25}{7} \Rightarrow \frac{5}{4}\leq x\leq \frac{25}{7}.$

But $16-x\in N$. So, x must be integer. Hence, $x=2, 3$.

23. Here $\left|\frac{1+x^2}{2x}\right|\leq 1$ and $1-x^2\geq 0$. The first inequality implies $x=1, -1$, which satisfy the second inequation.

24. $\frac{1-|x|}{2}\leq -1$ or $\frac{1-|x|}{2}\geq 1.$

$\frac{1-|x|}{2}\leq -1 \quad \Rightarrow |x|\geq 3 \quad \Rightarrow x\in(-\infty, -3]\cup[3, +\infty).$

$\frac{1-|x|}{2}\geq 1 \quad \Rightarrow |x|\leq -1$ (absurd). Also $0\leq \sec^{-1}\left\{\frac{1-|x|}{2}\right\}\leq \pi.$

25. $|\log_4 x^2|\leq 1 \quad \Rightarrow -1\leq \log_4 x^2\leq 1 \quad \Rightarrow \frac{1}{4}\leq x^2\leq 4$

$\Rightarrow \quad \frac{1}{2}\leq x\leq 2$ or $-2\leq x\leq -\frac{1}{2}.$

26. $x>0$ and $\log_e x\neq 0$

$\therefore \quad x>0$ and $x\neq 1 \quad \therefore$ domain $=(0, 1)\cup(1, +\infty)$.

27. $x^2-x-2\neq 0$ and $\left[x+\frac{1}{2}\right]>0, \left[x+\frac{1}{2}\right]\neq 1.$

$x^2-x-2\neq 0 \quad \Rightarrow x\neq -1, 2. \qquad \left[x+\frac{1}{2}\right]>0 \quad \Rightarrow x\geq \frac{1}{2}.$

$\left[x+\frac{1}{2}\right]\neq 1 \quad \Rightarrow x\notin\left[\frac{1}{2}, \frac{3}{2}\right)$. Thus $x\in\left[\frac{3}{2}, 2\right)\cup(2, +\infty)$.

28. $x-[x]>0$, or $x>[x]$. This is true for all $x\in R$ except integers.

29. $|x+[x]|\leq 1$ or $-1\leq x+[x]\leq 1$.

Let $x=n\in Z$. Then $x+[x]=n+n$.

So, we get $-1\leq 2n\leq 1 \quad \Rightarrow n=0$.

Let $x=n+k, n\in Z, 0<k<1$. Then $x+[x]=n+k+[n+k]=2n+k$.

$\therefore$ we get $-1\leq 2n+k\leq 1 \quad \Rightarrow n=0, 0<k<1$.

Therefore, $x\in[0, 1)$.

30. Domain of $f = D_1 = (-\infty, -1) \cup (-1, 0) \cup (0, 1) \cup (1, +\infty)$.

Domain of $g = D_2 = (0, 1) \cup (1, +\infty)$.

So, $D_1 \cap D_2 = (0, 1) \cup (1, +\infty)$.

In this set, $\log_{x^2} 25 = \frac{1}{2} \times 2\log_x 5 = \log_x 5$.

32. $f(x) = x^2 + \dfrac{1}{x^2+1} - 1 + 1 = 1 + x^2 - \dfrac{x^2}{1+x^2}$

$= 1 + x^2\left(1 - \dfrac{1}{1+x^2}\right) \geq 1$ for all $x \in R$.

The domain $f = R$. Clearly, as x increases $f(x)$ increases.

33. $\left|\dfrac{x^2}{1+x^2}\right| \leq 1$. This is true for all $x \in R$. So, the domain $= R$.

Now $\dfrac{x^2}{1+x^2} = 1 - \dfrac{1}{1+x^2}$.

$\therefore \quad 0 \leq \dfrac{x^2}{1+x^2} < 1.$

$\therefore \quad \cos^{-1} 0 \geq \cos^{-1} \dfrac{x^2}{1+x^2} > \cos^{-1} 1 \quad \Rightarrow \frac{\pi}{2} \geq \cos^{-1} \dfrac{x^2}{1+x^2} > 0.$

$\therefore \quad$ the range $= \left(0, \dfrac{\pi}{2}\right]$.

34. $9 - x^2 \geq 0 \;\Rightarrow -3 \leq x \leq 3$. As $9 - x^2$ is an even function, the value of $f(x)$ changes in $0 \leq x \leq 3$. Therefore,

$\max f(x) = \sqrt{9-0} = 3$ and $\min f(x) = \sqrt{9-9} = 0$.

35. If $x < 1$, $f(x) = -(x-1) - (x-2) = -2x + 3$. In this interval, $f(x)$ is decreasing.

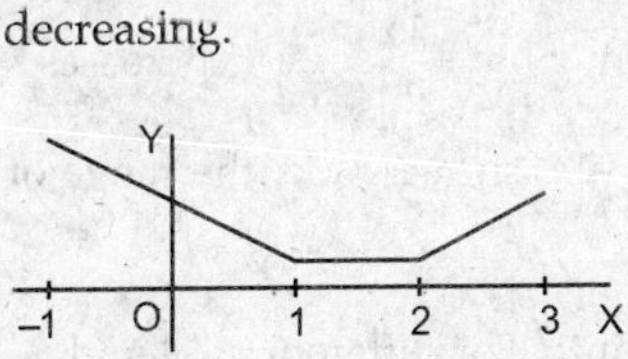

If $1 \leq x < 2, f(x) = x - 1 - (x-2) = 1$.

In this interval, $f(x)$ is constant.

If $2 \leq x \leq 3, f(x) = x - 1 + x - 2 = 2x - 3$.

In this interval, $f(x)$ is increasing.

$\therefore$ $\max f(x) =$ the greatest among $f(-1)$ and $f(3) = 5$, $\min f(x) = f(1) = 1$.

So, range $= [1, 5]$.

36. Here, $5+4x-x^2>0 \;\Rightarrow\; -1<x<5$.

Also, $5+4x-x^2=9-(x-2)^2$.

$\therefore \quad \max(5+4x-x^2)=9$, attained at $x=2 \in [-1, 5]$

$\min(5+4x-x^2)=0$, attained at $x=-1$ or 5.

So, the range $=(-\infty, \log_3 9]$.

37. $f(x) \neq b \quad \Rightarrow f(x)=a, f(y)=c, f(z)=b$ (uses (iii) also)

$f(x)=a, f(y)=b, f(z)=c$ (uses (ii) also)

$f(x)=c, f(y)=a, f(z)=b$ (uses (iii) also)

$f(x)=c, f(y)=b, f(z)=a$ (uses (ii) also)

$f(y)=b \quad \Rightarrow f(x)=a, f(y)=b, f(z)=c$ (uses (i) also)

$f(x)=c, f(y)=b, f(z)=a$ (uses (i) also)

$f(z) \neq a \quad \Rightarrow f(x)=a, f(y)=b, f(z)=c$ (uses (i) also)

$f(x)=a, f(y)=c, f(z)=b$ (uses (i) also)

$f(x)=b, f(y)=a, f(z)=c$ (possible)

$f(x)=c, f(y)=a, f(z)=b$ (uses (i) also)

38. $f(x)=(x-1)(x-2)(x-3)$

$\therefore f(1)=f(2)=f(3)=0$. So, the function is many-one.

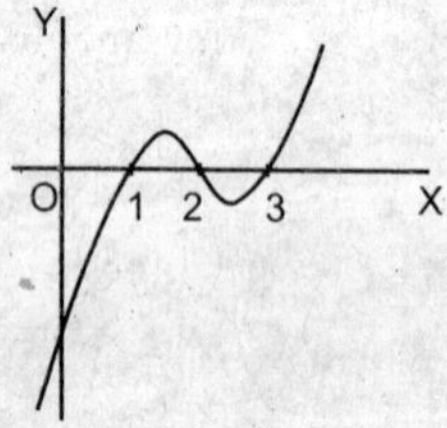

Clearly, for $x<0, f(x)<0$ and goes on decreasing as x decreases.

For $x>3, f(x)>0$ and goes on increasing as x increases.

$\therefore \quad f(x)$ can have all real values. So, f is onto.

39. $f(x)=x\left\{\left(x+\frac{1}{2}\right)^2+\frac{11}{4}\right\}+\sin x$. Clearly $x\left\{\left(x+\frac{1}{2}\right)^2+\frac{11}{4}\right\}$ increases with x and its value changes over R. Also, $-1 \le \sin x \le 1$. So, the range of $f=R$. Hence, f is onto and one-one.

40. $6^x>0, 6^{|x|}>0$ for all $x \in R$. So, f is into. For different x, 6^x and $6^{|x|}$ are different positive numbers. Clearly, f is one-one.

41. Clearly, $p \neq 0$. If $p=0$ then $f(x)=\sin x$ which is many-one.

43. $f(x) = f(-x)$. So, f is many-one. Also, $f(x) = 1 - \frac{5}{x^2+1} > 1 - 5 = -4$. So, f is into.

44. Clearly, $x^2 + x + k \geq 0$ for all x.

So, $D = 1 - 4k \leq 0 \quad \Rightarrow k \geq \frac{1}{4}$.

45. $f(x) = x + |x|$. Clearly, $f(-1) = f(-2) = \ldots = 0$. So, f is many-one.

Also, $f(x) \geq 0$ for all x. So, it is not surjective.

47. $g(x) = f(x+1) = |x-1| + |x-2| + |x-3|$.

If $x < 1$, $g(x) = -x + 1 - x + 2 - x + 3 = -3x + 6$.

If $1 \leq x < 2$, $g(x) = x - 1 - x + 2 - x + 3 = -x + 4$.

If $2 \leq x < 3$, $g(x) = x - 1 + x - 2 - x + 3 = x$.

If $x \geq 3$, $g(x) = x - 1 + x - 2 + x - 3 = 3x - 6$.

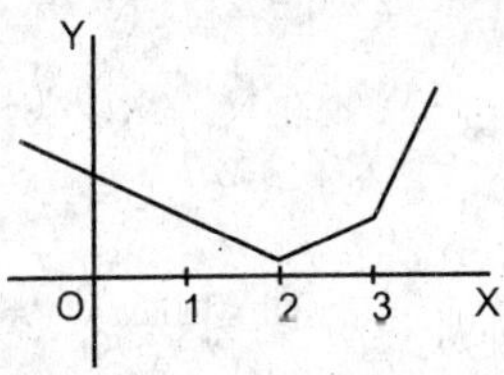

49. The function must be even.

$f(0+0) = f(0) + f(0) \quad \therefore \; f(0) = 0.$

$f(x - x) = f(x) + f(-x) \quad$ or $\quad 0 = f(x) + f(-x) \quad \therefore \; f(x)$ is odd.

Also, $\log_e (x + \sqrt{x^2+1})$ is odd because

$$\log_e (-x + \sqrt{(-x)^2 + 1}) = -\log_e (x + \sqrt{x^2+1}).$$

$f(x) = \cos x + \sin x = \sqrt{2} \cos\left(x - \frac{\pi}{4}\right)$. As $\cos\left(-x - \frac{\pi}{4}\right) \neq \cos\left(x - \frac{\pi}{4}\right)$ for every x, $\cos x + \sin x$ is not even.

50. A function $y = f(x)$ is symmetrical about the origin if (x, y) is on the graph implies that $(-x, -y)$ is also on the graph, i.e., $-y = f(-x)$. In other words, $f(-x) = -f(x)$. Now, see the hint to Q. No. 49.

51.

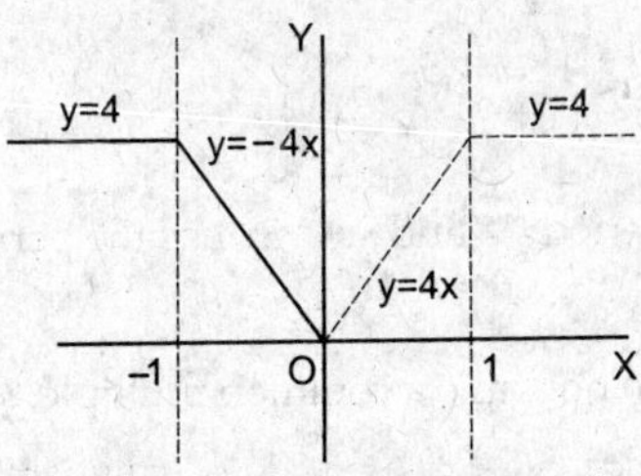

An even function is symmetrical about the y-axis.

Clearly from the graph, the definition given in option (a) is correct.

52. Here, $f(x) = x\,|x|, x \le -1$

$$x^2 \sin(\pi x/2), -1 < x < 1.$$

$$x\,|x|, x \ge 1.$$

Let $k \ge 0$. Then $f(-1-k) = (-1-k)\,|-1-k| = -(1+k)^2$

and $f(1+k) = (1+k)\,|1+k| = (1+k)^2$.

Therefore, $f(1+k) = -f(-\overline{1+k})$.

$$f(-1+k) = (-1+k)^2 \sin\frac{\pi}{2}(-1+k) = -(1-k)^2 \sin\frac{\pi}{2}(1-k) = -f(1-k)$$

$$= -f\{-(-1+k)\}$$

$\therefore$ $f(x) = -f(-x)$ for all x. Also, none of the pieces of definition are periodic.

53. $\left|\sin\frac{x}{2}\right|$ is a periodic function of period 2π and $|\cos x|$ is a periodic function of period π.

54. $f(x) = f(x+k) \Rightarrow f(kx+a) = f(kx+a+k) = f\{k(x+1)+a\}$.

So, the period is 1.

56. The period of $\sin\frac{\pi x}{3}$ is $\frac{2\pi}{\pi/3}$, i.e., 6. The period of $\cos\frac{\pi x}{4}$ is $\frac{2\pi}{\pi/4}$, i.e., 8.

LCM of 6 and 8 is 24. So, the period of $f(x) = 24$.

57. The period of $f(x) = \frac{2\pi}{\sqrt{p}} = \pi$ (from the question).

$\therefore \quad \sqrt{p} = 2 \quad$ or $\quad p = 4$

$\therefore \quad [a] = 4 \quad \therefore \quad 4 \le a < 5$.

58. $\cos 3x$ has the period $\frac{2\pi}{3}$ and $\sin\sqrt{3}x$ has the period $\frac{2\pi}{\sqrt{3}}$.

As $\frac{2\pi}{3}$ and $\frac{2\pi}{\sqrt{3}}$ do not have a common multiple, $f(x)$ is not periodic.

59. $\sin\frac{\pi x}{n!}$ has the period $\frac{2\pi}{\frac{\pi}{n!}}$, i.e., $2(n!)$ and $\cos\frac{\pi x}{(n+1)!}$ has the period

$\frac{2\pi}{\frac{\pi}{(n+1)!}}$, i.e., $2\{(n+1)!\}$. Their LCM $= 2\{(n+1)!\}$.

60. $x-[x]$ has the period 1 and $\cos x$ has the period 2π. Clearly, 1 and 2π do not have a common multiple.

61. $nx+n-[nx+n]$ has the period 1 and $\tan\frac{\pi x}{2}$ has the period $\frac{\pi}{\pi/2}$, i.e., 2.

LCM of 1, 2 is 2.

62. In $R-[0, 2]$, $f(x+2)=f(x)$.

The graph of $f(x)=x(2-x)$, $0\le x\le 2$ is shown.

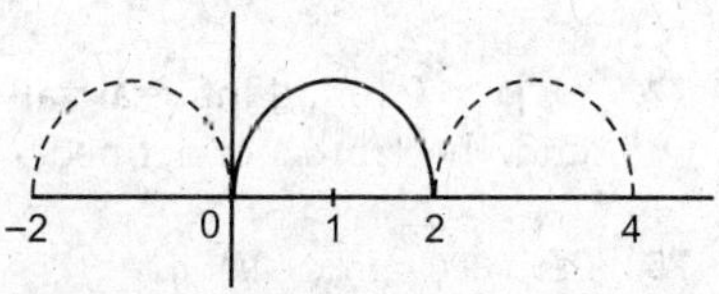

Clearly, f is a periodic function of period 2.

63. On simplification, $f(x)=\frac{5}{4}=$ constant. So, $g\{f(x)\}=g\left(\frac{5}{4}\right)=1$ (from the question).

64. $g\{f(x)\}=g(x^n)=ng(x)$. Also $\log x^n = n\log|x|$. So, $g(x)=\log|x|$ is possible.

65. Verify by trial.

66. $f\{f(x)\}=\frac{1}{1-f(x)}=\frac{1}{1-\frac{1}{1-x}}=\frac{1-x}{-x}=\frac{x-1}{x}$.

$\therefore\quad f\{f(f(x))\}=\frac{1}{1-f\{f(x)\}}=\frac{1}{1-\frac{x-1}{x}}=x$

$\therefore$ the graph has the equation $y=x$.

68. $f\{g(x)\}=f(|2x+5|)\quad\therefore\quad |2x+5|\in[-5, 7]$

$\therefore\quad -5\le|2x+5|\le 7\quad\Rightarrow |2x+5|\le 7$

$\Rightarrow\quad -7\le 2x+5\le 7\Rightarrow -6\le x\le 1.$

69. $y = x(2 - x) \quad \Rightarrow x^2 - 2x + y = 0 \quad \Rightarrow x = \dfrac{2 \pm 2\sqrt{1-y}}{2} = 1 \pm \sqrt{1-y}.$

But $x \le 1$. So, $x = 1 - \sqrt{1-y}$. Therefore, $f^{-1}(x) = 1 - \sqrt{1-x}$.

70. See the hint to Q. No. 69.

71. $y = 2^{x(x-1)} \quad \Rightarrow x(x-1) = \log_2 y \quad \Rightarrow x^2 - x - \log_2 y = 0$

$$\therefore \quad x = \frac{1 \pm \sqrt{1 + 4\log_2 y}}{2}.$$

But $x \ge 1$. $\quad \therefore \quad x = \dfrac{1 + \sqrt{1 + 4\log_2 y}}{2}$

$$\therefore \quad f^{-1}(x) = \frac{1}{2}\{1 + \sqrt{1 + 4\log_2 y}\}.$$

72. $x - [x] = 0$ for all integral values of x. Therefore, the function is many-one and, therefore, not defined.

75. Let $x + y = p, x - y = q$.

Then $f(p, q) = \dfrac{p+q}{2} \cdot \dfrac{p-q}{2} = \dfrac{p^2 - q^2}{4}$.

$$\therefore \quad f(x, y) = \frac{x^2 - y^2}{4} \quad \text{and} \quad f(y, x) = \frac{y^2 - x^2}{4}.$$

76. If $f(x) = f(-x)$, i.e., $f(0 + x) = f(0 - x)$, the graph is symmetrical about $x = 0$.

Therefore, the graph is symmetrical about $x = 2$ if $f(2 + x) = f(2 - x)$.

79. $\alpha + \alpha^3 = -f(a), \alpha \cdot \alpha^3 = a$.

Eliminating α, $a^{1/4} + a^{3/4} = -f(a)$.

$$\therefore \quad f(x) = -x^{1/4}(1 + x^{1/2})$$

80. As $f(x)$ is even and $f(x) = f\left(\dfrac{x+1}{x+2}\right)$, we get $\dfrac{x+1}{x+2} = \pm x$.

Solving these, we get the four values.

81. We have $[x + y] = [x] + [y + x - [x]]$ for all $x, y \in R$. It is proved as below.

If x is integer, $[x + y] = x + [y] = [x] + [y + x - [x]]$, because $x - [x] = 0$.

If $n < x < n + 1, \quad x = n + k, 0 < k < 1$. Then

$[x + y] = [n + k + y] = n + [y + k] = [x] + [y + x - [x]]$, because $x - [x] = k$.

82. By verification, $x = 16, 25$.

83. $f(\alpha) = (3\alpha)^3 + \frac{1}{\alpha^3} = \left(3\alpha + \frac{1}{\alpha}\right)^3 - 3 \cdot 3\alpha \cdot \frac{1}{\alpha}\left(3\alpha + \frac{1}{\alpha}\right) = 2^3 - 9 \times 2 = -10.$

Similarly, $f(\beta) = -10$.

85. $e^{f(x)} = e - e^x$ or $f(x) = \log_e (e - e^x)$.

For domain, $e - e^x > 0$ or $e > e^x \Rightarrow x < 1$.

Also, $e - e^x < e$. So, $f(x) < \log_e e = 1$.

86. For all $f(x)$, $\phi(x) = \frac{f(-x) + f(x)}{2} \Rightarrow \phi(x) = \phi(-x)$.

$\psi(x) = [\,|f(x)| + 1]$

$\Rightarrow \psi(-x) = [\,|f(-x)| + 1] = [\,|-f(x)| + 1] = [\,|f(x)| + 1] = \psi(x)$.

87. $\sec^{-1}[1 + \cos^2 x]$ is defined when $[1 + \cos^2 x] \geq 1$ or ≤ -1.

But $1 + \cos^2 x \geq 1$ and $1 + \cos^2 x \leq 2$. So, $[1 + \cos^2 x] = 1$ for all x except $x = 0, n\pi$, where $n \in Z$ and for these values of x, $[1 + \cos^2 x] = 2$.

89. $x^2 - \frac{\pi^2}{9} \geq 0 \Rightarrow x \geq \frac{\pi}{3}$ or $x \leq -\frac{\pi}{3}$.

91. $f(x) = [x]^2 + [x] + 1 - 3 = \{[x] + 2\}\{[x] - 1\}$.

So, $x = 1, 1.1, 1.2, \ldots \Rightarrow f(x) = 0$.

Only integral values will be attained.

92. g is decreasing $\Rightarrow g(2) \leq g(0)$. Therefore, f is increasing $\Rightarrow f\{g(0)\} \geq f\{g(2)\}$.

$\therefore \quad 0 \geq f\{g(2)\}$, from the question. But $f(x) \geq 0$, from the question.

$\therefore \quad f\{g(2)\} = 0$.

As $x \geq 0$, $g(x) \leq g(0)$. So, $f\{g(x)\} \leq f\{g(0)\}$.

As $x \geq 0$, $f(x) \geq f(0)$. So, $g\{f(x)\} \leq g\{f(0)\}$.

□

2

Differential Coefficient

• *Type 1* •

Choose the most appropriate option (a, b, c or d).

1. If $y = (1+x)(1+x^2)(1+x^4) \ldots (1+x^{2^n})$ then $\frac{dy}{dx}$ at $x = 0$ is

(a) 1 (b) -1

(c) 0 (d) none of these

2. If $y = |\cos x| + |\sin x|$ then $\frac{dy}{dx}$ at $x = \frac{2\pi}{3}$ is

(a) $\frac{1-\sqrt{3}}{2}$ (b) 0

(c) $\frac{1}{2}(\sqrt{3}-1)$ (d) none of these

3. The differential coefficient of $f(\log_e x)$ with respect to x, where $f(x) = \log_e x$, is

(a) $\frac{x}{\log_e x}$ (b) $\frac{1}{x}\log_e x$

(c) $\frac{1}{x\log_e x}$ (d) none of these

4. If $f(x) = \cos x \cdot \cos 2x \cdot \cos 4x \cdot \cos 8x \cdot \cos 16x$ then $f'\left(\frac{\pi}{4}\right)$ is

(a) $\sqrt{2}$ (b) $\frac{1}{\sqrt{2}}$

(c) 1 (d) none of these

5. If $y = \cos^{-1}(\cos x)$ then $\frac{dy}{dx}$ at $x = \frac{5\pi}{4}$ is equal to

(a) 1 (b) –1

(c) $\frac{1}{\sqrt{2}}$ (d) none of these

6. If $y = \sec(\tan^{1} x)$ then $\frac{dy}{dx}$ at $x = 1$ is equal to

(a) $\frac{1}{\sqrt{2}}$ (b) $-\frac{1}{\sqrt{2}}$

(c) 1 (d) none of these

7. If $x^y = e^{x-y}$ then $\frac{dy}{dx}$ at $x = 1$ is equal to

(a) 0 (b) –2

(c) 1 (d) none of these

8. If $xe^{xy} - y = \sin^2 x$ then $\frac{dy}{dx}$ at $x = 0$ is

(a) 0 (b) 1

(c) –1 (d) none of these

9. If $x = e^{y + e^{y + \dots \text{ to } \infty}}$ then $\frac{dy}{dx}$ is

(a) $\frac{x}{1+x}$ (b) $\frac{1}{x}$

(c) $\frac{1-x}{x}$ (d) none of these

10. If $y = \log_2(\log_e x)$ then $\frac{dy}{dx}$ is

(a) $\frac{1}{x}\log_2 e \cdot \log_x e$ (b) $\frac{1}{x}\log_2 x$

(c) $\frac{1}{x}\log_e x$ (d) none of these

11. If $f'(x) = \sqrt{2x^2 - 1}$ and $y = f(x^2)$ then $\frac{dy}{dx}$ at $x = 1$ is

(a) 2 (b) 1

(c) –2 (d) none of these

12. If $y = \cos^{-1}\left(\dfrac{5\cos x - 12\sin x}{13}\right), x \in \left(0, \dfrac{\pi}{2}\right)$, then $\dfrac{dy}{dx}$ is equal to

(a) 1 (b) –1

(c) 0 (d) none of these

13. If $y = \tan^{-1}\sqrt{\dfrac{x+1}{x-1}}$ then $\dfrac{dy}{dx}$ is equal to

(a) $\dfrac{-1}{2|x|\sqrt{x^2-1}}$ (b) $\dfrac{-1}{2x\sqrt{x^2-1}}$

(c) $\dfrac{1}{2x\sqrt{x^2-1}}$ (d) none of these

14. If $x^y \cdot y^x = 16$ then $\dfrac{dy}{dx}$ at (2, 2) is

(a) 1 (b) –1

(c) 0 (d) none of these

15. If $y = x^{1/x}$, the value of $\dfrac{dy}{dx}$ at $x = e$ is

(a) 1 (b) 0

(c) –1 (d) none of these

16. The derivative of $\tan^{-1}\dfrac{2x}{1-x^2}$ with respect to $\sin^{-1}\dfrac{2x}{1+x^2}$ is

(a) $\dfrac{1}{1+x^2}$ (b) $\dfrac{1}{1-x^2}$

(c) 0 (d) 1

17. Let the function $y = f(x)$ be given by $x = t^5 - 5t^3 - 20t + 7$ and $y = 4t^3 - 3t^2 - 18t + 3$, where $t \in (-2, 2)$. Then $f'(x)$ at $t = 1$ is

(a) $\dfrac{5}{2}$ (b) $\dfrac{2}{5}$

(c) $\dfrac{7}{5}$ (d) none of these

18. If $f(x) = \cos\left\{\dfrac{\pi}{2}[x] - x^3\right\}, 1 < x < 2$, and $[x]$ = the greatest integer $\leq x$, then $f'\left(\sqrt[3]{\dfrac{\pi}{2}}\right)$ is equal to

(a) 0 (b) $3\left(\frac{\pi}{2}\right)^{2/3}$

(c) $-3\left(\frac{\pi}{2}\right)^{3/2}$ (d) none of these

19. If $y = \sin x°$ and $z = \cos x$ then $\frac{dy}{dz}$ is equal to

(a) $-\text{cosec}\, x \cdot \cos x$ (b) $\frac{\pi}{180}\,\text{cosec}\,\frac{\pi x}{180} \cdot \cos x$

(c) $-\frac{\pi}{180}\,\text{cosec}\, x \cdot \cos\frac{\pi x}{180}$ (d) none of these

20. The derivative of $\tan^{-1}\frac{\sqrt{1+x^2}-1}{x}$ with respect to $\tan^{-1}x$ is

(a) $\frac{\sqrt{1+x^2}-1}{x^2}$ (b) 1

(c) $\frac{1}{1+x^2}$ (d) none of these

21. The differential coefficient of $\text{cosec}^{-1}\frac{1}{2x^2-1}$ with respect to $\sqrt{1-x^2}$ at $x = \frac{1}{2}$ is

(a) -4 (b) 4

(c) -1 (d) none of these

22. If the prime sign (′) represents differentiation w.r.t. x and $f'(x) = \sin x + \sin 4x \cdot \cos x$ then $f'\left(2x^2 + \frac{\pi}{2}\right)$ at $x = \sqrt{\frac{\pi}{2}}$ is equal to

(a) 0 (b) -1

(c) $-2\sqrt{2\pi}$ (d) none of these

23. If $u = f(x^3)$, $v = g(x^2)$, $f'(x) = \cos x$ and $g'(x) = \sin x$ then $\frac{du}{dv}$ is

(a) $\frac{3}{2}x \cdot \cos x^3 \cdot \text{cosec}\, x^2$ (b) $\frac{2}{3}\sin x^3 \cdot \sec x^2$

(c) $\tan x$ (d) none of these

24. If $y=f\left(\frac{2x-1}{x^2+1}\right)$ and $f'(x)=\sin x^2$ then $\frac{dy}{dx}$ is

(a) $\sin\left(\frac{2x-1}{x^2+1}\right)^2$

(b) $\frac{2(1+x-x^2)}{(x^2+1)^2}\sin\left(\frac{2x-1}{x^2+1}\right)^2$

(c) $\frac{2(2x-1)}{x^2+1}\sin\left(\frac{2x-1}{x^2+1}\right)^2$

(d) none of these

25. If $e^x=\frac{\sqrt{1+t}-\sqrt{1-t}}{\sqrt{1+t}+\sqrt{1-t}}$ and $\tan\frac{y}{2}=\sqrt{\frac{1-t}{1+t}}$ then $\frac{dy}{dx}$ at $t=\frac{1}{2}$ is

(a) $-\frac{1}{2}$

(b) $\frac{1}{2}$

(c) 0

(d) none of these

26. If $t(1+x^2)=x$ and $x^2+t^2=y$ then $\frac{dy}{dx}$ at $x=2$ is

(a) $\frac{88}{125}$

(b) $\frac{488}{125}$

(c) 1

(d) none of these

27. Let $f(x)$ be a polynomial function of the second degree. If $f(1)=f(-1)$ and a_1, a_2, a_3 are in AP then $f'(a_1), f'(a_2), f'(a_3)$ are in

(a) AP

(b) GP

(c) HP

(d) none of these

28. If $P(x)$ is a polynomial such that $P(x^2+1)=\{P(x)\}^2+1$ and $P(0)=0$ then $P'(0)$ is equal to

(a) 1

(b) 0

(c) –1

(d) none of these

29. If $5f(x)+3f\left(\frac{1}{x}\right)=x+2$ and $y=xf(x)$ then $\left(\frac{dy}{dx}\right)_{x=1}$ is equal to

(a) 14

(b) $\frac{7}{8}$

(c) 1

(d) none of these

30. If for all x, y the function f is defined by

$$f(x)+f(y)+f(x)\cdot f(y)=1 \text{ and } f(x)>0 \text{ then}$$

(a) $f'(x)$ does not exist

(b) $f'(x)=0$ for all x

(c) $f'(0)<f'(1)$

(d) none of these

31. There exists a function $f(x)$ satisfying $f(0)=1, f'(0)=-1, f(x)>0$ for all x and

(a) $f'(x)<0$ for all x (b) $-1<f''(x)<0$ for all x

(c) $-2<f''(x)\le -1$ for all x (d) $f''(x)\le -2$ for all x

32. If g is the inverse function of f and $f'(x)=\sin x$ then $g'(x)$ is

(a) cosec $\{g(x)\}$ (b) $\sin\{g(x)\}$

(c) $-\dfrac{1}{\sin\{g(x)\}}$ (d) none of these

33. Let $f(x)$ be a polynomial function of degree 2 and $f(x)>0$ for all $x\in R$. If $g(x)=f(x)+f'(x)+f''(x)$ then for any x

(a) $g(x)<0$ (b) $g(x)>0$

(c) $g(x)=0$ (d) $g(x)\ge 0$

34. If $y^2=P(x)=$ a polynomial of degree 3 then $2\dfrac{d}{dx}\left(y^3\dfrac{d^2y}{dx^2}\right)$ equals

(a) $P'''(x)+P'(x)$ (b) $P''(x)\cdot P'''(x)$

(c) $P(x)\cdot P'''(x)$ (d) none of these

35. Let $f(x)=\begin{vmatrix} x^3 & \sin x & \cos x \\ 6 & -1 & 0 \\ p & p^2 & p^3 \end{vmatrix}$,

where p is a constant. Then $\dfrac{d^3}{dx^3}\{f(x)\}$ at $x=0$ is

(a) p (b) $p+p^2$

(c) $p+p^3$ (d) independent of p

36. If $y=\sin 2x$ then $\dfrac{d^6y}{dx^6}$ at $x=\dfrac{\pi}{2}$ is equal to

(a) -64 (b) 0

(c) 64 (d) none of these

37. $x=t\cos t, y=t+\sin t$ then $\dfrac{d^2x}{dy^2}$ at $t=\dfrac{\pi}{2}$ is equal to

(a) $\dfrac{\pi+4}{2}$ (b) $-\dfrac{\pi+4}{2}$

(c) -2 (d) none of these

38. If $y = at^2, x = 2at$, where a is a constant, then $\frac{d^2y}{dx^2}$ at $x = \frac{1}{2}$ is

(a) $\frac{1}{2a}$ (b) 1

(c) $2a$ (d) none of these

• Type 2 •

Choose the correct options. One or more options may be correct.

39. Let $f(x) = (ax + b)\cos x + (cx + d)\sin x$ and $f'(x) = x\cos x$ be an identity in x. Then

(a) $a = 5$ (b) $b = 1$

(c) $c = 1$ (d) $d = -5$

40. Let $f(x) = 2\tan^{-1}x + \sin^{-1}\frac{2x}{1+x^2}\cdot$ Then

(a) $f'(2) = f'(3)$ (b) $f'(2) = 0$

(c) $f'\left(\frac{1}{2}\right) = \frac{16}{5}$ (d) $f'\left(\frac{1}{2}\right) = 0$

41. If $f(x) = x^3 + x^2 f'(1) + xf''(2) + f'''(3)$ for all $x \in R$ then

(a) $f(0) + f(2) = f(1)$ (b) $f(0) + f(3) = 0$

(c) $f(1) + f(3) = f(2)$ (d) none of these

42. If $f(x - y), f(x)\cdot f(y)$ and $f(x + y)$ are in AP for all x, y, and $f(0) \neq 0$, then

(a) $f(2) = f(-2)$ (b) $f(3) + f(-3) = 0$

(c) $f'(2) + f'(-2) = 0$ (d) $f'(3) = f'(-3)$

43. Let $f(-x) = f(x)$. Then $f'(x)$ must be

(a) an even function (b) an odd function

(c) a periodic function (d) neither even nor odd

44. If $f_n(x) = e^{f_{n-1}(x)}$ for all $n \in N$ and $f_0(x) = x$ then $\frac{d}{dx}\{f_n(x)\}$ is equal to

(a) $f_n(x)\cdot\frac{d}{dx}\{f_{n-1}(x)\}$ (b) $f_n(x)\cdot f_{n-1}(x)$

(c) $f_n(x)\cdot f_{n-1}(x)\cdot \ldots \cdot f_2(x)\cdot f_1(x)$ (d) none of these

45. Let $f(x) = x^2 + xg'(1) + g''(2)$ and $g(x) = f(1) \cdot x^2 + xf'(x) + f''(x)$ then

(a) $f'(1) + f'(2) = 0$

(b) $g'(2) = g'(1)$

(c) $g''(2) + f''(3) = 6$

(d) none of these

Answers

1. a	**2.** c	**3.** c	**4.** a	**5.** b
6. a	**7.** a	**8.** b	**9.** c	**10.** a
11. a	**12.** a	**13.** a	**14.** b	**15.** b
16. d	**17.** b	**18.** a	**19.** c	**20.** d
21. a	**22.** c	**23.** a	**24.** b	**25.** a
26. b	**27.** a	**28.** a	**29.** b	**30.** b
31. a	**32.** a	**33.** b	**34.** c	**35.** d
36. b	**37.** b	**38.** a	**39.** b, c	**40.** a, b, c
41. a, b, c	**42.** a, c	**43.** b	**44.** a, c	**45.** a, b, c

Hints and Solutions to Selected Questions

1. $y = \dfrac{(1-x)(1+x)(1+x^2)\ldots(1+x^{2^n})}{1-x} = \dfrac{1-x^{2^{n+1}}}{1-x}$

$\therefore \quad \dfrac{dy}{dx} = \dfrac{-2^{n+1}\cdot x^{2^{n+1}-1}\cdot(1-x)+1-x^{2^{n+1}}}{(1-x)^2};$

$\therefore \quad$ at $x=0, \dfrac{dy}{dx} = \dfrac{-2^{n+1}\cdot 0\cdot 1+1-0}{1^2} = 1.$

2. Around $x = \dfrac{2\pi}{3}$, $|\cos x| = -\cos x$ and $|\sin x| = \sin x$.

$\therefore \quad y = -\cos x + \sin x \qquad \therefore \quad \dfrac{dy}{dx} = \sin x + \cos x.$

At $x = \dfrac{2\pi}{3}$, $\dfrac{dy}{dx} = \sin\dfrac{2\pi}{3} + \cos\dfrac{2\pi}{3}.$

3. $f(\log_e x) = \log_e(\log_e x) \qquad \therefore \quad \dfrac{df(\log_e x)}{dx} = \dfrac{1}{\log_e x}\cdot\dfrac{1}{x}.$

4. $f(x) = \dfrac{2\sin x\cdot\cos x\cdot\cos 2x\cdot\cos 4x\cdot\cos 8x\cdot\cos 16x}{2\sin x} = \dfrac{\sin 32x}{2^5\sin x}$

$\therefore \quad f'(x) = \dfrac{1}{32}\cdot\dfrac{32\cos 32x\cdot\sin x - \cos x\cdot\sin 32x}{\sin^2 x}$

$\therefore \quad f'\left(\dfrac{\pi}{4}\right) = \dfrac{32\cdot\dfrac{1}{\sqrt{2}} - \dfrac{1}{\sqrt{2}}\cdot 0}{32\cdot\left(\dfrac{1}{\sqrt{2}}\right)^2}.$

5. $y = \cos^{-1}\cos(\pi + \overline{x-\pi}) = \cos^{-1}\cos(\pi - \overline{x-\pi})$. When x is around $\dfrac{5\pi}{4}$, $\pi - \overline{x-\pi}$ is in the second quadrant. So, $y = \pi - (x-\pi)$; $\therefore \quad \dfrac{dy}{dx} = -1.$

8. Differentiating w.r.t. x, $e^{xy} + x\cdot e^{xy}\left(y + x\dfrac{dy}{dx}\right) - \dfrac{dy}{dx} = 2\sin x\cos x$

or $\quad \dfrac{dy}{dx}\{x^2 \cdot e^{xy} - 1\} = \sin 2x - e^{xy} - xy \cdot e^{xy}.$

When $x = 0, 0 - y = 0$, i.e., $y = 0$.

$\therefore \quad \dfrac{dy}{dx} = \dfrac{0-1-0}{0-1} = 1.$

9. $x = e^{y+x} \quad \Rightarrow \log x = y + x \quad \therefore \ \dfrac{1}{x} = \dfrac{dy}{dx} + 1.$

11. $y = f(x^2) \quad \Rightarrow \dfrac{dy}{dx} = f'(x^2) \cdot 2x = 2x \cdot \sqrt{2(x^2)^2 - 1}.$

At $x = 1, \dfrac{dy}{dx} = 2 \cdot 1 \cdot \sqrt{2-1}.$

12. Let $\cos\alpha = \dfrac{5}{13}\cdot$ Then $\sin\alpha = \dfrac{12}{13}\cdot$ So, $y = \cos^{-1}\{\cos\alpha \cdot \cos x - \sin\alpha \cdot \sin x\}$.

$\therefore \ y = \cos^{-1}\{\cos(x+\alpha)\} = x + \alpha \quad (\because x + \alpha$ is in the first or the second quadrant).

13. Let $x = \sec\theta$.

Then $y = \tan^{-1}\sqrt{\dfrac{\sec\theta + 1}{\sec\theta - 1}} = \tan^{-1}\sqrt{\dfrac{1+\cos\theta}{1-\cos\theta}} = \tan^{-1}\left(\cot\dfrac{\theta}{2}\right).$

$\therefore \quad y = \tan^{-1}\left\{\tan\left(\dfrac{\pi}{2} - \dfrac{\theta}{2}\right)\right\} = \dfrac{\pi}{2} - \dfrac{1}{2}\sec^{-1}x$

$\therefore \quad \dfrac{dy}{dx} = -\dfrac{1}{2} \cdot \dfrac{1}{|x|\sqrt{x^2-1}}.$

18. Around $x = \sqrt[3]{\dfrac{\pi}{2}}$, $[x] = 1$. So, $f(x) = \cos\left\{\dfrac{\pi}{2} - x^3\right\} = \sin x^3$ around $x = \sqrt[3]{\dfrac{\pi}{2}}.$

$\therefore \quad f'(x) = 3x^2 \cos x^3$

$\therefore \quad f'\left(\sqrt[3]{\dfrac{\pi}{2}}\right) = 3 \cdot \left(\dfrac{\pi}{2}\right)^{2/3} \cdot \cos\dfrac{\pi}{2} = 0.$

20. Let $x = \tan\theta$.

Then $\tan^{-1}\dfrac{\sqrt{1+x^2} - 1}{x} = \tan^{-1}\dfrac{\sec\theta - 1}{\tan\theta}$

$= \tan^{-1}\dfrac{1-\cos\theta}{\sin\theta} = \dfrac{\theta}{2} = \dfrac{1}{2}\tan^{-1}x.$

$$\therefore \quad \frac{d\left(\tan^{-1}\frac{\sqrt{1+x^2}-1}{x}\right)}{d(\tan^{-1}x)}=\frac{1}{2}.$$

21. Let $x = \cos\theta$.

Then $y=\operatorname{cosec}^{-1}\dfrac{1}{2x^2-1}=\operatorname{cosec}^{-1}\sec 2\theta$

$$=\operatorname{cosec}^{-1}\left\{\operatorname{cosec}\left(\frac{\pi}{2}-2\theta\right)\right\}=\frac{\pi}{2}-2\theta.$$

$$z=\sqrt{1-\cos^2\theta}=\sin\theta$$

$$\therefore \quad \frac{dy}{dz}=\frac{\frac{dy}{d\theta}}{\frac{dz}{d\theta}}=\frac{-2}{\cos\theta}=\frac{-2}{x}.$$

22. $$f'\left(2x^2+\frac{\pi}{2}\right)=\frac{df\left(2x^2+\frac{\pi}{2}\right)}{dx}=\frac{df\left(2x^2+\frac{\pi}{2}\right)}{d\left(2x^2+\frac{\pi}{2}\right)}\cdot\frac{d\left(2x^2+\frac{\pi}{2}\right)}{dx}.$$

$\because \quad f'(x)=\dfrac{df(x)}{dx}=\sin x+\sin 4x\cdot\cos x$, we get

$$\frac{df\left(2x^2+\frac{\pi}{2}\right)}{d\left(2x^2+\frac{\pi}{2}\right)}=\sin\left(2x^2+\frac{\pi}{2}\right)+\sin 4\left(2x^2+\frac{\pi}{2}\right)\cdot\cos\left(2x^2+\frac{\pi}{2}\right)$$

$$=\cos 2x^2+\sin 8x^2\cdot(-\sin 2x^2)$$

$$\therefore \quad f'\left(2x^2+\frac{\pi}{2}\right)=(\cos 2x^2-\sin 8x^2\cdot\sin 2x^2)\cdot 4x$$

$$\therefore \quad \text{at } x=\sqrt{\frac{\pi}{2}},\ f'\left(2x^2+\frac{\pi}{2}\right)=(\cos\pi-\sin 4\pi\cdot\sin\pi)\cdot 4\cdot\sqrt{\frac{\pi}{2}}=-2\sqrt{2\pi}.$$

23. $$\frac{du}{dv}=\frac{\frac{du}{dx}}{\frac{dv}{dx}}=\frac{f'(x^3)\cdot 3x^2}{g'(x^2)\cdot 2x}=\frac{\cos x^3\cdot 3x^2}{\sin x^2\cdot 2x}=\frac{3}{2}x\cos x^3\cdot\operatorname{cosec} x^2.$$

24. $$\frac{dy}{dx}=\frac{df\left(\frac{2x-1}{x^2+1}\right)}{dx}=\frac{df\left(\frac{2x-1}{x^2+1}\right)}{d\left(\frac{2x-1}{x^2+1}\right)}\cdot\frac{d\left(\frac{2x-1}{x^2+1}\right)}{dx}$$

$$= \sin\left(\frac{2x-1}{x^2+1}\right)^2 \cdot \frac{2(x^2+1)-(2x-1)2x}{(x^2+1)^2} = \frac{2(1+x-x^2)}{(x^2+1)^2}\sin\left(\frac{2x-1}{x^2+1}\right)^2.$$

25. Let $t = \cos 2\theta$.

Then $e^x = \dfrac{\sqrt{1+\cos 2\theta} - \sqrt{1-\cos 2\theta}}{\sqrt{1+\cos 2\theta} + \sqrt{1-\cos 2\theta}} = \dfrac{\cos\theta - \sin\theta}{\cos\theta + \sin\theta}$

$$= \frac{1-\tan\theta}{1+\tan\theta} = \tan\left(\frac{\pi}{4} - \theta\right).$$

$\tan\dfrac{y}{2} = \sqrt{\dfrac{1-\cos 2\theta}{1+\cos 2\theta}} = \tan\theta.$

At $t = \frac{1}{2}$, $\cos 2\theta = \frac{1}{2} \Rightarrow \theta = \frac{\pi}{6}$.

Then $x = \log\tan\frac{\pi}{12}$, $y = \frac{\pi}{3}$.

Differentiating w.r.t. θ, $e^x \cdot \dfrac{dx}{d\theta} = -\sec^2\left(\dfrac{\pi}{4} - \theta\right)$ and $\dfrac{1}{2}\sec^2\dfrac{y}{2} \cdot \dfrac{dy}{d\theta} = \sec^2\theta$

$$\therefore \quad \frac{dy}{dx} = \frac{\frac{dy}{d\theta}}{\frac{dx}{d\theta}} = \frac{2\sec^2\theta \cdot \cos^2\frac{y}{2}}{-e^{-x} \cdot \sec^2\left(\frac{\pi}{4} - \theta\right)}.$$

At $t = \dfrac{1}{2}$, i.e., $\theta = \dfrac{\pi}{6}$, $\dfrac{dy}{dx} = \dfrac{2\sec^2\frac{\pi}{6} \cdot \cos^2\frac{\pi}{6}}{-e^{-\log\tan\pi/12} \cdot \sec^2\frac{\pi}{12}}$

$$\therefore \quad \frac{dy}{dx} = \frac{2}{-\cot\frac{\pi}{12} \cdot \sec^2\frac{\pi}{12}} = -2\sin\frac{\pi}{12} \cdot \cos\frac{\pi}{12} = -\sin\frac{\pi}{6} = -\frac{1}{2}.$$

26. Differentiating $x^2 + t^2 = y$ w.r.t. x, $\dfrac{dy}{dx} = 2x + 2t\dfrac{dt}{dx}$.

Now, $t = \dfrac{x}{1+x^2}$

$$\therefore \quad \frac{dt}{dx} = \frac{1+x^2 - x \cdot 2x}{(1+x^2)^2} = \frac{1-x^2}{(1+x^2)^2}$$

$$\therefore \quad \frac{dy}{dx} = 2x + 2 \cdot \frac{x}{1+x^2} \cdot \frac{1-x^2}{(1+x^2)^2}.$$

Now, put $x = 2$.

27. Let $f(x) = \lambda x^2 + \mu x + \nu$.

Then $f'(x) = 2\lambda x + \mu$.

Also, $f(1) = f(-1)$

$\Rightarrow \quad \lambda + \mu + \nu = \lambda - \mu + \nu \quad \Rightarrow \mu = 0 \quad \therefore\ f'(x) = 2\lambda x.$

$\therefore \quad f'(a_1) = 2\lambda a_1 \qquad f'(a_2) = 2\lambda a_2 \qquad f'(a_3) = 2\lambda a_3.$

As a_1, a_2, a_3 are in AP, $f'(a_1), f'(a_2), f'(a_3)$ are in AP.

28. $\because\ P(x^2+1) = \{P(x)\}^2 + 1$ and $P(x)$ is a polynomial, we have $P(x) = x$ which also satisfies $P(0) = 0$. Therefore, $P'(x) = 1$. So, $P'(0) = 1$.

29. $5f(x) + 3f\left(\frac{1}{x}\right) = x + 2 \Rightarrow 5f\left(\frac{1}{x}\right) + 3f(x) = \frac{1}{x} + 2.$

From these, $16f(x) = 5x - \frac{3}{x} + 4$.

$\therefore \quad 16f'(x) = 5 + \frac{3}{x^2}$

$\therefore \quad y = xf(x) \Rightarrow \frac{dy}{dx} = f(x) + xf'(x) = \frac{1}{16}\left(5x - \frac{3}{x} + 4\right) + x \cdot \frac{1}{16}\left(5 + \frac{3}{x^2}\right)$

$\therefore \quad$ at $x = 1, \frac{dy}{dx} = \frac{1}{16}(5 - 3 + 4) + \frac{1}{16}(5 + 3) = \frac{7}{8}$.

30. Putting $x = 0, y = 0$, we get $2f(0) + \{f(0)\}^2 = 1$

$\Rightarrow \quad f(0) = \sqrt{2} - 1 \qquad \{\because f(0) > 0\}.$

Putting $y = x$, $2f(x) + \{f(x)\}^2 = 1$.

Differentiating w.r.t. x,

$2f'(x) + 2f(x) \cdot f'(x) = 0 \quad$ or $\quad f'(x)\{1 + f(x)\} = 0$

$\Rightarrow \quad f'(x) = 0$, because $f(x) > 0$.

31. $f(x) = e^{-x}$ satisfies $f(0) = 1, f'(0) = -1, f(x) > 0$.

It also satisfies $f'(x) < 0$ for all x.

32. $f\{g(x)\} \Rightarrow \frac{df\{g(x)\}}{dg(x)} \cdot g'(x) = 1$

$\Rightarrow \quad \sin\{g(x)\} \cdot g'(x) = 1 \quad \Rightarrow g'(x) = \frac{1}{\sin\{g(x)\}}.$

33. Let $f(x) = ax^2 + bx + c$.

Then $g(x) = ax^2 + bx + c + 2ax + b + 2a = ax^2 + (2a + b)x + 2a + b + c$.

As $f(x) > 0$ for all x, $a > 0$ and $D = b^2 - 4ac < 0$.
$g(x) > 0$ for all x if $a > 0$ and $D = (2a+b)^2 - 4a(2a+b+c) < 0$.
Now, $(2a+b)^2 - 4a(2a+b+c) = -4a^2 + b^2 - 4ac < 0$.
Hence, $g(x) > 0$.

34. $y^2 = P(x) \quad \Rightarrow 2y\frac{dy}{dx} = P'(x) \quad \Rightarrow 2\left(\frac{dy}{dx}\right)^2 + 2y\frac{d^2y}{dx^2} = P''(x)$

$$\therefore \quad 2y = \frac{d^2y}{dx^2} = P''(x) - 2\left(\frac{dy}{dx}\right)^2$$

$$\text{or} \quad 2y^3\frac{d^2y}{dx^2} = y^2P''(x) - 2y^2\left(\frac{dy}{dx}\right)^2$$

$$\text{or} \quad 2y^3\frac{d^2y}{dx^2} = P(x)\cdot P''(x) - \frac{1}{2}\{P'(x)\}^2.$$

Differentiating w.r.t. x,

$$2\frac{d}{dx}\left(y^3\frac{d^2y}{dx^2}\right) = P'(x)\,P''(x) + P(x)\cdot P'''(x) - P'(x)\cdot P''(x) = P(x)\cdot P'''(x).$$

35. $f(x) = -p^3x^3 - 6p^3 \cdot \sin x + (6p^2 + p)\cos x$

$\therefore \quad f'(x) = -3p^3x^2 - 6p^3\cos x - (6p^2 + p)\sin x$

$\therefore \quad f''(x) = -6p^3x + 6p^3\sin x - (6p^2 + p)\cos x$

$\therefore \quad f'''(x) = -6p^3 + 6p^3\cos x + (6p^2 + p)\sin x$

$\therefore \quad f'''(0) = -6p^3 + 6p^3 \cdot 1 = 0 =$ independent of p.

36. $\frac{dy}{dx} = 2\cos 2x = 2\sin\left(2x + \frac{\pi}{2}\right) \quad \therefore \quad \frac{d^2y}{dx^2} = 2^2\cos\left(2x + \frac{\pi}{2}\right) = 2^2\sin\left(2x + 2\cdot\frac{\pi}{2}\right)$

....

$$\frac{d^6y}{dx^6} = 2^6 \cdot \sin\left(2x + 6\cdot\frac{\pi}{2}\right) = 2^6\sin(2x + 3\pi)$$

37. $\frac{dx}{dy} = \frac{\frac{dx}{dt}}{\frac{dy}{dt}} = \frac{\cos t - t\sin t}{1 + \cos t}$

$$\therefore \quad \frac{d^2x}{dy^2} = \frac{\frac{d}{dt}\left(\frac{dx}{dy}\right)}{\frac{dy}{dt}} = \frac{\frac{(-2\sin t - t\cos t)(1+\cos t) - (\cos t - t\sin t)(-\sin t)}{(1+\cos t)^2}}{1 + \cos t}$$

Now, put $t = \pi/2$.

39. $a\cos x - (ax+b)\sin x + c\sin x + (cx+d)\cos x \equiv x\cos x$

or $(a+cx+d)\cos x + (-ax-b+c)\sin x \equiv x\cos x$

$\Rightarrow$ $a+d=0, c=1, -ax-b+c=0$ for all $x \neq n\pi$.

$\therefore$ $a+d=0, c=1, -a=0, -b+c=0.$

40. If $x \le 1$, $\sin^{-1}\dfrac{2x}{1+x^2} = 2\tan^{-1}x.$

But, if $x > 1$, $\sin^{-1}\dfrac{2x}{1+x^2} = \pi - 2\tan^{-1}x$

$\therefore$ for $x < 1$, we have $f(x) = 4\tan^{-1}x$. So, $f'(x) = \dfrac{4}{1+x^2}\cdot$

$\therefore$ $f'\left(\dfrac{1}{2}\right) = \dfrac{4}{1+\frac{1}{4}} = \dfrac{16}{5}\cdot$

And for $x > 1$, we have $f(x) - \pi$ $\therefore$ $f'(x) = 0.$

Hence, $f'(2) = 0$ and $f'(2) = f'(3) = 0.$

41. Let $f(x) = x^3 + ax^2 + bx + c.$

Then $a = f'(1), b = f''(2), c = f'''(3).$

Now $f'(x) = 3x^2 + 2ax + b$ $\quad f''(x) = 6x + 2a$ $\quad f'''(x) = 6.$

$\therefore$ $f'(1) = a = 3 + 2a + b$ $\quad f''(2) = b = 12 + 2a$ $\quad f'''(3) = c = 6.$

Solving $3 + a + b = 0$ and $b = 12 + 2a$ we get $a = -5, b = 2.$

$\therefore$ $f(x) = x^3 - 5x^2 + 2x + 6$

$\therefore$ $f(0) = 6, f(1) = 4, f(2) = -2, f(3) = -6.$

42. $f(x+y) + f(x-y) = 2f(x)\cdot f(y)$ for all x, y ... (1)

Putting $x = 0, y = 0$ in (1), we get $f(0) + f(0) = 2f(0)\cdot f(0)$

$\Rightarrow$ $f(0) = 1$ ($\because f(0) \neq 0$).

Putting $x = 0, y = x$, we get $f(x) + f(-x) = 2f(0)\cdot f(x)$

or $f(x) = f(-x).$

$\therefore$ $f(2) = f(-2),\ f(3) = f(-3).$

Differentiating $f(x) = f(-x)$ w.r.t. x, $f'(x) + f'(-x) = 0.$

$\therefore$ $f'(2) + f'(-2) = 0, f'(3) + f'(-3) = 0.$

43. $f(-x) = f(x)$ $\Rightarrow -f'(-x) = f'(x)$ $\therefore$ $f'(x)$ is odd.

44. $\frac{d}{dx}\{f_n(x)\} = \frac{d}{dx}\{e^{f_{n-1}(x)}\} = e^{f_{n-1}(x)}\frac{d}{dx}\{f_{n-1}(x)\} = f_n(x)\cdot\frac{d}{dx}\{f_{n-1}(x)\}$

$$= f_n(x)\cdot\frac{d}{dx}\{e^{f_{n-2}(x)}\} = f_n(x)\cdot e^{f_{n-2}(x)}\cdot\frac{d}{dx}\{f_{n-2}(x)\}$$

$$= f_n(x)\cdot f_{n-1}(x)\cdot\frac{d}{dx}\{f_{n-2}(x)\}$$

$$= \dots = f_n(x)\cdot f_{n-1}(x)\dots f_2(x)\frac{d}{dx}\{f_1(x)\}$$

$$= f_n(x)\cdot f_{n-1}(x)\cdot\dots\cdot f_2(x)\frac{d}{dx}\{e^{f_0(x)}\}$$

$$= f_n(x)\cdot f_{n-1}(x)\cdot\dots\cdot f_2(x)\cdot e^{f_0(x)}\cdot\frac{d}{dx}\{f_0(x)\}.$$

Use $e^{f_0(x)} = f_1(x)$ and $f_0(x) = x$.

45. Let $g'(1) = a, g''(2) = b$. Then $f(x) = x^2 + ax + b$.

$\therefore\quad g(x) = (1 + a + b)x^2 + (2x + a)x + 2 = (3 + a + b)x^2 + ax + 2.$

$\therefore\quad g'(x) = 2(3 + a + b)x + a$. Hence,

$a = 2(3 + a + b)\cdot 1 + a$, i.e., $3 + a + b = 0$ and

$b = 2(3 + a + b)$, i.e., $b + 2a + 6 = 0$.

Hence $b = 0, a = -3$.

So, $f(x) = x^2 - 3x$ and $g(x) = -3x + 2$.

As a result, options *a*, *b*, *c* are correct.

□

3

Limits, Indeterminate Forms

• Type 1 •

Choose the most appropriate option (a, b, c or d).

1. $\lim_{x \to 1} \frac{(\sqrt{x}-1)(2x-3)}{2x^2+x-3}$ is equal to

(a) $\frac{1}{10}$ (b) $-\frac{1}{10}$

(c) 1 (d) none of these

2. $\lim_{x \to 2} \frac{\sqrt{x-2}+\sqrt{x}-\sqrt{2}}{\sqrt{x^2-4}}$ is equal to

(a) $\frac{1}{2}$ (b) 1

(c) 2 (d) none of these

3. $\lim_{x \to \infty} (\sqrt{x+\sqrt{x}}-\sqrt{x})$ is equal to

(a) 1 (b) 0

(c) $\frac{1}{2}$ (d) none of these

4. $\lim_{n \to \infty} \frac{a^n+b^n}{a^n-b^n}$, where $a > b > 1$, is equal to

(a) -1 (b) 1

(c) 0 (d) none of these

5. $\lim_{x \to 0} \frac{3^x - 1}{\sqrt{x+1} - 1}$ is equal to

(a) $\log_e 9$ (b) $\log_e 3$

(c) 0 (d) 1

6. $\lim_{n \to \infty} \frac{4^{1/n} - 1}{3^{1/n} - 1}$ is equal to

(a) $\log_4 3$ (b) 1

(c) $\log_3 4$ (d) none of these

7. $\lim_{x \to \infty} \frac{\cos x + \sin x}{x^2}$ is equal to

(a) 1 (b) 0

(c) ∞ (d) none of these

8. $\lim_{x \to 0} \frac{3^x - 2^x}{4^x - 3^x}$ is equal to

(a) 1 (b) –1

(c) 0 (d) none of these

9. $\lim_{h \to 0} \left\{ \frac{1}{h \cdot \sqrt[3]{8+h}} - \frac{1}{2h} \right\}$ is equal to

(a) $\frac{1}{12}$ (b) $-\frac{4}{3}$

(c) $-\frac{16}{3}$ (d) $-\frac{1}{48}$

10. $\lim_{n \to \infty} \frac{n^p \sin^2(n!)}{n+1}, 0 < p < 1$, is equal to

(a) 0 (b) ∞

(c) 1 (d) none of these

11. $\lim_{x \to 0} \frac{x\sqrt{y^2 - (y-x)^2}}{(\sqrt{8xy - 4x^2} + \sqrt{8xy})^3}$ is equal to

(a) $\frac{1}{4}$ (b) $\frac{1}{2}$

(c) $\frac{1}{2\sqrt{2}}$ (d) none of these

12. $\lim_{x \to 0} \frac{(1+x+x^2)-e^x}{x^2}$ is equal to

(a) 1 (b) 0

(c) $\frac{1}{2}$ (d) none of these

13. $\lim_{x \to 0} \frac{x\tan 2x - 2x\tan x}{(1-\cos 2x)^2}$ is equal to

(a) 2 (b) –2

(c) $\frac{1}{2}$ (d) $-\frac{1}{2}$

14. $\lim_{x \to 0} \left\{\frac{\log_e(1+x)}{x^2} + \frac{x-1}{x}\right\}$ is equal to

(a) $\frac{1}{2}$ (b) $-\frac{1}{2}$

(c) 1 (d) none of these

15. $\lim_{n \to \infty} (0.2)^{\log_{\sqrt{5}}(1/4 + 1/8 + 1/16 + \ldots \text{ to } n \text{ terms})}$ is equal to

(a) 2 (b) 4

(c) 8 (d) 0

16. Let $s_n = \frac{1}{1\cdot 4} + \frac{1}{4\cdot 7} + \frac{1}{7\cdot 10} + \ldots$ to n terms. Then $\lim_{n \to \infty} s_n$ is equal to

(a) $\frac{1}{3}$ (b) 3

(c) $\frac{1}{4}$ (d) ∞

17. Let the rth term, t_r, of a series is given by $t_r = \frac{r}{1+r^2+r^4}$. Then $\lim_{n \to \infty} \sum_{r=1}^{n} t_r$ is

(a) $\frac{1}{4}$ (b) 1

(c) $\frac{1}{2}$ (d) none of these

18. Let $a = \min\{x^2 + 2x + 3, x \in R\}$ and $b = \lim\limits_{\theta \to 0} \dfrac{1 - \cos\theta}{\theta^2}$. The value of $\sum\limits_{r=0}^{n} a^r \cdot b^{n-r}$ is

(a) $\dfrac{2^{n+1} - 1}{3 \cdot 2^n}$ (b) $\dfrac{2^{n+1} + 1}{3 \cdot 2^n}$

(c) $\dfrac{4^{n+1} - 1}{3 \cdot 2^n}$ (d) none of these

19. $\lim\limits_{x \to 1} \dfrac{\sum\limits_{r=1}^{n} x^r - n}{x - 1}$ is equal to

(a) $\dfrac{n}{2}$ (b) $\dfrac{n(n+1)}{2}$

(c) 1 (d) none of these

20. $\lim\limits_{x \to 0} \dfrac{x - \tan x}{x\tan x}$ is equal to

(a) 0 (b) 1

(c) $\dfrac{1}{2}$ (d) none of these

21. $\lim\limits_{x \to 0} \sqrt{\dfrac{x - \sin x}{x + \sin^2 x}}$ is equal to

(a) 1 (b) 0

(c) ∞ (d) none of these

22. $\lim\limits_{x \to 0} \dfrac{\log_e \cos x}{x^2}$ is equal to

(a) $-\dfrac{1}{2}$ (b) $\dfrac{1}{2}$

(c) 0 (d) none of these

23. $\lim\limits_{x \to 0} \dfrac{\sin^{-1}x - \tan^{-1}x}{x^2}$ is equal to

(a) $\dfrac{1}{2}$ (b) $-\dfrac{1}{2}$

(c) 0 (d) ∞

24. $\lim_{x \to 0} \frac{1 - \cos x}{x(2^x - 1)}$ is equal to

(a) $\frac{1}{2} \log_2 e$ (b) $\frac{1}{2} \log_e 2$

(c) 1 (d) none of these

25. If $f(4) = 4, f'(4) = 1$ then $\lim_{x \to 4} \frac{2 - \sqrt{f(x)}}{2 - \sqrt{x}}$ is equal to

(a) 0 (b) 1

(c) -1 (d) none of these

26. The graph of the function $y = f(x)$ has a unique tangent at the point $(a, 0)$ through which the graph passes. Then $\lim_{x \to a} \frac{\log_e \{1 + 6f(x)\}}{3f(x)}$ is

(a) 1 (b) 0

(c) 2 (d) none of these

27. Let $f(x)$ be a twice-differentiable function and $f''(0) = 2$ then

$\lim_{x \to 0} \frac{2f(x) - 3f(2x) + f(4x)}{x^2}$ is

(a) 6 (b) 3

(c) 12 (d) none of these

28. If $f(x), g(x)$ be differentiable functions and $f(1) = g(1) = 2$ then

$$\lim_{x \to 1} \frac{f(1)\,g(x) - f(x)\,g(1) - f(1) + g(1)}{g(x) - f(x)}$$ is equal to

(a) 0 (b) 1

(c) 2 (d) none of these

29. If $f(a) = 2, f'(a) = 1, g(a) = -1, g'(a) = 2$ then $\lim_{x \to a} \frac{g(x)f(a) - g(a)f(x)}{x - a}$ is

(a) -5 (b) $\frac{1}{5}$

(c) 5 (d) none of these

30. $\lim_{n \to \infty} \left(1 + \sin \frac{a}{n}\right)^n$ is equal to

(a) $e^{a/2}$ (b) e^a

(c) e (d) e^{2a}

31. $\lim_{x \to 0} \left(\frac{1+5x^2}{1+3x^2}\right)^{1/x^2}$ is equal to

(a) e (b) $e^{1/2}$

(c) e^{-2} (d) none of these

32. $\lim_{x \to 0} \left\{\tan\left(\frac{\pi}{4} - x\right)\right\}^{1/x}$ is equal to

(a) 1 (b) e

(c) e^2 (d) e^{-2}

33. $\lim_{x \to \pi/4} (2 - \tan x)^{\log \tan x}$ is equal to

(a) 0 (b) 1

(c) e (d) e^{-1}

34. $\lim_{x \to \infty} \left(\frac{x-1}{x+1}\right)^{x+2}$ is equal to

(a) e (b) e^{-1}

(c) e^{-2} (d) none of these

35. If $\lim_{x \to \infty} \left(1 + \frac{\lambda}{x} + \frac{\mu}{x^2}\right)^{2x} = e^2$ then

(a) $\lambda = 1, \mu = 2$ (b) $\lambda = 2, \mu = 1$

(c) $\lambda = 1, \mu =$ any real constant (d) $\lambda = \mu = 1$

36. $\lim_{\theta \to 0+} \frac{\sin\sqrt{\theta}}{\sqrt{\sin\theta}}$ is equal to

(a) 0 (b) 1

(c) -1 (d) none of these

37. Let $\{x\}$ denotes the fractional part of x. Then $\lim_{x \to 0} \frac{\{x\}}{\tan\{x\}}$ is equal to

(a) 1 (b) 0

(c) -1 (d) none of these

38. $\lim\limits_{x \to \infty} \dfrac{\log_e [x]}{x}$, where $[\cdot]$ denotes the greatest integer function, is

(a) 0 (b) 1

(c) –1 (d) nonexistent

39. $\lim\limits_{x \to 1} \dfrac{\sqrt{1 - \cos 2(x-1)}}{x-1}$

(a) exists and it is $\sqrt{2}$

(b) exists and it is $-\sqrt{2}$

(c) does not exist because $x - 1 \to 0$

(d) does not exist because LH lim $\neq$ RH lim

40. $\lim\limits_{x \to 1} \dfrac{x \sin \{x - [x]\}}{x-1}$, where $[\cdot]$ denotes the greatest integer function, is

(a) 0 (b) –1

(c) not existent (d) none of these

41. $\lim\limits_{x \to 0} \dfrac{x[x]}{\sin |x|}$, where $[\cdot]$ denotes the greatest integer function, is

(a) 0 (b) 1

(c) not existent (d) none of these

42. $\lim\limits_{x \to 2} \{[2 - x] + [x - 2] - x\}$

(a) is 0 (b) is 3

(c) is –3 (d) does not exist

43. $\lim\limits_{x \to -1} \{[x] + |x|\}$, where $[\cdot]$ denotes the greatest integer function,

(a) is 0 (b) is 1

(c) does not exist (d) none of these

44. If $f(x) = \dfrac{\sin [x]}{[x]}$, $[x] \neq 0$

0, $[x] = 0$, where $[\cdot]$ denotes the greatest integer function, then

$\lim\limits_{x \to 0} f(x)$ is equal to

(a) 1 (b) 0

(c) –1 (d) none of these

45. Let $f(x) = x^2 - 1, 0 < x < 2$
$2x + 3, 2 \le x < 3.$

The quadratic equation whose roots are $\lim_{x \to 2-0} f(x)$ and $\lim_{x \to 2+0} f(x)$ is

(a) $x^2 - 6x + 9 = 0$
(b) $x^2 - 10x + 21 = 0$
(c) $x^2 - 14x + 49 = 0$
(d) none of these

46. If $[\cdot]$ denotes the greatest integer function then $\lim_{n \to \infty} \frac{[x] + [2x] + \ldots + [nx]}{n^2}$ is

(a) 0
(b) x
(c) $\frac{x}{2}$
(d) $\frac{x^2}{2}$

47. $\lim_{x \to 0} \frac{\int_0^{x^2} \cos t^2 dt}{x \sin x}$ is equal to

(a) 1
(b) 2
(c) 0
(d) none of these

48. $\lim_{x \to a} \frac{x}{x-a} \cdot \int_a^x f(x)\, dx$ is equal to

(a) $f(a)$
(b) $af(a)$
(c) 0
(d) none of these

49. $\lim_{x \to 1+0} \frac{\int_1^x |t-1|\, dt}{\sin(x-1)}$ is equal to

(a) 0
(b) 1
(c) –1
(d) none of these

50. Let $f(x) = \sin x, x \ne n\pi$ and $g(x) = x^2 + 1, x \ne 2$
$2, x = n\pi$, where $n \in Z$, $\quad$ $3, \quad x = 2.$

Then $\lim_{x \to 0} g(f(x))$ is

(a) 0
(b) 1
(c) 3
(d) none of these

51. If $f(x)$ is continuous in $[0, 1]$ and $f\left(\frac{1}{3}\right)=1$ then $\lim\limits_{n\to\infty} f\left(\frac{n}{\sqrt{9n^2+1}}\right)$ is

(a) 1 (b) 0

(c) $\frac{1}{3}$ (d) none of these

52. If $f(x)$ is continuous and $f\left(\frac{9}{2}\right)=\frac{2}{9}$ then $\lim\limits_{x\to 0} f\left(\frac{1-\cos 3x}{x^2}\right)$ is equal to

(a) $\frac{9}{2}$ (b) $\frac{2}{9}$

(c) 0 (d) none of these

• Type 2 •

Choose the correct options. One or more options may be correct.

53. Let $f(x)=1+\frac{2x}{a}, 0\le x<1$

$ax,\ 1\le x<2.$

If $\lim\limits_{x\to 1} f(x)$ exists then a is

(a) 1 (b) –1

(c) 2 (d) –2

54. If α is a repeated root of $ax^2+bx+c=0$ then $\lim\limits_{x\to\alpha}\frac{\sin(ax^2+bx+c)}{(x-\alpha)^2}$ is

(a) 0 (b) a

(c) b (d) c

55. If $f(x)=|x-1|-[x]$, where $[x]$ = the greatest integer less than or equal to x, then

(a) $f(1+0)=-1, f(1-0)=0$ (b) $f(1+0)=0=f(1-0)$

(c) $\lim\limits_{x\to 1} f(x)$ exists (d) $\lim\limits_{x\to 1} f(x)$ does not exist

56. If $\lim\limits_{n \to \infty} \left(an - \dfrac{1+n^2}{1+n} \right) = b$, a finite number, then

(a) $a = 1$ (b) $a = 0$

(c) $b = 1$ (d) $b = -1$

57. Let $\tan \alpha \cdot x + \sin \alpha \cdot y = \alpha$ and $\alpha \operatorname{cosec} \alpha \cdot x + \cos \alpha \cdot y = 1$ be two variable straight lines, α being the parameter. Let P be the point of intersection of the lines. In the limiting position when $\alpha \to 0$, the point P lies on the line

(a) $x = 2$ (b) $x = -1$

(c) $y + 1 = 0$ (d) $y = 2$

Answers

1. b	**2.** a	**3.** c	**4.** b	**5.** a
6. c	**7.** b	**8.** d	**9.** d	**10.** a
11. a	**12.** c	**13.** c	**14.** a	**15.** b
16. a	**17.** c	**18.** c	**19.** b	**20.** a
21. b	**22.** a	**23.** c	**24.** a	**25.** b
26. c	**27.** a	**28.** c	**29.** c	**30.** b
31. d	**32.** d	**33.** b	**34.** c	**35.** c
36. b	**37.** a	**38.** a	**39.** d	**40.** c
41. c	**42.** c	**43.** c	**44.** d	**45.** b
46. c	**47.** a	**48.** b	**49.** a	**50.** b
51. a	**52.** b	**53.** b, c	**54.** b	**55.** a, d
56. a, c	**57.** a, c			

Hints and Solutions to Selected Questions

2. Limit $= \lim_{x \to 2} \left\{ \frac{1}{\sqrt{x+2}} + \frac{\sqrt{x} - \sqrt{2}}{\sqrt{x^2 - 4}} \right\} = \frac{1}{2} + \lim_{x \to 2} \frac{x-2}{\sqrt{x} + \sqrt{2}} \cdot \frac{1}{\sqrt{(x+2)(x-2)}}$

$$= \frac{1}{2} + \lim_{x \to 2} \frac{1}{\sqrt{x} + \sqrt{2}} \cdot \sqrt{\frac{x-2}{x+2}} = \frac{1}{2}.$$

3. Limit $= \lim_{x \to \infty} \frac{x + \sqrt{x} - x}{\sqrt{x + \sqrt{x}} + \sqrt{x}} = \lim_{x \to \infty} \frac{\sqrt{x}}{\sqrt{x} \left\{ \sqrt{1 + \frac{1}{\sqrt{x}}} + 1 \right\}} = \frac{1}{2}.$

4. Limit $= \lim_{n \to \infty} \frac{1 + \left(\frac{b}{a}\right)^n}{1 - \left(\frac{b}{a}\right)^n} = 1$ because $0 < \frac{b}{a} < 1$ implies $\left(\frac{b}{a}\right)^n \to 0$ as $n \to \infty$.

5. Limit $= \lim_{x \to 0} \frac{e^{x \log_e 3} - 1}{x} \cdot (\sqrt{x+1} + 1)$

$$= \lim_{x \to 0} \frac{\left\{ 1 + x \log_e 3 + \frac{(x \log_e 3)^2}{2!} + \dots \right\} - 1}{x} \cdot (\sqrt{x+1} + 1)$$

$$= (\log_e 3) \times 2 = \log_e 9.$$

6. Limit $= \lim_{x \to 0} \frac{4^x - 1}{3^x - 1} = \lim_{x \to 0} \frac{e^{x \log_e 4} - 1}{e^{x \log_e 3} - 1}$

$$= \lim_{x \to 0} \frac{\left\{ 1 + x \log_e 4 + \frac{(x \log_e 4)^2}{2!} + \dots \right\} - 1}{\left\{ 1 + x \log_e 3 + \frac{(x \log_e 3)^2}{2!} + \dots \right\} - 1} = \frac{\log_e 4}{\log_e 3}.$$

9. Limit $= \lim_{h \to 0} \frac{2 - \sqrt[3]{8+h}}{2h \cdot \sqrt[3]{8+h}}$

$$= \lim_{h \to 0} \frac{8 - (8+h)}{2h \cdot \sqrt[3]{8+h} \, \{8^{2/3} + 8^{1/3} \cdot (8+h)^{1/3} + (8+h)^{2/3}\}} = -\frac{1}{48}.$$

10. Limit $= \lim_{n \to \infty} \dfrac{\sin^2(n!)}{n^{1-p}\left(1+\dfrac{1}{n}\right)} = \dfrac{\text{some number between 0 and 1}}{\infty} = 0.$

11. Limit $= \lim_{x \to 0} \dfrac{x^{3/2}\sqrt{2y-x}}{x^{3/2}(\sqrt{8y-4x}+\sqrt{8y})} = \dfrac{\sqrt{2y}}{2\sqrt{8y}} = \dfrac{1}{4}.$

14. Limit $= \lim_{x \to 0} \dfrac{\log_e(1+x)+x^2-x}{x^2} = \lim_{x \to 0} \dfrac{\left(x-\dfrac{1}{2}x^2+\dfrac{1}{3}x^3-\ldots\right)+x^2-x}{x^2} = \dfrac{1}{2}.$

17. $t_r = \dfrac{1}{2} \cdot \dfrac{2r}{(r^2+1)^2-r^2} = \dfrac{1}{2}\left\{\dfrac{1}{r^2-r+1} - \dfrac{1}{r^2+r+1}\right\}$

$= \dfrac{1}{2}\left\{\dfrac{1}{r(r-1)+1} - \dfrac{1}{(r+1)r+1}\right\}$

$\therefore \quad \sum_{r=1}^{n} t_r = \sum_{r=1}^{n} \dfrac{1}{2}\{f(r)-f(r+1)\}$, where $f(r) = \dfrac{1}{r(r-1)+1}$

$= \dfrac{1}{2}\{f(1)-f(n+1)\} = \dfrac{1}{2}\left\{1-\dfrac{1}{(n+1)n+1}\right\} \to \dfrac{1}{2}$ as $n \to \infty$.

18. $x^2+2x+3 = (x+1)^2+2 \geq 2$. So, $a = 2$; $b = \lim_{\theta \to 0} \dfrac{2\sin^2\dfrac{\theta}{2}}{\theta^2} = \dfrac{1}{2}.$

$\therefore \quad \sum_{r=0}^{n} a^r b^{n-r} = b^n + ab^{n-1} + a^2b^{n-2} + \ldots + a^n = \dfrac{b^n\left\{1-\left(\dfrac{a}{b}\right)^{n+1}\right\}}{1-\dfrac{a}{b}}.$

$= \dfrac{\left(\dfrac{1}{2}\right)^n\{1-4^{n+1}\}}{1-4} = \dfrac{4^{n+1}-1}{3 \times 2^n}.$

19. Limit $= \lim_{x \to 1} \sum_{r=1}^{n} \dfrac{x^r - 1^r}{x-1} = \sum_{r=1}^{n} r \cdot 1^{r-1} = 1+2+3+\ldots+n = \dfrac{n(n+1)}{2}.$

20. Limit $= \lim_{x \to 0} \dfrac{1-\sec^2 x}{\tan x + x\sec^2 x}$ (using L'Hospital's rule)

$= \lim_{x \to 0} \dfrac{-2\sec^2 x \cdot \tan x}{2\sec^2 x + x \cdot 2\sec^2 x \cdot \tan x} = \dfrac{0}{2} = 0.$

21. Limit $= \sqrt{\lim_{x \to 0} \frac{x - \sin x}{x + \sin^2 x}} = \sqrt{\lim_{x \to 0} \frac{1 - \cos x}{1 + \sin 2x}} = 0.$

22. Limit $= \lim_{x \to 0} \frac{\frac{-\sin x}{\cos x}}{2x} = \lim_{x \to 0} -\frac{1}{2} \cdot \frac{\sin x}{x} \cdot \frac{1}{\cos x} = -\frac{1}{2}.$

23. Limit $= \lim_{x \to 0} \frac{\frac{1}{\sqrt{1-x^2}} - \frac{1}{1+x^2}}{2x} = \lim_{x \to 0} \frac{1 + x^2 - \sqrt{1-x^2}}{2x(1+x^2)\sqrt{1-x^2}}$

$$= \lim_{x \to 0} \frac{(1+x^2)^2 - (1-x^2)}{2x(1+x^2)\sqrt{1-x^2}\,\{1 + x^2 + \sqrt{1-x^2}\}}$$

$$= \lim_{x \to 0} \frac{x^4 + 3x^2}{2x(1+x^2)\sqrt{1-x^2}\,\{1 + x^2 + \sqrt{1-x^2}\}} = 0.$$

24. Limit $= \lim_{x \to 0} \frac{1 - \cos x}{x^2} \cdot \frac{x}{2^x - 1} = \lim_{x \to 0} \frac{2\sin^2 \frac{x}{2}}{4 \cdot \left(\frac{x}{2}\right)^2} \cdot \frac{x}{e^{x\log_e 2} - 1}$

$$= \lim_{x \to 0} \frac{1}{2} \cdot \left(\frac{\sin \frac{x}{2}}{\frac{x}{2}}\right)^2 \cdot \frac{x}{\left\{1 + x\log_e 2 + \frac{(x\log_e 2)^2}{n!} + \ldots\right\} - 1}$$

$$= \frac{1}{2} \cdot \frac{1}{\log_e 2} = \frac{1}{2} \log_2 e.$$

25. Limit $= \lim_{x \to 4} \frac{4 - f(x)}{4 - x} \cdot \frac{2 + \sqrt{x}}{2 + \sqrt{f(x)}} = \frac{2+2}{2+2} \cdot \lim_{x \to 4} \frac{4 - f(x)}{4 - x}$

$$= \lim_{x \to 4} \frac{-f'(x)}{-1} = f'(4) = 1.$$

26. From the question, $f(a) = 0$ and $f(x)$ is differentiable at $x = a$.

$\therefore \quad$ limit $= \lim_{x \to a} \frac{\frac{1}{1 + 6f(x)} \cdot 6f'(x)}{3f'(x)} = 2 \cdot \frac{1}{1 + 6f(a)} = 2.$

27. Limit $= \lim_{x \to 0} \frac{2f'(x) - 6f'(2x) + 4f'(4x)}{2x}$

$$= \lim_{x \to 0} \frac{2f''(x) - 12f''(2x) + 16f''(4x)}{2} = \frac{6f''(0)}{2} = 6.$$

28. Limit $= \lim_{x \to 1} \dfrac{f(1)\,g'(x) - f'(x)\,g(1)}{g'(x) - f'(x)} = \lim_{x \to 1} \dfrac{2\{g'(x) - f'(x)\}}{g'(x) - f'(x)} = \lim_{x \to 1} 2 = 2.$

30. Limit $= \lim_{n \to \infty} \left(1 + \dfrac{\sin \frac{a}{n}}{\frac{a}{n}} \cdot \dfrac{a}{n} \right)^n = \lim_{n \to \infty} \left(1 + \dfrac{a}{n} \right)^n = e^a.$

31. Limit $= \lim_{x \to 0} \left(1 + 2 \cdot \dfrac{x^2}{1 + 3x^2} \right)^{1/x^2} = \lim_{x \to 0} \left(1 + 2 \cdot \dfrac{1}{3 + \frac{1}{x^2}} \right)^{\left(3 + \frac{1}{x^2}\right) \cdot \frac{\frac{1}{x^2}}{3 + \frac{1}{x^2}}}$

$$= \left\{ \lim_{y \to \infty} \left(1 + \frac{2}{y} \right)^y \right\}^{\lim_{x \to 0} \frac{\frac{1}{x^2}}{3 + \frac{1}{x^2}}} = (e^2)^{\lim_{x \to 0} \frac{1}{3x^2 + 1}} = e^2.$$

32. Limit $= e^{\lim_{x \to 0} \frac{1}{x} \log \left\{ \tan \left(\frac{\pi}{4} - x \right) \right\}} = e^{\lim_{x \to 0} \frac{\cot \left(\frac{\pi}{4} - x \right) \cdot \sec^2 \left(\frac{\pi}{4} - x \right) \cdot (-1)}{1}} = e^{-2}.$

33. Limit $= (2 - 1)^{\log 1} = 1^0 = 1.$

34. Limit $= \lim_{x \to \infty} \left(1 + \dfrac{-2}{x + 1} \right)^{(x+1) \cdot \frac{x+2}{x+1}} = \left\{ \lim_{x \to 0} \left(1 + \dfrac{-2}{x + 1} \right)^{x+1} \right\}^{\lim_{x \to \infty} \frac{x+2}{x+1}}.$

$$= (e^{-2})^{\lim_{x \to \infty} \frac{1 + 2/x}{1 + 1/x}} = (e^{-2})^1 = e^{-2}.$$

35. Limit $= \lim_{x \to \infty} \left(1 + \dfrac{1}{\frac{x^2}{\lambda x + \mu}} \right)^{2x} = \lim_{x \to \infty} \left(1 + \dfrac{1}{\frac{x^2}{\lambda x + \mu}} \right)^{\frac{x^2}{\lambda x + \mu} \cdot \frac{2(\lambda x + \mu)}{x}}$

$$= \left\{ \lim_{x \to \infty} \left(1 + \frac{1}{\frac{x^2}{\lambda x + \mu}} \right)^{\frac{x^2}{\lambda x + \mu}} \right\}^{\lim_{x \to \infty} \frac{2(\lambda x + \mu)}{x}} = e^{\lim_{x \to \infty} \left(2\lambda + \frac{2\mu}{x} \right)} = e^{2\lambda}$$

$\therefore \quad e^{2\lambda} = e^2 \qquad \therefore \; \lambda = 1.$

36. Limit $= \lim_{h \to 0} \frac{\sin \sqrt{h}}{\sqrt{\sin h}} = \lim_{h \to 0} \left(\frac{\sin \sqrt{h}}{\sqrt{h}} \right) \sqrt{h} \left(\frac{1}{\sqrt{\frac{\sin h}{h}}} \right) \frac{1}{\sqrt{h}} = 1.$

37. When $x \to 0$, $\{x\} \to 0$.

$\therefore$ limit $= \lim_{h \to 0} \frac{h}{\tan h}$ or $\lim_{h \to 0} \frac{-h}{\tan(-h)} = 1.$

38. Let $x = n + k$, $n \in Z$, $0 \le k < 1$.

Limit $= \lim_{n \to \infty} \frac{\log n}{n + k} = \lim_{n \to \infty} \frac{\frac{1}{n}}{1} = 0.$

39. Limit $= \lim_{x \to 1} \frac{|\sin(x-1)|}{x-1} = \lim_{h \to 0} \frac{|\sin h|}{h}$ or $\lim_{h \to 0} \frac{|\sin h|}{-h}$

$\therefore$ RH limit $= \lim_{h \to 0} \frac{\sin h}{h} = 1$ and LH limit $= \lim_{h \to 0} \frac{\sin h}{-h} = -1.$

40. RH limit $= \lim_{h \to 0} \frac{(1+h)\sin\{1+h-[1+h]\}}{1+h-1} = \lim_{h \to 0} \frac{(1+h)\sin(1+h-1)}{h}$

$= \lim_{h \to 0} (1+h) \frac{\sin h}{h} = 1.$

LH limit $= \lim_{h \to 0} \frac{(1-h)\sin\{1-h-[1-h]\}}{1-h-1} = \lim_{h \to 0} \frac{(1-h)\sin(1-h)}{-h}$

$= \lim_{h \to 0} \frac{1-h}{-h} \cdot \sin(1-h) = -\infty \times \sin 1$

$\therefore$ RH limit $\ne$ LH limit.

41. RH limit $= \lim_{h \to 0} \frac{(0+h)[0+h]}{\sin|0+h|} = \lim_{h \to 0} \frac{0}{\sin h} = \lim_{h \to 0} 0 = 0.$

LH limit $= \lim_{h \to 0} \frac{(0-h)[0-h]}{\sin|0-h|} = \lim_{h \to 0} \frac{-h(-1)}{\sin h} = 1$

$\therefore$ RH limit $\ne$ LH limit.

42. RH limit $= \lim_{h \to 0} \{[2-(2+h)] + [(2+h)-2] - (2+h)\}$

$$= \lim_{h \to 0} \{[-h] + [h] - 2 - h\}$$

$$= \lim_{h \to 0} \{-1 + 0 - 2 - h\} = -3.$$

$$\text{LH limit} = \lim_{h \to 0} \{[2 - (2 - h)] + [(2 - h) - 2] - (2 - h)\}$$

$$= \lim_{h \to 0} \{[h] + [-h] - 2 + h\}$$

$$= \lim_{h \to 0} \{0 - 1 - 2 + h\} = -3.$$

44. $\text{RH limit} = \lim_{h \to 0} \dfrac{\sin[0 + h]}{[0 + h]} = \lim_{\theta \to 0} \dfrac{\sin \theta}{\theta} = 1.$

$\text{LH limit} = \lim_{k \to 0} \dfrac{\sin[0 - h]}{[0 - h]} = \lim_{h \to 0} \dfrac{\sin(-1)}{-1} = \sin 1.$

$\therefore$ RH limit $\neq$ LH limit.

45. $\lim\limits_{x \to 2-0} f(x) = \lim\limits_{x \to 2-0} (x^2 - 1) = 2^2 - 1 = 3$

$\lim\limits_{x \to 2+0} f(x) = \lim\limits_{x \to 2+0} (2x + 3) = 7.$

46. $nx - 1 < [nx] \le nx$. Putting $n = 1, 2, 3, \ldots n$ and adding them,

$x\Sigma n - n < \Sigma[nx] \le x\Sigma n$

$$\therefore \quad x \cdot \frac{\Sigma n}{n^2} - \frac{1}{n} < \frac{\Sigma[nx]}{n^2} \le x \cdot \frac{\Sigma n}{n^2} \qquad \ldots (1)$$

Now, $\lim\limits_{n \to \infty} \left\{x \cdot \dfrac{\Sigma n}{n^2} - \dfrac{1}{n}\right\} = x \cdot \lim\limits_{n \to \infty} \dfrac{\Sigma n}{n^2} - \lim\limits_{n \to \infty} \dfrac{1}{n} = \dfrac{x}{2}$

$$\lim_{n \to \infty} \left\{x \cdot \frac{\Sigma n}{n^2}\right\} = x \cdot \lim_{n \to \infty} \frac{\Sigma n}{n^2} = \frac{x}{2}.$$

As the two limits are equal, by (1), $\lim\limits_{n \to \infty} \dfrac{\Sigma[nx]}{n^2} = \dfrac{x}{2}$.

Note Here we have to select the correct option. In many cases we can adopt the method of elimination by trial. As the answer is to be correct for all x, take $x = 1$. Then

$$\lim_{n \to \infty} \frac{[x]+[2x]+\ldots+[nx]}{n} = \lim_{n \to \infty} \frac{1+2+\ldots+n}{n^2} = \frac{1}{2}.$$

When $x = 1$ is put in the options, we find that (a), (b) are wrong. Now, take $x = 2$. Then

$$\lim_{n \to \infty} \frac{[x]+[2x]+\ldots+[nx]}{n^2} = \lim_{n \to \infty} \frac{2+4+\ldots+2n}{n^2} = 1.$$

For $x = 2$, in the options (c), (d), we find that (d) is also wrong.

If the option (d) is given as "none of these" then this process of elimination becomes unsafe unless all other options are verified to be incorrect.

47. Limit $= \lim_{x \to 0} \dfrac{\dfrac{d}{dx}\displaystyle\int_0^{x^2} \cos t^2 dt}{\dfrac{d}{dx}(x\sin x)} = \lim_{x \to 0} \dfrac{\cos(x^2)^2 \cdot \dfrac{d(x^2)}{dx}}{\sin x + x\cos x}$

$$= \lim_{x \to 0} \frac{2x\cos x^4}{\sin x + x\cos x} = \lim_{x \to 0} \frac{2\cos x^4 - 8x^4 \sin x^4}{2\cos x - x\sin x} = 1.$$

48. Limit $= \lim_{x \to a} \dfrac{\dfrac{d}{dx}\left\{x\displaystyle\int_a^x f(x)\,dx\right\}}{\dfrac{d}{dx}(x-a)} = \lim_{x \to a} \dfrac{\displaystyle\int_a^x f(x)\,dx + xf(x)}{1} = af(a).$

49. Limit $= \lim_{h \to 0} \dfrac{\displaystyle\int_1^{1+h} |t-1|\,dt}{\sin(1+h-1)} = \lim_{h \to 0} \dfrac{\dfrac{d}{dh}\displaystyle\int_1^{1+h} |t-1|\,dt}{\dfrac{d}{dh}(\sin h)}$

$$= \lim_{h \to 0} \frac{|1+h-1| \cdot \dfrac{d(1+h)}{dh}}{\cos h} = \lim_{h \to 0} \frac{|h|}{\cos h} = 0.$$

50. $g\{f(x)\} = \{f(x)\}^2 + 1, f(x) \neq 2$

$3, f(x) = 2$

$\therefore \quad g\{f(x)\} = \sin^2 x + 1, x \neq n\pi$

$3, x = n\pi$

RH limit $= \lim_{x \to 0+0} g\{f(x)\} = \lim_{h \to 0} g\{f(0+h)\} = \lim_{h \to 0} \{\sin^2(0+h) + 1\} = 1.$

$$\text{LH limit} = \lim_{x \to 0-0} g\{f(x)\} = \lim_{h \to 0} g\{f(0-h)\} = \lim_{h \to 0} \{\sin^2(0-h)+1\} = 1.$$

51. $\lim_{n \to \infty} f\left(\frac{n}{\sqrt{9n^2+1}}\right) = f\left\{\lim_{n \to \infty} \frac{n}{\sqrt{9n^2+1}}\right\} = f\left(\frac{1}{3}\right) = 1.$

52. $\lim_{x \to 0} f\left(\frac{1-\cos 3x}{x^2}\right) = f\left\{\lim_{x \to 0} \frac{1-\cos 3x}{x^2}\right\} = f\left(\frac{9}{2}\right) = \frac{2}{9}.$

53. $\text{RH limit} = \lim_{h \to 0} f(1+h) = \lim_{h \to 0} a(1+h) = a$

$$\text{LH limit} = \lim_{h \to 0} f(1-h) = \lim_{h \to 0} \left\{1 + \frac{2}{a}(1-h)\right\} = 1 + \frac{2}{a}.$$

$$\lim_{x \to 1} f(x) \text{ exists} \Rightarrow \text{RH limit} = \text{LH limit} \Rightarrow a = 1 + \frac{2}{a} \Rightarrow a = 2, -1.$$

54. As α is a repeated root, $a\alpha^2 + b\alpha + c = 0$ and $2a\alpha + b = 0$.

$$\text{Limit} = \lim_{x \to \alpha} \frac{(2ax+b)\cos(ax^2+bx+c)}{2(x-\alpha)}$$

$$= \lim_{x \to \alpha} \frac{2a\cos(ax^2+bx+c) - (2ax+b)^2\sin(ax^2+bx+c)}{2}$$

$$= \frac{2a}{2} = a.$$

55. $f(1+0) = \lim_{h \to 0} \{|1+h-1| - [1+h]\} = \lim_{h \to 0} \{h-1\} = -1.$

$$f(1-0) = \lim_{h \to 0} \{|1-h-1| - [1-h]\} = \lim_{h \to 0} \{h-0\} = 0.$$

56. $\text{Limit} = \lim_{n \to \infty} \frac{an(1+n)-(1+n^2)}{1+n} = \lim_{n \to \infty} \frac{(a-1)n^2+an-1}{n+1}$

$= \infty$ if $a - 1 \neq 0$.

If $a - 1 = 0$, limit $= \lim_{n \to \infty} \frac{an-1}{n+1} = a = b$

$\therefore \quad a = b = 1.$

57. Solving the equations, $x = \dfrac{\alpha\cos\alpha - \sin\alpha}{\sin\alpha - \alpha}$ and $y = \dfrac{\alpha - x\tan\alpha}{\sin\alpha}$.

$$\therefore \quad \lim_{\alpha\to 0} x = \lim_{\alpha\to 0} \frac{\cos\alpha - \alpha\sin\alpha - \cos\alpha}{\cos\alpha - 1}$$

$$= \lim_{\alpha\to 0} \frac{\alpha\cdot\sin\alpha}{2\sin^2\dfrac{\alpha}{2}} = \lim_{\alpha\to 0} \frac{4\left(\dfrac{\alpha}{2}\right)^2\cdot\dfrac{\sin\alpha}{\alpha}}{\left(\sin\dfrac{\alpha}{2}\right)^2\cdot 2} = 2.$$

$$\lim_{\alpha\to 0} y = \lim_{\alpha\to 0} \frac{\alpha - x\tan\alpha}{\sin\alpha} = \lim_{\alpha\to 0}\left(\frac{\alpha}{\sin\alpha} - \frac{x}{\cos\alpha}\right) = 1 - 2 = -1.$$

$\therefore$ in the limiting position, P = (2, –1).

□

4

Continuity and Differentiability

• *Type 1* •

Choose the most appropriate option (a, b, c or d).

1. Let $f(x) = \log |x-1|, x \neq 1$. The value of $f'\left(\frac{1}{2}\right)$

(a) is –2 (b) is 2

(c) does not exist (d) none of these

2. Let $y = \left|\tan\left(\frac{\pi}{4} - x\right)\right|$. Then $\frac{dy}{dx}$ at $x = \frac{\pi}{4}$

(a) is 1 (b) is –1

(c) does not exist (d) none of these

3. Let $y = |x| + |x-2|$. Then $\frac{dy}{dx}$ at $x = 2$

(a) is 2 (b) is 0

(c) does not exist (d) none of these

4. Let $f(x) = \lambda + \mu|x| + \nu|x|^2$, where λ, μ, ν are real constants. Then $f'(0)$ exists if

(a) $\mu = 0$ (b) $\nu = 0$

(c) $\lambda = 0$ (d) $\mu = \nu$

5. If $f(x) = \frac{[x]}{|x|}, x \neq 0$ where $[\cdot]$ denotes the greatest integer function, then $f'(1)$ is

(a) -1 (b) ∞

(c) nonexistent (d) none of these

6. If $f(x) = |\cos 2x|$ then $f'\left(\frac{\pi}{4}+0\right)$ is equal to

(a) 2 (b) 0

(c) -2 (d) none of these

7. If $f(x) = \sin \pi[x]$ then $f'(1-0)$ is equal to

(a) -1 (b) 0

(c) 1 (d) none of these

8. Let $f(x) = [x^2] - [x]^2$, where $[\cdot]$ denotes the greatest integer function. Then

(a) $f(x)$ is discontinuous for all integral values of x

(b) $f(x)$ is discontinuous only at $x = 0, 1$

(c) $f(x)$ is continuous only at $x = 1$

(d) none of these

9. Let $f(x) = [\cos x + \sin x]$, $0 < x < 2\pi$ where $[x]$ denotes the greatest integer less than or equal to x. The number of points of discontinuity of $f(x)$ is

(a) 6 (b) 5

(c) 4 (d) 3

10. Let $f(x) = x - |x - x^2|$, $x \in [-1, 1]$. Then the number of points at which $f(x)$ is discontinuous is

(a) 1 (b) 2

(c) 0 (d) none of these

11. Let $f(x) = \sqrt{1+x^2}$, $x < \sqrt{3}$

$\sqrt{3}x - 1$, $\sqrt{3} \le x < 4$

$[x]$, $4 \le x < 5$, where $[x]$ is the greatest integer $\le x$

$|1 - x|$, $x \ge 5$

The number of points of discontinuity of $f(x)$ in R is

(a) 3 (b) 0

(c) infinite (d) none of these

12. Let $f(x)=\int_0^x t\sin\frac{1}{t}\,dt$. Then the number of points of discontinuity of the function $f(x)$ in the open interval $(0, \pi)$ is

(a) 0 (b) 1
(c) 2 (d) infinite

13. Let $f: [0, 1] \to [0, 1]$ be a continuous function. Then

(a) $f(x) = x$ for at least one $0 \le x \le 1$

(b) $f(x)$ will be differentiable in $[0, 1]$

(c) $f(x) + x = 0$ for at least one x such that $0 \le x \le 1$

(d) none of these

14. Let $f(x)$ be a continuous function defined for $1 \le x \le 3$. If $f(x)$ takes rational values for all x and $f(2) = 10$ then the value of $f(1.5)$ is

(a) 7.5 (b) 10
(c) 5 (d) none of these

15. If $f(x) = e^{-1/x^2}, x \ne 0$, and $f(0) = 0$ then $f'(0)$ is

(a) 0 (b) 1
(c) e (d) nonexistent

16. Let $f(x) = \sin x$, $g(x) = [x + 1]$ and $g\{f(x)\} = h(x)$, where $[\cdot]$ is the greatest integer function. Then $h'\left(\frac{\pi}{2}\right)$ is

(a) nonexistent (b) 1
(c) -1 (d) none of these

17. Let $f(x) = [x]$, $g(x) = |x|$ and $f\{g(x)\} = h(x)$, where $[\cdot]$ is the greatest integer function. Then $h'(-1)$ is

(a) 0 (b) $-\infty$
(c) nonexistent (d) none of these

18. The number of values of $x \in [0, 2]$ at which the real function $f(x) = \left|x - \frac{1}{2}\right| + |x - 1| + \tan x$ is not finitely differentiable is

(a) 2 (b) 3
(c) 1 (d) 0

19. Let $f(x) = [n + p\sin x]$, $x \in (0, \pi)$, $n \in \mathbb{Z}$, p is a prime number and $[x]$ = the greatest integer less than or equal to x. The number of points at which $f(x)$ is not differentiable is

(a) p (b) $p-1$
(c) $2p+1$ (d) $2p-1$

20. Let $f(x) = (x-1)^2 \cos \frac{1}{x-1} - |x|, x \neq 1$
$-1, x = 1.$
The set of points where $f(x)$ is not differentiable is
(a) {1} (b) {0, 1}
(c) {0} (d) none of these

21. Let $f(x) = \frac{1-\sin x}{\sin 2x}, x \neq \frac{\pi}{2}$. If $f(x)$ is continuous at $x = \frac{\pi}{2}$ then $f\left(\frac{\pi}{2}\right)$ should be
(a) 1 (b) 0
(c) $\frac{1}{2}$ (d) none of these

22. A function $f(x)$ is defined as below

$$f(x) = \frac{\cos(\sin x) - \cos x}{x^2}, x \neq 0 \text{ and } f(0) = a.$$

$f(x)$ is continuous at $x = 0$ if a equals
(a) 0 (b) 4
(c) 5 (d) 6

23. Let $f(x) = \frac{1-\tan x}{4x-\pi}, x \neq \frac{\pi}{4}$ and $x \in \left[0, \frac{\pi}{2}\right)$
$\lambda, x = \frac{\pi}{4}$

If $f(x)$ is continuous in $\left[0, \frac{\pi}{2}\right)$ then λ is
(a) 1 (b) $\frac{1}{2}$
(c) $-\frac{1}{2}$ (d) none of these

24. Let $f(x) = (\sin x)^{\frac{1}{\pi - 2x}}, x \neq \frac{\pi}{2}$.

If $f(x)$ is continuous at $x = \frac{\pi}{2}$ then $f\left(\frac{\pi}{2}\right)$ is
(a) e (b) 1
(c) 0 (d) none of these

25. Let $f(x) = \sin\frac{1}{x}$, $x \neq 0$. Then $f(x)$ can be continuous at $x = 0$

(a) if $f(0) = 1$ (b) if $f(0) = 0$

(c) if $f(0) = -1$ (d) for no value of $f(0)$

26. If $f(x) = px^2 - q, x \in [0, 1)$
$x + 1, x \in (1, 2]$

and $f(1) = 2$ then the value of the pair (p, q) for which $f(x)$ cannot be continuous at $x = 1$ is

(a) $(2, 0)$ (b) $(1, -1)$

(c) $(4, 2)$ (d) $(1, 1)$

27. If $f(x) = x, x \leq 1$, and $f(x) = x^2 + bx + c, x > 1$, and $f'(x)$ exists finitely for all $x \in R$ then

(a) $b = -1, c \in R$ (b) $c = 1, b \in R$

(c) $b = 1, c = -1$ (d) $b = -1, c = 1$

• *Type 2* •

Choose the correct options. One or more options may be correct.

28. If $f(x) = e^x, x < 2$
$a + bx, x \geq 2$

is differentiable for all $x \in R$ then

(a) $a + b = 0$ (b) $a + 2b = e^2$

(c) $b = e^2$ (d) none of these

29. If $f(x) = \cos^{-1}(\cos x)$ then $f(x)$ is

(a) continuous at $x = \pi$ (b) discontinuous at $x = -\pi$

(c) differentiable at $x = 0$ (d) nondifferentiable at $x = \pi$

30. Let $f(x) = x - |x|$. Then

(a) $f(x)$ is continuous everywhere

(b) $f(x)$ is differentiable everywhere

(c) $f(x)$ is discontinuous at $x = 0$

(d) $f(x)$ is not differentiable at $x = 0$

31. If $f(x) = [x] + \left[x + \frac{1}{2}\right]$, where $[\cdot]$ denotes the greatest integer function, then

(a) $f(x)$ is continuous at $x = \frac{1}{2}$ (b) $\lim_{x \to 1/2 + 0} f(x) = 1$

(c) $f(x)$ is discontinuous at $x = \frac{1}{2}$ (d) $\lim_{x \to 1/2 - 0} f(x) = 1$

32. If $f(x) = |2 - x| + (2 + x)$, where $(x) =$ the least integer greater than or equal to x, then

(a) $f(2 - 0) = f(2) = 4$

(b) $f(x)$ is continuous at $x = 2$

(c) $f(x)$ is nondifferentiable at $x = 2$

(d) $f(x)$ is differentiable but not continuous at $x = 2$

33. Let $h(x) = \min\{x, x^2\}$ for every real number x. Then

(a) h is continuous for all x

(b) h is differentiable for all x

(c) $h'(x) = 1$ for all $x > 1$

(d) h is not differentiable at two values of x

34. At $x = 0$, the function $y = e^{-|x|}$ is

(a) continuous

(b) continuous and differentiable

(c) differentiable with derivative $= 1$

(d) differentiable with derivative $= -1$

35. A function $f(x)$ is defined as follows:

$f(x) = -x^2, x \le 0$ $f(x) = 5x - 4, 0 < x \le 1$

$f(x) = 4x^2 - 3x, 1 < x \le 2$ $f(x) = 3x + 4, x > 2$

(a) $f(x)$ is not continuous at $x = 0$, but differentiable there

(b) $f(x)$ is continuous at $x = 1$, but not differentiable there

(c) $f(x)$ is continuous at $x = 2$, but not differentiable there

(d) none of the above

36. The function $f(x) = \frac{1}{x} - \frac{2}{e^{2x} - 1}$, $x \neq 0$, is continuous at $x = 0$. Then

(a) $f(0) = 1$

(b) $f(x)$ is differentiable at $x = 0$

(c) $f(x)$ is not differentiable at $x = 0$

(d) $f'(0) = \frac{1}{3}$

37. The function $f(x) = |x^2 - 3x + 2| + \cos|x|$ is not differentiable at $x =$

(a) -1 (b) 0

(c) 1 (d) 2

38. Let $f(x)$ be defined as follows:

$$f(x) = x^6, x^2 > 1$$
$$x^3, x^2 \leq 1.$$

Then $f(x)$ is

(a) continuous everywhere (b) differentiable everywhere

(c) discontinuous at $x = -1$ (d) not differentiable at $x = 1$

39. Let $f(x) = \sin x, x \geq 0$

$$-\sin x; x < 0.$$

Then $f(x)$ is

(a) continuous at $x = 0$ (b) differentiable at $x = 0$

(c) discontinuous at $x = 0$ (d) not differentiable at $x = 0$

40. If $f(x) = \sum_{n=0}^{n} a_n |x|^n$, where a_i's are real constants, then $f(x)$ is

(a) continuous at $x = 0$ for all a_i

(b) differentiable at $x = 0$ for all $a_i \in R$

(c) differentiable at $x = 0$ for all $a_{2k+1} = 0$

(d) none of these

41. Let $[x]$ denote the greatest integer less than or equal to x. Now $g(x)$ is defined as below:

$$g(x) = [f(x)],\ x \in \left(0, \frac{\pi}{2}\right) \cup \left(\frac{\pi}{2}, \pi\right)$$
$$= 3,\ x = \frac{\pi}{2}$$

where $f(x) = \dfrac{2(\sin x - \sin^n x) + |\sin x - \sin^n x|}{2(\sin x - \sin^n x) - |\sin x - \sin^n x|}$, $n \in R$. Then

(a) $g(x)$ is continuous and differentiable at $x = \frac{\pi}{2}$ when $n > 1$

(b) $g(x)$ is continuous and differentiable at $x = \frac{\pi}{2}$ when $0 < n < 1$

(c) $g(x)$ is continuous but not differentiable at $x = \frac{\pi}{2}$ when $n > 1$

(d) $g(x)$ is continuous but differentiable at $x = \frac{\pi}{2}$ when $0 < n < 1$

42. Let $f(x) = \phi(x) + \psi(x)$ and $\phi'(a)$, $\psi'(a)$ are finite and definite. Then

(a) $f(x)$ is continuous at $x = a$
(b) $f(x)$ is differentiable at $x = a$
(c) $f'(x)$ is continuous at $x = a$
(d) $f'(x)$ is differentiable at $x = a$

43. Let $f(x) = x + |x|$. Then $f(x)$ is

(a) differentiable at all x
(b) continuous at all x
(c) differentiable everywhere except at $x = 0$
(d) continuous everywhere except at $x = 0$

44. Let $f(x) = \lim\limits_{n \to \infty} \dfrac{1 - x^n}{1 + x^n}$. Then

(a) $f(x)$ is a constant in $0 < x < 1$
(b) $f(x)$ is continuous at $x = 1$
(c) $f(x)$ is not differentiable at $x = 1$
(d) none of these

45. Let $f(x) = 1 - |\cos x|$ for all $x \in R$. Then

(a) $f'\left(\frac{\pi}{2}\right)$ does not exist

(b) $f(x)$ is continuous everywhere

(c) $f(x)$ is not differentiable anywhere

(d) $\lim\limits_{x \to \pi/2+0} f(x) = 1$

46. Let $f(x) = [\tan^2 x]$, where $[\cdot]$ denotes the greatest integer function. Then

(a) $\lim\limits_{x \to 0} f(x)$ does not exist

(b) $f(x)$ is continuous at $x = 0$

(c) $f'(0) = 1$

(d) $f(x)$ is not differentiable at $x = 0$

47. If $f(x) = \dfrac{x}{\sqrt{x+1} - \sqrt{x}}$ be a real-valued function then

(a) $f(x)$ is continuous, but $f'(0)$ does not exist

(b) $f(x)$ is differentiable at $x = 0$

(c) $f(x)$ is not continuous at $x = 0$

(d) $f(x)$ is not differentiable at $x = 0$

48. A function $f(x)$ is defined in the interval $[1, 4)$ as follows:

$$f(x) = \log_e [x],\ 1 \le x < 3$$
$$|\log_e x|,\ 3 \le x < 4.$$

The graph of the function $f(x)$

(a) is broken at two points

(b) is broken at exactly one point

(c) does not have a definite tangent at two points

(d) does not have a definite tangent at more than two points

Answers

1. a	**2.** c	**3.** c	**4.** a	**5.** c
6. a	**7.** b	**8.** c	**9.** c	**10.** c
11. d	**12.** a	**13.** a	**14.** b	**15.** a
16. a	**17.** c	**18.** b	**19.** d	**20.** c
21. b	**22.** a	**23.** c	**24.** b	**25.** d
26. d	**27.** d	**28.** a, b, c	**29.** a, d	**30.** a, d
31. b, c	**32.** a, c	**33.** a, c, d	**34.** a	**35.** c
36. a, b	**37.** c, d	**38.** c, d	**39.** a, d	**40.** a, c
41. a	**42.** a, b	**43.** b, c	**44.** a, c	**45.** b, d
46. b	**47.** b	**48.** a, c		

Hints and Solutions to Selected Questions

1. RH derivative $= \lim\limits_{h \to 0} \dfrac{\log\left|\frac{1}{2}+h-1\right| - \log\left|\frac{1}{2}-1\right|}{h}$

$$= \lim_{h \to 0} \frac{\log\left|h-\frac{1}{2}\right| - \log\frac{1}{2}}{h}$$

$$= \lim_{h \to 0} \frac{\log\left(\frac{1}{2}-h\right) + \log 2}{h} = \lim_{h \to 0} \frac{\frac{1}{\frac{1}{2}-h}(-1)+0}{1} = -2.$$

LH derivative $= \lim\limits_{h \to 0} \dfrac{\log\left|\frac{1}{2}-h-1\right| - \log\left|\frac{1}{2}-1\right|}{-h} = \lim\limits_{h \to 0} \dfrac{\log\left(\frac{1}{2}+h\right) - \log\frac{1}{2}}{-h}$

$$= \lim_{h \to 0} \frac{\frac{1}{\frac{1}{2}+h} - 0}{-1} = -2.$$

Alternate method

Around $x = \frac{1}{2}$, $f(x) = \log(1-x)$. So, $f'(x) = \dfrac{-1}{1-x}$

$$\therefore \quad f'\left(\frac{1}{2}\right) = \frac{-1}{1-\frac{1}{2}} = -2.$$

2. RH derivative $= \lim\limits_{h \to 0} \dfrac{\left|\tan\left(\frac{\pi}{4}-\frac{\pi}{4}+h\right)\right| - |\tan 0|}{h} = \lim\limits_{h \to 0} \dfrac{\tan h}{h} = 1.$

LH derivative $= \lim\limits_{h \to 0} \dfrac{\left|\tan\left(\frac{\pi}{4}-\frac{\pi}{4}-h\right)\right| - |\tan 0|}{-h} = \lim\limits_{h \to 0} \dfrac{\tan h}{-h} = -1.$

4. RH derivative $= \lim\limits_{h \to 0} \dfrac{\lambda + \mu|0+h| + \nu|0+h|^2 - \lambda}{h} = \lim\limits_{h \to 0} \dfrac{\mu h + \nu h^2}{h} = \mu.$

$$\text{LH derivative} = \lim_{h \to 0} \frac{\lambda + \mu|0-h| + \nu|0-h|^2 - \lambda}{-h} = \lim_{h \to 0} \frac{\mu h + \nu h^2}{-h} = -\mu.$$

$f'(x)$ exists $\Rightarrow$ RH derivative = LH derivative $\Rightarrow$ $\mu = -\mu$.

5. $$f'(1+0) = \lim_{h \to 0} \frac{f(1+h) - f(1)}{h} = \lim_{h \to 0} \frac{\frac{[1+h]}{|1+h|} - \frac{[1]}{|1|}}{h}$$

$$= \lim_{h \to 0} \frac{\frac{1}{1+h} - 1}{h} = \lim_{h \to 0} \frac{-h}{h(1+h)} = -1.$$

$$f'(1-0) = \lim_{h \to 0} \frac{f(1-h) - f(1)}{-h} = \lim_{h \to 0} \frac{\frac{[1-h]}{|1-h|} - \frac{[1]}{|1|}}{-h} = \lim_{h \to 0} \frac{0-1}{-h} = \infty.$$

6. $$f'\left(\frac{\pi}{4} + 0\right) = \lim_{h \to 0} \frac{\left|\cos 2\left(\frac{\pi}{4} + h\right)\right| - \left|\cos 2 \cdot \frac{\pi}{4}\right|}{h} = \lim_{h \to 0} \frac{\sin 2h}{h} = 2.$$

7. $$f'(1-0) = \lim_{h \to 0} \frac{\sin \pi[1-h] - \sin \pi[1]}{-h} = \lim_{h \to 0} \frac{0}{-h} = \lim_{h \to 0} 0 = 0.$$

8. $$f(1+0) = \lim_{h \to 0} \{[(1+h)^2] - [1+h]^2\} = \lim_{h \to 0} \{1-1\} = 0.$$

$$f(1-0) = \lim_{h \to 0} \{[(1-h)^2] - [1-h]^2\} = \lim_{h \to 0} \{0-0\} = 0.$$

Also $f(1) = 0$. So, $f(x)$ is continuous at $x = 1$.

9. $\cos x + \sin x = \sqrt{2}\cos\left(x - \frac{\pi}{4}\right)$. So, $f(x) = \left[\sqrt{2}\cos\left(x - \frac{\pi}{4}\right)\right]$.

We know that $[x]$ is discontinuous at integral values of x.

Now, $\sqrt{2}\cos\left(x - \frac{\pi}{4}\right)$ is an integer at $x = \frac{\pi}{2}, \frac{\pi}{2} + \frac{\pi}{4}, \pi + \frac{\pi}{4}, \frac{3\pi}{2} + \frac{\pi}{4}$.

10. $f(x) = x - |x| \cdot |1-x|$. We know that x, $|x|$, $|1-x|$ are continuous everywhere. As the product and algebraic sum of continuous functions are continuous, $f(x)$ is continuous everywhere.

11. We know that $|1-x|$ is continuous everywhere, $[x]$ is continuous everywhere except at integers, $\sqrt{1+x^2}$ and $\sqrt{3x-1}$ are continuous in their respective intervals of definition. So, the only doubtful points of

continuity are $x = \sqrt{3}, 4, 5$.

$$f(\sqrt{3}-0) = \lim_{h\to 0} \sqrt{1+(\sqrt{3}-h)^2} = 2;$$

$$f(\sqrt{3}+0) = \lim_{h\to 0} \{\sqrt{3}(\sqrt{3}+h)-1\} = 2; \qquad f(\sqrt{3}) = 2.$$

$$f(4-0) = \lim_{h\to 0} \{\sqrt{3}(4-h)-1\} = 4\sqrt{3}-1; \qquad f(4+0) = \lim_{h\to 0} [4+h] = 4.$$

$$f(5-0) = \lim_{h\to 0} [5-h] = \lim_{h\to 0} 4 = 4;$$

$$f(5+0) = \lim_{h\to 0} |1-(5+h)| = 4; \quad f(5) = 4.$$

$\therefore \quad f(x)$ is discontinuous at $x = 4$ only.

12. $f'(x) = x\sin\frac{1}{x}\cdot$ At all points in $(0, \pi)$, $f'(x)$ has a definite finite value.

$\therefore$ $f(x)$ is differentiable finitely in $(0, \pi)$. As a finitely differentiable function is also continuous, we get $f(x)$ is continuous in $(0, \pi)$.

13.

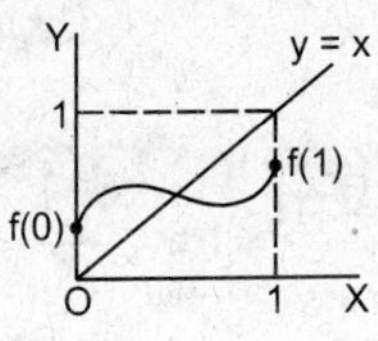

Clearly, $0 \le f(0) \le 1$ and $0 \le f(1) \le 1$. As $f(x)$ is continuous, $f(x)$ attains all values between $f(0)$ to $f(1)$, and the graph will have no breaks. So, the graph will cut the line $y = x$ at one point x at least where $0 \le x \le 1$. So, $f(x) = x$ at that point.

14. As $f(x)$ is continuous in $[1, 3]$, $f(x)$ will attain all values between $f(1)$ and $f(3)$. As $f(x)$ takes rational values for all x and there are innumerable irrational values between $f(1)$ and $f(3)$, $f(x)$ can take rational values for all x if $f(x)$ has a constant rational value at all points between $x = 1$ and $x = 3$. So, $f(2) = f(1.5) = 10$.

15. RH derivative $= \lim_{h\to 0} \dfrac{e^{-1/(0+h)^2} - 0}{h} = \lim_{h\to 0} \dfrac{1}{h}\cdot e^{-1/h^2} = \lim_{h\to 0} \dfrac{1/h}{e^{1/h^2}}$

$$= \lim_{h\to 0} \frac{-1/h^2}{e^{1/h^2}\cdot(-2/h^3)} = \lim_{h\to 0} \frac{h}{2e^{1/h^2}} = \frac{0}{\infty} = 0.$$

$$\text{LH derivative} = \lim_{h\to 0} \frac{e^{-1/(0-h)^2} - 0}{-h} = \lim_{h\to 0} -\frac{1}{h}\cdot e^{-1/h^2} = \lim_{h\to 0} \frac{-1/h}{e^{1/h^2}}$$

$$= \lim_{h\to 0} \frac{1/h^2}{e^{1/h^2}\cdot(-2/h^3)} = \lim_{h\to 0} \frac{h}{-2e^{1/h^2}} = \frac{0}{-\infty} = 0.$$

16. $h(x) = g\{f(x)\} = [f(x)+1] = [\sin x + 1]$.

$$h'\left(\frac{\pi}{2}+0\right) = \lim_{h\to 0} \frac{\left[\sin\left(\frac{\pi}{2}+h\right)+1\right] - \left[\sin\frac{\pi}{2}+1\right]}{h} = \lim_{h\to 0} \frac{[\cos h + 1] - [2]}{h}$$

$$= \lim_{h\to 0} \frac{1-2}{h} = \lim_{h\to 0} \frac{-1}{h} = -\infty.$$

$$h'\left(\frac{\pi}{2}-0\right) = \lim_{h\to 0} \frac{\left[\sin\left(\frac{\pi}{2}-h\right)+1\right] - \left[\sin\frac{\pi}{2}+1\right]}{-h}$$

$$= \lim_{h\to 0} \frac{[\cos h + 1] - [2]}{-h} = \lim_{h\to 0} \frac{1}{h} = +\infty.$$

17. $h(x) = f\{g(x)\} = [g(x)] = [\,|x|\,]$.

$$h'(-1+0) = \lim_{h\to 0} \frac{[\,|-1+h|\,] - [\,|-1|\,]}{h} = \lim_{h\to 0} \frac{[1-h]-[1]}{h} = \lim_{h\to 0} \frac{0-1}{h} = -\infty.$$

$$h'(-1-0) = \lim_{h\to 0} \frac{[\,|-1-h|\,] - [\,|-1|\,]}{-h} = \lim_{h\to 0} \frac{[1+h]-[1]}{-h}$$

$$= \lim_{h\to 0} \frac{1-1}{-h} = \lim_{h\to 0} = 0.$$

18. The doubtful points are $x = \frac{1}{2}, 1, \frac{\pi}{2}$.

$|x-1|$ is differentiable at $\frac{1}{2}, \frac{\pi}{2}$ but not differentiable at 1.

$\left|x-\frac{1}{2}\right|$ is differentiable at $1, \frac{\pi}{2}$ but not differentiable at $\frac{1}{2}$.

Remember $|x-a|$ is continuous everywhere, and differentiable everywhere except at $x = a$.

$\tan x$ is differentiable at $\frac{1}{2}$, 1 but not differentiable at $\frac{\pi}{2}$.

$\therefore \quad f(x)$ is not differentiable at $\frac{1}{2}, 1, \frac{\pi}{2}$.

19. $[n + p\sin x]$ is not differentiable at those points where $n + p\sin x$ is an integer. As p is a prime number, $n + p\sin x$ is an integer if $\sin x = 1, -1, \frac{r}{p}$, that is, $x = \frac{\pi}{2}, -\frac{\pi}{2}, \sin^{-1}\frac{r}{p}, \pi - \sin^{-1}\frac{r}{p}$, where $0 \le r \le p - 1$. But $x \ne -\frac{\pi}{2}, 0$.

$\therefore$ the function is not differentiable at $x = \frac{\pi}{2}, \sin^{-1}\frac{r}{p}, \pi - \sin^{-1}\frac{r}{p}$, where $0 < r < p - 1$. So, the required number of points $= 1 + 2(p - 1) = 2p - 1$.

20. The doubtful points are $x = 0, 1$ because these are the turning points of the definition. $|x|$ is not differentiable at $x = 0$, while $(x-1)^2 \cos\frac{1}{x-1}$ is differentiable. The algebraic sum of a differentiable function and a nondifferentiable function is nondifferentiable. So, $f(x)$ is not differentiable at $x = 0$. Now,

$$f'(1+0) = \lim_{h \to 0} \frac{(1+h-1)^2 \cos\frac{1}{1+h-1} - |1+h| - (-1)}{h}$$

$$= \lim_{h \to 0} \frac{h^2 \cos\frac{1}{h} - h}{h} = \lim_{h \to 0} \left(h\cos\frac{1}{h} - 1\right) = 0 - 1 = -1.$$

Similarly, $f'(1-0) = -1$. So, $f(x)$ is differentiable at $x = 1$.

21. Apply $\lim_{h \to 0} \frac{1 - \sin\left(\frac{\pi}{2} + h\right)}{\sin 2\left(\frac{\pi}{2} + h\right)} = \lim_{h \to 0} \frac{1 - \sin\left(\frac{\pi}{2} - h\right)}{\sin 2\left(\frac{\pi}{2} - h\right)} = f\left(\frac{\pi}{2}\right)$.

22. RH limit $= \lim_{h \to 0} \frac{\cos\{\sin(0+h)\} - \cos(0+h)}{(0+h)^2} = \lim_{h \to 0} \frac{\cos \sin h - \cos h}{h^2}$

$$= \lim_{h \to 0} \frac{-\sin(\sin h) \times \cos h + \sin h}{2h}$$

$$= \lim_{h \to 0} \frac{-\cos(\sin h) \times \cos^2 h + \sin(\sin h) \times \sin h + \cos h}{2} = 0.$$

Similarly, LH limit = 0.

As $f(x)$ is continuous at $x = 0, f(0) = 0$.

23. As $f(x)$ is continuous in $\left[0, \frac{\pi}{2}\right)$, it is continuous at $x = \frac{\pi}{4}$.

$$\therefore \quad \lim_{h \to 0} \frac{1 - \tan\left(\frac{\pi}{4} + h\right)}{4\left(\frac{\pi}{4} + h\right) - \pi} = \lim_{h \to 0} \frac{1 - \tan\left(\frac{\pi}{4} - h\right)}{4\left(\frac{\pi}{4} - h\right) - \pi} = f\left(\frac{\pi}{4}\right) = \lambda$$

$$\therefore \quad \lambda = \lim_{h \to 0} \frac{1 - \frac{1 + \tan h}{1 - \tan h}}{4h} = \lim_{h \to 0} \frac{-2\tan h}{(1 - \tan h)4h} = -\frac{1}{2}.$$

24. Here, $\lim_{h \to 0} \left\{\sin\left(\frac{\pi}{2} + h\right)\right\}^{\frac{1}{\pi - 2(\pi/2 + h)}} = \lim_{h \to 0} \left\{\sin\left(\frac{\pi}{2} - h\right)\right\}^{\frac{1}{\pi - 2(\pi/2 - h)}} = f\left(\frac{\pi}{2}\right)$.

Now, $\lim_{h \to 0} (\cos h)^{1/(-2h)} = e^{\lim_{h \to 0} 1/(-2h) \log \cos h} = e^{\lim_{h \to 0} \frac{(1/\cos h)(-\sin h)}{-2}} = e^0 = 1.$

$$\therefore \quad f\left(\frac{\pi}{2}\right) = 1.$$

25. For continuity at $x = 0$, $\lim_{h \to 0} \sin \frac{1}{0 + h} = \lim_{h \to 0} \sin \frac{1}{0 - h} = f(0)$.

But $\lim_{h \to 0} \sin \frac{1}{0 + h} = \lim_{h \to 0}$ {a number between 1 and −1} ≠ a definite number.

$\therefore \quad f(0)$ cannot be a definite number.

26. $f(x)$ is continuous at $x = 1$ if

$$\lim_{h \to 0} f(1 + h) = \lim_{h \to 0} f(1 - h) = f(1) = 2.$$

$$\lim_{h\to 0} f(1+h) = \lim_{h\to 0} \{1+h+1\} = 2$$

$$\lim_{h\to 0} f(1-h) = \lim_{h\to 0} \{p(1-h)^2 - q\} = p - q$$

$\therefore$ $f(x)$ is not continuous at $x = 1$ if $p - q \neq 2$.

27. $f(x)$ is differentiable at $x = 1$ also.

$$\therefore \quad \lim_{h\to 0} \frac{f(1+h)-f(1)}{h} = \lim_{h\to 0} \frac{f(1-h)-f(1)}{-h}.$$

Now, $\lim_{h\to 0} \dfrac{f(1+h)-f(1)}{h} = \lim_{h\to 0} \dfrac{(1+h)^2 + b(1+h) + c - 1}{h}$

$$= \lim_{h\to 0} \frac{h^2 + (2+b)h + b + c}{h}.$$

$$\lim_{h\to 0} \frac{f(1-h)-f(1)}{-h} = \lim_{h\to 0} \frac{1-h-1}{-h} = 1.$$

The two limits can be equal if $2 + b = 1$, $b + c = 0$.

28. $f(x)$ is differentiable at $x = 2$ also.

So, $\lim_{h\to 0} \dfrac{f(2+h)-f(2)}{h} = \lim_{h\to 0} \dfrac{f(2-h)-f(2)}{-h}.$

Now, $\lim_{h\to 0} \dfrac{f(2+h)-f(2)}{h} = \lim_{h\to 0} \dfrac{a + b(2+h) - (a+2b)}{h} = b.$

$$\lim_{h\to 0} \frac{f(2-h)-f(2)}{-h} = \lim_{h\to 0} \frac{e^{2-h} - (a+2b)}{-h}$$

$$= \lim_{h\to 0} \frac{e^{-h} - (a+2b)\,e^{-2}}{-h \cdot e^{-2}}$$

$$= \lim_{h\to 0} \frac{\left\{1 - h + \frac{h^2}{2!} - \dots\right\} - (a+2b)e^{-2}}{-he^{-2}}.$$

This can be equal to b if $1 - (a+2b)e^{-2} = 0$ and $e^{-2} = \dfrac{1}{b}$.

$$\therefore \quad 1 - (a+2b)\frac{1}{b} = 0 \quad \Rightarrow b - (a+2b) = 0$$

$$\Rightarrow a + b = 0.$$

29. Here, $f(x) = \cos^{-1}(\cos x) = 2\pi + x, -2\pi < x < -\pi$

$$-x, -\pi \le x < 0$$

$$x, 0 \le x \le \pi$$

$$2\pi - x, \pi < x < 2\pi.$$

Use these definitions for selecting the options.

30. Both the functions x and $|x|$ are continuous everywhere. So, their algebraic sum is also continuous everywhere. x is differentiable everywhere but $|x|$ is not differentiable at $x = 0$. So, $f(x)$ is not differentiable at $x = 0$.

31. $[x]$ is discontinuous at all integral values of x but continuous at all other values of x. $[x + 1/2]$ is discontinuous at all $x = n \pm 1/2, n \in Z$. So, $f(x) = [x] + [x + 1/2]$ is not continuous at $x = 1/2$.

$$\lim_{x \to \frac{1}{2}+0} f(x) = \lim_{h \to 0} f\left(\frac{1}{2} + h\right) = \lim_{h \to 0} \left\{\left[\frac{1}{2} + h\right] + \left[\frac{1}{2} + h + \frac{1}{2}\right]\right\}$$

$$= \lim_{h \to 0} \{0 + 1\} = 1.$$

$$\lim_{x \to \frac{1}{2}-0} f(x) = \lim_{h \to 0} f\left(\frac{1}{2} - h\right) = \lim_{h \to 0} \left\{\left[\frac{1}{2} - h\right] + \left[\frac{1}{2} - h + \frac{1}{2}\right]\right\}$$

$$= \lim_{h \to 0} \{0 + 0\} = 0.$$

32. $|2 - x|$ is continuous everywhere and $(2 + x)$ is discontinuous at all integral values of x. So, $f(x)$ is not continuous at $x = 2$. Hence, $f(x)$ is not differentiable at $x = 2$ because not continuous $\Rightarrow$ not differentiable.

$$f(2 - 0) = \lim_{h \to 0} \{|2 - (2 - h)| + (2 + 2 - h)\} = \lim_{h \to 0} \{h + 4\} = 4.$$

$f(2) = |2 - 2| + (2 + 2) = |0| + (4) = 0 + 4 = 4$. So, $f(2 - 0) = f(2) = 4$.

33. $h(x) = \min\{x, x^2\} = x, x \le 0$

$$x^2, 0 < x < 1$$

$$x, x \ge 1.$$

As x, x^2 are polynomial functions, they are continuous and differentiable in their respective intervals of definition. So, the only doubtful points are 0 and 1. Check continuity and differentiability at $x = 0, 1$.

34. RH limit $= \lim_{h \to 0} e^{-|0+h|} = \lim_{h \to 0} e^{-h} = 1.$

$$\text{LH limit} = \lim_{h \to 0} e^{-|0-h|} = \lim_{h \to 0} e^{-h} = 1.$$

So, the function is continuous at $x = 0$.

$$\text{RH derivative} = \lim_{h \to 0} \frac{e^{-|0+h|} - e^{-|0|}}{h} = \lim_{h \to 0} \frac{e^{-h} - 1}{h} = -1.$$

$$\text{LH derivative} = \lim_{h \to 0} \frac{e^{-|0-h|} - e^{-|0|}}{-h} = \lim_{h \to 0} \frac{e^{-h} - 1}{-h} = 1.$$

So, the function is not differentiable at $x = 0$.

36. $$\lim_{x \to 0+0} f(x) = \lim_{h \to 0} f(0+h) = \lim_{h \to 0} \left\{\frac{1}{0+h} - \frac{2}{e^{2(0+h)} - 1}\right\} = \lim_{h \to 0} \left\{\frac{1}{h} - \frac{2}{e^{2h} - 1}\right\}$$

$$= \lim_{h \to 0} \frac{e^{2h} - 1 - 2h}{h(e^{2h} - 1)} = \lim_{h \to 0} \frac{\left\{1 + 2h + \frac{(2h)^2}{2!} + \ldots\right\} - 1 - 2h}{h\left\{1 + 2h + \frac{(2h)^2}{2!} + \ldots\right\} - h} = 1.$$

$$\text{Similarly, } \lim_{x \to 0-0} f(x) = \lim_{h \to 0} \frac{e^{-2h} - 1 + 2h}{-h(e^{-2h} - 1)} = 1.$$

Hence, $f(0) = 1$.

$$\text{Now, } f'(0+0) = \lim_{h \to 0} \frac{\left\{\frac{1}{0+h} - \frac{2}{e^{2(0+h)} - 1}\right\} - 1}{h} = \lim_{h \to 0} \frac{\frac{1}{h} - \frac{2}{e^{2h} - 1} - 1}{h}$$

$$= \lim_{h \to 0} \frac{e^{2h} - 1 - 2h - h(e^{2h} - 1)}{h^2(e^{2h} - 1)}$$

$$= \lim_{h \to 0} \frac{\left\{1 + 2h + \frac{(2h)^2}{2!} + \ldots\right\} - 1 - 2h - h\left\{2h + \frac{(2h)^2}{2!} + \ldots\right\}}{h^2\left\{2h + \frac{(2h)^2}{2!} + \ldots\right\}}$$

$$= \lim_{h \to 0} \frac{\left(\frac{8}{3!} - \frac{4}{2!}\right)h^3 + \ldots}{h^3\left\{2 + \frac{4h}{2!} + \ldots\right\}} = \frac{\frac{8}{3!} - \frac{4}{2!}}{2} = -\frac{1}{3}.$$

$$\text{Similarly, } f'(0-h) = \lim_{h \to 0} \frac{\left\{\frac{1}{0-h} - \frac{2}{e^{2(0-h)} - 1}\right\} - 1}{-h} = -\frac{1}{3} \quad \text{(as above).}$$

37. $f(x) = |x-1|\ |x-2| + \cos|x|$. The only doubtful points are the turning points of the definition, namely, $x = 1, 2, 0$.

$$\lim_{h\to 0}\frac{f(0+h)-f(0)}{h} = \lim_{h\to 0}\frac{|h-1|\ |h-2| + \cos|h| - (2+1)}{h}$$

$$= \lim_{h\to 0}\frac{(1-h)(2-h)+\cos h - 3}{h}$$

$$= \lim_{h\to 0}\frac{-3+2h-\sin h}{1} = -3.$$

Similarly, $$\lim_{h\to 0}\frac{f(0-h)-f(0)}{-h} = \lim_{h\to 0}\frac{(1+h)(2+h)+\cos h - 3}{-h}$$

$$= \lim_{h\to 0}\frac{3+2h-\sin h}{-1} = -3.$$

$\therefore\ f(x)$ is differentiable at $x = 0$.

$$\lim_{h\to 0}\frac{f(1+h)-f(1)}{h} = \lim_{h\to 0}\frac{|h|\ |h-1| + \cos|1+h| - \cos 1}{h}$$

$$= \lim_{h\to 0}\frac{h(1-h)+\cos(1+h)-\cos 1}{h}$$

$$= \lim_{h\to 0}\frac{1-2h-\sin(1+h)}{1} = 1-\sin 1.$$

$$\lim_{h\to 0}\frac{f(1-h)-f(1)}{-h} = \lim_{h\to 0}\frac{|-h|\ |-1-h| + \cos|1-h| - \cos 1}{-h}$$

$$= \lim_{h\to 0}\frac{h(1+h)+\cos(1-h)-\cos 1}{-h}$$

$$= \lim_{h\to 0}\frac{1+2h+\sin(1-h)}{-h} = -1-\sin 1.$$

$\therefore\ f(x)$ is not differentiable at $x = 1$.

Similarly, $f(x)$ is not differentiable at $x = 2$.

38. $x^2 > 1 \Rightarrow x < -1$ or $x > 1$; $\quad x^2 \le 1 \Rightarrow -1 \le x \le 1$.

$\therefore\quad f(x) = x^6, x < -1$

$\qquad\qquad x^3, -1 \le x \le 1$

$\qquad\qquad x^6, x > 1.$

Use these definitions. The only doubtful points for continuity and differentiability are $x=-1, 1$.

Clearly, $f(-1-0)=1$ and $f(-1+0)=-1$.

So, $f(x)$ is not continuous at $x=-1$.

Also $f(1-0)=1, f(1+0)=1$ and $f(1)=1$. So, $f(x)$ is continuous at $x=1$.

As $f(x)$ is not continuous at $x=-1$, it is not differentiable there.

Now, $\lim\limits_{h\to 0}\dfrac{f(1+h)-f(1)}{h}=\lim\limits_{h\to 0}\dfrac{(1+h)^6-1}{(1+h)-1}=6\cdot 1^{6-1}=6.$

$\lim\limits_{h\to 0}\dfrac{f(1-h)-f(1)}{-h}=\lim\limits_{h\to 0}\dfrac{(1-h)^3-1}{(1-h)-1}=3\cdot 1^{3-1}=3.$

$\therefore$ $f(x)$ is not differentiable at $x=1$.

39. $f(0+0)=\sin 0=0, f(0-0)=-\sin 0=0$ and $f(0)=0$.

So, $f(x)$ is continuous at $x=0$.

$\lim\limits_{h\to 0}\dfrac{f(0+h)-f(0)}{h}=\lim\limits_{h\to 0}\dfrac{\sin(0+h)-0}{h}=1.$

$\lim\limits_{h\to 0}\dfrac{f(0-h)-f(0)}{-h}=\lim\limits_{h\to 0}\dfrac{-\sin(0-h)-0}{-h}=-1.$

So, $f(x)$ is not differentiable at $x=0$.

40. $|x|, |x|^2$, etc., are all continuous everywhere and the algebraic sum of continuous functions is also continuous. So, (a) is correct. $|x|, |x|^3$, etc., are not differentiable at $x=0$, whereas $|x|^2, |x|^4$, etc., are all differentiable at $x=0$. Hence, (c) is correct.

41. If $n>1, \sin x>\sin^n x$. If $0<n<1, \sin x<\sin^n x$.

$\therefore$ if $n>1, f(x)=\dfrac{2(\sin x-\sin^n x)+(\sin x-\sin^n x)}{2(\sin x-\sin^n x)-(\sin x-\sin^n x)}=3.$

If $0<n<1, f(x)=\dfrac{2(\sin x-\sin^n x)-(\sin x-\sin^n x)}{2(\sin x-\sin^n x)+(\sin x-\sin^n x)}=\dfrac{1}{3}.$

$\therefore$ if $n>1, g(x)=3, x\in(0,\pi)$

$\therefore$ $g(x)$ is continuous and differentiable at $x=\dfrac{\pi}{2}$.

If $0 < n < 1, g(x) = 0, x \in \left(0, \frac{\pi}{2}\right) \cup \left(\frac{\pi}{2}, \pi\right)$

$3, x = \frac{\pi}{2}\cdot$

Then $g\left(\frac{\pi}{2}+0\right) = 0, g\left(\frac{\pi}{2}-0\right) = 0, g\left(\frac{\pi}{2}\right) = 3$. So, $g(x)$ is not continuous at $x = \frac{\pi}{2}\cdot$ Hence, $g(x)$ is also not differentiable at $x = \frac{\pi}{2}\cdot$

43. x is continuous and differentiable everywhere while $|x|$ is continuous everywhere but not differentiable at $x = 0$ only.

44. If $0 < x < 1$ then $f(x) = \lim\limits_{x \to \infty} \frac{1 - x^n}{1 + x^n} = \frac{1-0}{1+0} = 1$. So, (a) is correct.

If $x = 1$ then $f(x) = \lim\limits_{n \to \infty} \frac{1-1}{1+1} = 0$.

If $x > 1$ then $f(x) = \lim\limits_{n \to \infty} \frac{\left(\frac{1}{x}\right)^n - 1}{\left(\frac{1}{x}\right)^n + 1} = \frac{-1}{1} = -1$.

$f(1-0) = 1, f(1+0) = -1$. So, $f(x)$ is not continuous at $x = 1$ and hence it is not differentiable at $x = 1$.

45. 1 as well as $|\cos x|$ are continuous everywhere. So, $f(x)$ is continuous everywhere.

1 is differentiable everywhere and $|\cos x|$ is not differentiable only at those x where $\cos x = 0$. So, $f'\left(\frac{\pi}{2}\right)$ does not exist.

$$\lim_{x \to \pi/2+0} f(x) = \lim_{h \to 0} f\left(\frac{\pi}{2} + h\right) = \lim_{h \to 0} \left\{1 - \left|\cos\left(\frac{\pi}{2} + h\right)\right|\right\} = \lim_{h \to 0} \{1 - \sin h\} = 1.$$

46. $\lim\limits_{x \to 0+0} f(x) = \lim\limits_{h \to 0} f(0+h) = \lim\limits_{h \to 0} [\tan^2 h] = \lim\limits_{h \to 0} 0 = 0$.

$\lim\limits_{x \to 0-0} f(x) = \lim\limits_{h \to 0} f(0-h) = \lim\limits_{h \to 0} [\tan^2(-h)] = \lim\limits_{h \to 0} 0 = 0; f(0) = 0$.

$\therefore \quad \lim\limits_{x \to 0} f(x) = f(0) = 0$

$\therefore \quad f(x)$ is continuous at $x = 0$.

$$\lim_{h\to 0}\frac{f(0+h)-f(0)}{h}=\lim_{h\to 0}\frac{[\tan^2 h]-0}{h}=\lim_{h\to 0}\frac{0}{h}=\lim_{h\to 0}0=0.$$

$$\lim_{h\to 0}\frac{f(0-h)-f(0)}{-h}=\lim_{h\to 0}\frac{[\tan^2(-h)]-0}{-h}=\lim_{h\to 0}\frac{0}{-h}=\lim_{h\to 0}0=0.\text{ So, }f'(0)=0.$$

47. $f(x)=\dfrac{x(\sqrt{x^2+1}+\sqrt{x})}{x+1-x}=x(\sqrt{x^2+1}+\sqrt{x})$. As $f(x)$ is real valued, $x\geq 0$.

$\therefore$ $f(x)$ is differentiable at $x=0$ if $\lim\limits_{h\to 0}\dfrac{f(0+h)-f(0)}{h}$ is finite and definite.

$$\lim_{h\to 0}\frac{f(0+h)-f(0)}{h}=\lim_{h\to 0}\frac{h(\sqrt{h^2+1}+\sqrt{h})}{h}=1.\text{ So, (b) is correct.}$$

A function which is finitely differentiable is also continuous.

48. Here, $f(x)=0,\ 1\leq x<2$

$\log_e 2,\ 2\leq x<3$

$\log_e x,\ 3\leq x<4.$

$\therefore$ $f(x)$ is continuous and differentiable everywhere except perhaps at $x=2, 3$.

$f(2-0)=0, f(2+0)=\log_e 2$ $\quad\Rightarrow$ $f(x)$ is not continuous at $x=2$.

$f(3-0)=\log_e 2, f(3+0)=\log_e 3$ $\quad\Rightarrow$ $f(x)$ is not continuous at $x=3$.

$\therefore$ $f(x)$ is not continuous at $x=2, 3$ (i.e., the graph is broken at $x=2, 3$)

$\Rightarrow$ $f(x)$ is not differentiable at $x=2, 3$ (i.e., the graph does not have a definite tangent at $x=2, 3$).

□

5

Application of dy/dx

• Type 1 •

Choose the most appropriate option (a, b, c or d).

1. If m be the slope of a tangent to the curve $e^y = 1 + x^2$ then

(a) $|m| > 1$ (b) $m < 1$

(c) $|m| < 1$ (d) $|m| \le 1$

2. If at each point of the curve $y = x^3 - ax^2 + x + 1$ the tangent is inclined at an acute angle with the positive direction of the x-axis then

(a) $a > 0$ (b) $a \le \sqrt{3}$

(c) $-\sqrt{3} \le a \le \sqrt{3}$ (d) none of these

3. The slope of the tangent to the curve $y = x^2 - x$ at the point where the line $y = 2$ cuts the curve in the first quadrant is

(a) 2 (b) 3

(c) –3 (d) none of these

4. The slope of the tangent to the curve $y = \sqrt{4 - x^2}$ at the point where the ordinate and the abscissa are equal, is

(a) –1 (b) 1

(c) 0 (d) none of these

5. The slope of the tangent to the locus $y = \cos^{-1}(\cos x)$ at $x = -\frac{\pi}{4}$ is

(a) 1 (b) 0

(c) 2 (d) –1

6. The slope of the tangent to the curve $y = \int_0^x \frac{dx}{1+x^3}$ at the point where $x = 1$ is

(a) $\frac{1}{2}$ (b) 1

(c) $\frac{1}{4}$ (d) none of these

7. The equation of a curve is given by $x = e^t \sin t$, $y = e^t \cos t$. The inclination of the tangent to the curve at the point $t = \frac{\pi}{4}$ is

(a) $\frac{\pi}{4}$ (b) $\frac{\pi}{3}$

(c) $\frac{\pi}{2}$ (d) 0

8. The curve given by $x + y = e^{xy}$ has a tangent parallel to the y-axis at the point

(a) (0, 1) (b) (1, 0)

(c) (1, 1) (d) none of these

9. P(2, 2) and Q$\left(\frac{1}{2}, -1\right)$ are two points on the parabola $y^2 = 2x$. The coordinates of the point R on the parabola, where the tangent to the curve is parallel to the chord PQ, is

(a) $\left(\frac{5}{4}, \sqrt{\frac{5}{2}}\right)$ (b) (2, –1)

(c) $\left(\frac{1}{8}, \frac{1}{2}\right)$ (d) none of these

10. The number of tangents to the curve $x^{3/2} + y^{3/2} = a^{3/2}$, where the tangents are equally inclined to the axes, is

(a) 2 (b) 1

(c) 0 (d) 4

11. The point on the curve $\sqrt{x} + \sqrt{y} = 2a^2$, where the tangent is equally inclined to the axes, is

(a) (a^4, a^4) (b) $(0, 4a^4)$

(c) $(4a^4, 0)$ (d) none of these

12. The parabolas $x^2 = 5 - 4y$ and $y = x^2$ cut at the point (1, 1) at an angle

(a) $\frac{\pi}{2}$ (b) $\frac{\pi}{4}$

(c) $\frac{\pi}{3}$ (d) none of these

13. The angle between two tangents to the ellipse $\frac{x^2}{16} + \frac{y^2}{9} = 1$ at the points where the line $y = 1$ cuts the curve is

(a) $\frac{\pi}{4}$ (b) $\tan^{-1}\frac{6\sqrt{2}}{7}$

(c) $\frac{\pi}{2}$ (d) none of these

14. The number of tangents to the curve $y^2 - 2x^3 - 4y + 8 = 0$ that pass through (1, 2) is

(a) 3 (b) 1

(c) 2 (d) 6

15. The equation of a curve is $y = f(x)$. The tangents at $\left(1, f(1)\right), \left(2, f(2)\right)$ and $\left(3, f(3)\right)$ make angles $\frac{\pi}{6}, \frac{\pi}{3}$ and $\frac{\pi}{4}$ respectively with the positive direction of the x-axis. Then the value of $\int_2^3 f'(x) f''(x) dx + \int_1^3 f''(x)\, dx$ is equal to

(a) $-\frac{1}{\sqrt{3}}$ (b) $\frac{1}{\sqrt{3}}$

(c) 0 (d) none of these

16. The equation of the tangent to the curve $y = e^{-|x|}$ at the point where the curve cuts the line $x = 1$ is

(a) $x + y = e$ (b) $e(x + y) = 1$

(c) $y + ex = 1$ (d) none of these

17. The equation of the tangent to the curve $y = be^{-x/a}$ where it cuts the y-axis is

(a) $\frac{x}{a}+\frac{y}{b}=1$ (b) $\frac{x}{a}+\frac{y}{b}=-1$

(c) $\frac{x}{a}-\frac{y}{b}=1$ (d) none of these

18. If the tangent to the curve $\sqrt{x}+\sqrt{y}=\sqrt{a}$ at any point on it cuts the axes OX and OY at P and Q respectively then $OP + OQ$ is

(a) $2a$ (b) a

(c) $\frac{1}{2}a$ (d) none of these

19. If the line joining the points (0, 3) and (5, –2) is a tangent to the curve $y=\frac{c}{x+1}$ then the value of c is

(a) 1 (b) –2

(c) 4 (d) none of these

20. The curve $\frac{x^n}{a^n}+\frac{y^n}{b^n}=2$ touches the line $\frac{x}{a}+\frac{y}{b}=2$ at the point

(a) (b, a) (b) (a, b)

(c) $(1, 1)$ (d) $\left(\frac{1}{b},\frac{1}{a}\right)$

21. The sum of the intercepts made on the axes of coordinates by any tangent to the curve $\sqrt{x}+\sqrt{y}=2$ is equal to

(a) 4 (b) 2

(c) 8 (d) none of these

22. The area bounded by the axes of reference and the normal to $y=\log_e x$ at the point (1, 0) is

(a) 1 unit^2 (b) 2 unit^2

(c) $\frac{1}{2}$ unit^2 (d) none of these

23. The normal to the curve $2x^2+y^2=12$ at the point (2, 2) cuts the curve again at

(a) $\left(-\frac{22}{9},-\frac{2}{9}\right)$ (b) $\left(\frac{22}{9},\frac{2}{9}\right)$

(c) $(-2, -2)$ (d) none of these

24. Two cyclists start from the junction of two perpendicular roads, their velocities being $3v$ metres/minute and $4v$ metres/minute. The rate at which the two cyclists are separating is

(a) $\frac{7}{2}v$ m/min (b) $5v$ m/min

(c) v m/min (d) none of these

25. A stick of length a cm rests against a vertical wall and the horizontal floor. If the foot of the stick slides with a constant velocity of b cm/s then the magnitude of the velocity of the middle point of the stick when it is equally inclined with the floor and the wall, is

(a) $\frac{b}{\sqrt{2}}$ cm/s (b) $\frac{b}{2}$ cm/s

(c) $\frac{ab}{2}$ cm/s (d) none of these

26. If $y = \int_0^x \frac{t^2}{\sqrt{t^2+1}}\,dt$ then the rate of change of y with respect to x when $x = 1$, is

(a) $\sqrt{2}$ (b) $\frac{1}{2}$

(c) $\frac{1}{\sqrt{2}}$ (d) none of these

27. On the curve $x^3 = 12y$ the abscissa changes at a faster rate than the ordinate. Then x belongs to the interval

(a) $(-2, 2)$ (b) $(-1, 1)$

(c) $(0, 2)$ (d) none of these

28. A balloon is pumped at the rate of a cm^3/minute. The rate of increase of its surface area when the radius is b cm, is

(a) $\frac{2a^2}{b^4}$ cm^2/min (b) $\frac{a}{2b}$ cm^2/min

(c) $\frac{2a}{b}$ cm^2/min (d) none of these

29. x and y are the sides of two squares such that $y = x - x^2$. The rate of change of the area of the second square with respect to that of the first square is

(a) $2(1-x^2)x$ (b) $2x^2 - 3x + 1$

(c) $2(2x^2 - 3x + 1)$ (d) none of these

30. Let the equation of a curve be $x = a(\theta + \sin\theta)$, $y = a(1 - \cos\theta)$. If θ changes at a constant rate k then the rate of change of the slope of the tangent to the curve at $\theta = \frac{\pi}{3}$ is

(a) $\frac{2k}{\sqrt{3}}$ (b) $\frac{k}{\sqrt{3}}$

(c) k (d) none of these

31. If there is an error of $k\%$ in measuring the edge of a cube then the per cent error in estimating its volume is

(a) k (b) $3k$

(c) $\frac{k}{3}$ (d) none of these

32. If $1° = \alpha$ radians then the approximate value of $\cos 60°1'$ is

(a) $\frac{1}{2} + \frac{\alpha\sqrt{3}}{120}$ (b) $\frac{1}{2} - \frac{\alpha}{120}$

(c) $\frac{1}{2} - \frac{\alpha\sqrt{3}}{120}$ (d) none of these

• Type 2 •

Choose the correct options. One or more options may be correct.

33. Let the parabolas $y = x^2 + ax + b$ and $y = x(c - x)$ touch each other at the point (1, 0). Then

(a) $a = -3$ (b) $b = 1$

(c) $c = 2$ (d) $b + c = 3$

34. Let $y = f(x)$ be the equation of a parabola which is touched by the line $y = x$ at the point where $x = 1$. Then

(a) $f'(0) = f'(1)$ (b) $f'(1) = 1$

(c) $f(0) + f'(0) + f''(0) = 1$ (d) $2f(0) = 1 - f'(0)$

35. A point on the ellipse $4x^2 + 9y^2 = 36$ where the tangent is equally inclined to the axes is

(a) $\left(\frac{9}{\sqrt{13}}, \frac{4}{\sqrt{13}}\right)$ (b) $\left(-\frac{9}{\sqrt{13}}, \frac{4}{\sqrt{13}}\right)$

(c) $\left(\frac{9}{\sqrt{13}}, -\frac{4}{\sqrt{13}}\right)$ (d) none of these

36. If the line $ax + by + c = 0$ is a normal to the rectangular hyperbola $xy = 1$ then

(a) $a > 0, b > 0$ (b) $a > 0, b < 0$

(c) $a < 0, b > 0$ (d) $a < 0, b < 0$

37. A tangent to the curve $y = \int_0^x |t|\, dt$, which is parallel to the line $y = x$, cuts off an intercept from the y-axis equal to

(a) 1 (b) $-\frac{1}{2}$

(c) $\frac{1}{2}$ (d) –1

Answers

1. d	**2.** c	**3.** b	**4.** a	**5.** d
6. a	**7.** d	**8.** b	**9.** c	**10.** b
11. a	**12.** a	**13.** b	**14.** c	**15.** a
16. d	**17.** a	**18.** b	**19.** c	**20.** b
21. a	**22.** c	**23.** a	**24.** b	**25.** a
26. c	**27.** a	**28.** c	**29.** b	**30.** d
31. b	**32.** c	**33.** a, d	**34.** b, d	**35.** a, b, c
36. b, c	**37.** b, c			

Hints and Solutions to Selected Questions

1. Differentiating w.r.t. x, $e^y \cdot \frac{dy}{dx} = 2x$

or $\quad \frac{dy}{dx} = \frac{2x}{1+x^2} \qquad (\because e^y = 1 + x^2)$

$\therefore \quad m = \frac{2x}{1+x^2}$ or $|m| = \frac{2|x|}{1+|x|^2}$.

But $1 + |x|^2 - 2|x| = (1 - |x|)^2 \geq 0$.

$\therefore \quad 1 + |x|^2 \geq 2|x|, \qquad \therefore \; |m| \leq 1$.

2. $\frac{dy}{dx} = 3x^2 - 2ax + 1$.

From the question, $\frac{dy}{dx} \geq 0 \qquad \Rightarrow 3x^2 - 2ax + 1 \geq 0$ for all x.

$\therefore \quad D \leq 0$ or $4a^2 - 12 \leq 0 \qquad \Rightarrow -\sqrt{3} \leq a \leq \sqrt{3}$.

4. Here $y > 0$. Putting $y = x$ in $y = \sqrt{4 - x^2}$, we get $x = \sqrt{2}, -\sqrt{2}$.

So, the point is $(\sqrt{2}, \sqrt{2})$. Differentiating $y^2 + x^2 = 4$ w.r.t. x,

$$2y\frac{dy}{dx} + 2x = 0 \quad \text{or} \quad \frac{dy}{dx} = -\frac{x}{y}$$

$\therefore \quad$ at $(\sqrt{2}, \sqrt{2})$, $\frac{dy}{dx} = -1$.

5. If $-\frac{\pi}{2} < x < 0$ then $y = \cos^{-1}(\cos x) = \cos^{-1}\{\cos(-x)\} = -x$.

$\therefore \quad \frac{dy}{dx} = -1 \quad \therefore \; \left.\frac{dy}{dx}\right)_{x=-\pi/4} = -1$.

6. $\frac{dy}{dx} = \frac{1}{1+x^3} \qquad \therefore \; \left.\frac{dy}{dx}\right)_{x=1} = \frac{1}{2}$.

7. $\frac{dy}{dx} = \frac{dy/dt}{dx/dt} = \frac{e^t(\cos t - \sin t)}{e^t(\sin t + \cos t)}$

$$\therefore \quad \left.\frac{dy}{dx}\right)_{t=\pi/4} = \frac{\frac{1}{\sqrt{2}} - \frac{1}{\sqrt{2}}}{\frac{1}{\sqrt{2}} + \frac{1}{\sqrt{2}}} = 0.$$

8. Differentiating w.r.t. x, $1 + \frac{dy}{dx} = e^{xy}\left(y + x\frac{dy}{dx}\right)$

or $\quad \frac{dy}{dx} = \frac{ye^{xy} - 1}{1 - xe^{xy}}.$

$$\frac{dy}{dx} = \infty \quad \Rightarrow 1 - xe^{xy} = 0 \quad \Rightarrow 1 - x(x+y) = 0.$$

This holds for $x = 1, y = 0$.

9. $y^2 = 2x \quad \Rightarrow 2y\frac{dy}{dx} = 2 \quad \Rightarrow \frac{dy}{dx} = \frac{1}{y}.$

The slope of PQ $= \frac{2+1}{2 - \frac{1}{2}} = 2.$

$$\therefore \quad \frac{1}{y} = 2 \quad \Rightarrow y = \frac{1}{2}$$

$$\therefore \quad x = \frac{y^2}{2} = \frac{\left(\frac{1}{2}\right)^2}{2} = \frac{1}{8}.$$

10. $x^{3/2} + y^{3/2} = a^{3/2}$

$$\Rightarrow \quad \frac{3}{2}\cdot x^{1/2} + \frac{3}{2}\cdot y^{1/2}\frac{dy}{dx} = 0 \quad \Rightarrow \frac{dy}{dx} = -\frac{x^{1/2}}{y^{1/2}}.$$

$$\therefore \quad \left.\frac{dy}{dx}\right)_{\alpha,\beta} = 1 \quad \Rightarrow \alpha^{1/2} + \beta^{1/2} = 0.$$

Also, $\alpha^{3/2} + \beta^{3/2} = a^{3/2}$ $\quad \{\because \ (\alpha, \beta)$ is on the curve$\}$.

These give no values of α, β.

Now, $\left.\frac{dy}{dx}\right)_{\alpha,\beta} = -1 \quad \Rightarrow \alpha^{1/2} = \beta^{1/2}.$

Also, $\alpha^{3/2} + \beta^{3/2} = a^{3/2} \quad \therefore \ \alpha = \beta = \frac{a}{2^{2/3}}.$

$\therefore$ there is only one point.

11. See the hint to Q. No. 10.

13. The two points are $\left(\pm\frac{8\sqrt{2}}{3}, 1\right)$. Differentiating, $\frac{x}{8}+\frac{2}{9}y\frac{dy}{dx}=0.$

$$\therefore \quad \frac{dy}{dx}=-\frac{9}{16}\cdot\frac{x}{y}$$

$$\therefore \quad m_1=\left(\frac{dy}{dx}\right)_{8\sqrt{2}/3,\,1}=-\frac{3}{2} \quad \text{and} \quad m_2=\left(\frac{dy}{dx}\right)_{-8\sqrt{2}/3,\,1}=\frac{3}{\sqrt{2}}$$

$$\therefore \quad \tan\phi=\left|\frac{m_1-m_2}{1+m_1m_2}\right|=\left|\frac{-3\sqrt{2}}{1-\frac{9}{2}}\right|=\frac{6\sqrt{2}}{7}.$$

14. Differentiating w.r.t. x,

$$2y\frac{dy}{dx}-6x^2-4\frac{dy}{dx}=0 \quad \Rightarrow \frac{dy}{dx}=\frac{3x^2}{y-2}$$

$\therefore$ the equation of the tangent at (α, β) is $y-\beta=\frac{3\alpha^2}{\beta-2}(x-\alpha)$.

It passes through (1, 2) if $2-\beta=\frac{3\alpha^2}{\beta-2}(1-\alpha)$

or $\quad (\beta-2)^2=3\alpha^2(\alpha-1)$.

Also, (α, β) satisfies the equation of the curve.

$\therefore \quad \beta^2-2\alpha^3-4\beta+8=0 \qquad$ or $\quad (\beta-2)^2=2\alpha^3-4$

$\therefore \quad (\beta-2)^2=3\alpha^2(\alpha-1)=2\alpha^3-4$

$\therefore \quad \alpha^3-3\alpha^2+4=0$

or $\quad (\alpha-2)(\alpha^2-\alpha-2)=0 \qquad$ or $\quad (\alpha-2)^2(\alpha+1)=0.$

When $\alpha=2, (\beta-2)^2=12 \qquad$ or $\qquad \beta=2\pm 2\sqrt{3}.$

When $\alpha=-1, (\beta-2)^2=-6 \qquad$ or $\qquad \beta=$ nonreal number.

$\therefore \quad (\alpha, \beta)$ has two values.

15. Here, $f'(1)=\tan\frac{\pi}{6}=\frac{1}{\sqrt{3}}, \quad f'(2)=\tan\frac{\pi}{3}=\sqrt{3}, \quad f'(3)=\tan\frac{\pi}{4}=1.$

Now, $\int_2^3 f'(x)\cdot f''(x)\,dx=\left[\frac{1}{2}\{f'(x)\}^2\right]_2^3=\frac{1}{2}[\{f'(3)\}^2-\{f'(2)\}^2]$

$$\int_1^3 f''(x)\,dx=\left[f'(x)\right]_1^3=f'(3)-f'(1).$$

$\therefore$ value $= \frac{1}{2}\{1^2 - (\sqrt{3})^2\} + 1 - \frac{1}{\sqrt{3}} = -\frac{1}{\sqrt{3}}$.

16. The point of intersection is $\left(1, \frac{1}{e}\right)$.

Around $x = 1$, $|x| = x$.

So, $y = e^{-x}$ $\quad\therefore\quad \frac{dy}{dx} = -e^{-x}$ $\quad\therefore\quad \left.\frac{dy}{dx}\right)_{x=1} = -e^{-1}$

$\therefore$ the equation of the tangent at $\left(1, \frac{1}{e}\right)$ is

$$y - \frac{1}{e} = -\frac{1}{e}(x - 1) \quad \text{or} \quad x + ey = 2.$$

19. The equation of the line is $y - 3 = \frac{3+2}{0-5}(x - 0)$, i.e., $x + y - 3 = 0$.

$$y = \frac{c}{x+1} \quad \Rightarrow \quad \frac{dy}{dx} = \frac{-c}{(x+1)^2}.$$

Let the line touches the curve at (α, β).

$$\therefore \quad \alpha + \beta - 3 = 0, \quad \left.\frac{dy}{dx}\right)_{\alpha, \beta} = \frac{-c}{(\alpha+1)^2} = -1 \quad \text{and} \quad \beta = \frac{c}{\alpha+1}$$

$$\therefore \quad \frac{c}{(c/\beta)^2} = 1 \quad \text{or} \quad \beta^2 = c \quad \text{or} \quad (3-\alpha)^2 = c = (\alpha+1)^2$$

$$\therefore \quad 3 - \alpha = \pm(\alpha + 1) \quad \text{or} \quad 3 - \alpha = \alpha + 1$$

$\therefore$ $\alpha = 1$. So, $c = (1+1)^2 = 4$.

20. Differentiating w.r.t. x, $\frac{nx^{n-1}}{a^n} + \frac{ny^{n-1}}{b^n} \cdot \frac{dy}{dx} = 0$

$$\text{or} \quad \frac{dy}{dx} = -\frac{b^n x^{n-1}}{a^n y^{n-1}}$$

$\therefore$ the equation of the tangent at (α, β) is

$$y - \beta = -\frac{b^n \alpha^{n-1}}{a^n \beta^{n-1}}(x - \alpha)$$

$$\text{or} \quad a^n \beta^{n-1} y + b^n \alpha^{n-1} x = a^n \beta^n + b^n \alpha^n.$$

But (α, β) is on the curve.

$$\therefore \quad a^n \beta^{n-1} y + b^n \alpha^{n-1} x = a^n b^n \cdot 2 \qquad \dots (1)$$

The line touches the curve at (α, β) if (1) and $\frac{x}{a}+\frac{y}{b}=2$ are identical.

$$\therefore \quad \frac{b^n\alpha^{n-1}}{\frac{1}{a}}=\frac{a^n\beta^{n-1}}{\frac{1}{b}}=\frac{2a^nb^n}{2} \qquad \Rightarrow \alpha^{n-1}=a^{n-1}, \beta^{n-1}=b^{n-1}$$

$$\therefore \quad (\alpha, \beta)=(a, b).$$

22. $y=\log_e x \quad \Rightarrow \frac{dy}{dx}=\frac{1}{x} \quad \therefore \left.\frac{dy}{dx}\right)_{1,0}=1.$

$\therefore$ the equation of the normal at (1, 0) is

$$y-0=-\frac{1}{1}(x-1) \quad \text{or} \quad x+y=1$$

$$\therefore \quad \text{area}=\frac{1}{2}\cdot 1\cdot 1.$$

24. At time t, the distance z between the cyclists is given by $z^2=(3vt)^2+(4vt)^2$

$$\therefore \quad z=5vt \quad \Rightarrow \frac{dz}{dt}=5v.$$

25. Here, $\frac{dx}{dt}=b$ cm/s, $\quad x^2+y^2=a^2.$

Differentiating w.r.t. t, $2x\frac{dx}{dt}+2y\frac{dy}{dt}=0.$

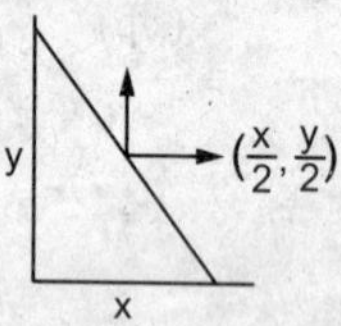

or $\quad 2bx+2y\frac{dy}{dt}=0 \quad \Rightarrow \frac{dy}{dt}=-\frac{bx}{y}.$

The velocity of the middle point at time t

$$=\sqrt{\left\{\frac{d(x/2)}{dt}\right\}^2+\left\{\frac{d(y/2)}{dt}\right\}^2}$$

$$=\frac{1}{2}\sqrt{\left(\frac{dx}{dt}\right)^2+\left(\frac{dy}{dt}\right)^2}=\frac{1}{2}\sqrt{b^2+\frac{b^2x^2}{y^2}}.$$

When the stick is equally inclined to the wall and to the floor, $x=y$.

$\therefore$ the required velocity $=\frac{1}{2}\sqrt{b^2+b^2}=\frac{b}{\sqrt{2}}.$

26. $\frac{dy}{dx}=\frac{x^2}{\sqrt{x^2+1}} \qquad \therefore \left.\frac{dy}{dx}\right)_{x=1}=\frac{1}{\sqrt{1+1}}=\frac{1}{\sqrt{2}}.$

27. From the question, $\left|\frac{dx}{dy}\right| > 1$. Differentiating $x^3 = 12y$ w.r.t. y,

$$3x^2 \frac{dx}{dy} = 12 \qquad \therefore \qquad \frac{dx}{dy} = \frac{4}{x^2}.$$

$$\therefore \quad \frac{4}{x^2} > 1 \quad \Rightarrow x^2 - 4 < 0 \quad \Rightarrow -2 < x < 2.$$

28. $$\frac{dV}{dt} = \frac{d\left(\frac{4}{3}\pi r^3\right)}{dt} = 4\pi r^2 \frac{dr}{dt} = a \text{ (given)}$$

$$\therefore \quad \frac{dr}{dt} = \frac{a}{4\pi r^2}.$$

$$\frac{dS}{dt} = \frac{d(4\pi r^2)}{dt} = 8\pi r \cdot \frac{dr}{dt} = 8\pi r \cdot \frac{a}{4\pi r^2} = \frac{2a}{r}$$

$$\therefore \quad \left.\frac{dS}{dt}\right)_{r=b} = \frac{2a}{b}.$$

29. We have to find $\frac{d(y^2)}{d(x^2)}$. Here, $\frac{dy}{dx} = 1 - 2x$.

$$\therefore \quad \frac{d(y^2)}{d(x^2)} = \frac{\frac{d(y^2)}{dx}}{\frac{d(x^2)}{dx}} = \frac{2y\frac{dy}{dx}}{2x} = \frac{y}{x}\cdot(1-2x) = \frac{(x-x^2)(1-2x)}{x}.$$

30. $$\frac{dy}{dx} = \frac{\frac{dy}{d\theta}}{\frac{dx}{d\theta}} = \frac{a\sin\theta}{a(1+\cos\theta)} = \tan\frac{\theta}{2}.$$

$\therefore$ the rate of change of the slope, i.e., $\frac{dy}{dx}$

$$= \frac{d\left(\frac{dy}{dx}\right)}{dt} = \frac{1}{2}\sec^2\frac{\theta}{2}\cdot\frac{d\theta}{dt} = \frac{k}{2}\sec^2\frac{\theta}{2}.$$

$\therefore$ the required rate $= \frac{k}{2}\cdot\sec^2\frac{\pi}{6} = \frac{k}{2}\cdot\left(\frac{2}{\sqrt{3}}\right)^2.$

31. $V = x^3$ and the per cent error in measuring $x = \frac{dx}{x}\times 100 = k$.

The per cent error in measuring volume $= \frac{dV}{V} \times 100.$

Now, $\frac{dV}{dx} = 3x^2$ $\quad \therefore \; dV = 3x^2 dx$ $\quad \therefore \; \frac{dV}{V} = \frac{3x^2 dx}{x^3} = 3\frac{dx}{x}$

$\therefore \quad \frac{dV}{V} \times 100 = 3\frac{dx}{x} \times 100 = 3k.$

32. Let $y = \cos x$ $\quad \therefore \; \frac{dy}{dx} = -\sin x.$

Now, $\cos 60°1' = \cos 60° + \delta y$

$$\delta y = \left.\frac{dy}{dt}\right)_{x=60°} \cdot \delta x = -\frac{\sqrt{3}}{2} \cdot \delta x = -\frac{\sqrt{3}}{2} \cdot 1' = -\frac{\sqrt{3}}{2} \times \frac{\alpha}{60}$$

$\therefore \quad \cos 60°1' = \frac{1}{2} - \frac{\alpha\sqrt{3}}{120}.$

33. $(1, 0)$ is on both the curves. So, $0 = 1 + a + b$ and $0 = c - 1.$

For the first parabola, $\frac{dy}{dx} = 2x + a$ $\quad \therefore \; \left.\frac{dy}{dx}\right)_{1,0} = 2 + a.$

For the second parabola, $\frac{dy}{dx} = c - 2x$ $\quad \therefore \; \left.\frac{dy}{dx}\right)_{1,0} = c - 2.$

$\therefore \quad 2 + a = c - 2$ and $0 = c - 1$

$\Rightarrow \quad c = 1, a = -3$ $\quad \therefore \; 0 = 1 + (-3) + b$ $\quad$ or $\; b = 2.$

34. Let $y = ax^2 + bx + c$ because it is a parabola. As it touches $y = x$ at $x = 1$,

$y = a + b + c$ and $y = 1$ $\quad$ or $\; a + b + c = 1.$

Also, $\frac{dy}{dx} = 2ax + b.$

$\therefore \quad \left.\frac{dy}{dx}\right)_{x=1} = 2a + b = 1$ $\quad$ ($\because$ from $y = x$, the slope $= 1$).

Now, $f(x) = ax^2 + bx + c$ $\quad \Rightarrow f'(x) = 2ax + b$ and $f''(x) = 2a$

$\therefore \quad f(0) = c, \quad f'(0) = b, \quad f''(0) = 2a, \quad f'(1) = 2a + b = 1.$

But $a + b + c = 1$ $\quad \Rightarrow 2a + 2b + 2c = 2$ $\quad \Rightarrow 1 + b + 2c = 2$ $\quad \Rightarrow 2c = 1 - b.$

35. Differentiating w.r.t. x, $8x + 18y\frac{dy}{dx} = 0$

or $\dfrac{dy}{dx} = -\dfrac{4x}{9y}$.

The tangent is equally inclined to the axes $\Rightarrow \dfrac{dy}{dx} = \pm 1$.

$\therefore \quad -\dfrac{4x}{9y} = \pm 1 \quad \Rightarrow y = -\dfrac{4}{9}x, \dfrac{4}{9}x.$

$\therefore \quad 4x^2 + 9 \cdot \dfrac{16}{81}x^2 = 36 \quad \Rightarrow \dfrac{52}{9}x^2 = 36 \quad \Rightarrow x = \pm \dfrac{9}{\sqrt{13}}$.

36. Differentiating w.r.t. x, $y + x\dfrac{dy}{dx} = 0 \qquad \therefore \quad \dfrac{dy}{dx} = -\dfrac{y}{x}$.

$\therefore$ the equation of the normal at (α, β) is

$$y - \beta = \frac{\alpha}{\beta}(x - \alpha) \text{ or } \alpha x - \beta y = \alpha^2 - \beta^2.$$

The given line is a normal at (α, β) if

$$\frac{\alpha}{a} = -\frac{\beta}{b} = \frac{\alpha^2 - \beta^2}{-c}$$

$$\Rightarrow \quad \frac{\alpha}{a} = \frac{\beta}{-b} = \frac{\sqrt{\alpha\beta}}{\sqrt{-ab}} = \frac{1}{\sqrt{-ab}} \qquad (\because \ \alpha\beta = 1)$$

$\therefore$ α, β are real if $ab < 0$, i.e., $a > 0, b < 0$ or $a < 0, b > 0$.

37. Differentiating w.r.t. x, $\dfrac{dy}{dx} = |x| = 1$ because the slope of $y = x$ is 1.

$\therefore$ at (α, β), $\left.\dfrac{dy}{dx}\right)_{\alpha, \beta} = 1 = |\alpha| \qquad \therefore \ \alpha = 1, -1.$

$\therefore$ when $\alpha = 1, \beta = \displaystyle\int_0^1 |t|\, dt = \int_0^1 t dt = \frac{1}{2}$ and

when $\alpha = -1, \beta = \displaystyle\int_0^{-1} |t| dt = -\int_{-1}^{0} |t| dt = -\int_{-1}^{0} -t dt = -\frac{1}{2}$.

$\therefore$ the points where the tangents are parallel to the line $y = x$ are $\left(1, \dfrac{1}{2}\right), \left(-1, -\dfrac{1}{2}\right)$.

The tangent at $\left(1, \dfrac{1}{2}\right)$ is $y - \dfrac{1}{2} = 1(x - 1)$, i.e., $2x - 2y = 1$.

The tangent at $\left(-1, -\dfrac{1}{2}\right)$ is $y + \dfrac{1}{2} = 1(x + 1)$, i.e., $2x - 2y + 1 = 0$.

□

6

Maxima and Minima

• Type 1 •

Choose the most appropriate option (a, b, c or d).

1. If $xy = a^2$ and $S = b^2x + c^2y$ where a, b and c are constants then the minimum value of S is

(a) abc (b) $bc\sqrt{a}$
(c) $2abc$ (d) none of these

2. The global minimum value of $f(x) = x^4 - x^2 - 2x + 6$ is

(a) 6 (b) 8
(c) 4 (d) nonexistent

3. The maximum value of $f(x) = 3\cos^2 x + 4\sin^2 x + \cos\frac{x}{2} + \sin\frac{x}{2}$ is

(a) 4 (b) $3 + \sqrt{2}$
(c) $4 + \sqrt{2}$ (d) none of these

4. If $a > b > 0$, the minimum value of $a\sec\theta - b\tan\theta$ is

(a) $b - a$ (b) $\sqrt{a^2 + b^2}$
(c) $\sqrt{a^2 - b^2}$ (d) $2\sqrt{a^2 - b^2}$

5. The number of values of x where the function $f(x) = \cos x + \cos(\sqrt{2}x)$ attains its maximum is

(a) 0 (b) 1
(c) 2 (d) infinite

6. The function $f(x)=\dfrac{x}{1+x\tan x}$ has

(a) one point of minimum in the interval $(0, \pi/2)$

(b) one point of maximum in the interval $(0, \pi/2)$

(c) no point of maximum, no point of minimum in the interval $(0, \pi/2)$

(d) two points of maxima in the interval $(0, \pi/2)$

7. Let $f(x)=e^x \sin x$ be the equation of a curve. If at $x=a$, $0 \le a \le 2\pi$, the slope of the tangent is the maximum then the value of a is

(a) $\pi/2$ (b) $3\pi/2$

(c) π (d) $\pi/4$

8. Let the tangent to the graph of $y=f(x)$ at the point $x=a$ be parallel to the x-axis, and let $f'(a-h)>0$ and $f'(a+h)<0$, where h is a very small positive number. Then the ordinate of the point is

(a) a maximum

(b) a minimum

(c) both a maximum and a minimum

(d) neither a maximum nor a minimum

9. Let $f(x)=x^3+3x^2-9x+2$. Then

(a) $f(x)$ has a maximum at $x=1$

(b) $f(x)$ has neither a minimum nor a maximum at $x=-3$

(c) $f(x)$ has a minimum at $x=1$

(d) none of these

10. If $f(x)=a\log_e|x|+bx^2+x$ has extremums at $x=1$ and $x=3$ then

(a) $a=-\dfrac{3}{4}, b=-\dfrac{1}{8}$ (b) $a=\dfrac{3}{4}, b=-\dfrac{1}{8}$

(c) $a=-\dfrac{3}{4}, b=\dfrac{1}{8}$ (d) none of these

11. The maximum value of $\left(\dfrac{1}{x}\right)^{2x^2}$ is

(a) e (b) $\sqrt[e]{e}$

(c) 1 (d) none of these

12. The maximum ordinate of a point on the graph of the function $f(x) = \sin x(1 + \cos x)$ is

(a) $\frac{2+\sqrt{3}}{4}$. (b) $\frac{3\sqrt{3}}{4}$

(c) 1 (d) none of these

13. If $\theta + \phi = \frac{\pi}{3}$ then $\sin\theta \cdot \sin\phi$ has a maximum value at $\theta =$

(a) $\frac{\pi}{6}$ (b) $\frac{2\pi}{3}$

(c) $\frac{\pi}{4}$ (d) none of these

14. Let $f(x) = x^3 - 6x^2 + 12x - 3$. Then at $x = 2, f(x)$ has

(a) a maximum

(b) a minimum

(c) both a maximum and a minimum

(d) neither a maximum nor a minimum

15. Let $f(x) = (x-p)^2 + (x-q)^2 + (x-r)^2$. Then $f(x)$ has a minimum at $x = \lambda$, where λ is equal to

(a) $\frac{p+q+r}{3}$ (b) $\sqrt[3]{pqr}$

(c) $\frac{3}{\frac{1}{p}+\frac{1}{q}+\frac{1}{r}}$ (d) none of these

16. Let $f(x) = 1 + 2x^2 + 2^2x^4 + \ldots + 2^{10}x^{20}$. Then $f(x)$ has

(a) more than one minimum (b) exactly one minimum

(c) at least one maximum (d) none of these

17. Let $f(x) = \frac{a}{x} + x^2$. If it has a maximum at $x = -3$ then a is

(a) -1 (b) 16

(c) 1 (d) none of these

18. Let $f(x)$ be a function such that $f'(a) \neq 0$. Then at $x = a, f(x)$

(a) cannot have a maximum

(b) cannot have a minimum

(c) must have neither a maximum nor a minimum

(d) none of these

19. Let the function $f(x)$ be defined as below.

$$f(x) = \sin^{-1}\lambda + x^2, 0 < x < 1$$
$$2x, x \geq 1$$

$f(x)$ can have a minimum at $x = 1$ if the value of λ is

(a) 1 (b) –1

(c) 0 (d) none of these

20. Let x be a number which exceeds its square by the greatest possible quantity. Then x is equal to

(a) $\frac{1}{2}$ (b) $\frac{1}{4}$

(c) $\frac{3}{4}$ (d) none of these

21. The sum of two nonzero numbers is 8. The minimum value of the sum of their reciprocals is

(a) $\frac{1}{4}$ (b) $\frac{1}{2}$

(c) $\frac{1}{8}$ (d) none of these

22. In a ΔABC, $\angle B = 90°$ and $b + a = 4$. The area of the triangle is the maximum when $\angle C$ is

(a) $\frac{\pi}{4}$ (b) $\frac{\pi}{6}$

(c) $\frac{\pi}{3}$ (d) none of these

23. The point (0, 3) is nearest to the curve $x^2 = 2y$ at

(a) $(2\sqrt{2}, 0)$ (b) (0, 0)

(c) (2, 2) (d) none of these

24. If λ, μ be real numbers such that $x^3 - \lambda x^2 + \mu x - 6 = 0$ has its roots real and positive then the minimum value of μ is

(a) $3 \times \sqrt[3]{36}$ (b) 11

(c) 0 (d) none of these

• Type 2 •

Choose the correct options. One or more options may be correct.

25. Let $f(x) = x + x^{-1}$. Then

(a) $f(x)$ has a maximum but no minimum

(b) $f(x)$ has no maximum but a minimum

(c) $f(x)$ has a maximum and a minimum

(d) $\max f(x) < \min f(x)$

26. Let $f(x) = ax^3 + bx^2 + cx + 1$ have extrema at $x = \alpha, \beta$ such that $\alpha\beta < 0$ and $f(\alpha) \cdot f(\beta) < 0$. Then the equation $f(x) = 0$ has

(a) three equal real roots

(b) three distinct real roots

(c) one positive root if $f(\alpha) < 0$ and $f(\beta) > 0$

(d) one negative root if $f(\alpha) > 0$ and $f(\beta) < 0$

27. The critical point(s) of $f(x) = \dfrac{|2-x|}{x^2}$ is(are)

(a) $x = 0$ (b) $x = 2$

(c) $x = 4$ (d) none of these

28. The value of x for which the function $f(x) = \int_0^x (1-t^2)e^{-t^2/2}dt$ has an extremum is

(a) 0 (b) 1

(c) –1 (d) none of these

29. Let $f(x) = x^3 + 3x^2 + 3x + 2$. Then, at $x = -1$

(a) $f(x)$ has a maximum (b) $f(x)$ has a minimum

(c) $f'(x)$ has a maximum (d) $f'(x)$ has a minimum

30. The function $f(x) = x^2 + \dfrac{\lambda}{x}$ has a

(a) minimum at $x = 2$ if $\lambda = 16$

(b) maximum at $x = 2$ if $\lambda = 16$

(c) maximum for no real value of λ

(d) point of inflection at $x = 1$ if $\lambda = -1$

31. Let $f(x) = (x-1)^4 \cdot (x-2)^n, n \in N$. Then $f(x)$ has

(a) a maximum at $x = 1$ if n is odd

(b) a maximum at $x = 1$ if n is even

(c) a minimum at $x = 2$ if n is even

(d) a maximum at $x = 2$ if n is odd

Answers

1. c	**2.** c	**3.** c	**4.** c	**5.** b
6. b	**7.** a	**8.** a	**9.** c	**10.** a
11. b	**12.** b	**13.** a	**14.** d	**15.** a
16. b	**17.** d	**18.** d	**19.** d	**20.** a
21. b	**22.** c	**23.** c	**24.** a	**25.** c, d
26. b, c, d	**27.** a, b, c	**28.** b, c	**29.** d	**30.** a, c, d
31. a, c				

Hints and Solutions to Selected Questions

1. $S = b^2 x + \dfrac{c^2 a^2}{x}$

$\therefore \quad \dfrac{dS}{dx} = b^2 - \dfrac{c^2 a^2}{x^2} = 0 \quad \Rightarrow x = \pm \dfrac{ca}{b}.$

$\dfrac{d^2 S}{dx^2} = 2c^2 a^2 \cdot \dfrac{1}{x^3} \qquad \therefore \quad \left.\dfrac{d^2 S}{dx^2}\right)_{x = ca/b} > 0$

$\therefore \quad \min S = b^2 \cdot \dfrac{ca}{b} + \dfrac{c^2 a^2}{\dfrac{ca}{b}} = 2abc.$

2. $f'(x) = 4x^3 - 2x - 2$

$\therefore \quad f'(x) = 0 \quad \Rightarrow 2x^3 - x - 1 = 0 \quad \Rightarrow (x-1)(2x^2 + 2x + 1) = 0$

$\therefore \quad x = 1.$

Now, $f''(x) = 12x^2 - 2 \qquad \therefore \ f''(1) = 12 - 2 > 0.$

Hence, global $\min f(x) = 1 - 1 - 2 + 6 = 4.$

3. $f(x) = 4 - \cos^2 x + \cos \dfrac{x}{2} + \sin \dfrac{x}{2}$

$\therefore \quad f'(x) = \sin 2x - \dfrac{1}{2}\left(\sin \dfrac{x}{2} - \cos \dfrac{x}{2}\right)$

$= 2\sin x \cdot \left(\cos^2 \dfrac{x}{2} - \sin^2 \dfrac{x}{2}\right) + \dfrac{1}{2}\left(\cos \dfrac{x}{2} - \sin \dfrac{x}{2}\right)$

$= \left(\cos \dfrac{x}{2} - \sin \dfrac{x}{2}\right)\left\{2\sin x \left(\cos \dfrac{x}{2} + \sin \dfrac{x}{2}\right) + \dfrac{1}{2}\right\}$

$= \left(\cos \dfrac{x}{2} - \sin \dfrac{x}{2}\right)\left\{2\sqrt{2}\sin x \cdot \sin\left(\dfrac{x}{2} + \dfrac{\pi}{4}\right) + \dfrac{1}{2}\right\}$

$\therefore \quad f'(x) = 0 \quad \Rightarrow \cos \dfrac{x}{2} - \sin \dfrac{x}{2} = 0$

$\Rightarrow x = \dfrac{\pi}{2}.$

$$f''(x) = 2\cos 2x - \frac{1}{2} \cdot \frac{1}{2}\left(\cos\frac{x}{2} + \sin\frac{x}{2}\right)$$

$$\therefore \quad f''\left(\frac{\pi}{2}\right) = -2 - \frac{1}{4}\left(\frac{1}{\sqrt{2}} + \frac{1}{\sqrt{2}}\right) < 0.$$

$$\therefore \quad \max f(x) = f\left(\frac{\pi}{2}\right) = 4 - 0 + \frac{1}{\sqrt{2}} + \frac{1}{\sqrt{2}} = 4 + \sqrt{2}.$$

4. $f(\theta) = a\sec\theta - b\tan\theta$

$$\therefore \quad f'(\theta) = a\sec\theta \cdot \tan\theta - b\sec^2\theta.$$

$$\therefore \quad f'(\theta) = 0 \quad \Rightarrow \sec\theta(a\tan\theta - b\sec\theta) = 0$$

$$\Rightarrow \sec\theta = 0 \text{ or } a\tan\theta = b\sec\theta.$$

But $\sec\theta \neq 0$.

$$\therefore \quad f'(\theta) = 0 \quad \Rightarrow a\tan\theta = b\sec\theta \quad \Rightarrow b = \frac{a\tan\theta}{\sec\theta} = a\sin\theta$$

$$\therefore \quad \sin\theta = \frac{b}{a}\cdot$$

$$f''(\theta) = a\sec\theta \cdot \tan^2\theta + a\sec^3\theta - 2b\sec^2\theta \cdot \tan\theta$$

$$= a \cdot \frac{a}{\sqrt{a^2 - b^2}} \cdot \frac{b^2}{(\sqrt{a^2 - b^2})^2} + a \cdot \left(\frac{a}{\sqrt{a^2 - b^2}}\right)^3 - 2b\left(\frac{a}{\sqrt{a^2 - b^2}}\right)^2 \cdot \frac{b}{\sqrt{a^2 - b^2}}$$

$$= \frac{1}{(a^2 - b^2)^{3/2}} \cdot \{a^2b^2 + a^4 - 2a^2b^2\} = \frac{a^2(a^2 - b^2)}{(a^2 - b^2)^{3/2}} = \frac{a^2}{\sqrt{a^2 - b^2}} > 0$$

$$\therefore \quad \min f(\theta) = a \cdot \frac{a}{\sqrt{a^2 - b^2}} - b \cdot \frac{b}{\sqrt{a^2 - b^2}} = \frac{a^2 - b^2}{\sqrt{a^2 - b^2}} = \sqrt{a^2 - b^2}.$$

5. $f(x) = 2 \cdot \cos\frac{\sqrt{2} + 1}{2}x \cdot \cos\frac{\sqrt{2} - 1}{2}x \leq 2,$

and it is 2 only when $\cos\frac{\sqrt{2} + 1}{2}x$ and $\cos\frac{\sqrt{2} - 1}{2}x$ are both equal to 1 for a value of x. This is possible only when $x = 0$.

6. $f'(x) = \dfrac{1 + x\tan x - x\{\tan x + x\sec^2 x\}}{(1 + x\tan x)^2} = \dfrac{1 - x^2\sec^2 x}{(1 + x\tan x)^2}.$

$f'(x)=0 \quad \Rightarrow 1-x^2\sec^2 x=0$

$\Rightarrow \cos x = \pm x$

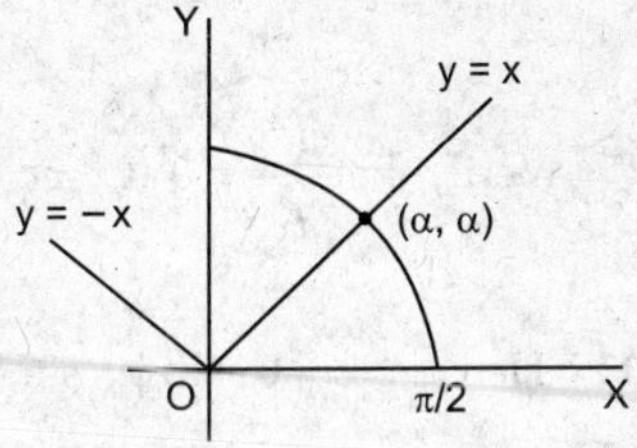

$\therefore$ in the interval $(0, \pi/2), f'(x)=0$ can hold for only one value of x, say α, where $\cos\alpha=\alpha$ (see the graph).

Verify $f''(x)<0$ at $x=\alpha$.

7. $f'(x)=e^x(\sin x+\cos x)$. Therefore, the slope m of the tangent is given by

$$m=e^x(\sin x+\cos x).$$

Now, $\dfrac{dm}{dx}=e^x\{\sin x+\cos x+\cos x-\sin x\}=2e^x\cdot\cos x$

and $\dfrac{d^2m}{dx^2}=2e^x(\cos x-\sin x)$.

At $x=a$, m is the maximum.

$$\therefore \quad \left.\frac{dm}{dx}\right)_{x=a}=0, \qquad \left.\frac{d^2m}{dx^2}\right)_{x=a}<0.$$

$\left.\dfrac{dm}{dx}\right)_{x=a}=0 \quad\Rightarrow\quad 2e^a\cos a=0 \quad \Rightarrow a=\dfrac{\pi}{2},\dfrac{3\pi}{2}$ because $0\le a\le 2\pi$.

$\left.\dfrac{d^2m}{dx^2}\right)_{x=a}<0 \quad\Rightarrow\quad 2e^a(\cos a-\sin a)<0$. This is satisfied by $a=\dfrac{\pi}{2}$.

10. Around $x=1,3$ we have $|x|=x$.

$$\therefore \quad f(x)=a\log_e x+bx^2+x$$

$$\therefore \quad f'(x)=\frac{a}{x}+2bx+1.$$

From the question, $f'(1)=0, \quad f'(3)=0$

$$\therefore \quad a+2b+1=0, \frac{a}{3}+6b+1=0.$$

11. $y=\left(\dfrac{1}{x}\right)^{2x^2} \Rightarrow \log y=2x^2\log\dfrac{1}{x}\cdot$

Differentiating w.r.t. x, $\dfrac{1}{y}\cdot\dfrac{dy}{dx}=4x\log\dfrac{1}{x}+2x^2\cdot\left(-\dfrac{1}{x}\right)$.

Then $\dfrac{dy}{dx}=0$ if $-4x\log x-2x=0 \quad$ or $\quad 2\log x+1=0$

$\therefore \quad x = e^{-1/2}.$

Verify that $\dfrac{d^2y}{dx^2} < 0$ at $x = e^{-1/2}$.

13. Here, $y = \sin\theta \cdot \sin\phi = \sin\theta \cdot \sin\left(\dfrac{\pi}{3} - \theta\right) = \dfrac{1}{2}\left\{\cos\left(2\theta - \dfrac{\pi}{3}\right) - \cos\dfrac{\pi}{3}\right\}$

$$= \frac{1}{2}\cos\left(2\theta - \frac{\pi}{3}\right) - \frac{1}{4}.$$

This has a maximum at $2\theta - \dfrac{\pi}{3} = 0$.

14. $f'(x) = 3x^2 - 12x + 12 = 3(x-2)^2$.

$\therefore \quad f'(2) = 0; \quad f'(2 - \in) = 3\in^2 > 0; \quad f'(2 + \in) = 3\in^2 > 0.$

Hence, $f(x)$ has neither a maximum nor a minimum at $x = 2$.

15. $f(x) = 3x^2 - 2(p+q+r)x + p^2 + q^2 + r^2$

$\therefore \quad f'(x) = 6x - 2(p+q+r).$

$\therefore \quad f'(x) = 0 \quad \Rightarrow x = \dfrac{p+q+r}{3} =$ AM of p, q, r.

Also, $f''(x) = 6 > 0$.

16. $f'(x) = x(4 + 4 \cdot 2^2 \cdot x^2 + \ldots + 20 \cdot 2^{10} \cdot x^{18})$

$\therefore \quad f'(x) = 0 \quad \Rightarrow x = 0$ only.

Also, $f''(0) > 0$.

18. Let $f(x) = |x|$.

$f'(0) \neq 0$ but $f(x)$ has a minimum at $x = 0$.

19. $\lim\limits_{h \to 0} \dfrac{f(1+h) - f(1)}{h} = \lim\limits_{h \to 0} \dfrac{2(1+h) - 2}{h} = \lim\limits_{h \to 0} 2 = 2.$

$$\lim_{h \to 0} \frac{f(1-h) - f(1)}{-h} = \lim_{h \to 0} \frac{\sin^{-1}\lambda + (1-h)^2 - 2}{-h}$$

$$= \lim_{h \to 0} \frac{\sin^{-1}\lambda - 1 - 2h + h^2}{-h} = 2 \text{ if } \sin^{-1}\lambda = 1, \text{ i.e., } \lambda = \sin 1.$$

$\therefore \quad f(x)$ is differentiable at $x = 1$ if $\lambda = \sin 1$. Then,

$f'(x) = 2x,\ 0 < x < 1$ and $f'(x) = 2,\ x \geq 1$.

$\therefore \quad f'(1-\in) = 2(1-\in) > 0$ and $f'(1+\in) = 2 > 0.$

So, $f(x)$ does not have a maximum or a minimum at $x = 1$.

20. $y = x - x^2$, where y is the greatest difference.

$$\frac{dy}{dx} = 1 - 2x$$

$$\therefore \quad \frac{dy}{dx} = 0 \Rightarrow x = \frac{1}{2}.$$

Again, $\dfrac{d^2y}{dx^2} = -2.$

$\therefore$ y is the maximum at $x = \dfrac{1}{2}.$

21. Here, $a + b = 8$. Let $y = \dfrac{1}{a} + \dfrac{1}{b} = \dfrac{1}{a} + \dfrac{1}{8-a} = \dfrac{8}{a(8-a)}$

$$\therefore \quad \frac{dy}{da} = \frac{-8}{a^2(8-a)^2}\{8-2a\}$$

$\therefore \quad \dfrac{dy}{da} = 0$ if $8 - 2a = 0$, i.e., $a = 4$.

Also, $\left.\dfrac{dy}{dx}\right)_{4-\in} = \dfrac{-8}{(4-\in)^2 \cdot (4+\in)^2}(2\in) < 0$

$$\left.\frac{dy}{dx}\right)_{4+\in} = \frac{-8}{(4+\in)^2 \cdot (4-\in)^2}(-2\in) > 0$$

$\therefore$ at $a = 4$, y is minimum. So, min $y = \dfrac{1}{4} + \dfrac{1}{8-4} = \dfrac{1}{2}.$

22. 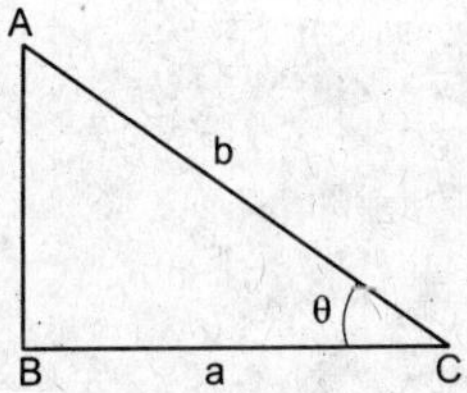

$$b\cos\theta = a$$

$$\therefore \quad b\cos\theta + b = 4$$

$$\text{or} \quad b = \frac{4}{1+\cos\theta}$$

$$\therefore \quad a = \frac{4\cos\theta}{1+\cos\theta}.$$

$$\therefore \quad \text{area} = \Delta = \frac{1}{2}ba\sin\theta = \frac{1}{2}\cdot\frac{4}{1+\cos\theta}\cdot\frac{4\cos\theta}{1+\cos\theta}\cdot\sin\theta = \frac{4\sin 2\theta}{(1+\cos\theta)^2}$$

$$\therefore \quad \frac{d\Delta}{d\theta} = 4\cdot\frac{2\cos 2\theta(1+\cos\theta)^2 + \sin 2\theta\cdot 2(1+\cos\theta)\sin\theta}{(1+\cos\theta)^4}$$

$\therefore \quad \dfrac{d\Delta}{d\theta} = 0 \quad \Rightarrow \cos 2\theta \cdot (1 + \cos\theta) + \sin 2\theta \cdot \sin\theta = 0$

or $\quad \cos 2\theta + \cos\theta = 0$

$\therefore \quad \cos 2\theta = -\cos\theta = \cos(\pi - \theta) \quad$ or $\quad \theta = \dfrac{\pi}{3}$.

$\therefore$ Δ is maximum when $\theta = \dfrac{\pi}{3}$. (It cannot be minimum, since in the given situation minimum is 0.)

23. Let $z^2 = (x - 0)^2 + (y - 3)^2 = x^2 + y^2 - 6y + 9 = y^2 - 4y + 9 \quad (\because x^2 = 2y)$

$\therefore \quad \dfrac{d(z^2)}{dy} = 2y - 4$

$\therefore \quad \dfrac{d(z^2)}{dy} = 0 \Rightarrow y = 2$ and then $x = 2, -2$.

As the question of a point being farthest does not arise (being ∞), we find (2, 2), as well as (–2, 2), are the nearest to the point (0, 3).

24. Let a, b, c be the roots. Then $ab + bc + ca = \mu$, $abc = 6$.

Dividing, $\dfrac{1}{a} + \dfrac{1}{b} + \dfrac{1}{c} = \dfrac{\mu}{6}$.

AM > GM $\quad \Rightarrow \dfrac{\frac{1}{a} + \frac{1}{b} + \frac{1}{c}}{3} \geq \sqrt[3]{\dfrac{1}{abc}}$

or $\quad \mu \geq 18 \cdot \sqrt[3]{\dfrac{1}{abc}} = 3 \cdot 6^{2/3}$

$\therefore$ the minimum value of $\mu = 3 \cdot 6^{2/3}$.

25. $f'(x) = 1 - \dfrac{1}{x^2}$

$\therefore \quad f'(x) = 0 \quad \Rightarrow x = \pm 1$.

$f''(x) = \dfrac{2}{x^3}$

$\therefore \quad f''(1) > 0$ and $f''(-1) < 0$.

$\therefore \quad f(x)$ has the maximum value $f(-1)$ and the minimum value $f(1)$.

$f(-1) = -1 - \dfrac{1}{1} = -2, \quad f(1) = 1 + \dfrac{1}{1} = 2$

$\therefore \quad \max f(x) < \min f(x)$.

26. As $f(x)$ has extrema at $x = \alpha, \beta, f'(x) = 0$ has two distinct real roots, α and β of which one is positive and other negative ($\because \alpha\beta < 0$).

Also, we know that between two real roots of $f(x) = 0$, at least one real root of $f'(x) = 0$ must lie. Hence, $f(x) = 0$ has at least two real roots, say λ, μ. So, the third root ν is also real because imaginary roots occur in pair. The third root ν cannot be equal to λ or μ because in that case $f'(x) = 0$ will also have a root λ or μ.

$\therefore$ only possible situation is

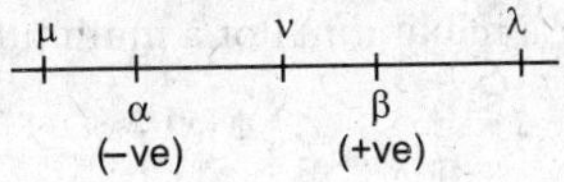

$\therefore$ one positive root will imply that ν is a negative root.

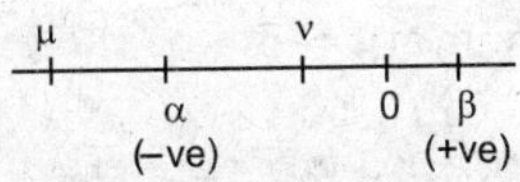

$\therefore \quad f(\alpha)$ and $f(0)$ must be of opposite signs.

But $f(0) = 1 > 0$.

$\therefore \quad f(\alpha) < 0, f(\beta) > 0 \quad \Rightarrow f(x) = 0$ has one positive root λ.

One negative root will imply that ν is a positive root.

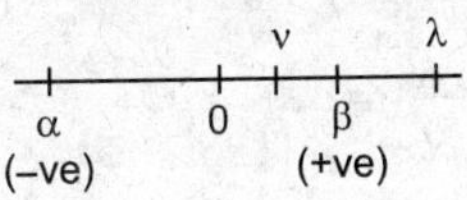

$\therefore \quad f(0)$ and $f(\beta)$ must be of opposite signs.

$\therefore \quad f(\alpha) > 0, f(\beta) < 0 \quad \Rightarrow f(x) = 0$ has one negative root μ.

27. Obviously, at $x = 0, f(x) = \infty$

$\therefore \quad f'(0)$ does not exist.

So, $x = 0$ is a critical point.

Now, $f(x) = \dfrac{2-x}{x^2}, 0 < x < 2$

$\dfrac{x-2}{x^2}, x \geq 2.$

At $x = 2, 4$ the function $f(x)$ is not differentiable. So, they are critical points.

28. $f'(x) = (1 - x^2)e^{-x^2/2}$.

For extremum, $(1 - x^2)e^{-x^2/2} = 0$, i.e., $x = 1, -1$.

29. $f(x) = (x+1)^3 + 1 \qquad \therefore \; f'(x) = 3(x+1)^2$.

$f'(x) = 0 \quad \Rightarrow x = -1$.

Now, $f'(-1 - \in) = 3(-\in)^2 > 0, f'(-1 + \in)^2 = 3\in^2 > 0$.

$\therefore \quad f(x)$ has neither a maximum nor a minimum at $x = -1$.

Let $f'(x) = \phi(x) = 3(x+1)^2 \qquad \therefore \; \phi'(x) = 6(x+1)$.

$\phi'(x) = 0 \quad \Rightarrow x = -1$

$\phi'(-1 - \in) = 6(-\in) < 0, \phi'(-1 + \in) = 6\in > 0$

$\therefore \quad \phi(x)$ has a minimum at $x = -1$

30. $f'(x) = 2x - \dfrac{\lambda}{x^2} \qquad \therefore \; f'(x) = 0 \quad \Rightarrow x = \left(\dfrac{\lambda}{2}\right)^{1/3}$.

If $\lambda = 16, x = 2$.

Now, $f''(x) = 2 + \dfrac{2\lambda}{x^3}$

$\therefore$ if $\lambda = 16, f''(x) > 0$, i.e., $f(x)$ has a minimum at $x = 2$.

Also, $f''\left\{\left(\dfrac{\lambda}{2}\right)^{1/3}\right\} = 2 + \dfrac{2\lambda}{\lambda/2} = 2 + 4 > 0$.

Hence, $f(x)$ has maximum for no real value of λ.

When $\lambda = -1, f''(x) = 0$ if $x = 1$. So, $f(x)$ has a point of inflection at $x = 1$.

31. $f'(x) = 4(x-1)^3 \cdot (x-2)^n + (x-1)^4 \cdot n(x-2)^{n-1}$

$= (x-1)^3(x-2)^{n-1}(4x - 8 + nx - n)$

$\therefore \quad f'(x) = 0$ if $x = 1, 2, \dfrac{8+n}{4+n}$.

$f'(1 - \in) = (-\in)^3 \cdot (-1 - \in)^{n-1} \cdot \{(4+n)(1 - \in) - 8 - n\}$

$= (-\text{ve}) \cdot (-\text{ve})^{n-1} \cdot \{-(4+n)\in - 4 - n\} = (-\text{ve})^{n-1}$

$f'(1 + \in) = \in^3 \cdot (-1 + \in)^{n-1} \cdot \{(4+n)(1 + \in) - 8 - n\}$

$$= (+\text{ve})(-\text{ve})^{n-1} \cdot \{(4+n)\in -4-n\} = (-\text{ve})^{n-1} \cdot (-\text{ve}) = (-\text{ve})^{n}$$

$\therefore \quad f'(1-\in) > 0, f'(1+\in) < 0$ if n is odd.

So, at $x = 1$ there is a maximum if n is odd.

$$f'(2-\in) = (1-\in)^{3} \cdot (-\in)^{n-1} \cdot \{(4+n)(2-\in) - 8 - n\}$$

$$= (+\text{ve})^{3} \cdot (-\text{ve})^{n-1} \cdot \{-(4+n)\in + n\}$$

$$= (-\text{ve})^{n-1} \cdot (+\text{ve}) = (-\text{ve})^{n-1}.$$

$$f'(2+\in) = (1+\in)^{3} \cdot (\in)^{n-1} \cdot \{(4+n)(2+\in) - 8 - n\}$$

$$= (+\text{ve})^{3} \cdot (+\text{ve})^{n-1} \cdot \{(4+n)\in + n\} = +\text{ve}.$$

$\therefore \quad f'(2-\in) > 0, f'(2+\in) < 0$ for no value of n.

$f'(2-\in) < 0, f'(2+\in) > 0$ for even value of n.

So, there is a minimum at $x = 2$ when n is even.

□

7

Monotonic Functions and Lagrange's Theorem

• *Type 1* •

Choose the most appropriate option (a, b, c or d).

1. If $f(x) = x^3 + 4x^2 + \lambda x + 1$ is a monotonically decreasing function of x in the largest possible interval $(-2, -2/3)$ then

(a) $\lambda = 4$ (b) $\lambda = 2$

(c) $\lambda = -1$ (d) λ has no real value

2. The function $f(x) = \sin^4 x + \cos^4 x$ increases if

(a) $0 < x < \frac{\pi}{8}$ (b) $\frac{\pi}{4} < x < \frac{3\pi}{8}$

(c) $\frac{3\pi}{4} < x < \frac{5\pi}{8}$ (d) $\frac{5\pi}{8} < x < \frac{3\pi}{4}$

3. If $f(x) = \frac{x}{\sin x}$ and $g(x) = \frac{x}{\tan x}$, where $0 < x \le 1$, then in the interval

(a) both $f(x)$ and $g(x)$ are increasing functions

(b) both $f(x)$ and $g(x)$ are decreasing functions

(c) $f(x)$ is an increasing function

(d) $g(x)$ is an increasing function

4. The function $f(x) = \tan^{-1} x - x$ is monotonically decreasing in the set

(a) R (b) $(0, +\infty)$

(c) $R - \{0\}$ (d) none of these

5. Let $f(x) = \tan^{-1}\{\phi(x)\}$, where $\phi(x)$ is m.i. for $0 < x < \pi/2$. Then $f(x)$ is

(a) increasing in $(0, \pi/2)$

(b) decreasing in $(0, \pi/2)$

(c) increasing in $(0, \pi/4)$ and decreasing in $(\pi/4, \pi/2)$

(d) none of these

6. If $f(x) = \dfrac{p^2 - 1}{p^2 + 1}x^3 - 3x + \log 2$ is a decreasing function of x in R then the set of possible values of p (independent of x) is

(a) $[-1, 1]$ (b) $[1, \infty)$

(c) $(-\infty, -1]$ (d) none of these

7. If $f(x) = (ab - b^2 - 2)x + \int_0^x (\cos^4\theta + \sin^4\theta)\, d\theta$ is a decreasing function of x for all $x \in R$ and $b \in R$, b being independent of x, then

(a) $a \in (0, \sqrt{6})$ (b) $a \in (-\sqrt{6}, \sqrt{6})$

(c) $a \in (-\sqrt{6}, 0)$ (d) none of these

8. Let $f(x) = \cos \pi x + 10x + 3x^2 + x^3, -2 \le x \le 3$. The absolute minimum value of $f(x)$ is

(a) 0 (b) -15

(c) $3 - 2\pi$ (d) none of these

9. If $x \in [-1, 1]$ then the minimum value of $f(x) = x^2 + x + 1$ is

(a) $-3/4$ (b) 1

(c) 3 (d) none of these

10. Let $f(x) = \dfrac{4}{3}x^3 - 4x, 0 \le x \le 2$. Then the global minimum value of the function is

(a) 0 (b) $-8/3$

(c) -4 (d) none of these

11. Let $f(x) = 6 - 12x + 9x^2 - 2x^3, 1 \le x \le 4$. Then the absolute maximum value of $f(x)$ in the interval is

(a) 2 (b) 1

(c) 4 (d) none of these

12. If $f(x)=\int_0^x (t^2+2t+2)dt,\ 2\le x\le 4$, then

(a) the maximum value of $f(x)$ is $\frac{136}{3}$

(b) the minimum value of $f(x)$ is 10

(c) the maximum value of $f(x)$ is 26

(d) none of these

13. Let the interval $I=[-1,4]$ and $f:I\to R$ be a function such that $f(x)=x^3-3x$. Then the range of the function is

(a) $[2,52]$ (b) $[-2,2]$

(c) $[-2,52]$ (d) none of these

14. The range of the function $f(x)=|2x+1|-2|x-1|,\ x\in R$, is

(a) $[-3,3]$ (b) $[0,6]$

(c) R (d) none of these

15. Let the function $f(x)$ be defined as follows:

$$f(x)=x^3+x^2-10x,\ -1\le x<0$$

$$\cos x,\ 0\le x<\frac{\pi}{2}$$

$$1+\sin x,\ \frac{\pi}{2}\le x\le\pi.$$

Then $f(x)$ has

(a) a local minimum at $x=\pi/2$

(b) a local maximum at $x=\pi/2$

(c) an absolute minimum at $x=-1$

(d) an absolute maximum at $x=\pi$

16. Let $f(x)$ be a function defined as below:

$$f(x)=\sin(x^2-3x),\ x\le 0$$

$$6x+5x^2,\ x>0.$$

Then at $x=0$, $f(x)$

(a) has a local maximum (b) has a local minimum

(c) is discontinuous (d) none of these

17. If θ is a positive acute angle then

(a) $\tan\theta < \theta < \sin\theta$ (b) $\theta < \sin\theta < \tan\theta$

(c) $\sin\theta < \tan\theta < \theta$ (d) none of these

18. Let $f: R \to R$ be a function such that $f(x) = ax + 3\sin x + 4\cos x$. Then $f(x)$ is invertible if

(a) $a \in (-5, 5)$ (b) $a \in (-\infty, -5)$ (c) $a \in (5, +\infty)$ (d) none of these

19. The function $f(x) = x^3 + \lambda x^2 + 5x + \sin 2x$ will be an invertible function if λ belongs to

(a) $(-\infty, -3)$ (b) $(-3, 3)$ (c) $(3, +\infty)$ (d) none of these

20. The value of c in Lagrange's theorem for the function $|x|$ in the interval $[-1, 1]$ is

(a) 0 (b) 1/2

(c) − 1/2 (d) nonexistent in the interval

21. The equation $\sin x + x\cos x = 0$ has at least one root in the interval

(a) $(-\pi/2, 0)$ (b) $(0, \pi)$ (c) $(-\pi/2, \pi/2)$ (d) none of these

22. If $4a + 2b + c = 0$ then the equation $3ax^2 + 2bx + c = 0$ has at least one real root lying between

(a) 0 and 1 (b) 1 and 2 (c) 0 and 2 (d) none of these

23. If the equation $a_n x^n + a_{n-1}x^{n-1} + \dots + a_1 x = 0$, $a_1 \neq 0$, $n \geq 2$, has a positive root α then the equation $na_n x^{n-1} + (n-1)a_{n-1}x^{x-2} + \dots + a_1 = 0$ has a positive root which is

(a) greater than α (b) smaller than α

(c) greater than or equal to α (d) equal to α

• Type 2 •

Choose the correct options. One or more options may be correct.

24. Let $h(x) = f(x) - \{f(x)\}^2 + \{f(x)\}^3$ for all real values of x. Then

(a) h is increasing whenever $f(x)$ is increasing

(b) h is increasing whenever $f'(x) < 0$

(c) h is decreasing whenever f is decreasing

(d) nothing can be said in general

25. If $f(x) = \sin x, -\pi/2 \le x \le \pi/2$, then

(a) $f(x)$ is increasing in the interval $[-\pi/2, \pi/2]$

(b) $f\{f(x)\}$ is increasing in the interval $[-\pi/2, \pi/2]$

(c) $f\{f(x)\}$ is decreasing in $[-\pi/2, 0]$ and increasing in $[0, \pi/2]$

(d) $f\{f(x)\}$ is invertible in $[-\pi/2, \pi/2]$

26. Let $f(x) = 2\sin^3 x - 3\sin^2 x + 12\sin x + 5, 0 \le x \le \pi/2$. Then $f(x)$ is

(a) decreasing in $[0, \pi/2]$

(b) increasing in $[0, \pi/2]$

(c) increasing in $[0, \pi/4]$ and decreasing in $[\pi/4, \pi/2]$

(d) none of these

27. Let $f(x) = -2x^3 + 21x^2 - 60x + 41$. Then

(a) $f(x)$ is m.d. in $(-\infty, 1)$

(b) $f(x)$ is m.i. in $(1, +\infty)$

(c) $f(x) > 0$ for $x < 1$

(d) $f(x) < 0$ for $x > 1$

28. Let $f(x) = \dfrac{x^2 + 1}{[x]}, 1 \le x \le 3.9$. $[\cdot]$ denotes the greatest integer function. Then

(a) $f(x)$ is m.d. in $[1, 3.9]$

(b) $f(x)$ is m.i. in $[1, 3.9]$

(c) The greatest value of $f(x)$ is $\dfrac{1}{3} \times 16.21$

(d) The least value of $f(x)$ is 2

29. Let $f(x) = |x^2 - 3x - 4|, -1 \le x \le 4$. Then

(a) $f(x)$ is m.i. in $[-1, 3/2)$

(b) $f(x)$ is m.d. in $(3/2, 4]$

(c) the maximum value of $f(x)$ is $\dfrac{25}{4}$

(d) the minimum value of $f(x)$ is 0

30. Let $f(x) = \phi(2 - x) + \phi(x)$ and $\phi''(x) < 0$ for $x \in [0, 2]$. Then

(a) $f(x)$ is m.i. in $[0, 1]$

(b) $f(x)$ is m.d. in $[0, 1]$

(c) $f(x)$ is m.i. in $[1, 2]$

(d) $f(x)$ is m.i. in $[1, 2]$

31. Let $f'(x) > 0$ and $g'(x) < 0$ for all $x \in R$. Then

(a) $f\{g(x)\} > f\{g(x+1)\}$ (b) $f\{g(x)\} > f\{g(x-1)\}$

(c) $g\{f(x)\} > g\{f(x+1)\}$ (d) $g\{f(x)\} > g\{f(x-1)\}$

32. Let $f(x) = 2x^2 - \log|x|, x \neq 0$. Then $f(x)$ is

(a) m.i. in $\left(-\frac{1}{2}, 0\right) \cup \left(\frac{1}{2}, +\infty\right)$ (b) m.d. in $\left(-\frac{1}{2}, 0\right) \cup \left(\frac{1}{2}, +\infty\right)$

(c) m.i. in $\left(-\infty, \frac{1}{2}\right) \cup \left(0, \frac{1}{2}\right)$ (d) m.d. in $\left(-\infty, -\frac{1}{2}\right) \cup \left(0, \frac{1}{2}\right)$

33. Let $f(x) = x^3 - 6x^2 + 15x + 3$. Then

(a) $f(x) > 0$ for all $x \in R$

(b) $f(x) > f(x+1)$ does not hold for any real x

(c) $f(x)$ is invertible

(d) $f(x)$ is a one-one function

34. Let $f(x) = a_5x^5 + a_4x^4 + a_3x^3 + a_2x^2 + a_1x$, where a_i's are real and $f(x) = 0$ has a positive root α_0. Then

(a) $f'(x) = 0$ has a root α_1 such that $0 < \alpha_1 < \alpha_0$

(b) $f'(x) = 0$ has at least one real root

(c) $f''(x) = 0$ has at least two real roots

(d) none of these

Answers

1. a	**2.** b	**3.** c	**4.** c	**5.** a
6. a	**7.** b	**8.** b	**9.** a	**10.** b
11. 2	**12.** a	**13.** c	**14.** a	**15.** b
16. b	**17.** d	**18.** a	**19.** b	**20.** d
21. b	**22.** c	**23.** b	**24.** a, c	**25.** a, b, d
26. b	**27.** a, c	**28.** c, d	**29.** a, b, c, d	**30.** a, d
31. a, c	**32.** a, d	**33.** b, c, d	**34.** a, b, c	

Hints and Solutions to Selected Questions

1. $f'(x) < 0 \quad \Rightarrow \ 3x^2 + 8x + \lambda < 0.$

So, λ should be such that $3x^2 + 8x + \lambda = 0$ has roots $= -2, -\frac{2}{3}\cdot$

Hence, $(-2)\left(-\frac{2}{3}\right) = \frac{\lambda}{3} \quad \therefore \ \lambda = 4.$

2. $f'(x) > 0 \qquad \Rightarrow 4\sin^3 x \cdot \cos x - 4\cos^3 x \cdot \sin x > 0$

$\Rightarrow \quad \sin x \cdot \cos x(\sin^2 x - \cos^2 x) > 0$

or $\quad \sin 2x \cdot \cos 2x < 0 \quad$ or $\quad \sin 4x < 0$

$\therefore \quad \pi < 4x < 2\pi \ \text{ or } \ 3\pi < 4x < 4\pi$

$\therefore \quad \frac{\pi}{4} < x < \frac{\pi}{2} \ \text{ or } \ \frac{3\pi}{4} < x < \pi.$

3. $f'(x) = \frac{\sin x - x\cos x}{\sin^2 x}, g'(x) = \frac{\tan x - x\sec^2 x}{\tan^2 x}.$

Now, $\frac{d}{dx}(\sin x - x\cos x) = x\sin x > 0$, when $0 < x \le 1$

$\therefore$ $\sin x - x\cos x$ is an increasing function. But at $x = 0$, $x\sin x$ is 0.

$\therefore$ in $0 < x \le 1$, $\sin x - x\cos x > 0$.

$\therefore$ $f'(x) > 0$ for $0 < x \le 1$. So, $f(x)$ is increasing in the interval.

Again, $\frac{d}{dx}(\tan x - x\sec^2 x) = -x \cdot 2\sec^2 x \cdot \tan x < 0$ for $0 < x \le 1$.

$\therefore$ $g(x)$ is decreasing in $0 < x \le 1$.

4. $f'(x) = \frac{1}{1+x^2} - 1 = \frac{-x^2}{1+x^2} < 0$ for all $x \ne 0$.

5. $f'(x) = \frac{\phi'(x)}{1 + \{\phi(x)\}^2} > 0$ for $0 < x < \frac{\pi}{2}$ because $\phi'(x) > 0$, $\phi(x)$ being m.i.

6. $f'(x) = \frac{p^2 - 1}{p^2 + 1} \cdot 3x^2 - 3 < 0$

$$\Rightarrow \quad \frac{p^2-1}{p^2+1}x^2 < 1 \qquad \Rightarrow (p^2-1)x^2-(p^2+1) < 0.$$

The inequation is satisfied for all $x \in R$ if $p^2 - 1 \le 0$, i.e., $-1 \le p \le 1$.

7. $f'(x) = ab - b^2 - 2 + \cos^4 x + \sin^4 x < 0$

$\Rightarrow \quad ab - b^2 - 2 + (\cos^2 x + \sin^2 x)^2 - 2\sin^2 x\cos^2 x < 0$

or $\quad ab - b^2 - 1 < \frac{1}{2}\sin^2 2x < \frac{1}{2}$

or $\quad 2ab - 2b^2 - 2 < 1$

or $\quad 2b^2 - 2ab + 3 > 0.$

This is true for any $b \in R$ if $D < 0$, that is, $4a^2 - 4 \cdot 2 \cdot 3 < 0$

or $\quad a^2 < 6 \qquad$ or $\quad -\sqrt{6} < a < \sqrt{6}.$

8. $f'(x) = -\pi\sin \pi x + 10 + 6x + 3x^2$

$\qquad = 3(x+1)^2 + 7 - \pi\sin \pi x > 0$ for all x.

$\therefore \quad f(x)$ is m.i. in $-2 \le x \le 3$. So, the absolute minimum $= f(-2)$.

9. $f'(x) = 2x + 1$. Therefore, for $-1 \le x < -\frac{1}{2}$, we get $f'(x) < 0$ and for $-\frac{1}{2} < x \le 1$ we get $f'(x) > 0$.

$\therefore \quad f(x)$ is m.d. in $\left[-1, -\frac{1}{2}\right]$ and m.i. in $\left(-\frac{1}{2}, 1\right]$.

$\therefore \quad \min f(x) = f\left(-\frac{1}{2}\right) = \left(-\frac{1}{2}\right)^2 + \left(-\frac{1}{2}\right) + 1 = \frac{3}{4}\cdot$

10. $f'(x) = 4(x^2 - 1)$. So, for $0 \le x < 1$, we get $f'(x) < 0$, i.e., $f(x)$ is m.d. and for $1 < x \le 2$ we get $f'(x) > 0$, i.e., $f(x)$ is m.i.

$\therefore \quad \min f(x) = f(1) = \frac{4}{3} - 4 = -\frac{8}{3}\cdot$

11. $f'(x) = -12 + 18x - 6x^2 = -6(x^2 - 3x + 2) = -6(x-1)(x-2).$

$\therefore \quad f'(x) > 0$ if $1 < x < 2$ and $f'(x) < 0$ if $2 < x \le 4$.

$\therefore \quad f(x)$ is m.i. in $1 < x < 2$ and m.d. in $2 < x \le 4$.

$\therefore \quad$ absolute maximum = the greatest among $\{f(1), f(2)\}$

$\qquad\qquad$ = the greatest among $\{1, 2\} = 2$.

12. $f'(x) = x^2 + 2x + 2 = (x+1)^2 + 1 > 0$ for all x. So, $f(x)$ is m.i. in [2, 4].

$\therefore$ $\min f(x) = f(2)$ and $\max f(x) = f(4)$.

$$\therefore \quad \min f(x) = \int_0^2 (t^2 + 2t + 2)\, dt = \left[\frac{t^3}{3} + t^2 + 2t\right]_0^2 = \frac{32}{3}.$$

$$\therefore \quad \max f(x) = \int_0^4 (t^2 + 2t + 2)\, dt = \left[\frac{t^3}{3} + t^2 + 2t\right]_0^4 = \frac{136}{3}.$$

13. $f'(x) = 3x^2 - 3 = 3(x^2 - 1)$.

So, $f'(x) \le 0$ if $-1 \le x \le 1$ and $f'(x) > 0$ if $1 < x \le 4$.

$\therefore$ $f(x)$ is decreasing in $-1 \le x \le 1$ and increasing in $1 < x \le 4$.

$\therefore$ $\min f(x) = f(1)$, $\max f(x) =$ the greatest among $\{f(-1), f(4)\}$.

Now, $f(1) = 1^3 - 3 \cdot 1 = -2$ $\qquad f(-1) = (-1)^3 - 3(-1) = 2$

$f(4) = 64 - 12 = 52$

$\therefore$ the range of the function $= [-2, 52]$.

14. Here, $f(x) = -(2x+1) - 2\{-(x-1)\} = -3,\ x < -\frac{1}{2}$

$$2x + 1 - 2\{-(x-1)\} = 4x - 1,\ -\frac{1}{2} \le x < 1$$

$$2x + 1 - 2(x-1) = 3,\ x \ge 1.$$

$f(x)$ is the constant -3 in $\left(-\infty, -\frac{1}{2}\right)$ and the constant 3 in $[1, +\infty)$.

In $-\frac{1}{2} \le x < 1$, $f'(x) = 4 > 0$ and so, $f(x)$ is m.i. in $\left[-\frac{1}{2}, 1\right)$.

In this interval, $f(x)$ increases from -3 to $3 - \in$.

As the function is continuous, range $= [-3, 3]$.

15. In $-1 \le x < 0$, $f'(x) = 3x^2 + 2x - 10 = 2x^2 + (x+1)^2 - 11 < 0$.

$\therefore$ $f(x)$ is m.d.

In $0 < x < \frac{\pi}{2}$, $f'(x) = -\sin x < 0$

$\therefore$ $f(x)$ is m.d.

In $\frac{\pi}{2} < x \le \pi$, $f'(x) = \cos x < 0$.

$\therefore \quad f(x)$ is m.d.

Displaying the trend of values of the function in different intervals, we get the adjoining graph.

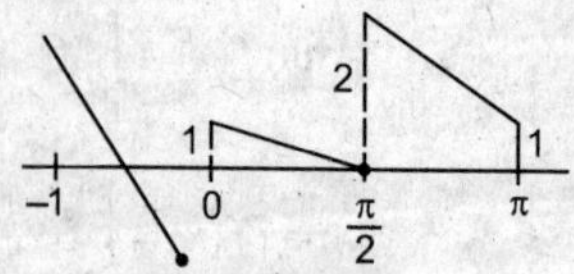

$\therefore \quad f\left(\frac{\pi}{2}-h\right)<2=f\left(\frac{\pi}{2}\right), \quad f\left(\frac{\pi}{2}+h\right)<f\left(\frac{\pi}{2}\right)$

$\therefore \quad f(x)$ has a local maximum at $x=\frac{\pi}{2}\cdot$

16. In $-\epsilon < x \le 0, f'(x) = \cos(x^2-3x)\cdot(2x-3) < 0$.

$\therefore \quad f(x)$ is decreasing.

In $x > 0, f'(x) = 6 + 10x > 0$.

$\therefore \quad f(x)$ is increasing. So, at $x=0, f(x)$ has a local minimum.

$\therefore \quad$ at $x=0, f(x)$ is continuous because $f(0-\epsilon)=f(0)=f(0+\epsilon)$.

17. Let $f(\theta)=\tan\theta-\theta$. Then $f'(\theta)=\sec^2\theta-1=\tan^2\theta>0$.

$\therefore \quad f(\theta)$ is increasing. Therefore, $f(\theta)>f(0) \quad \Rightarrow \tan\theta-\theta>0$.

Let $\phi(\theta)=\theta-\sin\theta$. Then $\phi'(\theta)=1-\cos\theta=2\sin^2\frac{\theta}{2}>0$.

$\therefore \quad \phi(\theta)$ is increasing. Therefore, $\phi(\theta)>\phi(0) \quad \Rightarrow \sin\theta-\theta>0$.

18. $f'(x)=a+3\cos x-4\sin x=a+5\cos(x+\alpha)$, where $\cos\alpha=\frac{3}{5}\cdot$

$\therefore \quad a-5\le f'(x)\le a+5$.

$\therefore \quad f'(x)>0$ if $a+5>0$, i.e., $a>-5$, and $f'(x)<0$ if $a-5<0$, i.e., $a<5$.

Hence, $f(x)$ is strictly monotonic if $a\in(-5,5)$ and hence it will be invertible.

19. $f'(x)=3x^2+2\lambda x+5+2\cos 2x\le 3x^2+2\lambda x+7 \quad (\because \max\cos 2x=1)$

$$\ge 3x^2+2\lambda x+3 \qquad (\because \min\cos 2x=-1)$$

But $3x^2+2\lambda x+7<0$ is not true for all $x\in R$.

$\therefore \quad f'(x)\ge 3x^2+2\lambda x+3>0$ for all x if $D<0$, i.e., $4\lambda^2-4\cdot 3\cdot 3<0$

or $\quad \lambda^2-9<0 \qquad \Rightarrow -3<\lambda<3$.

$\therefore$ if $-3<\lambda<3, f(x)$ is strictly m.i. and so $f(x)$ is invertible.

20. By Lagrange's theorem, $\frac{f(1)-f(-1)}{1-(-1)}=f'(c)$, i.e., $0=f'(c)$, where $-1<c<1$.

If $x > 0, f(x) = x$ and, therefore, $f'(x) = 1$, and

if $x < 0, f(x) = -x$ and, therefore, $f'(x) = -1$.

Hence, there is no $c \neq 0$ for which $0 = f'(c)$ holds.

Also at $x = 0, f(x)$ is not differentiable. Hence, c has no value.

Note $f(x)$ is not differentiable in $[-1, 1]$. So, Lagrange's theorem fails for the interval.

21. $\int (\sin x + x\cos x)\, dx = -\cos x + \int x\cos x\, dx = -\cos x + \{x\sin x - \int \sin x\, dx\}$

$$= -\cos x + x\sin x + \cos x = x\sin x.$$

Now, $x\sin x = 0$ has two roots $x = 0, \pi$.

$\therefore$ its derived equation $\sin x + x\cos x = 0$ has a root lying between 0 and π.

22. $3ax^2 + 2bx + c = 0 \quad \Rightarrow ax^3 + bx^2 + cx = 0$ (on integration).

Clearly $x = 0, 2$ are the roots of $ax^3 + bx^2 + cx = 0$ $(\because\ 4a + 2b + c = 0)$.

$\therefore$ the derived equation $3ax^2 + 2bx + c = 0$ will have a root lying between 0 and 2.

23. Two roots of the first equation are 0 and α. The second equation is the derived equation of the first. So, it has root in $(0, \alpha)$.

24. $h'(x) = f'(x) - 2f(x)f'(x) + 3\{f(x)\}^2 f'(x)$

$$= [1 - 2f(x) + 3\{f(x)\}^2] f'(x) = \left(3\left[\{f(x)\}^2 - \frac{2}{3}f(x) + \frac{1}{9}\right] + \frac{2}{3}\right) f'(x)$$

$$= \left[3\left\{f(x) - \frac{1}{3}\right\}^2 + \frac{2}{3}\right] f'(x).$$

If $f(x)$ is increasing, $f'(x) > 0$ and, therefore, $h'(x) > 0$, i.e., $h(x)$ is increasing.

If $f(x)$ is decreasing, $f'(x) < 0$ and, therefore, $h'(x) < 0$, i.e., $h(x)$ is decreasing.

25. We know that $\sin x$ is an increasing function of x in $\left[-\frac{\pi}{2}, \frac{\pi}{2}\right]$.

$f\{f(x)\} = \sin(\sin x)$; $\therefore \frac{d}{dx}\{f(f(x)\} = \cos(\sin x) \cdot \cos x \geq 0$ for $-\frac{\pi}{2} \leq x \leq \frac{\pi}{2}$.

$\therefore \quad f\{f(x)\}$ is increasing in $\left[-\frac{\pi}{2}, \frac{\pi}{2}\right]$.

As $f\{f(x)\}$ is strictly m.i. in $\left(-\frac{\pi}{2},\frac{\pi}{2}\right)$ and $f\left\{f\left(-\frac{\pi}{2}\right)\right\}=-\sin 1$, and $f\left\{f\left(\frac{\pi}{2}\right)\right\}=\sin 1$, and the values $\pm\sin 1$ are not attained at any point of $\left(-\frac{\pi}{2},\frac{\pi}{2}\right)$, we find that $f\{f(x)\}$ is invertible in $\left[-\frac{\pi}{2},\frac{\pi}{2}\right]$.

26. $f'(x)=6\sin^2 x\cos x-6\sin x\cos x+12\cos x$

$$=6\cos x\{\sin^2 x-\sin x+2\}=6\cos x\left\{\left(\sin x-\frac{1}{2}\right)^2+\frac{7}{4}\right\}.$$

$\therefore$ in $\left[0,\frac{\pi}{2}\right]$, $f'(x)\geq 0$. So, $f(x)$ is increasing in $\left[0,\frac{\pi}{2}\right]$.

27. $f'(x)=-6x^2+42x-60=-6(x^2-7x+10)=-6(x-2)(x-5)$.

If $x<1$, $f'(x)<0$, i.e., $f(x)$ is m.d.

$\therefore$ $f(x)>f(1)=0$.

28. Here, $f(x)=x^2+1,\ 1\leq x<2$

$\frac{x^2+1}{2},\ 2\leq x<3$

$\frac{x^2+1}{3},\ 3\leq x\leq 3.9.$

$f'(x)>0$ in each of the intervals and so $f(x)$ is increasing in each of the intervals.

$\therefore$ $2\leq f(x)<5$ in $1\leq x<2$; $\frac{5}{2}\leq f(x)<5$ in $2\leq x<3$;

$\frac{10}{3}\leq f(x)\leq\frac{1}{3}\times 16.21$ in $3\leq x\leq 3.9$.

Hence, the least value is 2 and the greatest value is $\frac{1}{3}\times 16.21$.

29. $x^2-3x-4=(x-4)(x+1)$. So, in $-1\leq x\leq 4$, $x^2-3x-4<0$.

$\therefore$ $f(x)=-(x^2-3x-4)=(4-x)(x+1)$

$\therefore$ $f'(x)=-(x+1)+4-x=3-2x$.

$\therefore$ in $-1\leq x<\frac{3}{2}$, $f'(x)>0$ and in $\frac{3}{2}<x\leq 4$, $f'(x)<0$.

$\therefore$ $f(x)$ is increasing in $-1\leq x<\frac{3}{2}$ and decreasing in $\frac{3}{2}<x\leq 4$.

As $f(x)$ is continuous at $x=\frac{3}{2}$, we find $f\left(\frac{3}{2}\right)$ is the maximum value.

The minimum value = the least among $\{f(-1), f(4)\}$.

30. $f'(x) = -\phi'(2-x) + \phi'(x)$. As $\phi''(x) < 0$, $\phi'(x)$ is m.d.

In $0 \le x \le 1, x \le 2-x$

$\therefore \quad \phi'(x) \ge \phi'(2-x)$

$\therefore \quad$ in $0 \le x \le 1, f'(x) > 0$. So, $f(x)$ is m.i.

In $1 \le x \le 2, x \ge 2-x$.

$\therefore \quad \phi'(2-x) \ge \phi'(x)$. So $f(x)$ is m.d.

31. $f'(x) > 0 \quad \Rightarrow f(x)$ is m.i. and $g'(x) < 0 \quad \Rightarrow g(x)$ is m.d.

$f(x)$ is m.i. $\quad \Rightarrow f(x) < f(x+1)$ and $f(x) > f(x-1)$

$\therefore \quad g\{f(x)\} > g\{f(x+1)\}$ and $g\{f(x)\} < g\{f(x-1)\}$ because $g(x)$ is m.d.

$g(x)$ is m.d. $\quad \rightarrow g(x) > g(x+1)$ and $g(x) < g(x-1)$

$\therefore \quad f\{g(x)\} > f\{g(x+1)\}$ and $f\{g(x)\} < f\{g(x-1)\}$ because $f(x)$ is m.i.

32.

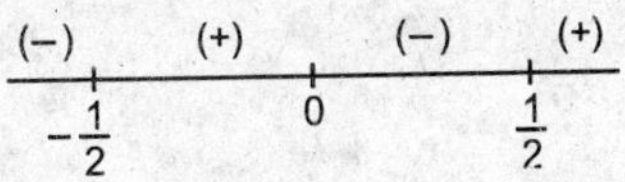

If $x > 0, f(x) = 2x^2 - \log x$

$\therefore \quad f'(x) = 4x - \frac{1}{x}$.

Now, $f'(x) > 0$ if $4x - \frac{1}{x} > 0$ or $\frac{4x^2-1}{x} > 0$

or $\quad x(4x^2-1) > 0$

$\therefore \quad f(x)$ is m.i. in $\left(\frac{1}{2}, +\infty\right)$.

Similarly $f'(x) < 0$, i.e., $f(x)$ is m.d. in $\left(0, \frac{1}{2}\right)$.

If $x > 0, f(x) = 2x^2 - \log(-x)$.

$\therefore \quad f'(x) = 4x - \frac{1}{-x}(-1) = 4x - \frac{1}{x}$.

Proceeding as before, $f(x)$ is m.d. in $\left(-\infty, -\frac{1}{2}\right)$ and m.i. in $\left(-\frac{1}{2}, 0\right)$.

33. $f'(x) = 3x^2 - 12x + 15 = 3(x^2 - 4x + 5) = 3\{(x-2)^2 + 1\} > 0$ for all $x \in R$.

$\therefore \quad f(x)$ is m.i. in R.

$\therefore \quad f(x) \le f(x+1)$.

As $f(x)$ is m.i., it is invertible and one-one.

34. Clearly, $f(0)=0$. So, $f(x)=0$ has two real roots $0, \alpha_0 (> 0)$.

$\therefore$ $f'(x)=0$ has a real root α_1 lying between 0 and α_0. So $0<\alpha_1<\alpha_0$.

Again, $f'(x)=0$ is a fourth-degree equation. As imaginary roots occur in conjugate pairs, $f'(x)=0$ will have another real root, α_2. Therefore, $f''(x)=0$ will have a real root lying between α_1 and α_2. As $f(x)=0$ is an equation of the fifth degree, it will have at least three real roots and so $f'(x)$ will have two real roots at least.

□

8

Indefinite Integration

• *Type 1* •

Choose the most appropriate option (a, b, c or d).

1. If $\phi(x) = \int \cot^4 x\, dx + \frac{1}{3}\cot^3 x - \cot x$ and $\phi\left(\frac{\pi}{2}\right) = \frac{\pi}{2}$ then $\phi(x)$ is

(a) $\pi - x$ (b) $x - \pi$

(c) $\frac{\pi}{2} - x$ (d) none of these

2. Integral of $f(x) = \sqrt{1 + x^2}$ with respect to x^2 is

(a) $\frac{2}{3}\frac{(1 + x^2)^{3/2}}{x} + k$ (b) $\frac{2}{3}(1 + x^2)^{3/2} + k$

(c) $\frac{2}{3}x(1 + x^2)^{3/2} + k$ (d) none of these

3. $\int \frac{d(x^2 + 1)}{\sqrt{x^2 + 2}}$ is equal to

(a) $2\sqrt{x^2 + 2} + k$ (b) $\sqrt{x^2 + 2} + k$

(c) $\frac{1}{(x^2 + 2)^{3/2}} + k$ (d) none of these

4. $\int \cos\left\{2\tan^{-1}\sqrt{\frac{1 - x}{1 + x}}\right\} dx$ is equal to

(a) $\frac{1}{8}(x^2-1)+k$ (b) $\frac{1}{2}x^2+k$

(c) $\frac{1}{2}x+k$ (d) none of these

5. $\int x^x(1+\log x)\,dx$ is equal to

(a) $x^x \log x + k$ (b) $e^{x^x}+k$

(c) x^x+k (d) none of these

6. Let the equation of a curve passing through the point (0, 1) be given by $y=\int x^2 \cdot e^{x^3}\,dx$. If the equation of the curve is written in the form $x=f(y)$ then $f(y)$ is

(a) $\sqrt{\log_e(3y-2)}$ (b) $\sqrt[3]{\log_e(3y-2)}$

(c) $\sqrt[3]{\log_e(2-3y)}$ (d) none of these

7. $\int \frac{x\,dx}{1+x^4}$ is equal to

(a) $\tan^{-1}x^2+k$ (b) $\frac{1}{2}\tan^{-1}x^2+k$

(c) $\log(1+x^4)+k$ (d) none of these

8. The antiderivative of $\frac{2^x}{\sqrt{1-4^x}}$ w.r.t. x is

(a) $\log_2 e \cdot \sin^{-1}(2^x)+k$ (b) $\sin^{-1}(2^x)+k$

(c) $\cos^{-1}(2^x)\cdot\frac{1}{\log_e 2}+k$ (d) none of these

9. $\int \frac{(1+x)^2}{x+x^3}\,dx$ is equal to

(a) $\log_e x+\log_e(1+x^2)+k$ (b) $\log_e x+\tan^{-1}x+k$

(c) $\log_e x+2\tan^{-1}x+k$ (d) none of these

10. $\int \frac{x^{5/2}}{\sqrt{1+x^7}}\,dx$ is

(a) $\frac{2}{7}\log(x^{7/2}+\sqrt{x^7+1})+c$ (b) $\frac{1}{2}\log\frac{x^7+1}{x^7-1}+c$

(c) $2\sqrt{1+x^7}+c$ (d) none of these

11. $\int\frac{dx}{x^{1/5}\left(1+x^{4/5}\right)^{1/2}}$ is

(a) $\sqrt{1+x^{4/5}}+k$ (b) $\frac{5}{2}\sqrt{1+x^{4/5}}+k$

(c) $x^{4/5}(1+x^{4/5})^{1/2}+k$ (d) none of these

12. The primitive of the function $x\,|\cos x|$ when $\frac{\pi}{2}<x<\pi$ is given by

(a) $\cos x+x\sin x$ (b) $-\cos x-x\sin x$

(c) $x\sin x-\cos x$ (d) none of these

13. $\int x\sec x^2 dx$ is equal to

(a) $\frac{1}{2}\log(\sec x^2+\tan x^2)+k$ (b) $\frac{x^2}{2}\log(\sec x^2+\tan x^2)+k$

(c) $2\log(\sec x^2+\tan x^2)+k$ (d) none of these

14. $\int\frac{\{f(x)\cdot\phi'(x)-f'(x)\cdot\phi(x)\}}{f(x)\cdot\phi(x)}\{\log\phi(x)-\log f(x)\}\,dx$ is equal to

(a) $\log\frac{\phi(x)}{f(x)}+k$ (b) $\frac{1}{2}\left\{\log\frac{\phi(x)}{f(x)}\right\}^2+k$

(c) $\frac{\phi(x)}{f(x)}\log\frac{\phi(x)}{f(x)}+k$ (d) none of these

15. $\int\sin 2x\cdot\log\cos x\,dx$ is equal to

(a) $\cos^2 x\left(\frac{1}{2}+\log\cos x\right)+k$ (b) $\cos^2 x\cdot\log\cos x+k$

(c) $\cos^2 x\left(\frac{1}{2}-\log\cos x\right)+k$ (d) none of these

16. $\int e^{-x}(1-\tan x)\sec x\,dx$ is equal to

(a) $e^{-x}\sec x+c$ (b) $e^{-x}\tan x+c$

(c) $-e^{-x}\tan x+c$ (d) none of these

17. $\int \frac{1+\sin x}{1+\cos x} \cdot e^x \, dx$ is equal to

(a) $e^x \tan\left(\frac{x}{2}\right) + k$
(b) $e^x \tan x + k$
(c) $\frac{1}{2} e^x \tan \frac{x}{2} + k$
(d) $e^x \sec^2 \frac{x}{2} + k$

18. Let $\int e^x \{f(x) - f'(x)\} \, dx = \phi(x)$. Then $\int e^x f(x) \, dx$ is

(a) $\phi(x) + e^x f(x)$
(b) $\phi(x) - e^x f(x)$
(c) $\frac{1}{2} \{\phi(x) + e^x f(x)\}$
(d) $\frac{1}{2} \{\phi(x) + e^x f'(x)\}$

19. If $f(0) = f'(0) = 0$ and $f''(x) = \tan^2 x$ then $f(x)$ is

(a) $\log \sec x - \frac{1}{2} x^2$
(b) $\log \cos x + \frac{1}{2} x^2$
(c) $\log \sec x + \frac{1}{2} x^2$
(d) none of these

20. Let $f(x) = \int \frac{x^2 dx}{(1+x^2)(1+\sqrt{1+x^2})}$ and $f(0) = 0$. Then $f(1)$ is

(a) $\log(1+\sqrt{2})$
(b) $\log(1+\sqrt{2}) - \frac{\pi}{4}$
(c) $\log(1+\sqrt{2}) + \frac{\pi}{4}$
(d) none of these

21. $\int \frac{dx}{\cos x + \sqrt{3} \sin x}$ is equal to

(a) $\log \tan\left(\frac{x}{2} + \frac{\pi}{3}\right) + k$
(b) $\log \tan\left(\frac{x}{2} - \frac{\pi}{3}\right) + k$
(c) $\frac{1}{2} \log \tan\left(\frac{x}{2} + \frac{\pi}{3}\right) + k$
(d) none of these

• Type 2 •

Choose the correct options. One or more options may be correct.

22. If $\int \tan^4 x\, dx = a\tan^3 x + b\tan x + \phi(x)$ then

(a) $a = \frac{1}{3}$ (b) $b = 1$

(c) $\phi(x) = x + c$ (d) $b = -1$

23. If $\int \frac{\sin x}{\sin(x - \alpha)} dx = Ax + B\log \sin(x - \alpha) + C$ then

(a) $A = \sin\alpha$ (b) $B = \cos\alpha$

(c) $A = \cos\alpha$ (d) $B = \sin\alpha$

24. If $\int \frac{4e^x + 6e^{-x}}{9e^x - 4e^{-x}} dx = Ax + B\log_e(9e^{2x} - 4) + C$ then

(a) $A = \frac{3}{2}$ (b) $B = \frac{35}{36}$

(c) C is indefinite (d) $A + B = -\frac{19}{36}$

25. If $\int x\log(1 + x^2)\, dx = \phi(x) \cdot \log(1 + x^2) + \psi(x) + c$ then

(a) $\phi(x) = \frac{1 + x^2}{2}$ (b) $\psi(x) = \frac{1 + x^2}{2}$

(c) $\psi(x) = -\frac{1 + x^2}{2}$ (d) $\phi(x) = -\frac{1 + x^2}{2}$

26. $\int \frac{dx}{(x + 1)(x - 2)} = A\log(x + 1) + B\log(x - 2) + C$, where

(a) $A + B = 0$ (b) $AB = -1$

(c) $A : B = -1$ (d) none of these

Answers

1. d	**2.** b	**3.** a	**4.** b	**5.** c
6. b	**7.** b	**8.** a	**9.** c	**10.** a
11. b	**12.** b	**13.** a	**14.** b	**15.** c
16. d	**17.** a	**18.** c	**19.** a	**20.** b
21. c	**22.** a, c, d	**23.** c, d	**24.** b, c, d	**25.** a, c
26. a, c				

Hints and Solutions to Selected Questions

1. $\int \cot^4 x\, dx = \int \cot^2 x \cdot (\operatorname{cosec}^2 x - 1)\, dx = \int -\cot^2 x\, d(\cot x) - \int (\operatorname{cosec}^2 x - 1)\, dx$

$$= -\frac{1}{3}\cot^3 x + \cot x + x + c$$

$\therefore \quad \phi(x) = -\frac{1}{3}\cot^3 x + \cot x + x + c + \frac{1}{3}\cot^3 x - \cot x = x + c$

$\therefore \quad \phi\left(\frac{\pi}{2}\right) = \frac{\pi}{2} + c$

$\therefore \quad c = 0 \qquad \therefore \quad \phi(x) = x.$

2. $\int \sqrt{1 + x^2}\, d(x^2) = \int \sqrt{1+t}\, dt$, where $x^2 = t$

$$= \frac{(1+t)^{3/2}}{\frac{3}{2}} + k = \frac{2}{3}(1 + x^2)^{3/2} + k.$$

3. $I = \int \frac{dt}{\sqrt{1+t}}$, where $x^2 + 1 = t$

$$= 2(1+t)^{1/2} + k = 2\sqrt{x^2 + 2} + k.$$

4. Put $x = \cos 2\theta$.

$\therefore \quad I = \int \cos\{2\tan^{-1}\tan\theta\}(-2\sin 2\theta)\, d\theta = -\int \sin 4\theta\, d\theta = \frac{1}{4}\cos 4\theta + c$

$$= \frac{1}{4}(2x^2 - 1) + c = \frac{1}{2}x^2 + k.$$

5. $\frac{d}{dx}(x^x) = x \cdot x^{x-1} + x^x \cdot \log x = x^x(1 + \log x).$

6. $y = \int \frac{1}{3} e^{x^3}\, d(x^3) = \frac{1}{3} e^{x^3} + c.$

It passes through (0, 1).

$\therefore \quad 1 = \frac{1}{3} e^0 + c$

$\therefore \quad c = \frac{2}{3} \cdot$ Hence, $y = \frac{1}{3}(e^{x^3} + 2)$.

$\therefore \quad e^{x^3} = 3y - 2 \qquad \therefore \; x = \sqrt[3]{\log_e(3y-2)}$.

7. $\int \frac{\frac{1}{2}d(x^2)}{1+(x^2)^2} = \frac{1}{2}\tan^{-1}(x^2) + k.$

8. $\int \frac{2^x}{\sqrt{1-4^x}}dx = \int \frac{1}{\log 2} \cdot \frac{d(2^x)}{\sqrt{1-(2^x)^2}} = \frac{1}{\log 2} \cdot \sin^{-1}(2^x) + k.$

9. $\int \frac{(1+x)^2}{x+x^3}dx = \int \frac{(1+x^2)+2x}{x(1+x^2)}dx = \int \left(\frac{1}{x} + \frac{2}{1+x^2}\right)dx = \log x + 2\tan^{-1}x + k.$

10. Put $x^{7/2} = t$.

$\therefore \quad \frac{7}{2}x^{5/2}dx = dt.$

$\therefore \quad I = \int \frac{2}{7}\frac{dt}{\sqrt{1+t^2}} = \frac{2}{7}\log\left(t + \sqrt{1+t^2}\right) + k = \frac{2}{7}\log(x^{7/2} + \sqrt{1+x^7}) + k.$

11. Put $1 + x^{4/5} = t$.

$\therefore \quad \frac{4}{5}x^{-1/5}dx = dt.$

$\therefore \quad I = \frac{5}{4}\int \frac{dt}{t^{1/2}} = \frac{5}{4} \cdot 2 \cdot \sqrt{t} + k = \frac{5}{2}\sqrt{1+x^{4/5}} + k.$

12. $f(x) = x\,|\cos x|, \frac{\pi}{2} < x < \pi$

$= -x\cos x$, because $\cos x$ is negative in $\left(\frac{\pi}{2}, \pi\right)$.

$\therefore$ the required primitive function $= \int -x\cos x\, dx$.

Now use integration by parts.

14. $I = \int \log\frac{\phi(x)}{f(x)}\, d\left\{\log\frac{\phi(x)}{f(x)}\right\} = \frac{1}{2}\left\{\log\frac{\phi(x)}{f(x)}\right\}^2 + k.$

15. $I = \int 2\sin x \cdot \cos x \cdot \log \cos x \, dx.$

Put $\log \cos x = z$

$\therefore \quad -\dfrac{\sin x}{\cos x} dx = dz$

$$\therefore \quad I = \int 2\sin x \cdot \cos x \cdot z \cdot \frac{\cos x}{-\sin x} dz = -2\int \cos^2 x \cdot z \, dz = -2\int ze^{2z} dz$$

$$= -2\left[z \cdot \frac{e^{2z}}{2} - \int \frac{e^{2z}}{2} \cdot dz\right] = -z \cdot e^{2z} + \frac{1}{2} e^{2z} + k$$

$$= e^{2z}\left(\frac{1}{2} - z\right) + k = \cos^2 x \cdot \left\{\frac{1}{2} - \log \cos x\right\} + k.$$

16. Putting $x = -z$, $I = \int -e^{z}(\sec z + \sec z \cdot \tan z) \, dz.$

$$\therefore \quad I = -\int e^{z}\left\{\sec z + \frac{d}{dz}(\sec z)\right\} dz = -\ e^{z}\sec z + k = -e^{-x}\sec x + k.$$

17. $I = \int e^{x} \cdot \dfrac{1 + \sin x}{2\cos^2 \dfrac{x}{2}} dx = \int e^{x}\left\{\dfrac{1}{2}\sec^2 \dfrac{x}{2} + \tan \dfrac{x}{2}\right\} dx = e^{x} \cdot \tan \dfrac{x}{2} + k.$

18. Here $\int e^{x}\{f(x) - f'(x)\} \, dx = \phi(x)$ and $\int e^{x}\{f(x) + f'(x)\} \, dx = e^{x} f(x).$

Adding, $2\int e^{x} f(x) \, dx = \phi(x) + e^{x} f(x).$

19. $f'(x) = \int \tan^2 x \, dx = \int (\sec^2 x - 1) \, dx = \tan x - x + k$

$\therefore \quad f'(0) = k = 0$

$\therefore \quad f'(x) = \tan x - x.$

$\therefore \quad f(x) = \int (\tan x - x) \, dx = \log \sec x - \dfrac{x^2}{2} + c$

$\therefore \quad f(0) = c = 0.$

20. $f(x) = \int \dfrac{x^2(\sqrt{1 + x^2} - 1)}{(1 + x^2)(1 + x^2 - 1)} dx = \int \dfrac{\sqrt{1 + x^2} - 1}{1 + x^2} dx = \int \dfrac{dx}{\sqrt{1 + x^2}} - \int \dfrac{dx}{1 + x^2}$

$= \log(x + \sqrt{1 + x^2}) - \tan^{-1} x + k$

$\therefore \quad f(0) = \log 1 - \tan^{-1} 0 + k = k = 0$

$\therefore \quad f(x) = \log(x + \sqrt{1 + x^2}) - \tan^{-1} x.$

$\therefore \quad f(1) = \log(1 + \sqrt{2}) - \frac{\pi}{4}.$

21. $I = \int \frac{dx}{2\cos\left(x - \frac{\pi}{3}\right)} = \frac{1}{2}\int \sec\left(x - \frac{\pi}{3}\right)dx = \frac{1}{2}\log\left\{\sec\left(x - \frac{\pi}{3}\right) + \tan\left(x - \frac{\pi}{3}\right)\right\} + k$

$= \frac{1}{2}\log\tan\left(\frac{\pi}{2} + \frac{x - \pi/3}{2}\right) + k = \frac{1}{2}\log\tan\left(\frac{x}{2} + \frac{\pi}{3}\right) + k.$

22. $\int \tan^4 x\, dx = \int \tan^2 x(\sec^2 x - 1)\, dx = \int \tan^2 x\, d(\tan x) - \int (\sec^2 x - 1)\, dx$

$= \frac{1}{3}\tan^3 x - \tan x + x + c.$

23. $I = \int \frac{\sin(\overline{x - \alpha} + \alpha)}{\sin(x - \alpha)}\, dx = \int \{\cos\alpha + \sin\alpha \cdot \cot(x - \alpha)\}\, dx$

$= x\cos\alpha + \sin\alpha \cdot \log\sin(x - \alpha) + C.$

24. $I = \int \frac{4e^{2x} + 6}{9e^{2x} - 4}\, dx.$

Put $9e^{2x} - 4 = z.$

Then $I = \int \frac{1}{z}\left\{4 \cdot \frac{z+4}{9} + 6\right\}\frac{1}{18}\frac{dz}{\frac{z+4}{9}} = \int \frac{1}{z(z+4)}\left\{\frac{2z+8}{9} + 3\right\} dz$

$= \frac{1}{9}\int \frac{2z + 35}{z(z+4)}\, dz = \frac{1}{9}\int \frac{2(z+4) + 27}{z(z+4)}\, dz$

$= \frac{2}{9}\int \frac{dz}{z} + \frac{3}{4}\int\left(\frac{1}{z} - \frac{1}{z+4}\right)dz = \left(\frac{2}{9} + \frac{3}{4}\right)\log z - \frac{3}{4}\log(z+4) + C$

$= \frac{35}{36}\log(9e^{2x} - 4) - \frac{3}{4}\log(e^{2x}) + C$

$= -\frac{3}{2}x + \frac{35}{36}\log(9e^{2x} - 4) - \frac{3}{2}\log 3 + C.$

25. $I = \log(1 + x^2) \times \frac{x^2}{2} - \int \frac{x^2}{2} \cdot \frac{2x}{1 + x^2}\, dx = \frac{x^2}{2}\log(1 + x^2) - \int\left(x - \frac{x}{1 + x^2}\right)dx$

$$= \frac{x^2}{2} \log(1 + x^2) - \frac{x^2}{2} + \frac{1}{2} \log(1 + x^2) + c'$$

$$= \frac{x^2 + 1}{2} \log(1 + x^2) - \frac{x^2 + 1}{2} + c.$$

26. $I = \frac{1}{3} \int \frac{(x+1) - (x-2)}{(x+1)(x-2)} dx = \frac{1}{3} \int \left(\frac{1}{x-2} - \frac{1}{x+1} \right) dx$

$$= \frac{1}{3} \log(x - 2) - \frac{1}{3} \log(x + 1) + C.$$

□

9

Definite Integration

• *Type 1* •

Choose the most appropriate option (a, b, c or d).

1. $\lim_{n \to \infty} \sum_{r=1}^{n} \frac{1}{\sqrt{nr}}$ is equal to

(a) 2 (b) 1

(c) 0 (d) none of these

2. $\lim_{n \to \infty} \sum_{r=0}^{n-1} \frac{1}{\sqrt{n^2 - r^2}}$ is

(a) π (b) $\pi/2$

(c) $\pi/4$ (d) none of these

3. $\lim_{n \to \infty} \sum_{r=1}^{n} \frac{1}{n} e^{r/n}$ is

(a) e (b) $e - 1$

(c) $1 - e$ (d) $e + 1$

4. $\lim_{n \to \infty} \sum_{r=1}^{n} \frac{1}{n} \sin \frac{r\pi}{2n}$ is

(a) $\pi/2$ (b) 2

(c) $2/\pi$ (d) none of these

5. $\lim_{n \to \infty} \sum_{r=1}^{4n} \frac{1}{n+r}$ is

(a) $\log_e 5$
(b) 0
(c) $\log_e 4$
(d) none of these

6. $\lim_{n \to \infty} \frac{1}{n} \sum_{r=1}^{2n} \frac{r}{\sqrt{n^2+r^2}}$ equals

(a) $1+\sqrt{5}$
(b) $-1+\sqrt{5}$
(c) $-1+\sqrt{2}$
(d) $1+\sqrt{2}$

7. $\lim_{n \to \infty} \left\{\frac{n!}{(kn)^n}\right\}^{1/n}$, where $k \neq 0$ is a constant and $n \in N$, is equal to

(a) ke
(b) $k^{-1} \cdot e$
(c) ke^{-1}
(d) $k^{-1} \cdot e^{-1}$

8. $\lim_{n \to \infty} \frac{2^k + 4^k + 6^k + \ldots + (2n)^k}{n^{k+1}}$, $k \neq -1$, is equal to

(a) 2^k
(b) $\frac{2^k}{k+1}$
(c) $\frac{1}{k+1}$
(d) none of these

9. Let (a, b) and (λ, μ) be two points on the curve $y = f(x)$. If the slope of the tangent to the curve at (x, y) be $\phi(x)$ then $\int_a^\lambda \phi(x)\, dx$ is

(a) $\lambda - a$
(b) $\mu - b$
(c) $\lambda + \mu - a - b$
(d) none of these

10. If $g(x) = \int_0^x \cos^4 t\, dt$ then $g(x + \pi)$ equals

(a) $g(x) + g(\pi)$
(b) $g(x) - g(\pi)$
(c) $g(x)g(\pi)$
(d) $g(x)/g(\pi)$

11. If $af(x) + bf\left(\frac{1}{x}\right) = \frac{1}{x} - 5$, $x \neq 0$, $a \neq b$, then $\int_1^2 f(x)\, dx$ equals

(a) $\dfrac{(\log 2-5)a+\dfrac{13}{2}b}{a^2-b^2}$ (b) $\dfrac{(\log 2-5)a+\dfrac{7b}{2}}{a^2-b^2}$

(c) $\dfrac{(5-\log 2)a+\dfrac{7b}{2}}{a^2-b^2}$ (d) none of these

12. $\int_0^2 \sqrt{\dfrac{2+x}{2-x}}\,dx$ is equal to

(a) $\pi+1$ (b) $1+\pi/2$

(c) $\pi+3/2$ (d) none of these

13. The value of $\int_0^{\pi} \sec^2 x\,dx$ is

(a) 0 (b) 2

(c) 1 (d) none of these

14. If $a_n = \int_0^{\pi/2} \dfrac{\sin^2 nx}{\sin x}\,dx$ then $a_2-a_1, a_3-a_2, a_4-a_3, \ldots$ are in

(a) AP (b) GP

(c) HP (d) none of these

15. Let $a_n = \int_0^{\pi/4} \tan^n x\,dx$. Then $a_2+a_4, a_3+a_5, a_4+a_6$ are in

(a) AP (b) GP

(c) HP (d) none of these

16. $\int_1^{e^{37}} \dfrac{\pi \sin(\pi \log_e x)}{x}\,dx$ is equal to

(a) 2 (b) –2

(c) $2/\pi$ (d) 2π

17. Let $\dfrac{d}{dx}F(x) = \dfrac{e^{\sin x}}{x}, x>0$. If $\int_1^4 \dfrac{2e^{\sin x^2}}{x}\,dx = F(k)-F(1)$ then one of the possible values of k is

(a) 4 (b) –4

(c) 16 (d) none of these

18. If $f(x)=\int_{-1}^{1}\frac{\sin x}{1+t^2}\,dt$ then $f'\left(\frac{\pi}{3}\right)$ is

(a) nonexistent (b) $\pi/4$

(c) $\pi\sqrt{3}/4$ (d) none of these

19. The value of $\int_{\pi^3/27}^{\pi^3/8}\sin x\,dt$, where $t=x^3$, is

(a) $\frac{\pi^2}{6}+(3-\sqrt{3})\pi-3$ (b) $\cos\frac{\pi^3}{27}-\cos\frac{\pi^3}{8}$

(c) $\frac{\pi^2}{6}$ (d) none of these

20. The value of $\int_{1}^{2}[f\{g(x)\}]^{-1}\cdot f'\{g(x)\}\cdot g'(x)\,dx$, where $g(1)=g(2)$, is equal to

(a) 1 (b) 2

(c) 0 (d) none of these

21. If $f\left(\frac{1}{x}\right)+x^2 f(x)=0$, $x>0$, and $I=\int_{1/x}^{x} f(z)dz$, $\frac{1}{2}\le x\le 2$, then I is

(a) $f(2)-f(1/2)$ (b) $f(1/2)-f(2)$

(c) 0 (d) none of these

22. $\int_{\pi/4}^{3\pi/4}\frac{dx}{1+\cos x}$ is equal to

(a) 2 (b) –2

(c) 1/2 (d) – 1/2

23. If $f(x)$ satisfies the conditions of Rolle's theorem in [1, 2] then $\int_{1}^{2} f'(x)\,dx$ is equal to

(a) 1 (b) 3

(c) 0 (d) none of these

24. The value of $\int_0^{\pi/2} \sin^8 x\, dx$ is

(a) $\frac{105\pi}{32(4\,!)}$ (b) $\frac{105\pi}{16(4\,!)}$

(c) $\frac{105}{16(4\,!)}$ (d) none of these

25. If $f(x)$ be a quadratic polynomial such that $f(0) = 2, f'(0) = -3$ and $f''(0) = 4$ then $\int_{-1}^{1} f(x)\, dx$ is equal to

(a) -3 (b) $16/3$

(c) 0 (d) none of these

• *Type 2* •

Choose the correct options. One or more options may be correct.

26. Let $I_n = \int_0^{\pi/2} \cos^n x\, dx, n \in N$. Then

(a) $I_{n-2} > I_n$ (b) $n(I_{n-2} - I_{n)} = I_{n-2}$

(c) $I_n : I_{n-1} = n : (n-1)$ (d) none of these

27. Let $I_n = \int_0^{\pi/4} \tan^n x\, dx, n \in N$. Then

(a) $I_1 = I_3 + 2I_5$ (b) $I_n + I_{n-2} = \frac{1}{n}$

(c) $I_n + I_{n-2} = \frac{1}{n-1}$ (d) none of these

Answers

1. a	**2.** b	**3.** b	**4.** c	**5.** a
6. b	**7.** d	**8.** b	**9.** b	**10.** a
11. b	**12.** d	**13.** d	**14.** c	**15.** c
16. a	**17.** c	**18.** b	**19.** a	**20.** c
21. c	**22.** a	**23.** c	**24.** a	**25.** b
26. a, b	**27.** a, c			

Hints and Solutions to Selected Questions

1. $$\lim_{n\to\infty}\sum_{r=1}^{n}\frac{1}{n}\cdot\frac{1}{\sqrt{\frac{r}{n}}}=\int_0^1\frac{1}{\sqrt{x}}\,dx=[2\sqrt{x}]_0^1=2.$$

2. $$\lim_{n\to\infty}\sum_{r=0}^{n-1}\frac{1}{n}\cdot\frac{1}{\sqrt{1-\left(\frac{r}{n}\right)^2}}=\int_0^1\frac{dx}{\sqrt{1-x^2}}=\left[\sin^{-1}x\right]_0^1=\frac{\pi}{2}.$$

3. $$\lim_{n\to\infty}\sum_{r=1}^{n}\frac{1}{n}e^{r/n}=\int_0^1 e^x dx=\left[e^x\right]_0^1=e-1.$$

4. $$\text{Limit}=\int_0^1\sin\frac{\pi x}{2}\,dx=\left[\frac{-\cos\frac{\pi x}{2}}{\frac{\pi}{2}}\right]_0^1=\frac{2}{\pi}\left\{-\cos\frac{\pi}{2}+\cos 0\right\}=\frac{2}{\pi}.$$

5. $$\text{Limit}=\lim_{n\to\infty}\sum_{r=1}^{4n}\frac{1}{n}\cdot\frac{1}{1+\frac{r}{n}}=\int_\alpha^\beta\frac{1}{1+x}\,dx$$

where $\alpha=\lim_{n\to\infty}\frac{r}{n}$, when $r=1$ $\Rightarrow \alpha=0$

$\beta=\lim_{n\to\infty}\frac{r}{n}$, when $r=4n$ $\Rightarrow \beta=4$

$$\therefore\quad \text{limit}=\int_0^4\frac{dx}{1+x}=[\log(1+x)]_0^4=\log 5.$$

6. $$\text{Limit}=\lim_{n\to\infty}\frac{1}{n}\sum_{r=1}^{2n}\frac{\frac{r}{n}}{\sqrt{1+\left(\frac{r}{n}\right)^2}}=\int_\alpha^\beta\frac{x}{\sqrt{1+x^2}}\,dx=\int_0^2\frac{x\,dx}{\sqrt{1+x^2}}$$

(See the hint to Q. No. 5.)

$$= \frac{1}{2}\int_0^2 \frac{d(1+x^2)}{\sqrt{1+x^2}} = \left[\frac{1}{2}\cdot 2\sqrt{1+x^2}\right]_0^2 = \sqrt{5}-1.$$

7. Limit $= \lim\limits_{n\to\infty} \frac{1}{k}\left\{\frac{1}{n}\cdot\frac{2}{n}\cdot\frac{3}{n}\cdot\ldots\cdot\frac{n}{n}\right\}^{1/n}$

$$= \frac{1}{k} e^{\lim\limits_{n\to\infty}\frac{1}{n}\sum\limits_{r=1}^{n}\log\frac{r}{n}} = \frac{1}{k} e^{\int_0^1 \log x\, dx} = \frac{1}{k} e^{[x\log x]_0^1 - \int_0^1 x\cdot\frac{1}{x}\, dx}$$

$$= \frac{1}{k} e^{-1} \quad \left(\because \lim_{x\to 0} x\log x = \lim_{x\to 0}\frac{\log x}{1/x} = \lim_{x\to 0}\frac{1/x}{-1/x^2} = \lim_{x\to 0}(-x) = 0\right).$$

8. Limit $= \lim\limits_{n\to\infty} 2^k\cdot\frac{1^k+2^k+3^k+\ldots+n^k}{n^{k+1}} = 2^k\cdot\lim\limits_{n\to\infty}\sum\frac{1}{n}\cdot\left(\frac{r}{n}\right)^k$

$$= 2^k\cdot\int_0^1 x^k dx = 2^k\cdot\left[\frac{x^{k+1}}{k+1}\right]_0^1 = \frac{2^k}{k+1}.$$

9. Here $f'(x) = \phi(x)$. So, $\int_a^\lambda \phi(x)\,dx = \int_a^\lambda f'(x)\,dx = \Big[f(x)\Big]_a^\lambda = f(\lambda) - f(a)$.

But $b = f(a)$, $\mu = f(\lambda)$. So, $\int_a^\lambda \phi(x)\,dx = \mu - b$.

10. $g(x) = \int_0^x \left(\frac{1+\cos 2t}{2}\right)^2 dt = \frac{1}{4}\int_0^x (1+2\cos 2t+\cos^2 2t)\,dt$

$$= \frac{1}{4}[t+\sin 2t]_0^x + \frac{1}{8}\int_0^x (1+\cos 4t)\,dt$$

$$= \frac{1}{4}(x+\sin 2x) + \frac{1}{8}\left[t+\frac{\sin 4t}{4}\right]_0^x$$

$$= \frac{1}{4}(x+\sin 2x) + \frac{1}{8}\left(x+\frac{\sin 4x}{4}\right) = \frac{3}{8}x + \frac{1}{4}\sin 2x + \frac{1}{32}\sin 4x.$$

$$\therefore \quad g(x+\pi) = \frac{3}{8}(x+\pi) + \frac{1}{4}\sin 2x + \frac{1}{32}\sin 4x = \frac{3}{8}\pi + g(x) = g(\pi) + g(x).$$

11. $af(x) + bf\left(\frac{1}{x}\right) = \frac{1}{x} - 5$

$$\Rightarrow \quad af\left(\frac{1}{x}\right) + bf(x) = x - 5.$$

From these, $(a^2 - b^2)f(x) = a\left(\frac{1}{x} - 5\right) - b(x - 5)$

$$\therefore \quad \int_1^2 f(x)\,dx = \int_1^2 \frac{1}{a^2 - b^2}\left\{\frac{a}{x} - bx - 5a + 5b\right\} dx$$

$$= \frac{1}{a^2 - b^2}\left[a\log x - b\frac{x^2}{2} + 5(b - a)x\right]_1^2$$

$$= \frac{a\log 2 - 2b + 10(b - a) + \frac{b}{2} - 5(b - a)}{a^2 - b^2}.$$

12. Put $x = 2\cos 2\theta$. Then $I = \int_{\pi/4}^{0} \cot\theta \cdot (-4\sin 2\theta)\,d\theta$.

$$\therefore \quad I = 8\int_0^{\pi/4} \cos^2\theta\,d\theta = 4\int_0^{\pi/4} (1 + \cos 2\theta)\,d\theta = 4\left[\theta + \frac{\sin 2\theta}{2}\right]_0^{\pi/4}$$

$$= 4\left[\frac{\pi}{4} + \frac{1}{2}\right] = \pi + 2.$$

13. $\int_0^{\pi} \sec^2 x\,dx = \lim_{\epsilon \to 0}\left(\int_0^{\pi/2 - \epsilon} \sec^2 x\,dx + \int_{\pi/2 + \epsilon}^{\pi} \sec^2 x\,dx\right)$

$$= \lim_{\epsilon \to 0}\left\{\tan\left(\frac{\pi}{2} - \epsilon\right) - \tan\left(\frac{\pi}{2} + \epsilon\right)\right\} = \lim_{\epsilon \to 0}(\cot\epsilon + \cot\epsilon) = \infty.$$

14. $a_n - a_{n-1} = \int_0^{\pi/2} \frac{\sin^2 nx - \sin^2(n-1)x}{\sin x}\,dx$

$$= \frac{1}{2}\int_0^{\pi/2} \frac{\cos 2(n-1)x - \cos 2nx}{\sin x}\,dx = \int_0^{\pi/2} \sin(2n-1)x\,dx$$

$$= \left[-\frac{\cos(2n-1)x}{2n-1}\right]_0^{\pi/2} = \frac{1}{2n-1}.$$

15. $a_n + a_{n+2} = \int_0^{\pi/4} (\tan^n x + \tan^{n+2} x)\,dx,\ n \geq 2$

$$= \int_0^{\pi/4} \tan^n x \sec^2 x\,dx = \left[\frac{\tan^{n+1} x}{n+1}\right]_0^{\pi/4} = \frac{1}{n+1}.$$

16. Put $\log_e x = z$. Then $I = \int_0^{37} \pi \cdot \sin \pi z \, dz = [-\cos \pi z]_0^{37}$

$$= -\cos 37\pi + \cos 0 = 1 + 1 = 2.$$

17. Put $x^2 = z$. Then $\int_1^4 \frac{2 \cdot e^{\sin x^2}}{x} dx = \int_1^{16} \frac{e^{\sin z}}{z} dz = \int_1^{16} \frac{d}{dz}\{F(z)\} dz$

$$= \Big[F(z)\Big]_1^{16} = F(16) - F(1).$$

$\therefore$ $F(16) = F(k)$, from the question.

Hence, one of the possible values of $k = 16$.

18. $f(x) = \sin x \cdot \int_{-1}^{1} \frac{dt}{1+t^2} = \sin x \cdot \left[\tan^{-1} t\right]_{-1}^{1} = \sin x \left\{\frac{\pi}{4} - \left(-\frac{\pi}{4}\right)\right\} = \frac{\pi}{2} \sin x$

$\therefore$ $f'(x) = \frac{\pi}{2} \cos x$

$\therefore$ $f'\left(\frac{\pi}{3}\right) = \frac{\pi}{2} \cdot \cos \frac{\pi}{3} = \frac{\pi}{4}.$

19. $I = \int_{\pi/3}^{\pi/2} \sin x \cdot d(x^3) = 3 \int_{\pi/3}^{\pi/2} x^2 \sin x \, dx.$

$\int x^2 \sin x \, dx = x^2(-\cos x) + \int 2x\cos x \, dx = -x^2 \cos x + 2\left\{x\sin x - \int \sin x \, dx\right\}$

$$= -x^2 \cos x + 2x\sin x + 2\cos x$$

$\therefore$ $I = 3\left[-x^2 \cos x + 2x\sin x + 2\cos x\right]_{\pi/3}^{\pi/2}$

$$= 3\pi - 3\left(-\frac{\pi^2}{9} \cdot \frac{1}{2} + 2 \cdot \frac{\pi}{3} \cdot \frac{\sqrt{3}}{2} + 2 \cdot \frac{1}{2}\right)$$

$$= 3\pi + \frac{\pi^2}{6} - \sqrt{3}\pi - 3 = \frac{\pi^2}{6} + (3 - \sqrt{3})\pi - 3.$$

20. Let $g(x) = z$. Then $I = \int_{g(1)}^{g(2)} \frac{1}{f(z)} \cdot f'(z) \, dz = \left[\log f(z)\right]_{g(1)}^{g(2)}$

$$= \log f\{g(2)\} - \log f\{g(1)\} = 0 \quad [\because g(1) = g(2)].$$

21. Put $z = \frac{1}{t}$ · Then $I = \int_x^{1/x} f\left(\frac{1}{t}\right) \cdot \frac{-1}{t^2} dt = \int_x^{1/x} \{-t^2 \cdot f(t)\} \cdot \frac{-1}{t^2} dt$

(from the question)

$$= \int_{x}^{1/x} f(t)\,dt = -\int_{1/x}^{x} f(t)\,dt = -I$$

$\therefore \quad 2I = 0.$

22. $I = \int_{\pi/4}^{3\pi/4} \frac{1}{2} \sec^2 \frac{x}{2}\,dx = \left[\tan \frac{x}{2}\right]_{\pi/4}^{3\pi/4} = \tan \frac{3\pi}{8} - \tan \frac{\pi}{8}$

$$= \tan\left(\frac{\pi}{2} - \frac{\pi}{8}\right) - \tan \frac{\pi}{8} = \cot \frac{\pi}{8} - \tan \frac{\pi}{8}$$

$$= \frac{\cos^2 \frac{\pi}{8} - \sin^2 \frac{\pi}{8}}{\sin \frac{\pi}{8} \cdot \cos \frac{\pi}{8}} = \frac{\cos \frac{\pi}{4}}{\frac{1}{2} \sin \frac{\pi}{4}} = 2.$$

23. As $f(x)$ satisfies the conditions of Rolle's theorem in [1, 2], $f(x)$ is continuous in the interval and $f(1) = f(2)$.

$\therefore \quad \int_1^2 f'(x)\,dx = [f(x)]_1^2 = f(2) - f(1) = 0.$

24. $I_n = \int_0^{\pi/2} \sin^n x\,dx = \int_0^{\pi/2} \sin^{n-1} x \cdot \sin x\,dx$

$$= \left[\sin^{n-1} x \cdot (-\cos x)\right]_0^{\pi/2} - \int_0^{\pi/2} -\cos x \cdot (n-1)\sin^{n-2} x \cdot \cos x\,dx$$

$$= (n-1)\int_0^{\pi/2} (\sin^{n-2} x - \sin^n x)\,dx$$

$\therefore \quad nI_n = (n-1)I_{n-2}$

$\therefore \quad I_n = \frac{n-1}{n} I_{n-2}$

$\therefore \quad I_8 = \frac{7}{8} I_6 = \frac{7}{8} \cdot \frac{5}{6} \cdot \frac{3}{4} \cdot \frac{1}{2} I_0 = \frac{1 \cdot 3 \cdot 5 \cdot 7}{1 \cdot 2 \cdot 3 \cdot 4} \cdot \frac{1}{2^4} \cdot \frac{\pi}{2}.$

25. Let $f(x) = ax^2 + bx + c$

$\therefore \quad f'(x) = 2ax + b, f''(x) = 2a.$

$\therefore \quad 2 = c, -3 = b, 4 = 2a$ (from the question)

$\therefore \quad f(x) = 2x^2 - 3x + 2$

$$\therefore \quad \int_{-1}^{1} f(x)\,dx = \int_{-1}^{1} (2x^2 - 3x + 2)\,dx = \left[\frac{2x^3}{3} - \frac{3x^2}{2} + 2x\right]_{-1}^{1}.$$

26. As in Q. No. 25, $I_n = \dfrac{n-1}{n} I_{n-2}$ and I_n, I_{n-1} are positive because $\cos^n x \geq 0$ in $[0, \pi/2]$. Therefore, $I_{n-2} > I_n$.

Also, $I_n = \dfrac{n-1}{n} I_{n-2} \quad \Rightarrow$ option (b).

27. $I_n = \displaystyle\int_0^{\pi/4} \tan^{n-2} x \cdot (\sec^2 x - 1)\,dx = \int_0^{\pi/4} \tan^{n-2} x\, d(\tan x) - I_{n-2}$

$$= \left[\frac{\tan^{n-1} x}{n-1}\right]_0^{\pi/4} - I_{n-2} = \frac{1}{n-1} - I_{n-2}$$

$$\therefore \quad I_5 = \frac{1}{4} - I_3 \text{ and } I_3 = \frac{1}{2} - I_1 \quad \Rightarrow I_1 = I_3 + 2I_5.$$

□

10

Properties and Applications of Definite Integrals

• *Type 1* •

Choose the most appropriate option (a, b, c or d).

1. $\displaystyle\int_0^{\pi/2} \frac{f(x)}{f(x)+f\left(\frac{\pi}{2}-x\right)}\,dx$, where $f(x) \neq -f\left(\frac{\pi}{2}-x\right)$ for $0 \le x \le \frac{\pi}{2}$, has the value

(a) $f(0)$

(b) $f\left(\frac{\pi}{2}\right)$

(c) $\frac{\pi}{2}$

(d) none of these

2. $\displaystyle\int_0^a \{f(x)+f(-x)\}\,dx$ is equal to

(a) $2\int_0^a f(x)\,dx$

(b) $\int_{-a}^a f(x)\,dx$

(c) 0

(d) $-\int_{-a}^a f(-x)\,dx$

3. $\displaystyle\int_{\pi/5}^{3\pi/10} \frac{\cos x}{\cos x+\sin x}\,dx$ is equal to

(a) π

(b) $\frac{\pi}{2}$

(c) $\frac{\pi}{4}$

(d) none of these

4. The equation $\int_{-\pi/4}^{\pi/4}\left(\lambda|\sin x| + \frac{\mu \sin x}{1+\cos x} + \nu\right)dx = 0$, where λ, μ, ν are constants, gives a relation between

(a) λ, μ and ν (b) λ and ν

(c) λ and μ (d) μ and ν

5. Let $f(x) = x - [x]$ for $x \in R$, where $[x]$ = the greatest integer $\le x$. Then $\int_{-2}^{2} f(x)\,dx$ is

(a) 4 (b) 2

(c) 0 (d) 1

6. $\int_0^{\pi} \sin^6 x \cdot \cos^5 x\,dx$ is equal to

(a) $2\int_0^{\pi/2} \sin^{50} x \cdot \cos^{47} x\,dx$ (b) 0

(c) 1 (d) none of these

7. $\int_{-2}^{2} |1 - x^2|\,dx$ is

(a) 4 (b) 2

(c) -2 (d) 0

8. $\int_0^{2\pi} \frac{x \sin^{2n} x}{\sin^{2n} x + \cos^{2n} x}\,dx$, $n > 0$, is equal to

(a) π (b) 2π

(c) π^2 (d) $\frac{1}{2}\pi^2$

9. $\int_{-\pi/4}^{\pi/4} \frac{e^x \cdot \sec^2 x\,dx}{e^{2x} - 1}$ is equal to

(a) 0 (b) 2

(c) e (d) none of these

10. If $[x]$ denotes the greatest integer less than or equal to x then $\int_0^{\infty}\left[\frac{2}{e^x}\right]dx$ is equal to

(a) $\log_e 2$ (b) e^2

(c) 0 (d) $\frac{2}{e}$

11. $\int_1^3 \left|(2-x)\log_e x\right| dx$ is equal to

(a) $\frac{3}{2}\log_e 3 + \frac{1}{2}$ (b) $\log_e \frac{16}{3\sqrt{2}} - \frac{1}{2}$

(c) $-\frac{3}{2}\log_e 3 - \frac{1}{2}$ (d) none of these

12. $\int_{-2}^{2} |x(x-1)|\, dx$ is

(a) $\frac{11}{3}$ (b) $\frac{13}{3}$

(c) $\frac{16}{3}$ (d) $\frac{17}{3}$

13. The value of $\int_{-2}^{2} \frac{\sin^2 x}{\left[\frac{x}{\pi}\right] + \frac{1}{2}} dx$, where $[x]$ = the greatest integer greater than or equal to x, is

(a) 1 (b) 0

(c) $4 - \sin 4$ (d) none of these

14. The value of $\int_0^{\pi} [\cos x]\, dx$, where $[\,\cdot\,]$ is the greatest integer function, is

(a) $\frac{\pi}{2}$ (b) 0

(c) π (d) $-\frac{\pi}{2}$

15. Let $a_n = \int_0^{\pi/2} \cos^n x \cdot \cos nx\, dx$. Then $a_n : a_{n+1}$ is equal to

(a) $3:1$ (b) $2:3$

(c) $2:1$ (d) $3:4$

16. If $\int_0^x f(t)\, dt = x + \int_x^1 t f(t)\, dt$ then the value of $f(1)$ is

(a) $\frac{1}{2}$ (b) 0

(c) 1 (d) $-\frac{1}{2}$

17. The value of $\int_{-1}^{1} \max\{2-x, 2, 1+x\}$ is

(a) 4 (b) $\frac{9}{2}$

(c) 2 (d) none of these

18. Let f be a positive function. If

$$I_1 = \int_{1-k}^{k} xf\{x(1-x)\}\,dx,\ I_2 = \int_{1-k}^{k} f\{x(1-x)\}\,dx,$$

where $2k-1>0$, then $\frac{I_1}{I_2}$ is

(a) 2 (b) k

(c) $\frac{1}{2}$ (d) 1

19. If $x \in (2n\pi, 2n\pi+\pi)$ then $\int_0^x [\sin x]\,dx$, where $[x]$ = greatest integer less than or equal to x, is equal to

(a) $-\pi$ (b) $-n\pi$

(c) 0 (d) none of these

20. The value of $\int_{-\pi/2}^{\pi/2} \frac{dx}{\sin^3 x + \sin x}$ is

(a) 0 (b) 2

(c) 1 (d) none of these

21. $\int_0^{\pi} \frac{dx}{1+3^{\cos x}}$ is equal to

(a) π (b) 0

(c) $\frac{\pi}{2}$ (d) none of these

22. If $[y]$ = the greatest integer less than or equal to y then $\int_{\pi/2}^{3\pi/2} [2\sin x]\,dx$ is

(a) $-\pi$ (b) 0

(c) $-\frac{\pi}{2}$ (d) $\frac{\pi}{2}$

23. The value of $\int_0^{2\pi} \frac{dx}{e^{\sin x}+1}$ is

(a) π (b) 0

(c) 2π (d) $\frac{\pi}{2}$

24. $\int_{a/4}^{3a/4} \frac{\sqrt{x}}{\sqrt{a-x}+\sqrt{x}}\, dx$ is equal to

(a) $\frac{a}{2}$ (b) a

(c) $-a$ (d) none of these

25. $\int_0^{\pi/4} \sin x \, d(x-[x])$ is equal to

(a) $\frac{1}{2}$ (b) $1-\frac{1}{\sqrt{2}}$

(c) 1 (d) none of these

26. If $f(x)=\int_0^{\sin x} \cos^{-1} t\, dt + \int_0^{\cos x} \sin^{-1} t\, dt, 0 < x < \frac{\pi}{2}$, then $f\left(\frac{\pi}{4}\right)$ is

(a) $\frac{\pi}{\sqrt{2}}$ (b) $1+\frac{\pi}{2\sqrt{2}}$

(c) 1 (d) none of these

27. The value of $\int_\alpha^\beta x\,|x|\, dx$, where $\alpha < 0 < \beta$, is

(a) $\frac{1}{2}(\alpha^2+\beta^2)$ (b) $\frac{1}{3}(\beta^2-\alpha^2)$

(c) $\frac{1}{3}(\alpha^3+\beta^3)$ (d) none of these

28. The value of $\int_0^{\pi/4} \log(1+\tan x)\, dx$ is equal to

(a) $\frac{\pi}{8}\log_e 2$ (b) $\frac{\pi}{4}\log_e 2$

(c) $\frac{\pi}{4}$ (d) none of these

29. $\int_{-1}^{1}(x-[2x])\,dx$ is equal to

(a) 1 (b) 0

(c) 2 (d) 4

30. If $f(x) = |x| + 1, -1 \le x < 0$

$1 + |x|^2, 0 \le x \le 1$

then $\int_{-1}^{1} f(x)\,dx$ is equal to

(a) $-\frac{1}{6}$ (b) $\frac{17}{6}$

(c) $-\frac{17}{6}$ (d) none of these

31. Let $f(x) = \text{maximum }\{x + |x|, x - [x]\}$, where $[x]$ = the greatest integer $\le x$.

Then $\int_{-2}^{2} f(x)\,dx$ is equal to

(a) 3 (b) 2

(c) 1 (d) none of these

32. $\int_{0}^{3}\left|x^3 - 3x^2 + 2x\right| dx$ is equal to

(a) $\frac{3}{4}$ (b) $\frac{7}{4}$

(c) $\frac{11}{4}$ (d) none of these

33. Let $f(x)$ be a continuous function such that $f(a-x)+f(x)=0$ for all $x \in [0, a]$. Then $\int_{0}^{a}\frac{dx}{1+e^{f(x)}}$ is equal to

(a) a (b) $\frac{a}{2}$

(c) $f(a)$ (d) $\frac{1}{2}f(a)$

34. $\int_{-a}^{a} \log_e (x+\sqrt{1+x^2})\, dx$ is equal to

(a) $2\log_e a$
(b) 0
(c) $\log_e 2 + \log a$
(d) none of these

35. Let $f(x) = \frac{e^x+1}{e^x-1}$ and $\int_0^1 \frac{e^x+1}{e^x-1} \cdot x\, dx = \lambda$. Then $\int_{-1}^{1} tf(t)\, dt$ is equal to

(a) 0
(b) 2λ
(c) λ
(d) none of these

36. Let $f(x)$ be a continuous function in R such that $f(x)+f(y)=f(x+y)$. If $\int_0^3 f(x)\, dx = k$ then $\int_{-3}^{3} f(x)\, dx$ is equal to

(a) $2k$
(b) 0
(c) $k/2$
(d) $-2k$

37. Let $f(x)$ be a continuous function such that $f(x)$ does not vanish for all $x \in R$. If $\int_2^3 f(x)\, dx = \int_{-2}^{3} f(x)\, dx$ then $f(x), x \in R$, is

(a) an even function
(b) an odd function
(c) a periodic function
(d) none of these

38. Let $I = \int_{-a}^{a} (p\tan^3 x + q\cos^2 x + r\sin x)\, dx$, where p, q, r are arbitrary constants. The numerical value of I depends on

(a) p, q, r, a
(b) q, r, a
(c) q, a
(d) p, r, a

39. Let $I_1 = \int_0^1 e^{-x^2} dx$, $I_2 = \int_0^1 e^{-x} \cos^2 x\, dx$ and $I_3 = \int_0^1 e^{-x^2} \cos^2 x\, dx$. Then

(a) $I_1 < I_2 < I_3$
(b) $I_3 < I_2 < I_1$
(c) $I_2 < I_1 < I_3$
(d) $I_2 < I_3 < I_1$

40. $\int_0^{100\pi} \sqrt{1+\cos 2x}\, dx$ is equal to

(a) 0
(b) $100\sqrt{2}$
(c) $200\sqrt{2}$
(d) 100

41. The value of $\int_0^1 (1+e^{-x^2})\,dx$ is

(a) -1 (b) 2

(c) $1+e^{-1}$ (d) none of these

42. If $\int_{-2}^{3} f(x)\,dx = 5$ and $\int_1^3 \{2-f(x)\}\,dx = 6$ then the value of $\int_{-2}^{1} f(x)\,dx$ is

(a) 7 (b) 3

(c) -7 (d) -3

43. Let $f(x)$ be a continuous function such that $\int_n^{n+1} f(x)\,dx = n^3, n \in Z$. Then the value of $\int_{-3}^{3} f(x)\,dx$ is

(a) 9 (b) -27

(c) -9 (d) none of these

44. If $I = \int_0^1 \frac{x\,dx}{8+x^3}$ then the smallest interval in which I lies is

(a) $\left(0, \frac{1}{8}\right)$ (b) $\left(0, \frac{1}{9}\right)$

(c) $\left(0, \frac{1}{10}\right)$ (d) $\left(0, \frac{1}{7}\right)$

45. If $\int_0^1 xe^{x^2}\,dx = \lambda \int_0^1 e^{x^2}\,dx$ then

(a) $\lambda = 0$ (b) $\lambda \in (0, 1)$

(c) $\lambda \in (-\infty, 0)$ (d) $\lambda \in (1, 2)$

46. $\int_{\log 1/2}^{\log 2} \sin\left(\frac{e^x-1}{e^x+1}\right) dx$ is equal to

(a) $\cos\frac{1}{3}$ (b) 0

(c) $2\cos 2$ (d) none of these

47. Let $\int_0^a f(x)\,dx = \lambda$ and $\int_0^a f(2a-x)\,dx = \mu$. Then $\int_0^{2a} f(x)\,dx$ is equal to

(a) $\lambda + \mu$ (b) $\lambda - \mu$

(c) $2\lambda - \mu$ (d) $\lambda - 2\mu$

48. If $f(-x) + f(x) = 0$ then $\int_a^x f(t)\, dt$ is

(a) an odd function (b) an even function

(c) a periodic function (d) none of these

49. If $f(x)$ and $g(x)$ be continuous functions over the closed interval $[0, a]$ such that $f(x) = f(a - x)$ and $g(x) + g(a - x) = 2$. Then $\int_0^a f(x) \cdot g(x)\, dx$ is equal to

(a) $\int_0^a f(x)\, dx$ (b) $\int_0^a g(x)\, dx$

(c) $2a$ (d) none of these

50. If $f(x) = f(a + x)$ and $\int_0^a f(x)\, dx = p$ then $\int_a^{na} f(x)\, dx$ is equal to

(a) np (b) $(n - 1)p$

(c) $(n + 1)p$ (d) none of these

51. Let $f(x)$ be a given integrable function such that $f(x + k) = f(x)$ for all $x \in R$. Then $\int_a^{a+k} f(x)\, dx$ depends for its value on

(a) a only (b) k only

(c) both a and k (d) neither a nor k

52. The value of $\int_{\pi/4}^{3\pi/4} \frac{x}{1 + \sin x}\, dx$ is equal to

(a) $(\sqrt{2} - 1)\pi$ (b) $(\sqrt{2} + 1)\pi$

(c) π (d) none of these

53. The value of $\int_0^{\pi/2} \frac{dx}{1 + \tan^3 x}$ is

(a) $\frac{\pi}{2}$ (b) $\frac{\pi}{4}$

(c) π (d) none of these

54. Let f and g be two continuous functions. Then

$$\int_{-\pi/2}^{\pi/2} \{f(x)+f(-x)\}\{g(x)-g(-x)\}\, dx$$

is equal to

(a) π (b) 1

(c) -1 (d) 0

55. Let $\int_a^b f(x)\, dx = p$ and $\int_a^b |f(x)|\, dx = q$. Then

(a) $|p| \le q$ (b) $p > q$

(c) $p + q = 0$ (d) none of these

56. If $f(x) = \int_0^x \log(1+t^2)\, dt$ then the value of $f''(1)$ is equal to

(a) 2 (b) 0

(c) 1 (d) none of these

57. If $f(x) = \int_x^{x^2} \frac{dt}{1+t^3}$ then $f'(2)$ is equal to

(a) $\frac{101}{585}$ (b) $\frac{-29}{585}$

(c) $\frac{-56}{585}$ (d) none of these

58. If $f(2a-x) = f(x)$ and $\int_0^a f(x)\, dx = \lambda$ then $\int_0^{2a} f(x)\, dx$ is

(a) 2λ (b) λ

(c) 0 (d) none of these

59. Let $f(x)$ be a continuous function such that the area bounded by the curve $y = f(x)$, the x-axis, and the lines $x = 0$ and $x = a$ is $1 + \frac{a^2}{2}\sin a$. Then

(a) $f\left(\frac{\pi}{2}\right) = 1 + \frac{\pi^2}{8}$ (b) $f(a) = 1 + \frac{a^2}{2}\sin a$

(c) $f(a) = a\sin a + \frac{1}{2}a^2\cos a$ (d) none of these

60. If $y=\int_0^x \sqrt{\sin x}\, dx$ then the value of $\frac{dy}{dx}$ at $x=\frac{\pi}{2}$ is

(a) 0 (b) 1

(c) –1 (d) none of these

61. If $y=\int_x^{x^2} \sqrt{5-t^2}\, dt$ then the value of $\frac{dy}{dx}$ at $x=\sqrt{2}$ is

(a) $1-\sqrt{3}$ (b) $\sqrt{3}(2\sqrt{6}-1)$

(c) $2\sqrt{2}-\sqrt{3}$ (d) none of these

62. The function $f(x)=\int_{-1}^{x} t(e^t-1)(t-2)^3(t-3)^5 dt$ has a local minimum at x which is equal to

(a) 0 (b) 1

(c) 2 (d) 3

63. Let $f(x)$ be a differentiable function and $f(1)=2$. If $\lim_{x\to 1}\int_2^{f(x)} \frac{2t}{x-1}\, dt=4$ then the value of $f'(1)$ is

(a) 1 (b) 2

(c) 4 (d) none of these

64. If $\phi(x)=\int_x^{x^2} (t-1)\, dt$, $1\le x\le 2$, then the greatest value of $\phi(x)$ is

(a) 2 (b) 4

(c) 8 (d) none of these

65. If $\int_0^1 (1+\sin^4 x)(ax^2+bx+c)\, dx=\int_0^2 (1+\sin^4 x)(ax^2+bx+c)\, dx$ then the quadratic equation $ax^2+bx+c=0$ has

(a) at least one root in (1, 2) (b) no root in (1, 2)

(c) two equal roots in (1, 2) (d) both roots imaginary

66. Let $f(x)$ be a function defined by $f(x)=\int_1^x x(x^2-3x+2)\, dx$, $1\le x\le 3$. Then the range of $f(x)$ is

(a) [0, 2] (b) $\left[-\frac{1}{4}, 4\right]$

(c) $\left[-\frac{1}{4}, 2\right]$ (d) none of these

• Type 2 •

Choose the correct options. One or more options may be correct.

67. If $A=\int_0^{\pi}\frac{\sin x}{\sin x+\cos x}\,dx$, $B=\int_0^{\pi}\frac{\sin x}{\sin x-\cos x}\,dx$ then

(a) $A+B=0$ (b) $A=B$

(c) $A=B=\pi/2$ (d) $A=-B=\pi$

68. Let $f(a)>0$, and let $f(x)$ be a nondecreasing continuous function in $[a, b]$. Then $\frac{1}{b-a}\int_a^b f(x)\,dx$ has the

(a) maximum value $f(b)$ (b) minimum value $f(a)$

(c) maximum value $bf(b)$ (d) minimum value $\frac{f(a)}{b-a}$

69. The value of $\int_0^{\pi}\frac{\sin nx}{\sin x}\,dx$, $n\in N$, is

(a) π if n is even (b) 0 if n is odd

(c) 0 if n is even (d) π for all $n\in N$

70. If $f(x)=\int_x^{x^2}\frac{dt}{(\log t)^2}$, $x\neq 0, x\neq 1$, then $f(x)$ is

(a) monotonically increasing in $(2, +\infty)$

(b) monotonically increasing in $(1, 2)$

(c) monotonically decreasing in $(2, +\infty)$

(d) monotonically decreasing in $(0, 1)$

71. Let $f(x)=ax^3+bx^2+cx$ have relative extrema $x=1$ and at $x=5$. If $\int_{-1}^{1}f(x)\,dx=6$ then

(a) $a=-1$ (b) $b=9$

(c) $c=15$ (d) $a=1$

72. Let $f(x)=\int_0^x |x-1|\,dx$, $x\geq 0$. Then $f'(x)$ is

(a) continuous at $x=1$ (b) continuous at $x=2$

(c) differentiable at $x=1$ (d) differentiable at $x=2$

Answers

1. d	**2.** b	**3.** d	**4.** b	**5.** b
6. b	**7.** a	**8.** c	**9.** a	**10.** a
11. b	**12.** d	**13.** b	**14.** d	**15.** c
16. a	**17.** b	**18.** c	**19.** b	**20.** a
21. c	**22.** c	**23.** a	**24.** a	**25.** b
26. b	**27.** c	**28.** a	**29.** a	**30.** b
31. d	**32.** c	**33.** b	**34.** b	**35.** b
36. b	**37.** d	**38.** c	**39.** d	**40.** c
41. d	**42.** a	**43.** b	**44.** b	**45.** b
46. b	**47.** a	**48.** b	**49.** a	**50.** b
51. b	**52.** a	**53.** b	**54.** d	**55.** a
56. c	**57.** b	**58.** a	**59.** c	**60.** b
61. c	**62.** d	**63.** a	**64.** b	**65.** a
66. c	**67.** b, c	**68.** a, b	**69.** c	**70.** a, d
71. a, b	**72.** a, b, d			

Hints and Solutions to Selected Questions

1. $I=\int_0^{\pi/2}\frac{f(x)\,dx}{f(x)+f\left(\frac{\pi}{2}-x\right)}=\int_0^{\pi/2}\frac{f\left(\frac{\pi}{2}-x\right)}{f\left(\frac{\pi}{2}-x\right)+f(x)}\,dx$

$$\left(\text{using}\int_0^a f(x)\,dx=\int_0^a f(a-x)\,dx\right)$$

$$\therefore\quad I+I=\int_0^{\pi/2}\frac{f(x)+f\left(\frac{\pi}{2}-x\right)}{f(x)+f\left(\frac{\pi}{2}-x\right)}\,dx=\int_0^{\pi/2}dx=[x]_0^{\pi/2}=\frac{\pi}{2}\cdot$$

$$\therefore\quad I=\frac{\pi}{4}\cdot$$

2. $I=\int_0^a f(x)\,dx+\int_0^a f(-x)\,dx=\int_0^a f(x)\,dx+\int_0^{-a} f(z)\,d(-z)$

$$=\int_0^a f(x)\,dx+\int_{-a}^0 f(z)\,dz=\int_0^a f(x)\,dx+\int_{-a}^0 f(x)\,dx=\int_{-a}^a f(x)\,dx.$$

3. Put $x=\frac{\pi}{2}-z$. Then $I=\int_{3\pi/10}^{\pi/5}\frac{\cos\left(\frac{\pi}{2}-z\right)}{\cos\left(\frac{\pi}{2}-z\right)+\sin\left(\frac{\pi}{2}-z\right)}\,d\left(\frac{\pi}{2}-z\right)$

$$\therefore\quad I=-\int_{3\pi/10}^{\pi/5}\frac{\sin z}{\sin z+\cos z}\,dz=\int_{\pi/5}^{3\pi/10}\frac{\sin x}{\sin x+\cos x}\,dx$$

$$\therefore\quad I+I=\int_{\pi/5}^{3\pi/10}\frac{\cos x}{\cos x+\sin x}\,dx+\int_{\pi/5}^{3\pi/10}\frac{\sin x}{\sin x+\cos x}\,dx$$

$$=\int_{\pi/5}^{3\pi/10}dx=\frac{3\pi}{10}-\frac{\pi}{5}=\frac{\pi}{10}\cdot$$

4. $\dfrac{\mu \sin x}{1+\cos x}$ is an odd function. So, $\displaystyle\int_{-\pi/4}^{\pi/4} \frac{\mu \sin x}{1+\cos x}\, dx = 0.$

$\therefore$ the equation is $\lambda \displaystyle\int_{-\pi/4}^{\pi/4} |\sin x|\, dx + \nu \int_{-\pi/4}^{\pi/4} dx = 0$

or $2\lambda \displaystyle\int_{0}^{\pi/4} |\sin x|\, dx + \nu \cdot \frac{\pi}{2} = 0$, which is a relation between λ, ν.

5. $I = \displaystyle\int_{-2}^{2} x\, dx - \int_{-2}^{2} [x]\, dx = 0 - \left\{ \int_{-2}^{-1} [x]\, dx + \int_{-1}^{0} [x]\, dx + \int_{0}^{1} [x]\, dx + \int_{1}^{2} [x]\, dx \right\}$

$= -\displaystyle\int_{-2}^{-1} -2\, dx - \int_{-1}^{0} -1\, dx - \int_{0}^{1} 0\, dx - \int_{1}^{2} 1\, dx$

$= 2[x]_{-2}^{-1} + [x]_{-1}^{0} - 0 - [x]_{1}^{2} = 2(-1+2) + (0+1) - (2-1) = 2.$

6. $I = \displaystyle\int_{0}^{\pi} \sin^6 x \cdot \cos^5 x\, dx = \int_{0}^{\pi} \sin^6(\pi - x) \cdot \cos^5(\pi - x)\, dx = -I$

$\therefore$ $I = 0.$

7. $|1 - x^2|$ is an even function.

$\therefore$ $I = 2\displaystyle\int_{0}^{2} |1 - x^2|\, dx$

$\therefore$ $I = 2\displaystyle\int_{0}^{1} (1 - x^2)\, dx + 2\int_{1}^{2} (x^2 - 1)\, dx = 2 \cdot \left[x - \frac{x^3}{3} \right]_0^1 + 2\left[\frac{x^3}{3} - x \right]_1^2$

$= 2\left\{ \left(1 - \dfrac{1}{3}\right) - 0 \right\} + 2\left\{ \left(\dfrac{8}{3} - 2\right) - \left(\dfrac{1}{3} - 1\right) \right\} = \dfrac{4}{3} + \dfrac{8}{3} = 4.$

8. $I = \displaystyle\int_{0}^{2\pi} \frac{(2\pi - x)\sin^{2n}(2\pi - x)}{\sin^{2n}(2\pi - x) + \cos^{2n}(2\pi - x)}\, dx$

$= 2\pi \displaystyle\int_{0}^{2\pi} \frac{\sin^{2n} x}{\sin^{2n} x + \cos^{2n} x}\, dx - I$

$\therefore$ $2I = 2\pi \displaystyle\int_{0}^{2\pi} \frac{\sin^{2n} x}{\sin^{2n} x + \cos^{2n} x}\, dx$

$$= 2\pi \cdot 2\int_0^{\pi} \frac{\sin^{2n} x}{\sin^{2n} x + \cos^{2n} x}\,dx,$$ because the integrand is a periodic function of period π.

$$I = 2\pi \int_{\pi/2}^{-\pi/2} \frac{\sin^{2n}\left(\frac{\pi}{2} - z\right)}{\sin^{2n}\left(\frac{\pi}{2} - z\right) + \cos^{2n}\left(\frac{\pi}{2} - z\right)}\,d\left(\frac{\pi}{2} - z\right)$$

$$= 2\pi \int_{-\pi/2}^{\pi/2} \frac{\cos^{2n} z}{\cos^{2n} z + \sin^{2n} z}\,dz = 4\pi \int_0^{\pi/2} \frac{\cos^{2n} z}{\cos^{2n} z + \sin^{2n} z}\,dz$$

$\therefore \quad I = 4\pi I'$, where $I' = \int_0^{\pi/2} \frac{\cos^{2n} z}{\cos^{2n} z + \sin^{2n} z}\,dz.$

$$\therefore \quad I' = \int_0^{\pi/2} \frac{\cos^{2n}\left(\frac{\pi}{2} - z\right)}{\cos^{2n}\left(\frac{\pi}{2} - z\right) + \sin^{2n}\left(\frac{\pi}{2} - z\right)}\,dz = \int_0^{\pi/2} \frac{\sin^{2n} z}{\sin^{2n} z + \cos^{2n} z}\,dz.$$

$$\therefore \quad I' + I' = \int_0^{\pi/2} dz = \frac{\pi}{2} \qquad \therefore \quad I' = \frac{\pi}{4}.$$

$$\therefore \quad I = 4\pi I' \quad \Rightarrow I = \pi^2.$$

9. Clearly, the integrand is an odd function. So, $I = 0$.

10. If $x = \log 2, \left[\frac{2}{e^x}\right] = 1$; if $x > \log 2, \left[\frac{2}{e^x}\right] = 0$

$$\therefore \quad I = \int_0^{\log 2} \left[\frac{2}{e^x}\right] dx + \int_{\log 2}^{\infty} \left[\frac{2}{e^x}\right] dx = \int_0^{\log 2} 1\,dx + \int_{\log 2}^{\infty} 0\,dx = \log 2.$$

11. $$\int_1^3 |2 - x|\,|\log_e x|\,dx = \int_1^2 |2 - x|\,|\log_e x|\,dx + \int_2^3 |2 - x|\,|\log_e x|\,dx$$

$$= \int_1^2 (2 - x)\log_e x\,dx + \int_2^3 -(2 - x)\log_e x\,dx.$$

Now, $\int (2 - x)\log_e x\,dx = \log_e x \cdot \left(2x - \frac{x^2}{2}\right) - \int \left(2x - \frac{x^2}{2}\right) \cdot \frac{1}{x}\,dx$

$$= \left(2x - \frac{x^2}{2}\right)\log_e x - \int\left(2 - \frac{x}{2}\right)dx = \left(2x - \frac{x^2}{2}\right)\log_e x - \left(2x - \frac{x^2}{4}\right)$$

$$\therefore \quad I = \left[\left(2x - \frac{x^2}{2}\right)\log_e x - \left(2x - \frac{x^2}{4}\right)\right]_1^2 + \left[-\left(2x - \frac{x^2}{2}\right)\log_e x + \left(2x - \frac{x^2}{4}\right)\right]_2^3$$

$$= (4-2)\log_e 2 - (4-1) + (2 - \frac{1}{4})$$

$$+ \left\{-\left(6 - \frac{9}{2}\right)\log_e 3 + \left(6 - \frac{9}{4}\right) + (4-2)\log_e 2 - (4-1)\right\}$$

$$= 2\log_e 2 - 3 + \frac{7}{4} - \frac{3}{2}\log_e 3 + \frac{15}{4} + 2\log_e 2 - 3$$

$$= 4\log_e 2 - \frac{3}{2}\log_e 3 - \frac{1}{2} = \log\frac{16}{3\sqrt{3}} - \frac{1}{2}.$$

12. $I = \int_{-2}^{0} |x(x-1)|\,dx + \int_0^1 |x(x-1)|\,dx + \int_1^2 |x(x-1)|\,dx$

$$= \int_{-2}^{0} x(x-1)\,dx + \int_0^1 -x(x-1)\,dx + \int_1^2 x(x-1)\,dx$$

$$= \left[\frac{x^3}{3} - \frac{x^2}{2}\right]_{-2}^{0} - \left[\frac{x^3}{3} - \frac{x^2}{2}\right]_0^1 + \left[\frac{x^3}{3} - \frac{x^2}{2}\right]_1^2$$

$$= 0 - \left(-\frac{8}{3} - 2\right) - \left(\frac{1}{3} - \frac{1}{2}\right) + \left(\frac{8}{3} - 2\right) - \left(\frac{1}{3} - \frac{1}{2}\right) = \frac{17}{3}.$$

13. $I = \int_{-2}^{0} \frac{\sin^2 x}{\left[\frac{x}{\pi}\right] + \frac{1}{2}}\,dx + \int_0^2 \frac{\sin^2 x}{\left[\frac{x}{\pi}\right] + \frac{1}{2}}\,dx = \int_{-2}^{0} \frac{\sin^2 x}{-1 + \frac{1}{2}}\,dx + \int_0^2 \frac{\sin^2 x}{0 + \frac{1}{2}}\,dx$

$$= 2\int_0^{-2} \sin^2 x\,dx + 2\int_0^2 \sin^2 x\,dx$$

$$= 2\int_0^2 \sin^2(-x)\,d(-x) + 2\int_0^2 \sin^2 x\,dx = 0.$$

14. $I = \int_0^{\pi/2} [\cos x]\,dx + \int_{\pi/2}^{\pi} [\cos x]\,dx$

$$= \int_0^{\pi/2} 0\,dx + \int_{\pi/2}^{\pi} (-1)\,dx = -\left(\pi - \frac{\pi}{2}\right) = -\frac{\pi}{2}.$$

15. $a_{n+1} - a_n = \int_0^{\pi/2} \{\cos^{n+1} x \cdot \cos(n+1)x - \cos^n x \cdot \cos nx] \, dx$

$$= \int_0^{\pi/2} \cos^n x \cdot \{\cos x \cdot \cos(n+1)x - \cos nx\} \, dx$$

$$= \int_0^{\pi/2} \cos^n x \cdot \{-\sin x \cdot \sin(n+1)x\} \, dx$$

$$= \left[\sin(n+1)x \cdot \frac{\cos^{n+1} x}{n+1}\right]_0^{\pi/2} - \int_0^{\pi/2} \frac{\cos^{n+1} x}{n+1} \cdot (n+1)\cos(n+1)x \, dx$$

$$= -a_{n+1}$$

$\therefore \quad 2a_{n+1} = a_n.$

16. $\int_0^x f(t)\, dt + \int_1^x tf(t) dt = x.$

Differentiating w.r.t. x,

$$f(x) + xf(x) = 1$$

$\therefore \quad f(x) = \dfrac{1}{1+x} \cdot$ So, $f(1) = \dfrac{1}{2} \cdot$

17. $I = \int_{-1}^0 \max\{2-x, 2, 1+x\} \, dx + \int_0^1 \max\{2-x, 2, 1+x\} \, dx$

$$= \int_{-1}^0 (2-x) \, dx + \int_0^1 2 \, dx = \left[2x - \frac{x^2}{2}\right]_{-1}^0 + 2[x]_0^1 = -\left(-2 - \frac{1}{2}\right) + 2 = \frac{9}{2} \cdot$$

18. $I_1 = \int_{1-k}^k xf\{x(1-x)\} \, dx$

Put $x = 1 - z$.

$$I_1 = \int_k^{1-k} (1-z) f\{(1-z)z\} d(1-z)$$

$$= \int_{1-k}^k f\{z(1-z)\} \, dz - \int_{1-k}^k zf\{z(1-z)\} \, dz$$

$$= I_2 - I_1$$

$\therefore \quad 2I_1 = I_2.$

19. Here $2n\pi < x < 2n\pi + \pi$.

$$\therefore \quad \int_0^x [\sin x]\, dx = \int_0^{2n\pi} [\sin x]\, dx + \int_{2n\pi}^x [\sin x]\, dx.$$

$$\int_0^{2n\pi} [\sin x]\, dx = n\int_0^{2\pi} [\sin x]\, dx \quad (\because\ [\sin \overline{x+2\pi}] = [\sin x])$$

$$= n\left\{\int_0^{\pi/2} [\sin x]\, dx + \int_{\pi/2}^{\pi} [\sin x]\, dx + \int_{\pi}^{3\pi/2} [\sin x]\, dx + \int_{3\pi/2}^{2\pi} [\sin x]\, dx\right\}$$

$$= n\left\{\int_0^{\pi/2} 0\, dx + \int 0\, dx + \int_{\pi}^{3\pi/2} -1\, dx + \int_{3\pi/2}^{2\pi} -1\, dx\right\}$$

$$= n\left\{-\left(\frac{3\pi}{2} - \pi\right) - \left(2\pi - \frac{3\pi}{2}\right)\right\} = -n\pi.$$

$$\int_{2n\pi}^{x} [\sin x]dx = \int_{2n\pi}^{x} 0\, dx = 0, \text{ if } 2n\pi < x < 2n\pi + \frac{\pi}{2}$$

$$\int_{2n\pi+\pi/2}^{x} 0\, dx = 0, \text{ if } 2n\pi + \frac{\pi}{2} \le x < 2n\pi + \pi$$

$$\therefore \quad I = -n\pi + 0 = -n\pi.$$

20. Here $f(x) = \dfrac{1}{\sin^3 x + \sin x}$

$\therefore \quad f(-x) = -f(x)$. So, $f(x)$ is odd. Hence, $I = 0$.

21. $I = \displaystyle\int_0^{\pi} \frac{dx}{1+3^{\cos(\pi - x)}} = \int_0^{\pi} \frac{dx}{1+3^{-\cos x}} = \int_0^{\pi} \frac{3^{\cos x}}{1+3^{\cos x}}\, dx.$

$$\therefore \quad I + I = \int_0^{\pi} \frac{dx}{1+3^{\cos x}} + \int_0^{\pi} \frac{3^{\cos x}}{1+3^{\cos x}}\, dx = \int_0^{\pi} 1\, dx = \pi.$$

22.

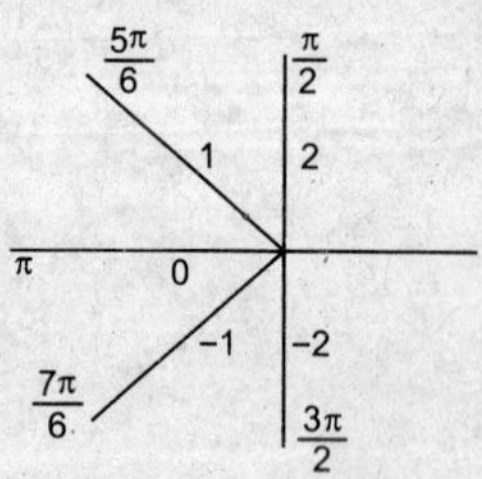

$$I=\int_{\pi/2}^{5\pi/6}[2\sin x]\,dx+\int_{5\pi/6}^{\pi}[2\sin x]\,dx+\int_{\pi}^{7\pi/6}[2\sin x]\,dx+\int_{7\pi/6}^{3\pi/2}[2\sin x]\,dx$$

$$=\int_{\pi/2}^{5\pi/6}1\,dx+\int_{5\pi/6}^{\pi}0\,dx+\int_{\pi}^{7\pi/6}-1\,dx+\int_{7\pi/6}^{3\pi/2}-2\,dx$$

$$=\frac{5\pi}{6}-\frac{\pi}{2}-\left(\frac{7\pi}{6}-\pi\right)-2\left(\frac{3\pi}{2}-\frac{7\pi}{6}\right)=-\frac{\pi}{2}\cdot$$

23. See the hint to Q. No. 21.

24. Put $a-x=z$. Then $I=\int_{3a/4}^{a/4}\frac{\sqrt{a-z}}{\sqrt{z}+\sqrt{a-z}}(-dz)=\int_{a/4}^{3a/4}\frac{\sqrt{a-x}}{\sqrt{x}+\sqrt{a-x}}dx$

$$\therefore\quad I+I=\int_{a/4}^{3a/4}\frac{\sqrt{x}}{\sqrt{a-x}+\sqrt{x}}dx+\int_{a/4}^{3a/4}\frac{\sqrt{a-x}}{\sqrt{x}+\sqrt{a-x}}dx$$

$$=\int_{a/4}^{3a/4}dx=\frac{3a}{4}-\frac{a}{4}=\frac{a}{2}\cdot$$

25. Let $x-[x]=z$. In $0\le x\le\frac{\pi}{4}$, $[x]=0$. So $x=z$.

$$\therefore\quad I=\int_0^{\pi/4}\sin z\,dz=[-\cos z]_0^{\pi/4}=1-\frac{1}{\sqrt{2}}\cdot$$

26. $f(x)=\int_0^{\sin x}\left(\frac{\pi}{2}-\sin^{-1}t\right)dt+\int_0^{\cos x}\left(\frac{\pi}{2}-\cos^{-1}t\right)dt$

$$=\frac{\pi}{2}[t]_0^{\sin x}-\int_0^{\sin x}\sin^{-1}t\,dt+\frac{\pi}{2}[t]_0^{\cos x}-\int_0^{\cos x}\cos^{-1}t\,dt$$

$$=\frac{\pi}{2}(\sin x+\cos x)-\int_0^{\sin x}\sin^{-1}t\,dt-\int_0^{\cos x}\cos^{-1}t\,dt$$

$$\int_0^{\sin x}\sin^{-1}t\,dt=\int_0^{x}\theta\cos\theta\,d\theta\qquad(\text{putting } t=\sin\theta)$$

$$=[\theta\sin\theta]_0^x-\int_0^x\sin\theta\,d\theta=x\sin x+[\cos\theta]_0^x$$

$$=x\sin x+\cos x-1$$

$$\int_0^{\cos x}\cos^{-1}t\,dt=\int_0^{x}-\phi\sin\phi\,d\phi\qquad(\text{putting } t=\cos\phi)$$

$$=[\phi\cos\phi]_0^x-\int_0^x\cos\phi\,d\phi=x\cos x-[\sin\phi]_0^x=x\cos x-\sin x$$

$$\therefore \quad f(x)=\frac{\pi}{2}(\sin x+\cos x)-x\sin x-\cos x+1-x\cos x+\sin x$$

$$\therefore \quad f\left(\frac{\pi}{4}\right)=\frac{\pi}{2}\cdot\sqrt{2}-\frac{\pi}{4}\cdot\frac{1}{\sqrt{2}}-\frac{1}{\sqrt{2}}+1-\frac{\pi}{4}\cdot\frac{1}{\sqrt{2}}+\frac{1}{\sqrt{2}}=1+\frac{\pi}{2\sqrt{2}}\cdot$$

27. $I=\int_{\alpha}^{0} x\,|x|\,dx+\int_{0}^{\beta} x\,|x|\,dx=\int_{\alpha}^{0} -x^2dx+\int_{0}^{\beta} x^2dx$

$$=\left[-\frac{x^3}{3}\right]_{\alpha}^{0}+\left[\frac{x^3}{3}\right]_{0}^{\beta}=\frac{\alpha^3}{3}+\frac{\beta^3}{3}\cdot$$

28. $I=\int_{0}^{\pi/4}\log(1+\tan x)\,dx=\int_{0}^{\pi/4}\log\left\{1+\tan\left(\frac{\pi}{4}-x\right)\right\}dx$

$$=\int_{0}^{\pi/4}\log\left\{1+\frac{1-\tan x}{1+\tan x}\right\}dx=\int_{0}^{\pi/4}\log\frac{2}{1+\tan x}\,dx=\int_{0}^{\pi/4}\log 2\,dx-I.$$

$$\therefore \quad I=\frac{1}{2}\int_{0}^{\pi/4}\log 2\,dx=\frac{1}{2}\cdot\log 2\cdot\frac{\pi}{4}\cdot$$

29. $I=\int_{-1}^{1} x\,dx-\int_{-1}^{1}[2x]\,dx$

$$=\left[\frac{x^2}{2}\right]_{-1}^{1}-\left\{\int_{-1}^{-1/2}[2x]\,dx+\int_{-1/2}^{0}[2x]\,dx+\int_{0}^{1/2}[2x]\,dx+\int_{1/2}^{1}[2x]\,dx\right\}$$

$$=0-\left\{\int_{-1}^{-1/2}-2\,dx+\int_{-1/2}^{0}-1\,dx+\int_{0}^{1/2}0\,dx+\int_{1/2}^{1}1\,dx\right\}$$

$$=2[x]_{-1}^{-1/2}-[x]_{-1/2}^{0}-[x]_{1/2}^{1}=2\left(-\frac{1}{2}+1\right)+\left(0+\frac{1}{2}\right)-\left(1-\frac{1}{2}\right)=1.$$

30. $I=\int_{-1}^{0} f(x)\,dx+\int_{0}^{1} f(x)\,dx=\int_{-1}^{0}(|x|+1)\,dx+\int_{0}^{1}(1+|x|^2)\,dx$

$$=\int_{-1}^{0}(-x+1)\,dx+\int_{0}^{1}(1+x^2)\,dx=\left[-\frac{x^2}{2}+x\right]_{-1}^{0}+\left[x+\frac{x^3}{3}\right]_{0}^{1}$$

$$=-\left(-\frac{1}{2}-1\right)+\left(1+\frac{1}{3}\right)=\frac{3}{2}+\frac{4}{3}=\frac{17}{6}\cdot$$

31. $|x|\geq-[x]$ for all positive x, and $|x|\leq-[x]$ for all negative x.

$$\therefore \quad \int_{-2}^{2} f(x)\,dx = \int_{-2}^{0} (x-[x])\,dx + \int_{0}^{2} (x+|x|)\,dx$$

$$= \int_{-2}^{0} x\,dx - \int_{-2}^{0} [x]\,dx + \int_{0}^{2} 2x\,dx$$

$$= \left[\frac{x^2}{2}\right]_{-2}^{0} - \left\{\int_{-2}^{-1} [x]\,dx + \int_{-1}^{0} [x]\,dx\right\} + [x^2]_0^2$$

$$= 0 - 2 - \int_{-2}^{-1} -2\,dx - \int_{-1}^{0} -1\,dx + 4$$

$$= -2 + 2[x]_{-2}^{-1} + [x]_{-1}^{0} + 4 = -2 + 2(-1+2) + (0+1) + 4 = 5.$$

32. $I = \int_0^3 |x|\,|x-1|\,|x-2|\,dx$

$$= \int_0^1 (x^3 - 3x^2 + 2x)\,dx + \int_1^2 -(x^3 - 3x^2 + 2x)\,dx + \int_2^3 (x^3 - 3x^2 + 2x)\,dx$$

$$= \left[\frac{x^4}{4} - x^3 + x^2\right]_0^1 - \left[\frac{x^4}{4} - x^3 + x^2\right]_1^2 + \left[\frac{x^4}{4} - x^3 + x^2\right]_2^3$$

$$= \frac{1}{4} - \left\{(4-8+4) - \frac{1}{4}\right\} + \left\{\left(\frac{81}{4} - 27 + 9\right) - (4-8+4)\right\} = \frac{11}{4}\cdot$$

33. $I = \int_0^a \frac{dx}{1+e^{f(x)}} = \int_0^a \frac{dx}{1+e^{f(a-x)}} = \int_0^a \frac{dx}{1+e^{-f(x)}}$ (from the question) .

$$= \int_0^a \frac{e^{f(x)}\,dx}{e^{f(x)}+1}\cdot$$

$$\therefore \quad I + I = \int_0^a \frac{dx}{1+e^{f(x)}} + \int_0^a \frac{e^{f(x)}}{1+e^{f(x)}}\,dx$$

$$\therefore \quad 2I = \int_0^a dx = a.$$

34. Let $f(x) = \log_e (x + \sqrt{1+x^2})$. Clearly, $f(-x) = -f(x)$. So, $f(x)$ is an odd function.

$\therefore \quad I = 0.$

35. $\phi(t) = tf(t) = t \cdot \frac{e^t+1}{e^t-1}\cdot$ So, $\phi(-t) = -t \cdot \frac{e^{-t}+1}{e^{-t}-1} = t \cdot \frac{e^t+1}{e^t-1}\cdot$

$\therefore \quad \phi(t) = \phi(-t)$. Hence, $\phi(t)$ is an even function.

$$\therefore \quad \int_{-1}^{1} \phi(t)\, dt = 2\int_{0}^{1} \phi(t)\, dt = 2\lambda.$$

36. Putting $x = 0, y = 0$, we get $f(0) + f(0) = f(0)$

$\therefore \quad f(0) = 0.$

Putting $y = -x$, $f(x) + f(-x) = f(x - x) = f(0) = 0$

$\therefore \quad f(-x) = -f(x).$

$\therefore \quad f(x)$ is an odd function. So, $\int_{-3}^{3} f(x)\, dx = 0.$

37. $\int_{-2}^{3} f(x)\, dx - \int_{2}^{3} f(x)\, dx = 0$

or $\int_{-2}^{2} f(x)\, dx = 0$ or $\int_{-2}^{0} f(x)\, dx + \int_{0}^{2} f(x)\, dx = 0$

or $\int_{0}^{2} \{f(x) + f(-x)\}\, dx = 0$, which may imply $f(-x) = -f(x)$ in $[-2, 2]$.

Nothing can be said for the whole of R.

38. $I = p\int_{-a}^{a} \tan^3 x\, dx + q\int_{-a}^{a} \cos^2 x\, dx + r\int_{-a}^{a} \sin x\, dx$

$= p \times 0 + 2q\int_{0}^{a} \cos^2 x\, dx + r \times 0$, because $\tan^3 x$ and $\sin x$ are odd functions, and $\cos^2 x$ is an even function.

39. In $(0, 1)$, $e^{-x^2} > e^{-x^2}\cos^2 x > 0 \qquad \therefore\ I_1 > I_3$

$e^{-x^2} > e^{-x} > 0 \qquad \therefore\ e^{-x^2}\cos^2 x > e^{-x}\cos^2 x > 0$

$\therefore \quad I_3 > I_2$

$\therefore \quad I_2 < I_3 < I_1.$

40. $I = \int_{0}^{100\pi} \sqrt{2}\,|\cos x|\, dx$. But $|\cos x|$ is a periodic function of period π.

$$\therefore \quad I = 100\int_{0}^{\pi} \sqrt{2}\,|\cos x|\, dx = 100\sqrt{2}\left\{\int_{0}^{\pi/2} |\cos x|\, dx + \int_{\pi/2}^{\pi} |\cos x|\, dx\right\}$$

$$= 100\sqrt{2}\left\{\int_{0}^{\pi/2} \cos x\, dx + \int_{\pi/2}^{\pi} -\cos x\, dx\right\}$$

$$= 100\sqrt{2}\left\{[\sin x]_0^{\pi/2} + [-\sin x]_{\pi/2}^{\pi}\right\}$$

$$= 100\sqrt{2}\{1 - 0 - (0 - 1)\} = 200\sqrt{2}.$$

41. $I = [x]_0^1 + \int_0^1 \frac{1}{e^{x^2}}\,dx = 1 + \int_0^1 \frac{1}{e^{x^2}}\,dx.$

In (0, 1), $\frac{1}{e} < \frac{1}{e^{x^2}} < 1$

$\therefore \quad \int_0^1 \frac{1}{e}\,dx < \int_0^1 \frac{1}{e^{x^2}}\,dx < \int_0^1 1\,dx$

$\therefore \quad \frac{1}{e} < \int_0^1 \frac{1}{e^{x^2}}\,dx < 1$

$\therefore \quad 1 + \frac{1}{e} < \int_0^1 (1 + e^{-x^2})\,dx < 1 + 1.$

42. $\int_1^3 \{2 - f(x)\}dx = 6 \quad \Rightarrow \quad 2(3-1) - 6 = \int_1^3 f(x)\,dx$

$\therefore \quad \int_1^3 f(x)\,dx = -2.$

Now, $\int_{-2}^1 f(x)\,dx = \int_{-2}^3 f(x)\,dx - \int_1^3 f(x)\,dx = 5 - (-2) = 7.$

43. $\int_{-3}^3 f(x)\,dx = \int_{-3}^{-3+1} f(x)\,dx + \int_{-2}^{-2+1} f(x)\,dx + \int_{-1}^{-1+1} f(x)\,dx$

$$+ \int_0^{0+1} f(x)\,dx + \int_1^{1+1} f(x)\,dx + \int_2^3 f(x)\,dx$$

$$= (-3)^3 + (-2)^3 + (-1)^3 + 0^3 + 1^3 + 2^3 = -27.$$

44. Let $f(x) = \frac{x}{8 + x^3}\cdot$ Then $f'(x) = \frac{1\cdot(8 + x^3) - x(3x^2)}{(8 + x^3)^2} = \frac{8 - 2x^3}{(8 + x^3)^2} > 0$ for $x \in [0, 1]$.

$\therefore \quad f(x)$ is m.i. in [0, 1]. So, $f(0) \le f(x) \le f(1)$.

$\therefore \quad 0 \le f(x) \le \frac{1}{8 + 1}$

$$\therefore \quad \int_0^1 0\,dx < \int_0^1 f(x)\,dx < \frac{1}{9}\int_0^1 1\,dx$$

$$\therefore \quad 0 < \int_0^1 f(x)\,dx < \frac{1}{9}.$$

45. Here $0 < x < 1$ and $e^{x^2} > 0$. So, $0 < xe^{x^2} < e^{x^2}$.

$$\therefore \quad \int_0^1 0\,dx < \int_0^1 xe^{x^2}\,dx < \int_0^1 e^{x^2}\,dx$$

$$\therefore \quad 0 < \lambda \int_0^1 e^{x^2}\,dx < \int_0^1 e^{x^2}\,dx \qquad \Rightarrow 0 < \lambda < 1.$$

46. Let $f(x) = \sin\left(\frac{e^x - 1}{e^x + 1}\right)$

$$\therefore \quad f(-x) = \sin\left(\frac{e^{-x} - 1}{e^{-x} + 1}\right) = \sin\left(\frac{1 - e^x}{1 + e^x}\right) = -\sin\left(\frac{e^x - 1}{e^x + 1}\right)$$

$$\therefore \quad f(-x) = -f(x)$$

$\therefore$ $f(x)$ is an odd function. Hence, $I = 0$.

47. $\int_0^{2a} f(x)\,dx = \int_0^a f(x)\,dx + \int_a^{2a} f(x)\,dx$

Putting $x = 2a - z$ for the second integral,

$$\int_0^{2a} f(x)\,dx = \int_0^a f(x)\,dx + \int_a^0 -f(2a - z)\,dz$$

$$= \int_0^a f(x)\,dx + \int_0^a f(2a - x)\,dx = \lambda + \mu.$$

48. Let $\phi(x) = \int_a^x f(t)\,dt$. Then $\phi(-x) = \int_a^{-x} f(t)\,dt$.

$$\therefore \quad \phi(-x) = \int_a^x f(t)\,dt + \int_x^{-x} f(t)\,dt = \phi(x) + \int_x^0 f(t)\,dt + \int_0^{-x} f(t)\,dt$$

$$= \phi(x) - \int_0^x f(t)\,dt + \int_0^x -f(-z)\,dz, \text{using } t = -z$$

$$= \phi(x) - \int_0^x \{f(t) + f(-t)\}\,dt = \phi(x)$$

($\because$ $f(t) + f(-t) = 0$, from the question)

$\therefore$ $\phi(x)$ is even.

49. $\int_0^a f(x)\cdot g(x)\,dx = \int_0^a f(a-x)\cdot g(a-x)\,dx$ (by property)

$$= \int_0^a f(x)\cdot \{2-g(x)\}\,dx \quad \text{(from the question)}$$

$$= 2\int_0^a f(x)\,dx - \int_0^a f(x)\cdot g(x)\,dx$$

$$\therefore \quad 2\int_0^a f(x)\cdot g(x)\,dx = 2\int_0^a f(x)\,dx.$$

50. $\int_a^{na} f(x)\,dx = \int_a^{2a} f(x)\,dx + \int_{2a}^{3a} f(x)\,dx + \ldots + \int_{(n-1)a}^{na} f(x)\,dx.$

Putting $x = a + z$ for the first integral,

$$\int_a^{2a} f(x)\,dx = \int_0^a f(a+z)\,dz = \int_0^a f(z)\,dz, \text{ because } f(a+x) = f(x).$$

Putting $x = 2a + z$ for the second integral,

$$\int_{2a}^{3a} f(x)\,dx = \int_0^a f(2a+z)\,dz = \int_0^a f(z)\,dz,$$

because $f(x) = f(a+x) = f(a + \overline{a+x})$.

Similarly, for the other integrals.

$$\therefore \quad \int_a^{na} f(x)\,dx = \int_0^a f(x)\,dx + \int_0^a f(x)\,dx + \ldots \text{ to } (n-1) \text{ times}$$

$$= (n-1)\int_0^a f(x)\,dx. \text{ (Remember this result)}$$

51. $\int_a^{a+k} f(x)\,dx = \int_0^{a+k} f(x)\,dx - \int_0^a f(x)\,dx$

$$= \int_0^k f(x)\,dx + \int_k^{a+k} f(x)\,dx - \int_0^a f(x)\,dx.$$

For the second integral, put $x = z + k$.

Then $\int_k^{a+k} f(x)\,dx = \int_0^a f(z+k)\,dz = \int_0^a f(z)\,dz.$

$$\therefore \quad \int_a^{a+k} f(x)\,dx = \int_0^k f(x)\,dx + \int_0^a f(x)\,dx - \int_0^a f(x)\,dx = \int_0^k f(x)\,dx,$$

which depends on k but not on a.

52. Put $x = \pi - z$. Then $\int_{\pi/4}^{3\pi/4} \frac{x}{1+\sin x}\, dx = \int_{3\pi/4}^{\pi/4} \frac{\pi - z}{1+\sin(\pi - z)}(-dz)$

$$\therefore \quad I = \int_{\pi/4}^{3\pi/4} \frac{\pi}{1+\sin z}\, dz - I$$

$$\therefore \quad 2I = \pi \int_{\pi/4}^{3\pi/4} \frac{dz}{1+\sin z}$$

$$\therefore \quad 2I = \pi \int_{\pi/4}^{3\pi/4} \frac{1-\sin z}{\cos^2 z}\, dz = \pi\,[\tan z - \sec z]_{\pi/4}^{3\pi/4}$$

$$= \pi\{(-1+\sqrt{2}) - (1-\sqrt{2})\} = 2\pi(\sqrt{2}-1).$$

53. $I = \int_0^{\pi/2} \frac{\cos^3 x}{\cos^3 x + \sin^3 x}\, dx = \int_0^{\pi/2} \frac{\cos^3\left(\frac{\pi}{2} - x\right)}{\cos^3\left(\frac{\pi}{2} - x\right) + \sin^3\left(\frac{\pi}{2} - \pi\right)}\, dx$

$$= \int_0^{\pi/2} \frac{\sin^3 x}{\sin^3 x + \cos^3 x}\, dx$$

$$\therefore \quad I + I = \int_0^{\pi/2} 1 dx = \frac{\pi}{2}$$

$$\therefore \quad I = \frac{\pi}{4}.$$

54. Clearly, $f(x) + f(-x)$ is an even function while $g(x) - g(-x)$ is an odd function.

$\therefore$ the integrand is an odd function. So, $I = 0$.

55. Use the property $\left|\int_a^b f(x)\, dx\right| \le \int_a^b |f(x)|\, dx$.

56. $f'(x) = \log(1+x^2)$ $\quad \therefore\ f''(x) = \frac{2x}{1+x^2}$ $\quad \therefore\ f''(1) = 1$.

57. $f(x) = \int_0^{x^2} \frac{dt}{1+t^3} - \int_0^{x} \frac{dt}{1+t^3}$

$$\therefore \quad f'(x) = \frac{d}{dx^2}\left\{\int_0^{x^2} \frac{dt}{1+t^3}\right\} \frac{d(x^2)}{dx} - \frac{1}{1+x^3}$$

$$\therefore \quad f'(x) = \frac{1}{1+x^6} \cdot 2x - \frac{1}{1+x^3}$$

$$\therefore \quad f'(2) = \frac{2 \times 2}{1+2^6} - \frac{1}{1+2^3} = \frac{4}{65} - \frac{1}{9} = \frac{-29}{585}.$$

58. $\int_0^{2a} f(x)\,dx = \int_0^a f(x)\,dx + \int_a^{2a} f(x)\,dx.$

Putting $x = 2a - z$ in the second integrand,

$$\int_0^{2a} f(x)\,dx = \int_0^a f(x)\,dx + \int_a^0 f(2a-z)\,(-dz) = \int_0^a f(x)\,dx + \int_0^a f(2a-z)\,dz$$

$$= \int_0^a f(x)\,dx + \int_0^a f(z)\,dz = \lambda + \lambda = 2\lambda.$$

59. Here $\int_0^u f(x)\,dx = 1 + \frac{a^2}{2}\sin a$. Differentiating w.r.t. a,

$$f(a) = a\sin a + \frac{a^2}{2}\cos a.$$

60. $\frac{dy}{dx} = \sqrt{\sin x} \qquad \therefore \quad \left.\frac{dy}{dx}\right)_{x=\frac{\pi}{2}} = 1.$

61. $y = \int_0^{x^2} \sqrt{5-t^2}\,dt - \int_0^x \sqrt{5-t^2}\,dt.$

Differentiating w.r.t. x,

$$\frac{dy}{dx} = \sqrt{5-(x^2)^2} \cdot 2x - \sqrt{5-x^2}.$$

Put $x = \sqrt{2}$.

62. $f'(x) = x(e^x - 1)(x-2)^3 \cdot (x-3)^5$

$f'(0-\in) = (0-\in)(e^{0-\in} - 1)(0-\in-2)^3(0-\in-3)^5 = (+\text{ve})$

$f'(0+\in) = (0+\in)(e^{0+\in} - 1)(0+\in-2)^3(0+\in-3)^5 = (+\text{ve})$

$f'(1-\in) = (1-\in)(e^{1-\in} - 1)(1-\in-2)^3(1-\in-3)^5 = (+\text{ve})$

$f'(1+\in) = (1+\in)(e^{1+\in} - 1)(1+\in-2)^3(1+\in-3)^5 = (+\text{ve})$

$f'(2-\in) = (2-\in)(e^{2-\in} - 1)(2-\in-2)^3(2-\in-3)^5 = (+\text{ve})$

$f'(2+\in) = (2+\in)(e^{2+\in} - 1)(2+\in-2)^3(2+\in-3)^5 = (-\text{ve});$

so, a local maximum at $x = 2$.

$f'(3-\in) = (3-\in)(e^{3-\in} - 1)(3-\in-2)^3(3-\in-3)^5 = (-\text{ve})$

$$f'(3+\in) = (3+\in)(e^{3+\in}-1)(3+\in-2)^3(3+\in-3)^5 = (+\text{ve});$$

so, a local minimum at $x = 3$.

63. Here, $4 = \lim\limits_{x\to 1} \dfrac{1}{x-1}\displaystyle\int_2^{f(x)} 2t\,dt = \lim\limits_{x\to 1} \dfrac{2f(x)\cdot f'(x)}{1}$ (using L'Hospital's rule)

$\therefore \quad 4 = 2f(1)\cdot f'(1)$

$\therefore \quad f'(1) = 1.$

64. $\phi(x) = \displaystyle\int_0^{x^2} (t-1)\,dt - \int_0^{x} (t-1)\,dt$

$\therefore \quad \phi'(x) = (x^2-1)\cdot 2x - (x-1) = 2x^3 - 3x + 1$

$\therefore \quad \phi'(x) = (x-1)(2x^2+2x-1) = 2(x-1)\left(x^2+x-\dfrac{1}{2}\right)$

$$= 2(x-1)\left\{\left(x+\frac{1}{2}\right)^2 - \frac{3}{4}\right\}.$$

In $1 \le x \le 2$, $\phi'(x) \ge 0$. So, $\phi(x)$ is m.i. in [1, 2].Hence, $\phi(2)$ is the greatest value. Now, $\phi(2) = \displaystyle\int_2^4 (t-1)\,dt = \left[\frac{t^2}{2} - t\right]_2^4 = (8-4)-(2-2) = 4.$

65. Clearly, $\displaystyle\int_1^2 (1+\sin^4 x)(ax^2+bx+c)\,dx = 0.$

As $1+\sin^4 x > 0$ for all $x \in [1, 2]$, the above can be true only if ax^2+bx+c changes sign in [1, 2]. So, $ax^2+bx+c = 0$ has at least one real root between 1 and 2.

66. $f'(x) = x(x^2-3x+2) = x(x-1)(x-2)$. The sign scheme for $f'(x)$ is as below.

(−) (+) (−) (+)

0 1 2

$\therefore \quad f'(x) \le 0$ in $1 \le x \le 2$ and $f'(x) \ge 0$ in $2 \le x \le 3$.

$\therefore \quad f(x)$ is m.d. in [1 , 2] and m.i. in [2, 3]

$\therefore \quad \min f(x) = f(2) = \displaystyle\int_1^2 x(x^2-3x+2)\,dx = \left[\frac{x^4}{4} - x^3 + x^2\right]_1^2 = -\frac{1}{4}.$

$\max f(x) =$ the greatest among $\{f(1), f(3)\}$

$$f(1)=\int_1^1 x(x^2-3x+2)\,dx=0$$

$$f(3)=\int_1^3 x(x^2-3x+2)\,dx=\left[\frac{x^4}{4}-x^3+x^2\right]_1^3=2.$$

$\therefore$ $\max f(x)=2$. So, the range $=\left[-\frac{1}{4},2\right]$.

67. $A=\int_0^\pi \frac{\sin(\pi-x)}{\sin(\pi-x)+\cos(\pi-x)}\,dx=\int_0^\pi \frac{\sin x}{\sin x-\cos x}\,dx=B.$

Now, $A+B=\int_0^\pi \frac{2\sin^2 x}{\sin^2 x-\cos^2 x}\,dx=\int_0^\pi \frac{1-\cos 2x}{-\cos 2x}\,dx=\int_0^\pi (1-\sec 2x)\,dx$

$$=\pi-\left[\frac{\log(\sec 2x+\tan 2x)}{2}\right]_0^\pi=\pi.\text{ Hence, } A=B=\frac{\pi}{2}\cdot$$

68. Here $f'(x)\geq 0$ in $[a,b]$. So, $f(x)$ is m.i. Hence,

$f(a)\leq f(x)\leq f(b)$.

Therefore, $\int_a^b f(a)\,dx\leq\int_a^b f(x)\,dx\leq\int_a^b f(b)\,dx$

or $f(a)\cdot(b-a)\leq\int_a^b f(x)\,dx\leq f(b)\cdot(b-a)$

$\therefore$ $f(a)\leq\frac{1}{b-a}\int_a^b f(x)\,dx\leq f(b).$

69. $I=\int_0^\pi \frac{\sin n(\pi-x)}{\sin(\pi-x)}\,dx=\int_0^\pi \frac{(-1)^{n-1}\sin nx}{\sin x}\,dx=(-1)^{n-1}I.$

$\therefore$ if n is even, $I=-I$ $\Rightarrow I=0$.

70. $f(x)=\int_0^{x^2}\frac{dt}{(\log t)^2}-\int_0^x \frac{dt}{(\log t)^2}$

$\therefore$ $f'(x)=\frac{1}{(\log x^2)^2}\cdot 2x-\frac{1}{(\log x)^2}$

$\therefore$ $f'(x)=\left(\frac{x}{2}-1\right)\frac{1}{(\log x)^2}$

$\therefore$ in $(2,+\infty)$, $f'(x)>0$. So, $f(x)$ is m.i. in $(2,+\infty)$.

And in (0, 1), $f'(x) < 0$. So, $f(x)$ is m.d. in (0, 1).

71. $f'(x) = 3ax^2 + 2bx + c$.

For extremum at $x = 1, 5$, we have

$f'(1) = 0 = 3a + 2b + c$ and $f'(5) = 0 = 75a + 10b + c$

$$\int_{-1}^{1} f(x)\,dx = \int_{-1}^{1} (ax^3 + bx^2 + cx)\,dx = \left[a\frac{x^4}{4} + b\frac{x^3}{3} + c\frac{x^2}{2}\right]_{-1}^{1}$$

$$= b\left(\frac{1}{3} - \frac{-1}{3}\right) = \frac{2b}{3} = 6$$

$\therefore \quad b = 9$.

Then, $0 = 3a + 18 + c$ and $0 = 75a + 90 + c$

$\therefore \quad 0 = 72a + 72$

$\therefore \quad a = -1$ and so $c = -15$.

72. $f'(x) = |x - 1|$, which is continuous everywhere and differentiable everywhere except at $x = 1$.

□

11

Area

• Type 1 •

Choose the most appropriate option (a, b, c or d).

1. $P(\alpha, f(\alpha))$ and $Q(\beta, f(\beta))$ are ends of an arc in the first quadrant. The area bounded by the arc, ordinates through P and Q, and the x-axis is

(a) $\int_{f(\alpha)}^{f(\beta)} f^{-1}(y)dy$ (b) $\int_{\alpha}^{\beta} f^{-1}(y)dy$

(c) $\int_{\alpha}^{\beta} f(x)dx$ (d) $\int_{f(\alpha)}^{f(\beta)} f(x)\, dx$

2. The area bounded by the lines $y = |x - 2|$, $|x| = 3$ and $y = 0$ is

(a) 13 unit^2 (b) 5 unit^2

(c) 9 unit^2 (d) 7 unit^2

3. The area bounded by the curve $y = \sqrt{4 - x^2}$ and the line $y = 0$ is

(a) 4π (b) 2π

(c) π (d) $\frac{\pi}{2}$

4. The area bounded by the curve $y^2 = 4x$ and the double ordinate $x = 2$, is

(a) $\frac{4}{3}\sqrt{2}$ unit^2 (b) $\frac{8}{3}\sqrt{2}$ unit^2

(c) $\frac{16}{3}\sqrt{2}$ unit^2 (d) $\frac{32}{3}\sqrt{2}$ unit^2

5. The area bounded by the curve $x^2 = ky, k > 0$ and the line $y = 3$ is 12 unit^2. Then k is

(a) 3 (b) $3\sqrt{3}$

(c) $\frac{3}{4}$ (d) none of these

6. The area bounded by the curve $y = 2^x$, the x-axis and the y-axis is

(a) $\log_e 2$ (b) $\log_e 4$

(c) $\log_4 e$ (d) $\log_2 e$

7. The area of the portion enclosed by the curve $\sqrt{x} + \sqrt{y} = \sqrt{a}$ and the axes of reference is

(a) $\frac{a^2}{6}$ (b) a^2

(c) $\frac{a^2}{2}$ (d) $\frac{a^2}{4}$

8. The area bounded by the curve $x = \cos^{-1} y$ and the lines $|x| = 1$ is

(a) $\sin 1$ (b) $\cos 1$

(c) $2\sin 1$ (d) $2\cos 1$

9. The area bounded by the curve $y = \sqrt{x}$, the line $2y + 3 = x$ and the x-axis in the first quadrant is

(a) 9 (b) $\frac{27}{4}$

(c) 36 (d) 18

10. The area of the region bounded by the pairs of lines $y = |x - 1|$ and $y = 3 - |x|$ is

(a) 3 unit^2 (b) 4 unit^2

(c) 6 unit^2 (d) 2 unit^2

11. The ratio in which the area bounded by the curves $y^2 = 4x$ and $x^2 = 4y$ is divided by the line $x = 1$ is

(a) 64 : 49 (b) 15 : 34

(c) 15 : 49 (d) none of these

12. Let $f(x)$ be a continuous function such that the area bounded by the curve $y = f(x)$, the x-axis and the two ordinates $x = 0$ and $x = a$ is $\frac{a^2}{2} + \frac{a}{2}\sin a + \frac{\pi}{2}\cos a$. Then $f\left(\frac{\pi}{2}\right)$ is

(a) $\frac{1}{2}$ (b) $\frac{\pi^2}{8}+\frac{\pi}{4}$

(c) $\frac{\pi+1}{2}$ (d) $\frac{\pi}{2}$

13. The area bounded by the curve $f(x)=ce^x (c>0)$, the x-axis and the two ordinates $x=p$ and $x=q$ is proportional to

(a) $f(p)\cdot f(q)$ (b) $| f(p)-f(q) |$

(c) $f(p)+f(q)$ (d) $\sqrt{f(p) f(q)}$

14. The ordinate $x=a$ divides the area bounded by the x-axis, the curve $y=1+\frac{8}{x^2}$ and the ordinates $x=2$ and $x=4$ into equal parts. Then a is

(a) 3 (b) $\frac{7}{2}$

(c) $2\sqrt{2}$ (d) $\frac{5}{2}$

15. The area of the ellipse $\frac{(x+1)^2}{4}+y^2=1$ falling in the first quadrant is

(a) $\frac{1}{6}(4\pi-3\sqrt{3})$ (b) $\frac{1}{12}(4\pi-3\sqrt{3})$

(c) $\frac{\sqrt{3}}{2}$ (d) $\frac{1}{3}(\pi-\sqrt{3})$

Answers

1. c	**2.** a	**3.** b	**4.** c	**5.** a
6. d	**7.** a	**8.** c	**9.** a	**10.** b
11. c	**12.** a	**13.** b	**14.** c	**15.** b

Hints and Solutions to Selected Questions

2. Area $= \frac{1}{2} \cdot 1 \cdot 1 + \frac{1}{2} \cdot 5 \cdot 5 \cdot \frac{26}{2} = 13.$

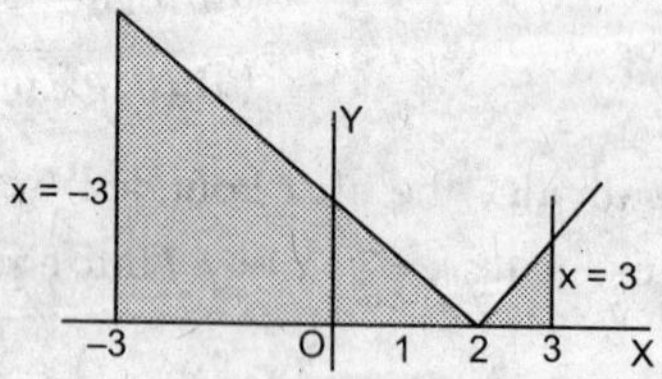

3. Area = area of the semicircle of radius 2.

5. Area $= 2\int_0^3 x\,dy = 2\int_0^3 \sqrt{ky}\,dy = \left[2\sqrt{k} \cdot \frac{y^{3/2}}{3/2}\right]_0^3$

$= \frac{4\sqrt{k}}{3} \cdot 3\sqrt{3}.$

$\therefore \quad 12 = 4\sqrt{k} \cdot \sqrt{3} \Rightarrow k = 3.$

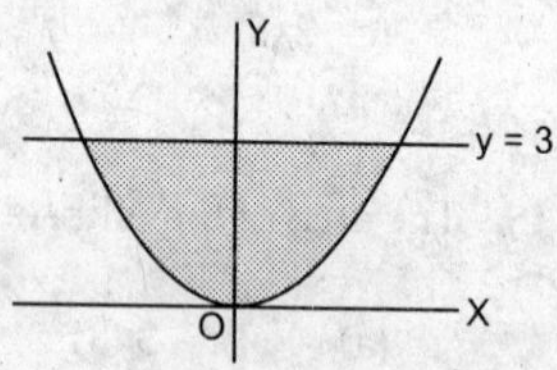

6. Area $= \int_{-\infty}^{0} 2^x\,dx = \left[\frac{2^x}{\log_e 2}\right]_{-\infty}^{0}$

$= \frac{1}{\log_e e} = \log_2 e.$

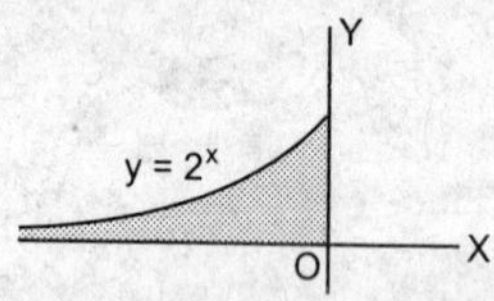

7. Area $= \int_0^a y\,dx$

$= \int_0^a (\sqrt{a} - \sqrt{x})^2 dx.$

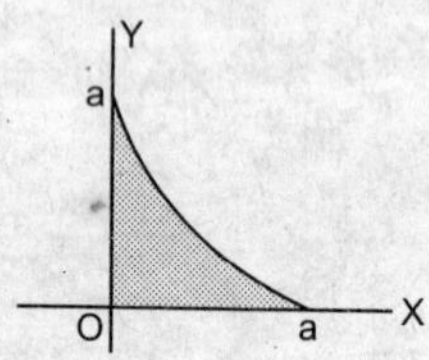

8. Area $= \int_{-1}^{1} \cos x\,dx = 2\int_0^1 \cos x\,dx$ (being an even function).

9. Solving $y = \sqrt{x}$ and $y = \frac{x}{2} - \frac{3}{2}$, $(x-1)(x-9) = 0.$

$\therefore \quad x = 9, 1.$ But y is positive, so $x \neq 1$.

Hence, $x = 9.$

$\therefore \quad$ Area $= \int_0^9 \sqrt{x}\,dx - \int_3^9 \left(\frac{1}{2}x - \frac{3}{2}\right)dx.$

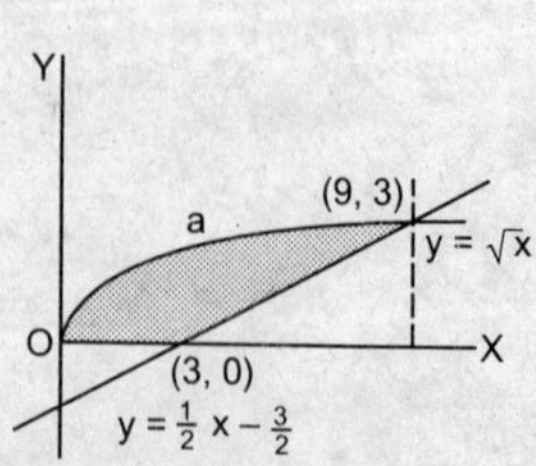

10. Area $= \sqrt{2} \times 2\sqrt{2}$.

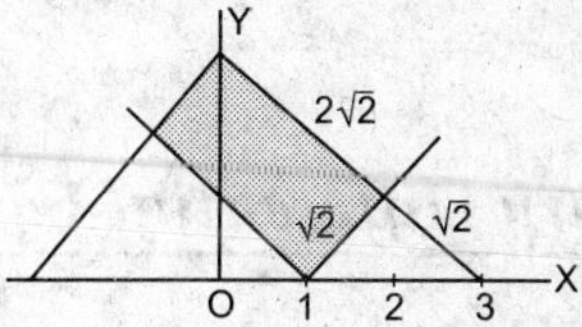

11. $A = \int_0^1 \left(\sqrt{4x} - \frac{x^2}{4}\right) dx = \frac{5}{4}$, $B = \int_1^4 \left(\sqrt{4x} - \frac{x^2}{4}\right) dx = \frac{49}{12}$.

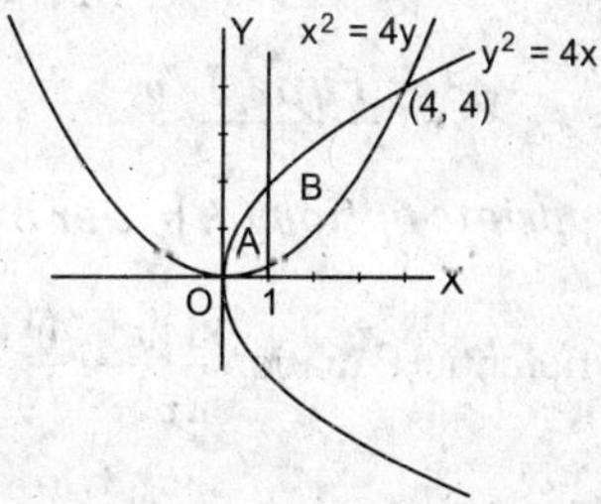

12. Here $\int_0^a f(x)\,dx = \frac{a^2}{2} + \frac{a}{2}\sin a + \frac{\pi}{2}\cos a$. Differentiating w.r.t. a,

$$f(a) = a + \frac{1}{2}(\sin a + a\cos a) - \frac{\pi}{2}\sin a. \text{ So, } f\left(\frac{\pi}{2}\right) = \frac{\pi}{2} + \frac{1}{2}(1+0) - \frac{\pi}{2} = \frac{1}{2}.$$

13. Area $= \int_q^p ce^x dx = \left[ce^x\right]_q^p = c(e^p - e^q) = f(p) - f(q)$.

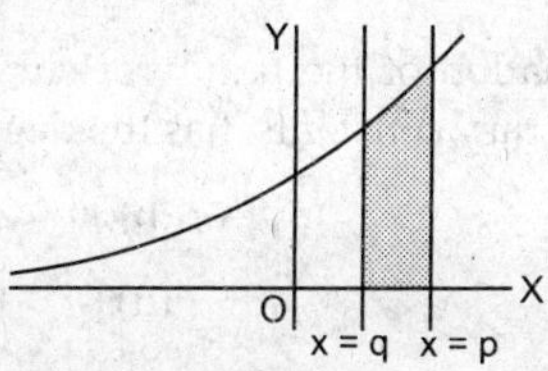

15. Area $= \int_0^1 y\,dx = \int_0^1 \sqrt{1 - \frac{(x+1)^2}{4}}\,dx = \frac{1}{2}\int_0^1 \sqrt{2^2 - (x+1)^2}\,dx$

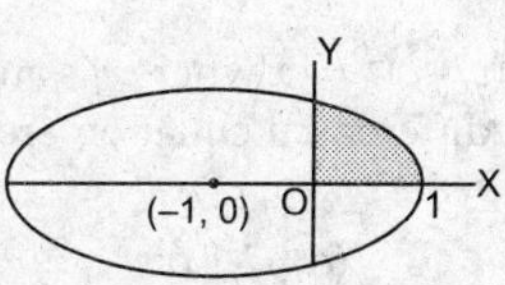

$$= \frac{1}{2}\left[\frac{2^2}{2}\sin^{-1}\frac{x+1}{2} + \frac{(x+1)\sqrt{2^2 - (x+1)^2}}{2}\right]_0^1$$

$$= \frac{1}{2}\left[\left(2 \cdot \frac{\pi}{2} + 0\right) - \left(2 \cdot \frac{\pi}{6} + \frac{1}{2} \cdot \sqrt{3}\right)\right]$$

□

12

Ordinary Differential Equations

• Type 1 •

Choose the most appropriate option (a, b, c or d).

1. The order of the differential equation $\frac{d^2y}{dx^2} - \left(\frac{dy}{dx}\right)^2 = 1$ is

(a) one (b) two
(c) four (d) zero

2. The degree of the differential equation $\sqrt{1 + \left(\frac{dy}{dx}\right)^2} = x^2$ is

(a) one (b) two
(c) half (d) four

3. The differential equation of the family of curves $y = e^x(A\cos x + B\sin x)$, where A, B are arbitrary constants, has the degree n and order m. Then

(a) $n = 2, m = 1$ (b) $n = 2, m = 2$
(c) $n = 1, m = 2$ (d) $n = 1, m = 1$

4. The general solution of a differential equation is $y = ae^{bx+c}$ where a, b, c are arbitrary constants. The order of the differential equation is

(a) 3 (b) 2
(c) 1 (d) none of these

5. The general solution of a differential equation is $(y + c)^2 = cx$ where c is an arbitrary constant. The order and degree of the differential equation are respectively

(a) 1, 2 (b) 2, 2 (c) 1, 1 (d) 2, 1

6. The degree and order of the differential equation of the family of all parabolas whose axis is the x-axis, are respectively

(a) 1, 2 (b) 3, 2

(c) 2, 3 (d) 2, 1

7. The order and degree of the differential equation of the family of circles touching the x-axis at the origin, are respectively

(a) 1, 1 (b) 1, 2

(c) 2, 1 (d) 2, 2

8. The order and degree of the differential equation of the family of ellipses having the same foci, are respectively

(a) 1, 1 (b) 2, 1

(c) 2, 2 (d) 1, 2

9. If $y(t)$ is a solution of the equation $(1+t)\frac{dy}{dt} - ty = 1$ and $y(0) = -1$ then $y(1)$ is

(a) $-\frac{1}{2}$ (b) $e + \frac{1}{2}$

(c) $e - \frac{1}{2}$ (d) $\frac{1}{2}$

10. The solution of $(x + \log y)dy + y\,dx = 0$ when $y(0) = 1$ is

(a) $y(x-1) + y\log y = 0$ (b) $y(x - 1 + \log y) + 1 = 0$

(c) $xy + y\log y + 1 = 0$ (d) none of these

11. The general solution of the equation $(1+y^2) + (x - e^{\tan^{-1}y})\frac{dy}{dx} = 0$ is

(a) $2xe^{\tan^{-1}y} = e^{2\tan^{-1}y} + k$ (b) $xe^{\tan^{-1}y} = \tan^{-1}y + k$

(c) $xe^{2\tan^{-1}y} = e^{\tan^{-1}y} + k$ (d) $x = 2 + ke^{-\tan^{-1}y}$

12. Let $\frac{df(x)}{dx} = \frac{e^{\sin x}}{x}$, $x > 0$. If $\int_1^4 \frac{3e^{\sin x^3}}{x}dx = f(k) - f(1)$ then one of the possible values of k is

(a) 16 (b) 63

(c) 64 (d) 15

13. If $x\frac{dy}{dx} + y = x \cdot \frac{f(x \cdot y)}{f'(x \cdot y)}$ then $f(x \cdot y)$ is equal to (k being an arbitrary constant)

(a) $ke^{x^2/2}$ (b) $ke^{y^2/2}$

(c) $ke^{xy/2}$ (d) none of these

14. The differential equation $\phi(x)dy = y\{\phi'(x) - y\}dx$ is changed in the form $df(x, y) = 0$. Then $f(x, y)$ is

(a) $\frac{1}{2}\phi(x) + y$ (b) $\frac{1}{y}\phi(x) - x$

(c) $\frac{1}{y}\phi(x) + x$ (d) $\frac{\phi(x)}{y}$

15. The solution of primitive integral equation $(x^2 + y^2)dy = xy \cdot dx$ is $y = y(x)$. If $y(1) = 1$ and $y(x_0) = e$ then x_0 is

(a) $\sqrt{2(e^2 - 1)}$ (b) $\sqrt{2(e^2 + 1)}$

(c) $\sqrt{3}e$ (d) $\sqrt{\frac{1}{2}(e^2 + 1)}$

Answers

1. b	**2.** b	**3.** c	**4.** b	**5.** a
6. a	**7.** a	**8.** d	**9.** a	**10.** b
11. a	**12.** c	**13.** a	**14.** b	**15.** c

Hints and Solutions to Selected Questions

2. On removal of the radical, $1+\left(\frac{dy}{dx}\right)^2 = x^4$. So, degree = 2.

3. $\frac{dy}{dx} = e^x(A\cos x + B\sin x) + e^x(-A\sin x + B\cos x) = y + e^x(-A\sin x + B\cos x)$.

$$\therefore \quad \frac{d^2y}{dx^2} = \frac{dy}{dx} + e^x(-A\sin x + B\cos x) + e^x(-A\cos x - B\sin x)$$

$$= \frac{dy}{dx} + \left(\frac{dy}{dx} - y\right) - y = 2\left(\frac{dy}{dx} - y\right)\cdot \text{ So, degree} = 1 \text{ and order} = 2.$$

4. $y = ae^{bx+c} = ae^c \cdot e^{bx} = ke^{bx}$. So, there are two arbitrary constants. Hence, the order of the differential equation will be 2.

6. The equation of the family is $y^2 = 4a(x-b)$ where a, b are arbitrary constants.

$$\therefore \quad 2y\frac{dy}{dx} = 4a \quad \text{or} \quad \left(\frac{dy}{dx}\right)^2 + y\frac{d^2y}{dx^2} = 0. \text{ So, degree} = 1 \text{ and order} = 2.$$

7. The equation of the family is $(x-c)^2 + y^2 = c^2$ or $x^2 + y^2 - 2cx = 0$

$$\text{or} \quad \frac{x^2+y^2}{x} = 2c \Rightarrow \frac{\left(2x + 2y\frac{dy}{dx}\right)x - (x^2+y^2)\cdot 1}{x^2} = 0.$$

So, degree = 1, order = 1.

8. The equation of the family is $\frac{x^2}{a^2+\lambda} + \frac{y^2}{b^2+\lambda} = 1$ because they have the same foci $(\pm\sqrt{a^2-b^2}, 0)$. On differentiation,

$$\frac{2x}{d^2+\lambda} + \frac{2y}{b^2+\lambda}\cdot\frac{dy}{dx} = 0 \quad \text{or} \quad \frac{x}{a^2+\lambda} + \frac{yp}{b^2+\lambda} = 0, \ \left(p = \frac{dy}{dx}\right)$$

$$\text{or} \quad x(b^2+\lambda) + yp(a^2+\lambda) = 0 \Rightarrow \lambda = \frac{-b^2x - a^2yp}{x+yp}$$

$$\therefore \quad \text{the differential equation is } \frac{x^2}{a^2 + \frac{-b^2x - a^2yp}{x+yp}} + \frac{y^2}{b^2 + \frac{-b^2x - a^2yp}{x+yp}} = 1$$

$$\text{or} \quad \frac{x(x+yp)}{a^2-b^2} + \frac{y(x+yp)}{(b^2-a^2)p} = 1. \text{ So, order} = 1 \text{ and degree} = 2.$$

9. $\frac{dy}{dx}-\frac{t}{1+t}y=\frac{1}{1+t}$ which is in the linear form.

$\text{IF}=e^{\int -\frac{t}{1+t}dt}=e^{-t+\log(1+t)}=(1+t)\,e^{-t}$

$\therefore\quad y(1+t)\,e^{-t}=\int\frac{1}{1+t}\cdot(1+t)\,e^{-t}dt=\int e^{-t}dt=-e^{-t}+c.$

When $t=0, y=-1$. So, $-1=-1+c$. Hence, $c=0$.

$\therefore\quad y(1+t)\,e^{-t}=-e^{-t}$. So, when $t=1$, $y\cdot 2\cdot e^{-1}=-e^{-1}$. $\therefore\quad y=-\frac{1}{2}\cdot$

11. $(1+y^2)\frac{dx}{dy}+x=e^{\tan^{-1}y}$

or $\frac{dx}{dy}+\frac{1}{1+y^2}\cdot x=\frac{e^{\tan^{-1}y}}{1+y^2}$ which is in the linear form.

$\text{IF}=e^{\int\frac{1}{1+y^2}dy}=e^{\tan^{-1}y}$. So, $x\cdot e^{\tan^{-1}y}=\int\frac{(e^{\tan^{-1}y})^2}{1+y^2}dy$

$\therefore\quad xe^{\tan^{-1}y}=\frac{(e^{\tan^{-1}x})^2}{2}+\frac{k}{2}$, (putting $e^{\tan^{-1}y}=z$).

12. $\frac{df(x)}{dx}=\frac{e^{\sin x}}{x}\Rightarrow\frac{df(x^3)}{d(x^3)}=\frac{e^{\sin x^3}}{x^3}\Rightarrow\frac{df(x^3)}{d(x^3)}\cdot\frac{d(x^3)}{dx}=\frac{e^{\sin x^3}}{x^3}\cdot 3x^2$

$\Rightarrow\quad\frac{df(x^3)}{dx}=\frac{3e^{\sin x^3}}{x}\Rightarrow\int_1^4\frac{3e^{\sin x^3}}{x}dx=\left[f(x^3)\right]_{x=1}^4=f(64)-f(1).$

13. $d(xy)=\frac{f(xy)}{f'(xy)}xdx$ or $\frac{f'(xy)}{f(xy)}d(xy)=x\,dx\Rightarrow\int\frac{f'(xy)}{f(xy)}d(xy)=\int x\,dx$

or $\log\{f(xy)\}=\frac{x^2}{2}+c.\quad\therefore\quad f(xy)=e^{x^2/2+c}=e^c\cdot e^{x^2/2}=ke^{x^2/2}.$

14. $\phi(x)dy=y\phi(x)dx-y^2dx$ or $\frac{y\phi'(x)dx-\phi(x)dy}{y^2}dy-dx=0$

or $d\left(\frac{\phi(x)}{y}\right)-dx=0$ or $d\left(\frac{\phi(x)}{y}-x\right)=0.$

15. Put $y=vx$. Solving, $\log y-\frac{1}{2}\cdot\frac{x^2}{y^2}=-\frac{1}{2}$, because at $x=1, y=1$.

For $x=x_0$, $y=e\Rightarrow 1-\frac{x_0^2}{2e^2}=-\frac{1}{2}\cdot$

□

Part 5

Vectors

1

Vectors and their Applications

• Type 1 •

Choose the most appropriate option (a, b, c or d).

1. ABCDEF is a regular hexagon where centre O is the origin. If the position vectors of A and B are $\vec{i}-\vec{j}+2\vec{k}$ and $2\vec{i}+\vec{j}+\vec{k}$ respectively then $\overrightarrow{BC}$ is equal to

(a) $\vec{i}-\vec{j}+2\vec{k}$ (b) $-\vec{i}+\vec{j}-2\vec{k}$

(c) $3\vec{i}+3\vec{j}-4\vec{k}$ (d) none of these

2. The position vectors of two vertices and the centroid of a triangle are $\vec{i}+\vec{j}$, $2\vec{i}-\vec{j}+\vec{k}$ and $\vec{k}$ respectively. The position vector of the third vertex of the triangle is

(a) $-3\vec{i}+2\vec{k}$ (b) $3\vec{i}-2\vec{k}$

(c) $\vec{i}+\frac{2}{3}\vec{k}$ (d) none of these

3. Let the position vectors of the points A, B, C be $\vec{i}+2\vec{j}+3\vec{k}$, $-\vec{i}-\vec{j}+8\vec{k}$ and $-4\vec{i}+4\vec{j}+6\vec{k}$ respectively. Then the ΔABC is

(a) right angled (b) equilateral

(c) isosceles (d) none of these

4. $\vec{a}, \vec{b}, \vec{c}$ are three vectors of which every pair is noncollinear. If the vector $\vec{a}+\vec{b}$ and $\vec{b}+\vec{c}$ are collinear with $\vec{c}$ and $\vec{a}$ respectively then $\vec{a}+\vec{b}+\vec{c}$ is

(a) a unit vector (b) the null vector

(c) equally inclined to $\vec{a}, \vec{b}, \vec{c}$ (d) none of these

5. If $\vec{r} = 3\vec{i} + 2\vec{j} - 5\vec{k}$, $\vec{a} = 2\vec{i} - \vec{j} + \vec{k}$, $\vec{b} = \vec{i} + 3\vec{j} - 2\vec{k}$ and $\vec{c} = -2\vec{i} + \vec{j} - 3\vec{k}$ such that $\vec{r} = \lambda\vec{a} + \mu\vec{b} + \nu\vec{c}$ then

(a) $\mu, \frac{\lambda}{2}, \nu$ are in AP (b) λ, μ, ν are in AP

(c) λ, μ, ν are in HP (d) μ, λ, ν are in GP

6. The position vectors of three points are $2\vec{a} - \vec{b} + 3\vec{c}$, $\vec{a} - 2\vec{b} + \lambda\vec{c}$ and $\mu\vec{a} - 5\vec{b}$ where $\vec{a}, \vec{b}, \vec{c}$ are noncoplanar vectors. The points are collinear when

(a) $\lambda = -2, \mu = \frac{9}{4}$ (b) $\lambda = -\frac{9}{4}, \mu = 2$

(c) $\lambda = \frac{9}{4}, \mu = -2$ (d) none of these

7. If $\vec{a} = \vec{i} + \vec{j} + \vec{k}$, $\vec{b} = 4\vec{i} + 3\vec{j} + 4\vec{k}$ and $\vec{c} = \vec{i} + \alpha\vec{j} + \beta\vec{k}$ are linearly dependent vectors and $|\vec{c}| = \sqrt{3}$ then

(a) $\alpha = 1, \beta = -1$ (b) $\alpha = 1, \beta = \pm 1$

(c) $\alpha = -1, \beta = \pm 1$ (d) $\alpha = \pm 1, \beta = 1$

8. Let $\vec{OA} = \vec{a}$ and $\vec{OB} = \vec{b}$. A vector along one of the bisectors of the angle $\angle AOB$ is

(a) $\vec{a} + \vec{b}$ (b) $\vec{a} - \vec{b}$

(c) $\frac{\vec{a}}{|\vec{a}|} + \frac{\vec{b}}{|\vec{b}|}$ (d) none of these

9. A vector has components $2p$ and 1 with respect to a rectangular cartesian system. The axes are rotated through an angle α about the origin in the anticlockwise sense. If the vector has components $p + 1$ and 1 with respect to the new system then

(a) $p = 1, -\frac{1}{3}$ (b) $p = 0$

(c) $p = -1, \frac{1}{3}$ (d) $p = 1, -1$

10. If $\vec{a}$ and $\vec{b}$ are two vectors of magnitude 2 inclined at an angle 60° then the angle between $\vec{a}$ and $\vec{a} + \vec{b}$ is

(a) 30° (b) 60°

(c) 45° (d) none of these

11. Let $|\vec{a}| = |\vec{b}| = |\vec{a}-\vec{b}| = 1$. Then the angle between $\vec{a}$ and $\vec{b}$ is

(a) $\frac{\pi}{6}$ (b) $\frac{\pi}{3}$

(c) $\frac{\pi}{4}$ (d) $\frac{\pi}{2}$

12. A vector of magnitude 4 which is equally inclined to the vectors $\vec{i}+\vec{j}, \vec{j}+\vec{k}$ and $\vec{k}+\vec{i}$ is

(a) $\frac{4}{\sqrt{3}}(\vec{i}-\vec{j}-\vec{k})$ (b) $\frac{4}{\sqrt{3}}(\vec{i}+\vec{j}-\vec{k})$

(c) $\frac{4}{\sqrt{3}}(\vec{i}+\vec{j}+\vec{k})$ (d) nonc of these

13. If $\vec{a}+\vec{b}=2\vec{i}$ and $2\vec{a}-\vec{b}=\vec{i}-\vec{j}$ then cosine of the angle between $\vec{a}$ and $\vec{b}$ is

(a) $\sin^{-1}\frac{4}{5}$ (b) $\cos^{-1}\frac{4}{5}$

(c) $\cos^{-1}\frac{3}{5}$ (d) none of these

14. Let $|\vec{a}| = 1, |\vec{b}| = \sqrt{2}, |\vec{c}| = \sqrt{3}$, and $\vec{a}\perp(\vec{b}+\vec{c}), \vec{b}\perp(\vec{c}+\vec{a})$ and $\vec{c}\perp(\vec{a}+\vec{b})$. Then $|\vec{a}+\vec{b}+\vec{c}|$ is

(a) $\sqrt{6}$ (b) 6

(c) $\sqrt{14}$ (d) none of these

15. $(\vec{a}\cdot\vec{i})\vec{i}+(\vec{a}\cdot\vec{j})\vec{j}+(\vec{a}\cdot\vec{k})\vec{k}$ is equal to

(a) $\vec{i}+\vec{j}+\vec{k}$ (b) $\vec{a}$

(c) $3\vec{a}$ (d) none of these

16. If $\hat{a}, \hat{b}$ are unit vectors such that $\hat{a}+\hat{b}$ is also a unit vector then the angle between the vectors $\hat{a}$ and $\hat{b}$ is

(a) $\frac{\pi}{6}$ (b) $\frac{\pi}{4}$

(c) $\frac{\pi}{3}$ (d) $\frac{2\pi}{3}$

17. If $\vec{a}\cdot\vec{i}=\vec{a}\cdot(\vec{i}+\vec{j})=\vec{a}\cdot(\vec{i}+\vec{j}+\vec{k})=1$ then $\vec{a}$ is

(a) $\vec{i}-\vec{j}$ (b) $\vec{i}$

(c) $\vec{j}$ (d) $\vec{k}$

18. $(\vec{a}\cdot\vec{i})^2+(\vec{a}\cdot\vec{j})^2+(\vec{a}\cdot\vec{k})^2$ is equal to

(a) $\vec{a}^2$ (b) 3

(c) $|\vec{a}\cdot(\vec{i}+\vec{j}+\vec{k})|^2$ (d) none of these

19. $|\vec{a}+\vec{b}|^2-|\vec{a}-\vec{b}|^2$ is equal to

(a) $4\vec{a}\cdot\vec{b}$ (b) 0

(c) $4|\vec{a}\cdot\vec{b}|$ (d) none of these

20. If a, b, c are the pth, qth, rth terms of an HP and

$$\vec{u}=(q-r)\vec{i}+(r-p)\vec{j}+(p-q)\vec{k},\ \vec{v}=\frac{\vec{i}}{a}+\frac{\vec{j}}{b}+\frac{\vec{k}}{c}$$

then

(a) $\vec{u}, \vec{v}$ are parallel vectors (b) $\vec{u}, \vec{v}$ are orthogonal vectors

(c) $\vec{u}\cdot\vec{v}=1$ (d) $\vec{u}\times\vec{v}=\vec{i}+\vec{j}+\vec{k}$

21. If $\vec{a}+\vec{b}\perp\vec{a}$ and $|\vec{b}|=\sqrt{2}\,|\vec{a}|$ then

(a) $(2\vec{a}+\vec{b})\parallel\vec{b}$ (b) $(2\vec{a}+\vec{b})\perp\vec{b}$

(c) $(2\vec{a}-\vec{b})\perp\vec{b}$ (d) $(2\vec{a}+\vec{b})\perp\vec{a}$

22. Let $\vec{a}=2\vec{i}+\vec{j}+\vec{k},\ \vec{b}=\vec{i}+2\vec{j}-\vec{k}$ and a unit vector $\vec{c}$ be coplanar. If $\vec{c}$ is perpendicular to $\vec{a}$ then $\vec{c}=$

(a) $\frac{1}{\sqrt{2}}(-\vec{j}+\vec{k})$ (b) $\frac{1}{\sqrt{3}}(-\vec{i}-\vec{j}-\vec{k})$

(c) $\frac{1}{\sqrt{5}}(\vec{i}-2\vec{j})$ (d) $\frac{1}{\sqrt{3}}(\vec{i}-\vec{j}-\vec{k})$

23. Let $\vec{\lambda}=\vec{a}\times(\vec{b}+\vec{c}), \vec{\mu}=\vec{b}\times(\vec{c}+\vec{a})$ and $\vec{\nu}=\vec{c}\times(\vec{a}+\vec{b})$. Then

(a) $\vec{\lambda}+\vec{\mu}=\vec{\nu}$ (b) $\vec{\lambda}, \vec{\mu}, \vec{\nu}$ are coplanar

(c) $\vec{\lambda}+\vec{v}=2\vec{\mu}$
(d) none of these

24. Let $\vec{a}, \vec{b}, \vec{c}$ be three unit vectors such that $|\vec{a}+\vec{b}+\vec{c}|=1$ and $\vec{a}\perp\vec{b}$. If $\vec{c}$ makes angles α, β with $\vec{a}, \vec{b}$ respectively then $\cos\alpha+\cos\beta$ is equal to

(a) $\frac{3}{2}$
(b) 1
(c) -1
(d) none of these

25. If $\vec{a}, \vec{b}, \vec{c}$ are three vectors of equal magnitude and the angle between each pair of vectors is $\frac{\pi}{3}$ such that $|\vec{a}+\vec{b}+\vec{c}|=\sqrt{6}$ then $|\vec{a}|$ is equal to

(a) 2
(b) -1
(c) 1
(d) $\frac{1}{3}\sqrt{6}$

26. If $|\vec{a}|=5$, $|\vec{a}-\vec{b}|=8$ and $|\vec{a}+\vec{b}|=10$ then $|\vec{b}|$ is

(a) 1
(b) $\sqrt{57}$
(c) 3
(d) none of these

27. If $\vec{a}$ and $\vec{b}$ are unit vectors and α is the angle between them then $\cos\frac{\alpha}{2}$ is equal to

(a) $\frac{1}{2}|\vec{a}+\vec{b}|$
(b) $\frac{1}{2}|\vec{a}-\vec{b}|$
(c) $|\vec{a}+\vec{b}|$
(d) none of these

28. If $|\vec{a}+\vec{b}|=|\vec{a}-\vec{b}|$ then

(a) $\vec{a}\parallel\vec{b}$
(b) $\vec{a}\perp\vec{b}$
(c) $|\vec{a}|=|\vec{b}|$
(d) none of these

29. Two vectors $\vec{a}=\vec{i}+\frac{\vec{j}}{\sqrt{3}}$ and $\vec{b}=\frac{\vec{i}}{\sqrt{3}}+\vec{j}$ are

(a) perpendicular to each other
(b) parallel to each other

(c) inclined to each other at an angle $\frac{\pi}{3}$

(d) inclined to each other at an angle $\frac{\pi}{6}$

30. Let $\vec{a} = 2\vec{i} - \vec{j} + \vec{k}$, $\vec{b} = \vec{i} + 2\vec{j} - \vec{k}$ and $\vec{c} = \vec{i} + \vec{j} - 2\vec{k}$. A vector in the plane of $\vec{b}$ and $\vec{c}$ whose projection on $\vec{a}$ has the magnitude $\sqrt{2/3}$ is

(a) $2\vec{i} + 3\vec{j} - 3\vec{k}$ (b) $2\vec{i} + 3\vec{j} + 3\vec{k}$

(c) $-2\vec{i} - \vec{j} + 5\vec{k}$ (d) $2\vec{i} + \vec{j} + 5\vec{k}$

31. ABC is an equilateral triangle of side a. The value of $\vec{AB} \cdot \vec{BC} + \vec{BC} \cdot \vec{CA} + \vec{CA} \cdot \vec{AB}$ is equal to

(a) $\frac{3a^2}{2}$ (b) $3a^2$

(c) $-\frac{3a^2}{2}$ (d) none of these

32. If $\vec{a} = \vec{i} + \vec{j}$, $\vec{b} = 2\vec{j} - \vec{k}$ and $\vec{r} \times \vec{a} = \vec{b} \times \vec{a}$, $\vec{r} \times \vec{b} = \vec{a} \times \vec{b}$ then $\frac{\vec{r}}{|\vec{r}|}$ is equal to

(a) $\frac{1}{\sqrt{11}}(\vec{i} + 3\vec{j} - \vec{k})$ (b) $\frac{1}{\sqrt{11}}(\vec{i} - 3\vec{j} + \vec{k})$

(c) $\frac{1}{\sqrt{3}}(\vec{i} - \vec{j} + \vec{k})$ (d) none of these

33. $(\vec{a} \times \vec{b})^2 + (\vec{a} \cdot \vec{b})^2$ is equal to

(a) 0 (b) $|\vec{a}|^2 |\vec{b}|^2$

(c) $(|\vec{a}| + |\vec{b}|)^2$ (d) 1

34. If $\vec{p}, \vec{q}$ are two noncollinear and nonzero vectors such that

$$(b - c)\, \vec{p} \times \vec{q} + (c - a)\, \vec{p} + (a - b)\, \vec{q} = 0,$$

where a, b, c are the lengths of the sides of a triangle, then the triangle is

(a) right angled (b) obtuse angled

(c) equilateral (d) isosceles

35. If $\vec{a}, \vec{b}, \vec{c}$ are any three vectors such that $(\vec{a} + \vec{b}) \cdot \vec{c} = (\vec{a} - \vec{b}) \cdot \vec{c} = 0$ then $(\vec{a} \times \vec{b}) \times \vec{c}$ is

(a) $\vec{0}$ (b) $\vec{a}$

(c) $\vec{b}$ (d) none of these

36. The unit vector perpendicular to both the vectors $\vec{a}=\vec{i}+\vec{j}+\vec{k}$ and $\vec{b}=2\vec{i}-\vec{j}+3\vec{k}$ and making an acute angle with the vector $\vec{k}$ is

(a) $-\frac{1}{\sqrt{26}}(4\vec{i}-\vec{j}-3\vec{k})$ (b) $\frac{1}{\sqrt{26}}(4\vec{i}-\vec{j}-3\vec{k})$

(c) $\frac{1}{\sqrt{26}}(4\vec{i}-\vec{j}+3\vec{k})$ (d) none of these

37. Let $\vec{a}=\vec{i}-2\vec{j}+3\vec{k}$, $\vec{b}=2\vec{i}+3\vec{j}-\vec{k}$ and $\vec{c}=\lambda\vec{i}+\vec{j}+(2\lambda-1)\vec{k}$. If $\vec{c}$ is parallel to the plane of the vectors $\vec{a}$ and $\vec{b}$ then λ is

(a) 1 (b) 0

(c) –1 (d) 2

38. Let $\vec{a}$ be a unit vector perpendicular to unit vectors $\vec{b}$ and $\vec{c}$ and if the angle between $\vec{b}$ and $\vec{c}$ be α then $\vec{b}\times\vec{c}$ is

(a) $\cos\alpha\,\vec{a}$ (b) $\operatorname{cosec}\alpha\,\vec{a}$

(c) $\sin\alpha\,\vec{a}$ (d) none of these

39. If $\vec{a}\cdot\vec{b}=0$ and $\vec{a}\times\vec{b}=\vec{0}$ then

(a) $\vec{a}\parallel\vec{b}$ (b) $\vec{a}\perp\vec{b}$

(c) $\vec{a}=\vec{0}$ or $\vec{b}=\vec{0}$ (d) none of these

40. The area of the parallelogram whose diagonals represent the vectors $3\vec{i}+\vec{j}-2\vec{k}$ and $\vec{i}-3\vec{j}+4\vec{k}$ is

(a) $10\sqrt{3}$ (b) $5\sqrt{3}$

(c) 8 (d) 4

41. $(\vec{r}\cdot\vec{i})(\vec{r}\times\vec{i})+(\vec{r}\cdot\vec{j})(\vec{r}\times\vec{j})+(\vec{r}\cdot\vec{k})(\vec{r}\times\vec{k})$ is equal to

(a) $3\vec{r}$ (b) $\vec{r}$

(c) $\vec{0}$ (d) none of these

42. Let $\vec{a}=\vec{i}+\vec{j}+\vec{k}$, $\vec{c}=\vec{j}-\vec{k}$. If $\vec{b}$ is a vector satisfying $\vec{a}\times\vec{b}=\vec{c}$ and $\vec{a}\cdot\vec{b}=3$ then $\vec{b}$ is

(a) $\frac{1}{3}(5\vec{i}+2\vec{j}+2\vec{k})$ (b) $\frac{1}{3}(5\vec{i}-2\vec{j}-2\vec{k})$

(c) $3\vec{i}-\vec{j}-\vec{k}$ (d) none of these

43. A unit vector perpendicular to the plane passing through the points whose position vectors are $\vec{i}-\vec{j}+2\vec{k}$, $2\vec{i}-\vec{k}$ and $2\vec{j}+\vec{k}$ is

(a) $2\vec{i}+\vec{j}+\vec{k}$ (b) $\frac{1}{\sqrt{6}}(2\vec{i}+\vec{j}+\vec{k})$

(c) $\frac{1}{\sqrt{6}}(\vec{i}+2\vec{j}+\vec{k})$ (d) none of these

44. Let $\vec{r}\times\vec{a}=\vec{b}\times\vec{a}$ and $\vec{r}\cdot\vec{c}=0$, where $\vec{a}\cdot\vec{b}\neq 0$. Then $\vec{r}$ is equal to

(a) $\vec{b}+t\,\vec{a}$ where t is a scalar (b) $\vec{b}-\frac{\vec{b}\cdot\vec{c}}{\vec{a}\cdot\vec{c}}\vec{a}$

(c) $\vec{a}-\vec{c}$ (d) none of these

45. For three vectors $\vec{u}, \vec{v}, \vec{w}$ which of the following expressions is not equal to any one of the remaining three options?

(a) $\vec{u}\cdot(\vec{v}\times\vec{w})$ (b) $(\vec{v}\times\vec{w})\cdot\vec{u}$

(c) $\vec{v}\cdot(\vec{u}\times\vec{w})$ (d) $(\vec{u}\times\vec{v})\cdot\vec{w}$

46. For three noncoplanar vectors $\vec{a}, \vec{b}, \vec{c}$ the relation

$$|\vec{a}\times\vec{b}\cdot\vec{c}| = |\vec{a}|\,|\vec{b}|\,|\vec{c}|$$

holds if and only if

(a) $\vec{b}\cdot\vec{c}=\vec{c}\cdot\vec{a}=0$ (b) $\vec{a}\cdot\vec{b}=\vec{b}\cdot\vec{c}=0$

(c) $\vec{a}\cdot\vec{b}=\vec{b}\cdot\vec{c}=\vec{c}\cdot\vec{a}=0$ (d) $\vec{c}\cdot\vec{a}=\vec{a}\cdot\vec{b}=0$

47. $[\vec{a}+\vec{b}\quad \vec{b}+\vec{c}\quad \vec{c}+\vec{a}]$ is equal to

(a) $2[\vec{a}\ \vec{b}\ \vec{c}]$ (b) $3[\vec{a}\ \vec{b}\ \vec{c}]$

(c) $[\vec{a}\ \vec{b}\ \vec{c}]$ (d) 0

48. $[\vec{a}-\vec{b}\quad \vec{b}-\vec{c}\quad \vec{c}-\vec{a}]$ is equal to

(a) $2[\vec{a}\ \vec{b}\ \vec{c}]$ (b) $[\vec{a}\ \vec{b}\ \vec{c}]$

(c) 0 (d) none of these

49. Let $\vec{a}, \vec{b}, \vec{c}$ be three unit vectors and $\vec{a} \cdot \vec{b} = \vec{a} \cdot \vec{c} = 0$. If the angle between $\vec{b}$ and $\vec{c}$ is $\frac{\pi}{3}$ then $|[\vec{a}\ \vec{b}\ \vec{c}]|$ is equal to

(a) $\frac{\sqrt{3}}{2}$ (b) $\frac{1}{2}$

(c) 1 (d) none of these

50. Let a, b, c be three distinct positive real numbers. If $\vec{p}, \vec{q}, \vec{r}$ lie in a plane, where $\vec{p} = a\vec{i} - a\vec{j} + b\vec{k}$, $\vec{q} = \vec{i} + \vec{k}$ and $\vec{r} = c\vec{i} + c\vec{j} + b\vec{k}$, then b is

(a) the AM of a, c (b) the GM of a, c

(c) the HM of a, c (d) equal to 0

51. Which of the following is not equal to $[\vec{a}\ \vec{b}\ \vec{c}]$?

(a) $\vec{a} \cdot \vec{b} \times \vec{c}$ (b) $\vec{c} \times \vec{a} \cdot \vec{b}$

(c) $\vec{b} \cdot \vec{a} \times \vec{c}$ (d) $\vec{c} \cdot \vec{a} \times \vec{b}$

52. If $[\vec{a}\ \vec{b}\ \vec{c}] = 1$ then $\dfrac{\vec{a} \cdot \vec{b} \times \vec{c}}{\vec{c} \times \vec{a} \cdot \vec{b}} + \dfrac{\vec{b} \cdot \vec{c} \times \vec{a}}{\vec{a} \times \vec{b} \cdot \vec{c}} + \dfrac{\vec{c} \cdot \vec{a} \times \vec{b}}{\vec{b} \times \vec{c} \cdot \vec{a}}$ is equal to

(a) 3 (b) 1

(c) –1 (d) none of these

53. $\vec{a}, \vec{b}, \vec{c}$ are noncoplanar vectors and $\vec{p}, \vec{q}, \vec{r}$ are defined as

$$\vec{p} = \frac{\vec{b} \times \vec{c}}{[\vec{b}\ \vec{c}\ \vec{a}]},\ \vec{q} = \frac{\vec{c} \times \vec{a}}{[\vec{c}\ \vec{a}\ \vec{b}]},\ \vec{r} = \frac{\vec{a} \times \vec{b}}{[\vec{a}\ \vec{b}\ \vec{c}]}.$$

$(\vec{a} + \vec{b}) \cdot \vec{p} + (\vec{b} + \vec{c}) \cdot \vec{q} + (\vec{c} + \vec{a}) \cdot \vec{r}$ is equal to

(a) 0 (b) 1

(c) 2 (d) 3

54. If $\vec{a}, \vec{b}, \vec{c}$ are three noncoplanar vectors represented by concurrent edges of a parallelepiped of volume 4 then

$$(\vec{a} + \vec{b}) \cdot (\vec{b} \times \vec{c}) + (\vec{b} + \vec{c}) \cdot (\vec{c} \times \vec{a}) + (\vec{c} + \vec{a}) \cdot (\vec{a} \times \vec{b})$$

is equal to

(a) 12 (b) 4

(c) ±12 (d) 0

55. If $\vec{a}, \vec{b}, \vec{c}$ are three noncoplanar nonzero vectors then

$$(\vec{a}\cdot\vec{a})\vec{b}\times\vec{c}+(\vec{a}\cdot\vec{b})\vec{c}\times\vec{a}+(\vec{a}\cdot\vec{c})\vec{a}\times\vec{b}$$

is equal to

(a) $[\vec{b}\ \vec{c}\ \vec{a}]\vec{a}$ (b) $[\vec{c}\ \vec{a}\ \vec{b}]\vec{b}$

(c) $[\vec{a}\ \vec{b}\ \vec{c}]\vec{c}$ (d) none of these

56. Let $\vec{r}$ be a vector perpendicular to $\vec{a}+\vec{b}+\vec{c}$, where $[\vec{a}\ \vec{b}\ \vec{c}]=2$. If $\vec{r}=l(\vec{b}\times\vec{c})+m(\vec{c}\times\vec{a})+n(\vec{a}\times\vec{b})$ then $l+m+n$ is

(a) 2 (b) 1

(c) 0 (d) none of these

57. If $\vec{a}, \vec{b}, \vec{c}$ are any three vectors in space then $(\vec{c}+\vec{b})\times(\vec{c}+\vec{a})\cdot(\vec{c}+\vec{b}+\vec{a})$ is equal to

(a) $3[\vec{a}\ \vec{b}\ \vec{c}]$ (b) 0

(c) $[\vec{a}\ \vec{b}\ \vec{c}]$ (d) none of these

58. If $\vec{a}, \vec{b}, \vec{c}$ are three noncoplanar vectors then $[\vec{a}+\vec{b}+\vec{c}\quad \vec{a}-\vec{c}\quad \vec{a}-\vec{b}]$ is equal to

(a) 0 (b) $[\vec{a}\ \vec{b}\ \vec{c}]$

(c) $-3[\vec{a}\ \vec{b}\ \vec{c}]$ (d) $2[\vec{a}\ \vec{b}\ \vec{c}]$

59. $[\vec{a}\quad \vec{b}+\vec{c}\quad \vec{a}+\vec{b}+\vec{c}]$ is equal to

(a) 0 (b) $2[\vec{a}\ \vec{b}\ \vec{c}]$

(c) $[\vec{a}\ \vec{b}\ \vec{c}]$ (d) none of these

60. If $\vec{a}, \vec{b}$ are nonzero and noncollinear vectors then

$$[\vec{a}\ \vec{b}\ \vec{i}]\vec{i}+[\vec{a}\ \vec{b}\ \vec{j}]\vec{j}+[\vec{a}\ \vec{b}\ \vec{k}]\vec{k}$$

is equal to

(a) $\vec{a}+\vec{b}$ (b) $\vec{a}\times\vec{b}$

(c) $\vec{a}-\vec{b}$ (d) $\vec{b}\times\vec{a}$

61. The three concurrent edges of a parallelepiped represent the vectors $\vec{a}, \vec{b}, \vec{c}$ such that $[\vec{a}\ \vec{b}\ \vec{c}]=\lambda$. Then the volume of the parallelepiped whose three concurrent edges are the three concurrent diagonals of three faces of the given parallelepiped is

(a) 2λ (b) 3λ

(c) λ (d) none of these

62. $\vec{i}\times(\vec{a}\times\vec{i})+\vec{j}\times(\vec{a}\times\vec{j})+\vec{k}\times(\vec{a}\times\vec{k})$ is equal to

(a) $2\vec{a}$ (b) $3\vec{a}$

(c) $\vec{0}$ (d) none of these

63. Let $\vec{a}, \vec{b}$ and $\vec{c}$ be three vectors having magnitudes 1, 1 and 2 respectively. If $\vec{a}\times(\vec{a}\times\vec{c})+\vec{b}=\vec{0}$, the acute angle between $\vec{a}$ and $\vec{c}$ is

(a) $\dfrac{\pi}{3}$ (b) $\dfrac{\pi}{4}$

(c) $\dfrac{\pi}{6}$ (d) none of these

64. If $\vec{b}$ is a unit vector then $(\vec{a}\cdot\vec{b})\vec{b}+\vec{b}\times(\vec{a}\times\vec{b})$ is equal to

(a) $\vec{a}^{\,2}\vec{b}$ (b) $(\vec{a}\cdot\vec{b})\vec{a}$

(c) $\vec{a}$ (d) none of these

65. Let $\vec{a}, \vec{b}, \vec{c}$ be three unit vectors such that $\vec{a}\times(\vec{b}\times\vec{c})=\dfrac{\vec{b}+\vec{c}}{\sqrt{2}}$ and the angles between $\vec{a}, \vec{c}$ and $\vec{a}, \vec{b}$ be α and β respectively then

(a) $\alpha=\dfrac{3\pi}{4}, \beta=\dfrac{\pi}{4}$ (b) $\alpha=\dfrac{\pi}{4}, \beta=\dfrac{7\pi}{4}$

(c) $\alpha=\dfrac{\pi}{4}, \beta=\dfrac{3\pi}{4}$ (d) none of these

66. Let $\vec{p}, \vec{q}, \vec{r}$ be three mutually perpendicular vectors of the same magnitude. If a vector $\vec{x}$ satisfies the equation

$$\vec{p}\times\{(\vec{x}-\vec{q})\times\vec{p}\}+\vec{q}\times\{(\vec{x}-\vec{r})\times\vec{q}\}+\vec{r}\times\{(\vec{x}-\vec{p})\times\vec{r}\}=\vec{0}$$

then $\vec{x}$ is given by

(a) $\dfrac{1}{2}(\vec{p}+\vec{q}-2\vec{r})$ (b) $\dfrac{1}{2}(\vec{p}+\vec{q}+\vec{r})$

(c) $\dfrac{1}{3}(\vec{p}+\vec{q}+\vec{r})$ (d) $\dfrac{1}{3}(2\vec{p}+\vec{q}+\vec{r})$

67. If $\cdot$ and $\times$ represent dot product and cross product respectively then which of the following is meaningless?

(a) $(\vec{a}\times\vec{b})\cdot(\vec{c}\times\vec{d})$ (b) $(\vec{a}\times\vec{b})\times(\vec{c}\times\vec{d})$

(c) $(\vec{a}\cdot\vec{b})(\vec{c}\times\vec{d})$ (d) $(\vec{a}\cdot\vec{b})\times(\vec{c}\times\vec{d})$

68. $(\vec{a}\times\vec{i})^2+(\vec{a}\times\vec{j})^2+(\vec{a}\times\vec{k})^2$ is equal to

(a) $\vec{a}^2$ (b) $3\vec{a}^2$

(c) $2\vec{a}^2$ (d) none of these

69. If $\vec{a}\parallel\vec{b}\times\vec{c}$ then $(\vec{a}\times\vec{b})\cdot(\vec{a}\times\vec{c})$ is equal to

(a) $\vec{a}^2(\vec{b}\cdot\vec{c})$ (b) $\vec{b}^2(\vec{a}\cdot\vec{c})$

(c) $\vec{c}^2(\vec{a}\cdot\vec{b})$ (d) none of these

70. If $\vec{a},\vec{b},\vec{c}$ are noncoplanar nonzero vectors then

$$(\vec{a}\times\vec{b})\times(\vec{a}\times\vec{c})+(\vec{b}\times\vec{c})\times(\vec{b}\times\vec{a})+(\vec{c}\times\vec{a})\times(\vec{c}\times\vec{b})$$

is equal to

(a) $[\vec{a}\ \vec{b}\ \vec{c}]^2(\vec{a}+\vec{b}+\vec{c})$ (b) $[\vec{a}\ \vec{b}\ \vec{c}](\vec{a}+\vec{b}+\vec{c})$

(c) $\vec{0}$ (d) none of these

71. If $\vec{a},\vec{b},\vec{c}$ are three noncoplanar nonzero vectors and $\vec{r}$ is any vector in space then $(\vec{a}\times\vec{b})\times(\vec{r}\times\vec{c})+(\vec{b}\times\vec{c})\times(\vec{r}\times\vec{a})+(\vec{c}\times\vec{a})\times(\vec{r}\times\vec{b})$ is equal to

(a) $2[\vec{a}\ \vec{b}\ \vec{c}]\vec{r}$ (b) $3[\vec{a}\ \vec{b}\ \vec{c}]\vec{r}$

(c) $[\vec{a}\ \vec{b}\ \vec{c}]\vec{r}$ (d) none of these

72. Let $\vec{a},\vec{b},\vec{c}$ be three unit vectors of which $\vec{b}$ and $\vec{c}$ are nonparallel. Let the angle between $\vec{a}$ and $\vec{b}$ be α and that between $\vec{a}$ and $\vec{c}$ be β. If $\vec{a}\times(\vec{b}\times\vec{c})=\frac{1}{2}\vec{b}$ then

(a) $\alpha=\frac{\pi}{3},\beta=\frac{\pi}{2}$ (b) $\alpha=\frac{\pi}{2},\beta=\frac{\pi}{3}$

(c) $\alpha=\frac{\pi}{6},\beta=\frac{\pi}{3}$ (d) none of these

73. Let $\vec{a} = 2\vec{i} + \vec{j} - 2\vec{k}$ and $\vec{b} = \vec{i} + \vec{j}$. If $\vec{c}$ is a vector such that $\vec{a} \cdot \vec{c} = |\vec{c}|$, $|\vec{c} - \vec{a}| = 2\sqrt{2}$ and the angle between $\vec{a} \times \vec{b}$ and $\vec{c}$ is 30° then $|(\vec{a} \times \vec{b}) \times \vec{c}|$ is equal to

(a) $\frac{2}{3}$ (b) $\frac{3}{2}$

(c) 2 (d) 3

74. Let $\vec{a}$ and $\vec{b}$ be two noncollinear unit vectors. If $\vec{u} = \vec{a} - (\vec{a} \cdot \vec{b})\vec{b}$ and $\vec{v} = \vec{a} \times \vec{b}$ then $|\vec{v}|$ is

(a) $|\vec{u}|$ (b) $|\vec{u}| + |\vec{u} \cdot \vec{a}|$

(c) $|\vec{u}| + |\vec{u} \cdot \vec{b}|$ (d) $|\vec{u}| + \vec{u} \cdot (\vec{a} + \vec{b})$

75. If $\vec{a}, \vec{b}, \vec{c}$ be three vectors such that $[\vec{a}\ \vec{b}\ \vec{c}] = 4$ then $[\vec{a} \times \vec{b}\ \ \vec{b} \times \vec{c}\ \ \vec{c} \times \vec{a}]$ is equal to

(a) 8 (b) 16

(c) 64 (d) none of these

76. If $\vec{d}$ is a unit vector such that $\vec{d} = \lambda \vec{b} \times \vec{c} + \mu \vec{c} \times \vec{a} + \nu \vec{a} \times \vec{b}$ then

$$|(\vec{d} \cdot \vec{a})(\vec{b} \times \vec{c}) + (\vec{d} \cdot \vec{b})(\vec{c} \times \vec{a}) + (\vec{d} \cdot \vec{c})(\vec{a} \times \vec{b})|$$

is equal to

(a) $|[\vec{a}\ \vec{b}\ \vec{c}]|$ (b) 1

(c) $3|[\vec{a}\ \vec{b}\ \vec{c}]|$ (d) none of these

77. $\vec{a} \times (\vec{b} \times \vec{c})$, $\vec{b} \times (\vec{c} \times \vec{a})$ and $\vec{c} \times (\vec{a} \times \vec{b})$ are

(a) linearly dependent (b) equal vectors

(c) parallel vectors (d) none of these

78. $[\vec{b}\ \vec{c}\ \vec{b} \times \vec{c}] + (\vec{b} \cdot \vec{c})^2$ is equal to

(a) $|\vec{b}|^2 |\vec{c}|^2$ (b) $(\vec{b} + \vec{c})^2$

(c) $|\vec{b}|^2 + |\vec{c}|^2$ (d) none of these

79. If the vectors $\vec{a}, \vec{b}, \vec{c}$ and $\vec{d}$ are coplanar then $(\vec{a} \times \vec{b}) \times (\vec{c} \times \vec{d})$ is equal to

(a) $\vec{a} + \vec{b} + \vec{c} + \vec{d}$ (b) $\vec{0}$

(c) $\vec{a} + \vec{b} = \vec{c} + \vec{d}$ (d) none of these

80. If $\vec{a} \times (\vec{a} \times \vec{b}) = \vec{b} \times (\vec{b} \times \vec{c})$ and $\vec{a} \cdot \vec{b} \neq 0$ then$[\vec{a}\ \vec{b}\ \vec{c}]$ is equal to

(a) 0 (b) 1

(c) 2 (d) none of these

81. Let $\overrightarrow{OA} = \vec{a}$, $\overrightarrow{OB} = 10\vec{a} + 2\vec{b}$ and $\overrightarrow{OC} = \vec{b}$, where O, A and C are noncollinear points. Let p denote the area of the quadrilateral OABC, and q denote the area of the parallelogram with OA and OC as adjacent sides. Then p/q is equal to

(a) 4 (b) 6

(c) $\frac{1}{2} \frac{|\vec{a} - \vec{b}|}{|\vec{a}|}$ (d) none of these

82. The position vectors of the vertices A, B, C of a triangle are $\vec{i} - \vec{j} - 3\vec{k}$, $2\vec{i} + \vec{j} - 2\vec{k}$ and $-5\vec{i} + 2\vec{j} - 6\vec{k}$ respectively. The length of the bisector AD of the angle BAC where D is on the line segment BC, is

(a) $\frac{15}{2}$ (b) $\frac{1}{4}$

(c) $\frac{11}{2}$ (d) none of these

83. P is a point on the line through the point A whose position vector is $\vec{a}$ and the line is parallel to the vector $\vec{b}$. If $PA = 6$, the position vector of P is

(a) $\vec{a} + 6\vec{b}$ (b) $\vec{a} + \frac{6}{|\vec{b}|} \vec{b}$

(c) $\vec{a} - 6\vec{b}$ (d) $\vec{b} + \frac{6}{|\vec{a}|} \vec{a}$

84. The coplanar points A, B, C, D are $(2 - x, 2, 2)$, $(2, 2 - y, 2)$, $(2, 2, 2 - z)$ and $(1, 1, 1)$ respectively. Then

(a) $\frac{1}{x} + \frac{1}{y} + \frac{1}{z} = 1$ (b) $x + y + z = 1$

(c) $\frac{1}{1-x} + \frac{1}{1-y} + \frac{1}{1-z} = 1$ (d) none of these

85. Let $\overrightarrow{AB} = 3\vec{i} + \vec{j} - \vec{k}$ and $\overrightarrow{AC} = \vec{i} - \vec{j} + 3\vec{k}$. If the point P on the line segment BC is equidistant from AB and AC then $\overrightarrow{AP}$ is

(a) $2\vec{i} - \vec{k}$ (b) $\vec{i} - 2\vec{k}$

(c) $2\vec{i}+\vec{k}$ (d) none of these

86. The cosine of the angle between two diagonals of a cube is

(a) $\frac{1}{3}$ (b) $\frac{2\sqrt{2}}{3}$

(c) $\frac{1}{2}$ (d) none of these

87. If $\vec{AB}=\vec{b}$ and $\vec{AC}=\vec{c}$ then the length of the perpendicular from A to the line BC is

(a) $\frac{|\vec{b}\times\vec{c}|}{|\vec{b}+\vec{c}|}$ (b) $\frac{|\vec{b}\times\vec{c}|}{|\vec{b}-\vec{c}|}$

(c) $\frac{1}{2}\frac{|\vec{b}\times\vec{c}|}{|\vec{b}-\vec{c}|}$ (d) none of these

88. The distance of the point (1, 1, 1) from the plane passing through the points (2, 1, 1), (1, 2, 1) and (1, 1, 2) is

(a) $\frac{1}{\sqrt{3}}$ (b) 1

(c) $\sqrt{3}$ (d) none of these

89. The projection of the vector $\vec{i}+\vec{j}+\vec{k}$ on the line whose vector equation is $\vec{r}=(3+t)\vec{i}+(2t-1)\vec{j}+3t\vec{k}$, t being the scalar parameter, is

(a) $\frac{1}{\sqrt{14}}$ (b) 6

(c) $\frac{6}{\sqrt{14}}$ (d) none of these

90. If the vertices of a tetrahedron have the position vectors $\vec{0}, \vec{i}+\vec{j}, 2\vec{j}-\vec{k}$ and $\vec{i}+\vec{k}$ then the volume of the tetrahedron is

(a) $\frac{1}{6}$ (b) 1

(c) 2 (d) none of these

• Type 2 •

Choose the correct options. One or more options may be correct.

91. A line passes through the points whose position vectors are $\vec{i}+\vec{j}-2\vec{k}$ and $\vec{i}-3\vec{j}+\vec{k}$. The position vector of a point on it at a unit distance from the first point is

(a) $\frac{1}{5}(6\vec{i}+\vec{j}-7\vec{k})$ (b) $\frac{1}{5}(4\vec{i}+9\vec{j}-13\vec{k})$

(c) $\vec{i}-4\vec{j}+3\vec{k}$ (d) none of these

92. A vector of magnitude 2 along a bisector of the angle between the two vectors $2\vec{i}-2\vec{j}+\vec{k}$ and $\vec{i}+2\vec{j}-2\vec{k}$ is

(a) $\frac{2}{\sqrt{10}}(3\vec{i}-\vec{k})$ (b) $\frac{1}{\sqrt{26}}(\vec{i}-4\vec{j}+3\vec{k})$

(c) $\frac{2}{\sqrt{26}}(\vec{i}-4\vec{j}+3\vec{k})$ (d) none of these

93. A unit vector coplanar with $\vec{i}+\vec{j}+2\vec{k}$ and $\vec{i}+2\vec{j}+\vec{k}$, and perpendicular to $\vec{i}+\vec{j}+\vec{k}$, is

(a) $\frac{1}{\sqrt{2}}(-\vec{j}+\vec{k})$ (b) $\frac{1}{\sqrt{2}}(\vec{k}-\vec{i})$

(c) $\frac{1}{\sqrt{2}}(\vec{i}-\vec{k})$ (d) $\frac{1}{\sqrt{2}}(\vec{j}-\vec{k})$

94. A unit vector which is equally inclined to the vectors $\vec{i}$, $\frac{-2\vec{i}+\vec{j}+2\vec{k}}{3}$ and $\frac{-4\vec{j}-3\vec{k}}{5}$ is

(a) $\frac{1}{\sqrt{51}}(-\vec{i}+5\vec{j}-5\vec{k})$ (b) $\frac{1}{\sqrt{51}}(\vec{i}+5\vec{j}-5\vec{k})$

(c) $\frac{1}{\sqrt{51}}(\vec{i}+5\vec{j}+5\vec{k})$ (d) $\frac{1}{\sqrt{51}}(\vec{i}-5\vec{j}+5\vec{k})$

95. If $|\vec{a}|=4$, $|\vec{b}|=2$ and the angle between $\vec{a}$ and $\vec{b}$ is $\frac{\pi}{6}$ then $(\vec{a}\times\vec{b})^2$ is equal to

(a) 48 (b) 16

(c) $\vec{a}^2$ (d) none of these

96. Three points whose position vectors are $\vec{a}, \vec{b}, \vec{c}$ will be collinear if

(a) $\lambda\vec{a} + \mu\vec{b} = (\lambda + \mu)\vec{c}$ (b) $\vec{a} \times \vec{b} + \vec{b} \times \vec{c} + \vec{c} \times \vec{a} = \vec{0}$

(c) $[\vec{a}\ \vec{b}\ \vec{c}] = 0$ (d) none of these

97. Let $\vec{b} = 4\vec{i} + 3\vec{j}$. Let $\vec{c}$ be a vector perpendicular to $\vec{b}$ and it lies in the x-y plane. A vector in the x-y plane having projections 1 and 2 along $\vec{b}$ and $\vec{c}$ is

(a) $2\vec{i} - \vec{j}$ (b) $\vec{i} - 2\vec{j}$

(c) $\frac{1}{5}(-2\vec{i} + 11\vec{j})$ (d) none of these

98. If $\vec{a}, \vec{b}, \vec{c}$ are noncoplanar nonzero vectors and $\vec{r}$ is any vector in space then $[\vec{b}\ \vec{c}\ \vec{r}]\vec{a} + [\vec{c}\ \vec{a}\ \vec{r}]\vec{b} + [\vec{a}\ \vec{b}\ \vec{r}]\vec{c}$ is equal to

(a) $3[\vec{a}\ \vec{b}\ \vec{c}]\vec{r}$ (b) $[\vec{a}\ \vec{b}\ \vec{c}]\vec{r}$

(c) $[\vec{b}\ \vec{c}\ \vec{a}]\vec{r}$ (d) none of these

99. If $\vec{a}, \vec{b}, \vec{c}$ are noncoplanar vectors such that $\vec{b} \times \vec{c} = \vec{a}$, $\vec{a} \times \vec{b} = \vec{c}$ and $\vec{c} \times \vec{a} = \vec{b}$ then

(a) $|\vec{a}| = 1$ (b) $|\vec{b}| = 1$

(c) $|\vec{a}| + |\vec{b}| + |\vec{c}| = 3$ (d) none of these

100. Let $\vec{a}, \vec{b}, \vec{c}$ be noncoplanar vectors and $\vec{p} = \dfrac{\vec{b} \times \vec{c}}{[\vec{a}\ \vec{b}\ \vec{c}]}$, $\vec{q} = \dfrac{\vec{c} \times \vec{a}}{[\vec{b}\ \vec{c}\ \vec{a}]}$, $\vec{r} = \dfrac{\vec{a} \times \vec{b}}{[\vec{c}\ \vec{a}\ \vec{b}]}$ then

(a) $\vec{p} \cdot \vec{a} = 1$ (b) $\vec{p} \cdot \vec{a} + \vec{q} \cdot \vec{b} + \vec{r} \cdot \vec{c} = 3$

(c) $\vec{p} \cdot \vec{a} + \vec{q} \cdot \vec{b} + \vec{r} \cdot \vec{c} = 0$ (d) none of these

101. If $\vec{a}, \vec{b}, \vec{c}$ are any three vectors then $(\vec{a} \times \vec{b}) \times \vec{c}$ is a vector

(a) perpendicular to $\vec{a} \times \vec{b}$ (b) coplanar with $\vec{a}$ and $\vec{b}$

(c) parallel to $\vec{c}$ (d) parallel to either $\vec{a}$ or $\vec{b}$

102. If $\vec{c}=\vec{a}\times\vec{b}$ and $\vec{b}=\vec{c}\times\vec{a}$ then

(a) $\vec{a}\cdot\vec{b}=\vec{c}^{\,2}$ (b) $\vec{c}\cdot\vec{a}=\vec{b}^{\,2}$

(c) $\vec{a}\perp\vec{b}$ (d) $\vec{a}\parallel\vec{b}\times\vec{c}$

103. If $\vec{x}\times\vec{b}=\vec{c}\times\vec{b}$ and $\vec{x}\perp\vec{a}$ then $\vec{x}$ is equal to

(a) $\dfrac{\vec{b}\times(\vec{a}\times\vec{c})}{\vec{b}\cdot\vec{c}}$ (b) $\dfrac{(\vec{b}\times\vec{c})\times\vec{a}}{\vec{b}\cdot\vec{a}}$

(c) $\dfrac{\vec{a}\times(\vec{c}\times\vec{b})}{\vec{a}\cdot\vec{b}}$ (d) none of these

104. The resolved part of the vector $\vec{a}$ along the vector $\vec{b}$ is $\vec{\lambda}$ and that perpendicular to $\vec{b}$ is $\vec{\mu}$. Then

(a) $\vec{\lambda}=\dfrac{(\vec{a}\cdot\vec{b})\vec{a}}{\vec{a}^{\,2}}$ (b) $\vec{\lambda}=\dfrac{(\vec{a}\cdot\vec{b})\vec{b}}{\vec{b}^{\,2}}$

(c) $\vec{\mu}=\dfrac{(\vec{b}\cdot\vec{b})\vec{a}-(\vec{a}\cdot\vec{b})\vec{b}}{\vec{b}^{\,2}}$ (d) $\vec{\mu}=\dfrac{\vec{b}\times(\vec{a}\times\vec{b})}{\vec{b}^{\,2}}$

105. $(\vec{a}\times\vec{b})\cdot(\vec{c}\times\vec{d})$ is equal to

(a) $\vec{a}\cdot\{\vec{b}\times(\vec{c}\times\vec{d})\}$ (b) $(\vec{a}\cdot\vec{c})(\vec{b}\cdot\vec{d})-(\vec{a}\cdot\vec{d})(\vec{b}\cdot\vec{c})$

(c) $\{(\vec{a}\times\vec{b})\times\vec{c}\}\cdot\vec{d}$ (d) $(\vec{d}\times\vec{c})\cdot(\vec{b}\times\vec{a})$

106. If $\vec{a},\vec{b},\vec{c},\vec{d}$ are any four vectors then $(\vec{a}\times\vec{b})\times(\vec{c}\times\vec{d})$ is a vector

(a) perpendicular to $\vec{a},\vec{b},\vec{c},\vec{d}$

(b) along the line of intersection of two planes, one containing $\vec{a},\vec{b}$ and the other containing $\vec{c},\vec{d}$

(c) equally inclined to both $\vec{a}\times\vec{b}$ and $\vec{c}\times\vec{d}$

(d) none of these

Answers

1. b	**2.** a	**3.** b	**4.** b	**5.** a
6. c	**7.** d	**8.** c	**9.** a	**10.** a
11. b	**12.** c	**13.** b	**14.** a	**15.** b
16. d	**17.** b	**18.** a	**19.** a	**20.** b
21. b	**22.** a	**23.** b	**24.** c	**25.** c
26. b	**27.** a	**28.** b	**29.** d	**30.** c
31. c	**32.** a	**33.** b	**34.** c	**35.** a
36. a	**37.** b	**38.** c	**39.** c	**40.** b
41. c	**42.** a	**43.** b	**44.** b	**45.** c
46. c	**47.** a	**48.** c	**49.** a	**50.** c
51. c	**52.** a	**53.** d	**54.** c	**55.** a
56. c	**57.** c	**58.** c	**59.** a	**60.** b
61. a	**62.** a	**63.** c	**64.** c	**65.** c
66. b	**67.** d	**68.** c	**69.** a	**70.** b
71. a	**72.** b	**73.** b	**74.** a	**75.** b
76. a	**77.** a	**78.** a	**79.** b	**80.** a
81. b	**82.** a	**83.** b	**84.** a	**85.** c
86. a	**87.** b	**88.** a	**89.** c	**90.** a
91. a, b	**92.** a, c	**93.** a, d	**94.** a, d	**95.** b, c
96. a, b	**97.** a, c	**98.** b, c	**99.** a, b, c	**100.** a, b
101. a, b	**102.** c, d	**103.** b, c	**104.** b, c, d	**105.** a, b, c, d
106. b, c				

Hints and Solutions to Selected Questions

1. 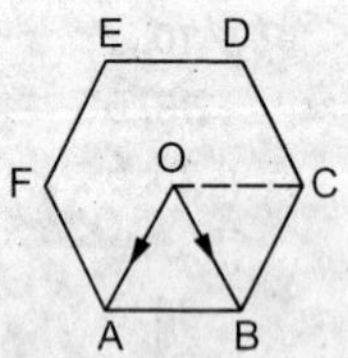

$\overrightarrow{OA} = \vec{i} - \vec{j} + 2\vec{k}$ $\quad\quad \overrightarrow{OB} = 2\vec{i} + \vec{j} + \vec{k}$

$\therefore \quad \overrightarrow{OC} = \overrightarrow{AB} = \overrightarrow{OB} - \overrightarrow{OA} = \vec{i} + 2\vec{j} - \vec{k}$

$\therefore \quad \overrightarrow{BC} = \overrightarrow{OC} - \overrightarrow{OB} = (\vec{i} + 2\vec{j} - \vec{k}) - (2\vec{i} + \vec{j} + \vec{k})$

$= -\vec{i} + \vec{j} - 2\vec{k}.$

2. The positive vector of the centroid $= \dfrac{\vec{a} + \vec{b} + \vec{c}}{3}$, where $\vec{a}, \vec{b}, \vec{c}$ are the position vectors of the vertices.

$\therefore \quad \vec{k} = \dfrac{\vec{i} + \vec{j} + 2\vec{i} - \vec{j} + \vec{k} + \vec{a}}{3}$ $\quad\quad \therefore \quad \vec{a} = -3\vec{i} + 2\vec{k}.$

3. $\overrightarrow{AB} = \overrightarrow{OB} - \overrightarrow{OA} = -\vec{i} - \vec{j} + 8\vec{k} - (\vec{i} + 2\vec{j} + 3\vec{k}) = -2\vec{i} - 3\vec{j} + 5\vec{k}.$

Similarly, find $\overrightarrow{BC}$ and $\overrightarrow{CA}$. Obtain $|\overrightarrow{AB}| = |\overrightarrow{BC}| = |\overrightarrow{CA}|$.

4. Here, $\vec{a} + \vec{b} = t\vec{c}$, $\vec{b} + \vec{c} = s\vec{a}$. Subtracting,

$\vec{a} - \vec{c} = t\vec{c} - s\vec{a}$

or $\quad (1 + s)\,\vec{a} = (1 + t)\,\vec{c}.$

But $\vec{a}, \vec{c}$ are noncollinear.

$\therefore \quad 1 + s = 0,\ 1 + t = 0$. Hence, $\vec{a} + \vec{b} = -\vec{c}.$

5. Here, $3\vec{i} + 2\vec{j} - 5\vec{k} = \lambda(2\vec{i} - \vec{j} + \vec{k}) + \mu(\vec{i} + 3\vec{j} - 2\vec{k}) + \nu(-2\vec{i} + \vec{j} - 3\vec{k})$

$\Rightarrow \quad 3 = 2\lambda + \mu - 2\nu, \quad 2 = -\lambda + 3\mu + \nu, \quad -5 = \lambda - 2\mu - 3\nu.$

Solving these, $\mu = 1$, $\nu = 2$, $\lambda = 3$. Hence, $\mu, \dfrac{\lambda}{2}, \nu$ are in AP.

6. If the points be A, B and C respectively then

$\overrightarrow{AB} = \overrightarrow{OB} - \overrightarrow{OA} = \vec{a} - 2\vec{b} + \lambda\vec{c} - (2\vec{a} - \vec{b} + 3\vec{c}) = -\vec{a} - \vec{b} + (\lambda - 3)\vec{c}$

$\overrightarrow{AC} = \overrightarrow{OC} - \overrightarrow{OA} = \mu\vec{a} - 5\vec{b} - (2\vec{a} - \vec{b} + 3\vec{c}) = (\mu - 2)\vec{a} - 4\vec{b} - 3\vec{c}.$

The points are collinear if $\vec{AB} = t\vec{AC}$

$\Rightarrow \quad -1 = t(\mu - 2), \quad -1 = -4t, \quad \lambda - 3 = -3t.$

$\therefore \quad t = \frac{1}{4}$, and $-1 = \frac{1}{4}(\mu - 2), \lambda - 3 = -\frac{3}{4}$.

7. Here, $\vec{c} = t\vec{a} + s\vec{b}$ and $|\vec{c}| = \sqrt{3} = \sqrt{1 + \alpha^2 + \beta^2}$.

Now $\vec{i} + \alpha\vec{j} + \beta\vec{k} = t(\vec{i} + \vec{j} + \vec{k}) + s(4\vec{i} + 3\vec{j} + 4\vec{k})$

$\Rightarrow \quad 1 = t + 4s, \quad \alpha = t + 3s, \quad \beta = t + 4s$

$\therefore$ we get $3 = 1 + \alpha^2 + \beta^2 \quad \Rightarrow 2 = (t + 3s)^2 + (t + 4s)^2$

or $\quad 2 = (1 - 4s + 3s)^2 + (1 - 4s + 4s)^2 \quad$ or $\quad 1 = (1 - s)^2$

$\therefore \quad 1 - s = \pm 1 \quad \therefore \quad s = 1 \mp 1 = 0, 2$

$\therefore \quad t = 1 - 4s = 1, -7$

$\therefore \quad \alpha = t + 3s = 1 + 3 \cdot 0, -7 + 3 \cdot 2 = 1, -1$ and

$\beta = t + 4s = 1 + 4 \cdot 0, -7 + 4 \cdot 2 = 1, 1.$

9. Here, $\vec{a} = 2p\vec{i} + \vec{j}$. After rotation, let the vector be $\vec{b}$ and let the unit vectors along the new axes be $\vec{i'}, \vec{j'}$.

Then $\vec{b} = (p + 1)\vec{i'} + \vec{j'}$. But the magnitude of a vector does not change with rotation of axes.

$\therefore \quad |\vec{a}| = |\vec{b}| \quad \Rightarrow \sqrt{(2p)^2 + 1^2} = \sqrt{(p + 1)^2 + 1^2}$

$\Rightarrow \quad 4p^2 + 1 = p^2 + 2p + 2 \quad$ or $\quad 3p^2 - 2p - 1 = 0$

$\therefore \quad p = 1, -\frac{1}{3}$.

10. Here, $|\vec{a}| = 2 = |\vec{b}|$ and $\vec{a} \cdot \vec{b} = 2 \cdot 2\cos 60° = 2$.

If the angle between $\vec{a}$ and $(\vec{a} + \vec{b})$ be θ then

$$\vec{a} \cdot (\vec{a} + \vec{b}) = |\vec{a}|\,|\vec{a} + \vec{b}|\cos\theta$$

or $\quad |\vec{a}|^2 + \vec{a} \cdot \vec{b} = 2|\vec{a} + \vec{b}|\cos\theta$

or $\quad 4 + 2 = 2|\vec{a} + \vec{b}|\cos\theta$

$\therefore \quad \cos\theta = \frac{3}{|\vec{a} + \vec{b}|}$.

Now, $|\vec{a}+\vec{b}|^2=(\vec{a}+\vec{b})^2=\vec{a}^2+\vec{b}^2+2\vec{a}\cdot\vec{b}$

$=4+4+2\cdot 2=12$

$\therefore \quad |\vec{a}+\vec{b}|=2\sqrt{3}.$

$\therefore \quad \cos\theta=\dfrac{3}{2\sqrt{3}}=\dfrac{\sqrt{3}}{2} \quad \Rightarrow \theta=30°.$

11. $1=|\vec{a}-\vec{b}|^2=(\vec{a}-\vec{b})^2=\vec{a}^2+\vec{b}^2-2\vec{a}\cdot\vec{b}=1+1-2\vec{a}\cdot\vec{b}$

$\therefore \quad \vec{a}\cdot\vec{b}=\dfrac{1}{2}$

or $\quad |\vec{a}|\,|\vec{b}|\cos\theta=\dfrac{1}{2} \quad$ or $\cos\theta=\dfrac{1}{2} \quad \therefore\ \theta=\dfrac{\pi}{3}.$

12. Let $\vec{a}=x\vec{i}+y\vec{j}+z\vec{k}$. Then $|\vec{a}|=4=\sqrt{x^2+y^2+z^2}$.

Now, $\dfrac{\vec{a}\cdot(\vec{i}+\vec{j})}{|\vec{a}|\,|\vec{i}+\vec{j}|}=\dfrac{\vec{a}\cdot(\vec{j}+\vec{k})}{|\vec{a}|\,|\vec{j}+\vec{k}|}=\dfrac{\vec{a}\cdot(\vec{k}+\vec{i})}{|\vec{a}|\,|\vec{k}+\vec{i}|}$ (from the question)

or $\quad x+y=y+z=z+x=t$ (say).

Adding, $2(x+y+z)=3t \quad$ or $x+y+z=\dfrac{3t}{2}$

$\therefore \quad x=y=z=\dfrac{t}{2}$

$\therefore \quad 4=\sqrt{x^2+y^2+z^2} \Rightarrow 16=3\cdot\left(\dfrac{t}{2}\right)^2$

$\therefore \quad t=\pm\dfrac{8}{\sqrt{3}}$. So, $\vec{a}=\pm\dfrac{8}{2\sqrt{3}}(\vec{i}+\vec{j}+\vec{k})$.

13. $\vec{a}+\vec{b}+2\vec{a}-\vec{b}=2\vec{i}+\vec{i}-\vec{j} \quad$ or $3\vec{a}=3\vec{i}-\vec{j}$

$\therefore \quad \vec{a}=\vec{i}-\dfrac{1}{3}\vec{j}.$

$\therefore \quad \vec{b}=2\vec{i}-\vec{a}=\vec{i}+\dfrac{1}{3}\vec{j}.$

So, $\cos\theta=\dfrac{\vec{a}\cdot\vec{b}}{|\vec{a}|\,|\vec{b}|}=\dfrac{1-\dfrac{1}{9}}{\sqrt{1+\dfrac{1}{9}}\sqrt{1+\dfrac{1}{9}}}=\dfrac{8}{10}=\dfrac{4}{5}.$

14. $\vec{a}\cdot(\vec{b}+\vec{c})=0 \quad$ or $\vec{a}\cdot\vec{b}+\vec{a}\cdot\vec{c}=0.$

Similarly, we get $\vec{b}\cdot\vec{c}+\vec{b}\cdot\vec{a}=0$ and $\vec{c}\cdot\vec{a}+\vec{c}\cdot\vec{b}=0$.

Adding these, $\vec{a}\cdot\vec{b}+\vec{b}\cdot\vec{c}+\vec{c}\cdot\vec{a}=0$.

$$|\vec{a}+\vec{b}+\vec{c}|^2=(\vec{a}+\vec{b}+\vec{c})^2$$

$$=|\vec{a}|^2+|\vec{b}|^2+|\vec{c}|^2+2(\vec{a}\cdot\vec{b}+\vec{b}\cdot\vec{c}+\vec{c}\cdot\vec{a})$$

$$=1+2+3+0=6$$

$\therefore \quad |\vec{a}+\vec{b}+\vec{c}|=\sqrt{6}$.

15. Let $\vec{a}=x\vec{i}+y\vec{j}+z\vec{k}$. Then $\vec{a}\cdot\vec{i}=x, \vec{a}\cdot\vec{j}=y, \vec{a}\cdot\vec{k}=z$.

16. See the hint to Q. No. 11.

17. $\vec{a}\cdot\vec{i}=1, \vec{a}\cdot\vec{j}=0, \vec{a}\cdot\vec{k}=0$. These imply $\vec{a}=\vec{i}$.

18. See the hint to Q. No. 15. As $\vec{u}=(\vec{a}\cdot\vec{i})\vec{i}+(\vec{a}\cdot\vec{j})\vec{j}+(\vec{a}\cdot\vec{k})\vec{k}$,

$$|\vec{a}|^2=(\vec{a}\cdot\vec{i})^2+(\vec{a}\cdot\vec{j})^2+(\vec{a}\cdot\vec{k})^2 \text{ and } |\vec{a}|^2=\vec{a}^2.$$

20. $\vec{u}\cdot\vec{v}=\dfrac{q-r}{a}+\dfrac{r-p}{b}+\dfrac{p-q}{c}$ ... (1)

From the question, $\dfrac{1}{a}=x+(p-1)y, \dfrac{1}{b}=x+(q-1)y, \dfrac{1}{c}=x+(r-1)y$.

Putting these in (1) and simplifying, $\vec{u}\cdot\vec{v}=0$.

Clearly, $(\vec{i}+\vec{j}+\vec{k})\cdot\vec{v}=\dfrac{1}{a}+\dfrac{1}{b}+\dfrac{1}{c}\neq 0$. So, $\vec{u}\times\vec{v}\neq\vec{i}+\vec{j}+\vec{k}$.

21. Here, $\vec{a}\cdot(\vec{a}+\vec{b})=0$ or $|\vec{a}|^2+\vec{a}\cdot\vec{b}=0$

or $\dfrac{1}{2}|\vec{b}|^2+\vec{a}\cdot\vec{b}=0$.

$$(2\vec{a}+\vec{b})\cdot\vec{b}=2\vec{a}\cdot\vec{b}+|\vec{b}|^2=2\left\{\vec{a}\cdot\vec{b}+\frac{1}{2}|\vec{b}|^2\right\}=0$$

$\therefore \quad (2\vec{a}+\vec{b})\perp\vec{b}$.

$(2\vec{a}-\vec{b})\cdot\vec{b}=2\vec{a}\cdot\vec{b}-|\vec{b}|^2=2\vec{a}\cdot\vec{b}+2\vec{a}\cdot\vec{b}=4\vec{a}\cdot\vec{b}\neq 0$.

$\vec{a}\cdot(2\vec{a}+\vec{b})=2|\vec{a}|^2+\vec{a}\cdot\vec{b}=|\vec{a}|^2+(|\vec{a}|^2+\vec{a}\cdot\vec{b})=|\vec{a}|^2\neq 0$.

22. $\vec{c}=t\vec{a}+s\vec{b}$. Also $|\vec{c}|=1$ (given).

Now, $\vec{c}=t(2\vec{i}+\vec{j}+\vec{k})+s(\vec{i}+2\vec{j}-\vec{k})=(2t+s)\vec{i}+(t+2s)\vec{j}+(t-s)\vec{k}$

$\therefore \quad |\vec{c}| = 1 \quad \therefore \quad (2t+s)^2 + (t+2s)^2 + (t-s)^2 = 1 \qquad \dots (1)$

and $\vec{a} \cdot \vec{c} = 0 \quad \Rightarrow \quad 2(2t+s) + 1(t+2s) + 1(t-s) = 0$

or $\quad 6t + 3s = 0 \quad$ or $\quad 2t + s = 0.$

$\therefore \quad$ from (1), $0^2 + (t-4t)^2 + (t+2t)^2 = 1$

or $\quad 18t^2 = 1 \qquad \therefore \quad t = \pm \dfrac{1}{3\sqrt{2}}$

$\therefore \quad s = -2t = \mp \dfrac{\sqrt{2}}{3}.$

$\therefore \quad \vec{c} = (2t+s)\vec{i} + (t+2s)\vec{j} + (t-s)\vec{k}$

$$= 0\vec{i} - 3t\vec{j} + 3t\vec{k} = -3t(\vec{j} - \vec{k}) = -3 \cdot \frac{\pm 1}{3\sqrt{2}}(\vec{j} - \vec{k}).$$

23. Adding and simplifying, $\vec{\lambda} + \vec{\mu} + \vec{\nu} = 0$

$\therefore$ there is a linear relation between $\vec{\lambda}, \vec{\mu}, \vec{\nu}$ and, therefore, they are coplanar.

24. $|\vec{a}| = |\vec{b}| = |\vec{c}| = 1, \vec{a} \cdot \vec{b} = 0, \vec{c} \cdot \vec{a} = \cos\alpha, \vec{c} \cdot \vec{b} = \cos\beta.$

Now, see the hint to Q. No. 14.

25. Let $|\vec{a}| = |\vec{b}| = |\vec{c}| = k$. Then

$\cos\dfrac{\pi}{3} = \dfrac{\vec{a} \cdot \vec{b}}{|\vec{a}||\vec{b}|} = \dfrac{\vec{a} \cdot \vec{b}}{k^2} \qquad$ or $\quad \vec{a} \cdot \vec{b} = \dfrac{k^2}{2}.$

Similarly, $\vec{b} \cdot \vec{c} = \vec{c} \cdot \vec{a} = \dfrac{k^2}{2}$. So, $\vec{a} \cdot \vec{b} + \vec{b} \cdot \vec{c} + \vec{c} \cdot \vec{a} = \dfrac{3k^2}{2}.$

Now, $6 = |\vec{a} + \vec{b} + \vec{c}|^2 = |\vec{a}|^2 + |\vec{b}|^2 + |\vec{c}|^2 + 2(\vec{a} \cdot \vec{b} + \vec{b} \cdot \vec{c} + \vec{c} \cdot \vec{a})$

$\therefore \quad 6 = 3k^2 + 2 \cdot \dfrac{3k^2}{2}$

$\therefore \quad k = \pm 1$. But $|\vec{a}|$ is positive. So, $|\vec{a}| = 1$.

26. $\because \quad |\vec{a} + \vec{b}|^2 = |\vec{a} - \vec{b}|^2 = 4\vec{a} \cdot \vec{b} \qquad \therefore \quad 4\vec{a} \cdot \vec{b} = 10^2 - 8^2 = 36$

$\therefore \quad \vec{a} \cdot \vec{b} = 9.$

Now, $|\vec{a} - \vec{b}|^2 = |\vec{a}|^2 + |\vec{b}|^2 - 2\vec{a} \cdot \vec{b} = 5^2 + |\vec{b}|^2 - 18$

$\therefore \quad 8^2 = 7 + |\vec{b}|^2 \qquad \therefore \quad |\vec{b}| = \sqrt{57}.$

27. Here $\cos\alpha = \vec{a}\cdot\vec{b}$ or $2\cos^2\frac{\alpha}{2} - 1 = \vec{a}\cdot\vec{b}$ or $4\cos^2\frac{\alpha}{2} = 2 + 2\vec{a}\cdot\vec{b}$

$$\therefore \quad \left(2\cos\frac{\alpha}{2}\right)^2 = \vec{a}^2 + \vec{b}^2 + 2\vec{a}\cdot\vec{b} = (\vec{a}+\vec{b})^2 = |\vec{a}+\vec{b}|^2$$

or $\quad 2\cos\frac{\alpha}{2} = |\vec{a}+\vec{b}| \qquad \therefore \cos\frac{\alpha}{2} = \frac{1}{2}|\vec{a}+\vec{b}|.$

29. $\vec{a}\cdot\vec{b} = \sqrt{1+\frac{1}{3}}\cdot\sqrt{1+\frac{1}{3}}\cos\theta$ or $\frac{1}{\sqrt{3}} + \frac{1}{\sqrt{3}} = \frac{2}{\sqrt{3}}\cdot\frac{2}{\sqrt{3}}\cos\theta$

or $\quad \cos\theta = \frac{\sqrt{3}}{2} \qquad \therefore \quad \theta = 30°.$

30. Let $\vec{p} = \lambda\vec{b} + \mu\vec{c}$.

The projection of $\vec{p}$ on $\vec{a} = \frac{\vec{p}\cdot\vec{a}}{|\vec{a}|} = \sqrt{\frac{2}{3}}$ ... (1)

$\because \quad \vec{p} = \lambda(\vec{i} + 2\vec{j} - \vec{k}) + \mu(\vec{i} + \vec{j} - 2\vec{k})$

$\therefore \quad \vec{p}\cdot\vec{a} = 2(\lambda+\mu) - 1(2\lambda+\mu) + 1(-\lambda - 2\mu) = -\lambda - \mu$

$\therefore \quad$ (1) gives $\frac{-\lambda-\mu}{\sqrt{2^2+1^2+1^2}} = \sqrt{\frac{2}{3}}$ or $\lambda + \mu = -2.$

$\therefore \quad \vec{p} = \lambda(\vec{i} + 2\vec{j} - \vec{k}) + \mu(\vec{i} + \vec{j} - 2\vec{k})$

$= (\lambda+\mu)\vec{i} + (2\lambda+\mu)\vec{j} - (\lambda+2\mu)\vec{k} = -2\vec{i} + (\lambda-2)\vec{j} + (\lambda+4)\vec{k},$

where λ is a scalar parameter.

When $\lambda = 1$, $\vec{p} = -2\vec{i} - \vec{j} + 5\vec{k}$. Other options hold for no real λ.

31. $(\vec{AB} + \vec{BC} + \vec{CA})^2 = |AB|^2 + |\vec{BC}|^2 + |\vec{CA}|^2$

$+ 2(\vec{AB}\cdot\vec{BC} + \vec{BC}\cdot\vec{CA} + \vec{CA}\cdot\vec{AB})$

or $\quad 0 = a^2 + a^2 + a^2 + 2(\vec{AB}\cdot\vec{BC} + \vec{BC}\cdot\vec{CA} + \vec{CA}\cdot\vec{AB}).$

32. Here, $\vec{r}\times\vec{a} + \vec{r}\times\vec{b} = 0$ or $\vec{r}\times(\vec{a}+\vec{b}) = 0$ $\therefore \vec{r} \parallel (\vec{a}+\vec{b})$

$\therefore \quad \vec{r} = t(\vec{a}+\vec{b}) = t(\vec{i} + 3\vec{j} - \vec{k})$

$\therefore \quad |\vec{r}| = t\cdot\sqrt{1^2+3^2+1^2} = t\cdot\sqrt{11}$

$$\therefore \quad \frac{\vec{r}}{|\vec{r}|} = \frac{1}{\sqrt{11}}(\vec{i} + 3\vec{j} - \vec{k}).$$

33. $(\vec{a} \times \vec{b})^2 = (|\vec{a}||\vec{b}| \sin\theta\, \hat{n})^2 = |\vec{a}|^2 |\vec{b}|^2 \sin^2\theta$

$(\vec{a} \cdot \vec{b})^2 = (|\vec{a}||\vec{b}| \cos\theta)^2 = |\vec{a}|^2 |\vec{b}|^2 \cos^2\theta$. Add these.

34. $\vec{p} \times \vec{q}$ is perpendicular to $\vec{p}$ and $\vec{q}$. So $\vec{p}, \vec{q}, \vec{p} \times \vec{q}$ are noncoplanar. Therefore, the given linear relation is possible only when $b - c = 0,\ c - a = 0,\ a - b = 0$. So, $a = b = c$.

35. $(\vec{a} + \vec{b}) \cdot \vec{c} = (\vec{a} - \vec{b}) \cdot \vec{c} = 0 \quad \Rightarrow \vec{a} \cdot \vec{c} = \vec{b} \cdot \vec{c} = 0.$

$\therefore \quad \vec{c} \perp \vec{a}$ as well as $\vec{b}$.

$\therefore \quad \vec{c} \parallel \vec{a} \times \vec{b}$. Hence, $(\vec{a} \times \vec{b}) \times \vec{a} = \vec{0}$.

36. The unit vector $= \pm \dfrac{\vec{a} \times \vec{b}}{|\vec{a} \times \vec{b}|} = \pm \dfrac{4\vec{i} - \vec{j} - 3\vec{k}}{\sqrt{4^2 + (-1)^2 + (-3)^2}} = \pm \dfrac{1}{\sqrt{26}}(4\vec{i} - \vec{j} - 3\vec{k})$.

Now, $\vec{k} \cdot \dfrac{\pm 1}{\sqrt{26}}(4\vec{i} - \vec{j} - 3\vec{k}) = \dfrac{\pm 1}{\sqrt{26}}(-3)$.

$\therefore \quad$ for acute angle the unit vector $= -\dfrac{1}{\sqrt{26}}(4\vec{i} - \vec{j} - 3\vec{k})$.

37. $\vec{c} \parallel$ the plane of $\vec{a}$ and $\vec{b} \quad \Rightarrow \vec{c} \perp \vec{a} \times \vec{b}$.

$\therefore \quad \vec{a} \times \vec{b} \cdot \vec{c} = 0$

or $\quad (\vec{i} - 2\vec{j} + 3\vec{k}) \times (2\vec{i} + 3\vec{j} - \vec{k}) \cdot \{\lambda\vec{i} + \vec{j} + (2\lambda - 1)\vec{k}\} = 0$

or $\quad (-7\vec{i} + 7\vec{j} + 7\vec{k}) \cdot \{\lambda\vec{i} + \vec{j} + (2\lambda - 1)\vec{k}\} = 0$

or $\quad -7\lambda + 7 + 7(2\lambda - 1) = 0 \qquad$ or $\lambda = 0$.

38. $|\vec{b} \times \vec{c}| = |\vec{b}||\vec{c}| \sin\alpha = \sin\alpha$.

Now, $\dfrac{\vec{b} \times \vec{c}}{|\vec{b} \times \vec{c}|} = \vec{a}$ (given) $\qquad \therefore \vec{b} \times \vec{c} = \sin\alpha\, \vec{a}$.

41. $\vec{r} = (\vec{r} \cdot \vec{i})\vec{i} + (\vec{r} \cdot \vec{j})\vec{j} + (\vec{r} \cdot \vec{k})\vec{k} \qquad$ (See the hint to Q. No.15.)

$\therefore \quad$ the given vector $= \vec{r} \times \vec{r} = \vec{0}$.

42. Let $\vec{b} = x\vec{i} + y\vec{j} + z\vec{k}$.

$\vec{a} \cdot \vec{b} = 3 \quad \Rightarrow x + y + z = 3$

$\vec{a} \times \vec{b} = \vec{c} \quad \Rightarrow (\vec{i} + \vec{j} + \vec{k}) \times (x\vec{i} + y\vec{j} + z\vec{k}) = \vec{j} - \vec{k}$

or $\quad (z - y)\,\vec{i} + (x - z)\,\vec{j} + (y - x)\,\vec{k} = \vec{j} - \vec{k}$

$\Rightarrow \quad z - y = 0, x - z = 1, y - x = -1.$

Solving the four equations in x, y, z, we get $x = \frac{5}{3}, y = \frac{2}{3}, z = \frac{2}{3}$.

43. Let the points be A, B, C. Then $\vec{OA} = \vec{i} - \vec{j} + 2\vec{k}, \vec{OB} = 2\vec{i} - \vec{k}, \vec{OC} = 2\vec{j} + \vec{k}$.

The required vector $= \dfrac{\vec{AB} \times \vec{AC}}{|\vec{AB} \times \vec{AC}|}$. Calculate its value.

44. Here $(\vec{r} - \vec{b}) \times \vec{a} = 0$. So, $(\vec{r} - \vec{b}) \parallel \vec{a}$.

$\therefore \quad \vec{r} - \vec{b} = t\vec{a} \qquad$ or $\vec{r} = \vec{b} + t\vec{a}$.

But $\vec{r} \cdot \vec{c} = 0$.

$\therefore \quad 0 = \vec{b} \cdot \vec{c} + t\vec{a} \cdot \vec{c} \qquad \therefore \; t = -\dfrac{\vec{b} \cdot \vec{c}}{\vec{a} \cdot \vec{c}}$.

$\therefore \quad \vec{r} = \vec{b} - \dfrac{\vec{b} \cdot \vec{c}}{\vec{a} \cdot \vec{c}}\,\vec{a}$.

46. $|\vec{a} \times \vec{b} \cdot \vec{c}|$ = volume of the parallelepiped $= |\vec{a}|\,|\vec{b}|\,|\vec{c}|$

$\Rightarrow$ it is a rectangular parallelepiped, i.e., concurrent edges are perpendicular to each other. So, $\vec{a} \cdot \vec{b} = 0, \vec{b} \cdot \vec{c} = 0, \vec{c} \cdot \vec{a} = 0$.

47. $(\vec{a} + \vec{b}) \times (\vec{b} + \vec{c}) = \vec{a} \times \vec{b} + \vec{a} \times \vec{c} + \vec{b} \times \vec{c}$

$\therefore \quad [\vec{a} + \vec{b} \quad \vec{b} + \vec{c} \quad \vec{c} + \vec{a}] = (\vec{a} \times \vec{b} + \vec{a} \times \vec{c} + \vec{b} \times \vec{c}) \cdot (\vec{c} + \vec{a})$

$= [\vec{a}\ \vec{b}\ \vec{c}] + [\vec{b}\ \vec{c}\ \vec{a}]$ ($\because$ $[\vec{a}\ \vec{b}\ \vec{c}] = 0$ if two vectors are equal)

$= [\vec{a}\ \vec{b}\ \vec{c}] + [\vec{a}\ \vec{b}\ \vec{c}] = 2[\vec{a}\ \vec{b}\ \vec{c}]$.

48. $\{(\vec{a} - \vec{b}) \times (\vec{b} - \vec{c})\} \cdot (\vec{c} - \vec{a}) = (\vec{a} \times \vec{b} - \vec{a} \times \vec{c} + \vec{b} \times \vec{c}) \cdot (\vec{c} - \vec{a})$

$= [\vec{a}\ \vec{b}\ \vec{c}] - [\vec{b}\ \vec{c}\ \vec{a}] = [\vec{a}\ \vec{b}\ \vec{c}] - [\vec{a}\ \vec{b}\ \vec{c}] = 0.$

49. $\vec{a} \perp \vec{b}$, $\vec{a} \perp \vec{c}$, i.e., $\vec{a} \parallel (\vec{b} \times \vec{c})$ and $|\vec{b} \times \vec{c}| = |\vec{b}|\,|\vec{c}| \sin\frac{\pi}{3} = \frac{\sqrt{3}}{2}$.

$$|[\vec{a}\ \vec{b}\ \vec{c}]| = |\vec{a} \cdot (\vec{b} \times \vec{c})| = |\vec{a}|\,|\vec{b} \times \vec{c}| \cos 0° = 1 \cdot \frac{\sqrt{3}}{2} \cdot 1 = \frac{\sqrt{3}}{2}.$$

50. Here $[\vec{p}\ \vec{q}\ \vec{r}] = 0$. So, $\begin{vmatrix} a & -a & b \\ 1 & 0 & 1 \\ c & c & b \end{vmatrix} = 0$

or $\quad a(0 - c) + a(b - c) + b(c - 0) = 0 \quad \Rightarrow ab + bc = 2ac$

$\therefore \quad b = \dfrac{2ac}{a + c}$.

So, b is the HM of a, c.

53. $$\text{Expression} = \frac{(\vec{a}+\vec{b}) \cdot (\vec{b} \times \vec{c})}{[\vec{a}\ \vec{b}\ \vec{c}]} + \frac{(\vec{b}+\vec{c}) \cdot (\vec{c} \times \vec{a})}{[\vec{a}\ \vec{b}\ \vec{c}]} + \frac{(\vec{c}+\vec{a}) \cdot (\vec{a} \times \vec{b})}{[\vec{a}\ \vec{b}\ \vec{c}]}$$

$$= \frac{[\vec{a}\ \vec{b}\ \vec{c}]}{[\vec{a}\ \vec{b}\ \vec{c}]} + \frac{[\vec{b}\ \vec{c}\ \vec{a}]}{[\vec{a}\ \vec{b}\ \vec{c}]} + \frac{[\vec{c}\ \vec{a}\ \vec{b}]}{[\vec{a}\ \vec{b}\ \vec{c}]} = 1 + 1 + 1 = 3.$$

54. Value $= [\vec{a}\ \vec{b}\ \vec{c}] + [\vec{b}\ \vec{c}\ \vec{a}] + [\vec{c}\ \vec{a}\ \vec{b}] = 3[\vec{a}\ \vec{b}\ \vec{c}] = 3(\pm 4) = \pm 12$.

55. As $\vec{a}, \vec{b}, \vec{c}$ are noncoplanar, $\vec{b} \times \vec{c}$, $\vec{c} \times \vec{a}$, $\vec{a} \times \vec{b}$ are also noncoplanar.

So, any vector can be expressed as a linear combination of these vectors.

Let $\vec{a} = \lambda\, \vec{b} \times \vec{c} + \mu\, \vec{c} \times \vec{a} + \nu\, \vec{a} \times \vec{b}$

$\therefore \quad \vec{a} \cdot \vec{a} = \lambda[\vec{b}\ \vec{c}\ \vec{a}]$, $\vec{a} \cdot \vec{b} = \mu[\vec{c}\ \vec{a}\ \vec{b}]$, $\vec{a} \cdot \vec{c} = \nu[\vec{a}\ \vec{b}\ \vec{c}]$

$$\therefore \quad \vec{a} = \frac{(\vec{a} \cdot \vec{a})\, \vec{b} \times \vec{c}}{[\vec{b}\ \vec{c}\ \vec{a}]} + \frac{(\vec{a} \cdot \vec{b})\, \vec{c} \times \vec{a}}{[\vec{c}\ \vec{a}\ \vec{b}]} + \frac{(\vec{a} \cdot \vec{c})\, \vec{a} \times \vec{b}}{[\vec{a}\ \vec{b}\ \vec{c}]}.$$

56. Here $\vec{r} \cdot (\vec{a} + \vec{b} + \vec{c}) = 0 \qquad \therefore \ \vec{r} \cdot \vec{a} + \vec{r} \cdot \vec{b} + \vec{r} \cdot \vec{c} = 0$.

Now, $\vec{r} \cdot \vec{a} = l[\vec{b}\ \vec{c}\ \vec{a}] \qquad \vec{r} \cdot \vec{b} = m[\vec{c}\ \vec{a}\ \vec{b}] \qquad \vec{r} \cdot \vec{c} = n[\vec{a}\ \vec{b}\ \vec{c}]$

$\therefore \quad l[\vec{b}\ \vec{c}\ \vec{a}] + m[\vec{c}\ \vec{a}\ \vec{b}] + n[\vec{a}\ \vec{b}\ \vec{c}] = 0$

$\Rightarrow \quad l + m + n = 0 \quad (\because\ [\vec{a}\ \vec{b}\ \vec{c}] = 2 \neq 0)$.

57. $(\vec{c} + \vec{b}) \times (\vec{c} + \vec{a}) \cdot (\vec{c} + \vec{b} + \vec{a}) = (\vec{c} \times \vec{a} + \vec{b} \times \vec{c} + \vec{b} \times \vec{a}) \cdot (\vec{c} + \vec{b} + \vec{a})$

$= [\vec{c}\ \vec{a}\ \vec{b}] + [\vec{b}\ \vec{c}\ \vec{a}] + [\vec{b}\ \vec{a}\ \vec{c}]$

$$= [\vec{a}\ \vec{b}\ \vec{c}] + [\vec{a}\ \vec{b}\ \vec{c}] - [\vec{a}\ \vec{b}\ \vec{c}] = [\vec{a}\ \vec{b}\ \vec{c}].$$

58. $\{(\vec{a}+\vec{b}+\vec{c})\times(\vec{a}-\vec{c})\}\cdot(\vec{a}-\vec{b}) = \{-\vec{a}\times\vec{c}+\vec{b}\times\vec{a}-\vec{b}\times\vec{c}+\vec{c}\times\vec{a}\}\cdot(\vec{a}-\vec{b})$

$$= [\vec{a}\ \vec{c}\ \vec{b}] - [\vec{b}\ \vec{c}\ \vec{a}] - [\vec{c}\ \vec{a}\ \vec{b}]$$

$$= -[\vec{a}\ \vec{b}\ \vec{c}] - [\vec{a}\ \vec{b}\ \vec{c}] - [\vec{a}\ \vec{b}\ \vec{c}] = -3[\vec{a}\ \vec{b}\ \vec{c}].$$

59. $\{\vec{a}\times(\vec{b}+\vec{c})\}\cdot(\vec{a}+\vec{b}+\vec{c}) = (\vec{a}\times\vec{b}+\vec{a}\times\vec{c})\cdot(\vec{a}+\vec{b}+\vec{c})$

$$= [\vec{a}\ \vec{b}\ \vec{c}] + [\vec{a}\ \vec{c}\ \vec{b}] = [\vec{a}\ \vec{b}\ \vec{c}] - [\vec{a}\ \vec{b}\ \vec{c}] = 0.$$

60. Let $\vec{a}\times\vec{b} = x\vec{i}+y\vec{j}+z\vec{k}.$

$\therefore\quad \vec{a}\times\vec{b}\cdot\vec{i} = x \qquad \vec{a}\times\vec{b}\cdot\vec{j} = y \qquad \vec{a}\times\vec{b}\cdot\vec{k} = z$

$\therefore\quad \vec{a}\times\vec{b} = [\vec{a}\ \vec{b}\ \vec{i}]\vec{i} + [\vec{a}\ \vec{b}\ \vec{j}]\vec{j} + [\vec{a}\ \vec{b}\ \vec{k}]\vec{k}.$

61. The required volume of the parallelepiped

$= [\vec{b}+\vec{c}\ \ \vec{c}+\vec{a}\ \ \vec{a}+\vec{b}] = 2[\vec{a}\ \vec{b}\ \vec{c}] = 2\lambda.$ (See the hint to Q. No.47.)

62. Expression $= (\vec{i}\cdot\vec{i})\vec{a} - (\vec{i}\cdot\vec{a})\vec{i} + (\vec{j}\cdot\vec{j})\vec{a} - (\vec{j}\cdot\vec{a})\vec{j} + (\vec{k}\cdot\vec{k})\vec{a} - (\vec{k}\cdot\vec{a})\vec{k}$

$$= 3\vec{a} - \{(\vec{i}\cdot\vec{a})\vec{i} + (\vec{j}\cdot\vec{a})\vec{j} + (\vec{k}\cdot\vec{a})\vec{k}\}$$

$= 3\vec{a} - \vec{a} = 2\vec{a}.$ (See the hint to Q. No.15.)

63. Here $|\vec{a}| = 1,\ |\vec{b}| = 1,\ |\vec{c}| = 2.$

Now, $\vec{a}\times(\vec{a}\times\vec{c}) + \vec{b} = \vec{0}$ or $(\vec{a}\cdot\vec{c})\vec{a} - (\vec{a}\cdot\vec{a})\vec{c} + \vec{b} = \vec{0}$

or $2\cos\theta\,\vec{a} - \vec{c} + \vec{b} = \vec{0}$ or $(\vec{c} - 2\cos\theta\,\vec{a})^2 = \vec{b}^{\,2}$

or $|\vec{c}|^2 + 4\cos^2\theta\,|\vec{a}|^2 - 4\cos\theta\,\vec{c}\cdot\vec{a} = |\vec{b}|^2$

or $4 + 4\cos^2\theta - 4\cos\theta\cdot 2\cos\theta = 1$ or $4\cos^2\theta = 3$

$\therefore\quad \theta = \dfrac{\pi}{6}.$

64. Expression

$= (\vec{a}\cdot\vec{b})\vec{b} + (\vec{b}\cdot\vec{b})\vec{a} - (\vec{b}\cdot\vec{a})\vec{b} = \vec{b}^{\,2}\vec{a} = \vec{a}$ ($\because$ $\vec{b}$ is a unit vector).

65. $\vec{a}\times(\vec{b}\times\vec{c}) = \dfrac{\vec{b}+\vec{c}}{\sqrt{2}} \quad\Rightarrow\quad (\vec{a}\cdot\vec{c})\vec{b} - (\vec{a}\cdot\vec{b})\vec{c} = \dfrac{\vec{b}+\vec{c}}{\sqrt{2}}$

$$\Rightarrow \quad \vec{a}\cdot\vec{c}=\frac{1}{\sqrt{2}}, \quad \vec{a}\cdot\vec{b}=-\frac{1}{\sqrt{2}}$$

$$\Rightarrow \quad |\vec{a}|\,|\vec{c}|\cos\alpha=\frac{1}{\sqrt{2}}, \quad |\vec{a}|\,|\vec{b}|\cos\beta=-\frac{1}{\sqrt{2}}$$

$$\therefore \quad \cos\alpha=\frac{1}{\sqrt{2}}, \quad \cos\beta=-\frac{1}{\sqrt{2}}$$

$$\therefore \quad \alpha=\frac{\pi}{4}, \beta=\frac{3\pi}{4}.$$

66. Here, $(\vec{p}\cdot\vec{p})(\vec{x}-\vec{q})-\{\vec{p}\cdot(\vec{x}-\vec{q})\}\vec{p}+(\vec{q}\cdot\vec{q})(\vec{x}-\vec{r})-\{\vec{q}\cdot(\vec{x}-\vec{r})\}\vec{q}$

$$+(\vec{r}\cdot\vec{r})(\vec{x}-\vec{p})-\{\vec{r}\cdot(\vec{x}-\vec{p})\}\vec{r}=0$$

or $\quad \lambda^2(\vec{x}-\vec{q}+\vec{x}-\vec{r}+\vec{x}-\vec{p})$

$$=\{\vec{p}\cdot(\vec{x}-\vec{q})\}\vec{p}+\{\vec{q}\cdot(\vec{x}-\vec{r})\}\vec{q}+\{\vec{r}\cdot(\vec{x}-\vec{p})\}\vec{r},$$

where $|\vec{p}|=|\vec{q}|=|\vec{r}|=\lambda$

or $\quad \lambda^2\{3\vec{x}-(\vec{p}+\vec{q}+\vec{r})\}=(\vec{p}\cdot\vec{x})\vec{p}+(\vec{q}\cdot\vec{x})\vec{q}+(\vec{r}\cdot\vec{x})\vec{r},$

because $\vec{p}, \vec{q}, \vec{r}$ are mutually perpendicular.

Let $\vec{x}=\alpha\vec{p}+\beta\vec{q}+\gamma\vec{r}$. Then

$$\vec{p}\cdot\vec{x}=\alpha|\vec{p}|^2=\alpha\lambda^2, \quad \vec{q}\cdot\vec{x}=\beta|\vec{q}|^2=\beta\lambda^2$$

$$\vec{r}\cdot\vec{x}=\gamma|\vec{r}|^2=\gamma\lambda^2.$$

$$\therefore \quad \lambda^2\{3\vec{x}-(\vec{p}+\vec{q}+\vec{r})\}=\lambda^2\vec{x} \qquad \text{or} \quad 3\vec{x}-(\vec{p}+\vec{q}+\vec{r})=\vec{x}.$$

67. $\vec{a}\cdot\vec{b}$ is a scalar and $\vec{a}\times\vec{b}$ is a vector. Dot or cross product is applicable for vectors only. Also, a vector may have a scalar multiple.

68. $(\vec{a}\times\vec{i})^2=(\vec{a}\times\vec{i})\cdot(\vec{a}\times\vec{i})=\begin{vmatrix}\vec{a}\cdot\vec{a} & \vec{a}\cdot\vec{i}\\ \vec{i}\cdot\vec{a} & \vec{i}\cdot\vec{i}\end{vmatrix}=\vec{a}^2-(\vec{a}\cdot\vec{i})^2.$

Similarly, $(\vec{a}\times\vec{j})^2=\vec{a}^2-(\vec{a}\cdot\vec{j})^2$, and $(\vec{a}\times\vec{k})^2=\vec{a}^2-(\vec{a}\cdot\vec{k})^2$.

$\therefore$ the expression $=3\vec{a}^2-\{(\vec{a}\cdot\vec{i})^2+(\vec{a}\cdot\vec{j})^2+(\vec{a}\cdot\vec{k})^2\}$

$$=3\vec{a}^2-\vec{a}^2=2\vec{a}^2. \quad \text{(See the hint to Q. No. 18.)}$$

69. $\vec{a}\parallel(\vec{b}\times\vec{c}) \qquad \Rightarrow \vec{a}\times(\vec{b}\times\vec{c})=0 \qquad \Rightarrow (\vec{a}\cdot\vec{c})\vec{b}=(\vec{a}\cdot\vec{b})\vec{c}.$

But $\vec{b}$ is not collinear with $\vec{c}$. So, $\vec{a}\cdot\vec{c}=0$ and $\vec{a}\cdot\vec{b}=0$.

Now, $(\vec{a}\times\vec{b})\cdot(\vec{a}\times\vec{c})=\begin{vmatrix}\vec{a}\cdot\vec{a} & \vec{a}\cdot\vec{c}\\ \vec{b}\cdot\vec{a} & \vec{b}\cdot\vec{c}\end{vmatrix}=\begin{vmatrix}\vec{a}\cdot\vec{a} & 0\\ 0 & \vec{b}\cdot\vec{c}\end{vmatrix}$

$=\vec{a}^{\,2}\,(\vec{b}\cdot\vec{c}).$

70. $(\vec{u}\times\vec{b})\times(\vec{a}\times\vec{c})=(\vec{a}\times\vec{b}\cdot\vec{c})\,\vec{a}-(\vec{a}\times\vec{b}\cdot\vec{a})\,\vec{c}$

$=[\vec{a}\ \vec{b}\ \vec{c}]\,\vec{a}-[\vec{a}\ \vec{b}\ \vec{a}]\,\vec{c}=[\vec{a}\ \vec{b}\ \vec{c}]\,\vec{a}.$

Similarly, $(\vec{b}\times\vec{c})\times(\vec{b}\times\vec{a})=[\vec{b}\ \vec{c}\ \vec{a}]\,\vec{b}$

$(\vec{c}\times\vec{a})\times(\vec{c}\times\vec{b})=[\vec{c}\ \vec{a}\ \vec{b}]\,\vec{c}$

$\therefore$ the expression $=[\vec{a}\ \vec{b}\ \vec{c}]\,(\vec{a}+\vec{b}+\vec{c}).$

71. $(\vec{a}\times\vec{b})\times(\vec{r}\times\vec{c})=[\vec{a}\ \vec{b}\ \vec{c}]\,\vec{r}-[\vec{a}\ \vec{b}\ \vec{r}]\,\vec{c}$ ... (1)

$(\vec{b}\times\vec{c})\times(\vec{r}\times\vec{a})=[\vec{b}\ \vec{c}\ \vec{a}]\,\vec{r}-[\vec{b}\ \vec{c}\ \vec{r}]\,\vec{a}$

$(\vec{c}\times\vec{a})\times(\vec{r}\times\vec{b})=[\vec{c}\ \vec{a}\ \vec{b}]\,\vec{r}-[\vec{c}\ \vec{a}\ \vec{r}]\,\vec{b}$

$\therefore$ the expression

$=3[\vec{a}\ \vec{b}\ \vec{c}]\,\vec{r}-\{[\vec{a}\ \vec{b}\ \vec{r}]\,\vec{c}+[\vec{b}\ \vec{c}\ \vec{r}]\,\vec{a}+[\vec{c}\ \vec{a}\ \vec{r}]\,\vec{b}$ ... (2)

Again, $(\vec{a}\times\vec{b})\times(\vec{r}\times\vec{c})=[\vec{a}\ \vec{r}\ \vec{c}]\,\vec{b}-[\vec{b}\ \vec{r}\ \vec{c}]\,\vec{a}$ ... (3)

From (1) and (3), $[\vec{a}\ \vec{b}\ \vec{c}]\,\vec{r}-[\vec{a}\ \vec{b}\ \vec{r}]\,\vec{c}=[\vec{a}\ \vec{r}\ \vec{c}]\,\vec{b}-[\vec{b}\ \vec{r}\ \vec{c}]\,\vec{a}$

$\therefore\quad [\vec{a}\ \vec{b}\ \vec{c}]\,\vec{r}=[\vec{a}\ \vec{b}\ \vec{r}]\,\vec{c}+[\vec{c}\ \vec{a}\ \vec{r}]\,\vec{b}+[\vec{b}\ \vec{c}\ \vec{r}]\,\vec{a}$

$\therefore$ from (2), the expression $=3[\vec{a}\ \vec{b}\ \vec{c}]\,\vec{r}-[\vec{a}\ \vec{b}\ \vec{c}]\,\vec{r}=2[\vec{a}\ \vec{b}\ \vec{c}]\,\vec{r}.$

72. $\vec{a}\times(\vec{b}\times\vec{c})=\frac{1}{2}\vec{b}\quad\Rightarrow(\vec{a}\cdot\vec{c})\,\vec{b}-(\vec{a}\cdot\vec{b})\,\vec{c}=\frac{1}{2}\vec{b}$

$\therefore\quad \vec{a}\cdot\vec{c}=\frac{1}{2},\quad \vec{a}\cdot\vec{b}=0$

$\therefore\quad |\vec{a}|\,|\vec{c}|\cos\beta=\frac{1}{2},\quad |\vec{a}|\,|\vec{b}|\cos\alpha=0$

or $\cos\beta=\frac{1}{2},\quad \cos\alpha=0\quad\Rightarrow\beta=\frac{\pi}{3},\ \alpha-\frac{\pi}{2}.$

73. $\vec{a}\times\vec{b}=(2\vec{i}+\vec{j}-2\vec{k})\times(\vec{i}+\vec{j})=2\vec{i}-2\vec{j}+\vec{k}.$

$\therefore\quad |(\vec{a}\times\vec{b})\times\vec{c}|=|\vec{a}\times\vec{b}|\,|\vec{b}|\sin 30°$

$=\sqrt{2^2+(-2)^2+1^2}\cdot|\vec{c}|\cdot\frac{1}{2}=\frac{3}{2}|\vec{c}|.$

Now, $|\vec{c}-\vec{a}| = 2\sqrt{2} \qquad \Rightarrow (\vec{c}-\vec{a})^2 = 8$

or $\quad |\vec{c}|^2 + |\vec{a}|^2 - 2\vec{c}\cdot\vec{a} = 8$

or $\quad |\vec{c}|^2 + (\sqrt{2^2+1^2+(-2)^2})^2 - 2|\vec{c}| = 8$

or $\quad |\vec{c}|^2 - 2|\vec{c}| + 1 = 0$

$\therefore \quad |\vec{c}| = 1$

$\therefore \quad |(\vec{a}\times\vec{b})\times\vec{c}| = |\vec{a}\times\vec{b}|\,|\vec{c}| \sin 30° = \frac{3}{2}.$

74. $(\vec{a}\times\vec{b})\times\vec{b} = (\vec{a}\cdot\vec{b})\vec{b} - (\vec{b}\cdot\vec{b})\vec{a}$

$\therefore \quad (\vec{a}\cdot\vec{b})\vec{b} = (\vec{a}\times\vec{b})\times\vec{b} + \vec{a}$

$\therefore \quad \vec{u} = \vec{a} - \{(\vec{a}\times\vec{b})\times\vec{b} + \vec{a}\} = \vec{b}\times(\vec{a}\times\vec{b})$

$\therefore \quad \vec{u} = \vec{b}\times\vec{v} \qquad \Rightarrow |\vec{u}| = |\vec{b}|\,|\vec{v}|$ because $\vec{b}\cdot\vec{v} = 0$, i.e., $\vec{b}\perp\vec{v}$

$\Rightarrow \quad |\vec{u}| = |\vec{v}|$ because $|\vec{b}| = 1$.

75. $[\vec{a}\times\vec{b} \;\; \vec{b}\times\vec{c} \;\; \vec{c}\times\vec{a}] = (\vec{a}\times\vec{b})\times(\vec{b}\times\vec{c})\cdot\vec{c}\times\vec{a}$

$= \{(\vec{a}\times\vec{b}\cdot\vec{c})\vec{b} - (\vec{a}\times\vec{b}\cdot\vec{b})\vec{c}\}\cdot\vec{c}\times\vec{a}$

$= [\vec{a}\;\vec{b}\;\vec{c}][\vec{b}\;\vec{c}\;\vec{a}] = [\vec{a}\;\vec{b}\;\vec{c}]^2 = 4^2 = 16.$

76. $\vec{d}\cdot\vec{a} = \lambda[\vec{b}\;\vec{c}\;\vec{a}] + \mu[\vec{c}\;\vec{a}\;\vec{a}] + \nu[\vec{a}\;\vec{b}\;\vec{a}] = \lambda[\vec{a}\;\vec{b}\;\vec{c}].$

Similarly, $\vec{d}\cdot\vec{b} = \mu[\vec{c}\;\vec{a}\;\vec{b}]$ and $\vec{d}\cdot\vec{c} = \nu[\vec{a}\;\vec{b}\;\vec{c}]$

$\therefore \quad \lambda = \frac{\vec{d}\cdot\vec{a}}{[\vec{a}\;\vec{b}\;\vec{c}]}, \quad \mu = \frac{\vec{d}\cdot\vec{b}}{[\vec{a}\;\vec{b}\;\vec{c}]}, \quad \nu = \frac{\vec{d}\cdot\vec{c}}{[\vec{a}\;\vec{b}\;\vec{c}]}$

$\therefore \quad \vec{d} = \frac{\vec{d}\cdot\vec{a}}{[\vec{a}\;\vec{b}\;\vec{c}]}\vec{b}\times\vec{c} + \frac{\vec{d}\cdot\vec{b}}{[\vec{a}\;\vec{b}\;\vec{c}]}\vec{c}\times\vec{a} + \frac{\vec{d}\cdot\vec{c}}{[\vec{a}\;\vec{b}\;\vec{c}]}\vec{a}\times\vec{b}$

$\therefore$ the expression $= |[\vec{a}\;\vec{b}\;\vec{c}]\vec{d}| = |[\vec{a}\;\vec{b}\;\vec{c}]|\,|\vec{d}| = |[\vec{a}\;\vec{b}\;\vec{c}]|.$

77. $\vec{a}\times(\vec{b}\times\vec{c}) + \vec{b}\times(\vec{c}\times\vec{a}) + \vec{c}\times(\vec{a}\times\vec{b})$

$= (\vec{a}\cdot\vec{c})\vec{b} - (\vec{a}\cdot\vec{b})\vec{c} + (\vec{b}\cdot\vec{a})\vec{c} - (\vec{b}\cdot\vec{c})\vec{a} + (\vec{c}\cdot\vec{b})\vec{a} - (\vec{c}\cdot\vec{a})\vec{b} = \vec{0}$

$\therefore$ the given vectors are linearly dependent.

78. $[\vec{b}\ \vec{c}\ \vec{b}\times\vec{c}] = \vec{b}\cdot\vec{c}\times(\vec{b}\times\vec{c}) = \vec{b}\cdot\{(\vec{c}\cdot\vec{c})\vec{b} - (\vec{c}\cdot\vec{b})\vec{c}\}$

$$= (\vec{c}\cdot\vec{c})(\vec{b}\cdot\vec{b}) - (\vec{b}\cdot\vec{c})^2.$$

$\therefore \quad [\vec{b}\ \vec{c}\ \vec{b}\times\vec{c}] + (\vec{b}\cdot\vec{c})^2 = |\vec{c}|^2\,|\vec{b}|^2.$

79. Let their plane be π. $\vec{a}\times\vec{b}$ is perpendicular to π and $\vec{c}\times\vec{d}$ is also perpendicular to π. So, $\vec{a}\times\vec{b} \parallel \vec{c}\times\vec{d}$. Hence, $(\vec{a}\times\vec{b})\times(\vec{c}\times\vec{d}) = \vec{0}$.

80. $(\vec{a}\cdot\vec{b})\vec{a} - (\vec{a}\cdot\vec{a})\vec{b} = (\vec{b}\cdot\vec{c})\vec{b} - (\vec{b}\cdot\vec{b})\vec{c}$. Taking dot product with $\vec{b}\times\vec{c}$,

$$(\vec{a}\cdot\vec{b})[\vec{a}\ \vec{b}\ \vec{c}] = 0 \quad \Rightarrow [\vec{a}\ \vec{b}\ \vec{c}] = 0.$$

81. Here $|\vec{a}\times\vec{b}| = q$ and

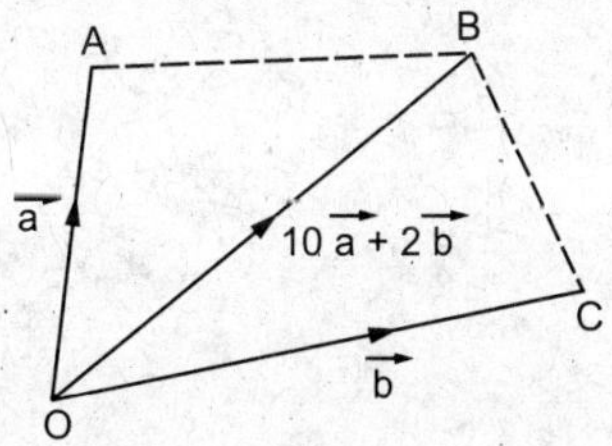

$$\frac{1}{2}|\vec{b}\times(10\vec{a}+2\vec{b})| + \frac{1}{2}|\vec{a}\times(10\vec{a}+2\vec{b})| = p$$

$\therefore \quad 5|\vec{b}\times\vec{a}| + |\vec{a}\times\vec{b}| = p$

or $\quad 6q = p \qquad \therefore \ \frac{p}{q} = 6.$

82. $\vec{AB} = \vec{OB} - \vec{OA} = (2\vec{i}+\vec{j}-2\vec{k}) - (\vec{i}-\vec{j}-3\vec{k}) = \vec{i}+2\vec{j}+\vec{k}.$

$\vec{AC} = \vec{OC} - \vec{OA} = (-5\vec{i}+2\vec{j}-6\vec{k}) - (\vec{i}-\vec{j}-3\vec{k}) = -6\vec{i}+3\vec{j}-3\vec{k}.$

A vector along the bisector of the angle BAC

$$= \frac{\vec{AB}}{|\vec{AB}|} + \frac{\vec{AC}}{|\vec{AC}|} = \frac{\vec{i}+2\vec{j}+\vec{k}}{\sqrt{1^2+2^2+1^2}} + \frac{-6\vec{i}+3\vec{j}-3\vec{k}}{\sqrt{(-6)^2+3^2+(-3)^2}}$$

$$= \frac{1}{\sqrt{6}}(\vec{i}+2\vec{j}+\vec{k}) + \frac{1}{3\sqrt{6}}(-6\vec{i}+3\vec{j}-3\vec{k}) = \frac{1}{3\sqrt{6}}(-3\vec{i}+9\vec{j}) = \frac{-\vec{i}+3\vec{j}}{\sqrt{6}}.$$

$\therefore \quad$ the unit vector along AD $= \dfrac{-\vec{i}+3\vec{j}}{\sqrt{10}}.$

$\therefore \quad \vec{AD} = \dfrac{-\vec{i}+3\vec{j}}{10}AD.$

As D is on BC, $\vec{BD} = t\vec{BC}$.

$\therefore \quad \vec{BA} + \vec{AD} = t(\vec{BA} + \vec{AC})$

or $\quad -\vec{i}-2\vec{j}-\vec{k}+\frac{-\vec{i}+3\vec{j}}{10}AD = t\{-\vec{i}-2\vec{j}-\vec{k}-6\vec{i}+3\vec{j}-3\vec{k}\}$

$$= t(-7\vec{i}+\vec{j}-4\vec{k})$$

$$\Rightarrow \quad -1-\frac{AD}{10} = -7t, \quad -2+\frac{3}{10}AD = t, \quad -1 = -4t.$$

$$\therefore \quad t = \frac{1}{4}$$

$$\therefore \quad -1-\frac{AD}{10} = -\frac{7}{4} \quad \text{or} \quad \frac{AD}{10} = \frac{3}{4}$$

$$\therefore \quad AD = \frac{15}{2}.$$

83. 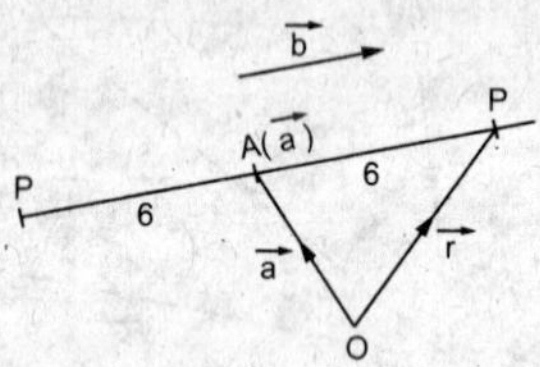

Clearly, $\vec{r} = \vec{a} + \vec{AP} = \vec{a} \pm 6\frac{\vec{b}}{|\vec{b}|}$.

84. $\vec{AB} = \vec{OB} - \vec{OA} = \{2\vec{i} + (2-y)\vec{j} + 2\vec{k}\} - \{(2-x)\vec{i} + 2\vec{j} + 2\vec{k}\} = x\vec{i} - y\vec{j}$

$\vec{AC} = \vec{OC} - \vec{OA} = \{2\vec{i} + 2\vec{j} + (2-z)\vec{k}\} - \{(2-x)\vec{i} + 2\vec{j} + 2\vec{k}\} = x\vec{i} - z\vec{k}$

$\vec{AD} = \vec{OD} - \vec{OA} = \{\vec{i} + \vec{j} + \vec{k}\} - \{(2-x)\vec{i} + 2\vec{j} + 2\vec{k}\} = (x-1)\vec{i} - \vec{j} - \vec{k}.$

As these vectors are coplanar, $\begin{vmatrix} x & -y & 0 \\ x & 0 & -z \\ x-1 & -1 & -1 \end{vmatrix} = 0.$

On simplification, $\frac{1}{x} + \frac{1}{y} + \frac{1}{z} = 1.$

85. A point equidistant from AB and AC is on the bisectors of the angle BAC.

A vector along the internal bisector of the angle BAC $= \frac{\vec{AB}}{|\vec{AB}|} + \frac{\vec{AC}}{|\vec{AC}|}$

$$= \frac{3\vec{i}+\vec{j}-\vec{k}}{\sqrt{3^2+1^2+(-1)^2}} + \frac{\vec{i}-\vec{j}+3\vec{k}}{\sqrt{1^2+(-1)^2+3^2}}$$

$$= \frac{1}{\sqrt{11}}(4\vec{i}+2\vec{k})$$

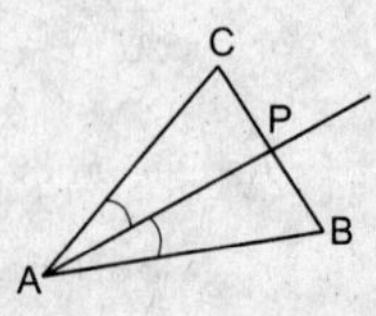

$\therefore \quad \vec{AP} = t(2\vec{i} + \vec{k})$

$\therefore \quad \vec{BP} = \vec{AP} - \vec{AB} = t(2\vec{i} + \vec{k}) - (3\vec{i} + \vec{j} - \vec{k})$

$\therefore \quad \vec{BP} = (2t - 3)\,\vec{i} - \vec{j} + (t+1)\,\vec{k}.$

Also, $\vec{BC} = \vec{AC} - \vec{AB} = (\vec{i} - \vec{j} + 3\vec{k}) - (3\vec{i} + \vec{j} - \vec{k}) = -2\vec{i} - 2\vec{j} + 4\vec{k}.$

But $\vec{BP} = s\vec{BC}$

$\therefore \quad (2t-3)\,\vec{i} - \vec{j} + (t+1)\,\vec{k} = s(-2\vec{i} - 2\vec{j} + 4\vec{k})$

or $\quad 2t - 3 = -2s, \quad -1 = -2s, \quad t + 1 = 4s$

$\therefore \quad s = \frac{1}{2}$ and $t = 1$

$\therefore \quad \vec{AP} = 2\vec{i} + \vec{k}.$

87. $\text{ar}(\Delta ABC) = \frac{1}{2}|\vec{b} \times \vec{c}| = \frac{1}{2}|\vec{b} - \vec{c}|\,h,$

where h = the length of the perpendicular.

$\therefore \quad h = \dfrac{|\vec{b} \times \vec{c}|}{|\vec{b} - \vec{c}|}.$

88.

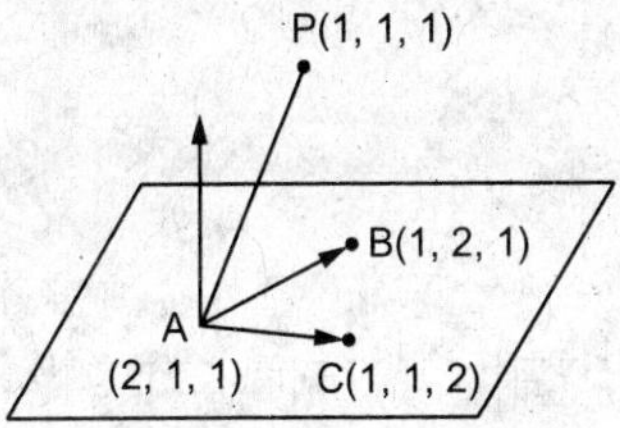

$\vec{AP} = \vec{OP} - \vec{OA}$

$= (\vec{i} + \vec{j} + \vec{k}) - (2\vec{i} + \vec{j} + \vec{k}) = -\vec{i}.$

$\vec{AB} = \vec{OB} - \vec{OA}$

$= (\vec{i} + 2\vec{j} + \vec{k}) - (2\vec{i} + \vec{j} + \vec{k}) = -\vec{i} + \vec{j}.$

$\vec{AC} = \vec{OC} - \vec{OA}$

$= (\vec{i} + \vec{j} + 2\vec{k}) - (2\vec{i} + \vec{j} + \vec{k}) = -\vec{i} + \vec{k}.$

$\vec{AC} \times \vec{AB} = (-\vec{i} + \vec{k}) \times (-\vec{i} + \vec{j}) = -\vec{k} - \vec{j} - \vec{i}$

$\therefore$ unit vector perpendicular to the plane ABC $= \dfrac{-\vec{k} - \vec{j} - \vec{i}}{\sqrt{3}}.$

The required distance $= \vec{AP} \cdot \dfrac{-\vec{k} - \vec{j} - \vec{i}}{\sqrt{3}} = (-\vec{i}) \cdot \dfrac{-\vec{k} - \vec{j} - \vec{i}}{\sqrt{3}} = \dfrac{1}{\sqrt{3}}.$

89. $\vec{r} = 3\vec{i} - \vec{j} + t(\vec{i} + 2\vec{j} + 3\vec{k})$. So, a vector parallel to the line is $\vec{b} = \vec{i} + 2\vec{j} + 3\vec{k}$

$\therefore$ unit vector along the line is $\dfrac{\vec{i} + 2\vec{j} + 3\vec{k}}{\sqrt{1^2 + 2^2 + 3^2}}$, i.e., $\dfrac{\vec{i} + 2\vec{j} + 3\vec{k}}{\sqrt{14}}.$

$\therefore$ the projection $= (\vec{i}+\vec{j}+\vec{k}) \cdot \dfrac{\vec{i}+2\vec{j}+3\vec{k}}{\sqrt{14}} = \dfrac{1+2+3}{\sqrt{14}} = \dfrac{6}{\sqrt{14}}$.

90.

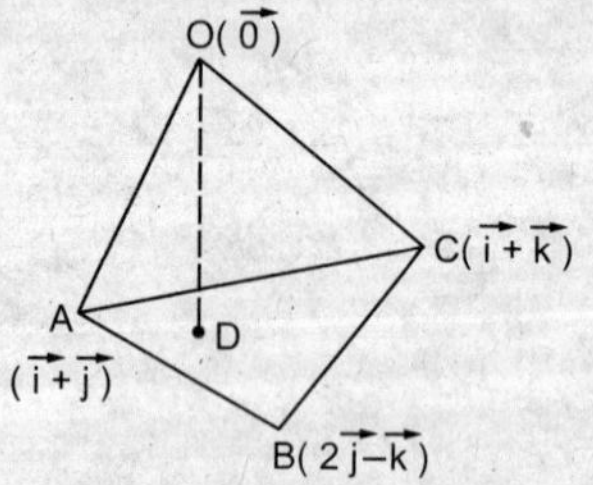

Volume of the tetrahedron

$$= V = \left|\frac{1}{6}[\vec{AB}\ \ \vec{AC}\ \ \vec{AO}]\right|.$$

Now,

$$\vec{AB} = \vec{OB} - \vec{OA} = 2\vec{j} - \vec{k} - (\vec{i}+\vec{j})$$

$$= -\vec{i}+\vec{j}-\vec{k}.$$

$$\vec{AC} = \vec{OC} - \vec{OA} = \vec{i}+\vec{k} - (\vec{i}+\vec{j}) = \vec{k}-\vec{j}.$$

$$\therefore \quad V = \left|\frac{1}{6}[-\vec{i}+\vec{j}-\vec{k}\quad \vec{k}-\vec{j}\quad -\vec{i}-\vec{j}]\right| = \left|\frac{1}{6}\begin{vmatrix}-1 & 1 & -1\\ 0 & -1 & 1\\ -1 & -1 & 0\end{vmatrix}\right| = \frac{1}{6}.$$

91. $\vec{OP} = \vec{r} = \vec{OA} \pm 1 \cdot \dfrac{\vec{AB}}{|\vec{AB}|}$

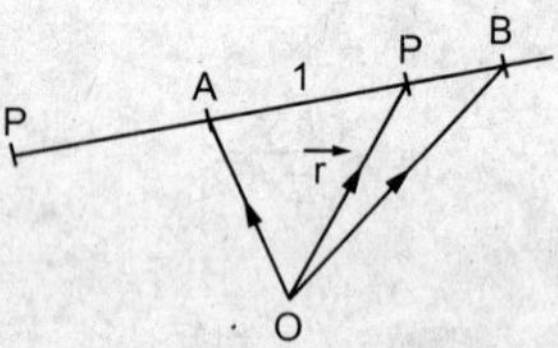

$$= \vec{i}+\vec{j}-2\vec{k} \pm \frac{\vec{OB}-\vec{OA}}{|\vec{OB}-\vec{OA}|}$$

$$= \vec{i}+\vec{j}-2\vec{k} \pm \frac{-4\vec{j}+3\vec{k}}{|-4\vec{j}+3\vec{k}|}$$

$$= \vec{i}+\vec{j}-2\vec{k} \pm \frac{1}{5}(\vec{i}-4\vec{j}+3\vec{k})$$

$$= \left(1+\frac{1}{5}\right)\vec{i}+\left(1-\frac{4}{5}\right)\vec{j}+\left(-2+\frac{3}{5}\right)\vec{k},\ \left(1-\frac{1}{5}\right)\vec{i}+\left(1+\frac{4}{5}\right)\vec{j}+\left(-2-\frac{3}{5}\right)\vec{k}.$$

92. $\vec{a} = 2\vec{i}-2\vec{j}+\vec{k},\ \vec{b} = \vec{i}+2\vec{j}-2\vec{k}$

$\therefore \quad |\vec{a}| = 3, \quad |\vec{b}| = 3$

$\therefore$ a vector along bisectors

$$= \frac{\vec{a}}{|\vec{a}|} \pm \frac{\vec{b}}{|\vec{b}|} = \frac{2\vec{i}-2\vec{j}+\vec{k}}{3} \pm \frac{\vec{i}+2\vec{j}-2\vec{k}}{3}$$

$$= \vec{i} - \frac{1}{3}\vec{k},\quad \frac{1}{3}\vec{i} - \frac{4}{3}\vec{j} + \vec{k}$$

$\therefore$ the required vector $= 2 \cdot \dfrac{\vec{i} - \frac{1}{3}\vec{k}}{\sqrt{1^2 + \left(-\frac{1}{3}\right)^2}}, 2 \cdot \dfrac{\frac{1}{3}\vec{i} - \frac{4}{3}\vec{j} + \vec{k}}{\sqrt{\left(\frac{1}{3}\right)^2 + \left(-\frac{4}{3}\right)^2 + 1^2}}$.

93. $\vec{a} = t(\vec{i} + \vec{j} + 2\vec{k}) + s(\vec{i} + 2\vec{j} + \vec{k})$. Here, $\vec{a} \cdot (\vec{i} + \vec{j} + \vec{k}) = 0$

and $t + s + t + 2s + 2t + s = 0 \quad \Rightarrow t + s = 0$.

$\therefore \quad \vec{a} = t(\vec{i} + \vec{j} + 2\vec{k} - \vec{i} - 2\vec{j} - \vec{k}) = t(-\vec{j} + \vec{k})$.

But $|\vec{a}| = 1$. So, $1 = 2t^2$

$\therefore \quad t = \pm \dfrac{1}{\sqrt{2}}$ · Therefore, $\vec{a} = \pm \dfrac{-\vec{j} + \vec{k}}{\sqrt{2}}$.

94. See the hint to Q. No. 12.

95. $(\vec{a} \times \vec{b})^2 = (\vec{a} \times \vec{b}) \cdot (\vec{a} \times \vec{b}) = \begin{vmatrix} \vec{a} \cdot \vec{a} & \vec{a} \cdot \vec{b} \\ \vec{b} \cdot \vec{a} & \vec{b} \cdot \vec{b} \end{vmatrix}$

$$= \begin{vmatrix} 16 & 4 \cdot 2 \cdot \frac{\sqrt{3}}{2} \\ 4 \cdot 2 \cdot \frac{\sqrt{3}}{2} & 4 \end{vmatrix} = 64 - 48 = 16 = |\vec{a}|^2.$$

96. Points A, B, C are collinear if $\lambda\vec{CA} = \mu\vec{BC}$ or $\vec{CA} \times \vec{CB} = 0$.

97. Let $\vec{c} = \lambda \vec{i} + \mu \vec{j}$. But $\vec{b} \cdot \vec{c} = 0$.

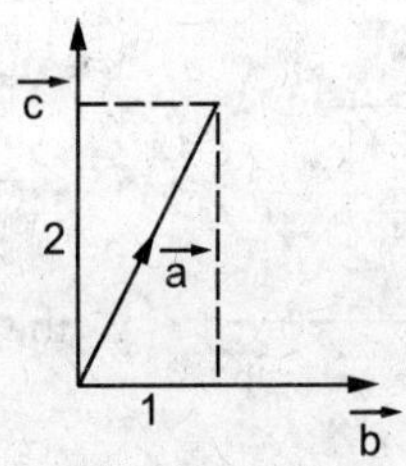

$\therefore \quad 4\lambda + 3\mu = 0$.

$\therefore \quad \dfrac{\lambda}{3} = \dfrac{\mu}{-4} = t$.

$\therefore \quad \vec{c} = t(3\vec{i} - 4\vec{j})$.

$\therefore \quad \dfrac{\vec{c}}{|\vec{c}|} = \pm \dfrac{3\vec{i} - 4\vec{j}}{5}$.

From the question, $\vec{a} = 1 \dfrac{\vec{b}}{|\vec{b}|} + 2 \dfrac{\vec{c}}{|\vec{c}|} = \dfrac{4\vec{i} + 3\vec{j}}{5} \pm 2 \dfrac{3\vec{i} - 4\vec{j}}{5}$.

98. $(\vec{a}\times\vec{b})\times(\vec{c}\times\vec{r})=[\vec{a}\ \vec{b}\ \vec{r}]\vec{c}-[\vec{a}\ \vec{b}\ \vec{c}]\vec{r}.$

Also, $(\vec{a}\times\vec{b})\times(\vec{c}\times\vec{r})=[\vec{a}\ \vec{c}\ \vec{r}]\vec{b}-[\vec{b}\ \vec{c}\ \vec{r}]\vec{a}$

$\therefore\quad [\vec{a}\ \vec{b}\ \vec{r}]\vec{c}-[\vec{a}\ \vec{b}\ \vec{c}]\vec{r}=[\vec{a}\ \vec{c}\ \vec{r}]\vec{b}-[\vec{b}\ \vec{c}\ \vec{r}]\vec{a}.$

99. $\vec{b}\times\vec{c}=\vec{a}\quad\Rightarrow [\vec{a}\ \vec{b}\ \vec{c}]=\vec{a}\cdot\vec{a}=|\vec{a}|^2.$

Similarly, $[\vec{a}\ \vec{b}\ \vec{c}]=|\vec{b}|^2=|\vec{c}|^2$

$\therefore\quad |\vec{a}|=|\vec{b}|=|\vec{c}|=\sqrt{[\vec{a}\ \vec{b}\ \vec{c}]}.$

Now, $[\vec{b}\times\vec{c}\quad \vec{c}\times\vec{a}\quad \vec{a}\times\vec{b}]=[\vec{a}\ \vec{b}\ \vec{c}]^2$ (see the hint to Q. No.75)

$\therefore\quad [\vec{a}\ \vec{b}\ \vec{c}]=[\vec{a}\ \vec{b}\ \vec{c}]^2$ (from the question).

But $[\vec{a}\ \vec{b}\ \vec{c}]\neq 0$. So, $[\vec{a}\ \vec{b}\ \vec{c}]=1$. Hence, $|\vec{a}|=|\vec{b}|=|\vec{c}|=1$.

102. $\because\ \vec{b}=\vec{c}\times\vec{a}$, by definition of cross product $\vec{a}\perp\vec{b}$.

$\because\ \vec{c}=\vec{a}\times\vec{b}$, by definition of cross product $\vec{a}\perp\vec{c}$. So, $\vec{a}\parallel\vec{b}\times\vec{c}$.

103. $\vec{x}\times\vec{b}=\vec{c}\times\vec{b}\quad\Rightarrow\ \vec{a}\times(\vec{x}\times\vec{b})=\vec{a}\times(\vec{c}\times\vec{b})$

or $\quad (\vec{a}\cdot\vec{b})\vec{x}-(\vec{a}\cdot\vec{x})\vec{b}=\vec{a}\times(\vec{c}\times\vec{b})$

or $\quad \vec{x}=\dfrac{\vec{a}\times(\vec{c}\times\vec{b})}{\vec{a}\cdot\vec{b}}\qquad (\because\ \vec{a}\perp\vec{x}).$

Also, $\vec{a}\times(\vec{c}\times\vec{b})=-(\vec{c}\times\vec{b})\times\vec{a}=(\vec{b}\times\vec{c})\times\vec{a}.$

104.

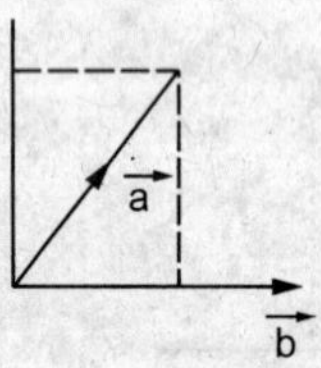

The magnitude of the resolved part along the vector $\vec{b}$

$$=\vec{a}\cdot\frac{\vec{b}}{|\vec{b}|}$$

$\therefore$ the resolved part along the vector $\vec{b}$

$$=\left(\frac{\vec{a}\cdot\vec{b}}{|\vec{b}|}\right)\frac{\vec{b}}{|\vec{b}|}=\frac{(\vec{a}\cdot\vec{b})\vec{b}}{\vec{b}^2}$$

$\therefore$ resolved part perpendicular to $\vec{b}=\vec{a}-\left(\dfrac{\vec{a}\cdot\vec{b}}{|\vec{b}|}\right)\dfrac{\vec{b}}{|\vec{b}|}$

$$=\frac{(\vec{b}\cdot\vec{b})\vec{a}-(\vec{a}\cdot\vec{b})\vec{b}}{\vec{b}^2}=\frac{\vec{b}\times(\vec{a}\times\vec{b})}{\vec{b}^2}.$$

105. $(\vec{a}\times\vec{b})\cdot(\vec{c}\times\vec{d})=$ scalar triple product of $\vec{a},\ \vec{b},\ \vec{c}\times\vec{d}=\vec{a}\cdot\{\vec{b}\times(\vec{c}\times\vec{d})\}$

$(\vec{a}\times\vec{b})\cdot(\vec{c}\times\vec{d})=$ scalar triple product of $\vec{a}\times\vec{b},\ \vec{c},\ \vec{d}=\{(\vec{a}\times\vec{b})\times\vec{c}\}\cdot\vec{d}$

$$(\vec{a}\times\vec{b})\cdot(\vec{c}\times\vec{d})=\vec{a}\cdot\{\vec{b}\times(\vec{c}\times\vec{d})\}=\vec{a}\cdot\{(\vec{b}\cdot\vec{d})\vec{c}-(\vec{b}\cdot\vec{c})\vec{d}\}$$

$$=(\vec{b}\cdot\vec{d})(\vec{a}\cdot\vec{c})-(\vec{b}\cdot\vec{c})(\vec{a}\cdot\vec{d}).$$

106.

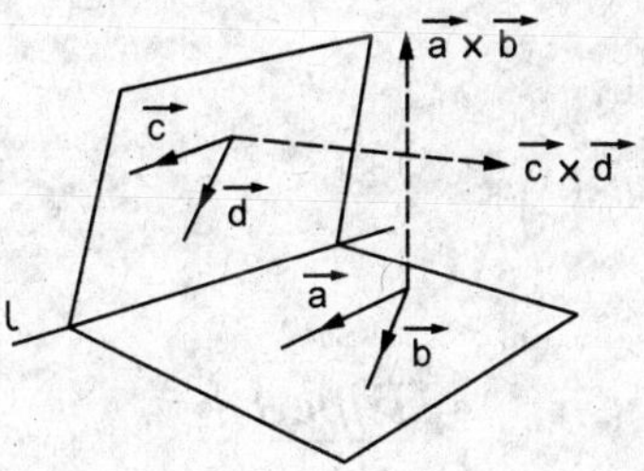

$(\vec{a}\times\vec{b})\times(\vec{c}\times\vec{d})=$ a vector perpendicular to $\vec{a}\times\vec{b}$ and $\vec{c}\times\vec{d}$.

If l is the line of intersection of the two planes of $\vec{a},\vec{b}$ and $\vec{c},\vec{d}$ then $l\perp\vec{a}\times\vec{b}$ and $l\perp\vec{c}\times\vec{d}$. So, $(\vec{a}\times\vec{b})\times(\vec{c}\times\vec{d})$ is along l which is equally inclined to $\vec{a}\times\vec{b}$ and $\vec{c}\times\vec{d}$.

□

2

Application of Vectors in Mechanics

• *Type 1* •

Choose the most appropriate option (a, b, c or d).

1. P is a point in the plane of the $\triangle ABC$ whose orthocentre is H and the circumcentre is O. Forces $\overrightarrow{AP}, \overrightarrow{BP}, \overrightarrow{CP}$ and $\overrightarrow{PH}$ act at P. The force that will keep the given forces in equilibrium is

(a) $2\overrightarrow{OP}$ (b) $3\overrightarrow{PO}$

(c) $2\overrightarrow{PO}$ (d) none of these

2. Three forces $\overrightarrow{P}, \overrightarrow{Q}$ and $\overrightarrow{R}$, each of 15 units, act along AB, BC and CA respectively. The position vectors of A, B and C are $\overrightarrow{OA} = 2\vec{i} - \vec{j} + 3\vec{k}$, $\overrightarrow{OB} = 5\vec{i} + 3\vec{j} - 2\vec{k}$ and $\overrightarrow{OC} = -2\vec{i} + 2\vec{j} + 3\vec{k}$ respectively. The resultant force vector is

(a) $\left(12 + \frac{9}{\sqrt{2}} - 7\sqrt{3}\right)\vec{i} - (9 - 6\sqrt{2} + \sqrt{3})\,\vec{j} + \left(5\sqrt{3} - \frac{15}{\sqrt{2}}\right)\vec{k}$

(b) $\left(12 + \frac{9}{\sqrt{2}} - 7\sqrt{3}\right)\vec{i} + (9 - 6\sqrt{3} + \sqrt{3})\,\vec{j} + \left(\frac{15}{\sqrt{2}} - 5\sqrt{3}\right)\vec{k}$

(c) $75\vec{i} + 60\vec{j} + 60\vec{k}$

(d) none of these

3. A ship is sailing towards north at a speed of 1.25 m/s. The current is taking it towards east at the rate of 1 m/s. A sailor is climbing a vertical pole on the ship at the rate of 0.5 m/s. The magnitude of the velocity of the sailor in space is

(a) 2.75 m/s (b) $\frac{3\sqrt{5}}{4}$ m/s

(c) $\frac{3\sqrt{5}}{2}$ m/s (d) none of these

4. A force $10\vec{i} - 5\vec{j} + 7\vec{k}$ displaces a particle from the point A to the point B. The position vectors of A and B are $3\vec{i} - \vec{j} + 2\vec{k}$ and $\vec{i} + 3\vec{j} + 2\vec{k}$ respectively. Then the work done is

(a) 40 (b) 20

(c) 60 (d) none of these

5. Constant forces $\vec{P} = \vec{i} - 2\vec{j} + 3\vec{k}$, $\vec{Q} = -\vec{i} + 3\vec{j} - \vec{k}$ and $\vec{R} = 2\vec{i} - 4\vec{j} + 3\vec{k}$ act on a particle. The work done when the particle is displaced from a point A with position vector $4\vec{i} - 3\vec{j} - 2\vec{k}$ to the point B with position vector $6\vec{i} + \vec{j} - 3\vec{k}$ is

(a) 15 (b) 13

(c) $\sqrt{13}$ (d) none of these

6. The vertices of a triangle ABC are A(–1, 0, 2), B(1, 2, 0) and C(2, 3, 4). The moment of a force of magnitude 10 acting at A along AB about C is

(a) $\frac{50\sqrt{6}}{3}$ (b) $20\sqrt{6}$

(c) $\frac{50}{\sqrt{3}}$ (d) none of these

7. The vector moment about the point $\vec{i} + 2\vec{j} + 3\vec{k}$ of the resultant of the forces $\vec{i} - 2\vec{j} + 5\vec{k}$ and $3\vec{j} - 4\vec{k}$ acting at the point $-2\vec{i} + 3\vec{j} - \vec{k}$ is

(a) $5\vec{i} + \vec{j} - 4\vec{k}$ (b) $5\vec{i} - \vec{j} - 4\vec{k}$

(c) $3\vec{i} + \vec{j} - 4\vec{k}$ (d) none of these

8. A rigid body is rotating at 5 radians per second about an axis AB, where A and B are points whose position vectors are $2\vec{i} + \vec{j} + \vec{k}$ and $8\vec{i} - 2\vec{j} + 3\vec{k}$ respectively. The velocity of the particle of the body at the point whose position vector is $5\vec{i} + \vec{j} - \vec{k}$ is

(a) $\frac{15}{7}(2\vec{i} + 6\vec{j} + 3\vec{k})$ (b) $\frac{5}{7}(4\vec{i} - 6\vec{j} - 3\vec{k})$

(c) $\frac{15}{7}(4\vec{i} - 6\vec{j} - 3\vec{k})$ (d) none of these

• Type 2 •

Choose the correct options. One or more options may be correct.

9. A particle is in equilibrium when the forces $\vec{F_1} = -10\vec{k}$,

$\vec{F_2} = \frac{u}{13}(4\vec{i} - 12\vec{j} + 3\vec{k})$, $\vec{F_3} = \frac{v}{13}(-4\vec{i} - 12\vec{j} + 3\vec{k})$ and $\vec{F} = w(\cos\theta\,\vec{i} + \sin\theta\,\vec{j})$

act on it. Then

(a) $u = 65(1 - 3\cot\theta)$

(b) $v = \frac{65}{3} + 65\cot\theta$

(c) $w = 40\operatorname{cosec}\theta$

(d) none of these

10. The resolved parts of the force vector $5\vec{i} + 4\vec{j} + 2\vec{k}$ along and perpendicular to the vector $3\vec{i} + 4\vec{j} - 5\vec{k}$ are $\vec{u}$ and $\vec{v}$ respectively. Then

(a) $\vec{u} = \frac{21}{50}(3\vec{i} + 4\vec{j} - 5\vec{k})$

(b) $\vec{u} = \frac{21}{5\sqrt{2}}(3\vec{i} + 4\vec{j} - 5\vec{k})$

(c) $\vec{v} = \frac{1}{50}(187\vec{i} + 116\vec{j} + 205\vec{k})$

(d) $\frac{1}{5\sqrt{2}}(\vec{i} - 2\vec{j} - \vec{k})$

Answers

1. c	**2.** a	**3.** b	**4.** a	**5.** b
6. a	**7.** b	**8.** a	**9.** b, c	**10.** a, c

Hints and Solutions to Selected Questions

1. $\vec{AP} + \vec{PH} = \vec{AH}$, $\vec{BP} + \vec{CP} = 2\vec{MP}$

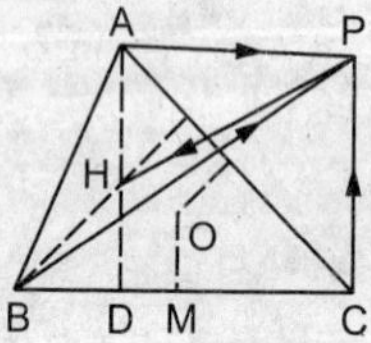

$\therefore \quad \vec{AP} + \vec{BP} + \vec{CP} + \vec{PH} = \vec{AH} + 2(\vec{MO} + \vec{OP})$

$= 2\vec{OM} + 2\vec{MO} + 2\vec{OP} \quad (\because \; AH = 2OM)$

$= 2\vec{OP}.$

$\therefore$ the required force $= 2\vec{PO}$.

2. $\vec{AB} = \vec{OB} - \vec{OA} = 3\vec{i} + 4\vec{j} - 5\vec{k}$

$\therefore \quad \dfrac{\vec{AB}}{|\vec{AB}|} = \dfrac{1}{5\sqrt{2}}(3\vec{i} + 4\vec{j} - 5\vec{k})$

$\therefore \quad \vec{P} = 15 \cdot \dfrac{1}{5\sqrt{2}}(3\vec{i} + 4\vec{j} - 5\vec{k}) = \dfrac{3}{\sqrt{2}}(3\vec{i} + 4\vec{j} - 5\vec{k}).$

Similarly, $\vec{Q} = 15 \cdot \dfrac{\vec{BC}}{|\vec{BC}|} = 15 \cdot \dfrac{-7\vec{i} - \vec{j} + 5\vec{k}}{5\sqrt{3}} = \sqrt{3}(-7\vec{i} - \vec{j} + 5\vec{k})$, and

$\vec{R} = 15 \cdot \dfrac{\vec{CA}}{|\vec{CA}|} = 15 \cdot \dfrac{4\vec{i} - 3\vec{j}}{5} = 3(4\vec{i} - 3\vec{j}).$

$\therefore$ the resultant

$$= \vec{P} + \vec{Q} + \vec{R} = \left(12 + \frac{9}{\sqrt{2}} - 7\sqrt{3}\right)\vec{i} + (-9 + 6\sqrt{2} - \sqrt{3})\,\vec{j} + \left(5\sqrt{3} - \frac{15}{\sqrt{2}}\right)\vec{k}.$$

3.

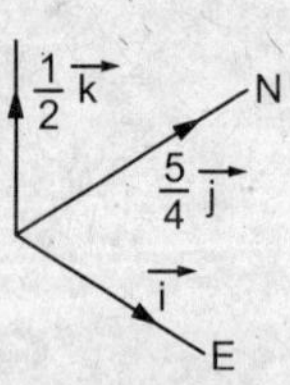

The velocity vector $= \vec{i} + \dfrac{5}{4}\vec{j} + \dfrac{1}{2}\vec{k}.$

$\therefore$ velocity $= \left|\vec{i} + \dfrac{5}{4}\vec{j} + \dfrac{1}{2}\vec{k}\right|$

$= \sqrt{1^2 + \left(\dfrac{5}{4}\right)^2 + \left(\dfrac{1}{2}\right)^2} = \dfrac{3\sqrt{5}}{4}.$

4. Work done $= |\vec{F} \cdot \vec{AB}| = |(10\vec{i} - 5\vec{j} + 7\vec{k}) \cdot (-2\vec{i} + 4\vec{j})| = |-20 - 20| = 40.$

5. Work done $= |(\vec{P} + \vec{Q} + \vec{R}) \cdot \vec{AB}| = |(2\vec{i} - 3\vec{j} + 5\vec{k}) \cdot (2\vec{i} + 4\vec{j} - \vec{k})| = 13.$

6. $\vec{AB} = \vec{OB} - \vec{OA} = \vec{i} + 2\vec{j} - (-\vec{i} + 2\vec{k}) = 2\vec{i} + 2\vec{j} - 2\vec{k}.$

$\therefore$ vector moment about C $= \vec{CA} \times \dfrac{10(2\vec{i} + 2\vec{j} - 2\vec{k})}{\sqrt{12}}$

$$= -(3\vec{i} + 3\vec{j} + 2\vec{k}) \times \frac{10}{\sqrt{3}}(\vec{i} + \vec{j} - \vec{k}) = \frac{10}{\sqrt{3}}(5\vec{i} - 5\vec{j}).$$

$\therefore$ moment $= \dfrac{10}{\sqrt{3}} \cdot \sqrt{5^2 + 5^2} = \dfrac{50\sqrt{2}}{\sqrt{3}} = \dfrac{50\sqrt{6}}{3}.$

7. Vector moment $= \vec{AB} \times (\vec{F_1} + \vec{F_2}) = (\vec{OB} - \vec{OA}) \times (\vec{F_1} + \vec{F_2})$

$$= \{(-2\vec{i} + 3\vec{j} - \vec{k}) - (\vec{i} + 2\vec{j} + 3\vec{k})\} \times (\vec{i} - 2\vec{j} + 5\vec{k} + 3\vec{j} - 4\vec{k})$$

$$= (-3\vec{i} + \vec{j} - 4\vec{k}) \times (\vec{i} + \vec{j} + \vec{k}) = 5\vec{i} - \vec{j} - 4\vec{k}.$$

8.

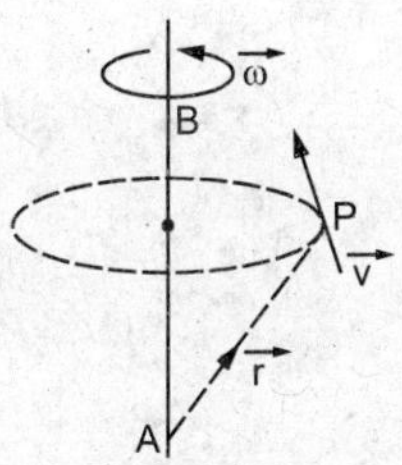

$\vec{v} = \vec{\omega} \times \vec{r}.$

Here, $\vec{\omega} = 5\dfrac{\vec{AB}}{|\vec{AB}|}$ and $\vec{r} = \vec{AP}$.

Now, $\vec{AB} = (8\vec{i} - 2\vec{j} + 3\vec{k}) - (2\vec{i} + \vec{j} + \vec{k})$

$$= 6\vec{i} - 3\vec{j} + 2\vec{k}$$

$\therefore \quad \vec{\omega} = 5 \cdot \dfrac{6\vec{i} - 3\vec{j} + 2\vec{k}}{\sqrt{6^2 + 3^2 + 2^2}} = \dfrac{5}{7}(6\vec{i} - 3\vec{j} + 2\vec{k})$

$\therefore \quad \vec{v} = \dfrac{5}{7}(6\vec{i} - 3\vec{j} + 2\vec{k}) \times (5\vec{i} + \vec{j} - \vec{k} - 2\vec{i} - \vec{j} - \vec{k})$

$$= \frac{5}{7}(6\vec{i} - 3\vec{j} + 2\vec{k}) \times (3\vec{i} - 2\vec{k}).$$

9. $\vec{F_1} + \vec{F_2} + \vec{F_3} + \vec{F_4} = 0.$

$\therefore \quad \dfrac{4u}{13} - \dfrac{4v}{13} + w\cos\theta = 0, \qquad -\dfrac{12}{13}u - \dfrac{12}{13}v + w\sin\theta = 0,$

$$-10 + \frac{3u}{13} + \frac{3v}{13} = 0.$$

Solving the three equations in u, v, w, we get

$$u = \frac{65}{3}(1 - 3\cot\theta), \quad v = \frac{65}{3}(1 + 3\cot\theta), \quad w = 40\operatorname{cosec}\theta.$$

10. Unit vector along $3\vec{i}+4\vec{j}-5\vec{k}=\dfrac{3\vec{i}+4\vec{j}-5\vec{k}}{5\sqrt{2}}$.

$$\therefore \quad \vec{u}=\left\{(5\vec{i}+4\vec{j}+2\vec{k})\cdot\frac{3\vec{i}+4\vec{j}-5\vec{k}}{5\sqrt{2}}\right\}\frac{3\vec{i}+4\vec{j}-5\vec{k}}{5\sqrt{2}}$$

$$=\frac{21}{50}(3\vec{i}+4\vec{j}-5\vec{k}).$$

$$\vec{v}=(5\vec{i}+4\vec{j}+2\vec{k})-\vec{u}.$$

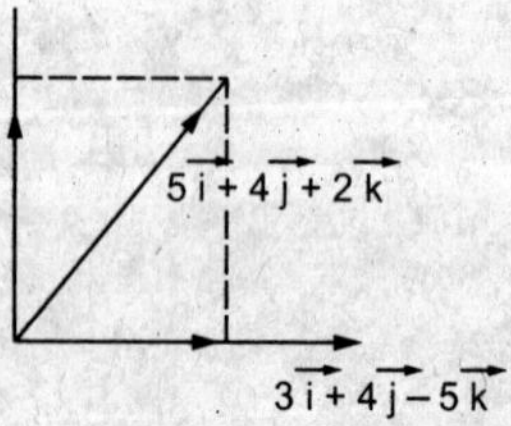

□

Part 6

Probability

1

Probability

• Type 1 •

Choose the most appropriate option (a, b, c or d).

1. Five boys and three girls are seated at random in a row. The probability that no boy sits between two girls is

(a) $\frac{1}{56}$ (b) $\frac{1}{8}$

(c) $\frac{3}{28}$ (d) none of these

2. In a convex hexagon two diagonals are drawn at random. The probability that the diagonals intersect at an interior point of the hexagon is

(a) $\frac{5}{12}$ (b) $\frac{7}{12}$

(c) $\frac{2}{5}$ (d) none of these

3. 4 five-rupee coins, 3 two-rupee coins and 2 one-rupee coins are stacked together in a column at random. The probability that the coins of the same denomination are consecutive is

(a) $\frac{13}{9!}$ (b) $\frac{1}{210}$

(c) $\frac{1}{35}$ (d) none of these

4. Two cards are drawn at random from a pack of 52 cards. The probability of getting at least a spade and an ace is

(a) $\frac{1}{34}$ (b) $\frac{8}{221}$

(c) $\frac{1}{26}$ (d) $\frac{2}{51}$

5. A five-digit number is written down at random. The probability that the number is divisible by 5 and no two consecutive digits are identical, is

(a) $\frac{1}{5}$ (b) $\frac{1}{5}\cdot\left(\frac{9}{10}\right)^3$

(c) $\left(\frac{3}{5}\right)^4$ (d) none of these

6. If the letters of the word ATTEMPT are written down at random, the chance that all *T*s are consecutive is

(a) $\frac{1}{42}$ (b) $\frac{6}{7}$

(c) $\frac{1}{7}$ (d) none of these

7. In a single cast with two dice the odds against drawing 7 is

(a) $\frac{1}{6}$ (b) $\frac{1}{12}$

(c) $5:1$ (d) $1:5$

8. 7 white balls and 3 black balls are placed in a row at random. The probability that no two black balls are adjacent is

(a) $\frac{1}{2}$ (b) $\frac{7}{15}$

(c) $\frac{2}{15}$ (d) $\frac{1}{3}$

9. 10 apples are distributed at random among 6 persons. The probability that at least one of them will receive none is

(a) $\frac{6}{143}$ (b) $\frac{^{14}C_4}{^{15}C_5}$

(c) $\frac{137}{143}$ (d) none of these

10. 4 gentlemen and 4 ladies take seats at random round a table. The probability that they are sitting alternately is

(a) $\frac{4}{35}$ (b) $\frac{1}{70}$

(c) $\frac{2}{35}$ (d) $\frac{1}{35}$

11. Let $x = 33^n$. The index n is given a positive integral value at random. The probability that the value of x will have 3 in the units place is

(a) $\frac{1}{4}$ (b) $\frac{1}{2}$

(c) $\frac{1}{3}$ (d) none of these

12. Three dice are thrown simultaneously. The probability of getting a sum of 15 is

(a) $\frac{1}{72}$ (b) $\frac{5}{36}$

(c) $\frac{5}{72}$ (d) none of these

13. Three dice are thrown. The probability of getting a sum which is a perfect square is

(a) $\frac{2}{5}$ (b) $\frac{9}{20}$

(c) $\frac{1}{4}$ (d) none of these

14. The probability of getting a sum of 12 in four throws of an ordinary dice is

(a) $\frac{1}{6}\left(\frac{5}{6}\right)^3$ (b) $\left(\frac{5}{6}\right)^4$

(c) $\frac{1}{36}\left(\frac{5}{6}\right)^2$ (d) none of these

15. Three different numbers are selected at random from the set $A = \{1, 2, 3, \ldots, 10\}$. The probability that the product of two of the numbers is equal to the third is

(a) $\frac{3}{4}$ (b) $\frac{1}{40}$

(c) $\frac{1}{8}$ (d) none of these

16. There are 7 seats in a row. Three persons take seats at random. The probability that the middle seat is always occupied and no two persons are consecutive is

(a) $\frac{9}{70}$ (b) $\frac{9}{35}$

(c) $\frac{4}{35}$ (d) none of these

17. A second-order determinant is written down at random using the numbers 1, –1 as elements. The probability that the value of the determinant is nonzero is

(a) $\frac{1}{2}$ (b) $\frac{3}{8}$

(c) $\frac{5}{8}$ (d) $\frac{1}{3}$

18. $x_1, x_2, x_3, \ldots x_{50}$ are fifty real numbers such that $x_r < x_{r+1}$ for $r = 1, 2, 3, \ldots, 49$. Five numbers out of these are picked up at random. The probability that the five numbers have x_{20} as the middle number is

(a) $\frac{{}^{20}C_2 \times {}^{30}C_2}{{}^{50}C_5}$ (b) $\frac{{}^{30}C_2 \times {}^{19}C_2}{{}^{50}C_5}$

(c) $\frac{{}^{19}C_2 \times {}^{31}C_3}{{}^{50}C_5}$ (d) none of these

19. Numbers 1, 2, 3, … , 100 are written down on each of the cards A, B and C. One number is selected at random from each of the cards. The probability that the numbers so selected can be the measures (in cm) of three sides of right-angled triangles no two of which are similar, is

(a) $\frac{4}{100^3}$ (b) $\frac{3}{50^3}$

(c) $\frac{3!}{100^3}$ (d) none of these

20. Three numbers are chosen at random without replacement from the set $A = \{x \mid 1 \le x \le 10, x \in N\}$. The probability that the minimum of the chosen numbers is 3 and maximum is 7, is

(a) $\frac{1}{12}$ (b) $\frac{1}{15}$

(c) $\frac{1}{40}$ (d) none of these

21. Three natural numbers are taken at random from the set $A = \{x \mid 1 \le x \le 100, x \in N\}$. The probability that the AM of the numbers taken is 25 is

(a) $\frac{^{77}C_2}{^{100}C_3}$ (b) $\frac{^{25}C_2}{^{100}C_3}$

(c) $\frac{^{74}C_{72}}{^{100}C_{97}}$ (d) none of these

22. Let S be the universal set and $n(X) = k$. The probability of selecting two subsets A and B of the set X such that $B = \overline{A}$ is

(a) $\frac{1}{2}$ (b) $\frac{1}{2^k - 1}$

(c) $\frac{1}{2^k}$ (d) $\frac{1}{3^k}$

23. From a group of 10 persons consisting of 5 lawyers, 3 doctors and 2 engineers, four persons are selected at random. The probability that the selection contains at least one of each category is

(a) $\frac{1}{2}$ (b) $\frac{1}{3}$

(c) $\frac{2}{3}$ (d) none of these

24. 10 different books and 2 different pens are given to 3 boys so that each gets equal number of things. The probability that the same boy does not receive both the pens is

(a) $\frac{5}{11}$ (b) $\frac{7}{11}$

(c) $\frac{2}{3}$ (d) $\frac{6}{11}$

25. Two distinct numbers are selected at random from the first twelve natural numbers. The probability that the sum will be divisible by 3 is

(a) $\frac{1}{3}$ (b) $\frac{23}{66}$

(c) $\frac{1}{2}$ (d) none of these

26. The probability of a number n showing in a throw of a dice marked 1 to 6 is proportional to n. Then the probability of the number 3 showing in a throw is

(a) $\frac{1}{2}$ (b) $\frac{1}{6}$

(c) $\frac{1}{7}$ (d) $\frac{1}{21}$

27. The probability that out of 10 persons, all born in April, at least two have the same birthday is

(a) $\frac{^{30}C_{10}}{(30)^{10}}$ (b) $1-\frac{^{30}C_{10}}{30\,!}$

(c) $\frac{(30)^{10}-{}^{30}C_{10}}{(30)^{10}}$ (d) none of these

28. If one ball is drawn at random from each of the three boxes containing 3 white and 1 black, 2 white and 2 black, 1 white and 3 black balls then the probability that 2 white and 1 black balls will be drawn is

(a) $\frac{13}{32}$ (b) $\frac{1}{4}$

(c) $\frac{1}{32}$ (d) $\frac{3}{16}$

29. A draws two cards at random from a pack of 52 cards. After returning them to the pack and shuffling it, B draws two cards at random. The probability that their draws contain exactly one common card is

(a) $\frac{25}{546}$ (b) $\frac{50}{663}$

(c) $\frac{25}{663}$ (d) none of these

30. A and B draw two cards each, one after another, from a pack of well-shuffled pack of 52 cards. The probability that all the four cards drawn are of the same suit is

(a) $\frac{44}{85\times 49}$ (b) $\frac{11}{85\times 49}$

(c) $\frac{13\times 24}{17\times 25\times 49}$ (d) none of these

31. Three numbers are chosen at random without replacement from 1, 2, 3,..., 10. The probability that the minimum of the chosen numbers is 4 or their maximum is 8, is

(a) $\frac{11}{40}$ (b) $\frac{3}{10}$

(c) $\frac{1}{40}$ (d) none of these

32. A man draws a card from a pack of 52 cards and then replaces it. After shuffling the pack, he again draws a card. This he repeats a number of times. The probability that he will draw a heart for the first time in the third draw is

(a) $\frac{9}{64}$ (b) $\frac{27}{64}$

(c) $\frac{1}{4}\times\frac{^{39}C_2}{^{52}C_2}$ (d) none of these

33. A fair coin is tossed repeatedly. The probability of getting a result in the fifth toss different from those obtained in the first four tosses is

(a) $\frac{1}{2}$ (b) $\frac{1}{32}$

(c) $\frac{31}{32}$ (d) $\frac{1}{16}$

34. If the integers m and n are chosen at random between 1 and 100 then the probability that a number of the form $7^m + 7^n$ is divisible by 5 is

(a) $\frac{1}{5}$ (b) $\frac{1}{7}$

(c) $\frac{1}{4}$ (d) $\frac{1}{49}$

35. It has been found that if A and B play a game 12 times, A wins 6 times, B wins 4 times and they draw twice. A and B take part in a series of 3 games. The probability that they will win alternately is

(a) $\frac{5}{72}$ (b) $\frac{5}{36}$

(c) $\frac{19}{27}$ (d) none of these

36. If the probability of A to fail in an examination is $\frac{1}{5}$ and that of B is $\frac{3}{10}$ then the probability that either A or B fails is

(a) $\frac{1}{2}$ (b) $\frac{11}{25}$

(c) $\frac{19}{50}$ (d) none of these

37. A and B are two events where $P(A) = 0.25$ and $P(B) = 0.5$. The probability of both happening together is 0.14. The probability of both A and B not happening is

(a) 0.39 (b) 0.25

(c) 0.11 (d) none of these

38. Three faces of an ordinary dice are yellow, two faces are red and one face is blue. The dice is tossed 3 times. The probability that yellow, red and blue faces appear in the first, second and third tosses respectively is

(a) $\frac{1}{36}$ (b) $\frac{1}{6}$

(c) $\frac{1}{30}$ (d) none of these

39. India play two matches each with West Indies and Australia. In any match the probabilities of India getting 0, 1 and 2 points are 0.45, 0.05 and 0.50 respectively. Assuming that the outcomes are independent, the probability of India getting at least 7 points is

(a) 0.0875 (b) $\frac{1}{16}$

(c) 0.1125 (d) none of these

40. Let A and B be two independent events such that $P(A) = \frac{1}{5}$, $P(A \cup B) = \frac{7}{10}$. Then $P(\overline{B})$ is equal to

(a) $\frac{3}{8}$ (b) $\frac{2}{7}$

(c) $\frac{7}{9}$ (d) none of these

41. Let A and B be two independent events such that their probabilities are $\frac{3}{10}$ and $\frac{2}{5}$. The probability of exactly one of the events happening is

(a) $\frac{23}{50}$ (b) $\frac{1}{2}$

(c) $\frac{31}{50}$ (d) none of these

42. The probability that at least one of the events A and B occurs is $\frac{3}{5}$. If A and B occur simultaneously with probability $\frac{1}{5}$ then $P(A') + P(B')$ is

(a) $\frac{2}{5}$
(b) $\frac{4}{5}$
(c) $\frac{6}{5}$
(d) $\frac{7}{5}$

43. A, B, C are three events for which $P(A) = 0.6$, $P(B) = 0.4$, $P(C) = 0.5$, $P(A \cup B) = 0.8$, $P(A \cap C) = 0.3$ and $P(A \cap B \cap C) = 0.2$.
If $P(A \cup B \cup C) \geq 0.85$ then the interval of values of $P(B \cap C)$ is

(a) [0.2, 0.35]
(b) [0.55, 0.7]
(c) [0.2, 0.55]
(d) none of these

44. A coin is tossed $2n$ times. The chance that the number of times one gets head is not equal to the number of times one gets tail is

(a) $\frac{(2n)!}{(n!)^2} \cdot \left(\frac{1}{2}\right)^{2n}$
(b) $1 - \frac{(2n)!}{(n!)^2}$
(c) $1 - \frac{(2n)!}{(n!)^2} \cdot \frac{1}{4^n}$
(d) none of these

45. A coin is tossed n times. The probability of getting at least one head is greater than that of getting at least two tails by $\frac{5}{32}$. Then n is

(a) 5
(b) 10
(c) 15
(d) none of these

46. A coin is tossed 7 times. Each time a man calls head. The probability that he wins the toss on more occasions is

(a) $\frac{1}{4}$
(b) $\frac{5}{8}$
(c) $\frac{1}{2}$
(d) none of these

47. A bag contains 14 balls of two colours, the number of balls of each colour being the same. 7 balls are drawn at random one by one. The ball in hand is returned to the bag before each new draw. If the probability that at least 3 balls of each colour are drawn is p then

(a) $p > \frac{1}{2}$
(b) $p = \frac{1}{2}$

(c) $p<1$ (d) $p<\frac{1}{2}$

48. From a box containing 20 tickets of value 1 to 20, four tickets are drawn one by one. After each draw, the ticket is replaced. The probability that the largest value of tickets drawn is 15 is

(a) $\left(\frac{3}{4}\right)^4$ (b) $\frac{27}{320}$

(c) $\frac{27}{1280}$ (d) none of these

49. A dice is thrown $2n+1$ times, $n \in N$. The probability that faces with even numbers show odd number of times is

(a) $\frac{2n+1}{4n+3}$ (b) less than $\frac{1}{2}$

(c) greater than $\frac{1}{2}$ (d) none of these

50. 6 ordinary dice are rolled. The probability that at least half of them will show at least 3 is

(a) $41\times\frac{2^4}{3^6}$ (b) $\frac{2^4}{3^6}$

(c) $20\times\frac{2^4}{3^6}$ (d) none of these

51. An ordinary dice is rolled a certain number of times. The probability of getting an odd number 2 times is equal to the probability of getting an even number 3 times. Then the probability of getting an odd number an odd number of times is

(a) $\frac{1}{32}$ (b) $\frac{5}{16}$

(c) $\frac{1}{2}$ (d) none of these

52. A card is drawn from a pack. The card is replaced and the pack is reshuffled. If this is done six times, the probability that 2 hearts, 2 diamonds and 2 black cards are drawn is

(a) $90\cdot\left(\frac{1}{4}\right)^6$ (b) $\frac{45}{2}\cdot\left(\frac{3}{4}\right)^4$

(c) $\frac{90}{2^{10}}$ (d) none of these

53. A man firing at a distant target has 10% chance of hitting the target in one shot. The number of times he must fire at the target to have about 50% chance of hitting the target is

(a) 11 (b) 9

(c) 7 (d) 5

54. There are four machines and it is known that exactly two of them are faulty. They are tested one by one in a random order till both the faulty machines are identified. Then the probability that only two tests will be required is

(a) $\frac{1}{3}$ (b) $\frac{1}{6}$

(c) $\frac{1}{2}$ (d) $\frac{1}{4}$

55. Let $A = \{2, 3, 4, \ldots, 20, 21\}$. A number is chosen at random from the set A and it is found to be a prime number. The probability that it is more than 10 is

(a) $\frac{9}{10}$ (b) $\frac{1}{10}$

(c) $\frac{1}{5}$ (d) none of these

56. All the spades are taken out from a pack of cards. From these cards, cards are drawn one by one without replacement till the ace of spades comes. The probability that the ace comes in the 4th draw is

(a) $\frac{1}{13}$ (b) $\frac{12}{13}$

(c) $\frac{4}{13}$ (d) none of these

57. A point is selected at random from the interior of a circle. The probability that the point is closer to the centre than the boundary of the circle is

(a) $\frac{3}{4}$ (b) $\frac{1}{2}$

(c) $\frac{1}{4}$ (d) none of these

58. A, B and C are contesting the election for the post of secretary of a club which does not allow ladies to become members. The probabilities of A,

B and C winning the election are $\frac{1}{3}$, $\frac{2}{9}$ and $\frac{4}{9}$ respectively. The probabilities of introducing the clause of admitting lady members to the club by A, B, and C are 0.6, 0.7 and 0.5 respectively. The probability that ladies will be taken as members in the club after the election is

(a) $\frac{26}{45}$ (b) $\frac{5}{9}$

(c) $\frac{19}{45}$ (d) none of these

59. There are 4 white and 3 black balls in a box. In another box there are 3 white and 4 black balls. An unbiased dice is rolled. If it shows a number less than or equal to 3 then a ball is drawn from the first box but if it shows a number more than 3 then a ball is drawn from the second box. If the ball drawn is black then the probability that the ball was drawn from the first box is

(a) $\frac{1}{2}$ (b) $\frac{6}{7}$

(c) $\frac{4}{7}$ (d) $\frac{3}{7}$

• Type 2 •

Choose the correct options. One or more options may be correct.

60. The probabilities that a student passes in mathematics, physics and chemistry are m, p and c respectively. Of these subjects, a student has a 75% chance of passing in at least one, a 50% chance of passing in at least two, and a 40% chance of passing in exactly two subjects. Which of the following relations are true?

(a) $p+m+c=\frac{19}{20}$ (b) $p+m+c=\frac{27}{20}$

(c) $pmc=\frac{1}{10}$ (d) $pmc=\frac{1}{4}$

61. If E and F are two events with $P(E) \leq P(F) > 0$ then

(a) occurrence of E $\Rightarrow$ occurrence of F

(b) occurrence of F $\Rightarrow$ occurrence of E

(c) nonoccurrence of E $\Rightarrow$ nonoccurrence of F

(d) none of the above implications hold

62. If A and B are two events such that $P(A \cup B) \geq \frac{3}{4}$ and $\frac{1}{8} \leq P(A \cap B) \leq \frac{3}{8}$ then

(a) $P(A) + P(B) \leq \frac{11}{8}$ (b) $P(A) \cdot P(B) \leq \frac{3}{8}$

(c) $P(A) + P(B) \geq \frac{7}{8}$ (d) none of these

63. If $\overline{E}$ and $\overline{F}$ are the complementary events of the events E and F respectively then

(a) $P(E/F) + P(\overline{E}/F) = 1$ (b) $P(E/F) + P(E/\overline{F}) = 1$

(c) $P(\overline{E}/F) + P(E/\overline{F}) = 1$ (d) $P(E/\overline{F}) + P(\overline{E}/\overline{F}) = 1$

64. Given that $x \in [0, 1]$ and $y \in [0, 1]$. Let A be the event of (x, y) satisfying $y^2 \leq x$ and B be the event of (x, y) satisfying $x^2 \leq y$. Then

(a) $P(A \cap B) = \frac{1}{3}$ (b) A, B are exhaustive

(c) A, B are mutually exclusive (d) A, B are independent

65. Let A and B be two events such that $P(A \cap B) = \frac{1}{3}$, $P(A \cup B) = \frac{5}{6}$ and $P(\overline{A}) = \frac{1}{2}$. Then

(a) A, B are independent (b) A, B are mutually exclusive

(c) $P(A) = P(B)$ (d) $P(B) \leq P(A)$

66. The probability that exactly one of the independent events A and B occurs is equal to

(a) $P(A) + P(B) - 2P(A \cap B)$ (b) $P(A) + P(B) - P(A \cap B)$

(c) $P(\overline{A}) + P(\overline{B}) - 2P(\overline{A} \cap \overline{B})$ (d) none of these

67. If A and B are independent events such that $0 < P(A) < 1, 0 < P(B) < 1$ then

(a) A, B are mutually exclusive

(b) A and $\overline{B}$ are independent

(c) $\overline{A}, \overline{B}$ are independent

(d) $P(A/B) + P(\overline{A}/B) = 1$

68. For any two events A and B

(a) $P(A \cap B) \geq P(A) + P(B) - 1$

(b) $P(A \cap B) \geq P(A) + P(B)$

(c) $P(A \cap B) = P(A) + P(B) - P(A \cup B)$

(d) $P(A \cap B) = P(A) + P(B) + P(A \cup B)$

69. A coin is tossed repeatedly. A and B call alternately for winning a prize of Rs 30. One who calls correctly first wins the prize. A starts the call. Then the expectation of

(a) A is Rs 10 (b) B is Rs 10

(c) A is Rs 20 (d) B is Rs 20

Answers

1. c	**2.** a	**3.** b	**4.** c	**5.** c
6. c	**7.** c	**8.** b	**9.** c	**10.** d
11. a	**12.** d	**13.** d	**14.** a	**15.** b
16. c	**17.** a	**18.** b	**19.** d	**20.** c
21. c	**22.** b	**23.** a	**24.** d	**25.** a
26. c	**27.** c	**28.** a	**29.** b	**30.** a
31. a	**32.** a	**33.** d	**34.** a	**35.** b
36. c	**37.** a	**38.** a	**39.** a	**40.** a
41. a	**42.** c	**43.** a	**44.** c	**45.** a
46. c	**47.** a	**48.** b	**49.** d	**50.** a
51. c	**52.** c	**53.** c	**54.** b	**55.** c
56. a	**57.** c	**58.** a	**59.** d	**60.** b, c
61. d	**62.** a, c	**63.** a, d	**64.** a	**65.** a
66. a, c	**67.** b, c, d	**68.** a, c	**69.** b, c	

Hints and Solutions to Selected Questions

1. $n(S) = 8\,!$

$n(E)$ = the number of arrangements of 5 boys and 3 girls when the 3 girls are consecutive $= 6\,! \times 3\,!$.

$\therefore$ the required probability $= \dfrac{6\,! \times 3\,!}{8\,!} = \dfrac{3}{28}$.

2. $n(S)$ = total number of sections of two diagonals

$= {}^{{}^6C_2 - 6}C_2 = {}^9C_2 = \dfrac{9 \times 8}{2} = 36.$

$n(E)$ = the number of selections of two diagonals which intersect at an interior point

= the number of selections of four vertices $= {}^6C_4 = 15$.

$\therefore$ the required probability $= \dfrac{15}{36} = \dfrac{5}{12}$.

3. $n(S) = \dfrac{9\,!}{4\,!\,3\,!\,2\,!}$ and $n(E) = 3\,!$.

$\therefore \quad P(E) = \dfrac{(3\,!)\,(4\,!)\,(3\,!)\,(2\,!)}{9\,!} = \dfrac{1}{210}$.

4. $n(S) = {}^{52}C_2$ and

$n(E)$ = the number of selections of 1 spade, 1 ace from 3 aces or selections of the ace of spade and 1 other spade

$= {}^{13}C_1 \times {}^3C_1 + {}^{12}C_1 \times {}^1C_1 = 51$

$\therefore \quad P(E) = \dfrac{51}{{}^{52}C_2} = \dfrac{1}{26}$.

5. $P(E) = \dfrac{n(E)}{n(S)} = \dfrac{2 \times 9 \times 9 \times 9 \times 8}{9 \times 10 \times 10 \times 10 \times 10} = \dfrac{9 \times 9}{5 \times 5 \times 5 \times 5} = \left(\dfrac{3}{5}\right)^4$.

6. $n(S) = \dfrac{7\,!}{3\,!}$, $n\,(E) = 5\,!$. So, $P(E) = \dfrac{5\,!}{7\,!/3\,!}$.

7. $E = \{(1, 6), (2, 5), (3, 4), (4, 3), (5, 2), (6, 1)\}.$

$\therefore \quad P(E) = \dfrac{6}{6 \times 6} = \dfrac{1}{6}$. So, the odds against drawing $7 = \dfrac{P(\overline{E})}{P(E)} = \dfrac{1 - \dfrac{1}{6}}{\dfrac{1}{6}} = \dfrac{5}{1}$.

8.

```
 B   B   B   B              B   D   B
| W | W | W |------------| W | W |
```

$n(S) = \dfrac{10\,!}{(7\,!)\,(3\,!)}$

$n(E) = {}^8C_3 = \dfrac{8\,!}{(3\,!)\,(5\,!)}$, because there are 8 places for 3 black balls.

$$\therefore \quad P(E) = \frac{\dfrac{18}{(3\,!)\,(5\,!)}}{\dfrac{10\,!}{(7\,!)\,(3\,!)}} = \frac{(8\,!)\,(7\,!)}{(10\,!)\,(5\,!)} = \frac{7 \cdot 6}{10 \cdot 9} = \frac{7}{15}.$$

9. The required probability $= 1 -$ probability of each receiving at least one

$$= 1 - \frac{n(E)}{n(S)}.$$

Now, the number of integral solutions of $x_1 + x_2 + x_3 + x_4 + x_5 + x_6 = 10$ such that $x_1 \geq 1, x_2 \geq 1, \ldots, x_6 \geq 1$ gives $n(E)$ and the number of integral solutions of $x_1 + x_2 + \ldots + x_5 + x_6 = 10$ such that $x_1 \geq 0, x_2 \geq 0, \ldots, x_6 \geq 0$ gives $n(S)$.

$$\therefore \quad \text{the required probability} = 1 - \frac{{}^{10-1}C_{6-1}}{{}^{10+6-1}C_{6-1}} = 1 - \frac{{}^9C_5}{{}^{15}C_5} = \frac{137}{143}.$$

10. $n(S) = 7\,!$, $n(E) = (3\,!) \times (4\,!)$ ($\because$ after making 4 gentlemen sit in $3\,!$ ways, 4 ladies can sit in $4\,!$ ways in between the gentlemen)

$$\therefore \quad P(E) = \frac{(3\,!) \times (4\,!)}{7\,!} = \frac{6}{7 \times 6 \times 5} = \frac{1}{35}.$$

11. $n(S) =$ the number of ways $(33)^n$ can have a digit in the units place

$= 4$ ($\because$ it is 3, 9, 7 or 1)

$n(E) = 1$. So, $P(E) = \dfrac{1}{4}$.

12. $n(S) = 6 \times 6 \times 6$.

$n(E) =$ the number of solutions of $x + y + z = 15$,

where $1 \leq x \leq 6, 1 \leq y \leq 6, 1 \leq z \leq 6$

$=$ coefficient of x^{15} in $(x + x^2 + \ldots + x^6)^3$

$= \text{coefficient of } x^{12} \text{ in } (1 + x + \ldots + x^5)^3$

$= \text{coefficient of } x^{12} \text{ in } \left(\dfrac{1-x^6}{1-x}\right)^3$

$= \text{coefficient of } x^{12} \text{ in } (1 - 3 \cdot x^6 + 3 \cdot x^{12} - x^{18}) \cdot (1-x)^{-3}$

$= \text{coefficient of } x^{12} \text{ in } (1 - 3x^6 + 3x^{12} - x^{18})({}^2C_0 + {}^3C_1 x + {}^4C_2 x^2 + \ldots)$

$= {}^{14}C_{12} - 3 \times {}^8C_6 + 3 \times {}^2C_0 = 10.$

$\therefore \quad P(E) = \dfrac{10}{6 \times 6 \times 6} = \dfrac{5}{108}.$

13. $n(S) = 6 \times 6 \times 6$. Clearly, the sum varies from 3 to 18, and among these 4, 9, 16 are perfect squares.

The number of ways to get the sum 4

$= \text{the number of integral solutions of } x_1 + x_2 + x_3 = 4,$

where $1 \le x_1 \le 6, 1 \le x_2 \le 6, 1 \le x_3 \le 6$

$= \text{coefficient of } x^4 \text{ in } (x + x^2 + \ldots + x^6)^3$

$= \text{coefficient of } x \text{ in } \left(\dfrac{1-x^6}{1-x}\right)^3$

$= \text{coefficient of } x \text{ in } (1-x^6)^3 \cdot (1-x)^{-3} = {}^3C_1.$

Similarly, the number of ways to get the sum 9

$= \text{coefficient of } x^6 \text{ in } (1-x^6)^3 \cdot (1-x)^{-3}$

$= -3 \times 1 + {}^8C_6 = 28 - 3 = 25.$

The number of ways to get the sum 16

$= \text{coefficient of } x^{13} \text{ in } (1-x^6)^3 \cdot (1-x)^{-3}$

$= \text{coefficient of } x^{13} \text{ in } (1 - 3x^6 + 3x^{12} - x^{18})({}^2C_0 + {}^3C_1 x + {}^4C_2 x^2 + \ldots)$

$= {}^{15}C_{13} - 3 \times {}^9C_7 + 3 \times {}^3C_1 = 105 - 108 + 9 = 6.$

$\therefore \quad n(E) = 3 + 25 + 6 = 34.$ So, $P(E) = \dfrac{34}{6 \times 6 \times 6} = \dfrac{17}{108}.$

14. $n(S) = 6 \times 6 \times 6 \times 6.$

$n(E) = \text{the number of integral solutions of } x_1 + x_2 + x_3 + x_4 = 12,$

where $1 \le x_1 \le 6, \ldots, 1 \le x_4 \le 6$

$= \text{coefficient of } x^{12} \text{ in } (x + x^2 + \ldots + x^6)^4$

$$= \text{coefficient of } x^8 \text{ in } \left(\frac{1-x^6}{1-x}\right)^4$$

$$= \text{coefficient of } x^8 \text{ in } (1-x^6)^4 \cdot ({}^3C_0 + {}^4C_1 x + {}^5C_2 x^2 + \ldots)$$

$$= {}^{11}C_8 - 4 \cdot {}^5C_2 = 125.$$

$$\therefore \quad P(E) = \frac{125}{6 \times 6 \times 6 \times 6}.$$

15. $n(S) = {}^{10}C_3$. Clearly, $E = \{(10, 5, 2), (8, 2, 4), (6, 2, 3)\}$. So, $n(E) = 3$.

$$\therefore \quad P(E) = \frac{n(E)}{n(S)} = \frac{3}{{}^{10}C_3} = \frac{1}{40}.$$

16. $n(S) = {}^7C_3 \times 3\,! = \dfrac{7 \cdot 6 \cdot 5}{6} \cdot 6 = 210.$

$n(E) = {}^2C_1 \times {}^2C_1 \times {}^1C_1 \times 3\,!,$

because one has to sit at any one of the two marked seats on the left and the other has to sit at any one of the two marked seats on the right.

$$\therefore \quad P(E) = \frac{n(E)}{n(S)} = \frac{2 \times 2 \times 6}{210} = \frac{4}{35}.$$

17. $n(S) = 2 \times 2 \times 2 \times 2 = 16,$

because each of the four places can be filled in 2 ways.

The zero determinants are $\begin{vmatrix} 1 & 1 \\ -1 & -1 \end{vmatrix}, \begin{vmatrix} -1 & -1 \\ 1 & 1 \end{vmatrix}, \begin{vmatrix} 1 & 1 \\ 1 & 1 \end{vmatrix}, \begin{vmatrix} -1 & -1 \\ -1 & -1 \end{vmatrix},$

$\begin{vmatrix} -1 & 1 \\ -1 & 1 \end{vmatrix}, \begin{vmatrix} 1 & -1 \\ 1 & -1 \end{vmatrix}, \begin{vmatrix} -1 & 1 \\ 1 & -1 \end{vmatrix}, \begin{vmatrix} 1 & -1 \\ -1 & 1 \end{vmatrix}.$

$$\therefore \quad P(E) = \frac{8}{16} = \frac{1}{2}.$$

18. $n(S) = {}^{50}C_5$, $n(E) = {}^{30}C_2 \times {}^{19}C_2$.

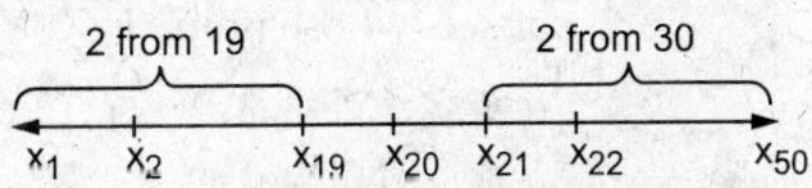

$$\therefore \quad P(E) = \frac{{}^{30}C_2 \times {}^{19}C_2}{{}^{50}C_5}.$$

19. $n(S) = 100 \times 100 \times 100.$

We know that $(2n+1)^2 + (2n^2 + 2n)^2 = (2n^2 + 2n + 1)^2$ for all $n \in N$.

$\therefore$ for $n = 1, 2, 3, 4, 5, 6$, we get lengths of the three sides of a right-angled triangle whose longest side ≤ 100.

For example, when $n=1$, sides are 3, 4, 5; when $n=2$, sides are 5, 12, 13 and so on.

The number of selections of 3, 4, 5 from the three cards by taking one from each is 3 !.

$\therefore \quad n(E)=6(3\,!)$. Hence, $P(E)=\dfrac{6(3\,!)}{100\times100\times100}=\dfrac{1}{100}\left(\dfrac{3}{50}\right)^2$.

20. $n(S)={}^{10}C_3$ and

$n(E)={}^{3}C_1$, because on selecting 3, 7 and we have to select one from 4, 5 and 6.

$$\therefore \quad P(E)=\frac{{}^{3}C_1}{{}^{10}C_3}=\frac{1}{40}.$$

21. $n(S)={}^{100}C_3$.

As the AM of three numbers is 25, their sum $=75$.

$\therefore \quad n(E)=$ the number of integral solutions of $x_1+x_2+x_3=75$,

where $x_1\ge1, x_2\ge1, x_3\ge1$

$=$ coefficient of x^{75} in $(x+x^2+x^3+\ldots)^3$

$=$ coefficient of x^{72} in $\left(\dfrac{1}{1-x}\right)^3$

$=$ coefficient of x^{72} in $(1-x)^{-3}$

$={}^{74}C_{72}$.

$$\therefore \quad P(E)=\frac{{}^{74}C_{72}}{{}^{100}C_3}=\frac{{}^{74}C_{72}}{{}^{100}C_{97}}.$$

22. The total number of subsets of X is 2^k. So, $n(S)={}^{2^k}C_2$.

$n(E)=$ the number of selections of two nonintersecting subsets whose union is X

$=\dfrac{1}{2}({}^kC_0+{}^kC_1+{}^kC_2+\ldots)$

($\because$ the number of selections in which one subset has r elements and the rest are in the other subset $={}^kC_r$ and every selection appears twice in the total number of selections).

$$\therefore \quad P(E)=\frac{(1/2)\cdot2^k}{{}^{2^k}C_2}=\frac{2^{k-1}}{\dfrac{2^k(2^k-1)}{2}}=\frac{1}{2^k-1}.$$

23. $n(S) = {}^{10}C_4 = 210.$

$n(E) = {}^5C_2 \times {}^3C_1 \times {}^2C_1 + {}^5C_1 \times {}^3C_2 \times {}^2C_1 + {}^5C_1 \times {}^3C_1 \times {}^2C_2 = 105$

$\therefore \quad P(E) = \frac{105}{210} = \frac{1}{2}.$

24. $n(S) = {}^{12}C_4 \times {}^8C_4 \times {}^4C_4.$

$n(E) = n(S)$ – the number of ways in which one boy gets both the pens

$= n(S) - {}^{10}C_2 \times {}^8C_4 \times {}^4C_4 \times (3\,!)$

$\therefore \quad P(E) = 1 - \frac{{}^{10}C_2 \times {}^8C_4 \times {}^4C_4 \times (3\,!)}{{}^{12}C_4 \times {}^8C_4 \times {}^4C_4} = 1 - \frac{6}{11} = \frac{5}{11}.$

25. Let E_3 = the event of the sum being 3. Similarly, E_6, E_9, E_{12}, E_{15}, E_{18}, E_{21}.

$n(S) = {}^{12}C_2$

$n(E_3) = 1, n(E_6) = 2, n(E_9) = 4, n(E_{12}) = 5, n(E_{15}) = 5, n(E_{18}) = 3, n(E_{21}) = 2.$

$\therefore \quad P(E) = P(E_3) + \ldots + P(E_{21}) = \frac{1+2+4+5+5+3+2}{{}^{12}C_2} = \frac{22 \times 2}{12 \times 11} = \frac{1}{3}.$

Alternatively The sum of the numbers for every selection is divisible by 3 or leaves the remainder 1 or leaves the remainder 2. These are equally probable. So, the required probability $= \frac{1}{3}$, because the sum of the three probabilities is 1.

26. $P(E_n)$ = probability of the dice showing $n = kn$.

Clearly, $P(E_1) + P(E_2) + \ldots + P(E_6) = 1$

$\therefore \quad k(1+2+3+4+5+6) = 1$

$\therefore \quad k = \frac{1}{\frac{6 \cdot 7}{2}} = \frac{1}{21}$. So, $P(E_3) = k \cdot 3 = \frac{1}{21} \times 3 = \frac{1}{7}.$

27. There are 30 days in April.

$n(S)$ = the number of ways in which 10 persons can have birthdays in the month of April

$= 30 \times 30 \times \ldots$ to 10 times $= 30^{10}$

(because each person can have birthday in 30 ways).

$n(E) = n(S)$ – the number of ways in which 10 persons can have different birthdays

$$= 30^{10} - {}^{30}C_{10}.$$

$$\therefore \quad P(E) = \frac{30^{10} - {}^{30}C_{10}}{30^{10}}.$$

28. Let E_1 = the event of drawing a white ball from the first box.

Similarly, E_2 and E_3.

Here $P(E_1) = \frac{3}{4}$, $\quad P(E_2) = \frac{1}{2}$, $\quad P(E_3) = \frac{1}{4}$.

The required probability $= P(E_1E_2\overline{E}_3) + P(E_1\overline{E}_2E_3) + P(\overline{E}_1E_2E_3)$

$$= P(E_1) \cdot P(E_2) \cdot P(\overline{E}_3) + P(E_1) \cdot P(\overline{E}_2) \cdot P(E_3) + P(\overline{E}_1) \cdot P(E_2) \cdot P(E_3)$$

$$= \frac{3}{4} \cdot \frac{1}{2} \cdot \frac{3}{4} + \frac{3}{4} \cdot \frac{1}{2} \cdot \frac{1}{4} + \frac{1}{4} \cdot \frac{1}{2} \cdot \frac{1}{4} = \frac{13}{32}.$$

29. The probability of both drawing the common card x

$= P(x) =$ (probability of A drawing the card x and any other card y) $\times$
(probability of B drawing the card x and a card other than y)

$$\therefore \quad P(x) = \frac{{}^{51}C_1}{{}^{52}C_2} \times \frac{{}^{50}C_1}{{}^{52}C_2} \text{ for all } x, \text{ where } x \text{ has 52 values.}$$

$$\therefore \quad \text{the required probability} = \Sigma P(x) = 52 \times \frac{51 \times 50 \times 4}{52 \times 51 \times 52 \times 51} = \frac{50}{663}.$$

30. The probability of the four cards being spades $= \frac{{}^{13}C_2}{{}^{52}C_2} \times \frac{{}^{11}C_2}{{}^{50}C_2}$.

Similarly, for other suits.

$$\therefore \text{ the required probability} = 4 \times \frac{{}^{13}C_2 \times {}^{11}C_2}{{}^{52}C_2 \times {}^{50}C_2} = \frac{44}{85 \times 49}.$$

31. The probability of 4 being the minimum number $= \frac{{}^{6}C_2}{{}^{10}C_3}$

(because, after selecting 4 any two can be selected from 5, 6, 7, 8, 9, 10).

The probability of 8 being the maximum number $= \frac{{}^{7}C_2}{{}^{10}C_3}$.

The probability of 4 being the minimum number and 8 being the maximum number $= \frac{3}{{}^{10}C_3}$

$$\therefore \text{ the required probability} = P(A \cup B) = P(A) + P(B) - P(A \cap B)$$

$$= \frac{{}^6C_2}{{}^{10}C_3} + \frac{{}^7C_2}{{}^{10}C_3} - \frac{3}{{}^{10}C_3} = \frac{11}{40}.$$

32. Let E = the event of drawing a heart. Clearly, $P(E) = \frac{{}^{13}C_1}{{}^{52}C_1} = \frac{1}{4}$.

$\therefore$ the required probability $= P(\overline{E}\,\overline{E}\,E) = P(\overline{E}) \cdot P(\overline{E}) \cdot P(E)$

$$= \frac{3}{4} \cdot \frac{3}{4} \cdot \frac{1}{4} = \frac{9}{64}.$$

33. Required probability

$$= P(E\,E\,E\,E\,\overline{E}) + P(\overline{E}\,\overline{E}\,\overline{E}\,\overline{E}\,E) = \{P(E)\}^4 \cdot P(\overline{E}) + \{P(\overline{E})\}^4 \cdot P(E).$$

34. We know $7^k, k \in N$, has 1, 3, 9, 7 at the units place for $k = 4p, 4p-1, 4p-2, 4p-3$ respectively, where $p = 1, 2, 3, \ldots$.

Clearly, $7^m + 7^n$ will be divisible by 5 if 7^m has 3 or 7 in the units place and 7^n has 7 or 3 in the units place or 7^m has 1 or 9 in the units place and 7^n has 9 or 1 in the units place.

$\therefore$ for any choice of m, n the digit in the units place of $7^m + 7^n$ is 2, 4, 6, 0 or 8. It is divisible by 5 only when this digit is 0.

$\therefore$ the required probability $= \frac{1}{5}$.

35. The probability of A winning in a game $= P(A) = \frac{6}{12} = \frac{1}{2}$.

The probability of B winning in a game $= P(B) = \frac{4}{12} = \frac{1}{3}$.

The required probability $= P(A \cap B \cap A) + P(B \cap A \cap B)$

$$= P(A) \cdot P(B) \cdot P(A) + P(B) \cdot P(A) \cdot P(B) = \frac{1}{2} \cdot \frac{1}{3} \cdot \frac{1}{2} + \frac{1}{3} \cdot \frac{1}{2} \cdot \frac{1}{3} = \frac{5}{36}.$$

36. The probability of A failing $= P(A) = \frac{1}{5}$

and the probability of B failing $= P(B) = \frac{3}{10}$.

The required probability $= P(A\overline{B} \cup \overline{A}B) = P(A\overline{B}) + P(\overline{A}B)$

$$= P(A) \cdot P(\overline{B}) + P(\overline{A}) \cdot P(B) = \frac{1}{5}\left(1 - \frac{3}{10}\right) + \left(1 - \frac{1}{5}\right) \cdot \frac{3}{10} = \frac{19}{50}.$$

37. $P(\overline{A}\,\overline{B}) = P(\overline{A \cup B}) = 1 - P(A \cup B) = 1 - \{P(A) + P(B) - P(A \cap B)\}$

$= 1 - \{0.25 + 0.50 - 0.14\} = 0.39.$

38. $P(Y) = \frac{3}{6} = \frac{1}{2}, \quad P(R) = \frac{2}{6} = \frac{1}{3}, \quad P(B) = \frac{1}{6}$

$\therefore \quad P(Y \cap R \cap B) = P(Y) \cdot P(R) \cdot P(B) = \frac{1}{2} \cdot \frac{2}{6} \cdot \frac{1}{6} = \frac{1}{36} \cdot$

39. $P(E_0) = 0.45, \quad P(E_1) = 0.05, \quad P(E_2) = 0.50.$

For 7 points there should be at least three instances of 2 points and one of 1 or 2 points.

$\therefore$ the required probability

$$= P(E_2E_2E_2E_2) + P(E_2E_2E_2E_1) + P(E_2E_2E_1E_2) + P(E_2E_1E_2E_2) + P(E_1E_2E_2E_2)$$

$$= \{P(E_2)\}^4 + 4\{P(E_2)\}^3 \cdot P(E_1)$$

$$= (0.50)^4 + 4(0.50)^3 \cdot (0.05) = 0.0875.$$

40. Let $P(\overline{B}) = x$. Then $P(B) = 1 - x$.

$\therefore \quad P(A \cap B) = P(A)\, P(B) = \frac{1}{5}(1 - x).$

But $P(A \cup B) = P(A) + P(B) - P(A \cap B) \quad \Rightarrow \quad \frac{7}{10} = \frac{1}{5} + (1 - x) - \frac{1}{5}(1 - x)$

or $\quad \frac{1}{2} = \frac{4}{5}(1 - x)$

$\therefore \quad 1 - x = \frac{5}{8} \qquad \therefore \quad x = \frac{3}{8} \cdot$

41. The required probability $= P(A\overline{B} \cup \overline{A}B) = P(A)P(\overline{B}) + P(\overline{A})P(B)$

$$= \frac{3}{10}\left(1 - \frac{2}{5}\right) + \left(1 - \frac{3}{10}\right)\frac{2}{5} = \frac{23}{50} \cdot$$

42. Here, $P(A \cup B) = \frac{3}{5}$ and $P(A \cap B) = \frac{1}{5} \cdot$ So, from the addition theorem,

$$\frac{3}{5} = P(A) + P(B) - \frac{1}{5}$$

or $\quad \frac{4}{5} = 1 - P(A') + 1 - P(B')$

$\therefore \quad P(A') + P(B') = 2 - \frac{4}{5} = \frac{6}{5} \cdot$

43. $P(A \cup B \cup C) = P(A) + P(B) + P(C) - P(A \cap B) - P(B \cap C)$

$$- P(C \cap A) + P(A \cap B \cap C)$$

$$= 0.6 + 0.4 + 0.5 - 0.2 - P(B \cap C) - 0.3 + 0.2 = 1.2 - P(B \cap C),$$

because $P(A \cup B) = P(A) + P(B) - P(A \cap B) \Rightarrow 0.8 = 0.6 + 0.4 - P(A \cap B)$.

But $0.85 \le P(A \cup B \cup C) \le 1$

$\therefore \quad 0.85 \le 1.2 - P(B \cap C) \le 1 \quad \Rightarrow \quad 0.2 \le P(B \cap C) \le 0.35.$

44. The required probability

$= 1 -$ probability of equal number of heads and tails

$$= 1 - {}^{2n}C_n \cdot \left(\frac{1}{2}\right)^n \cdot \left(\frac{1}{2}\right)^{2n-n} = 1 - \frac{(2n)!}{(n!)(n!)}\left(\frac{1}{4}\right)^n.$$

45. The probability of getting at least one head

$$= 1 - \text{probability of getting no heads} = 1 - {}^nC_0\left(\frac{1}{2}\right)^0 \cdot \left(\frac{1}{2}\right)^n = 1 - \frac{1}{2^n}.$$

The probability of getting at least two tails

$= 1 -$ probability of getting no tails $-$ probability of getting 2 tails

$$= 1 - {}^nC_n\left(\frac{1}{2}\right)^n \cdot \left(\frac{1}{2}\right)^0 - {}^nC_{n-1}\left(\frac{1}{2}\right)^{n-1} \cdot \frac{1}{2} = 1 - \frac{1}{2^n} - n\frac{1}{2^n}.$$

From the question, $\left(1 - \frac{1}{2^n}\right) - \left(1 - \frac{1+n}{2^n}\right) = \frac{5}{32}$

or $\quad \frac{n+1}{2^n} - \frac{1}{2^n} = \frac{5}{32} \qquad$ or $\quad \frac{n}{2^n} = \frac{5}{32} \qquad \Rightarrow n = 5.$

46. The man has to win at least 4 times.

$\therefore$ the required probability

$$= {}^7C_4\left(\frac{1}{2}\right)^4 \cdot \left(\frac{1}{2}\right)^3 + {}^7C_5 \cdot \left(\frac{1}{2}\right)^5 \cdot \left(\frac{1}{2}\right)^2 + {}^7C_6\left(\frac{1}{2}\right)^6 \cdot \frac{1}{2} + {}^7C_7\left(\frac{1}{2}\right)^7$$

$$= ({}^7C_4 + {}^7C_5 + {}^7C_6 + {}^7C_7) \cdot \frac{1}{2^7} = \frac{64}{2^7} = \frac{1}{2}.$$

47. A ball drawn is of either colour. If drawing a ball of one colour is a success then drawing ball of the other colour is a failure.

$\therefore$ the probability of success in one draw $= \frac{7}{14} = \frac{1}{2}$ and

the probability of failure in one draw $= \frac{7}{14} = \frac{1}{2}$.

$\therefore$ the probability that at least 3 balls of each colour is drawn

$$= {}^7C_3 \cdot \left(\frac{1}{2}\right)^3 \cdot \left(\frac{1}{2}\right)^4 + {}^7C_4 \left(\frac{1}{2}\right)^4 \cdot \left(\frac{1}{2}\right)^3 = 2 \cdot \frac{7\,!}{(3\,!)\,(4\,!)} \cdot \frac{1}{2^7}$$

$$= 2 \cdot \frac{7 \cdot 6 \cdot 5}{6} \cdot \frac{1}{2^7} = \frac{35}{64} > \frac{1}{2}.$$

48. The probability of drawing a number less than or equal to 15 in a draw $= \frac{15}{20} = \frac{3}{4}$.

The probability of drawing the ticket of value 15 in a draw $= \frac{1}{20}$.

$\therefore$ the required probability $= {}^4C_1 \cdot \frac{1}{20} \cdot \left(\frac{3}{4}\right)^3$.

49. The probability of showing an even number in a throw $= \frac{3}{6} = \frac{1}{2}$.

$\therefore$ the required probability

$$= {}^{2n+1}C_1 \cdot \frac{1}{2} \cdot \left(\frac{1}{2}\right)^{2n} + {}^{2n+1}C_3 \cdot \left(\frac{1}{2}\right)^3 \cdot \left(\frac{1}{2}\right)^{2n-2} + \ldots + {}^{2n+1}C_{2n+1} \left(\frac{1}{2}\right)^{2n+1}$$

$$= \left({}^{2n+1}C_1 + {}^{2n+1}C_3 + \ldots + {}^{2n+1}C_{2n+1}\right)\left(\frac{1}{2}\right)^{2n+1}$$

$$= 2^{2n} \cdot \left(\frac{1}{2}\right)^{2n+1} = \frac{1}{2}.$$

50. The probability of getting at least 3 in a throw $= \frac{4}{6} = \frac{2}{3}$.

$\therefore$ the required probability

$$= {}^6C_3 \cdot \left(\frac{2}{3}\right)^3 \cdot \left(\frac{1}{3}\right)^3 + {}^6C_4 \cdot \left(\frac{2}{3}\right)^4 \left(\frac{1}{3}\right)^2 + {}^6C_5 \cdot \left(\frac{2}{3}\right)^5 \cdot \frac{1}{3} + {}^6C_6 \cdot \left(\frac{2}{3}\right)^6.$$

51. The probability of getting an odd number in a throw $= \frac{3}{6} = \frac{1}{2}$.

The probability of getting an odd number 2 times in n trials

$$= {}^nC_2 \cdot \left(\frac{1}{2}\right)^2 \cdot \left(\frac{1}{2}\right)^{n-2}.$$

Similarly, the probability of getting an even number 3 times in n trials

$$= {}^nC_3 \cdot \left(\frac{1}{2}\right)^3 \cdot \left(\frac{1}{2}\right)^{n-3}.$$

From the question, ${}^nC_2 \cdot \left(\frac{1}{2}\right)^n = {}^nC_3 \cdot \left(\frac{1}{2}\right)^n \quad \Rightarrow n = 5.$

$\therefore$ the required probability

$$= {}^5C_1 \cdot \left(\frac{1}{2}\right) \cdot \left(\frac{1}{2}\right)^4 + {}^5C_3 \cdot \left(\frac{1}{2}\right)^3 \cdot \left(\frac{1}{2}\right)^2 + {}^5C_5 \left(\frac{1}{2}\right)^5$$

$$= (5 + 10 + 1) \cdot \frac{1}{2^5} = \frac{16}{32} = \frac{1}{2}.$$

52. The probability of getting a heart in one draw $= \frac{13}{52} = \frac{1}{4}.$

Similarly for a diamond, the probability $= \frac{1}{4}.$

The probability of getting a black card in one draw $= \frac{26}{52} = \frac{1}{2}.$

$\therefore$ the required probability

$$= {}^6C_2 \times {}^4C_2 \times \left(\frac{1}{4}\right)^2 \times \left(\frac{1}{4}\right)^2 \times \left(\frac{1}{2}\right)^2,$$

because in 6 draws, 2 draws will be for hearts, 2 for diamonds and 2 for black cards, and this selection can be done in ${}^6C_2 \times {}^4C_2 \times {}^2C_2$ ways.

53. The probability of hitting in one shot $= \frac{10}{100} = \frac{1}{10}.$

If he fires n shots, the probability of hitting at least once

$$= 1 - \left(1 - \frac{1}{10}\right)^n = 1 - \left(\frac{9}{10}\right)^n = \frac{1}{2} \quad \text{(from the question)}$$

$$\therefore \quad \left(\frac{9}{10}\right)^n = \frac{1}{2}$$

$$\therefore \quad n\{2\log_{10} 3 - 1\} = -\log_{10} 2$$

$$\therefore \quad n = \frac{\log_{10} 2}{1 - 2\log_{10} 3} = \frac{0.3010}{1 - 2 \times 0.4771} = 6.5 \text{ (nearly)}$$

$\therefore$ for 6 shots, the probability is about 53% while for 7 shots it is nearly 48%.

54. The required probability = (probability of the first being faulty) $\times$ (probability of the second being faulty when the first is faulty)

$$= \frac{2}{4} \times \frac{1}{3} = \frac{1}{6}.$$

55. The required probability = (probability of drawing a prime number) $\times$ (probability of drawing more than 10 from the prime numbers)

$$= \frac{8}{20} \times \frac{4}{8} = \frac{1}{5}.$$

56. The probability of not drawing the ace in the first draw, in the second draw and in the third draw are $\frac{12}{13}, \frac{11}{12}$ and $\frac{10}{11}$ respectively.

The probability of drawing ace of spades in the fourth draw $= \frac{1}{10}$.

$\therefore$ the required probability $= \frac{12}{13} \times \frac{11}{12} \times \frac{10}{11} \times \frac{1}{10} = \frac{1}{13}$.

57. $n(S)$ = the area of the circle of radius r.

$n(E)$ = the area of the circle of radius $\frac{r}{2}$.

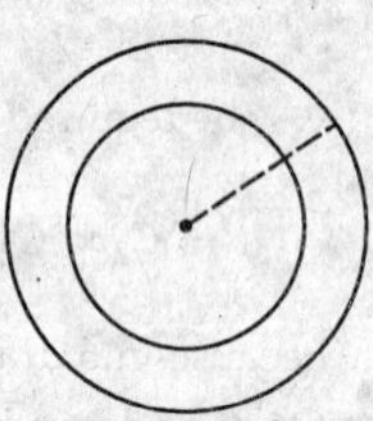

$\therefore$ the required probability

$$= \frac{n(E)}{n(S)} = \frac{\pi \cdot \left(\frac{r}{2}\right)^2}{\pi r^2} = \frac{1}{4}.$$

58. Let E_A = the event of A becoming secretary. Similarly, E_B and E_C.

E_L = the event of admitting lady members.

Here, $P(E_A) = \frac{1}{3}, P(E_B) = \frac{2}{9}, P(E_C) = \frac{4}{9}$.

Clearly, E_A, E_B, E_C are mutually exclusive and exhaustive.

Also, $P\left(\frac{E_L}{E_A}\right) = 0.6, \quad P\left(\frac{E_L}{E_B}\right) = 0.7, \quad P\left(\frac{E_L}{E_C}\right) = 0.5.$

$\therefore$ the required probability

$$= P(E_A) \cdot P\left(\frac{E_L}{E_A}\right) + P(E_B) \cdot P\left(\frac{E_L}{E_B}\right) + P(E_C) \cdot P\left(\frac{E_L}{E_C}\right)$$

$$= \frac{1}{3} \times \frac{3}{5} + \frac{2}{9} \times \frac{7}{10} + \frac{4}{9} \times \frac{5}{10} = \frac{26}{45}.$$

59. E_1 = the event of drawing from the first box.

E_2 = the event of drawing from the second box.

$P(E_1) = \frac{3}{6} = \frac{1}{2}$ and $P(E_2) = \frac{3}{6} = \frac{1}{2}$.

Clearly, E_1 and E_2 are mutually exclusive and exhaustive.

Now, $P\left(\frac{B}{E_1}\right) = \frac{3}{7}$ and $P\left(\frac{B}{E_2}\right) = \frac{4}{7}$.

We have to find $P\left(\frac{E_1}{B}\right)$.

By Bayes' theorem,

$$P\left(\frac{E_1}{B}\right) = \frac{P(E_1) \cdot P\left(\frac{B}{E_1}\right)}{P(E_1) \cdot P\left(\frac{B}{E_1}\right) + P(E_2) \cdot P\left(\frac{B}{E_2}\right)} = \frac{\frac{1}{2} \cdot \frac{3}{7}}{\frac{1}{2} \cdot \frac{3}{7} + \frac{1}{2} \cdot \frac{4}{7}} = \frac{3}{7}.$$

60. Here, $P(M) = m$, $P(P) = p$, $P(C) = c$.

The probability of passing in at least one subject $= 1 - P(\overline{M}\,\overline{P}\,\overline{C})$.

$$\therefore \quad \frac{75}{100} = 1 - (1-m)(1-p)(1-c)$$

$$\text{or} \quad \frac{3}{4} = m + p + c - mp - pc - mc + mpc \qquad \text{... (1)}$$

The probability of passing in at least two subjects

$$= P(MPC) + P(MP\overline{C}) + P(M\overline{P}C) + P(\overline{M}PC)$$

$$\therefore \quad \frac{1}{2} = mpc + mp(1-c) + m(1-p)c + (1-m)pc$$

$$\therefore \quad 2mpc = mp + mc + pc - \frac{1}{2} \qquad \text{... (2)}$$

The probability of passing in exactly two subjects

$$= P(MP\overline{C}) + P(M\overline{P}C) + P(\overline{M}PC).$$

$$\therefore \quad \frac{2}{5} = mp(1-c) + m(1-p)c + (1-m)pc = mp + mc + pc - 3mpc \qquad \text{... (3)}$$

(2) and (3) $\Rightarrow 2mpc + \frac{1}{2} = \frac{2}{5} + 3mpc$ or $mpc = \frac{1}{2} - \frac{2}{5} = \frac{1}{10}$

$\therefore \quad (2) \Rightarrow \frac{1}{5} = mp + mc + pc - \frac{1}{2}$ or $mp + mc + pc = \frac{1}{5} + \frac{1}{2} = \frac{7}{10}$.

$\therefore \quad (1) \Rightarrow \frac{3}{4} = m + p + c - \frac{7}{10} + \frac{1}{10}$

$\therefore \quad m + p + c = \frac{27}{20}$.

61. $P(E) \le P(F) \Rightarrow \frac{n(E)}{n(S)} \le \frac{n(F)}{n(S)} \Rightarrow n(E) \le n(F)$.

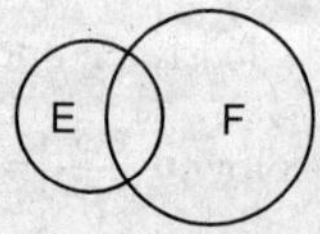

$P(E \cap F) > 0 \quad \Rightarrow E \cap F \ne \phi$.

Clearly, $x \in E \not\Rightarrow x \in F,\ x \in F \not\Rightarrow x \in E,$

$x \notin E \not\Rightarrow x \notin F$.

62. $P(A \cup B) = P(A) + P(B) - P(A \cap B)$.

$\therefore \quad 1 \ge P(A) + P(B) - P(A \cap B) \ge \frac{3}{4}$.

As the minimum value of $P(A \cap B) = \frac{1}{8}$, we get

$P(A) + P(B) - \frac{1}{8} \ge \frac{3}{4} \quad \Rightarrow P(A) + P(B) \ge \frac{1}{8} + \frac{3}{4} = \frac{7}{8}$.

As the maximum value of $P(A \cap B) = \frac{3}{8}$, we get

$1 \ge P(A) + P(B) - \frac{3}{8} \quad \Rightarrow P(A) + P(B) \le 1 + \frac{3}{8} = \frac{11}{8}$.

63. $P\left(\frac{E}{F}\right) = \frac{P(E \cap F)}{P(F)},\ P\left(\frac{\overline{E}}{F}\right) = \frac{P(\overline{E} \cap F)}{P(F)}$.

$\therefore \quad P\left(\frac{E}{F}\right) + P\left(\frac{\overline{E}}{F}\right) = \frac{P(E \cap F) + P(\overline{E} \cap F)}{P(F)} = \frac{P\{(E \cap F) \cup (\overline{E} \cap F)\}}{P(F)}$

(because $E \cap F, \overline{E} \cap F$ are nonintersecting)

$= \frac{P(F)}{P(F)} = 1$.

Similarly, $P\left(\frac{E}{\overline{F}}\right) + P\left(\frac{\overline{E}}{\overline{F}}\right) = \frac{P(\overline{F})}{P(\overline{F})} = 1$.

64. A = the event of (x, y) belonging to the area OTQPO

B = the event of (x, y) belonging to the area OSQRO

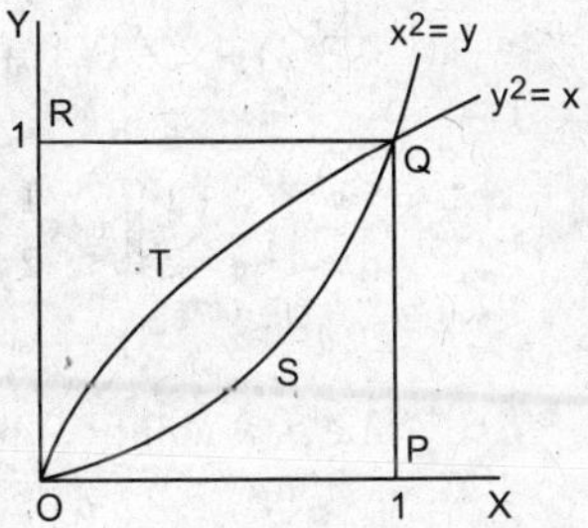

$$P(A) = \frac{\text{ar}(OTQPO)}{\text{ar}(OPQRO)} = \frac{\int_0^1 \sqrt{x}\,dx}{1 \times 1}$$

$$= \left[\frac{2}{3}x^{3/2}\right]_0^1 = \frac{2}{3}.$$

$$P(B) = \frac{\text{ar}(OSQRO)}{\text{ar}(OPQRO)} = \frac{\int_0^1 \sqrt{y}\,dy}{1 \times 1} = \frac{2}{3}.$$

$$P(A \cap B) = \frac{\text{ar}(OTQS)}{\text{ar}(OPQRO)} = \frac{\int_0^1 \sqrt{x}\,dx - \int_0^1 x^2 dx}{1 \times 1} = \frac{2}{3} - \frac{1}{3} = \frac{1}{3}.$$

$P(A) + P(B) = \frac{2}{3} + \frac{2}{3} \neq 1$. So, A and B are not exhaustive.

$P(A) \cdot P(B) = \frac{2}{3} \cdot \frac{2}{3} = \frac{4}{9} \neq P(A \cap B)$. So, A and B are not independent.

$P(A \cup B) = 1$, $P(A) + P(B) = \frac{2}{3} + \frac{2}{3} \neq P(A \cup B)$. So A and B are not mutually exclusive.

65. $P(A \cup B) = P(A) + P(B) - P(A \cap B)$

$$\Rightarrow \quad \frac{5}{6} = \left(1 - \frac{1}{2}\right) + P(B) - \frac{1}{3} \quad \Rightarrow P(B) = \frac{2}{3}.$$

$P(A) \cdot P(B) = \frac{1}{2} \cdot \frac{2}{3} = \frac{1}{3} = P(A \cap B)$. So, A and B are independent and therefore, not mutually exclusive.

66. The probability of exactly one of A and B to occur

$$= P(A\bar{B}) + P(\bar{A}B)$$
$$= P(A) \cdot P(\bar{B}) + P(\bar{A}) \cdot P(B) = P(A) \cdot \{1 - P(B)\} + \{1 - P(A)\}P(B)$$
$$= P(A) + P(B) - 2P(A) \cdot P(B) = P(A) + P(B) - 2P(A \cap B)$$
$$= 1 - P(\bar{A}) + 1 - P(\bar{B}) - 2\{1 - P(\bar{A})\}\{1 - P(\bar{B})\}.$$
$$= P(\bar{A}) + P(\bar{B}) - 2P(\bar{A}) \cdot P(\bar{B}).$$

67. See the hint to Q. No. 63.

68. Use $P(A \cup B) = P(A) + P(B) - P(A \cap B)$ and each probability lies between 0 and 1.

69. The probability of A winning

$$= P(A) + P(\overline{A}\,\overline{B}\,A) + P(\overline{A}\,\overline{B}\,\overline{A}\,\overline{B}\,A) + \ldots$$

$$= \frac{1}{2} + \left(1 - \frac{1}{2}\right)\left(1 - \frac{1}{2}\right)\frac{1}{2} + \ldots = \frac{1}{2} + \left(\frac{1}{2}\right)^3 + \ldots = \frac{\frac{1}{2}}{1 - \frac{1}{4}} = \frac{2}{3}$$

$\therefore$ the probability of B winning $= 1 - \frac{2}{3} = \frac{1}{3}$

$\therefore$ the expectation of A = Rs $30 \times \frac{2}{3}$ = Rs 20, and

the expectation of B = Rs $30 \times \frac{1}{3}$ = Rs 10.

□

Part 7

Coordinate Geometry of Three Dimensions

1

Points, Direction Cosines and Direction Ratios

• *Type 1* •

Choose the most appropriate option (a, b, c or d).

1. A line makes angles α, β, γ with the positive directions of the axes of reference. The value of $\cos 2\alpha + \cos 2\beta + \cos 2\gamma$ is

(a) 1 (b) 2

(c) –1 (d) 0

2. If a line makes angles of 60° and 45° with the positive directions of the x-axis and y-axis respectively then the acute angle between the line and the z-axis is

(a) 60° (b) 45°

(c) 75° (d) 15°

3. If θ is an angle given by $\cos\theta = \dfrac{\cos^2\alpha + \cos^2\beta + \cos^2\gamma}{\sin^2\alpha + \sin^2\beta + \sin^2\gamma}$ where α, β, γ are the angles made by a line with the positive directions of the axes of reference then the measure of θ is

(a) $\dfrac{\pi}{4}$ (b) $\dfrac{\pi}{6}$

(c) $\dfrac{\pi}{2}$ (d) $\dfrac{\pi}{3}$

4. If the direction cosines of the line joining the origin and a point at unit distance from the origin are $\dfrac{1}{\sqrt{3}}, -\dfrac{1}{2}, \lambda$ then the point has coordinates

(a) $\left(\frac{-1}{\sqrt{3}}, \frac{1}{2}, \frac{-1}{2} \cdot \sqrt{\frac{5}{4}}\right)$ (b) $\left(\frac{2}{\sqrt{3}}, -1, \sqrt{\frac{5}{4}}\right)$

(c) $\left(\frac{-2}{\sqrt{3}}, 1, -\sqrt{\frac{5}{4}}\right)$ (d) none of these

5. The direction ratios of two parallel lines are 4, –3, –1 and $\lambda + \mu, 1 + \mu, 2$. The value of the pair (λ, μ) is

(a) (1, 7) (b) (–1, –7)

(c) (7, 1) (d) no fixed value

6. The direction ratios of two perpendicular lines are 1, –3, 5 and $\lambda, 1 + \lambda, 2 + \lambda$. Then λ is

(a) $-\frac{7}{3}$ (b) $\frac{7}{2}$

(c) $-\frac{1}{4}$ (d) $\frac{1}{2}$

7. The direction cosines of the line joining the points (2, 3, –1) and (3, –2, 1) are

(a) –1, 5, –2 (b) $\frac{1}{\sqrt{30}}, -\sqrt{\frac{5}{6}}, \sqrt{\frac{2}{15}}$

(c) $\frac{-1}{30}, \frac{1}{6}, -\frac{1}{15}$ (d) none of these

8. If $A = (1, 2, 3)$, $B = (-2, 4, \lambda)$ and $\angle AOB = \pi/2$ where O is the origin then λ is

(a) 6 (b) –6

(c) 0 (d) –2

9. Let $A = (2, 3, -1)$, $B = (1, -1, 2)$. A point C on the y-axis such that $AB \perp BC$ has the coordinates

(a) $\left(-\frac{9}{4}, 0, \frac{-9}{4}\right)$ (b) $\left(0, -\frac{4}{9}, 0\right)$

(c) $\left(0, \frac{-9}{4}, 0\right)$ (d) $\left(0, \frac{9}{4}, 0\right)$

10. If the points $A(1, 2, -1)$, $B(2, 6, 2)$ and $C(\lambda, -2, -4)$ are collinear then λ is

(a) 0 (b) 2 (c) –2 (d) 1

11. If $A = (1, 1, 1)$, $B = (2, -1, 3)$, $C = (0, 4, -2)$, $D = (1, 2, \lambda)$ and AB, AC and AD are coplanar then λ is

(a) 1 (b) 0

(c) –1 (d) 3

12. The angle between the lines joining the points (1, 0, –3), (2, –1, 2), and (1, 1, 1), (3, 2, 0) is

(a) $\cos^{-1}\frac{2}{9}$ (b) 90°

(c) 0° (d) $\cos^{-1}\frac{2\sqrt{2}}{9}$

13. The angle between the lines joining the points $(1, 1, 0)$, $(-3, \sqrt{3}+1, 3)$, and $(0, -1, 0)$, $(-1, \sqrt{3}-1, \lambda)$ is $\cos^{-1}\left(\frac{1}{8}\right)$. If λ is an integer then its value is

(a) 1 (b) 0

(c) –1 (d) 2

14. $L_r, m_r, n_r; r = 1, 2, 3;$ are the direction cosines of three mutually perpendicular lines. The direction cosines of the line equally inclined to them are

(a) $l_1 + l_2 + l_3, m_1 + m_2 + m_3, n_1 + n_2 + n_3$

(b) $\frac{l_1 + l_2 + l_3}{3}, \frac{m_1 + m_2 + m_3}{3}, \frac{n_1 + n_2 + n_3}{3}$

(c) $\frac{1}{\sqrt{3}}(l_1 + l_2 + l_3), \frac{1}{\sqrt{3}}(m_1 + m_2 + m_3), \frac{1}{\sqrt{3}}(n_1 + n_2 + n_3)$

(d) $\frac{l_1 l_2 l_3}{\sqrt{3}}, \frac{m_1 m_2 m_3}{\sqrt{3}}, \frac{n_1 n_2 n_3}{\sqrt{3}}$

15. Let $A = (1, 2, 2)$, $B = (2, 3, 6)$ and $C = (3, 4, 12)$. The direction cosines of a line equally inclined with OA, OB and OC where O is the origin, are

(a) $\frac{1}{\sqrt{2}}, \frac{-1}{\sqrt{2}}, 0$ (b) $\frac{1}{\sqrt{2}}, \frac{1}{\sqrt{2}}, 0$

(c) $\frac{1}{\sqrt{3}}, -\frac{1}{\sqrt{3}}, \frac{1}{\sqrt{3}}$ (d) $\frac{1}{\sqrt{3}}, -\frac{1}{\sqrt{3}}, -\frac{1}{\sqrt{3}}$

16. The angle between the lines whose direction cosines satisfy the equations $l + m + n = 0$ and $l^2 = m^2 + n^2$ is

(a) $\frac{\pi}{6}$ (b) $\frac{\pi}{2}$

(c) $\frac{\pi}{3}$ (d) $\frac{\pi}{4}$

17. The angle between two lines whose direction cosines satisfy the equations $n = l + m$ and $m = 2l + 3n$ is

(a) 0° (b) 90°

(c) 60° (d) none of these

18. The value of λ for which the triangle ABC whose vertices are $A(6, 10, 10)$, $B(1, 0, -5)$ and $C(6, -10, \lambda)$ is right-angled at B, is

(a) 0 (b) 30

(c) $\frac{70}{3}$ (d) $\frac{3}{70}$

19. Let $A = (1, 2, 3)$, $B = (-1, -2, -1)$, $C = (2, 3, 2)$ and $D = (4, 7, 6)$. Then $ABCD$ is a

(a) rectangle (b) square

(c) parallelogram (d) none of these

20. The points (0, 0, 0), (0, 2, 0), (1, 0, 0), (0, 0, 4) are

(a) coplanar (b) vertices of a parallelogram

(c) vertices of a rectangle (d) on a sphere

21. If $A = (5, -1, 1)$, $B = (7, -4, 7)$, $C = (1, -6, 10)$, $D = (-1, -3, 4)$ then $ABCD$ is a

(a) square (b) rectangle

(c) rhombus (d) none of these

22. If $A = (0, 0, 2)$, $B = (\sqrt{2}, \sqrt{2}, 2)$, $C = (\sqrt{2}, \sqrt{2}, 0)$ and

$$D = \left(\frac{8\sqrt{2} - 20}{17}, \frac{12\sqrt{2} + 4}{17}, \frac{20 - 8\sqrt{2}}{17}\right)$$

then $ABCD$ is a

(a) rhombus (b) square

(c) parallelogram (d) none of these

23. If the vertices of a triangle are $(-1, 6, -4)$, $(2, 1, 1)$ and $(5, -1, 0)$ then the centroid of the triangle is

(a) $(6, 6, -3)$ (b) $(2, 2, -1)$

(c) $\left(3, 3, -\frac{3}{2}\right)$ (d) none of these

24. If two vertices of a triangle ABC are $A(-1, 2, 4)$ and $B(2, -3, 0)$, and the centroid is $(2, 0, 2)$ then the vertex C has the coordinates

(a) $(5, 1, 2)$ (b) $\left(1, -\frac{1}{3}, \frac{7}{3}\right)$

(c) $\left(3, -\frac{2}{3}, \frac{5}{3}\right)$ (d) none of these

25. Four vertices of a tetrahedron are $(0, 0, 0)$, $(4, 0, 0)$, $(0, -8, 0)$ and $(0, 0, 12)$. Its centroid has the coordinates

(a) $\left(\frac{4}{3}, -\frac{8}{3}, 4\right)$ (b) $(2, -4, 6)$

(c) $(1, -2, 3)$ (d) none of these

26. Three vertices of a tetrahedron are $(0, 0, 0)$, $(6, -5, -1)$ and $(-4, 1, 3)$. If the centroid of the tetrahedron be $(1, -2, 5)$ then the fourth vertex is

(a) $(2, -4, 18)$ (b) $(2, -4, -18)$

(c) $\left(\frac{3}{4}, \frac{-3}{2}, \frac{7}{4}\right)$ (d) none of these

27. The points $A(1, 2, -1)$, $B(2, 5, -2)$, $C(4, 4, -3)$ and $D(3, 1, -2)$ are

(a) collinear (b) vertices of a rectangle

(c) vertices of a square (d) vertices of a rhombus

28. The projection of a line segment on the axes of reference are 3, 4 and 12 respectively. The length of the line segment is

(a) 19 (b) $\frac{19}{3}$

(c) 5 (d) 13

29. ABC is a triangle where $A = (2, 3, 5)$, $B = (-1, 3, 2)$ and $C = (\lambda, 5, \mu)$. If the median through A is equally inclined with the axes then

(a) $\lambda = 14, \mu = 20$ (b) $\lambda = 7, \mu = 10$

(c) $\lambda = \frac{7}{2}, \mu = 5$ (d) $\lambda = 10, \mu = 7$

30. The volume of the tetrahedron whose vertices are $(0, 1, 2)$, $(4, 3, 6)$, $(2, 3, 2)$ and $(3, 0, 1)$ is (in unit^3)

(a) 0 (b) 1

(c) 6 (d) 3

31. If $(1, -1, 0)$, $(-2, 1, 8)$ and $(-1, 2, 7)$ are three consecutive vertices of a parallelogram then the fourth vertex is

(a) $(2, 0, -1)$ (b) $(1, 0, -1)$

(c) $(1, -2, 0)$ (d) $(0, -2, 1)$

32. Let $P_r(x_r, y_r, z_r)$; $r = 1, 2, 3$; be three points where x_1, x_2, x_3; y_1, y_2, y_3 and z_1, z_2, z_3 are each in GP with the same common ratio. Then P_1, P_2, P_3 are

(a) coplanar points

(b) collinear points

(c) vertices of an equilateral triangle

(d) none of these

• Type 2 •

Choose the correct options. One or more options may be correct.

33. A point Q at a distance 3 from the point $P(1, 1, 1)$ lying on the line joining the points $A(0, -1, 3)$ and P, has the coodinates

(a) $(2, 3, -1)$ (b) $(4, 7, -5)$

(c) $(0, -1, 3)$ (d) $(-2, -5, 7)$

34. If $A = (2, -3, 7)$, $B = (-1, 4, -5)$ and P is a point on the line AB such that $AP : BP = 3 : 2$ then P has the coordinates

(a) $\left(\frac{4}{5}, \frac{-1}{5}, \frac{11}{5}\right)$ (b) $\left(\frac{1}{5}, \frac{6}{5}, -\frac{1}{5}\right)$

(c) $\left(\frac{7}{5}, \frac{-18}{5}, \frac{29}{5}\right)$ (d) $(-7, 18, -29)$

35. If the direction ratios of a line are $1 + \lambda, 1 - \lambda, 2$, and it makes an angle of $60°$ with the y-axis then λ is

(a) $1 + \sqrt{3}$ (b) $2 + \sqrt{5}$ (c) $1 - \sqrt{3}$ (d) $2 - \sqrt{5}$

Answers

1. c	**2.** a	**3.** d	**4.** a	**5.** b
6. a	**7.** b	**8.** d	**9.** c	**10.** a
11. b	**12.** d	**13.** b	**14.** c	**15.** d
16. c	**17.** a	**18.** a	**19.** c	**20.** d
21. c	**22.** d	**23.** b	**24.** a	**25.** c
26. a	**27.** b	**28.** d	**29.** b	**30.** c
31. a	**32.** b	**33.** a, c	**34.** b, d	**35.** b, d

Hints and Solutions to Selected Questions

10. Direction ratios of the line AB are $2-1, 6-2, 2-(-1)$, i.e., $1, 4, 3$. Direction ratios of the line AC are $\lambda-1, -2-2, -4-(-1)$, i.e., $\lambda-1, -4, -3$. For collinearity, $AB \parallel AC$ $\Rightarrow \dfrac{\lambda-1}{1}=\dfrac{-4}{4}=\dfrac{-3}{3}$ $\Rightarrow \lambda=0$.

11. Direction ratios of AB, AC and AD are respectively $1, -2, 2; -1, 3, -3; 0, 1, \lambda-1$. They are coplanar $\Rightarrow$ there is a line with direction ratios l, m, n which is perpendicular to AB, AC and AD.

$\therefore$ $l-2m+2n=0, -l+3m-3n=0, 0.l+m+(\lambda-1)n=0.$

Eliminating l, m, n we get $\begin{vmatrix} 1 & -2 & 2 \\ -1 & 3 & -3 \\ 0 & 1 & \lambda-1 \end{vmatrix}=0 \Rightarrow \lambda=0$

14. Let $OA=OB=OC=1$ and their direction cosines are $l_1, m_1, n_1; l_2, m_2, n_2$ and l_3, m_3, n_3 respectively.

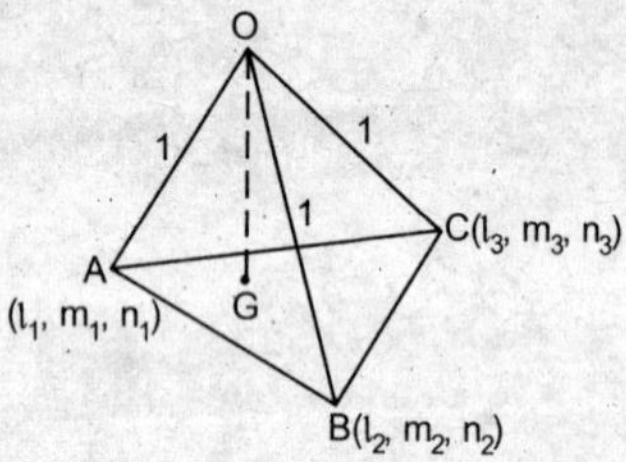

Then $A=(l_1, m_1, n_1), B=(l_2, m_2, n_2)$,

$C=(l_3, m_3, n_3)$

and $OA \perp OB, OB \perp OC, OC \perp OA$.

Clearly, ABC is equilateral, and OG is equally inclined with OA, OB, OC where G is the centroid (also circumcentre) of the ΔABC.

$$G=\left(\frac{l_1+l_2+l_3}{3}, \frac{m_1+m_2+m_3}{3}, \frac{n_1+n_2+n_3}{3}\right).$$

$\therefore$ direction ratios of $OG = l_1+l_2+l_3, m_1+m_2+m_3, n_1+n_2+n_3$.

Also, $(l_1+l_2+l_3)^2+(m_1+m_2+m_3)^2+(n_1+n_2+n_3)^2$

$=\Sigma l_1^2+\Sigma l_2^2+\Sigma l_3^2+2\Sigma l_1l_2+2\Sigma l_2l_3+2\Sigma l_3l_1=1+1+1+0=3,$

from the conditions of the problem.

$\therefore$ direction cosines of OG are $\dfrac{l_1+l_2+l_3}{\sqrt{3}}, \dfrac{m_1+m_2+m_3}{\sqrt{3}}, \dfrac{n_1+n_2+n_3}{\sqrt{3}}$.

15. Let l, m, n be the required direction cosines. Then

$$l\cdot\frac{1}{3}+m\cdot\frac{2}{3}+n\cdot\frac{2}{3}=l\cdot\frac{2}{7}+m\cdot\frac{3}{7}+n\cdot\frac{6}{7}=l\cdot\frac{3}{13}+m\cdot\frac{4}{13}+n\cdot\frac{12}{13}$$

$$\Rightarrow\quad\left(\frac{1}{3}-\frac{2}{7}\right)l+\left(\frac{2}{3}-\frac{3}{7}\right)m+\left(\frac{2}{3}-\frac{6}{7}\right)n=0\qquad\Rightarrow\ l+5m-4n=0$$

$$\left(\frac{1}{3}-\frac{3}{13}\right)l+\left(\frac{2}{3}-\frac{4}{13}\right)m+\left(\frac{2}{3}-\frac{12}{13}\right)n=0\qquad\Rightarrow\ 4l+14m-10n=0$$

$$\Rightarrow\ 2l+7m-5n=0$$

$$\therefore\quad\frac{l}{3}=\frac{m}{-3}=\frac{n}{-3}\quad\text{or}\quad\frac{l}{1}=\frac{m}{-1}=\frac{n}{-1}.\qquad\therefore\ l=\frac{1}{\sqrt{3}},\ m=\frac{-1}{\sqrt{3}},\ n=\frac{-1}{\sqrt{3}}.$$

16. $\{-(m+n)\}^2=m^2+n^2 \quad\Rightarrow\ mn=0.$

When $m=0, l=-n$ direction ratios are $-1, 0, 1$.

When $n=0, l=-m$ direction ratios are $-1, 1, 0$.

$$\therefore\quad\cos\theta=\frac{-1}{\sqrt{2}}\cdot\frac{-1}{\sqrt{2}}+\frac{0}{\sqrt{2}}\cdot\frac{1}{\sqrt{2}}+\frac{1}{\sqrt{2}}\cdot\frac{0}{\sqrt{2}}=\frac{1}{2}.$$

20. From the location of the points it is clear that they are not on the same plane. Through four noncoplanar points one and only one sphere passes.

22. The points are not on the same plane.

26. If the fourth vertex is (α, β, γ) then centroid

$$\text{centroid}=\left(\frac{0+6-4+\alpha}{4},\frac{0-5+1+\beta}{4},\frac{0-1+3+\gamma}{4}\right)$$

$$=(1,-2,5).$$

28. If the direction cosines be $\cos\alpha, \cos\beta, \cos\gamma$ and the length is l then

$$l\cos\alpha=3,\ l\cos\beta=4,\ l\cos\gamma=12.$$

Squaring and adding, $l^2(\cos^2\alpha+\cos^2\beta+\cos^2\gamma)=3^2+4^2+12^2=13^2.$

$\therefore\quad l^2=13^2.$

29. Centroid $G=\left(\frac{2-1+\lambda}{3},\frac{3+3+5}{3},\frac{5+2+\mu}{3}\right).$

Direction ratios of AG are $2-\frac{1+\lambda}{3}, 3-\frac{11}{3}, 5-\frac{7+\mu}{3}$, i.e., $\frac{5-\lambda}{3},\frac{-2}{3},\frac{8-\mu}{3}.$

As AG is equally inclined with the axes, the direction ratios are 1, 1, 1 also.

$$\therefore \quad \frac{\frac{5-\lambda}{3}}{1}=\frac{-\frac{2}{3}}{1}=\frac{\frac{8-\mu}{3}}{1} \quad \Rightarrow 5-\lambda=-2=8-\mu.$$

32. $P_1=(x_1, rx_1, r^2x_1), P_2=(y_1, ry_1, r^2y_1), P_3=(z_1, rz_1, r^2z_1)$.

Direction ratios of P_1P_2 are $y_1-x_1, r(y_1-x_1), r^2(y_1-x_1)$.

Direction ratios of P_2P_3 are $z_1-y_1, r(z_1-y_1), r^2(z_1-y_1)$.

As $\frac{y_1-x_1}{z_1-y_1}=\frac{r(y_1-x_1)}{r(z_1-y_1)}=\frac{r^2(y_1-x_1)}{r^2(z_1-y_1)}$, $P_1P_2 \parallel P_2P_3$.

So P_1, P_2, P_3 are collinear.

33. If Q is (α, β, γ) then (α, β, γ), $(1, 1, 1)$ and $(0, -1, 3)$ are collinear.

$$\therefore \quad \frac{\alpha-1}{1-0}=\frac{\beta-1}{1-(-1)}=\frac{\gamma-1}{1-3}. \quad \text{Also, } PQ=3$$

$$\Rightarrow \quad (\alpha-1)^2+(\beta-1)^2+(\gamma-1)^2=9.$$

Solve these equations for α, β, γ.

□

2

Straight Lines

• *Type 1* •

Choose the most appropriate option (a, b, c or d).

1. The direction cosines of a line whose equations are $\frac{x-1}{2}=\frac{y+3}{4}=\frac{z-2}{-3}$ are

(a) $\frac{1}{\sqrt{14}}, \frac{-3}{\sqrt{14}}, \frac{2}{\sqrt{14}}$ (b) $\frac{2}{\sqrt{29}}, \frac{4}{\sqrt{29}}, \frac{-3}{\sqrt{29}}$

(c) $\frac{1}{\sqrt{29}}, \frac{-3}{\sqrt{29}}, \frac{2}{\sqrt{29}}$ (d) $2, 4, -3$

2. The equations of the line passing through the point (1, 2, 3) having the direction ratios 3, 2, 1 are

(a) $\frac{x-1}{1}=\frac{y-2}{2}=\frac{z-3}{3}$ (b) $\frac{x}{3}=\frac{y}{6}=\frac{z}{9}$

(c) $\frac{x+2}{3}=\frac{y}{2}=\frac{z-2}{1}$ (d) none of these

3. The equations of the line passing through the points $(-2, 1, 0)$ and $(3, 4, -1)$ are

(a) $\frac{x+7}{5}=\frac{y+2}{3}=\frac{z-1}{-1}$ (b) $\frac{x+2}{-1}=\frac{y-1}{3}=\frac{z}{5}$

(c) $\frac{x+3}{5}=\frac{y+4}{3}=\frac{z-1}{-1}$ (d) none of these

4. The coordinates of a point on the line $\frac{x-1}{2}=\frac{y-2}{3}=\frac{z}{\sqrt{3}}$ at a distance 1 unit from the point $(-1, -1, -\sqrt{3})$ are

(a) $\left(-\frac{3}{2}, -\frac{7}{4}, -\frac{5}{4}\sqrt{3}\right)$ (b) $\left(\frac{3}{2}, \frac{11}{4}, \frac{\sqrt{3}}{4}\right)$

(c) $\left(\frac{1}{2}, \frac{5}{4}, -\frac{\sqrt{3}}{4}\right)$ (d) none of these

5. The equation of the locus of the point $\left(1+\frac{r}{4}, -1+\frac{r}{3}, 2\right)$, where $r \in R$, is given by

(a) $\frac{x-1}{4}=\frac{y+1}{3}=\frac{z-2}{0}$ (b) $\frac{x-1}{3}=\frac{y+1}{4}=\frac{z-2}{0}$

(c) $4x-3y=7$ (d) $z=2$

6. The lines $\frac{x+3}{-2}=\frac{y}{1}=\frac{z-4}{3}$ and $\frac{x}{\lambda}=\frac{y-1}{\lambda+1}=\frac{z}{\lambda+2}$ are perpendicular to each other. Then λ is equal to

(a) $-\frac{5}{3}$ (b) 4

(c) $-\frac{1}{4}$ (d) -4

7. If the lines $\frac{x+2}{4\lambda+1}=\frac{y-1}{4}=\frac{z}{-18}$ and $\frac{x}{-3}=\frac{y+1}{5\mu-3}=\frac{z-1}{6}$ are parallel to each other then the value of the pair (λ, μ) is

(a) $\left(-2, \frac{1}{3}\right)$ (b) $\left(2, -\frac{1}{3}\right)$

(c) $\left(2, \frac{1}{3}\right)$ (d) cannot be found

8. The angle between the lines $\frac{x+2}{1}=\frac{y+3}{-2}=\frac{z-4}{1}$ and $\frac{x-1}{-1}=\frac{y}{1}=\frac{z}{0}$ is

(a) $\frac{\pi}{6}$ (b) $\frac{\pi}{3}$

(c) $\frac{\pi}{2}$ (d) 0

9. If $\frac{\alpha}{\alpha'}, \frac{\beta}{\beta'}, \frac{\gamma}{\gamma'}$ are not all equal, the point of intersection of the lines $\frac{x-\alpha'}{\alpha}=\frac{y-\beta'}{\beta}=\frac{z-\gamma'}{\gamma}$ and $\frac{x-\alpha}{\alpha'}=\frac{y-\beta}{\beta'}=\frac{z-\gamma}{\gamma'}$ is

(a) $(\alpha-\alpha', \beta-\beta', \gamma-\gamma')$

(b) $(\alpha+\alpha', \beta+\beta', \gamma+\gamma')$

(c) $(\alpha\alpha', \beta\beta', \gamma\gamma')$

(d) none because they are nonintersecting

10. The equation of the straight line passing through the origin and perpendicular to the lines $\frac{x+1}{-3}=\frac{y-2}{2}=\frac{z}{1}$ and $\frac{x-1}{1}=\frac{y}{-3}=\frac{z+1}{2}$ has the equation

(a) $x=y=z$

(b) $\frac{x}{4}=\frac{y}{3}=\frac{z}{6}$

(c) $\frac{x}{3}=\frac{y}{1}=\frac{z}{0}$

(d) none of these

11. The shortest distance between the line $\frac{x-3}{3}=\frac{y}{0}=\frac{z}{-4}$ and the y-axis is

(a) $\frac{1}{5}$

(b) 1

(c) 0

(d) $\frac{12}{5}$

12. The equations of the line of shortest distance between the lines $\frac{x+4}{4}=\frac{y-2}{-2}=\frac{z-3}{0}$ and $\frac{x-5}{5}=\frac{y-3}{3}=\frac{z}{0}$ are

(a) $\frac{x+4}{0}=\frac{y-2}{0}=\frac{z-3}{1}$

(b) $\frac{x-5}{0}=\frac{y-3}{0}=\frac{z}{1}$

(c) $\frac{x}{0}=\frac{y}{0}=\frac{z-3}{1}$

(d) none of these

13. The projection of the line segment joining the point $(6, -2, 1)$ and the origin on the line $\frac{x-2}{4}=\frac{y+1}{-3}=\frac{z-1}{0}$ is

(a) 30

(b) 6

(c) 5

(d) none of these

14. If $A=(p, q, r)$ and $B=(p', q', r')$ are two points on the line $\lambda x=\mu y=\nu z$ such that $OA=a$, $OB=b$ then $pp'+qq'+rr'$ is equal to

(a) $a+b$

(b) ab

(c) $\sqrt{a^2+b^2}$

(d) none of these

15. The number of real values of k for which the lines $\frac{x-k}{4}=\frac{y-1}{2}=\frac{z+1}{1}$ and $\frac{x-(k+1)}{1}=\frac{y}{-1}=\frac{z-1}{2}$ are intersecting, is

(a) 0 (b) 2 (c) 1 (d) infinite

16. The distance between the lines $\frac{x-4}{2}=\frac{y+1}{-3}=\frac{z}{6}$ and $\frac{x}{-1}=\frac{y-1}{3/2}=\frac{z+1}{-3}$ is

(a) $\sqrt{\frac{629}{7}}$ (b) $\sqrt{\frac{39}{7}}$

(c) $\frac{\sqrt{629}}{7}$ (d) none of these

17. The point $A(3,-2,4)$ is shifted parallel to the line $\frac{x}{\sqrt{3}}=\frac{y-1}{2}=\frac{z+1}{3}$ by a distance 1. The coordinates of P in the new position are

(a) $\left(\frac{12-\sqrt{3}}{4},-\frac{5}{2},\frac{13}{4}\right)$ (b) $(3+\sqrt{3},3,2)$

(c) $(3-\sqrt{3},-1,-4)$ (d) none of these

18. The image of the origin in the line $\frac{x+1}{2}=\frac{y-2}{3}=\frac{z}{\sqrt{3}}$ is

(a) $\left(-1,\frac{11}{2},\frac{\sqrt{3}}{2}\right)$ (b) $\left(3,\frac{-5}{2},\frac{\sqrt{3}}{2}\right)$

(c) $\left(-3,\frac{5}{2},-\frac{\sqrt{3}}{2}\right)$ (d) $\left(1,-\frac{11}{2},-\frac{\sqrt{3}}{2}\right)$

19. The distance of the point $(1,2,\lambda)$ from the line $\frac{x}{3}=\frac{y}{0}=\frac{z}{4}$ is 2. Then λ is

(a) $\frac{4}{3}$ (b) $\frac{3}{4}$

(c) 1 (d) nonexistent

20. If the lines $\frac{x}{1}=\frac{y}{2}=\frac{z}{3}$, $\frac{x-1}{3}=\frac{y-2}{-1}=\frac{z-3}{4}$ and $\frac{x+k}{3}=\frac{y-1}{2}=\frac{z-2}{h}$ are concurrent then

(a) $h=-2, k=-6$ (b) $h=\frac{1}{2}, k=2$

(c) $h=6, k=2$ (d) $h=2, k=\frac{1}{2}$

21. The number of real values of k for which the lines $\frac{x-1}{4}=\frac{y+1}{3}=\frac{z}{k}$ and $\frac{x}{1}=\frac{y-k}{3}=\frac{z-1}{-2}$ are coplanar, is

(a) 2 (b) 1

(c) 3 (d) 0

• Type 2 •

Choose the correct options. One or more options may be correct.

22. A point on the line $\frac{x-1}{1}=\frac{y-2}{2}=\frac{z+1}{3}$ at a distance $\sqrt{6}$ from the origin is

(a) $\left(\frac{-5}{7},\frac{-10}{7},\frac{13}{7}\right)$ (b) $(1, 2, -1)$

(c) $\left(\frac{5}{7},\frac{10}{7},\frac{-13}{7}\right)$ (d) $(-1, -2, 1)$

23. The direction cosines of a line passing through the origin and cutting the line $\frac{x+2}{1}=\frac{y-1}{2}=\frac{z}{-1}$ at $\cos^{-1}\sqrt{\frac{6}{11}}$ are

(a) $\frac{-1}{\sqrt{11}},\frac{3}{\sqrt{11}},\frac{-1}{\sqrt{11}}$ (b) $\frac{1}{\sqrt{11}},\frac{3}{\sqrt{11}},\frac{1}{\sqrt{11}}$

(c) $\frac{-1}{\sqrt{6}},\frac{2}{\sqrt{6}},\frac{1}{\sqrt{6}}$ (d) $\frac{-3}{\sqrt{11}},\frac{-1}{\sqrt{11}},\frac{1}{\sqrt{11}}$

Answers

1. b	**2.** c	**3.** a	**4.** a	**5.** b
6. d	**7.** c	**8.** a	**9.** b	**10.** a
11. d	**12.** c	**13.** b	**14.** b	**15.** d
16. c	**17.** a	**18.** c	**19.** a	**20.** d
21. a	**22.** b, c	**23.** a, d		

Hints and Solutions to Selected Questions

5. Let $(\alpha, \beta, \gamma) = \left(1 + \frac{r}{4}, -1 + \frac{r}{3}, 2\right)$. Then $\alpha = 1 + \frac{r}{4}, \beta = -1 + \frac{r}{3}, \gamma = 2$.

$\therefore \quad 4(\alpha - 1) = 3(\beta + 1) = r$ and $\gamma = 2 \quad \Rightarrow \frac{\alpha - 1}{3} = \frac{\beta + 1}{4} = \frac{\gamma - 2}{0}$.

7. Direction ratios of the lines are $4\lambda + 1, 4, -18$, and $-3, 5\mu - 3, 6$.

They are parallel $\Rightarrow \frac{4\lambda + 1}{-3} = \frac{4}{5\mu - 3} = \frac{-18}{6}$

$\Rightarrow \frac{4\lambda + 1}{-3} = -3$ and $\frac{4}{5\mu - 3} = -3$

9. From the question, the lines are not parallel.

Any point on the first line is $(\alpha' + \alpha r, \beta' + \beta r, \gamma' + \gamma r)$. It is on the other line if

$$\frac{\alpha' + \alpha r - \alpha}{\alpha'} = \frac{\beta' + \beta r - \beta}{\beta'} = \frac{\gamma' + \gamma r - \gamma}{\gamma'}$$

$$\Rightarrow \quad 1 + \frac{\alpha(r-1)}{\alpha'} = 1 + \frac{\beta(r-1)}{\beta'} = 1 + \frac{\gamma(r-1)}{\gamma'}$$

$$\Rightarrow \quad \frac{\alpha(r-1)}{\alpha'} = \frac{\beta(r-1)}{\beta'} = \frac{\gamma(r-1)}{\gamma'} \quad \Rightarrow r - 1 = 0 \ \left(\because \frac{\alpha}{\alpha'} = \frac{\beta}{\beta'} = \frac{\gamma}{\gamma'} \text{ is not true}\right)$$

$\therefore$ the point of intersection is $(\alpha' + \alpha \cdot 1, \beta' + \beta \cdot 1, \gamma' + \gamma \cdot 1)$.

14. Here, $\lambda p = \mu q = \nu r$ and $\lambda p' = \mu q' = \nu r'$. So,

$$\frac{p'}{p} = \frac{q'}{q} = \frac{r'}{r} = \frac{\sqrt{p'^2 + q'^2 + r'^2}}{\sqrt{p^2 + q^2 + r^2}} = \frac{\sqrt{b^2}}{\sqrt{a^2}} = \frac{b}{a}.$$

$$\therefore \quad pp' + qq' + rr' = \frac{b}{a}(p^2 + q^2 + r^2) = \frac{b}{a} \cdot a^2 = ab.$$

15. Any point on the first line is $(4r + k, 2r + 1, r - 1)$, and any point on the second line is $(r' + k + 1, -r', 2r' + 1)$. The lines are intersecting if

$4r + k = r' + k + 1, 2r + 1 = -r', r - 1 = 2r' + 1$ for some r and r'

$\Rightarrow \quad 4r - r' = 1, 2r + r' = -1, r - 2r' = 2.$

Now, $4r - r' = 1, 2r + r' = -1 \quad \Rightarrow r = 0, r' = -1$ which satisfy $r - 2r' = 2$.

This is true for all k.

16. The lines are parallel because $\dfrac{2}{-1}=\dfrac{-3}{3/2}=\dfrac{6}{-3}$.

$\therefore$ the distance = the distance of (4, –1, 0) from the line

$$\frac{x}{-1}=\frac{y-1}{3/2}=\frac{z+1}{-3}.$$

17. Direction cosines of the line are $\dfrac{\sqrt{3}}{4},\dfrac{2}{4},\dfrac{3}{4}$. So,

$$P=\left(3\pm1\cdot\frac{\sqrt{3}}{4},-2\pm1\cdot\frac{2}{4},4\pm1\cdot\frac{3}{4}\right).$$

$$\therefore \quad P=\left(3\pm\frac{\sqrt{3}}{4},-2\pm\frac{1}{2},4\pm\frac{3}{4}\right).$$

20. Any point on the first line is $(r, 2r, 3r)$, and any point on the second line is $(1+3r', 2-r', 3+4r')$. For the point of intersection,

$$r=1+3r', 2r=2-r', 3r=3+4r' \quad \Rightarrow r=1 \text{ and } r'=0.$$

So, the point of intersection of the first two lines is (1, 2, 3). It is on the third line. So, $\dfrac{1+k}{3}=\dfrac{2-1}{2}=\dfrac{3-2}{h}$.

21. The lines are coplanar if they intersect.

Any point on the first line is $(1+4r, -1+3r, kr)$, and any point on the second line is $(r', k+3r', 1-2r')$. They interesect if

$$1+4r=r', -1+3r=k+3r', 1-2r'=kr$$

$$\Rightarrow \quad 9r+4+k=0 \quad \text{and} \quad 8r+1+kr=0 \quad \Rightarrow r=\frac{-4-k}{9}=\frac{-1}{8+k}$$

$\Rightarrow \quad k^2+12k+23=0$ which gives two real values of k.

23. If l, m, n be the direction cosines of the line then

$$l\cdot\frac{1}{\sqrt{6}}+m\cdot\frac{2}{\sqrt{6}}+n\cdot\frac{-1}{\sqrt{6}}=\cos\theta=\frac{\sqrt{6}}{11}$$

or $\quad l+2m-n=\dfrac{6}{\sqrt{11}} \quad$ and $\quad l^2+m^2+n^2=1.$

Both the equations are satisfied by (a) and (d).

□

3

Planes

• Type 1 •

Choose the most appropriate option (a, b, c or d).

1. The sum of thc intercepts made hy the plane $ax + by + cz = d$ on the three axes of reference is

(a) $a + b + c$ (b) $\frac{1}{a} + \frac{1}{b} + \frac{1}{c}$

(c) $d\left(\frac{1}{a} + \frac{1}{b} + \frac{1}{c}\right)$ (d) $\frac{1}{d}\sqrt{a^2 + b^2 + c^2}$

2. If the sum of the reciprocals of the intercepts made by the plane $ax + by + cz = 1$ on the three axes is 1 then the plane always passes through the point

(a) $(2, -1, 0)$ (b) $(1, 1, 1)$

(c) $(-1, -1, -1)$ (d) $\left(\frac{1}{2}, -1, \frac{1}{2}\right)$

3. The direction cosines of the perpendicular from the origin to the plane $3x - y + 4z = 5$ are

(a) $4, -1, 3$ (b) $3, -1, 4$

(c) $\frac{3}{\sqrt{26}}, \frac{-1}{\sqrt{26}}, \frac{4}{\sqrt{26}}$ (d) $\frac{4}{\sqrt{26}}, \frac{-1}{\sqrt{26}}, \frac{3}{\sqrt{26}}$

4. The length of the perpendicular from the origin to the plane $2x + 3y + \lambda z = 1$ $(\lambda > 0)$ is $\frac{1}{5}$. Then λ is

(a) $2\sqrt{3}$ (b) $3\sqrt{2}$ (c) 0 (d) 1

5. The direction cosines of the normal to the plane $5(x-2)=3(y-z)$ are

(a) $5, -3, 3$ (b) $\frac{5}{\sqrt{43}}, \frac{-3}{\sqrt{43}}, \frac{3}{\sqrt{43}}$

(c) $\frac{1}{2}, \frac{-3}{10}, \frac{3}{10}$ (d) $1, \frac{-3}{5}, \frac{3}{5}$

6. A plane passing through the line joining the points $A(1,-3,5)$ and $B(4,1,-1)$ is turned about AB till it passes through the origin. The equation of the plane in the new position is

(a) $3x+4y-6z=0$ (b) $2x-21y+13z=0$

(c) $2x-21y-13z=0$ (d) none of these

7. The equations of a line passing through the point $(-1,0,3)$ and perpendicular to the plane $4x+3y-5z=12$ are

(a) $\frac{x-1}{4}=\frac{y}{3}=\frac{z+3}{-5}$ (b) $5(3x-1)=20(y-1)=-4(z-4)$

(c) $\frac{x+1}{-5}=\frac{y}{3}=\frac{z-3}{4}$ (d) none of these

8. The equation of the plane passing through the line $\frac{x-1}{2}=\frac{y+1}{-1}=\frac{z}{3}$ and parallel to the direction whose direction numbers are 3, 4, 2 is

(a) $14x-5y-11z=19$ (b) $3x+4y+2z+1=0$

(c) $2x-y+3z=3$ (d) none of these

9. If the line $\frac{x-1}{1}=\frac{y+1}{-2}=\frac{z+1}{\lambda}$ lies in the plane $3x-2y+5z=0$ then λ is

(a) 1 (b) $-\frac{7}{5}$

(c) $\frac{5}{7}$ (d) no possible value

10. The equations of the line of intersection of the planes $x+y+z=2$ and $3x-y+2z=5$ in symmetric form are

(a) $\frac{x-\frac{7}{4}}{4}=\frac{y-\frac{1}{4}}{-1}=\frac{z}{-3}$ (b) $\frac{x}{3}=\frac{y+\frac{1}{3}}{1}=\frac{z-\frac{7}{3}}{-4}$

(c) $\frac{x}{1}=\frac{3y+1}{1}=\frac{3z-7}{-4}$ (d) none of these

11. The direction cosines of a line parallel to the planes $3x + 4y + z = 0$ and $x - 2y - 3z = 5$ are

(a) $(-1, 1, -1)$ (b) $\left(-\frac{1}{\sqrt{3}}, -\frac{1}{\sqrt{3}}, \frac{1}{\sqrt{3}}\right)$

(c) $\left(-\frac{1}{\sqrt{3}}, \frac{1}{\sqrt{3}}, \frac{-1}{\sqrt{3}}\right)$ (d) no line possible

12. If $(3, \lambda, \mu)$ is a point on the line $2x + y + z - 3 = 0 = x - 2y + z - 1$ then

(a) $\lambda = \frac{-8}{3}, \mu = -\frac{1}{3}$ (b) $\lambda = -\frac{1}{3}, \mu = -\frac{8}{3}$

(c) $\lambda = -1, \mu = -5$ (d) $\lambda = -5, \mu = -1$

13. The equations of the perpendicular from the point $(-2, 4, 1)$ to the plane $7x - 2y + 3z = 1$ are

(a) $\frac{x-5}{7} = \frac{y-2}{-2} = \frac{z-4}{3}$ (b) $\frac{x-2}{7} = \frac{y+4}{-2} = \frac{z+1}{3}$

(c) $\frac{x+2}{1} = \frac{y-4}{4} = \frac{z-1}{-2}$ (d) none of these

14. P is a point on the y–z plane, making equal angles with the y-axis and z-axis and at a distance 2 from the origin. M is the foot of the perpendicular from P to the plane $3x + y - \sqrt{2}z = 2\sqrt{2}$. The coordinates of M are

(a) $\left(1, \frac{5}{3}, \frac{\sqrt{2}}{3}\right)$ (b) $(1, -3, -2)$

(c) $\left(\frac{1}{\sqrt{2}}, \frac{5}{3\sqrt{2}}, \frac{1}{3}\right)$ (d) none of these

15. The distance of the point $(2, 0, -3)$ from the plane $5x - 12y = 0$ is

(a) $\frac{10}{13}$ (b) $\frac{46}{13}$

(c) $\frac{36}{13}$ (d) none of these

16. The image of the point $P(\alpha, \beta, \gamma)$ by the plane $lx + my + nz = 0$ is the point $Q(\alpha', \beta', \gamma')$. Then

(a) $\alpha^2 + \beta^2 + \gamma^2 = l^2 + m^2 + n^2$

(b) $\alpha^2 + \beta^2 + \gamma^2 = \alpha'^2 + \beta'^2 + \gamma'^2$

(c) $\alpha\alpha' + \beta\beta' + \gamma\gamma' = 0$

(d) $l(\alpha - \alpha') + m(\beta - \beta') + n(\gamma - \gamma') = 0$

17. The image of the point (2, –1, 1) by the plane $3x + 4y - 5z = 0$ is

(a) $(-2, 1, -1)$ (b) $\left(\frac{2}{3}, \frac{-1}{4}, \frac{-1}{5}\right)$

(c) $\left(\frac{59}{25}, \frac{-13}{25}, \frac{2}{5}\right)$ (d) none of these

18. If the image of the point (1, 1, 1) by a plane be (3, –1, 5) then the equation of the plane is

(a) $x - y + 2z = 8$ (b) $x - y + 2z = 16$

(c) $x - y + 2z = 14$ (d) none of these

19. The angle between the line $x = y = z$ and the plane $4x - 3y + 5z = 2$ is

(a) $\cos^{-1}\frac{\sqrt{6}}{5}$ (b) $\sin^{-1}\sqrt{\frac{6}{5}}$

(c) $\frac{\pi}{2}$ (d) $\sin^{-1}\frac{1}{\sqrt{6}}$

20. The equation of the plane passing through the origin and containing the line of intersection of the planes $5x + y - 3z = 2$ and $x + 2y + 3z = 1$ is

(a) $2x + y = 1$ (b) $x - y - 3z = 0$

(c) $4x - y - 6z = 0$ (d) $7x + 5y + 3z = 0$

21. What is the equation of the plane passing through the line of intersection of the planes $x - y + 3z = 4$ and $2x + y + 3z = 5$ and parallel to the plane $x + y + z = 1$?

(a) $x + y + z = 2$ (b) $x + y + z + 2 = 0$

(c) $2x = y + z$ (d) no plane exists

22. What is the equation of the plane passing through the line $3x + y - 5z = 2 = x - 2y + 3z$ and perpendicular to the plane $x - y + z = 3$?

(a) $2x + 3y + z = 2$ (b) $3x + 2y - z = 2$

(c) $7(x - z) = 6$ (d) no plane exists

23. The equation of the plane passing through the line $x + y - 2 = 0 = x - y - 2z$ and at a distance 1 from the point (0, 1, 1) is

(a) $2x+y-z-3=\sqrt{3}(2-x-y)$ (b) $x-y-2z+\sqrt{3}(x+y-2)=0$

(c) $x+y-2=\sqrt{3}(x-y-2z)$ (d) none of these

24. The angle between the planes $x+y+z=0$ and $3x-4y+5z=0$ is

(a) $\cos^{-1}\left(\frac{1}{5}\sqrt{\frac{2}{3}}\right)$ (b) $\frac{\pi}{2}$

(c) $\frac{\pi}{3}$ (d) $\cos^{-1}\left(\frac{2}{5}\cdot\sqrt{\frac{2}{3}}\right)$

25. The variable plane $(2\lambda+1)x+(3-\lambda)y+z=4$ always passes through the line

(a) $\frac{x}{0}=\frac{y}{0}=\frac{z+4}{1}$ (b) $\frac{x}{1}=\frac{y}{2}=\frac{z}{-3}$

(c) $\frac{x}{1}=\frac{y}{2}=\frac{z-4}{-7}$ (d) none of these

26. The distance between the planes $4x-5y+3z=5$ and $4x-5y+3z+2=0$ is

(a) $\frac{7}{2\sqrt{5}}$ (b) 7

(c) $\frac{7}{5\sqrt{2}}$ (d) 3

27. The distance between the planes $x+2y-3z-4=0$ and $2x+4y-6z=1$ along the line $\frac{x}{1}=\frac{y}{-3}=\frac{z}{2}$ is

(a) $\frac{19}{22}$ (b) $\frac{3}{22}$

(c) 5 (d) none of these

28. The shortest distance between the lines $x-y=0=2x+z$ and $x+y-2=0=3x-y+z-1$ is

(a) $\frac{1}{\sqrt{3}}$ (b) $\frac{1}{2\sqrt{3}}$

(c) $\frac{1}{2}$ (d) 1

29. Which of the following planes is equally inclined to the planes $4x+3y-5z=0$ and $5x-12y+13z=0$?

(a) $11x-3y=0$ (b) $3x+11y=0$

(c) $3x + 11y = 65z$ (d) none of these

30. The equation of the plane bisecting the angle between the planes $3x + 4y = 4$ and $6x - 2y + 3z + 5 = 0$ that contains the origin, is

(a) $9x - 38y + 15z + 43 = 0$ (b) $51x + 18y + 15z = 13$

(c) $9x + 2y + 3z + 1 = 0$ (d) none of these

31. The equation of the plane bisecting the acute angle between the planes $x - y + z - 1 = 0$ and $x + y + z = 2$ is

(a) $x + z = \frac{3}{2}$ (b) $2y = 1$

(c) $x - y - z = 3$ (d) none of these

32. The direction cosines of the projection of the line $\frac{x}{-2} = \frac{y-1}{1} = \frac{z+1}{-1}$ on the plane $2x + y - 3z = 5$ are

(a) $2, -1, 1$ (b) $\frac{2}{7}, \frac{-1}{7}, \frac{1}{7}$

(c) $\frac{-2}{\sqrt{6}}, \frac{1}{\sqrt{6}}, \frac{-1}{\sqrt{6}}$ (d) $\frac{2}{\sqrt{6}}, \frac{-1}{\sqrt{6}}, \frac{1}{\sqrt{6}}$

33. Two systems of rectangular axes have the same origin. If a plane cuts them at distances a, b, c and a', b', c' from the origin then

(a) $a^{-2} + b^{-2} - c^{-2} + a'^{-2} + b'^{-2} - c'^{-2} = 0$

(b) $a^{-2} - b^{-2} - c^{-2} + a'^{-2} - b'^{-2} - c'^{-2} = 0$

(c) $a^{-2} + b^{-2} + c^{-2} - a'^{-2} - b'^{-2} - c'^{-2} = 0$

(d) none of these

34. The lines $x = ay + b, z = cy + d$
and $x = a'y + b', z = c'y + d'$
will be perpendicular if and only if

(a) $aa' + bb' + cc' = 0$ (b) $(a + a')(b + b') + c + c' = 0$

(c) $aa' + cc' + 1 = 0$ (d) $aa' + bb' + cc' + 1 = 0$

35. Which of the following planes intersects the planes $x - y + 2z = 3$ and $4x + 3y - z = 1$ along the same line?

(a) $11x + 10y - 5z = 0$ (b) $7x + 7y - 4z = 0$

(c) $5x + 2y + z = 2$ (d) none of these

36. The line $\frac{x-1}{2}=\frac{y}{-1}=\frac{z+2}{2}$ cuts the plane $x+y+z=1$ at P. If the foot of the perpendicular from P to a plane be $(3,-4,1)$ then the equation of the plane is

(a) $3x-2y-z=0$ (b) $2x-y+2z=12$

(c) $2x-10y+5z=51$ (d) none of these

37. A variable plane at a distance of 1 unit from the origin cuts the coordinate axes at A, B and C. If the centroid $D(x, y, z)$ satisfies the relation $x^{-2}+y^{-2}+z^{-2}=k$ then the value of k is

(a) 3 (b) 1

(c) $\frac{1}{3}$ (d) 9

• *Type 2* •

Choose the correct options. One or more options may be correct.

38. A plane through the line $\frac{x-1}{1}=\frac{y+1}{-2}=\frac{z}{1}$ has the equation

(a) $x+y+z=0$ (b) $3x+2y-z=1$

(c) $4x+y-2z=3$ (d) $3x+2y+z=0$

39. The equation of a plane is $2x-y-3z=5$ and $A(1,1,1)$, $B(2,1,-3)$, $C(1,-2,-2)$ and $D(-3,1,2)$ are four points. Which of the following line segments are intersected by the plane?

(a) AD (b) AB

(c) AC (d) BC

Answers

1. c	**2.** b	**3.** c	**4.** a	**5.** b
6. c	**7.** b	**8.** a	**9.** b	**10.** b
11. c	**12.** b	**13.** a	**14.** d	**15.** a
16. b	**17.** c	**18.** a	**19.** b	**20.** b
21. a	**22.** c	**23.** a	**24.** d	**25.** c
26. c	**27.** d	**28.** b	**29.** a	**30.** b
31. a	**32.** d	**33.** c	**34.** c	**35.** a
36. c	**37.** d	**38.** a, c	**39.** b, c	

Hints and Solutions to Selected Questions

8. The point (1, –1, 0) is on the line. So, the equation of the plane is

$a(x-1)+b(y+1)+cz=0$ where

$a\cdot 2+b(-1)+c\cdot 3=0$ and $a\cdot 3+b\cdot 4+c\cdot 2=0.$

$\therefore$ eliminating a, b, c from these three, the equation of the plane is

$$\begin{vmatrix} x-1 & y+1 & z \\ 2 & -1 & 3 \\ 3 & 4 & 2 \end{vmatrix}=0.$$

14. $P=(0, 2\cos 45°, 2\sin 45°)=\left(0, \frac{1}{\sqrt{2}}, \frac{1}{\sqrt{2}}\right)$. If $M=(\alpha, \beta, \gamma)$ then

$$3\alpha+\beta-\sqrt{2}\gamma=2\sqrt{2} \text{ and } \frac{\alpha-0}{3}=\frac{\beta-\frac{1}{\sqrt{2}}}{1}=\frac{\gamma-\frac{1}{\sqrt{2}}}{-\sqrt{2}}.$$

$$\therefore \quad \frac{\alpha}{3}=\frac{\beta-\frac{1}{\sqrt{2}}}{1}=\frac{\sqrt{2}\gamma-1}{-2}=\frac{3\alpha+\beta-\frac{1}{\sqrt{2}}-(\sqrt{2}\gamma-1)}{9+1-(-2)}$$

$$=\frac{2\sqrt{2}-\frac{1}{\sqrt{2}}+1}{12}=\frac{3+\sqrt{2}}{12\sqrt{2}}.$$

$$\therefore \quad \alpha=\frac{3+\sqrt{2}}{4\sqrt{2}},\ \beta=\frac{1}{\sqrt{2}}+\frac{3+\sqrt{2}}{12\sqrt{2}}=\frac{15+\sqrt{2}}{12\sqrt{2}},$$

$$\gamma=\frac{1}{\sqrt{2}}\left\{1-\frac{6+2\sqrt{2}}{12\sqrt{2}}\right\}=\frac{10\sqrt{2}-6}{24}.$$

16. $lx+my+nz=0$ is a plane through the origin. So, $OP=OQ$.

25. The plane is $x+3y+z-4+\lambda(2x-y)=0$.

This always passes through the intersection of the planes $x+3y+z-4=0$ and $2x-y=0$, which is a line.

Now, $2x-y=0 \quad \Rightarrow \frac{x}{1}=\frac{y}{2}.$

$\therefore \quad x+3y+z-4=0 \quad \Rightarrow x+3\cdot 2x+z-4=0 \quad \Rightarrow 7x+z-4=0.$

$\therefore \quad 7x = -(z-4) \quad \Rightarrow \quad \frac{x}{1} = \frac{z-4}{-7}$.

$\therefore$ the line is $\frac{x}{1} = \frac{y}{2} = \frac{z-4}{-7}$.

27. Any point on $\frac{x}{1} = \frac{y}{-3} = \frac{z}{2}$ is $(r, -3r, 2r)$. It is on the first plane if

$$r + 2(-3r) - 3(2r) - 4 = 0 \quad \text{or} \quad -11r - 4 = 0 \quad \text{or} \quad r = -\frac{4}{11}.$$

$\therefore$ the point is $\left(-\frac{4}{11}, \frac{12}{11}, -\frac{8}{11}\right)$.

$(r, -3r, 2r)$ is on the second plane if

$$2r + 4(-3r) - 6(2r) = 1 \quad \text{or} \quad -22r = 1 \quad \text{or} \quad r = -\frac{1}{22}.$$

$\therefore$ the point is $\left(-\frac{1}{22}, \frac{3}{22}, -\frac{1}{11}\right)$.

$\therefore$ the distance $= \sqrt{\left(-\frac{4}{11} + \frac{1}{22}\right)^2 + \left(\frac{12}{11} - \frac{3}{22}\right)^2 + \left(-\frac{8}{11} + \frac{1}{11}\right)^2}$

$$= \sqrt{\frac{49}{(22)^2} + \frac{441}{(22)^2} + \frac{49}{(11)^2}} = \frac{1}{22}\sqrt{49 + 441 + 196}.$$

28. Equations of the lines in symmetric form are

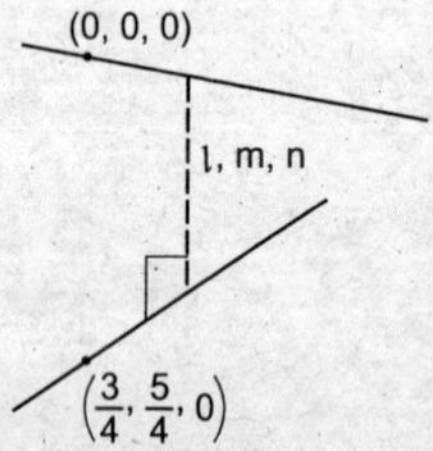

$$\frac{x}{1} = \frac{y}{1} = \frac{z}{-2} \quad \text{and} \quad \frac{x - \frac{3}{4}}{1} = \frac{y - \frac{5}{4}}{-1} = \frac{z}{-4}.$$

Now, $l + m - 2n = 0$

$l - m - 4n = 0$

Solving, $\frac{l}{-2} = \frac{m}{2} = \frac{n}{-2} \quad \Rightarrow \quad l = \frac{-1}{\sqrt{3}}, m = \frac{1}{\sqrt{3}}, n = \frac{-1}{\sqrt{3}}$.

$\therefore$ shortest distance $= \left(\frac{3}{4} - 0\right)\frac{-1}{\sqrt{3}} + \left(\frac{5}{4} - 0\right)\frac{1}{\sqrt{3}} + (0-0)\frac{-1}{\sqrt{3}}$.

31. Rewriting the equations $-x + y - z + 1 = 0$

$-x - y - z + 2 = 0$

Here, $d_1 = 1 > 0, d_2 = 2 > 0$ and

$$a_1a_2 + b_1b_2 + c_1c_2 = (-1)(-1) + 1(-1) + (-1)(-1) = 1 > 0.$$

$\therefore$ the bisector of the angle containing the origin is not the bisector of the acute angle.

$\therefore$ the required bisector has the equation

$$\frac{-x+y-z+1}{\sqrt{(-1)^2+1^2+(-1)^2}}=-\frac{-x-y-z+2}{\sqrt{(-1)^2+(-1)^2+(-1)^2}}.$$

32. Let $P=(\alpha,\beta,\gamma)$. Then

$$2\alpha+\beta-3\gamma=5 \qquad \ldots (1)$$

and $\dfrac{\alpha-0}{2}=\dfrac{\beta-1}{1}=\dfrac{\gamma+1}{-3}$. $\qquad \ldots (2)$

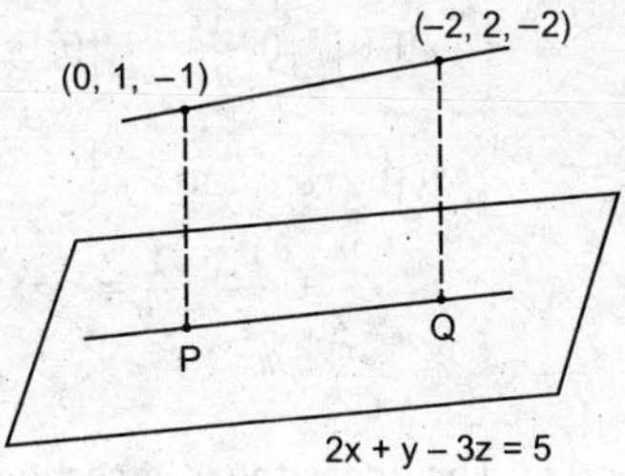

Solving (1) and (2), $\alpha=\dfrac{1}{7}, \beta=\dfrac{15}{14}, \gamma=\dfrac{-17}{14}$

Let $Q=(\alpha',\beta',\gamma')$. Then

$2\alpha'+\beta'-3\gamma'=5$ ... (3) and $\dfrac{\alpha'+2}{2}=\dfrac{\beta'-2}{1}=\dfrac{\gamma'+2}{-3}$. $\qquad \ldots (4)$

Solving (3) and (4), $\alpha'=\dfrac{-13}{7}, \beta'=\dfrac{29}{14}, \gamma'=\dfrac{-31}{14}$.

$\therefore$ direction ratios of PQ are $\dfrac{1}{7}-\left(-\dfrac{13}{7}\right), \dfrac{15}{14}-\dfrac{29}{14}, \dfrac{-17}{14}-\left(\dfrac{-31}{14}\right)$

i.e., $2,-1,1$. So, direction cosines are $\dfrac{2}{\sqrt{6}}, \dfrac{-1}{\sqrt{6}}, \dfrac{1}{\sqrt{6}}$.

33. The length of the perpendicular to the plane $\dfrac{x}{a}+\dfrac{y}{b}+\dfrac{z}{c}=1$ from the origin

$$=\frac{1}{\sqrt{\dfrac{1}{a^2}+\dfrac{1}{b^2}+\dfrac{1}{c^2}}}.$$

The length of the perpendicular to the plane $\dfrac{x'}{a'}+\dfrac{y'}{b'}+\dfrac{z'}{c'}=1$ from the origin (the same point) $=\dfrac{1}{\sqrt{\dfrac{1}{a'^2}+\dfrac{1}{b'^2}+\dfrac{1}{c'^2}}}$.

These are equal. So, $\dfrac{1}{a^2}+\dfrac{1}{b^2}+\dfrac{1}{c^2}=\dfrac{1}{a'^2}+\dfrac{1}{b'^2}+\dfrac{1}{c'^2}$.

34. For the first line, $x-b=ay, z-d=cy \quad \Rightarrow \dfrac{x-b}{a}=\dfrac{y}{1}=\dfrac{z-d}{c}$.

For the second line $x-b'=a'y, z-d'=c'y \quad \Rightarrow \dfrac{x-b'}{a'}=\dfrac{y}{1}=\dfrac{z-d'}{c'}$.

These lines are perpendicular $\quad \Rightarrow a\cdot a'+1\cdot 1+c\cdot c'=0$.

37. The plane is $lx+my+nz=1$ where $l^2+m^2+n^2=1$. It cuts axes at $\left(\frac{1}{l},0,0\right), \left(0,\frac{1}{m},0\right), \left(0,0,\frac{1}{n}\right)$.

$\therefore$ the centroid $=\left(\frac{1}{3l},\frac{1}{3m},\frac{1}{3n}\right)$. It satisfies

$$\frac{1}{x^2}+\frac{1}{y^2}+\frac{1}{z^2}=k \Rightarrow 9(l^2+m^2+n^2)=k \Rightarrow 9=k.$$

38. The line is the intersection of the planes $\dfrac{x-1}{1}=\dfrac{y+1}{-2}$ and $\dfrac{y+1}{-2}=\dfrac{z}{1}$,

i.e., $2x+y-1=0$ and $y+2z+1=0$.

$\therefore$ any plane through the line is $2x+y-1+\lambda(y+2z+1)=0$.

For $\lambda=1$, it is $x+y+z=0$. For $\lambda=-\frac{1}{2}$, it is $4x+y-2z=3$.

39. For $A(1,1,1)$, $2x-y-3z-5=2-1-3-5<0$.

For $B(2,1,-3)$, $2x-y-3z-5=4-1+9-5>0$.

For $C(1,-2,-2)$, $2x-y-3z-5=2+2+6-5>0$.

For $D(-3,1,2)$, $2x-y-3z-5=-6-1-6-5<0$.

$\therefore$ A, D are on one side of the plane and B, C are on the other side.

$\therefore$ the line segments AB, AC, BD, CD intersect the plane.

□

Part 8

Set Theory

1

Set Operations, Relations and Mappings

• Type 1 •

Choose the most appropriate option (a, b, c or d).

1. Let A and B be two sets such that $A \cup B = A$. Then $A \cap B$ is equal to

(a) ϕ (b) B

(c) A (d) none of these

2. Let A and B be two sets. Then $(A \cup B)' \cup (A' \cap B)$ is equal to

(a) A' (b) A

(c) B' (d) none of these

3. Let U be the universal set and $A \cup B \cup C = U$. Then

$$\{(A - B) \cup (B - C) \cup (C - A)\}'$$

is equal to

(a) $A \cup B \cup C$ (b) $A \cup (B \cap C)$

(c) $A \cap B \cap C$ (d) $A \cap (B \cup C)$

4. If A and B are two sets then $(A - B) \cup (B - A) \cup (A \cap B)$ is equal to

(a) $A \cup B$ (b) $A \cap B$

(c) A (d) B'

5. 20 teachers of a school either teach mathematics or physics. 12 of them teach mathematics while 4 teach both the subjects. Then the number of teachers teaching only physics is

(a) 12 (b) 8

(c) 16 (d) none of these

6. Of the members of three athletic teams in a school, 21 are in the cricket team, 26 are in the hockey team and 29 are in the football team. Among them, 14 play hockey and cricket, 15 play hockey and football, and 12 play football and cricket. Eight play all the three games. The total number of members in the three athletic teams is

(a) 43 (b) 76
(c) 49 (d) none of these

7. The relation "congruence modulo m" is

(a) reflexive only (b) transitive only
(c) symmetric only (d) an equivalence relation

8. R is a relation over the set of real numbers and it is given by $mn \geq 0$. Then R is

(a) symmetric and transitive (b) reflexive and symmetric
(c) a partial-order relation (d) an equivalence relation

9. R is a relation over the set of integers and it is given by $(x, y) \in R \Leftrightarrow |x - y| \leq 1$. Then R is

(a) reflexive and transitive (b) reflexive and symmetric
(c) symmetric and transitive (d) an equivalence relation

10. Let r be a relation over the set $N \times N$ and it is defined by $(a, b)\ r\ (c, d) \quad \Rightarrow\ a + d = b + c$. Then r is

(a) reflexive only (b) symmetric only
(c) transitive only (d) an equivalence relation

11. Let $A = \{1, 2, 3\}$. The total number of distinct relations that can be defined over A is

(a) 2^9 (b) 6
(c) 8 (d) none of these

12. On the set $A = \{1, 2, 3, 4\}$, a relation is $R = \{(1, 3,), (4, 2), (2, 4), (2, 3), (3, 1)\}$. The relation R is

(a) a function (b) transitive
(c) not symmetric (d) reflexive

13. Let $R = \{(3, 3), (6, 6), (9, 9), (12, 12), (6, 12), (3, 9), (3, 12), (3, 6)\}$ be a relation on the set $A = \{3, 6, 9, 12\}$. The relation is

(a) reflexive only (b) reflexive and transitive only
(c) reflexive and symmetric only (d) an equivalence relation

14. The range of the function $f(x) = {}^{7-x}P_{x-3}$ is $\left(\text{where } {}^nP_r = \dfrac{n!}{(n-r)!}\right)$

(a) $\{1, 2, 3\}$ (b) $\{1, 2, 3, 4, 5, 6\}$

(c) $\{1, 2, 3, 4\}$ (d) $\{1, 3\}$

15. Let R = set of real numbers and R_c = set of real angles in radian measure. If $f: R_c \to R$ be a mapping such that $f(x) = \sin x$, $x \in R_c$, then f is

(a) one-one and into (b) one-one and onto

(c) many-one and onto (d) many-one and into

16. Let $f: R \to R$ such that $f(x) = \dfrac{1}{1+x^2}$, $x \in R$. Then f is

(a) injective (b) surjective

(c) bijective (d) none of these

17. $f: R \times R \to R$ such that $f(x + iy) = +\sqrt{x^2 + y^2}$. Then f is

(a) many-one and into (b) one-one and onto

(c) many-one and onto (d) one-one and into

18. Let $A = \{x \mid -1 < x < 1\} = B$. If $f: A \to B$ be bijective then a possible definition of $f(x)$ is

(a) $|x|$ (b) $x|x|$

(c) $\sin \pi x$ (d) none of these

19. Let $A = \{1, 2, 3\}$ and $B = \{a, b\}$. Which of the following subsets of $A \times B$ is a mapping from A to B?

(a) $\{(1, a), (3, b), (2, a), (2, b)\}$ (b) $\{(1, b), (2, a), (3, a)\}$

(c) $\{(1, a), (2, b)\}$ (d) none of these

• Type 2 •

Choose the correct options. One or more options may be correct.

20. Let R be the relation over the set of straight lines of a plane such that $l_1 R l_2 \Leftrightarrow l_1 \perp l_2$. Then R is

(a) symmetric (b) reflexive

(c) transitive (d) an equivalence relation

21. Let R be the relation over the set of integers such that $m\,R\,n$ if and only if m is a multiple of n. Then R is

(a) reflexive
(b) symmetric
(c) transitive
(d) an equivalence relation

22. Let $A = \{1, 2, 3, 4\}$ and R be a relation in A given by

$$R = \{(1, 1), (2, 2), (3, 3), (4, 4), (1, 2), (2, 1), (3, 1), (1, 3)\}.$$

Then R is

(a) reflexive
(b) symmetric
(c) transitive
(d) an equivalence relation

23. Let $f : R \to R$ be a mapping such that $f(x) = \dfrac{x^2}{1+x^2}$. Then f is

(a) many-one
(b) one-one
(c) into
(d) onto

24. Let $A = \{1, 2, 3\}$ and $B = \{a, b, c\}$. If f is a function from A to B and g is a one-one function from A to B then the maximum number of definitions of

(a) f is 9
(b) g is 9
(c) f is 27
(d) g is 6

Note For more problems on mapping (function), see the chapter on functions in the calculus part of this book.

Answers

1. b	**2.** a	**3.** c	**4.** a	**5.** b
6. a	**7.** d	**8.** d	**9.** b	**10.** d
11. a	**12.** c	**13.** a	**14.** a	**15.** d
16. d	**17.** a	**18.** d	**19.** b	**20.** a
21. a, c	**22.** a, b	**23.** a, c	**24.** c, d	

Hints and Solutions to Selected Questions

1. $A \cup B = A \quad \Rightarrow B \subseteq A \quad \Rightarrow A \cap B = B.$

2. $(A \cup B)' \cup (A' \cap B) = (A' \cap B') \cup (A' \cap B)$

$$= (A' \cup A') \cap (A' \cup B) \cap (B' \cup A') \cap (B' \cup B)$$

$$= A' \cap \{A' \cup (B \cap B')\} \cap U = A' \cap (A' \cup \phi) \cap U$$

$$= A' \cap A' \cap U = A' \cap U = A'.$$

3.

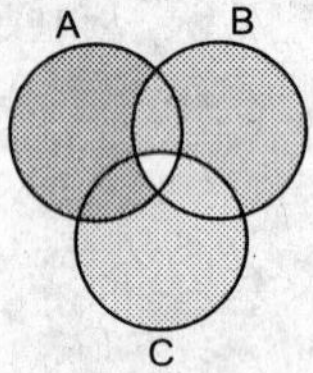

$(A - B) \cup (B - C) \cup (C - A)$ is represented by the shaded portion in the figure. The unshaded portion is $A \cap B \cap C$.

$\therefore \quad \{(A - B) \cup (B - C) \cup (C - A)\}' = A \cap B \cap C.$

4. Draw the Venn diagram. From the Venn diagram,

$$(A - B) \cup (B - A) \cup (A \cap B) = A \cup B.$$

5. $n(M \cup P) = 20, \quad n(M) = 12, \quad n(M \cap P) = 4.$

$n(M \cup P) = n(M) + n(P) - n(M \cap P) \Rightarrow 20 = 12 + n(P) - 4$

$\therefore \quad n(P) = 12$. So, the required number $= n(P) - n(M \cap P) = 12 - 4 = 8.$

6. $n(C) = 21, n(H) = 26, n(F) = 29, n(H \cap C) = 14, n(H \cap F) = 15, n(F \cap C) = 12,$ $n(C \cap H \cap F) = 8.$

$$n(C \cup H \cup F) = n(C) + n(H) + n(F) - n(H \cap C) - n(H \cap F) - n(F \cap C) + n(C \cap H \cap F)$$

$$= 21 + 26 + 29 - 14 - 15 - 12 + 8 = 43.$$

7. If R be the relation, $x R y \Leftrightarrow x - y$ is divisible by m.

$x R x$ because $x - x$ is divisible by m. So, R is reflexive.

$x R y \Rightarrow y R x$. So, R is symmetric.

$x R y$ and $y R z \Rightarrow x - y = k_1 m, y - z = k_2 m$

$\therefore \quad x - z = (k_1 + k_2)m$. So, R is transitive.

As R is reflexive, symmetric and transitive, it is an equivalence relation.

8. R is reflexive, symmetric and transitive. So, the most appropriate option is (d).

10. $(a, b)\, r\, (a, b)$ because $a + b = b + a$. So, r is reflexive.

$(a, b)\, r\, (c, d) \quad \Rightarrow a + d = b + c \quad \Rightarrow c + b = d + a \quad \Rightarrow (c, d)\, r\, (a, b)$.

So, r is symmetric.

$(a, b)\, r\, (c, d)$ and $(c, d)\, r\, (e, f) \quad \Rightarrow a + d = b + c, c + f = d + e$

Adding, $a + d + c + f = b + c + d + e \quad \Rightarrow a + f = b + e \quad \Rightarrow (a, b)\, r\, (e, f)$.

$\therefore \quad r$ is transitive.

11. $n(A \times A) = n(A) \cdot n(A) = 3^2 = 9$. So, the total number of subsets of $A \times A$ is 2^9 and a subset of $A \times A$ is a relation over the set A.

18. $f(x) = |x|$ is many-one and into.

$f(x) = x|x|$ is one-one but into as $f(x)$ will have only rational values.

$f(x) = \sin \pi x$ is onto but many-one $\quad \left[\because\ f\left(\frac{3}{4}\right) = f\left(\frac{1}{4}\right)\right]$.

24. Every element of A can have an image in B in 3 ways. So, the total number of ways in which 3 elements of A can have images in B = maximum number of definitions of $f = 3 \times 3 \times 3$.

The number of ways of arranging 1, 2, 3 in places of a, b, c is $3! =$ the maximum number of definitions of one-one function g.

□

Part 9

Miscellaneous Questions

1

Miscellaneous Questions—1

• Type 1 •

Choose the most appropriate option (a, b, c or d).

1. If y is a function of x defined by $a^{x+y} = a^x + a^y$ where a is a real constant $(a > 1)$ then the domain of $y(x)$ is

(a) $(0, +\infty)$ (b) $(-\infty, 0)$

(c) $(-1, +\infty)$ (d) $(-\infty, 1)$

2. If y is a function of x given by $2\log y - \log x - \log(y-1) = 0$ then the

(a) domain $= [4, +\infty)$, range $= (1, +\infty)$

(b) domain $= [4, +\infty)$, range $= (2, +\infty)$

(c) domain $= (2, +\infty)$, range $= (2, +\infty)$

(d) none of these

3. If the latus rectum of the parabola $2x^2 - ky + 3 = 0$ be 2 then the vertex is

(a) $\left(0, \frac{3}{4}\right)$ (b) $\left(0, \frac{3}{2}\right)$

(c) $\left(\frac{3}{4}, 0\right)$ (d) $(0, 0)$

4. The ends of a quadrant of a circle have the coordinates (1, 3) and (3, 1). Then the centre of such a circle is

(a) (2, 2) (b) (1, 1)

(c) (4, 4) (d) (2, 6)

5. If the function $f(x) = x^2 - ax + 4$ is monotonic increasing in the open interval $(2, +\infty)$ then a is

(a) 2 (b) 4

(c) 1 (d) 1/2

6. If $\begin{vmatrix} x & 4 & 7 & 9 \\ 0 & 1 & 5 & 8 \\ 0 & 0 & 2 & 6 \\ 0 & 0 & 0 & 3 \end{vmatrix}$ = the total number of arrangements of 4 different things, taking at least one at a time, then x is equal to

(a) $\frac{11}{8}$ (b) 4

(c) $\frac{11}{2}$ (d) $\frac{32}{3}$

7. Let $A_r; r = 1, 2, 3, \ldots;$ be points on the number line such that $OA_1, OA_2, OA_3, \ldots$ are in GP where O is the origin, and the common ratio of the GP be a positive proper fraction. Let M_r be the middle point of the line segment $A_{r+1}A_r$.

Then the value of $\sum_{r=1}^{\infty} OM_r$ is equal to

(a) $\frac{OA_1(OA_1 - OA_2)}{2(OA_1 + OA_2)}$ (b) $\frac{OA_1(OA_1 + OA_2)}{2(OA_1 - OA_2)}$

(c) $\frac{OA_1}{2(OA_1 - OA_2)}$ (d) ∞

8. If $x^2 - px + q, q \neq 0$, is a factor of $x^3 - ax^2 + bx - c, c \neq 0$, then

(a) $pq + c = aq$ (b) $q(ac - b + 1) = c^2$

(c) $q^2 + c^2 = pqc$ (d) $a^2 + b^2 + c^2 = p^2 + q^2$

9. The shortest distance between the lines whose equations are $\vec{r} = t(\vec{i} + \vec{j} + \vec{k})$ and $\vec{r} = \vec{k} + s(\vec{i} - 2\vec{j} + 3\vec{k})$ is

(a) 3 (b) $\frac{3}{\sqrt{38}}$

(c) $\sqrt{\frac{3}{14}}$ (d) $\frac{2}{\sqrt{13}}$

10. Four cards are drawn at random from a pack of 52 playing cards. The probability that the draw contains exactly one pair (i.e., having the same number) is

(a) $\dfrac{13 \times {}^{12}C_2 \times 4^2}{{}^{52}C_4}$ (b) $\dfrac{13 \times {}^4C_2 \times {}^{12}C_2 \times 4^2}{{}^{52}C_4}$

(c) $\dfrac{13 \times {}^4C_2 \times {}^{48}C_2}{{}^{52}C_4}$ (d) none of these

11. If $\frac{1}{\sqrt{2}} < x < 1$ then $\cos^{-1} x + \cos^{-1}\left(\frac{x + \sqrt{1 - x^2}}{\sqrt{2}}\right)$ is equal to

(a) $2\cos^{-1} x - \frac{\pi}{4}$ (b) $2\cos^{-1} x$

(c) $\frac{\pi}{4}$ (d) 0

12. Two circles $x^2 + y^2 - 2kx = 0$ and $x^2 + y^2 - 4x - 8y + 16 = 0$ touch each other externally. Then k is

(a) 4 (b) 1

(c) 2 (d) –4

13. ABC is an equilateral triangle of side 4 cm. If R, r, h be the circumradius, inradius and altitude respectively then $\frac{R + r}{h}$ is equal to

(a) 4 (b) 2

(c) 1 (d) 3

14. If $y = 2$ be the directrix and (0, 1) be the vertex of the parabola $x^2 + \lambda y + \mu = 0$ then (λ, μ) is equal to

(a) (8, –8) (b) (8, 8)

(c) (–4, 4) (d) (4, –4)

15. The real value of a for which the value of m satisfying the equation $(a^2 - 1)m^2 - (2a - 3)m + a = 0$ gives the slope of a line parallel to the y-axis is

(a) $\frac{3}{2}$ (b) 0

(c) 1 (d) ± 1

16. The number of values of θ satisfying $4\cos\theta + 3\sin\theta = 5$ as well as $3\cos\theta + 4\sin\theta = 5$ is

(a) one (b) two (c) zero (d) none of these

17. Four boys and eight girls sit at random in a row. The probability that each boy sits between two girls is

(a) $\dfrac{7!\times 8!}{3!\times 12!}$ (b) $\dfrac{{}^7C_3}{12!}$

(c) $\dfrac{{}^7C_3\times 8!}{12!}$ (d) none of these

18. $\lim_{x\to\frac{1}{2}} \dfrac{\sin 3[x]}{[x]}$, where $[x]$ denotes the greatest integer $\le x$, is equal to

(a) $\dfrac{3}{2}$ (b) 3

(c) 1 (d) none of these

19. If $\int_{a-n}^{n} e^{x(a-x)}\,dx = \lambda$ then the value of $\int_{a-n}^{n} xe^{x(a-x)}\,dx,\ a\ne 2n$, is

(a) $\dfrac{a\lambda}{2}$ (b) $a\lambda$

(c) $2a\lambda$ (d) none of these

20. The value of $\tan^{-1}2+\tan^{-1}3$ is equal to

(a) $-\dfrac{\pi}{4}$ (b) $\dfrac{3\pi}{4}$

(c) $\tan^{-1}5$ (d) $\dfrac{\pi}{4}$

21. If in the ΔABC, AC is double of AB then the value of $\cot\dfrac{A}{2}\cdot\cot\dfrac{B-C}{2}$ is equal to

(a) $\dfrac{1}{3}$ (b) $-\dfrac{1}{3}$

(c) 3 (d) $\dfrac{1}{2}$

22. Let $C_1, C_2, \ldots, C_n, \ldots$ be a sequence of concentric circles. The nth circle has the radius n, and it has n openings. A point P starts travelling on the smallest circle C_1 and leaves it at an opening along the normal at the point of opening to reach the next circle C_2. Then it moves on the second circle C_2 and leaves it likewise to reach the third circle C_3 and so on. The total number of different paths in which the point can come out of the nth circle is

(a) $2^n \cdot n!$ (b) $2^{n-1} \cdot n!$

(c) $n!$ (d) $2^{n-1} \cdot (n-1)!$

23. If $\vec{a}, \vec{b}$ are orthogonal unit vectors then for a vector $\vec{r}$, noncoplanar with $\vec{a}$ and $\vec{b}$, the vector $\vec{r} \times \vec{a}$ is equal to

(a) $[\vec{r}\ \vec{a}\ \vec{b}]\,\vec{b} + (\vec{r} \cdot \vec{b})\,\vec{b} \times \vec{a}$ (b) $[\vec{r}\ \vec{a}\ \vec{b}](\vec{a} + \vec{b})$

(c) $[\vec{r}\ \vec{a}\ \vec{b}]\,\vec{a} + (\vec{r} \cdot \vec{a})\,\vec{a} \times \vec{b}$ (d) none of these

24. If the rank of the matrix $\begin{bmatrix} 4 & 2 & 1-x \\ 5 & k & 1 \\ 6 & 3 & 1+x \end{bmatrix}$ is 2 then

(a) $k = \frac{5}{2}, x = \frac{1}{5}$ (b) $k = \frac{5}{2}, x \neq \frac{1}{5}$

(c) $k = \frac{1}{5}, x = \frac{5}{2}$ (d) none of these

25. The number of solutions of the equation $\int_{-2}^{x} |\cos x|\, dx = 0, 0 < x < \frac{\pi}{2}$, is

(a) 0 (b) 1

(c) 2 (d) 4

26. If the line $ax + by = 2$ is a normal to the circle $x^2 + y^2 - 4x - 4y = 0$ and a tangent to the circle $x^2 + y^2 = 1$ then

(a) $a = \frac{1}{2}, b = \frac{1}{2}$ (b) $a = \frac{1 + \sqrt{7}}{2}, b = \frac{1 - \sqrt{7}}{2}$

(c) $a = \frac{1}{4}, b = \frac{3}{4}$ (d) $a = 1, b = \sqrt{3}$

27. If A and B are two independent events then which of the following is not equal to any of the remaining?

(a) $P(A' \cap B') - P(A \cap B)$ (b) $P(A') + P(B') - 1$

(c) $P(B) - P(A')$ (d) $P(B') - P(A)$

28. An event X can take place in conjunction with any one of the mutually exclusive and exhaustive events A, B and C. If A, B, C are equiprobable and the probability of X is 5/12, and the probability of X taking place when A has happened is 3/8 while it is 1/4 when B has taken place then the probability of X taking place in conjunction with C is

(a) $\frac{5}{8}$ (b) $\frac{3}{8}$

(c) $\frac{5}{24}$ (d) none of these

29. The largest value of a for which the circle $x^2 + y^2 = a^2$ falls totally in the interior of the parabola $y^2 = 4(x+4)$ is

(a) $4\sqrt{3}$ (b) 4

(c) $4\sqrt{\frac{6}{7}}$ (d) $2\sqrt{3}$

30. Let $\vec{f}(t) = [t]\,\vec{i} + \{t - [t]\}\,\vec{j} + [t+1]\,\vec{k}$, where $[\cdot]$ denotes the greatest integer function. Then the vectors $\vec{f}\left(\frac{5}{4}\right)$ and $\vec{f}(t)$, $0 < t < 1$, are

(a) perpendicular to each other

(b) parallel to each other

(c) inclined at an angle $\cos^{-1}\frac{2}{\sqrt{5(1+t^2)}}$

(d) inclined at $\cos^{-1}\frac{8+t}{9\sqrt{1+t^2}}$

31. The equation of the diagonal OB of the rectangle given in the diagram is

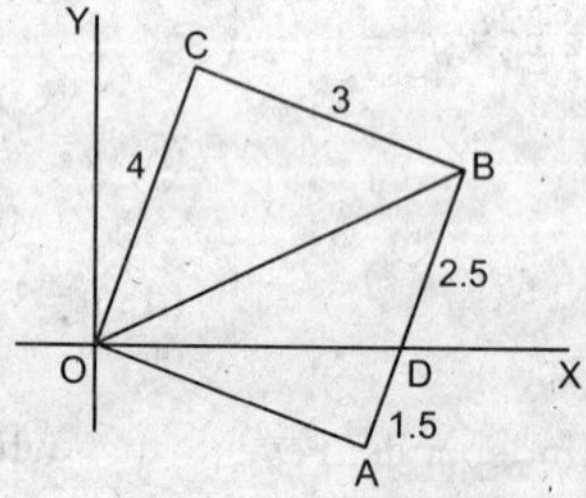

(a) $y = 2x$ (b) $2y = x$

(c) $3y = x$ (d) none of these

32. Let $f(x)$ be a continuous function whose range is $[2, 6.5]$. If $h(x) = \left[\frac{\cos x + f(x)}{\lambda}\right]$, $\lambda \in N$, be continuous, where $[\cdot]$ denotes the greatest integer function, then the least value of λ is

(a) 6 (b) 7
(c) 8 (d) none of these

33. For the equation $1+\sin x = ax$, the number of solutions is

(a) 1 if $a > 1$ (b) at least 3 if $0 < a < 1$
(c) 2 if $a = 1$ (d) none of these

34. If the point $(0, \lambda)$ is a point of the region bounded by the semicircle $x = \sqrt{9 - y^2}$ and the line $x = 0$ then the equation $x^2 - 2x - \lambda = 0$ has

(a) at least one real root in $[2, 3]$ (b) at least one real root in $[-1, 0]$
(c) two real roots in $[-1, 3]$ (d) no real roots in $(0, 2)$

35. Let E be an event which is neither a certainty nor an impossibility. Its probability is such that $P(E) = 1 + \lambda + \lambda^2$ and $P(E') = (1+\lambda)^2$ in terms of an unknown λ. Then $P(E)$ is equal to

(a) 1 (b) $\frac{3}{4}$
(c) $\frac{1}{4}$ (d) none of these

36. A word of at least 5 letters is made at random from 3 vowels and 3 consonants, all the letters being different. The probability that no consonant falls between any two vowels in the word is

(a) $\frac{9}{20}$ (b) $\frac{9}{10}$
(c) $\frac{7}{10}$ (d) $\frac{11}{20}$

37. A number is selected at random from the first twenty-five natural numbers. If it is a composite number then it is divided by 5. But if it is not a composite number, it is divided by 2. The probability that there will be no remainder in the division is

(a) $\frac{11}{30}$ (b) 0.4
(c) 0.2 (d) none of these

38. If $P_r = (x_r, y_r)$; $r = 1, 2, 3$; are three points on the parabola $x^2 = 8y$, where x_1, x_2, x_3 are in GP, then y_1, y_2, y_3 are in

(a) AP (b) GP
(c) HP (d) none of these

39. Let $f(x)$ be an even function in R. If $f(x)$ is monotonic increasing in [2, 6] then

(a) $f(3) > f(-5)$ (b) $f(-2) < f(2)$

(c) $f(-2) > f(2)$ (d) $f(-3) < f(5)$

40. If $f(2) = g(2)$ and $f'(2) = a, g'(2) = 2a$ then $\lim\limits_{x \to 2} \dfrac{f(x) - g(x)}{x - 2}$ is equal to

(a) 1 (b) a

(c) $-a$ (d) none of these

41. If $\vec{a}, \vec{b}, \vec{c}$ are three vectors such that $|\vec{a}| = |\vec{c}| = 1$, $|\vec{b}| = 4$, $|\vec{b} \times \vec{c}| = 2$ and $2\vec{b} = \vec{c} + \lambda\vec{a}$ then λ is

(a) $\sqrt{65 - 8\sqrt{3}}$ (b) $\sqrt{17}$

(c) $\sqrt{3}$ (d) $\sqrt{\dfrac{17}{2}(2 + \sqrt{3})}$

42. If $f(x) = \int_0^x \sin [2x]\, dx$ where $[x]$ = greatest integer less than or equal to x then $f(\pi/2)$ is

(a) $\dfrac{1}{2}\{\sin 1 + (\pi - 2)\sin 2\}$ (b) $\dfrac{1}{2}\{\sin 1 + \sin 2 + (\pi - 3)\sin 3\}$

(c) 0 (d) $\sin 1 + \left(\dfrac{\pi}{2} - 2\right)\sin 2$

43. $\int_0^2 e^{x + [x]}\, d(x - [x])$, where $[x]$ = greatest integer $\leq x$, is equal to

(a) $(e^2 - 1)(e + 1)$ (b) $e - 1$

(c) $(e^2 + 1)(e - 1)$ (d) $e^2 + 1$

44. Let $f(x) = \min\{x, 1 - x\}$ for all $x \in R$. Then the value of $\int_0^2 f(x)\, dx$ is

(a) $-\dfrac{1}{4}$ (b) $\dfrac{1}{4}$

(c) 2 (d) 0

45. A dice is thrown six times, it being known that each time a different digit shows up. The probability that a sum of 12 will be obtained in the first three throws is

(a) $\frac{5}{24}$ (b) $\frac{25}{216}$

(c) $\frac{3}{20}$ (d) $\frac{1}{12}$

46. A composite number is selected at random from the first 30 natural numbers and it is divided by 5. The probability that there will be a remainder is

(a) $\frac{14}{19}$ (b) $\frac{5}{19}$

(c) $\frac{5}{6}$ (d) $\frac{7}{15}$

47. In a ΔABC the sides b, c and the angle B are given such that a has two values a_1, a_2. Then $|a_1 - a_2|$ is equal to

(a) $\sqrt{b^2 - c^2\sin^2 B}$ (b) $2\sqrt{b^2 - c^2\sin^2 B}$

(c) $\sqrt{c^2 - b^2\sin^2 B}$ (d) $2\sqrt{c^2 - b^2\sin^2 B}$

48. If $f: (3, 4) \to (0, 1)$ is defined by $f(x) = x - [x]$ where $[x]$ denotes the greatest integer function then $f^{-1}(x)$ is

(a) $\frac{1}{x - [x]}$ (b) $[x] - x$

(c) $x - 3$ (d) $x + 3$

49. Let $f(x) = [4 + 3\cos x]$, $x \in \left(-\frac{\pi}{2}, \frac{\pi}{2}\right)$, where $[x]$ = greatest integer less than or equal to x. The number of points of discontinuity of $f(x)$ is

(a) 2 (b) 3

(c) 5 (d) none of these

50. Let $f(x) = 1 - (1 - x)^2, x \le 1$. The number of solutions of the equation $f\{f(x)\} = x$ is

(a) one (b) two

(c) zero (d) none of these

51. The system of inequations $x - y \le -1, y \le x, x \ge 0, y \ge 0$ has

(a) one solution (b) infinite number of solutions

(c) finite number of solutions (d) no solution

52. If $p^x = q^w = a$ and $q^y = p^z = b$ then

(a) $xy = wz$ (b) $xz = yw$

(c) $x + z = y + w$ (d) $x + y = z + w$

53. If $2\sin\theta - 1 = 0$ and $2\cos\theta + \sqrt{3} = 0$ then the general values of θ are given by

(a) $2n\pi + \frac{\pi}{6}$ (b) $2n\pi + \frac{\pi}{4}$

(c) $2n\pi + \frac{7\pi}{6}$ (d) $2n\pi + \frac{5\pi}{6}$

54. The angle between the straight lines joining the point $(-1, 0)$ to the common points of $3x^2 + 5xy - 3y^2 + 8x + 8y + 5 = 0$ and $3x - 2y + 2 = 0$ is

(a) $\frac{\pi}{2}$ (b) $\frac{\pi}{4}$

(c) $\frac{\pi}{3}$ (d) none of these

55. If $x\cos\alpha + y\sin\alpha = p$ is a tangent to the circle $x^2 + y^2 = 2q(x\cos\alpha + y\sin\alpha)$ then the set of possible values of p is

(a) $\{q\}$ (b) $\{0, q\}$

(c) $\{0, 2q\}$ (d) $\{q, 2q\}$

56. f and g are two functions such that $(fg)(x) = (gf)(x)$ for all x. Then f and g may be defined as

(a) $f(x) = \sqrt{x}, g(x) = \cos x$

(b) $f(x) = x^3, g(x) = x + 1$

(c) $f(x) = x - 1, g(x) = x^2 + 1$

(d) $f(x) = x^m, g(x) = x^n$ where m, n are unequal integers

57. If the parabola $y = ax^2 - 6x + b$ passes through $(0, 2)$ and has its tangent at $x = \frac{3}{2}$ parallel to the x-axis then

(a) $a = 2, b = -2$ (b) $a = 2, b = 2$

(c) $a = -2, b = 2$ (d) $a = -2, b = -2$

58. If $f(x) = \sqrt{1 + \sqrt{x}}, x > 0$, then $f(x) \cdot f'(x)$ is equal to

(a) $\frac{1}{2\sqrt{x}}$ (b) $\frac{1}{2}$

(c) $\frac{1}{4\sqrt{x}}$ (d) $\frac{2\sqrt{x}+1}{4\sqrt{x}}$

59. If in a $\triangle ABC$, $4\tan A + 3 = 0$ then $\sin\frac{A}{2}$ and $\cos\frac{A}{2}$ are the roots of

(a) $10x^2 - 4\sqrt{10}x + 3 = 0$ (b) $10x^2 + 4\sqrt{10}x + 3 = 0$

(c) $10x^2 - 4\sqrt{10}x - 3 = 0$ (d) $10x^2 + 4\sqrt{10}x - 3 = 0$

60. If $\sin^6\theta + \cos^6\theta + k\cos^2 2\theta = 1$ then k is equal to

(a) $\frac{1}{2}\tan^2 2\theta$ (b) $\frac{1}{4}\tan^2 2\theta$

(c) $4\cot^2 2\theta$ (d) $\frac{3}{4}\tan^2 2\theta$

61. If a, b, c are in AP and one root of the equation $ax^2 + bx + c = 0$ is 2 then the other root is

(a) $\frac{3}{4}$ (b) $\frac{-3}{4}$

(c) $-\frac{5}{4}$ (d) $\frac{-5}{2}$

62. The number of solutions of $[\cos x] + |\sin x| = 1$ in $\pi \le x \le 3\pi$ is

(a) 3 (b) 4

(c) 2 (d) 1

63. A variable line $\frac{x}{a} + \frac{y}{b} = 1$ is drawn through the point $(k, 2k)$ so as to form a triangle of area A. If a, b are of the same sign then the least value of A is

(a) k^2 (b) $2k^2$

(c) $4k^2$ (d) $8k^2$

64. If the set $A = \left\{x \mid \cos\frac{\pi}{x} > 0\right\}$ then which of the following intervals is a subset of A?

(a) $\left(\frac{2}{5}, \frac{2}{3}\right)$ (b) $\left(\frac{2}{7}, \frac{2}{5}\right)$

(c) $\left(\frac{2}{7}, \frac{2}{3}\right)$ (d) $\left(\frac{2}{9}, \frac{2}{3}\right)$

65. If $\theta \in \left(0, \frac{\pi}{2}\right)$ then the value of $\cos\left(\theta - \frac{\pi}{4}\right)$ lies in the interval

(a) $\left(\frac{1}{2}, 1\right)$ (b) $\left(\frac{1}{\sqrt{2}}, 1\right]$

(c) $\left(-\frac{1}{\sqrt{2}}, 1\right]$ (d) $(0, 1)$

66. In ΔABC, the median AD divides $\angle BAC$ such that $\angle BAD : \angle CAD = 2 : 1$. Then $\cos \frac{A}{3}$ is equal to

(a) $\frac{\sin B}{2\sin C}$ (b) $\frac{\sin C}{2\sin B}$

(c) $\frac{2\sin B}{\sin C}$ (d) none of these

67. If the coordinates of two vertices and the orthocentre of a triangle are integers then the coordinates of the third vertex must be

(a) integers (b) rational numbers

(c) irrational numbers (d) none of these

68. The number of triangles that the four lines $y = x + 3$, $y = 2x + 3$, $y = 3x + 2$ and $y + x = 3$ form is

(a) 4 (b) 2

(c) 3 (d) 1

69. Let PQR be a right-angled triangle, right angled at P(2, 1). If the equation of the line QR is $2x + y = 3$ and $Q = (2, -1)$ then the centroid of the ΔPQR is

(a) $\left(\frac{5}{3}, \frac{1}{3}\right)$ (b) $\left(2, \frac{1}{3}\right)$

(c) $(2, 0)$ (d) $\left(\frac{1}{3}, \frac{5}{3}\right)$

70. If n is a multiple of 3 then the coefficient of x^n in the expansion of $\log_e(1 + x + x^2)$, $|x| < 1$, is

(a) $-\frac{1}{n}$ (b) $\frac{2}{n}$

(c) $-\frac{2}{n}$ (d) $\frac{1}{n}$

71. The possible number of different orders that a matrix can have when it has 24 elements, is

(a) 8 (b) 16

(c) 4 (d) none of these

72. If in the expansion of $(1+x)^m \cdot (1-x)^n$, the coefficients of x and x^2 are 3 and –6 respectively then

(a) $m=6, n=15$ (b) $m=9, n=12$

(c) $m=12, n=9$ (d) $m=24, n=21$

73. The coefficient of x^{28} in the expansion of $(1+x^3-x^6)^{30}$ is

(a) 1 (b) 0

(c) ${}^{30}C_6$ (d) ${}^{30}C_3$

74. $(n+2) \cdot {}^nC_0 \cdot 2^{n+1} - (n+1) \cdot {}^nC_1 \cdot 2^n + n \cdot {}^nC_2 \cdot 2^{n-1} - \dots$ is equal to

(a) 4 (b) $4n$

(c) $4(n+1)$ (d) $2(n+2)$

75. If $u_n = 2\cos n\theta$ then $u_1 u_n - u_{n-1}$ is equal to

(a) u_{n-2} (b) u_{n+1}

(c) 0 (d) none of these

76. The value of $\tan 9° - \tan 27° - \tan 63° + \tan 81°$ is

(a) 4 (b) 3

(c) 2 (d) 1

77. $\tan \dfrac{10\pi}{3}, \tan \dfrac{9\pi}{4}, \tan \dfrac{7\pi}{6}$ are in

(a) AP (b) GP

(c) HP (d) none of these

78. The number of real solutions of the equation $\cos^7 x + \sin^5 x = 1, 0 \le x \le 2\pi$, is

(a) 1 (b) 2

(c) 3 (d) infinite

79. The sum of $\cot^{-1}3 + \cot^{-1}7 + \cot^{-1}13 + \cot^{-1}21 + \cot^{-1}31$ is

(a) $\cot^{-1}\dfrac{7}{5}$ (b) $\cot^{-1}\dfrac{5}{7}$ (c) $\dfrac{\pi}{4}$ (d) $\dfrac{\pi}{2}$

80. One root of the equation $2\cos x + 1 = 2x$ lies in the interval

(a) $\left[-\frac{\pi}{2}, 0\right]$ (b) $\left[0, \frac{\pi}{2}\right]$

(c) $\left[\frac{\pi}{2}, \pi\right]$ (d) none of these

81. If 2, $\log_{3^x-4} 4$, $\log_{3^x+\frac{7}{2}} 4$ are in HP then x is equal to

(a) 1 (b) 2

(c) 4 (d) 0

82. The value of $\sum_{r=1}^{\infty} \tan^{-1} \frac{2r}{2+r^2+r^4}$ is equal to

(a) $\frac{\pi}{4}$ (b) $\frac{\pi}{3}$

(c) $\frac{\pi}{2}$ (d) π

83. In a ΔPQR, $\angle R = \frac{\pi}{2}$. If $\tan\frac{P}{2}$ and $\tan\frac{Q}{2}$ are the roots of the equation $ax^2 + bx + c = 0$ then

(a) $a + b = c$ (b) $b + c = a$

(c) $a + c = b$ (d) $b = c$

84. A particle moving on a curve has the position given by $x = f'(t)\sin t + f''(t)\cos t$, $y = f'(t)\cos t - f''(t)\sin t$ at time t where f is a thrice-differentiable function. Then the velocity of the particle at time t is

(a) $f'''(t)$ (b) $f'(t) + f'''(t)$

(c) $f'(t) + f''(t)$ (d) $f'(t) - f'''(t)$

85. If $f(x) = \cos^{-1}\frac{x - x^{-1}}{x + x^{-1}}$ then $f'(-2)$ is

(a) $\frac{2}{5}$ (b) $\frac{-2}{5}$

(c) $-\frac{1}{5}$ (d) none of these

86. The slope of the tangent to the curve $\tan y = \frac{\sqrt{1+x} - \sqrt{1-x}}{\sqrt{1+x} + \sqrt{1-x}}$ at $x = \frac{1}{2}$ is

(a) $\frac{1}{\sqrt{3}}$ (b) $\sqrt{3}$

(c) 1 (d) $\frac{1}{2}$

87. If $\frac{3x^3 + 5x - 4}{(x^2 - 1)(x^2 - 4)} \doteq \frac{A}{x+1} + \frac{B}{x-1} + \frac{C}{x+2} + \frac{D}{x-2}$ then $A + B + C + D$ is

(a) 4 (b) 3

(c) 0 (d) 2

88. The distance between the plane whose equation is $\vec{r} \cdot (2\vec{i} + \vec{j} - 3\vec{k}) = 5$ and the line whose equation is $\vec{r} = \vec{i} + \lambda(2\vec{i} + 5\vec{j} + 3\vec{k})$, is

(a) $\frac{3}{\sqrt{14}}$ (b) $\frac{5}{\sqrt{14}}$

(c) 5 (d) 0

89. The number of ways in which 10 red balls, 10 green balls and a white ball can be arranged in a row with the white ball in the middle, and the arrangement of colours on balls being symmetrical about the white ball, is

(a) $\frac{2(10\,!)}{(5\,!)^2}$ (b) $\frac{10\,!}{(5\,!)^2}$

(c) $\frac{20\,!}{(10\,!)^2}$ (d) none of these

90. A flag is to be coloured in four stripes by using 6 different colours, no two consecutive stripes being of the same colour. This can be done in

(a) 1500 ways (b) 750 ways

(c) 6^4 ways (d) none of these

91. Two vertices of an equilateral triangle in the Argand plane represent the nonreal cube roots of unity. The area of the triangle is

(a) $\frac{3\sqrt{3}}{4}$ (b) $\frac{3\sqrt{3}}{2}$

(c) $3\sqrt{3}$ (d) none of these

92. If $2x - y + 1 = 0, x + yz = 2, x > 0, y > 0$ then

(a) $z \in [0, 2)$ (b) $z \in \left(-\frac{1}{2}, 0\right]$

(c) $z \in [0, 1]$ (d) $z \in \left(-\frac{1}{2}, 2\right)$

93. Let $\Delta = \begin{vmatrix} a_{11} & a_{12} & a_{13} \\ a_{21} & a_{22} & a_{23} \\ a_{31} & a_{32} & a_{33} \end{vmatrix}$ and $a_{pq} = i^{p+q}$ where $i = \sqrt{-1}$. The value of Δ is

(a) real and positive (b) real and negative

(c) 0 (d) imaginary

94. A piece of paper is in the shape of a square of side 1 metre. It is cut at the four corners to make a regular polygon of eight sides (octagon). The area of the polygon is

(a) $2(\sqrt{2} - 1)\ \text{m}^2$ (b) $(\sqrt{2} - 1)\ \text{m}^2$

(c) $\frac{1}{\sqrt{2}}\ \text{m}^2$ (d) none of these

95. The sum of the first n terms of the series $3 + \frac{5}{2} + \frac{9}{4} + \frac{17}{8} + \ldots$ is

(a) $2n + 2^{1-n}$ (b) $2\{n + 1 - 2^{-n}\}$

(c) $n + 2^n$ (d) none of these

96. The number of solutions of $||x - 4| - 1| = 2$ is

(a) four (b) two

(c) three (d) one

97. If $1 \in (\alpha, \beta)$ where α, β are roots of $x^2 - a(x + 1) + 3 = 0$ then

(a) $a > 2$ (b) $a < -6$ or $a > 2$

(c) $-6 < a < 2$ (d) $a < 2$

98. The number of ways of selecting two numbers from the first 15 natural numbers such that their sum is a multiple of 5, is

(a) 20 (b) 36

(c) 21 (d) 16

99. If $\left|z - \frac{2}{z}\right| = 1$ then the greatest value of $|z|$ is

(a) 2 (b) 1

(c) 4 (d) 3

100. If $|z_1 - z_0| = |z_2 - z_0| = a$ and amp $\dfrac{z_2 - z_0}{z_0 - z_1} = \dfrac{\pi}{2}$ then z_0 is equal to

(a) $\dfrac{1}{2}\{(1+i)z_1 + (1-i)z_2\}$ (b) $\dfrac{1}{2}\{(1-i)z_1 + (1+i)z_2\}$

(c) $\dfrac{1}{2}\{z_1 + z_2\}$ (d) none of these

101. The fractional part of the number $\dfrac{3^{33}}{25}$ is

(a) 3/25 (b) 23/25

(c) 22/25 (d) 8/25

102. A boy goes to a place with a speed of a km/h and comes back with a speed of b km/h. The constant speed with which he should travel to cover the entire distance in the same total time is

(a) $\dfrac{a+b}{2}$ km/h (b) $\sqrt{ab}$ km/h

(c) $\dfrac{2ab}{a+b}$ km/h (d) none of these

103. If α, β be unequal real roots of the equation $ax^2 + bx + c = 0$ where a, b, c are real and γ is the solution of $2ax + b = 0$ then

(a) $\gamma > \alpha, \gamma > \beta$ (b) $\gamma < \alpha, \gamma < \beta$

(c) $\alpha < \gamma < \beta$ or $\beta < \gamma < \alpha$ (d) none of these

104. If f be a function defined as $f(x) = x^3 - 3x, -1 \le x \le 3$ then the range of f is

(a) $[0, 24]$ (b) $[-2, 2]$

(c) $[-2, 18]$ (d) none of these

105. Let $\hat{a}, \hat{b}, \hat{c}$ be unit vectors and α, β, γ are angles between the vectors $\hat{a}, \hat{b}$; $\hat{b}, \hat{c}$ and $\hat{c}, \hat{a}$ respectively. If $\hat{a} + \hat{b} + \hat{c}$ is also a unit vector then $\cos\alpha + \cos\beta + \cos\gamma$ is equal to

(a) -1 (b) 3

(c) -3 (d) 1

106. If $\cos^{-1}x + \cos^{-1}y + \cos^{-1}z = 3\pi$ then the value of $xy + yz + zx$ is

(a) -3 (b) 1

(c) 3 (d) none of these

107. Let $f_n(x) = \dfrac{n+1}{(n-1)!} x^n$. The value of $\int_0^1 \left(\sum_{n=1}^{\infty} f_n(x) \right) dx$ is

(a) e (b) 0
(c) $2e$ (d) none of these

108. Let three positive numbers a, b, c be in GP such that $a, b+8, c$ are in AP, while $a, b+8, c+64$ are in GP. Then the AM of a, b, c is

(a) $\dfrac{52}{3}$ (b) $\dfrac{28}{9}$
(c) 26 (d) 14

109. The number of ways in which the five digits 1, 2, 3, 4, 5 can be arranged to make a number greater than 10000 such that the odd digits are in the ascending order, is

(a) 10 (b) 20
(c) 120 (d) 60

110. The number of real solutions of the equation $\left(\dfrac{3}{5}\right)^x + \dfrac{7}{5} = 2^x$ is

(a) 2 (b) 1
(c) 0 (d) infinite

111. If $f(x) = x^3 - 6x^2 + 6x$ then $f(\sqrt[3]{2} + \sqrt[3]{4} + 2)$ has the value

(a) 2 (b) 4
(c) 6 (d) 8

112. Let $f(x) = |x+1|$. The number of values of $x \in [-2, 2]$ for which $f(x-3), f(x-1), f(x+1)$ are in AP is

(a) 1 (b) 2
(c) 0 (d) infinite

113. In a college of 300 students, every student reads 5 newspapers and every newspaper is read by 60 students. The number of newspapers is

(a) at least 30 (b) at most 20
(c) exactly 25 (d) none of these

114. If $(x+1)(x^{1/2}+1)(x^{1/2^2}+1) \ldots (x^{1/2^n}+1) = (x^2-1) f(x)$ then $f(x)$ is

(a) $\dfrac{1}{x^{\frac{1}{2^{n-1}}} - 1}$ (b) $\dfrac{1}{x^{\frac{1}{2^n}} - 1}$

(c) $x^{\frac{1}{2^n}} - 1$ (d) none of these

115. The minimum positive integral value of x such that $(1073)^{71} - x$ is divisible by 10, is

(a) 1 (b) 3

(c) 7 (d) 9

116. The minimum value of $|z - 1| + |z|$ for complex values of z is

(a) 2 (b) $\frac{1}{2}$

(c) 0 (d) 1

117. If in ΔABC, $a^2 + b^2 > c^2$ then the Δ is

(a) acute angled at C (b) obtuse angled at C

(c) an acute-angled triangle (d) none of these

118. In the ΔABC, BC is produced to D and $\angle ACD = \frac{3\pi}{4}$ and $\tan A, \tan B$ are roots of the equation $x^2 - \lambda x + \mu = 0$. Then

(a) $\lambda^2 - \mu^2 = 1 + 6\mu$ (b) $\lambda^2 - \mu^2 = 1$

(c) $\lambda = \mu - 1$ (d) none of these

119. If α, β are the roots of $x^2 + px + q = 0$ and γ, δ are the roots of $x^2 + px - r = 0$ then $\frac{\alpha - \gamma}{\beta - \gamma} \cdot \frac{\alpha - \delta}{\beta - \delta}$ is equal to

(a) 1 (b) -1

(c) $\frac{q + r}{p + r}$ (d) $\frac{q - r}{p - r}$

120. If a body starts with a velocity u in a straight line with uniform acceleration f and covers a distance s in time t seconds, and s_t denotes the distance covered by it in the tth second, then s_2, s_4, s_6 are in

(a) AP (b) GP

(c) the ratio 3 : 7 : 15 (d) the ratio 1 : 3 : 7

121. The value of $\sum_{r=1}^{n} r({}^nC_r + {}^rP_r)$ is

(a) $n \cdot 2^{n-1} - 1$ (b) $n \cdot 2^{n-1} + (n + 1)\,!$

(c) $n \cdot 2^{n-1} + (n+1)! - 1$ (d) $n^2 + n + 5$

122. For a positive integer n, let $a(n) = 1 + \frac{1}{2} + \frac{1}{3} + \frac{1}{4} + \ldots + \frac{1}{2^n - 1}$. Then

(a) $a(100) \le 100$ (b) $a(100) > 100$

(c) $a(200) \le 100$ (d) $a(200) > 100$

123. If α is a nonreal root of $x^5 + 1 = 0$ then $\alpha^{10n+2} + \alpha^{5n+2} + \alpha^{5n}$, where n is an odd positive integer, has the value

(a) 1 (b) 0

(c) –1 (d) none of these

124. If the 6th, 11th and 16th terms of a GP are a, b, c respectively then

(a) $a + c = 2b$ (b) $b^2 = ac$

(c) $a^2 + c^2 = b^2$ (d) none of these

125. Let $f(x) = x^3 - 3x + t$ and $f(0), f(1)$ are of opposite signs. Then the set of values of t is

(a) $(0, 2)$ (b) $(-\infty, 0)$

(c) $(2, +\infty)$ (d) none of these

126. If $z = 1 + \cos\frac{2\pi}{3} + i\sin\frac{2\pi}{3}$ then

(a) $\text{Re}(z^5) = \frac{\sqrt{3}}{2}$ (b) $\text{Re}(z^5) = \frac{1}{2}$

(c) $\text{Im}(z^5) = \frac{1}{2}$ (d) $\text{Im}(z^5) = \frac{\sqrt{3}}{2}$

127. If $f(x) = \sum_{\lambda=1}^{n} \left(x - \frac{1}{\lambda}\right)\left(x - \frac{1}{\lambda+1}\right)$ then $\lim_{n \to \infty} f(0)$ is

(a) 0 (b) –1

(c) 1 (d) none of these

128. If a, b, c, d are non-negative real numbers where $a + b + c + d = 1$ then the maximum value of $ab + bc + cd$ is

(a) $\frac{1}{4}$ (b) 3

(c) 4 (d) none of these

129. If $x^3 - mx^2 - 3x + 2 = 0$ has two roots equal in magnitude but opposite in sign then m is

(a) $\frac{3}{2}$ (b) $\frac{2}{3}$

(c) $-\frac{2}{3}$ (d) none of these

130. If $u_i = 1 - \frac{1}{i}$ then $u_2 \cdot u_3 \cdot \ldots \cdot u_n$ is equal to

(a) $\frac{1}{n}$ (b) $\frac{1}{n!}$

(c) 1 (d) none of these

131. The value of $\sum_{j=1}^{n} ({}^{n+1}C_j - {}^nC_j)$ is equal to

(a) 2^n (b) $2^n + 1$

(c) $3 \cdot 2^n$ (d) $2^n - 1$

132. If in the triangle ABC the equation of the side BC is $x + 2y = 3$, the vertex A is (1, 2) and the abscissa of the orthocentre is 3 then the ordinate of the orthocentre is

(a) 6 (b) 4

(c) 2 (d) 0

133. If the circles $x^2 + y^2 - 2x = 0$ and $x^2 + y^2 + 2\lambda y = 4$ have only one common tangent then λ is

(a) 1 (b) –1

(c) 0 (d) 2

134. The least number of sides of a polygon in which the number of diagonals is at least 10 more than the number of sides, is

(a) 8 (b) 10

(c) 11 (d) 25

135. If the algebraic sum of the distances of the points (1, 2), (–3, 1) and (2, –5) from a straight line be equal to zero then the straight line must pass through

(a) (0, –2) (b) (0, 0)

(c) $\left(0, \frac{-2}{3}\right)$ (d) at least one of the given points

136. Let $ABCDEF$ be a regular hexagon in the x-y plane and $\overrightarrow{AB} = 4\vec{i}$. Then $\overrightarrow{CD}$ is equal to

(a) $6\vec{i} + 2\sqrt{3}\vec{j}$ (b) $2(-\vec{i} + \sqrt{3}\vec{j})$

(c) $2(\vec{i} + \sqrt{3}\vec{j})$ (d) $2(\vec{i} - \sqrt{3}\vec{j})$

137. In a ΔABC, $\angle B = \frac{\pi}{3}$ · The range of values of x, where $x = \sin A \cdot \sin C$, is the interval

(a) $\left[-\frac{1}{4}, \frac{3}{4}\right]$ (b) $\left(0, \frac{3}{4}\right)$

(c) $\left(0, \frac{3}{4}\right]$ (d) $\left[\frac{1}{4}, \frac{3}{4}\right]$

138. If $\phi(x) = \dfrac{1}{1 + e^{-x}}$ then the value of

$$\phi(5) + \phi(4) + \phi(3) + \ldots + \phi(-3) + \phi(-4) + \phi(-5) \text{ is}$$

(a) 5 (b) $\frac{9}{2}$

(c) $\frac{11}{2}$ (d) none of these

139. If $f(x+2) = \frac{1}{2}\left\{f(x+1) + \frac{4}{f(x)}\right\}$ and $f(x) > 0$ for all $x \in R$ then $\lim\limits_{x \to \infty} f(x)$ is

(a) 1 (b) 2

(c) –2 (d) 0

140. If $I = \int_{1/\pi}^{\pi} \frac{1}{x} \cdot \sin\left(x - \frac{1}{x}\right) dx$ then I is equal to

(a) 0 (b) π

(c) $\pi - \frac{1}{\pi}$ (d) $\pi + \frac{1}{\pi}$

141. If the equations $4x^2 - x - 1 = 0$ and $3x^2 + (\lambda + \mu)x + \lambda - \mu = 0$ have a common root then the rational values of λ and μ are

(a) $\lambda = 0, \mu - \frac{-3}{4}$ (b) $\lambda = \frac{-3}{4}, \mu = 0$

(c) $\lambda = \frac{-3}{4}, \mu = \frac{3}{4}$ (d) $\lambda = \frac{-3}{4}, \mu = \frac{1}{4}$

142. If $ax^2 + bx + 1 = 0$, $a \in R$, $b \in R$, does not have distinct real roots then the least value of $3a - 2b$ is

(a) $-\frac{2}{3}$ (b) -1

(c) $\frac{1}{3}$ (d) $-\frac{4}{3}$

143. A square is inscribed in the circle $x^2 + y^2 - 2x - 4y = 0$ whose sides are parallel to the axes of reference. A vertex of the square is

(a) $(3, 1)$ (b) $(-2, 3)$

(c) $\left(\frac{\sqrt{2}-\sqrt{5}}{\sqrt{2}}, \frac{2\sqrt{2}-\sqrt{5}}{\sqrt{2}}\right)$ (d) $\left(\frac{\sqrt{5}+\sqrt{2}}{\sqrt{2}}, -\frac{2\sqrt{2}+\sqrt{5}}{\sqrt{2}}\right)$

144. Let $f(x) = ax^2 - bx + c^2$, $b \neq 0$ and $f(x) \neq 0$ for all $x \in R$. Then

(a) $a + c^2 < b$ (b) $4a + c^2 > 2b$

(c) $9a - 3b + c^2 < 0$ (d) none of these

145. Let $f(3) = 4$ and $f'(3) = 5$. Then $\lim_{x \to 3} [f(x)]$, where $[\cdot]$ denotes the greatest integer function, is

(a) 3 (b) 4

(c) 5 (d) nonexistent

146. If $[x + [x]] \leq 2$ where $[x]$ denotes the greatest integer $\leq x$, then x lies in the interval,

(a) $(-\infty, 1]$ (b) $(-\infty, 2)$

(c) $(-\infty, 2]$ (d) $(-\infty, 1)$

147. Let A = (3, 4) and B is a variable point on the lines $|x| = 6$. If $AB \leq 4$ then the number of positions of B with integral coordinates is

(a) 5 (b) 4

(c) 6 (d) 10

148. Let A = (3, –4), B = (1, 2). P = $(2k - 1, 2k + 1)$ is a variable point such that $PA + PB$ is the minimum. Then k is

(a) $\frac{7}{9}$ (b) 0

(c) $\frac{7}{8}$ (d) none of these

149. Let $f(x)$ be a function which can be expressed as a power series such that $f(0)=p, f'(0)=pq, f''(0)=pq^2, \ldots, f^n(0)=pq^n, \ldots$ where

$$f^n(0)=\left\{\frac{d^n f(x)}{dx^n}\right\}_{x=0}.$$

Then $\lim\limits_{x\to p} f(x)$ is equal to

(a) p (b) q

(c) pe^{pq} (d) qe^{pq}

150. If the angle between the trangents from the point $(\lambda, 1)$ to the parabola $y^2=16x$ be $\frac{\pi}{2}$ then λ is

(a) 4 (b) –4

(c) –1 (d) 2

151. The number of ways in which 9 flowers, of which 5 are identical and white, and the other four are of different colours, can be set on a garland so that no two of the coloured flowers are consecutive, is

(a) 60 (b) 120

(c) 180 (d) none of these

152. If r, r_0 be the inradius and an exradius respectively of an equilateral triangle then $r : r_0$ is equal to

(a) $1:2$ (b) $3:1$

(c) $1:\sqrt{3}$ (d) $1:3$

153. The value of $\int_0^{\pi/4} \frac{\sec x}{(\sec x+\tan x)^2}\,dx$ is

(a) $1+\sqrt{2}$ (b) $-(1+\sqrt{2})$

(c) $-\sqrt{2}$ (d) none of these

154. If the two circles $(x-1)^2+(y-3)^2=r^2$ and $x^2+y^2-8x+2y+8=0$ intersect in two distinct points then

(a) $r<2$ (b) $r=2$

(c) $r>2$ (d) $2<r<8$

155. The number of points with integral coordinates, exactly in the interior of the triangle with vertices (0, 0), (0, 21), (21, 0), is

(a) 133 (b) 190 (c) 233 (d) 105

156. If $\lim_{x \to 0} \frac{\{(a-n)nx - \tan x\}\sin nx}{x^2} = 0,\ n \neq 0$ then a is equal to

(a) 0 (b) $1 + \frac{1}{n}$

(c) n (d) $n + \frac{1}{n}$

157. A point on the line $\frac{x-1}{0} = \frac{y}{1} = \frac{z+1}{2}$ at a distance 3 from the plane $x + y + z = 3$ is

(a) $(1, 1, 1)$ (b) $(1 + \sqrt{3}, 1, 2\sqrt{3} + 1)$

(c) $(1, \sqrt{3} + 1, 2\sqrt{3} + 1)$ (d) $(1 - \sqrt{3}, 1, 2\sqrt{3} - 1)$

158. Two numbers are selected one by one without replacement at random from the set $S = \{1, 2, 3, 4, 5, 6\}$. The probability that the smaller of the two numbers is less than 4 is

(a) $\frac{1}{15}$ (b) $\frac{14}{15}$

(c) $\frac{1}{5}$ (d) $\frac{4}{5}$

159. Given $f'(x) = 6$ and $f'(1) = 4$, $\lim_{h \to 0} \frac{f(2h + 2 + h^2) - f(2)}{f(h - h^2 + 1) - f(1)}$ is

(a) 3 (b) $-\frac{3}{2}$

(c) $\frac{3}{2}$ (d) does not exist

160. The domain of the function $f(x) = \sqrt{\sin^{-1}(2x) + \frac{\pi}{6}}$ for real x, is

(a) $\left[-\frac{1}{4}, \frac{1}{2}\right]$ (b) $\left[-\frac{1}{2}, \frac{1}{2}\right]$

(c) $\left(-\frac{1}{2}, \frac{1}{9}\right)$ (d) $\left[-\frac{1}{4}, \frac{1}{4}\right]$

161. If $f(x) = x^2 + 2bx + 2c^2$ and $g(x) = -x^2 - 2cx + b^2$ such that $\min f(x) > \max g(x)$ then the relation between b and c is

(a) $0 < c < b\sqrt{2}$ (b) $|c| < \sqrt{2}\,|b|$

(c) $|c| > \sqrt{2}\,|b|$ (d) no real b and c

162. The coefficient of x^{24} in $(1+x^2)^{12} \cdot (1+x^{12}) \cdot (1+x^{24})$ is

(a) ${}^{12}C_6 + 3$ (b) ${}^{12}C_6 + 1$

(c) ${}^{12}C_6$ (d) ${}^{12}C_6 + 2$

163. If the sum of the roots of the quadratic equation $ax^2 + bx + c = 0$ is equal to the sum of the squares of their reciprocals then $\frac{a}{c}, \frac{b}{a}, \frac{c}{b}$ are in

(a) GP (b) HP.

(c) AP

(d) arithmetico-geometric progression

164. Let $\vec{u} = \vec{i} + \vec{j}$, $\vec{v} = \vec{i} - \vec{j}$ and $\vec{w} = \vec{i} + 2\vec{j} + 3\vec{k}$. If $\hat{n}$ is a unit vector such that $\vec{u} \cdot \hat{n} = 0$ and $\vec{v} \cdot \hat{n} = 0$ then $|\vec{w} \cdot \hat{n}|$ is equal to

(a) 1 (b) 2

(c) 3 (d) 0

165. Let $f'(x) = f(x)$ where $f(0) = 1$. If $f(x) + g(x) = x^2$ then $\int_0^1 f(x) g(x)\, dx$ is equal to

(a) $\frac{e^2}{2} + e - \frac{3}{2}$ (b) $-\frac{e^2}{2} + e - \frac{3}{2}$

(c) $\frac{e^2}{2} + e + \frac{5}{2}$ (d) $-\frac{e^2}{2} + e - \frac{5}{2}$

166. Let a, b, c be distinct real numbers such that $a^2 - b = b^2 - c = c^2 - a$. Then $(a+b)(b+c)(c+a)$ equals

(a) 0 (b) 1

(c) –1 (d) none of these

167. The number of elements in the set $S = \{i^n + i^{-n} : n \text{ is an integer}\}$ where $i = \sqrt{-1}$, is

(a) 3 (b) 1

(c) greater than 4 but finite (d) infinite

168. If $a_0 = 1, a_1 = 1$ and $a_n = a_{n-1} \cdot a_{n-2} + 1, n > 2$, then

(a) a_{165} is odd, a_{166} is even (b) a_{165} is even, a_{166} is odd

(c) both a_{165}, a_{166} are odd (d) both a_{165}, a_{166} are even

169. If $P(B)=\frac{3}{4}$, $P(A\cap B\cap \overline{C})=\frac{1}{3}$ and $P(\overline{A}\cap B\cap \overline{C})=\frac{1}{3}$ then $P(B\cap C)$ is

(a) $\frac{1}{12}$ (b) $\frac{1}{6}$

(c) $\frac{1}{15}$ (d) $\frac{1}{5}$

170. If $f(x)=\int_{x^2}^{x^2+1} e^{-t^2}\, dt$ then $f(x)$ increases in

(a) $(0, \infty)$ (b) $(-\infty, 0)$

(c) $(2, 2)$ (d) no value of x

171. Events A, B, C are mutually exclusive events such that $P(A)=\frac{3x+1}{3}$, $P(B)=\frac{1-x}{4}$ and $P(C)=\frac{1-2x}{2}$. The set of possible values of x is in the interval

(a) $\left[\frac{1}{3}, \frac{2}{3}\right]$ (b) $\left[\frac{1}{3}, \frac{13}{3}\right]$

(c) $[0, 1]$ (d) $\left[\frac{1}{3}, \frac{1}{2}\right]$

172. If in a $\triangle ABC$, $b=3$ cm, $c=4$ cm and the length of the perpendicular from A to the side BC is 2 cm then the number of solutions of the triangle is

(a) 0 (b) 1

(c) 2 (d) 3

173. The solution set of the equation $[2x]=[x]+3, 0\le x\le 4$ where $[x]=$ the greatest integer less than or equal to x, is

(a) $\{3\}$ (b) $\left\{3, \frac{5}{2}\right\}$

(c) $\left[\frac{5}{2}, \frac{7}{2}\right)$ (d) $[2, 3]$

174. The ratio in which the line segment joining the points whose position vectors are $2\vec{i}-4\vec{j}-7\vec{k}$ and $-3\vec{i}+5\vec{j}-8\vec{k}$ is divided by the plane whose equation is $\vec{r}\cdot(\vec{i}-2\vec{j}+3\vec{k})=13$, is

(a) 13 : 12 internally (b) 12 : 25 externally

(c) 13 : 25 internally (d) 37 : 25 internally

175. The value of $\sum_{r=0}^{n} {}^{n+r}C_r$ is equal to

(a) ${}^{2n+1}C_n$ (b) ${}^{2n}C_{n-1}$

(c) ${}^{2n}C_{n+1}$ (d) ${}^{2n+1}C_{n-1}$

176. The value of p for which one root of the quadratic equation $(p^2 - 5p + 3)x^2 + (3p - 1)x + 2 = 0$ is twice as large as the other, is

(a) $-\frac{2}{3}$ (b) $\frac{1}{3}$

(c) $-\frac{1}{3}$ (d) $\frac{2}{3}$

177. If $f(a + b - x) = f(x)$ then $\int_a^b xf(x)\,dx$ is equal to

(a) $\frac{a+b}{2}\int_a^b f(x)\,dx$ (b) $\frac{b-a}{2}\int_a^b f(x)\,dx$

(c) $\frac{a+b}{2}\int_a^b f(a+b-x)\,dx$ (d) $\frac{a+b}{2}\int_a^b f(b-x)\,dx$

178. The number of integral terms in the expansion of $(\sqrt{3} + \sqrt[5]{5})^{256}$ is

(a) 25 (b) 26

(c) 24 (d) none of these

179. ${}^nC_{r+1} + {}^nC_{r-1} + 2 \times {}^nC_r$ is equal to

(a) ${}^{n+2}C_{r+1}$ (b) ${}^{n+1}C_r$

(c) ${}^{n+1}C_{r+1}$ (d) ${}^{n+2}C_r$

180. A focal chord of the parabola $y^2 = 16x$ touches the circle $(x - 6)^2 + y^2 = 2$. Then the possible values of the slope of the chord are

(a) $\{-1, 1\}$ (b) $\{-2, 2\}$

(c) $\left\{-2, \frac{1}{2}\right\}$ (d) $\left\{2, -\frac{1}{2}\right\}$

181. The value of λ so that the volume of the parallelopiped formed by the vectors $\vec{i} + \lambda\vec{j} + \vec{k}$, $\vec{j} + \lambda\vec{k}$ and $\lambda\vec{i} + \vec{k}$ becomes minimum, is

(a) -3 (b) 3 (c) $\frac{1}{\sqrt{3}}$ (d) $\sqrt{3}$

182. The sum of the radii of inscribed and circumscribed circles for an n-sided regular polygon of side a, is

(a) $\frac{a}{2}\cot\frac{\pi}{2n}$ (b) $a\cot\frac{\pi}{2n}$

(c) $\frac{a}{4}\cot\frac{\pi}{2n}$ (d) $a\cot\frac{\pi}{n}$

183. If $|z| = 1$ and $w = \frac{z-1}{z+1}$ $(z \neq -1)$ then Re(w) is

(a) 0 (b) $\frac{-1}{|z+1|^2}$

(c) $\frac{1}{|z+1|^2}$ (d) $\frac{\sqrt{2}}{|z+1|^2}$

184. If z and w are two nonzero complex numbers such that $|zw| = 1$ and arg $z - \arg w = \frac{\pi}{2}$ then $\bar{z}w$ is equal to

(a) –1 (b) i (c) $-i$ (d) 1

185. If the system of equations $x + ay = az + y = ax + z = 0$ has infinite solutions then a is

(a) –1 (b) 1

(c) 0 (d) no real value

186. The origin, z_1 and z_2 are the vertices of an equilateral triangle where z_1, z_2 are roots of the equation $z^2 + az + b = 0$. Then

(a) $a^2 = 2b$ (b) $a^2 = 3b$

(c) $a^2 = 4b$ (d) $a^2 = b$

187. If $I(m, n) = \int_0^1 t^m(1+t)^n\,dt$ then $I(m, n)$ is equal to

(a) $\frac{2^n}{m+1} - \frac{n}{m+1}I(m+1, n-1)$ (b) $\frac{n}{m+1}I(m+1, n-1)$

(c) $\frac{2^n+n}{m+1}I(m+1, n-1)$ (d) $\frac{m}{m+1}I(m+1, n-1)$

188. Let $f(x) = \sqrt{x^2+x} + \frac{\tan^2\alpha}{\sqrt{x^2+x}}$, $\alpha \in \left(0, \frac{\pi}{2}\right)$. The value of $f(x)$ is always greater than or equal to

(a) $2\tan\alpha$ (b) 1 (c) 2 (d) $\sec^2\alpha$

189. A function $f: N \to Z$ is defined by

$$f(n) = \frac{n-1}{2}, \text{ when } n \text{ is odd}$$
$$= -\frac{n}{2}, \text{ when } n \text{ is even.}$$

The function f is

(a) onto but not one-one (b) onto and one-one

(c) neither onto nor one-one (d) one-one but not onto

190. If $f: [0, \infty) \to [0, \infty)$ and $f(x) = \dfrac{x}{x+1}$ then f is

(a) one-one and onto (b) one-one but not onto

(c) onto but not one-one (d) neither one-one nor onto

191. In [0, 1] Lagrange's Mean-Value theorem is not applicable to

(a) $f(x) = \frac{1}{2} - x, x < \frac{1}{2}$

$\left(\frac{1}{2} - x\right)^2, x \geq \frac{1}{2}$

(b) $f(x) = \frac{\sin x}{x}, x \neq 0$

$1, x = 0$

(c) $f(x) = x|x|$ (d) $f(x) = |x|$

192. If $\vec{u}, \vec{v}, \vec{w}$ are three noncoplanar vectors then $[\vec{u} + \vec{v} - \vec{w} \;\; \vec{u} - \vec{v} \;\; \vec{v} - \vec{w}]$ equals

(a) $[\vec{u} \; \vec{v} \; \vec{w}]$ (b) $[\vec{u} \; \vec{w} \; \vec{v}]$

(c) $3[\vec{u} \; \vec{v} \; \vec{w}]$ (d) 0

193. If the system of linear equations

$$x + 2ay + az = 0$$
$$x + 3by + bz = 0$$
$$x + 4cy + cz = 0$$

has a nonzero solution then a, b, c

(a) are in GP (b) are in HP

(c) are in AP (d) satisfy $a + 2b + 3c = 0$

194. If the equation $\lambda x^2 - 2x + 3 = 0$ has positive roots for some real λ then

(a) $\lambda > 0$ (b) $\lambda > \frac{1}{3}$ (c) $0 \leq \lambda \leq \frac{1}{3}$ (d) $\lambda \leq \frac{1}{3}$

195. If ${}^{2n}C_6$ is the greatest coefficient in the expansion of $(1+x)^{2n}$, $n \in N$, then n is

(a) 6 (b) 7

(c) 5 (d) 8

196. The lines $\frac{x-2}{1}=\frac{y-3}{1}=\frac{z-4}{-k}$ and $\frac{x-1}{k}=\frac{y-4}{2}=\frac{z-5}{1}$ are coplanar if k is

(a) 1 or –1 (b) 0 or –3

(c) 3 or –3 (d) 0 or –1

197. The equation of the plane bisecting the obtuse angle between the planes $x+y+z=1$ and $x+2y-4z=5$ is

(a) $(\sqrt{7}-1)x+(\sqrt{7}-2)y+(\sqrt{7}+4)z+5-\sqrt{7}=0$

(b) $(\sqrt{7}+1)x+(\sqrt{7}+2)y+(\sqrt{7}+4)z+5-\sqrt{7}=0$

(c) $(\sqrt{7}+1)x+(\sqrt{7}+2)y+(\sqrt{7}-4)z=5+\sqrt{7}$

(d) none of these

198. Two gas companies G_1 and G_2, situated at (40, 0) and (0, 30) respectively (unit of distance = 1 km), offer to install equally priced gas ovens at a buyer's house. The company G_1 adds a charge of Rs 40/km of aerial distance between the company and the buyer's house while it is Rs 60 in case of the company G_2. Then the region where it is cheaper to install the oven from the gas company G_1 is

(a) the interior of the circle $(x+32)^2+(y-54)^2=3600$

(b) the interior of the circle $(x-54)^2+(y+32)^2=3600$

(c) the exterior of the circle $(x+32)^2+(y-54)^2=3600$

(d) the exterior of the circle $(x-54)^2+(y+32)^2=3600$

Answers

1. a	**2.** a	**3.** a	**4.** b	**5.** b
6. d	**7.** b	**8.** a	**9.** b	**10.** b
11. c	**12.** c	**13.** c	**14.** d	**15.** d
16. c	**17.** a	**18.** b	**19.** a	**20.** b
21. c	**22.** a	**23.** a	**24.** b	**25.** a
26. b	**27.** c	**28.** a	**29.** d	**30.** d

31. b	**32.** c	**33.** a	**34.** c	**35.** b
36. d	**37.** c	**38.** b	**39.** d	**40.** c
41. a	**42.** b	**43.** c	**44.** a	**45.** c
46. a	**47.** b	**48.** d	**49.** c	**50.** b
51. d	**52.** a	**53.** d	**54.** a	**55.** c
56. d	**57.** b	**58.** c	**59.** a	**60.** d
61. c	**62.** a	**63.** c	**64.** a	**65.** b
66. a	**67.** b	**68.** c	**69.** a	**70.** c
71. a	**72.** c	**73.** b	**74.** c	**75.** b
76. a	**77.** b	**78.** c	**79.** a	**80.** b
81. b	**82.** a	**83.** a	**84.** b	**85.** b
86. a	**87.** b	**88.** a	**89.** b	**90.** b
91. a	**92.** d	**93.** c	**94.** a	**95.** b
96. b	**97.** a	**98.** c	**99.** a	**100.** a
101. b	**102.** c	**103.** c	**104.** c	**105.** a
106. c	**107.** a	**108.** a	**109.** b	**110.** b
111. a	**112.** b	**113.** c	**114.** b	**115.** c
116. d	**117.** a	**118.** c	**119.** a	**120.** a
121. c	**122.** d	**123.** c	**124.** b	**125.** a
126. b	**127.** c	**128.** a	**129.** b	**130.** a
131. d	**132.** a	**133.** c	**134.** a	**135.** c
136. b	**137.** d	**138.** c	**139.** b	**140.** a
141. b	**142.** d	**143.** d	**144.** b	**145.** d
146. b	**147.** a	**148.** c	**149.** c	**150.** b
151. d	**152.** d	**153.** d	**154.** d	**155.** b
156. d	**157.** c	**158.** d	**159.** a	**160.** a
161. c	**162.** d	**163.** b	**164.** c	**165.** b
166. b	**167.** a	**168.** c	**169.** a	**170.** b
171. d	**172.** c	**173.** c	**174.** b	**175.** a
176. d	**177.** a	**178.** b	**179.** a	**180.** a
181. c	**182.** a	**183.** a	**184.** c	**185.** a
186. b	**187.** a	**188.** a	**189.** b	**190.** b
191. a	**192.** a	**193.** b	**194.** c	**195.** a
196. b	**197.** c	**198.** c		

Hints and Solutions to Selected Questions

1. $a^y(a^x - 1) = a^x$ or $a^y = \dfrac{a^x}{a^x - 1} > 0$

$\Rightarrow \quad a^x > 1 = a^0 \quad \Rightarrow x > 0.$

2. Here, $\log \dfrac{y^2}{x(y-1)} = 0. \quad \therefore \quad \dfrac{y^2}{x(y-1)} = 1$

or $\quad y^2 = xy - x \quad$ or $\quad y^2 - xy + x = 0. \quad \therefore \; y = \dfrac{x \pm \sqrt{x^2 - 4x}}{2}.$

So, y is real if $x^2 - 4x \geq 0$ or $x(x-4) >= 0.$

$\therefore \quad x \leq 0$ or $x \geq 4$. But $x > 0$ for $\log x$ to be defined.

So $x \geq 4$.

Now, $x = \dfrac{y^2}{y-1} \geq 4$ or $\dfrac{y^2 - 4y + 4}{y-1} \geq 0$

or $\quad (y-2)^2 \cdot (y-1) \geq 0 \quad \Rightarrow y \geq 1$. But $y > 1$ for $\log(y-1)$ to be real.

3. $2x^2 - ky + 3 = 0$ or $x^2 = \dfrac{k}{2}\left(y - \dfrac{3}{k}\right).$

$\therefore \quad$ latus rectum $= \dfrac{k}{2} = 2. \quad \therefore \; k = 4.$

The vertex $= \left(0, \dfrac{3}{k}\right) = \left(0, \dfrac{3}{4}\right).$

4. If (α, β) is the centre then $(\alpha - 1)^2 + (\beta - 3)^2 = (\alpha - 3)^2 + (\beta - 1)^2 \quad \ldots (1)$

and $\dfrac{\beta - 3}{\alpha - 1} \cdot \dfrac{\beta - 1}{\alpha - 3} = -1$ or $(\alpha - 1)(\alpha - 3) + (\beta - 1)(\beta - 3) = 0. \quad \ldots (2)$

(1) $\Rightarrow 4\alpha - 4\beta = 0. \quad \therefore \; \alpha = \beta.$

(2) $\Rightarrow 2(\alpha - 1)(\alpha - 3) = 0. \quad \therefore \; \alpha = 1, 3.$

$\therefore \quad (\alpha, \beta) = (1, 1), (3, 3).$

5. $f'(x) = 2x - a > 0$ or $x > \dfrac{a}{2}. \quad \therefore$ m.i. in $\left(\dfrac{a}{2}, +\infty\right).$

$\therefore \quad \frac{a}{2} = 2 \quad$ or $\quad a = 4.$

6. Use the fact that

$^4P_1 + {}^4P_2 + {}^4P_3 + {}^4P_4 =$ the value of the determinant $= x \cdot 1 \cdot 2 \cdot 3.$

7. $OM_r = OA_r + \frac{OA_{r+1} - OA_r}{2} = \frac{OA_r + OA_{r+1}}{2} = \frac{1}{2}\{OA_1 \cdot k^{r-1} + OA_1 \cdot k^r\}$

$$= \frac{OA_1}{2}(1+k)k^{r-1}.$$

$$\therefore \quad \sum_{r=1}^{\infty} OM_r = \frac{OA_1}{2}(1+k) \cdot \sum_{r=1}^{\infty} k^{r-1} = \frac{OA_1}{2}(1+k) \cdot \frac{1}{1-k}$$

$$= \frac{OA_1}{2} \cdot \frac{1 + \frac{OA_2}{OA_1}}{1 - \frac{OA_2}{OA_1}} = \frac{OA_1(OA_1 + OA_2)}{2(OA_1 - OA_2)}.$$

8. $x^3 - ax^2 + bx - c \equiv (x^2 - px + q)(x - \lambda).$

Equating the coefficients of x^2, x, x^0 on both sides,

$$-a = -\lambda - p, b = p\lambda + q, -c = -q\lambda.$$

$\therefore \quad \lambda = a - p = \frac{b-q}{p} = \frac{c}{q}$ which gives the result.

9. s.d. $= \left| \frac{(\vec{i}+\vec{j}+\vec{k}) \times (\vec{i} - 2\vec{j} + 3\vec{k})}{|(\vec{i}+\vec{j}+\vec{k}) \times (\vec{i} - 2\vec{j} + 3\vec{k})|} \cdot \vec{k} \right|$

$$= \left| \frac{5\vec{i} - 2\vec{j} - 3\vec{k}}{\sqrt{5^2 + (-2)^2 + (-3)^2}} \cdot \vec{k} \right| = \frac{3}{\sqrt{38}}.$$

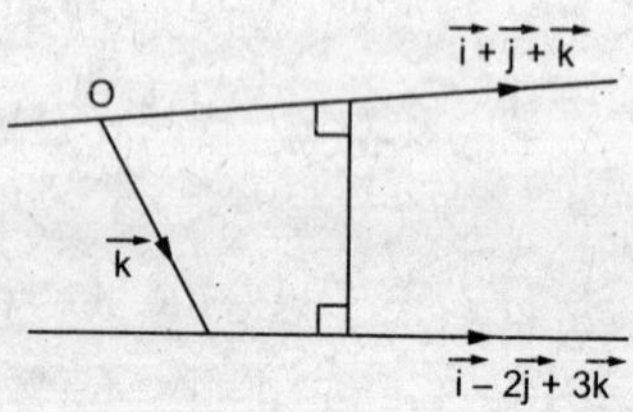

10. $n(S) = {}^{52}C_4.$

$n(E) =$ the number of ways to select a pair and two others different

$= {}^{13}C_1 \times {}^4C_2 \times {}^{12}C_2 \times 4^2$

because there are 13 different denominations and there are four cards of each denomination, and after that there are 12 cards of different denominations from which 2 cards of different denominations are to be selected, there being 4 cards of each of the remaining denominations.

11. Expression $= \cos^{-1}x + \cos^{-1}\left\{x\cdot\frac{1}{\sqrt{2}} + \sqrt{1-x^2}\cdot\sqrt{1-\left(\frac{1}{\sqrt{2}}\right)^2}\right\}$

$$= \cos^{-1}x + \cos^{-1}\frac{1}{\sqrt{2}} - \cos^{-1}x \left\{\because \frac{1}{\sqrt{2}} < x \Rightarrow \cos^{-1}\frac{1}{\sqrt{2}} > \cos^{-1}x\right\}$$

$$= \cos^{-1}\frac{1}{\sqrt{2}} = \frac{\pi}{4}.$$

12. Here, $r_1 + r_2 = d$. So, $k + \sqrt{2^2 + 4^2 - 16}$ = distance between $(k, 0)$ and $(2, 4)$

or $\quad k + 2 = \sqrt{(k-2)^2 + 4^2}$ $\quad$ or $\quad (k+2)^2 = k^2 - 4k + 20$

or $\quad 8k = 16. \quad \therefore\ k = 2.$

13. $R = \dfrac{abc}{4\Delta} = \dfrac{4\cdot 4\cdot 4}{4\cdot\frac{1}{2}\cdot 4\cdot 4\cdot \sin 60°} = \dfrac{4}{\sqrt{3}}$

$$r = \frac{\Delta}{s} = \frac{\frac{1}{2}\cdot 4\cdot 4\cdot \sin 60°}{\frac{1}{2}\cdot(4+4+4)} = \frac{4}{3}\cdot\frac{\sqrt{3}}{2} = \frac{2}{\sqrt{3}}$$

$h = \sqrt{4^2 - 2^2} = \sqrt{12} = 2\sqrt{3}. \quad \therefore\ \dfrac{R+r}{h} = \dfrac{\frac{4}{\sqrt{3}} + \frac{2}{\sqrt{3}}}{2\sqrt{3}} = 1.$

14. Here, $x^2 = -\lambda\left(y + \frac{\mu}{\lambda}\right)\cdot$

$\therefore\quad$ vertex $= \left(0, -\frac{\mu}{\lambda}\right)$ and the directrix is $\left(y + \frac{\mu}{\lambda}\right) + \frac{-\lambda}{4} = 0.$

Comparing with the given values, $-\frac{\mu}{\lambda} = 1$ and $\frac{\mu}{\lambda} - \frac{\lambda}{4} = -2.$

$\therefore\quad -1 - \frac{\lambda}{4} = -2 \quad$ or $\quad \lambda = 4$ and so, $\mu = -4.$

15. A line parallel to the y-axis has the slope ∞.

$\therefore\quad (a^2-1)m^2 - (2a-3)m + a = 0$ must have one infinite root.

$\therefore\quad a^2 - 1 = 0$ ($\because$ both roots cannot be ∞ for then $a^2 - 1 = 0$, $2a - 3 = 0$ are to hold for the same a).

16. Adding, $7(\cos\theta + \sin\theta) = 10 \quad$ or $\quad \cos\theta + \sin\theta = \frac{10}{7}\cdot$

But maximum value of $\cos\theta+\sin\theta=\sqrt{1^2+1^2}=\sqrt{2}<\frac{10}{7}$.

So, no θ is possible.

17.

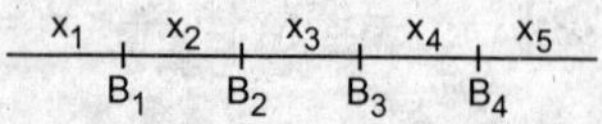

B_1, etc., indicate places for boys and x_1, etc., indicate the numbers of girls in different gaps.

Now, $x_1+x_2+x_3+x_4+x_5=8;\ x_1\geq 1, \ldots, x_5>1$.

The number of solutions = coefficient of x^8 in $(x+x^2+x^3+\ldots)^5$

$= \text{coefficient of } x^3 \text{ in } (1+x+x^2+\ldots)^5$

$= \text{coefficient of } x^3 \text{ in } (1-x)^{-5}$

$= \text{coefficient of } x^3 \text{ in } \{{}^4C_0+{}^5C_1x+{}^6C_2x^2+\ldots\}={}^7C_3.$

$\therefore \quad n(S)=12\,!$ and $n(E)={}^7C_3\times 8\,!\times 4\,!.$

$$\therefore \quad P(E)=\frac{{}^7C_3\times 8\,!\times 4\,!}{12\,!}=\frac{7\,!\times 8\,!}{3\,!\,12\,!}.$$

Alternatively

There are 7 places in between 8 girls. So, they can be seated in $8\,!\times{}^7C_4\times 4\,!$ ways.

18. $$\text{RH limit}=\lim_{h\to 0}\frac{\sin 3\left[\frac{1}{2}+h\right]}{\left[\frac{1}{2}+h\right]}=\lim_{\theta\to 0}\frac{\sin 3\theta}{\theta}=3.$$

$$\text{LH limit}=\lim_{h\to 0}\frac{\sin 3\left[\frac{1}{2}-h\right]}{\left[\frac{1}{2}-h\right]}=\lim_{\theta\to 0}\frac{\sin 3\theta}{\theta}=3.$$

19. Put $a-x=z$. Then $I=\int_{a-n}^{n} xe^{x(a-x)}\,dx=\int_{n}^{a-n}(a-z)e^{(a-z)z}(-dz).$

$$\therefore \quad I=\int_{a-n}^{n}(a-z)e^{(a-z)z}\,dz=a\int_{a-n}^{n}e^{x(a-x)}\,dx-\int_{a-n}^{n}xe^{x(a-x)}\,dx.$$

$$\therefore \quad 2I=a\int_{a-n}^{n}e^{x(a-x)}\,dx=a\lambda.$$

20. $\tan^{-1}2 + \tan^{-1}3 = \pi + \tan^{-1}\dfrac{2+3}{1-2\cdot 3}$

($\because$ $\tan^{-1}2 + \tan^{-1}3 \in$ first or second quadrant)

$$= \pi + \tan^{-1}(-1) = \pi - \frac{\pi}{4}\cdot$$

21. Here, $b = 2c$. Now,

$$\tan\frac{B-C}{2} = \frac{b-c}{b+c}\cot\frac{A}{2} \quad \text{or} \quad \cot\frac{A}{2}\cdot\cot\frac{B-C}{2} = \frac{b+c}{b-c} = \frac{3c}{c} = 3.$$

22. There are r openings to come out of the rth circle and after that it can travel clockwise or anticlockwise to enter the $(r+1)$th circle.

Thus, the number of paths

$= (1\cdot 2)\times(2\cdot 2)\times(3\cdot 2)\times\ldots\times\{(n-1)\cdot 2\}\times\{n\cdot 2\}$

$= n!\cdot 2^n$.

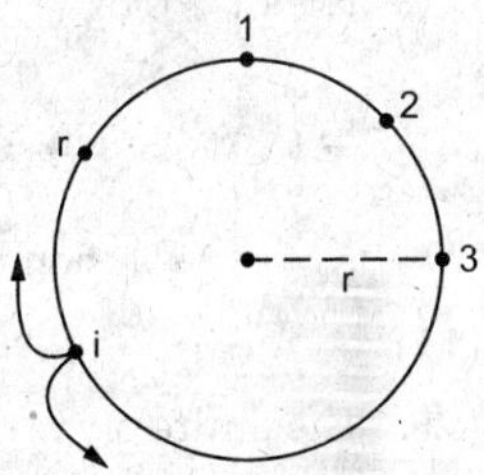

23. $\vec{r}\times\vec{a} = \lambda\vec{a} + \mu\vec{b} + \gamma\vec{a}\times\vec{b}$.

$\therefore \quad [\vec{r}\ \vec{a}\ \vec{a}] = \lambda\vec{a}\cdot\vec{a} + \mu\vec{b}\cdot\vec{a} + \gamma[\vec{a}\ \vec{b}\ \vec{a}]$.

$\therefore \quad 0 = \lambda \quad (\because\ |\vec{a}| = 1, \vec{a}\cdot\vec{b} = 0)$.

Also, $[\vec{r}\ \vec{a}\ \vec{b}] = \lambda\vec{a}\cdot\vec{b} + \mu\vec{b}\cdot\vec{b} + \gamma[\vec{a}\ \vec{b}\ \vec{b}] = \mu$.

Also, $(\vec{r}\times\vec{a})\times\vec{b} = \gamma(\vec{a}\times\vec{b})\times\vec{b}$

or $\quad (\vec{r}\cdot\vec{b})\,\vec{a} - (\vec{a}\cdot\vec{b})\,\vec{r} = \gamma\{(\vec{a}\cdot\vec{b})\,\vec{b} - (\vec{b}\cdot\vec{b})\,\vec{a}\}$

or $\quad (\vec{r}\cdot\vec{b})\,\vec{a} = -\gamma\vec{a}. \qquad \therefore\ \gamma = -(\vec{r}\cdot\vec{b})$.

24. The rank is 2 if $\begin{vmatrix} 4 & 2 & 1-x \\ 5 & k & 1 \\ 6 & 3 & 1+x \end{vmatrix} = 0.$

$$\therefore \quad \begin{vmatrix} 0 & 2 & 1-x \\ 5-2k & k & 1 \\ 0 & 3 & 1+x \end{vmatrix} = 0 \quad \text{or} \quad -(5-2k)\begin{vmatrix} 2 & 1-x \\ 3 & 1+x \end{vmatrix} = 0$$

or $\quad (5-2k)(5x-1) = 0. \quad \therefore\ k = \dfrac{5}{2} \quad$ or $\quad x = \dfrac{1}{5}\cdot$

If $k = \dfrac{5}{2}, x \neq \dfrac{1}{5}$ then $\begin{vmatrix} 4 & 2 & 1-x \\ 5 & k & 1 \\ 6 & 3 & 1+x \end{vmatrix} = \dfrac{1}{2}\begin{vmatrix} 4 & 4 & 1-x \\ 5 & 5 & 1 \\ 6 & 6 & 1+x \end{vmatrix} = 0$

and $\begin{vmatrix} 4 & 1-x \\ 5 & 1 \end{vmatrix} = 4 - 5 + 5x = 5x - 1 \neq 0$. So, the rank = 2.

25. If $0 < x < \frac{\pi}{2}$, $\int_{-2}^{x} |\cos x|\, dx = \int_{-2}^{-\pi/2} |\cos x|\, dx + \int_{-\pi/2}^{x} |\cos x|\, dx$

$$= \int_{-2}^{-\pi/2} -\cos x\, dx + \int_{-\pi/2}^{x} \cos x\, dx = -\sin x\Big]_{-2}^{-\pi/2} + \sin x\Big]_{-\pi/2}^{x}$$

$$= 1 - \sin 2 + \sin x + 1.$$

$\therefore \quad \int_{-2}^{x} |\cos x|\, dx = 0 \quad \Rightarrow \quad 1 - \sin 2 + \sin x + 1 = 0$

$\Rightarrow \quad \sin x = \sin 2 - 2 < -1.$

$\therefore$ no solution in $\left(0, \frac{\pi}{2}\right)$.

26. The centre of $x^2 + y^2 - 4x - 4y = 0$ is $(2, 2)$. It will be on $ax + by = 2$.

$\therefore \quad 2a + 2b = 2 \quad$ or $\quad a + b = 1$.

$ax + by = 2$ touches $x^2 + y^2 = 1$. So, $1 = \left|\frac{-2}{\sqrt{a^2 + b^2}}\right|$.

$\therefore \quad a^2 + b^2 = 4 \quad$ or $\quad a^2 + (1 - a)^2 = 4$

or $\quad 2a^2 - 2a - 3 = 0. \quad \therefore \; a = \frac{2 \pm \sqrt{4 + 24}}{4} = \frac{1 \pm \sqrt{7}}{2}$.

$\therefore \quad b = 1 - a = 1 - \frac{1 \pm \sqrt{7}}{2} = \frac{1 \mp \sqrt{7}}{2}$.

27. $P(A' \cap B') - P(A \cap B) = P(A') \cdot P(B') - P(A) \cdot P(B)$

$$= \{1 - P(A)\}\{1 - P(B)\} - P(A) \cdot P(B) = 1 - P(A) - P(B).$$

$$P(A') + P(B') - 1 = 1 - P(A) + 1 - P(B) - 1 = 1 - P(A) - P(B)$$

$$P(B) - P(A') = P(B) - \{1 - P(A)\} = P(A) + P(B) - 1$$

$$P(B') - P(A) = 1 - P(B) - P(A).$$

28. $P(A) = P(B) = P(C)$ and $P(A) + P(B) + P(C) = 1$.

$\therefore \quad P(A) = P(B) = P(C) = \frac{1}{3}$. Also, $P(X) = \frac{5}{12}$, $P(X/A) = \frac{3}{8}$, $P(X/B) = \frac{1}{4}$.

We have $P(X) = P(A) \cdot P(X/A) + P(B) \cdot P(X/B) + P(C) \cdot P(X/C)$.

$$\therefore \quad \frac{5}{12} = \frac{1}{3}\left\{\frac{3}{8} + \frac{1}{4} + P(X/C)\right\}\cdot$$

29. Solving the equations, $x^2 + 4(x+4) = a^2$.

$D = 0 \quad \Rightarrow 16 - 4\cdot(16 - a^2) = 0 \quad \Rightarrow a = \pm 2\sqrt{3}$.

Also, (0, 0) is an interior point of $y^2 = 4(x+4)$ for $0^2 - 4\cdot 0 - 16 < 0$.

30. $\vec{f}\left(\frac{5}{4}\right) = \left[\frac{5}{4}\right]\vec{i} + \left\{\frac{5}{4} - \left[\frac{5}{4}\right]\right\}\vec{j} + \left[\frac{5}{4} + 1\right]\vec{k} = \vec{i} + \left(\frac{5}{4} - 1\right)\vec{j} + 2\vec{k}$

$$= \vec{i} + \frac{1}{4}\vec{j} + 2\vec{k}.$$

When $0 < t < 1$, $\vec{f}(t) = 0\vec{i} + \{t - 0\}\vec{j} + \vec{k} = t\vec{j} + \vec{k}$.

$\vec{f}\left(\frac{5}{4}\right)\cdot\vec{f}(t) = 2 + \frac{t}{4}\cdot$ So, $\cos\theta = \dfrac{2 + \frac{t}{4}}{\left|\vec{i} + \frac{1}{4}\vec{j} + 2\vec{k}\right| \; |t\vec{j} + \vec{k}|}$

$$= \frac{2 + \frac{t}{4}}{\sqrt{1 + \frac{1}{16} + 4}\;\sqrt{1 + t^2}} = \frac{8 + t}{9\sqrt{1 + t^2}}\cdot$$

31. $OB = 5, OA = 3$. So, $\dfrac{OA}{OB} = \dfrac{3}{5} = \dfrac{1.5}{2.5} = \dfrac{AD}{DB}\cdot$

$\therefore$ OD bisects $\angle AOB$. Let $\angle BOC = \theta$. Then $\tan\theta = \frac{3}{4}\cdot$

$\angle BOD = \frac{1}{2}\angle AOB = \frac{1}{2}(90° - \theta)$.

$\therefore$ the slope of OB $= \tan\frac{1}{2}(90° - \theta) = \dfrac{\tan 45° - \tan\frac{\theta}{2}}{1 + \tan 45°\cdot\tan\frac{\theta}{2}}\cdot$

Now, $\tan\theta = \frac{3}{4} \quad \Rightarrow \quad \dfrac{2\tan\frac{\theta}{2}}{1 - \tan^2\frac{\theta}{2}} = \dfrac{3}{4}$

or $\quad 3 - 3\tan^2\frac{\theta}{2} = 8\tan\frac{\theta}{2} \quad$ or $\quad 3\tan^2\frac{\theta}{2} + 8\tan\frac{\theta}{2} - 3 = 0$

or $\quad \left(3\tan\frac{\theta}{2} - 1\right)\left(\tan\frac{\theta}{2} + 3\right) = 0. \quad$ So, $\tan\frac{\theta}{2} = \frac{1}{3} \qquad \left(\because \frac{\theta}{2} \text{ is acute}\right)\cdot$

$\therefore$ the slope of OB $= \dfrac{1 - \frac{1}{3}}{1 + \frac{1}{3}} = \dfrac{1}{2}$.

$\therefore$ the equation of OB is $y = \dfrac{1}{2}x$.

32. As $2 \le f(x) \le 6.5$ and $-1 \le \cos x \le 1$, we get $1 \le \cos x + f(x) \le 7.5$.

$\therefore \quad \dfrac{1}{\lambda} \le \dfrac{\cos x + f(x)}{\lambda} \le \dfrac{7.5}{\lambda}$.

Clearly, $h(x) = \left[\dfrac{\cos x + f(x)}{\lambda}\right]$ cannot be continuous if

$\left[\dfrac{1}{\lambda}, \dfrac{7.5}{\lambda}\right]$ contains any integer. So $\lambda = 8, 9, \ldots$.

33. Here, $\sin x = ax - 1$. The graphs of $y = \sin x$ and $y = ax - 1$ show that option (a) is correct.

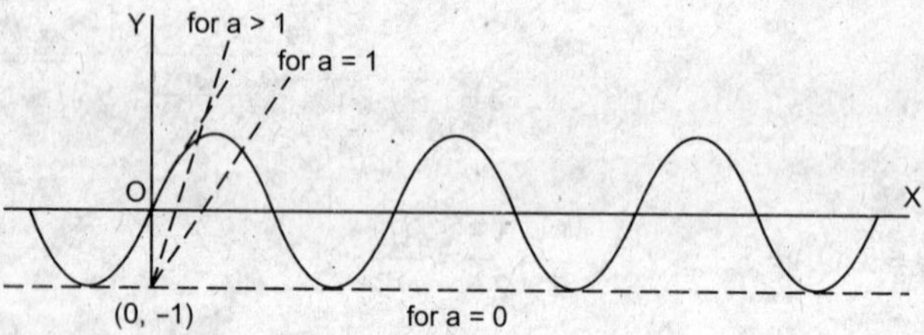

34. Here, $x^2 + y^2 = 9$. The point $(0, \lambda)$ is a point of the given region implies

$\lambda^2 - 9 \le 0 \quad$ or $\quad -3 \le \lambda \le 3$.

Roots of $x^2 - 2x - \lambda = 0$ are $1 \pm \sqrt{1 + \lambda}$.

$\therefore$ they are real for $\lambda \in [-1, 3]$.

If $\lambda = -1$, roots are 1 ± 0.

If $\lambda = 3$, roots are 1 ± 2, i.e., $3, -1$.

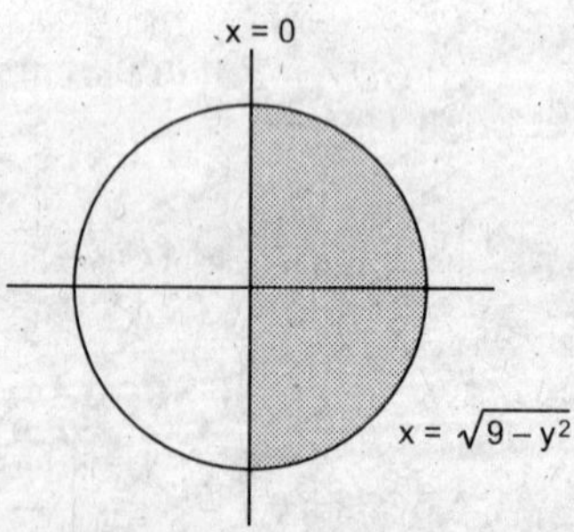

35. $P(E) + P(E') = 1 = 1 + \lambda + \lambda^2 + (1 + \lambda)^2$

or $\quad 2\lambda^2 + 3\lambda + 1 = 0 \quad$ or $\quad (2\lambda + 1)(\lambda + 1) = 0. \quad \therefore \lambda = -1, -\dfrac{1}{2}$.

Then $P(E) = 1 + (-1) + (-1)^2 = 1$ (not possible)

or $\quad P(E) = 1 + \left(-\frac{1}{2}\right) + \left(-\frac{1}{2}\right)^2 = \frac{1}{2} + \frac{1}{4} = \frac{3}{4}$.

36. E_5 = the event of making words of 5 letters

E_6 = the event of making words of 6 letters

$P(E_6) = \frac{4!\times 3!}{6!}$, arrangement being done by considering all the vowels together for calculating $n(E_6)$

$P(E_5) = \frac{{}^3C_2 \cdot 4! \cdot 2! + {}^3C_2 \cdot 3! \cdot 3!}{{}^6P_5}$ {∵ there are 2 vowels, 3 consonants or 3 vowels, 2 consonants}

∴ the required probability

$$= P(E_6) + P(E_5) = \frac{4!\cdot 3! + 3\cdot 4!\cdot 2! + 3\cdot 3!\cdot 3!}{6!}$$

$$= \frac{144 + 144 + 108}{720} = \frac{396}{720} = \frac{11}{20}.$$

37. Let C = the event of the selected number being composite,

E = the event of there being no remainder.

$P(C) = \frac{n(C)}{n(S)} = \frac{15}{25} = \frac{3}{5}$, $P(\overline{C}) = \frac{2}{5}$.

$P(E/C) = \frac{4}{15}$, $P(E/\overline{C}) = \frac{1}{10}$

$P(E) = P(C)\cdot P(E/C) + P(\overline{C})\cdot P(E/\overline{C})$

$$= \frac{3}{5}\cdot\frac{4}{15} + \frac{2}{5}\cdot\frac{1}{10} = \frac{4}{25} + \frac{1}{25} = \frac{1}{5} = 0.2.$$

38. Here, $x_r^2 = 8y_r$. As x_1, x_2, x_3 are in GP x_1^2, x_2^2, x_3^2 are also in GP.

∴ $\quad 8y_1, 8y_2, 8y_3$ are in GP. So y_1, y_2, y_3 are in GP.

39. As $f(x)$ is even, $f(-2) = f(2)$.

As $f(x)$ is m.i. in $[2, 6]$, $f(3) < f(5) = f(-5)$.

Also, $f(3) < f(5) \quad \Rightarrow \quad f(-3) < f(5)$.

40. $\lim\limits_{x \to 2} \dfrac{f(x) - g(x)}{x - 2} = \lim\limits_{x \to 2} \dfrac{f'(x) - g'(x)}{1}$ (by L 'Hospital's rule)

$$= f'(2) - g'(2) = a - 2a = -a.$$

41. $|\vec{b} \times \vec{c}| = 2 = |\vec{b}|\,|\vec{c}| \sin\theta$

or $\quad 2 = 4 \cdot 1 \cdot \sin\theta. \quad \therefore \; \theta = 30°.$

In ΔACD, $AC = |\vec{c}| = 1$, $AD = |2\vec{b}| = 2|\vec{b}| = 8$

$CD = AB = |\lambda\vec{a}| = \lambda|\vec{a}| = \lambda.$

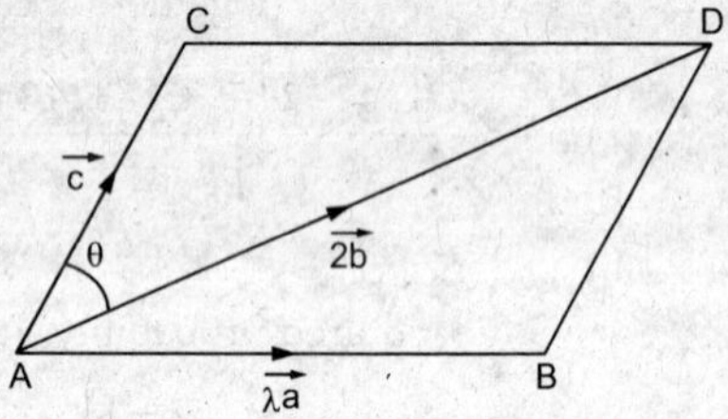

$\therefore \quad \cos\theta = \dfrac{AC^2 + AD^2 - CD^2}{2 \cdot AC \cdot AD}$

gives $\quad \dfrac{\sqrt{3}}{2} = \dfrac{1^2 + 8^2 - \lambda^2}{2 \cdot 1 \cdot 8}$

or $\quad 8\sqrt{3} = 65 - \lambda^2$

or $\quad \lambda = \pm\sqrt{65 - 8\sqrt{3}}.$

42. $f\left(\dfrac{\pi}{2}\right) = \displaystyle\int_0^{\pi/2} \sin[2x]\,dx = \int_0^{1/2} \sin[2x]\,dx + \int_{1/2}^{1} \sin[2x]\,dx + \int_1^{3/2} \sin[2x]\,dx + \int_{3/2}^{\pi/2} \sin[2x]\,dx$

$$= \int_0^{1/2} \sin 0\,dx + \int_{1/2}^{1} \sin 1\,dx + \int_1^{3/2} \sin 2\,dx + \int_{3/2}^{\pi/2} \sin 3\,dx$$

$$= (\sin 1) \cdot \frac{1}{2} + (\sin 2) \cdot \frac{1}{2} + (\sin 3) \cdot \left(\frac{\pi}{2} - \frac{3}{2}\right)$$

$$= \frac{1}{2}\{\sin 1 + \sin 2 + (\pi - 3)\sin 3\}.$$

43. Value $= \displaystyle\int_0^1 e^{x + [x]}\, d(x - [x]) + \int_1^2 e^{x + [x]}\, d(x - [x])$

$$= \int_0^1 e^x\,dx + \int_1^2 e^{x+1}\,d(x - 1) = \left[e^x\right]_0^1 + \int_1^2 e^{x+1}\,dx$$

$$= e - 1 + \left[e^{x+1}\right]_1^2 = e - 1 + e^3 - e^2 = (e^2 + 1)(e - 1).$$

44. $x < 1 - x$ if $x < \dfrac{1}{2}$ and $x \geq 1 - x$ if $x \geq \dfrac{1}{2}\cdot$

$\therefore \quad f(x) = x \quad \text{if } x < \frac{1}{2}$

$1 - x \text{ if } x \geq \frac{1}{2}.$

$$\therefore \quad \int_0^2 f(x)\, dx = \int_0^{1/2} f(x)\, dx + \int_{1/2}^2 f(x)\, dx$$

$$= \int_0^{1/2} x\, dx + \int_{1/2}^2 (1-x)\, dx = \left[\frac{x^2}{2}\right]_0^{1/2} + \left[x - \frac{x^2}{2}\right]_{1/2}^2$$

$$= \frac{1}{2}\left\{\frac{1}{4} - 0\right\} + \left\{\left(2 - \frac{4}{2}\right) - \left(\frac{1}{2} - \frac{1}{2}\cdot\frac{1}{4}\right)\right\}$$

$$= \frac{1}{8} - \frac{3}{8} = -\frac{1}{4}.$$

45. The sum is 12 in the first three throws if they are (1, 5, 6) in any order or (2, 4, 6) in any order or (3, 4, 5) in any order.

$\therefore$ the required probability

$$= \frac{1}{6} \times \frac{1}{5} \times \frac{1}{4} \times 3\,! + \frac{1}{6} \times \frac{1}{5} \times \frac{1}{4} \times 3\,! + \frac{1}{6} \times \frac{1}{5} \times \frac{1}{4} \times 3\,! = \frac{3}{20}$$

(because after throwing 1, it cannot come up in the next throw, etc.).

46. $n(S)$ = the number of composite numbers in 1 to 30 = 19

$n(E)$ = the number of composite numbers when divided by 5 leaves a remainder

= 14

$\therefore$ the required probability $= \frac{14}{19}.$

47. $\cos B = \dfrac{c^2 + a^2 - b^2}{2ca}$ or $a^2 - 2c \cdot \cos B \cdot a + c^2 - b^2 = 0$

$\therefore \quad a_1 + a_2 = 2c \cdot \cos B,\ a_1 a_2 = c^2 - b^2.$

$$\therefore \quad (a_1 - a_2)^2 = (a_1 + a_2)^2 - 4a_1 a_2 = 4c^2 \cos^2 B - 4(c^2 - b^2)$$

$$= 4b^2 - 4c^2 \sin^2 B = 4(b^2 - c^2 \sin^2 B).$$

$$\therefore \quad |a_1 - a_2| = 2\sqrt{b^2 - c^2 \sin^2 B}.$$

48. Let $x = 3 + k, 0 < k < 1$. Then $y = f(x) = x - [x] = 3 + k - 3 = k = x - 3$

$\therefore \quad x = y + 3. \qquad \therefore \quad f^{-1}(y) = y + 3.$

$\therefore \quad f^{-1}(x) = x + 3.$

49. $4 \le 4 + 3\cos x \le 7$ for $x \in \left(-\frac{\pi}{2}, \frac{\pi}{2}\right)$.

$f(x) = [4 + 3\cos x]$ is discontinuous at those points where $4 + 3\cos x$ is an integer.

Now $4 + 3\cos x = 4$ if $\cos x = 0$. So, $x = \frac{-\pi}{2}, \frac{\pi}{2}$ (not possible).

$4 + 3\cos x = 5$ if $\cos x = \frac{1}{3}$. So, x has two values $\cos^{-1}\frac{1}{3}, -\cos^{-1}\frac{1}{3}$.

$4 + 3\cos x = 6$ if $\cos x = \frac{2}{3}$. So, x has two values $\cos^{-1}\frac{2}{3}, -\cos^{-1}\frac{2}{3}$.

$4 + 3\cos x = 7$ if $\cos x = 1$. So, $x = 0$.

$\therefore$ the number of values of $x = 2 + 2 + 1 = 5$.

50. Here, $f(x) = 2x - x^2, x \le 1$.

$\therefore$ if $x \le 1$, $f(f(x)) = f(2x - x^2) = 2(2x - x^2) - (2x - x^2)^2$ when $2x - x^2 \le 1$

$= (2x - x^2)(2 - 2x + x^2)$ when $x^2 - 2x + 1 \ge 0$

$= x(2 - x)(2 - 2x + x^2)$ for all x when $x \le 1$.

$\therefore$ here, $x(2 - x)(2 - 2x + x^2) = x$ when $x \le 1$.

$\therefore$ when $x \le 1$, $x\{(2 - x)(2 - 2x + x^2) - 1\} = 0$

or $x\{-x^3 + 4x^2 - 6x + 3\} = 0$ or $x(x - 1)(-x^2 + 3x - 3) = 0$.

$\therefore$ $x = 0, 1, \frac{-3 \pm \sqrt{9 - 12}}{-2}$. Thus $x = 0, 1$ satisfy $x \le 1$.

51. $x - y \le -1, y \le x$. Adding, $x \le x - 1$ or $1 \le 0$ (absurd).

So, there is no value of x.

52. $p^x = q^w = a \Rightarrow p = a^{1/x}, q = a^{1/w}$.

$q^y = p^z = b \Rightarrow p = b^{1/z}, q = b^{1/y}$.

$\therefore$ $p = a^{1/x} = b^{1/z}$ and $q = a^{1/w} = b^{1/y}$. So, $b = a^{z/x} = a^{y/w} \Rightarrow \frac{z}{x} = \frac{y}{w}$.

53. $\sin\theta = \frac{1}{2}$ and $\cos\theta = \frac{-\sqrt{3}}{2}$. So, θ must belong to the second quadrant.

$\therefore$ $\theta = 2n\pi + \left(\pi - \frac{\pi}{6}\right)$.

54. Let $x+1=X, y=Y$. So, $(-1, 0)_{x, y}=(0, 0)_{x, y}$.

Changed equations, when $(-1, 0)$ is taken as the origin, are

$$3(X-1)^2+5(X-1)Y-3Y^2+8(X-1)+8Y+5=0$$

and $3(X-1)-2Y+2=0$

i.e., $3X^2+5XY-3Y^2+2X+3Y=0$ and $3X-2Y=1$.

Making the first equation homogeneous of the second degree with the help of the second equation,

$$3X^2+5XY-3Y^2+(2X+3Y)(3X-2Y)=0$$

or $\quad 9X^2+10XY-9Y^2=0.$ ... (1)

$\therefore$ the required angle = the angle between the pair of lines given by (1)

$$=\frac{\pi}{2} \quad (\because \ a+b=9-9=0).$$

55. Here, centre $=(q\cos\alpha, q\sin\alpha)$ and radius $=\sqrt{q^2\cos^2\alpha+q^2\sin^2\alpha}=q$.

$\therefore$ the line touches the circle if $q=\left|\dfrac{q\cos\alpha\cdot\cos\alpha+q\sin\alpha\cdot\sin\alpha-p}{\sqrt{\cos^2\alpha+\sin^2\alpha}}\right|$

or $\quad q=|q-p| \quad$ or $\quad q-p=\pm q$

$\therefore \quad p=0 \quad$ or $\quad p=2q.$

56. $(fg)(x)=f(g(x))=f(x^n)=(x^n)^m=x^{nm}$

$(gf)(x)=g(f(x))=g(x^m)=(x^m)^n=x^{mn}$. So, (*d*) is correct.

57. Here, $2=a\cdot0^2-6\cdot0+b. \quad \therefore \ b=2.$

Also, $\dfrac{dy}{dx}=2ax-6. \quad \therefore \ \left(\dfrac{dy}{dx}\right)_{x=\frac{3}{2}}=2a\cdot\dfrac{3}{2}-6=3a-6=\tan 0°=0.$

$\therefore \quad a=2.$

58. $f(x)^2=1+\sqrt{x}. \quad \therefore \ 2f(x)\cdot f'(x)=0+\dfrac{1}{2}\cdot\dfrac{1}{\sqrt{x}}\cdot$

59. Here, $\tan A=-\dfrac{3}{4}\cdot \quad \therefore \ \sin A=\dfrac{3}{5}$ because $0°<A<180°$.

$\therefore \ 1+\sin A=\dfrac{8}{5}$ or $\left(\sin\dfrac{A}{2}+\cos\dfrac{A}{2}\right)^2=\dfrac{8}{5}\cdot \quad \therefore \ \sin\dfrac{A}{2}+\cos\dfrac{A}{2}=\sqrt{\dfrac{8}{5}}\cdot$

Again, $\sin A=\dfrac{3}{5}$ or $\sin\dfrac{A}{2}\cdot\cos\dfrac{A}{2}=\dfrac{3}{10}\cdot$

$\therefore$ the required equation is

$$x^2 - \sqrt{\frac{8}{5}}\,x + \frac{3}{10} = 0 \quad \text{or} \quad 10x^2 - 4\sqrt{10}x + 3 = 0.$$

60. $(\sin^2\theta + \cos^2\theta)(\sin^4\theta - \sin^2\theta \cdot \cos^2\theta + \cos^4\theta) + k\cos^2 2\theta = 1$

or $\sin^4\theta + \cos^4\theta - \sin^2\theta \cdot \cos^2\theta + k\cos^2 2\theta = 1$

or $(\sin^2\theta + \cos^2\theta)^2 - 3\sin^2\theta \cdot \cos^2\theta + k\cos^2 2\theta = 1$

or $3\sin^2\theta \cdot \cos^2\theta = k\cos^2 2\theta$ or $\frac{3}{4}\sin^2 2\theta = k\cos^2 2\theta.$

$\therefore$ $k = \frac{3}{4}\tan^2 2\theta.$

61. Let α be the other root. Here, $a \cdot 2^2 + b \cdot 2 + c = 0$ and $2b = a + c$.

$\therefore$ $5a + 2c = 0.$ $\therefore$ $\frac{c}{a} = \frac{-5}{2}.$

Now, $2 \times \alpha = \frac{c}{a} = -\frac{5}{2}.$ $\therefore$ $\alpha = \frac{-5}{4}.$

62. Clearly, $[\cos x] = 0, 1, -1$.

If $[\cos x] = -1$, the equation is $-1 + |\sin x| = 1 \Rightarrow |\sin x| = 2$ (absurd).

If $[\cos x] = 0$, the equation is $|\sin x| = 1$.

This is possible only when $x = \frac{3\pi}{2}, \frac{5\pi}{2}.$

If $[\cos x] = 1$, the equation is $|\sin x| = 0$. Hence, $x = 2\pi$.

63. Here, $\frac{k}{a} + \frac{2k}{b} = 1$ or $\frac{1}{a} = \frac{1}{k} - \frac{2}{b} = \frac{b - 2k}{bk}.$

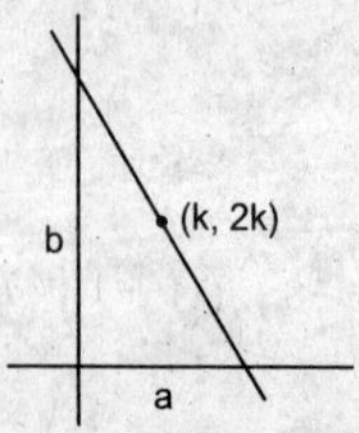

Now, $A = \frac{1}{2}ab = \frac{1}{2} \cdot \frac{bk}{b - 2k} \cdot b.$

$\therefore$ $\frac{dA}{db} = \frac{k}{2} \cdot \frac{2b(b - 2k) - b^2 \cdot 1}{(b - 2k)^2} = \frac{k}{2} \cdot \frac{b^2 - 4bk}{(b - 2k)^2}.$

$\therefore$ $\frac{dA}{db} = 0 \Rightarrow b = 0, 4k$. As no question of maximum arises (max. $= \infty$),

A is minimum when $b = 4k$.

$\therefore$ $A_{\text{least}} = \frac{k}{2} \cdot \frac{(4k)^2}{4k - 2k} = \frac{k}{2} \cdot \frac{16k^2}{2k} = 4k^2.$

64. $\cos\frac{\pi}{x} > 0 \quad \Rightarrow \quad \frac{\pi}{x} \in \left(2n\pi - \frac{\pi}{2}, 2n\pi + \frac{\pi}{2}\right)$.

$\therefore \quad 2n\pi - \frac{\pi}{2} < \frac{\pi}{x} < 2n\pi + \frac{\pi}{2} \quad$ or $\quad 2n - \frac{1}{2} < \frac{1}{x} < 2n + \frac{1}{2}$

or $\quad \frac{4n-1}{2} < \frac{1}{x} < \frac{4n+1}{2}$. If $x > 0$ then

$\frac{2}{4n+1} < x < \frac{2}{4n-1}$ when $n \geq 1$ and n is an integer.

When $n = 1$, $\frac{2}{5} < x < \frac{2}{3}$.

65. When $\theta = \frac{\pi}{4}$, $\cos\left(\theta - \frac{\pi}{4}\right) = \cos 0 = 1 =$ maximum value.

$0 < \theta < \frac{\pi}{2} \quad \Rightarrow \quad -\frac{\pi}{4} < \theta - \frac{\pi}{4} < \frac{\pi}{4}$. So, minimum $\cos\left(\theta - \frac{\pi}{4}\right) > \cos\frac{\pm\pi}{4} = \frac{1}{\sqrt{2}}$.

$\therefore \quad \frac{1}{\sqrt{2}} < \cos\left(\theta - \frac{\pi}{4}\right) \leq 1.$

66.

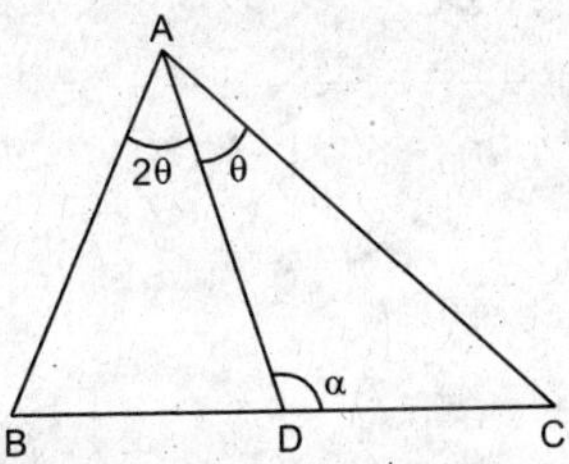

$\frac{A}{3} = \angle CAD = \theta.$

Now, $(1+1)\cot\alpha = 1 \cdot \cot 2\theta - 1 \cdot \cot\theta$

or $\quad 2\cot(B + 2\theta) = \cot 2\theta - \cot\theta$

or $\quad \cot(B + 2\theta) + \cot\theta = \cot 2\theta - \cot(B + 2\theta)$

or $\quad \frac{\sin(B+3\theta)}{\sin(B+2\theta)\cdot\sin\theta} = \frac{\sin B}{\sin(B+2\theta)\cdot\sin 2\theta}$

or $\quad \frac{\sin(B+A)}{\sin\theta} = \frac{\sin B}{\sin 2\theta} \quad$ or $\quad \sin C = \frac{\sin B}{2\cos\theta} \quad$ or $\quad \cos\frac{A}{3} = \frac{\sin B}{2\sin C}$.

67. Let two vertices be $(x_1, y_1), (x_2, y_2)$ and the orthocentre $= (\alpha, \beta)$, all the coordinates being integers.

Let (x_3, y_3) be the third vertex. Then

$$\frac{y_2 - y_1}{x_2 - x_1} \cdot \frac{\beta - y_3}{\alpha - x_3} = -1 \text{ which is of the form } ax_3 + by_3 + c = 0 \quad \ldots (1)$$

where a, b, c are integers.

Also, $$\frac{y_2 - y_3}{x_2 - x_3} \cdot \frac{y_1 - \beta}{x_1 - \alpha} = -1 \text{ which is of the form } a'x_3 + b'y_3 + c' = 0 \quad \ldots (2)$$

where a', b', c' are integers.

The solution of (1), (2) must give rational values of x_3, y_3.

68. A rough sketch of the lines is given below.

There are 3 triangles, namely, ABC, ABD, ACD.

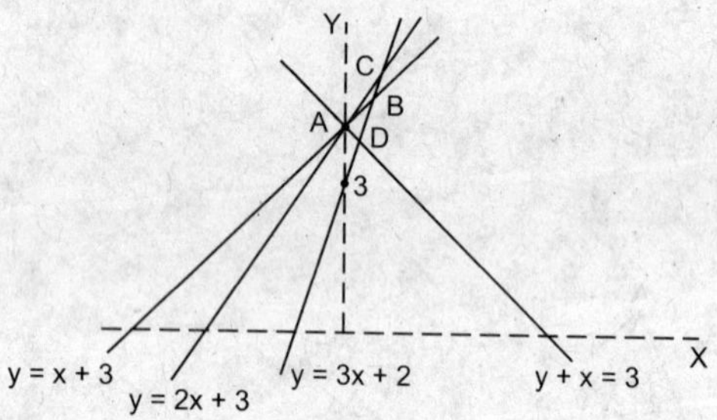

69.

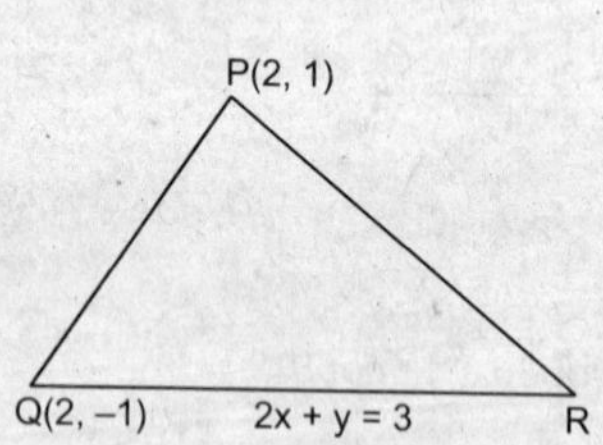

The equation of PR is $y = \lambda$ (being perpendicular to PQ whose slope $= \infty$).

But it passes through P(2, 1). So $\lambda = 1$.

$\therefore$ PR is $y = 1$. Hence, $R = (1, 1)$.

$\therefore$ centroid $= \left(\frac{2+2+1}{3}, \frac{1-1+1}{3}\right) = \left(\frac{5}{3}, \frac{1}{3}\right)$.

70. $\log_e(1+x+x^2) = \log_e \dfrac{1-x^3}{1-x} = \log_e(1-x^3) - \log_e(1-x)$

$$= \left\{-x^3 - \frac{1}{2}(x^3)^2 - \frac{1}{3}(x^3)^3 - \ldots\right\} - \left\{-x - \frac{1}{2}x^2 - \frac{1}{3}x^3 - \ldots\right\}.$$

Let $n = 3m$, the coefficient of x^n = coefficient of x^{3m}

$$= -\frac{1}{m} + \frac{1}{3m} = \frac{-2}{3m} = \frac{-2}{n}.$$

71. The matrix can be of the order

$24 \times 1, 1 \times 24, 12 \times 2, 2 \times 12, 3 \times 8, 8 \times 3, 4 \times 6, 6 \times 4.$

72. $(1+x)^m \cdot (1-x)^n = ({}^mC_0 + {}^mC_1 x + {}^mC_2 x^2 + \ldots)({}^nC_0 - {}^nC_1 x + {}^nC_2 x^2 - \ldots).$

$\therefore$ coefficient of $x = {}^mC_1 \cdot {}^nC_0 - {}^mC_0 \, {}^nC_1 = m - n = 3$ ... (1)

coefficient of $x^2 = {}^mC_0 \cdot {}^nC_2 - {}^mC_1 \cdot {}^nC_1 + {}^mC_2 \cdot {}^nC_0$

$$= \frac{n(n-1)}{2} - mn + \frac{m(m-1)}{2} = \frac{m^2 + n^2 - m - n - 2mn}{2} = -6. \quad \ldots (2)$$

$\therefore$ $(m-n)^2 - (m+n) = -12.$

$\therefore$ using (1), $9 - (m+n) = -12$ or $m + n = 21.$ ... (3)

(1) and (3) $\Rightarrow$ $m = 12, n = 9.$

73. $(1 + x^3 - x^6)^{30} = \{1 + x^3(1 - x^3)\}^{30}$

$$= {}^{30}C_0 + {}^{30}C_1 x^3(1-x^3) + {}^{30}C_2 x^6(1-x^3)^2 + \ldots.$$

Obviously, each term will contain x^{3m}, $m \in N$. But 28 is not divisible by 3.

$\therefore$ there will be no term containing x^{28}.

74. $t_{r+1} = (-1)^r \cdot (n - r + 2)\ {}^nC_r \cdot 2^{n-r+1}$

$$= (n+2) \cdot 2^{n+1} \cdot (-1)^r \cdot {}^nC_r \cdot \left(\frac{1}{2}\right)^r - 2^{n+1} \cdot (-1)^r \cdot r \cdot {}^nC_r \cdot \left(\frac{1}{2}\right)^r$$

$$= (n+2) \cdot 2^{n+1} \cdot (-1)^r \cdot {}^nC_r \cdot \left(\frac{1}{2}\right)^r - 2^n \cdot n \cdot (-1)^r \cdot {}^{n-1}C_{r-1}\left(\frac{1}{2}\right)^{r-1}.$$

$\therefore$ $\text{sum} = (n+2) \cdot 2^{n+1} \cdot \left\{{}^nC_0 - {}^nC_1 \cdot \frac{1}{2} + {}^nC_2 \cdot \left(\frac{1}{2}\right)^2 - \ldots\right\}$

$$- n \cdot 2^n \left\{-{}^{n-1}C_0 + {}^{n-1}C_1 \cdot \frac{1}{2} - {}^{n-1}C_2 \cdot \left(\frac{1}{2}\right)^2 + \ldots\right\}$$

$$= (n+2)\cdot 2^{n+1}\cdot\left(1-\frac{1}{2}\right)^{n} + n\cdot 2^{n}\cdot\left(1-\frac{1}{2}\right)^{n-1}$$

$$= 2(n+2) + 2n = 4n + 4.$$

75. $u_1u_n - u_{n-1} = 2\cos\theta\cdot 2\cos n\theta - 2\cos(n-1)\theta$

$$= 2\{\cos(n-1)\theta + \cos(n+1)\theta\} - 2\cos(n-1)\theta = 2\cos(n+1)\theta = u_{n+1}.$$

76. Value $= \tan 9^\circ + \tan 81^\circ - (\tan 27^\circ + \tan 63^\circ)$

$$= \frac{\sin 90^\circ}{\cos 9^\circ\cdot\cos 81^\circ} - \frac{\sin 90^\circ}{\cos 27^\circ\cdot\cos 63^\circ}$$

$$= \frac{1}{\cos 9^\circ\cdot\sin 9^\circ} - \frac{1}{\cos 27^\circ\cdot\sin 27^\circ}$$

$$= \frac{2}{\sin 18^\circ} - \frac{2}{\sin 54^\circ} = 2\cdot\frac{\sin 54^\circ - \sin 18^\circ}{\sin 18^\circ\cdot\cos 36^\circ}$$

$$= 2\cdot\frac{2\cos 36^\circ\cdot\sin 18^\circ}{\sin 18^\circ\cdot\cos 36^\circ} = 4.$$

77. $\tan\frac{10\pi}{3} = \tan\left(3\pi+\frac{\pi}{3}\right) = \tan\frac{\pi}{3} = \sqrt{3}$, $\tan\frac{9\pi}{4} = \tan\left(2\pi+\frac{\pi}{4}\right) = \tan\frac{\pi}{4} = 1$,

$\tan\frac{7\pi}{6} = \tan\left(\pi+\frac{\pi}{6}\right) = \tan\frac{\pi}{6} = \frac{1}{\sqrt{3}}$. Clearly, $\sqrt{3}, 1, \frac{1}{\sqrt{3}}$ are in GP.

78. Clearly, $x = 0, \frac{\pi}{2}, 2\pi$ satisfy the equation.

For other values, $\cos^7 x < \cos^2 x$ or $\sin^5 x < \sin^2 x$.

But $\cos^2 x + \sin^2 x = 1$ and so $\cos^7 x + \sin^5 x \neq 1$.

79. $t_n = \cot^{-1}(1+n+n^2) = \tan^{-1}\frac{1}{1+n(n+1)}$

$$= \tan^{-1}\frac{(n+1)-n}{1+(n+1)n} = \tan^{-1}(n+1) - \tan^{-1}n.$$

$\therefore$ sum $= t_1 + t_2 + t_3 + t_4 + t_5$

$$= (\tan^{-1}2 - \tan^{-1}1) + (\tan^{-1}3 - \tan^{-1}2) + \ldots + (\tan^{-1}6 - \tan^{-1}5)$$

$$= \tan^{-1}6 - \tan^{-1}1 = \tan^{-1}\frac{6-1}{1+6\cdot 1} = \tan^{-1}\frac{5}{7} = \cot^{-1}\frac{7}{5}.$$

80. Here, $\cos x = x - \frac{1}{2}\cdot$ Consider the graphs of $y = \cos x$ and $y = x - \frac{1}{2}\cdot$

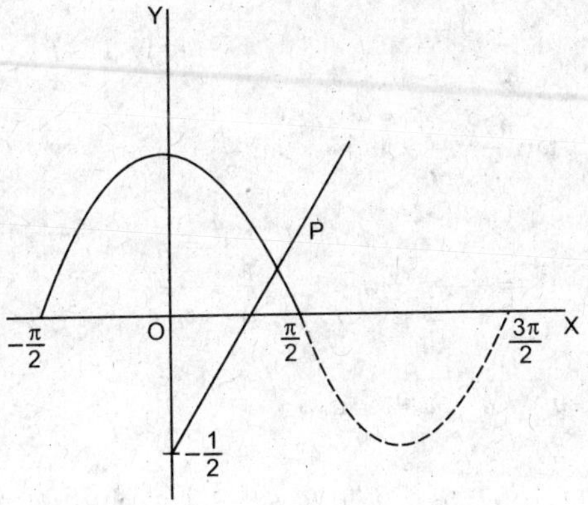

Clearly, they cut at P whose x-coordinate lies in $\left[0, \frac{\pi}{2}\right]\cdot$

81. $\frac{2\log 2}{\log 2}, \frac{\log 4}{\log(3^x - 4)}, \frac{\log 4}{\log\left(3^x + \frac{7}{2}\right)}$ are in HP

$\therefore$ $\log 2, \log(3^x - 4), \log\left(3^x + \frac{7}{2}\right)$ are in AP.

$\therefore$ $2\log(3^x - 4) = \log 2 + \log\left(3^x + \frac{7}{2}\right)$

or $(3^x - 4)^2 = 2\left(3^x + \frac{7}{2}\right)$ or $(3^x)^2 - 10\cdot 3^x + 9 = 0$

or $(3^x - 9)(3^x - 1) = 0$. $\therefore$ $x = 2, 0$.

But $3^x - 4 = 1 - 4 = -3 < 0$ when $x = 0$. So, $x = 2$.

82. $t_r = \tan^{-1}\frac{2r}{2 + r^2 + r^4} = \tan^{-1}\frac{2r}{1 + (r^4 + r + 1)} = \tan^{-1}\frac{2r}{1 + (r^2 + 1)^2 - r^2}$

$= \tan^{-1}\frac{(r^2 + r + 1) - (r^2 - r + 1)}{1 + (r^2 + r + 1)(r^2 - r + 1)} = \tan^{-1}(r^2 + r + 1) - \tan^{-1}(r^2 - r + 1)$

$t_{r+1} = \tan^{-1}\{(r + 1)^2 + r + 1 + 1\} - \tan^{-1}\{(r + 1)^2 - (r + 1) + 1\}$

$= \tan^{-1}(r^2 + 3r + 3) - \tan^{-1}(r^2 + r + 1)$

$$\therefore \quad \sum_{1}^{n} t_n = \tan^{-1}(n^2+n+1) - \tan^{-1}1 = \tan^{-1}\frac{n^2+n}{2+n+n^2}$$

$$\to \tan^{-1}1 \text{ as } n \to \infty.$$

83. Here, $\tan\frac{P}{2} + \tan\frac{Q}{2} = \frac{-b}{a}$, $\tan\frac{P}{2}\cdot\tan\frac{Q}{2} = \frac{c}{a}$.

$$\therefore \quad \tan\left(\frac{P}{2}+\frac{Q}{2}\right) = \frac{-\frac{b}{a}}{1-\frac{c}{a}} \quad \text{or} \quad \tan\frac{\pi}{4} = \frac{-b}{a-c}.$$

$$\therefore \quad a-c=-b \quad \text{or} \quad a+b=c.$$

84. $\frac{dx}{dt} = f''(t)\sin t + f'(t)\cos t + f'''(t)\cos t - f''(t)\sin t$

$= \{f'(t) + f'''(t)\}\cos t$

$\frac{dy}{dt} = f''(t)\cos t - f'(t)\sin t - f'''(t)\sin t - f''(t)\cos t$

$= -\{f'(t)+f'''(t)\}\sin t$

$$\therefore \quad \text{velocity} = \sqrt{\left(\frac{dx}{dt}\right)^2 + \left(\frac{dy}{dt}\right)^2} = f'(t) + f'''(t).$$

85. Here, $f(x) = \cos^{-1}\frac{1-\frac{1}{x^2}}{1+\frac{1}{x^2}}$. At $x=-2$, $\frac{1}{x^2} < 1$.

So, $f(x) = 2\tan^{-1}\frac{1}{x}$. $\quad \therefore \quad f'(x) = 2\cdot\frac{1}{1+\frac{1}{x^2}}\cdot\frac{-1}{x^2}$.

$$\therefore \quad f'(-2) = \frac{-2}{\left(1+\frac{1}{4}\right)\cdot 4} = \frac{-2}{\frac{5}{4}\cdot 4} = \frac{-2}{5}.$$

86. $\tan y = \frac{(\sqrt{1+x}-\sqrt{1-x})^2}{1+x-(1-x)} = \frac{1+x+1-x-2\sqrt{1-x^2}}{2x} = \frac{1-\sqrt{1-x^2}}{x}$.

$$\therefore \quad y = \tan^{-1}\frac{1-\sqrt{1-x^2}}{x}.$$

Put $x = \sin 2\theta$. Then

$$y = \tan^{-1} \frac{1 - \cos 2\theta}{\sin 2\theta} = \tan^{-1} \frac{2\sin^2\theta}{2\sin\theta \cdot \cos\theta} = \tan^{-1} \tan\theta = \theta.$$

$$\therefore \quad \frac{dy}{dx} = \frac{dy/d\theta}{dx/d\theta} = \frac{1}{2\cos 2\theta} = \frac{1}{2\sqrt{1-x^2}}.$$

$$\therefore \quad \text{at } x = \frac{1}{2}, \frac{dy}{dx} = \frac{1}{2 \cdot \sqrt{1 - \frac{1}{4}}} = \frac{1}{\sqrt{3}}.$$

87. Here, $3x^3 + 5x - 4 = A(x-1)(x^2-4) + B(x+1)(x^2-4)$

$$+ C(x^2-1)(x-2) + D(x^2-1)(x+2).$$

Put $x = -1, 1, -2, 2$ respectively. Then

$$-3 - 5 - 4 = A(-2)(-3), \quad \text{i.e.,} \quad A = -2$$

$$3 + 5 - 4 = B(2)(-3), \quad \text{i.e.,} \quad B = -\frac{2}{3}$$

$$-24 - 10 - 4 = C(3)(-4), \quad \text{i.e.,} \quad C = \frac{19}{6}$$

$$24 + 10 - 4 = D(3)(4), \quad \text{i.e.,} \quad D = \frac{5}{2}$$

$$\therefore \quad A + B + C + D = -2 - \frac{2}{3} + \frac{19}{6} + \frac{5}{2} = 3.$$

88. If the line intersects the plane then for some λ,

$$\{\vec{i} + \lambda(2\vec{i} + 5\vec{j} + 3\vec{k})\} \cdot (2\vec{i} + \vec{j} - 3\vec{k}) = 5$$

or $\quad 2 + 4\lambda + 5\lambda - 9\lambda = 5$, which is false.

Hence, the line is parallel to the plane.

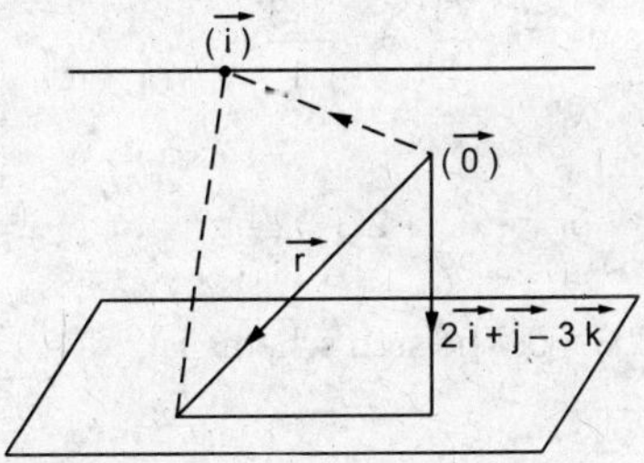

The distance = projection of $(\vec{r} - \vec{i})$ on $\dfrac{2\vec{i} + \vec{j} - 3\vec{k}}{\sqrt{14}}$

$$= \frac{\vec{r} \cdot (2\vec{i} + \vec{j} - 3\vec{k})}{\sqrt{14}} - \frac{2}{\sqrt{14}} = \frac{5}{\sqrt{14}} - \frac{2}{\sqrt{14}} = \frac{3}{\sqrt{14}}.$$

89. There should be 5 red and 5 green balls on each side of the white ball for symmetry. Also, corresponding to any arrangement on one side, the number of arrangement on the other side is fixed.

$\therefore$ the required number of ways = the number of arrangements of 5 red and 5 green balls

$$= \frac{10!}{5!\,5!}.$$

90. 

The first stripe can be coloured in 6 ways.

The next in 5 ways ($\because$ the colour used for the first stripe cannot be used for the second).

The next in 5 ways also and so on.

$\therefore$ the required number of ways $= 6 \times 5 \times 5 \times 5$.

91.

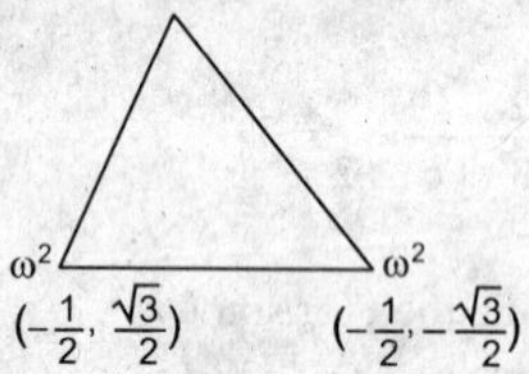

A side $= |\omega - \omega^2| = \sqrt{\left(-\frac{1}{2} + \frac{1}{2}\right)^2 + \left(\frac{\sqrt{3}}{2} + \frac{\sqrt{3}}{2}\right)^2} = \sqrt{3}$

$\therefore$ area $= \dfrac{\sqrt{3}}{4}(\text{side})^2 = \dfrac{\sqrt{3}}{4} \cdot 3 = \dfrac{3\sqrt{3}}{4}$.

92. $x + (2x + 1)z = 2$ or $x(1 + 2z) = 2 - z$ or $x = \dfrac{2 - z}{1 + 2z} > 0$

or $(2 - z)(1 + 2z) > 0$. The sign scheme is

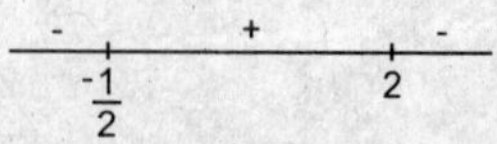

$\therefore$ $-\dfrac{1}{2} < z < 2$.

93. $\Delta = \begin{vmatrix} i^2 & i^3 & i^4 \\ i^3 & i^4 & i^5 \\ i^4 & i^5 & i^6 \end{vmatrix} = i^2 \cdot i^3 \cdot i^4 \begin{vmatrix} 1 & i & i^2 \\ 1 & i & i^2 \\ 1 & i & i^2 \end{vmatrix} = 0.$

94. Clearly, $x + \sqrt{2}x + x = 1$. So, $x = \dfrac{1}{2+\sqrt{2}}$.

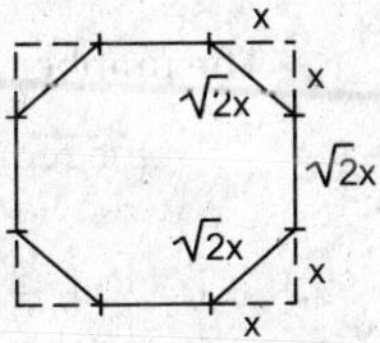

The required area $= \left(1^2 - 4 \times \frac{1}{2} x^2\right)$ metre^2

$= \left\{1 - 2 \cdot \dfrac{1}{(2+\sqrt{2})^2}\right\}$ metre^2

$= \left\{1 - \dfrac{1}{(\sqrt{2}+1)^2}\right\}$ metre^2 $= \{1 - (\sqrt{2}-1)^2\}$ metre^2

$= 2(\sqrt{2}-1)$ metre^2.

95. $t_n = \dfrac{3 + 2 + 2^2 + 2^3 + \dots \text{ to } n \text{ terms}}{2^{n-1}} = \dfrac{2 + \dfrac{2^n - 1}{2-1}}{2^{n-1}} = \dfrac{2^n + 1}{2^{n-1}} = 2 + \dfrac{1}{2^{n-1}}.$

$\therefore \quad \Sigma t_n = 2\Sigma 1 + \Sigma \dfrac{1}{2^{n-1}} = 2n + \left(1 + \dfrac{1}{2} + \dfrac{1}{2^2} + \dots \text{ to } n \text{ terms}\right).$

$= 2n + \dfrac{1 - \left(\frac{1}{2}\right)^n}{1 - \frac{1}{2}} = 2n + 2\left\{1 - \dfrac{1}{2^n}\right\} = 2\left\{n + 1 - \dfrac{1}{2^n}\right\}.$

96. $|x-4| - 1 = \pm 2$ or $|x-4| = 3$ $\{\because\ |x-4|$ is non-negative$\}$.

$\therefore \quad x - 4 = \pm 3.$

97. Here, $\alpha < 1 < \beta$. Now, $\alpha - 1 < 0$ and $\beta - 1 > 0$.

$\therefore \quad (\alpha-1)(\beta-1) < 0$ or $\alpha\beta - (\alpha+\beta) + 1 < 0$

or $3 - a - a + 1 < 0$ or $2a > 4$. $\therefore\ a > 2$.

98. $x + y = 5, 10, 15, 20, 25, 30$ where $x \neq y$, $(x, y) = (y, x)$.

The number of solutions of $x + y = 5$ is $\dfrac{4}{2!}$.

The number of solutions of $x + y = 10$ is $\dfrac{9-1}{2!}$.

The number of solutions of $x + y = 15$ is $\dfrac{14}{2!}$.

The number of solutions of $x+y=20$ is $\frac{11-1}{2\,!}$ $\{\because\ x=5, 6, \ldots, 15\}$.

The number of solutions of $x+y=25$ is $\frac{6}{2\,!}$ $\{\because\ x=10, 11, \ldots, 15\}$.

The number of solutions of $x+y=30$ is 0 $\{\because\ x=y=15\}$.

$\therefore$ the required number of selections $=2+4+7+5+3+0=21$.

99. $|z|=\left|\left(z-\frac{2}{z}\right)+\frac{2}{z}\right|\le\left|z-\frac{2}{z}\right|+\left|\frac{2}{z}\right|=1+\frac{2}{|z|}$

or $|z|^2-|z|-2\le 0$ or $(|z|-2)(|z|+1)\le 0$.

$\therefore$ $|z|\le 2$. So, the greatest value of $|z|$ is 2.

100. Here, $\left|\frac{z_2-z_0}{z_0-z_1}\right|=1$ and amp $\frac{z_2-z_0}{z_0-z_1}=\frac{\pi}{2}$.

$\therefore$ $\frac{z_2-z_0}{z_0-z_1}=1\left\{\cos\frac{\pi}{2}+i\sin\frac{\pi}{2}\right\}=i$

or $z_2-z_0=iz_0-iz_1$ or $z_2+iz_1=(i+1)z_0$.

$\therefore$ $z_0=\frac{z_2+iz_1}{1+i}=\frac{(z_2+iz_1)(1-i)}{1^2+1^2}=\frac{1}{2}\{(i-i^2)z_1+(1-i)z_2\}$.

101. $3^{33}=3.3^{32}=3.9^{16}=3.(10-1)^{16}$

$=3\{{}^{16}C_0\cdot 10^{16}-{}^{16}C_1\cdot 10^{15}+\ldots-{}^{16}C_{15}\cdot 10+{}^{16}C_{16}\}$

$\therefore$ when 3^{33} is divided by 25, the remainder is $3(-{}^{16}C_{15}\cdot 10+{}^{16}C_{16})$,

i.e., $3(-160+1)$, i.e., -477, i.e., $-477+25\times 20$, i.e., 23.

$\therefore$ fractional part $=\frac{23}{25}$.

102. The total time taken in the first case $=\left(\frac{x}{a}+\frac{x}{b}\right)$ hours.

If the required speed is v km/h then the time taken in the second case

$=\frac{2x}{v}$ hours.

$\therefore$ $\frac{x}{a}+\frac{x}{b}=\frac{2x}{v}$ or $\frac{2}{v}=\frac{1}{a}+\frac{1}{b}$.

$\therefore$ v is the HM between a and b. So, $v=\frac{2ab}{a+b}$ km/h.

103. Let $f(x) = ax^2 + bx + c$. Here, $f(\alpha) = 0$ and $f(\beta) = 0$.

Also, $f'(x) = 2ax + b$. So, $f'(\gamma) = 0$.

By Lagrange's theorem, γ lies between α and β.

104. $f'(x) = 3x^2 - 3 = 3(x^2 - 1)$.

$\therefore \quad f'(x) \geq 0$ if $x \leq -1$ or $x \geq 1$

$f'(x) \leq 0$ if $-1 \leq x \leq 1$.

$\therefore$ in $[-1, 1]$ it is monotonically decreasing and in $[1, 3]$ it is monotonically increasing

$\therefore \quad \max f(x) =$ greatest among $\{f(-1), f(3)\} = f(3) = 18$.

$\min f(x) = f(1) = -2$. So, the range of $f = [-2, 18]$.

105. $\hat{a} \cdot \hat{b} = \cos \alpha, \hat{b} \cdot \hat{c} = \cos \beta, \hat{c} \cdot \hat{a} = \cos \gamma$.

Now, $|\hat{a} + \hat{b} + \hat{c}| = 1. \quad \therefore \quad (\hat{a} + \hat{b} + \hat{c})^2 = 1$

or $\quad \hat{a}^2 + \hat{b}^2 + \hat{c}^2 + 2(\hat{a} \cdot \hat{b} + \hat{b} \cdot \hat{c} + \hat{c} \cdot \hat{a}) = 1$

or $\quad 1 + 1 + 1 + 2(\cos \alpha + \cos \beta + \cos \gamma) = 1$.

106. The greatest value of each of $\cos^{-1} x, \cos^{-1} y, \cos^{-1} z$ is π.

$\therefore \quad \cos^{-1} x + \cos^{-1} y + \cos^{-1} z = 3\pi \Rightarrow \cos^{-1} x = \pi = \cos^{-1} y = \cos^{-1} z$.

$\therefore \quad x = y = z = -1$.

107. $f_n(x) = \dfrac{n - 1 + 2}{(n-1)!} x^n = \left(\dfrac{1}{(n-2)!} + \dfrac{2}{(n-1)!}\right) x^n$.

$$\therefore \quad \sum_{n=1}^{\infty} f_n(x) = \sum_{n=1}^{\infty} \frac{1}{(n-2)!} x^n + \sum_{n=1}^{\infty} \frac{2}{(n-1)!} x^n = x^2 \cdot e^x + 2xe^x.$$

$$\therefore \quad \int_0^1 \left(\sum_{n=1}^{\infty} f_n(x)\right) dx = \int_0^1 (x^2 + 2x)e^x \, dx$$

$$= \int_0^1 e^x \left\{x^2 + \frac{d}{dx}(x^2)\right\} dx = [e^x \cdot x^2]_0^1 = e.$$

108. Here, $b^2 = ac$, $2(b + 8) = a + c$, $(b + 8)^2 = a(c + 64)$.

$\therefore \quad b^2 + 16b + 64 = ac + 64a \quad$ or $\quad b + 4 = 4a. \quad \therefore \; b = 4a - 4$.

$\therefore \quad 2(b + 8) = a + c \Rightarrow 2(4a + 4) = a + c. \quad \therefore \; c = 7a + 8$.

$\therefore \quad b^2 = ac \Rightarrow (4a - 4)^2 = a(7a + 8)$

or $\quad 9a^2 - 40a + 16 = 0 \quad$ or $\quad (9a-4)(a-4) = 0.$

$\therefore \quad a = \frac{4}{9}$ or 4. But for $a = \frac{4}{9}$, $b < 0$.

So, $\quad \frac{a+b+c}{3} = \frac{a+4a-4+7a+8}{3} = 4a + \frac{4}{3}$.

$\therefore \quad \frac{a+b+c}{3} = 4{\cdot}4 + \frac{4}{3} = \frac{52}{3}$

109.

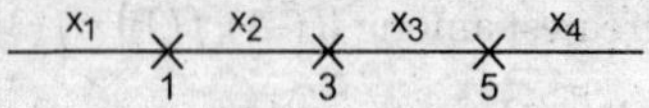

$x_1 + x_2 + x_3 + x_4 = 2$ where $x_i \geq 0$.

The number of solutions = the number of ways of placing 2, 4

$= \text{coefficient of } x^2 \text{ in } (x^0 + x + x^2 + \ldots)^4$

$= \text{coefficient of } x^2 \text{ in } (1-x)^{-4} = {}^5C_2.$

But 2, 4 can be arranged among themselves in 2 ! ways.

$\therefore$ the required number of arrangements $= {}^5C_2 \cdot (2\,!) = 20.$

110. Clearly, $x = 1$ satisfies it because $\frac{3}{5} + \frac{7}{5} = 2.$

If $x > 1$, $\left(\frac{3}{5}\right)^x < \frac{3}{5}$ and $2^x > 2$. So, $x > 1$ cannot satisfy it.

If $x < 1$, $\left(\frac{3}{5}\right)^x > \frac{3}{5}$ and $2^x < 2$. So $x < 1$ cannot satisfy it.

$\therefore \quad x = 1$ is the only solution.

111. $x = \sqrt[3]{2} + \sqrt[3]{4} + 2 \quad \Rightarrow \quad (x-2)^3 = 2 + 4 + 3 \cdot \sqrt[3]{2} \cdot \sqrt[3]{4}\,(\sqrt[3]{2} + \sqrt[3]{4})$

or $\quad (x-2)^3 = 6 + 6(x-2) \quad$ or $\quad x^3 - 6x^2 + 6x - 2 = 0.$

$\therefore \quad f(x) = x^3 - 6x^2 + 6x = 2.$

112.

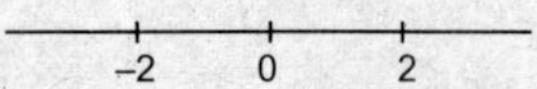

$f(x-3) = |x-3+1| = |x-2|,$

$f(x-1) = |x-1+1| = |x|$ and

$f(x+1) = |x+1+1| = |x+2|.$

These are in AP if $2|x| = |x-2| + |x+2|.$

If $x < -2$, $-2x = -(x-2) - (x+2)$

or $\quad -2x = -2x$. So x has all values. Hence, $x < -2$.

If $-2 \le x < 0, -2x = -(x-2) + (x+2)$. $\quad \therefore \; x = -2$.

If $0 \le x < 2, 2x = -(x-2) + (x+2)$. $\quad \therefore \; x = 2$ (absurd).

If $2 \le x, 2x = x - 2 + x + 2$. $\quad \therefore \; 2x = 2x$.

$\therefore$ x has all values. Hence, $x \ge 2$.

But $x \in [-2, 2]$. Hence, $x = -2, 2$. So, there are two values.

113. Let the number of newspapers be n.

The total number of times newspapers are read $= 300 \times 5 = n \times 60$.

$\therefore \quad n = 25$.

114. $(x+1)(x^{1/2}+1)(x^{(1/2)^2}+1) \ldots (x^{(1/2)^n}+1)(x^{(1/2)^n}-1) = (x^2-1)f(x) \cdot (x^{(1/2)^n}-1)$

or $\quad (x+1)(x^{1/2}+1)(x^{(1/2)^2}+1) \ldots (x^{(1/2)^{n-1}}+1) = (x^2-1)f(x) \cdot (x^{(1/2)^n}-1)$

or $\quad (x^2-1) = (x^2-1)f(x) \cdot (x^{(1/2)^n}-1)$. $\quad \therefore \; f(x) = \dfrac{1}{x^{(1/2)^n}-1}$.

115. The digit in the units place of $(1073)^{71}$ = the digit in the units place of 3^{71}

= the digit in the units place of $(3^4)^{17} \cdot 3^3$

= the digit in the units place of 3^3 $\quad (\because \; 3^4$ has 1 in the units place)

$= 7$.

$\therefore \quad x$ must be 7 so that $(1073)^{71} - x$ has 0 in the units place.

116. $|z-1| + |z|$ = the sum of the distances of z from $z = 1$ and $z = 0$.

$\therefore \quad |z-1| + |z|$ is minimum if z is collinear with $z = 1, z = 0$.

Then $|z-1| + |z|$ = the distance between $z = 1$ and $z = 0$

$= 1$.

117. $\cos C = \dfrac{a^2+b^2-c^2}{2ab} > 0$. $\quad \therefore \; 0° < C < 90°$.

118. $\tan A + \tan B = \lambda$, $\tan A \cdot \tan B = \mu$.

Also, $A + B = \dfrac{3\pi}{4}$.

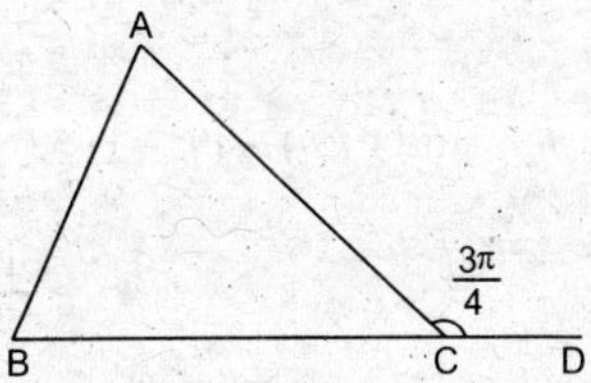

$\therefore \quad \tan \dfrac{3\pi}{4} = \tan(A+B) = \dfrac{\lambda}{1-\mu}$

or $\quad -1 = \dfrac{\lambda}{1-\mu}$. $\quad \therefore \; \mu - 1 = \lambda$.

119. $$\frac{\alpha-\gamma}{\beta-\gamma}\cdot\frac{\alpha-\delta}{\beta-\delta}=\frac{\alpha^2-\alpha(\gamma+\delta)+\gamma\delta}{\beta^2-\beta(\gamma+\delta)+\gamma\delta}=\frac{\alpha^2-\alpha\cdot(-p)-r}{\beta^2-\beta\cdot(-p)-r}$$

$$=\frac{\alpha^2+p\alpha-r}{\beta^2+p\beta-r}=\frac{-q-r}{-q-r}=1 \qquad (\because\ \alpha,\beta \text{ satisfy } x^2+px+q=0).$$

120. $\dfrac{d^2s}{dt^2}=f.\quad\therefore\quad \dfrac{ds}{dt}=\int f\,dt=ft+c.$

At $t=0$, $\dfrac{ds}{dt}=u.\quad\therefore\quad u=c.$

$\therefore\quad \dfrac{ds}{dt}=u+ft.\quad\therefore\quad \int_{t-1}^{t}ds=\int_{t-1}^{t}(u+ft)\,dt$

or $\quad s_t=u+\dfrac{f}{2}\{t^2-(t-1)^2\}=u+\dfrac{f}{2}(2t-1).$

$\therefore\quad s_2=u+\dfrac{3}{2}f,\ s_4=u+\dfrac{7}{2}f,\ s_6=u+\dfrac{11}{2}f.$

$\therefore\quad s_2, s_4, s_6$ are in AP.

121. $t_r=r({}^nC_r+{}^nP_r)$

$=r\cdot{}^nC_r+r\cdot{}^nP_r=n\cdot{}^{n-1}C_{r-1}+r\cdot(r\,!)$

$=n\cdot{}^{n-1}C_{r-1}+(r+1)\,!-r\,!.$

$\therefore\quad \text{sum}=n\{{}^{n-1}C_0+{}^{n-1}C_1+\ldots+{}^{n-1}C_{n-1}\}+\{(2\,!-1\,!)+(3\,!-2\,!)+\ldots+((n+1)\,!-n\,!)$

$=n\cdot2^{n-1}+(n+1)\,!-1\,!=n\cdot2^{n-1}-1+(n+1)\,!.$

122. Here, $a(n)-a(n-1)=\left(1+\dfrac{1}{2}+\dfrac{1}{3}+\ldots+\dfrac{1}{2^n-1}\right)-\left(1+\dfrac{1}{2}+\dfrac{1}{3}+\ldots+\dfrac{1}{2^{n-1}-1}\right)$

$$=\frac{1}{2^{n-1}}+\frac{1}{2^{n-1}+1}+\ldots+\frac{1}{2^n-1}.$$

$\therefore\quad a(n)-a(n-1)<\dfrac{1}{2^{n-1}}+\dfrac{1}{2^{n-1}}+\ldots+\dfrac{1}{2^{n-1}}=\dfrac{1}{2^{n-1}}(2^n-2^{n-1})=1$

and $\quad a(n)-a(n-1)>\dfrac{1}{2^n-1}+\dfrac{1}{2^n-1}+\ldots+\dfrac{1}{2^n-1}=\dfrac{1}{2^n-1}(2^n-2^{n-1})$

$$>\frac{1}{2^n}(2^n-2^{n-1})=\frac{1}{2}.$$

Thus, $\frac{1}{2} = a(n) - a(n-1) < 1$.

Putting $n = 2, 3, \ldots, n$ and adding, we get

$$(n-1) \cdot \frac{1}{2} < a(n) - a(1) < n - 1.$$

When $n = 100$, $\frac{99}{2} < a(100) - a(1) < 99$

or $\quad \frac{99}{2} < a(100) - 1 < 99. \quad \therefore \quad \frac{101}{2} < a(100) - a(1) < 100.$

When $n = 200$, $\frac{199}{2} < a(200) - a(1) < 199$

or $\quad \frac{199}{2} < a(200) - 1 < 199. \quad \therefore \quad \frac{201}{2} < a(200) < 200.$

123. $\alpha^{10n+2} + \alpha^{5n+2} + \alpha^{5n} = (\alpha^5)^{2n} \cdot \alpha^2 + (\alpha^5)^n \cdot \alpha^2 + (\alpha^5)^n$

$$= (-1)^{2n} \cdot \alpha^2 + (-1)^n \alpha^2 + (-1)^n = \alpha^2 - \alpha^2 - 1 = -1.$$

124. $a = xy^5,\ b = xy^{10},\ c = xy^{15}$.

$\therefore \quad ac = xy^5 \cdot xy^{15} = x^2 \cdot y^{20} = (xy^{10})^2 = b^2.$

125. $f(0) = t, f(1) = t - 2$.

Either $f(0) > 0, f(1) < 0 \quad$ or $\quad f(0) < 0, f(1) > 0$

$\Rightarrow \quad t > 0, t - 2 < 0 \quad$ or $\quad t < 0, t - 2 > 0$ (absurd)

$\Rightarrow \quad t \in (0, 2)$.

126. $z = 1 + \cos\frac{2\pi}{3} + i\sin\frac{2\pi}{3} = 2\cos^2\frac{\pi}{3} + i \cdot 2\sin\frac{\pi}{3} \cdot \cos\frac{\pi}{3}$

$$= 2\cos\frac{\pi}{3}\left(\cos\frac{\pi}{3} + i\sin\frac{\pi}{3}\right) = \cos\frac{\pi}{3} + i\sin\frac{\pi}{3}.$$

$\therefore \quad z^5 = \left(\cos\frac{\pi}{3} + i\sin\frac{\pi}{3}\right)^5 = \cos\frac{5\pi}{3} + i\sin\frac{5\pi}{3}$

$$= \cos\left(-\frac{\pi}{3}\right) + i\sin\left(-\frac{\pi}{3}\right) = \frac{1}{2} - \frac{\sqrt{3}}{2}i.$$

127. $f(0) = \sum_{\lambda=1}^{n} \frac{1}{n} \cdot \frac{1}{\lambda+1} = \sum_{\lambda=1}^{n} \left(\frac{1}{\lambda} - \frac{1}{\lambda+1}\right) = 1 - \frac{1}{n+1} = \frac{n}{n+1}.$

$\therefore \quad \lim_{n \to \infty} f(0) = 1.$

128. $\dfrac{(a+c)+(b+d)}{2} \geq \sqrt{(a+c)(b+d)}$

or $\left(\dfrac{1}{2}\right)^2 \geq (a+c)(b+d) = ab + bc + cd + ad.$

$\therefore \quad \dfrac{1}{4} \geq (ab + bc + cd) + ad.$

$\therefore \quad \dfrac{1}{4} \geq ab + bc + cd$, equality holding when $a + c = b + d$ and $ad = 0$.

129. Let α, $-\alpha$ be roots and the third root $= \beta$.

Then $\alpha + (-\alpha) + \beta = m$. $\quad \therefore \; \beta = m.$

$\therefore \quad m^3 - m \cdot m^2 - 3m + 2 = 0 \quad$ or $\quad m = \dfrac{2}{3}.$

130. $u_2 \cdot u_3 \cdot \ldots u_n = \left(1 - \dfrac{1}{2}\right)\left(1 - \dfrac{1}{3}\right) \ldots \left(1 - \dfrac{1}{n}\right)$

$= \dfrac{1}{2} \cdot \dfrac{2}{3} \cdot \dfrac{3}{4} \cdot \ldots \cdot \dfrac{n-1}{n} = \dfrac{1}{n}.$

131. ${}^{n+1}C_j - {}^nC_j = ({}^nC_j + {}^nC_{j-1}) - {}^nC_j = {}^nC_{j-1}.$

$\therefore \quad \text{sum} = \sum_{j=1}^{n} {}^nC_{j-1} = {}^nC_0 + {}^nC_1 + {}^nC_2 + \ldots + {}^nC_{n-1} = 2^n - {}^nC_n = 2^n - 1.$

132. Here, $\dfrac{2-\lambda}{1-3} \cdot \dfrac{-1}{2} = -1.$

$\therefore \quad 2 - \lambda = -4. \quad \therefore \; \lambda = 6.$

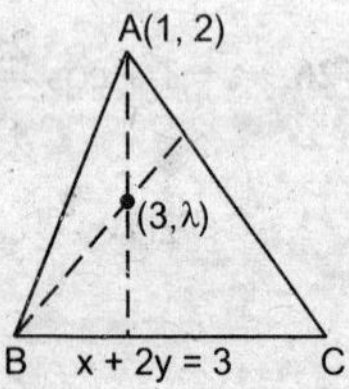

133. They must touch internally.

Here, $r_1 = 1$, $r_2 = \sqrt{\lambda^2 + 4}$, $d = \sqrt{1 + \lambda^2}$.

Now, $|r_1 - r_2| = d$

$\Rightarrow \quad \sqrt{\lambda^2 + 4} - 1 = \sqrt{1 + \lambda^2}$

or $\quad \sqrt{\lambda^2 + 4} - \sqrt{1 + \lambda^2} = 1$

or $\lambda^2+4-(1+\lambda^2)=\sqrt{\lambda^2+4}+\sqrt{1+\lambda^2}$.

Adding, $2\sqrt{\lambda^2+4}=1+3=4$. $\therefore\ \lambda^2+4=4$. $\therefore\ \lambda=0$.

134. If n = the number of sides then ${}^nC_2-n$ = number of diagonals.

$\therefore\quad ({}^nC_2-n)-n\ge 10$ or $\dfrac{n(n-1)}{2}-2n\ge 10$ or $n^2-5n\ge 20$

or $n^2-5n-20\ge 0$.

Roots of corresponding equation are $n=\dfrac{5\pm\sqrt{25+80}}{2}=\dfrac{5\pm\sqrt{105}}{2}$.

$\therefore$ 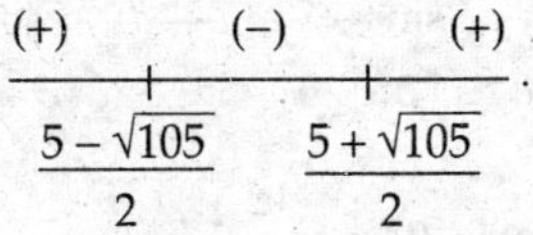

$\therefore\quad n\in\left[\dfrac{5+\sqrt{105}}{2},+\infty\right)$, because $n>0$. But n is a positive integer.

So, $n\ge 8$.

135. The line must pass through the centroid of the triangle formed by the points.

136. 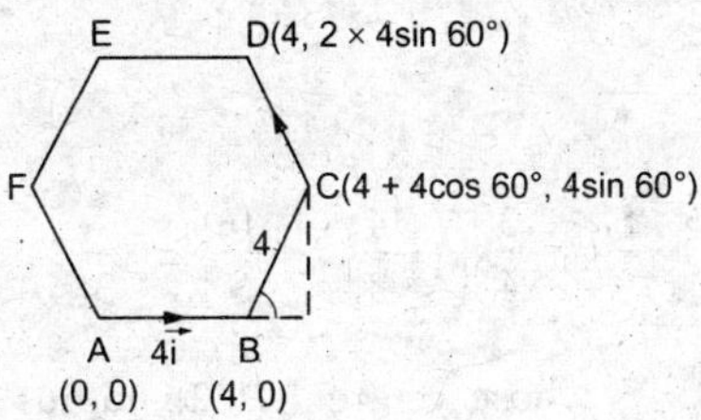

$$\vec{CD}=\vec{AD}-\vec{AC}$$
$$=\left(4\vec{i}+8\cdot\frac{\sqrt{3}}{2}\vec{j}\right)-(6\vec{i}+2\sqrt{3}\vec{j})$$
$$=-2\vec{i}+2\sqrt{3}\vec{j}$$
$$=2(-\vec{i}+\sqrt{3}\vec{j}).$$

137. $x=\dfrac{1}{2}\cdot\{\cos(A-C)-\cos(A+C)\}=\dfrac{1}{2}\left\{\cos(A-C)+\dfrac{1}{2}\right\}$.

But $0\le\cos(A-C)\le 1$. So $\dfrac{1}{2}\left(0+\dfrac{1}{2}\right)\le x\le\dfrac{1}{2}\left(1+\dfrac{1}{2}\right)$.

138. Here, $\phi(-x)=\dfrac{1}{1+e^x}$. So $\phi(x)+\phi(-x)=\dfrac{1}{1+e^{-x}}+\dfrac{1}{1+e^x}$.

$$\therefore\quad \phi(x)+\phi(-x)=\frac{e^x}{e^x+1}+\frac{1}{1+e^x}=\frac{e^x+1}{e^x+1}=1.$$

$$\therefore\quad \text{sum}=\{\phi(5)+\phi(-5)\}+\dots+\{\phi(1)+\phi(-1)\}+\phi(0)$$
$$=1+1+1+1+1+\phi(0)=5+\frac{1}{1+e^0}=5+\frac{1}{2}=\frac{11}{2}.$$

139. Clearly, $\lim_{x \to \infty} f(x+2) = \lim_{x \to \infty} f(x+1) = \lim_{x \to \infty} f(x) = l$, (say).

Then taking limit, $l = \frac{1}{2}\left(l + \frac{4}{l}\right)$

or $\frac{1}{2} l = \frac{2}{l}$ or $l^2 = 4$. $\therefore$ $l = 2$ [$\because f(x) > 0$ for all x]

140. Put $\frac{1}{x} = z$. Then $\frac{-1}{x^2} dx = dz$.

$$\therefore \quad I = \int_{\pi}^{1/\pi} z \cdot \sin\left(\frac{1}{z} - z\right) \cdot (-x^2)\, dz = \int_{\pi}^{1/\pi} z \cdot \sin\left(\frac{1}{z} - z\right) \cdot \frac{-1}{z^2}\, dz$$

$$= \int_{\pi}^{1/\pi} \frac{1}{z} \sin\left(z - \frac{1}{z}\right) dz = -I. \quad \therefore \quad I = 0.$$

141. Roots of $4x^2 - x - 1 = 0$ are irrational. So, one common root implies both roots are common.

$$\therefore \quad \frac{4}{3} = \frac{-1}{\lambda + \mu} = \frac{-1}{\lambda - \mu} = \frac{-2}{2\lambda} = \frac{0}{2\mu} \cdot \text{So, } \lambda = \frac{-3}{4},\ \mu = 0.$$

142. Here, $b^2 - 4a \le 0$ or $a \ge \frac{b^2}{4}$.

$$\therefore \quad 3a - 2b \ge 3 \cdot \frac{b^2}{4} - 2b = \frac{3}{4}\left(b^2 - \frac{8}{3} b\right) = \frac{3}{4}\left\{\left(b - \frac{4}{3}\right)^2 - \frac{16}{9}\right\} \ge -\frac{4}{3}.$$

143. The circle is $(x-1)^2 + (y-2)^2 = 5$ whose centre $= (1, 2)$, radius $= \sqrt{5}$. Changing the origin to the centre (1, 2), the four vertices of the square are (a, a), $(-a, a)$, $(-a, -a)$, $(a, -a)$. So, the diagonal of the square is $2\sqrt{2}a =$ diameter $= 2\sqrt{5}$. So, $a = \sqrt{5/2}$.

$\therefore$ vertices in old coordinates are $(\pm(1 + \sqrt{5/2}), \pm(2 + \sqrt{5/2}))$.

144. Here, $ax^2 - bx + c^2 = 0$ does not have real roots.
So, $D < 0 \Rightarrow b^2 - 4ac^2 < 0 \Rightarrow a > 0$.

$\therefore$ $f(x)$ is always positive. So, $f(2) > 0 \Rightarrow 4a - 2b + c^2 > 0$.

145. $f'(3) = \lim_{h \to 0} \frac{f(3+h) - f(3)}{h} = \lim_{h \to 0} \frac{f(3-h) - f(3)}{-h} = 5$.

$\therefore$ $f(3+h) - f(3) = 5h + h\epsilon_1$ where $\epsilon_1 \to 0$ when $h \to 0$

and $f(3-h)-f(3)=-5h+h\in_2$ were $\in_2 \to 0$ when $h \to 0$.

$\therefore \quad f(3+h)=f(3)+5h+h\in_1 = 4+5h+h\in_1$.

$\therefore \quad [f(3+h)]=[4+5h+h\in_1]=4$ when $h>0$ and h is very small.

$\therefore \quad \lim\limits_{h\to 0} [f(3+h)]=4.$

Similarly, $[f(3-h)]=[4-5h+h\in_2]=[4-h(5-\in_2)]=3$
when $h>0$ and h is very small implying $\in_2$ is very small.

$\therefore \quad \lim\limits_{h\to 0} [f(3-h)]=3\neq 4=\lim\limits_{h\to 0} [f(3+h)].$

$\therefore \quad \lim\limits_{x\to 3} [f(x)]$ does not exist.

146. $[x+[x]]\le 2 \Rightarrow [x]+[x]\le 2$ because $[x]$ = integer
or $\quad [x]\le 1. \quad \therefore \quad x<2.$

147. $B=(\pm 6, y)$. So, $AB\le 4 \Rightarrow (3\mp 6)^2+(y-4)^2\le 16.$

$\therefore \quad 9+(y-4)^2\le 16, \{\because\ 81+(y-4)^2\le 16$ is absurd$\}$.

$\therefore \quad y^2-8y+9\le 0 \Rightarrow$ + $\quad$ – $\quad$ + (sign scheme with critical points $4-\sqrt{7}$, $4+\sqrt{7}$)

$\therefore \quad 4-\sqrt{7}\le y\le 4+\sqrt{7}$. But y is an integer.

$\therefore \quad y=2,3,4,5,6.$

148. $PA+PB\ge AB$ (by triangle inequality). So, $PA+PB$ is the minimum if $PA+PB=AB$, i.e., A, P, B are collinear.

$$\therefore \quad \begin{vmatrix} 3 & -4 & 1 \\ 1 & 2 & 1 \\ 2k-1 & 2k+1 & 1 \end{vmatrix}=0$$

or $\quad 3(2-2k-1)+4(1-2k+1)+1(2k+1-4k+2)=0$

or $\quad 3-6k+8-8k+3-2k=0$

or $\quad 14-16k=0. \quad \therefore \quad k=\dfrac{7}{8}$.

149. Let $f(x)=a_0+a_1x+a_2x^2+\ldots+a_nx^n+\ldots$

Then $f(0)=a_0,\ f'(0)=a_1,\ f''(0)=a_2\cdot \underline{|2},\ldots,f^n(0)=a_n\ \underline{|n},\ldots$, etc.

$$\therefore \quad f(x)=f(0)+\frac{f'(0)}{\underline{|1}}x+\frac{f''(0)}{\underline{|2}}x^2+\ldots+\frac{f^n(0)}{\underline{|n}}x^n+\ldots$$

$$= p + \frac{pq}{\lfloor 1}x + \frac{pq^2}{\lfloor 2}x^2 + \ldots + \frac{pq^n}{\lfloor n}x^n + \ldots$$

$$= p\left\{1 + \frac{qx}{\lfloor 1} + \frac{(qx)^2}{\lfloor 2} + \ldots + \frac{(qx)^n}{\lfloor n} + \ldots\right\} = p \cdot e^{qx}.$$

$$\therefore \quad \lim_{x \to p} f(x) = \lim_{x \to p} (p \cdot e^{qx}) = p \cdot e^{pq}.$$

150. The equation of pair of tangents is $S \cdot S_1 = T^2$

or $\quad (y^2 - 16x)(1 - 16\lambda) = \{y - 8(x + \lambda)\}^2$

or $\quad 64x^2 - 16xy + 16\lambda y^2 + \ldots = 0.$

The pair is at an angle $\frac{\pi}{2}$ if $a + b = 0$, i.e., $64 + 16\lambda = 0 \Rightarrow \lambda = -4$.

151. 5 white flowers can be set in only 1 way (treating them as identical things). There are 5 places in between them where the flowers of other colours are to be placed. The number of selections of 4 places is 1 because the white flowers are identical.

$\therefore$ the required number of ways $= \frac{\lfloor 4}{2}$, treating clockwise and anticlockwise arrangements identical.

152. $r = \frac{\Delta}{s} = \frac{\Delta}{\frac{3a}{2}}, \; r_0 = \frac{\Delta}{s - a} = \frac{\Delta}{\frac{3a}{2} - a} = \frac{\Delta}{\frac{a}{2}}.$

153. Let $\sec x + \tan x = z$. Then $\sec x(\tan x + \sec x)dx = dz$.

$$\therefore \quad I = \int_1^{\sqrt{2}+1} \frac{dz}{z^3} = \left[\frac{-z^{-2}}{2}\right]_1^{\sqrt{2}+1} = -\frac{1}{2}\left\{\frac{1}{(\sqrt{2}+1)^2} - 1\right\}$$

$$= -\frac{1}{2}\{(\sqrt{2}-1)^2 - 1\} = -\frac{1}{2}(2 - 2\sqrt{2}) = \sqrt{2} - 1.$$

154. The centre of the second circle $= (4, -1)$ and radius $= \sqrt{4^2 + 1^2 - 8} = 3$.

The centre of the first circle $= (1, 3)$ and radius $= r$.

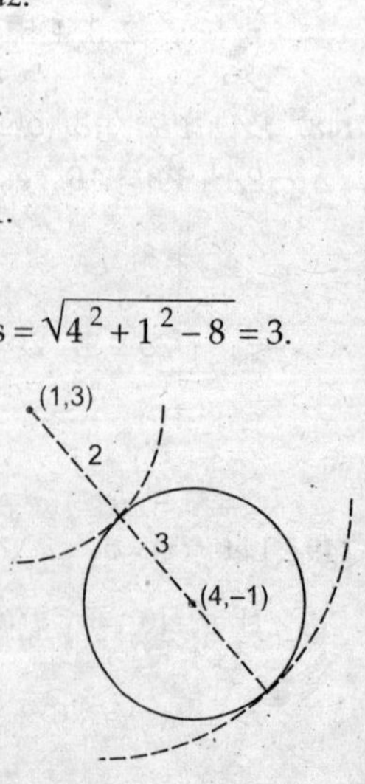

Clearly, $(1, 3)$ is outside the second circle because

$1^2 + 3^2 - 8 \cdot 1 + 2 \cdot 3 + 8 > 0.$

Also, the distance between the centres

$= \sqrt{(1-4)^2 + (3+1)^2} = 5.$

So, for distinct intersection,

$r > 5 - 3$ and $r < 2 + 3 + 3 \Rightarrow 2 < r < 8.$

155. Total number of interior points in the square $= 20 \times 20 = 400$.

The equation of AB is $x + y = 21$. Clearly, $(1, 20)$, $(2, 19)$, ... , $(20, 1)$ satisfy the equation. So there are 20 points on AB.

$\therefore$ required number of interior points

$$= \frac{400 - 20}{2} = 190.$$

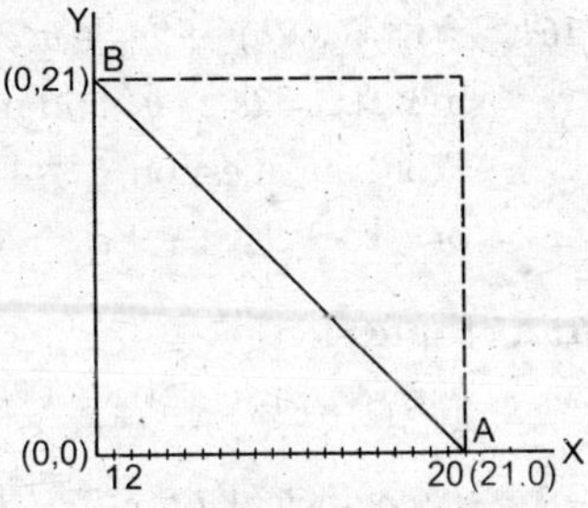

156. Limit $= \lim\limits_{x \to 0} \dfrac{(a-n)nx - \tan x}{x} \cdot n\left(\dfrac{\sin nx}{nx}\right) = n \lim\limits_{x \to 0} \dfrac{(a-n)nx - \tan x}{x}$

$$= n\left\{\lim_{x \to 0} \frac{(a-n)nx}{x} - \lim_{x \to 0} \frac{\tan x}{x}\right\}$$

$$= n\{(a-n)n - 1\} = 0 \quad \text{(from the question)}$$

157. Any point on the line is $(1, r, 2r - 1)$. The distance of it from the plane is 3

$$\Rightarrow \quad \left|\frac{1 + r + 2r - 1 - 3}{\sqrt{1^2 + 1^2 + 1^2}}\right| = 3 \quad \text{or} \quad \left|\frac{3(r-1)}{\sqrt{3}}\right| = 3$$

or $\quad |r - 1| = \sqrt{3}$. $\qquad \therefore \quad r = 1 \pm \sqrt{3}$.

$\therefore \quad$ the point is $(1, 1 \pm \sqrt{3}, 1 \pm 2\sqrt{3})$.

158. Both the numbers are not to be selected from 4, 5, 6.

Probability of selecting both from 4, 5, 6 $= \dfrac{3}{6} \times \dfrac{2}{5}$.

$\therefore \quad$ the required probability $= 1 - \dfrac{3}{6} \times \dfrac{2}{5} = 1 - \dfrac{1}{5} = \dfrac{4}{5}$.

159. Use L' Hospital's rule.

160. $\sin^{-1}(2x) + \dfrac{\pi}{6} \ge 0$ and $|2x| \le 1$

$$\Rightarrow \quad \frac{\pi}{2} \ge \sin^{-1}(2x) \ge -\frac{\pi}{6} \quad \text{and} \quad -\frac{1}{2} \le x \le \frac{1}{2}$$

$$\Rightarrow \quad 1 \ge 2x \ge -\frac{1}{2}, \text{ i.e., } \frac{1}{2} \ge x \ge -\frac{1}{4} \text{ and } -\frac{1}{2} \le x \le \frac{1}{2} \Rightarrow -\frac{1}{4} \le x \le \frac{1}{2}.$$

161. $f(x) = (x+b)^2 + 2c^2 - b^2, g(x) = b^2 + c^2 - (x+c)^2$
$\min f(x) = 2c^2 - b^2$ and $\max g(x) = b^2 + c^2$.
From the question, $2c^2 - b^2 > b^2 + c^2$.
or $c^2 > 2b^2 \Rightarrow |c| > \sqrt{2}|b|$.

162. Expression
$= (1 + x^{12} + x^{24} + x^{36})\{{}^{12}C_0 + {}^{12}C_1(x^2) + {}^{12}C_2(x^2)^2 + \dots + {}^{12}C_{12}(x^2)^{12}\}$.
$\therefore$ coefficient of $x^{24} = {}^{12}C_{12} + {}^{12}C_6 + {}^{12}C_0 = 2 + {}^{12}C_6$.

163. $\alpha + B = \frac{1}{\alpha^2} + \frac{1}{\beta^2} \Rightarrow -\frac{b}{a} = \frac{b^2 - 2ca}{c^2} \Rightarrow 2ca^2 = ab^2 + bc^2 \Rightarrow 2 \cdot \frac{a}{b} = \frac{b}{c} + \frac{c}{a}$
$\therefore$ $\frac{c}{a}, \frac{a}{b}, \frac{b}{c}$ are in AP.

164. $\hat{n} \parallel \vec{u} \times \vec{v} = -2\vec{k}$. $\therefore$ $|\vec{w} \cdot \hat{n}| = |(\vec{i} + 2\vec{j} + 3\vec{k}) \cdot \vec{k}| = 3$.

165. $f'(x) = f(x) \Rightarrow \frac{f'(x)}{f(x)} = 1 \Rightarrow \log f(x) = x + c$.
$\therefore$ $f(0) = 1 \Rightarrow \log 1 = 0 + c$, $\therefore c = 0$. Hence, $\log f(x) = x \Rightarrow f(x) = e^x$.
$\therefore$ $g(x) = x^2 - e^x$.
Hence, $\int_0^1 f(x)g(x)dx = \int_0^1 e^x(x^2 - e^x)dx = \int_0^1 x^2 \cdot e^x dx - \int_0^1 e^{2x} dx$.

166. $a^2 - b = b^2 - c \Rightarrow a^2 - b^2 = b - c \Rightarrow a + b = \frac{b-c}{a-b}$. Similarly find $b + c$ and $c + a$.

167. $i^n + i^{-n} = i^n + \frac{1}{i^n} = \frac{i^{2n} + 1}{i^n} = \frac{(-1)^n + 1}{i^n}$
If $n = 4m$, value $= \frac{1+1}{1} = 2$; if $n = 4m + 1$, value $= \frac{-1+1}{i} = 0$;
if $n = 4m + 2$, value $= \frac{1+1}{-1} = -2$; if $n = 4m + 3$, value $= \frac{-1+1}{i^3} = 0$.
$\therefore$ different values are 2, 0, –2.

168. $a_1 = 1 = \text{odd}$; $a_2 = a_1 a_0 + 1 = 2 = \text{even}$; $a_3 = a_2 a_1 + 1 = \text{even} + 1 = \text{odd}$;
$a_4 = a_3 a_2 + 1 = \text{even} + 1 = \text{odd}$; $a_3 = a_4 a_3 + 1 = \text{odd} + 1 = \text{even}$, etc.
$\therefore$ the numbers $a_1, a_2, a_3, a_4, a_5, \dots$ in the sequence are in the order odd, | even, odd, odd, | even, odd, odd, | even, ...
$165 = 1 + 164 = (1) + (3 \times 52) + 2$. So, a_{165} is odd.
$166 = 1 + 165 = (1) + (3 \times 52) + 3$, So, a_{166} is odd.

169. $A \cap (B \cap \bar{C})$ and $\bar{A} \cap (B \cap \bar{C})$ are mutually exclusive, having no common outcome.

$\therefore \quad P[\{A \cap (B \cap \bar{C})\} \cup \{\bar{A} \cap (B \cap \bar{C}\}]$

$$= P\{A \cap (B \cap \bar{C})\} + P\{\bar{A} \cap (B \cap \bar{C})\} = \frac{1}{3} + \frac{1}{3}$$

or $\quad P\{(A \cup \bar{A}) \cap (B \cap \bar{C})\} = \frac{2}{3} \Rightarrow P(B \cap \bar{C}) = \frac{2}{3}$

Again, $P(B) = P\{B \cap (C \cup \bar{C})\} = P\{(B \cap C) \cup (B \cap \bar{C})\}$

$$= P(B \cap C) + P(B \cap \bar{C})$$

$B \cap C$ and $B \cap \bar{C}$ being mutually exclusive.

$\therefore \quad \frac{3}{4} = P(B \cap C) + \frac{2}{3} \Rightarrow P(B \cap C) = \frac{1}{12}$.

170. $f(x) = \int_{\alpha}^{x^2+1} e^{-t^2} dt - \int_{\alpha}^{x^2} e^{-t^2} dt \Rightarrow \frac{df}{dx} = e^{-(x^2+1)^2} \cdot (2x + 0) - e^{-x^4} \cdot 2x$

$\Rightarrow \quad f'(x) = 2x\left(e^{-(x^2+1)^2} - e^{-x^4}\right) > 0$ (for increasing function)

$\Rightarrow \quad x\left\{e^{x^4 - (x^2+1)^2} - 1\right\} > 0$

$\Rightarrow \quad x > 0$ and $e^{-2x^2-1} - 1 > 0$ ... (1) or $x < 0$ and $e^{-2x^2-1} - 1 < 0$. ... (2)

(1) cannot hold for any x. But (2) holds for all $x < 0$.

171. Clearly, $0 \le \frac{3x+1}{3} \le 1,\ 0 \le \frac{1-x}{4} \le 1, 0 \le \frac{1-2x}{2} \le 1$

$\Rightarrow \quad -\frac{1}{3} \le x \le \frac{2}{3},\ -3 \le x \le 1,\ -\frac{1}{2} \le x \le \frac{1}{2} \Rightarrow -\frac{1}{3} \le x \le \frac{1}{2}$.

Also, $P(A) + P(B) + P(C) = P(A \cup B \cup C)$ and $0 \le P(A \cup B \cup C) \le 1$.

$\therefore \quad 0 \le \frac{3x+1}{3} + \frac{1-x}{4} + \frac{1-2x}{2} \le 1$

or $\quad 0 \le 12x + 4 + 3 - 3x + 6 - 12x \le 12$ or $\frac{1}{3} \le x \le \frac{13}{3}$.

$\therefore \quad \frac{1}{3} \le x \le \frac{1}{2}$.

172. $CD = \sqrt{3^2 - 2^2} = \sqrt{5},\ BD = \sqrt{4^2 - 2^2} = 2\sqrt{3}$.

$\therefore \quad BC = \sqrt{5} + 2\sqrt{3}$.

Now, $ar(\Delta ABC) = \frac{1}{2} \cdot 4 \cdot 3 \cdot \sin A = \frac{1}{2}(\sqrt{5} + 2\sqrt{3}) \cdot 2$.

$\therefore \quad \sin A = \frac{\sqrt{5} + 2\sqrt{3}}{6} < 1$. Also $a > b, a > c$.

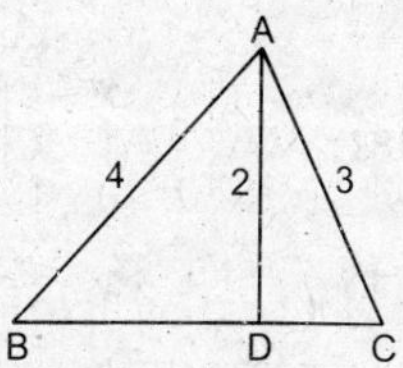

$\therefore$ A can have two values, one acute and the other obtuse.

173. $0 \quad \frac{1}{2} \quad 1 \quad \frac{3}{2} \quad 2 \quad \frac{5}{2} \quad 3 \quad \frac{7}{2} \quad 4$

Verify the solutions at $x = 0, \frac{1}{2}, 1, \frac{3}{2}, \ldots, 4$.

Also, solve for (i) $0 < x < \frac{1}{2}$, (ii) $\frac{1}{2} < x < 1, \ldots$, (viii) $\frac{7}{2} < x < 4$.

175. Value $= {}^nC_0 + {}^{n+1}C_1 + {}^{n+2}C_2 + \ldots + {}^{2n}C_n$

$= {}^nC_n + {}^{n+1}C_n + {}^{n+2}C_n + \ldots + {}^{2n}C_n$

$=$ coefficient of x^n in

$$\{(1+x)^n + (1+x)^{n+1} + (1+x)^{n+2} + \ldots + (1+x)^{2n}\}$$

$=$ coefficient of x^n in $\dfrac{(1+x)^n \{1-(1+x)^{n+1}\}}{1-(1+x)}$

$=$ coefficient of x^{n+1} in $\{(1+x)^{2n+1} - (1+x)^n\}$

$= {}^{2n+1}C_{n+1} = {}^{2n+1}C_n$.

177. $I = \int_a^b x f(a+b-x)dx = \int_a^b (a+b-z) f(z)dz$ (put $a+b-x = z$)

$= (a+b)\int_a^b f(z)\,dz - \int_a^b zf(z)\,dz \Rightarrow 2I = (a+b)\int_a^b f(x)\,dx.$

178. $t_{r+1} = {}^{256}C_r\, 3^{\frac{256-r}{2}} \cdot 5^{\frac{r}{5}}$. For integral index, r should be an even multiple of 5. So, $r = 0, 10, 20, \ldots 250$.

179. Value $= ({}^nC_{r+1} + {}^nC_r) + ({}^nC_{r-1} + {}^nC_r) = {}^{n+1}C_{r+1} + {}^{n+1}C_r = {}^{n+2}C_{r+1}$.

180. A focal chord is $y = m(x-4)$. It touches the circle $\Rightarrow \left|\dfrac{m(6-4)}{\sqrt{1+m^2}}\right| = \sqrt{2}$.

181. $V = \begin{vmatrix} 1 & \lambda & 1 \\ 0 & 1 & \lambda \\ \lambda & 0 & 1 \end{vmatrix} = \lambda^3 - \lambda + 1$. So $\dfrac{dV}{d\lambda} = 0 \Rightarrow 3\lambda^2 - 1 = 0 \Rightarrow \lambda = \pm\dfrac{1}{\sqrt{3}}$.

$\dfrac{d^2V}{dx^2} > 0$ at $\lambda = \dfrac{1}{\sqrt{3}}$.

182. Here, $\frac{\frac{a}{2}}{R} = \sin\frac{\pi}{n} \Rightarrow R = \frac{a}{2\sin\frac{\pi}{n}}$

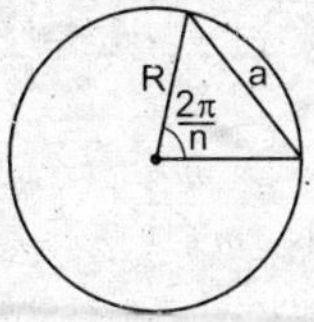

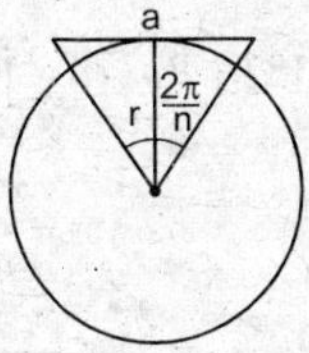

$$\frac{\frac{a}{2}}{r} = \tan\frac{\pi}{n} \Rightarrow r = \frac{a\cos\frac{\pi}{n}}{2\sin\frac{\pi}{n}}.$$

$$\therefore \quad R + r = \frac{a}{2}\cdot\frac{1+\cos\frac{\pi}{n}}{\sin\frac{\pi}{n}} = \frac{a}{2}\cdot\frac{2\cos^2\frac{\pi}{2n}}{2\sin\frac{\pi}{2n}\cdot\cos\frac{\pi}{2n}}.$$

183. $w = \frac{\cos\theta + i\sin\theta - 1}{\cos\theta + i\sin\theta + 1} = \frac{(\cos\theta - 1 + i\sin\theta)(\cos\theta + 1 - i\sin\theta)}{(\cos\theta+1)^2 + \sin^2\theta} = 0 + iy$

184. Let $w = r(\cos\theta + i\sin\theta)$. Then $z = \frac{1}{r}\left\{\cos\left(\theta+\frac{\pi}{2}\right) + i\sin\left(\theta+\frac{\pi}{2}\right)\right\}$.

$\therefore \quad \bar{z}w = (-\sin\theta - i\cos\theta)(\cos\theta + i\sin\theta) = -i.$

185. For homogeneous equations infinite solutions
$\Rightarrow$ existence of nontrivial solution.

186. Let $z_1 = r(\cos\theta + i\sin\theta)$. Then $z_2 = r\left\{\cos\left(\theta+\frac{\pi}{3}\right) + i\sin\left(\theta+\frac{\pi}{3}\right)\right\}$.

$$\therefore \quad z_1 + z_2 = r\left[\cos\theta + \cos\left(\theta+\frac{\pi}{3}\right) + i\left\{\sin\theta + \sin\left(\theta+\frac{\pi}{3}\right)\right\}\right]$$

$$= r\sqrt{3}\left\{\cos\left(\theta+\frac{\pi}{6}\right) + i\sin\left(\theta+\frac{\pi}{6}\right)\right\} = -a.$$

$$z_1z_2 = r^2(\cos\theta + i\sin\theta)\left\{\cos\left(\theta+\frac{\pi}{3}\right) + i\sin\left(\theta+\frac{\pi}{3}\right)\right\}$$

$$= r^2\left\{\cos 2\left(\theta+\frac{\pi}{6}\right) + i\sin 2\left(\theta+\frac{\pi}{6}\right)\right\}$$

$$= \left[r\left\{\cos\left(\theta+\frac{\pi}{6}\right) + i\sin\left(\theta+\frac{\pi}{6}\right)\right\}\right]^2 = b.$$

$$\therefore \quad \left(-\frac{a}{\sqrt{3}}\right)^2 = b.$$

187. $I(m, n) = \int_0^1 t^m \cdot (1+t)^n dt = \left[(1+t)^n \cdot \frac{t^{m+1}}{m+1}\right]_0^1 - \int_0^1 \frac{t^{m+1}}{m+1} \cdot n(1+t)^{n-1} dt$

$$= \frac{2^n}{m+1} - \frac{n}{m+1}\int_0^1 t^{m+1} \cdot (1+t)^{n-1} dt$$

$$= \frac{2^n}{m+1} - \frac{n}{m+1} I(m+1, n-1).$$

188. For positive quantities AM $\geq$ GM $\Rightarrow \dfrac{\sqrt{x^2+x} + \dfrac{\tan^2\alpha}{\sqrt{x^2+x}}}{2} \geq \sqrt{\tan^2\alpha}$.

189. $f(n) = \frac{n-1}{2}$ when n is odd gives the images 0, 1, 2, 3, ... successively.

$f(n) = -\frac{n}{2}$ when n is even gives the images $-1, -2, -3, \ldots$ successively.

So, f is one-one and onto.

190. $f(x_1) = f(x_2) \Rightarrow \frac{x_1}{x_1+1} = \frac{x_2}{x_2+1} \Rightarrow x_1 = x_2$. And $f(x) = \frac{x}{x+1} < 1$ for all $x \in [0, \infty)$.

$\therefore$ one-one and into function.

191. The function in (*a*) is not differentiable at $x = \frac{1}{2}$ and $0 < \frac{1}{2} < 1$. So, LMVT is not applicable. Others are continuous in [0, 1] and differentiable in (0, 1).

192. $(\vec{u} - \vec{v}) \times (\vec{v} - \vec{w}) = \vec{u} \times \vec{v} - \vec{u} \times \vec{w} + \vec{v} \times \vec{w}$

$\therefore$ value $= (\vec{u} + \vec{v} - \vec{w}) \cdot (\vec{u} \times \vec{v} - \vec{u} \times \vec{w} + \vec{v} \times \vec{w})$

$$= [\vec{u}\ \vec{v}\ \vec{w}] - [\vec{v}\ \vec{u}\ \vec{w}] - [\vec{w}\ \vec{u}\ \vec{v}]$$

$$= [\vec{u}\ \vec{v}\ \vec{w}] + [\vec{u}\ \vec{v}\ \vec{w}] - [\vec{u}\ \vec{v}\ \vec{w}] = [\vec{u}\ \vec{v}\ \vec{w}].$$

196. Condition is $\begin{vmatrix} x_1 - x_2 & y_1 - y_2 & z_1 - z_2 \\ l_1 & m_1 & n_1 \\ l_2 & m_2 & n_2 \end{vmatrix} = 0.$

Here, $\begin{vmatrix} 2-1 & 3-4 & 4-5 \\ 1 & 1 & -k \\ k & 2 & 1 \end{vmatrix} = 0.$

197. Planes are $-x-y-z+1=0$, $-x-2y+4z+5=0$. So, $d_1>0$, $d_2>0$ and $a_1a_2+b_1b_2+c_1c_2=(-1)(-1)+(-1)(-2)+(-1)4<0$.

So, the bisector of the angle containing the origin is the bisector of the acute angle.

$\therefore$ bisector of the obtuse angle is

$$\frac{-x-y-z+1}{\sqrt{(-1)^2+(-1)^2+(-1)^2}}=-\frac{-x-2y+4z+5}{\sqrt{(-1)^2+(-2)^2+4^2}}.$$

198. Here $40\sqrt{(x-40)^2+y^2}<60\sqrt{x^2+(y-30)^2}$

$\Rightarrow 4[x^2+y^2-80x+1600]<9[x^2+y^2-60y+900]$

$\Rightarrow (x+32)^2+(y-54)^2>3600.$

□

2

Miscellaneous Questions—2

• Type 2 •

Choose the correct options. One or more options may be correct.

1. If in a ΔABC, $\begin{vmatrix} \sin A & \sin B & \sin C \\ \sin B & \sin C & \sin A \\ \sin C & \sin A & \sin B \end{vmatrix} = 0$ then

(a) ΔABC is equilateral

(b) $a^3 + b^3 + c^3 = 3abc$

(c) $\Sigma \sin A = 0$

(d) $\Sigma \sin^3 A - 3\Pi \sin A = 0$

2. α, β are the roots of $x^2 - px + q = 0$ and α, β, γ are the roots of $x^3 - ax^2 + bx - c = 0$. Then

(a) $pq + c = aq$

(b) $p + \gamma = a$

(c) $p\gamma + q = b$

(d) $\gamma q = c$

3. Let $\vec{c} = \lambda \vec{a} + \mu \vec{b} + \gamma \vec{a} \times \vec{b}$ where $\vec{a}$ and $\vec{b}$ are unit vectors such that $\vec{a} \cdot \vec{b} = 0$. Then

(a) $\mu = \vec{b} \cdot \vec{c}$

(b) $\lambda = |\vec{a} \times \vec{c}|$

(c) $\gamma = [\vec{a}\ \vec{b}\ \vec{c}]$

(d) $\lambda + \mu + \gamma = (\vec{a} + \vec{b} + \vec{a} \times \vec{b}) \cdot \vec{c}$

4. ABC is an isosceles triangle whose vertices A and C have the coordinates (1, 1) and (2, –1) respectively, and AC is one of the two equal sides. If the equation of the side AB is $x + y = 2$ then B has the coordinates

(a) (4, –2)

(b) (2, 0)

(c) $\left(1 - \sqrt{\frac{5}{2}},\ 1 + \sqrt{\frac{5}{2}}\right)$

(d) $\left(1 + \frac{1}{\sqrt{2}},\ 1 - \frac{1}{\sqrt{2}}\right)$

5. A vertical, rectangular, thin wall of length $= a$ metres, height $= b$ metres runs from east to west. A vertical flagstaff of height h metres stands on the top edge of the wall, dividing the wall in the ratio $p : q$ $(p > q)$. A man starts walking towards the east from the eastern end of the wall. After walking a distance x metres he can just see the top of the flagstaff. Instead, if he walks a distance x metres towards the west from the western end then he can see two-thirds of the flagstaff. Then

(a) $hx = \frac{3}{4}ab$ (b) $hx = \frac{4}{3}ab$

(c) $p = 2q$ (d) $p = 3q$

6. Which of the following is/are equal to $(\vec{u} \times \vec{v}) \cdot \vec{w}$?

(a) $\vec{u} \cdot (\vec{v} \times \vec{w})$ (b) $(\vec{v} \times \vec{w}) \cdot \vec{u}$

(c) $\vec{w} \cdot (\vec{u} \times \vec{v})$ (d) $\vec{v} \times (\vec{w} \cdot \vec{u})$

7. For the vectors $\vec{a}$, $\vec{b}$, $\vec{c}$ and $\vec{d}$ which of the following is/are meaningful if $\cdot$ and $\times$ represent scalar and vector products respectively?

(a) $(\vec{a} \times \vec{b}) \cdot (\vec{c} \times \vec{d})$ (b) $\{\vec{a} \times (\vec{b} \times \vec{c})\} \cdot \vec{d}$

(c) $\vec{a} \times \{(\vec{b} \times \vec{c}) \cdot \vec{d}\}$ (d) $\{(\vec{a} \cdot \vec{b}) \times \vec{c}\} \cdot \vec{d}$

8. If a, b, c are three consecutive integers then the determinant

$$\begin{vmatrix} -a^2 & ab & ac \\ ab & -b^2 & bc \\ ac & bc & -c^2 \end{vmatrix} \text{ is divisible by}$$

(a) 36 (b) 144

(c) 48 (d) none of these

9. Let P, A, B, C be four collinear points in order, the distances of A, B, C from P being a, b, c respectively. If the equation $(b-a)x^2 + (a-c)x + c - b = 0$ has one root double the other then

(a) B divides AC in the ratio 2 : 1 internally

(b) C divides AB in the ratio 2 : 1 externally

(c) B divides AC in the ratio 1 : 2 internally

(d) none of these

10. If $x^2 + px + q$ is an integer for all integral values of x then

(a) p is an integer (b) q is an integer

(c) $\frac{p+q}{2}$ is an integer (d) $\frac{p-q}{2}$ is an integer

11. The equation $px^3 - (p+1)x^2 + px = 0$ has real roots and p is any positive integer. Then the equation has

(a) exactly one root less than 1 (b) more than one root ≤ 1

(c) all roots greater than 1 (d) none of these

12. If $4x^2 + 4x + 12\lambda^2 \geq 47$ for all $x \in R$ then λ can have the value

(a) $\frac{3}{2}$ (b) 2

(c) -2 (d) 1

13. Let $f(x) = 5a^2 + 4(x-1) - x^2, x \in R$ and a is a real constant. If the maximum value of $f(x)$ is 20 then a is

(a) -2 (b) 4

(c) 2 (d) none of these

14. If $A = \int_0^1 \sin^{-1}x\, dx$ and $B = \int_0^1 \cos^{-1}x\, dx$ then $\frac{\pi}{2}$ is equal to

(a) $A - B + 2$ (b) $A - 2B$

(c) $A + 1$ (d) $A + B$

15. The equation of a parabola is $x^2 + y = 0$. The line $x = 1$, after getting reflected from the parabola, passes

(a) along the line $3x + 4y + 1 = 0$ (b) through the point $(-1, 1)$

(c) through the point $\left(0, -\frac{1}{4}\right)$ (d) through the point $\left(\frac{1}{3}, \frac{-1}{2}\right)$

16. If $f(x) = \{x - [x]\} |\cos \pi x|$ where $[\cdot]$ is the greatest integer function then $f(x)$ is

(a) an even function (b) an odd function

(c) a periodic function (d) a many-one function

17. Let $f(x)$ be a continuous even periodic function of period a and $f(0) = 0$. Then

(a) $f(x)$ is monotone in $\left[0, \frac{a}{2}\right]$ (b) $f\left(\frac{a}{2} - x\right) = f\left(\frac{a}{2} + x\right)$ for all x

(c) $f(x)$ is differentiable in $\left[0, \frac{a}{2}\right]$ (d) $f(x)$ has a maximum value

18. If in a ΔABC, $a = \sqrt{3}$, $c = 3$ and $A = 30°$ then the angle B has the measure

(a) 90° (b) 60° (c) 30° (d) 120°

19. There are $n+1$ identical balls on which the numbers ${}^nC_0, {}^nC_1, {}^nC_2, \dots, {}^nC_n$ are written, each ball carrying a different number. The total number of ways in which the balls can be arranged in a row so that the values of numbers on the first and the last are equal, is

(a) $n!$ when n is even (b) $(n^2-1)\cdot(n-2)!$ when n is odd

(c) $\frac{1}{2}\cdot n!$ when n is even (d) $\frac{1}{2}(n+1)\cdot(n-1)!$ when n is odd

20. The figure given below can be the rough sketch of the function given by

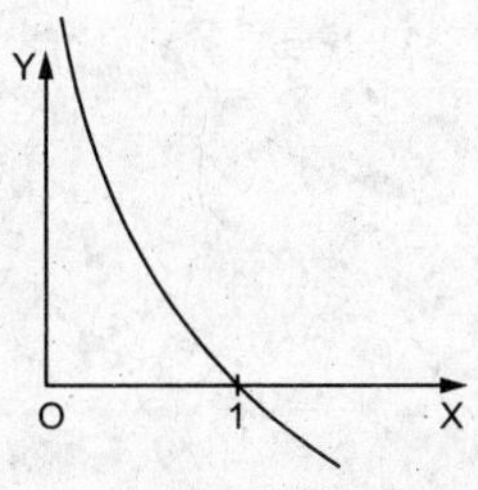

(a) $y = \log_2 x$ (b) $x \cdot 2^y = 1$

(c) $x = \log_{0.5} y$ (d) $y = \log_{0.5} x$

21. If $px^2 + qx + r = 0$ has no real roots for real values of p, q, r and $4p + 2q + r > 0$ then

(a) $r > 0$ (b) $p + q + r > 0$

(c) $p + q + r < 0$ (d) $r \geq 0$

22. If $f(x) = 3x^2 + 12x - 1, x \in [-1, 2]$

$37 - x, \quad x \in (2, 3]$ then

(a) $f(x)$ is monotonic increasing in $[-1, 2]$

(b) $f(x)$ is continuous in $[-1, 3]$

(c) $f(x)$ has the maximum value at $x = 2$

(d) $f'(2)$ does not exist

23. P and Q are two points on the curve $x^2 + y^2 = a^2$, where $\angle POQ = \frac{\pi}{2}$, O being the origin. The equation of the tangent to the curve at P is $x - \sqrt{3}y + 2a = 0$. The equation of the tangent to the curve at Q is

(a) $x + \sqrt{3}y - 2a = 0$ (b) $\sqrt{3}x + y - 2a = 0$

(c) $x + 2y - 2a = 0$ (d) $\sqrt{3}x + y + 2a = 0$

Answers

1. a, b, d	**2.** a, b, c, d	**3.** a, c, d	**4.** a, c	**5.** a, d
6. a, b, c	**7.** a, b	**8.** a, b, c	**9.** a, c	**10.** a, b
11. a, b	**12.** b, c	**13.** a, c	**14.** a, c, d	**15.** a, c, d
16. c, d	**17.** b, d	**18.** a, c	**19.** a, b	**20.** b, d
21. a, b	**22.** a, b, c, d	**23.** b, d		

Hints and Solutions to Selected Questions

1. Here, $\begin{vmatrix} a & b & c \\ b & c & a \\ c & a & b \end{vmatrix} = 0 \quad (\because \ a = 2R\sin A, \text{ etc.})$

or $\quad (a+b+c)\begin{vmatrix} 1 & 1 & 1 \\ b & c & a \\ c & a & b \end{vmatrix} = 0 \quad$ or $\quad (a+b+c)\begin{vmatrix} 1 & 0 & 0 \\ b & c-b & a-b \\ c & a-c & b-c \end{vmatrix} = 0$

or $\quad (a+b+c)\{-(b-c)^2 - (a-b)(a-c)\} = 0$

or $\quad (a+b+c)(a^2+b^2+c^2-bc-ca-ab) = 0. \qquad \ldots (1)$

But in a ΔABC, $a+b+c \neq 0 \quad (\therefore \ \sin A + \sin B + \sin C \neq 0)$.

$\therefore \quad (1) \Rightarrow \dfrac{1}{2}\{(a-b)^2 + (b-c)^2 + (c-a)^2\} = 0 \quad \Rightarrow a = b = c.$

$\therefore \quad \Delta$ABC is equilateral.

Also, (1) $\Rightarrow a^3 + b^3 + c^3 - 3abc = 0$. This implies (d) is also true.

2. Use $\alpha + \beta = p, \alpha\beta = q, \alpha + \beta + \gamma = a, \alpha\beta + \beta\gamma + \gamma\alpha = b, \alpha\beta\gamma = c$.

3. $\vec{c} = \lambda\vec{a} + \mu\vec{b} + \gamma\vec{a} \times \vec{b}$.

$\therefore \quad \vec{b} \cdot \vec{c} = \vec{b} \cdot (\lambda\vec{a} + \mu\vec{b} + \gamma\vec{a} \times \vec{b}) = \mu \quad \{\because \ \vec{a} \cdot \vec{b} = 0, \vec{b} \cdot \vec{b} = 1, [\vec{b}\ \vec{a}\ \vec{b}] = 0\}$

$\vec{c} \cdot \vec{c} = \vec{c}^{\,2} = \lambda^2 \vec{a}^{\,2} + \mu^2 \vec{b}^{\,2} + \gamma^2 (\vec{a} \times \vec{b})^2 + 2\lambda\mu\, \vec{a} \cdot \vec{b} + 2\lambda\gamma[\vec{a}\ \vec{a}\ \vec{b}]$

$+ 2\mu\gamma[\vec{b}\ \vec{a}\ \vec{b}]$

or $\quad \vec{c} \cdot \vec{c} = \lambda^2 + \mu^2 + \gamma^2. \qquad \ldots (1)$

Also, $\vec{c} \cdot \vec{c} = \vec{c} \cdot (\lambda\vec{a} + \mu\vec{b} + \gamma\vec{a} \times \vec{b})$

$= \lambda\vec{c} \cdot \vec{a} + \mu\vec{c} \cdot \vec{b} + \gamma[\vec{c}\ \vec{a}\ \vec{b}]$

$= \lambda(\lambda\vec{a} + \mu\vec{b} + \gamma\vec{a} \times \vec{b}) \cdot \vec{a} + \mu\mu + \gamma[\vec{a}\ \vec{b}\ \vec{c}]$

$= \lambda^2 + \mu^2 + \gamma[\vec{a}\ \vec{b}\ \vec{c}]. \qquad \ldots (2)$

$\therefore \quad (1), (2) \Rightarrow \gamma^2 = \gamma[\vec{a}\ \vec{b}\ \vec{c}]. \qquad \therefore \ \gamma = [\vec{a}\ \vec{b}\ \vec{c}].$

Also, $(\vec{a} + \vec{b} + \vec{a} \times \vec{b}) \cdot \vec{c} = \vec{a} \cdot \vec{c} + \vec{b} \cdot \vec{c} + [\vec{a}\ \vec{b}\ \vec{c}]$

$= \lambda + \mu + \gamma.$

4.

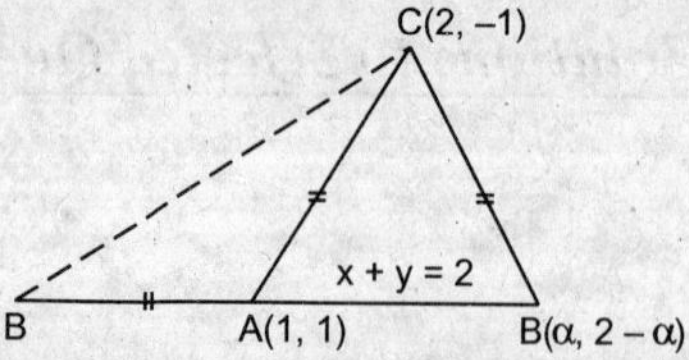

Here, $(\alpha-2)^2+(3-\alpha)^2=1^2+2^2$ when $AC=BC$ (first case).

$\therefore \quad 2\alpha^2-10\alpha+8=0 \quad$ or $\quad \alpha^2-5\alpha+4=0.$

$\therefore \quad \alpha=1,4. \quad \therefore\ 2-\alpha=1,-2. \quad \therefore\ B=(4,-2).$

Also, $(\alpha-1)^2+(1-\alpha)^2=1^2+2^2$ when $AC=AB$ (second case).

$\therefore \quad (\alpha-1)^2=\frac{5}{2}. \quad \therefore\ \alpha=1\pm\sqrt{\frac{5}{2}}. \quad \therefore\ 2-\alpha=1\mp\sqrt{\frac{5}{2}}.$

$\therefore \quad B=\left(1+\sqrt{\frac{5}{2}},1-\sqrt{\frac{5}{2}}\right),\left(1-\sqrt{\frac{5}{2}},1+\sqrt{\frac{5}{2}}\right).$

5.

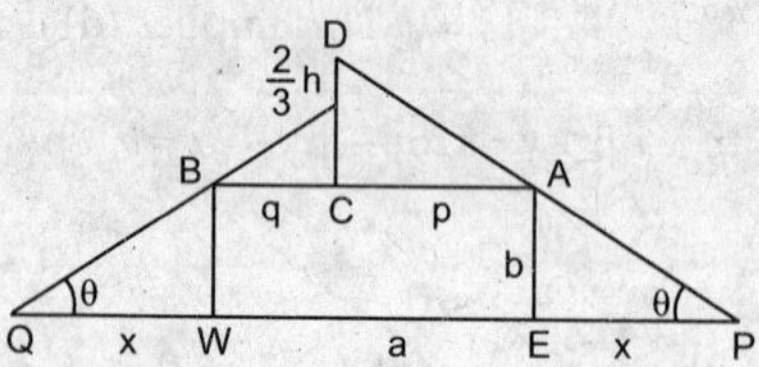

$$\tan\theta=\frac{b}{x}=\frac{h}{AC}=\frac{h}{\frac{pa}{p+q}}=\frac{h(p+q)}{pa}; \quad \therefore\ pab=xh(p+q). \qquad \ldots (1)$$

$$\text{Also, } \tan\theta=\frac{b}{x}=\frac{\frac{1}{3}h}{\frac{qa}{p+q}}=\frac{\frac{1}{3}h(p+q)}{qa}; \quad \therefore\ qab=\frac{1}{3}xh(p+q). \qquad \ldots (2)$$

$\therefore \quad (1), (2) \Rightarrow \frac{p}{q}=3. \quad \therefore\ 3qab=xh(3q+q); \quad \therefore\ xh=\frac{3}{4}ab.$

8. $\Delta=abc\begin{vmatrix}-a & b & c\\ a & -b & c\\ a & b & -c\end{vmatrix}=abc\begin{vmatrix}-a & b & c\\ 0 & 0 & 2c\\ 0 & 2b & 0\end{vmatrix}$

$=abc\cdot 4abc=4(abc)^2.$

But abc is divisible by 3 ! as a, b, c are three consecutive integers.

$\therefore$ Δ is divisible by $4(3\,!)^2$, i.e., 144 and so by 36, 48.

9. As $b-a+a-c+c-b=0$, $x=1$ is a root. So, the other root is 2 or $\frac{1}{2}$.

$$\therefore \quad 1\cdot 2=\frac{c-b}{b-a} \quad \text{or} \quad 1\cdot\frac{1}{2}=\frac{c-b}{b-a}.$$

$$\therefore \quad 2b-2a=c-b \quad \text{or} \quad b-a=2c-2b.$$

$$\therefore \quad 3b=2a+c \quad \text{or} \quad 3b=a+2c.$$

$$\therefore \quad b=\frac{2a+c}{2+1} \quad \text{or} \quad b=\frac{a+2c}{1+2}.$$

$\therefore$ B divides AC in the ratio 2 : 1 or 1 : 2 internally.

10. For $x=0$, x^2+px+q is equal to q which is an integer.

For $x=1$, $1+p+q$ is an integer. But q is an integer. So, p is an integer.

11. Roots are given by $x=0$ and $px^2-(p+1)x+p=0$, i.e.,

$$x=0, \frac{p+1\pm\sqrt{(p+1)^2-4p^2}}{2p} \quad \text{or} \quad x=0, \frac{p+1\pm\sqrt{(3p+1)(1-p)}}{2p}.$$

As p is a positive integer, x is real if $p=1$ only. Then $x=\frac{2\pm 0}{2}=1, 1$.

12. $4x^2+4x+12\lambda^2-47\geq 0$ for all $x\in R$.

So, $D\leq 0$ (as $a=4>0$).

$$\therefore \quad 4^2-4\cdot 4\cdot(12\lambda^2-47)\leq 0 \quad \text{or} \quad 1-12\lambda^2+47\leq 0$$

or $\quad 12\lambda^2\geq 48. \quad \therefore \ \lambda^2\geq 4.$ So, $\lambda\geq 2$ or $\lambda\leq -2$.

13. $f'(x)=4-2x=0$ if $x=2$. So, $f(x)$ has an extremum at $x=2$.

Now $f''(2)=-2<0$.

$\therefore$ $f(x)$ has the maximum at $x=2$.

$$\therefore \quad 5a^2+4(2-1)-2^2=20 \quad \text{or} \quad 5a^2=20. \quad \therefore \ a=2, -2.$$

14. $A+B=\int_0^1 (\sin^{-1}x+\cos^{-1}x)\,dx=\int_0^1 \frac{\pi}{2}\,dx=\frac{\pi}{2}.$

$$A-B=\int_0^1 (\sin^{-1}x-\cos^{-1}x)\,dx=\int_0^1\left(\frac{\pi}{2}-2\cos^{-1}x\right)dx=\frac{\pi}{2}-2\int_0^1 \cos^{-1}x\,dx$$

$$=\frac{\pi}{2}-\left[\{(\cos^{-1}x)x\}_0^1-\int_0^1 x\cdot\frac{-1}{\sqrt{1-x^2}}dx\right]$$

$$=\frac{\pi}{2}+\int_0^1\frac{-2x}{\sqrt{1-x^2}}dx=\frac{\pi}{2}+\int_1^0\frac{dz}{\sqrt{z}}=\frac{\pi}{2}+\left[\frac{z^{1/2}}{1/2}\right]_1^0=\frac{\pi}{2}-2.$$

$\therefore \quad A-B=A+B-2 \quad \therefore\ B=1.$

$\therefore \quad A-B+2=\frac{\pi}{2}$ or $A-1+2=\frac{\pi}{2}\cdot \quad \therefore\ A+1=\frac{\pi}{2}\cdot$

15.

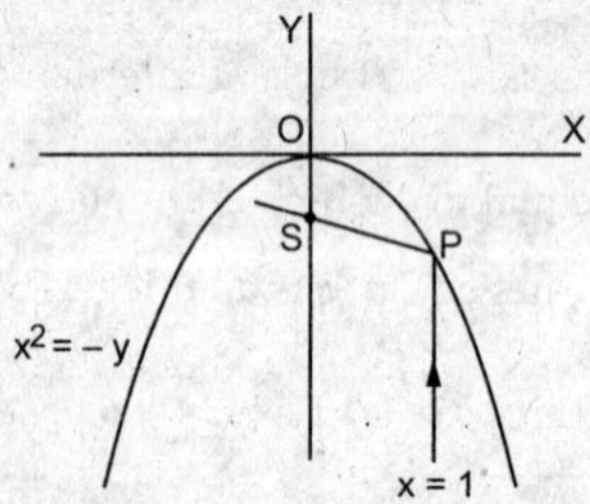

Clearly, the line $x=1$ is parallel to the axis $x=0$ of the parabola. So, after reflection it will pass through the focus $\left(0,-\frac{1}{4}\right)\cdot$

Also, P is $(1,-1)$, obtained by solving $x=1, x^2=-y$.

$\therefore$ the equation of PS is $y+\frac{1}{4}=-\frac{3}{4}x$ and it is satisfied by $\left(\frac{1}{3},-\frac{1}{2}\right)\cdot$

16. $f(x+1)=\{x+1-[x+1]\}\,|\cos\pi(x+1)|=\{x+1-[x]-1\}\,|-\cos\pi x|$

$=\{x-[x]\}\,|\cos\pi x|=f(x)$. So, $f(x)$ is periodic.

Also, periodic functions are many-one functions and so $f(x)$ is many-one.

17. Here, $f(x)=f(-x)=f(x+a)$. Replacing x by $x-\frac{a}{2}$,

$f\left\{-\left(x-\frac{a}{2}\right)\right\}=f\left\{\left(x-\frac{a}{2}\right)+a\right\}$ or $f\left(\frac{a}{2}-x\right)=f\left(\frac{a}{2}+x\right)\cdot$

$f(0+a)=f(0)=0 \quad [\because\ f(x)$ is periodic of period $a]$.

Also, $f(-a)=f(a)=0$. Thus, $f(na)=0$ for all $n\in\mathbb{Z}$.

Graphically,

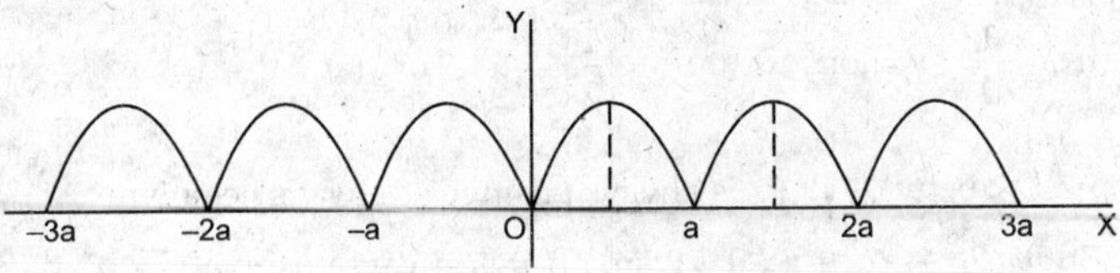

$f(x)$ may or may not be monotone in $\left(0, \frac{a}{2}\right)$ and $f(x)$ has a maximum in $(0, a)$. $f(x)$ is not differentiable at $x = 0$.

18. $\cos A = \dfrac{b^2 + c^2 - a^2}{2bc} \quad \Rightarrow \cos 30° = \dfrac{b^2 + 9 - 3}{2 \cdot b \cdot 3}$

or $\quad 3\sqrt{3}b = b^2 + 6 \quad \Rightarrow b = 2\sqrt{3}, \sqrt{3}.$

$\dfrac{a}{\sin A} = \dfrac{b}{\sin B} \quad \Rightarrow \dfrac{\sqrt{3}}{\sin 30°} = \dfrac{2\sqrt{3}}{\sin B}$ or $\dfrac{\sqrt{3}}{\sin B}\cdot$

$\therefore \quad \sin B = 1, \frac{1}{2}\cdot \qquad \therefore B = 90°, 30° \qquad (\because B \neq 150° \text{ as } A = 30°).$

19. When n is even, let $n = 2m$. So, the numbers are

${}^{2m}C_0, {}^{2m}C_1, {}^{2m}C_2, \ldots, {}^{2m}C_{m-1}, {}^{2m}C_m, {}^{2m}C_{m+1}, \ldots, {}^{2m}C_{2m}.$

The first place can be filled by any one of ${}^{2m}C_0, {}^{2m}C_1, \ldots, {}^{2m}C_{m-1}$ or ${}^{2m}C_{2m}, {}^{2m}C_{2m-1}, \ldots, {}^{2m}C_{m+1}$ and the last place is filled accordingly.

$\therefore$ the first place and the last place can be filled in $2m$ ways. The remaining $2m - 1$ places can be filled by the remaining $2m - 1$ numbers in $(2m - 1)!$ ways.

$\therefore$ the total number of ways to fill according to the condition

$$= 2m \cdot (2m - 1)! = n \cdot (n - 1)! = n!.$$

When n is odd, let $n = 2m + 1$. So, the numbers are

${}^{2m+1}C_0, {}^{2m+1}C_1, \ldots, {}^{2m+1}C_m, {}^{2m+1}C_{m+1}, \ldots, {}^{2m+1}C_{2m+1}.$

As above, the first and last places can be filled in $2(m + 1)$ ways and the rest $2m$ numbers can be arranged in $2m!$ ways.

$\therefore$ the total number of ways to fill according to the condition

$$= 2(m + 1) \cdot 2m! = (n + 1) \cdot (n - 1)! = (n^2 - 1) \cdot (n - 2)!.$$

20. It is the graph of $y = \log_a x$ where $0 < a < 1$.

If $a = \frac{1}{2}$, $y = \log_{1/2} x$ or $\left(\frac{1}{2}\right)^y = x$ or $x \cdot 2^y = 1$.

21. As $D < 0$, $px^2 + qx + r$ is always positive or always negative.

But $p \cdot 2^2 + q \cdot 2 + r > 0$ (given). So it is always positive.

Hence, $px^2 + qx + r$ is positive for $x = 0$ as well as 1.

22. In $[-1, 2]$, $f'(x) = 6x + 12 > 0$. So, $f(x)$ is m.i.

As $\lim\limits_{h \to 0} f(2 - h) = \lim\limits_{h \to 0} f(2 - h) = f(2) = 35$, $f(x)$ is continuous at $x = 2$ also.

So, $f(x)$, being a polynomial function, is continuous in $[-1, 3]$.

As $f'(2 - 0) \neq f'(2 + 0)$, $f(x)$ is not differentiable at $x = 2$.

Also $f(x)$ is m.i. in $[-1, 2]$ and m.d. in $(2, 3]$, being continuous at $x = 2$. So, it has a maximum at $x = 2$.

□

3

Miscellaneous Questions—3

• Type 1 •

Choose the most appropriate option (a, b, c or d).

1. $\lim_{n \to \infty} \left[\frac{1}{n^2} \sec^2 \frac{1}{n^2} + \frac{2}{n^2} \sec^2 \frac{4}{n^2} + \frac{3}{n^2} \sec^2 \frac{9}{n^2} + \dots + \frac{1}{n} \sec^2 1 \right]$ equals

(a) $\frac{1}{2}$ cosec 1 (b) $\frac{1}{2}$ sec 1

(c) $\frac{1}{2}$ tan 1 (d) tan 1

2. If A be a 3×3 matrix and I be the unit matrix of that order such that $A = A^2 + I$ then A^{-1} is equal to

(a) A (b) $A + I$

(c) $I - A$ (d) $A - I$

3. If $w = \frac{z}{z - \frac{1}{3} i}$ and $|w| = 1$ then z lies on

(a) a circle (b) an ellipse

(c) a parabola (d) a straight line

4. The value of $\binom{30}{0}\binom{30}{10} - \binom{30}{1}\binom{30}{11} + \binom{30}{2}\binom{30}{12} - \dots + \binom{30}{20}\binom{30}{30}$ is, where $\binom{n}{r} = {}^nC_r$

(a) $\binom{30}{10}$ (b) $\binom{30}{15}$

(c) $\binom{60}{30}$ (d) $\binom{31}{10}$

5. If a, b, c are in AP and x, y, z are finite numbers such that $x = \sum_{n=0}^{\infty} a^n$, $y = \sum_{n=0}^{\infty} b^n$ and $z = \sum_{n=0}^{\infty} c^n$ then x, y, z are in

(a) AP

(b) GP

(c) HP

(d) Arithmetico-geometric progression

6. In a triangle ABC, $a^2 + b^2 = c^2$. If R and r are its circumradius and inradius respectively then $R + r$ is equal to

(a) AM of a and b (b) AM of b and c

(c) AM of c and a (d) perimeter

7. Suppose $f(x)$ is differentiable at $x = 1$ and $\lim_{h \to 0} \frac{f(1+h)}{h} = 5$. Then $f'(1)$ equals

(a) 4 (b) 3

(c) 6 (d) 5

8. If f is a real-valued differentiable function satisfying $|f(x) - f(y)| \le (x-y)^2$ for all $x, y \in R$ and $f(0) = 0$ then $f(1)$ equals

(a) 0 (b) –1

(c) 1 (d) 2

9. If $\int_{\sin x}^{1} t^2 \cdot f(t)dt = 1 - \sin x$ then $f\left(\frac{1}{\sqrt{3}}\right)$ is equal to

(a) $\frac{1}{3}$ (b) –3

(c) 3 (d) $\sqrt{3}$

10. If $f(x)$ is a twice differentiable function for which $f(1) = 1, f(2) = 4$ and $f(3) = 9$ then

(a) $f''(x) = 2$ for all $x \in (1, 3)$

(b) $f''(x) = f'(x) = 5$ for some $x \in (2, 3)$

(c) $f''(x) = 3$ for all $x \in (2, 3)$

(d) $f''(x) = 2$ for some $x \in (1, 3)$

11. Let f be differentiable for all x. If $f(1) = -2$ and $f'(x) \ge 2$ for $x \in [1, 6]$ then the value of $f(6)$ is

(a) less than 8 (b) greater than or equal to 8

(c) equal to 5 (d) less than 5

12. The value of $\int_1^e \left\{\frac{\log x - 1}{1+(\log x)^2}\right\}^2 dx$ is equal to

(a) $\frac{1}{2}(e-2)$ (b) $-\frac{(e-1)^2}{2(1+e^2)}$

(c) $\frac{1}{2}(e-1)$ (d) $\frac{1}{2}e$

13. Let A and B be two events such that $P(\overline{A \cup B}) = \frac{1}{6}$, $P(A \cap B) = \frac{1}{4}$ and $P(\overline{A}) = \frac{1}{4}$ where $\overline{A}$ = complementary of event A. Then A and B are

(a) equally likely but not indepenent

(b) equally likely and mutually exclusive

(c) mutually exclusive and independent

(d) independent but not equally likely

14. Let $\vec{a} = \vec{i} - \vec{k}$, $\vec{b} = x\,\vec{i} + \vec{j} + (1-x)\,\vec{k}$ and $\vec{c} = y\,\vec{i} + x\,\vec{j} + (1-x-y)\,\vec{k}$. Then $[\vec{a}\ \vec{b}\ \vec{c}]$ depends on

(a) only x (b) only y

(c) neither x nor y (d) both x and y

15. If $\vec{a}$, $\vec{b}$ and $\vec{c}$ are three nonzero noncoplanar vectors and $\vec{p} = \vec{b} - \frac{\vec{b} \cdot \vec{a}}{|\vec{a}|^2}\vec{a}$,

$\vec{q} = \vec{b} + \frac{\vec{b} \cdot \vec{a}}{|\vec{a}|^2}\vec{a}$, $\vec{r} = \vec{c} - \frac{\vec{c} \cdot \vec{a}}{|\vec{a}|^2}\vec{a} + \frac{\vec{b} \cdot \vec{c}}{|\vec{c}|^2}\vec{p}$, $\vec{s} = \vec{c} - \frac{\vec{c} \cdot \vec{a}}{|\vec{a}|^2}\vec{a} - \frac{\vec{p} \cdot \vec{c}}{|\vec{p}|^2}\vec{p}$,

$\vec{t} = \vec{c} - \frac{\vec{c} \cdot \vec{a}}{|\vec{a}|^2}\vec{a} + \frac{\vec{b} \cdot \vec{c}}{|\vec{c}|^2}\vec{p}$, $\vec{u} = \vec{c} - \frac{\vec{c} \cdot \vec{a}}{|\vec{a}|^2}\vec{a} - \frac{\vec{b} \cdot \vec{c}}{|\vec{b}|^2}\vec{p}$

then the set of orthogonal vectors is

(a) $(\vec{a}, \vec{p}, \vec{t})$ (b) $(\vec{a}, \vec{p}, \vec{s})$

(c) $(\vec{a}, \vec{p}, \vec{u})$ (d) $(\vec{a}, \vec{q}, \vec{s})$

16. A six faced fair dice is thrown until 1 comes. The probability that 1 comes in even number of trials, is

(a) $\frac{5}{11}$ (b) $\frac{5}{6}$

(c) $\frac{6}{11}$ (d) $\frac{1}{6}$

17. Four flats are available in Krishna Apartment. Four persons visit the flats and each gives his choice to the promoter without consulting others. The probability that all will give different choices, is

(a) $\frac{1}{8}$ (b) $\frac{1}{64}$

(c) $\frac{3}{32}$ (d) $\frac{3}{8}$

18. The area of the greatest rectangle that can be inscribed in the ellipse $\frac{x^2}{a^2}+\frac{y^2}{b^2}=1$ is

(a) ab (b) $2ab$

(c) $\frac{a}{b}$ (d) $\sqrt{ab}$

19. The value of $(\vec{a}\times\vec{i})^2+(\vec{a}\times\vec{j})^2+(\vec{a}\times\vec{k})^2$ is equal to

(a) $\vec{a}^2$ (b) $3\vec{a}^2$

(c) $4\vec{a}^2$ (d) $2\vec{a}^2$

20. The minimum area of the triangle formed by a tangent to the ellipse $\frac{x^2}{a^2}+\frac{y^2}{b^2}=1$ and the coordinate axes, is

(a) ab (b) $\frac{a^2+b^2}{2}$

(c) $\frac{(a+b)^2}{2}$ (d) $\frac{a^2+ab+b^2}{3}$

21. If y is a function of x given by $ydx+y^2dy=xdy$ where $y>0$ and $y(1)=1$ then $y(-3)$ is

(a) 3 (b) 2

(c) 1 (d) 5

22. The function $y=||x|-1|$ is differentiable for all $x\in R$ except for the values

(a) $0, 1, -1$ (b) ± 1

(c) 1 (d) -1

23. 3 coins of radii 1 unit each are kept inside an equilateral triangle so that they touch each other and also the sides of the triangle. The area of the triangle is

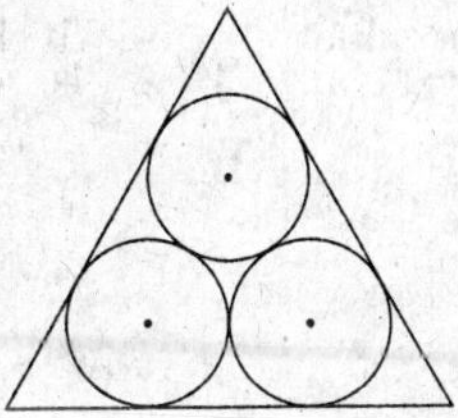

(a) $(4+2\sqrt{3})$ unit^2

(b) $(6+4\sqrt{3})$ unit^2

(c) $\left(12+\frac{7\sqrt{3}}{4}\right)$ unit^2

(d) $\left(3+\frac{7\sqrt{3}}{4}\right)$ unit^2

24. If a, b, c are integers not all equal and $w = \sqrt[3]{1}, (w \neq 1)$ then the minimum value of $|a + bw + cw^2|$ is

(a) 0
(b) 1
(c) $\frac{\sqrt{3}}{2}$
(d) $\frac{1}{2}$

25. If both the roots of the quadratic equation $x^2 - 2kx + k^2 + k - 5 = 0$ are less than 5 then k lies in the interval

(a) $(6, \infty)$
(b) $(5, 6]$
(c) $[4, 5]$
(d) $(-\infty, 4)$

26. If $a_1, a_2, a_3, \dots, a_n, \dots$ are in GP then the determinant

$$\Delta = \begin{vmatrix} \log a_n & \log a_{n+1} & \log a_{n+2} \\ \log a_{n+3} & \log a_{n+4} & \log a_{n+5} \\ \log a_{n+6} & \log a_{n+7} & \log a_{n+8} \end{vmatrix}$$

is equal to

(a) 0
(b) 1
(c) 2
(d) 4

27. A real valued function $f(x)$ satisfies the functional equation $f(x-y) = f(x)f(y) - f(a-x)(a+y)$ where a is a given constant and $f(0) = 1$. Then $f(2a - x)$ equals

(a) $f(x)$
(b) $-f(x)$
(c) $f(-x)$
(d) $f(a) + f(a-x)$

28. Let $f : (-1, 1) \to B$ be a function defined by $f(x) = \tan^{-1}\frac{2x}{1-x^2}$. Then f is both one-one and onto function when B is the interval

(a) $\left[0, \frac{\pi}{2}\right)$
(b) $\left(0, \frac{\pi}{2}\right)$
(c) $\left(-\frac{\pi}{2}, \frac{\pi}{2}\right)$
(d) $\left[-\frac{\pi}{2}, \frac{\pi}{2}\right]$

29. The value of a for which the sum of the squares of the roots of the equation $x^2 - (a-2)x - a - 1 = 0$ assume the least value is

(a) 0 (b) 1

(c) 2 (d) 3

30. The system of equations $\alpha x + y + z = \alpha - 1$, $x + \alpha y + z = \alpha - 1$, $x + y + \alpha z = \alpha - 1$ has no solution if α is

(a) either –2 or 1 (b) –2

(c) 1 (d) 2

31. A circle is given by $x^2 + (y-1)^2 = 1$. Then the locus of the centre of the circle touching the given circle and the x-axis, is

(a) $\{(x, y) \mid x^2 = 4y\} \cup \{(x, y) \mid y \le 0\}$

(b) $\{(x, y) \mid x^2 + (y-1)^2 = 4\} \cup \{(x, y) \mid y \le 0\}$

(c) $\{(x, y) \mid x^2 = 4y\}$

(d) $\{(x, y) \mid x^2 = 4y\} \cup \{(0, y) \mid y \le 0\}$

32. Let $\cos(\alpha - \beta) = 1$ and $\cos(\alpha + \beta) = \frac{1}{e}$, where $\alpha, \beta \in [-\pi, \pi]$ and e is the exponential number. The number of values of the pair (α, β) satisfying both the equations, is

(a) 1 (b) 2

(c) 3 (d) 4

33. The value of $\int_{-2}^{0} \{x^3 + 3x^2 + 3x + 3 + (x+1)\cos(x+1)\}dx$ is

(a) –4 (b) 0

(c) 4 (d) 6

34. Let α, β be roots of the equation $ax^2 + bx + c = 0$ and $\Delta = b^2 - 4ac$. If $\alpha + \beta, \alpha^2 + \beta^2, \alpha^3 + \beta^3$ are in GP then

(a) $\Delta \neq 0$ (b) $b \cdot \Delta = 0$

(c) $c \cdot \Delta = 0$ (d) $\Delta = 0$

35. The number of measures of the angle A for which $\cos A$, $\cos 2A$ and $\cos 3A$ are in GP, is

(a) 0 (b) 2

(c) 4 (d) infinite

36. If z_1 and z_2 are two nonzero complex numbers such that $|z_1 + z_2| = |z_1| + |z_2|$ then $\arg z_1 - \arg z_2$ is equal to

(a) $-\pi$ (b) $\frac{\pi}{2}$

(c) $-\frac{\pi}{2}$ (d) 0

37. If $a^2 + b^2 + c^2 + 2 = 0$ and $f(x) = \begin{vmatrix} 1 + a^2x & (1+b^2)x & (1+c^2)x \\ (1+a^2)x & 1+b^2x & (1+c^2)x \\ (1+a^2)x & (1+b^2)x & 1+c^2x \end{vmatrix}$ then $f(x)$ is a polynomial of degree

(a) 0 (b) 1

(c) 2 (d) 3

38. Let α and β be the distinct roots of $ax^2 + bx + c = 0$ then

$$\lim_{n \to \alpha} \frac{1 - \cos(ax^2 + bx + c)}{(x - \alpha)^2}$$

is equal to

(a) 0 (b) $\frac{1}{2}a^2(\alpha - \beta)^2$

(c) $\frac{1}{2}(\alpha - \beta)^2$ (d) $-\frac{1}{2}a^2(\alpha - \beta)^2$

39. If $I_1 = \int_0^1 2^{x^2} dx$, $I_2 = \int_0^1 2^{x^3} dx$, $I_3 = \int_1^2 2^{x^2} dx$ and $I_4 = \int_1^2 2^{x^3} dx$ then

(a) $I_1 > I_2$ (b) $I_2 > I_1$

(c) $I_1 = I_2$ (d) $I_3 > I_4$

40. The angle between the lines $2x = 3y = -z$ and $6x = -y = -4z$ is

(a) $\frac{\pi}{2}$ (b) 0

(c) $\frac{\pi}{6}$ (d) $\frac{\pi}{4}$

41. If the angle θ between the line $\frac{x+1}{1} = \frac{y-1}{2} = \frac{z-2}{2}$ and the plane $2x - y + \sqrt{\lambda} z + 4 = 0$ is such that $\sin\theta = \frac{1}{3}$ then the value of λ is

(a) $-\frac{3}{5}$ (b) $\frac{5}{3}$

(c) $-\frac{4}{3}$ (d) $\frac{3}{4}$

42. The tangent to the curve $y = x^2 + 6$ at $(1, 7)$ touches the circle $x^2 + y^2 + 16x + 12y + c = 0$ at the point

(a) $(-6, -11)$ (b) $(-9, -13)$

(c) $(-10, -15)$ (d) $(-6, -7)$

43. The distance between the line $\vec{r} = 2\vec{i} - 2\vec{j} + 3\vec{k} + \lambda(\vec{i} - \vec{j} + 4\vec{k})$ and the plane $\vec{r} \cdot (\vec{i} + 5\vec{j} + \vec{k}) = 5$ is

(a) $\frac{10}{3\sqrt{3}}$ (b) $\frac{10}{9}$

(c) $\frac{10}{3}$ (d) 0

44. If $f(x)$ is continuous and differentiable function and $f\left(\frac{1}{n}\right) = 0$ for all $n \geq 1$ and $n \in I$ then

(a) $f(x) = 0, x \in (0, 1]$

(b) $f(0) = 0, f'(0) = 0$

(c) $f'(0) = 0 = f''(0), x \in (0, 1]$

(d) $f(0) = 0$ and $f'(0)$ need not be zero

45. If $f(x)$ and $g(x)$ are real functions defined by

$f(x) = 0$, x is rational; x, x is irrational and $g(x) = 0$, x is irrational; x, x is rational

then $(f - g)(x)$ is

(a) one-one and onto (b) neither one-one nor onto

(c) one-one but not onto (d) onto but not one-one

46. The value of $\int_{-\pi}^{\pi} \frac{\cos^2 x}{1 + a^x} dx, a > 0$ is

(a) $\frac{\pi}{2}$ (b) $a\pi$

(c) 2π (d) $\frac{\pi}{a}$

47. If the pair of lines $ax^2 + 2(a+b)xy + by^2 = 0$ lie along two diameters of the circle $x^2 + y^2 = 4$ such that the circle is divided into four sectors in which one sector has double the area of another sector then

(a) $a = 2b$ (b) $2a = b$

(c) $a = b$ (d) $a + b + 1 = 0$

48. A rectangle with side lengths $2m - 1$ and $2n - 1$ is divided into unit squares by drawing parallel lines as shown in the diagram. Then the number of rectangles possible with odd side lengths is

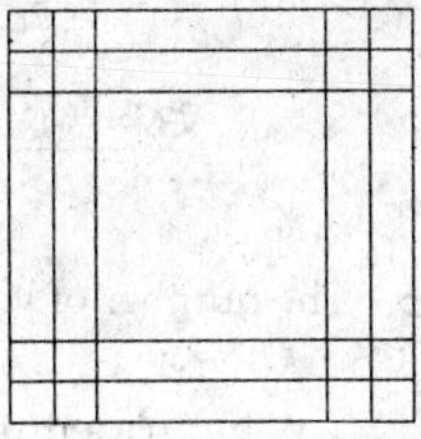

(a) $(m + n + 1)^2$

(b) 4^{m+n-1}

(c) $m^2 n^2$

(d) $mn(m+1)(n+1)$

49. Let $A = \{x \mid x \in N, x \leq 15\}$. If $3^m + 3^n$ is divisible by 5 where $(m, n) \in A \times A$ then the number of possible values of the ordered pair (m, n) is

(a) 32 (b) 56

(c) 28 (d) 64

50. If z_1, z_2 be two nth roots of unity such that they represent two points A, B in the Argand plane where $\angle AOB = 60°$ and O is the origin than the positive integer n is of the form

(a) $4k, k \in N$ (b) $4k + 3, k \in N$

(c) $6k, k \in N$ (d) $6k + 5, k \in N$

51. If the circle passing through the points (0, 0), (1, 0) and (0, 1) also passes through the point $\left(\lambda, \frac{1}{\lambda}\right)$ then λ is

(a) 1 (b) –1

(c) 2 (d) –2

52. If in a ΔABC, $\tan A + \tan B + \tan C$ has the value 9 then the value of $\cot A \cdot \cot B \cdot \cot C$ is equal to

(a) 9 (b) 1

(c) $\frac{1}{9}$ (d) none of these

53. The number of negative integral solutions of the equation $x_1 + x_2 + x_3 + x_4 + 15 = 0$ is

(a) 364 (b) ${}^{15}C_3$

(c) ${}^{14}C_4$ (d) none of these

54. Which of the following is equal to $\sqrt[3]{-1}$?

(a) $\dfrac{\sqrt{3}+\sqrt{-1}}{2}$ (b) $\dfrac{-\sqrt{3}+\sqrt{-1}}{\sqrt{-4}}$

(c) $\dfrac{\sqrt{3}-\sqrt{-1}}{\sqrt{-4}}$ (d) $-\sqrt{-1}$

55. The number of values of a for which

$$(x^2 + x + 1)a^2 - (x^2 + 2)a - 3(2x^2 + 3x + 1) = 0$$

will be an identity in x is

(a) one (b) zero

(c) three (d) two

56. If α, β are two values of θ satisfying the equation $\sec^2\theta + p\tan\theta + q = 1$ then

(a) $\tan(\alpha+\beta) = \dfrac{p}{q-1}$ (b) $\tan(\alpha-\beta) = \dfrac{p}{q-1}$

(c) $\tan(\alpha+\beta) = \dfrac{q}{p-1}$ (d) $\tan(\alpha-\beta) = \dfrac{q}{p-1}$

57. The set of values of a for which the graph of the function $f(x) = a(x^2 + 1) + x$ is always above the x-axis is

(a) $\left(\frac{1}{2}, +\infty\right)$ (b) $\left(-\infty, -\frac{1}{2}\right)$

(c) $\left(-\frac{1}{2}, \frac{1}{2}\right)$ (d) $(0, +\infty)$

58. $\lim\limits_{x \to -\infty} \dfrac{x^3 - x}{\sqrt{1+x^6}}$ is equal to

(a) -1 (b) 1

(c) 0 (d) 3

59. Three particles A, B and C start to move from three vertices of a triangle with the same uniform speed along the sides of the triangle at random. The probability that they will collide is

(a) $\frac{3}{8}$ (b) $\frac{1}{4}$

(c) $\frac{3}{4}$ (d) $\frac{1}{2}$

60. $\{(\vec{a} \times \vec{b}) \times \vec{a}\} \times \vec{b} + \vec{a} \times \{\vec{b} \times (\vec{a} \times \vec{b})\}$ is equal to

(a) $2(\vec{a} \cdot \vec{b})\vec{b} \times \vec{a}$ (b) $\vec{0}$

(c) $2\vec{a} \times \vec{b}$ (d) $(\vec{a} \times \vec{b})^2$

61. The four common tangents to the ellipses $\frac{x^2}{9} + \frac{y^2}{4} = 1$ and $\frac{x^2}{4} + \frac{y^2}{9} = 1$ form

(a) a rectangle of area 13 unit^2

(b) a square of area 26 unit^2

(c) a parallelogram which is neither a square nor a rectangle

(d) a rhombus

62. A bag contains 12 tickets numbered 1, 2, 3, ..., 12. Five of them are drawn at random and arranged in the ascending order of their numbers. What is the probability that the third in the order is 5?

(a) $\frac{7}{44}$ (b) $\frac{5}{12}$

(c) $\frac{2}{3}$ (d) $\frac{1}{2}$

63. The lines $\frac{x}{0} = \frac{y}{1} = \frac{z}{1}, \frac{x}{2} = \frac{y}{5} = \frac{z}{7}$ and $\frac{x}{1} = \frac{y-1}{2} = \frac{z-1}{3}$

(a) are three concurrent lines

(b) are three lines of which at least two are skew lines

(c) are coplanar lines forming a triangle of area $2\sqrt{3}$ unit^2

(d) form a triangle of area $\sqrt{3}$ unit^2

64. If $(1-y)^m \cdot (1+y)^n = 1 + a_1 y + a_2 y^2 + \ldots + a_{m+n} y^{m+n}$, where $m \in N$, $n \in N$ and $a_1 = a_2 = 10$, then (m, n) is

(a) (20, 45) (b) (35, 20)

(c) (45, 35) (d) (35, 45)

65. If (a, a^2) falls inside the angle made by the lines $y = \frac{1}{2}x\ (x > 0)$ and $y = 3x\ (x > 0)$ then a belongs to

(a) $\left(0, \frac{1}{2}\right)$ (b) $(3, +\infty)$

(c) $\left(\frac{1}{2}, 3\right)$ (d) $\left(-3, -\frac{1}{2}\right)$

66. If $\frac{1}{(1-ax)(1-bx)} = a_0 + a_1x + a_2x^2 + \ldots$ then a_n equals

(a) $\frac{b^n - a^n}{b - a}$ (b) $\frac{a^n - b^n}{b - a}$

(c) $\frac{a^{n+1} - b^{n+1}}{b - a}$ (d) $\frac{b^{n+1} - a^{n+1}}{b - a}$

67. The locus of vertices of the family of parabolas $y = \frac{a^3x^2}{3} + \frac{a^2x}{2} - 2a$, where a is a parameter, is

(a) a circle (b) an ellipse

(c) a parabola (d) a rectangular hyperbola

68. Let E^c denote the complement of an event. Let E, F, G be pairwise independent events with $P(G) > 0$ and $P(E \cap F \cap G) = 0$. Then $P(E^c \cap F^c \mid G)$ equals

(a) $P(E^c) + P(F^c)$ (b) $P(E^c) - P(F^c)$

(c) $P(E^c) - P(F)$ (d) $P(E) - P(F^c)$

Answers

1. c	**2.** c	**3.** d	**4.** a	**5.** c
6. a	**7.** d	**8.** a	**9.** c	**10.** d
11. b	**12.** a	**13.** d	**14.** a	**15.** b
16. a	**17.** c	**18.** b	**19.** d	**20.** a
21. a	**22.** a	**23.** b	**24.** b	**25.** d
26. a	**27.** b	**28.** c	**29.** b	**30.** b
31. d	**32.** c	**33.** c	**34.** c	**35.** a
36. d	**37.** c	**38.** b	**39.** a	**40.** a
41. b	**42.** d	**43.** a	**44.** b	**45.** a
46. a	**47.** c	**48.** c	**49.** b	**50.** c
51. a	**52.** c	**53.** a	**54.** b	**55.** a
56. a	**57.** a	**58.** a	**59.** c	**60.** a
61. b	**62.** a	**63.** d	**64.** d	**65.** c
66. d	**67.** d	**68.** c		

Hints and Solutions to Selected Questions

1. The limit $= \lim_{n \to \infty} \sum_{r=1}^{n} \frac{r}{n^2} \sec^2 \frac{r^2}{n^2} = \lim_{n \to \infty} \sum_{r=1}^{n} \frac{1}{n} \left\{ \frac{r}{n} \cdot \sec^2 \left(\frac{r}{n} \right)^2 \right\}$

$$= \int_0^1 x \sec^2 x^2 \, dx = \frac{1}{2} \int_0^1 \sec^2 z \, dz \qquad \text{(putting } x^2 = z).$$

2. $A = A^2 + I \Rightarrow A^{-1} A = A^{-1}(A^2 + I) \Rightarrow I = A^{-1}(AA) + A^{-1} I$

$\Rightarrow I = (A^{-1}A) A + A^{-1} = IA + A^{-1} = A + A^{-1} \Rightarrow A^{-1} = I - A.$

3. $|w| = 1 = \left| \frac{z}{z - \frac{1}{3} i} \right| = \frac{|z|}{\left| z - \frac{1}{3} i \right|} \Rightarrow \left| z - \frac{1}{3} i \right| = |z|$

$\Rightarrow \left| x + i\left(y - \frac{1}{3} \right) \right| = |x + iy| \Rightarrow x^2 + \left(y - \frac{1}{3} \right)^2 = x^2 + y^2$

$\Rightarrow -2 \cdot \frac{1}{3} y + \frac{1}{9} = 0 \Rightarrow y = \frac{1}{6}.$

4. $(1 + x)^{30} = {}^{30}C_0 + {}^{30}C_1 x + {}^{30}C_2 x^2 + \ldots + {}^{30}C_{30} x^{30}$

$(x - 1)^{30} = {}^{30}C_0 x^{30} - {}^{30}C_1 x^{29} + {}^{30}C_2 x^{28} - \ldots + {}^{30}C_{30}$

Multiplying these and equating coefficient of x^{20} on both sides,

series = coefficient of x^{20} in $(1 + x)^{30} \cdot (x - 1)^{30}$, i.e., $(1 - x^2)^{30}$

$$= {}^{30}C_{10} (-1)^{10} = \binom{30}{10}.$$

5. $x = \frac{1}{1 - a}, \; y = \frac{1}{1 - b}, \; z = \frac{1}{1 - c}$. So, $\frac{1}{x} = 1 - a, \; \frac{1}{y} = 1 - b, \; \frac{1}{z} = 1 - c.$

a, b, c are in AP $\Rightarrow -a, -b, c$ are in AP $\Rightarrow 1 - a, 1 - b, 1 - c$ are in AP.

6. Clearly, $\angle C = 90°$, $2R = \frac{c}{\sin C} \Rightarrow R = \frac{c}{2 \sin 90°} = \frac{c}{2}.$

Again, $r = (s - c) \tan \frac{C}{2} = \left(\frac{a + b + c}{2} - c \right) \tan 45° = \frac{a + b - c}{2}.$

$\therefore \quad R + r = \frac{a + b}{2}.$

7. $f'(1) = \lim_{h \to n} \dfrac{f(1+h) - f(1)}{h}$ is finite $\Rightarrow \lim_{h \to 0} \dfrac{f(1+h)}{h} - \lim_{h \to 0} \dfrac{f(1)}{h}$ is finite

$\Rightarrow \quad 5 - \lim_{h \to 0} \dfrac{f(1)}{h}$ is finite $\Rightarrow f(1) = 0$, otherwise $5 - \infty$ is finite (absurd).

$\therefore \quad f'(1) = \lim_{h \to 0} \dfrac{f(1+h)}{h} = 5.$

8. $|f(x) - f(y)| < |x - y|^2 \Rightarrow \lim_{x \to y} \left| \dfrac{f(x) - f(y)}{x - y} \right| \le \lim_{x \to y} |x - y| \to 0$

$\therefore \quad \left| \lim_{x \to y} \dfrac{f(x) - f(y)}{x - y} \right| = 0 \Rightarrow |f'(y)| = 0 \Rightarrow f'(y) = 0$ for all real y.

$\therefore \quad f(y) = c$. So, $f(0) = 0 = c$. $\quad \therefore \quad f(y) = 0 \quad \therefore \quad f(1) = 0.$

9. Taking $\sin x = y$, $-\int_1^y t^2 \cdot f(t)\, dt = 1 - y$. Differentiating w.r.t. y,

$-t^2 \cdot f(t) \Big]_{t=y} = -1 \Rightarrow y^2 \cdot f(y) = 1 \Rightarrow f(y) = \dfrac{1}{y^2}$. So, $f\left(\dfrac{1}{\sqrt{3}}\right) = 3.$

10. Here, $f(1) = 1^2, f(2) = 2^2, f(3) = 3^2$. Observing these, take $\phi(x) = f(x) - x^2$. Then $\phi(1) = \phi(2) = \phi(3) = 0$. Also $\phi(x)$ is twice differentiable.

Using Rolle's theorem on $\phi(x)$ in $[1, 2]$, we get $\phi'(a) = 0$ for some $a \in (1, 2)$.

Using Rolle's theorem on $\phi(x)$ in $[2, 3]$, we get $\phi'(b) = 0$ for some $b \in (2, 3)$.

Thus $\phi'(a) = \phi'(b) = 0$. Also, $\phi'(x)$ is differentiable in $[a, b]$.

$\therefore$ by using Rolle's theorem on $\phi'(x)$ in $[a, b]$ we get $\phi''(c) = 0$ for some $c \in (a, b)$.

$\therefore \ \phi''(x) = 0$ for some $x \in (a, b) \subset (1, 3) \Rightarrow f''(x) - 2 = 0$

$\Rightarrow f''(x) = 2$ for some $x \in (1, 3)$.

11. Using Lagrange's theorem on $f(x)$ in $[1, 6]$,

$\dfrac{f(6) - f(1)}{6 - 1} = f'(c)$, where $c \in (1, 6)$.

$\therefore \quad \dfrac{f(6) - (-2)}{5} = f'(c) \ge 2 \Rightarrow f(6) + 2 \ge 10.$

12. $I = \int_0^1 \left(\dfrac{z - 1}{1 + z^2}\right)^2 \cdot e^z\, dz$ (substituting $\log x = z$).

$$= \int_0^1 \left\{ \frac{1}{1+z^2} - \frac{2z}{(1+z^2)^2} \right\} e^z \, dz$$

$$= \int_0^1 \left\{ \frac{1}{1+z^2} + \frac{d}{dx}\left(\frac{1}{1+z^2} \right) \right\} e^z \, dz = \frac{e^z}{1+z^2} \Bigg]_0^1 = \frac{e}{2} - 1.$$

13. $P(A \cup B) = P(A) + P(B) - P(A \cap B) \Rightarrow \left(1 - \frac{1}{6}\right) = \left(1 - \frac{1}{4}\right) + P(B) - \frac{1}{4}$

$$\Rightarrow P(B) = \frac{1}{3}.$$

$\therefore \; P(A) \neq P(B)$ and $P(A \cap B) = P(A) \cdot P(B)$.

15. $\vec{a} \cdot \vec{p} = \vec{a} \cdot \vec{b} - \frac{\vec{b} \cdot \vec{a}}{|\vec{a}|^2} \vec{a} \cdot \vec{a} = 0$ and $\vec{a} \cdot \vec{q} = 2\,\vec{a} \cdot \vec{b} \neq 0$

$$\vec{a} \cdot \vec{t} = 0, \;\; \vec{p} \cdot \vec{t} = \vec{p} \cdot \vec{c} - \frac{\vec{c} \cdot \vec{a}}{|\vec{a}|} \vec{p} \cdot \vec{a} + \frac{\vec{b} \cdot \vec{c}}{|\vec{c}|^2} \vec{p} \cdot \vec{p} = \vec{p} \cdot \vec{c} + \frac{\vec{b} \cdot \vec{c}}{|\vec{c}|^2} \vec{p}^{\,2} \neq 0,$$

$$\vec{a} \cdot \vec{s} = \vec{a} \cdot \vec{c} - \frac{\vec{c} \cdot \vec{a}}{|\vec{a}|^2} \vec{a} \cdot \vec{a} - \frac{\vec{p} \cdot \vec{c}}{|\vec{p}|^2} \vec{p} \cdot \vec{a} = 0,$$

$\vec{p} \cdot \vec{s} = \vec{p} \cdot \vec{c} - \frac{\vec{c} \cdot \vec{a}}{|\vec{a}|} \vec{p} \cdot \vec{a} - \frac{\vec{p} \cdot \vec{c}}{|\vec{p}|^2} \vec{p} \cdot \vec{p} = 0.$ So, $(\vec{a}, \vec{p}, \vec{s})$ are orthogonal vectors.

16. E = event of throwing 1. Then $P(E) = \frac{1}{6}$, $P(\bar{E}) = \frac{5}{6}$.

Required probability $= P(\bar{E}E) + P(\bar{E}\,\bar{E}\,\bar{E}\,E) + P(\bar{E}\,\bar{E}\,\bar{E}\,\bar{E}\,\bar{E}\,E) + \ldots$

$$= P(\bar{E}) \cdot P(E) + \{P(\bar{E})\}^3 \cdot P(E) + \{P(\bar{E})\}^5 \cdot P(E) + \ldots \text{ to } \infty$$

$$= \frac{5}{6} \cdot \frac{1}{6} + \left(\frac{5}{6}\right)^3 \cdot \frac{1}{6} + \left(\frac{5}{6}\right)^5 \cdot \frac{1}{6} + \ldots \text{ to } \infty = \frac{5}{36} \cdot \frac{1}{1 - \left(\frac{5}{6}\right)^2}.$$

17. Probability $= \frac{n(E)}{n(S)} = \frac{4 \times 3 \times 2 \times 1}{4 \times 4 \times 4 \times 4}$.

18. $A = 2a\cos\phi \cdot 2b\sin\phi$

$= 2ab\sin 2\phi$

$\therefore \; A_{\max} = 2ab$

($\because$ max $\sin 2\phi = 1$).

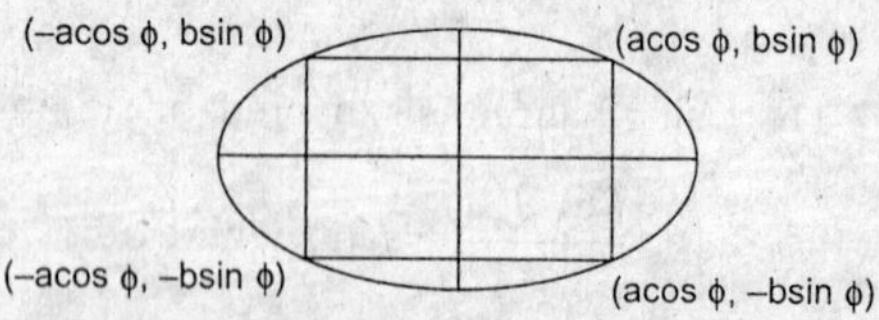

19. Let $\vec{a} = a_1\vec{i} + a_2\vec{j} + a_3\vec{k}$. Then $\vec{a} \times \vec{i} = -a_2\vec{k} + a_3\vec{j}$. So, $(\vec{a} \times \vec{i})^2 = a_2^2 + a_3^2$.

Similarly, $(\vec{a} \times \vec{j})^2 = a_1^2 + a_3^2$, $(\vec{a} \times \vec{k})^2 = a_1^2 + a_2^2$.

$\therefore$ value $= 2(a_1^2 + a_2^2 + a_3^2) = 2\,\vec{a}^{\,2}$.

20. The equation of any tangent to $\frac{x^2}{a^2}+\frac{y^2}{b^2}=1$ is $\frac{x\cos\phi}{a}+\frac{y\sin\phi}{b}=1.$

The intercepts of this on the axes are $a\sec\phi$ and $b\operatorname{cosec}\phi$.

$\therefore$ the area of the triangle $=\frac{1}{2}a\sec\phi\cdot b\operatorname{cosec}\phi=\frac{ab}{\sin 2\phi}\geq ab$

(because $|\sin 2\phi|\leq 1$).

21. Here, $\frac{y\,dx-x\,dy}{y^2}+dy=0 \Rightarrow d\left(\frac{x}{y}\right)+dy=0 \Rightarrow \frac{x}{y}+y=e.$

Now, $y(1)=1 \Rightarrow \frac{1}{1}+1=c \Rightarrow c=2.$ So, $\frac{x}{y}+y=2.$

When $x=-3$, $\frac{-3}{y}+y=2 \Rightarrow y^2-2y-3=0 \Rightarrow y=3,-1.$

But $y>0$. So $y(-3)=3$.

22. The turning points of definition are $|x|=0$ (i.e., $x=0$) and $|x|=1$ (i.e., $x=\pm1$).

$\therefore$ for $x<-1$, $|x|>1$ and so $y=|x|-1=-x-1$;

for $-1\leq x<0$, $|x|<1$ and so $y=1-|x|=1-(-x)=1+x$;

for $0\leq x<1$, $|x|<1$ and so $y=1-|x|=1-x$;

for $x\geq 1$, $|x|\geq 1$ and so $y=|x|-1=x-1$.

So, the graph is as follows.

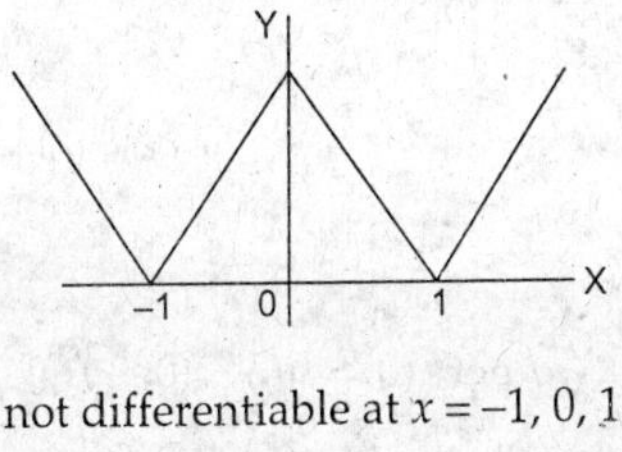

So, the function is not differentiable at $x=-1, 0, 1$.

23. Here, each side $=2x+2=2\sqrt{3}+2$ $\left(\because \frac{1}{x}=\tan 30^\circ\right)$.

$\therefore$ area of the equilateral triangle $=\frac{\sqrt{3}}{4}(2\sqrt{3}+2)^2.$

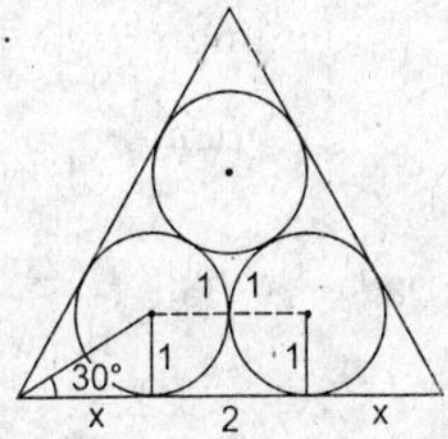

24. $|a + b\omega + c\omega^2| = \left|a + b \cdot \frac{-1+\sqrt{3}i}{2} + c \cdot \frac{-1-\sqrt{3}i}{2}\right|$

$$= \sqrt{\left(a - \frac{b}{2} - \frac{c}{2}\right)^2 + \frac{3}{4}(b-c)^2}$$

$$= \sqrt{\{a^2 + b^2 + c^2 - bc - ca - ab\}}$$

$$= \sqrt{\frac{1}{2}\{(a-b)^2 + (b-c)^2 + (c-a)^2}$$

If $a = b = c$ then the value is 0. But, here all cannot be equal.

If $b = c$, $|a + b\omega + c\omega^2| = \sqrt{\frac{1}{2}\{(a-b)^2 + (b-a)^2\}} = |a - b| \geq 1$ became a and b are unequal integers.

25. Here, $D \geq 0$, $\alpha - 5 < 0$, $\beta - 5 < 0$, where α, β are roots of the equation.

$D \geq 0 \Rightarrow 4k^2 - 4(k^2 + k - 5) \geq 0 \Rightarrow k \leq 5$.

$\alpha - 5 < 0$ and $\beta - 5 < 0 \Rightarrow \alpha + \beta - 10 < 0$ and $\alpha\beta - 5(\alpha + \beta) + 25 > 0$

$\Rightarrow 2k - 10 < 0$ and $k^2 + k - 5 - 10k + 25 > 0$,

i.e., $k^2 - 9k + 20 > 0$

$= k < 5$ and $(k-4)(k-5) > 0$, i.e., $k < 4$ or $k > 5$.

Hence, $k < 4$.

26. $\Delta = \begin{vmatrix} \log a_n & \log r & \log r \\ \log a_{n+3} & \log r & \log r \\ \log a_{n+6} & \log r & \log r \end{vmatrix}$ $\begin{pmatrix} C_3 \to C_3 - C_2 \\ C_2 \to C_2 - C_1 \end{pmatrix}$,

where r is the common ratio of the GP

$= 0$ $(\because \ C_2 \equiv C_3)$.

27. Putting $x = 0, y = 0$ we get $f(0) = f(0) \cdot f(0) - f(a) \cdot f(a) \Rightarrow f(a) = 0$ because $f(0) = 1$.

Putting $x = a, y = x - a$ we get $f(2a - x) = f(a) \cdot f(x - a) - f(0) \cdot f(x)$

$\therefore \quad f(2a - x) = 0 \cdot f(x - a) - 1 \cdot f(x) = -f(x)$.

28. $f(x) = \tan^{-1}\frac{2x}{1-x^2} = 2\tan^{-1}x$ because $|x| < 1$. We know that $\tan^{-1}x$ is one-to-one and onto in $\left(-\frac{\pi}{2}, \frac{\pi}{2}\right)$.

29. Let the roots be α, β. Then $\alpha^2 + \beta^2 = (\alpha+\beta)^2 - 2\alpha\beta = (a-2)^2 - 2\cdot(-a-1)$

$\therefore\ \alpha^2+\beta^2 = a^2 - 2a + 6 = (a-1)^2 + 5 \geq 5$, equality holding when $a = 1$.

30. $\Delta = \begin{vmatrix} \alpha & 1 & 1 \\ 1 & \alpha & 1 \\ 1 & 1 & \alpha \end{vmatrix} = (\alpha-1)^2 \cdot (\alpha+2)$. For no solution, $\Delta = 0$. So, $\alpha = 1$ or -2.

When $\alpha = 1$, all the equations give $x+y+z=0$ which has infinite number of solutions.

When $\alpha = -2$, ading all the equations we get $0\cdot x + 0\cdot y + 0\cdot z = -9$, which cannot hold for any real values of x, y, z.

31. Let the circle be $(x-\alpha)^2 + (y-\beta)^2 = r^2$. Then, from the condition of tangency with given circle, $r + 1 = \sqrt{(\alpha-0)^2 + (\beta-1)^2}$ and from the condition of tangency with x-axis, $|\beta| = r$.

$\therefore\ (|\beta|+1)^2 = \alpha^2 + (\beta-1)^2 \Rightarrow \alpha^2 = 2(\beta + |\beta|)$.

$\therefore$ if $y > 0$, the locus is $x^2 = 4y$

and if $y \leq 0$, the locus is $x = 0$.

As the locus is of real points, we have the locus is the set of points (x, y) given by

$$\{(x,y) \mid x^2 = 4y\} \cup \{(x,y) \mid x = 0, y \leq 0\}.$$

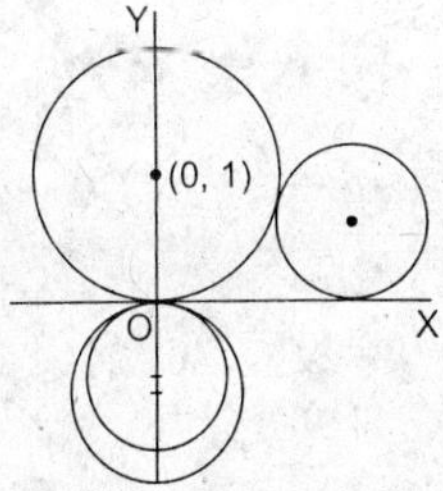

32. $\cos(\alpha-\beta) = 1 \Rightarrow \alpha - \beta = 0, 2\pi$ because $\alpha, \beta \Rightarrow [-\pi, \pi]$.
When $\alpha - \beta = 0$ we get $\alpha = \beta$

$$\therefore\ \cos(\alpha+\beta) = \frac{1}{e} \Rightarrow \cos 2\alpha = \frac{1}{e} \Rightarrow 2\alpha = 2n\pi \pm \cos^{-1}\frac{1}{e}$$

$$\Rightarrow \alpha = n\pi \pm \frac{1}{2}\cos^{-1}\frac{1}{e}$$

$$= \frac{1}{2}\cos^{-1}\frac{1}{e}, -\frac{1}{2}\cos^{-1}\frac{1}{e}, \pi - \frac{1}{2}\cos^{-1}\frac{1}{e}.$$

When $\alpha - \beta = 2\pi$ we get $\alpha = \pi$ and $\beta = -\pi$ because $\alpha, \beta \in [-\pi, \pi]$.

$$\therefore\ \cos(\alpha+\beta) = \frac{1}{e} \Rightarrow \cos(\pi - \pi) = \frac{1}{c} \Rightarrow e = 1 \text{ (absurd)}.$$

$\therefore$ the number of values of (α, β) is 3.

33. Put $x + 1 = z$. Then $I = \int_{-1}^{1} (z^3 + 2 + z\cos z)\,dz = \int_{-1}^{1} 2\,dz + \int_{-1}^{1} (z^3 + z\cos z)\,dz$

$$= 2\cdot z\Big]_{-1}^{1} + 0 \qquad (z^3 + z\cos z \text{ being an odd function}).$$

34. $(\alpha^2+\beta^2)^2=(\alpha+\beta)(\alpha^3+\beta^3) \Rightarrow 2\alpha^2\beta^2=\alpha\beta^3+\beta\alpha^3=\alpha\beta(\alpha^2+\beta^2)$

$\Rightarrow \alpha\beta(\alpha-\beta)^2=0 \Rightarrow \alpha\beta=0$ or $\alpha=\beta \Rightarrow c=0$ or $\Delta=0 \Rightarrow c\Delta=0$.

35. $\cos^2 2A=\cos A\cdot\cos 3A \Rightarrow 2\cos^2 2A=\cos 2A+\cos 4A$

$$\Rightarrow \cos 2A=1 \Rightarrow 2A=2n\pi \Rightarrow A=n\pi.$$

When $A=n\pi$, at least two of the given members are equal. So, there is no value of A.

36. We know that $|z_1+z_2|\le|z_1|+|z_2|$, equality holding when $0, z_1, z_2$ are represented by collinear points with 0 at an end. So, amp $z_1=$ amp z_2.

37. $f(x)=\begin{vmatrix}1+a^2x & (1+b^2)x & (1+c^2)x\\ x-1 & 1-x & 0\\ 0 & x-1 & 1-x\end{vmatrix}$ $(R_3\to R_3-R_2, R_2\to R_2-R_1)$

$$=(x-1)^2\begin{vmatrix}1+a^2x & (1+b^2)x & (1+c^2)x\\ 1 & -1 & 0\\ 0 & 1 & -1\end{vmatrix}$$

$$=(x-1)^2\begin{vmatrix}1+(a^2+b^2+1)x & (1+b^2)x & (1+c^2)x\\ 0 & -1 & 0\\ 1 & 1 & -1\end{vmatrix} \quad (C_1\to C_1+C_2)$$

$$=(x-1)^2(-1)\begin{vmatrix}1+(a^2+b^2+1)x & (1+c^2)x\\ 1 & -1\end{vmatrix}$$

$$=-(x-1)^2\begin{vmatrix}1+(a^2+b^2+c^2+2)x & (1+c^2)x\\ 0 & -1\end{vmatrix} \quad (C_1\to C_1+C_2)$$

$$=-(x-1)^2\begin{vmatrix}1 & (1+c^2)x\\ 0 & -1\end{vmatrix}=(x-1)^2.$$

38. $$\text{Limit}=\lim_{x\to\alpha}\frac{2\sin^2\dfrac{a(x-\alpha)(x-\beta)}{2}}{\dfrac{a^2}{4}(x-\alpha)^2(x-\beta)^2}\cdot\frac{a^2}{4}(x-\beta)^2$$

$$=\lim_{x\to\alpha}\frac{a^2}{2}(x-\beta)^2\cdot\left\{\frac{\sin\dfrac{a(x-\alpha)(x-\beta)}{2}}{\dfrac{a(x-\alpha)(x-\beta)}{2}}\right\}^2$$

$$=\frac{a^2}{2}(\alpha-\beta)^2\cdot 1^2.$$

39. In the interval (0, 1), $2^{x^2} > 2^{x^3} > 0$. So, $\int_0^1 2^{x^2} dx > \int_0^1 2^{x^3} dx$.

In the interval (1, 2), $0 < 2^{x^2} > 2^{x^3}$. So, $\int_1^2 2^{x^2} dx < \int_1^2 2^{x^3} dx$.

40. The lines are $\frac{x}{3} = \frac{y}{2} = \frac{z}{-6}$ and $\frac{x}{2} = \frac{y}{-12} = \frac{z}{-3}\cdot$ So, their d, r are 3, 2, –6 and 2, –12, –3 and $3 \cdot 2 + 2(-12) + (-6)(-3) = 0$.

41. $\sin\theta = \frac{1}{3} \Rightarrow \cos(90° - \theta) = \frac{1}{3}\cdot$ So, the angle between the line and the normal to the plane is ϕ, where $\cos\phi = \frac{1}{3}\cdot$

D.r. of the line = 1, 2, 2 and *d.r.* of the normal to the plane = $2, -1, \sqrt{\lambda}$.

$$\therefore\ \cos\phi = \frac{1}{3} = \frac{1}{3}\cdot\frac{2}{\sqrt{5+\lambda}} + \frac{2}{3}\cdot\frac{-1}{\sqrt{5+\lambda}} + \frac{2}{3}\cdot\frac{\sqrt{\lambda}}{\sqrt{5+\lambda}} = \frac{2\sqrt{\lambda}}{3\sqrt{\lambda+5}}$$

$$\Rightarrow\ \lambda + 5 = 4\lambda \quad \Rightarrow\ \lambda = \frac{5}{3}\cdot$$

42. The tangent at (1, 7) is $\frac{y+7}{2} = 1\cdot x + 6$ or $y = 2x + 5$. ... (1)

The centre of the circle = (–8, –6). The line perpendicular to (1) passing through the centre is $y + 6 = -\frac{1}{2}(x+8)$ or $x + 2y + 20 = 0$. ... (2)

Solving (1) and (2) we get coordinates of *Q*. So, $Q = (-6, -7)$.

43. The line passes through (2, –2, 3) having *d.r.* = 1, –1, 4 and the plane has the equation $x + 5y + z = 5$.

The normal to the plane has *d.r.* = 1, 5, 1.

$\because\ 1\cdot 1 + (-1)\cdot 5 + 4\cdot 1 = 0$, the line is parallel to the plane.

$\therefore$ the required distance = distance of (2, –2, 3) from the plane

$$= \left|\frac{2 + 5(-2) + 3 - 5}{\sqrt{1^2 + 5^2 + 1^2}}\right|.$$

44. As $f(x)$ is continuous $f\left(\frac{1}{n}\right) = \lim_{n\to\infty} f\left(\frac{1}{n}\right) = f\left(\lim_{n\to\infty}\frac{1}{n}\right) = f(0)$. So, $f(0) = 0$.

$$f'(0) = \lim_{h\to 0}\frac{f(h) - f(0)}{h} = \lim_{n\to\infty}\frac{f\left(\frac{1}{n}\right) - 0}{\frac{1}{n}} = \lim_{n\to\infty}\frac{0-0}{\frac{1}{n}} = \lim_{n\to\infty} 0 = 0.$$

45. $(f-g)(x) = f(x) - g(x)$. When x is rational, $f(x) - g(x) = 0 - x$ and when x is irrational, $f(x) - g(x) = x - 0$. So,

$$f(x) - g(x) = \begin{cases} -x & \text{when } x \text{ is rational} \\ x & \text{when } x \text{ is irrational.} \end{cases}$$

As x as well as $-x$ are one-to-one and onto functions for their domain, $f(x) - g(x)$ is one-to-one and onto.

46. $$I = \int_{-\pi}^{0} \frac{\cos^2 x}{1+a^x}\,dx + \int_{0}^{\pi} \frac{\cos^2 x}{1+a^x}\,dx$$

$$= \int_{0}^{\pi} \frac{\cos^2 x}{1+a^{-x}}\,dx + \int_{0}^{\pi} \frac{\cos^2 x}{1+a^x}\,dx \qquad \text{(using } -x \text{ for } x \text{ in the first)}$$

$$= \int_{0}^{\pi} \cos^2 x \left(\frac{a^x}{1+a^x} + \frac{1}{1+a^x}\right) dx$$

$$= \int_{0}^{\pi} \cos^2 x\,dx = \frac{1}{2}\int_{0}^{\pi} (1+\cos 2x)\,dx$$

$$= \frac{1}{2}(\pi - 0).$$

47. If the angle betwen the pair be θ then two adjacent sectors will have angles θ and 2θ, where $\theta + 2\theta = 180°$.

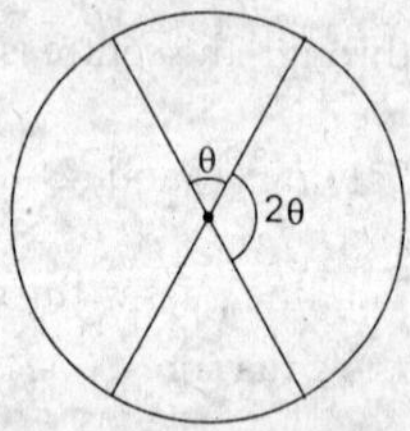

$$\therefore \quad \tan\theta = \left|\frac{2\sqrt{h^2 - ab}}{a+b}\right|$$

$$\Rightarrow \quad \tan 60° = \left|\frac{2\sqrt{(a+b)^2 - ab}}{a+b}\right|$$

$$\therefore \quad 3(a+b)^2 = 4(a^2+b^2+ab) \Rightarrow a^2+b^2-2ab = 0 \Rightarrow (a-b)^2 = 0 \Rightarrow a = b.$$

48. There are $2m$ horizontal lines numbered 1, 2, 3, ..., $2m$ and $2n$ vertical lines numbered 1, 2, 3, ..., $2n$.

As rectangles must have odd side lengths our selection should be as below:

(i) Of the two lines from horizontal set one should be odd numbered (like 1, 3, 5, ..., $2m-1$) and one should be even numbered like 2, 4, 6, ..., $2m$ and

(ii) Of the two lines from vertical set one should be odd numbered and one should be even numbered.

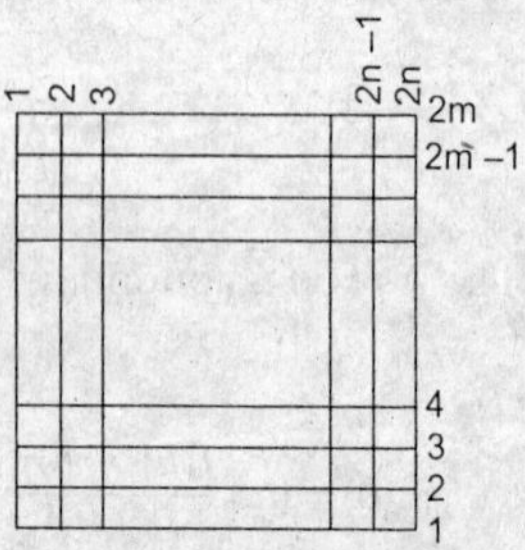

As there are m odd numbered and m even numbered horizontal lines, the number of selections in (i) $= {}^mC_1 \cdot {}^mC_1 = m^2$.

As there are n odd numbered and n even numbered vertical lines, the number of selection in (ii) $= {}^nC_1 \cdot {}^nC_1 = n^2$.

$\therefore$ the required number of rectangles $= m^2 \times n^2$.

49. $3^p, p \in N$ has the digit 3, 9, 7, 1 in units place according as p is of the form $4k+1, 4k+2, 4k+3, 4k$.

$\therefore$ 3^m has 3 in units place if $m = 1, 5, 9, 13$ and

3^n has 7 in units place if $n = 3, 7, 11, 15$

$\therefore$ $3^m + n^n$ will have 0 in units place if $m \in \{1, 5, 9, 3\}$ and $n \in \{3, 7, 11, 15\}$

$\therefore$ (m, n) will have 4×4 value. Also m, n can be interchanged. So, (m, n) will have 32 values in this case.

Again, 3^m has 9 in units place if $m = 2, 6, 10, 14$ and 3^n has 1 in units place if $n = 4, 8, 12$ $(\because n \neq 0)$.

$\therefore$ $3^m + 3^n$ will have 0 in unity place if $m \in \{2, 6, 10, 14\}$ and $n \in \{4, 8, 12\}$.

$\therefore$ (m, n) will have 4×3 values. Also m, n can be interchanged.

So, (m, n) will have 24 values in this case.

$\therefore$ (m, n) will have $32 + 24$ values.

50. $z = 1^{1/n} = 1, \alpha, \alpha^2, \alpha^3, \ldots, \alpha^{n-1}$ where $\alpha = \cos\frac{2\pi}{n} + i\sin\frac{2\pi}{n}$.

Let $z_1 = \alpha^r, z_2 = \alpha^s$. So, $\angle AOB = \text{amp } z_1 - \text{amp } z_2$.

$$\text{Clearly, amp } z_1 = \text{amp}\left(\cos\frac{2\pi}{n} + i\sin\frac{2\pi}{n}\right)^r$$

$$= \text{amp}\left(\cos\frac{2\pi r}{n} + i\sin\frac{2\pi r}{n}\right) = \frac{2\pi r}{n}.$$

Similarly, amp $z_2 = \frac{2\pi s}{n}$. So, $\angle AOB = \frac{2\pi r}{n} - \frac{2\pi s}{n}$

or $\frac{\pi}{3} = \frac{2\pi r}{n} - \frac{2\pi s}{n} \Rightarrow n = 6(r-s) = \text{multiple of } 6$.

51. The equation of the circle passing through (0, 0), (1, 0) and (0, 1) is $x^2 + y^2 - x - y = 0$. The point $\left(\lambda, \frac{1}{\lambda}\right)$ is on it.

$\therefore \quad \lambda^2+\frac{1}{\lambda^2}-\lambda-\frac{1}{\lambda}=0 \qquad$ or $\quad \left(\lambda+\frac{1}{\lambda}\right)^2-2-\left(\lambda+\frac{1}{\lambda}\right)=0$

or $\quad \left(\lambda+\frac{1}{\lambda}-2\right)\left(\lambda+\frac{1}{\lambda}+1\right)=0 \quad$ or $\quad (\lambda^2-2\lambda+1)(\lambda^2+\lambda+1)=0$

or $\quad (\lambda-1)^2(\lambda^2+\lambda+1)=0 \quad \Rightarrow \quad \lambda=1.$

52. In ΔABC, $\tan A+\tan B+\tan C=\tan A\tan B\tan C=9.$

$\therefore \quad \cot A\cot B\cot C=\frac{1}{9}.$

53. Let $x_i=-y_i$. Then $y_1+y_2+y_3+y_4=15$. The required number of negative integral solutions is the number of positive integral solutions of

$$y_1+y_2+y_3+y_4=15 \qquad (y_i\in N),$$

which is equal to the coefficient of x^{15} in $(x+x^2+x^3+\ldots)^4$

$= $ coefficient of x^{11} in $(1+x+x^2+\ldots)^4$, i.e., $(1-x)^{-4}$

$= $ coefficient of x^{11} in $\{{}^3C_0+{}^4C_1x+{}^5C_2x^2+\ldots\}$

$={}^{14}C_{11}=\frac{14\cdot13\cdot12}{6}.$

54. $\sqrt[3]{-1}=(\cos\pi+i\sin\pi)^{1/3}=\cos\frac{(2r+1)\pi}{3}+i\sin\frac{(2r+1)\pi}{3}$, where $r=0,1,2$

$=\cos\frac{\pi}{3}+i\sin\frac{\pi}{3},\ \cos\pi+i\sin\pi,\ \cos\frac{5\pi}{3}+i\sin\frac{5\pi}{3}$

$=\frac{1}{2}+i\frac{\sqrt{3}}{2},\ -1,\ \cos\left(-\frac{\pi}{3}\right)+i\sin\left(-\frac{\pi}{3}\right)=\frac{1+\sqrt{3}i}{2},\ -1,\ \frac{1-\sqrt{3}i}{2}$

$=\frac{-\sqrt{3}+i}{2i},\ -1,\ \frac{\sqrt{3}+i}{2i}=\frac{-\sqrt{3}+\sqrt{-1}}{\sqrt{-4}},\ -1,\ \frac{\sqrt{3}+\sqrt{-1}}{\sqrt{-4}}.$

55. Here, $(a^2-a-6)x^2+(a^2-9)x+a^2-2a-3=0.$

It is an identity in x if $a^2-a-6=0,\ a^2-9=0,\ a^2-2a-3=0.$

$a^2-a-6=0 \quad \Rightarrow \quad a=3,-2;$

$a^2-9=0 \quad \Rightarrow \quad a=3,-3;$

$a^2-2a-3=0 \quad \Rightarrow \quad a=3,-1.$

Hence, $a=3$ is the only common value.

56. Here, $\tan^2\theta + p\tan\theta + q = 0$, whose roots are $\tan\alpha$ and $\tan\beta$.

$$\therefore \quad \tan(\alpha+\beta) = \frac{\tan\alpha + \tan\beta}{1 - \tan\alpha\cdot\tan\beta} = \frac{-p}{1-q} = \frac{p}{q-1}$$

$$\text{and} \quad \tan(\alpha-\beta) = \frac{\tan\alpha - \tan\beta}{1 + \tan\alpha\cdot\tan\beta} = \frac{\sqrt{(-p)^2 - 4\cdot q}}{1+q}.$$

57. Here, $a(x^2+1) + x > 0$ for all $x \in R$.

$\therefore \quad ax^2 + x + a > 0$. So, $a > 0$ and $D = 1 - 4a^2 < 0$.

$\therefore \quad a > 0$ and $a < -\frac{1}{2}$ or $a > \frac{1}{2}$.

58. $$\lim_{x\to-\infty} \frac{x^3 - x}{\sqrt{1+x^6}} = \lim_{y\to\infty} \frac{-y^3 + y}{\sqrt{1+y^6}} = \lim_{y\to\infty} \frac{y^3\left(-1+\frac{1}{y^2}\right)}{y^3\sqrt{\frac{1}{y^6}+1}} = \frac{-1}{1} = -1.$$

59. Total number of ways for the three particles to move from the vertices $= 2\times2\times2 = 8$.

The number of ways not to collide = 2.

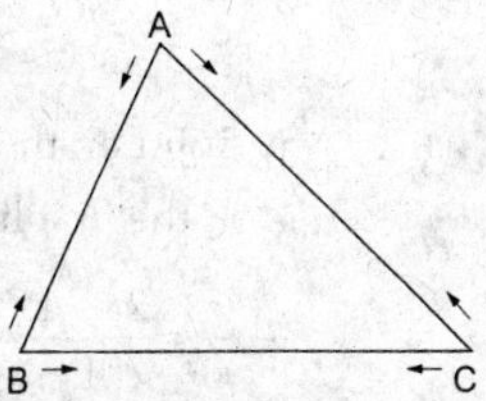

$\therefore$ the probability that they will not collide

$= \frac{2}{8} = \frac{1}{4}$.

$\therefore$ the required probability $= 1 - \frac{1}{4} = \frac{3}{4}$.

60. Expression $= \{(\vec{a}\times\vec{b})\cdot\vec{b}\}\vec{a} - (\vec{b}\cdot\vec{a})(\vec{a}\times\vec{b}) + \{\vec{a}\cdot(\vec{a}\times\vec{b})\}\vec{b} - (\vec{a}\cdot\vec{b})(\vec{a}\times\vec{b})$

$$= [\vec{a}\ \vec{b}\ \vec{b}]\vec{a} - (\vec{a}\cdot\vec{b})\,\vec{a}\times\vec{b} + [\vec{a}\ \vec{a}\ \vec{b}]\vec{b} - (\vec{a}\cdot\vec{b})\vec{a}\times\vec{b}$$

$$= -2(\vec{a}\cdot\vec{b})\vec{a}\times\vec{b} = 2(\vec{a}\cdot\vec{b})\vec{b}\times\vec{a}$$

61. Any tangent to $\frac{x^2}{9} + \frac{y^2}{4} = 1$ is $y = mx + \sqrt{9m^2+4}$. ... (1)

It touches $\frac{x^2}{4} + \frac{y^2}{9} = 1$. So, (1) is of the form $x = m'y + \sqrt{9m'^2+4}$. ... (2)

Comparing (1) and (2),

$$\frac{m}{-1} = \frac{-1}{m'} = \frac{\sqrt{9m^2+4}}{\sqrt{9m'^2+4}}$$

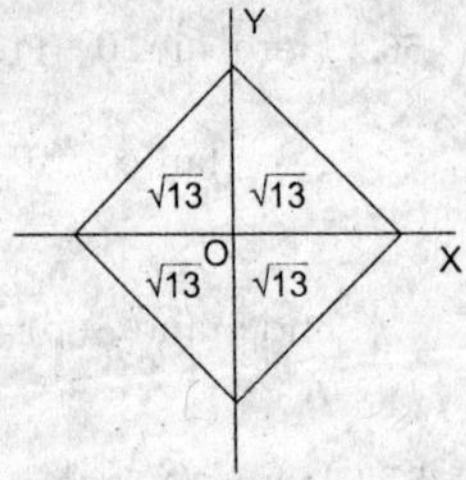

$$\Rightarrow \quad mm' = 1 \text{ and } m^2 = \frac{9m^2+4}{9m'^2+4}$$

$$\Rightarrow 9 + 4m^2 = 9m^2 + 4 \quad (\because mm' = 1)$$

$$\Rightarrow \quad m^2 = 1 \quad \Rightarrow m = 1, -1.$$

$\therefore$ the common tangents are $y = x \pm \sqrt{13}$ and $y = -x \pm \sqrt{13}$.

62. $\underbrace{* < *}_{1,2,3,4} < 5 < \underbrace{* < *}_{6,7,8,9,10,11,12}$

Total number of ways $= n(S) = {}^{12}C_5$. (No question of arrangement)

The number of favourable ways $= n(E) = {}^4C_2 \times {}^7C_2$.

$\therefore$ the required probability $= \dfrac{{}^4C_2 \times {}^7C_2}{{}^{12}C_5}$

$$= \frac{6 \times 21}{11 \times 9 \times 8} = \frac{7}{44}.$$

63. Any point on the third line is $(r, 2r+1, 3r+1)$.
It is on the first line if

$$\frac{r}{0} = \frac{2r+1}{1} = \frac{3r+1}{1} \quad \Rightarrow r = 0.$$

$\therefore$ the third line cuts the first line at $(0, 1, 1)$.
Again, $(r, 2r+1, 3r+1)$ is on the second line if

$$\frac{r}{2} = \frac{2r+1}{5} = \frac{3r+1}{7} \quad \Rightarrow r = 2.$$

$\therefore$ the third line cuts the second line at $(2, 5, 7)$
Clearly, the first two lines pass through $(0, 0, 0)$.

$\therefore$ they form a triangle of area $\frac{1}{2}|\vec{a} \times \vec{b}|$

$$= \frac{1}{2}|(\vec{j} + \vec{k}) \times (2\vec{i} + 5\vec{j} + 7\vec{k})|$$

$$= \frac{1}{2}|2\vec{i} + 2\vec{j} - 2\vec{k}| = \sqrt{3}$$

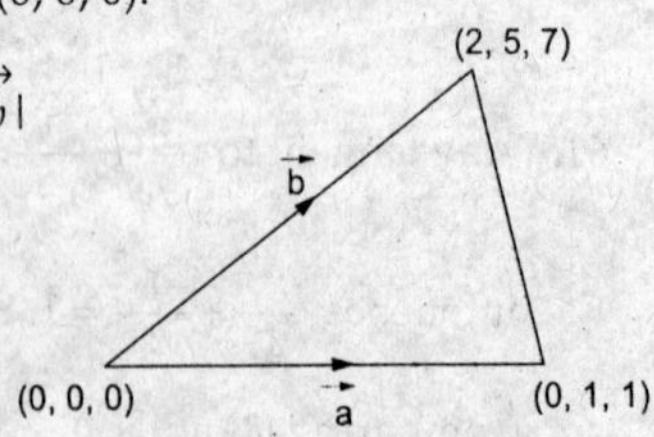

64. Differentiating w.r.t. y,

$$m(1-y)^{m-1} \cdot (-1) \cdot (1+y)^n + n(1+y)^{n-1} \cdot (1-y)^m = a_1 + 2a_2y + 3a_3y^2 + \dots.$$

Putting $y = 0$,

$$-m + n = a_1 = 10. \quad \ldots (1)$$

Differentiating again w.r.t. y,

$$-m(m-1)(1-y)^{m-2} \cdot (-1) \cdot (1+y)^n - mn(1-y)^{m-1}(1+y)^{n-1}$$
$$+ n(n-1)(1+y)^{n-2} \cdot (1-y)^m + nm(1+y)^{n-1} \cdot (1-y)^{m-1} \cdot (-1)$$
$$= 2a_2 + 6a_3 y + \ldots .$$

Putting $y = 0$,

$$m(m-1) - mn + n(n-1) - mn = 2a_2 = 2 \times 10 = 20$$

or $\quad m^2 + n^2 - 2mn - m - n = 20$

or $\quad (m-n)^2 - (m+n) = 20 \quad$ or $\quad 10^2 - (m+n) = 20$

or $\quad m + n = 80. \quad \ldots (2)$

Solving (1) and (2), $n = 45$ and $m = 35$. So, $(m, n) = (35, 45)$.

65. Here $P(a, a^2)$ and $Q(1, 0)$ are on the opposite sides of $x - 2y = 0$.

$\therefore \quad a - 2a^2 < 0$ because $(1, 0)$ makes $x - 2y$ positive.

But P and Q are on the same side of $3x - y = 0$.

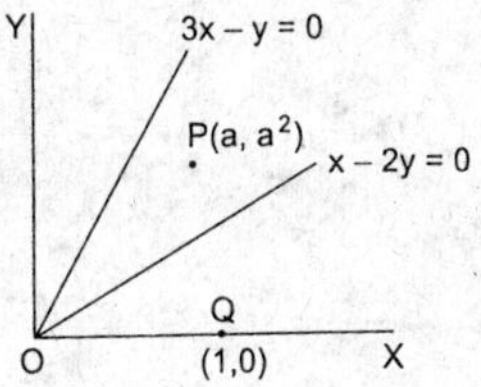

$\therefore \quad 3a - a^2 > 0$

Now, $a - 2a^2 < 0 \quad \Rightarrow a(1 - 2a) < 0$

$\Rightarrow \quad a < 0$ or $a > \frac{1}{2}$.

Again, $3a - a^2 > 0 \quad \Rightarrow a(3 - a) > 0 \quad \Rightarrow 0 < a < 3$.

$\therefore \quad$ the common values of a are given by $\frac{1}{2} < a < 3$.

66. $a_0 + a_1 x + a_2 x^2 + \ldots = \dfrac{1}{(1-ax)(1-bx)} = \dfrac{b(1-ax) - a(1-bx)}{(1-ax)(1-bx)} \cdot \dfrac{1}{b-a}$

$$= \frac{1}{b-a}\left\{\frac{b}{1-bx} - \frac{a}{1-ax}\right\}$$

$$= \frac{1}{b-a}[b\{1 + bx + (bx)^2 + \ldots\} - a\{1 + ax + (ax)^2 + \ldots\}]$$

$\therefore \quad$ coefficient of $x^n = a_n = \dfrac{1}{b-a}(b \cdot b^n - a \cdot a^n) = \dfrac{b^{n+1} - a^{n+1}}{b-a}$.

67. $y=\frac{a^3}{3}x^2+\frac{a^2}{2}x-2a$ or $6(y+2a)=2a^3x^2+3a^2x$

or $\frac{3}{a^3}(y+2a)=x^2+\frac{3}{2a}x=\left(x+\frac{3}{4a}\right)^2-\frac{9}{16a^2}$

or $\left(x+\frac{3}{4a}\right)^2=\frac{3}{a^3}\left(y+2a+\frac{3a}{16}\right)$ or $\left(x+\frac{3}{4a}\right)^2=\frac{3}{a^3}\left(y+\frac{35}{16}a\right)$.

If (α, β) be the vertex,

$$\alpha=\frac{-3}{4a} \text{ and } \beta=\frac{-35}{16}a.$$

$\therefore$ $\alpha\beta=\frac{105}{64}$. So, the locus of vertices is $xy=\frac{105}{64}$, which is a rectangular hyperbola.

68. $P(E\cap F\cap G)=0 \Rightarrow E\cap F\cap G=\phi$

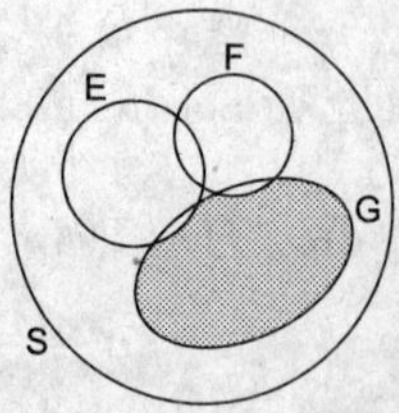

Now, $P(E^c\cap F^c \mid G)=\frac{P\{(E^c\cap F^c)\cap G\}}{P(G)}$

$$=\frac{n\{(E^c\cap F^c)\cap G\}}{n(G)}$$

$$=\frac{n(G)-n(E\cap G)-n(F\cap G)}{n(G)}, \quad \text{(see figure)}$$

$$=1-\frac{n(E\cap G)}{n(G)}-\frac{n(F\cap G)}{n(G)}$$

$$=1-\frac{P(E\cap G)}{P(G)}-\frac{P(F\cap G)}{P(G)}$$

$$=1-P(E)-P(F)=P(E^c)-P(F)$$

because $P(E\cap G)=P(E)\cdot P(G)$, E and G being independent

$P(F\cap G)=P(F)\cdot P(G)$, F and G being independent

□

4

Miscellaneous Questions—4

Comprehension-Type Questions

- $2^m 3^n 5^p$ is a divisor of $2^{10} \cdot 3^8 \cdot 5^7$ if m, n and p are all whole numbers such that $0 \le m \le 10, 0 \le n \le 8$ and $0 \le p \le 7$. Also,

$$(2^0 + 2^1 + 2^2 + \dots + 2^{10})(3^0 + 3^1 + 3^2 + \dots + 3^8)(5^0 + 5^1 + 5^2 + \dots + 5^7)$$
$$= \text{sum of all the divisors of } 2^{10} \cdot 3^8 \cdot 5^7.$$

1. The number of proper divisors of 16200 is

(a) 60 (b) 24

(c) 58 (d) 22

2. The sum of proper divisors of 360 is

(a) 1170 (b) 810

(c) 1169 (d) 809

3. The sum of odd divisors ($\ne 1$) of 10800 is

(a) 1239 (b) 1240

(c) 1238 (d) 2479

- We know that the number of positive integral solutions of the equation $x_1 + x_2 + x_3 + \dots + x_m = n$ ($n \in N$) is equal to the coefficient of x^n in the expansion of $(x + x^2 + x^3 + \dots \text{ to } \infty)^m$ when $|x| < 1$. Also, we have the expansion

$$(1 - x)^{-n} = {}^{n-1}C_0 + {}^nC_1 x + {}^{n+1}C_2 x^2 + \dots + {}^{n+r}C_{r+1} x^{r+1} + \dots \text{ to } \infty,$$
where $|x| < 1$.

4. The number of positive integral solutions of the equation $x_1 + x_2 + x_3 + x_4 = 10$ is

(a) ${}^9C_6$ (b) ${}^{10}C_4$

(c) ${}^9C_4$ (d) ${}^{10}C_5$

5. The number of negative integral solutions of the equation $x_1 + x_2 + x_3 + x_4 + x_5 + 10 = 0$ is

(a) ${}^{10}C_4$ (b) ${}^{10}C_5$

(c) ${}^9C_5$ (d) ${}^9C_6$

6. The number of ways in which 10 identical things can be distributed among 3 persons so that each gets at least one thing is

(a) ${}^8C_3$ (b) ${}^{10}C_4$

(c) ${}^9C_3$ (d) ${}^9C_2$

- r is called the rank of the matrix A if there exists at least one nonzero minor of order r and every minor of order $r+1$ of the matrix equals zero. Again if A is a square matrix and I is the unit matrix of the same order then the equation $|A - xI| = 0$ is called the characteristic equation and the roots of the characteristic equation are called eigenvalues of the matrix A.

Let $A = \begin{bmatrix} 8 & -6 & 2 \\ -6 & 7 & -4 \\ 2 & -4 & 3 \end{bmatrix}$

7. The rank of A is

(a) 3 (b) 2

(c) 1 (d) none of these

8. The characteristic equation of A is

(a) $x^3 - 18x^2 + 45x = 0$ (b) $x^3 - 6x^2 + 9x - 4 = 0$

(c) $x^3 + 18x^2 - 45x = 0$ (d) $x^3 + 6x^2 - 9x + 4 = 0$

9. One of the eigenvalues of A is

(a) 1 (b) 2

(c) 3 (d) 5

• If $\alpha_1, \alpha_2, \alpha_3, \ldots, \alpha_n$ are in an AP whose common difference is λ then we have

$$\sum_{i=1}^{n} \sin \alpha_i = \frac{\sin \frac{n\lambda}{2}}{\sin \frac{\lambda}{2}} \sin \frac{\alpha_1 + \alpha_n}{2}.$$

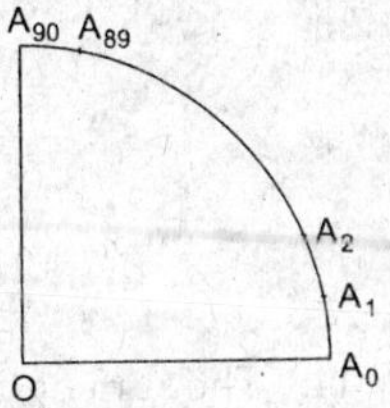

A quadrant $A_0 A_{90}$ of a unit circle is divided into 90 equal parts by inserting the points $A_1, A_2, A_3, \ldots, A_{89}$ on the arc of the circle. Let $\cot 0.5° = a$.

10. $\sum_{i=1}^{90} \sin \angle A_0 OA_i$ is equal to

(a) $\frac{a+1}{\sqrt{2}}$ (b) $\frac{a-1}{2\sqrt{2}}$

(c) $\frac{a+1}{2a}$ (d) $\frac{a+1}{2}$

11. $\sum_{i=1}^{90} \text{ar}(\Delta A_0OA_i)$ is equal to

(a) $\frac{a+1}{4}$ (b) $\frac{a+1}{2\sqrt{2}}$

(c) $\frac{a-1}{2\sqrt{2}}$ (d) $\frac{a+1}{4a}$

12. $\sum_{i=1}^{90} \sqrt{1 + \cos 2\angle A_0OA_i}$ is equal to

(a) $\frac{a+1}{2\sqrt{2}}$ (b) $\frac{a-1}{\sqrt{2}}$

(c) $\frac{a-1}{\sqrt{2}a}$ (d) $\frac{a-1}{2}$

• We know from Gregory's series that

$$\theta = \tan\theta - \frac{1}{3}\tan^3\theta + \frac{1}{5}\tan^5\theta - \ldots \text{ to } \infty, \qquad \text{where } |\theta| \le \frac{\pi}{4}.$$

13. The sum of the series $1 - \frac{1}{3} + \frac{1}{5} - \frac{1}{7} + \ldots$ to ∞ is

(a) $\frac{\pi}{2}$ (b) $\frac{\pi}{4}$

(c) 1 (d) $\frac{\pi}{8}$

14. The sum of the series $\frac{1}{1 \cdot 3} + \frac{1}{5 \cdot 7} + \frac{1}{9 \cdot 11} + \ldots$ to ∞ is

(a) $\frac{\pi}{2}$ (b) $\frac{\pi}{4}$

(c) 1 (d) $\frac{\pi}{8}$

15. The sum of the series $1 - \frac{1}{3 \cdot 3} + \frac{1}{5 \cdot 3^2} - \frac{1}{7 \cdot 3^3} + \ldots$ to ∞ is

(a) $\frac{\pi}{2}$ (b) $\frac{\pi}{\sqrt{3}}$

(c) $\frac{\pi}{2\sqrt{3}}$ (d) $\frac{\sqrt{3}\pi}{2}$

- A second-degree curve inscribing the ΔABC, where the sides AB, BC and CA have the equations $L_1 = 0$, $L_2 = 0$ and $L_3 = 0$, respectively, has the equation $L_1L_2 + \lambda L_2L_3 + \mu L_3L_1 = 0$, where λ and μ are parameters. Let the equations of AB, BC and CA be $y = 2x$, $y + 2x = 0$ and $x = 1$, respectively.

16. The circumcentre of the ΔABC is

(a) $\left(\frac{5}{4}, \frac{5}{8}\right)$ (b) $\left(\frac{5}{2}, 0\right)$

(c) $\left(0, \frac{5}{4}\right)$ (d) $\left(\frac{5}{4}, -\frac{5}{8}\right)$

17. The parabola circumscribing ΔABC and passing through the point (4, 4) has the focus

(a) (0, 1) (b) (4, 0)

(c) (1, 0) (d) (0, 4)

18. The parabola circumscribing ΔABC and passing through the point (4, 4) has the latus rectum

(a) 4 (b) 1

(c) 16 (d) 8

- The director circle of a circle is the locus of points from which tangents to the circle are at right angles. $C_1, C_2, C_3, \ldots$ is a sequence of circles such that C_{r+1} is the director circle of C_r. Let the equation of C_1 be $x^2 + y^2 = 4$.

19. The length of a chord of C_{11} which touches C_{10} is

(a) $64\sqrt{2}$ (b) 64

(c) $32\sqrt{2}$ (d) 32

20. The distance of the chord of contact of tangents of a point on C_{11} with respect to C_{10} from the centre of C_{10} is

(a) $64\sqrt{2}$ (b) $32\sqrt{2}$

(c) 64 (d) 32

21. The locus of the poles of the tangents to C_9 with respect to C_{10} is

(a) $x^2 + y^2 = 2 \times 32^2$ (b) $x^2 + y^2 = 32^2$

(c) $x^2 + y^2 = 64^2$ (d) $x^2 + y^2 = 2 \times 16^2$

- The circle drawn on the major axis of an ellipse as a diameter is the auxiliary circle of the ellipse. Let an ellipse have the equation $\frac{x^2}{a^2} + \frac{y^2}{b^2} = 1$. Let P be a point on the ellipse and the perpendicular to the major axis from the point P when produced towards P meets the auxiliary circle at Q.

22. The equation of the locus of the middle point R of PQ is

(a) $\frac{x^2}{a^2} + \frac{9y^2}{(a+b)^2} = 1$ (b) $\frac{x^2}{4a^2} + \frac{4y^2}{b^2} = 1$

(c) $\frac{x^2}{a^2} + \frac{4y^2}{(a+b)^2} = 1$ (d) $\frac{x^2}{a^2} + \frac{y^2}{(a+b)^2} = \frac{1}{4}$

23. The area enclosed by the locus of R is

(a) $\frac{\pi}{2}a(a+b)$ (b) $2\pi ab$

(c) $\frac{\pi}{3}a(a+b)$ (d) $\frac{\pi}{4}a(a+b)$

24. The point of intersection of the tangents at P and Q (P and Q being different points) to the corresponding curves lies on

(a) $b^2x^2 + a^2y^2 - 2b^2x = a^2 - b^2$ (b) $y = 0$

(c) $b^2x^2 - a^2y^2 + 2b^2x = a^2 - b^2$ (d) $x = 0$

- Two diameters AB and CD of an ellipse are called conjugate diameters if chords parallel to one of them are bisected by the other diameter. It is found that the difference of eccentric angles of A and C is $\frac{\pi}{2}$ · Let the equation of the ellipse be $9x^2 + 16y^2 = 144$.

25. If $A = \left(2, \frac{3\sqrt{3}}{2}\right)$ then coordinates of C are

(a) $\left(-2, \frac{3\sqrt{3}}{2}\right)$ (b) $\left(\sqrt{2}-\sqrt{6}, \frac{3(\sqrt{3}+1)}{2\sqrt{2}}\right)$

(c) $\left(-2\sqrt{3}, \frac{3}{2}\right)$ (d) $(0, 3)$

26. If the equation of the diameter AB is $x = y$ then the equation of the conjugate diameter CD will be

(a) $9x + 16y = 0$ (b) $x + y = 0$

(c) $16x + 9y = 0$ (d) $9x + 16y = 7$

27. The area of the quadrilateral formed by the tangents at the ends of conjugate diameters AB and CD, where $A = \left(\frac{12}{5}, \frac{12}{5}\right)$ is

(a) 96 (b) 24

(c) 12 (d) 48

- The quadratic equation $ax^2 + bx + c = 0$ will have two roots equal to zero if $b = 0, c = 0$. Also, roots of $cx^2 + bx + a = 0$ are reciprocal to the roots of $ax^2 + bx + c = 0$.

  A straight line $y = mx + c$ will be an asymptote to the curve $f(x, y) = 0$ if the line touches the curve at infinity. As a result, $f(x, mx + c) = 0$ should have two infinite roots.

28. The value of λ for which the equation $\lambda^2 - 4\lambda + 3 + (\lambda^2 + \lambda - 2)x + 6x^2 - 5x^3 = 0$ will have two roots equal to zero is

(a) 3 (b) −2

(c) 0 (d) 1

29. The value of (λ, μ) for which $(\lambda - \mu)x^3 - (\lambda - 2)x^2 - 3x + 7 = 0$ will have two infinite roots is

(a) (0, 0) (b) (2, 2)

(c) (1, 1) (d) (2, 0)

30. The equations of asymptotes to the curve $x^2 + 4xy + 3y^2 + 4x - 3y + 1 = 0$ are

(a) $2x + 2y = 7$ and $x + 3y + 15 = 0$

(b) $x + y - 7 = 0$ and $x + 3y + 15 = 0$

(c) $2(x + y) - 7 = 0$ and $2(x + 3y) + 15 = 0$

(d) $x + y = 7$ and $2(x + 3y) + 15 = 0$

- Perpendiculars AP, AQ and AR are drawn to the x-, y- and z-axes, respectively, from the point $A(1, -1, 2)$.

31. The AM of AP^2, AQ^2 and AR^2 is

(a) 4 (b) 5

(c) 3 (d) 2

32. The equation of the plane PQR is

(a) $2x + 2y - z = 2$ (b) $2x - 2y + z = 2$

(c) $2x - y + z = 2$ (d) $x - 2y + z = 2$

33. The volume of the tetrahedron (A, PQR) is

(a) 4 (b) $\frac{4}{3}$

(c) $\frac{2}{3}$ (d) $\frac{1}{3}$

- A ray of light emancipating from the point source $P(1, -3, 2)$ and travelling parallel to the line $\frac{x-2}{1} = \frac{y}{3} = \frac{z+1}{2}$ is incident on the plane $x + y - 3z = 0$ at the point Q. After reflecting from the plane the ray travels along the line QR. It is also known that the incident ray, reflected ray and the normal to the plane at the point of incidence are in the same plane.

34. The point Q is

(a) $(3, 15, 6)$ (b) $(3, 6, 3)$

(c) $(-3, -6, -3)$ (d) $(-3, -15, -6)$

35. The equations of the line QR are

(a) $\frac{x-12}{15} = \frac{y-22}{37} = \frac{z-4}{10}$ (b) $\frac{x-3}{3} = \frac{y-15}{7} = \frac{z-6}{2}$

(c) $\frac{x-3}{15} = \frac{y-6}{37} = \frac{z-3}{10}$ (d) $\frac{x+3}{3} = \frac{y+6}{7} = \frac{z+3}{2}$

36. The equation of the plane PQR is

(a) $5x+2y-z+3=0$ (b) $11x-5y+2z=30$

(c) $5x-y-z=6$ (d) $x-y+z=6$

- The sphere whose centre $=(\alpha, \beta, \gamma)$ and radius $=a$, has the equation $(x-\alpha)^2+(y-\beta)^2+(z-\gamma)^2=a^2$. Again, the intersection of a sphere by a plane is a circle.

37. The distance of the centre of the sphere $x^2+y^2+z^2-2x-4y=0$ from the origin is

(a) 5 (b) $\sqrt{5}$

(c) $2\sqrt{5}$ (d) $\frac{5}{2}$

38. The radius of the circle of intersection of the sphere $x^2+y^2+z^2=9$ by the plane $3x+4y+5z=5$ is

(a) $\sqrt{\frac{17}{2}}$ (b) 3

(c) $\sqrt{34}$ (d) $\frac{1}{\sqrt{2}}$

39. The area of the circle on the sphere $x^2+y^2+z^2=25$ whose centre is $(1, 1, 1)$ is

(a) 13π (b) 3π

(c) 22π (d) 25π

- Two one-parameter families of curves are called orthogonal trajectories of one another if each member of a family cuts all the members of the other family at right angles.

  It is found that the orthogonal trajectories of the family of curves $f\left(x, y, \frac{dy}{dx}\right)=0$ is the family of curves $f\left(x, y, -\frac{dx}{dy}\right)=0$.

40. The family of parabolas with a common vertex at the origin whose foci are on the x-axis and the directrices are parallel to the y-axis can have the equation (a being a parameter)

(a) $y^2=3ax$ (b) $x^2=4ay$

(c) $y^2=4a(x+2)$ (d) $(y-1)^2=4ax$

41. The orthogonal trajectory of the family of parabolas $y^2=4ax$ is

(a) $x^2+y^2=c^2$ (b) $x^2+2y^2=c^2$

(c) $2x^2 + y^2 = c^2$ (d) $y^2 - x^2 = c^2$

42. The curve passing through the point (1, 2) that cuts each member of the family of parabolas $y^2 = 4ax$ orthogonally is

(a) $2x^2 + y^2 = 6$ (b) $x^2 + y^2 = 5$

(c) $x^2 + 2y^2 = 9$ (d) $y^2 - x^2 = 3$

- If any point of the closed interval $[\alpha, \beta]$ is a sample point and any point of the closed interval $[a, b] \subseteq [\alpha, \beta]$ be a favourable point for the event E then the probability of the event E is defined by

$$P(E) = \frac{\text{length of } [a, b]}{\text{length of } [\alpha, \beta]}.$$

43. The set of values of $p \in R$ for which $x^2 + px + \frac{1}{4}(p+2) \geq 0$ for all $x \in R$ is

(a) $(-2, 1)$ (b) $(2, +\infty)$

(c) $[-1, 2]$ (d) $[1, 2]$

44. The set of values of $p \in R$ for which the equation $x^2 + px + \frac{1}{4}(p+2) = 0$ will have real roots is

(a) $[2, +\infty)$ (b) $(2, +\infty)$

(c) $(-\infty, 2)$ (d) $R - (-1, 2)$

45. If p is chosen at random from the interval [0, 6] then the probability that the roots of the equation $x^2 + px + \frac{1}{4}(p+2) = 0$ will be real is

(a) $\frac{3}{5}$ (b) $\frac{1}{2}$

(c) $\frac{3}{4}$ (d) $\frac{2}{3}$

- In a series of n independent trials for an event of constant probability p, the most probable number r of successes is given by $(n+1)p - 1 < r < (n+1)p$. Hence, the most probable number of successes is the integral part of $(n+1)p$. But if $(n+1)p$ is an integer, the chance of r successes is equal to that of $r+1$ successes and both $r, r+1$ are most probable numbers of successes.

A bag contains 2 white balls and 1 black ball. A ball is drawn at random and returned to the bag.

46. The experiment is done 10 times. The probability that a white ball is drawn exactly 5 times is

(a) $\frac{10!}{5!5!}\left(\frac{2}{3}\right)^5$ (b) $\frac{10!}{5!5!}\left(\frac{1}{3}\right)^5$

(c) $\frac{10!}{(5!)^2}\left(\frac{2}{9}\right)^5$ (d) $\frac{10!}{5!}\left(\frac{2}{9}\right)^5$

47. The experiment is done 5 times. The probability that a white ball is drawn most of the time is

(a) $\frac{8}{81}$ (b) $9\cdot\left(\frac{2}{3}\right)^6$

(c) $\left(\frac{2}{3}\right)^6$ (d) $\frac{8}{27}$

48. The experiment is done 10 times. The most probable number of successes is

(a) 7 (b) 6

(c) 8 (d) 5

- The envelope of a family of curves $f(x, y, c) = 0$ is a curve whose equation is obtained by eliminating the parameter c from

$$f(x, y, c) = 0 \text{ and } \frac{\partial f}{\partial c} = 0,$$

where $\frac{\partial f}{\partial c}$ is the differential coefficient of f with respect to c, treating x and y as constants.

Moreover, the envelope of the family of normals to a curve is known as the envolute of the curve.

49. The envelope of the family of straight lines whose sum of intercepts on the axes is 4 is

(a) $(x+y)^2 - 8(x-y) + 16 = 0$ (b) $(x-y)^2 - 8(x+y) + 16 = 0$

(c) $(x-y)^2 = 4(x+y)$ (d) $(x+y)^2 = 4(x-y)$

50. The envelope of the family of tangents to the curve $y^2 = x$ is

(a) $x + y^2 = 0$ (b) $x^2 = y$

(c) $x^2 + y = 0$ (d) $y^2 - x = 0$

51. The evolute of the curve $x^2 = 4y$ is

(a) $27x^2 = 4(y-2)^3$ (b) $4y^2 = 27(x-2)^3$

(c) $27x^2 = 4y^3$ (d) none of these

- The arc rate of turning of tangent to a curve at a point is called the curvature of the curve at that point. The angle that the tangent to the curve at the point (x, y) makes with the positive direction of the x-axis is denoted by ψ. Thus, curvature $=\frac{d\psi}{ds}$, where s is the arc distance of the point (x, y) from a fixed point on the curve.

  Also, for a curve, $(ds)^2=(dx)^2+(dy)^2$ and the reciprocal of the curvature at a point is known as the radius of curvature at that point.

52. The radius of curvature of the curve $y=f(x)$ at (x, y) is equal to

(a) $\dfrac{\frac{d^2y}{dx^2}}{\left\{1+\left(\frac{dy}{dx}\right)^2\right\}^{3/2}}$

(b) $\dfrac{\sqrt{1+\left(\frac{dy}{dx}\right)^2}}{\frac{d^2y}{dx^2}}$

(c) $\dfrac{\left\{1+\left(\frac{dy}{dx}\right)^2\right\}^{3/2}}{\frac{d^2y}{dx^2}}$

(d) $\dfrac{\frac{d^2y}{dx^2}}{\sqrt{1+\left(\frac{dy}{dx}\right)^2}}$

53. The radius of curvature of the curve $y^2=4x$ at the point $(1, 2)$ is

(a) $4\sqrt{2}$

(b) $2\sqrt{2}$

(c) $\frac{1}{\sqrt{2}}$

(d) $\sqrt{2}$

54. The radius of curvature of $9x^2+4y^2=36$ is the greatest at the point

(a) $(0, 2)$

(b) $(2, 0)$

(c) $(0, 3)$

(d) $\left(\sqrt{3}, \frac{3}{2}\right)$

- It is known that $3205=3\times10^3+2\times10^2+0\times10^1+5$. This method of representing a number is in the denary scale (scale of 10) of notation and it is a denary number. Similarly, in a scale of r,

$$3205=3\times r^3+2\times r^2+0\times r^1+5.$$

More generally,

$$a_n\,a_{n-1}\,a_{n-2}\ldots a_0=a_n r^n+a_{n-1}r^{n-1}+\ldots+a_1r+a_0,$$

where $a_0, a_1, \ldots, a_n$ are digits whose values range from 0 to $r-1$. Thus, in the scale of 7, the digits will be 0, 1, 2, 3, 4, 5, 6.

Conversion of a denary number to the scale of 5

$$\begin{array}{r|l}5 & 268 \\ \hline 5 & \ \ 53 \quad \ldots 3 \\ \hline 5 & \ \ 10 \quad \ldots 3 \\ \hline & \ \ \ \ 2 \quad \ldots 0\end{array} \qquad \therefore \ 268 = 2033 \text{ in the scale of } 5$$

Addition of numbers in the scale of 5

$$\begin{array}{r}34 \\ +22 \\ \hline 111\end{array} \qquad \begin{array}{l}34 = 3 \times 5 + 4 \\ 22 = 2 \times 5 + 2 \\ \hline \text{sum} = 5 \times 5 + 6 = 1 \times 5^2 + 1 \times 5 + 1\end{array}$$

Multiplication of numbers in the scale of 5

$$\begin{array}{r}23 \\ \times 4 \\ \hline 202\end{array} \qquad \begin{array}{l}3 \times 4 = 12 = 2 \times 5 + 2 \\ 2 \times 4 + 2 = 10 = 2 \times 5 + 0\end{array}$$

On the basis of the above discussions, answer the following.

55. The denary number 43125 in the scale of 6 will be represented by

(a) 353135 (b) 531353

(c) 515313 (d) 55453

56. The sum of the numbers 2053 and 412 in the scale of seven is

(a) 2465 (b) 3020

(c) 3004 (d) none of these

57. The product of the numbers 5623 and 6 in the scale of eight is

(a) 41672 (b) 33738

(c) 42562 (d) 45262

- Z is the set of integers and N is the set of natural numbers. If $b \in Z$, $c \in Z$ and $n \in N$ are such that when b and c are divided by n, we get the same remainder then b and c are said to be congruent with respect to modulo n. Symbolically it is written as $b \equiv c \pmod{n}$

58. Which of the following is true?

(a) $4 \equiv 7 \pmod 5$ (b) $118 \equiv 18 \pmod 5$

(c) $110 \equiv 93 \pmod{12}$ (d) $63 \equiv 9 \pmod{11}$

59. Let $a \equiv b \pmod{n}$, $a' \equiv b' \pmod{n}$ and $d, m \in N - \{1\}$. Then which of the following need not be true?

(a) $a + a' \equiv b + b' \pmod{n}$ (b) $aa' \equiv bb' \pmod{n}$

(c) $a^m \equiv b^m \pmod{n}$ (d) $\frac{a}{d} \equiv \frac{b}{d} \pmod{n}$

60. If $5x \equiv 2 \pmod{7}$ then

(a) $x \equiv 6 \pmod{9}$ (b) $x \equiv 6 \pmod{7}$

(c) $x \equiv 7 \pmod{6}$ (d) $x \equiv 2 \pmod{7}$

- The sum of an infinite series $u_1 + u_2 + u_3 + \ldots$ will be the finite number l if $\lim\limits_{n \to \infty} S_n = l$, where $S_n = u_1 + u_2 + u_3 + \ldots u_n$. In such cases it is said that the infinite series is convergent to the sum l.

  Moreover, a series of positive terms $u_1 + u_2 + u_3 + \ldots$ up to ∞, is found to be convergent if $\lim\limits_{n \to \infty} \frac{u_n}{u_{n+1}} > 1$.

61. The series $\frac{1}{1 \cdot 2} + \frac{1}{2 \cdot 3} + \frac{1}{3 \cdot 4} + \ldots$ to ∞ has the sum

(a) 1 (b) $\frac{2}{3}$

(c) ∞ (d) 2

62. The series $\frac{x}{1 \cdot 2} + \frac{x^2}{2 \cdot 3} + \frac{x^3}{3 \cdot 4} + \ldots$ $(x > 0)$ is not convergent where x is

(a) $\frac{1}{2}$ (b) 1

(c) 2 (d) $\frac{2}{3}$

63. Let $f(n) = \frac{1}{n(n^2 - 1)}$. The series $\sum\limits_{n=1}^{\infty} f(n+1)$ is convergent to the sum

(a) $\frac{1}{2}$ (b) $\frac{1}{4}$

(c) 1 (d) $\frac{3}{4}$

- The number of points of intersection of the graphs of the functions $y = f(x)$ and $y = \phi(x)$ give the number of solutions of the equation $f(x) - \phi(x) = 0$. Let $f(x) = ke^x$ and $\phi(x) = x$, where k is a real constant.

64. The line $y = x$ meets the curve $y = ke^x$ for $k \le 0$ at

(a) no point (b) one point

(c) two points (d) more than two points

65. The positive value of k for which $ke^x - x = 0$ has only one real solution is

(a) $\frac{1}{e}$ (b) 1 (c) e (d) $\log_e 2$

66. For $k > 0$, the set of values of k for which $ke^x - x = 0$ has two distinct real roots is

(a) $\left(0, \frac{1}{e}\right)$ (b) $\left(\frac{1}{e}, 1\right)$ (c) $\left(\frac{1}{e}, \infty\right)$ (d) (0, 1)

Answers

1. c	**2.** d	**3.** a	**4.** a	**5.** c
6. d	**7.** b	**8.** a	**9.** c	**10.** d
11. a	**12.** b	**13.** b	**14.** d	**15.** c
16. b	**17.** c	**18.** a	**19.** a	**20.** d
21. c	**22.** c	**23.** a	**24.** b	**25.** c
26. a	**27.** d	**28.** d	**29.** b	**30.** c
31. a	**32.** b	**33.** c	**34.** d	**35.** a
36. b	**37.** b	**38.** a	**39.** c	**40.** a
41. c	**42.** a	**43.** c	**44.** d	**45.** d
46. c	**47.** b	**48.** a	**49.** b	**50.** d
51. a	**52.** c	**53.** a	**54.** b	**55.** b
56. a	**57.** c	**58.** b	**59.** d	**60.** b
61. a	**62.** c	**63.** b	**64.** b	**65.** a
66. a				

Hints and Solutions to Selected Questions

1. $16200 = 2^3 \times 3^4 \times 5^2$.

The number of divisors = number of selections from 2, 2, 2, 3, 3, 3, 3, 5, 5

$$= (3+1)\cdot(4+1)\cdot(2+1) = 4 \times 5 \times 3 = 60.$$

$\therefore$ the number of proper divisors = 60 – 2 (excluding 1 and the number).

2. $360 = 2^3 \times 3^2 \times 5$.

The sum of all divisors of $2^3 \times 3^2 \times 5$

$= (2^0 + 2^1 + 2^2 + 2^3)(3^0 + 3^1 + 3^2)(5^0 + 5^1)$

$= 15 \times 13 \times 6 = 1170.$

$\therefore$ the sum of proper divisors $= 1170 - 1 - 2^3 \times 3^2 \times 5 = 1170 - 361 = 809.$

3. $10800 = 2^4 \times 3^3 \times 5^2$.

The sum of odd divisors $= (3^0 + 3^1 + 3^2 + 3^3)(5^0 + 5^1 + 5^2) - 1$

$$= 40 \times 31 - 1 = 1239.$$

4. Required number = coefficient of x^{10} in $(x + x^2 + x^3 + \dots \text{ to } \infty)^4$

$= \text{coefficient of } x^6 \text{ in } (1 + x + x^2 + \dots \text{ to } \infty)^4$

$= \text{coefficient of } x^6 \text{ in } \left(\dfrac{1}{1-x}\right)^4, \text{ i.e., } (1-x)^{-4}$

$= \text{coefficient of } x^6 \text{ in } ({}^3C_0 + {}^4C_1 x + {}^5C_2 x^2 + \dots \text{ to } \infty)$

$= {}^9C_6.$

5. The number of negative integral solutions of the equation

= the number of positive integral solutions of $(y_1 + y_2 + {}_3 + y_4 + y_5 = 10)$

= coefficient of x^{10} in $(x + x^2 + x^3 + \dots \text{ to } \infty)^5$

= coefficient of x^5 in $\left(\dfrac{1}{1-x}\right)^5$, i.e., $(1-x)^{-5} = {}^9C_5.$

6. Required number of ways

= the number of positive integral solutions of the equation $(x_1 + x_2 + x_3 = 10)$.

$= \text{coefficient of } x^{10} \text{ in } (x + x^2 + x^3 + \ldots \text{ to } \infty)^3$

$= \text{coefficient of } x^7 \text{ in } (1 - x)^{-3} = {}^9C_7 = {}^9C_2.$

7. $|A| = \begin{vmatrix} 8 & -6 & 2 \\ -6 & 7 & -4 \\ 2 & -4 & 3 \end{vmatrix} = 0$. But $\begin{vmatrix} 8 & -6 \\ -6 & 7 \end{vmatrix} \neq 0$. So, rank of $A = 2$.

8. $A - xI = \begin{bmatrix} 8 & -6 & 2 \\ -6 & 7 & -4 \\ 2 & -4 & 3 \end{bmatrix} - \begin{bmatrix} x & 0 & 0 \\ 0 & x & 0 \\ 0 & 0 & x \end{bmatrix} = \begin{bmatrix} 8-x & -6 & 2 \\ -6 & 7-x & -4 \\ 2 & -4 & 3-x \end{bmatrix}.$

$\therefore \quad |A - xI| = \begin{vmatrix} 8-x & -6 & 2 \\ -6 & 7-x & -4 \\ 2 & -4 & 3-x \end{vmatrix} = x^3 - 18x^2 + 45x.$

$\therefore$ the characteristic equation is $|A - xI| = 0$ or $x^3 - 18x^2 + 45x = 0$.

9. The characteristic equation is

$$x^3 - 18x^2 + 45x = 0$$

or $x(x^2 - 18x + 45) = 0$ or $x(x-3)(x-15) = 0$.

$\therefore \; x = 0, 3, 15.$

$\therefore$ eigenvalues are 0, 3, 15.

10. $\sin 1^\circ + \sin 2^\circ + \sin 3^\circ + \ldots + \sin 90^\circ = \dfrac{\sin 90 \times \dfrac{1^\circ}{2}}{\sin \dfrac{1^\circ}{2}} \times \sin \dfrac{91^\circ}{2}$

$$= \frac{\dfrac{1}{\sqrt{2}}}{\sin \dfrac{1^\circ}{2}} \left(\sin 45^\circ \cos \frac{1^\circ}{2} + \cos 45^\circ \sin \frac{1^\circ}{2} \right)$$

$$= \frac{1}{\sqrt{2}} \left(\frac{1}{\sqrt{2}} \cot \frac{1^\circ}{2} + \frac{1}{\sqrt{2}} \right) = \frac{a+1}{2}.$$

11. $\dfrac{1}{2} \cdot 1 \cdot 1 \cdot (\sin 1^\circ + \sin 2^\circ + \sin 3^\circ + \ldots + \sin 90^\circ) = \dfrac{1}{2} \cdot \dfrac{a+1}{2}.$

12. $\sqrt{2}(\cos 1^\circ + \cos 2^\circ + \cos 3^\circ + \ldots + \cos 89^\circ + \cos 90^\circ)$

$= \sqrt{2}(\sin 0^\circ + \sin 1^\circ + \sin 2^\circ + \ldots + \sin 89^\circ)$

$$= \sqrt{2} \cdot \frac{\sin 89 \cdot \dfrac{1^\circ}{2}}{\sin \dfrac{1^\circ}{2}} \cdot \sin \frac{89^\circ + 1^\circ}{2} = \sqrt{2} \sin 45^\circ \cdot \frac{\sin \left(45^\circ - \dfrac{1^\circ}{2} \right)}{\sin \dfrac{1^\circ}{2}}$$

$$= \sqrt{2} \cdot \frac{1}{\sqrt{2}} \cdot \left(\frac{1}{\sqrt{2}} \cot \frac{1^\circ}{2} - \frac{1}{\sqrt{2}} \right) = \frac{1}{\sqrt{2}}(a - 1).$$

13. Putting $\theta = \frac{\pi}{4}$ in Gregory's series, $\frac{\pi}{4} = 1 - \frac{1}{3} + \frac{1}{5} - \frac{1}{7} + \dots$ to ∞.

14. Series $= \frac{1}{2}\left(\frac{3-1}{1.3} + \frac{7-5}{5.7} + \frac{11-9}{9.11} + \dots \text{ to } \infty \right)$

$$= \frac{1}{2}\left(1 - \frac{1}{3} + \frac{1}{5} - \frac{1}{7} + \frac{1}{9} - \frac{1}{11} + \dots \text{ to } \infty \right)$$

$= \frac{1}{2} \cdot \frac{\pi}{4}$ (using the result in **13** above).

15. Series $= \sqrt{3}\left\{ \frac{1}{\sqrt{3}} - \frac{1}{3} \cdot \left(\frac{1}{\sqrt{3}} \right)^3 + \frac{1}{5} \cdot \left(\frac{1}{\sqrt{3}} \right)^5 - \dots \text{ to } \infty \right\}$

$= \sqrt{3} \cdot \frac{\pi}{6}$ (putting $\theta = \frac{\pi}{6}$ in Gregory's series).

16. The curve circumscribing the ΔABC is

$(y - 2x)(y + 2x) + \lambda(y + 2x)(x - 1) + \mu(x - 1)(y - 2x) = 0$, where $(\lambda \neq 0, \mu \neq 0)$,

or $(-4 + 2\lambda - 2\mu)x^2 + (\lambda + \mu)xy + y^2 + (-2\lambda + 2\mu)x + (-\lambda - \mu)y = 0.$ ... (1)

It is a circle if $\lambda + \mu = 0, -4 + 2\lambda - 2\mu = 1$.

$\therefore$ $-4 + 4\lambda = 1$ or $\lambda = \frac{5}{4}$ and $\mu = -\frac{5}{4}$.

$\therefore$ the circumcircle is $x^2 + y^2 - 5x = 0$, whose centre is $\left(\frac{5}{2}, 0 \right)$.

17. In **16**, (1) is a parabola if $(-4 + 2\lambda - 2\mu) \cdot 1 = \frac{(\lambda + \mu)^2}{4}$. ... (2)

Also, $(-4 + 2\lambda - 2\mu) \cdot 16 + (\lambda + \mu) \cdot 16 + 16 + (-2\lambda + 2\mu) \cdot 4 + (-\lambda - \mu) \cdot 4 = 0$

or $36\lambda - 12\mu - 48 = 0$ or $3\lambda - \mu - 4 = 0$.

$\therefore$ (2) $\Rightarrow \{-4 + 2\lambda - 2(3\lambda - 4)\} \cdot 4 = (\lambda + 3\lambda - 4)^2 \Rightarrow (-4\lambda + 4) \cdot 4 = (4\lambda - 4)^2$

or $-\lambda + 1 = (\lambda - 1)^2$ or $(\lambda - 1)(\lambda - 1 + 1) = 0 \Rightarrow \lambda = 1, 0$.

But $\lambda \neq 0, \mu \neq 0$. So, $\lambda = 1$ and $\mu = -1$.

$\therefore$ the parabola is $(-4 + 2 + 2)x^2 + (1 - 1)xy + y^2 - 4x = 0$ or $y^2 - 4x = 0$.

So, focus $= (1, 0)$.

18. From **17,** latus rectum = 4.

19. C_2 is $x^2+y^2=(2\sqrt{2})^2$; C_3 is $x^2+y^2=\{\sqrt{2}\,(2\sqrt{2})\}^2$; etc.

Thus the radii of $C_1, C_2, C_3, \ldots$ are $2, 2\sqrt{2}, 2(\sqrt{2})^2, \ldots$, respectively.

$\therefore$ C_{10} is $x^2+y^2=\{2(\sqrt{2})^9\}^2$ or $x^2+y^2=(32\sqrt{2})^2$ and

C_{11} is $x^2+y^2=64^2$.

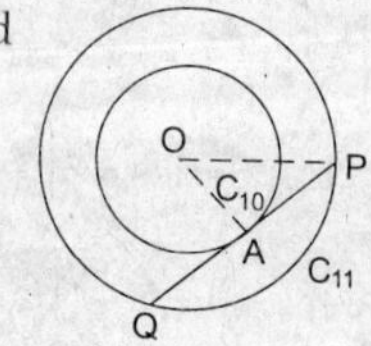

$PQ=2AP=2\cdot\sqrt{OP^2-OA^2}=2\cdot\sqrt{64^2-(32\sqrt{2})^2}$

$=2\cdot\sqrt{32^2(2^2-2)}=2\cdot 32\sqrt{2}=64\sqrt{2}.$

20. A point on C_{11} is $(64\cos\theta, 64\sin\theta)$. Its chord of contact w.r.t. C_{10}, i.e., $x^2+y^2=(32\sqrt{2})^2$, is

$$(64\cos\theta\cdot x)+(64\sin\theta)\cdot y=(32\sqrt{2})^2$$

or $\quad x\cos\theta+y\sin\theta=32$.

Its distance from (0, 0) $=\left|\dfrac{32}{\sqrt{\cos^2\theta+\sin^2\theta}}\right|=32$.

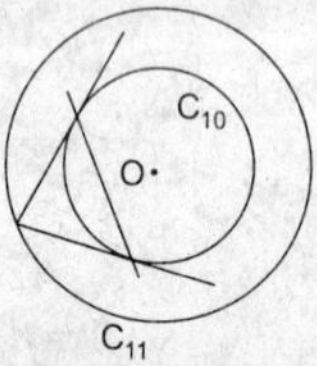

21. Let (α, β) be the pole. Then the polar of (α, β) with respect to C_{10} is

$$x\alpha+y\beta=(32\sqrt{2})^2.$$

It touches C_9. So, $\left|\dfrac{-(32\sqrt{2})^2}{\sqrt{\alpha^2+\beta^2}}\right|=32$ because C_9 has the equation $x^2+y^2=32^2$.

$\therefore$ $\sqrt{\alpha^2+\beta^2}=64$. So, the locus of poles is $x^2+y^2=64^2$.

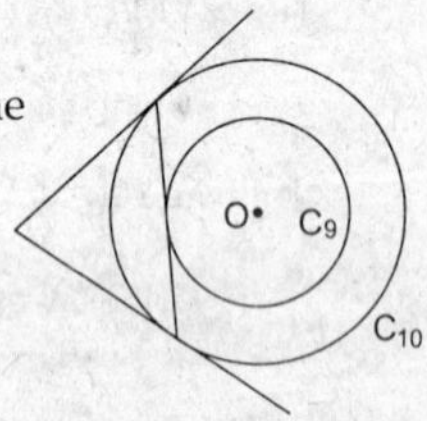

22. $P=(a\cos\phi, b\sin\phi)$ and $Q=(a\cos\phi, a\sin\phi)$.

$\therefore$ $R=\left(a\cos\phi, \dfrac{b+a}{2}\sin\phi\right)$. The locus of R is $\dfrac{x^2}{a^2}+\dfrac{y^2}{\left(\dfrac{a+b}{2}\right)^2}=1.$

23. Area $=\pi\cdot$ (semimajor axis) $\cdot$ (semiminor axis) $=\pi\cdot a\cdot\dfrac{a+b}{2}$.

24. The tangent at P to the ellipse $\dfrac{x^2}{a^2}+\dfrac{y^2}{b^2}=1$ is

$$\frac{x\cos\phi}{a}+\frac{y\sin\phi}{b}=1. \qquad \ldots (1)$$

and the tangent at Q to the circle $x^2+y^2=a^2$ is

$$x\cos\phi+y\sin\phi=a. \qquad \ldots (2)$$

$(1)\times a-(2)\Rightarrow \left(\frac{a}{b}-1\right)y\sin\phi=0 \Rightarrow y=0$ (when P, Q are different points).

25. Let the ecentric angle be ϕ at A. Then $A=(a\cos\phi, b\sin\phi)$ for the ellipse $\frac{x^2}{a^2}+\frac{y^2}{b^2}=1$. Here, $a^2=16, b^2=9$.

$$\therefore \quad A=(4\cos\phi, 3\sin\phi)=\left(2, \frac{3\sqrt{3}}{2}\right) \Rightarrow \cos\phi=\frac{1}{2}, \sin\phi=\frac{\sqrt{3}}{2} \Rightarrow \phi=\frac{\pi}{3}.$$

$$\therefore \quad C=\left(4\cos\left(\phi+\frac{\pi}{2}\right), 3\sin\left(\phi+\frac{\pi}{2}\right)\right)=\left(4\cos\frac{5\pi}{6}, 3\sin\frac{5\pi}{6}\right)=\left(-2\sqrt{3}, \frac{3}{2}\right).$$

26. Let $A=(4\cos\phi, 3\sin\phi)$. Then A is on $x=y$. So,

$$4\cos\phi=3\sin\phi \Rightarrow \tan\theta=\frac{4}{3}.$$

$$\therefore \quad C=\left(4\cos\left(\phi+\frac{\pi}{2}\right), 3\sin\left(\phi+\frac{\pi}{2}\right)\right)=(-4\sin\phi, 3\cos\phi)=\left(-4\cdot\frac{4}{5}, 3\cdot\frac{3}{5}\right).$$

$\therefore$ the equation of the diameter CD is

$$y-0=\frac{\frac{9}{5}-0}{-\frac{16}{5}-0}(x-0)$$

(because CD passes through the centre (0, 0) of the ellipse)

or $\quad 9x+16y=0.$

27. $A=(4\cos\phi, 3\sin\phi)=\left(\frac{12}{5}, \frac{12}{5}\right) \Rightarrow \cos\phi=\frac{3}{5}, \sin\phi=\frac{4}{5}.$

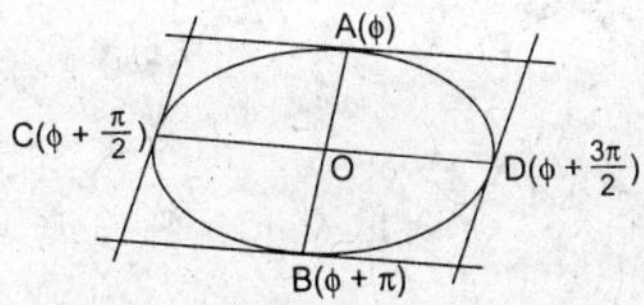

The tangent at A is $9x\cdot\frac{12}{5}+16y\cdot\frac{12}{5}=144$ or $9x+16y=60$.

$B=\left(-\frac{12}{5}, -\frac{12}{5}\right)$. So, the tangent at B is $9x+16y=-60$.

$$C=\left(4\cos\left(\phi+\frac{\pi}{2}\right), 3\sin\left(\phi+\frac{\pi}{2}\right)\right)=(-4\sin\phi, 3\cos\phi)=\left(-\frac{16}{5}, \frac{9}{5}\right).$$

So, the tangent at C is $9x \cdot \left(-\frac{16}{5}\right) + 16y \cdot \frac{9}{5} = 144$ or $y - x = 5$.

$D = \left(\frac{16}{5}, -\frac{9}{5}\right)$. So, the tangent at D is $9x \cdot \frac{16}{5} + 16y \cdot \left(-\frac{9}{5}\right) = 144$ or $x - y = 5$.

$\therefore$ the tangents make a parallelogram whose sides are

$9x + 16y = \pm 60,\ x - y = \pm 5.$

Solving equations, we find three vertices as

$\left(-\frac{4}{5}, \frac{21}{5}\right), \left(\frac{4}{5}, -\frac{21}{5}\right)$ and $\left(\frac{28}{5}, \frac{3}{5}\right)$.

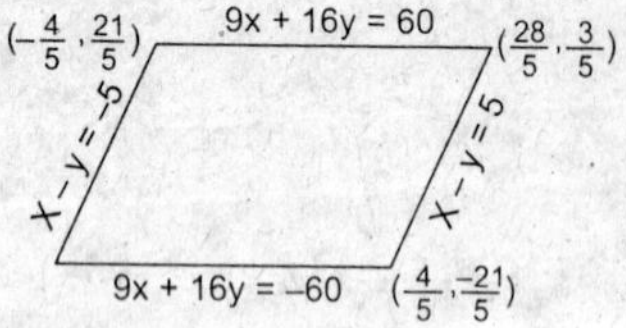

$$\therefore \quad \text{area} = 2 \times \left| \frac{1}{2} \begin{vmatrix} -\frac{4}{5} & \frac{21}{5} & 1 \\ \frac{4}{5} & -\frac{21}{5} & 1 \\ \frac{28}{5} & \frac{3}{5} & 1 \end{vmatrix} \right| = 48.$$

28. Two roots will be zero if $\lambda^2 - 4\lambda + 3 = 0, \lambda^2 + \lambda - 2 = 0$

$\Rightarrow (\lambda - 3)(\lambda - 1) = 0, (\lambda + 2)(\lambda - 1) = 0 \Rightarrow \lambda = 1.$

29. $(\lambda - \mu)x^3 - (\lambda - 2)x^2 - 3x + 7 = 0$ will have two infinite roots if $7x^3 - 3x^2 - (\lambda - 2)x + \lambda - \mu = 0$ has two zero roots. For that,

$$\lambda - 2 = 0, \lambda - \mu = 0 \Rightarrow \lambda = \mu = 2.$$

30. Let $y = mx + c$ be an asymptote. Then

$$x^2 + 4x(mx + c) + 3(mx + c)^2 + 4x - 3(mx + c) + 1 = 0$$

or $\quad (1 + 4m + 3m^2)x^2 + (4c + 6mc + 4 - 3m)x + 3c^2 - 3c + 1 = 0$

will have two infinite roots.

$\therefore \quad 1 + 4m + 3m^2 = 0$ and $4c + 6mc + 4 - 3m = 0$

$\Rightarrow \quad (1 + 3m)(1 + m) = 0$

$\Rightarrow \quad m = -1, -\frac{1}{3}$. $\quad \Big| \quad \Rightarrow c = \frac{3m - 4}{4 + 6m} = \frac{-7}{-2}, \frac{-5}{2}$

$\therefore$ the asymptotes are $y = -x + \frac{7}{2}$ and $y = -\frac{x}{3} - \frac{5}{2}$

$$\Rightarrow \quad 2(x + y) = 7 \text{ and } 2x + 6y + 15 = 0.$$

31. $AQ^2 = 1^2 + 2^2 = 5$.Similarly,

$AP^2 = (-1)^2 + 2^2 = 5, AR^2 = 1^2 + (-1)^2 = 2.$

$\therefore \quad AM = \dfrac{5+5+2}{3} = 4.$

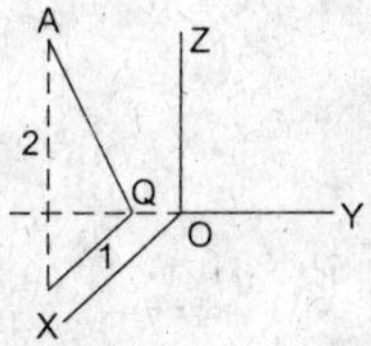

32. $P = (1, 0, 0), Q = (0, -1, 0), R = (0, 0, 2).$

Any plane through P is $a(x-1) + b(y-0) + c(z-0) = 0.$

It passes through $Q \Rightarrow a(0-1) + b(-1-0) + c(0-0) = 0.$

It passes through $R \Rightarrow a(0-1) + b(0-0) + c(2-0) = 0.$

$\therefore$ the equation of the plane PQR is $\begin{vmatrix} x-1 & y & z \\ -1 & -1 & 0 \\ -1 & 0 & 2 \end{vmatrix} = 0$

$\Rightarrow \quad -2(x-1) - y(-2) + z(0-1) = 0 \quad$ or $\quad 2x - 2y + z = 2.$

33. Volume of the tetrahedron $(A, PQR) = \dfrac{1}{6}\begin{vmatrix} 1-1 & 0+1 & 0-2 \\ 0 & 1 & -1+1 & 0-2 \\ 0-1 & 0+1 & 2-2 \end{vmatrix}$

$$= \frac{1}{6}\begin{vmatrix} 0 & 1 & -2 \\ -1 & 0 & -2 \\ -1 & 1 & 0 \end{vmatrix} = \frac{1}{6}\begin{vmatrix} 0 & 1 & -2 \\ 0 & -1 & -2 \\ -1 & 1 & 0 \end{vmatrix}$$

$$= \frac{1}{6}\cdot(-1)(-2-2) = \frac{4}{6} = \frac{2}{3}.$$

34. The equations of PQ are $\dfrac{x-1}{1} = \dfrac{y+3}{3} = \dfrac{z-2}{2}.$

Any point on PQ has coordinates $(1+r, -3+3r, 2+2r).$

It is Q if $1 + r + (-3+3r) - 3(2+2r) = 0 \Rightarrow r = -4.$

$\therefore \quad Q = (-3, -15, -6).$

35. Let $P' = (\alpha, \beta, \gamma)$ be the image of P in the plane. Then the middle point of PP' is $\left(\dfrac{1+\alpha}{2}, \dfrac{-3+\beta}{2}, \dfrac{\gamma+2}{2}\right)$ which is on the plane.

So, $\dfrac{1+\alpha}{2} + \dfrac{-3+\beta}{2} - 3\cdot\dfrac{\gamma+2}{2} = 0 \Rightarrow \alpha + \beta - 3\gamma - 8.$

Also, PP' is perpendicular to the plane. So,

$$\frac{\alpha-1}{1} = \frac{\beta+3}{1} = \frac{\gamma-2}{-3} = \frac{\alpha-1+\beta+3-3(\gamma-2)}{1+1-3(-3)} = \frac{16}{11}.$$

$\therefore \quad \alpha = \dfrac{27}{11}, \beta = -\dfrac{17}{11}, \gamma = -\dfrac{26}{11}$. So, $P' = \left(\dfrac{27}{11}, -\dfrac{17}{11}, -\dfrac{26}{11}\right).$

Clearly, P', Q, R are on the same line. So, QR has the equations

$$\frac{x+3}{15}=\frac{y+15}{37}=\frac{z+6}{10} \Rightarrow \frac{x+3-15}{15}=\frac{y+15-37}{37}=\frac{z+6-10}{10}.$$

36. Three points on the plane are $P=(1,-3,2)$, $Q=(-3,-15,-6)$ and $P'=\left(\frac{27}{11},-\frac{17}{11},-\frac{26}{11}\right)$. So, the equation of the plane is

$$\begin{vmatrix} x-1 & y+3 & z-2 \\ -3-1 & -15+3 & -6-2 \\ \frac{27}{11}-1 & -\frac{17}{11}+3 & -\frac{26}{11}-2 \end{vmatrix}=0 \Rightarrow \begin{vmatrix} x-1 & y+3 & z-2 \\ -4 & -12 & -8 \\ 16 & 16 & -48 \end{vmatrix}=0$$

or $\quad 11x-5y+2z=30.$

37. Here, $x^2-2x+1+y^2-4y+4+z^2=5$ or $(x-1)^2+(y-2)^2+z^2=5.$

$\therefore \quad$ centre $=(1,2,0)$. So, distance $=\sqrt{(1-0)^2+(2-0)^2+(0-0)^2}=\sqrt{5}.$

38. $d=\left|\frac{-5}{\sqrt{3^2+4^2+5^2}}\right|=\frac{5}{5\sqrt{2}}=\frac{1}{\sqrt{2}}$

$\therefore \quad r=\sqrt{3^2-d^2}=\sqrt{9-\frac{1}{2}}=\sqrt{\frac{17}{2}}.$

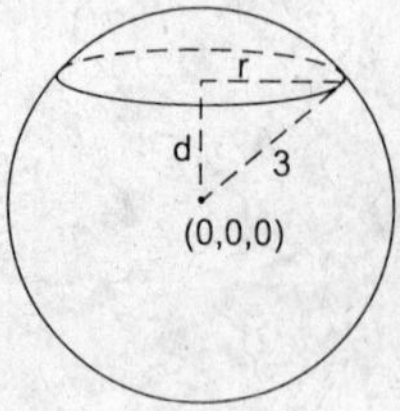

39. $d=\sqrt{1^2+1^2+1^2}=\sqrt{3}$

$r=\sqrt{5^2-d^2}=\sqrt{25-3}=\sqrt{22}$

$\therefore \quad$ area $=\pi r^2=\pi\cdot 22=22\pi.$

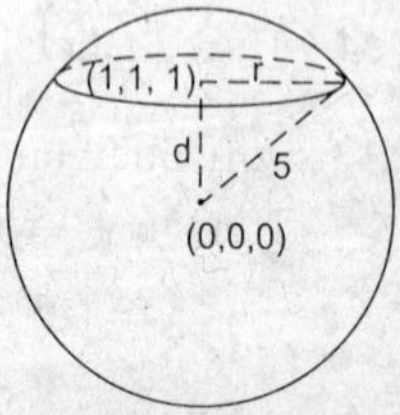

40. The equation of the family of parabolas is

$$(x-c)^2+(y-0)^2=1\cdot\left(\frac{x+c}{\sqrt{1}}\right)^2$$

or $\quad y^2=4cx=3\cdot\frac{4c}{3}x$ or $y=3ax$ $\qquad \left(\text{putting } \frac{4c}{3}=a\right).$

41. Differential equation of the family of parabolas is

$$2y\frac{dy}{dx}=4a \quad \text{or} \quad 2y\frac{dy}{dx}=\frac{y^2}{x} \qquad (\because\ y^2=4ax)$$

or $\quad 2x\dfrac{dy}{dx} = y$. Orthogonal trajectories are

$$2x\left(-\frac{dx}{dy}\right) = y \quad \text{or} \quad y\,dy + 2x\,dx = 0 \quad \text{or} \quad \frac{y^2}{2} + x^2 = \frac{c^2}{2}.$$

$\therefore \quad 2x^2 + y^2 = c^2.$

42. From **41**,

$2x^2 + y^2 = c^2$ passes through $(1, 2) \Rightarrow 2\cdot 1^2 + 2^2 = c^2 \Rightarrow c^2 = 6.$

$\therefore$ the curve is $2x^2 + y^2 = 6$.

43. Here, $a = 1 > 0$. So, the condition is $D \le 0 \quad \Rightarrow p^2 - 4\cdot\frac{1}{4}(p+2) \le 0$

$\Rightarrow \quad p^2 - p - 2 \le 0 \quad \Rightarrow -1 \le p \le 2$.

44. Roots real $\quad \Rightarrow D \ge 0 \quad \Rightarrow p^2 - p - 2 \ge 0 \quad \Rightarrow p \le -1$ or $p \ge 2$.

45. Here, the length of $[\alpha, \beta] = 6 - 0 = 6$.

$\therefore$ from **44**, $[a, b] = \{(-\infty, -1] \cup [2, +\infty)\} \cap [0, 6] = [2, 6]$.

$\therefore$ the length of $[a, b] = 6 - 2 = 4$. So, $P(E) = \frac{4}{6} = \frac{2}{3}$.

46. Probability $= {}^{10}C_5\left(\frac{2}{3}\right)^5\cdot\left(\frac{1}{3}\right)^5 \qquad \left(\because \text{ here, } n = 10, r = 5, p = \frac{2}{3}, q = \frac{1}{3}\right)$

$$= \frac{10!}{(5!)^2}\cdot\left(\frac{2}{3\times 3}\right)^5.$$

47. Probability $= {}^5C_3\left(\frac{2}{3}\right)^3\cdot\left(\frac{1}{3}\right)^2 + {}^5C_4\left(\frac{2}{3}\right)^4\cdot\left(\frac{1}{3}\right)^1 + {}^5C_5\left(\frac{2}{3}\right)^5$

$$= \frac{2^3}{3^5}\{{}^5C_3 + {}^5C_4\cdot 2 + {}^5C_5\cdot 2^2\} = \frac{2^3}{3^5}\cdot 24 = 9\cdot\frac{2^6}{3^6}.$$

48. Here, $(n+1)p = (10+1)\cdot\frac{2}{3} = \frac{22}{3} = 7\frac{1}{3}$. So, the integral part of $(n+1)p$ is 7.

$\therefore$ most probable number of successes = 7.

49. The equation of the family of straight lines is $\frac{x}{\lambda} + \frac{y}{4-\lambda} = 1$.

$\therefore \quad f(x, y, \lambda) \equiv (4-\lambda)x + \lambda y - \lambda(4-\lambda) = 0 \qquad \text{... (1)}$

and $\dfrac{\partial f}{\partial \lambda} \equiv -x + y - 4 + 2\lambda = 0.$ ... (2)

(2) $\Rightarrow$ $\lambda = \dfrac{4 + x - y}{2}\cdot$ Putting in (1),

$$\left(4 - \frac{4 + x - y}{2}\right)\cdot x + \frac{4 + x - y}{2}\cdot y - \frac{4 + x - y}{2}\left(4 - \frac{4 + x - y}{2}\right) = 0$$

or $\dfrac{4 - x + y}{2}\cdot x + \dfrac{4 + x - y}{2}\cdot y - \dfrac{4 + x - y}{2}\cdot\dfrac{4 - x + y}{2} = 0$

or $2x(4 - x + y) + 2y(4 + x - y) - (4 + x - y)(4 - x + y) = 0$

or $8x - 2x^2 + 2xy + 8y + 2xy - 2y^2 - \{16 - (x - y)^2\} = 0$

or $-x^2 - y^2 + 2xy + 8x + 8y - 16 = 0$ or $x^2 - 2xy + y^2 - 8x - 8y + 16 = 0.$

50. The tangent to $y^2 = x$ at (t^2, t) is $yt = \dfrac{x + t^2}{2}\cdot$

$\therefore$ the family of tangents to $y^2 = x$ is $f(x, y, t) \equiv x - 2yt + t^2 = 0.$ ... (1)

$\therefore$ $\dfrac{\partial f}{\partial t} \equiv -2y + 2t = 0.$ ... (2)

(2) $\Rightarrow$ $t = y$. So, from (1), $x - 2y\cdot y + y^2 = 0$ or $y^2 = x.$

51. The normal to $x^2 = 4y$ at any point $(2t, t^2)$ has the equation

$$y - t^2 = \frac{-1}{\left.\dfrac{dy}{dx}\right]_{(2t,\, t^2)}}\cdot(x - 2t) \text{ or } y - t^2 = \frac{-1}{t}(x - 2t) \text{ or } x + yt - t^3 - 2t = 0$$

$\therefore$ $f(x, y, t) \equiv x + yt - t^3 - 2t = 0$ ... (1)

and $\dfrac{\partial f}{\partial t} \equiv y - 3t^2 - 2 = 0 \Rightarrow t^2 = \dfrac{y - 2}{3}\cdot$ ... (2)

Putting from (2) in (1), $x + (y - 2)t - t\cdot\dfrac{y - 2}{3} = 0$ or $x + \dfrac{2}{3}(y - 2)t = 0$

or $x^2 = \dfrac{4}{9}(y - 2)^2\cdot t^2$ or $9x^2 = 4(y - 2)^2\cdot\dfrac{y - 2}{3}$ or $27x^2 = 4(y - 2)^3.$

52. Radius of curvature $= \dfrac{1}{\dfrac{d\psi}{ds}} = \dfrac{ds}{d\psi} = \dfrac{ds}{dx}\cdot\dfrac{dx}{d\psi}\cdot$ ... (1)

$\dfrac{dy}{dx} = \tan\psi \Rightarrow \dfrac{d^2y}{dx^2} = \sec^2\psi\cdot\dfrac{d\psi}{dx} \Rightarrow \dfrac{d\psi}{dx} = \dfrac{\dfrac{d^2y}{dx^2}}{1 + \left(\dfrac{dy}{dx}\right)^2}\cdot$ ... (2)

$$(ds)^2 = (dx)^2 + (dy)^2 \Rightarrow \left(\frac{ds}{dx}\right)^2 = 1 + \left(\frac{dy}{dx}\right)^2 \Rightarrow \frac{ds}{dx} = \sqrt{1 + \left(\frac{dy}{dx}\right)^2} \quad \text{... (3)}$$

Putting in (1) from (2) and (3),

$$\text{radius of curvature} = \sqrt{1 + \left(\frac{dy}{dx}\right)^2} \cdot \frac{1 + \left(\frac{dy}{dx}\right)^2}{\frac{d^2y}{dx^2}}$$

53. $y^2 = 4x \Rightarrow 2y\frac{dy}{dx} = 4 \Rightarrow \frac{dy}{dx} = \frac{2}{y} \Rightarrow \frac{d^2y}{dx^2} = \frac{-2}{y^2}\frac{dy}{dx} = \frac{-2}{y^2} \cdot \frac{2}{y}.$

$\therefore$ at (1, 2), $\frac{dy}{dx} = \frac{2}{2} = 1$ and $\frac{d^2y}{dx^2} = \frac{-4}{2^3} = -\frac{1}{2}.$

$\therefore$ radius of curvature $= \left|\frac{(1+1^2)^{3/2}}{-\frac{1}{2}}\right| = 4\sqrt{2}.$

54. The arc rate of turning of tangent is the minimum at the ends of minor axis. So, the radius of curvature is the maximum at the ends of minor axis.

55.

6	43125	
6	7187	... 3
6	1197	... 5
6	199	... 3
6	33	... 1
	5	... 3

$\therefore$ $43125 = 5 \times 6^5 + 3 \times 6^4 + 1 \times 6^3 + 3 \times 6^2 + 5 \times 6^1 + 3.$

$\therefore$ in the scale of 6, the number is 531353.

56.

$$\begin{array}{r} 2053 \\ 412 \\ \hline 2465 \end{array}$$

or $2053 = 2 \times 7^3 + 0 \times 7^2 + 5 \times 7^1 + 3$

$412 = 4 \times 7^2 + 1 \times 7^1 + 2$

sum $= 2 \times 7^3 + 4 \times 7^2 + 6 \times 7^1 + 5 = 2465$

57.

$$\begin{array}{r} 5623 \\ 6 \\ \hline 42562 \end{array}$$

$3 \times 6 = 18 = 2 \times 8 + 2$

$2 \times 6 + 2 = 14 = 1 \times 8 + 6$

$6 \times 6 + 1 = 37 = 4 \times 8 + 5$

$5 \times 6 + 4 = 34 = 4 \times 8 + 2$

58. $4 - 7 = -3$, which is not divisible by 5.

$118 - 18 = 100$, which is divisible by 5.

$110 - 93 = 17$, which is not divisible by 12.

$63 - 9 = 54$, which is not divisible by 11.

59. $a \equiv b \pmod{n}$ and $a' \equiv b' \pmod{n} \Rightarrow a - b = kn$ and $a' - b' = k'n$

$\Rightarrow \quad a - b + a' - b' = (k + k')n \Rightarrow (a + a') - (b + b') = pn$

$\Rightarrow \quad a + a' \equiv b + b' \pmod{n}$.

Again, $aa' - bb' = (a - b)a' + (a' - b')b = kna' + k'nb = (ka' + k'b)n = p'n$

$\Rightarrow \quad aa' \equiv bb' \pmod{n}$.

Also, $a \equiv b \pmod{n}$ and $a \equiv b \pmod{n} \Rightarrow aa \equiv bb \pmod{n}$

$\Rightarrow \quad a^2 \equiv b^2 \pmod{n}$.

Using the above result repeatedly, $a^m \equiv b^m$.

Again, $a \equiv b \pmod{n} \Rightarrow a - b = kn \Rightarrow \frac{a}{d} - \frac{b}{d} = \frac{k}{d} n$.

Since $\frac{k}{d}$ need not be integer, $\frac{a}{d} \equiv \frac{b}{d} \pmod{n}$ need not be true.

60. $x \equiv 6 \pmod{9} \Rightarrow x = 6 + 9k \Rightarrow 5x - 2 = 5(6 + 9k) - 2 = 45k + 28$.

This is not divisible by 7 for all $k \in Z$.

$x \equiv 6 \pmod{7} \Rightarrow x = 6 + 7k \Rightarrow 5x - 2 = 5(6 + 7k) - 2 = 35k + 28$

$= 7(5k + 4)$, which is divisible by 7.

$x \equiv 7 \pmod{6} \Rightarrow x = 7 + 6k \Rightarrow 5x - 2 = 5(7 + 6k) - 2 = 30k + 33$.

This is not divisible by 7 for all $k \in Z$.

$x \equiv 2 \pmod{7} \Rightarrow x = 2 + 7k \Rightarrow 5x - 2 = 5(2 + 7k) - 2 = 35k + 8$.

This is not divisible by 7.

61. $S_n = \frac{1}{1.2} + \frac{1}{2.3} + \ldots + \frac{1}{n(n+1)} = \left(1 - \frac{1}{2}\right) + \left(\frac{1}{2} - \frac{1}{3}\right) + \ldots + \left(\frac{1}{n} - \frac{1}{n+1}\right)$

$= 1 - \frac{1}{n+1}$.

So, $\lim_{n \to \infty} S_n = \lim_{n \to \infty} \left(1 - \frac{1}{n+1}\right) = 1$.

62. $u_n = \frac{x^n}{n(n+1)}$ and $u_{n+1} = \frac{x^{n+1}}{(n+1)(n+2)}$.

$\therefore \quad \lim_{n \to \infty} \frac{u_n}{u_{n+1}} = \lim_{n \to \infty} \frac{x^n}{n(n+1)} \cdot \frac{(n+1)(n+2)}{x^{n+1}} = \lim_{n \to \infty} \frac{n+2}{n} \cdot \frac{1}{x} = \frac{1}{x}$.

$\therefore \quad$ the series is convergent when $\frac{1}{x} > 1$, that is, $x < 1$.

When $x = 1$, we get the series in the above solution of **61**—which is convergent to the sum 1. So, when $x = 2 > 1$, the series is not convergent.

63. $\sum_{n=1}^{\infty} f(n+1) = f(2) + f(3) + f(4) + \ldots = \frac{1}{2(2^2 - 1)} + \frac{1}{3(3^2 - 1)} + \frac{1}{4(4^2 - 1)} + \ldots$

$$= \frac{1}{1\cdot 2\cdot 3} + \frac{1}{2\cdot 3\cdot 4} + \frac{1}{3\cdot 4\cdot 5} + \dots.$$

Now, $\frac{1}{n(n+1)(n+2)} = \frac{1}{n(n+1)} - \frac{1}{n(n+2)} = \frac{1}{n} - \frac{1}{n+1} - \frac{1}{2}\left(\frac{1}{n} - \frac{.1}{n+2}\right)$

$$= \frac{1}{n} - \frac{1}{n+1} - \frac{1}{2}\left\{\left(\frac{1}{n} - \frac{1}{n+1}\right) + \left(\frac{1}{n+1} - \frac{1}{n+2}\right)\right\}$$

$$= \frac{1}{2}\left(\frac{1}{n} - \frac{1}{n+1}\right) - \frac{1}{2}\left(\frac{1}{n+1} - \frac{1}{n+2}\right).$$

$\therefore \quad S_n = \frac{1}{2}\left(1 - \frac{1}{n+1}\right) - \frac{1}{2}\left(\frac{1}{2} - \frac{1}{n+2}\right) = \frac{1}{4} - \frac{1}{2(n+1)} + \frac{1}{2(n+2)}.$

$\therefore \quad \lim_{n \to \infty} S_n = \frac{1}{4}.$

64.

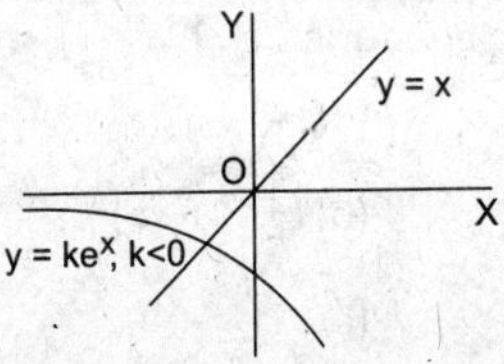

65. For one real solution, the curve $y = ke^x$ should touch the line $y = x$. (**Note** the slope of tangent to $y = ke^x$ is increasing.)

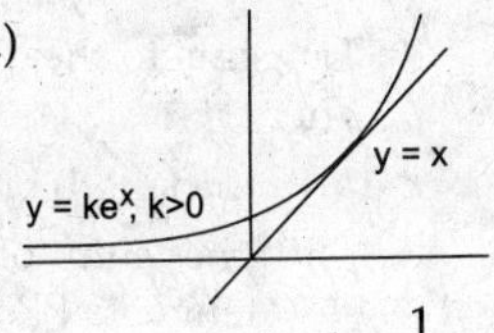

For the curve, $\frac{dy}{dx} = ke^x$ and $\frac{dy}{dx} = 1$ (for the line)

$\therefore$ we must have $ke^x = 1$ or $x = \log_e \frac{1}{k}$

$\therefore \quad y = ke^x = ke^{\log_e \frac{1}{k}} = k \cdot \frac{1}{k} = 1$. As the point is on $y = x$, we get $1 = \log_e \frac{1}{k}$

or $e = \frac{1}{k}$ or $k = \frac{1}{e}$.

66. For one solution, the curve $y = ke^x$ touches the line $y = x$ and then the curve makes an intercept $\frac{1}{e}(= k)$. For two roots, the curve should cross the line $y = x$. (**Note** The slope of tangent to $y = ke^x$ is increasing.) In that case the intercept k on y-axis should be less than $\frac{1}{e}$(= intercept in case of one solution). So, $0 < k < \frac{1}{e}$.

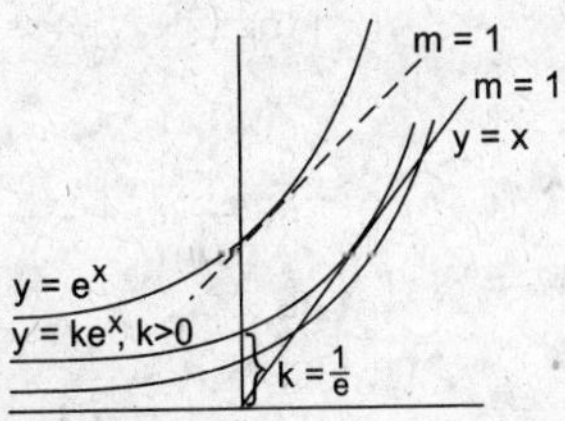

□

5

Miscellaneous Questions—5

Matching Questions

1. $\lambda x^2 + 2(\lambda - 1)xy + y^2 + 3x + 5y + 1 = 0$ is the equation of

(i) a circle if λ is	(a) 2
(ii) a parabola if λ is	(b) $\frac{1}{2}(3+\sqrt{5})$
(iii) an ellipse if λ is	(c) 3
(iv) a hyperbola if λ is	(d) 1

2. Tangents to the parabola $y^2 = 4x$ at the points P and Q meet at $R(-1, 0)$. Then

(i) the length of the chord PQ is	(a) $4\sqrt{2}$
(ii) the sum of the lengths of the tangents PR and QR is	(b) 2
(iii) the inradius of the ΔPQR is	(c) 4
(iv) the circumradius of the ΔPQR is	(d) $2\sqrt{2} - 2$

3.

(i) $\int_{-2}^{2} \log(x + \sqrt{1+x^2})\, dx$ is equal to	(a) 1
(ii) $\lim_{n \to \infty} \frac{1}{\sqrt{n}} \left\{1 + \frac{1}{\sqrt{2}} + \frac{1}{\sqrt{3}} + \ldots + \frac{1}{\sqrt{n}}\right\}$ is equal to	(b) 2
(iii) If $\lim_{n \to \infty} \left(\frac{n^2+1}{n+1} - an\right) + 1 = 0$ then a is	(c) 0

(iv) If $y = |\cos x|$ then $\left.\frac{dy}{dx}\right]_{x = 3\pi/4}$ is (d) $\frac{1}{\sqrt{2}}$

4. (i) The maximum value of $(2\sin x)(\sqrt{5}\cos x + 2\sin x) - 6$ is (a) $-\frac{3}{8}$

(ii) $\int_{-2}^{2} |x|[x]\, dx$ is equal to (b) $-\frac{1}{2}$

(iii) $\lim_{x \to 0} \frac{\tan^{-1}x - \sin^{-1}x}{x^3}$ is (c) -2

(iv) If $x^2y + y^3 = 2$ then $\left.\frac{d^2y}{dx^2}\right]_{(1, 1)}$ is (d) -1

5. (i) If $x^2 - ax + 1 - 2a^2 \geq 0$ for all $x \in R$ then (a) $-5 < a < 3$

(ii) If the difference between the roots of $x^2 + ax + 1 = 0$ is less than $\sqrt{5}$ then (b) $-3 \leq a \leq -2$

(iii) If the complex number $a + 1 + 3i$ for all $a \in R$ is represented by an interior point of the circle $|z| = 5$ then (c) $-\frac{2}{3} \leq a \leq \frac{2}{3}$

(iv) If $\cos^4 x - (a + 2)\cos^2 x - (a + 3) = 0$ has real solutions for x then (d) $|a| < 3$

6. (i) If $\cos\theta - \sin\theta = \cos\alpha - \sin\alpha$ then the minimum value of $|\theta + \alpha|$ is (a) $\frac{\pi}{4}$

(ii) In a ΔABC, if $A = 90°$ then $\tan^{-1}\frac{b}{c + a} + \tan^{-1}\frac{c}{a + b}$ is (b) $\frac{\pi}{2}$

(iii) If $x \in [0, \pi]$ and

$$\log_2(\tan x) + \log_2(\tan 2x) = 0$$

then x is (c) $\frac{\pi}{3}$

(iv) In a ΔABC, if

$$\frac{\tan A - \tan B}{\tan A + \tan B} = \frac{c - b}{c}$$

then A is (d) $\frac{\pi}{6}$

7. (i) $f(x) = x|x-2|$

(ii) $g(x) = x^2[x]$

(iii) $h(x) = \dfrac{x}{4+|x|}$

(iv) $\phi(x) = \begin{cases} [x-1] & \text{for } x \le 1 \\ x^3 & \text{for } 1 < x \le 2 \\ 4|x-2| + 2x^2 & \text{for } x > 2 \end{cases}$

(a) neither continuous nor differentiable at $x = 2$

(b) differentiable and continuous for all x

(c) continuous but not differentiable at $x = 2$

(d) differentiable and continuous at $x = 2$

8. (i) If ${}^nC_{r-1} = 36, {}^nC_r = 84$ and ${}^nC_{r+1} = 126$ then r is

(ii) The number of real solutions of the equation $2^x + x^2 = 1$ is

(iii) If the focus of the parabola $x^2 - ky + 3 = 0$ $(k \ne 2)$ is $(0, 2)$ then k is

(iv) The number of different ways in which the first twelve natural numbers can be divided in three equal groups such that the numbers in each group are in AP is

(a) 6

(b) 2

(c) 4

(d) 3

9. (i) If z_1, z_2, z_3 are unimodular complex numbers such that $|z_1 - z_2 + z_3| = 4$

then $\left|\dfrac{1}{z_1} - \dfrac{1}{z_2} + \dfrac{1}{z_3}\right|$ is

(ii) The planes $2x - 3y - 7z = 0$, $\lambda x - 14y - 13z = 0$ and $8x - 31y - 33z = 0$ pass through the same line if λ is

(iii) If $\tan^{-1}\dfrac{\sqrt{1-x^2}-1}{x} : \tan^{-1}x = a : 1$

then a is

(a) $\dfrac{1}{2}$

(b) 3

(c) 2

(iv) In the ΔABC, the median $AD = \dfrac{1}{\sqrt{11-6\sqrt{3}}}$ and the median divides $\angle A$ into angles of $30°$ and $45°$. The length of BC is — (d) 4

10. (i) The line $\dfrac{x-1}{l} = \dfrac{y+1}{4} = \dfrac{z-2}{2}$ lies on the plane $2x - y + z - 5 = 0$ if l is — (a) 3

(ii) The latus rectum of the rectangular hyperbola $x^2 - y^2 = 1$ is — (b) 1

(iii) The line cutting a positive intercept 1 on the y-axis will be a tangent to the curve $y^2 = 12x$ if its slope is — (c) 4

(iv) The lines $\dfrac{x-5}{4} = \dfrac{y-7}{m} = \dfrac{z+3}{-5}$ and $\dfrac{x-8}{7} = \dfrac{y-4}{1} = \dfrac{z-5}{3}$ are intersecting lines if m is — (d) 2

11. (i) $\displaystyle\int_0^4 \frac{f(t)}{f(t)+f(4-t)}\,dt$ is equal to — (a) 2

(ii) If $f(x)$ is an even function then $\displaystyle\int_{-2}^{2} \{x^3 f(x) + xf''(x) + 2\}\,dx$ is equal to — (b) 1

(iii) If $f(x) = \displaystyle\int_0^x |\cos t|\,dt$ then $f'\left(\dfrac{\pi}{2}\right)$ equals — (c) 0

(iv) The value of b for which the area bounded by the parabolas $y = x - bx^2$ and $y = \dfrac{1}{b}x^2$ $(b > 0)$ is the maximum is — (d) 8

12. (i) If $f: [2, +\infty) \to X$ be a bijective, where $f(x) = 5 - 4x + x^2$, then X is — (a) $R - [-2, 2]$

(ii) The domain of

$$\phi(x) = \frac{x-3}{(x-1)\sqrt{x^2-4}}$$

is — (b) $[1, +\infty)$

(iii) The interval in which $f(x) = 2x^2 - \log x$ is monotonic increasing is — (c) $(-2, -1)$

(iv) $4x^3 + 9x^2 + 6x + 3 = 0$ has a real root lying in — (d) $\left(\frac{1}{2}, +\infty\right)$

13. (i) The focus of the parabola $y^2 - 2y + 8x - 23 = 0$ is — (a) $(1, -2)$

(ii) The centre of the ellipse $9x^2 + 5y^2 - 30y = 0$ is — (b) $\left(\frac{3}{2}, -\frac{3}{2}\right)$

(iii) The vertex of the parabola $x^2 + 2xy + x = 3 + y - y^2$ is — (c) $(1, 1)$

(iv) The tangents at the ends of the double ordinate through the focus of $x^2 = 2(x + 2y) + 3$ meet at — (d) $(0, 3)$

14. (i) If α is a nonreal root of $x^3 - 3x^2 + 3x - 2 = 0$ then $\alpha^{6n} + \alpha^{-6n}$ is equal to — (a) 3

(ii) The order of the differential equation whose general solution is $y = (c_1x + c_2)e^{ax} + c_3e^{bx + c_4}$, where c_1, c_2, c_3, c_4 are arbitrary constants and a, b are fixed constants, is — (b) 4

(iii) Let $f(x + y) = f(x) + f(y)$ and $f(1) = 1$. If $\phi(x) = \int_0^x f(2x)\, dx$ then $\phi'(2)$ is — (c) $\frac{4}{3}$

(iv) If $\vec{a}, \vec{b}, \vec{c}$ are three mutually perpendicular vectors where $|\vec{a}| = |\vec{b}| = 2$, $|\vec{c}| = 1$ then $\frac{1}{12}[\vec{a} \times \vec{b}\ \ \vec{b} \times \vec{c}\ \ \vec{c} \times \vec{a}]$ is — (d) 2

15. (i) If $|z - 2i| \le 3$ then the maximum value of $|iz + 3|$ is — (a) 3

(ii) If the greatest coefficient in the expansion of $(1+x)^n$ $(n \in N)$ is $\frac{1}{24}n(n-1)(n-2)(n-3)$ then the value of n is	(b) 5
(iii) If $\vec{\lambda} = \vec{a} + 2\vec{b} - \vec{c}$, $\vec{\mu} = \vec{a} + \vec{b} + 3\vec{c}$, $\vec{\gamma} = 2\vec{a} + 3\vec{b} + \vec{c}$ and $[\vec{a}\ \vec{b}\ \vec{c}] = 3$ then $[\vec{\lambda}\ \vec{\mu}\ \vec{\nu}]$ is	(c) 4
(iv) Let $\lvert f(x)\rvert \le 1$ for $x \in [0, 1)$ and $\lvert f(x)\rvert \le 2$ for $x \in [1, 3]$ then the greatest value of $\int_0^3 f(x)\,dx$ is	(d) 8

16. In the parabola $y^2 + 4 = 4x$, a chord passing through the point (2, 0) cuts the parabola at P and Q. If $P = (5, 4)$ and the tangents at P and Q meet at R then match the following.

(i) The focus is	(a) $\left(0, \frac{3}{2}\right)$
(ii) The centroid of the ΔPQR is	(b) (2, 0)
(iii) The circumcentre of the ΔPQR is	(c) $\left(\frac{25}{12}, \frac{3}{2}\right)$
(iv) The orthocentre of the ΔPQR is	(d) $\left(\frac{25}{8}, \frac{3}{2}\right)$

17.

(i) $[\vec{a}\ \vec{b}\ \vec{i}]\vec{i} + [\vec{a}\ \vec{b}\ \vec{j}]\vec{j} + [\vec{a}\ \vec{b}\ \vec{k}]\vec{k}$ equals	(a) $\vec{a} + \vec{b}$
(ii) If $\vec{a} \times (\vec{a} \times \vec{b}) = \vec{b} \times (\vec{b} \times \vec{c})$ and $\vec{a} \cdot \vec{b} \ne 0$ then $[\vec{a}\ \vec{b}\ \vec{c}]$ equals	(b) $\vec{a} \times \vec{b}$
(iii) If $\lvert\vec{a}\rvert = \lvert\vec{b}\rvert = 1$, $\lvert\vec{a} + \vec{b}\rvert = \sqrt{2}$ and $\vec{c} = \vec{a} - \vec{b} + \vec{a} \times \vec{b}$ then $(\vec{a} \times \vec{b}) \times \vec{c}$ equals	(c) 0
(iv) If $\vec{a}, \vec{b}$ are orthogonal unit vectors then $[\vec{a} + \vec{a} \times \vec{b}\ \ \vec{b} + \vec{a} \times \vec{b}\ \ \vec{a} \times \vec{b}]$ equals	(d) 1

18. (i) Given $a^2 + a + 1$ is divisible by 3. When a is divided by 3 the remainder will be — (a) 2

(ii) $\lim\limits_{n \to \infty} \dfrac{\sqrt[3]{n^2} \cdot \sin(n!)}{n+1}$ is — (b) 3

(iii) If $f(n+1) = \dfrac{1}{2}\left\{f(n) + \dfrac{9}{f(n)}\right\}$ for all $n \in N$ and $f(n) > 0$ for all $n \in N$ then $\lim\limits_{n \to \infty} f(n)$ is — (c) 0

(iv) If $|\vec{a}| = 1$ then $|\vec{a} \times \vec{i}|^2 + |\vec{a} \times \vec{j}|^2 + |\vec{a} \times \vec{k}|^2$ equals — (d) 1

19. (i) An unbiased dice is thrown and the number shown is put for p in $x^2 + px + 2 = 0$. The probability of the equation to have real roots is — (a) $\dfrac{2}{3}$

(ii) $\vec{a}, \vec{b}$ are two unit vectors inclined at $\dfrac{\pi}{3}$. Then $[\vec{a} \;\; \vec{b} + \vec{a} \times \vec{b} \;\; \vec{b}]$ is — (b) $\dfrac{2}{5}$

(iii) If $f(x)$ is differentiable and $\displaystyle\int_0^{t^2} xf(x)\, dx = \frac{2}{5}t^5$ then $f\left(\dfrac{4}{25}\right)$ equals — (c) -1

(iv) Three lines through the origin having direction ratios $1, a, a^2$; $1, b, b^2$; $1, c, c^2$ are noncoplanar. But the lines through the origin having direction ratios $a, a^2, 1 + a^3$; $b, b^2, 1 + b^3$; $c, c^2, 1 + c^3$ are coplanar then abc equals — (d) $-\dfrac{3}{4}$

20. (i) Let

$$f(x) = 2\sin^2\beta + 4\cos(x+\beta) \cdot \sin x \cdot \sin\beta + \cos 2(x+\beta).$$

Then the value of $\{f(\alpha)\}^2 + \left\{f\left(\dfrac{\pi}{4} - \alpha\right)\right\}^2$ is — (a) 0

(ii) n is given a value from the set $\{1, 2, 3, \ldots, 12\}$ at random. The probability of the value of $(1 + i)^n$ being real is

(iii) If $[\vec{x}\ \vec{y}\ \vec{a}] = [\vec{x}\ \vec{y}\ \vec{b}] = [\vec{a}\ \vec{b}\ \vec{c}] = 0$, each of the vectors being a nonzero vector, then $[\vec{x}\ \vec{y}\ \vec{c}]$ equals

(iv) If $\frac{2 + \sin x}{y + 1} \frac{dy}{dx} + \cos x = 0$, where $y(0) = 1$, then $y\left(\frac{\pi}{2}\right)$ equals

(b) $\frac{1}{3}$

(c) 1

(d) $\frac{1}{4}$

21. (i) If the vertices of a triangle are represented by the complex numbers 0, z_1 and z_2 such that $\frac{z_1}{z_2} + \frac{z_2}{z_1} = 1$ then the triangle is

(ii) If a vertex, the circumcentre and the centroid of a triangle be (0, 0), (3, 4) and (6, 8), respectively, then the triangle is

(iii) If in a ΔABC, $\sin A \cdot \cos B = 1 - \frac{1}{\sqrt{2}}$ and $\tan A \cdot \cot B = \sqrt{2} - 1$ then the triangleis

(iv) The triangle formed by the lines $\frac{x}{3} = \frac{y}{0} = \frac{z}{4}$, $\frac{x}{0} = \frac{y}{5} = \frac{z}{0}$ and $\frac{x - 3}{3} = \frac{y}{-5} = \frac{z - 4}{4}$ is

(a) right-angled

(b) equilateral

(c) isosceles

(d) isosceles right-angled

22. (i) If $x^2 - 4x - 2\pi < \sin^{-1}(\sin 5)$ then x lies in

(ii) If $f(x) = x^2 - 4$ and $g(x) = f(x) + 5$ such that $f\{g(x)\} > 0$ then x lies in

(iii) If $3\sin\theta + 4\cos\theta = \frac{5(2a - 3)}{2 - 3a}$ has a real solution for θ then a lies in

(a) $R - [-1, 1]$

(b) $R - (-1, 1)$

(c) $[-4, 4]$

(iv) If $4x^4 + 9y^4 = 64$ and $z = \sqrt{3}xy$ then z lies in — (d) $[-1, 5]$

23. (i) The number of sides of the quadrilateral formed by the lines $x^2y^2 + 1 = x^2 + y^2$ that touch the circle $x^2 + y^2 - 2x = 0$ is — (a) 1

(ii) $\vec{b}, \vec{c}$ are orthogonal unit vectors and $\vec{b} \times \vec{c} = \vec{a}$. Then $[\vec{a} + \vec{b} + \vec{c} \;\; \vec{a} + \vec{b} \;\; \vec{b} + \vec{c}]$ equals — (b) 0

(iii) A line having direction ratios $1, -1, 5$ is perpendicular to the plane OPQ, where $P = (\lambda, 2, 1)$, $Q = (-2, \mu, 1)$ and O is the origin. Then $\lambda + \mu$ equals — (c) 2

(iv) If the projection of the vector $\vec{i} + \vec{j} + \vec{k}$ on the vector $a\vec{i} + 4\vec{j} - 5\vec{k}$ be $\frac{1}{3\sqrt{5}}$ then a can have the value — (d) 3

24. (i) Let $a + b + c + d = 8$ and $abcd = 16$, where a, b, c, d are positive numbers. The value of $a + b$ is — (a) 5

(ii) The total number of ways of selecting an odd number of things from $(2n + 1)$ different things is 1024. Then n is — (b) 3

(iii) If $\sqrt{3}\sin x + \cos x - 2 = (y - 1)^2$ for $0 \le x \le 8\pi$ then the number of values of the pair (x, y) is — (c) 6

(iv) $f(x) = 2x^3 - 3ax^2 + \frac{4}{3}a^2x + 1$ $(a > 0)$ attains its maximum and minimum at $x = p$ and $x = p^2$, respectively. Then a equals — (d) 4

25. (i) If $|z_1| = 12$ and $|z_2 - 3 - 4i| = 5$ then the minimum value of $|z_1 - z_2|$ is — (a) 3

(ii) If $\lim\limits_{x \to 0} \dfrac{(1-\cos x)(e^x - \cos x)}{x^n}$ is a nonzero finite number then the integer n is — (b) 1

(iii) If $f(x) = \dfrac{1-x}{1+x}$ for $x > 0$ then the minimum value of $f\{f(x)\} + f\left\{f\left(\dfrac{1}{x}\right)\right\}$ is — (c) 2

(iv) If z is a complex number satisfying $z\bar{z} - 2(z + \bar{z}) + 3 = 0$ then the greatest value of $|z|$ is — (d) 4

26. (i) The domain of the real valued function $f(x)$ for which $4^{f(x)} + 4^{1-f(x)} = 4^x$ is — (a) $(1, \infty)$

(ii) A point is moving along the curve $y^3 = 27x$. The interval of values of x in which the ordinate changes faster than the abscissa is — (b) $\left[-\dfrac{1}{2}, \dfrac{1}{2}\right]$

(iii) The range of values of x for which $x^2 - 1, 2x + 1$ and $x^2 + x + 1$ will be the lengths of three sides of a triangleis — (c) $[1, \infty]$

(iv) The equation $\sin^{-1}x - 3\sin^{-1}a = 0$ has real solutions for x if a belongs to — (d) $(-1, 1)$

27. (i) The total number of positive integral solutions of $xyz = 18$ is — (a) 18

(ii) In a ΔABC, the minimum value of $\operatorname{cosec}\dfrac{A}{2} + \operatorname{cosec}\dfrac{B}{2} + \operatorname{cosec}\dfrac{C}{2}$ is — (b) $\dfrac{3}{2}$

(iii) If $x^2 + x + 1 = 0$ the value of $\sum\limits_{n=1}^{9}\left(x^n + \dfrac{1}{x^n}\right)^2$ is — (c) 15

(iv) The value of $\displaystyle\int_3^6 \frac{\sqrt{x}}{\sqrt{9-x} + \sqrt{x}}\,dx$ is — (d) 6

28. (i) If ${}^{2n+1}C_{n+1} + {}^{2n+1}C_{n+2} + {}^{2n+1}C_{n+3} + \ldots + {}^{2n+1}C_{2n+1}$ equals 1024 then n is — (a) 16

(ii) Let $A = \begin{bmatrix} 0 & 2y & z \\ x & y & -z \\ x & -y & z \end{bmatrix}$ and $A^T A = I$. Then the value of $x^2 + y^2 + z^2$ is — (b) 1

(iii) In the expansion of $(1+x)^{47}$ the coefficients of $(2r+1)$th term and the 16th term are equal. Then r is — (c) 15

(iv) Let $\frac{dF(x)}{dx} = \frac{e^{\sin x}}{x}$ for $x > 0$. If $\int_1^4 \frac{e^{\sin x^2}}{x} dx$ equals $\frac{1}{2}\{F(k) - F(1)\}$ then k is — (d) 5

29. (i) If $|\vec{a}| = |\vec{b}| = \sqrt{2}$ then $[\vec{a} \times \vec{b} \;\; \vec{a} \;\; \vec{b}] + (\vec{a} \cdot \vec{b})^2$ equals — (a) 2

(ii) The number of points $(a+1, a)$, where $a \in Z$, lying inside the region bounded by the circles $x^2 + y^2 - 2x - 1 = 0$ and $x^2 + y^2 - 2x - 17 = 0$ is — (b) 4

(iii) The straight line joining the points (0, 3) and (5, –2) is a tangent to the curve $y = \frac{c}{x+1}$. Then c is — (c) 3

(iv) A bag contains 8 balls of two colours: red and green. 3 balls are taken out at random. The probability of getting more red balls than green is $\frac{15}{28}$ and that of getting more green balls than red is $\frac{15}{56}$. The number of green balls is — (d) 5

30. (i) If exactly two real common tangents can be drawn to the circles $x^2 + y^2 - 2x - 2y = 0$ and $x^2 + y^2 - 8x - 8y + 6\lambda = 0$ for $\lambda \in Z$ then the greatest possible value of λ equals

(a) 2

(ii) $\int_{-2\pi}^{2\pi} \{|\sin x| + |\cos x|\} \, dx$ equals

(b) 3

(iii) The slope of a curve at (x, y) is $-\frac{x+y}{x}$ and it passes through the points $\left(3, -\frac{1}{6}\right)$ and $(\lambda, 1)$ then one of the values of λ is

(c) 1

(iv) $(\lambda, 6, 2)$ is a point on the plane passing through the line $\frac{x+1}{2} = \frac{y-1}{-1} = \frac{z+3}{1}$ and parallel to the line of intersection of the planes $x - y - 5z = 6$ and $3x + 5y + 3z = 4$. Then λ is

(d) 16

Answers

1.	(i) ↔ (d)	(ii) ↔ (b)	(iii) ↔ (c)	(iv) ↔ (a)
2.	(i) ↔ (c)	(ii) ↔ (a)	(iii) ↔ (d)	(iv) ↔ (b)
3.	(i) ↔ (c)	(ii) ↔ (b)	(iii) ↔ (a)	(iv) ↔ (d)
4.	(i) ↔ (d)	(ii) ↔ (c)	(iii) ↔ (b)	(iv) ↔ (a)
5.	(i) ↔ (c)	(ii) ↔ (d)	(iii) ↔ (a)	(iv) ↔ (b)
6.	(i) ↔ (b)	(ii) ↔ (a)	(iii) ↔ (d)	(iv) ↔ (c)
7.	(i) ↔ (c)	(ii) ↔ (a)	(iii) ↔ (b)	(iv) ↔ (d)
8.	(i) ↔ (d)	(ii) ↔ (b)	(iii) ↔ (a)	(iv) ↔ (c)
9.	(i) ↔ (d)	(ii) ↔ (b)	(iii) ↔ (a)	(iv) ↔ (c)
10.	(i) ↔ (b)	(ii) ↔ (d)	(iii) ↔ (a)	(iv) ↔ (c)
11.	(i) ↔ (a)	(ii) ↔ (d)	(iii) ↔ (c)	(iv) ↔ (b)
12.	(i) ↔ (b)	(ii) ↔ (a)	(iii) ↔ (d)	(iv) ↔ (c)
13.	(i) ↔ (c)	(ii) ↔ (d)	(iii) ↔ (b)	(iv) ↔ (a)
14.	(i) ↔ (d)	(ii) ↔ (a)	(iii) ↔ (b)	(iv) ↔ (c)
15.	(i) ↔ (c)	(ii) ↔ (d)	(iii) ↔ (a)	(iv) ↔ (b)
16.	(i) ↔ (b)	(ii) ↔ (c)	(iii) ↔ (d)	(iv) ↔ (a)
17.	(i) ↔ (b)	(ii) ↔ (c)	(iii) ↔ (a)	(iv) ↔ (d)
18.	(i) ↔ (d)	(ii) ↔ (c)	(iii) ↔ (b)	(iv) ↔ (a)
19.	(i) ↔ (a)	(ii) ↔ (d)	(iii) ↔ (b)	(iv) ↔ (c)
20.	(i) ↔ (c)	(ii) ↔ (d)	(iii) ↔ (a)	(iv) ↔ (b)
21.	(i) ↔ (b)	(ii) ↔ (c)	(iii) ↔ (a)	(iv) ↔ (d)
22.	(i) ↔ (d)	(ii) ↔ (a)	(iii) ↔ (b)	(iv) ↔ (c)
23.	(i) ↔ (c)	(ii) ↔ (a)	(iii) ↔ (b)	(iv) ↔ (c)
24.	(i) ↔ (d)	(ii) ↔ (a)	(iii) ↔ (d)	(iv) ↔ (c)
25.	(i) ↔ (c)	(ii) ↔ (a)	(iii) ↔ (c)	(iv) ↔ (a)
26.	(i) ↔ (c)	(ii) ↔ (d)	(iii) ↔ (a)	(iv) ↔ (b)
27.	(i) ↔ (a)	(ii) ↔ (d)	(iii) ↔ (a)	(iv) ↔ (b)
28.	(i) ↔ (d)	(ii) ↔ (b)	(iii) ↔ (a)	(iv) ↔ (a)
29.	(i) ↔ (b)	(ii) ↔ (a)	(iii) ↔ (b)	(iv) ↔ (c)
30.	(i) ↔ (b)	(ii) ↔ (d)	(iii) ↔ (a)	(iv) ↔ (b)

Hints and Solutions to Selected Questions

1. $a=b, h=0 \Rightarrow \lambda=1, \lambda-1=0 \Rightarrow \lambda=1.$

$ab-h^2=\lambda\cdot 1-(\lambda-1)^2=-\lambda^2+3\lambda-1.$

$$\therefore \quad ab-h^2=0 \Rightarrow \lambda^2-3\lambda+1=0 \Rightarrow \lambda=\frac{3\pm\sqrt{9-4}}{2}=\frac{3\pm\sqrt{5}}{2}.$$

$$ab-h^2>0 \Rightarrow \lambda^2-3\lambda+1>0 \Rightarrow \lambda<\frac{3-\sqrt{5}}{2} \text{ or } \lambda>\frac{3+\sqrt{5}}{2}.$$

$$ab-h^2<0 \Rightarrow \lambda^2-3\lambda+1<0 \Rightarrow \frac{3-\sqrt{5}}{2}<\lambda<\frac{3+\sqrt{5}}{2}.$$

2. The equation of the polar line PQ of $R(-1, 0)$ is $y\cdot 0=2(x-1)$, i.e., $x-1=0$.

Solving $y^2=4x$ and $x-1=0$ we get $P=(1, 2)$ and $Q=(1, -2)$.

$\therefore \quad PQ=4, PR=QR=2\sqrt{2}$. So, $\angle PRQ=90°$.

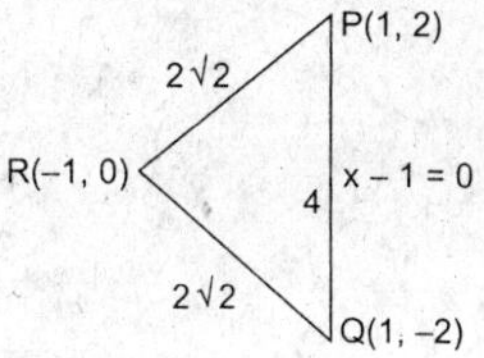

$$\text{Inradius}=\frac{\Delta}{s}=\frac{\frac{1}{2}\cdot 2\sqrt{2}\cdot 2\sqrt{2}}{\frac{1}{2}(2\sqrt{2}+2\sqrt{2}+4)}=2\sqrt{2}-2.$$

By sine rule, $\dfrac{4}{\sin 90°}=2R \Rightarrow R=2$.

3. (i) $$f(x)=\log(x+\sqrt{1+x^2})=\log\frac{(1+x^2)-x^2}{\sqrt{1+x^2}-x}=\log\frac{1}{\sqrt{1+x^2}-x}$$

$$=-\log\{-x+\sqrt{1=(-x)^2}\}=-f(-x). \text{ So, } f(x) \text{ is an odd function.}$$

$$\therefore \int_{-2}^{2} f(x)\,dx-0.$$

(ii) $$\lim_{n\to\infty}\frac{1}{\sqrt{n}}\left\{1+\frac{1}{\sqrt{2}}+\frac{1}{\sqrt{3}}+\ldots+\frac{1}{\sqrt{n}}\right\}=\lim_{n\to\infty}\sum_{r=1}^{n}\frac{1}{\sqrt{n}}\cdot\frac{1}{\sqrt{r}}$$

$$=\lim_{n\to\infty}\sum_{r+1}^{n}\frac{1}{n}\cdot\sqrt{\frac{n}{r}}$$

$$= \lim_{n \to \infty} \sum_{r=1}^{n} \frac{1}{n} \cdot \frac{1}{\sqrt{\frac{r}{n}}}$$

$$= \int_0^1 \frac{1}{\sqrt{x}}\, dx = 2\sqrt{x}\Big]_0^1 = 2.$$

(iii) $\lim\limits_{n \to \infty} \left(\frac{n^2+1}{n+1} - an \right) = \lim\limits_{n \to \infty} \frac{(1-a)n^2 - an + 1}{n+1} = \infty$ unless $1 - a = 0$.

$\therefore$ for the limit to be finite (i.e., –1), $1 - a = 0$.

(iv) Around $x = \frac{3\pi}{4}$, $\cos x < 0$. So, $y = -\cos x$.

$\therefore \frac{dy}{dx} = \sin x$. $\quad \therefore \frac{dy}{dx}\Big]_{x = 3\pi/4} = \sin \frac{3\pi}{4} = \frac{1}{\sqrt{2}}$.

4. (i) Expression $= \sqrt{5}\sin 2x + 4\sin^2 x - 6 = \sqrt{5}\sin 2x - 2\cos 2x - 4$.

$\therefore$ maximum value $= \sqrt{(\sqrt{5})^2 + (-2)^2} - 4 = -1$.

(ii) $\int_{-2}^{2} |x|[x]\, dx = \int_{-2}^{-1} |x|[x]\, dx + \int_{-1}^{0} |x|[x]\, dx + \int_{0}^{1} |x|[x]\, dx + \int_{1}^{2} |x|[x]\, dx$

$$= \int_{-2}^{-1} -x(-2)\, dx + \int_{-1}^{0} -x(-1)\, dx + \int_{0}^{1} x(0)\, dx + \int_{1}^{2} x(1)\, dx$$

$$= x^2\Big]_{-2}^{-1} + \frac{x^2}{2}\Big]_{-1}^{0} + \frac{x^2}{2}\Big]_{1}^{2} = (1-4) + \left(0 - \frac{1}{2}\right) + \left(2 - \frac{1}{2}\right) = -2.$$

(iii) Limit $= \lim\limits_{x \to 0} \dfrac{\frac{1}{1+x^2} - \frac{1}{\sqrt{1-x^2}}}{3x^2} = \lim\limits_{x \to 0} \dfrac{\sqrt{1-x^2} - (1+x^2)}{3x^2 \cdot (1+x^2)\sqrt{1-x^2}}$

$$= \lim_{x \to 0} \frac{1 - x^2 - (1+x^2)^2}{3x^2(1+x^2)\sqrt{1-x^2}\{\sqrt{1-x^2} + (1+x^2)\}}$$

$$= \lim_{x \to 0} \frac{-x^2 - 3}{3(1+x^2)\sqrt{1-x^2}\{\sqrt{1-x^2} + (1+x^2)\}} = \frac{-3}{3 \cdot 1 \cdot 1 \cdot 2} = -\frac{1}{2}.$$

(iv) $x^2 y + y^3 = 2 \Rightarrow 2xy + x^2 \frac{dy}{dx} + 3y^2 \frac{dy}{dx} = 0$ ... (1)

$\Rightarrow \frac{dy}{dx} = \frac{-2xy}{x^2 + 3y^2} \Rightarrow \frac{dy}{dx}\Big]_{(1,1)} = \frac{-2}{1+3} = -\frac{1}{2}$.

Differentiating (1) w.r.t. x,

$$2y + 2x\frac{dy}{dx} + 2x\frac{dy}{dx} + x^2\frac{d^2y}{dx^2} + 6y\left(\frac{dy}{dx}\right)^2 + 3y^2\frac{d^2y}{dx^2} = 0$$

$$\therefore \text{ at } (1, 1),\ 2 + 2\left(-\frac{1}{2}\right) + 2\left(-\frac{1}{2}\right) + 1\cdot\left.\frac{d^2y}{dx^2}\right]_{(1,1)} + 6\cdot 1\cdot\left(-\frac{1}{2}\right)^2 + 3\cdot 1\cdot\left.\frac{d^2y}{dx^2}\right]_{(1,1)} = 0$$

or $4\cdot\left.\dfrac{d^2y}{dx^2}\right]_{(1,1)} = -\dfrac{3}{2}$. $\quad\therefore \left.\dfrac{d^2y}{dx^2}\right]_{(1,1)} = -\dfrac{3}{8}$.

5. (i) Here, coefficient of $x^2 = 1 > 0$. So, the condition is $D \le 0$.

$\therefore\ a^2 - 4(1 - 2a^2) < 0$

or $9a^2 - 4 \le 0 \Rightarrow a^2 \le \left(\dfrac{2}{3}\right)^2 \Rightarrow -\dfrac{2}{3} \le a \le \dfrac{2}{3}$.

(ii) If α, β be roots of $x^2 + ax + 1 = 0$ then $\alpha + \beta = -a$, $\alpha\beta = 1$ and $|\alpha - \beta| < \sqrt{5}$.

$\therefore\ (\alpha - \beta)^2 < 5$ or $(\alpha + \beta)^2 - 4\alpha\beta < 5$ or $a^2 - 4 < 5$ or $a^2 < 3^2$

or $-3 < a < 3$.

(iii) $(a + 1) + 3i$ is an interior point of $|z| = 5 \Rightarrow |a + 1 + 3i| < 5$

$\Rightarrow (a + 1)^2 + 3^2 < 5^2 \Rightarrow a^2 + 2a - 15 < 0$

$\Rightarrow (a + 5)(a - 3) < 0 \Rightarrow -5 < a < 3$.

(iv) $\cos^4 x - (a + 2)\cos^2 x - (a + 3) = 0$

$$\Rightarrow \cos^2 x = \frac{a + 2 \pm \sqrt{(a + 2)^2 + 4(a + 3)}}{2}$$

$$\text{or } \cos^2 x = \frac{a + 2 \pm \sqrt{a^2 + 8a + 16}}{2} = \frac{a + 2 \pm (a + 4)}{2} = a + 3, -1.$$

But $0 \le \cos^2 x \le 1$. So, $0 \le a + 3 \le 1 \Rightarrow -3 \le a \le -2$.

6. (i) $\sqrt{2}\cos\left(\dfrac{\pi}{4} + \theta\right) = \sqrt{2}\cos\left(\dfrac{\pi}{4} + \alpha\right) \Rightarrow \dfrac{\pi}{4} + \theta = 2n\pi \pm \left(\dfrac{\pi}{4} + \alpha\right)$.

Taking the negative sign, $\theta + \alpha = 2n\pi - \dfrac{\pi}{2}$.

So, $|\theta + \alpha| = \left|2n\pi - \dfrac{\pi}{2}\right|$ for $n \in Z$.

$\therefore$ the minimum value of $|\theta + \alpha| = \left|-\frac{\pi}{2}\right| = \frac{\pi}{2}$.

(ii) $\tan^{-1}\frac{b}{c+a} + \tan^{-1}\frac{c}{a+b} = \tan^{-1}\frac{\frac{b}{c+a}+\frac{c}{a+b}}{1-\frac{bc}{(c+a)(a+b)}}$

$$= \tan^{-1}\frac{ab+ca+b^2+c^2}{ca+ab+a^2}$$

$$= \tan^{-1}1 = \frac{\pi}{4} \qquad (\because\ b^2+c^2=a^2).$$

(iii) $\log_2(\tan x) + \log_2(\tan 2x) = 0 \quad \Rightarrow \log_2(\tan x \cdot \tan 2x) = 0$

$\Rightarrow \tan x \cdot \tan 2x = 2^0 = 1 \quad \Rightarrow \frac{2\tan^2 x}{1-\tan^2 x} = 1$

$\Rightarrow \tan^2 x = \frac{1}{3} \Rightarrow x = \frac{\pi}{6} \qquad (\because\ \tan x > 0, \tan 2x > 0).$

(iv) $\frac{\sin(A-B)}{\sin(A+B)} = \frac{\sin C - \sin B}{\sin C}$ or $\sin(A-B) = \sin(A+B) - \sin B$

or $\sin B = 2\cos A \sin B$. $\therefore \cos A = \frac{1}{2}$. $\therefore A = \frac{\pi}{3}$.

7. (i) x is continuous and differentiable everywhere but $|x-2|$ is not differentiable at $x = 2$. So, $x|x-2|$ is continuous but not differentiable at $x = 2$.

(ii) x^2 is continuous and differentiable everywhere but $[x]$ is neither continuous nor differentiable at integers. So, $x^2[x]$ is neither continuous nor differentiable at $x = 2$.

(iii) As x and $4+|x|$ are continuous everywhere, $h(x)$ is continuous everywhere. x is differentiable everywhere and $|x|$ is differentiable everywhere except at $x = 0$. Now,

$$h'(0^+) = \lim_{k\to 0}\frac{h(0+k)-h(0)}{k} = \lim_{k\to 0}\frac{\frac{k}{4+k}-0}{k} = \frac{1}{4} \text{ and}$$

$$h'(0^-) = \lim_{k\to 0}\frac{h(0-k)-h(0)}{-k} = \lim_{k\to 0}\frac{\frac{-k}{4+k}-0}{-k} = \frac{1}{4}.$$

$\therefore$ $h(x)$ is continuous and differentiable at all x.

(iv) $\phi(x) = \begin{cases} x^3 & \text{for } 1 < x \le 2 \\ 4(x-2) + 2x^2 & \text{for } x > 2 \end{cases}$

because when $x > 2$, $|x-2| = x-2$.

Clearly, $\phi'(2^+) = \phi'(2^-) = 12$.

$\therefore$ $\phi(x)$ is differentiable and continuous at $x = 2$.

8. (i) $\dfrac{{}^nC_{r-1}}{{}^nC_r} = \dfrac{36}{84} \Rightarrow \dfrac{r}{n-r+1} = \dfrac{3}{7} \Rightarrow 10r = 3n + 3.$ ... (1)

$\dfrac{{}^nC_r}{{}^nC_{r+1}} = \dfrac{84}{126} \Rightarrow \dfrac{r+1}{n-(r+1)+1} = \dfrac{2}{3} \Rightarrow 5r = 2n - 3.$ ... (2)

Solving (1) and (2), $r = 3$.

(ii) Here, $2^x + x^2 = 1$ or $2^x = 1 - x^2$.

The graphs of the curves $y = 2^x$ and $y = 1 - x^2$ cut at two points.

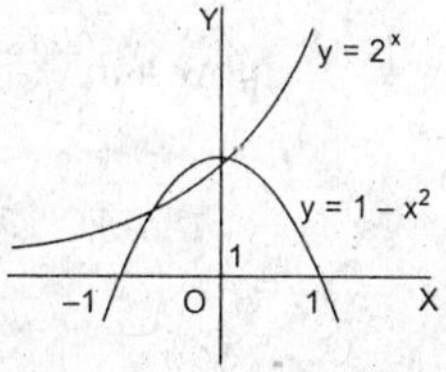

$\therefore$ the number of real solutions is two.

(iii) $x^2 = ky - 3 = k\left(y - \dfrac{3}{k}\right)$. Putting $x = Y, y - \dfrac{3}{k} = X$, we get

$Y^2 = kX = 4 \cdot \dfrac{k}{4} \cdot X$. So, focus $= \left(\dfrac{k}{4}, 0\right)_{X,Y} = \left(0, \dfrac{3}{k} + \dfrac{k}{4}\right) = (0, 2)$.

$\therefore$ $\dfrac{3}{k} + \dfrac{k}{4} = 2$ or $12 + k^2 = 8k$ or $(k-6)(k-2) = 0 \Rightarrow k = 6$ ($\because$ $k \neq 2$).

(iv) No group can have the common difference ≥ 4 as there is no number ≥ 13.

If the group containing 1 has the c.d. = 1, i.e., if one group is (1, 2, 3, 4), there will be two ways to group others—(5, 6, 7, 8), (9, 10, 11, 12) and (5, 7, 9, 11), (6, 8, 10, 12).

If the group containing 1 has the c.d. =2, i.e., if one group is (1, 3, 5,7), there will be one way to group others—(2, 4, 6, 8), (9, 10, 11, 12).

If the group containing 1 has the c.d. = 3, i.e., if one group is (1, 4, 7, 10), there will be one way to group others—(2, 5, 8, 11), (3, 6, 9, 12).

$\therefore$ the number of ways = $2 + 1 + 1 = 4$.

9. (i) $\left|\dfrac{1}{z_1} - \dfrac{1}{z_2} + \dfrac{1}{z_3}\right| = \left|\dfrac{z_1\bar{z}_1}{z_1} - \dfrac{z_2\bar{z}_2}{z_2} + \dfrac{z_3\bar{z}_3}{z_3}\right|$

(because $|z_1| = 1 \Rightarrow z_1\bar{z}_1 = |z_1|^2 = 1$, etc.)

$= |\bar{z}_1 - \bar{z}_2 + \bar{z}_3| = |\overline{z_1 - z_2 + z_3}| = |z_1 - z_2 + z_3| = 4.$

(ii) Any plane through the line of intersection of $2x - 3y - 7z = 0$ and $8x - 31y - 33z = 0$ is $2x - 3y - 7z + \mu(8x - 31y - 33z) = 0$. For some μ, it should be the plane $\lambda x - 14y - 13z = 0$.

$$\therefore \quad \frac{2+8\mu}{\lambda} = \frac{-3-31\mu}{-14} = \frac{-7-33\mu}{-13} = \frac{2\mu+4}{-1}.$$

So, $-3 - 31\mu = 14(2\mu + 4)$ or $\mu = -1$.

Then $\dfrac{2-8}{\lambda} = \dfrac{-2+4}{-1}$ or $-6 = -2\lambda$. $\therefore$ $\lambda = 3$.

(iii) Let $x = \tan\theta$. Then $\tan^{-1}\dfrac{\sec\theta - 1}{\tan\theta} : \theta = \lambda : 1$.

Now, $$\tan^{-1}\frac{\sec\theta-1}{\tan\theta} = \tan^{-1}\frac{1-\cos\theta}{\sin\theta} = \tan^{-1}\frac{2\sin^2\frac{\theta}{2}}{2\sin\frac{\theta}{2}\cos\frac{\theta}{2}}$$

$$= \tan^{-1}\tan\frac{\theta}{2} = \frac{\theta}{2}.$$

$$\therefore \quad \lambda : 1 = \frac{\theta}{2} : \theta \Rightarrow \lambda = \frac{1}{2}.$$

(iv) Using $(m+n)\cot\theta = n\cot\beta - m\cot\alpha$, we get

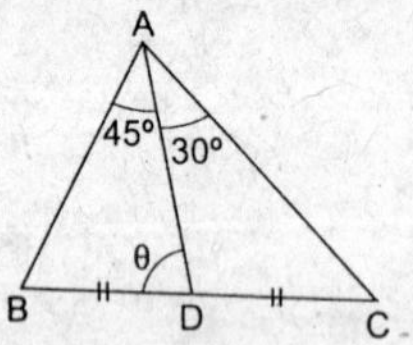

$$(1+1)\cot\theta = 1\cdot\cot 30^\circ - 1\cdot\cot 45^\circ$$

$$\Rightarrow \quad \cot\theta = \frac{\sqrt{3}-1}{2}.$$

From $\triangle ABD$, $\dfrac{BD}{\sin 45^\circ} = \dfrac{AD}{\sin(45^\circ+\theta)}$.

$$\therefore \quad BD = \frac{1}{\sqrt{2}}\cdot\frac{1}{\sqrt{11-6\sqrt{3}}}\cdot\frac{1}{\frac{1}{\sqrt{2}}(\cos\theta+\sin\theta)} = \frac{1}{\sqrt{11-6\sqrt{3}}}\cdot\frac{\operatorname{cosec}\theta}{1+\cot\theta}.$$

$$\therefore \quad BC = \frac{2}{\sqrt{11-6\sqrt{3}}}\cdot\frac{\sqrt{1+\left(\frac{\sqrt{3}-1}{2}\right)^2}}{1+\frac{\sqrt{3}-1}{2}}$$

$$= \frac{4}{\sqrt{11-6\sqrt{3}}}\cdot\frac{1}{\sqrt{3}+1}\cdot\sqrt{1+\frac{4-2\sqrt{3}}{4}}$$

$$= \frac{2}{\sqrt{11-6\sqrt{3}}}\cdot\frac{\sqrt{8-2\sqrt{3}}}{\sqrt{3}+1} = 2\cdot\frac{\sqrt{8-2\sqrt{3}}}{\sqrt{(11-6\sqrt{3})(4+2\sqrt{3})}} = 2.$$

10. (i) The point $(1, -1, 2)$ of the line is on the plane.

So, the line lies on the plane $\Rightarrow\ l \cdot 2 + 4(-1) + 2 \cdot 1 = 0 \ \Rightarrow\ l = 1.$

(ii) Latus rectum $= \dfrac{2b^2}{a} = \dfrac{2 \cdot 1^2}{1} = 2.$

(iii) Any tangent to $y^2 = 12x$ or $y^2 = 4 \cdot 3 \cdot x$ is $y = mx + \dfrac{3}{m}$.

$\therefore\ \dfrac{3}{m} = 1 \quad$ or $\quad m = 3 =$ slope of the tangent.

(iv) Any point on the first line is $(5 + 4r, 7 + mr, -3 - 5r)$.

Any point on the second line is $(8 + 7s, 4 + s, 5 + 3s)$.

The lines will be intersecting if

$$5 + 4r = 8 + 7s,\ 7 + mr = 4 + s,\ -3 - 5r = 5 + 3s.$$

Solving the first and the third equations, $r = -1, s = -1$.

$\therefore\ 7 + m(-1) = 4 + (-1) \quad$ or $\quad m = 4.$

11. (i) $I = \displaystyle\int_0^4 \frac{f(t)}{f(t) + f(4-t)}\, dt = \int_0^4 \frac{f(4-t)}{f(4-t) + f\{4 - (4-t)\}}\, dt = \int_0^4 \frac{f(4-t)}{f(t) + f(4-t)}\, dt.$

$$\therefore\ 2I = I + I = \int_0^4 \left\{\frac{f(t)}{f(t) + f(4-t)} + \frac{f(4-t)}{f(t) + f(4-t)}\right\} dt = \int_0^4 dt = 4.$$

$\therefore\ I = \dfrac{4}{2} = 2.$

(ii) $f(-x) = f(x)$. So, $\phi(x) = x^3 f(x)$ is odd because

$$\phi(-x) = (-x)^3 f(-x) = -x^3 f(x) = -\phi(x).$$

Again, $\psi(x) = xf''(x)$.

$f(-x) = f(x) \ \Rightarrow\ f'(-x) \cdot (-1) = f'(x) \ \Rightarrow\ f''(-x) \cdot (-1)^2 = f''(x)$

$\Rightarrow\ f''(-x) = f''(x).$

$\therefore\quad \psi(-x) = -xf''(-x) = -xf''(x) = -\psi(x)$. So, $\psi(x)$ is odd.

$$\therefore\quad I = \int_{-2}^{2} x^3 f(x)\, dx + \int_{-2}^{2} xf''(x)\, dx + \int_{-2}^{2} 2\, dx = 0 + 0 + 2x\Big]_{-2}^{2} = 8.$$

(iii) $f'(x) = |\cos t|\Big]_{t=x} = |\cos x|$. So, $f'\left(\frac{\pi}{2}\right) = \left|\cos\frac{\pi}{2}\right| = 0.$

(iv) Solving $y = 0$ and $y = x - bx^2$, the parabola cuts the x-axis at (0, 0) and $\left(\frac{1}{b}, 0\right)$.

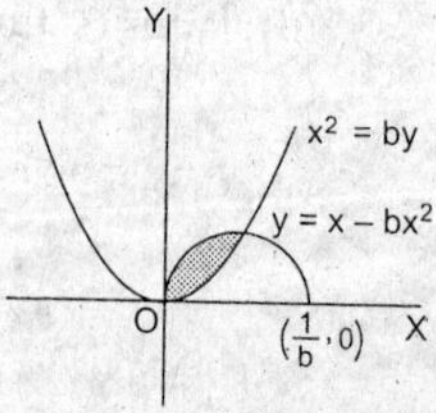

Solving $x^2 = by$ and $y = x - bx^2$,

$$x^2 = b(x - bx^2) \text{ or } x = 0, \frac{b}{1+b^2}.$$

$\therefore$ the area enclosed $A = \int_0^{b/(1+b^2)} \left(x - bx^2 - \frac{x^2}{b}\right)dx = \frac{1}{6}\cdot\frac{b^2}{(1+b^2)^2}.$

$$\therefore \quad \frac{dA}{db} = \frac{1}{6}\cdot\frac{2b(1+b^2)^2 - b^2\cdot 2(1+b^2)\cdot 2b}{(1+b^2)^2} = 0$$

$\Rightarrow 2b(1+b^2)(1+b^2-2b^2) = 0$

$\Rightarrow b = 0, 1, -1$. But $b > 0$. So, $b = 1$.

12. (i) $f(x) = 1 + (x^2 - 4x + 4) = 1 + (x-2)^2 \geq 1$ and $f(x)$ gets different values for different values of $x \geq 2$.. So, $X = [1, +\infty)$.

(ii) $\phi(x)$ is real for all $x \in R$ except those for which $x - 1 = 0$ or $x^2 - 4 \geq 0$. So, $x \neq 1$ and $x \notin [-2, 2]$.

Therefore, $x \in R - [-2, 2]$.

(iii) $f'(x) = 4x - \frac{1}{x} > 0 \Rightarrow \frac{4x^2 - 1}{x} > 0 \Rightarrow 4x^2 - 1 > 0 \quad (\because x > 0).$

$\therefore x < -\frac{1}{2}$ or $x > \frac{1}{2}$. But $x > 0$. So, $x > \frac{1}{2}$.

(iv) Consider the equation $x^4 + 3x^3 + 3x^2 + 3x + 2 = 0$ (integrating the given equation and adjusting the constant of integration). It has two roots $x = -1, -2$ (by trial). So, the derived equation will have a real root between –2 and –1.

13. (i) $(y-1)^2 = 8(3-x)$. Putting $y - 1 = Y, 3 - x = X$, we get

$$Y^2 = 8x = 4\cdot 2\cdot X.$$

$\therefore$ focus $= (2, 0)_{X,Y} = (1, 1)$.

(ii) $9x^2 + 5(y-3)^2 = 45$ or $\dfrac{x^2}{5} + \dfrac{(y-3)^2}{9} = 1$. Putting $y - 3 = X$, $x = Y$, we get

$$\frac{X^2}{3^2} + \frac{Y^2}{(\sqrt{5})^2} = 1.$$

So, centre $= (0, 0)_{X, Y} = (0, 3)$.

(iii) $(x+y)^2 = 3 + y - x$. Putting $\dfrac{x+y}{\sqrt{2}} = Y$, $\dfrac{y-x+3}{\sqrt{2}} = X$, we get

$$2Y^2 = \sqrt{2}X \quad \text{or} \quad Y^2 = \frac{1}{\sqrt{2}}X.$$

So, vertex $= (0, 0)_{X, Y}$.

When $X = 0$ and $Y = 0$, we have

$$x + y = 0 \text{ and } y - x + 3 = 0 \Rightarrow x = \frac{3}{2} \text{ and } y = -\frac{3}{2}.$$

$\therefore$ vertex $= \left(\dfrac{3}{2}, -\dfrac{3}{2}\right)$.

(iv) $(x-1)^2 = 4(y+1)$. Putting $x - 1 = Y$, $y + 1 = X$, we get $Y^2 = 4X$.

The ends of the double ordinate through the focus are $(1, 2)_{X, Y}$ and $(1, -2)_{X, Y}$. Tangents meet at $(a t_1 t_2, a(t_1 + t_2))_{X, Y}$, where $a = 1$, $t_1 = 1$, $t_2 = -1$.

$\therefore$ tangents meet at $(-1, 0)_{X, Y}$, i.e., $(1, -2)$.

14. (i) $(x-1)^3 = 1$ or $x - 1 = 1, \omega, \omega^2$. $\therefore$ $x = 2, 1 + \omega, 1 + \omega^2 = 2, -\omega^2, -\omega$.

$\therefore$ $\alpha = \omega, -\omega^2$ or $\alpha^3 = -1$.

So, $\alpha^{6n} + \alpha^{-6n} = (-1)^{2n} + (-1)^{-2n} = 2$.

(ii) $y = (c_1 x + c_2) e^{ax} + (c_3 e^{c_4}) e^{bx} = (c_1 x + c_2) e^{ax} + c_3' e^{bx}$.

There are only three arbitrary constants. So, the order is 3.

(iii) $\phi'(x) = f(2x) = f(x+x) = f(x) + f(x) = 2f(x)$.

$\therefore$ $\phi'(2) = 2f(2) = 2f(1+1) = 2\{f(1) + f(1)\} = 2(1+1) = 4$.

(iv) $[\vec{a}\ \vec{b}\ \vec{c}] = [2\hat{a}\ 2\hat{b}\ \hat{c}] = 4[\hat{a}\ \hat{b}\ \hat{c}] = 4$

($\because$ $\hat{a}, \hat{b}, \hat{c}$ are three mutually perpendicular unit vectors).

Now, $\dfrac{1}{12}[\vec{a} \times \vec{b}\ \ \vec{b} \times \vec{c}\ \ \vec{c} \times \vec{a}] = \dfrac{1}{12}[\vec{a}\ \vec{b}\ \vec{c}]^2 = \dfrac{1}{12} \cdot 4^2 = \dfrac{4}{3}$.

15. (i) $|iz + 3| = |i(z - 2i) + 1| \le |i(z - 2i)| + 1 \le |i|\,|z - 2i| + 1 \le 1 \cdot 3 + 1 = 4$.

(ii) $\frac{1}{24}n(n-1)(n-2)(n-3) = \frac{n\,!}{4\,!\,(n-4)\,!} = {}^nC_4$

$= \text{greatest coefficient} = \text{middle coefficient.}$

$\therefore$ the 5th term is the middle term. So, there will be 9 terms, i.e., $n + 1 = 9$.

(iii) $[\vec{\lambda}\ \vec{\mu}\ \vec{\nu}] = \begin{vmatrix} 1 & 2 & -1 \\ 1 & 1 & 3 \\ 2 & 3 & 1 \end{vmatrix} [\vec{a}\ \vec{b}\ \vec{c}] = \{1(-8) - 2(-5) + (-1)\cdot 1\}\cdot 3 = 3.$

(iv) $\int_0^3 f(x)\,dx = \int_0^1 f(x)\,dx + \int_1^3 f(x)\,dx \le \int_0^1 |f(x)|\,dx + \int_1^3 |f(x)|\,dx$

$$\le \int_0^1 dx + \int_1^3 2\,dx = 1 + 2\cdot 2 = 5.$$

16. (i) $y^2 = 4(x-1)$. Putting $x - 1 = X$ and $y = Y$, we get

$Y^2 = 4X = 4\cdot 1\cdot X.$

$\therefore$ focus $= (1, 0)_{X,\,Y} = (2, 0).$

(ii) PQ has the equation $y - 0 = \frac{4-0}{5-2}(x-2)$ or $4x - 3y - 8 = 0$.

If $R = (\alpha, \beta)$, the polar line PQ should be

$y\beta = 2(x+\alpha) - 4$

or $2x - \beta y + 2\alpha - 4 = 0$.

$\therefore \frac{4}{2} = \frac{-3}{-\beta} = \frac{-8}{2\alpha - 4} \Rightarrow \beta = \frac{3}{2}$ and $\alpha = 0$. So, $R = \left(0, \frac{3}{2}\right)$.

Solving $4x - 3y - 8 = 0$ and $y^2 = 4(x-1)$, we get

$\left(\frac{4x-8}{3}\right)^2 = 4(x-1)$ or $x = 5, \frac{5}{4}$. So, $Q = \left(\frac{5}{4}, -1\right)$.

$\therefore$ centroid, $G = \left(\frac{5 + \frac{5}{4} + 0}{3}, \frac{4 - 1 + \frac{3}{2}}{3}\right) = \left(\frac{25}{12}, \frac{3}{2}\right)$.

(iii) (α, β) is the circumcentre if

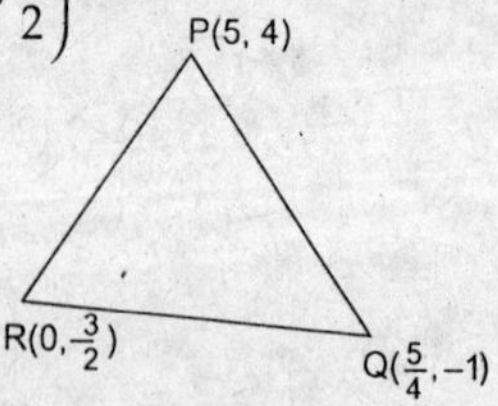

$$(\alpha - 0)^2 + \left(\beta - \frac{3}{2}\right)^2 = (\alpha - 5)^2 + (\beta - 4)^2$$

$$= \left(\alpha - \frac{5}{4}\right)^2 + (\beta + 1)^2$$

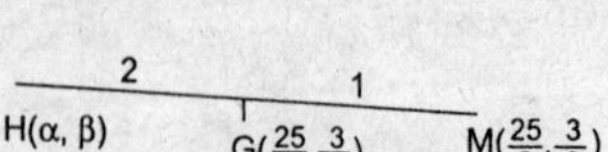

$$\Rightarrow -3\beta + \frac{9}{4} = -10\alpha - 8\beta + 41$$

$$= -\frac{5}{2}\alpha + 2\beta + \frac{41}{16}.$$

Solving, $\alpha = \frac{25}{8}, \beta = \frac{3}{2}\cdot$ So, circumcentre, $M = \left(\frac{25}{8}, \frac{3}{2}\right)\cdot$

(iv) (α, β) is the orthocentre if

$$\frac{\alpha + \frac{25}{4}}{3} = \frac{25}{12} \text{ and } \frac{\beta + 3}{3} = \frac{3}{2}\cdot$$

$\therefore\ \alpha = 0, \beta = \frac{3}{2}\cdot$ So, orthocentre $= \left(0, \frac{3}{2}\right)\cdot$

17. (i) We know that $\vec{r} = (\vec{r}\cdot\vec{i})\,\vec{i} + (\vec{r}\cdot\vec{j})\,\vec{j} + (\vec{r}\cdot\vec{k})\,\vec{k}$.

Putting $\vec{r} = \vec{a}\times\vec{b}$, we obtain $\vec{a}\times\vec{b} = [\vec{a}\ \ \vec{b}\ \ \vec{i}\,]\,\vec{i} + [\vec{a}\ \ \vec{b}\ \ \vec{j}\,]\,\vec{j} + [\vec{a}\ \ \vec{b}\ \ \vec{k}\,]\,\vec{k}$.

(ii) $(\vec{a}\cdot\vec{b})\,\vec{a} - (\vec{a}\cdot\vec{a})\,\vec{b} = (\vec{b}\cdot\vec{c})\,\vec{b} - (\vec{b}\cdot\vec{b})\,\vec{c}$

$\Rightarrow\ (\vec{a}\cdot\vec{b})\,\vec{a} - \{(\vec{a}\cdot\vec{a}) + (\vec{b}\cdot\vec{c})\}\,\vec{b} + (\vec{b}\cdot\vec{b})\,\vec{c} = 0.$

Taking dot product by $\vec{b}\times\vec{c}$, $(\vec{a}\cdot\vec{b})\,[\vec{a}\ \ \vec{b}\ \ \vec{c}] = 0 \ \Rightarrow\ [\vec{a}\ \ \vec{b}\ \ \vec{c}] = 0.$

(iii) $|\vec{a}+\vec{b}|^2 = 2 \Rightarrow (\vec{a}+\vec{b})^2 = 2 \Rightarrow |\vec{a}|^2 + |\vec{b}|^2 + 2\vec{a}\cdot\vec{b} = 2 \Rightarrow \vec{a}\cdot\vec{b} = 0.$

$(\vec{a}\times\vec{b})\times\vec{c} = (\vec{a}\times\vec{b})\times\{\vec{a} - \vec{b} + (\vec{a}\times\vec{b})\}$

$= (\vec{a}\times\vec{b})\times(\vec{a}-\vec{b}) = \{\vec{a}\cdot(\vec{a}-\vec{b})\}\vec{b} - \{\vec{b}\cdot(\vec{a}-\vec{b})\}\vec{a}$

$= \vec{b} + \vec{a}.$

(iv) Expression $= [\vec{a}\ \ \vec{b}\ \ \vec{a}\times\vec{b}] = (\vec{a}\times\vec{b})\cdot(\vec{a}\times\vec{b})$

$= |\vec{a}\times\vec{b}|^2 = (1\cdot 1\cdot \sin 90^\circ)^2 = 1.$

18. (i) Let $a = 3n$. Then $a^2 + a + 1 = (3n)^2 + 3n + 1 = 3(3n^2 + n) + 1$, which is not divisible by 3.

Let $a = 3n + 1$. Then $a^2 + a + 1 = (3n+1)^2 + (3n+1) + 1$

$= 9n^2 + 9n + 3 = 3(3n^2 + 3n + 1),$

which is divisible by 3. So, when a is divided by 3, the remainder is 1.

(ii) $$\lim_{n\to\infty} \frac{n^{2/3}\cdot\sin(n!)}{n\left(1+\frac{1}{n}\right)} = \lim_{n\to\infty} \frac{1}{n^{1/3}}\cdot\frac{\sin(n!)}{1+\frac{1}{n}}$$

$$= \frac{1}{\infty}\cdot\frac{\text{a number between 1 and } -1}{1}$$

$$= 0.$$

(iii) $\lim_{n\to\infty} f(n+1) = \lim_{n\to\infty} f(n) = l$ (say).

$\therefore$ when $n \to \infty$, $l = \frac{1}{2}\left(l + \frac{9}{l}\right)$ or $2l^2 = l^2 + 9$

or $l^2 = 9$ or $l = 3$ (because $f(n) > 0$ for all $n \in N$).

(iv) Let $\vec{a} = l\vec{i} + m\vec{j} + n\vec{k}$. As $|\vec{a}| = 1$, l, m and n are direction cosines of $\vec{a}$. So, $l^2 + m^2 + n^2 = 1$. Let $l = \cos\alpha$, $m = \cos\beta$, $n = \cos\gamma$.

$|\vec{a}\times\vec{i}|^2 = \sin^2\alpha$, $|\vec{a}\times\vec{j}|^2 = \sin^2\beta$, $|\vec{a}\times\vec{k}|^2 = \sin^2\gamma$

$\therefore$ value $= \sin^2\alpha + \sin^2\beta + \sin^2\gamma = 3 - (\cos^2\alpha + \cos^2\beta + \cos^2\gamma)$

$= 3 - 1 = 2$.

19. (i) $p^2 - 8 \geq 0 \Rightarrow p = 3, 4, 5, 6$. So, $n(E) = 4$ and $n(S) = 6$.
So, probability $= \frac{4}{6} = \frac{2}{3}$.

(ii) $[\vec{a}\ \ \vec{b} + \vec{a}\times\vec{b}\ \ \vec{b}] = [\vec{a}\ \ \vec{a}\times\vec{b}\ \ \vec{b}] = (\vec{b}\times\vec{a})\cdot(\vec{a}\times\vec{b}) = -|\vec{a}\times\vec{b}|^2$

$= -\left(\sin\frac{\pi}{3}\right)^2 = -\frac{3}{4}$.

(iii) $\int_0^{t^2} xf(x)\,dx = \frac{2}{5}t^5$. Differentiating both sides w.r.t. t,

$\{xf(x)\}\Big]_{x=t^2} \times \frac{d(t^2)}{dt} = \frac{2}{5}\cdot 5t^4$ or $t^2\cdot f(t^2)\cdot 2t = 2t^4$.

$\therefore f(t^2) = t$. So, $f\left\{\left(\frac{2}{5}\right)^2\right\} = \frac{2}{5}$.

(iv) Here, $\begin{vmatrix} 1 & a & a^2 \\ 1 & b & b^2 \\ 1 & c & c^2 \end{vmatrix} \neq 0$ and $\begin{vmatrix} a & a^2 & 1+a^3 \\ b & b^2 & 1+b^3 \\ c & c^2 & 1+c^3 \end{vmatrix} = 0$.

$\Rightarrow (1+abc)\begin{vmatrix} 1 & a & a^2 \\ 1 & b & b^2 \\ 1 & c & c^2 \end{vmatrix} = 0 \Rightarrow 1 + abc = 0$.

20. (i) $f(x) = 2\sin^2\beta + 2\cos(x+\beta)\{\cos(x-\beta) - \cos(x+\beta)\} + \cos 2(x+\beta)$

$= 2\sin^2\beta + 2\cos(x+\beta)\cos(x-\beta) - 2\cos^2(x+\beta) + 2\cos^2(x+\beta) - 1$

$= 2\sin^2\beta + 2(\cos^2 x - \sin^2\beta) - 1 = \cos 2x$.

$$\therefore\ \{f(\alpha)\}^2 + \left\{f\left(\frac{\pi}{4} - \alpha\right)\right\}^2 = \cos^2 2\alpha + \cos^2\left(\frac{\pi}{2} - 2\alpha\right) = 1.$$

(ii) $(1+i)^n = \left\{\sqrt{2}\left(\cos\frac{\pi}{4} + i\sin\frac{\pi}{4}\right)\right\}^n = 2^{n/2}\left(\cos\frac{n\pi}{4} + i\sin\frac{n\pi}{4}\right).$

$\sin\frac{n\pi}{4} = 0$ if $n = 4, 8, 12$. Hence, $n(E) = 3$ and $n(S) = 12$.

$\therefore$ probability $= \frac{3}{12} = \frac{1}{4}$.

(iii) $[\vec{a}\ \vec{b}\ \vec{c}] = 0 \Rightarrow \vec{a}, \vec{b}, \vec{c}$ are coplanar.

$\therefore\ \vec{c} = \lambda\vec{a} + \mu\vec{b}$. So, $\vec{x}\times\vec{y}\cdot\vec{c} = \vec{x}\times\vec{y}\cdot(\lambda\vec{a} + \mu\vec{b})$.

$\therefore\ [\vec{x}\ \vec{y}\ \vec{c}] = \lambda[\vec{x}\ \vec{y}\ \vec{a}] + \mu[\vec{x}\ \vec{y}\ \vec{b}] = \lambda\cdot 0 + \mu\cdot 0 = 0.$

(iv) $\frac{dy}{y+1} + \frac{\cos x\, dx}{2+\sin x} = 0 \Rightarrow \log(y+1) + \log(2+\sin x) = \log c$

$\Rightarrow (y+1)(2+\sin x) = c$. As $y(0) = 1$, $(1+1)(2+\sin 0) = c$ or $c = 4$.

$\therefore\ (y+1)(2+\sin x) = 4$. So, $\left\{y\left(\frac{\pi}{2}\right) + 1\right\}\left(2 + \sin\frac{\pi}{2}\right) = 4$. $\therefore\ y\left(\frac{\pi}{2}\right) = \frac{1}{3}$.

21. (i) $z_1^2 + z_2^2 = z_1 z_2$ or $\left(\frac{z_1}{z_2}\right)^2 - \frac{z_1}{z_2} + 1 = 0$ or $\frac{z_1}{z_2} = \frac{1 \pm \sqrt{3}i}{2} = -\omega, -\omega^2$

or $z_1 = -\omega z_2, -\omega^2 z_2$.

$\therefore\ |z_1| = |-\omega z_2|, |-\omega^2 z_2| = |z_2|$

and $|z_1 - z_2| = |-(1+\omega)z_2|, |-(1+\omega^2)z_2|$

$= |\omega^2 z_2|, |\omega z_2| = |z_2|$.

So, $|z_1| = |z_2| = |z_1 - z_2|$.

$\therefore$ the triangle is equilateral.

(ii) (3, 4) is the midpoint of the line segment joining (0, 0) and (6, 8).

Hence, a vertex, the centroid and the circumcentre are collinear.

$\therefore$ the circumcentre is on a median. But the circumcentre lies on the perpendicular bisector of the opposite side. Hence, the median is the perpendicular bisector of the base. So, the triangle must be, at least, isosceles.

(iii) $\dfrac{\sin A}{\cos A}\cdot\dfrac{\cos B}{\sin B}=\sqrt{2}-1 \;\Rightarrow\; \dfrac{1-\dfrac{1}{\sqrt{2}}}{\cos A\sin B}=\sqrt{2}-1$

$$\Rightarrow \cos A\sin B=\frac{1}{\sqrt{2}}.$$

$$\therefore\; \sin(A+B)=\sin A\cos B+\cos A\sin B=\frac{\sqrt{2}-1}{\sqrt{2}}+\frac{1}{\sqrt{2}}=1.$$

$\therefore\; A+B=90^\circ \qquad \therefore\; C=90^\circ.$

(iv) Any point on the first line $=(3r, 0, 4r)$; any point on the second line $=(0, 5s, 0)$; any point on the third line $=(3t+3, -5t, 4t+4)$.

$(3r, 0, 4r)=(0, 5s, 0) \Rightarrow s=0, r=0$. So, the point of intersection, P, of the first two lines $=(0, 0, 0)$.

$(3r, 0, 4r)=(3t+3, -5t, 4t+4)\Rightarrow t=0, r=1$. So, the point of intersection, Q, of the first and the third lines $=(3, 0, 4)$.

$(0, 5s, 0)=(3t+3, -5t, 4t+4)\Rightarrow t=-1, s=1$. So, the point of intersection, R, of the second and the third lines $=(0, 5, 0)$.

$\therefore\; PQ=5, QR=2\sqrt{5}, PR=5$. Hence, $QR^2=PQ^2+PR^2$ and $PQ=PR$.

$\therefore$ the triangle is isosceles right-angled.

22. (i) $x^2=-4x-2\pi\le\sin^{-1}\sin(5-2\pi)$

($5-2\pi$ is an angle in the 4th quadrant)

or $x^2-4x-2\pi\le 5-2\pi \;\Rightarrow\; x^2-4x-5\le 0 \;\Rightarrow\; (x-5)(x+1)\le 0$

$\Rightarrow\; x\in[-1, 5]$.

(ii) $f\{g(x)\}=f(x^2-4+5)=f(x^2+1)=(x^2+1)^2-4=(x^2+3)(x^2-1)$.

$\therefore\; f\{g(x)\}>0 \;\Rightarrow\; x^2-1>0 \;\Rightarrow\; x<-1$ or $x>1 \;\Rightarrow\; x\in R-[-1, 1]$.

(iii) $-5\le 3\sin\theta+4\cos\theta\le 5$. So, $-5\le\dfrac{5(2a-3)}{2-3a}\le 5$.

$$\therefore\; -1\le\frac{2a-3}{2-3a}\le 1 \;\Rightarrow\; \frac{2a-3+2-3a}{2-3a}\ge 0 \text{ and } \frac{2a-3-2+3a}{2-3a}\le 0$$

$$\Rightarrow\; (1+a)(2-3a)\le 0 \text{ and } (a-1)(2-3a)\le 0$$

$$\Rightarrow\; a\le -1 \text{ or } a\ge\frac{2}{3} \text{ and } \frac{2}{3}\le a \text{ or } a\ge 1$$

$$\Rightarrow\; a\le -1 \text{ or } a\ge 1 \;\Rightarrow\; a\in R-(-1, 1).$$

(iv) Let $x^2 = 4\cos\theta$ and $y^2 = \frac{8}{3}\sin\theta$, which satisfy $4x^4 + 9y^4 = 64$.

$\therefore\ z^2 = 3x^2y^2 = 3 \cdot 4\cos\theta \cdot \frac{8}{3}\sin\theta = 16\sin 2\theta \le 16.$

$\therefore\ z^2 - 16 \le 0 \ \Rightarrow\ -4 \le z \le 4.$

23. (i) $x^2(y^2-1) + 1 - y^2 = 0$ or $(y^2-1)(x^2-1) = 0.$

So, the sides of the quadrilateral are $x = 1, x = -1, y = 1, y = -1.$

The centre of the circle = (1, 0) and radius = 1.

The distance of (1, 0) from each of $y = 1, y = -1$ is equal to 1, i.e., the radius.

$x = 1, x = -1$ do not touch the circle. So, two sides touch the circle.

(ii) $[\vec{a}+\vec{b}+\vec{c}\ \ \vec{a}+\vec{b}\ \ \vec{b}+\vec{c}] = [\vec{c}\ \ \vec{a}+\vec{b}\ \ \vec{b}+\vec{c}] = [\vec{c}\ \ \vec{a}\ \ \vec{b}]$

$= [\vec{b}\ \ \vec{c}\ \ \vec{a}] = \vec{b}\times\vec{c}\cdot\vec{a} = \vec{a}\cdot\vec{a} = |\vec{a}|^2 = |\vec{b}\times\vec{c}|^2$

$= (|\vec{b}|\,|\vec{c}|\sin 90°)^2 = 1.$

(iii) Direction ratios of *OP* and *OQ* are λ, 2, 1 and –2, μ, 1, respectively.

$\therefore$ perpendicularity $\Rightarrow 1\cdot\lambda + (-1)\cdot 2 + 5\cdot 1 = 0,\ 1(-2) + (-1)\mu + 5\cdot 1 = 0$

$\Rightarrow \lambda = -3, \mu = 3 \Rightarrow \lambda + \mu = 0.$

(iv) The projection $= (\vec{i}+\vec{j}+\vec{k}) \cdot \dfrac{a\vec{i}+4\vec{j}-5\vec{k}}{\sqrt{41+a^2}} = \dfrac{1}{3\sqrt{5}}$

or $\dfrac{a-1}{\sqrt{41+a^2}} = \dfrac{1}{3\sqrt{5}} \ \Rightarrow\ 41 + a^2 = 45(a^2 - 2a + 1)$

$\Rightarrow\ 44a^2 - 90a + 4 = 0 \ \Rightarrow\ 22a^2 - 45a + 2 = 0 \ \Rightarrow\ a = \frac{1}{22}, 2.$

24. (i) AM $= \dfrac{a+b+c+d}{4} = \dfrac{8}{4} = 2$ and GM $= \sqrt[4]{abcd} = \sqrt[4]{16} = 2.$

$\therefore$ AM = GM. So, the numbers are equal, i.e., $a = b = c = d = \frac{8}{4} = 2.$

$\therefore\ a + b = 4.$

(ii) ${}^{2n+1}C_1 + {}^{2n+1}C_3 + \ldots + {}^{2n+1}C_{2n+1} = 1024$ or $2^{2n+1-1} = 1024$

or $2^{2n} = 2^{10} \ \Rightarrow\ n = 5.$

(iii) $\max(\sqrt{3}\sin x + \cos x) = \sqrt{(\sqrt{3})^2 + 1} = 2$ and $(y-1)^2 \ge 0$.

So, the equation $\Rightarrow \sqrt{3}\sin x + \cos x - 2 = 0 = (y-1)^2$

$$\Rightarrow y = 1 \text{ and } \frac{\sqrt{3}}{2}\sin x + \frac{1}{2}\cos x = 1$$

$$\Rightarrow \cos\left(x - \frac{\pi}{3}\right) = 1 \Rightarrow x - \frac{\pi}{3} = 2n\pi \Rightarrow x = 2n\pi + \frac{\pi}{3}.$$

$$\therefore 0 \le 2n\pi + \frac{\pi}{3} \le 8\pi \Rightarrow n = 0, 1, 2, 3.$$

$\therefore$ there are four values of x and, hence, four pairs (x, y).

(iv) $f'(x) = 0 \Rightarrow 6x^2 - 6ax + \frac{4}{3}a^2 = 0 \Rightarrow 9x^2 - 9ax + 2a^2 = 0$

$$\Rightarrow (3x - a)(3x - 2a) = 0 \Rightarrow x = \frac{a}{3}, \frac{2a}{3}.$$

$f''(x) = 12x - 6a$. So, $f''\left(\frac{a}{3}\right) = -2a < 0$ and $f''\left(\frac{2a}{3}\right) = 2a > 0$.

$$\therefore p = \frac{a}{3} \text{ and } p^2 = \frac{2a}{3} \Rightarrow \frac{a^2}{9} = \frac{2a}{3} \Rightarrow a = 6.$$

25. (i) $|z_1 - z_2| \ge \big||z_1| - |z_2|\big|$. Taking $z_2 + z_3$ for z_2,

$|z_1 - z_2 - z_3| \ge \big||z_1| - |z_2 + z_3|\big| \ge \big||z_1| - |z_2| - |z_3|\big|$

$(\because |z_2 + z_3| \le |z_2| + |z_3|)$.

$\therefore |z_1 - (z_2 - 3 - 4i) - (3 + 4i)| \ge \big||z_1| - |z_2 - 3 - 4i| - |3 + 4i|\big|$

or $|z_1 - z_2| \ge |12 - 5 - 5| = 2$.

So, the minimum value of $|z_1 - z_2|$ is 2.

(ii) $$\text{Limit} = \lim_{x \to 0} \frac{2\sin^2 \frac{x}{2}}{4 \cdot \left(\frac{x}{2}\right)^2} \cdot \frac{e^x - \cos x}{x^{n-2}}$$

$$= \frac{1}{2} \lim_{x \to 0} \frac{e^x - \cos x}{x^{n-2}} \quad (= 0 \text{ unless } n \ge 3)$$

$$= \frac{1}{2} \lim_{x \to 0} \frac{e^x + \sin x}{(n-2)x^{n-3}} \quad (= \infty \text{ unless } n = 3).$$

$\therefore \quad n = 3.$

(iii) $f\{f(x)\} = \dfrac{1-f(x)}{1+f(x)} = \dfrac{1-\dfrac{1-x}{1+x}}{1+\dfrac{1-x}{1+x}} = \dfrac{2x}{2} = x.$

$$f\left\{f\left(\frac{1}{x}\right)\right\} = \frac{1-f\left(\frac{1}{x}\right)}{1+f\left(\frac{1}{x}\right)} = \frac{1-\dfrac{1-\frac{1}{x}}{1+\frac{1}{x}}}{1+\dfrac{1-\frac{1}{x}}{1+\frac{1}{x}}} = \frac{\frac{2}{x}}{2} = \frac{1}{x}.$$

$\therefore$ expression $= x + \dfrac{1}{x} \geq 2\sqrt{x \cdot \dfrac{1}{x}} = 2$ (using AM $\geq$ GM).

(iv) Let $x = x + iy$. Then $|z| = \sqrt{x^2 + y^2}$ and $\bar{z} = x - iy$.

$\therefore$ $z\bar{z} - 2(z + \bar{z}) + 3 = 0 \Rightarrow x^2 + y^2 - 4x + 3 = 0$

$\Rightarrow y^2 = -(x^2 - 4x + 3).$

As y is real, $x^2 - 4x + 3 \leq 0$ or $(x-1)(x-3) \leq 0$. So, $1 \leq x \leq 3$.

Again, $x^2 + y^2 - 4x + 3 = 0 \Rightarrow |z|^2 = 4x - 3$.

$\therefore$ $|z|$ is maximum when x is maximum, i.e., 3.

So, $\max|z| = \sqrt{4x-3} = \sqrt{4 \times 3 - 3} = 3$.

26. (i) $\{4^{f(x)}\}^2 + 4 = 4^x \cdot 4^{f(x)}$ or $\{4^{f(x)}\}^2 - 4^x \cdot 4^{f(x)} + 4 = 0$.

or $4^{f(x)} = \dfrac{4^x \pm \sqrt{4^{2x} - 4^2}}{2}$, which is real and positive.

$\therefore$ $4^{2x} - 4^2 \geq 0 \Rightarrow 4^{2x} \geq 4^2 \Rightarrow 2x \geq 2$. So, $x \in [1, \infty)$.

(ii) $\dfrac{dy}{dt} > \dfrac{dx}{dt}$.

Differentiating $y^3 = 27x$ w.r.t. t, we obtain $3y^2 \dfrac{dy}{dt} = 27\dfrac{dx}{dt}$.

So, $3y^2 < 27$ or $y^2 < 9 \Rightarrow -3 < y < 3 \Rightarrow -27 < y^3 < 27$

$\Rightarrow -27 < 27x < 27 \Rightarrow -1 < x < 1$. So, $x \in (-1, 1)$.

(iii) We must have $x^2 - 1 + 2x + 1 > x^2 + x + 1,$ ... (1)

$x^2 - 1 + x^2 + x + 1 > 2x + 1,$ ... (2)

$2x + 1 + x^2 + x + 1 > x^2 - 1.$ ... (3)

$(1) \Rightarrow \quad x > 1.$

$(2) \Rightarrow \quad 2x^2 - x - 1 > 0$

$\Rightarrow \quad (2x+1)(x-1) > 0 \quad \Rightarrow \quad x < -\frac{1}{2} \quad \text{or} \quad x > 1.$

$(3) \Rightarrow \quad 3x > -3 \quad \Rightarrow \quad x > -1.$

$\therefore$ all are satisfied if $x > 1$. So, $x \in (1, \infty)$.

(iv) As x has real solutions, $\sin^{-1} x \in \left[-\frac{\pi}{2}, \frac{\pi}{2}\right]$.

$$\therefore \quad -\frac{\pi}{2} \le 3\sin^{-1} a \le \frac{\pi}{2} \quad \Rightarrow \quad -\frac{\pi}{6} < \sin^{-1} a \le \frac{\pi}{6}$$

$$\Rightarrow \quad \sin^{-1}\left(-\frac{1}{2}\right) \le \sin^{-1} a \le \sin^{-1}\left(\frac{1}{2}\right) \quad \Rightarrow \quad -\frac{1}{2} \le a \le \frac{1}{2}.$$

27. (i) $18 = 1 \cdot 1 \cdot 18; 1 \cdot 2 \cdot 9; 1 \cdot 3 \cdot 6; 2 \cdot 3 \cdot 3.$

$\therefore$ the required number of positive integral solutions

$$= \frac{3\,!}{2\,!} + 3\,! + 3\,! + \frac{3\,!}{2\,!} = 3 + 6 + 6 + 3 = 18.$$

(ii) AM $\ge$ GM $\Rightarrow$

$$\frac{\operatorname{cosec}\frac{A}{2} + \operatorname{cosec}\frac{B}{2} + \operatorname{cosec}\frac{C}{2}}{3} \ge \sqrt[3]{\frac{1}{\sin\frac{A}{2}\sin\frac{B}{2}\sin\frac{C}{2}}}$$

But in a ΔABC, $\sin\frac{A}{2}\sin\frac{B}{2}\sin\frac{C}{2} \le \frac{1}{8}$.

$$\therefore \quad \frac{\operatorname{cosec}\frac{A}{2} + \operatorname{cosec}\frac{B}{2} + \operatorname{cosec}\frac{C}{2}}{3} > \sqrt[3]{8} = 2.$$

(iii) Here, $x = \omega, -\omega$.

$$\therefore \quad \left(x^n + \frac{1}{x^n}\right)^2 = \left(\omega^n + \frac{1}{\omega^n}\right)^2, \ (-1)^{2n}\left(\omega^n + \frac{1}{\omega^n}\right)^2$$

$= (-1)^2$ if n is not a multiple of 3 and

2^2 if n is a multiple of 3.

$\therefore$ sum $= 1 + 1 + 2^2 + 1 + 1 + 2^2 + 1 + 1 + 2^2 = 18.$

(iv) $$I = \int_3^6 \frac{\sqrt{x}}{\sqrt{9-x} + \sqrt{x}}\,dx = \int_3^6 \frac{\sqrt{9-z}}{\sqrt{z} + \sqrt{9-z}}\,dz$$ (putting $9 - x = z$).

$$\therefore \quad I + I = \int_3^6 dx = 3. \qquad \therefore \quad I = \frac{3}{2}.$$

28. (i) Expression $= {}^{2n+1}C_0 + {}^{2n+1}C_1 + \ldots + {}^{2n+1}C_n$ $\quad (\because {}^mC_r = {}^mC_{m-r})$

$$= \frac{1}{2}\left\{{}^{2n+1}C_0 + {}^{2n+1}C_1 + \ldots + {}^{2n+1}C_{2n+1}\right\} = 2^{2n} = 2^{10}$$

$\Rightarrow n = 5.$

(ii) $$A^T A = \begin{bmatrix} 0 & x & x \\ 2y & y & -y \\ z & -z & z \end{bmatrix}\begin{bmatrix} 0 & 2y & z \\ x & y & -z \\ x & -y & z \end{bmatrix} = \begin{bmatrix} 2x^2 & 0 & 0 \\ 0 & 6y^2 & 0 \\ 0 & 0 & 3z^2 \end{bmatrix} = I$$

$$\therefore\ 2x^2 = 1, 6y^2 = 1, 3z^2 = 1 \ \Rightarrow\ x^2 + y^2 + z^2 = \frac{1}{2} + \frac{1}{6} + \frac{1}{3} = 1.$$

(iii) ${}^{47}C_{2r} = {}^{47}C_{15} \ \Rightarrow\ 2r + 15 = 47 \ \Rightarrow\ r = 16.$

(iv) Put $z^2 = z$. Then $\displaystyle\int_1^4 \frac{e^{\sin x^2}}{x}\,dx = \int_1^{16} \frac{e^{\sin z}}{z}\cdot\frac{1}{2}\,dz = \frac{1}{2}\int_1^{16} \frac{e^{\sin x}}{x}\,dx$

$$= \frac{1}{2}\int_1^{16} \frac{dF(x)}{dx}\,dx = \frac{1}{2}\int_1^{16} dF(x) = \frac{1}{2}F(x)\Big]_1^{16}$$

$$= \frac{1}{2}\{F(16) - F(1)\}.\ \text{So, } k = 16.$$

29. (i) Expression $= (\vec{a} \times \vec{b}) \cdot (\vec{a} \times \vec{b}) + (\vec{a} \cdot \vec{b})^2$

$$= |\vec{a}|^2 |\vec{b}|^2 \sin^2\theta + |\vec{a}|^2 |\vec{b}|^2 \cos^2\theta$$
$$= 4(\sin^2\theta + \cos^2\theta) = 4.$$

(ii) Two circles are concentric. So,

$(a+1)^2 + a^2 - 2(a+1) - 1 > 0 \ \Rightarrow\ a^2 - 1 > 0 \ \Rightarrow\ a < -1$ or $a > 1.$

$(a+1)^2 + a^2 - 2(a+1) - 17 < 0 \ \Rightarrow\ a^2 - 9 < 0 \ \Rightarrow\ -3 < a < 3.$

$\therefore\ a = -2, 2.$

(iii) The equation of the line is $x + y = 3$. Solving it with $y = \dfrac{c}{x+1}$, we get

$x^2 - 2x + c - 3 = 0$. To touch, $D = 0 \ \Rightarrow\ 4 - 4(c-3) = 0 \ \Rightarrow\ c = 4.$

(iv) Let the number of green balls $= n$. Then

$$\frac{15}{28} = \frac{{}^{8-n}C_2 \cdot {}^nC_1}{{}^8C_3} \text{ and } \frac{15}{56} = \frac{{}^nC_2 \cdot {}^{8-n}C_1}{{}^8C_3} \ \Rightarrow\ \frac{7-n}{n-1} = 2 \ \Rightarrow\ n = 3.$$

30. (i) The two circles must be intersecting. Solving them,

$$6x + 6y - 6\lambda = 0 \ \text{ or } \ x + y - \lambda = 0.$$

Putting in the first equation,

$$x^2+(\lambda-x)^2-2x-2(\lambda-x)=0 \text{ or } 2x^2-2\lambda x+\lambda^2-2\lambda=0.$$

Roots must be real and unequal.

$\therefore\ 4\lambda^2-4\cdot 2(\lambda^2-2\lambda)>0$ or $-\lambda^2+4\lambda>0$ or $\lambda(\lambda-4)<0$.

$\therefore\ 0<\lambda<4$. But $\lambda\in Z$. So, $\lambda=1, 2, 3$.

(ii) $I=2\int_0^{2\pi}\{|\sin x|+|\cos x|\}\,dx=2\times 2\int_0^{\pi}\{|\sin x|+|\cos x|\}\,dx$

$$=4\int_0^{\pi}\sin x\,dx+4\int_0^{\pi/2}\cos x\,dx+4\int_{\pi/2}^{\pi}(-\cos x)\,dx$$

$$=4\Big[-\cos x\Big]_0^{\pi}+4\Big[\sin x\Big]_0^{\pi/2}+4\Big[-\sin x\Big]_{\pi/2}^{\pi}=16.$$

(iii) $\dfrac{dy}{dx}=-\dfrac{x+y}{x}$ or $x\,dy+y\,dx+x\,dx=0$ or $d(xy)+x\,dx=0$

or $xy+\dfrac{x^2}{2}=c$. As $\left(3-\dfrac{1}{6}\right)$ is on it, $-\dfrac{1}{2}+\dfrac{9}{2}=c$. So, $xy+\dfrac{x^2}{2}=4$.

As $(\lambda, 1)$ is on it, $\lambda+\dfrac{\lambda^2}{2}=4$ or $\lambda^2+2\lambda-8=0 \Rightarrow \lambda=-4, 2$.

(iv) The line of intersection of $x-y-5z=6$ and $3x+5y+3z=4$ is

$$\frac{x-\frac{17}{4}}{11}=\frac{y+\frac{7}{4}}{-9}=\frac{z}{4}. \qquad \text{... (1)}$$

Any plane passing through the line $\dfrac{x+1}{2}=\dfrac{y-1}{-1}=\dfrac{z+3}{1}$ is

$$a(x+1)+b(y-1)+c(z+3)=0, \text{ where } 2a-b+c=0.$$

It is parallel to (1) if $11a-9b+4c=0$.

$\therefore\ \dfrac{a}{5}=\dfrac{b}{3}=\dfrac{c}{-7}$. So, the plane is $5(x+1)+3(y-1)-7(z+3)=0$.

Now, since $(\lambda, 6, 2)$ is on it, we get $5(\lambda+1)+15-35=0 \Rightarrow \lambda=3$.

□

6

Miscellaneous Questions—6

Fill-in/Completion Questions

1. If $(1+x+x^2+x^3)^7 = \sum_{r=0}^{21} a_r x^r$ then the value of $a_0 + a_4 + a_8 + a_{12} + a_{16} + a_{20}$ is

2. The number of different ways in which 24 apples can be distributed among 3 boys and 3 girls such that the total number of apples received by boys is double that received by the girls is

3. If $|\vec{a}| = 3$, $|\vec{b}| = 4$, $|\vec{c}| = 5$, $\vec{a} \perp (\vec{b} + \vec{c})$, $\vec{b} \perp (\vec{c} + \vec{a})$ and $\vec{c} \perp (\vec{a} + \vec{b})$ then $\sqrt{2}\,|\vec{a} + \vec{b} + \vec{c}|$ is equal to

4. Three times the least value of the area bounded by the line $y = mx + 1$ and the parabola $y = x^2 + 2x - 3$, m being a parameter, is

5. If $f(x)$ and $\phi(x)$ are continuous functions in $[0, 4]$ satisfying $f(x) = f(4-x)$, $\phi(x) + \phi(4-x) = 3$ and $\int_0^4 f(x)\, dx = 2$ then the value of $\int_0^4 f(x) \cdot \phi(x)\, dx$ is

6. $\int_{10}^{100} (x - [x])\, dx$ is equal to

7. If $f(x) = \int_a^x \frac{1}{f(x)} dx$ $(x > 0)$ and $\int_a^1 \frac{1}{f(x)} dx = \sqrt{2}$ then $f(50)$ is

8. Consider the closed figure C made by the lines $|x| + |y| = \sqrt{2}$. Let S be the region inside the figure C such that any point in it is nearer to the side $x + y = \sqrt{2}$ than the origin. Then three times the area of S is

9. If $a_i \geq 0$, where $i = 1, 2, 3, \ldots, n$, and $\sum_{i=1}^{n-1} a_i = 20$ then the greatest possible value of $\sum_{i=1}^{n} a_i a_{i+1}$ is

10. Let $2^{a_1} + 3^{a_2} + 5^{a_3} + 7^{a_4} + 9^{a_5}$ be divisible by 4, where a_1, a_2, a_3, a_4, a_5 are digits. The largest possible number of values of $(a_1, a_2, a_3, a_4, a_5)$ is

11. Let $f(x) = ax^4 + bx^3 + cx^2 + dx + \lambda$, where $\frac{f(1)}{1} = \frac{f(2)}{2} = \frac{f(3)}{3}$ and $\lim_{x \to \infty} \frac{f(x)}{x^4} = 1$. If $f(i) + f(\bar{i}) + k = 0$, where $i = \sqrt{-1}$, then k is

12. A plane exists which is perpendicular to the line $\frac{x}{3} = \frac{y-1}{5} = \frac{z+1}{7}$ and passes through the line $\frac{x+3}{10} = \frac{y-1}{-20} = \frac{z+1}{k}$. Then k is

Answers

1. 4096	**2.** 6885	**3.** 10	**4.** 32	**5.** 3
6. 45	**7.** 10	**8.** 4	**9.** 100	**10.** 45000
11. 20	**12.** 10			

Note If the answers of completion questions are whole numbers then the answers may be given by shading properly in a matrix of digits as shown below.

For example, the answer '4096', i.e., '04096', of **Q.1** can be shown as below by shading:

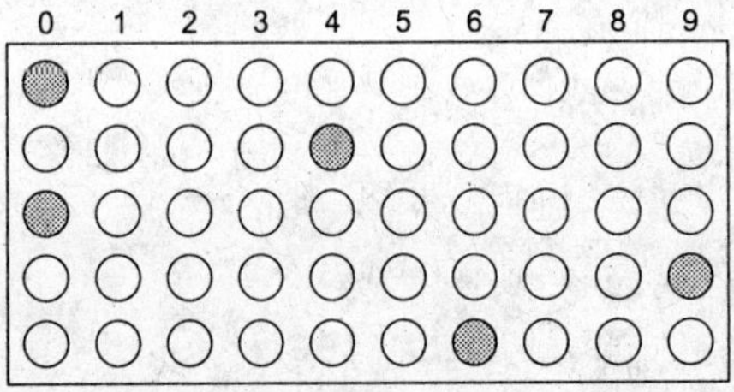

The answer '10', i.e., '00010', of **Q.12** can be shown as below by shading:

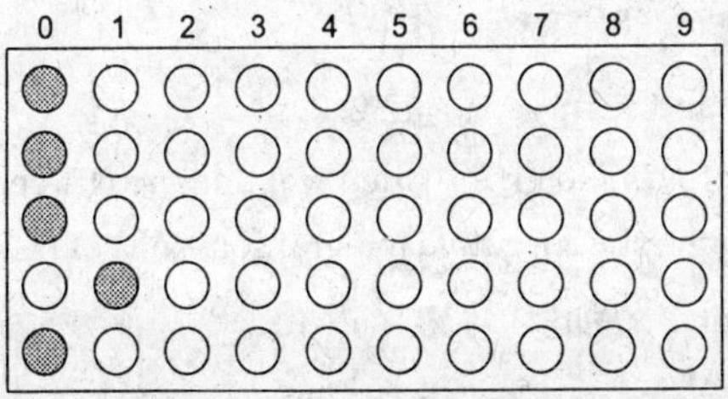

Hints and Solutions to Selected Questions

1. Putting $x = i$,

$$(1 + i + i^2 + i^3)^7 = a_0 + a_1 i + a_2 i^2 + \dots + a_{21} i^{21}$$

$\Rightarrow \quad 0 = (a_0 - a_2 + a_4 - \dots) + i(a_1 - a_3 + a_5 - \dots).$

$\therefore \quad a_0 - a_2 + a_4 - \dots = 0.$ ... (1)

Putting $x = 1, -1$, we get

$$(1 + 1 + 1^2 + 1^3)^7 = a_0 + a_1 + a_2 + a_3 + \dots$$

and $(1 - 1 + 1 - 1)^7 = a_0 - a_1 + a_2 - a_3 + \dots$.

Adding, $4^7 = 2(a_0 + a_2 + a_4 + \dots)$ or $a_0 + a_2 + a_4 + \dots = 2 \times 4^6$ ... (2)

So, (1) + (2) $\Rightarrow$ $2(a_0 + a_4 + a_8 + \dots) = 2 \times 4^6$

$$\Rightarrow a_0 + a_4 + a_8 + \dots + a_{20} = 4^6 = 4096.$$

2. Let the boys get x_1, x_2, x_3 and girls get y_1, y_2, y_3. Then

$x_1 + x_2 + x_3 = 16$ ... (1)

and $y_1 + y_2 + y_3 = 8,$ ... (2)

where $x_i \in W$ and $y_i \in W$ for $i = 1, 2, 3$.

The number of ways to give 16 identical things to three boys

= the number of nonegative integral solutions of (1)

= coefficient of x^{16} in $(x^0 + x^1 + x^2 + \dots)^3$

= coefficient of x^{16} in $(1 - x)^{-3} = {}^{18}C_{16}$.

Similarly, the number of ways to give 8 identical things to three girls = ${}^{10}C_8$.

$\therefore$ the required number of ways $= {}^{18}C_{16} \times {}^{10}C_8 = \dfrac{18 \times 17}{2} \times \dfrac{10 \times 9}{2} = 6885.$

3. $|\vec{a} + \vec{b} + \vec{c}|^2 = (\vec{a} + \vec{b} + \vec{c})^2 = \vec{a}^2 + \vec{b}^2 + \vec{c}^2 + 2(\vec{a} \cdot \vec{b} + \vec{b} \cdot \vec{c} + \vec{c} \cdot \vec{a}).$

$\vec{a} \cdot (\vec{b} + \vec{c}) = 0 \Rightarrow \vec{a} \cdot \vec{b} + \vec{a} \cdot \vec{c} = 0.$

$\vec{b}\cdot(\vec{c}+\vec{a})=0 \Rightarrow \vec{b}\cdot\vec{c}+\vec{b}\cdot\vec{a}=0.$

$\vec{c}\cdot(\vec{a}+\vec{b})=0 \Rightarrow \vec{c}\cdot\vec{a}+\vec{c}\cdot\vec{b}=0.$

Addingthese, $\vec{a}\cdot\vec{b}+\vec{b}\cdot\vec{c}+\vec{c}\cdot\vec{a}=0.$

$\therefore \quad |\vec{a}+\vec{b}+\vec{c}|^2=3^2+4^2+5^2=50 \Rightarrow \sqrt{2}\,|\vec{a}+\vec{b}+\vec{c}|=10.$

4. The parabola is $(x+1)^2=y+4$.

Solving with $y=0$, we get $x=1,-3$.

Solving $y=mx+1$ and $y=x^2+2x-3$,

we get $x^2+(2-m)x-4=0$.

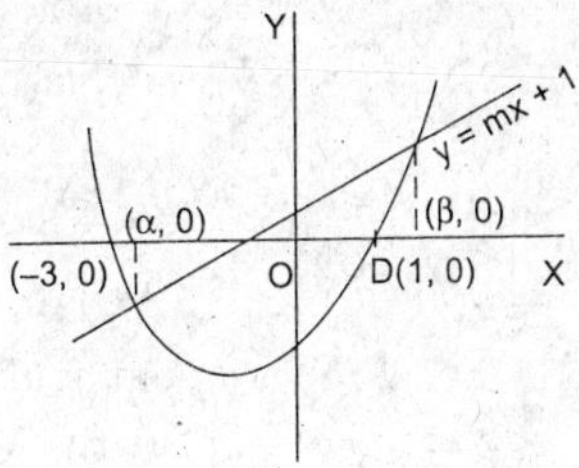

If α, β are the roots,

$\alpha+\beta=m-2, \alpha\beta=-4.$

Let the area bounded be Δ. Then

$$\Delta=\int_{\alpha}^{\beta}\{(mx+1)-(x^2+2x-3)\}dx=\sqrt{m^2-4m+20}\left\{\frac{1}{6}(m-2)^2+\frac{8}{3}\right\}.$$

For maximum or minimum Δ,

$$\frac{d\Delta}{dm}=0 \Rightarrow (m-2)(m^2-4m+20)=0 \Rightarrow m=2 \text{ (only real value)}.$$

$\therefore \quad \Delta_{max}$ is infinity. So, $\Delta_{min}=\sqrt{2^2-4\times 2+20}\left\{\frac{1}{6}(2-2)^2+\frac{8}{3}\right\}=\frac{32}{3}.$

5. $I=\int_0^4 f(4-x)\cdot\phi(4-x)\,dx=\int_0^4 f(x)\cdot\{3-\phi(x)\}\,dx=3\int_0^4 f(x)\,dx-I.$

$\therefore \quad 2I=3\times 2 \Rightarrow I=3.$

6. $f(x)=x-[x]$ is a periodic function of the period 1.

$$\therefore \quad \int_{10}^{100}(x-[x])\,dx=(100-10)\int_0^1(x-[x])\,dx=90\int_0^1 x\,dx-90\int_0^1[x]\,dx$$

$$=90\left[\frac{x^2}{2}\right]_0^1-90\int_0^1 0\,dx=45.$$

7. $f'(x)=\frac{1}{f(x)}$ or $f(x)\cdot f'(x)=1$ or $\frac{\{f(x)\}^2}{2}=x+c.$

Also, $f(1) = \int_a^1 \frac{1}{f(x)}\, dx = \sqrt{2}$. So, $\frac{\{f(1)\}^2}{2} = 1 + c \Rightarrow 1 = 1 + c \Rightarrow c = 0.$

$\therefore \quad \{f(x)\}^2 = 2x. \qquad \therefore \quad f(x) = \sqrt{2x}. \qquad \therefore \quad f(50) = \sqrt{100} = 10.$

8. If $(x, y) \in S$ then $\sqrt{x^2 + y^2} > \left|\frac{x + y - \sqrt{2}}{\sqrt{2}}\right|$ or $2(x^2 + y^2) > (x + y - \sqrt{2})^2$

$\Rightarrow \quad (x - y)^2 + 2\sqrt{2}(x + y) - 2 > 0. \qquad \text{... (1)}$

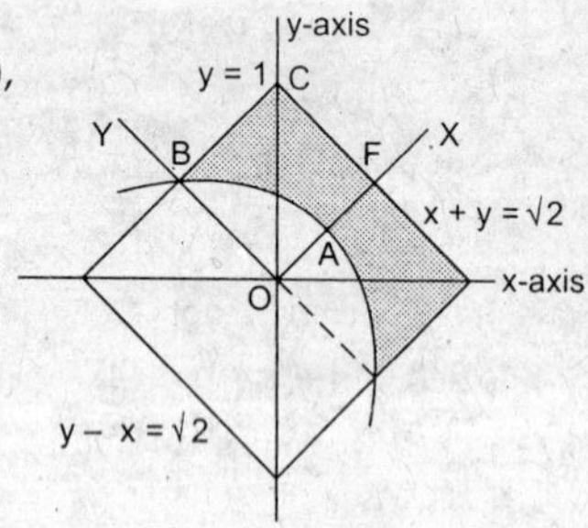

Substituting $\frac{x - y}{\sqrt{2}} = Y$ and $\frac{x + y}{\sqrt{2}} = X$ in (1),

$2Y^2 + 4X - 2 > 0$

or $\quad Y^2 + 2X - 1 > 0.$

So, (X, Y) is a point in the exterior of

$Y^2 = -2\left(X - \frac{1}{2}\right)$.

$x + y = \sqrt{2} \Rightarrow X = 1.$

$y - x = \sqrt{2} \Rightarrow Y = 1.$

Now, $A = \left(\frac{1}{2}, 0\right)_{X, Y}$ and $B = (0, 1)_{X, Y}$.

$\therefore \quad$ area of $S = 2 \times \text{ar}(BAFCB) = 2\{1 \times 1 - \text{ar}(BAOB)\}$

$$= 2 - 2\int_0^{1/2} Y\, dX = 2 - 2\int_0^{1/2} \sqrt{1 - 2X}\, dX = 2 - \frac{2}{3} = \frac{4}{3}.$$

9. $A \geq G$

$$\Rightarrow \frac{(a_1 + a_3 + a_5 + \dots) + (a_2 + a_4 + a_6 + \dots)}{2} \geq \sqrt{(a_1 + a_3 + a_5 + \dots)(a_2 + a_4 + a_6 + \dots)}$$

or $\quad \frac{20}{2} \geq \sqrt{\sum_{i=1}^{n-1} a_i a_{i+1} + k}$, for $k \geq 0$.

$\therefore \quad 100 \geq \sum_{i=1}^{n-1} a_i a_{i+1} + k \geq \sum_{i=1}^{n-1} a_i a_{i+1}.$

Hence, the greatest possible value = 100.

10. Expression $= 2^{a_1} + (4 - 1)^{a_2} + (4 + 1)^{a_3} + (8 - 1)^{a_4} + (8 + 1)^{a_5}$

$= 2^{a_1} + \{4m_1 + (-1)^{a_2}\} + \{4m_2 + 1^{a_3}\} + \{8m_3 + (-1)^{a_4}\} + \{8m_4 + 1^{a_5}\}$

(using binomial expansion)

$= 2^{a_1} + (-1)^{a_2} + 1^{a_3} + (-1)^{a_4} + 1^{a_5} + 4m$

$\therefore$ $2^{a_1} + (-1)^{a_2} + 1 + (-1)^{a_4} + 1$ must be divisible by 4. Clearly, $a_1 \neq 0$.

$\therefore$ $(a_1, a_2, a_3, a_4, a_5)$

= (1, odd digit, any digit, even digit, any digit)

or (1, even digit, any digit, odd digit, any digit)

or (any digit ≥ 2, even digit, any digit, even digit, any digit)

or (any digit ≥ 2, odd digit, any digit, odd digit, any digit)

$\therefore$ total number of ways to form $(a_1, a_2, a_3, a_4, a_5)$

$= (1 \times 5 \times 10 \times 5 \times 10) + (1 \times 5 \times 10 \times 5 \times 10) + (8 \times 5 \times 10 \times 5 \times 10) + (8 \times 5 \times 10 \times 5 \times 10)$

$= 72 \times 5^4 = 45000.$

11. $\lim\limits_{x \to \infty} \dfrac{f(x)}{x^4} = \lim\limits_{x \to \infty} \dfrac{ax^4 + bx^3 + cx^2 + dx + \lambda}{x^4} = a = 1.$

$$\frac{f(1)}{1} = \frac{f(2)}{2} = \frac{f(3)}{3} \Rightarrow \frac{1 + b + c + d + \lambda}{1} = \frac{16 + 8b + 4c + 2d + \lambda}{2} = \frac{81 + 27b + 9c + 3d + \lambda}{3}. \quad \text{... (1)}$$

Now, $f(i) + f(\overline{i}) = f(i) + f(-i) = a - bi - c + di + \lambda + a + bi - c - di + \lambda$

$= 2a - 2c + 2\lambda = 2 - 2c + 2\lambda.$

Eliminating b and d from (1), we obtain $22 - 2c + 2\lambda = 0$.

$\therefore$ $f(i) + f(\overline{i}) = 2 - 22 = -20$. So, $k = 20$.

12. Here, $\dfrac{x+3}{10} = \dfrac{y-1}{-20} = \dfrac{z+1}{k} \Rightarrow 2x + y + 5 = 0 = kx - 10z + 3k - 10.$

Any plane through the line has the equation

$kx - 10z + 3k - 10 + \lambda(2x + y + 5) = 0$

or $(k + 2\lambda)x + \lambda y - 10z + 3k - 10 + 5\lambda = 0.$

$\therefore$ $\dfrac{x}{3} = \dfrac{y-1}{5} = \dfrac{z+1}{7}$ is perpendicular to it. So, $\dfrac{k + 2\lambda}{3} = \dfrac{\lambda}{5} = \dfrac{-10}{7}$.

Solving these, we have $k = 10$.

□

7

Miscellaneous Questions—7

Assertion-Reason-Type Questions

- [Each question contains STATEMENT-1 (Assertion) and STATEMENT-2 (Reason). Each question has 4 choices out of which only one is correct.]

1. The curve C is the locus of a point whose sum of distances from the points $S(\sqrt{3}, 0)$ and $S'(-\sqrt{3}, 0)$ is 4.

   STATEMENT-1: The curve C cuts off intercept $2\sqrt{3}$ from the line $2y - 1 = 0$.

   <u>*because*</u>

   STATEMENT-2: The equation of the curve C is $\frac{x^2}{4} + 2y^2 = \frac{5}{4}$.

   (a) Statement-1 is True, Statement-2 is True; Statement-2 is a correct explanation for Statement-1

   (b) Statement-1 is True, Statement-2 is True; Statement-2 is *Not* a correct explanation for Statement-1

   (c) Statement-1 is True, Statement-2 is False

   (d) Statement-1 is False, Statement-2 is True

2. STATEMENT-1: $\int_0^{\pi/2} \frac{dx}{1 + \tan^3 x} = \frac{\pi}{4}$.

   <u>*because*</u>

   STATEMENT-2: $\int_0^a f(x)dx = \int_0^a f(x + a)dx$ where $f(x)$ is an integrable function.

   (a) Statement-1 is True, Statement-2 is True; Statement-2 is a correct explanation for Statement-1

(b) Statement-1 is True, Statement-2 is True; Statement-2 is *Not* a correct explanation for Statement-1

(c) Statement-1 is True, Statement-2 is False

(d) Statement-1 is False, Statement-2 is True

3. Let $F(x) = \int_1^{x^2} \cos\sqrt{t}\, dt$.

STATEMENT-1: $F'(x) = \cos x$.

because

STATEMENT-2: If $f(x) = \int_a^x \phi(t)dt$ then $f'(x) = \phi(x)$.

(a) Statement-1 is True, Statement-2 is True; Statement-2 is a correct explanation for Statement-1

(b) Statement-1 is True, Statement-2 is True; Statement-2 is *Not* a correct explanation for Statement-1

(c) Statement-1 is True, Statement-2 is False

(d) Statement-1 is False, Statement-2 is True

4. $a_1, a_2, a_3, \ldots, a_n$ are in AP.

STATEMENT-1:

$$\frac{1}{a_1 a_n} + \frac{1}{a_2 a_{n-1}} + \frac{1}{a_3 a_{n-2}} + \ldots + \frac{1}{a_n a_1} = \frac{2}{a_1 + a_n}\left(\frac{1}{a_1} + \frac{1}{a_2} + \frac{1}{a_3} + \ldots + \frac{1}{a_n}\right).$$

because

STATEMENT-2: $a_1 + a_n = a_r + a_{n-r+1}$ for $1 \le r \le n$.

(a) Statement-1 is True, Statement-2 is True; Statement-2 is a correct explanation for Statement-1

(b) Statement-1 is True, Statement-2 is True; Statement-2 is *Not* a correct explanation for Statement-1

(c) Statement-1 is True, Statement-2 is False

(d) Statement-1 is False, Statement-2 is True

5. STATEMENT-1: $1 \cdot 3 \cdot 5 \cdot \ldots \cdot (2n-1) > n^n, n \in N$.

because

STATEMENT-2: The sum of the first n odd natural numbers is equal to n^2.

(a) Statement-1 is True, Statement-2 is True; Statement-2 is a correct explanation for Statement-1

(b) Statement-1 is True, Statement-2 is True; Statement-2 is *Not* a correct explanation for Statement-1

(c) Statement-1 is True, Statement-2 is False

(d) Statement-1 is False, Statement-2 is True

6. *ABCD* is a quadrilateral in which a circle is inscribed.

STATEMENT-1: The lengths of the sides of the quadrilateral can be in AP.

because

STATEMENT-2: The lengths of tangents from an external point to a circle are equal.

(a) Statement-1 is True, Statement-2 is True; Statement-2 is a correct explanation for Statement-1

(b) Statement-1 is True, Statement-2 is True; Statement-2 is *Not* a correct explanation for Statement-1

(c) Statement-1 is True, Statement-2 is False

(d) Statement-1 is False, Statement-2 is True

7. a, b, c are three unequal positive numbers.

STATEMENT-1: The product of their sum and the sum of their reciprocals exceeds 9.

because

STATEMENT-2: AM of n positive numbers exceeds their HM.

(a) Statement-1 is True, Statement-2 is True; Statement-2 is a correct explanation for Statement-1

(b) Statement-1 is True, Statement-2 is True; Statement-2 is *Not* a correct explanation for Statement-1

(c) Statement-1 is True, Statement-2 is False

(d) Statement-1 is False, Statement-2 is True

8. z is a unimodular complex number.

STATEMENT-1: $\arg(z^2 + \bar{z}) = \arg z$.

because

STATEMENT-2: $\bar{z} = \cos(\arg z) - i \sin(\arg z)$

(a) Statement-1 is True, Statement-2 is True; Statement-2 is a correct explanation for Statement-1

(b) Statement-1 is True, Statement-2 is True; Statement-2 is *Not* a correct explanation for Statement-1

(c) Statement-1 is True, Statement-2 is False

(d) Statement-1 is False, Statement-2 is True

9. Let $(1 + x)^{36} = a_0 + a_1x + a_2x^2 + \ldots + a_{36}x^{36}$.

STATEMENT-1: $a_0 + a_3 + a_6 + \ldots + a_{36} = \frac{2}{3}(2^{35} + 1)$

<u>*because*</u>

STATEMENT-2: $a_0 + a_1 + a_2 + \ldots + a_{36} = 2^{36}$ and

$$a_0 + a_2 + a_4 + \ldots + a_{36} = 2^{35}$$

(a) Statement-1 is True, Statement-2 is True; Statement-2 is a correct explanation for Statement-1

(b) Statement-1 is True, Statement-2 is True; Statement-2 is *Not* a correct explanation for Statement-1

(c) Statement-1 is True, Statement-2 is False

(d) Statement-1 is False, Statement-2 is True

10. Let a, b, c be real such that $ax^2 + bx + c = 0$ and $x^2 + x + 1 = 0$ have a common root.

STATEMENT-1: $a = b = c$.

<u>*because*</u>

STATEMENT-2: Two quadratic equations with real coefficients cannot have only one imaginary root common.

(a) Statement-1 is True, Statement-2 is True; Statement-2 is a correct explanation for Statement-1

(b) Statement-1 is True, Statement-2 is True; Statement-2 is *Not* a correct explanation for Statement-1

(c) Statement-1 is True, Statement-2 is False

(d) Statement-1 is False, Statement-2 is True

11. Given in ΔABC, $a : b : c = \cos A : \cos B : \cos C$.

STATEMENT-1: ΔABC is equilateral.

<u>*because*</u>

STATEMENT-2: $\cos A = \dfrac{b^2 + c^2 - a^2}{2bc}$, $\cos B = \dfrac{c^2 + a^2 - b^2}{2ca}$, $\cos C = \dfrac{a^2 + b^2 - c^2}{2ab}$.

(a) Statement-1 is True, Statement-2 is True; Statement-2 is a correct explanation for Statement-1

(b) Statement-1 is True, Statement-2 is True; Statement-2 is *Not* a correct explanation for Statement-1

(c) Statement-1 is True, Statement-2 is False

(d) Statement-1 is False, Statement-2 is True

12. STATEMENT-1: $[\vec{b} \times \vec{c} \;\; \vec{c} \times \vec{a} \;\; \vec{a} \times \vec{b}] = 0$.

<u>*because*</u>

STATEMENT-2: If $\vec{p}$, $\vec{q}$, $\vec{r}$ are linearly dependent vectors then they are coplanar.

(a) Statement-1 is True, Statement-2 is True; Statement-2 is a correct explanation for Statement-1
(b) Statement-1 is True, Statement-2 is True; Statement-2 is *Not* a correct explanation for Statement-1
(c) Statement-1 is True, Statement-2 is False
(d) Statement-1 is False, Statement-2 is True

13. STATEMENT-1: The probability of getting a tail most of the time in 10 tosses of a unbiased coin is $\frac{1}{2}\left\{1 - \frac{10\,!}{2^{10}\,5\,!\,5\,!}\right\}$.

because

STATEMENT-2: ${}^{2n}C_0 + {}^{2n}C_1 + {}^{2n}C_2 + \ldots + {}^{2n}C_n = 2^{2n-1}$, $n \in N$.

(a) Statement-1 is True, Statement-2 is True; Statement-2 is a correct explanation for Statement-1
(b) Statement-1 is True, Statement-2 is True; Statement-2 is *Not* a correct explanation for Statement-1
(c) Statement-1 is True, Statement-2 is False
(d) Statement-1 is False, Statement-2 is True

14. STATEMENT-1: The probability of solving a new problem by 3 students are $\frac{1}{2}, \frac{1}{3}$ and $\frac{1}{4}$ respectively. The probability that the problem will be solved by them is $\frac{1}{24}$.

because

STATEMENT-2: If A, B and C are three independent events then the probability of at least one of them happenning $= 1 - P(\overline{A})\, P(\overline{B})\, P(\overline{C})$.

(a) Statement-1 is True, Statement-2 is True; Statement-2 is a correct explanation for Statement-1
(b) Statement-1 is True, Statement-2 is True; Statement-2 is *Not* a correct explanation for Statement-1
(c) Statement-1 is True, Statement-2 is False
(d) Statement-1 is False, Statement-2 is True

15. Let A and B are two independent events.

STATEMENT-1: If $P(A) = 0.3$ and $P(A \cup \overline{B}) = 0.8$ then $P(B)$ is $\frac{2}{7}$.

because

STATEMENT-2: $P(\overline{E}) = 1 - P(E)$ where E is any event.

(a) Statement-1 is True, Statement-2 is True; Statement-2 is a correct explanation for Statement-1

(b) Statement-1 is True, Statement-2 is True; Statement-2 is *Not* a correct explanation for Statement-1

(c) Statement-1 is True, Statement-2 is False

(d) Statement-1 is False, Statement-2 is True

16. The equations of two straight lines are

$$\frac{x-1}{2}=\frac{y+3}{1}=\frac{z-2}{-3} \text{ and } \frac{x-2}{1}=\frac{y-1}{-3}=\frac{z+3}{2}.$$

STATEMENT-1: The given lines are coplanar.

because

STATEMENT-2: The equations

$$2r-s=1$$
$$r+3s=4$$
$$3r+2s=5 \text{ are consistent.}$$

(a) Statement-1 is True, Statement-2 is True; Statement-2 is a correct explanation for Statement-1

(b) Statement-1 is True, Statement-2 is True; Statement-2 is *Not* a correct explanation for Statement-1

(c) Statement-1 is True, Statement-2 is False

(d) Statement-1 is False, Statement-2 is True

17. Given two straight lines whose equations are

$$\frac{x-3}{1}=\frac{y-5}{-2}=\frac{z-7}{1} \text{ and } \frac{x+1}{7}=\frac{y+1}{-6}=\frac{z+1}{1}$$

STATEMENT-1: The line of shortest distance between the given lines is perpendicular to the plane $x+3y+5z=0$.

because

STATEMENT-2: The direction ratios of the normal to the plane $ax+by+cz+d=0$ are $\frac{a}{d}, \frac{b}{d}, \frac{c}{d}$.

(a) Statement-1 is True, Statement-2 is True; Statement-2 is a correct explanation for Statement-1

(b) Statement-1 is True, Statement-2 is True; Statement-2 is *Not* a correct explanation for Statement-1

(c) Statement-1 is True, Statement-2 is False

(d) Statement-1 is False, Statement-2 is True

18. Let $I_n = \int_0^{\pi/4} \tan^n x \, dx$ where $n \in N$.

STATEMENT-1: $\int_0^{\pi/4} \tan^4 x \, dx = \frac{3\pi - 8}{12}$.

<u>*because*</u>

STATEMENT-2: $I_n + I_{n-2} = \frac{1}{n-1}$

(a) Statement-1 is True, Statement-2 is True; Statement-2 is a correct explanation for Statement-1

(b) Statement-1 is True, Statement-2 is True; Statement-2 is *Not* a correct explanation for Statement-1

(c) Statement-1 is True, Statement-2 is False

(d) Statement-1 is False, Statement-2 is True

19. STATEMENT-1: The cubic equation $4x^3 - 15x^2 + 14x - 5 = 0$ has a root in the interval (2, 3).

<u>*because*</u>

STATEMENT-2: If $f(x) = 0$ is a polynomial equation which has two real roots α, β ($\alpha < \beta$) then $f'(x) = 0$ will have a root γ such that $\alpha < \gamma < \beta$.

(a) Statement-1 is True, Statement-2 is True; Statement-2 is a correct explanation for Statement-1

(b) Statement-1 is True, Statement-2 is True; Statement-2 is *Not* a correct explanation for Statement-1

(c) Statement-1 is True, Statement-2 is False

(d) Statement-1 is False, Statement-2 is True

Answers

1. c	**2.** c	**3.** d	**4.** a	**5.** d
6. a	**7.** a	**8.** d	**9.** b	**10.** a
11. a	**12.** d	**13.** c	**14.** d	**15.** a
16. a	**17.** d	**18.** a	**19.** a	

Hints and Solutions to Selected Questions

1. Equation of C : $\sqrt{(x+\sqrt{3})^2+y^2}+\sqrt{(x-\sqrt{3})^2+y^2}=4$ ... (1)

$\Rightarrow \quad (x+\sqrt{3})^2+y^2-(x-\sqrt{3})^2-y^2=4\left\{\sqrt{(x+\sqrt{3})^2+y^2}-\sqrt{(x-\sqrt{3})^2+y^2}\right\}$

or . $\sqrt{(x+\sqrt{3})^2+y^2}-\sqrt{(x-\sqrt{3})^2+y^2}=\sqrt{3}x$ ... (2)

$\therefore$ adding (1) and (2), $2\sqrt{(x+\sqrt{3})^2+y^2}=4+\sqrt{3}x$

or $\quad 4(x^2+y^2+2\sqrt{3}x+3)=16+8\sqrt{3}x+3x^2$

or $\quad x^2+4y^2=4$ or $\dfrac{x^2}{4}+y^2=1$. So, Statement-2 is False.

Solving $2y-1=0$ with $\dfrac{x^2}{4}+y^2=1$ we get $(\pm\sqrt{3}, 0)$. So, the intercept $=2\sqrt{3}$.

So, Statement-1 is True.

2. $\displaystyle\int_0^a f(x+a)\,dx=\int_a^{2a} f(z)\,dz$ (putting $a+x=z$)

$$=\int_a^{2a} f(x)\,dx \neq \int_0^a f(x)\,dx. \text{ So, Statement-2 is False.}$$

$$I=\int_0^{\pi/2}\frac{dx}{1+\tan^3 x}=\int_0^{\pi/2}\frac{\cos^3 x}{\sin^3 x+\cos^3 x}\,dx$$

$$=\int_0^{\pi/2}\frac{\cos^3\left(\frac{\pi}{2}-x\right)dx}{\sin^3\left(\frac{\pi}{2}-x\right)+\cos^3\left(\frac{\pi}{2}-x\right)}=\int_0^{\pi/2}\frac{\sin^3 x\,dx}{\cos^3 x+\sin^3 x}$$

$$\therefore \quad 2I=I+I=\int_0^{\pi/2}\frac{\cos^3 x}{\sin^3 x+\cos^3 x}\,dx+\int_0^{\pi/2}\frac{\sin^3 x}{\sin^3 x+\cos^3 x}\,dx$$

$$=\int_0^{\pi/2} 1\,dx=x\Big]_0^{\pi/2}=\frac{\pi}{2}\cdot$$

$\therefore \quad I=\dfrac{\pi}{4}\cdot$ So, Statement-1 is True.

3. $\dfrac{dF(x)}{d(x^2)}=\cos\sqrt{t}\Big]_{t=x^2}=\cos x. \quad \therefore \quad \dfrac{dF(x)}{dx}=\dfrac{dF(x)}{d(x^2)}\times\dfrac{d(x^2)}{dx}=2x\cos x.$

$\therefore$ Statement-1 is False. But Statement-2 is True.

4. $$\text{LHS} = \frac{1}{a_1 + a_n}\left\{\frac{a_1 + a_n}{a_1 a_n} + \frac{a_2 + a_{n-1}}{a_2 a_{n-1}} + \frac{a_3 + a_{n-2}}{a_3 a_{n-2}} + \ldots + \frac{a_n + a_1}{a_n a_1}\right\}$$

$$= \frac{1}{a_1 + a_n}\left\{\left(\frac{1}{a_n} + \frac{1}{a_1}\right) + \left(\frac{1}{a_{n-1}} + \frac{1}{a_2}\right) + \left(\frac{1}{a_{n-2}} + \frac{1}{a_3}\right) + \ldots + \left(\frac{1}{a_1} + \frac{1}{a_n}\right)\right\}$$

$$= \frac{2}{a_1 + a_n}\left(\frac{1}{a_1} + \frac{1}{a_2} + \frac{1}{a_3} + \ldots + \frac{1}{a_n}\right) \cdot \text{ So, Statement-1 is True.}$$

$$a_r + a_{n-r+1} = \{a_1 + (r-1)d\} + \{a_1 + (n - r + 1 - 1)d\}$$
$$= a_1 + (r-1)d + a_1 + (n-r)d$$
$$= a_1 + a_1 + (r - 1 + n - r)d$$
$$= a_1 + \{a_1 + (n-1)d\} = a_1 + a_n.$$

So, Statement-2 is True and it is a correct explanation for Statement-1.

5. $$\frac{1 + 3 + 5 + \ldots + (2n-1)}{n} \geq \sqrt[n]{1 \cdot 3 \cdot 5 \cdot \ldots \cdot (2n-1)}$$

($\because$ $AM \geq GM$ for positive numbers)

or $\frac{n^2}{n} \geq \{\sqrt{1 \cdot 3 \cdot 5 \cdot \ldots \cdot (2n-1)}\}^{(1/n)}$ or $n^n \geq 1 \cdot 3 \cdot 5 \cdot \ldots \cdot (2n-1)$.

So, Statement-1 is False.

But $1 + 3 + 5 + \ldots$ to n terms $= n^2$. So, Statement-2 is True.

6. Clearly $AB + DC = BC + AD$ ... (1)

$\therefore$ $CD = \lambda$, $AB = \lambda + 3\mu$, $BC = \lambda + \mu$, $AD = \lambda + 2\mu$ is possible.

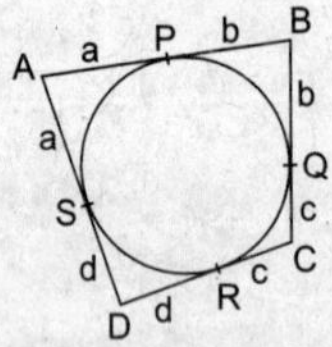

$\therefore$ lengths of sides of $ABCD$ can be in AP.

So, Statement-1 is True.

Also, Statement-2 is True by geometry.

(1) is established using Statement-2.

So Statement-2 is a correct explanation for Statement-1.

7. $\text{AM} = \frac{a+b+c}{3}$, $\text{HM} = \dfrac{1}{\dfrac{\frac{1}{a} + \frac{1}{b} + \frac{1}{c}}{3}}$.

Statement-2 is true. Using it

$$\frac{a+b+c}{3} > \frac{3}{\frac{1}{a} + \frac{1}{b} + \frac{1}{c}} \Rightarrow (a+b+c)\left(\frac{1}{a} + \frac{1}{b} + \frac{1}{c}\right) > 9.$$

$\therefore$ Statement-1 is True and Statement-2 is a correct explanation of Statement-1.

8. $|z| = 1 \Rightarrow z = \cos\theta + i\sin\theta$ where $\theta = \arg z$. Therefore, $\bar{z} = \cos\theta - i\sin\theta$. So, Statement-2 is True.

$\therefore\ \arg(z^2 + \bar{z}) = \arg\{\cos 2\theta + i\sin 2\theta + \cos\theta - i\sin\theta\} = \tan^{-1}\dfrac{\sin 2\theta - \sin\theta}{\cos 2\theta + \cos\theta}$

$$= \tan^{-1}\frac{2\cos\frac{3\theta}{2}\cdot\sin\frac{\theta}{2}}{2\cos\frac{3\theta}{2}\cdot\cos\frac{\theta}{2}} = \tan^{-1}\tan\frac{\theta}{2} = \frac{1}{2}\theta = \frac{1}{2}\arg z.$$

So, Statement-1 is False.

9. Putting $x = 1, \omega, \omega^2$ and adding,

$$3(a_0 + a_3 + a_6 + \ldots + a_{36}) = 2^{36} + (-\omega^2)^{36} + (-\omega)^{36} = 2^{36} + 1 + 1.$$

$\therefore\quad a_0 + a_3 + a_6 + \ldots + a_{36} = \frac{2}{3}(2^{35} + 1)$. So, Statement-1 is True.

Putting $x = 1$, $a_0 + a_1 + a_2 + \ldots + a_{36} = 2^{36}$.

Putting $x = -1$, $a_0 - a_1 + a_2 - \ldots = 0 \Rightarrow a_0 + a_2 + a_4 + \ldots + a_{36} = \frac{1}{2}\cdot 2^{36} = 2^{35}$.

$\therefore$ Statement-2 is True. But Statement-2 is not a correct explanation for Statement-1.

10. Roots of $x^2 + x + 1 = 0$ are imaginary complex numbers. Also, imaginary roots occur in conjugate pair. So, one imaginary root common implies both roots must be common. Thus Statement-2 is True.

When both roots are common, the two quadratic equations are identical.

$\therefore\ \frac{a}{1} = \frac{b}{1} = \frac{c}{1}\cdot$ So, Statement-1 is True and Statement-2 is a correct explanation for Statement-1.

11. Statement-2 is True. Then $\dfrac{a}{\frac{b^2 + c^2 - a^2}{2bc}} = \dfrac{b}{\frac{c^2 + a^2 - b^2}{2ca}} = \dfrac{c}{\frac{a^2 + b^2 - c^2}{2ab}}$

$\therefore\quad b^2 + c^2 - a^2 = c^2 + a^2 - b^2 = a^2 + b^2 - c^2$

$\Rightarrow\quad b^2 - a^2 = a^2 - b^2$ and $c^2 - b^2 = b^2 - c^2 \Rightarrow a^2 = b^2 = c^2 \Rightarrow a = b = c.$

$\therefore$ Statement-1 is True. Thus, Statement-2 is a correct explanation for Statement-1.

12. Statement-2 is True. $[\vec{b}\times\vec{c}\ \ \vec{c}\times\vec{a}\ \ \vec{a}\times\vec{b}] = (\vec{b}\times\vec{c})\cdot(\vec{c}\times\vec{a})\times(\vec{a}\times\vec{b})$

$= (\vec{b}\times\vec{c})\cdot\{[\vec{c}\ \vec{a}\ \vec{b}]\vec{a} - [\vec{c}\ \vec{a}\ \vec{a}]\vec{b}\} = [\vec{a}\ \vec{b}\ \vec{c}]^2 \neq 0$ unless $\vec{a}, \vec{b}, \vec{c}$, are coplanar.

$\therefore$ Statement-1 is False.

13. Stated probabiliy $= {}^{10}C_6\left(\frac{1}{2}\right)^6 \cdot \left(\frac{1}{2}\right)^4 + {}^{10}C_7\left(\frac{1}{2}\right)^7 \cdot \left(\frac{1}{2}\right)^3 + \ldots + {}^{10}C_{10}\left(\frac{1}{2}\right)^{10}$

$$= \left(\frac{1}{2}\right)^{10} \{{}^{10}C_0 + {}^{10}C_1 + \ldots + {}^{10}C_4\}$$

$$= \left(\frac{1}{2}\right)^{10} \left\{\left({}^{10}C_0 + {}^{10}C_1 + \ldots + {}^{10}C_4 + \frac{1}{2}\,{}^{10}C_5\right) - \frac{1}{2} \cdot {}^{10}C_5\right\}$$

$$= \left(\frac{1}{2}\right)^{10} \left\{2^9 - \frac{1}{2}\frac{10\,!}{5\,!\,5\,!}\right\} = \frac{1}{2}\left\{1 - \frac{10\,!}{2^{10}\,5\,!\,5!}\right\}.$$

$\therefore$ Statement-1 is True. Statament-2 is False because

$${}^{2n}C_0 + {}^{2n}C_1 + {}^{2n}C_2 + \ldots + {}^{2n}C_{n-1} + \frac{1}{2}\,{}^{2n}C_n = 2^{2n-1}.$$

14. Statement-2 is True because the probability of none happening $= P(\overline{A} \cap \overline{B} \cap \overline{C})$

$= P(\overline{A}) \cdot P(\overline{B}) \cdot P(\overline{C})$

The probability that the problem is solved

$= 1 -$ probability of all not solving

$$= 1 - \left(1 - \frac{1}{2}\right)\left(1 - \frac{1}{3}\right)\left(1 - \frac{1}{4}\right) = 1 - \frac{1}{2} \cdot \frac{2}{3} \cdot \frac{3}{4} = \frac{3}{4}.$$

$\therefore$ Statement-1 is False.

15. $P(A \cup \overline{B}) = 1 - P(\overline{A \cup \overline{B}}) = 1 - P(\overline{A} \cap B) = 1 - P(\overline{A}) \cdot P(B)$

$\therefore \quad 0.8 = 1 - (1 - 0.3) \cdot P(B) \quad \therefore \; 0.7P(B) = 0.2 \quad \therefore \; P(B) = \frac{2}{7}$

$\therefore$ Statement-1 is True. Also, Statement-2 is True and it is a correct explanation for Statement-1.

16. Any point on the first line is $(2r + 1, r - 3, -3r + 2)$.

Any point on the second line is $(s + 2, -3s + 1, 2s - 3)$.

Two lines will intersect (be coplanar) if $2r + 1 = s + 2$, $r - 3 = -3s + 1$ and $-3r + 2 = 2s - 3$ are consistent, i.e., $2r - s = 1$, $r + 3s = 4$ and $3r + 2s = 5$ are consistent. Condition for this is

$$\begin{vmatrix} 2 & -1 & -1 \\ 1 & 3 & -4 \\ 3 & 2 & -5 \end{vmatrix} = 0, \text{ which is True.}$$

So, Statement-1 and Statement-2 are True and Statement-2 is a correct explanation for Statement-1.

17. The d.r. of the normal to the plane $x + 3y + 5z = 0$ are 1, 3, 5.

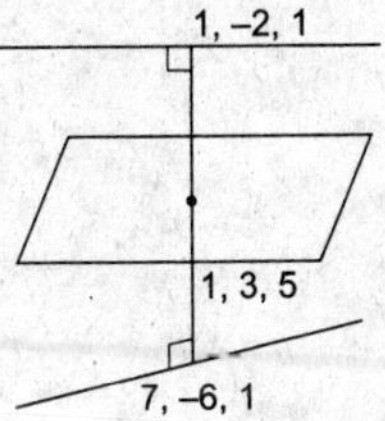

$\therefore$ d.r. of the s.d. will be 1, 3, 5.

By perpendicularity, $1 \times 1 + (-2)3 + 1 \times 5 = 0$ (correct)

and $1 \times 7 + 3(-6) + 5 \times 1 = 0$ (wrong).

$\therefore$ Statement-1 is False.

Statement-2 is True because d.r. of the normal are a, b, c and numbers proportional to a, b, c will also be d.r.

18. $I_n = \int_0^{\pi/4} \tan^{n-2}x\,(\sec^2 x - 1)\,dx = \int_0^{\pi/4} \tan^{n-2}x \sec^2 x\,dx - I_{n-2}$

$$\therefore \quad I_n + I_{n-2} = \int_0^{\pi/4} \tan^{n-2}x\,d(\tan x) = \left.\frac{\tan^{n-1}x}{n-1}\right]_0^{\pi/4} = \frac{1}{n-1}$$

$\therefore$ Statement-2 is True.

$$\therefore \quad I_4 + I_2 = \frac{1}{4-1} \text{ and } I_2 + I_0 = \frac{1}{2-1}$$

$$\therefore \quad I_4 = \frac{1}{3} - I_2 = \frac{1}{3} - (1 - I_0) = I_0 - \frac{2}{3} = \int_0^{\pi/4} \tan^0 x\,dx - \frac{2}{3} = \frac{\pi}{4} - \frac{2}{3}$$

$\therefore$ Statement-1 is True and Statement-2 is a correct explanation for Statment-1.

19. Consider the equation $f(x) = x^4 - 5x^3 + 7x^2 - 5x + 6 = 0$. Clearly, $x = 2$ and $x = 3$ satisfy it. So, $f(x) = 0$ has two real roots 2 and 3.

$\therefore f'(x) = 0$ has a root in (2, 3), i.e., $4x^3 - 15x^2 + 14x - 5 = 0$ has a root in (2, 3).

$\therefore$ Statement-2 and Statement-1 are True and Statement-2 is a correct explanation for Statement-1.

□

Part 10

Practice Worksheets

Practice Worksheets

In the following pages, four practice worksheets are provided with the following objectives

(a) to assess your preparation,

(b) to develop your skills of time management, and

(c) to familiarize you with the actual examination hall situation.

• *Instructions* •

The time for answering each worksheet is 2 hours. You may attempt questions in any order within this time limit. Use a separate blank sheet handy for your rough work. Do not use calculators, log tables or slide rules.

Each worksheet contains 38 questions. There is negative marking. For every two wrong answers, the marks of one correctly answered question will be deducted. However, there is no negative marking for question numbers 36 and 37.

Practice Worksheet—1

One-Option-Correct Questions

1. The value of $\sum_{j=0}^{n}\left({}^{4n+1}C_j + {}^{4n+1}C_{2n-j}\right)$ is

   (a) $2^{4n} + {}^{4n+1}C_n$ (b) 2^{4n+1}

   (c) $2^{4n+1} + {}^{4n+1}C_n$ (d) 2^{4n}

2. If $A = \{(x, y) \mid x^2 + y^2 \le 4\}$ and $B = \{(x, y) \mid (x-3)^2 + y^2 \le 4\}$ and the point $P\left(a, a - \frac{1}{2}\right)$ belongs to the set $B - A$ then the set of possible real values of a is

   (a) $\left(\frac{1+\sqrt{31}}{4}, \frac{7+\sqrt{7}}{4}\right]$ (b) $\left[\frac{7-\sqrt{7}}{4}, \frac{1+\sqrt{31}}{4}\right)$

   (c) $\left(\frac{1-\sqrt{31}}{4}, \frac{7-\sqrt{7}}{4}\right]$ (d) none of these

3. The graph of the curve $x^2 + y^2 - 2xy - 8x - 8y + 32 = 0$ falls wholly in

   (a) the first quadrant (b) the second quadrant

   (c) the third quadrant (d) none of these

4. If α, β be two angles belonging to the interval $[0, 4\pi]$ such that $\sin\alpha + \cos\beta = 2$ then the number of possible values of the pair (α, β) is

   (a) 5 (b) 2 (c) 4 (d) 6

5. If $f(x)$ is a monotonic increasing function and $a \in R$ then

   (a) $f(a^2 + a + 1) > f(1)$ (b) $f(a^2 + a + 1) < f(a^2 + a)$

   (c) $f(a^2 + a + 1) > f\left(\frac{3}{4}\right)$ (d) $f(a^2 + a + 1) < f(a + 1)$

6. The number of divisors of the form $2n - 1$ $(n \ge 2)$ of the number $2^p 3^q 4^r 5^s$, where p, q, r, s belong to N, is

(a) $qs+q+s+1$ (b) $(p+1)(q+1)(r+1)(s+1)-1$

(c) $qs+q+s$ (d) qs

7. If the vertices of a triangle are such that the coordinates of the orthocentre as well as the circumcentre are integers then the coordinates of the centroid must be

(a) integers (b) rational numbers

(c) irrational numbers (d) none of these

8. The number of sides of the quadrilateral whose joint equation is $x^2y^2+1=x^2+y^2$ and which are touched by the circle $x^2+y^2=2x$ is

(a) 4 (b) 3

(c) 2 (d) 1

9. $\sin^8\theta-4\sin^6\theta+6\sin^4\theta-4\sin^2\theta+1$ is equal to

(a) $\cos^8\theta$ (b) $\sin 8\theta$

(c) $\cos 8\theta$ (d) $\sin^4\theta\cdot\cos^4\theta$

10. A vertical wall running from south to north has the height a. A policeman of height $b(>a)$ is standing in front of the wall at a distance c from it on the eastern side. What should be the maximum distance of a crawling thief from the wall so that the thief can hide from the view of the policeman if the thief is on the other side of the wall in the west of the policeman?

(a) $\dfrac{ac}{b-a}$ (b) $\dfrac{bc}{b-a}$

(c) $\dfrac{a+b}{b-a}\cdot c$ (d) $\dfrac{ac}{a+b}$

11. If $\dfrac{\sin x}{x}=\sum\limits_{r=0}^{\infty}a_rx^r$ then

(a) $a_0=1, a_1=0$ (b) $a_0=-1, a_1=0$

(c) $a_0=0, a_1=1$ (d) $a_0=0, a_1=0$

12. A batsman can score 0, 1, 2, 3, 4 or 6 runs from a ball. The number of different sequences in which he can score exactly 30 runs in an over of six balls is

(a) 4 (b) 72

(c) 56 (d) 71

13. If $z = 1 + i\cot\alpha, -\frac{\pi}{2} < \alpha < 0$ then $|z|$ is equal to

(a) $\operatorname{cosec}\alpha$ (b) $-\operatorname{cosec}\alpha$

(c) $\operatorname{cosec}\alpha$ or $-\operatorname{cosec}\alpha$ (d) none of these

14. $C_1, C_2, C_3, \ldots$ is a sequence of circles such that C_{n+1} is the director circle of C_n. If the radius of C_1 is a then the area bounded by the circles C_n and C_{n+1} is

(a) $\pi \cdot 2^n \cdot a^2$ (b) $\pi \cdot 2^{2n-2} \cdot a^2$

(c) $\pi \cdot 2^{n-1} \cdot a^2$ (d) none of these

15. The value of $\int_0^{3\pi/4} \sqrt{1 - \sin 2x}\, dx$ is

(a) $2\sqrt{2} - 1$ (b) -1

(c) 1 (d) none of these

16. $y^2 = 4x$ is a curve and P, Q, R are three points on it where $P = (1, 2), Q = \left(\frac{1}{4}, 1\right)$ and the tangent to the curve at R is parallel to the chord PQ of the curve. Then the coordinates of R are

(a) $\left(\frac{5}{8}, \sqrt{\frac{5}{2}}\right)$ (b) $\left(\frac{9}{16}, \frac{3}{2}\right)$

(c) $\left(\frac{5}{8}, -\sqrt{\frac{5}{2}}\right)$ (d) $\left(\frac{9}{16}, -\frac{3}{2}\right)$

17. If the real-valued function $f(x) = x^3 + 3(a^2 - 1)x + 1$ be invertible then the set of possible real values of a is

(a) $(-\infty, -1) \cup (1, +\infty)$ (b) $(-1, 1)$

(c) $[-1, 1]$ (d) $(-\infty, -1] \cup [1, +\infty)$

18. Let $P(4, 1, \lambda)$ and $Q(2, -1, \lambda)$ be two points. A line having direction ratios $1, -1, 6$ is perpendicular to the plane passing through the origin, P and Q. Then λ equals

(a) $-\frac{1}{2}$ (b) $\frac{1}{2}$

(c) 1 (d) none of these

19. The slope of the tangent to the curve $x = \sec y$ at the point $y = \sec^{-1}(-2)$ is

(a) $\frac{1}{2\sqrt{3}}$ (b) $\frac{-1}{2\sqrt{3}}$

(c) $\frac{1}{2\sqrt{3}}$ or $\frac{-1}{2\sqrt{3}}$ (d) none of these

20. $\lim_{n \to \infty} \frac{1}{n}\{ \sqrt[n]{e} + \sqrt[n]{e^2} + \sqrt[n]{e^3} + \ldots + \sqrt[n]{e^n} \}$ is equal to

(a) $e + 1$ (b) $e - 1$

(c) e (d) none of these

21. $\int_0^1 \cos^{-1} x \, dx$ is equal to

(a) 1 (b) $\frac{\pi}{2}$

(c) -1 (d) π

22. If $\vec{a}, \vec{b}$ and $\vec{c}$ are unit vectors and the angles between $\vec{a}, \vec{b}$; $\vec{b}, \vec{c}$ and $\vec{c}, \vec{a}$ are $\frac{\pi}{6}, \frac{\pi}{3}$ and $\frac{\pi}{4}$ respectively then $|\vec{a} + \vec{b} + \vec{c}|$ is

(a) 3 (b) $4 + \sqrt{3} + \sqrt{2}$

(c) $\sqrt{4 + \sqrt{3} + \sqrt{2}}$ (d) none of these

23. $\int_0^a f(x)\, dx + \int_0^a f(2a - x)\, dx$ is equal to

(a) $2\int_0^a f(x)\, dx$ (b) $\int_0^{2a} f(x)\, dx$

(c) $\int_{-a}^{a} f(x)\, dx$ (d) $\int_0^a f(a + x)\, dx$

24. Five horses are in a race. A man selects two of them at random and bets on them. All the horses have equal chances of winning. The probability that the man selected the winning horse is

(a) $\frac{3}{5}$ (b) $\frac{1}{5}$

(c) $\frac{2}{5}$ (d) $\frac{4}{25}$

25. If $\sin x + \cos y = \frac{1}{3}$ and $\cos x + \sin y = \frac{3}{4}$ then $\tan \frac{x-y}{2}$ is equal to

(a) $\frac{5}{13}$ (b) $\frac{12}{13}$

(c) $-\frac{5}{13}$ (d) none of these

26. The term independent of a in the expansion of $\left(1 + \sqrt{a} + \frac{1}{\sqrt{a} - 1}\right)^{-30}$ is

(a) ${}^{30}C_{20}$ (b) 0

(c) ${}^{30}C_{10}$ (d) none of these

One-or-More-Options-Correct Questions

27. If $\sum_{p=1}^{n} \left\{ \sum_{m=1}^{p} \left(\sum_{k=1}^{m} k \right) \right\} = a_0 n^4 + a_1 n^3 + a_2 n^2 + a_3 n + a_4$ then

(a) $a_0 = \frac{1}{6}$ (b) $a_1 = \frac{1}{4}$

(c) $a_3 = \frac{1}{4}$ (d) $a_4 = 0$

28. If $\vec{a} + \vec{b} + \vec{c} = 0$ then $\vec{a} \times \vec{b}$ is equal to

(a) $\vec{c} \times \vec{a}$ (b) $\vec{c}$

(c) $\vec{b} \times \vec{c}$ (d) $\vec{a} \times \vec{c}$

29. A and B are two independent events such that $P(A' \cap B) = \frac{2}{15}$ and $P(A \cap B') = \frac{1}{6}$. Then $P(B)$ is equal to

(a) $\frac{4}{5}$ (b) $\frac{1}{6}$

(c) $\frac{1}{5}$ (d) $\frac{5}{6}$

30. The centre of a circle which touches the line $2x - y = 1$ at (1, 1) and also touches the line $2x + y = 4$ is

(a) $\left(0, \frac{3}{2}\right)$ (b) $\left(0, -\frac{3}{2}\right)$

(c) $\left(-\frac{5}{4}, -\frac{7}{8}\right)$ (d) $\left(\frac{5}{4}, \frac{7}{8}\right)$

Comprehension-Type Questions

The general solution of a differential equation of the form

$$\frac{d^n y}{dx^n} + a_1 \frac{d^{n-1} y}{dx^{n-1}} + a_2 \frac{d^{n-2} y}{dx^{n-2}} + \ldots + a_n y = 0 \qquad (a_i\text{s being constants})$$

is given by the following rules.

(i) If the roots of the corresponding auxiliary equation

$$D^n + a_1 D^{n-1} + a_2 D^{n-2} + \ldots + a_n = 0$$

in D has unequal real roots $\alpha_1, \alpha_2, \alpha_3, \ldots, \alpha_n$ then the general solution will be

$$y = c_1 e^{\alpha_1 x} + c_2 e^{\alpha_2 x} + \ldots + c_n e^{\alpha_n x}, \text{ where } c_i\text{s are arbitrary constants.}$$

(ii) If the roots of the corresponding auxiliary equation has two equal roots (say, $\alpha_1 = \alpha_2$) and the rest are unequal real roots then the general solution will be

$$y = (c_1 x + c_2)\, e^{\alpha_1 x} + c_3 e^{\alpha_3 x} + \ldots + c_n e^{\alpha_n x}.$$

(iii) If two roots of the corresponding auxiliary equation are $\alpha_1 = \alpha + i\beta$, $\alpha_2 = \alpha - i\beta$ and the rest are real and unequal then the general solution will be

$$y = (A\cos \beta x + B\sin \beta x)\, e^{\alpha x} + c_3 e^{\alpha_3 x} + \ldots + c_n e^{\alpha_n x}.$$

31. The general solution of the equation $\frac{d^3 y}{dx^3} - 7\frac{d^2 y}{dx^2} + 16\frac{dy}{dx} - 12y = 0$ is

(a) $c_1 e^{2x} + c_2 e^{-2x} + c_3 e^{-3x}$ (b) $(c_1 x + c_2)\, e^{2x} + c_3 e^{3x}$

(c) $(c_1 x + c_2)\, e^{-3x} + c_3 e^{2x}$ (d) $(A\cos x + B\sin x)\, e^{2x} + c_3 e^{3x}$

32. $y = (c_1 \cos x + c_2 \sin x)\, e^{-x} + c_3 e^{x}$ is the general solution of the equation

(a) $\frac{d^3 y}{dx^3} - \frac{d^2 y}{dx^2} + 2y = 0$ (b) $\frac{d^3 y}{dx^3} - y = 0$

(c) $\frac{d^3 y}{dx^3} + \frac{d^2 y}{dx^2} - 2y = 0$ (d) $\frac{d^2 y}{dx^2} - y = 0$

33. The general solution of the equation $\dfrac{d^3y}{dx^3} + y = 0$ is

(a) $ce^{-x} + e^{x/2}\left(a\cos\dfrac{\sqrt{3}}{2}x + b\sin\dfrac{\sqrt{3}}{2}x\right)$

(b) $c_1e^{\sqrt{3}x/2} + c_2e^{x/2} + c_3e^{-x}$

(c) $c_1e^{-x} + c_2e^{-\omega x} + c_3e^{-\omega^2 x}$, where $\omega = \dfrac{-1+\sqrt{3}i}{2}$

(d) none of these

Matching Questions

One or more may match with the same.

34. (i) If $f(x)$ is continuous in R and $f(x) + f(x+1) = 1$, where $0 \le x \le 1$, then the value of $\int_0^2 f(x)\,dx$ is — (a) 3

(ii) In a triangle ABC, the scalar triple product $[(\tan A)\vec{i} + \vec{j} + \vec{k} \quad \vec{i} + (\tan B)\vec{j} + \vec{k} \quad \vec{i} + \vec{j} + (\tan C)\vec{k}]$ is equal to — (b) 4

(iii) The intercept on the x-axis cut off by the normal to the parabola $x^2 = 4y$ passing through $(1, 2)$ is — (c) 1

(iv) If $f(x)$ is twice differentiable and $\lambda^2 f(x) - 2\lambda f'(x) + f''(x) = 0$ provides two equal values of λ for all x, and $f(0) = 1$, $f'(0) = 2$ then $f(\log_e 2)$ is equal to — (d) 2

35. (i) Let $\int_0^{f(x)} t \cdot \cos t\, dt = \int_0^{x^2} \sqrt{z} \cdot \sin\sqrt{z}\, dz$ and $f(0) = 0$. Then $4x\sin x + 4\cos x - 2x^2\cos x - f(x)\sin f(x) - \cos f(x)$ equals — (a) 1

(ii) The number of real roots of the equation $x + e^x = 0$ is — (b) 3

(iii) If $\sum_{r=1}^{n} \dfrac{4r+1}{(4r-1)^2(4r+3)^2} > \dfrac{1}{75}$ then the least possible value of n is — (c) 6

(iv) From a bag containing 2 white balls and 1 black ball a ball is drawn at random and returned to the bag. It is done n times. If the probability of drawing a white ball three times is $\frac{160}{729}$ then n is

(d) 4

Complete the following statements.

36. If $(1+x)^n = {}^nC_0 + {}^nC_1 x + {}^nC_2 x^2 + \ldots + {}^nC_n x^n$ then the value of $\sum_{i=0}^{8} \sum_{j=1}^{8} {}^8C_j \, {}^jC_i$ for all $i \le j$ is

37. The area enclosed by the tangents from $P(6, 8)$ to the circle $x^2 + y^2 = r^2$ and the chord of contact of tangents is the maximum when r is

Assertion-Reason Type

38. Let the vectors $\overrightarrow{PQ}, \overrightarrow{QR}, \overrightarrow{RS}, \overrightarrow{ST}, \overrightarrow{TU}$ and $\overrightarrow{UP}$ represent the sides of a regular hexagon.

STATEMENT-1: $\overrightarrow{PQ} \times (\overrightarrow{RS} + \overrightarrow{ST}) \neq \vec{0}$

because

STATEMENT-2: $\overrightarrow{PQ} \times \overrightarrow{RS} = \vec{0}$ and $\overrightarrow{PQ} \times \overrightarrow{ST} \neq \vec{0}$.

(a) Statement-1 is True, Statement-2 is True; Statement-2 is a correct explanation for Statement-1

(b) Statement-1 is True, Statement-2 is True; Statement-2 is *Not* a correct explanation for Statement-1

(c) Statement-1 is True, Statement-2 is False

(d) Statement-1 is False, Statement-2 is True

Answers

1. a	**2.** a	**3.** a	**4.** d	**5.** c
6. c	**7.** b	**8.** c	**9.** a	**10.** a
11. a	**12.** d	**13.** b	**14.** c	**15.** a
16. b	**17.** d	**18.** a	**19.** a	**20.** b
21. a	**22.** c	**23.** b	**24.** c	**25.** c
26. b	**27.** b, c, d	**28.** a, c	**29.** a, b	**30.** a, d
31. b	**32.** c	**33.** a		

34. (i) $\leftrightarrow$ c (ii) $\leftrightarrow$ c (iii) $\leftrightarrow$ a (iv) $\leftrightarrow$ b

35. (i) $\leftrightarrow$ b (ii) $\leftrightarrow$ a (iii) $\leftrightarrow$ d (iv) $\leftrightarrow$ c

36. 6560 **37.** 5 **38.** c

Hints and Solutions to Selected Questions

1. Value $= ({}^{4n+1}C_0 + {}^{4n+1}C_1 + \ldots + {}^{4n+1}C_n) + ({}^{4n+1}C_{2n} + {}^{4n+1}C_{2n-1} + \ldots + {}^{4n+1}C_n)$

$= ({}^{4n+1}C_0 + {}^{4n+1}C_1 + \ldots + {}^{4n+1}C_{2n}) + {}^{4n+1}C_n = 2^{4n} + {}^{4n+1}C_n.$

2. $P \in B$ but $P \notin A$. So, $a^2 + \left(a - \frac{1}{2}\right)^2 > 4$ and $(a-3)^2 + \left(a - \frac{1}{2}\right)^2 \le 4.$

$\therefore$ we get $8a^2 - 4a - 15 > 0$ ... (1)

$8a^2 - 28a + 21 \le 0$ ... (2)

Solving (1) and (2), we get the set of possible values of a.

3. Here $x^2 - (8+2y)x + y^2 - 8y + 32 = 0$. $\therefore$ $x = 4 + y \pm 2\sqrt{y-1}$.

For real x, $y \ge 1$ and then $x = 4 + y \pm 2\sqrt{y-1} = 4 + (\sqrt{y-1} \pm 1)^2 > 0$.

Similarly, for real y, $x \ge 1$ and then $y = 4 + x \pm 2\sqrt{x-1} > 0$.

Hence, the points on the graph will have their coordinates positive.

Alternatively Here $(x-y)^2 = 8(x+y-4)$ is a parabola whose axis is $x - y = 0$ and the tangent at the vertex is $x + y - 4 = 0$.

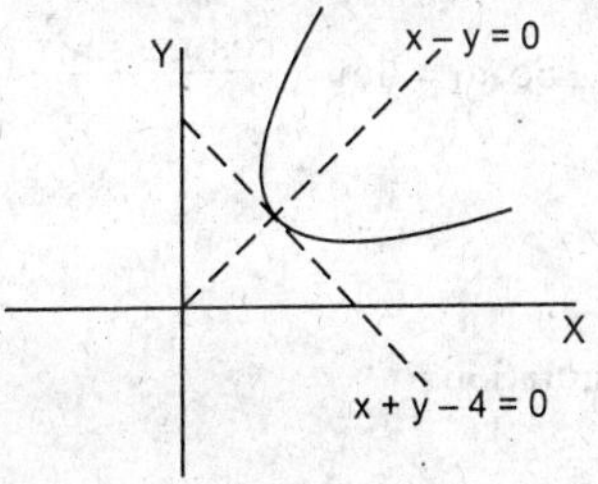

Also, when $y = 0$ we have $x^2 - 8x + 32 = 0$ which gives no real values of x.

When $x = 0$ we have $y^2 - 8y + 32 = 0$ which gives no real values of y.

So, the parabola does not intersect the axes. Hence, the graph is as shown in the figure.

4. $\sin\alpha + \cos\beta = 2$ $\Rightarrow$ $\sin\alpha = 1$, $\cos\beta = 1$ because the maximum value of each is 1.

$\therefore$ $\alpha = \frac{\pi}{2}, \frac{5\pi}{2}$ and $\beta = 0, 2\pi, 4\pi$. Hence (α, β) has 2×3 possible values.

5. $a^2 + a + 1 = \left(a + \frac{1}{2}\right)^2 + \frac{3}{4} > \frac{3}{4}$; $a^2 + a + 1 < 1$ for $a \in (-1, 0)$.

$a^2 + a + 1 > a^2 + a$ for all a; $a^2 + a + 1 \ge a + 1$ for all a.

As $f(x)$ is m.i., $f(a^2+a+1) > f\left(\frac{3}{4}\right)$.

6. Required number of divisors = the number of selections of at least one from q 3s, s 5s

$$= (q+1)(s+1) - 1.$$

7. The centroid divides the line segment joining the circumcentre and the orthocentre in the ratio 1 : 2. Use this fact.

8. Sides are $(x^2-1)(y^2-1)=0$ or $x+1=0, x-1=0, y+1=0, y-1=0$.
Clearly, only $y \pm 1 = 0$ touch the circle.

9. Expression $= {}^4C_0 \cdot (\sin^2\theta)^4 - {}^4C_1(\sin^2\theta)^3 + {}^4C_2(\sin^2\theta)^2 - {}^4C_3(\sin^2\theta) + {}^4C_4$

$$= (1-\sin^2\theta)^4 = (\cos^2\theta)^4 = \cos^8\theta.$$

10. From similar triangles ABT and PQT,

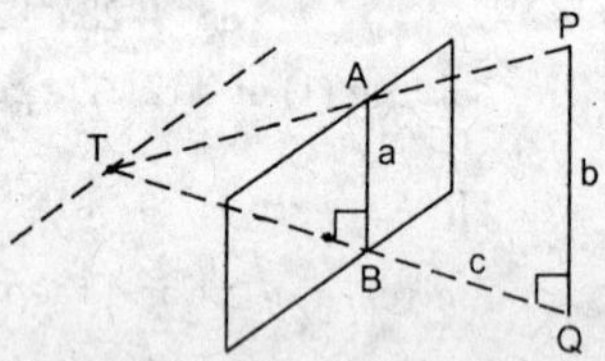

$$\frac{AB}{PQ} = \frac{BT}{QT} \quad \text{or} \quad \frac{a}{b} = \frac{x}{x+c}$$

or $x(b-a) = ac$.

11. $\dfrac{\sin x}{x} = a_0 + a_1 x + a_2 x^2 + \ldots$ ... (1)

$$\therefore \quad a_0 = \lim_{x \to 0} \frac{\sin x}{x} = 1.$$

Differentinting (1) w.r.t. x,

$$\frac{x\cos x - \sin x}{x^2} = a_1 + 2a_2 x + \ldots . \qquad \text{So, } a_1 = \lim_{x\to 0} \frac{x\cos x - \sin x}{x^2}.$$

$$\therefore \quad a_1 = \lim_{x \to 0} \frac{\cos x - x\sin x - \cos x}{2x} = 0.$$

12.

Possibilities	**No. of permutations**
6, 6, 6, 6, 6, 0	$\frac{6!}{5!} = 6$
6, 6, 6, 6, 4, 2	$\frac{6!}{4!} = 30$
6, 6, 6, 6, 3, 3	$\frac{6!}{4!2!} = 15$
6, 6, 6, 4, 4, 4	$\frac{6!}{3!3!} = 20$

$\therefore$ required number of sequences = 71.

13. $|z| = |\text{cosec}\,\alpha| = -\text{cosec}\,\alpha$ because cosec α is negative in $\left(-\frac{\pi}{2}, 0\right)$.

14. The sequence of radii is $a, \sqrt{2}a, 2a, 2\sqrt{2}a, \ldots$.

$\therefore$ required area $= \pi \cdot t_{n+1}^2 - \pi t_n^2 = \pi\{a \cdot (\sqrt{2})^n\}^2 - \pi\{a(\sqrt{2})^{n-1}\}^2$

$$= \pi a^2(2^n - 2^{n-1}) = \pi a^2 \cdot 2^{n-1}.$$

15. $I = \int_0^{3\pi/4} |\cos x - \sin x| dx = \int_0^{\pi/4} |\cos x - \sin x| dx + \int_{\pi/4}^{3\pi/4} |\cos x - \sin x| dx$

$= \int_0^{\pi/4} (\cos x - \sin x)\, dx + \int_{\pi/4}^{3\pi/4} (\sin x - \cos x)\, dx$

$= \left[\sin x + \cos x\right]_0^{\pi/4} + \left[-\cos x - \sin x\right]_{\pi/4}^{3\pi/4}$

$= (\sqrt{2} - 1) + \sqrt{2} = 2\sqrt{2} - 1.$

16. Let $R = (t^2, 2t)$. Then m of the tangent at R is $\left.\frac{dy}{dx}\right]_{(t^2, 2t)}$, i.e., $\frac{1}{t}$.

The slope of $PQ = \dfrac{2-1}{1-\frac{1}{4}} = \frac{4}{3}$. So, $\frac{1}{t} = \frac{4}{3}$.

$\therefore \quad t = \frac{3}{4}$. $\quad \therefore \quad R = \left(\frac{9}{16}, 2 \cdot \frac{3}{4}\right) = \left(\frac{9}{16}, \frac{3}{2}\right)$.

17. $f'(x) = 3x^2 + 3(a^2 - 1)$

$\therefore$ $f(x)$ is m.i. if $x^2 + a^2 - 1 > 0$ for all $x \in R$ and this happens if $a^2 - 1 > 0$.

Clearly, $f(x)$ cannot be m.d. because the coefficient of x^2 in $f'(x) = 3 \nless 0$.

Now, $a^2 - 1 > 0 \quad \Rightarrow a \in (-\infty, -1) \cup (1, +\infty)$.

If $a = \pm 1$, $y = f(x) = x^3 + 1$ or $x = (y-1)^{1/3}$ which is real for all $y \in R$.

Hence, $a \in (-\infty, -1] \cup [1, +\infty)$.

18. The equation of the plane is $1(x - 0) + (-1)(y - 0) + 6(z - 0) = 0$.

It passes through P and $Q \Rightarrow 4 - 1 + 6\lambda = 0, 2 + 1 + 6\lambda = 0 \Rightarrow \lambda = -\frac{1}{2}$.

19. Here, $y = \sec^{-1} x$. So, $\dfrac{dy}{dx} = \dfrac{1}{|x|\sqrt{x^2 - 1}}$.

$\therefore$ at $x = -2$, $\dfrac{dy}{dx} = \dfrac{1}{|-2|\sqrt{(-2)^2 - 1}} = \dfrac{1}{2\sqrt{3}}$.

20. Limit $= \lim_{n \to \infty} \sum_{r=1}^{n} \frac{1}{n} \cdot e^{r/n} = \int_0^1 e^x\, dx = e^x \Big]_0^1 = e - 1.$

21. $I = \left[\cos^{-1} x \cdot x\right]_0^1 - \int_0^1 x \cdot \frac{-1}{\sqrt{1-x^2}}\, dx = \int_0^1 \frac{x}{\sqrt{1-x^2}}\, dx$

$$= \frac{-1}{2}\int_0^1 \frac{-2x}{(1-x^2)^{1/2}}\, dx = -\frac{1}{2}\int_1^0 \frac{dz}{\sqrt{z}} = \frac{-1}{2}\left[\frac{z^{-\frac{1}{2}+1}}{-\frac{1}{2}+1}\right]_1^0 = -(0-1) = 1.$$

22. $\vec{a} \cdot \vec{b} = \cos\frac{\pi}{6} = \frac{\sqrt{3}}{2}, \quad \vec{b} \cdot \vec{c} = \cos\frac{\pi}{3} = \frac{1}{2}, \quad \vec{c} \cdot \vec{a} = \cos\frac{\pi}{4} = \frac{1}{\sqrt{2}}.$

$$|\vec{a} + \vec{b} + \vec{c}|^2 = (\vec{a} + \vec{b} + \vec{c})^2 = \vec{a}^2 + \vec{b}^2 + \vec{c}^2 + 2\vec{a} \cdot \vec{b} + 2\vec{b} \cdot \vec{c} + 2\vec{c} \cdot \vec{a}$$

$$= 1 + 1 + 1 + \sqrt{3} + 1 + \sqrt{2} = 4 + \sqrt{2} + \sqrt{3}.$$

23. $\int_0^a f(x)\, dx + \int_0^a f(2a - x)\, dx = \int_0^a f(x)\, dx + \int_{2a}^a f(z)(-dz)$ (putting $2a - x = z$)

$$= \int_0^a f(x)\, dx + \int_a^{2a} f(z)\, dz = \int_0^a f(x)\, dx + \int_a^{2a} f(x)\, dx = \int_0^{2a} f(x)\, dx.$$

24. Required probability

$$= \frac{\text{number of ways to select the winning horse and any other}}{\text{number of ways to select 2 from 5}}$$

$$= \frac{1 \times {}^4C_1}{{}^5C_2} = \frac{4}{10} = \frac{2}{5}.$$

25. $\sin x + \sin\left(\frac{\pi}{2} - y\right) = \frac{1}{3}$ or $2\sin\left(\frac{\pi}{4} + \frac{x-y}{2}\right) \cdot \cos\left(\frac{x+y}{2} - \frac{\pi}{4}\right) = \frac{1}{3}.$

$\cos x + \cos\left(\frac{\pi}{2} - y\right) = \frac{3}{4}$ or $2\cos\left(\frac{\pi}{4} + \frac{x-y}{2}\right) \cdot \cos\left(\frac{x+y}{2} - \frac{\pi}{4}\right) = \frac{3}{4}.$

Dividing, $\tan\left(\frac{\pi}{4} + \frac{x-y}{2}\right) = \frac{4}{9}$ or $\frac{1 + \tan\frac{x-y}{2}}{1 - \tan\frac{x-y}{2}} = \frac{4}{9}.$

Using componendo–dividendo we get the result.

26. $\left(1+\sqrt{a}+\dfrac{1}{\sqrt{a}-1}\right)^{-30}=\left(\dfrac{a}{\sqrt{a}-1}\right)^{-30}=\left(\dfrac{\sqrt{a}-1}{a}\right)^{30}=\dfrac{1}{a^{30}}(1-\sqrt{a})^{30}$

$$=\frac{1}{a^{30}}\{{}^{30}C_0-{}^{30}C_1\sqrt{a}+\ldots+{}^{30}C_{30}(\sqrt{a})^{30}\}.$$

There is no term independent of a.

27. $\displaystyle\sum_{k=1}^{m}k=\frac{m(m+1)}{2}=\frac{1}{2}m^2+\frac{1}{2}m.$

$$\therefore\quad \sum_{m=1}^{p}\left(\sum_{k=1}^{m}k\right)=\frac{1}{2}\sum_{m=1}^{p}m^2+\frac{1}{2}\sum_{m=1}^{p}m.$$

$$\therefore\quad \sum_{m=1}^{p}\left(\sum_{k=1}^{m}k\right)=\frac{m(m+1)(2m+1)}{12}+\frac{m(m+1)}{4}=\frac{1}{6}(m^3+3m^2+2m).$$

$$\therefore\quad a_0n^4+a_1n^3+a_2n^2+a_3n+a_4=\sum_{m=1}^{n}\frac{1}{6}(m^3+3m^2+2m)$$

$$=\frac{1}{6}\cdot\frac{n^2(n+1)^2}{4}+\frac{1}{2}\cdot\frac{n(n+1)(2n+1)}{6}+\frac{1}{3}\cdot\frac{n(n+1)}{2}$$

$$=\frac{n(n+1)}{24}\{n(n+1)+2(2n+1)+4\}=\frac{n(n+1)(n^2+5n+6)}{24}$$

$$=\frac{1}{24}(n^4+6n^3+11n^2+6n).$$

$$\therefore\quad a_0=\frac{1}{24},\ a_1=\frac{1}{4},\ a_2=\frac{11}{24},\ a_3=\frac{1}{4},\ a_4=0.$$

28. $\vec{a}\times(\vec{a}+\vec{b}+\vec{c})=\vec{0}\quad\Rightarrow\ \vec{a}\times\vec{b}+\vec{a}\times\vec{c}=\vec{0}\quad\Rightarrow\ \vec{a}\times\vec{b}=\vec{c}\times\vec{a}$

Also, $\vec{b}\times(\vec{a}+\vec{b}+\vec{c})=\vec{0}\quad\Rightarrow\ \vec{b}\times\vec{a}+\vec{b}\times\vec{c}=\vec{0}\quad\Rightarrow\ \vec{b}\times\vec{c}=\vec{a}\times\vec{b}$

29. $P(A\cap B')=P(A)\cdot P(B')=\dfrac{1}{6}$ and $P(A'\cap B)=P(A')\cdot P(B)=\dfrac{2}{15}$.

If $P(B)=x$ then $P(A)\cdot(1-x)=\dfrac{1}{6}$ and $P(A')\cdot x=\dfrac{2}{15}$.

$$\therefore\quad \frac{1}{6(1-x)}+\frac{2}{15x}=P(A)+P(A')=1\quad\Rightarrow\ 5x+4(1-x)=30x(1-x)$$

or $\quad 30x^2-29x+4=0\quad$ or $\quad(6x-1)(5x-4)=0\quad$ or $\quad x=\dfrac{1}{6},\dfrac{4}{5}$.

30. The centre will be on the bisectors of the angle between the given lines. So, it will lie on

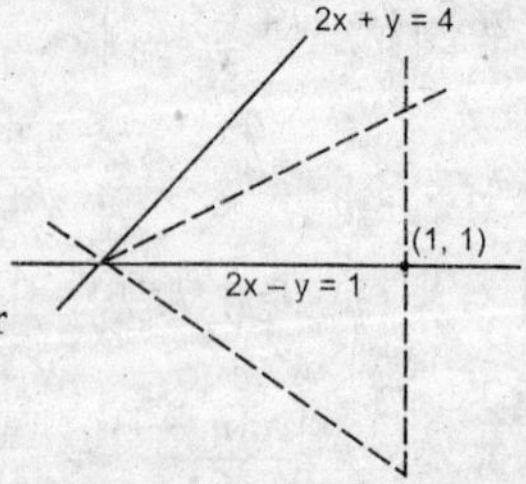

$$\frac{2x-y-1}{\sqrt{5}} = \pm\frac{2x+y-4}{\sqrt{5}}$$

$\Rightarrow \quad 2y = 3$ and $4x = 5$

The centre should also lie on the perpendicular to $2x - y = 1$ at $(1, 1)$ whose equation is

$x + 2y = k$, where $1 + 2 = k$, i.e., $k = 3$.

$\therefore$ the centre also lies on $x + 2y = 3$.

Solving $2y = 3$ and $x + 2y = 3$, the centre $= \left(0, \frac{3}{2}\right)$.

Solving $4x = 5$ and $x + 2y = 3$, the centre $= \left(\frac{5}{4}, \frac{7}{8}\right)$.

31. The auxiliary equation is $D^3 - 7D^2 + 16D - 12 = 0$

or $\quad (D-3)(D^2 - 4D + 4) = 0 \quad$ or $\quad D = 3, 2, 2$.

$\therefore$ by rule (ii), $y = (c_1x + c_2)\, e^{2x} + c_3e^{3x}$.

32. The general solution tells that the auxiliary equation has roots $-1 + i, -1 - i, 1$. So, the auxiliary equation is

$(D + 1 - i)(D + 1 + i)(D - 1) = 0 \quad$ or $\quad \{(D+1)^2 + 1\}(D-1) = 0$

or $\quad (D^2 + 2D + 2)(D - 1) = 0 \quad$ or $\quad D^3 + D^2 - 2 = 0.$

$\therefore$ the differential equation is $\frac{d^3y}{dx^3} + \frac{d^2y}{dx^2} - 2y = 0.$

33. The auxiliary equation is $D^3 + 1 = 0$, whose roots are $-1, \frac{1-\sqrt{3}i}{2}, \frac{1+\sqrt{3}i}{2}$.

$\therefore$ the general solution is $y = ce^{-x} + \left(a\cos\frac{\sqrt{3}}{2}x + b\sin\frac{\sqrt{3}}{2}x\right)e^{x/2}$

34. (i) $\int_0^2 f(x)\,dx = \int_0^1 f(x)\,dx + \int_1^2 f(x)\,dx$

$= \int_0^1 f(x)\,dx + \int_0^1 f(z+1)\,dz \qquad$ (putting $x = z + 1$)

$= \int_0^1 \{f(x) + f(x+1)\}\,dx = \int_0^1 dx = 1.$

(ii) Value $= \begin{vmatrix} \tan A & 1 & 1 \\ 1 & \tan B & 1 \\ 1 & 1 & \tan C \end{vmatrix}$

$= (\tan A)(\tan B \tan C - 1) - (\tan C - 1) + (1 - \tan B)$

$= \tan A \tan B \tan C + 1 - (\tan A + \tan B + \tan C) = 1.$

(iii) the normal at $(2t, t^2)$ is $y - t^2 = \dfrac{-1}{\left.\dfrac{x}{2}\right]_{x=2t}}(x - 2t)$

or $\quad y - t^2 = -\dfrac{1}{t}(x - 2t)$. It passes through (1, 2).

$\therefore \quad 2 - t^2 = -\dfrac{1}{t}(1 - 2t) \quad \Rightarrow t^3 = 1 \quad \Rightarrow t = 1$

$\therefore$ the normal is $y - 1 = -(x - 2)$ or $x + y = 3$.
So, the intercept = 3.

(iv) Here $4\{f'(x)\}^2 - 4f(x) \cdot f''(x) = 0 \quad \Rightarrow \dfrac{f''(x) \cdot f(x) - f'(x) \cdot f'(x)}{\{f(x)\}^2} = 0$

$\Rightarrow \quad \dfrac{d}{dx}\left\{\dfrac{f'(x)}{f(x)}\right\} = 0 \quad \Rightarrow \dfrac{f'(x)}{f(x)} = c$. So, $2 = c$.

$\therefore \quad \log f(x) = 2x + c'$. So, $0 = c'$.

$\therefore \quad f(x) = e^{2x}$.

$\therefore \quad f(\log_e 2) = e^{2\log_e 2} = 4.$

35. (i) Differentiating w.r.t. x,

$f(x) \cos f(x) \cdot f'(x) = x\sin x \cdot 2x$ or $f(x) \cos f(x)\, df(x) = 2x^2 \sin x\, dx$

$\Rightarrow \quad \int f(x) \cos f(x)\, df(x) = 2\int x^2 \sin x\, dx$

$\Rightarrow \quad f(x) \sin f(x) - \int \sin f(x) \cdot f'(x)\, dx = -2x^2 \cos x + 4\int x\cos x\, dx$

$\Rightarrow \quad f(x) \sin f(x) - \int \sin f(x) \cdot df(x) = -2x^2 \cos x + 4\{x\sin x - \int \sin x\, dx\}$

$\Rightarrow \quad f(x) \sin f(x) + \cos f(x) + c = -2x^2 \cos x + 4x\sin x + 4\cos x$

$\Rightarrow \quad 4x\sin x + 4\cos x - 2x^2 \cos x - f(x) \sin f(x) - \cos f(x) = c.$

Since $f(0) = 0$, $\quad 4 - 1 = c$.

(ii) Here $e^x = -x$. So, we take graphs of $y = e^x$ and $y = -x$.

Clearly, there is only one solution.

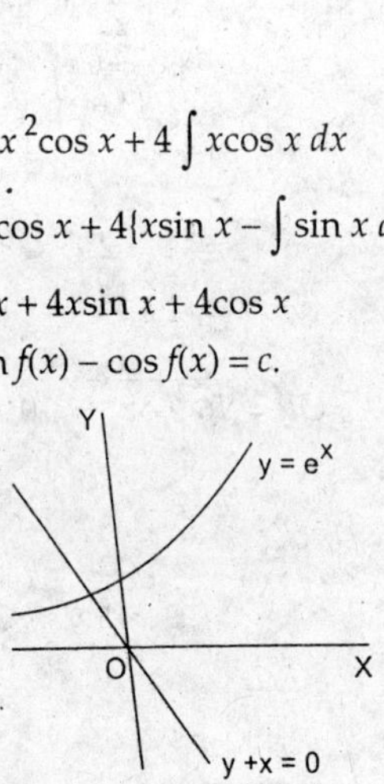

(iii) $\dfrac{4r+1}{(4r-1)^2(4r+3)^2}=\dfrac{1}{8}\cdot\dfrac{(4r+3)^2-(4r-1)^2}{(4r-1)^2(4r+3)^2}=\dfrac{1}{8}\left\{\dfrac{1}{(4r-1)^2}-\dfrac{1}{(4r+3)^2}\right\}$

$$\therefore\quad \sum_{r=1}^{n}\frac{4r+1}{(4r-1)^2(4r+3)^2}$$

$$=\frac{1}{8}\left[\left(\frac{1}{3^2}-\frac{1}{7^2}\right)+\left(\frac{1}{7^2}-\frac{1}{11^2}\right)+\ldots+\left\{\frac{1}{(4n-1)^2}-\frac{1}{(4n+3)^2}\right\}\right]$$

$$=\frac{1}{8}\left\{\frac{1}{9}-\frac{1}{(4n+3)^2}\right\}>\frac{1}{75}.$$

$$\therefore\quad \frac{1}{72}-\frac{1}{75}>\frac{1}{8}\cdot\frac{1}{(4n+3)^2}\quad\text{or}\quad\frac{3}{9\times75}>\frac{1}{(4n+3)^2}$$

or $(4n+3)^2>225=15^2.$

$\therefore$ $4n+3>15$ $\Rightarrow$ $n>3$. So, the least possible value of n is 4.

(iv) ${}^nC_3\left(\dfrac{2}{3}\right)^3\left(\dfrac{1}{3}\right)^{n-3}=\dfrac{160}{729}.$ $\quad\therefore\quad \dfrac{n(n-1)(n-2)}{6}\cdot\dfrac{8}{3^n}=\dfrac{160}{3^6}$

$\Rightarrow$ $n(n-1)(n-2)=40\times3^{n-5}=5\times4\times2\times3^{n-5}=6\times5\times4\times3^{n-6}.$

As the LHS is the product of three consecutive natural numbers, $n=6$.

36. Value $=\displaystyle\sum_{i=0}^{8}({}^8C_1\,{}^1C_i+{}^8C_2\,{}^2C_i+{}^8C_3\,{}^3C_i+\ldots+{}^8C_8\,{}^8C_i)$

$$={}^8C_1\sum_{i=0}^{8}{}^1C_i+{}^8C_2\cdot\sum_{i=0}^{8}{}^2C_i+{}^8C_3\cdot\sum_{i=0}^{8}{}^3C_i+\ldots+{}^8C_8\sum_{i=0}^{8}{}^8C_i$$

$$={}^8C_1({}^1C_0+{}^1C_1)+{}^8C_2({}^2C_0+{}^2C_1+{}^2C_2)+\ldots$$
$$+\,{}^8C_8({}^8C_0+{}^8C_1+\ldots+{}^8C_8)$$

$$={}^8C_1\cdot2+{}^8C_2\cdot2^2+{}^8C_3\cdot2^3+\ldots+{}^8C_8\cdot2^8$$
$$=(1+2)^8-1=3^8-1=6560$$

37. $OP=\sqrt{6^2+8^2}=10$; so, $AP=BP=\sqrt{10^2-r^2}$ and $\sin\theta=\dfrac{r}{10}.$

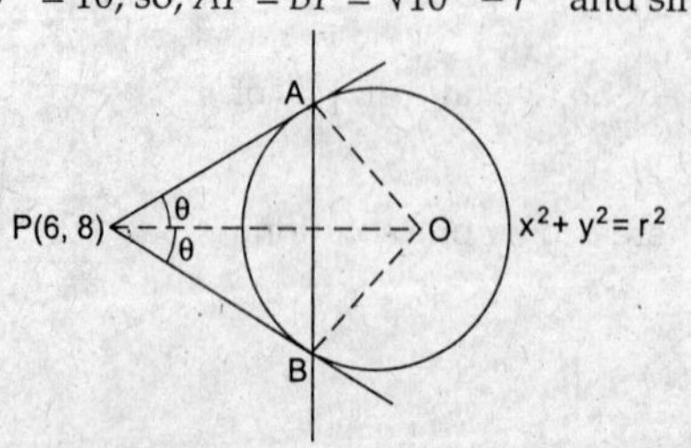

$$\therefore \quad \Delta = \frac{1}{2} AP \cdot BP \cdot \sin 2\theta = \frac{100 - r^2}{2} \cdot 2 \cdot \frac{r}{10} \cdot \sqrt{1 - \frac{r^2}{10^2}}$$

$$= \frac{r(100 - r^2)}{100} \cdot \sqrt{100 - r^2} = \frac{1}{100} r(100 - r^2)^{3/2}.$$

$$\therefore \quad \frac{d\Delta}{dr} = \frac{1}{100} \left\{ (100 - r^2)^{3/2} + r \cdot \frac{3}{2} (100 - r^2)^{1/2} \cdot (-2r) \right\}.$$

$$\therefore \quad \frac{d\Delta}{dr} = 0 \text{ if } (100 - r^2)^{3/2} - 3r^2 (100 - r^2)^{1/2} = 0$$

$$\Rightarrow \quad \sqrt{100 - r^2}\,(100 - r^2 - 3r^2) = 0.$$

$$\therefore \quad r = 10, 5.$$

When $r = 10$, the circle passes through P and Δ is minimum.

$\therefore$ for maximum area, $r = 5$.

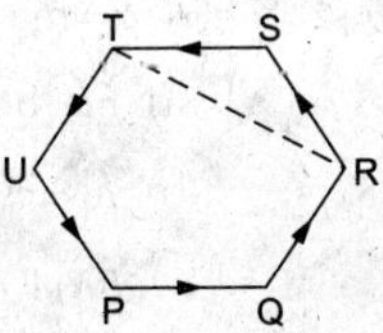

38. $\overrightarrow{RS} + \overrightarrow{ST} = \overrightarrow{RT} \nparallel \overrightarrow{PQ} \Rightarrow \overrightarrow{PQ} \times (\overrightarrow{RS} + \overrightarrow{ST}) \neq \overrightarrow{0}.$

So Statement-1 is True.

$\overrightarrow{PQ} \nparallel \overrightarrow{RS}$. So, $\overrightarrow{PQ} \times \overrightarrow{RS} \neq \overrightarrow{0}$.

$\overrightarrow{PQ} \parallel \overrightarrow{ST}$. So, $\overrightarrow{PQ} \times \overrightarrow{ST} = 0$.

So Statement-2 is False.

□

Practice Worksheet—2

One-Option-Correct-Questions

1. If $f(a+x) = f(2a)$ for all $x \in R$ and a is a positive constant then $f(x)$ is

(a) a periodic function of the period a

(b) a periodic function of indeterminate period

(c) a nonperiodic function

(d) a monotonic function

2. Let $f(x) = \max\{x, 2-x\}$ for all $x \in R$. Then

(a) $f(x)$ is not continuous everywhere

(b) $f(x)$ is differentiable everywhere

(c) $f(x)$ is continuous at $x = 1$ but not differentiable there

(d) $f(x)$ is neither continuous nor differentiable at $x = 1$

3. If $y = \int_0^{x^2} \sin\sqrt{t}\, dt$ then $\frac{dy}{dx}$ is equal to

(a) $\sin x$ (b) $\frac{\cos x}{2x}$

(c) $2x \sin x$ (d) none of these

4. The coefficient of $x^{2^{n+1}}$ in $\frac{1}{(1+x)(1+x^2)(1+x^4)\ldots(1+x^{2^n})}$, where $|x| < 1$, is

(a) -1 (b) 1

(c) 0 (d) none of these

5. If $f(x) = (x^2 - x + 2)(x^2 - 4)$ then the set $\{x \mid f(x) > 0, x \in R\}$ is equal to

(a) $(-\infty, -2)$ (b) $(2, +\infty)$

(c) $(-2, 2)$ (d) $(-\infty, -2) \cup (2, +\infty)$

6. If $\Delta_r = \begin{vmatrix} 2^n + 1 & 2^{n+1} & n(n+2) \\ 1 & 1 & -1 \\ {}^nC_r & 2^r & 2r+1 \end{vmatrix}$ then the value of $\sum_{r=0}^{n} \Delta_r$ is

(a) 2 (b) 0

(c) a function of n (d) 1

7. Let P be a point whose coordinates differ by unity and the point does not lie on any of the axes of reference. If the parabola $y^2 = 4x + 1$ passes through P then the ordinate of P may be

(a) 3 (b) –1

(c) 7 (d) 1

8. Four vertices of a square are $(\pm 3, \pm 3)$. If (a, a) is a point inside the square but falls outside the circle $x^2 + y^2 = 1$ then the set of possible values of a is

(a) $\left(\frac{1}{\sqrt{2}}, 3\right)$ (b) $\left(-3, -\frac{1}{\sqrt{2}}\right)$

(c) $\left(-3, -\frac{1}{\sqrt{2}}\right) \cup \left(\frac{1}{\sqrt{2}}, 3\right)$ (d) $(-3, 3)$

9. In a right-angled isosceles triangle the ratio of the circumradius and inradius is

(a) $2(\sqrt{2} + 1) : 1$ (b) $(\sqrt{2} + 1) : 1$

(c) $2 : 1$ (d) $\sqrt{2} : 1$

10. If $\log_{\sin x} \cos x = 1 + \log_{\sin x} \sin 2x$ then the values of x are (n being an integer)

(a) $n\pi + \frac{\pi}{4}$ (b) $2n\pi + \frac{\pi}{4}$

(c) $n\pi - \frac{\pi}{4}$ (d) $(2n+1)\pi - \frac{\pi}{4}$

11. If $\vec{a}$ and $\vec{b}$ are two perpendicular unit vectors then $\vec{a} \times (\vec{a} \times \vec{b})$ is equal to

(a) $-\vec{a}$ (b) $\vec{a}$

(c) $\vec{b}$ (d) $-\vec{b}$

12. If z is a nonreal complex number then which of the following may not be real?

(a) $z+\bar{z}$ (b) $z^3+\bar{z}^3$

(c) $(z-\bar{z})^2$ (d) z^2+z^{-2}

13. If the difference of the number of arrangements of 3 things from a certain number of dissimilar things and the number of selections of the same number of things from them exceeds 100 then the least number of dissimilar things is

(a) 6 (b) 7

(c) 5 (d) 8

14. If $z=\omega, \omega^2$, where ω is a nonreal complex cube root of unity, are two vertices of an equilateral triangle in the Argand plane then the third vertex may be represented by

(a) $z=2$ (b) $z=0$

(c) $z=-2$ (d) $z=-1$

15. The order and degree of the differential equation of the family of circles with centres on the *y*-axis are respectively

(a) 2, 1 (b) 2, 2

(c) 1, 2 (d) 1, 1

16. The value of $\int_0^{\pi/2} \frac{x}{\sqrt{1+\sin 2x}}\,dx$ is equal to

(a) 0 (b) $\frac{\pi}{4\sqrt{2}}\log_e(\sqrt{2}+1)$

(c) $\frac{\pi}{2\sqrt{2}}\log_e(\sqrt{2}+1)$ (d) $\frac{\pi}{2}\log_e 2$

17. Let $f(x)$ be a periodic function of the period 1 and $f\left(\frac{1}{2}\right)=\frac{1}{2}\cdot$ Let $\phi(x)=\int_0^x f(x+n)\,dx$ for all $n\in N$. Then $\phi'\left(\frac{3}{2}\right)$ is equal to

(a) $\frac{3}{2}$ (b) $\frac{1}{2}$

(c) 2 (d) 1

18. If $f(x)=(x-1)+(x+1)$ and $g(x)=f\{f(x)\}$ then $g'(3)$

(a) equals 1 (b) equals 0

(c) equals 3 (d) does not exist

19. The equation $\cos^4\theta+\sin^2\theta+\lambda=0$ admits of real solutions for θ if

(a) $\lambda \geq -1$ (b) $\lambda \leq -1$

(c) $-\frac{3}{2} \leq \lambda \leq -\frac{1}{2}$ (d) $-1 \leq \lambda \leq -\frac{3}{4}$

20. The value of $\tan^{-1}\left(\frac{\sin 2 - 1}{\cos 2}\right)$ is equal to

(a) $\frac{\pi}{2} - 1$ (b) $2 - \frac{\pi}{2}$

(c) $1 - \frac{\pi}{4}$ (d) $\frac{\pi}{4} - 1$

21. If $A + B = \frac{\pi}{3}$ and $\cos A + \cos B = 1$ then $\cos(A - B)$ is

(a) $-\frac{1}{3}$ (b) $\frac{1}{3}$ (c) $\frac{1}{4}$ (d) $-\frac{1}{4}$

22. If $f(x) = \int_0^x \frac{t^2}{e^t + e^{-t}}\, dt$ then $f(x)$ is

(a) even (b) odd

(c) periodic (d) none of these

23. The number of tangents to the curve $y = e^{|x|}$ at the point (0, 1) is

(a) 2 (b) 1

(c) 4 (d) 0

24. The function $f(x) = ax^3 - 3x^2 - 12x + 1$ is monotonic decreasing in the open interval (–1, 2). Then one value of a is

(a) 2 (b) 12

(c) 1 (d) none of these

25. The circles $x^2 + y^2 - 2\lambda x + c^2 = 0$ and $x^2 + y^2 + 2\lambda y = 0$ will have three common tangents if $|c : \lambda|$ is

(a) $\sqrt{\sqrt{2} - 1} : 1$ (b) $\sqrt{2\sqrt{2} - 2} : 1$

(c) $3 : 2$ (d) none of these

26. Let the function $f(x)$ be defined as follows

$$f(x) = \begin{cases} |x| & \text{for } -2 \leq x \leq 2 \\ [x] & \text{for } 2 < x \leq 4, \end{cases}$$

where [ ] represents the greatest integer function. Then

(a) $f(x)$ is continuous at $x = 2$ (b) $f(x)$ is differentiable at $x = 2$

(c) $f(x)$ is differentiable in $[-2, 4]$ (d) $f(x)$ is continuous in $[-2, 4]$

One-or-More-Options-Correct Questions

27. A plane parallel to $x + y + z = 3$ and at distance $\frac{4}{\sqrt{3}}$ from it has the equation

(a) $x - 2y - z = 0$ (b) $x + y + z + 1 = 0$

(c) $2x - y - z = \frac{3\sqrt{3} + 4}{\sqrt{3}}$ (d) $x + y + z = 7$

28. Let $(1 + x + x^2 + x^3 + x^4)^{10} = a_0 + a_1x + a_2x^2 + \ldots + a_{40}x^{40}$. Then

(a) $a_1 = 20$ (b) $a_2 = 110$

(c) $a_2 = 55$ (d) $a_1 + a_2 = 65$

29. In the ΔABC, $b : c = 2 : 1$ and $\sin(B - C) = \frac{3}{5}$ · Then

(a) ΔABC is right-angled (b) ΔABC is obtuse-angled

(c) $a : c = 3 : 1$ (d) $a : c = \sqrt{5} : 1$

30. If $f(x) = \log \frac{1 + x}{1 - x}$ then

(a) $f(x)$ is odd (b) $f(x)$ is even

(c) $f(x_1) + f(x_2) = f\left(\frac{x_1 + x_2}{1 + x_1x_2}\right)$ (d) $f(x_1) \cdot f(x_2) = f(x_1 + x_2)$

Comprehension-Type Questions

It is known that $\frac{x^2}{a^2} - \frac{y^2}{b^2} = -1$ is the equation of the conjugate hyperbola of the hyperbola $\frac{x^2}{a^2} - \frac{y^2}{b^2} = 1$. Also, the pair of lines $\frac{x^2}{a^2} - \frac{y^2}{b^2} = 0$ are the asymptotes of the hyperbola $\frac{x^2}{a^2} - \frac{y^2}{b^2} = 1$.

Let the equation of a hyperbola be $x^2 - 3y^2 - 2x - 8 = 0$.

31. The equation of its conjugate hyperbola is

(a) $x^2 - 3y^2 - 2x + 8 = 0$ (b) $3x^2 - y^2 - 2y - 8 = 0$

(c) $x^2 - 3y^2 - 2x + 10 = 0$ (d) $x^2 - 3y^2 = -8$

32. The equation of the asymptotes of the hyperbola is

(a) $x^2 - 3y^2 - 2x = 0$ (b) $x^2 - 3y^2 = 0$

(c) $3x^2 - y^2 - 2y = 0$ (d) $x^2 - 3y^2 - 2x + 1 = 0$

33. If $(a-1, 2)$ is a point interior to the hyperbola but exterior to its conjugate hyperbola then the number of positive integral values of a is

(a) 3 (b) 2

(c) 1 (d) 4

Matching Questions

One or more may match with the same.

34. (i) In an acute-angled ΔABC, the minimum value of $\sec A + \sec B + \sec C$ is — (a) 3

(ii) $\lim\limits_{\substack{x \to 0 \\ y \to 0}} \dfrac{x^2 + \sin y}{y^2 + \sin x^2}$ when $(x, y) \to (0, 0)$ along the curve $y = x^2$ equals — (b) 1

(iii) If $\vec{c} = \dfrac{\vec{a} \times \vec{b} - \vec{a} \times (\vec{a} \times \vec{b})}{(\vec{a} \times \vec{b})^2}$ then $[\vec{c}\ \vec{a}\ \vec{b}]$ equals — (c) 6

(iv) A plane exists which is perpendicular to the line $\dfrac{x}{3} = \dfrac{y-1}{5} = \dfrac{z+1}{7}$ and passes through the line $\dfrac{x+3}{1} = \dfrac{y-1}{-2} = \dfrac{z+1}{k}$. Then k is — (d) 2

35. (i) If $f(x) = \sin x + \cos x$ and $g(x) = x^2 - 1$ then $g\{f(x)\}$ is invertible in the domain — (a) $\left[0, \dfrac{\pi}{2}\right]$

(ii) If $x^2 + 2ax + 10 > 3a$ for all $x \in R$ then a belongs to — (b) $(-5, 2)$

(iii) Let $\phi(x) = \int_0^x \cos^3 x \, dx$. Then $\phi(x)$ is monotonic increasing in — (c) $\left[-\frac{\pi}{4}, \frac{\pi}{4}\right]$

(iv) The vertex of the parabola $y^2 = 2(2y + x)$ is — (d) $(-2, 2)$

Complete the following statements.

36. The number of ways in which 25 identical things can be distributed among five persons when each gets odd number of things is

37. A card from a pack of 52 cards is lost. From the remaining cards of the pack two cards are drawn at random and are found to be spades. The probability of the missing card being a spade is p. Then $50p$ is equal to

Assertion-Reason Type

38. Let $z_1 = r_1 e^{i\theta_1}$ and $z_2 = r_2 e^{i\theta_2}$, where $r_1 > 1, r_2 > 1$.

STATEMENT-1: $\left|1 - \bar{z}_1 z_2\right| > \left|z_1 - z_2\right|$.

<u>because</u>

STATEMENT-2: $\left|z_1 - z_2\right| < |z_1| + |z_2|$.

(a) Statement-1 is True, Statement-2 is True; Statement-2 is a correct explanation for Statement-1

(b) Statement-1 is True, Statement-2 is True; Statement-2 is *Not* a correct explanation for Statement-1

(c) Statement-1 is True, Statement-2 is False

(d) Statement-1 is False, Statement-2 is True

Answers

1. b	**2.** c	**3.** c	**4.** b	**5.** d
6. b	**7.** a	**8.** c	**9.** b	**10.** b
11. d	**12.** d	**13.** b	**14.** c	**15.** a
16. c	**17.** b	**18.** d	**19.** d	**20.** c
21. a	**22.** b	**23.** a	**24.** a	**25.** b
26. a	**27.** b, d	**28.** c, d	**29.** a, d	**30.** a, c
31. c	**32.** d	**33.** b		

34. (i) ↔ c (ii) ↔ d (iii) ↔ b (iv) ↔ b

35. (i) ↔ c (ii) ↔ b (iii) ↔ a (iv) ↔ d

36. 1001	**37.** 11	**38.** b

Hints and Solutions to Selected Questions

1. $f(a+x)=f(2a) \Rightarrow f(a+x)=f(a+\overline{a+x})=f\left(a+\overline{x-\frac{a}{2}}\right)=\ldots=f(2a)$.

Also, $f(a+x)=f(2a) \Rightarrow f(a+\overline{x-a})=f(2a)$. So, $f(a+x)=f(x)$.

So, $f(x)$ is periodic but period is indeterminate.

Also $f(a+x) \ngtr f(x)$ or $f(a+x) \nless f(x)$. So, $f(x)$ is not monotonic.

2. $x \geq 2-x \Rightarrow x \geq 1$ and $x < 2-x \Rightarrow x < 1$.

$$\therefore \quad f(x)=\begin{cases} 2-x & \text{for } x<1 \\ x & \text{for } x \geq 1. \end{cases}$$

$\therefore \quad f(1-0)=f(1+0)=f(1)$. So, $f(x)$ is continuous at $x=1$.

As polynomial functions are continuous and differentiable everywhere, $f(x)$ is continuous in $(-\infty, 1)$ and $[1, +\infty)$.

As $f'(1-0)=-1, f'(1+0)=1$ we get $f(x)$ is not differentiable at $x=1$.

3. $$\frac{dy}{d(x^2)}=\sin\sqrt{t}\Big]_{t=x^2}=\sin x.$$

$$\therefore \quad \frac{dy}{dx}=\frac{dy}{d(x^2)}\cdot\frac{d(x^2)}{dx}=\sin x \cdot 2x.$$

4. $$\frac{1}{(1+x)(1+x^2)(1+x^4)\ldots(1+x^{2^n})}$$

$$=\frac{1-x}{(1-x)(1+x)(1+x^2)\ldots(1+x^{2^n})}=\frac{1-x}{1-x^{2^{n+1}}}=(1-x)(1-x^{2^{n+1}})^{-1}$$

$=(1-x)\{1+x^{2^{n+1}}+(x^{2^{n+1}})^2+\ldots\}$. So, the coefficient of $x^{2^{n+1}}=1$.

5. Here, $(x^2-x+2)(x^2-4)>0$.

But $x^2-x+2>0$ for all $x \in R$ $\quad (\because D<0, a>0)$.

$\therefore \quad x^2-4>0 \Rightarrow x<-2$ or $x>2$.

6. $\sum_{r=0}^{n} \Delta_r = \begin{vmatrix} 2^n+1 & 2^{n+1} & n(n+2) \\ 1 & 1 & -1 \\ \sum_{r=0}^{n} {}^nC_r & \sum_{r=0}^{n} 2^r & \sum_{r=0}^{n} (2r+1) \end{vmatrix} = \begin{vmatrix} 2^n+1 & 2^{n+1} & n(n+2) \\ 1 & 1 & -1 \\ 2^n & 2^{n+1}-1 & (n+1)^2 \end{vmatrix}$

$= \begin{vmatrix} 0 & 0 & 0 \\ 1 & 1 & -1 \\ 2^n & 2^{n+1}-1 & (n+1)^2 \end{vmatrix}$ $(R_1 \to R_1 - R_2 - R_3)$

$= 0.$

7. $P = (\alpha, \alpha+1)$, where $\alpha \neq 0, -1$, or $P = (\alpha, \alpha - 1)$, where $\alpha \neq 0, 1$.

$(\alpha, \alpha+1)$ is on $y^2 = 4x + 1$ $\Rightarrow (\alpha+1)^2 = 4\alpha + 1$

$\Rightarrow \alpha^2 - 2\alpha = 0 \Rightarrow \alpha = 2$ $(\because \alpha \neq 0)$.

$\therefore$ ordinate of $P = 3$.

$(\alpha, \alpha-1)$ is on $y^2 = 4x + 1$ $\Rightarrow (\alpha-1)^2 = 4\alpha + 1$

$\Rightarrow \alpha^2 - 6\alpha = 0 \Rightarrow \alpha = 6$ $(\because \alpha \neq 0)$.

$\therefore$ ordinate of $P = 5$.

8. (a, a) is inside the square $\Rightarrow |a| < 3 \Rightarrow -3 < a < 3$.

(a, a) is outside the circle $x^2 + y^2 - 1 = 0$ $\Rightarrow a^2 + a^2 - 1 > 0$.

$\therefore 2a^2 - 1 > 0 \Rightarrow a < \frac{-1}{\sqrt{2}}$ or $a > \frac{1}{\sqrt{2}}$.

Thus $a \in \left(-3, -\frac{1}{\sqrt{2}}\right) \cup \left(\frac{1}{\sqrt{2}}, 3\right)$.

9. 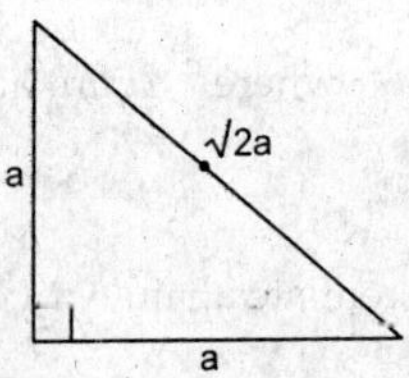

Here, $R = \frac{\sqrt{2}a}{2} = \frac{a}{\sqrt{2}}$.

$r = \frac{\Delta}{s} = \frac{\frac{1}{2}a^2}{\frac{1}{2}(a + a + \sqrt{2}a)} = \frac{a}{2+\sqrt{2}}$

$\therefore \frac{R}{r} = \frac{a}{\sqrt{2}} \cdot \frac{2+\sqrt{2}}{a} = \sqrt{2} + 1$.

10. $\log_{\sin x} \cos x = 1 + \log_{\sin x}(2 \sin x \cdot \cos x)$

or $\log_{\sin x} \cos x = 1 + \log_{\sin x}(2 \cos x) + \log_{\sin x} \sin x$

or $2 + \log_{\sin x} 2 = 0 \Rightarrow 2 = (\sin x)^{-2} \Rightarrow \operatorname{cosec} x = \pm\sqrt{2}$.

But $\cos x > 0$, $\sin 2x > 0$ and $\sin x > 0$, $\sin x \neq 1$.

Hence, $x \in$ the first quadrant.

$$\therefore \quad \text{cosec}\, x = \sqrt{2} \quad \text{or} \quad \sin x = \frac{1}{\sqrt{2}} \quad \therefore \quad x = 2n\pi + \frac{\pi}{4}.$$

11. $\vec{a} \times (\vec{a} \times \vec{b}) = (\vec{a} \cdot \vec{b})\,\vec{a} - (\vec{a} \cdot \vec{a})\,\vec{b} = 0\vec{a} - 1\vec{b} = -\vec{b}.$

12. Let $z = x + iy$. Then $\bar{z} = x - iy$.

$$\therefore \quad z + \bar{z} = 2x, \; (z - \bar{z})^2 = (2iy)^2 = -4y^2$$

$$z^3 + \bar{z}^3 = (x + iy)^3 + (x - iy)^3 = 2(x^3 + 3xi^2y^2) = 2(x^3 - 3xy^2)$$

$$z^2 + z^{-2} = (x + iy)^2 + \frac{1}{(x + iy)^2}$$

$$= x^2 + 2ixy - y^2 + \frac{1}{(x^2 + y^2)^2}(x^2 - 2ixy - y^2).$$

13. Here, ${}^nP_3 - {}^nC_3 > 100$ or $\dfrac{n!}{(n-3)!} - \dfrac{n!}{3!\,(n-3)!} > 100$

or $\dfrac{5}{6} \cdot n(n-1)(n-2) > 100$

or $n(n-1)(n-2) > 120 = 6 \cdot 5 \cdot 4$.

$\therefore$ by trial, $n = 7, 8, \ldots$.

14. Clearly, we have to find it for real z. Let $z = x$. Then

$$|x - \omega| = |x - \omega^2| = |\omega - \omega^2|$$

$$\Rightarrow \quad \left(x + \frac{1}{2}\right)^2 + \frac{3}{4} = \left|\frac{-1 + \sqrt{3}i}{2} - \frac{-1 - \sqrt{3}i}{2}\right|^2 = 3 \;\Rightarrow\; x + \frac{1}{2} = \pm\frac{3}{2}$$

$$\Rightarrow \quad x = 1, -2.$$

15. The equation of the family is $x^2 + (y - c)^2 = \alpha^2$, where c, α are arbitrary constants. Differentiating w.r.t. x,

$$2x + 2y\frac{dy}{dx} - 2c\frac{dy}{dx} = 0 \quad \text{or} \quad \frac{x + y\dfrac{dy}{dx}}{\dfrac{dy}{dx}} = c.$$ Differentiate again w.r.t. x.

16. $$I = \int_0^{\pi/2} \frac{x}{\sqrt{1 + \sin 2x}}\,dx = \int_0^{\pi/2} \frac{x}{|\sin x + \cos x|}\,dx = \int_0^{\pi/2} \frac{x}{\sin x + \cos x}\,dx$$

$$= \int_0^{\pi/2} \frac{\frac{\pi}{2} - x}{\cos x + \sin x}\,dx = \frac{\pi}{2}\int_0^{\pi/2} \frac{dx}{\cos x + \sin x} - I.$$

$$\therefore \quad 2I = \frac{\pi}{2}\int_0^{\pi/2} \frac{dx}{\sqrt{2}\cos\left(x-\frac{\pi}{4}\right)} = \frac{\pi}{2\sqrt{2}}\int_0^{\pi/2} \sec\left(x-\frac{\pi}{4}\right)dx.$$

$$\therefore \quad I = \frac{\pi}{4\sqrt{2}}\cdot\left[\log\left\{\sec\left(x-\frac{\pi}{4}\right)+\tan\left(x-\frac{\pi}{4}\right)\right\}\right]_0^{\pi/2}$$

$$= \frac{\pi}{4\sqrt{2}}\left\{\log\left(\sec\frac{\pi}{4}+\tan\frac{\pi}{4}\right)-\log\left(\sec\frac{-\pi}{4}+\tan\frac{-\pi}{4}\right)\right\}$$

$$= \frac{\pi}{4\sqrt{2}}\{\log(\sqrt{2}+1)-\log(\sqrt{2}-1)\} = \frac{\pi}{4\sqrt{2}}\log\frac{\sqrt{2}+1}{\sqrt{2}-1}$$

$$= \frac{\pi}{4\sqrt{2}}\log(\sqrt{2}+1)^2 = \frac{\pi}{2\sqrt{2}}\log(\sqrt{2}+1).$$

17. Here, $\phi'(x) = f(x+n) = f(x+\overline{n-1}+1) = f(x+\overline{n-1}) = \ldots = f(x+1) = f(x)$.

$$\therefore \quad \phi'\left(\frac{3}{2}\right) = f\left(\frac{3}{2}\right) = f\left(\frac{1}{2}+1\right) = f\left(\frac{1}{2}\right) = \frac{1}{2}.$$

18. $f(x) = [x-1]+[x+1] = [x]-1+[x]+1 = 2[x]$.

Now, $\dfrac{g(3+h)-g(3)}{h} = \dfrac{f\{f(3+h)\}-f\{f(3)\}}{h} = \dfrac{f(6)-f(6)}{h} = \dfrac{0}{h}$.

$$\therefore \quad g'(3+0) = \lim_{h\to 0}\frac{0}{h} = \lim_{h\to 0} 0 = 0.$$

$$\frac{g(3-h)-g(3)}{-h} = \frac{f\{f(3-h)\}-f\{f(3)\}}{-h} = \frac{f(4)-f(6)}{-h} = \frac{8-12}{-h} \to \infty.$$

$\therefore \quad g'(3)$ does not exist.

19. $\cos^4\theta - \cos^2\theta + 1 + \lambda = 0$. $\quad \therefore \quad \cos^2\theta = \dfrac{1 \pm \sqrt{1-4(1+\lambda)}}{2} = \dfrac{1 \pm \sqrt{-4\lambda-3}}{2}$

As $0 \le \cos^2\theta \le 1$, we get $0 \le \dfrac{1\pm\sqrt{-4\lambda-3}}{2} \le 1$ or $0 \le 1 \pm \sqrt{-4\lambda-3} \le 2$.

Now, $1 \pm \sqrt{-4\lambda-3} \le 2 \quad \Rightarrow \pm\sqrt{-4\lambda-3} \le 1$.

$\therefore \quad \sqrt{-4\lambda-3} \le 1 \quad$ or $\quad -4\lambda-3 \le 1 \quad$ or $\quad -4\lambda \le 4. \quad \therefore \lambda \ge -1$.

Also, $-4\lambda-3 \ge 0 \quad$ or $\quad -4\lambda \ge 3 \quad$ or $\quad \lambda \le \dfrac{-3}{4}$.

$$\therefore \quad -1 \le \lambda \le \frac{-3}{4}.$$

20. $\dfrac{\sin 2-1}{\cos 2}=-\dfrac{(\cos 1-\sin 1)^2}{\cos^2 1-\sin^2 1}=-\dfrac{\cos 1-\sin 1}{\cos 1+\sin 1}=-\dfrac{1-\tan 1}{1+\tan 1}$

$$=-\tan\left(\frac{\pi}{4}-1\right)=\tan\left(1-\frac{\pi}{4}\right) \qquad \left(\because\ 1-\frac{\pi}{4}\in \text{ the first quadrant}\right)$$

$$\therefore \quad \tan^{-1}\frac{\sin 2-1}{\cos 2}=\tan^{-1}\tan\left(1-\frac{\pi}{4}\right)=1-\frac{\pi}{4}.$$

21. $\cos A+\cos B=1 \quad \Rightarrow \ 2\cos\dfrac{A+B}{2}\cdot\cos\dfrac{A-B}{2}=1$

or $\quad 2\cos\dfrac{\pi}{6}\cdot\cos\dfrac{A-B}{2}=1 \quad \therefore\ \cos\dfrac{A-B}{2}=\dfrac{1}{\sqrt{3}}$

$\therefore \quad \cos(A-B)=2\cos^2\dfrac{A-B}{2}-1=2\cdot\dfrac{1}{3}-1=-\dfrac{1}{3}.$

22. $f(-x)=\displaystyle\int_0^{-x}\frac{t^2}{e^t+e^{-t}}\,dt=\int_0^{x}\frac{(-z)^2}{e^{-z}+e^{z}}\,d(-z)=-\int_0^{x}\frac{z^2}{e^{-z}+e^{z}}\,dz$

$$=-\int_0^{x}\frac{t^2}{e^{-t}+e^{t}}\,dt=-f(x).$$

23. $\left.\dfrac{dy}{dx}\right]_{(0,\,1)}$ has two values because $f'(0+h)$ and $f'(0-h)$ are different.

$f'(0+h)=\lim\limits_{h\to 0}\dfrac{f(0+h)-f(0)}{h}=\lim\limits_{h\to 0}\dfrac{e^h-1}{h}=1$ and

$f'(0-h)=\lim\limits_{h\to 0}\dfrac{f(0-h)-f(0)}{-h}=\lim\limits_{h\to 0}\dfrac{e^h-1}{-h}=-1.$

24. $f'(x)=3ax^2-6x-12<0$ for $-1<x<2$.

The corresponding equation is $3ax^2-6x-12=0$

or $\quad ax^2-2x-4=0$ or $\quad x=\dfrac{2\pm\sqrt{4+16a}}{2a}=\dfrac{1\pm\sqrt{1+4a}}{a}.$

If $a>0$ then $f(x)$ is m.d. in $\left(\dfrac{1-\sqrt{1+4a}}{a},\ \dfrac{1+\sqrt{1+4a}}{a}\right).$

For $a=2$, the interval is $(-1, 2)$.

25.

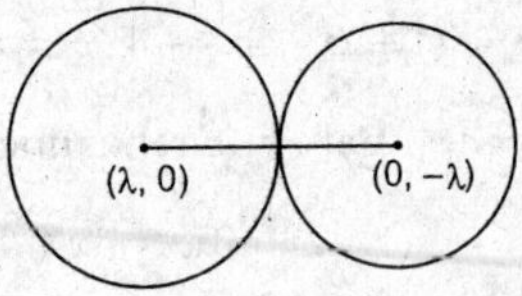

For three common tangents, the circles should touch externally.

$\therefore \quad r_1 + r_2 = d \quad$ or $\quad \sqrt{\lambda^2 - c^2} + \sqrt{\lambda^2} = \sqrt{\lambda^2 + \lambda^2}$

or $\quad \sqrt{\lambda^2 - c^2} = (\sqrt{2} - 1)\lambda \quad$ or $\quad \lambda^2 - c^2 = (\sqrt{2} - 1)^2 \lambda^2$

or $\quad \{1 - (\sqrt{2} - 1)^2\}\lambda^2 = c^2 \quad$ or $\quad \dfrac{c^2}{\lambda^2} = 2\sqrt{2} - 2. \quad \therefore \quad \left|\dfrac{c}{\lambda}\right| = \sqrt{2\sqrt{2} - 2}.$

26. $|x|$ is continuous everywhere in its domain but not differentiable at $x = 0$.

$[x]$ is discontinuous at $x = 3$ but continuous at other points of $(2, 4)$

At the turning point of definition $x = 2$,

$$\lim_{h \to 0} f(2 - h) = \lim_{h \to 0} |2 - h| = 2,$$

$$\lim_{h \to 0} f(2 + h) = \lim_{h \to 0} [2 + h] = 2 \quad \text{and} \quad f(2) = |2| = 2.$$

$\therefore \quad f(x)$ is continuous at $x = 2$.

$$\lim_{h \to 0} \frac{f(2 + h) - f(2)}{h} = \lim_{h \to 0} \frac{[2 + h] - 2}{h} = \lim_{h \to 0} \frac{2 - 2}{h} = 0 \quad \text{and}$$

$$\lim_{h \to 0} \frac{f(2 - h) - f(2)}{-h} = \lim_{h \to 0} \frac{|2 - h| - 2}{-h} = \lim_{h \to 0} \frac{2 - h - 2}{-h} = \lim_{h \to 0} \frac{-h}{-h} = 1.$$

So, $f'(2)$ does not exist.

One-or-More-Options-Correct Questions

27. A plane parallel to $x + y + z = 3$ is $x + y + z = k$.

A point on $x + y + z = 3$ is $(1, 1, 1)$. So, $\left|\dfrac{1 + 1 + 1 - k}{\sqrt{3}}\right| = \dfrac{4}{\sqrt{3}}$.

or $\quad |3 - k| = 4 \quad$ or $\quad 3 - k = \pm 4 \quad$ or $\quad k = -1, 7$

$\therefore \quad$ the planes are $x + y + z = -1$ and $x + y + z = 7$.

28. Differentiating,

$$10(1+x+x^2+x^3+x^4)^9(1+2x+3x^2+4x^3)=a_1+2a_2x+3a_3x^2+\dots.$$

Putting $x=0$ we obtain $10=a_1$. Again differentiating,

$$90(1+x+x^2+x^3+x^4)^8\cdot(1+2x+3x^2+4x^3)^2$$
$$+10(1+x+x^2+x^3+x^4)^9\cdot(2+6x+12x^2)=2a_2+6a_3x+\dots$$

Putting $x=0$, $90+10\cdot 2=2a_2 \quad \Rightarrow a_2=55 \quad \therefore\ a_1+a_2=65.$

29. $\sin(B-C)=\dfrac{3}{5} \quad \Rightarrow \cos(B-C)=\dfrac{4}{5}$ or $\dfrac{1-t^2}{1+t^2}=\dfrac{4}{5}$, where $t=\tan\dfrac{B-C}{2}$.

$$\therefore\quad t^2=\frac{1}{9} \quad \therefore\ \tan\frac{B-C}{2}=\pm\frac{1}{3}=\frac{b-c}{b+c}\cot\frac{A}{2}=\frac{1}{3}\cot\frac{A}{2}.$$

$$\therefore\quad \cot\frac{A}{2}=1,-1\text{ (absurd)} \quad \therefore\ \frac{A}{2}=\frac{\pi}{4} \quad \therefore\ A=\frac{\pi}{2}.$$

$$\therefore\quad a=\sqrt{(2k)^2+k^2}=\sqrt{5}k \quad \therefore\ a:c=\sqrt{5}:1$$

30. $f(-x)=\log\dfrac{1-x}{1+x}=-\log\dfrac{1+x}{1-x}=-f(x)$. So, $f(x)$ is odd.

$$f(x_1)+f(x_2)=\log\left(\frac{1+x_1}{1-x_1}\cdot\frac{1+x_2}{1-x_2}\right)=\log\frac{1+x_1+x_2+x_1x_2}{1-x_1-x_2+x_1x_2}=\log\frac{1+\dfrac{x_1+x_2}{1+x_1x_2}}{1-\dfrac{x_1+x_2}{1+x_1x_2}}$$

$$=f\left(\frac{x_1+x_2}{1-x_1x_2}\right).$$

31. $x^2-2x+1-3y^2=9$ or $\dfrac{(x-1)^2}{3^2}-\dfrac{y^2}{3}=1.$

Putting $x-1=X$ and $y=Y$, the equation becomes $\dfrac{X^2}{3^2}-\dfrac{Y^2}{(\sqrt{3})^2}=1.$

$\therefore$ the equation of the conjugate hyperbola is $\dfrac{X^2}{3^2}-\dfrac{Y^2}{3}=-1$

or $\dfrac{(x-1)^2}{9}-\dfrac{y^2}{3}+1=0$ or $x^2-2x+1-3y^2+9=0$

or $x^2-3y^2-2x+10=0.$

32. Equation of asymptotes is $\frac{X^2}{3^2} - \frac{Y^2}{3} = 0$

or $\quad \frac{(x-1)^2}{9} - \frac{y^2}{3} = 0 \quad$ or $\quad (x-1)^2 - 3y^2 = 0$

or $\quad x^2 - 3y^2 - 2x + 1 = 0.$

33. $(a-1, 2)$ is interior to the hyperbola $x^2 - 3y^2 - 2x - 8 = 0$. So,

$$(a-1)^2 - 3 \cdot 2^2 - 2 \cdot 2 - 8 < 0 \quad \text{or} \quad a^2 - 2a - 23 < 0. \qquad \dots (1)$$

$(a-1, 2)$ is exterior to the conjugate hyperbola $x^2 - 3y^2 - 2x + 10 = 0$. So,

$$(a-1)^2 - 3 \cdot 2^2 - 2 \cdot 2 + 10 > 0 \quad \text{or} \quad a^2 - 2a - 5 > 0. \qquad \dots (2)$$

$(1) \Rightarrow (a-1)^2 < 24 \quad \Rightarrow -2\sqrt{6} < a - 1 < 2\sqrt{6} \quad \Rightarrow 1 - 2\sqrt{6} < a < 1 + 2\sqrt{6}$

$(2) \Rightarrow (a-1)^2 > 6 \quad \Rightarrow a - 1 > \sqrt{6} \quad \text{or} \quad a - 1 < -\sqrt{6}$

$\qquad \Rightarrow a < 1 - \sqrt{6} \quad \text{or} \quad a > 1 + \sqrt{6}$

$1 - 2\sqrt{6} \qquad 1 - \sqrt{6} \qquad 1 + \sqrt{6} \qquad 1 + 2\sqrt{6}$

$-3 \cdot xx \qquad -1 \cdot xx \qquad 3 \cdot xx \qquad 5 \cdot xx$

$\therefore \quad a = 4, 5.$

34. (i) AM $\geq$ GM $\Rightarrow \frac{\sec A + \sec B + \sec C}{3} \geq \sqrt[3]{\sec A \cdot \sec B \cdot \sec C} \qquad \dots (1)$

and $\cos A \cdot \cos B \cdot \cos C = \frac{1}{2} \cos A\{\cos(B-C) + \cos(B+C)\}$

$$= \frac{1}{2} \cos A \cos(B-C) - \frac{1}{2} \cos^2 A$$

$$\leq \frac{1}{2} \cos A - \frac{1}{2} \cos^2 A.$$

$$\therefore \quad \cos A \cos B \cos C \leq \frac{1}{2}\left\{\frac{1}{4} - \left(\cos^2 A - \cos A + \frac{1}{4}\right)\right\}$$

$$= \frac{1}{2}\left\{\frac{1}{4} - \left(\cos A - \frac{1}{2}\right)^2\right\} \leq \frac{1}{8}.$$

$\therefore \quad \sec A \sec B \sec C \geq 8.$

$\therefore \quad (1) \Rightarrow \sec A + \sec B + \sec C \geq 3 \cdot \sqrt[3]{\sec A \sec A \sec C} \geq 3 \cdot \sqrt[3]{8} = 6$

(ii) $\text{Limit} = \lim_{y \to 0} \frac{y + \sin y}{y^2 + \sin y} = \lim_{y \to 0} \frac{1 + \frac{\sin y}{y}}{y + \frac{\sin y}{y}} = \frac{1+1}{0+1} = 2.$

(iii) $\vec{c}\cdot(\vec{a}\times\vec{b}) = \dfrac{(\vec{a}\times\vec{b})\cdot(\vec{a}\times\vec{b}) - \vec{a}\times(\vec{a}\times\vec{b})\cdot(\vec{a}\times\vec{b})}{(\vec{a}\times\vec{b})^2} = \dfrac{(\vec{a}\times\vec{b})^2}{(\vec{a}\times\vec{b})^2} = 1.$

(iv) $\dfrac{x+3}{1} = \dfrac{y-1}{-2} = \dfrac{z+1}{k} \quad \Rightarrow \; 2x + y + 5 = 0$ and $kx - z + 3k - 1 = 0.$

$\therefore$ any plane passing through the line is

$$kx - z + 3k - 1 + \lambda(2x + y + 5) = 0$$

or $(k + 2\lambda)x + \lambda y - z + 3k + 5\lambda - 1 = 0$, which is perpendicular to the line $\dfrac{x}{3} = \dfrac{y-1}{5} = \dfrac{z+1}{7}$.

$\therefore \quad \dfrac{k+2\lambda}{3} = \dfrac{\lambda}{5} = \dfrac{-1}{7} = \dfrac{k}{3-10} \quad \Rightarrow \; k = 1.$

35. (i) $g\{f(x)\} = (\sin x + \cos x)^2 - 1 = \sin 2x$. So, $2x = \sin^{-1} g\{f(x)\} \in \left[-\dfrac{\pi}{2}, \dfrac{\pi}{2}\right]$

$\therefore$ it is invertible if $x \in \left[-\dfrac{\pi}{4}, \dfrac{\pi}{4}\right]$.

(ii) $x^2 + 2ax + 10 - 3a > 0$ for all $x \in R \;\Rightarrow\; D < 0$

($\because$ coefficient of $x^2 = 1 > 0$).

$\therefore \quad 4a^2 - 4(10 - 3a) < 0 \quad$ or $\quad a^2 + 3a - 10 < 0$

$\Rightarrow \quad -5 < a < 2$. So, $a \in (-5, 2)$.

(iii) $\phi'(x) = \cos^3 x > 0$ in $\left[0, \dfrac{\pi}{2}\right]$.

(iv) Here, $y^2 - 4y + 4 = 2x + 4 \quad$ or $\quad (y-2)^2 = 2(x+2)$

Putting $Y = y - 2$ and $X = x + 2$, the equation becomes $Y^2 = 2X$.

$\therefore \quad$ vertex $= (0, 0)_{X, Y} = (-2, 2)$.

36. Take the equation $x_1 + x_2 + x_3 + x_4 + x_5 = 25$, where x_1, x_2, x_3, x_4, x_5 are odd integers.

The required number of ways to distribute

$= $ coefficient of x^{25} in $(x + x^3 + x^5 + \ldots)^5$

$= $ coefficient of x^{20} in $(1 - x^2)^{-5}$

$= $ coefficient of x^{20} in ${}^4C_0 + {}^5C_1 x^2 + {}^6C_2 (x^2)^2 + \ldots$

$= {}^{14}C_{10} = 1001.$

37. Let

E_s = event of the lost card being spade and

E = event of two cards drawn being spades.

$$\therefore \quad P(E_s) = \frac{^{13}C_1}{^{52}C_1} = \frac{1}{4}, \quad P(\overline{E}_s) = \frac{3}{4},$$

$$P\left(\frac{E}{E_s}\right) = \frac{^{12}C_2}{^{51}C_2} = \frac{22}{425}, \quad P\left(\frac{E}{\overline{E}_s}\right) = \frac{^{13}C_2}{^{51}C_2} = \frac{78}{1275}.$$

Now, $$P\left(\frac{E_s}{E}\right) = \frac{P(E_s)\cdot P\left(\frac{E}{E_s}\right)}{P(E_s)\cdot P\left(\frac{E}{E_s}\right) + P(\overline{E}_s)\cdot P\left(\frac{E}{\overline{E}_s}\right)}$$

$$\therefore \quad p = \frac{\frac{1}{4}\cdot\frac{22}{425}}{\frac{1}{4}\cdot\frac{22}{425} + \frac{3}{4}\cdot\frac{78}{1275}} = \frac{22\times 3}{22\times 3 + 3\times 78} = \frac{22\times 3}{3\times 100} = \frac{11}{50}.$$

$\therefore \quad 50p = 11.$

38. Statement-2 is True because

$$|z_1 - z_2| = |z_1 + (-z_2)| < |z_1| + |-z_2| = |z_1| + |z_2|$$

$$|1 - \overline{z}_1 z_2|^2 - |z_1 - z_2|^2 = (1 - \overline{z}_1 z_2)(1 - z_1\overline{z}_2) - (z_1 - z_2)(\overline{z}_1 - \overline{z}_2)$$

$$= 1 - z_1\overline{z}_2 - \overline{z}_1 z_2 + z_1\overline{z}_1 z_2\overline{z}_2 - z_1\overline{z}_1 + z_1\overline{z}_2 + \overline{z}_1 z_2 - z_2\overline{z}_2$$

$$= 1 + |z_1|^2|z_2|^2 - |z_1|^2 - |z_2|^2 = 1 + r_1^2 r_2^2 - r_1^2 - r_2^2$$

$$= (1 - r_1^2)(1 - r_2^2) = (-)(-) = +$$

$$\therefore \quad |1 - \overline{z}_1 z_2|^2 > |z_1 - z_2|^2$$

$\therefore \quad |1 - \overline{z}_1 z_2| > |z_1 - z_2|$. So, Statement-1 is True and Statement-2 has not been used to establish Statement-1.

□

Practice Worksheet—3

One-Option-Correct Questions

1. If the polynomial function $f(x) = ax^3 + bx^2 + cx + d$ has extrema at $x = \alpha$ and $x = 2\alpha$ then

   (a) a, b, c are in AP
   (b) a, b, c are in GP
   (c) $6a, 4b, 9c$ are in AP
   (d) $6a, 4b, 9c$ are in GP

2. If $f(x) = |x| + |\cos x|$ then

   (a) $f'\left(\frac{\pi}{2}\right) = 2$
   (b) $f'\left(\frac{\pi}{2}\right) = 0$
   (c) $f'\left(\frac{\pi}{2}\right) = 1$
   (d) $f'\left(\frac{\pi}{2}\right)$ does not exist

3. If the perpendicular bisectors of the sides AB and AC of the ΔABC meet at the middle point of the median AD then $\cos B$ is equal to

   (a) $\frac{a}{2c}$
   (b) $\frac{b}{2a}$
   (c) $\frac{a}{b}$
   (d) $\frac{2b}{a}$

4. If $A = \begin{bmatrix} a & b \\ b & a \end{bmatrix}$ and $A^2 = \begin{bmatrix} \alpha & \beta \\ \beta & \alpha \end{bmatrix}$ then

   (a) $\alpha = a^2 + b^2,\ \beta = a^2 - b^2$
   (b) $\alpha = 2ab,\ \beta = a^2 + b^2$
   (c) $\alpha = a^2 + b^2,\ \beta = 2ab$
   (d) $\alpha = a^2 + b^2,\ \beta = ab$

5. Let AB be a chord of the parabola $y^2 = 4ax$. If the pole of AB with respect to the parabola be $(2a, 3a)$ then the length of AB is

   (a) $\sqrt{13}a$
   (b) $4a$
   (c) $5a$
   (d) $2\sqrt{3}a$

6. The equation of a diameter of a circle passing through the origin is $x + y = 1$ and the greatest distance of any point of the circle from the diameter is $\sqrt{5}$. Then an equation of the circle is

(a) $x^2+y^2-2x+4y=0$ (b) $x^2+y^2-4x+2y=0$

(c) $x^2+y^2+4x-2y=0$ (d) $x^2+y^2+2x+4y=0$

7. The area of the region in the Argand plane in which z can belong, satisfying the inequalities $1 \le |z-1| \le 4$ and $z+\bar{z} \ge 2$, is

(a) 15π (b) $\frac{15}{2}\pi$

(c) π (d) none of these

8. Let $\phi(x)=\int_0^x \cos^3 x\, dx$. The function $\phi(x)$ is

(a) monotonic increasing in $[0, \pi]$

(b) monotonic increasing in $\left[0, \frac{\pi}{2}\right]$

(c) monotonic decreasing in $[0, 2\pi]$

(d) monotonic decreasing in $\left[-\frac{\pi}{2}, \frac{3\pi}{2}\right]$

9. Let $A=\{1, 2, 3, \ldots, 50\}$ and $B=\{2, 4, 6, \ldots, 100\}$. The number of elements $(x, y) \in A \times B$ such that $x+y=50$ is

(a) 24 (b) 25 (c) 50 (d) 75

10. A word has 4 identical letters and some different letters. If the total number of words that can be made with the letters of the word be 210 then the number of different letters in the word is

(a) 3 (b) 5

(c) 4 (d) 7

11. $ax^2+bx+c=0$ is an equation whose sum of roots exceeds 2 but the product of roots is less than 2. Then

(a) $a(b+c)<0$ (b) $b+c<0$

(c) $a+b+c>0$ (d) $a+b<0$

12. If α, β be the positive roots of a quadratic equation such that $|\alpha-\beta|=1$ and $\frac{\alpha}{\beta}+\frac{\beta}{\alpha}=\frac{5}{2}$ then the quadratic equation is

(a) $x^2+3x+2=0$ (b) $x^2-3x+2=0$

(c) $x^2-3x-4=0$ (d) $x^2-x+2=0$

13. If $\lambda > 0$ such that the system of equations

$$x + y - 2z = 0,$$
$$\lambda x - 2y + z = 0,$$
$$x + 3y - \lambda z = 0$$

has infinite number of solutions then λ is

(a) $\sqrt{10} + 2$ (b) $\sqrt{10} - 2$

(c) 2 (d) 1

14. The value of $\sum_{r=1}^{n} r^2 \cdot \frac{{}^nC_r}{{}^nC_{r-1}}$ is

(a) $\frac{(n+1)(n+2)}{12}$ (b) $\frac{n^2(n+2)}{6}$

(c) $\frac{n(n+1)}{2}$ (d) $\frac{n(n+1)(n+2)}{6}$

15. If the centroid of an equilateral triangle ABC in the Argand plane be at $z = 1$, the vertex C is at $z = \omega = \frac{-1 - \sqrt{3}i}{2}$ and A, B, C are in the anticlockwise sense then the vertex A is at

(a) $z = -2\omega$ (b) $z = 3 + \omega$

(c) $z = 0$ (d) $z = \omega^2$

16. If $f(x)$ is not differentiable finitely at $x = 2$ and $f(x) + g(x)$ is differentiable finitely at $x = 2$ then

(a) $g(x)$ must be differentiable finitely at $x = 2$

(b) $g(x)$ must be continuous at $x = 2$

(c) $g(x)$ may not be differentiable at $x = 2$

(d) $g(x)$ cannot be differentiable at $x = 2$

17. If the equation of a family of curves be $y = a\cos(x + b)$, where a, b are arbitrary constants, and $y' = \frac{dy}{dx}$, $y'' = \frac{d^2y}{dx^2}$, etc., then their differential equation is

(a) $y'' = y$ (b) $y'' = yy'$

(c) $y'' + y = 0$ (d) $y'' = y + y'$

18. A function $y = f(x)$ is invertible only when

(a) $f(x)$ is monotonic increasing (b) $f(x)$ is bijective

(c) $f(x)$ is monotonic decreasing (d) $f(x)$ is injective

19. The function $f(x) = |\sin x|$ $(-2\pi \le x \le 2\pi)$ is

(a) continuous everywhere (b) differentiable everywhere

(c) monotonic increasing (d) invertible

20. The distance through which the point (1, 2, 3) is to be translated parallel to the line $\frac{x}{2} = \frac{y-1}{3} = \frac{z+1}{4}$ to reach the plane $x + y + z = 2$ is

(a) $\frac{4}{9}\sqrt{29}$ (b) $\frac{1}{2}\sqrt{29}$

(c) $\frac{4}{9}$ (d) $\frac{2}{9}\sqrt{29}$

21. If the point $(a, a + 1)$ belongs to the interior of the region bounded by the curve $y = \sqrt{5 - x^2}$ and the x-axis then a belongs to

(a) (–2, 1) (b) (–1, 1)

(c) (–2, 0) (d) (0, 1)

22. The vertex of the parabola $x^2 = 2(2x + y)$ is

(a) (–2, 2) (b) (–1, 1)

(c) (–2, 0) (d) (2, –2)

23. The value of $\int_0^1 \frac{x^2}{1 + x^6}\, dx$ is equal to

(a) $\frac{\pi}{12}$ (b) $\frac{\pi}{8}$

(c) π (d) $\frac{\pi}{4}$

24. In a ΔABC, the value of

$\sin A \cos B \cos C + \sin B \cos C \cos A + \sin C \cos A \cos B$ is equal to

(a) $\cos A \cos B \cos C$ (b) $\sin A \sin B \sin C$

(c) 0 (d) none of these

25. If the lines $ax + by + c = 0$ and $x + y + 1 = 0$ meet at the point $(\cos\theta, \sin\theta)$ for some θ then $|b - c|$, $|c - a|$ and $|a - b|$ are sides of a triangle which is

(a) isosceles (b) scalene

(c) equilateral (d) right-angled

26. Let ABC be an equilateral triangle whose orthocentre is the origin O. If $\overrightarrow{OA} = \vec{a}$, $\overrightarrow{OB} = \vec{b}$ then $\overrightarrow{OC}$ is

(a) $\vec{a} + \vec{b}$ (b) $\frac{1}{2}(\vec{a} + \vec{b})$

(c) $-(\vec{a} + \vec{b})$ (d) $-2(\vec{a} + \vec{b})$

One-or-More-Options-Correct Questions

27. The equation of a common tangent to the circle $x^2 + y^2 = 2a^2$ and the parabola $y^2 = 8ax$ is

(a) $y = x - 2a$ (b) $y + x + 2a = 0$

(c) $y = x + 2a$ (d) $y + x = 2a$

28. If amp $\dfrac{z_1 - z_2}{z_1 - z_3} = \dfrac{\pi}{2}$, where $|z_1| \neq |z_2| \neq |z_3|$ then the points representing z_1, z_2 and z_3 are

(a) concyclic

(b) collinear

(c) vertices of a right-angled triangle

(d) vertices of an isosceles right-angled triangle

29. If $f(x) = \begin{cases} |x - 3| & \text{for } x \geq 1 \\ \dfrac{x^2}{4} - \dfrac{3x}{2} + \dfrac{13}{4} & \text{for } x < 1 \end{cases}$

then $f(x)$ is

(a) continuous at $x = 1$ (b) differentiable at $x = 1$

(c) discontinuous at $x = 3$ (d) nondifferentiable at $x = 3$

30. In a triangle ABC in which $C > \dfrac{\pi}{2}$,

(a) $\tan A \tan B > 1$ (b) $\cos A + \cos B + \cos C > 1$

(c) $\tan A \tan B < 1$ (d) $\cos A + \cos B + \cos C < 1$

Comprehension-Type Questions

According to Fermat's theorem, $n^p - n = M(p)$ = a multiple of p when p is a prime number and $n \in N$.

It is also known that the highest index of power of a prime number p which divides n ! is equal to $\left[\frac{n}{p}\right]+\left[\frac{n}{p^2}\right]+\left[\frac{n}{p^3}\right]+\ldots$ where $[x]$ = greatest integer less than or equal to x.

31. If $n(>2)$ is a prime number which is coprime with a, b and $a+b$ then $a^{n-2}\cdot b - a^{n-3}\cdot b^2 + a^{n-4}\cdot b^3 - \ldots + ab^{n-2}$ is equal to

(a) $M(n)-1$ (b) $M(n)+1$

(c) $M(n)$ (d) none of these

32. $17^{22}-1$ is a multiple of

(a) 16 (b) 44

(c) 46 (d) 27

33. The highest power of 7 which divides 1000 ! is

(a) 7^{164} (b) 7^{162}

(c) 7^{167} (d) 7^{142}

Matching Questions

One or more may match with the same.

34. Three lines OA, OB and OC through the origin O have direction ratios $l, 2, 1$; $1, m, 2$ and $2, 1, n$ respectively.

(i) $OB \perp OC$ if — (a) $lmn + 9 = 2(l+m+n)$

(ii) OC is perpendicular to the plane AOB if — (b) $m + 2n = -2$

(iii) OA lies in the plane BOC if — (c) $l = m = n = -\frac{2}{3}$

(iv) OA, OB and OC form three mutually perpendicular lines if — (d) $n = -2(l+1) = -\frac{m+2}{2}$

35. (i) If $\alpha = \sqrt[5]{1}$ and α is nonreal then the value of $\alpha^{5n+1} + \alpha^{5n+7} + \alpha^{5n+13} + \alpha^{5n+19}$ is equal to — (a) 1

(ii) In ΔABC, if $\sin A$, $\sin B$, $\sin C$ are in AP then $3\tan\frac{A}{2}\cdot\tan\frac{C}{2}$ is equal to — (b) 4

(iii) Let $f: R \to R$ such that $|f(x) - f(y)| \le |x = y|^3$ for $x, y \in R$. Also, $f(2) = 4$. Then $f(4)$ equals (c) -1

(iv) The greatest value of $f(x) = \dfrac{x^4 - x^2 - 2(x-3)}{x^4 - x^2 - 2(x-2)}$ for $x \in R$ is (d) 2

Complete the following statements.

36. Normals are drawn from the point P with slopes m_1, m_2, m_3 to the parabola $y^2 = 4x$. If the locus of P when $m_1 m_2 = k$ is part of the parabola itself then the value of k is

37. Let $F(x) = f(x) \cdot g(x) \cdot h(x)$. At some point x_0, it is given that $F'(x_0) = 21F(x_0)$, $f'(x_0) = 4f(x_0)$, $g'(x_0) = -7g(x_0)$ and $h'(x_0) = kh(x_0)$. Then k is

Assertion-Reason Type

38. Two tangents to the parabola $x^2 = 6y$ meet at the point $\left(-1, -\dfrac{3}{2}\right)$.

STATEMENT-1: The tangents are perpendicular to each other.

because

STATEMENT-2: Mutually perpendicular tangents to the parabola meet on the line $2y + 3 = 0$.

(a) Statement-1 is True, Statement-2 is True; Statement-2 is a correct explanation for Statement-1

(b) Statement-1 is True, Statement-2 is True; Statement-2 is *Not* a correct explanation for Statement-1

(c) Statement-1 is True, Statement-2 is False

(d) Statement-1 is False, Statement-2 is True

Answers

1. d	**2.** d	**3.** a	**4.** c	**5.** a
6. b	**7.** b	**8.** b	**9.** a	**10.** c
11. a	**12.** b	**13.** a	**14.** d	**15.** b
16. d	**17.** c	**18.** b	**19.** a	**20.** a
21. b	**22.** d	**23.** a	**24.** b	**25.** d
26. c	**27.** b, c	**28.** a, c	**29.** a, b, d	**30.** b, c
31. b	**32.** c	**33.** a		

34. (i) ↔ b (ii) ↔ d (iii) ↔ a (iv) ↔ c

35. (i) ↔ c (ii) ↔ a (iii) ↔ b (iv) ↔ d

36. 2	**37.** 24	**38.** a

Hints and Solutions to Selected Questions

1. $f'(x) = 3ax^2 + 2bx + c = 0$ for extrema.

$\therefore$ α, 2α are the roots of $3ax^2 + 2bx + c = 0$.

$$\therefore \quad \alpha + 2\alpha = \frac{-2b}{3a}, \quad \alpha \cdot 2\alpha = \frac{c}{3a} \quad \Rightarrow 2 \cdot \left(\frac{-2b}{9a}\right)^2 = \frac{c}{3a}$$

$$\Rightarrow \quad 8b^2 = 27ac \quad \Rightarrow (4b)^2 = 6a \cdot 9c \quad \Rightarrow 6a, 4b, 9c \text{ are in GP.}$$

2. $$f\left(\frac{\pi}{2} - h\right) = \frac{\pi}{2} - h + \cos\left(\frac{\pi}{2} - h\right) = \frac{\pi}{2} - h + \sin h$$

$$f\left(\frac{\pi}{2} + h\right) = \frac{\pi}{2} + h - \cos\left(\frac{\pi}{2} + h\right) = \frac{\pi}{2} + h + \sin h$$

$$\therefore \quad f'\left(\frac{\pi}{2} + 0\right) = \lim_{h \to 0} \frac{f\left(\frac{\pi}{2} + h\right) - f\left(\frac{\pi}{2}\right)}{h} = \lim_{h \to 0} \frac{\frac{\pi}{2} + h + \sin h - \frac{\pi}{2}}{h}$$

$$= \lim_{h \to 0} \frac{h + \sin h}{h} = \lim_{h \to 0} \left(1 + \frac{\sin h}{h}\right) = 2.$$

$$f'\left(\frac{\pi}{2} - 0\right) = \lim_{h \to 0} \frac{f\left(\frac{\pi}{2} - h\right) - f\left(\frac{\pi}{2}\right)}{-h} = \lim_{h \to 0} \frac{\frac{\pi}{2} - h + \sin h - \frac{\pi}{2}}{-h} = 0.$$

$\therefore$ $f'\left(\frac{\pi}{2}\right)$ does not exist.

Note $|x|$ is differentiable everywhere except at $x = 0$. Also, $|\cos x|$ is not differentiable at $x = \frac{\pi}{2}$, the graph having two tangents at $x = \frac{\pi}{2}$.

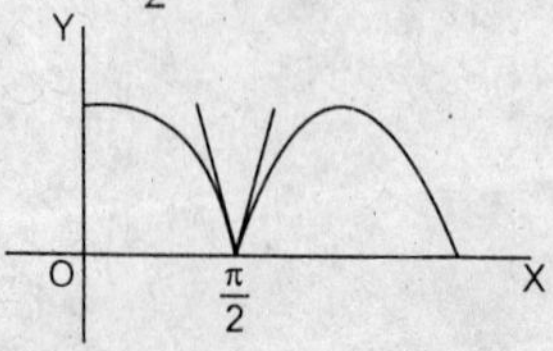

3.

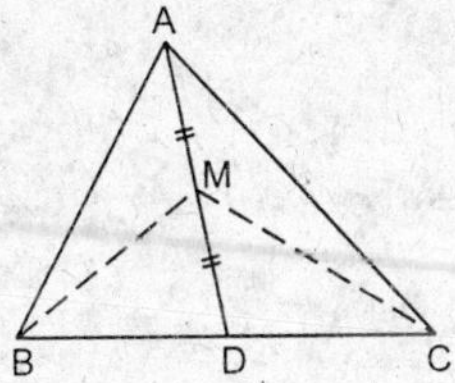

As the circumcentre M is on the median AD, $MD \perp BC$ and MD bisects BC.

$\therefore \quad AB = AC$, i.e., $c = b$.

$$\therefore \quad \cos B = \frac{c^2 + a^2 - b^2}{2ca} = \frac{a^2}{2ca} = \frac{a}{2c}.$$

4. $A^2 = \begin{bmatrix} a & b \\ b & a \end{bmatrix}\begin{bmatrix} a & b \\ b & a \end{bmatrix}$

$= \begin{bmatrix} a^2 + b^2 & ab + ba \\ ba + ab & b^2 + a^2 \end{bmatrix}$

$= \begin{bmatrix} \alpha & \beta \\ \beta & \alpha \end{bmatrix}$

$\therefore \quad \alpha = a^2 + b^2,\ \beta = 2ab.$

5. Let $A = (at_1^2, 2at_1)$ and $B = (at_2^2, 2at_2)$. The point of intersection of the tangents at A and B is the pole of AB with respect to the parabola.

$\therefore \quad \left(at_1t_2, a(t_1 + t_2)\right) = (2a, 3a)$

$\Rightarrow \quad t_1t_2 = 2,\ t_1 + t_2 = 3 \quad \Rightarrow \quad t_1 = 2, t_2 = 1 \quad$ or $\quad t_1 = 1, t_2 = 2.$

$\therefore \quad AB =$ the distance between $(4a, 4a)$ and $(a, 2a)$

$$= \sqrt{(4a - a)^2 + (4a - 2a)^2} = \sqrt{13} \cdot a.$$

6. The centre $= (t, 1 - t)$ and radius $= \sqrt{5}$.

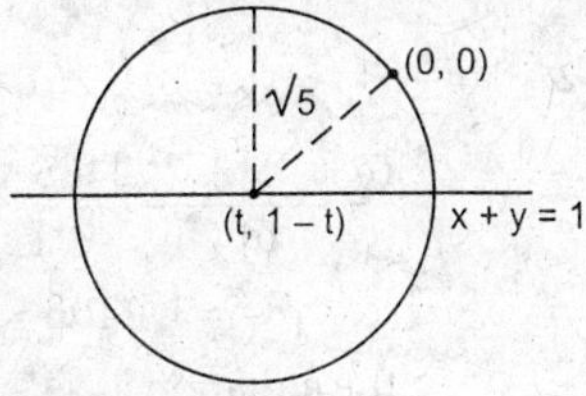

$\therefore \quad t^2 + (1 - t)^2 = (\sqrt{5})^2$

or $\quad 2t^2 - 2t - 4 = 0$

or $\quad t^2 - t - 2 = 0$

or $\quad t = 2, -1$

$\therefore \quad$ the equation of the circle is

$(x - 2)^2 + (y + 1)^2 = 5 \quad$ or $\quad x^2 + y^2 - 4x + 2y = 0.$

7. $z + \bar{z} \geq 2 \quad \Rightarrow \quad x \geq 1$, where $z = x + iy$.

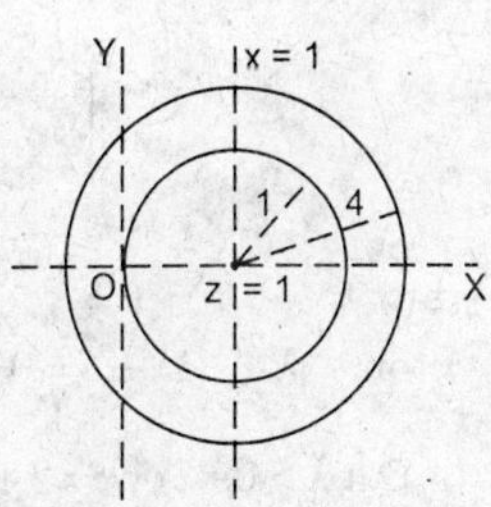

$|z - 1| \leq 4 \quad \Rightarrow \quad z \in$ interior or on the circle whose centre is $z = 1$ and radius $= 4$.

$|z - 1| \geq 1 \quad \Rightarrow \quad z \in$ exterior or on the circle whose centre is $z = 1$ and the radius $= 1$.

Clearly, the area $= \frac{1}{2}(\pi \cdot 4^2 - \pi \cdot 1^2) = \frac{15\pi}{2}$.

8. $\phi'(x) = \cos^3 x$. So, $\phi'(x) \geq 0 \Rightarrow \cos^3 x \geq 0 \Rightarrow \cos x \geq 0$.

$\therefore$ $\phi(x)$ is m.i. in $\left[-\frac{\pi}{2}, \frac{\pi}{2}\right]$. So, (b) is true.

$\phi(x)$ is m.d. if $\cos x \leq 0$, i.e., $\frac{\pi}{2} \leq x \leq \frac{3\pi}{2}$.

9. Here $x + y = 50$. But y is even. So x must also be even.

Let $x = 2n, y = 2m$. So, $n + m = 25$, where m, n are positive integers.

$\therefore$ $(n, m) = (1, 24), (2, 23), \ldots, (24, 1)$. So, the required number = 24.

10. Let the number of distinct letters other than the identical letters be n.

Then $\frac{(n+4)!}{4!} = 210$ or $(n+4)(n+3)\ldots 6 \cdot 5 = 7 \cdot 6 \cdot 5 \Rightarrow n + 4 = 7$

$\therefore$ $n = 3$.

$\therefore$ the required number of distinct letters = 3 + 1 = 4.

11. $\alpha + \beta > 2,\ \alpha\beta < 2 \Rightarrow -\frac{b}{a} > 2, \frac{c}{a} < 2$

$\Rightarrow \frac{b}{a} < -2, \frac{c}{a} < 2 \Rightarrow \frac{b}{a} + \frac{c}{a} < 0 \Rightarrow \frac{(b+c)a^2}{a} < 0.$

12. $|\alpha - \beta|^2 = (\alpha - \beta)^2 = (\alpha + \beta)^2 - 4\alpha\beta = 1$ and $\frac{\alpha^2 + \beta^2}{\alpha\beta} = \frac{5}{2}$.

$\therefore$ $\frac{(\alpha+\beta)^2 - 2\alpha\beta}{\alpha\beta} = \frac{5}{2}$ or $\frac{1 + 4\alpha\beta - 2\alpha\beta}{\alpha\beta} = \frac{5}{2}$ $\therefore$ $\alpha\beta = 2$

$\therefore$ $(\alpha + \beta)^2 = 1 + 4\alpha\beta = 9$ $\therefore$ $\alpha + \beta = 3, -3$. But α, β are positive.

$\therefore$ the equation is $x^2 - 3x + 2 = 0$.

13. The system of homogeneous linear equations will have infinite number of solutions if it has nontrivial solutions. So,

$$\begin{vmatrix} 1 & 1 & -2 \\ \lambda & -2 & 1 \\ 1 & 3 & -\lambda \end{vmatrix} = 0$$

or $1(2\lambda - 3) - 1(-\lambda^2 - 1) - 2(3\lambda + 2) = 0$

or $\lambda^2 - 4\lambda - 6 = 0$ $\therefore$ $\lambda = \frac{4 \pm \sqrt{16 + 24}}{2} = 2 \pm \sqrt{10}$.

But $\lambda > 0$. So, $\lambda = 2 + \sqrt{10}$.

14. $t_r = r^2 \cdot \dfrac{{}^nC_r}{{}^nC_{r-1}} = r^2 \cdot \dfrac{n!}{r!\,(n-r)!} \cdot \dfrac{(r-1)!\,(n-r+1)!}{n!}$

$$= r(n+1-r) = (n+1)r - r^2.$$

$\therefore \quad \text{sum} = (n+1)\sum_1^n r - \sum_1^n r^2 = (n+1)\dfrac{n(n+1)}{2} - \dfrac{n(n+1)(2n+1)}{6}$

$$= \frac{n(n+1)}{6}\{3(n+1)-(2n+1)\} = \frac{n(n+1)(n+2)}{6}.$$

15. Here, $|z-1| = |\omega - 1|$ ($\because$ in an equilateral Δ, the centroid is also the circumcentre).

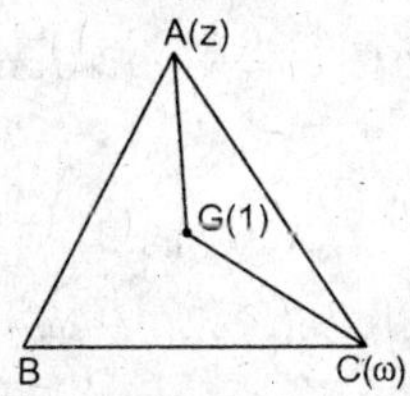

$\text{amp}\,\dfrac{z-1}{\omega-1} = \dfrac{2\pi}{3}.$

$\therefore \quad \dfrac{z-1}{\omega-1} = 1\left(\cos\dfrac{2\pi}{3} + i\sin\dfrac{2\pi}{3}\right) = \dfrac{-1+\sqrt{3}i}{2} = \omega^2$

$\therefore \quad z - 1 = \omega^3 - \omega^2 = 1 - \omega^2$

or $\quad z = 2 - \omega^2 = 2 + 1 + \omega = 3 + \omega.$

16. As $f(x)$ is not differentiable at $x = 2$, $f(x) + g(x)$ can be differentiable at $x = 2$ only if $g(x)$ is also not differentiable at $x = 2$.

17. Differentiating w.r.t. x, $\quad y' = -a\sin(x+b).$

Squaring and adding, $\quad y^2 + y'^2 = a^2.$

Differentiating again w.r.t. x, $\quad 2yy' + 2y' \cdot y'' = 0.$

But $y' \neq 0$ since then y = constant. So, $\quad y + y'' = 0.$

18. $f(x)$ may be invertible when $f(x)$ is m.d. or m.i. $f(x)$ is invertible if $f(x)$ is strictly monotonic.

The necessary and sufficient condition for a function $f(x)$ to be invertible is that $f(x)$ is one-to-one and onto (i.e., bijective).

19. $|\sin x|$ is not differentiable at $-2\pi, -\pi, 0, \pi, 2\pi$ as in each case the graph of $y = |\sin x|$ has two tangents. It is continuous at all points. $|\sin x|$ is even and so many-to-one. So, it is not monotonic and invertible.

20. The direction cosines of the line PQ are $\frac{2}{\sqrt{29}}, \frac{3}{\sqrt{29}}, \frac{4}{\sqrt{29}}$.

If $PQ = r$ then $Q = \left(1+\frac{2r}{\sqrt{29}}, 2+\frac{3r}{\sqrt{29}}, 3+\frac{4r}{\sqrt{29}}\right)$.

As Q is on the plane $x+y+z=2$,

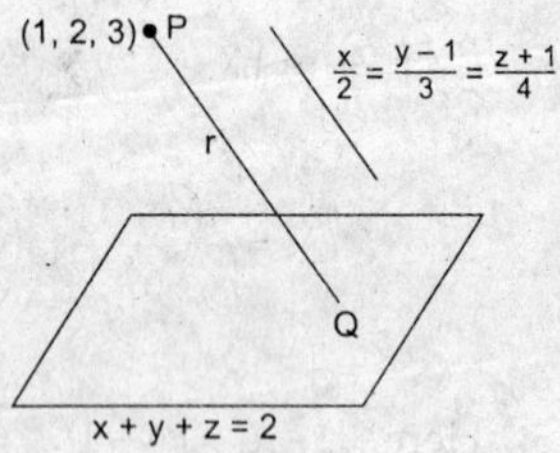

$$1+\frac{2r}{\sqrt{29}}+2+\frac{3r}{\sqrt{29}}+3+\frac{4r}{\sqrt{29}}=2$$

or $\quad r = -\frac{4\sqrt{29}}{9}$.

$\therefore \quad PQ = \left|-\frac{4\sqrt{29}}{9}\right| = \frac{4\sqrt{29}}{9}$.

21. Clearly, $y > 0$ and $-\sqrt{5} < x < \sqrt{5}, y < \sqrt{5}$

($\because$ the curve is the upper half of the circle $x^2+y^2-5=0$).

Now, $(a+1)^2 + a^2 - 5 < 0$

or $\quad 2a^2+2a-4<0 \quad$ or $\quad a^2+a-2<0$

or $\quad (a+2)(a-1)<0 \quad \Rightarrow \quad -2<a<1$.

But $a+1>0$. So, $a>-1$. Then $-1<a<1$. So, $a \in (-1, 1)$.

22. Here, $x^2-4x+4=2(y+2) \quad$ or $\quad (x-2)^2 = 2(y+2)$.

So, the vertex is $(2, -2)$.

23. Put $x^3 = z$. Then $3x^2dx = dz$ and $z = 0, 1$ according as $x = 0, 1$.

$\therefore \quad I = \int_0^1 \frac{1}{3}\cdot\frac{dz}{1+z^2} = \frac{1}{3}\tan^{-1}z\Big]_0^1 = \frac{1}{3}\cdot\frac{\pi}{4}$.

24. Value $= (\cos C)(\sin A\cdot\cos B + \sin B\cdot\cos A) + \sin C\cdot\cos A\cdot\cos B$

$= \cos C\cdot\sin(A+B) + \sin C\cdot\cos A\cdot\cos B$

$= \cos C\cdot\sin C + \sin C\cdot\cos A\cdot\cos B = (\sin C)(\cos C + \cos A\cdot\cos B)$

$= (\sin C)\cdot\{-\cos(A+B) + \cos A\cdot\cos B\} = \sin C\cdot\sin A\cdot\sin B$.

25. Here, $a\cos\theta + b\sin\theta + c = 0 \quad$ and $\quad \cos\theta + \sin\theta + 1 = 0$.

$$\therefore \quad \frac{\cos\theta}{b-c} = \frac{\sin\theta}{c-a} = \frac{1}{a-b} = \frac{\sqrt{\cos^2\theta+\sin^2\theta}}{\sqrt{(b-c)^2+(c-a)^2}}$$

$\therefore \quad (a-b)^2 = (b-c)^2 + (c-a)^2$

$\Rightarrow \quad |b-c|, |c-a|, |a-b|$ are sides of a right-angled triangle.

26. 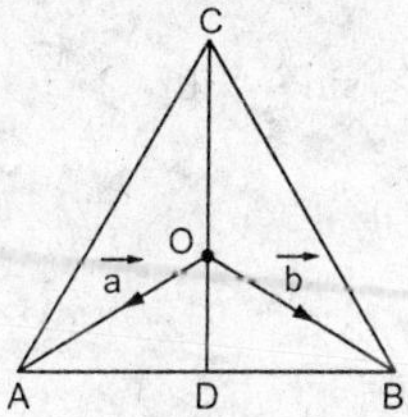

$\vec{OD} = \frac{1}{2}(\vec{a} + \vec{b})$. But $\vec{OC} = 2\vec{DO}$

($\because$ in an equilateral triangle, the orthocentre and centroid coincide).

$\therefore \quad \vec{OC} = -2 \cdot \frac{1}{2}(\vec{a} + \vec{b})$.

27. $y = mx + \frac{2a}{m}$ is any tangent to the parabola $y^2 = 8ax$.

Any tangent to $x^2 + y^2 = 2a^2$ is $y = \lambda x + \sqrt{2}a\sqrt{1+\lambda^2}$.

For a common tangent the two tangents should be identical.

So, $m = \lambda$ and $\frac{2a}{m} = \sqrt{2}a\sqrt{1+\lambda^2}$.

$\therefore \quad \frac{\sqrt{2}}{m} = \sqrt{1+m^2} \quad \Rightarrow \quad (1+m^2)m^2 = 2 \quad \Rightarrow \quad (m^2+2)(m^2-1) = 0$

$\therefore \quad m = 1, -1$. So, the common tangents are $y = x + 2a$ and $y = -x - 2a$.

28. Here, amp $\frac{z_2 - z_1}{z_3 - z_1} = \frac{\pi}{2} \quad \Rightarrow \quad \angle Z_2Z_1Z_3 = \frac{\pi}{2}$.

$\therefore \quad \Delta Z_2Z_1Z_3$ is right-angled. Hence, the points are concyclic (not collinear)

29. Here, $f(x) = \begin{cases} \frac{x^2}{4} - \frac{3x}{2} + \frac{13}{4} & \text{for } x < 1 \\ 3 - x & \text{for } 1 \le x < 3 \\ x - 3 & \text{for } x \ge 3. \end{cases}$

$\therefore \quad f(1+0) = f(1-0) = f(1) = 2,\ f'(1-0) = f'(1+0),$

$f(3-0) = f(3+0) = f(3) = 0,\ f'(3-0) \ne f'(3+0).$

$\therefore \quad f(x)$ is continuous at $x = 1, 3$ and differentiable at $x = 1$ but not differentiable at $x = 3$.

30. $1 - \tan A \tan B = 1 - \frac{\sin A \sin B}{\cos A \cos B} = \frac{\cos(A+B)}{\cos A \cos B} = \frac{-\cos C}{\cos A \cos B} > 0$

($\because \cos A > 0, \cos B > 0, \cos C < 0$).

$\therefore \quad \tan A \tan B < 1.$

Also, $\cos A + \cos B + \cos C = 1 + 4\sin\frac{A}{2}\cdot\sin\frac{B}{2}\cdot\sin\frac{C}{2} > 1.$

31. Adding, expression $= \dfrac{a^{n-2}\cdot b\left\{1-\left(-\frac{b}{a}\right)^{n-2}\right\}}{1-\left(-\frac{b}{a}\right)}$

$$= \frac{a^{n-1}\cdot b}{a+b}\cdot(a^{n-2}+b^{n-2}).$$

Since n is a prime number and a, b are prime to n, we have

$$a^n - a = M_1(n) \text{ and } b^n - b = M_2(n)$$

$$\Rightarrow \quad a^n + b^n - (a+b) = M_1(n) + M_2(n) = \text{a multiple of } n$$

$$= n\cdot k \text{ (say)}.$$

$$\therefore \quad \frac{a^n+b^n}{a+b} = 1 + \frac{nk}{a+b}$$

$$\Rightarrow \quad a^{n-1} - a^{n-2}b + \ldots - ab^{n-2} + b^{n-1} = 1 + n\cdot\frac{k}{a+b}.$$

As the LHS is an integer and $n, a+b$ are coprime, k is a multiple of $a+b$.

$$\therefore \quad a^{n-2}b - a^{n-3}\cdot b^2 + \ldots + ab^{n-2} = a^{n-1} + b^{n-1} - 1 + M_3(n)$$

$$= (a^{n-1}-1) + (b^{n-1}-1) + 1 + M_3(n)$$

$= M_4(n) + M_5(n) + 1 + M_3(n)$ $\quad$ ($\because$ $a^n - a = M(n) \Rightarrow a^{n-1} - 1$ is a multiple of n because a, n are coprime)

$$= 1 + M(n).$$

32. $n^p - n = M(p)$ if p is prime.

$\therefore \quad n^{p-1} - 1 = M(p)$ when n, p are coprime.

Take $n = 17, p = 23$. Then $17^{22} - 1 = M(23)$.

But 17^{22} is odd. So, $17^{22} - 1$ is even. Hence, $17^{22} - 1 = M(23\times 2)$.

33. The required index of power

$$= \left[\frac{1000}{7}\right] + \left[\frac{1000}{7^2}\right] + \left[\frac{1000}{7^3}\right] = \left[142\frac{6}{7}\right] + \left[20\frac{20}{49}\right] + \left[2\frac{314}{343}\right]$$

$$= 142 + 20 + 2 = 164.$$

$\therefore$ $\quad$ the highest power of 7 which divides $1000! = 7^{164}$.

34. (i) $OB \perp OC \quad \Rightarrow \quad 1\cdot 2 + m\cdot 1 + 2\cdot n = 0 \quad$ or $\quad m + 2n + 2 = 0.$

(ii) $OC \perp$ plane $AOB \quad \Rightarrow \quad OC \perp OA,\ OC \perp OB$

$\Rightarrow \quad l \cdot 2 + 2 \cdot 1 + 1 \cdot n = 0, \ 1 \cdot 2 + m \cdot 1 + 2 \cdot n = 0$

$\Rightarrow \quad 2l + n + 2 = 0, \ m + 2n + 2 = 0.$

(iii) If λ, μ, ν are d.r. of the normal to the plane then

$$\lambda \cdot l + \mu \cdot 2 + \nu \cdot 1 = 0, \ \lambda \cdot 1 + \mu \cdot m + \gamma \cdot 2 = 0, \ \lambda \cdot 2 + \mu \cdot 1 + \nu \cdot n = 0$$

$$\Rightarrow \quad \begin{vmatrix} l & 2 & 1 \\ 1 & m & 2 \\ 2 & 1 & n \end{vmatrix} = 0 \ \Rightarrow \ lmn = 2(l + m + n) - 9.$$

(iv) $OA \perp OB \ \Rightarrow \ l + 2m + 2 = 0.$

$OA \perp OC \ \Rightarrow \ 2l + 2 + n = 0.$

$OB \perp OC \ \Rightarrow \ 2 + m + 2n = 0.$

Solving, $l = m = n = -\dfrac{2}{3}$.

35. (i) Expression $= \alpha + \alpha^2 + \alpha^3 + \alpha^4$ $\quad (\because \ \alpha^5 = 1)$

$$= -1 + (1 + \alpha + \alpha^2 + \alpha^3 + \alpha^4)$$

$$= -1 + \frac{1 - \alpha^5}{1 - \alpha} = -1 \qquad (\because \alpha \neq 1).$$

(ii) Here a, b, c are in AP. So, $a + c = 2b \ \Rightarrow \ 2s = 3b$.

$$\therefore \quad 3\tan\frac{A}{2} \cdot \tan\frac{C}{2} = 3 \cdot \frac{\Delta}{s(s-a)} \cdot \frac{\Delta}{s(s-c)}$$

$$= 3 \cdot \frac{s(s-a)(s-b)(s-c)}{s^2 \cdot (s-a)(s-c)} = \frac{3(s-b)}{s}$$

$$= 3 \cdot \frac{\frac{1}{2}(3b - 2b)}{\frac{3b}{2}} = 1.$$

(iii) $f'(x) = \lim\limits_{y \to x} \dfrac{f(y) - f(x)}{y - x}$. $\therefore \ |f'(x)| = \lim\limits_{y \to x} \left| \dfrac{f(y) - f(x)}{y - x} \right| \leq \lim\limits_{y \to x} |x - y|^2$

(from the question)

$\therefore \quad |f'(x)| = 0 \quad \Rightarrow f'(x) = 0 \quad \Rightarrow f(x) = \text{constant}$

$\Rightarrow \quad f(4) = f(2) = 4.$

(iv) $f(x) = 1 + \dfrac{2}{x^4 - x^2 - 2x + 4}$. Let $\phi(x) = x^4 - x^2 - 2x + 4$.

$\therefore \quad \phi'(x) = 4x^3 - 2x - 2 = (x - 1)(4x^2 + 4x + 2).$

$\therefore \quad \phi'(x) = 0 \quad \Rightarrow x = 1 \quad \text{or} \quad 2x^2 + 2x + 1 = 0$ (imaginary roots)

Also, $\phi''(x) = 12x^2 - 2 > 0$ when $x = 1$.

So, for $x = 1$, $\phi(x)$ has the minimum value.

$\therefore \quad f(x)$ has the greatest value at $x = 1$.

$\therefore$ the greatest $f(x) = 1 + \dfrac{2}{1-1-2+4} = 2.$

36. Let $P = (\alpha, \beta)$. Any normal to the parabola $y^2 = 4x$ is $y + tx = 2t + t^3$

or $y = -tx + 2t + t^3$

$\therefore$ slope $m = -t$. So, the normal is

$y = mx - 2am - am^3$, where m is the slope.

It passes through P. So, $\beta = m\alpha - 2m - m^3$ or $m^3 + (2-\alpha)m + \beta = 0$.

From the question, its roots are m_1, m_2, m_3, where $m_1 m_2 = k$.

$\therefore$ $m_1 + m_2 + m_3 = 0,\ m_1 m_2 m_3 = -\beta \quad \Rightarrow km_3 = -\beta \quad \Rightarrow m_3 = -\dfrac{\beta}{k}.$

Now, $\beta = m_3\alpha - 2m_3 - m_3^3 \quad \Rightarrow \beta = -\dfrac{\beta}{k}\alpha + \dfrac{2\beta}{k} + \dfrac{\beta^3}{k^3}$

($\because$ the locus is a part of the parabola and $\beta \neq 0$)

or $\beta^2 = (\alpha - 2)k^2 + k^3$. So, the locus of P is

$y^2 = (x-2)k^2 + k^3$ or $y^2 = k^2 x + (k^3 - 2k^2)$. ... (1)

From the question, (1) is identical with $y^2 = 4x$.

$\therefore$ $k^2 = 4,\ k^3 - 2k^2 = 0 \quad \Rightarrow k = \pm 2,\ k = 0, 2.$

The common value of k is 2.

37. $F'(x) = f'(x)\cdot g(x)\cdot h(x) + f(x)\cdot g'(x)\cdot h(x) + f(x)\cdot g(x)\cdot h'(x)$

$\therefore$ $F'(0) = 21F(x_0)$ implies

$$f'(x_0)\cdot g(x_0)\cdot h(x_0) + f(x_0)\cdot g'(x_0)\cdot h(x_0) + f(x_0)\cdot g(x_0)\cdot h'(x_0) = 21f(x_0)\cdot g(x_0)\cdot h(x_0)$$

$\therefore$
$$4f(x_0)\cdot g(x_0)\cdot h(x_0) + f(x_0)\cdot\{-7g(x_0)\} + f(x_0)\cdot g(x_0)\cdot kh(x_0) = 21f(x_0)\cdot g(x_0)\cdot h(x_0)$$

$\Rightarrow$ $4 - 7 + k = 21 \quad \Rightarrow k = 24.$

38. The equation of the directrix of $x^2 = 6y$ or $x^2 = 4\cdot\dfrac{3}{2}y$ is $y + \dfrac{3}{2} = 0$ or $2y + 3 = 0$.

$\therefore$ $\left(-1, -\dfrac{3}{2}\right)$ is a point on the directirix. We know that the mutually perpendicular tangents to a parabola meet on its directrix. Hence, the Statement-1 is True as well as Statement-2 is True. Moreover, Statement-2 is a correct explanation for Statement-1.

□

Practice Worksheet—4

One-Option-Correct Questions

1. Let z, ω be complex numbers such that $\bar{z} + i\bar{\omega} = 0$ and $\arg(z\omega) = \pi$. Then $\arg z$ equals

(a) $\frac{\pi}{4}$ (b) $\frac{\pi}{2}$ (c) $\frac{3\pi}{4}$ (d) $\frac{5\pi}{4}$

2. If $|z^2 - 1| = |z|^2 + 1$ then z lies on

(a) the real axis (b) the imaginary axis
(c) a circle (d) an ellipse

3. Let $A = \begin{bmatrix} 0 & 0 & -1 \\ 0 & -1 & 0 \\ -1 & 0 & 0 \end{bmatrix}$. The only correct statement about the matrix A is

(a) A is a zero matrix (b) $A = (-1)I_3$
(c) A^{-1} does not exist (d) $A^2 = I$

4. If $(x + p - 1)$ is a factor of the expression $x^2 + px + 1 - p$ then the roots of the equation $x^2 + px + 1 = p$ are

(a) 0, 1 (b) −1, 1 (c) 0, −1 (d) −1, 2

5. If $f(x) = \sqrt{a^2\cos^2 x + b^2\sin^2 x} + \sqrt{a^2\sin^2 x + b^2\cos^2 x}$ then the difference between the maximum and minimum values of $\{f(x)\}^2$ is given by

(a) $2(a^2 + b^2)$ (b) $2\sqrt{a^2 + b^2}$
(c) $(a + b)^2$ (d) $(a - b)^2$

6. A point on the parabola $y^2 = 18x$ at which the ordinate increases at twice the rate of the abscissa is

(a) (2, 4) (b) (2, −4) (c) $\left(-\frac{9}{8}, \frac{9}{2}\right)$ (d) $\left(\frac{9}{8}, \frac{9}{2}\right)$

7. The normal to the curve $x = a(1 + \cos\theta)$, $y = a\sin\theta$ at the point 'θ' always passes through the fixed point

(a) $(a, 0)$ (b) $(0, a)$ (c) $(0, 0)$ (d) (a, a)

8. If $\int_0^{\pi} xf(\sin x)\, dx = A \cdot \int_0^{\pi/2} f(\sin x)\, dx$ then A is

(a) 0 (b) π (c) $\pi/4$ (d) 2π

9. If $f(x) = \dfrac{e^x}{1+e^x}$, $I_1 = \int_{f(-a)}^{f(a)} x \cdot g\{x(1-x)\}\, dx$ and $I_2 = \int_{f(-a)}^{f(a)} g\{x(1-x)\}\, dx$

then the value of I_2/I_1 is

(a) 2 (b) –3 (c) –1 (d) 1

10. The differential equation for the family of curves $x^2 + y^2 - 2ay = 0$, where a is an arbitrary constant, is

(a) $2(x^2 - y^2)y' = xy$ (b) $2(x^2 + y^2)y' = xy$

(c) $(x^2 - y^2)y' = 2xy$ (d) $(x^2 + y^2)y' = 2xy$

11. The intercept on the line $y = x$ by the circle $x^2 + y^2 - 2x = 0$, is AB. The equation of the circle on AB as a diameter, is

(a) $x^2 + y^2 = x + y$ (b) $x^2 + y^2 - x + y = 0$

(c) $x^2 + y^2 + x + y = 0$ (d) $x^2 + y^2 + x = y$

12. If a circle passes through the point (a, b) and cuts the circle $x^2 + y^2 = 4$ orthogonally then the locus of its centre is

(a) $2ax + 2by + a^2 + b^2 + 4 = 0$ (b) $2ax + 2by - (a^2 + b^2 + 4) = 0$

(c) $2ax - 2by + a^2 + b^2 + 4 = 0$ (d) $2ax - 2by - (a^2 + b^2 + 4) = 0$

13. The eccentricity of an ellipse with its centre at the origin is $\frac{1}{2}$. If one of the directrices is $x = 4$ then the equation of the ellipse is

(a) $3x^2 + 4y^2 = 1$ (b) $3x^2 + 4y^2 = 12$

(c) $4x^2 + 3y^2 = 12$ (d) $4x^2 + 3y^2 = 1$

14. A line with direction ratios 2, 1, 2 meets the lines $x = y + a = z$ and $x + a = 2y = 2z$ at P and Q, respectively. Then the length of PQ is

(a) $2\sqrt{3}a$ (b) $3a$ (c) $\sqrt{6}a$ (d) $2\sqrt{7}a$

15. The distance between the planes $2x + y + 2z = 8$ and $4x + 2y + 4z + 5 = 0$ is

(a) $\frac{3}{2}$ (b) $\frac{5}{2}$ (c) $\frac{7}{2}$ (d) $\frac{9}{2}$

16. If $\vec{a} \cdot \vec{b} = 1$ and $\vec{a} \times \vec{b} = \vec{j} - \vec{k}$, where $\vec{a} = \vec{i} + \vec{j} + \vec{k}$, then $\vec{b}$ is

(a) $\vec{i} - \vec{j} + \vec{k}$ (b) $2\vec{j} - \vec{k}$

(c) $\vec{i}$ (d) $2\vec{i}$

17. The unit vector which is orthogonal to the vector $\vec{a} = 3\vec{i} + 2\vec{j} + 6\vec{k}$ and is coplanar with the vectors $\vec{b} = 2\vec{i} + \vec{j} + \vec{k}$ and $\vec{c} = \vec{i} - \vec{j} + \vec{k}$ is

(a) $\frac{1}{\sqrt{41}}(2\vec{i} - 6\vec{j} + \vec{k})$ (b) $\frac{2\vec{i} - 3\vec{j}}{\sqrt{13}}$

(c) $\frac{3\vec{j} - \vec{k}}{\sqrt{10}}$ (d) $\frac{4\vec{i} + 3\vec{j} - 3\vec{k}}{\sqrt{14}}$

18. If $f(x) = \sin x + \cos x$ and $g(x) = x^2 - 1$ then $g\{f(x)\}$ is invertible in the domain

(a) $\left[0, \frac{\pi}{2}\right]$ (b) $\left[-\frac{\pi}{4}, \frac{\pi}{4}\right]$

(c) $\left[-\frac{\pi}{2}, \frac{\pi}{2}\right]$ (d) $[0, \pi]$

19. For all $x \in R$, $x^2 + 2ax + 10 - 3a > 0$. The interval in which a lies is

(a) $a < -5$ (b) $-5 < a < 2$

(c) $a > 5$ (d) $2 < a < 5$

20. If $a_1, a_2, a_3, \ldots$ are in GP then the value of the determinant

$$\begin{vmatrix} \log a_1 & \log a_{n+1} & \log a_{2n+1} \\ \log a_2 & \log a_{n+2} & \log a_{2n+2} \\ \log a_3 & \log a_{n+3} & \log a_{2n+3} \end{vmatrix} \text{ is}$$

(a) 0 (b) 1

(c) $\log a_1$ (d) $\log a_2 - \log a_1$

21. If the sides of a triangle are in the ratio $1 : \sqrt{3} : 2$ then the angles of the triangle are in the ratio

(a) $1 : 3 : 5$ (b) $2 : 3 : 4$

(c) $3 : 2 : 1$ (d) $1 : 2 : 3$

22. A question paper contains 10 questions divided in two equal groups. A candidate has to attempt 6 questions selecting not more than 4 questions from one group. The probability that the candidate will answer by selecting an equal number of questions from the two groups is

(a) $\frac{1}{2}$ (b) $\frac{2}{3}$ (c) $\frac{10}{21}$ (d) $\frac{11}{21}$

23. If $f(x) = \begin{cases} xe^{-(1/|x| + 1/x)} & \text{for } x \neq 0 \\ 0 & \text{for } x = 0 \end{cases}$

then $f(x)$ is

(a) continuous for all x but not differentiable at $x = 0$

(b) neither differentiable nor continuous at $x = 0$

(c) discontinuous everywhere

(d) continuous and differentiable for all x

24. If $f: R \to R$ satisfies $f(x + y) = f(x) + f(y)$ for all $x, y \in R$ and $f(1) = 7$ then $\sum_{r=1}^{n} f(r)$ is

(a) $\frac{7(n+1)}{2}$ (b) $7n(n+1)$

(c) $\frac{7n(n+1)}{2}$ (d) $\frac{7n}{2}$

25. If $\lim_{x \to a} \frac{x \cdot \int_a^x f(x)\, dx}{x - a} = a^3 + a^2$ $(a \neq 0)$ then $f(2a)$ is

(a) $a(2a + 1)$ (b) $2a(2a + 1)$

(c) $4a^2 + 1$ (d) $2a(a + 1)$

26. The number of positive integral solutions of (x, y) when $x^2 - 4xy + 6y^2 - 2x - 20y = 29$ is

(a) six (b) three

(c) two (d) zero

One-or-More-Options-Correct Questions

27. The equation of a line which is parallel to $x - y = 0$ and touches the hyperbola $x^2 - 2y^2 = 18$ is

(a) $y - x = 3$ (b) $x - y = 3\sqrt{2}$

(c) $y - x = 3\sqrt{2}$ (d) $y + 3 = x$

28. Let $f(x) = 1 + |\sin x|$. Then

(a) $f(x)$ is continuous nowhere (b) $f(x)$ is continuous everywhere

(c) $f(x)$ is differentiable nowhere (d) $f'(0)$ does not exist

29. If $|z_1| = |z_2| = 1$ and amp z_1 + amp $z_2 = 0$ then

(a) $z_1 z_2 = 1$ (b) $z_1 + z_2 = 0$

(c) $z_1 = \bar{z}_2$ (d) $z_1 = z_2$

30. A perpendicular PQ is dropped from the point $P(1, 2, 3)$ to the plane $x + y + z = 3$, Q being the foot. Then

(a) $PQ = 3$ (b) $PQ = \sqrt{3}$

(c) $Q = (0, 1, 2)$ (d) $Q = (2, 1, 0)$

Comprehension-Type Questions

Let $f(x, y) = 0$ be the equation of a curve. The length of the perpendicular OP from the point O to the tangent to the curve at $P(x, y)$ is p and the distance of the point $P(x, y)$ from O is r. Then the functional relation between p and r is called the pedal equation of the curve with regard to O. Also, the locus of the foot P of the perpendicular from O is called the first positive pedal curve of the curve $f(x, y) = 0$ with regard to O.

31. The pedal equation of a parabola $y^2 = 4x$ with regard to the focus is

(a) $r^2 = p$ (b) $rp = 1$

(c) $p^2 = r$ (d) $p^2 = 1 + r^2$

32. The pedal equation of the circle $x^2 + y^2 = 4$ with regard to the point $(2, 0)$ is

(a) $r^2 = 4$ (b) $4p = r^2$

(c) $p^2 = 4r$ (d) $p = 4r^2$

33. The equation of the first positive pedal curve of the ellipse $3x^2 + 4y^2 = 12$ with regard to the focus whose abscissa is positive is

(a) $x^2 + y^2 - 2x + 1 = 0$ (b) $x^2 + y^2 + 2x - 1 = 0$

(c) $x^2 + y^2 = 3$ (d) $x^2 + y^2 = 4$

Matching Questions

One or more may match with the same.

34. (i) Let $f : R \to R$ such that $f(x+y) = f(x) + f(y)$ for all x, y. If $\sum_{n=1}^{5}\left(\sum_{r=1}^{n} rf(r)\right) = 315$ then $f\left(\frac{7}{3}\right)$ equals — (a) 4

(ii) A line $\frac{x+2}{1} = \frac{y-3}{2} = \frac{2z-k}{6}$ cuts the yz-plane and the xy-plane at A and B respectively. If AB subtends a right angle at the origin then k is — (b) 8

(iii) If $2\log_2\log_2 x + \log_{1/2}\log_2 2\sqrt{2}x = 1$ for all $x \in R$ then x equals — (c) 7

(iv) Let a, b and c be real numbers such that $a + 2b + c = 4$. Then the greatest value of $ab + bc + ca$ is — (d) 9

35. (i) If $\vec{a} \times (\vec{b} \times \vec{c}) + (\vec{a} \cdot \vec{b})\,\vec{b} = (4 - 2\lambda - \sin\alpha)\,\vec{b} + (\lambda^2 - 1)\,\vec{c}$ and $(\vec{c} \cdot \vec{c})\,\vec{a} = \vec{c}$, where $\vec{b}, \vec{c}$ are noncollinear, then λ is — (a) 8

(ii) A coin whose two faces are marked 3 and 5 is tossed 4 times. The probability of getting a sum less than 15 is λ. Then 16λ is equal to — (b) 1

(iii) $\lim_{x \to 2} \frac{2^x + 2^{3-x} - 6}{2^{-x/2} - 2^{1-x}}$ is — (c) 3

(iv) The greatest value of $f(x) = \left|\frac{x}{2} + 1\right| - \left|\frac{x}{2} - 1\right| + 1$ is — (d) 5

Complete the following statements.

36. The area bounded by the curve $y = x(3 - x)^2$, the x-axis and the lines of ordinates of the maximum and minimum values of the curve is

37. Let $f(x)$ be a continuous function such that $f(x+4) = f(x+2) - f(x)$ and $\phi(x) = \int_x^{x+12} f(x)\, dx$. If $\phi(3) = 5$ then $\int_4^{16} f(x)\, dx$ equals

Assertion-Reason Type

38. Let $0 < \alpha < \beta < \frac{\pi}{2}$.

STATEMENT-1: $\frac{\tan \beta}{\tan \alpha} > \frac{\alpha}{\beta}$.

<u>*because*</u>

STATEMENT-2: If $f''(x) > 0$, $x_1 > x_2$ then $f(x_1) > f(x_2)$.

(a) Statement-1 is True, Statement-2 is True; Statement-2 is a correct explanation for Statement-1

(b) Statement-1 is True, Statement-2 is True; Statement-2 is *Not* a correct explanation for Statement-1

(c) Statement-1 is True, Statement-2 is False

(d) Statement-1 is False, Statement-2 is True

Answers

1. c	**2.** b	**3.** d	**4.** c	**5.** d
6. d	**7.** a	**8.** b	**9.** a	**10.** c
11. a	**12.** b	**13.** b	**14.** b	**15.** c
16. c	**17.** c	**18.** b	**19.** b	**20.** a
21. d	**22.** a	**23.** a	**24.** c	**25.** b
26. a	**27.** a, d	**28.** b, d	**29.** a, c	**30.** b, c
31. c	**32.** b	**33.** d		

34. (i) $\leftrightarrow$ c (ii) $\leftrightarrow$ d (iii) $\leftrightarrow$ b (iv) $\leftrightarrow$ a

35. (i) $\leftrightarrow$ b (ii) $\leftrightarrow$ d (iii) $\leftrightarrow$ a (iv) $\leftrightarrow$ c

36. 4	**37.** 5	**38.** c

Hints and Solutions to Selected Questions

1. $\bar{z} + i\bar{\omega} = 0 \Rightarrow z + \bar{i}\omega = 0 \Rightarrow z = i\omega \Rightarrow \omega = -iz$

$\therefore \quad \arg(z\omega) = \pi \Rightarrow \arg(-iz^2) = \pi \Rightarrow \arg(-i) + \arg(z^2) = \pi$

$\Rightarrow \quad -\frac{\pi}{2} + 2\arg z = \pi \Rightarrow \arg z = \frac{3\pi}{4}.$

2. $|x^2 - y^2 + 2ixy - 1| = x^2 + y^2 + 1 \Rightarrow (x^2 - y^2 - 1)^2 + 4x^2y^2 = (x^2 + y^2 + 1)^2$

$\Rightarrow \quad 4x^2y^2 = \{(x^2 + y^2 + 1) - (x^2 - y^2 - 1)\}\{(x^2 + y^2 + 1) + (x^2 - y^2 - 1)\}$

$= (2y^2 + 2)(2x^2) = 4x^2y^2 + 4x^2 \Rightarrow x = 0.$

3. $A = (-1)\begin{bmatrix} 0 & 0 & 1 \\ 0 & 1 & 0 \\ 1 & 0 & 0 \end{bmatrix}$. But $\begin{bmatrix} 0 & 0 & 1 \\ 0 & 1 & 0 \\ 1 & 0 & 0 \end{bmatrix}$ is neither a zero matrix nor a unit matrix.

Also $|A| = 1 \neq 0$. So (a), (b), (c) are not correct. Check (d).

4. Here, $(1 - p)$ is a root of $x^2 + px + 1 - p = 0$. If the other root is α,

$\alpha + 1 - p = -p$ and $\alpha(1 - p) = 1 - p$

$\therefore \quad \alpha = -1$ and $-1 + p = 1 - p$, i.e., $p = 1$. So, $1 - p = 0$.

5. $\{f(x)\}^2 = a^2 + b^2 + 2\sqrt{(a^4 + b^4)\sin^2 x \cos^2 x + a^2b^2(\cos^4 x + \sin^4 x)}$

$= a^2 + b^2 + \sqrt{(a^4 + b^4)\sin^2 2x + 4a^2b^2(1 - 2\sin^2 x \cos^2 x)}$

$= a^2 + b^2 + \sqrt{(a^4 + b^4)\sin^2 2x + 4a^2b^2 - 2a^2b^2\sin^2 2x}$

$= a^2 + b^2 + \sqrt{(a^4 + b^4)\sin^2 2x + 4a^2b^2 - 2a^2b^2\sin^2 2x}$

$= a^2 + b^2 + \sqrt{(a^2 - b^2)^2 \sin^2 2x + 4a^2b^2}.$

$\therefore \quad \max\{f(x)\}^2 = a^2 + b^2 + \sqrt{(a^2 - b^2)^2 + 4a^2b^2},$

$= a^2 + b^2 + (a^2 + b^2).$

$\min\{f(x)\}^2 = a^2 + b^2 + \sqrt{0 + 4a^2b^2}$

$= a^2 + b^2 + 2ab.$

$\therefore \quad$ difference $= 2(a^2 + b^2) - (a^2 + b^2 + 2ab) = (a - b)^2.$

6. $\dfrac{dy}{dt} = 2\dfrac{dx}{dt}$ (given) $\Rightarrow \dfrac{dy}{dx} = 2.$

$$y^2 = 18x \quad \Rightarrow 2y\frac{dy}{dx} = 18 \quad \Rightarrow 4y = 18 \quad \Rightarrow y = \frac{9}{2}$$

$$\therefore \quad x = \frac{y^2}{18} = \frac{\left(\frac{9}{2}\right)^2}{18} = \frac{9}{8}.$$

7. $\dfrac{dy}{dx} = \dfrac{a\cos\theta}{-a\sin\theta} = -\cot\theta$. So, the equation of the normal at 'θ' is

$$y - a\sin\theta = -\frac{1}{-\cot\theta}\{x - a(1+\cos\theta)\} \qquad \Rightarrow y = (x-a)\tan\theta,$$

which passes through $(a, 0)$.

8. $I = \displaystyle\int_0^{\pi} xf(\sin x)\,dx = \int_{\pi}^{0} (\pi - z)f(\sin z)\,(-dz) = \int_0^{\pi} (\pi - x)f(\sin x)\,dx$

$$\therefore \quad 2I = \pi\int_0^{\pi} f(\sin x)\,dx = \pi\left\{\int_0^{\pi/2} f(\sin x)\,dx + \int_{\pi/2}^{\pi} f(\sin x)\,dx\right\}$$

$$\int_{\pi/2}^{\pi} f(\sin x)\,dx = \int_{\pi/2}^{0} f\{\sin(\pi - z)\}\,(-dz) = \int_0^{\pi/2} f(\sin z)\,dz$$

$$\therefore \quad 2I = \pi\cdot 2\int_0^{\pi/2} f(\sin x)\,dx.$$

9. Here, $f(a) + f(-a) = 1$ because $\dfrac{e^a}{1+e^a} + \dfrac{e^{-a}}{1+e^{-a}} = 1$

$$I_1 = \int_{1-f(-a)}^{1-f(a)} (1-z)g\{(1-z)z\}\,(-dz), \quad \text{(putting } 1 - x = z)$$

$$= \int_{f(a)}^{f(-a)} -(1-x)g\{x(1-x)\}\,dx = \int_{f(-a)}^{f(a)} (1-x)g\{x(1-x)\}\,dx$$

$= I_2 - I_1$. So, $2I_1 = I_2$.

12. If (α, β) is the centre then radius $= \sqrt{(\alpha - a)^2 + (\beta - b)^2}$.

$\therefore$ the circle is $(x-\alpha)^2 + (y-\beta)^2 = (a-\alpha)^2 + (b-\beta)^2$

or $x^2 + y^2 - 2\alpha x - 2\beta y + (2a\alpha + 2b\beta - a^2 - b^2) = 0$. For orthogonality,

$c_1 + c_2 = 2g_1 g_2 + 2f_1 f_2 \Rightarrow 2a\alpha + 2b\beta - a^2 - b^2 - 4 = 0.$

$\therefore$ the locus is $2ax + 2by - (a^2 + b^2 + 4) = 0$.

13. Here, $e = \frac{1}{2}$ and $\frac{a}{e} = 4$, and the equation in standard form is $\frac{x^2}{a^2} + \frac{y^2}{b^2} = 1$.

$\therefore \quad \frac{a}{\frac{1}{2}} = 4 \Rightarrow a = 2$. So, $b^2 = a^2(1 - e^2) \Rightarrow b^2 = 4\left(1 - \frac{1}{4}\right) = 3$

$\therefore$ the equation is $\frac{x^2}{4} + \frac{y^2}{3} = 1$.

14. Any point on the line $\frac{x}{1} = \frac{y + a}{1} = \frac{z}{1}$ is $(t, -a + t, t)$ and on the line $\frac{x + a}{2} = \frac{y}{1} = \frac{z}{1}$ is $(-a + 2s, s, s)$.

From the question, $\frac{t - (-a + 2s)}{2} = \frac{-a + t - s}{1} = \frac{t - s}{2}$

$$= \frac{-a + t - s - (t - s)}{1 - 2}$$

$$= \frac{t + a - 2s - (-a + t - s)}{2 - 1}.$$

$\therefore \quad \frac{t - s}{2} = a = \frac{2a - s}{1}$. So, $s = a$ and $t = 3a$.

$\therefore \quad P = (3a, 2a, 3a)$ and $Q = (a, a, a)$. So, $PQ = 3a$.

15. A point on $2x + y + 2z = 8$ is $(0, 8, 0)$. Its distance from the other plane is the required distance.

16. Let $\vec{b} = x\vec{i} + y\vec{j} + z\vec{k}$. Then $\vec{a} \cdot \vec{b} = 1 \Rightarrow x + y + z = 1$.

$\vec{a} \times \vec{b} = \vec{j} - \vec{k} \Rightarrow (z - y)\vec{i} + (x - z)\vec{j} + (y - x)\vec{k} = \vec{j} - \vec{k}$

$$\Rightarrow z - y = 0,\ x - z = 1,\ y - x = -1.$$

Solving $x + y + z = 1$, $z = y$, $y - x = -1$, we get $x = 1$, $y = 0$, $z = 0$.

17. A vector lying in the plane of $\vec{b}$, $\vec{c}$ and perpendicular to $\vec{a}$ is

$$\vec{a} \times (\vec{b} \times \vec{c}) = (\vec{a} \cdot \vec{c})\,\vec{b} - (\vec{a} \cdot \vec{b})\,\vec{c}$$

$$= 7(2\vec{i} + \vec{j} + \vec{k}) - 14(\vec{i} - \vec{j} + \vec{k}) = 7(3\vec{j} - \vec{k}).$$

18. $y = g(f(x)) = \{f(x)\}^2 - 1 = (\sin x + \cos x)^2 - 1 = \sin 2x$

$\Rightarrow \quad x = \frac{1}{2}\sin^{-1} y.$

As $\frac{-\pi}{2} \le \sin^{-1} y \le \frac{\pi}{2}, \quad -\frac{\pi}{4} \le x \le \frac{\pi}{4}.$

19. $ax^2 + bx + c > 0$ for all $x \in R$ if $a > 0,\ D < 0$.

Here, the condition is $(2a)^2 - 4 \cdot 1 \cdot (10 - 3a) < 0$ or $a^2 + 3a - 10 < 0$

or $\quad (a+5)(a-2) < 0 \Rightarrow -5 < a < 2$ (by sign scheme).

20. Use $C_3 \to C_3 - C_2, C_2 \to C_2 - C_1$. Then $C_2 \equiv C_3$.

21. $a = k,\ b = \sqrt{3}k,\ c = 2k.$

$$\cos C = \frac{k^2 + 3k^2 - 4k^2}{2 \cdot k \cdot \sqrt{3}k} = 0 \Rightarrow C = 90°;$$

$$\cos A = \frac{3k^2 + 4k^2 - k^2}{2\sqrt{3}k \cdot 2k} = \frac{6k^2}{4\sqrt{3}k^2} = \frac{\sqrt{3}}{2}.$$

$\therefore \quad A = 30°$ and $B = 60°. \qquad \therefore \quad A : B : C = 1 : 2 : 3.$

22. Selections may contain (4, 2), (3, 3) and (2, 4) questions from the two groups.

$\therefore \quad n(S) = {}^5C_4 \times {}^5C_2 + {}^5C_3 \times {}^5C_3 + {}^5C_2 \times {}^5C_4 = 200$ and

$n(E) = {}^5C_3 \times {}^5C_3 = 100.$

23. Here,

$$f(x) = \begin{cases} xe^{-2/x} & \text{for } x > 0 \\ 0 & \text{for } x = 0 \\ x & \text{for } x < 0. \end{cases}$$

$\therefore \quad \lim_{h \to 0} f(0+h) = \lim_{h \to 0} he^{-2/h} = 0 \times e^{-\infty} = 0 \times 0 = 0$ and

$\lim_{h \to 0} f(0-h) = \lim_{h \to 0} (0-h) = 0,\ f(0) = 0.$

So, $f(x)$ is continuous at $x = 0$.

Also, $xe^{-2/x}$, x, 0 are continuous in their respective intervals.

$\therefore \quad f(x)$ is continuous everywhere.

Again, $\lim_{h \to 0} \dfrac{f(0+h)-f(0)}{h} = \lim_{h \to 0} \dfrac{he^{-2/h} - 0}{h} = 0$ and

$$\lim_{h \to 0} \frac{f(0-h)-f(0)}{-h} = \lim_{h \to 0} \frac{-h-0}{-h} = 1.$$

So, $f(x)$ is not differentiable at $x = 0$.

24. $f(0+0) = f(0) + f(0) \Rightarrow f(0) = 0.$

$f(r) = f(1 + \overline{r-1}) = f(1) + f(r-1) = f(1) + \{f(1 + \overline{r-2})\} = f(1) + f(1) + f(r-2)$

$= 2f(1) + f(r-2) = \dots = rf(1) + f(r-r) = rf(1) + 0 = 7r.$

$$\therefore \quad \sum_{r=1}^{n} f(r) = 7 \cdot \sum_{r=1}^{n} r = 7 \cdot \frac{n(n+1)}{2}.$$

25. Using L'Hospital's rule, $\lim_{x \to a} \dfrac{\int_a^x f(x)\,dx + x \cdot f(x)}{1} = af(a) = a^3 + a^2$

$\Rightarrow \quad f(a) = a^2 + a \Rightarrow f(2a) = (2a)^2 + 2a = 2a(2a+1).$

26. Solving $x^2 - 2(1+2y)x + 6y^2 - 20y - 29 = 0$, we get

$x = 2y + 1 \pm \sqrt{102 - 2(y-6)^2}$. As x, y are positive integers, $102 - 2(y-6)^2$ must be a perfect square of an integer. So, $(y-6)^2 = 1, 49$

$\therefore \quad y - 6 = \pm 1, \pm 7 \quad \Rightarrow \quad y = 7, 5, 13, -1$ (absurd)

$\therefore \quad y$ has the values 7, 5, 13. For each of these values we get two positive integral values of x. So (x, y) will have 3×2, i.e., 6 values.

27. Here, $\dfrac{x^2}{18} - \dfrac{y^2}{9} = 1$. Any tangent to it has the equation

$y = mx \pm \sqrt{a^2m^2 - b^2}$ or $y = mx \pm \sqrt{18m^2 - 9}$.

It is parallel to $x - y = 0$ if $m = 1$. Then the tangent is $y = x \pm 3$.

28. $|x|$ is continuous for all x and differentiable for all x except at $x = 0$.

$\therefore \quad |\sin x|$ is continuous for all x and differentiable for all x except at $\sin x = 0$, i.e., $x = 0, \pm\pi, \pm 2\pi, \dots\dots$. So, $f(x) = 1 + |\sin x|$ is continuous everywhere and $f'(0)$ does not exist.

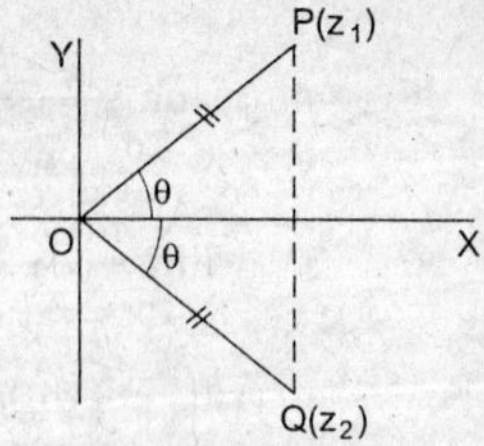

29. If amp $z_1 = \theta$ then amp $z_2 = -\text{amp } z_1 = -\theta$.

Also, $OP = |z_1| = |z_2| = OQ$. Hence, Q is the image of P by the real axis. So,

$$z_2 = \bar{z}_1 \Rightarrow \bar{z}_2 = z_1$$

$$\Rightarrow z_2\bar{z}_2 = z_2 z_1 \Rightarrow z_1 z_2 = |z_2|^2 = 1.$$

30. $PQ = \left| \dfrac{1+2+3-3}{\sqrt{1^2+1^2+1^2}} \right| = \sqrt{3}$. Let $Q = (\alpha, \beta, \gamma)$. Then $\alpha + \beta + \gamma = 3$ and

d.r. of $PQ = \alpha - 1, \beta - 2, \gamma - 3$, which is perpendicular to the plane.

$$\therefore \quad \frac{\alpha-1}{1} = \frac{\beta-2}{1} = \frac{\gamma-3}{1} = \frac{\alpha+\beta+\gamma-6}{3} = \frac{3-6}{3} = -1$$

$\therefore \quad \alpha = 0, \beta = 1, \gamma = 2$. So, $Q = (0, 1, 2)$.

31. The tangent at $(t^2, 2t)$ is $yt = x + t^2$ and $r^2 = (t^2-1)^2 + 4t^2 = (t^2+1)^2$.

$\therefore \quad r = t^2 + 1$.

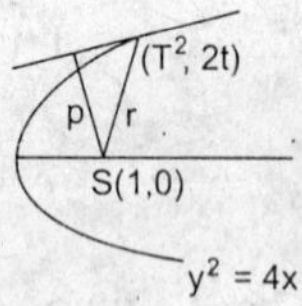

Also, $p = \left| \dfrac{1+t^2}{\sqrt{1+t^2}} \right| = \sqrt{1+t^2}$. $\quad \therefore \; p^2 = 1 + t^2$.

$\therefore \quad p^2 = r$.

32. Clearly, $\dfrac{p}{r} = \sin\theta = \dfrac{r}{4}$.

So, $r^2 = 4p$.

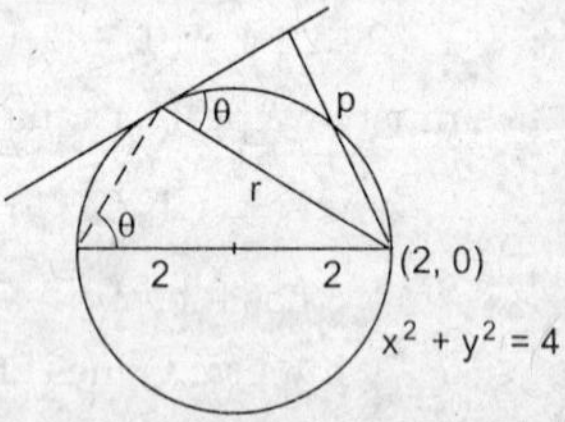

33. Here, $\dfrac{x^2}{4} + \dfrac{y^2}{3} = 1$. So, $a^2 = 4$ and $b^2 = 3$.

$\therefore \quad b^2 = a^2(1 - e^2)$

$\Rightarrow \quad 3 = 4(1 - e^2) \quad \Rightarrow \quad e = \dfrac{1}{2}$.

$\therefore \quad$ focus $S = (ae, 0) = (1, 0)$.

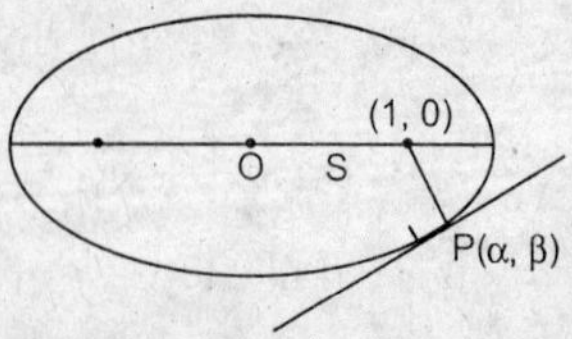

Any tangent to the ellipse is $y = mx + \sqrt{a^2m^2 + b^2}$

or $\quad y = mx + \sqrt{4m^2 + 3}$.

Let $P(\alpha, \beta)$ be the foot of the perpendicular to the tangent from the focus S.

Then $\frac{\beta}{\alpha-1}\cdot m=-1$ and $\beta=m\alpha+\sqrt{4m^2+3}$

Eliminating m, $\beta=\frac{1-\alpha}{\beta}\cdot\alpha+\sqrt{4\cdot\left(\frac{1-\alpha}{\beta}\right)^2+3}$

or $\quad \beta^2=\alpha-\alpha^2+\sqrt{4(1-\alpha)^2+3\beta^2}$

or $\quad \{(\alpha^2+\beta^2)-\alpha\}^2=4(1-\alpha)^2+3\beta^2$

or $\quad \lambda^2-2\lambda\alpha=3\lambda-8\alpha+4 \qquad$ (where $\alpha^2+\beta^2=\lambda$)

or $\quad \lambda^2-\lambda(2\alpha+3)+8\alpha-4=0 \quad$ or $\quad (\lambda-4)(\lambda-2\alpha+1)=0.$

$\therefore$ the locus of P is $x^2+y^2-4=0$ or $x^2+y^2-2x+1=0$

But (2, 0) is a position of P and it does not satisfy $x^2+y^2-2x+1=0$.

34. (i) Here, $\sum_{n=1}^{5}\{f(1)+2f(2)+3f(3)+\ldots+nf(n)\}=315$

Now, $f(x+y)=f(x)+f(y)$.

$\therefore \quad f(2)=f(1)+f(1)=2f(1),\ f(3)=f(2)+f(1)=3f(1)$, etc.

$\therefore \quad \sum_{n=1}^{5}(1^2+2^2+3^2+\ldots+n^2)\,f(1)=315$

$$\Rightarrow \quad f(1)\sum_{n=1}^{5}\frac{n(n+1)(2n+1)}{6}=315$$

$$\Rightarrow \quad f(1)\sum_{n=1}^{5}\left(\frac{1}{3}n^3+\frac{1}{2}n^2+\frac{1}{6}n\right)=315$$

$$\Rightarrow \quad f(1)\left[\frac{1}{3}\cdot\left\{\frac{5(5+1)}{2}\right\}^2+\frac{1}{2}\cdot\frac{5(5+1)(10+1)}{6}+\frac{1}{6}\cdot\frac{5(5+1)}{2}\right]=315$$

$\Rightarrow \quad f(1)=3.$

Now, $f(7)=f\left(\frac{7}{3}+\frac{7}{3}+\frac{7}{3}\right)=3f\left(\frac{7}{3}\right) \qquad (\because\ f(x+y)=f(x)+f(y))$

or $\quad 7f(1)=3f\left(\frac{7}{3}\right). \quad \therefore\ f\left(\frac{7}{3}\right)=\frac{7}{3}f(1)=\frac{7}{3}\times3=7.$

(ii) Any point on the line $=\left(r-2, 2r+3, 3r+\dfrac{k}{2}\right)$

The yz-plane is $x=0$ and the xy-plane is $z=0$.

$\therefore \quad A=\left(0, 7, \dfrac{12+k}{2}\right)$ and $B=\left(-\dfrac{12+k}{6}, \dfrac{9-k}{3}, 0\right)$

$\therefore \quad OA \perp OB \quad \Rightarrow 0\cdot\left(-\dfrac{12+k}{6}\right)+7\cdot\dfrac{9-k}{3}+0=0 \quad \Rightarrow k=9.$

(iii) Let $\log_2 x = y$. Then $2\log_2 y + \dfrac{\log_2\log_2(2^{3/2}\cdot x)}{\log_2 \frac{1}{2}} = 1$

$\Rightarrow \quad 2\log_2 y + \dfrac{\log_2\left(\frac{3}{2}+\log_2 x\right)}{-1} = 1 \quad \Rightarrow \log_2 y^2 - \log_2\left(\dfrac{3}{2}+y\right) = 1$

$\Rightarrow \quad \log_2 \dfrac{y^2}{\frac{3}{2}+y} = 1 \quad \Rightarrow \dfrac{2y^2}{3+2y} = 2 \quad \Rightarrow y^2-2y-3=0$

$\Rightarrow \quad (y-3)(y+1)=0 \quad \Rightarrow \log_2 x = 3, -1.$

But $\log_2 x > 0$ for $\log_2\log_2 x$ to be real.

$\therefore \quad \log_2 x = 3 \quad \Rightarrow x = 2^3 = 8.$

(iv) Let $a+b=4\cos^2\theta$, $b+c=4\sin^2\theta$. $\Rightarrow (a+b)(b+c)=16\cos^2\theta\cdot\sin^2\theta$

$\Rightarrow \quad ab+bc+ca+b^2=4\sin^2 2\theta \quad \Rightarrow ab+bc+ca \le 4\sin^2 2\theta \le 4.$

35. (i) Here, $(\vec{a}\cdot\vec{c})\vec{b}-(\vec{a}\cdot\vec{b})\vec{c}+(\vec{a}\cdot\vec{b})\vec{b}=(4-2\lambda-\sin\alpha)\vec{b}+(\lambda^2-1)\vec{c}$

$\Rightarrow \quad \vec{a}\cdot\vec{c}+\vec{a}\cdot\vec{b}=4-2\lambda-\sin\alpha, -\vec{a}\cdot\vec{b}=\lambda^2-1.$

Adding, $\vec{a}\cdot\vec{c}=\lambda^2-1+4-2\lambda-\sin\alpha.$

Now, $(\vec{c}\cdot\vec{c})\vec{a}=\vec{c} \quad \Rightarrow (\vec{c}\cdot\vec{c})(\vec{a}\cdot\vec{c})=\vec{c}\cdot\vec{c} \quad \Rightarrow \vec{a}\cdot\vec{c}=1.$

$\therefore \quad 1=\lambda^2+3-2\lambda-\sin\alpha \quad \Rightarrow (\lambda^2-2\lambda+1)+(1-\sin\alpha)=0$

$\Rightarrow \quad (\lambda-1)^2+\left(\cos\dfrac{\alpha}{2}-\sin\dfrac{\alpha}{2}\right)^2=0 \quad \Rightarrow \lambda=1, \cos\dfrac{\alpha}{2}=\sin\dfrac{\alpha}{2}.$

(ii) Let throwing a 3 in one trial be a success and throwing a 5 is a failure. Then $p=\dfrac{1}{2}, q=1-\dfrac{1}{2}=\dfrac{1}{2}$.

$3+3+3+3<15, \ 3+3+3+5<15.$

So, a sum less than 15 is obtained when there are 4 successes or 3 successes.

$\therefore$ the probability $= {}^4C_3 p^3 q + {}^4C_4 p^4 = {}^4C_3 \left(\frac{1}{2}\right)^3 \cdot \frac{1}{2} + {}^4C_4 \left(\frac{1}{2}\right)^4$.

$\therefore \quad \lambda = 4 \cdot \left(\frac{1}{2}\right)^4 + \left(\frac{1}{2}\right)^4 = 5 \cdot \frac{1}{16}. \quad \therefore \quad 16\lambda = 5.$

(iii) $\text{Limit} = \lim_{x \to 2} \frac{(2^x)^2 - 6(2^x) + 8}{\sqrt{2^x} - 2} = \lim_{x \to 2} \frac{(2^x - 4)(2^x - 2)}{\sqrt{2^x} - 2}$

$= \lim_{x \to 2} \{(\sqrt{2^x} + 2)(2^x - 2)\} = (2 + 2)(4 - 2) = 8.$

(iv) $f(x) = \frac{1}{2}|x + 2| - \frac{1}{2}|x - 2| + 1.$

$$\therefore \quad f(x) = \begin{cases} -\frac{1}{2}(x+2) + \frac{1}{2}(x-2) + 1 & \text{for } x < -2 \\ \frac{1}{2}(x+2) + \frac{1}{2}(x-2) + 1 & \text{for } -2 \le x < 2 \\ \frac{1}{2}(x+2) - \frac{1}{2}(x-2) + 1 & \text{for } x \ge 2 \end{cases}$$

$$\Rightarrow \quad f(x) = \begin{cases} -1 & \text{for } x < -2 \\ x + 1 & \text{for } -2 \le x < 2 \\ 3 & \text{for } x \ge 2 \end{cases}$$

$\therefore \quad \max\{f(x)\} = 3.$

36. $y = x(3 - x)^2 \quad \Rightarrow \quad \frac{dy}{dx} = (3 - x)^2 + 2x(3 - x)(-1).$

$\therefore$ for maximum or minimum y,

$\frac{dy}{dx} = 0 = (3 - x)(3 - x - 2x) \quad \Rightarrow \quad x = 3, 1.$

$\frac{d^2y}{dx^2} = -(3 - 3x) + (3 - x)(-3) = 6x - 12$

$\therefore \quad \left.\frac{d^2y}{dx^2}\right]_{x=3} > 0 \text{ and } \left.\frac{d^2y}{dx^2}\right]_{x=1} < 0.$

So, y is maximum at $x = 1$ and minimum at $x = 3$.

$\therefore \quad \text{area} = \int_1^3 y\,dx = \int_1^3 x(3 - x)^2\,dx$

$= \int_2^0 (3 - z)z^2(-dz)$ (putting $3 - x = z$)

$= \int_0^2 (3z^2 - z^3)\,dz = \left[z^3 - \frac{z^4}{4}\right]_0^2 = 8 - \frac{16}{4} = 4.$

37. $\phi(x) = \int_{x}^{x+12} f(x)\,dx = \int_{0}^{x+12} f(x)\,dx - \int_{0}^{x} f(x)\,dx$

$\therefore \quad \phi'(x) = f(x+12) - f(x).$

As $f(x+4) = f(x+2) - f(x)$,

$$f(x+12) = f(x+10) - f(x+8)$$
$$= \{f(x+8) - f(x+6)\} - f(x+8) = -f(x+6)$$
$$= -\{f(x+4) - f(x+2)\} = f(x).$$

$\therefore \quad \phi'(x) = f(x) - f(x) = 0 \quad \Rightarrow \phi(x) = \text{constant} = 5 \qquad (\because \ \phi(3) = 5).$

$\therefore \quad \int_{4}^{16} f(x)\,dx = \phi(4) = 5.$

38. Let $f(x) = x\tan x$. Then $f'(x_1) = \tan x + x\sec^2 x > 0$ in $\left(0, \frac{\pi}{2}\right)$.

$\therefore$ $f(x)$ is m.i. So, $\beta > \alpha \Rightarrow f(\beta) > f(\alpha)$; $\therefore$ $\beta\tan\beta > \alpha\tan\alpha \Rightarrow \frac{\tan\beta}{\tan\alpha} > \frac{\alpha}{\beta}$.

$\therefore$ Statement-1 is True.

$f''(x) > 0 \Rightarrow f'(x)$ is m.i. So, $f'(x_1) > f'(x_2)$. It does not imply $f(x)$ is m.i.

$\therefore$ Statement-2 is False.

□

IIT Questions—1

One-Option-Correct Questions

1. For all $x \in (0, 1)$,

   (a) $e^x < 1 + x$ (b) $\log_e(1 + x) < x$

   (c) $\sin x > x$ (d) $\log_e x > x$

2. The incentre of the triangle with vertices $(1, \sqrt{3})$, $(0, 0)$ and $(2, 0)$ is

   (a) $\left(1, \frac{\sqrt{3}}{2}\right)$ (b) $\left(\frac{2}{3}, \frac{1}{\sqrt{3}}\right)$

   (c) $\left(\frac{2}{3}, \frac{\sqrt{3}}{2}\right)$ (d) $\left(1, \frac{1}{\sqrt{3}}\right)$

3. The line $x - 1 = 0$ is the directrix of the parabola $y^2 - kx + 8 = 0$. Then one of the values of k is

   (a) $\frac{1}{8}$ (b) 8

   (c) 4 (d) $\frac{1}{4}$

4. The domain of definition of the function $y(x)$ given by $2^x + 2^y = 2$ is

   (a) $(0, 1]$ (b) $[0, 1]$

   (c) $(-\infty, 0]$ (d) $(-\infty, 1)$

5. The triangle PQR is inscribed in the circle $x^2 + y^2 = 25$. If Q and R have coordinates $(3, 4)$ and $(-4, 3)$, respectively then $\angle QPR$ is equal to

   (a) $\frac{\pi}{2}$ (b) $\frac{\pi}{3}$

   (c) $\frac{\pi}{4}$ (d) $\frac{\pi}{6}$

6. Let $f(x) = \int e^x \cdot (x - 1)(x - 2)\, dx$. Then f decreases in the interval

   (a) $(-\infty, -2)$ (b) $(-2, -1)$

(c) $(1, 2)$ (d) $(2, +\infty)$

7. Let $g(x) = \int_0^x f(t)\,dt$, where $\frac{1}{2} \le f(t) \le 1$ for $t \in [0, 1]$ and $0 \le f(t) \le \frac{1}{2}$ for $t \in (1, 2]$. Then

(a) $-\frac{3}{2} \le g(2) < \frac{1}{2}$ (b) $0 \le g(2) < 2$

(c) $\frac{3}{2} < g(2) \le \frac{5}{2}$ (d) $2 < g(2) < 4$

8. If $x^2 + y^2 = 1$ then $\left(\text{where } y' = \frac{dy}{dx}, y'' = \frac{d^2y}{dx^2}\right)$

(a) $yy'' - 2y'^2 + 1 = 0$ (b) $yy'' + y'^2 + 1 = 0$

(c) $yy'' - y'^2 - 1 = 0$ (d) $yy'' + 2y'^2 + 1 = 0$

9. If the vectors $\vec{a}$, $\vec{b}$ and $\vec{c}$ are represented by the sides BC, CA and AB, respectively, of the ΔABC then

(a) $\vec{a} \cdot \vec{b} + \vec{b} \cdot \vec{c} + \vec{c} \cdot \vec{a} = 0$ (b) $\vec{a} \times \vec{b} = \vec{b} \times \vec{c} = \vec{c} \times \vec{a}$

(c) $\vec{a} \cdot \vec{b} = \vec{b} \cdot \vec{c} = \vec{c} \cdot \vec{a}$ (d) $\vec{a} \times \vec{b} + \vec{b} \times \vec{c} + \vec{c} \times \vec{a} = \vec{0}$

10. A pole stands vertically inside a triangular park ABC. If the angle of elevation of the top of the pole from each corner of the park is the same then in the ΔABC, the foot of the pole is at the

(a) centroid (b) circumcentre

(c) incentre (d) orthocentre

11. Let $f(\theta) = (\sin\theta)(\sin\theta + \sin 3\theta)$. Then

(a) $f(\theta) \ge 0$ only when $\theta \ge 0$ (b) $f(\theta) \le 0$ for all real θ

(c) $f(\theta) \ge 0$ for all real θ (d) $f(\theta) \le 0$ only when $\theta \le 0$

12. If $x + y = k$ is a normal to the parabola $y^2 = 12x$ then k is

(a) 3 (b) 9

(c) –9 (d) –3

13. If z_1, z_2, z_3 are complex numbers such that $|z_1| = |z_2| = |z_3| = \left|\frac{1}{z_1} + \frac{1}{z_2} + \frac{1}{z_3}\right| = 1$ then $|z_1 + z_2 + z_3|$ is

(a) equal to 1 (b) less than 1

(c) greater than 3 (d) equal to 3

14. If $\arg z < 0$ then $\arg(-z) - \arg z$ is equal to

(a) π (b) $-\pi$

(c) $-\frac{\pi}{2}$ (d) $\frac{\pi}{2}$

15. If a, b, c, d are positive real numbers such that $a+b+c+d=2$ then $M=(a+b)(c+d)$ satisfies the relation

(a) $0 < M \leq 1$ (b) $1 \leq M \leq 2$

(c) $2 \leq M \leq 3$ (d) $3 \leq M \leq 4$

16. If the circles $x^2+y^2+2x+2ky+6=0$ and $x^2+y^2+2ky+k=0$ intersect orthogonally then k is

(a) 2 or $-\frac{3}{2}$ (b) -2 or $-\frac{3}{2}$

(c) 2 or $\frac{3}{2}$ (d) -2 or $\frac{3}{2}$

17. Let PS be the median of the triangle with vertices $P(2, 2)$, $Q(6, -1)$ and $R(7, 3)$. The equation of the line passing through $(1, -1)$ and parallel to PS is

(a) $2x-9y-7=0$ (b) $2x-9y-11=0$

(c) $2x+9y-11=0$ (d) $2x+9y+7=0$

18. If $\vec{a}$, $\vec{b}$ and $\vec{c}$ are unit coplanar vectors then the scalar triple product $[2\vec{a}-\vec{b}\ \ 2\vec{b}-\vec{c}\ \ 2\vec{c}-\vec{a}]$ is equal to

(a) 0 (b) 1

(c) $-\sqrt{3}$ (d) $\sqrt{3}$

19. Let the vectors $\vec{a}$, $\vec{b}$, $\vec{c}$ and $\vec{d}$ be such that $(\vec{a}\times\vec{b})\times(\vec{c}\times\vec{d})=\vec{0}$. Let P_1 and P_2 be planes determined by pairs of vectors $\vec{a}$, $\vec{b}$ and $\vec{c}$, $\vec{d}$, respectively. Then the angle between P_1 and P_2 is

(a) 0 (b) $\frac{\pi}{4}$

(c) $\frac{\pi}{3}$ (d) $\frac{\pi}{2}$

20. In a ΔABC, $2ac \cdot \sin \dfrac{A-B+C}{2}$ is equal to

(a) $a^2+b^2-c^2$ (b) $c^2+a^2-b^2$
(c) $b^2-c^2-a^2$ (d) $c^2-a^2-b^2$

21. In a ΔABC, let $\angle C = \dfrac{\pi}{2}$. If r and R are the inradius and the circumradius, respectively, of the triangle then $2(r+R)$ is equal to

(a) $a+b$ (b) $b+c$
(c) $c+a$ (d) $a+b+c$

22. If

$$f(x) = \begin{cases} e^{\cos x} \cdot \sin x & \text{for } |x| \le 2 \\ 2 & \text{otherwise} \end{cases}$$

then $\int_{-2}^{3} f(x)\, dx$ is equal to

(a) 0 (b) 1
(c) 2 (d) 3

23. The value of $\int_{e^{-1}}^{e^2} \left| \dfrac{\log_e x}{x} \right| dx$ is

(a) $\dfrac{3}{2}$ (b) $\dfrac{5}{2}$
(c) 3 (d) 5

24. If the system of equations $x-ky-z=0$, $kx-y-z=0$ and $x+y-z=0$ has a nonzero solution then the possible values of k are

(a) $-1, 2$ (b) $1, 2$
(c) $0, 1$ (d) $-1, 1$

25. Consider an infinite geometric series with the first term a and the common ratio r. If its sum is 4 and the second term is $\dfrac{3}{4}$ then (a, r) has the value

(a) $\left(\dfrac{7}{4}, \dfrac{3}{7}\right)$ (b) $\left(2, \dfrac{3}{8}\right)$
(c) $\left(\dfrac{3}{2}, \dfrac{1}{2}\right)$ (d) $\left(3, \dfrac{1}{4}\right)$

26. For $x \in R$, $\lim\limits_{x \to \infty} \left(\dfrac{x-3}{x+2}\right)^x$ is equal to

(a) e (b) e^{-1}

(c) e^{5} (d) e^{5}

27. If the normal to the curve $y = f(x)$ at the point (3, 4) makes an angle $\dfrac{3\pi}{4}$ with the positive x-axis then $f'(3)$ is equal to

(a) -1 (b) $-\dfrac{3}{4}$

(c) $\dfrac{4}{3}$ (d) 1

28. For the equation $3x^2 + px + 3 = 0$ $(p > 0)$, if one of the roots is square of the other then p is equal to

(a) $\dfrac{1}{3}$ (b) 1

(c) 3 (d) $\dfrac{2}{3}$

29. If α and β $(\alpha < \beta)$ are the roots of the equation $x^2 + bx + c = 0$, where $c < 0 < b$ then

(a) $0 < \alpha < \beta$ (b) $\alpha < 0 < \beta < |\alpha|$

(c) $\alpha < \beta < 0$ (d) $\alpha < 0 < |\alpha| < \beta$

30. Let $f: R \to R$ be a function. Define $g: R \to R$ by $g(x) = |f(x)|$ for all x. Then g is

(a) onto if f is onto (b) one-to-one if f is one-to-one

(c) continuous if f is continuous (d) differentiable if f is differentiable

31. For $2 \le r \le n$, $\binom{n}{r} + 2\binom{n}{r-1} + \binom{n}{r-2}$ is equal to

(a) $\binom{n+1}{r-1}$ (b) $2\binom{n+1}{r+1}$

(c) $2\binom{n+2}{r}$ (d) $\binom{n+2}{r}$

32. If $b > a$ then the equation $(x - a)(x - b) = 1$ has
    (a) both roots in $[a, b]$
    (b) both roots in $(-\infty, a)$
    (c) both roots in $(b, +\infty)$
    (d) one root in $(-\infty, a)$ and the other in $(b, +\infty)$

33. Let

$$f(x) = \begin{cases} |x| & \text{for } 0 < |x| \le 2 \\ 1 & \text{for } x = 0. \end{cases}$$

Then at $x = 0$, f has
    (a) a local maximum (b) no local maximum
    (c) a local minimum (d) no extremum

34. How many different nine-digit numbers can be formed from the digits of the number 223355888 by rearrangement of the digits so that the odd digits occupy even places?
    (a) 16 (b) 36
    (c) 60 (d) 180

35. Consider the following statements S and R.

S: Both $\sin x$ and $\cos x$ are decreasing functions in $\left(\frac{\pi}{2}, \pi\right)$.

R: If a differentiable function decreases in (a, b) then its derivative also decreases in (a, b).

Which of the following is true?
    (a) Both S and R are wrong.
    (b) Both S and R are correct but R is not the correct explanation for S.
    (c) S is correct and R is the correct explanation for S.
    (d) S is correct and R is wrong.

36. For $x > 0$, $\lim_{x \to 0} \left\{(\sin x)^{1/x} + \left(\frac{1}{x}\right)^{\sin x}\right\}$ is
    (a) 0 (b) –1
    (c) 1 (d) 2

37. Given an isosceles triangle whose inradius is $\sqrt{3}$. If one of the angles of the triangle is 120° then the area of the triangle in square units is
    (a) 4π (b) $12 + 7\sqrt{3}$
    (c) $12 - 7\sqrt{3}$ (d) $7 + 12\sqrt{3}$

38. Let $\vec{a} = \vec{i} + 2\vec{j} + \vec{k}$, $\vec{b} = \vec{i} - \vec{j} + \vec{k}$ and $\vec{c} = \vec{i} + \vec{j} - \vec{k}$. A vector in the plane of $\vec{a}$ and $\vec{b}$ whose projection on $\vec{c}$ is $\frac{1}{\sqrt{3}}$ is

(a) $3\vec{i} + \vec{j} - 3\vec{k}$ (b) $4\vec{i} - \vec{j} + 4\vec{k}$

(c) $4\vec{i} + \vec{j} - 4\vec{k}$ (d) $2\vec{i} + \vec{j} - 2\vec{k}$

39. A plane passes through $(1, -2, 1)$ and is perpendicular to the planes $2x - 2y + z = 0$ and $x - y + 2z = 4$. Then the distance of that plane from the point $(1, 2, 2)$ is

(a) 0 (b) 1 (c) $\sqrt{2}$ (d) $2\sqrt{2}$

40. The locus of Z representing the complex number z and lying in the shaded region is best represented by

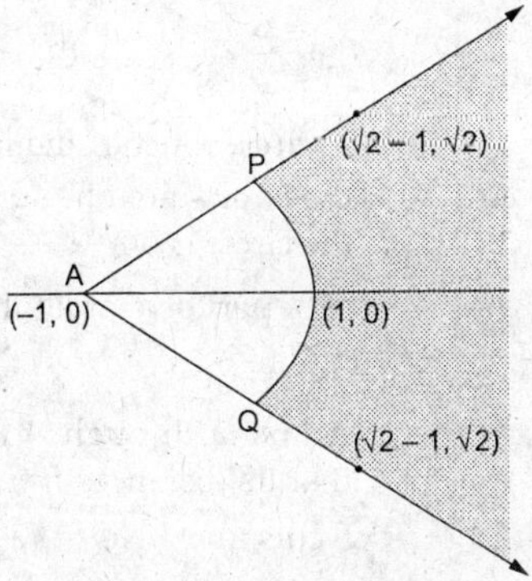

(a) $z : |z + 1| > 2,\ |\arg(z + 1)| < \frac{\pi}{4}$

(b) $z : |z - 1| > 2,\ |\arg(z - 1)| < \frac{\pi}{4}$

(c) $z : |z + 1| < 2,\ |\arg(z + 1)| < \frac{\pi}{2}$

(d) $z : |z - 1| < 2,\ |\arg(z - 1)| < \frac{\pi}{2}$

One-or-More-Options-Correct Questions

41. If $f(x) = \min\{1, x^2, x^3\}$ then

(a) $f(x)$ is continuous $\forall\ x \in R$

(b) $f'(x) > 0\ \ \forall\ x > 1$

(c) $f(x)$ is not differentiable but continuous $\forall\ x \in R$

(d) $f(x)$ is not differentiable for two values of x

42. If a hyperbola passes through the focus of the ellipse $\frac{x^2}{25} + \frac{y^2}{16} = 1$ and its transverse and conjugate axes coincide with the major and minor axes of the ellipse, the product of their eccentricities being 1 then

(a) the equation of the hyperbola is $\frac{x^2}{9} - \frac{y^2}{16} = 1$

(b) the equation of the hyperbola is $\frac{x^2}{9} - \frac{y^2}{25} = 1$

(c) the focus of the hyperbola is (5, 0)

(d) the focus of the hyperbola is $(5\sqrt{3}, 0)$

Comprehension-Type Questions

43. Let $ABCD$ be a square of sides of length 2 units. C_1 is the circle touching all the sides of the square and C_2 is the circle circumscribing the square. l is a line through A.

(i) If P is a point on C_1 and Q is a point on C_2 then $\dfrac{PA^2+PB^2+PC^2+PD^2}{QA^2+QB^2+QC^2+QD^2}$ equals

(a) $\frac{3}{4}$ (b) $\frac{5}{4}$ (c) 1 (d) $\frac{1}{2}$

(ii) A circle touches the line l and the circle C_1 externally such that both the circles are on the same side of the line. The locus of the centre of the circle is

(a) part of a straight line (b) parabola

(c) ellipse (d) hyperbola

(iii) A line m through A is drawn parallel to BD. A point S moves such that its distances from BD and the vertex A are equal. If the locus of S cuts the line m at T_2 and T_3, and AC at T_1 then ar($\Delta T_1T_2T_3$) in square units is

(a) $\frac{1}{2}$ (b) $\frac{2}{3}$ (c) 1 (d) 2

44. Let $A = \begin{bmatrix} 1 & 0 & 0 \\ 2 & 1 & 0 \\ 3 & 2 & 1 \end{bmatrix}$ and U_1, U_2, U_3 are three column matrices such that

$$AU_1 = \begin{bmatrix} 1 \\ 0 \\ 0 \end{bmatrix}, \ AU_2 = \begin{bmatrix} 2 \\ 3 \\ 0 \end{bmatrix} \text{ and } AU_3 = \begin{bmatrix} 2 \\ 3 \\ 1 \end{bmatrix}.$$

U is a 3×3 matrix whose first column is the column in U_1, the second column is the column in U_2 and the third column is the column in U_3. On the basis of this, answer the following questions.

(i) The value of $|U|$ is

(a) 3 (b) –3 (c) $\frac{3}{2}$ (d) 2

(ii) The sum of the elements of U^{-1} is

(a) 0 (b) –1 (c) 1 (d) 3

(iii) The sum of the elements of the product $[3\ 2\ 0] \cdot U \cdot \begin{bmatrix} 3 \\ 2 \\ 0 \end{bmatrix}$ is

(a) $\frac{5}{2}$ (b) $\frac{3}{2}$ (c) 4 (d) 5

Matching Questions

More than one may match with the same.

45. (i) $\int_0^{\pi/2} (\sin x)^{\cos x} \cdot \{\cos x \cot x - \log(\sin x)^{\sin x}\}\, dx$ (a) 1

(ii) The area bounded by $-4y^2 = x$ and $x - 1 = -5y^2$ is (b) 0

(iii) Cosine of the angle of intersection of curves $y = 3^{x-1} \log x$ and $y = x^x - 1$ is (c) 6log 2

(iv) $\lim\limits_{x \to \infty} \dfrac{\left(\int_0^x e^{x^2} dx\right)^2}{\int_0^x e^{2x^2} dx}$ is equal to (d) $\frac{4}{3}$

Complete the following statements.

46. If the roots of the equation $x^2 - 10cx - 11d = 0$ are a, b and those of $x^2 - 10ax - 11b = 0$ are c, d then the positive value of $a + b + c + d$ is (a, b, c and d being distinct numbers)

47. If $a_n = \frac{3}{4} - \left(\frac{3}{4}\right)^2 + \left(\frac{3}{4}\right)^3 - \ldots + (-1)^{n-1} \cdot \left(\frac{3}{4}\right)^n$ and $b_n = 1 - a_n$ then the smallest natural number n_0, such that $b_n > a_n\ \forall\ n > n_0$, is

Answers

1. b	**2.** d	**3.** c	**4.** d	**5.** c
6. c	**7.** b	**8.** b	**9.** b	**10.** b
11. c	**12.** b	**13.** a	**14.** a	**15.** a
16. a	**17.** d	**18.** a	**19.** a	**20.** b
21. a	**22.** c	**23.** b	**24.** d	**25.** d
26. c	**27.** d	**28.** c	**29.** b	**30.** c
31. d	**32.** d	**33.** a	**34.** c	**35.** d
36. c	**37.** b	**38.** b	**39.** d	**40.** a
41. a, c	**42.** a, c			

43. (i) a (ii) b (iii) c   **44.** (i) a (ii) a (iii) d

45. (i) $\leftrightarrow$ a (ii) $\leftrightarrow$ d (iii) $\leftrightarrow$ a (iv) $\leftrightarrow$ b

46. 1210   **47.** 5

Hints and Solutions to Selected Questions

1. Both e^x and $1+x$ are increasing and $\sqrt{e} > 1 + \frac{1}{2}$ because $\sqrt{e} = 1.65$ nearly.

So the answer (a) is not correct.

$\sin\frac{\pi}{6} < \frac{\pi}{6}$ because $\frac{1}{2} < \frac{22}{42}$ · So the answer (c) is not correct.

$\log_e \frac{1}{2} < \frac{1}{2}$ because $\log_e \frac{1}{2}$ is negative. So, the answer (d) is not correct.

Thus, by elimination the answer (b) is correct.

2.

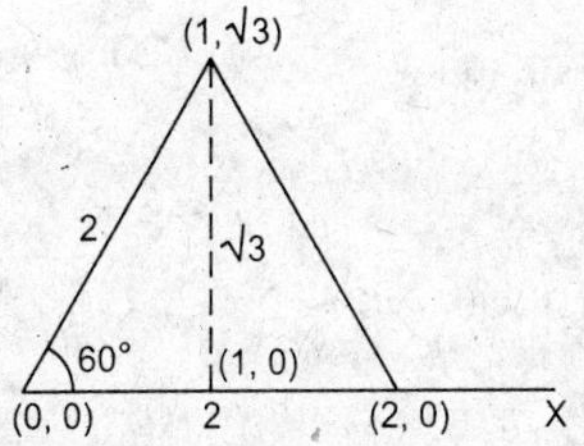

Clearly, the triangle is equilateral.

So, the incentre is the same as the centroid.

$\therefore$ incentre $= \left(\frac{1+0+2}{3}, \frac{\sqrt{3}+0+0}{3}\right) = \left(1, \frac{1}{\sqrt{3}}\right)$.

3. The parabola is $y^2 = 4 \cdot \frac{k}{4} \cdot \left(x - \frac{8}{k}\right)$ · Putting $y = Y,\ x - \frac{8}{k} = X$,

the equation is $Y^2 = 4 \cdot \frac{k}{4} \cdot X$.

$\therefore$ the directrix is $X + \frac{k}{4} = 0$, i.e., $x - \frac{8}{k} + \frac{k}{4} = 0$.

But $x - 1 = 0$ is the directrix. So, $\frac{8}{k} - \frac{k}{4} = 1 \quad \Rightarrow k = -8, 4$.

4. $2^y = 2 - 2^x$; y is real if $2 - 2^x \geq 0$ or $2 > 2^x$. So, $1 > x$.

5. Here, the centre $= (0, 0)$. So m of OQ is $\frac{4}{3}$ and m of OR is $-\frac{3}{4}$ ·

$\therefore \quad \angle QOR = \frac{\pi}{2}$. Hence, $\angle QPR = \frac{1}{2} \times \frac{\pi}{2}$.

6. $f'(x) < 0$ if $e^x(x-1)(x-2) < 0$. Clearly $e^x > 0$.

$\therefore \quad (x-1)(x-2) < 0 \quad \Rightarrow 1 < x < 2$ (using sign scheme).

7. $g(2) = \int_0^2 f(t)\,dt = \int_0^1 f(t)\,dt + \int_1^2 f(t)\,dt$.

As $\frac{1}{2} \le f(t) \le 1$ for $0 \le t \le 1$,

$\int_0^1 \frac{1}{2}\,dt \le \int_0^1 f(t)\,dt \le \int_0^1 1\,dt \quad$ or $\quad \frac{1}{2} \le \int_0^1 f(t)\,dt \le 1$ ... (1)

As $0 \le f(t) \le \frac{1}{2}$ for $1 < t \le 2$, $\int_1^2 0\,dt \le \int_1^2 f(t)\,dt \le \int_1^2 \frac{1}{2}\,dt$

$\therefore \quad 0 \le \int_1^2 f(t)\,dt \le \frac{1}{2}$ ... (2)

Adding (1) and (2), $\frac{1}{2} \le g(2) \le \frac{3}{2}$

$\therefore \quad g(2)$ satisfies the inequality $0 \le g(2) < 2$.

8. Differentiating w.r.t. x,

$2x + 2yy' = 0 \quad$ or $\quad x + yy' = 0$.

Differentiating again w.r.t. x,

$1 + y'^2 + yy'' = 0$.

9. Here, $\vec{a} + \vec{b} + \vec{c} = 0$. Take cross products with $\vec{a}$ and $\vec{b}$ by turn.

10. If P be the foot then $PA = PB = PC$, i.e., P is the circumcentre.

11. $f(\theta) = \sin^2\theta\{1 + (3 - 4\sin^2\theta)\} = 4\sin^2\theta \cdot \cos^2\theta \ge 0$ for all θ.

12. Any normal is $y + tx = 6t + 3t^3$. It is identical with $x + y = k$ if

$\frac{t}{1} = \frac{1}{1} = \frac{6t + 3t^3}{k}$. $\quad \therefore \ t = 1$ and $1 = \frac{6+3}{k}$.

$\therefore \quad k = 9$.

13. $1 = \left|\frac{1}{z_1} + \frac{1}{z_2} + \frac{1}{z_3}\right| = \left|\frac{z_1\bar{z}_1}{z_1} + \frac{z_2\bar{z}_2}{z_2} + \frac{z_3\bar{z}_3}{z_3}\right|$ $\quad (\because \ |z_1|^2 = 1 = z_1\bar{z}_1$, etc.)

$= |\bar{z}_1 + \bar{z}_2 + \bar{z}_3| = |\overline{z_1 + z_2 + z_3}| = |z_1 + z_2 + z_3|$ $\quad (\because \ |\bar{z}_1| = |z_1|)$.

14. Clearly from the diagram,

$\arg(-z) - \arg z = \pi.$

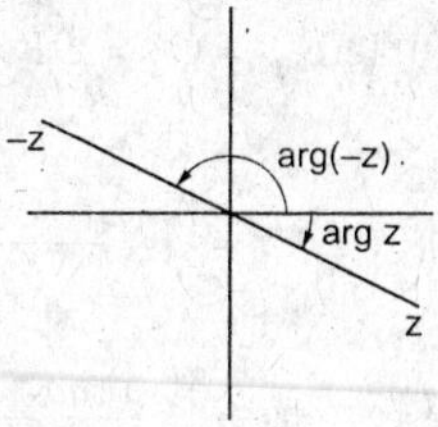

15. $\dfrac{(a+b)+(c+d)}{2} \geq \sqrt{(a+b)(c+d)}$ or $\dfrac{2}{2} > \sqrt{M}$ or $1 \geq M.$

Also, $M > 0$. So, $0 < M \leq 1$.

16. $2gg' + 2ff' = c + c'$, i.e., $2{\cdot}1{\cdot}0 + 2{\cdot}k{\cdot}k = 6 + k$

or $2k^2 - k - 6 = 0$ or $(2k+3)(k-2) = 0$ $\therefore$ $k = 2, -\dfrac{3}{2}$.

17. $S =$ midpoint of $QR = \left(\dfrac{6+7}{2}, \dfrac{-1+3}{2}\right) = \left(\dfrac{13}{2}, 1\right)$

$\therefore$ m of $PS = \dfrac{2-1}{2-\dfrac{13}{2}} = -\dfrac{2}{9}$.

$\therefore$ the required equation is $y + 1 = -\dfrac{2}{9}(x-1)$.

18. Here $[\vec{a}\ \vec{b}\ \vec{c}\,] = 0$. The given scalar triple product $= k[\vec{a}\ \vec{b}\ \vec{c}\,] = 0$.

19. A vector perpendicular to the plane P_1 of $\vec{a}, \vec{b}$ is $\vec{a} \times \vec{b}$.

A vector perpendicular to the plane P_2 of $\vec{c}, \vec{d}$ is $\vec{c} \times \vec{d}$.

$(\vec{a} \times \vec{b}) \times (\vec{c} \times \vec{d}) = \vec{0}$ $\Rightarrow$ $\vec{a} \times \vec{b} \parallel \vec{c} \times \vec{d}$

$\Rightarrow$ the angle between the planes is $0°$.

20. $2ac \cdot \sin \dfrac{A-B+C}{2} = 2ac \cdot \sin \dfrac{\pi - 2B}{2} = 2ac \cdot \cos B$

$= 2ac \cdot \dfrac{c^2 + a^2 - b^2}{2ca} = c^2 + a^2 - b^2.$

21. Here, $R = OA = OB = OC = \frac{1}{2} AB = \frac{c}{2}$.

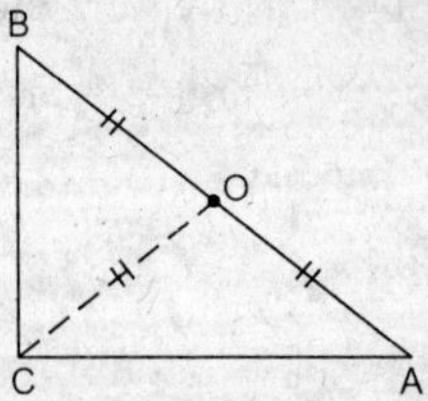

$$\therefore \quad r = \frac{\Delta}{s} = \frac{\frac{1}{2}ab}{\frac{1}{2}(a+b+c)} = \frac{ab}{a+b+c}.$$

$$\therefore \quad r + R = \frac{ab}{a+b+c} + \frac{c}{2} = \frac{2ab + c(a+b+c)}{2(a+b+c)}$$

$$= \frac{2ab + ca + bc + a^2 + b^2}{2(a+b+c)} \qquad (\because\ c^2 = a^2 + b^2)$$

$$= \frac{(a+b)^2 + c(a+b)}{2(a+b+c)} = \frac{(a+b)(a+b+c)}{2(a+b+c)}.$$

$$\therefore \quad 2(r+R) = a + b.$$

22. $\int_{-2}^{3} f(x)\,dx = \int_{-2}^{2} f(x)\,dx + \int_{2}^{3} f(x)\,dx$

$$= \int_{-2}^{2} e^{\cos x} \cdot \sin x\,dx + \int_{2}^{3} 2\,dx = 0 + 2(3-2)$$

(because $e^{\cos x} \cdot \sin x$ is an odd function)

$$= 2.$$

23. $\int_{e^{-1}}^{e^2} \left|\frac{\log_e x}{x}\right| dx = \int_{e^{-1}}^{1} \left|\frac{\log_e x}{x}\right| dx + \int_{1}^{e^2} \left|\frac{\log_e x}{x}\right| dx$

$$= \int_{e^{-1}}^{1} -\frac{\log_e x}{x}\,dx + \int_{1}^{e^2} \frac{\log_e x}{x}\,dx$$

$$= \int_{-1}^{0} -z\,dz + \int_{0}^{2} z\,dz \qquad (\text{putting } \log_e x = z)$$

$$= \left[-\frac{z^2}{2}\right]_{-1}^{0} + \left[\frac{z^2}{2}\right]_{0}^{2} = \frac{1}{2} + 2 = \frac{5}{2}.$$

24. The condition is $\begin{vmatrix} 1 & -k & -1 \\ k & -1 & -1 \\ 1 & 1 & -1 \end{vmatrix} = 0$

or $\quad 1(1+1) + k(-k+1) - 1(k+1) = 0 \quad$ or $\quad -k^2 + 1 = 0 \quad \therefore\ k = \pm 1.$

25. Here, $\frac{a}{1-r} = 4$ and $ar = \frac{3}{4}$. Dividing these,

$$r(1-r) = \frac{3}{16} \quad \text{or} \quad 16r^2 - 16r + 3 = 0 \quad \text{or} \quad (4r-3)(4r-1) = 0.$$

$\therefore \quad r = \frac{1}{4}, \frac{3}{4}$ and $a = \frac{3}{1}, \frac{3}{3} = 3, 1$. So $(a, r) = \left(3, \frac{1}{4}\right), \left(1, \frac{3}{4}\right)$.

26. Limit $= \lim\limits_{x \to \infty} \left(1 + \frac{-5}{x+2}\right)^{x+2} \cdot \frac{1}{\left(1 + \frac{-5}{x+2}\right)^2} = e^{-5} \cdot 1 = e^{-5}$.

27. m of the normal $= \frac{-1}{f'(x)\big]_{(3,4)}} = \tan\frac{3\pi}{4} = -1$.

$\therefore \quad f'(x)\big]_{(3,4)} = 1. \quad \therefore f'(3) = 1.$

28. $\alpha + \alpha^2 = -\frac{p}{3}$ and $\alpha \cdot \alpha^2 = 1$. So $\alpha = 1, \omega, \omega^2$. If $\alpha = 1, p < 0$.

If $\alpha = \omega$ or ω^2, we have $\omega + \omega^2 = -\frac{p}{3}$. $\quad \therefore \; -1 = -\frac{p}{3}$. $\quad \therefore \; p = 3$.

29. Here $D = b^2 - 4c > 0$ because $c < 0 < b$. So, roots are real and unequal.

Now, $\alpha + \beta = -b < 0$ and $\alpha\beta = c < 0$.

$\therefore$ one root is positive and the other negative, the negative root being numerically bigger. As $\alpha < \beta$, α is the negative root while β is the positive root. So, $|\alpha| > \beta$ and $\alpha < 0 < \beta$.

30. $g(x) = |f(x)| \geq 0$. So $g(x)$ cannot be onto.

If $f(x)$ is one-to-one and $f(x_1) = -f(x_2)$ then $g(x_1) = g(x_2)$. So, "$f(x)$ is one-to-one" does not ensure that $g(x)$ is one-to-one.

If $f(x)$ is continuous for $x \in R$, $|f(x)|$ is also continuous for $x \in R$. This is obvious from the following graphical consideration.

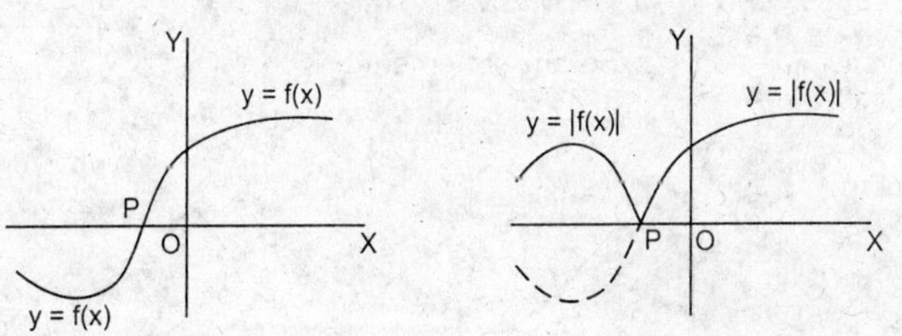

So the answer (c) is correct. The fourth answer (d) is not correct from the above graphs. $y = f(x)$ is differentiable at P while $y = |f(x)|$ has two tangents at P, i.e., not differentiable at P.

31. Expression $= {}^nC_r + 2 \cdot {}^nC_{r-1} + {}^nC_{r-2}$

$$= ({}^nC_r + {}^nC_{r-1}) + ({}^nC_{r-1} + {}^nC_{r-2}) = {}^{n+1}C_r + {}^{n+1}C_{r-1} = {}^{n+2}C_r.$$

32. The equation is $x^2 - (a+b)x + ab - 1 = 0$.

$\therefore$ discriminant $= (a+b)^2 - 4(ab-1) = (b-a)^2 + 4 > 0$.

$\therefore$ both roots are real. Let them be α, β, where

$$\alpha = \frac{a+b-\sqrt{(b-a)^2+4}}{2} \text{ and } \beta = \frac{a+b+\sqrt{(b-a)^2+4}}{2}.$$

Clearly, $\alpha < \frac{a+b-\sqrt{(b-a)^2}}{2} = \frac{a+b-(b-a)}{2} = a \quad (\because \; b > a)$

and $\quad \beta > \frac{a+b+\sqrt{(b-a)^2}}{2} = \frac{a+b+b-a}{2} = b.$

Hence, one root α is less than a and the other root β is greater than b.

33. The graph of the function is as given below.

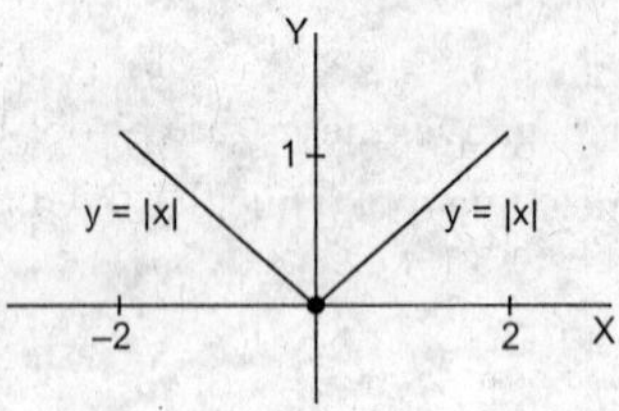

$\because \quad f(0) = 1, f(0-h) < 1, f(0+h) < 1,$

$\therefore \quad f(x)$ has a maximum at $x = 0$.

34. There are four even places for the four odd digits 3, 3, 5, 5.

$\therefore$ the required number of ways $= \frac{4!}{2!2!} \cdot \frac{5!}{2!3!} = 60$.

35. From the trend of values of $\sin x$ and $\cos x$ we know $\sin x$ and $\cos x$ decrease in $\frac{\pi}{2} < x < \pi$. So, the statement S is correct.

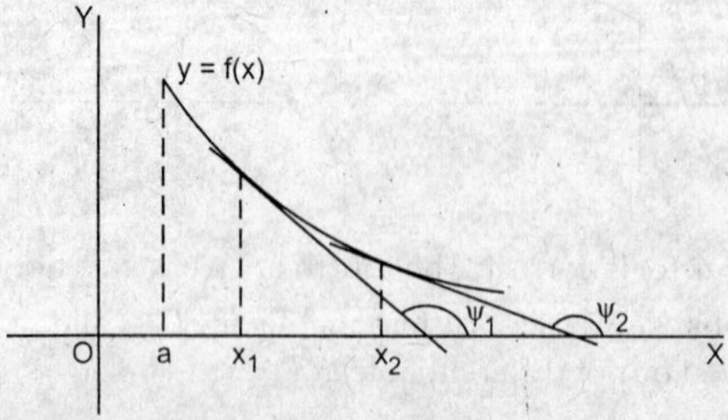

The statement R is incorrect, which is clear from the given graph.

Clearly, $f(x)$ is differentiable (as there is unique tangent at all points) in (a, b).

Also, $a < x_1 < x_2 < b$.

But $f'(x_1) = \tan \psi_1 < \tan \psi_2 = f'(x_2)$. (For example, $\psi_2 = 150°, \psi_1 = 135°$.)

36. Limit $= \lim\limits_{x \to 0} (\sin x)^{1/x} + \lim\limits_{x \to 0} \left(\dfrac{1}{x}\right)^{\sin x} = 0 + e^{\lim\limits_{x\to 0} \sin x \cdot \log^{(1/x)}}$ $(\because\ x > 0)$

$$= e^{\lim\limits_{x\to 0} \frac{-\log x}{\operatorname{cosec} x}} = e^{\lim\limits_{x\to 0} \frac{-1/x}{-\operatorname{cosec} x \cdot \cot x}} = e^{\lim\limits_{x\to 0} \frac{\sin x}{x} \cdot \tan x} = e^{1 \cdot 0} = 1.$$

37. Let $\angle A$ be 120° in the ΔABC. Then $\angle B = \angle C = 30°$ in the isosceles triangle.

$$\therefore \quad \frac{\sin 120°}{a} = \frac{\sin 30°}{b} \quad \Rightarrow a = \sqrt{3}b \quad \text{and} \quad b = c.$$

Here, $r = \sqrt{3} = \dfrac{\Delta}{s}$ and $s = \dfrac{1}{2}(a + b + c) = \dfrac{1}{2}(a + 2b)$.

$$\therefore \quad \Delta = \sqrt{3}s = \frac{\sqrt{3}}{2}(a + 2b) = \frac{\sqrt{3}}{2}\left(a + 2 \cdot \frac{a}{\sqrt{3}}\right) = \frac{1}{2}(2 + \sqrt{3})a.$$

Also, $\Delta = \dfrac{1}{2}ab \sin 30° = \dfrac{1}{2}a \cdot \dfrac{a}{\sqrt{3}} \cdot \dfrac{1}{2} = \dfrac{a^2}{4\sqrt{3}}$.

$$\therefore \quad \frac{1}{2}(2 + \sqrt{3})a = \frac{a^2}{4\sqrt{3}} \quad \text{or} \quad a = 2\sqrt{3}(2 + \sqrt{3}).$$

$$\therefore \quad \Delta = \frac{a^2}{4\sqrt{3}} = \frac{1}{4\sqrt{3}} \cdot \{2\sqrt{3}(2 + \sqrt{3})\}^2 = \sqrt{3}(7 + 4\sqrt{3}) = 12 + 7\sqrt{3}.$$

38. A vector in the plane of $\vec{a}$ and $\vec{b}$ is $\lambda\vec{a} + \mu\vec{b}$.

$$\therefore \quad (\lambda\vec{a} + \mu\vec{b}) \cdot \frac{\vec{c}}{|\vec{c}|} = \frac{1}{\sqrt{3}} \quad \text{or} \quad (\lambda\vec{a} + \mu\vec{b}) \cdot \frac{\vec{i} + \vec{j} - \vec{k}}{\sqrt{3}} = \frac{1}{\sqrt{3}}$$

$$\text{or} \quad \{\lambda(\vec{i} + 2\vec{j} + \vec{k}) + \mu(\vec{i} - \vec{j} + \vec{k})\} \cdot (\vec{i} + \vec{j} - \vec{k}) = 1$$

$$\text{or} \quad \{(\lambda + \mu)\vec{i} + (2\lambda - \mu)\vec{j} + (\lambda + \mu)\vec{k}\} \cdot (\vec{i} + \vec{j} - \vec{k}) = 1$$

$$\text{or} \quad \lambda + \mu + 2\lambda - \mu - (\lambda + \mu) = 1 \quad \text{or} \quad 2\lambda - \mu = 1.$$

$$\therefore \quad \lambda\vec{a} + \mu\vec{b} = \lambda(\vec{i} + 2\vec{j} + \vec{k}) + (2\lambda - 1)(\vec{i} - \vec{j} + \vec{k})$$

$$= (3\lambda - 1)\vec{i} + \vec{j} + (3\lambda - 1)\vec{k}$$

Taking $\lambda = -1$, the vector is $-4\vec{i} + \vec{j} - 4\vec{k}$ which is negative of $4\vec{i} - \vec{j} + 4\vec{k}$.

39. The plane has the equation $a(x-1) + b(y+2) + c(z-1) = 0$.

It is perpendicular to the given planes. So,

$$\left.\begin{array}{l} a \cdot 2 + b(-2) + c \cdot 1 = 0 \\ \text{and} \\ a \cdot 1 + b(-1) + c \cdot 2 = 0 \end{array}\right\} \Rightarrow \frac{a}{-3} = \frac{b}{-3} = \frac{c}{0} \quad \text{or} \quad \frac{a}{1} = \frac{b}{1} = \frac{c}{0}$$

$\therefore$ the plane has the equation $x - 1 + y + 2 = 0$ or $x + y + 1 = 0$

$\therefore$ the required distance $= \dfrac{1+2+1}{\sqrt{1^2+1^2}} = 2\sqrt{2}$.

40. The shaded portion is outside the circle whose centre is $z = -1$ and the radius is $PA = QA = 2$. So, $|z+1| > 2$.

Also, the point falls in the angle PAQ whose arms make $\frac{\pi}{4}$ and $-\frac{\pi}{4}$ with the real axis. If A is the origin then $|\arg z| < \frac{\pi}{4}$. But A is $z = -1$.

So, $|\arg\{z - (-1)\}| < \frac{\pi}{4}$ or $|\arg(z+1)| < \frac{\pi}{4}$.

Hence, z is such that $|z+1| > 2$, $|\arg(z+1)| < \frac{\pi}{4}$.

41. Here, $f(x) = \begin{cases} x^3 & \text{for } x < 0 \\ x^3 & \text{for } 0 \le x < 1 \\ 1 & \text{for } x \ge 1. \end{cases}$

Clearly, $f(0-0) = f(0+0) = 0$ and $f(1-0) = f(1+0) = f(1)$.

Also, x^3 and 1 are polynomial functions. So, $f(x)$ is continuous $\forall\, x$.

$f'(0-0) = f'(0+0) = 0$ but $f'(1-0) = 3, f'(1+0) = 0$.

So, $f(x)$ is not differentiable at $x = 1$ only.

Also, $f'(x) = 0$ for all $x > 1$.

42. The equation of the hyperbola is $\dfrac{x^2}{a^2} - \dfrac{y^2}{b^2} = 1$. ... (1)

The eccentricity of the ellipse is given by $16 = 25(1 - e^2)$ or $e = \frac{3}{5}$ and the focus $= \left(\pm 5 \cdot \frac{3}{5}, 0\right) = (\pm 3, 0)$.

(1) passes through (±3, 0). So, $\frac{9}{a^2}=1$ or $a^2=9$.

Also for the hyperbola $e=\frac{5}{3}$. So, $b^2=a^2(e^2-1)$ or $b^2=9\left(\frac{25}{9}-1\right)$

$\therefore$ $b^2=16$. Hence, the hyperbola has the equation $\frac{x^2}{9}-\frac{y^2}{16}=1$

and its focus $=(\pm ae, 0)=\left(\pm 3\cdot\frac{5}{3}, 0\right)=(\pm 5, 0)$.

43. (i) The equation of C_1 is $x^2+y^2=1$ and that of C_2 is $x^2+y^2=2$.

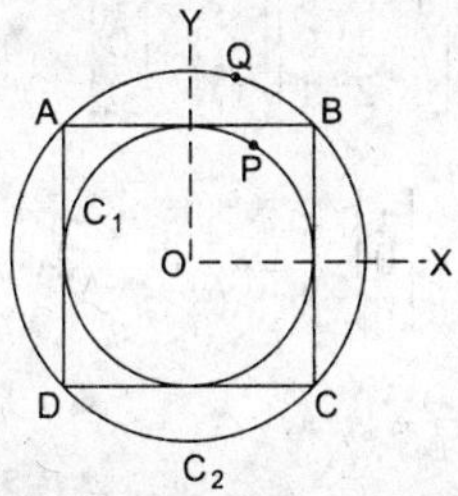

Let $P=(x_1, y_1)$ and $Q=(x_2, y_2)$. Then $x_1^2+y_1^2=1$, $x_2^2+y_2^2=2$.
Also, $A=(-1,1)$, $B=(1,1)$, $C=(1,-1)$, $D=(-1,-1)$.

$$\begin{aligned}\therefore\quad PA^2+PB^2+PC^2+PD^2 &= (x_1+1)^2+(y_1-1)^2\\ &\quad+(x_1-1)^2+(y_1-1)^2\\ &\quad+(x_1-1)^2+(y_1+1)^2\\ &\quad+(x_1+1)^2+(y_1+1)^2\\ &=2\{(x_1+1)^2+(x_1-1)^2\}+2\{(y_1-1)^2+(y_1+1)^2\}\\ &=4(x_1^2+1)+4(y_1^2+1)=4(x_1^2+y_1^2)+8=4+8=12.\end{aligned}$$

Similarly, $QA^2+QB^2+QC^2+QD^2=4(x_2^2+y_2^2)+8=4\times 2+8=16$.

$\therefore$ value $=\frac{12}{16}=\frac{3}{4}$.

(ii) The circle C always touches the line l and the fixed circle C_1, whose radius is 1. Take a line $m \parallel l$ at a distance 1 from l not cutting C, whose centre is G. Clearly, the distance of G from the centre O of C_1 is equal to the distance of G from the line p. Thus, the centre G of the circle C moves keeping equal distances from the fixed point O and the fixed line p. Hence, the locus is a parabola whose focus is O and the directrix is the line p.

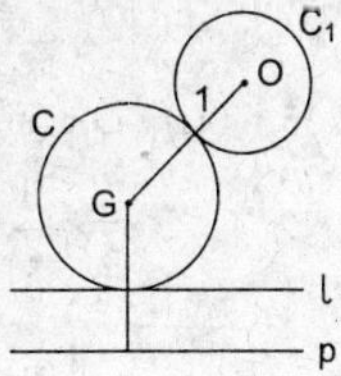

(iii) $AT_1 = T_1O = \frac{1}{2}AO = \frac{1}{4}AC = \frac{1}{4} \cdot 2\sqrt{2} = \frac{1}{\sqrt{2}}$.

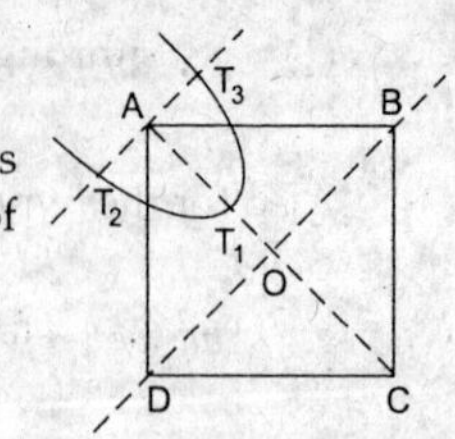

Clearly, the locus of S is a parabola whose focus is A and BD is the directrix. So, OA is the line of axis and T_2T_3 is the latus rectum.

$$\therefore \quad T_2T_3 = 4 \times AT_1 = 4 \times \frac{1}{\sqrt{2}} = 2\sqrt{2}$$

$$\therefore \quad \text{ar}(\Delta T_1T_2T_3) = \frac{1}{2} \cdot T_2T_3 \cdot AT_1 = \frac{1}{2} \cdot 2\sqrt{2} \cdot \frac{1}{\sqrt{2}} = 1.$$

44. (i) Let $U_1 = \begin{bmatrix} x_1 \\ y_1 \\ z_1 \end{bmatrix}$. Then $\begin{bmatrix} 1 & 0 & 0 \\ 2 & 1 & 0 \\ 3 & 2 & 1 \end{bmatrix}\begin{bmatrix} x_1 \\ y_1 \\ z_1 \end{bmatrix} = \begin{bmatrix} 1 \\ 0 \\ 0 \end{bmatrix}$

or $\begin{bmatrix} x_1 \\ 2x_1 + y_1 \\ 3x_1 + 2y_1 + z_1 \end{bmatrix} = \begin{bmatrix} 1 \\ 0 \\ 0 \end{bmatrix} \Rightarrow x_1 = 1, 2x_1 + y_1 = 0, 3x_1 + 2y_1 + z_1 = 0.$

$\therefore \; x_1 = 1, y_1 = -2, z_1 = 1.$

Similarly, $U_2 = \begin{bmatrix} 2 \\ -1 \\ -4 \end{bmatrix}$ and $U_3 = \begin{bmatrix} 2 \\ -1 \\ -3 \end{bmatrix}$.

$\therefore \; U = \begin{bmatrix} 1 & 2 & 2 \\ -2 & -1 & -1 \\ 1 & -4 & -3 \end{bmatrix}$. $\quad \therefore \; |U| = \begin{vmatrix} 1 & 2 & 2 \\ -2 & -1 & -1 \\ 1 & -4 & -3 \end{vmatrix} = 3.$

(ii) $U^{-1} = \frac{1}{3}\begin{bmatrix} -1 & -2 & 0 \\ -7 & -5 & -3 \\ 9 & 6 & 3 \end{bmatrix}$. So, the sum of elements $= \frac{0}{3} = 0.$

(iii) The product $= [3 \;\; 2 \;\; 0]\begin{bmatrix} 1 & 2 & 2 \\ -2 & -1 & -1 \\ 1 & -4 & -3 \end{bmatrix}\begin{bmatrix} 3 \\ 2 \\ 0 \end{bmatrix}$

$$= [-1 \;\; 4 \;\; 4]\begin{bmatrix} 3 \\ 2 \\ 0 \end{bmatrix} = [5].$$

45. (i) Let $(\sin x)^{\cos x} = y \Rightarrow \log y = \cos x \cdot \log \sin x$

$$\Rightarrow \frac{1}{y}\frac{dy}{dx} = \cos x \cdot \cot y - \log(\sin x)^{\sin x}$$

$$\therefore \; I = \int_0^1 y \cdot \frac{1}{y}\, dy = \int_0^1 dy = 1.$$

(ii) Parabolas are $x = -4y^2$ and $x = 1 - 5y^2$.

Solving, their points of intersection are (–4, 1) and (–4, –1).

$\therefore$ the area bounded by the parabolas

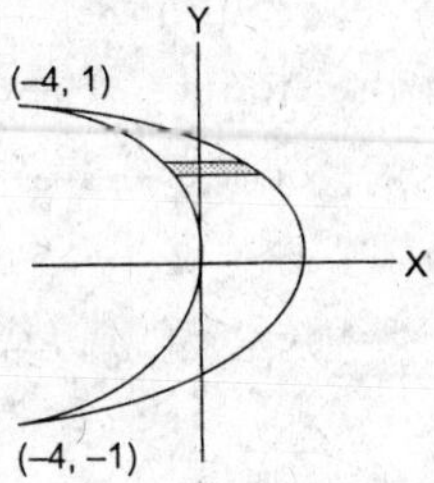

$$= \left|2\left\{\int_0^1 (x_1 - x_2)\,dy\right\}\right|$$

$$= \left|2\left[\int_0^1 \{(1-5y^2) - (-4y^2)\}\,dy\right]\right|$$

$$= \left|2\int_0^1 (1-y^2)\,dy\right| = 2\left(1 - \frac{1}{3}\right) = \frac{4}{3}.$$

(iii) Solving $y = 3^{x-1}\log x$ and $y = x^x - 1$ we get $3^{x-1}\log x = x^x - 1$. By trial, $x = 1$ satisfies it.

Then $y = 1^1 - 1 = 0$. So, the point of intersection is (1, 0).

Now, $y = 3^{x-1}\log x \Rightarrow \dfrac{dy}{dx} = 3^{x-1}\cdot\dfrac{1}{x} + 3^{x-1}\cdot\log 3\cdot\log x$

$$y = x^x - 1 \Rightarrow \frac{dy}{dx} = x\cdot x^{x-1} + x^x\log x = x^x(1 + \log x)$$

$\therefore$ at (1, 0), $m_1 = \left.\dfrac{dy}{dx}\right]_{(1,0)} = 1$ and $m_2 = \left.\dfrac{dy}{dx}\right]_{(1,0)} = 1$

$\therefore$ the curves touch at (1, 0). So, $\cos\theta = \cos 0° = 1$.

(iv) $$\text{Limit} = \lim_{n\to\infty}\frac{\left(\int_0^x e^{x^2}dx\right)^2}{\int_0^x e^{2x^2}dx} \quad \left(\text{from } \frac{\infty}{\infty}\right)$$

$$= \lim_{n\to\infty}\frac{\left(2\cdot\int_0^x e^{x^2}dx\right)e^{x^2}}{e^{2x^2}} \quad \text{(by L'Hospital's rule)}$$

$$= \lim_{n\to\infty}\frac{2\cdot\int_0^x e^{x^2}dx}{e^{x^2}} = \lim_{n\to\infty}\frac{2\cdot e^{x^2}}{e^{x^2}\cdot 2x} = \lim_{n\to\infty}\frac{1}{x} = 0.$$

46. Here, $a + b = 10c$, $ab = -11d$, $c + d = 10a$, $cd = -11b$.

$\therefore$ $abcd = (-11d)(-11b)$ or $ac = 121$.

Also, $a+b+c+d=10c+10a$ or $b+d=9(a+c)$.

Again, $a^2-10ca-11d=0$ and $c^2-10ac-11b=0$

(by meaning of a root).

Adding, $a^2+c^2-20ac-11(b+d)=0$

or $(a+c)^2-22\times 121-11\times 9(a+c)=0$

or $(a+c-121)(a+c+22)=0$. $\therefore$ $a+c=121, -22$.

$\therefore$ $a+b+c+d=10(a+c)=1210, -220$.

47. $a_n=\dfrac{\frac{3}{4}\left\{1-\left(-\frac{3}{4}\right)^n\right\}}{1-\left(-\frac{3}{4}\right)}=\dfrac{3}{7}\left\{1-\left(-\frac{3}{4}\right)^n\right\}$

$\therefore$ $b_n>a_n \Rightarrow 1-a_n>a_n \Rightarrow 2a_n<1$.

$\Rightarrow \dfrac{6}{7}\left\{1-\left(-\frac{3}{4}\right)^n\right\}<1 \Rightarrow 1-\left(-\frac{3}{4}\right)^n<\dfrac{7}{6} \Rightarrow -\dfrac{1}{6}<\left(-\frac{3}{4}\right)^n$

This is not true for $n=1, 3, 5$ and is true for $n=2, 4, 6, 7, 8, 9, \ldots$.

$\therefore$ the minimum value of n_0 is 5 such that $b_n>a_n \ \forall\ n>n_0$.

□

IIT Questions—2

One-Option-Correct Questions

1. Let α, β be roots of $x^2 - x + p = 0$ and γ, δ be the roots of $x^2 - 4x + q = 0$. If $\alpha, \beta, \gamma, \delta$ are in GP then the integral values of p and q, respectively, are

(a) $-2, -32$ (b) $-2, 3$
(c) $-6, 3$ (d) $-6, -32$

2. In the binomial expansion of $(a-b)^n$ $(n \geq 5)$, the sum of the 5th and 6th terms is zero. Then $\frac{a}{b}$ is equal to

(a) $\frac{1}{6}(n-5)$ (b) $\frac{1}{5}(n-4)$
(c) $\frac{5}{n-4}$ (d) $\frac{6}{n-5}$

3. Let $f(x) = (1+b^2)x^2 + 2bx + 1$ and $m(b)$ the minimum value of $f(x)$ for a given b. As b varies, the range of $m(b)$ is

(a) $[0, 1]$ (b) $\left(0, \frac{1}{2}\right]$
(c) $\left[\frac{1}{2}, 1\right]$ (d) $(0, 1]$

4. The number of distinct real roots of

$$\begin{vmatrix} \sin x & \cos x & \cos x \\ \cos x & \sin x & \cos x \\ \cos x & \cos x & \sin x \end{vmatrix} = 0$$

in the interval $-\frac{\pi}{4} \leq x \leq \frac{\pi}{4}$ is

(a) 0 (b) 2
(c) 1 (d) 3

5. Let $E = \{1, 2, 3, 4\}$ and $F = \{1, 2\}$. Then the number of onto functions from E to F is

(a) 14 (b) 16
(c) 12 (d) 8

6. Let T_n denote the number of triangles which can be formed using the vertices of a regular polygon on n sides. If $T_{n+1} - T_n = 21$ then n equals

(a) 5 (b) 7
(c) 6 (d) 4

7. The complex numbers z_1, z_2 and z_3 satisfying $\frac{z_1 - z_3}{z_2 - z_3} = \frac{1 - \sqrt{3}i}{2}$ are the vertices of a triangle which is

(a) of area = 0 (b) right-angled isoceles
(c) equilateral (d) obtuse-angled isoceles

8. If the sum of the first $2n$ terms of 2, 5, 8, ... is equal to the sum of the first n terms of 57, 59, 61, ... then n is equal to

(a) 10 (b) 12
(c) 11 (d) 13

9. Let z_1 and z_2 be nth roots of unity which are ends of a line segment that subtend a right angle at the origin. Then n must be of the form

(a) $4k + 1$ (b) $4k + 2$
(c) $4k + 3$ (d) $4k$

10. Let the positive numbers a, b, c, d be in AP. Then abc, abd, acd, bcd are

(a) not in AP/GP/HP (b) in AP
(c) in GP (d) in HP

11. Let $f(x) = \frac{\alpha x}{x+1}$, where $x \neq -1$. Then for what value of α is $f\{(f(x)\} = x$ true?

(a) $\sqrt{2}$ (b) $-\sqrt{2}$
(c) 1 (d) -1

12. Let $f: R \to R$ be a function defined by $f(x) = \max\{x, x^3\}$. The set of all points where $f(x)$ is not differentiable is

(a) $\{-1, 1\}$ (b) $\{-1, 0\}$
(c) $\{0, 1\}$ (d) $\{-1, 0, 1\}$

13. Let $f: (0, +\infty) \to R$ and $F(x) = \int_0^x f(t)\, dt$. If $F(x^2) = x^2(1 + x)$ then $f(4)$ equals

(a) $\frac{5}{4}$ (b) 7

(c) 4 (d) 2

14. The left-hand derivative of $f(x) = [x]\sin \pi x$ at $x = k$, where k is an integer and $[x]$ = greatest integer not greater than x, is

(a) $(-1)^k(k-1)\pi$ (b) $(-1)^{k-1} \cdot (k-1)\pi$

(c) $(-1)^k k\pi$ (d) $(-1)^{k-1} k\pi$

15. If $f(x) = xe^{x(1-x)}$ then $f(x)$ is

(a) increasing on $\left[-\frac{1}{2}, 1\right]$ (b) decreasing on R

(c) increasing on R (d) decreasing on $\left[-\frac{1}{2}, 1\right]$

16. $\lim\limits_{x \to 0} \dfrac{\sin(\pi \cos^2 x)}{x^2}$ equals

(a) $-\pi$ (b) π

(c) $\frac{\pi}{2}$ (d) 1

17. The triangle formed by the tangent to the curve $f(x) = x^2 + bx - b$ at the point (1, 1) and the coordinate axes, lies in the first quadrant. If its area is 2 then the value of b is

(a) –1 (b) 3

(c) –3 (d) 1

18. Let $g(x) = 1 + x - [x]$ and

$$f(x) = \begin{cases} -1 & \text{for } x < 0 \\ 0 & \text{for } x = 0 \\ 1 & \text{for } x > 0. \end{cases}$$

Then for all x, $f\{g(x)\}$ is equal to

(a) x (b) 1

(c) $f(x)$ (d) $g(x)$

19. If $f: [1, +\infty) \to [2, +\infty)$ is given by $f(x) = x + \frac{1}{x}$ then f^{-1} equals

(a) $\dfrac{x + \sqrt{x^2 - 4}}{2}$ (b) $\dfrac{x}{1 + x^2}$

(c) $\dfrac{x - \sqrt{x^2 - 4}}{2}$ (d) $1 + \sqrt{x^2 - 4}$

20. The domain of $f(x) = \dfrac{\log_2(x+3)}{x^2+3x+2}$ is

(a) $R - \{-1, -2\}$ (b) $(-2, +\infty)$

(c) $R - \{-1, -2, -3\}$ (d) $(-3, +\infty) - \{-1, -2\}$

21. The value of $\int_{-\pi}^{\pi} \dfrac{\cos^2 x}{1+a^x}\, dx \ (a > 0)$ is

(a) π (b) $a\pi$

(c) $\dfrac{\pi}{2}$ (d) 2π

22. The equation of the common tangent touching the circle $(x-3)^2 + y^2 = 9$ and the parabola $y^2 = 4x$ above the x-axis, is

(a) $\sqrt{3}y = 3x + 1$ (b) $\sqrt{3}y = -(x+3)$

(c) $\sqrt{3}y = x + 3$ (d) $\sqrt{3}y = -(3x+1)$

23. Let AB be a chord of the circle $x^2 + y^2 = r^2$ subtending a right angle at the centre. Then the locus of the centroid of the ΔPAB as P moves on the circle is

(a) a parabola (b) a circle

(c) an ellipse (d) a pair of straight lines

24. The number of integral values of m, for which the x-coordinate of the point of intersection of the lines $3x + 4y = 9$ and $y = mx + 1$ is also an integer, is

(a) 2 (b) 0

(c) 4 (d) 1

25. The equation of the directrix of the parabola $y^2 + 4y + 4x + 2 = 0$ is

(a) $x = -1$ (b) $x = 1$

(c) $x = -\dfrac{3}{2}$ (d) $x = \dfrac{3}{2}$

26. If $\vec{a}, \vec{b}$ and $\vec{c}$ are unit vectors then $|\vec{a} - \vec{b}|^2 + |\vec{b} - \vec{c}|^2 + |\vec{c} - \vec{a}|^2$ does not exceed

(a) 4 (b) 9

(c) 8 (d) 6

27. Which of the following functions is differentiable at $x = 0$?

(a) $\cos|x| + |x|$ (b) $\cos|x| - |x|$

(c) $\sin|x| + |x|$ (d) $\sin|x| - |x|$

28. The number of solutions of $\log_4(x-1) = \log_2(x-3)$ is

(a) 3 (b) 1

(c) 2 (d) 0

29. Let $\vec{a} = \vec{i} - \vec{k}$, $\vec{b} = x\,\vec{i} + \vec{j} + (1-x)\,\vec{k}$ and $\vec{c} = y\,\vec{i} + x\,\vec{j} + (1 + x - y)\,\vec{k}$. Then $[\vec{a}\ \vec{b}\ \vec{c}]$ depends on

(a) only x (b) only y

(c) neither x nor y (d) both x and y

30. The area of the parallelogram formed by the lines $y = mx$, $y = mx + 1$, $y = nx$ and $y = nx + 1$ equals

(a) $\dfrac{|m+n|}{(m-n)^2}$ (b) $\dfrac{2}{|m+n|}$

(c) $\dfrac{1}{|m+n|}$ (d) $\dfrac{1}{|m-n|}$

31. Let PQ and RS be tangents at the extremities of the diameter PR of a circle of radius r. If PS and RQ intersect at a point X on the circumference of the circle then $2r$ equals

(a) $\sqrt{PQ \cdot RS}$ (b) $\dfrac{PQ + RS}{2}$

(c) $\dfrac{2PQ \cdot RS}{PQ + RS}$ (d) $\sqrt{\dfrac{PQ^2 + RS^2}{2}}$

32. A man from the top of a 100-metre-high tower sees a car moving towards the tower at an angle of depression of 30°. After some time, the angle of depression becomes 60°. The distance (in metres) travelled by the car during the time is

(a) $100\sqrt{3}$ (b) $\dfrac{200\sqrt{3}}{3}$

(c) $\dfrac{100\sqrt{3}}{3}$ (d) $200\sqrt{3}$

33. If $\alpha+\beta=\frac{\pi}{2}$ and $\beta+\gamma=\alpha$ then $\tan\alpha$ equals

(a) $2(\tan\beta+\tan\gamma)$ (b) $\tan\beta+\tan\gamma$

(c) $\tan\beta+2\tan\gamma$ (d) $2\tan\beta+\tan\gamma$

34. If $\sin^{-1}\left(x-\frac{x^2}{2}+\frac{x^3}{4}-\ldots\right)+\cos^{-1}\left(x^2-\frac{x^4}{2}+\frac{x^6}{4}-\ldots\right)=\frac{\pi}{2}$

for $0<|x|<\sqrt{2}$ then x equals

(a) $\frac{1}{2}$ (b) 1

(c) $-\frac{1}{2}$ (d) -1

35. The maximum value of $\cos\alpha_1\cdot\cos\alpha_2\cdot\ldots\cdot\cos\alpha_n$, under the restrictions $0\le\alpha_1,\alpha_2,\ldots,\alpha_n\le\frac{\pi}{2}$ and $\cot\alpha_1\cdot\cot\alpha_2\cdot\ldots\cdot\cot\alpha_n=1$, is

(a) $\frac{1}{2^{n/2}}$ (b) $\frac{1}{2^n}$

(c) $\frac{1}{2n}$ (d) 1

36. $\int\frac{x^2-1}{x^3\cdot\sqrt{2x^4-2x^2+1}}dx$ is equal to

(a) $\frac{1}{x^2}\cdot\sqrt{2x^4-2x^2+1}+c$ (b) $\frac{1}{x^3}\cdot\sqrt{2x^4-2x^2+1}+c$

(c) $\frac{1}{x}\cdot\sqrt{2x^4-2x^2+1}+c$ (d) $\frac{1}{2x^2}\cdot\sqrt{2x^4-2x^2+1}+c$

37. Let a, b, c be the lengths of sides of a scalene triangle. If the roots of the equation $x^2+2(a+b+c)x+3\lambda(ab+bc+ca)=0$ $(\lambda\in R)$ are real then

(a) $\lambda<\frac{4}{3}$ (b) $\lambda>\frac{5}{3}$

(c) $\lambda\in\left(\frac{1}{3},\frac{5}{3}\right)$ (d) $\lambda\in\left(\frac{4}{3},\frac{5}{3}\right)$

38. If $A=\begin{bmatrix}1&0&0\\0&1&1\\0&-2&4\end{bmatrix}$ and $A^{-1}=\frac{1}{6}(A^2+cA+dI)$, where I is the identity matrix, then the value of (c, d) is

(a) (–6, –11) (b) (6, 11)
(c) (–6, 11) (d) (6, –11)

39. Let $f''(x) = -f(x)$ and $g(x) = f'(x)$. If $F(x) = \left\{f\left(\frac{x}{2}\right)\right\}^2 + \left\{g\left(\frac{x}{2}\right)\right\}^2$ and $F(5) = 5$ then $F(10)$ is equal to

(a) 5 (b) 10
(c) 0 (d) 15

40. Let $\theta \in \left(0, \frac{\pi}{4}\right)$ and $t_1 = (\tan\theta)^{\tan\theta}$, $t_2 = (\tan\theta)^{\cot\theta}$, $t_3 = (\cot\theta)^{\tan\theta}$ and $t_4 = (\cot\theta)^{\cot\theta}$ then

(a) $t_1 > t_2 > t_3 > t_4$ (b) $t_4 > t_3 > t_1 > t_2$
(c) $t_3 > t_1 > t_2 > t_4$ (d) $t_2 > t_3 > t_1 > t_4$

One-or-More-Options-Correct Questions

41. Let $\vec{A}$ be a vector parallel to the line of intersection of the planes P_1 and P_2 through the origin. The plane P_1 is parallel to the vectors $2\vec{j} + 3\vec{k}$ and $4\vec{j} - 3\vec{k}$ while the plane P_2 is parallel to the vectors $\vec{j} - \vec{k}$ and $3\vec{i} + 3\vec{j}$. The angle between $\vec{A}$ and $2\vec{i} + \vec{j} - 2\vec{k}$ is

(a) $\frac{\pi}{2}$ (b) $\frac{\pi}{4}$ (c) $\frac{\pi}{6}$ (d) $\frac{3\pi}{4}$

42. The equations of common tangents to the parabolas $y = x^2$ and $y = -(x-2)^2$ are

(a) $y = 0$ (b) $y = 4(x-1)$
(c) $y + 4(x-1) = 0$ (d) $y + 30x + 50 = 0$

43. Let

$$f(x) = \begin{cases} e^x & \text{for } 0 \le x \le 1 \\ 2 - e^{x-1} & \text{for } 1 < x \le 2 \\ x - e & \text{for } 2 < x \le 3 \end{cases}$$

and $g(x) = \int_0^x f(t)\,dt$, $1 \le x \le 3$. Then $g(x)$ has

(a) a local maximum at $x = 1 + \log 2$ and a local minimum at $x = e$

(b) a local maximum at $x = e$ and a local minimum at $x = 1 + \log 2$

(c) no local maxima

(d) no local minima

Comprehension-Type Questions

44. There are n urns each containing $n+1$ balls such that the ith urn contains i white balls and $(n+1-i)$ red balls. Let u_i be the event of selecting ith urn and w denote the event of getting a white ball.

(i) If $P(u_i) \infty i$ where $i = 1, 2, 3, \ldots, n$ then $\lim\limits_{n \to \infty} P(w)$ is equal to

(a) $\frac{2}{3}$ (b) $\frac{3}{4}$

(c) $\frac{1}{4}$ (d) 1

(ii) If $P(u_i) = c$, a constant then, $P\left(\frac{u_n}{w}\right)$ is equal to

(a) $\frac{1}{n+1}$ (b) $\frac{2}{n+1}$ (c) $\frac{n}{n+1}$ (d) $\frac{1}{2}$

(iii) If n is even and E denotes the event of choosing even-numbered urn and u_is are equiprobable then the value of $P\left(\frac{w}{E}\right)$ is

(a) $\frac{1}{n+1}$ (b) $\frac{n}{n+1}$ (c) $\frac{n+2}{2(n+1)}$ (d) $\frac{n+2}{2n+1}$

Matching Questions

More than one may match with the same.

45. Normals are drawn at points P, Q and R lying on the parabola $y^2 = 4x$ and they intersect at $(3, 0)$. Then

(i) Area of ΔPQR	(a) 2
(ii) Circumradius of ΔPQR	(b) $\frac{5}{2}$
(iii) Centroid of ΔPQR	(c) $\left(\frac{5}{2}, 0\right)$
(iv) Circumcentre of ΔPQR	(d) $\left(\frac{2}{3}, 0\right)$

46. (i) Two rays in the first quadrant are $x + y = |a|$ and $ax - y = 1$ which intersect each other if $a \in (a_0, \infty)$. The value of a_0 is

(ii) The point (α, β, γ) lies on the plane $x + y + z = 2$. If $\vec{a} = \alpha\vec{i} + \beta\vec{j} + \gamma\vec{k}$ and $\vec{k} \times (\vec{k} \times \vec{a}) = 0$ then γ is

(iii) $\left|\int_0^1 (1 - y^2)\, dy\right| + \left|\int_1^0 (y^2 - 1)\, dy\right|$ equals

(iv) If in a ΔABC,
$\sin A \sin B \sin C + \cos A \cos B = 1$
then the value of $\sin C$ is

(a) 2

(b) $\frac{4}{3}$

(c) $\left|\int_0^1 \sqrt{1 - x}\, dx\right| + \left|\int_{-1}^0 \sqrt{1 + x}\, dx\right|$

(d) 1

Complete the following statement.

47. The value of $5050 \times \dfrac{\int_0^1 (1 - x^{50})^{100}\, dx}{\int_0^1 (1 - x^{50})^{101}\, dx}$ is

Answers

1. a	**2.** b	**3.** d	**4.** c	**5.** a
6. b	**7.** c	**8.** c	**9.** d	**10.** d
11. d	**12.** d	**13.** c	**14.** a	**15.** a
16. b	**17.** c	**18.** b	**19.** a	**20.** d
21. c	**22.** c	**23.** b	**24.** a	**25.** d
26. b	**27.** d	**28.** b	**29.** c	**30.** d
31. a	**32.** b	**33.** c	**34.** b	**35.** a
36. d	**37.** a	**38.** c	**39.** a	**40.** b
41. b, d	**42.** a, b	**43.** a		

44. (i) a (ii) b (iii) c

45. (i) $\leftrightarrow$ a (ii) $\leftrightarrow$ b (iii) $\leftrightarrow$ d (iv) $\leftrightarrow$ c

46. (i) $\leftrightarrow$ d (ii) $\leftrightarrow$ a (iii) $\leftrightarrow$ b, c (iv) $\leftrightarrow$ d

47. 5051

Hints and Solutions to Selected Questions

1. $\alpha+\beta=1, \alpha\beta=p, \gamma+\delta=4, \gamma\delta=q.$

As $\alpha, \beta, \gamma, \delta$ are in GP, $\beta=\alpha r, \gamma=\alpha r^2, \delta=\alpha r^3$.

$\therefore \quad p=\alpha\beta=\alpha^2 r, q=\gamma\delta=\alpha r^2\cdot\alpha r^3=\alpha^2 r^5.$

$\therefore \quad \frac{q}{p}=\frac{\alpha^2 r^5}{\alpha^2 r}=r^4.$

Also $\alpha+\alpha r=1, \alpha r^2+\alpha r^3=4.$

$\therefore \quad \alpha(1+r)=1, \alpha r^2(1+r)=4 \Rightarrow r^2=4.$

$\therefore \quad \frac{q}{p}=r^4=4^2=16$ or $16p=q$. So, (a) is true.

2. $t_{r+1}={}^nC_r\, a^{n-r}\cdot(-b)^r$. So, $t_5+t_6=0$ gives

$${}^nC_4\, a^{n-4}\cdot(-b)^4+{}^nC_5\, a^{n-5}\cdot(-b)^5=0$$

or $$\frac{n(n-1)(n-2)(n-3)}{24}\cdot a^{n-4}\cdot b^4 - \frac{n(n-1)(n-2)(n-3)(n-4)}{120}a^{n-5}\cdot b^5=0$$

or $$\frac{n(n-1)(n-2)(n-3)}{120}\cdot a^{n-5}\cdot b^4\{5a-(n-4)\cdot b\}=0$$

$\Rightarrow \quad 5a=(n-4)b \qquad \therefore \quad \frac{a}{b}=\frac{n-4}{5}.$

3. $$f(x)=(1+b^2)\left\{x^2+\frac{2b}{1+b^2}x+\frac{b^2}{(1+b^2)^2}\right\}-\frac{b^2}{1+b^2}+1$$

$$=(1+b^2)\left(x+\frac{b}{1+b^2}\right)^2+\frac{1}{1+b^2}\geq\frac{1}{1+b^2}$$

$\therefore \quad m(b)=\frac{1}{1+b^2}\cdot$ So, range of $m(b)=(0,1]$.

4. Here, $(2\cos x+\sin x)\begin{vmatrix}1 & \cos x & \cos x\\ 1 & \sin x & \cos x\\ 1 & \cos x & \sin x\end{vmatrix}=0$

$$\text{or} \quad (2\cos x + \sin x)\begin{vmatrix} 1 & \cos x & \cos x \\ 0 & \sin x - \cos x & 0 \\ 0 & 0 & \sin x - \cos x \end{vmatrix} = 0$$

or $\quad (2\cos x + \sin x)(\sin x - \cos x)^2 = 0.$

$\therefore \quad \tan x = -2, 1.$ But $\tan x \neq -2$ in $\left[-\frac{\pi}{4}, \frac{\pi}{4}\right]$.

$\therefore \quad \tan x = 1.$ So, $x = \frac{\pi}{4}$.

5. The number of ways to give images to each element of E in F is 2.

$\therefore$ the total number of ways to give images to the elements of E in $F = 2 \times 2 \times 2 \times 2$.

But in two of them all the elements of E have the same image 1 or the same image 2 (the mapping being into in these two cases).

$\therefore$ the number of onto functions $= 2^4 - 2 = 14$.

6. Clearly, ${}^nC_3 = T_n$. So, ${}^{n+1}C_3 - {}^nC_3 = 21$ or $({}^nC_3 + {}^nC_2) - {}^nC_3 = 21$

$\therefore \quad {}^nC_2 = 21 \quad$ or $\quad n(n-1) = 42 = 7 \cdot 6. \quad \therefore \quad n = 7.$

7. $\left|\frac{z_1 - z_3}{z_2 - z_3}\right| = \left|\frac{1}{2} - i\frac{\sqrt{3}}{2}\right| = \sqrt{\frac{1}{4} + \frac{3}{4}} = 1.$ So, $|z_1 - z_3| = |z_2 - z_3|$.

Also, $\text{amp}\,\frac{z_1 - z_3}{z_2 - z_3} = \tan^{-1}\frac{-\frac{\sqrt{3}}{2}}{\frac{1}{2}} = \tan^{-1}(-\sqrt{3}) = -\frac{\pi}{3}$

or $\quad \text{amp}\,\frac{z_2 - z_3}{z_1 - z_3} = \frac{\pi}{3}$.

$\therefore \quad \angle Z_2Z_3Z_1 = 60°.$

$\therefore$ the triangle has two sides equal and the angle between the equal sides $= 60°$. So, it is equilateral.

8. $\frac{2n}{2}\{2 \cdot 2 + (2n-1) \cdot 3\} = \frac{n}{2}\{2 \cdot 57 + (n-1) \cdot 2\}$

or $\quad 2(6n+1) = 112 + 2n \quad$ or $\quad 10n = 110.$

$\therefore \quad n = 11.$

9. $1^{1/n} = \cos\dfrac{2r\pi}{n} + i\sin\dfrac{2r\pi}{n}$.

Let $z_1 = \cos\dfrac{2r_1\pi}{n} + i\sin\dfrac{2r_1\pi}{n}$ and $z_2 = \cos\dfrac{2r_2\pi}{n} + i\sin\dfrac{2r_2\pi}{n}$.

Then $\angle Z_1OZ_2 = \text{amp}\,\dfrac{z_1}{z_2} = \text{amp}\, z_1 - \text{amp}\, z_2 = \dfrac{2(r_1 - r_2)\pi}{n} = \dfrac{\pi}{2}$ (given)

$\therefore$ $n = 4(r_1 - r_2) = 4 \times$ integer. So, n is of the form $4k$.

10. a, b, c, d are in AP $\Rightarrow \dfrac{a}{abcd}, \dfrac{b}{abcd}, \dfrac{c}{abcd}, \dfrac{d}{abcd}$ are in AP.

$\therefore$ $\dfrac{1}{bcd}, \dfrac{1}{acd}, \dfrac{1}{abd}, \dfrac{1}{abc}$ are in AP.

$\therefore$ bcd, acd, abd, abc are in HP.

$\therefore$ in reverse order, abc, abd, acd, bcd are in HP.

11. $$f\{f(x)\} = \frac{\alpha f(x)}{f(x)+1} = \frac{\alpha \cdot \dfrac{\alpha x}{x+1}}{\dfrac{\alpha x}{x+1} + 1} = \frac{\alpha^2 \cdot x}{\alpha x + x + 1}$$

$\therefore$ $x = \dfrac{\alpha^2 \cdot x}{(\alpha+1)x+1}$ or $x\{(\alpha+1)x + 1 - \alpha^2\} = 0$

or $(\alpha+1)x^2 + (1-\alpha^2)x = 0$. This should hold for all x.

$\Rightarrow$ $\alpha + 1 = 0, 1 - \alpha^2 = 0$. $\therefore$ $\alpha = -1$.

12. If $x < -1$ then $x > x^3$. So, $f(x) = x$.
If $x = -1$ then $x = x^3$. So, $f(x) = x$.
If $-1 < x < 0$ then $x < x^3$. So, $f(x) = x^3$.
If $x = 0$ then $x = x^3$. So, $f(x) = x^3$.
If $0 < x < 1$ then $x > x^3$. So, $f(x) = x$.
If $x = 1$ then $x = x^3$. So, $f(x) = x$.
If $x > 1$ then $x < x^3$. So, $f(x) = x^3$.
Thus,

$$f(x) = \begin{cases} x & \text{for } x \le -1 \\ x^3 & \text{for } -1 < x \le 0 \\ x & \text{for } 0 < x \le 1 \\ x^3 & \text{for } x > 1. \end{cases}$$

Clearly, $f(x)$ is not differentiable at $x = -1, 0, 1$.

13. $x^2(1+x) = \int_0^{x^2} f(t)\,dt$. Differentiating w.r.t x,

$$2x(1+x) + x^2 = f(x^2)\cdot 2x \quad \text{or} \quad f(x^2) = 1 + x + \frac{x}{2} \qquad (\text{for } x > 0).$$

Putting $x = 2$, $f(4) = 1 + 2 + \dfrac{2}{2} = 4$.

14. $f'(k-0) = \lim\limits_{h\to 0} \dfrac{[k-h]\sin \pi(k-h) - [k]\sin \pi k}{-h}$

$$= \lim_{h\to 0} \frac{(-1)^{k-1}(k-1)\sin \pi h - (k\times 0)}{-h} = \lim_{h\to 0} \frac{(-1)^{k-1}(k-1)\sin \pi h}{-h}$$

$$= (-1)^k \cdot (k-1)\pi.$$

15. $f'(x) = e^{x(1-x)} + x\cdot e^{x(1-x)}\cdot(1-2x)$

$$= e^{x(1-x)}\{1 + x(1-2x)\} = e^{x(1-x)}\cdot(-2x^2 + x + 1).$$

Now, the sign scheme for $-2x^2 + x + 1$ is as follows:

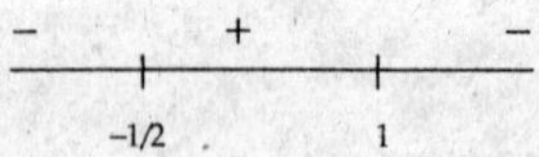

$\therefore \quad f'(x) \geq 0$ if $x \in \left[-\dfrac{1}{2}, 1\right]$ because $e^{x(1-x)}$ is always positive. So, $f(x)$ is increasing on $\left[-\dfrac{1}{2}, 1\right]$.

16. Limit $= \lim\limits_{x\to 0} \dfrac{\cos(\pi\cos^2 x)\cdot \pi \cdot 2\cos x(-\sin x)}{2x}$

$$= \lim_{x\to 0} \pi\cos(\pi\cos^2 x)\cdot \cos x \cdot \frac{-\sin x}{x} = \pi(-1)\cdot 1\cdot(-1) = \pi.$$

17.

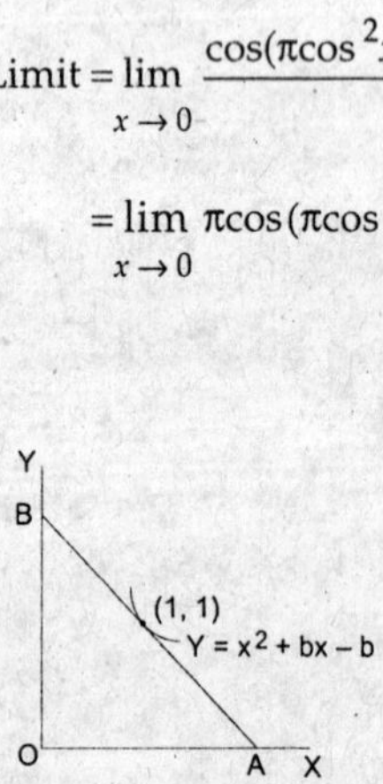

$$\frac{dy}{dx} = 2x + b.$$

So, the equation of the tangent at (1, 1) is

$$y - 1 = \left.\frac{dy}{dx}\right]_{(1,1)} \cdot (x-1) = (2+b)(x-1).$$

or $\quad (2+b)x - y = 1 + b.$

So, $\quad OA = \dfrac{1+b}{2+b}$ and $OB = -(1+b)$.

Being in the first quadrant, $\frac{1+b}{2+b} > 0$ and $1+b < 0$.

$\therefore \quad 1+b<0, 2+b<0 \Rightarrow b<-2 \qquad \ldots (1)$

Now, $\text{ar}(\Delta AOB) = 2 = \frac{1}{2} \cdot \frac{1+b}{2+b} \cdot \{-(1+b)\}$

or $\quad 4(2+b)+(1+b)^2 = 0 \quad$ or $\quad b^2+6b+9=0 \quad$ or $\quad (b+3)^2 = 0.$

$\therefore \quad b = -3$, which satisfies (1).

18. Here,

$$g(x) = \begin{cases} 1+n-n = 1 & \text{for } x = n \in Z \\ 1+n+k-n = 1+k & \text{for } x = n+k, \end{cases}$$

where $n \in Z$ and $0 < k < 1$.

Now,

$$f\{g(x)\} = \begin{cases} -1 & \text{for } g(x) < 0 \\ 0 & \text{for } g(x) = 0 \\ 1 & \text{for } g(x) > 0. \end{cases}$$

Clearly, $g(x) > 0$ for all x. So, $f\{g(x)\} = 1$ for all x.

19. Here, $y = x + \frac{1}{x}$, where $x \geq 1$ and $y \geq 2$.

Now, $x^2 - yx + 1 = 0$ or $x = \frac{y \pm \sqrt{y^2-4}}{2}$.

$\therefore \quad f^{-1}(x) = \frac{x \pm \sqrt{x^2-4}}{2} \geq 1$ when $x \geq 2$.

But, for $x \geq 2$, $\frac{x - \sqrt{x^2-4}}{2} \ngeq 1$.

$\therefore \quad f^{-1}(x) = \frac{x + \sqrt{x^2-4}}{2}$.

20. Here, $x+3 > 0$ and $x^2+3x+2 \neq 0$

$\therefore \quad x > -3$ and $(x+1)(x+2) \neq 0$, i.e., $x \neq -1, -2$.

$\therefore \quad$ the domain is $(-3, +\infty) - \{-1, -2\}$.

21. $I = \int_{-\pi}^{\pi} \frac{\cos^2 x}{1+a^x} dx = \int_{\pi}^{-\pi} \frac{\cos^2 x}{1+a^{-x}} (-dx) \qquad$ (putting $-x$ for x)

$$= \int_{-\pi}^{\pi} \frac{\cos^2 x}{1+a^{-x}} dx$$

$$\therefore \quad I + I = \int_{-\pi}^{\pi} \cos^2 x \left(\frac{1}{1+a^x} + \frac{1}{1+a^{-x}} \right) dx = \int_{-\pi}^{\pi} \cos^2 x \, dx$$

$$= 2\int_0^{\pi} \cos^2 x \cdot dx = \int_0^{\pi} (1 + \cos 2x)\, dx = \left. x \right]_0^{\pi} + \left. \frac{\sin 2x}{2} \right]_0^{\pi}.$$

$\therefore \quad 2I = \pi. \qquad \therefore \quad I = \frac{\pi}{2}.$

22. Any tangent to $y^2 = 4x$ is $y = mx + \frac{1}{m}$.

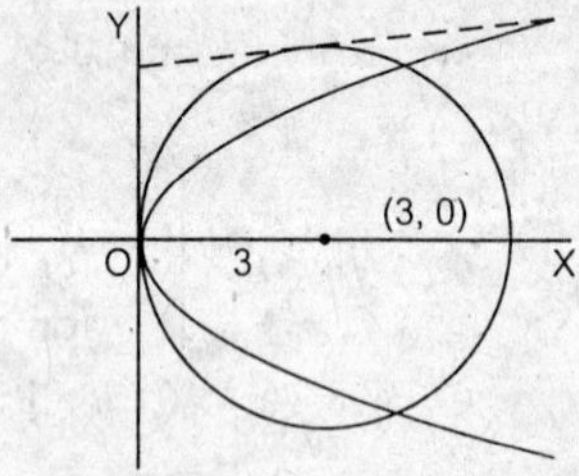

It touches the circle if

$$3 = \left| \frac{3m + \frac{1}{m}}{\sqrt{1+m^2}} \right|$$

or $\quad 9(1 + m^2) = \left(3m + \frac{1}{m}\right)^2$

or $\quad \frac{1}{m^2} = 3 \qquad \therefore \quad m = \pm \frac{1}{\sqrt{3}}.$

For the common tangent to be above the x-axis, $m = \frac{1}{\sqrt{3}}$.

$\therefore$ the common tangent is $y = \frac{1}{\sqrt{3}} x + \sqrt{3}$, i.e., $\sqrt{3}y = x + 3$.

23. Let the centroid = (α, β). Then

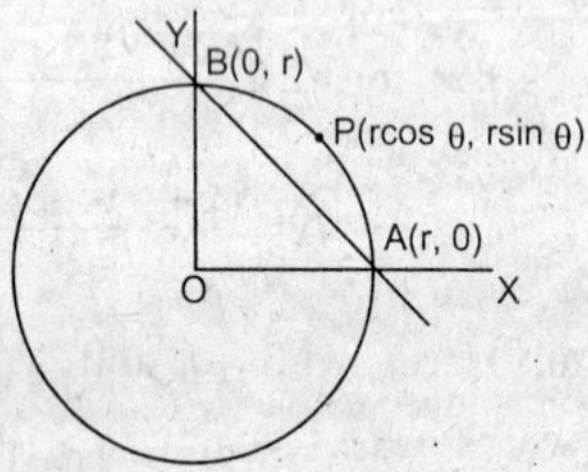

$$\alpha = \frac{r + r\cos\theta}{3}, \quad \beta = \frac{r + r\sin\theta}{3}$$

or $\quad \left(\alpha - \frac{r}{3}\right)^2 + \left(\beta - \frac{r}{3}\right)^2 = \frac{r^2}{9}$

$\therefore$ the locus is $\left(x - \frac{r}{3}\right)^2 + \left(y - \frac{r}{3}\right)^2 = \left(\frac{r}{3}\right)^2$, which is a circle.

24. Solving $3x + 4y = 9$ and $y = mx + 1$, we get $x = \frac{5}{3+4m}$.

x is an integer if $3 + 4m = 1, -1, 5, -5$.

$\therefore \quad m = -\frac{2}{4}, -\frac{4}{4}, \frac{2}{4}, -\frac{8}{4}.$ So, m has two integral values.

25. Here, $y^2 + 4y + 4 + 4x - 2 = 0$ or $(y+2)^2 = -4\left(x - \frac{1}{2}\right)$.

Let $y + 2 = Y$, $\frac{1}{2} - x = X$. Then the parabola is $Y^2 = 4X$.

$\therefore$ the directrix is $X + 1 = 0$ or $\frac{1}{2} - x + 1 = 0$. $\therefore$ $x = \frac{3}{2}$.

26. $|\vec{a} - \vec{b}|^2 + |\vec{b} - \vec{c}|^2 + |\vec{c} - \vec{a}|^2$

$= 2(\vec{a}^2 + \vec{b}^2 + \vec{c}^2) - 2(\vec{a}\cdot\vec{b} + \vec{b}\cdot\vec{c} + \vec{c}\cdot\vec{a})$

$= 2 \times 3 - 2(\vec{a}\cdot\vec{b} + \vec{b}\cdot\vec{c} + \vec{c}\cdot\vec{a})$

$= 6 - \left\{(\vec{a} + \vec{b} + \vec{c})^2 - \vec{a}^2 - \vec{b}^2 - \vec{c}^2\right\}$

$= 9 - |\vec{a} + \vec{b} + \vec{c}|^2 \le 9.$

27. $\cos|x| = \cos x$, which is differentiable everywhere and $|x|$ is not differentiable at $x = 0$. So, both $\cos|x| \pm |x|$ are not differentiable at $x = 0$.

For $x < 0$, $\sin|x| + |x| = \sin(-x) - x = -\sin x - x$.

For $x > 0$, $\sin|x| + |x| = \sin x + x$.

$\therefore$ its LH derivative at $x = 0$ is $\left[-\cos x - 1\right]_{x=0} = -2$

and its RH derivative at $x = 0$ is $\left[\cos x + 1\right]_{x=0} = 2$.

$\therefore$ $\sin|x| + |x|$ is not differentiable at $x = 0$.

For $x < 0$, $\sin|x| - |x| = \sin(-x) - (-x) = -\sin x + x$.

For $x > 0$, $\sin|x| - |x| = \sin x - x$

$\therefore$ its LH derivative at $x = 0$ is $\left[-\cos x + 1\right]_{x=0} = 0$

and its RH derivative at $x = 0$ is $\left[\cos x - 1\right]_{x=0} = 0$

$\therefore$ $\sin|x| - |x|$ is differentiable at $x = 0$.

28. $\log_4(x-1) = \log_2(x-3) = \log_{2^2}(x-3)^2 = \log_4(x-3)^2$.

$\therefore$ $x - 1 = (x-3)^2$ or $x^2 - 7x + 10 = 0$ or $(x-5)(x-2) = 0$.

$\therefore$ $x = 5, 2$. But $x - 3 < 0$ when $x = 2$.

$\therefore$ the only solution is $x = 5$.

29. $[\vec{a}\ \vec{b}\ \vec{c}] = \begin{vmatrix} 1 & 0 & -1 \\ x & 1 & 1-x \\ y & x & 1+x-y \end{vmatrix} = \begin{vmatrix} 1 & 0 & 0 \\ x & 1 & 1 \\ y & x & 1+x \end{vmatrix} = \begin{vmatrix} 1 & 1 \\ x & 1+x \end{vmatrix} = 1.$

30. Solving $y = nx$ and $y = mx + 1$, we get

$$P = \left(\frac{1}{n-m}, \frac{n}{n-m}\right).$$

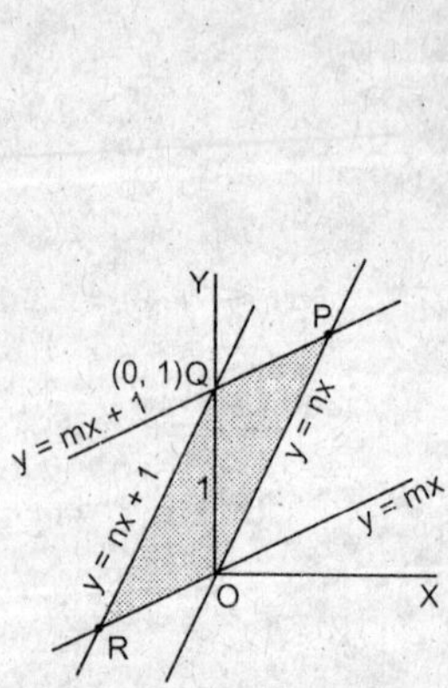

$\therefore$ $\quad ar(\square OPQR) = 2 \times ar(\Delta POQ)$

$$= 2 \times \left|\frac{1}{2} \times OQ \times \frac{1}{n-m}\right|$$

$$= \frac{1}{|n-m|} = \frac{1}{|m-n|}.$$

31.

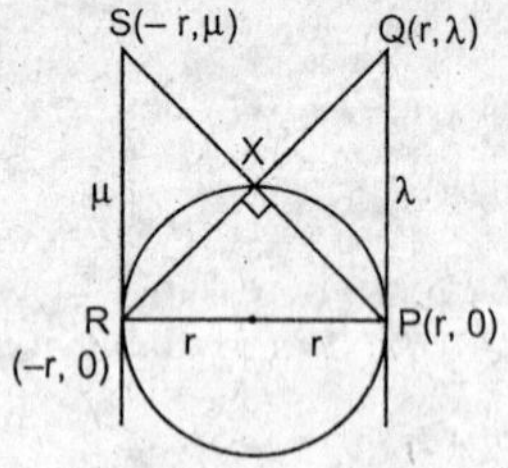

Clearly, from similar triangles,

$$\frac{\lambda}{2r} = \frac{PX}{RX} \text{ and } \frac{\mu}{2r} = \frac{RX}{PX}.$$

$$\therefore \quad \frac{\lambda}{2r} \cdot \frac{\mu}{2r} = 1 \quad \text{or} \quad 4r^2 = \lambda\mu.$$

$$\therefore \quad 2r = \sqrt{PQ \cdot RS}.$$

32. $x = QR - QS = 100 \cot 30° - 100 \cot 60°$

$$= 100\sqrt{3} - \frac{100}{\sqrt{3}} = \frac{200}{\sqrt{3}} = \frac{200\sqrt{3}}{\sqrt{3}}.$$

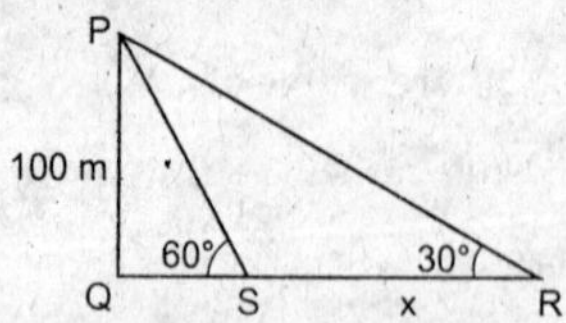

33. $\beta + \gamma = \frac{\pi}{2} - \beta = \alpha$. So, $2\beta + \gamma = \frac{\pi}{2}$ and $\beta + \gamma = \alpha$.

Solving, $\beta = \frac{\pi}{2} - \alpha$ and $\gamma = 2\alpha - \frac{\pi}{2}$.

$\therefore$ $\quad \tan\beta = \cot\alpha$ and $\tan\gamma = \tan\left(2\alpha - \frac{\pi}{2}\right) = -\cot 2\alpha$.

But $\tan\alpha = \cot\alpha - 2\cot 2\alpha$.

$\therefore$ $\quad \tan\alpha = \tan\beta + 2\tan\gamma$.

34. We know that $\sin^{-1} y + \cos^{-1} y = \frac{\pi}{2}$, where $|y| \le 1$.

$\therefore$ the question implies $x - \frac{x^2}{2} + \frac{x^3}{4} - \dots = x^2 - \frac{x^4}{2} + \frac{x^6}{4} - \dots$

or $\frac{x}{1+\frac{x}{2}} = \frac{x^2}{1+\frac{x^2}{2}}$ ($\because$ $0 < |x| < \sqrt{2}$ given)

or $\frac{x}{2+x} = \frac{x^2}{2+x^2}$ or $2x + x^3 = 2x^2 + x^3$ or $x = x^2$

$\therefore$ $x = 0, 1$. But $x \neq 0$. So, $x = 1$.

35. Here, $(\cot \alpha_1) \cdot (\cot \alpha_2) \cdot \dots \cdot (\cot \alpha_n) = 1$.

$\therefore$ $\cos \alpha_1 \cdot \cos \alpha_2 \cdot \dots \cdot \cos \alpha_n = \sin \alpha_1 \cdot \sin \alpha_2 \cdot \dots \cdot \sin \alpha_n$.

Now, $(\cos \alpha_1 \cdot \cos \alpha_2 \cdot \dots \cdot \cos \alpha_n)^2$

$= (\cos \alpha_1 \cdot \cos \alpha_2 \cdot \dots \cdot \cos \alpha_n)(\cos \alpha_1 \cdot \cos \alpha_2 \cdot \dots \cdot \cos \alpha_n)$

$= (\cos \alpha_1 \cdot \cos \alpha_2 \cdot \dots \cos \alpha_n)(\sin \alpha_1 \cdot \sin \alpha_2 \cdot \dots \cdot \sin \alpha_n)$

$= \frac{1}{2^n} \sin 2\alpha_1 \cdot \sin 2\alpha_2 \cdot \dots \cdot \sin 2\alpha_n$.

But each of $\sin 2\alpha_i \leq 1$.

$\therefore$ $(\cos \alpha_1 \cdot \cos \alpha_2 \cdot \dots \cdot \cos \alpha_n)^2 \leq \frac{1}{2^n}$ · But each of $\cos \alpha_i$ is positive.

$\therefore$ $\cos \alpha_1 \cdot \cos \alpha_2 \cdot \dots \cdot \cos \alpha_n \leq \sqrt{\frac{1}{2^n}} = \frac{1}{2^{n/2}}$.

36. $I = \int \frac{x^2 - 1}{x^5 \sqrt{2 - \frac{2}{x^2} + \frac{1}{x^4}}} dx = \int \frac{\frac{1}{x^3} - \frac{1}{x^5}}{\sqrt{2 - \frac{2}{x^2} + \frac{1}{x^4}}} dx$

Put $2 - \frac{2}{x^2} + \frac{1}{x^4} = z^2$. Then $\left(\frac{4}{x^3} - \frac{4}{x^5}\right) dx = 2z\, dz$.

$\therefore$ $I = \int \frac{\frac{1}{2} z\, dz}{z} = \frac{1}{2} \int dz = \frac{1}{2} z + c,$ where c = constant of integration

$= \frac{1}{2} \sqrt{2 - \frac{2}{x^2} + \frac{1}{x^4}} + c$

$= \frac{1}{2x^2} \sqrt{2x^4 - 2x^2 + 1} + c.$

37. Roots are real $\Rightarrow D \ge 0 \Rightarrow 4(a+b+c)^2 - 12\lambda(ab+bc+ca) \ge 0$

$$\Rightarrow \lambda \le \frac{(a+b+c)^2}{3(ab+bc+ca)} = \frac{1}{3}\left(\frac{a^2+b^2+c^2}{ab+bc+ca}+2\right). \qquad \text{... (1)}$$

In the scalene triangle, $|a-b| < c \Rightarrow (a-b)^2 < c^2$.

Similarly, $(b-c)^2 < a^2$ and $(c-a)^2 < b^2$.

Adding, $2(a^2+b^2+c^2-ab-bc-ca) < a^2+b^2+c^2$

or $a^2+b^2+c^2 < 2(ab+bc+ca)$ or $\dfrac{a^2+b^2+c^2}{ab+bc+ca} < 2$.

$\therefore$ from (1), $\lambda \le \frac{1}{3}(2+2) = \frac{4}{3}$.

38. Here, $A^2 = \begin{bmatrix}1 & 0 & 0\\ 0 & 1 & 1\\ 0 & -2 & 4\end{bmatrix}\begin{bmatrix}1 & 0 & 0\\ 0 & 1 & 1\\ 0 & -2 & 4\end{bmatrix} = \begin{bmatrix}1 & 0 & 0\\ 0 & -1 & 5\\ 0 & -10 & 14\end{bmatrix}$; $|A| = \begin{vmatrix}1 & 0 & 0\\ 0 & 1 & 1\\ 0 & -2 & 4\end{vmatrix} = 6$.

$$\therefore \quad A^{-1} = \frac{1}{6}\begin{bmatrix}6 & 0 & 0\\ 0 & 4 & -1\\ 0 & 2 & 1\end{bmatrix} = \frac{1}{6}\left\{\begin{bmatrix}1 & 0 & 0\\ 0 & -1 & 5\\ 0 & -10 & 14\end{bmatrix} + \begin{bmatrix}c & 0 & 0\\ 0 & c & c\\ 0 & -2c & 4c\end{bmatrix} + \begin{bmatrix}d & 0 & 0\\ 0 & d & 0\\ 0 & 0 & d\end{bmatrix}\right\}$$

$\Rightarrow \quad 6 = 1+c+d,\ 4 = -1+c+d,\ -1 = 5+c,\ 2 = -10-2c,\ 1 = 14+4c+d$

$\Rightarrow \quad c = -6$ and $d = 11 \qquad \therefore (c, d) = (-6, 11)$.

39. $f''(x) + f(x) = 0 \Rightarrow 2f'(x)\cdot f''(x) + 2f(x)\cdot f'(x) = 0$

$\Rightarrow \quad d[\{f'(x)\}^2] + d[\{f(x)\}^2] = 0 \Rightarrow \{f'(x)\}^2 + \{f(x)\}^2 = c$

$\Rightarrow \quad \{g(x)\}^2 + \{f(x)\}^2 = c \Rightarrow \left\{g\left(\frac{x}{2}\right)\right\}^2 + \left\{f\left(\frac{x}{2}\right)\right\}^2 = c$

$\Rightarrow \quad F(x) = c. \qquad \therefore F(10) = F(5) = 5$.

40. As $\theta \in \left(0, \frac{\pi}{4}\right)$ where $0 < \tan\theta < 1$, we get $\cot\theta > 1$.

$\therefore \quad t_2 < t_1,\ t_3 > t_1$ and $t_4 > t_3$. So, $t_4 > t_3 > t_1 > t_2$.

41. A vector along the line of intersection of two planes, one of them being parallel to the vectors $\vec{a}, \vec{b}$ and the other plane being parallel to the vectors $\vec{c}, \vec{d}$, is $(\vec{a}\times\vec{b})\times(\vec{c}\times\vec{d})$.

$\therefore \quad \vec{A} \parallel \{(2\vec{j}+3\vec{k})\times(4\vec{j}-3\vec{k})\}\times\{(\vec{j}-\vec{k})\times(3\vec{i}+3\vec{j})\}$

$= -18\vec{i} \times (3\vec{i} - 3\vec{j} - 3\vec{k}) = 54(\vec{k} - \vec{j})$. If θ be an acute angle then

$$\cos\theta = \left| \frac{54(\vec{k} - \vec{j})}{54\sqrt{2}} \cdot (2\vec{i} + \vec{j} - 2\vec{k}) \right| = \left| -\frac{1}{\sqrt{2}} \right| = \frac{1}{\sqrt{2}} \cdot \text{ So, angle} = \frac{\pi}{4}, \pi - \frac{\pi}{4} \cdot$$

42. The parabola is $x^2 = y$. The equation of any tangent to it is $y = mx + c$. Then $x^2 = mx + c$ will have equal roots.

$$\therefore \quad m^2 + 4c = 0 \quad \text{or} \quad c = -\frac{m^2}{4} \cdot$$

$$\therefore \quad y = mx - \frac{m^2}{4} \text{ is any tangent to } x^2 = y.$$

Similarly, any tangent to $y = -(x-2)^2$ is $y = \lambda(x-2) + \dfrac{\lambda^2}{4} \cdot$

They are the same if $m = \lambda$, $-\dfrac{m^2}{4} = \dfrac{\lambda^2}{4} - 2\lambda$.

$$\therefore \quad -\frac{m^2}{4} = \frac{m^2}{4} - 2m \quad \text{or} \quad \frac{m^2}{2} - 2m = 0. \qquad \therefore \quad m = 0, 4.$$

$$\therefore \quad \text{common tangents are } y = mx - \frac{m^2}{4}, \text{ where } m = 0, 4.$$

43. $g(x) = \int_0^x f(t)\,dt \quad \Rightarrow \quad g'(x) = f(x)$

For maximum or minimum of $g(x)$, $g'(x) = 0 \quad \Rightarrow \quad f(x) = 0$.

Now, $f(x) = 0$ if $2 - e^{x-1} = 0$ or $x - e = 0$ $\qquad (\because\ e^x \neq 0 \text{ when } 0 \leq x \leq 1)$

$$\Rightarrow \quad x = 1 + \log 2 \quad \text{or} \quad e.$$

Now,

$$g''(x) = f'(x) = \begin{cases} -e^{x-1} & \text{for } 1 < x \leq 2 \\ 1 & \text{for } 2 < x \leq 3. \end{cases}$$

$\therefore \quad g''(x) < 0$ when $x = 1 + \log 2$ and $g''(x) > 0$ when $x = e$.

$\therefore \quad x = 1 + \log 2$ is a point of local maximum and $x = e$ is a point of local minimum.

44. (i) Here, $P(u_i) = k \cdot i$. Also, we have $\sum_{i=1}^{n} P(u_i) = 1$.

$$\therefore \quad k(1 + 2 + 3 + \ldots + n) = 1 \quad \text{or} \quad k = \frac{2}{n(n+1)} \cdot \text{ So, } P(u_i) = \frac{2i}{n(n+1)} \cdot$$

$$\therefore \quad P(w) = \sum_{i=1}^{n} P(u_i) \cdot \frac{{}^{i}C_1}{{}^{n+1}C_1} = \sum_{i=1}^{n} \frac{2i}{n(n+1)} \cdot \frac{i}{n+1} = \frac{2}{n(n+1)^2} \sum_{i=1}^{n} i^2$$

$$= \frac{2}{n(n+1)^2} \cdot (1^2 + 2^2 + 3^2 + \ldots + n^2)$$

$$= \frac{2}{n(n+1)^2} \cdot \frac{n(n+1)(2n+1)}{6} = \frac{2n+1}{3(n+1)}.$$

$$\therefore \quad \lim_{n \to \infty} P(w) = \lim_{n \to \infty} \frac{2n+1}{3(n+1)} = \frac{2}{3}.$$

(ii) The events $u_1, u_2, \ldots, u_n$ are equiprobable.

$$\therefore \quad P\left(\frac{u_n}{w}\right) = \frac{P(u_n) \cdot P\left(\frac{w}{u_n}\right)}{P(u_1) \cdot P\left(\frac{w}{u_1}\right) + \ldots + P(u_n) \cdot P\left(\frac{w}{u_n}\right)} = \frac{P\left(\frac{w}{u_n}\right)}{P\left(\frac{w}{u_1}\right) + \ldots + P\left(\frac{w}{u_n}\right)}$$

$$= \frac{\frac{n}{n+1}}{\frac{1}{n+1} + \frac{2}{n+1} + \frac{3}{n+1} + \ldots + \frac{n}{n+1}} = \frac{n}{\frac{n(n+1)}{2}} = \frac{2}{n+1}.$$

(iii) $P(E) \cdot P\left(\frac{w}{E}\right) = P(u_2) \cdot P\left(\frac{w}{u_2}\right) + P(u_4) \cdot P\left(\frac{w}{u_4}\right) + \ldots + P(u_n) \cdot P\left(\frac{w}{u_n}\right)$

$$= \frac{1}{n} \cdot \frac{{}^{2}C_1}{{}^{n+1}C_1} + \frac{1}{n} \cdot \frac{{}^{4}C_1}{{}^{n+1}C_1} + \ldots + \frac{1}{n} \cdot \frac{{}^{n}C_1}{{}^{n+1}C_1} = \frac{2+4+6+\ldots+n}{n(n+1)}$$

$$\therefore \quad \frac{1}{2} \cdot P\left(\frac{w}{E}\right) = \frac{2}{n(n+1)}\left(1 + 2 + 3 + \ldots + \frac{n}{2}\right)$$

$$= \frac{2}{n(n+1)} \frac{\frac{n}{2}\left(\frac{n}{2}+1\right)}{2} = \frac{n+2}{4(n+1)}$$

$$\Rightarrow \quad P\left(\frac{w}{E}\right) = \frac{n+2}{2(n+1)}.$$

45. (i) Normal at $(t^2, 2t)$ to $y^2 = 4x$ is $y + tx = 2t + t^3$.

It passes through $(3, 0)$. So, $3t = 2t + t^3$ or $t(t^2 - 1) = 0$.

$\therefore \; t = 0, \pm 1$. So, the vertices P, Q and R are $(0, 0)$, $(1, 2)$, $(1, -2)$.

$\therefore \; \text{ar}(\Delta PQR) = \frac{1}{2} \cdot 4 \cdot 1 = 2.$

(ii) In ΔPQR, $\dfrac{QR}{\sin \angle QPR} = 2 \times$ circumradius.

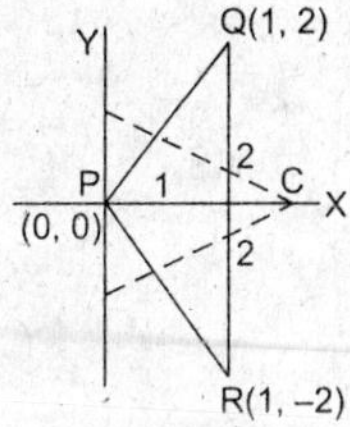

$\therefore$ circumradius $= \dfrac{4}{2\sin 2\theta}$, where $\tan\theta = 2$

$$= \frac{1}{\sin\theta\cos\theta} = \frac{1}{\frac{2}{\sqrt{5}}\cdot\frac{1}{\sqrt{5}}} = \frac{5}{2}.$$

(iii) Centroid $= \left(\dfrac{0+1+1}{3}, \dfrac{0+2-2}{3}\right) = \left(\dfrac{2}{3}, 0\right)$.

(iv) As $PQ = PR$, the circumcentre C is on the x-axis and PC = circumradius.

$\therefore\ C = \left(\dfrac{5}{2}, 0\right)$.

46. (i) Solving, $x = \dfrac{1+|a|}{1+a}$ and $y = \dfrac{a|a|-1}{1+a}$. It should be in the first quadrant. So, $\dfrac{1+|a|}{1+a} > 0$ and $\dfrac{a|a|-1}{1+a} > 0$

$\Rightarrow\ 1 + a > 0$ and $a|a| - 1 > 0$. ... (1)

Now, $a|a| - 1 > 0 \quad \Rightarrow\ a > \dfrac{1}{|a|} > 0$. So, $a = |a|$.

$\therefore$ (1) $\quad\Rightarrow a^2 - 1 > 0 \quad \Rightarrow a > 1$. As $a \in (a_0, \infty)$, $a_0 = 1$.

(ii) $\alpha + \beta + \gamma = 2$. Now, $\vec{k} \times (\vec{k} \times \vec{a}) = \vec{0} \quad \Rightarrow (\vec{k}\cdot\vec{a})\ \vec{k} - \vec{a} = \vec{0}$

or $\gamma\vec{k} - (\alpha\vec{i} + \beta\vec{j} + \gamma\vec{k}) = \vec{0} \quad \Rightarrow \alpha\vec{i} + \beta\vec{j} = \vec{0} \quad \Rightarrow \alpha = 0, \beta = 0$.

$\therefore\ \alpha + \beta + \gamma \quad \Rightarrow \gamma = 2$.

(iii) $$\left|\int_0^1 (1-y^2)\,dy\right| + \left|\int_1^0 (y^2-1)\,dy\right| = 2\left|\int_0^1 (1-y^2)\,dy\right|$$

$$= 2\left|\left[y - \frac{y^3}{3}\right]_0^1\right| = 2\left(1 - \frac{1}{3}\right) = \frac{4}{3}.$$

$$\left|\int_0^1 \sqrt{1-x}\,dx\right| + \left|\int_{-1}^0 \sqrt{1+x}\,dx\right|$$

$$= 2\left|\int_0^1 \sqrt{1-x}\,dx\right| \quad \text{(putting } x = -z \text{ in the second)}$$

$$= 2\left|\left[\frac{(1-x)^{3/2}}{-\frac{3}{2}}\right]_0^1\right| = \frac{4}{3}\,|0-1| = \frac{4}{3}.$$

(iv) $\sin A \sin B \sin C + \cos A \cos B \leq \sin A \sin B + \cos A \cos B$

(equality holding when $\sin C = 1$)

$= \cos(A-B) \leq 1.$

$\therefore$ $\sin A \sin B \sin C + \cos A \cos B = 1$ only when $\sin C = 1$.

47. $I_{101} = \int_0^1 (1-x^{50})^{101}\,dx = \int_0^1 (1-x^{50}) \cdot (1-x^{50})^{100}\,dx$

$$= \int_0^1 (1-x^{50})^{100}\,dx - \int_0^1 x^{50} \cdot (1-x^{50})^{100}\,dx.$$

$$\therefore \quad I_{101} = I_{100} + \int_0^1 \frac{x}{50} \cdot (1-x^{50})^{100}\,d(1-x^{50})$$

$$= I_{100} + x \cdot \left.\frac{(1-x^{50})^{101}}{50 \times 101}\right]_0^1 - \int_0^1 \frac{(1-x^{50})^{101}}{50 \times 101}\,dx.$$

$$\therefore \quad I_{101} = I_{100} - \frac{I_{101}}{5050} \quad \Rightarrow \quad \frac{5051}{5050} I_{101} = I_{100}.$$

$$\therefore \quad 5050 \cdot \frac{I_{100}}{I_{101}} = 5051.$$

□

IIT Questions—3

One-Option-Correct Questions

1. Let $\omega = -\frac{1}{2} + i\frac{\sqrt{3}}{2}$. Then the value of the determinant $\begin{vmatrix} 1 & 1 & 1 \\ 1 & -1-\omega^2 & \omega^2 \\ 1 & \omega^2 & \omega^4 \end{vmatrix}$ is

   (a) 3ω (b) $3\omega(\omega - 1)$

   (c) $3\omega^2$ (d) $3\omega(1 - \omega)$

2. For all complex numbers z_1, z_2 satisfying $|z_1| = 12$ and $|z_2 - 3 - 4i| = 5$, the minimum value of $|z_1 - z_2|$ is

   (a) 0 (b) 2 (c) 7 (d) 17

3. If $a_1, a_2, ..., a_n$ are positive real numbers whose product is a fixed number c, the minimum value of $a_1 + a_2 + \ldots + a_{n-1} + 2a_n$ is

   (a) $n(2c)^{1/n}$ (b) $(n+1)c^{1/n}$ (c) $2nc^{1/n}$ (d) $(n+1)(2c)^{1/n}$

4. Suppose a, b, c are in AP and a^2, b^2, c^2 are in GP. If $a < b < c$ and $a + b + c = \frac{3}{2}$ then the value of a is

   (a) $\frac{1}{2\sqrt{2}}$ (b) $\frac{1}{2\sqrt{3}}$ (c) $\frac{1}{2} - \frac{1}{\sqrt{3}}$ (d) $\frac{1}{2} - \frac{1}{\sqrt{2}}$

5. The number of arrangements of the letters of the word *BANANA* in which two *N*s do not appear adjacently is

   (a) 40 (b) 60 (c) 80 (d) 100

6. The sum $\sum_{i=0}^{m} \binom{10}{i}\binom{20}{m-i}$, where $\binom{p}{q} = 0$ if $p < q$, is the maximum when m is

   (a) 5 (b) 10 (c) 15 (d) 20

7. The number of values of k for which the system of equations

   $$(k+1)\,x + 8y = 4k,$$

   $$k \cdot x + (k+3)y = 3k - 1$$

has infinitely many solutions is

(a) 0 (b) 1 (c) 2 (d) infinite

8. The set of all real x for which $x^2 - |x+2| + x > 0$ is

(a) $(-\infty, -2) \cup (2, \infty)$ (b) $(-\infty, -\sqrt{2}) \cup (\sqrt{2}, \infty)$

(c) $(-\infty, -1) \cup (1, \infty)$ (d) $(\sqrt{2}, \infty)$

9. The length of the longest interval in which the function $3\sin x - 4\sin^3 x$ is increasing is

(a) $\frac{\pi}{3}$ (b) $\frac{\pi}{2}$ (c) $\frac{3\pi}{2}$ (d) π

10. Which of the following pieces of data does not uniquely determine an acute-angled ΔABC (R = circumradius)?

(a) $a, \sin A, \sin B$ (b) a, b, c

(c) $a, \sin B, R$ (d) $a, \sin A, R$

11. The number of integral values of k for which the equation $7\cos x + 5\sin x = 2k + 1$ has a solution is

(a) 4 (b) 8 (c) 10 (d) 12

12. Let $0 < \alpha < \frac{\pi}{2}$ be a fixed angle. If $P = (\cos\theta, \sin\theta)$ and $Q = \left(\cos(\alpha - \theta), \sin(\alpha - \theta)\right)$ then Q is obtained from P by

(a) clockwise rotation around origin through an angle α

(b) anticlockwise rotation around origin through an angle α

(c) reflection in the line through origin with slope $\tan\alpha$

(d) reflection in the line through origin with slope $\tan\frac{\alpha}{2}$

13. Let $P = (-1, 0)$, $Q = (0, 0)$ and $R = (3, 3\sqrt{3})$ be three points. Then the equation of the bisector of the angle PQR is

(a) $\frac{\sqrt{3}}{2}x + y = 0$ (b) $x + \sqrt{3}y = 0$

(c) $\sqrt{3}x + y = 0$ (d) $x + \frac{\sqrt{3}}{2}y = 0$

14. A straight line through the origin O meets the parallel lines $4x + 2y = 9$ and $2x + y + 6 = 0$ at points P and Q, respectively. Then the point O

divides the segment PQ in the ratio

(a) 1 : 2 (b) 3 : 4 (c) 2 : 1 (d) 4 : 3

15. If the tangent at the point P on the circle $x^2+y^2+6x+6y=2$ meets the straight line $5x-2y+6=0$ at a point Q on the y-axis then the length of PQ is

(a) 4 (b) $2\sqrt{5}$ (c) 5 (d) $3\sqrt{5}$

16. If $a>2b>0$ then the positive value of m for which $y=mx-b\sqrt{1+m^2}$ is a common tangent to $x^2+y^2=b^2$ and $(x-a)^2+y^2=b^2$ is

(a) $\dfrac{2b}{\sqrt{a^2-4b^2}}$ (b) $\dfrac{\sqrt{a^2-4b^2}}{2b}$

(c) $\dfrac{2b}{a-2b}$ (d) $\dfrac{b}{a-2b}$

17. The locus of the midpoint of the line segment joining the focus to a moving point on the parabola $y^2=4ax$ is another parabola with the directrix

(a) $x=-a$ (b) $x=-\frac{a}{2}$ (c) $x=0$ (d) $x=\frac{a}{2}$

18. The area bounded by the curves $y=|x|-1$ and $y=-|x|+1$ is

(a) 1 (b) 2 (c) $2\sqrt{2}$ (d) 4

19. Suppose $f(x)=(x+1)^2$ for $x\geq -1$ If $g(x)$ is the function whose graph is the reflection of the graph of $f(x)$ with respect to the line $y=x$ then $g(x)$ equals

(a) $-\sqrt{x}-1$ for $x\geq 0$ (b) $\dfrac{1}{(x+1)^2}$ for $x>-1$

(c) $\sqrt{x+1}$ for $x\geq -1$ (d) $\sqrt{x}-1$ for $x\geq 0$

20. Let the function $f: R\to R$ be defined by $f(x)=2x+\sin x$ $(x\in R)$. Then f is

(a) one-to-one and onto (b) one-to-one but not onto

(c) onto but not one-to-one (d) neither one-to-one nor onto

21. The domain of the derivative of the function

$$f(x)=\begin{cases}\tan^{-1}x & \text{for } |x|\leq 1\\ \frac{1}{2}(|x|-1) & \text{for } |x|>1\end{cases}$$

is

(a) $R - \{0\}$ (b) $R - \{1\}$ (c) $R - \{-1\}$ (d) $R - \{-1, 1\}$

22. The integer n for which $\lim_{x \to 0} \frac{(\cos x - 1)(\cos x - e^x)}{x^n}$ is a finite nonzero number is

(a) 1 (b) 2 (c) 3 (d) 4

23. Let $f: R \to R$ be such that $f(1) = 3$ and $f'(1) = 6$. Then $\lim_{x \to 0} \left\{\frac{f(1+x)}{f(1)}\right\}^{1/x}$ equals

(a) 1 (b) $e^{1/2}$ (c) e^2 (d) e^3

24. The point(s) on the curve $y^3 + 3x^2 = 12y$, where the tangent is vertical (i.e., parallel to the y-axis), is/are

(a) $\left(\pm\frac{4}{\sqrt{3}}, -2\right)$ (b) $\left(\pm\frac{\sqrt{11}}{3}, 1\right)$

(c) $(0, 0)$ (d) $\left(\pm\frac{4}{\sqrt{3}}, 2\right)$

25. The equation of the common tangent to the curves $y^2 = 8x$ and $xy = -1$ is

(a) $3y = 9x + 2$ (b) $y = 2x + 1$

(c) $2y = x + 8$ (d) $y = x + 2$

26. Let $f(x) = \int_1^x \sqrt{2 - t^2}\, dt$. Then real roots of the equation $x^2 - f'(x) = 0$ are

(a) ± 1 (b) $\pm\frac{1}{\sqrt{2}}$ (c) $\pm\frac{1}{2}$ (d) 0 and 1

27. Let $T > 0$ be a fixed number. Suppose f is a continuous function such that for all $x \in R$, $f(x + T) = f(x)$. If $I = \int_0^T f(x)\, dx$ then the value of $\int_3^{3+3T} f(2x)\, dx$ is

(a) $\frac{3}{2}I$ (b) $2I$ (c) $3I$ (d) $6I$

28. The integral $\int_{-1/2}^{1/2}\left\{[x]+\log\left(\frac{1+x}{1-x}\right)\right\}dx$ equals

(a) $-\frac{1}{2}$ (b) 0

(c) 1 (d) $2\log\frac{1}{2}$

29. If $\vec{a}$ and $\vec{b}$ are two unit vectors such that $\vec{a}+2\vec{b}$ and $5\vec{a}-4\vec{b}$ are perpendicular to each other then the angle between $\vec{a}$ and $\vec{b}$ is

(a) 45° (b) 60°

(c) $\cos^{-1}\left(\frac{1}{3}\right)$ (d) $\cos^{-1}\left(\frac{2}{7}\right)$

30. Let $\vec{V}=2\vec{i}+\vec{j}-\vec{k}$ and $\vec{W}=\vec{i}+3\vec{k}$. If $\vec{U}$ is a unit vector then the maximum value of the scalar triple product $[\vec{U}\ \vec{V}\ \vec{W}]$ is

(a) -1 (b) $\sqrt{10}+\sqrt{6}$

(c) $\sqrt{59}$ (d) $\sqrt{60}$

31. If $\theta \in (0, 2\pi)$ then the interval of values of θ for which $2\sin^2\theta - 5\sin\theta + 2 > 0$ is

(a) $\left(\frac{\pi}{8}, \frac{5\pi}{6}\right)$ (b) $\left(\frac{41\pi}{48}, \pi\right)$

(c) $\left(0, \frac{\pi}{6}\right)\cup\left(\frac{5\pi}{6}, 2\pi\right)$ (d) $\left(0, \frac{\pi}{8}\right)\cup\left(\frac{\pi}{6}, \frac{5\pi}{6}\right)$

32. If w is a nonreal complex number such that $\frac{w-\overline{w}z}{1-z}$ is purely real then the set of values of z is

(a) $\{z: z=\overline{z}\}$ (b) $\{z: |z|=1\}$

(c) $\{z: z\neq 1\}$ (d) $\{z: |z|=1, z\neq 1\}$

33. If r, s and t are prime numbers and p, q are positive integers such that the LCM of p, q is $r^2t^4s^2$ then the number of ordered pair (p, q) is

(a) 254 (b) 252

(c) 225 (d) 224

34. The axis of a parabola is along the line $y = x$ and the distance of the origin from its vertex is $\sqrt{2}$ and that from its focus is $2\sqrt{2}$. If the vertex and focus both lie in the first quadrant then the equation of the parabola is

(a) $(x+y)^2 = x - y - 2$ (b) $(x-y)^2 = x + y - 2$

(c) $(x-y)^2 = 4(x+y-2)$ (d) $(x-y)^2 = 8(x+y-2)$

One-or-More-Options-Correct Questions

35. A tangent to the curve $y = f(x)$ at $P(x, y)$ cuts the x-axis and the y-axis at A and B, respectively, such that $BP : AP = 3 : 1$. If $f(1) = 1$ then

(a) the equation of the curve is $x\frac{dy}{dx} + 3y = 0$

(b) the curve passes through $\left(2, \frac{1}{8}\right)$

(c) the equation of the curve is $x\frac{dy}{dx} - 3y = 0$

(d) the normal at (1, 1) is $x + 3y = 4$

36. The internal bisector of the angle A of the ΔABC meets BC at D. A line drawn through D perpendicular to AD intersects the side AC at E and the side AB produced at F. Then

(a) HM of b and c is equal to AE (b) the ΔAEF is isosceles

(c) $AD = \frac{2bc}{b+c}\cos\frac{A}{2}$ (d) $EF = \frac{4bc}{b+c}\sin\frac{A}{2}$

37. $f(x)$ is a polynomial of the third degree which has a local maximum at $x = -1$. If $f(1) = -1$, $f(2) = 18$ and $f'(x)$ has a local minimum at $x = 0$ then

(a) $f(0) = 5$

(b) $f(x)$ has a local minimum at $x = 1$

(c) $f(x)$ is increasing in $[1, 2\sqrt{5}]$

(d) the distance between $(-1, 2)$ and $(a, f(a))$, where a is a point of local minimum, is $2\sqrt{5}$

Comprehension-Type Questions

38. Suppose we define definite integral using the formula $\int_a^b f(x)\,dx = \frac{b-a}{2}\{f(a) + f(b)\}$. For more accurate result, we have

$$\int_a^b f(x)\,dx = \frac{b-a}{4}\{f(a) + f(b) + 2f(c)\} \text{ when } c = \frac{a+b}{2}.$$ Also, let

$$F(c) = \frac{c-a}{2}\{f(a) + f(c)\} + \frac{b-c}{2}\{f(b) + f(c)\} \text{ when } c \in (a, b).$$

(i) $\int_0^{\pi/2} \sin x\, dx$ equals

(a) $\frac{\pi}{8}(1 + \sqrt{2})$ (b) $\frac{\pi}{4}(1 + \sqrt{2})$

(c) $\frac{\pi}{8\sqrt{2}}$ (d) $\frac{\pi}{4\sqrt{2}}$

(ii) $\int_a^b f(x)\, dx$ is equal to

(a) $F\left(\frac{a-b}{2}\right)$ (b) $F\left(\frac{a+b}{2}\right)$

(c) $F\left(\frac{a-b}{4}\right)$ (d) $F\left(\frac{a+b}{2}\right)$

(iii) If $f''(x) < 0 \ \forall \ x \in (a, b)$ and $(c, f(c))$ is a point lying on the curve $y = f(x)$, where $a < c < b$ and for that value of c, $F(c)$ has a maximum, then $f'(c)$ equals

(a) $\frac{f(b) - f(a)}{b - a}$ (b) $\frac{2}{b-a}\{f(b) - f(a)\}$

(c) $\frac{2f(b) - f(a)}{2b - a}$ (d) 0

Matching Questions

More than one may match with the same.

39. (i) If $\sum_{n=1}^{\infty} \tan^{-1} \frac{1}{2n^2} = t$ then $\tan t$ equals (a) 0

(ii) If in the ΔABC the sides a, b, c are in AP and $\cos \theta_1 = \frac{a}{b+c}$, $\cos \theta_2 = \frac{b}{a+c}$ and $\cos \theta_3 = \frac{c}{a+b}$ then $\tan^2 \frac{\theta_1}{2} + \tan^2 \frac{\theta_3}{2}$ equals (b) 1

(iii) A line is perpendicular to $x + 2y + 2z = 0$ and passes through (0, 1, 0). The perpendicular distance of the line from the origin is — (c) $\frac{\sqrt{5}}{3}$

(iv) $\int_0^1 e^{x^2} dx + \int_1^e (-1 + \sqrt{\log x})\, dx$ has the value — (d) $\frac{2}{3}$

Complete the following statement.

40. If $f(x)$ is a twice differentiable function such that $f(a) = 0, f(b) = 2, f(c) = -1, f(d) = 2$ and $f(e) = 0$, where $a < b < c < d < e$, then the minimum number of zeros of $g(x) = \{f'(x)\}^2 + f''(x) \cdot f(x)$ in the interval $[a, e]$ is.

Answers

1. b	**2.** b	**3.** a	**4.** d	**5.** a
6. a	**7.** b	**8.** b	**9.** a	**10.** d
11. b	**12.** d	**13.** c	**14.** b	**15.** c
16. a	**17.** c	**18.** b	**19.** d	**20.** a
21. c	**22.** c	**23.** c	**24.** d	**25.** d
26. a	**27.** c	**28.** a	**29.** b	**30.** c
31. c	**32.** d	**33.** d	**34.** d	**35.** a, b
36. a, b, c, d	**37.** b, c			

38. (i) a (ii) b (iii) a

39. (i) $\leftrightarrow$ b (ii) $\leftrightarrow$ d (iii) $\leftrightarrow$ c (iv) $\leftrightarrow$ b

40. 6

Hints and Solutions of Selected Questions

1. $\Delta = \begin{vmatrix} 3 & 1 & 1 \\ 0 & -1-\omega^2 & \omega^2 \\ 0 & \omega^2 & \omega \end{vmatrix}$ $(C_1 \to C_1 + C_2 + C_3)$

$= 3\{\omega \cdot \omega - \omega^4\} = 3(\omega^2 - \omega) = 3\omega(\omega - 1)$

2. $\min|z_1 - z_2|$ = minimum distance between the circles
$= 12 - 5 - 5 = 2.$

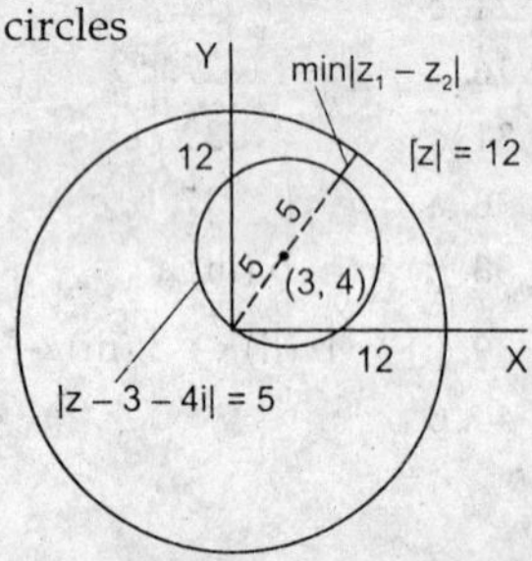

3. $\text{AM} \ge \text{GM} \Rightarrow \dfrac{a_1 + a_2 + \ldots + a_{n-1} + 2a_n}{n} \ge (a_1 \cdot a_2, \ldots a_{n-1} \cdot 2a_n)^{1/n} = (2c)^{1/n}$

4. $b = a + d,\ c = a + 2d$, where $d > 0$.

Now, $a^2, (a+d)^2, (a+2d)^2$ are in GP.

$\therefore \quad (a+d)^4 = a^2(a+2d)^2$

or $\quad (a+d)^2 = \pm a(a+2d)$

or $\quad a^2 + d^2 + 2ad = \pm(a^2 + 2ad).$

Taking + sign, $d = 0$ (not possible as $a < b < c$).

Taking − sign, $2a^2 + 4ad + d^2 = 0$. Also, $a + b + c = \dfrac{3}{2} \Rightarrow a + d = \dfrac{1}{2}\cdot$

$$\therefore \quad 2a^2 + 4a\left(\frac{1}{2} - a\right) + \left(\frac{1}{2} - a\right)^2 = 0 \text{ or } 4a^2 - 4a - 1 = 0.$$

$\therefore \quad a = \dfrac{1}{2} \pm \dfrac{1}{\sqrt{2}}\cdot$ Here $d = \dfrac{1}{2} - a > 0$. So, $a < \dfrac{1}{2}\cdot$ Hence $a = \dfrac{1}{2} - \dfrac{1}{\sqrt{2}}\cdot$

5. Required number of arrangements
= (total number of arrangements) − (number of arrangements in which *N*s are together)

$$= \frac{6!}{2! \times 3!} - \frac{5!}{3!} = 60 - 20 = 40.$$

6. Expression $= {}^{10}C_0 \cdot {}^{20}C_m + {}^{10}C_1 \cdot {}^{20}C_{m-1} + {}^{10}C_2 \cdot {}^{20}C_{m-2} + \ldots + {}^{10}C_m \cdot {}^{20}C_0$.

$(1+x)^{10} = {}^{10}C_0 + {}^{10}C_1 x + {}^{10}C_2 x^2 + \ldots + {}^{10}C_m x^m + \ldots + {}^{10}C_{10} x^{10}$.

$(x+1)^{20} = {}^{20}C_0 x^{20} + {}^{20}C_1 x^{19} + {}^{20}C_2 x^{18} + \cdots + {}^{20}C_m x^{20-m} + \cdots + {}^{20}C_{20}$.

Multiply and equate coefficient of x^{20-m} on both sides.
Sum = coefficient of x^{20-m} in $(1+x)^{30}$.
The middle coefficient of $(1+x)^{30}$ is the greatest coefficient , i.e., the coefficient of x^{15}. So, $20 - m = 15$ or $m = 5$.

7. Condition is $\dfrac{k+1}{k} = \dfrac{8}{k+3} = \dfrac{4k}{3k-1} \Rightarrow k^2 - 4k + 3 = 0$ and $k^2 - 3k + 2 = 0$

$\Rightarrow k = 1, 3$ and $k = 1, 2$. One common value.

8. $x < -2 \Rightarrow x^2 + x + 2 + x > 0$ or $x^2 + 2x + 2 > 0$

or $(x+1)^2 + 1 > 0$ which is true for all x.

$\therefore \quad x < -2.$

$x \geq -2 \Rightarrow x^2 - x - 2 + x > 0$ or $x^2 - 2 > 0$ or $x < -\sqrt{2}$ or $x > \sqrt{2}$.

$\therefore \quad x \in [-2, -\sqrt{2}) \cup (\sqrt{2}, +\infty)$.

$\therefore$ the solution is $x \in (-\infty, -2) \cup [-2, -\sqrt{2}) \cup (\sqrt{2}, +\infty)$,

i.e., $(-\infty, -\sqrt{2}) \cup (\sqrt{2}, +\infty)$.

9. $3\sin x - 4\sin^3 x = \sin 3x$. It is increasing when $-\dfrac{\pi}{2} \leq 3x \leq \dfrac{\pi}{2}$,

i.e., $-\dfrac{\pi}{6} \leq x \leq \dfrac{\pi}{6}$. The length of this interval $= \left|\dfrac{\pi}{6} - \left(-\dfrac{\pi}{6}\right)\right| = \dfrac{\pi}{3}$.

10. $\dfrac{a}{\sin A} = 2R$ and $b = 2R\sin B$. So, two sides and two angles are known.
So, $\angle C$ is known. Therefore, two sides and the included angle is known.
So, Δ is uniquely known in case (a).
If a, b, c are known, Δ is uniquely known in case (b).
$b = 2R\sin B$, $\sin A = \dfrac{a}{2R}$ · So, sides a, b and angles A, B are known. So, $\angle C$ is known. Therefore, two sides and included angle is known. So, Δ is uniquely known in case (c).
$\dfrac{a}{\sin A} = R$. So, only a side and an angle is known. So, Δ is not uniquely known in case (d).

11. $-\sqrt{7^2+5^2} \le 7\cos x + 5\sin x \le \sqrt{7^2+5^2}$.

So, for solution, $-\sqrt{74} \le 2k+1 \le \sqrt{74}$ or $-8.6 \le 2k+1 \le 8.6$

or $-9.6 \le 2k \le 7.6$ or $-4.8 \le k \le 3.8$.

So, integral values of k are $-4, -3, -2, -1, 0, 1, 2, 3$ (eight values).

12. Clearly, (d) is correct.

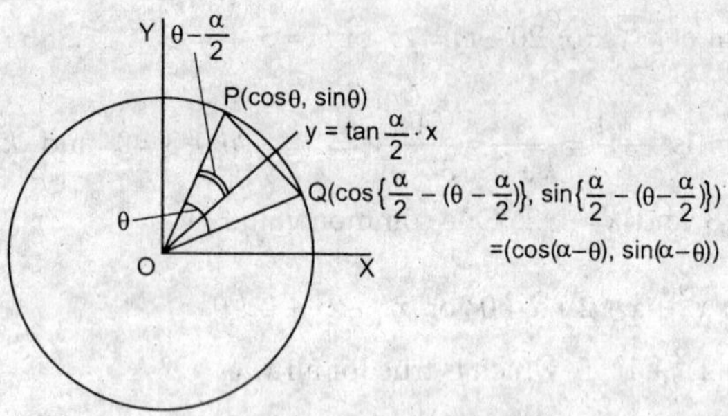

13. The equation of the bisector of $\angle PQR$ is

$$y = (\tan 120°)x$$

or $y = -\sqrt{3}x$.

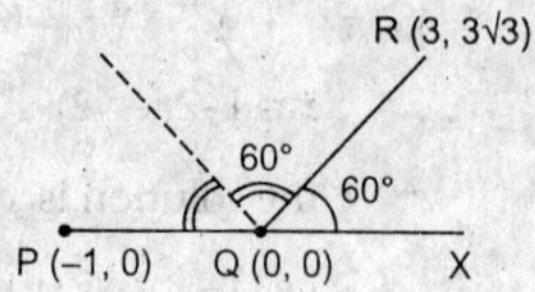

14. $OP : OQ = OR : OS$.

The equation of SR is $y = \frac{1}{2}x$.

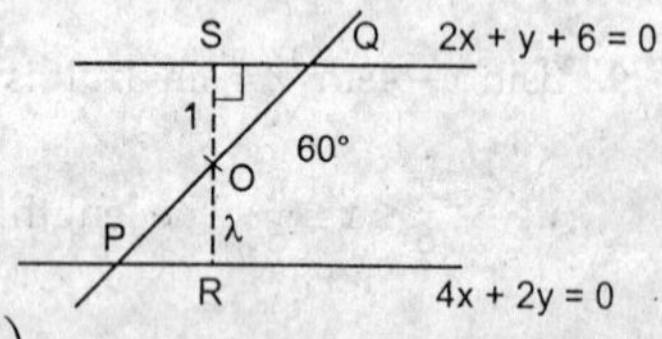

$\therefore \quad R = \left(\frac{9}{5}, \frac{9}{10}\right)$ and $S = \left(-\frac{12}{5}, -\frac{6}{5}\right)$.

$$\therefore \quad O = (0, 0) = \left(\frac{-\frac{12}{5}\lambda + \frac{9}{5}}{\lambda+1}, \frac{-\frac{6}{5}\lambda + \frac{9}{10}}{\lambda+1}\right).$$

$\therefore \quad -\frac{12}{5}\lambda + \frac{9}{5} = 0, -\frac{6}{5}\lambda + \frac{9}{10} = 0 \Rightarrow 12\lambda = 9. \therefore \lambda = \frac{3}{4}$. So, $OP : OQ = 3 : 4$.

15. Let $P = (x_1, y_1)$.

The tangent at P is $xx_1 + yy_1 + 3(x + x_1) + 3(y + y_1) - 2 = 0$. ... (1)

Coordinates of Q satisfy (1), $5x - 2y + 6 = 0$, $x = 0$.

So, $3x_1 + 6y_1 + 7 = 0$ and $Q = (0, 3)$.

$\therefore \quad PQ^2 = x_1^2 + (y_1 - 3)^2 = x_1^2 + y_1^2 - 6y_1 + 9$

$= 11 - 6x_1 - 12y_1 \qquad (\because x_1^2 + y_1^2 + 6x_1 + 6y_1 - 2 = 0)$

$= 11 - 2(3x_1 + 6y_1) = 11 - 2(-7) = 25$. So, $PQ = 5$.

16. Any tangent to $x^2 + y^2 = b^2$ is $y = mx - b\sqrt{1+m^2}$.

It touches $(x-a)^2 + y^2 = b^2$ if $\dfrac{ma - b\sqrt{1+m^2}}{\sqrt{m^2+1}} = b$

or $ma = 2b\sqrt{1+m^2}$ or $m^2a^2 = 4b^2 + 4b^2m^2$.

$\therefore \quad m = \pm\dfrac{2b}{\sqrt{a^2 - 4b^2}}$.

17. $\alpha = \dfrac{at^2 + a}{2}, \beta = \dfrac{2at + 0}{2} \Rightarrow 2\alpha = at^2 + a, \beta = at.$

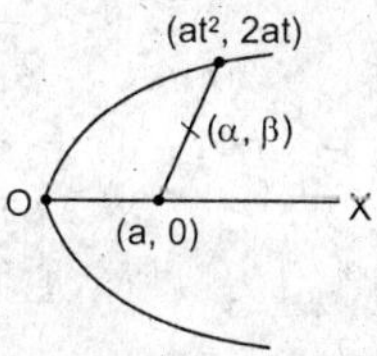

$\therefore \quad 2\alpha = a \cdot \dfrac{\beta^2}{a^2} + a$ or $2a\alpha = \beta^2 + a^2$.

$\therefore$ the locus is $y^2 = \dfrac{4a}{2}\left(x - \dfrac{a}{2}\right)$

$= 4b(x - b), \left(b = \dfrac{a}{2}\right)$.

$\therefore$ the directrix is $(x - b) + b = 0$ or $x = 0$.

18. The lines are

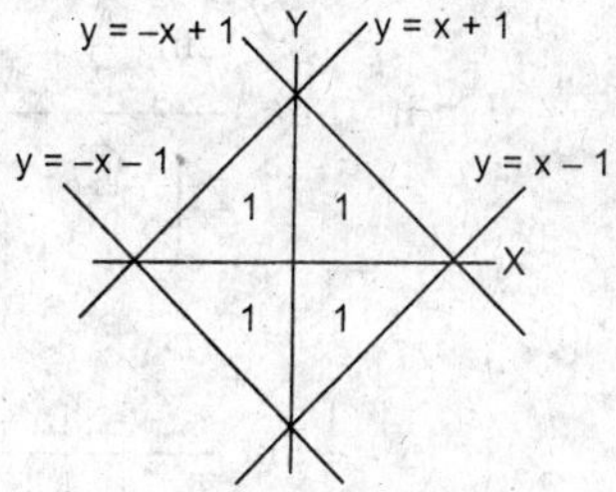

$y = x - 1 \quad$ for $x \geq 0$,

$y = -x - 1 \quad$ for $x < 0$,

$y = -x + 1 \quad$ for $x \geq 0$,

$y = x + 1 \quad$ for $x < 0$.

$\therefore \quad$ area $= 4 \times \left(\dfrac{1}{2} \times 1 \times 1\right) = 2$.

19. The graph of $f(x)$ has the equation $y = (x+1)^2$, where $x \geq -1$. The reflection of the graph of $f(x)$ is obtained by interchanging x, y. So, the graph of $g(x)$ has the equation $x = (y+1)^2$, where $y \geq -1$

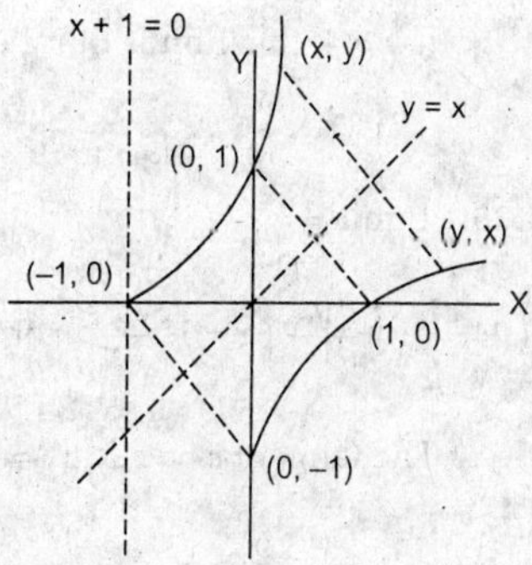

$\therefore \quad y = \sqrt{x} - 1 \qquad$ (because $y \geq -1$)

$\therefore \quad \phi(x) = \sqrt{x} - 1, \qquad$ where $x \geq 0$.

20. $f'(x) = 2 + \cos x > 0$. So, $f(x)$ is strictly m.i. So, $f(x)$ is one-to-one and onto.

21. $f(x) = \begin{cases} \frac{1}{2}(-x-1) & \text{for } x < -1 \\ \tan^{-1}x & \text{for } -1 \le x \le 1 \\ \frac{1}{2}(x+1) & \text{for } x > 1. \end{cases}$

$$\therefore \quad f'(x) = \begin{cases} -\frac{1}{2} & \text{for } x < -1 \\ \frac{1}{1+x^2} & \text{for } -1 < x < 1 \\ \frac{1}{2} & \text{for } x > 1. \end{cases}$$

$$f'(-1+0) = \frac{1}{1+(-1+0)^2} = \frac{1}{2}, f'(-1-0) = -\frac{1}{2},$$

$$f'(1-0) = \frac{1}{1+(1-0)^2} = \frac{1}{2}, f'(1+0) = \frac{1}{2}.$$

$\therefore \quad f'(-1)$ does not exist.

$\therefore \quad$ domain of $f'(x) = R - \{-1\}$.

22. n cannot be a negative integer for then the limit = 0.

$$\text{Limit} = \lim_{x \to 0} \frac{2\sin^2 \frac{x}{2}}{2^2\left(\frac{x}{2}\right)^2} \cdot \frac{e^x - \cos x}{x^{n-2}} = \frac{1}{2} \lim_{x \to 0} \frac{e^x - \cos x}{x^{n-2}}$$

($n \ne 1$ for then the limit = 0)

$= \frac{1}{2} \lim_{x \to 0} \frac{e^x + \sin x}{(n-2)\, x^{n-3}}$. So, if $n = 3$, the limit is $\frac{1}{2(n-2)}$, which is finite.

If $n = 4$, the limit is infinite.

23. Limit $= e^{\lim_{x \to 0} \frac{1}{x}\{\log f(1+x) - \log f(1)\}} = e^{\lim_{x \to 0} \frac{f'(1+x)/f(1+x)}{1}} = e^{\frac{f'(1)}{f(1)}} = e^{6/3} = e^2.$

24. $3y^2 \frac{dy}{dx} + 6x = 12 \frac{dy}{dx}$ or $\frac{dy}{dx} = \frac{6x}{12 - 3y^2}$.

The tangent is vertical $\Rightarrow \frac{dy}{dx} = \infty \Rightarrow 12 - 3y^2 = 0,$ where $x \ne 0$.

$\therefore \quad y = 2, -2.$

If $y = 2$ then $8 + 3x^2 = 24$.

or $x^2 = \frac{16}{3}$. $\therefore x = \pm\frac{4}{\sqrt{3}}$

If $y = -2$, $-8 + 3x^2 = -24$ or $x^2 = -\frac{16}{3}$ (absurd).

So, the point are $\left(\pm\frac{4}{\sqrt{3}}, 2\right)$.

25. Any point on $y^2 = 8x$ is $(2t^2, 4t)$, where the tangent is $yt = x + 2t^2$.
Solving it with $xy = -1$,

$$y(yt - 2t^2) = -1 \quad \text{or} \quad ty^2 - 2t^2y + 1 = 0.$$

For common tangent, it should have equal roots.

$\therefore$ $4t^4 - 4t = 0$. $\therefore t = 0, 1$.

$\therefore$ the common tangent is $y = x + 2$

(when $t = 0$, it is $x = 0$, which can touch $xy = -1$ at infinity only).

26. $f'(x) = \sqrt{2 - x^2}$.

$\therefore$ $x^2 - \sqrt{2 - x^2} = 0$

or $x^4 + x^2 - 2 = 0$ or $(x^2 + 2)(x^2 - 1) = 0$.

$\therefore$ $x^2 - 1 = 0$. $\therefore x = \pm 1$.

27. Put $2x = z$. Then $\int_3^{3+3T} f(2x)\,dx = \frac{1}{2}\int_6^{6+6T} f(z)\,dz$

$$= \frac{1}{2}\left[\int_6^0 f(x)\,dx + \int_0^T f(x)\,dx + \int_T^{6+6T} f(x)\,dx\right]$$

$$= \frac{1}{2}\left[-\int_0^6 f(x)\,dx + I + \int_0^{6+5T} f(T+z)\,dz\right]$$

$$= \frac{1}{2}\left[-\int_0^6 f(x)\,dx + I + \int_0^T f(z)\,dz + \int_T^{6+5T} f(z)\,dz\right]$$

$$= \frac{1}{2}\left[-\int_0^6 f(x)\,dx + I + I + \int_T^{6+5T} f(x)\,dx\right]$$

$$= \ldots\ldots = \frac{1}{2}\left[-\int_0^6 f(x)\,dx + I + 5I + \int_T^{6+T} f(x)\,dx\right]$$

$$= \frac{1}{2}\left[-\int_0^6 f(x)\,dx + 6I + \int_0^6 f(z+T)\,dz\right]$$

$$= \frac{1}{2}\left[-\int_0^6 f(x)\,dx + 6I + \int_0^6 f(z)\,dz\right] = \frac{1}{2} \times 6I = 3I.$$

28. $I = \int_{-1/2}^{1/2} [x]\,dx + \int_{-1/2}^{1/2} \log \frac{1+x}{1-x}\,dx.$

If $f(x) = \log \frac{1+x}{1-x}$ then $f(-x) = \log \frac{1-x}{1+x} = -\log \frac{1+x}{1-x} = -f(x)$.

$\therefore \quad I = \int_{-1/2}^{1/2} [x]\,dx + 0$ (being integral of an odd function)

$$= \int_{-1/2}^{0} -1\,dx + \int_0^{1/2} 0\,dx = -1 \cdot x\Big]_{-1/2}^{0} = -\frac{1}{2}$$

29. $(\vec{a} + 2\vec{b}) \cdot (5\vec{a} - 4\vec{b}) = 0$

or $5\vec{a}^{\,2} + 6\vec{a}\cdot\vec{b} - 8\vec{b}^{\,2} = 0$ or $6\vec{a}\cdot\vec{b} = 3$ $\left(\because \vec{a}^{\,2} = 1, \vec{b}^{\,2} = 1\right)$

$\therefore \quad \vec{a}\cdot\vec{b} = \frac{1}{2}$ or $|\vec{a}|\,|\vec{b}|\cos\theta = \frac{1}{2}$ $\therefore \cos\theta = \frac{1}{2}$ $\therefore \theta = 60°$.

30. $\vec{V} \times \vec{W} = \begin{vmatrix} \vec{i} & \vec{j} & \vec{k} \\ 2 & 1 & -1 \\ 1 & 0 & 3 \end{vmatrix} = 3\vec{i} - 7\vec{j} - \vec{k}$

$\therefore \quad \vec{U} \cdot (\vec{V} \times \vec{W})$ is maximum when $\vec{U} \parallel (\vec{V} \times \vec{W})$. But $\vec{U}$ is a unit vector.

$\therefore \quad \vec{U} = \frac{3\vec{i} - 7\vec{j} - \vec{k}}{\sqrt{59}}$. So, $[\vec{U}\ \vec{V}\ \vec{W}] = \frac{3^2 + 7^2 + 1^2}{\sqrt{59}} = \sqrt{59}$.

31. Here, $(2\sin\theta - 1)(\sin\theta - 2) > 0$. But $\sin\theta - 2 < 0$ for all θ.

$\therefore \quad 2\sin\theta - 1 < 0$ or $\sin\theta < \frac{1}{2} = \sin\frac{\pi}{6}$.

$\therefore \quad$ by value scheme,

$0 < \theta < \frac{\pi}{6}$ or $\frac{5\pi}{6} < \theta < 2\pi$.

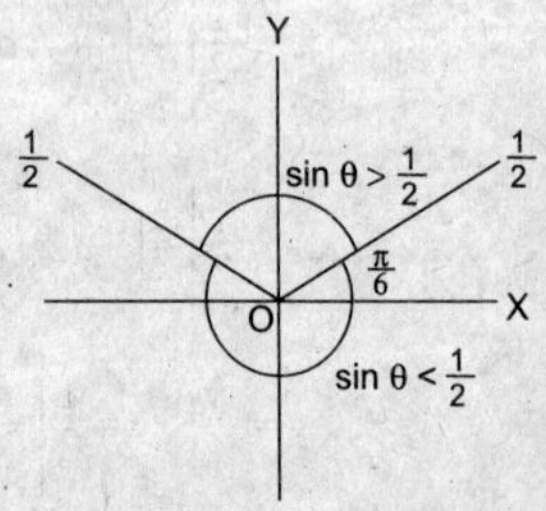

32. $\dfrac{w-\overline{w}z}{1-z}$ is purely real $\Rightarrow 1-z\neq 0$ and $\dfrac{w-\overline{w}z}{1-z}=\overline{\left(\dfrac{w-\overline{w}z}{1-z}\right)}$.

$\therefore \quad z\neq 1$ and $\dfrac{w-\overline{w}z}{1-z}=\dfrac{\overline{w}-w\overline{z}}{1-\overline{z}}$

$\Rightarrow \quad w-w\overline{z}-\overline{w}z+\overline{w}z\overline{z}=\overline{w}-w\overline{z}-\overline{w}z+wz\overline{z}$

$\Rightarrow \quad w+\overline{w}|z|^2=\overline{w}+w|z|^2 \quad \Rightarrow \quad (\overline{w}-w)(|z|^2-1)=0.$

But w is a nonreal complex number. So, $w\neq\overline{w}$. Hence, $|z|^2-1=0$.

$\therefore \quad |z|=1,$ where $z\neq 1$.

33. LCM of $p, q=(r\cdot r)(t\cdot t\cdot t\cdot t)(s\cdot s)$

There are two rs. So, p, q can have powers of r as factors in $(2\times 3-1)$ ways, namely (0, 2), (1, 2), (2, 0), (2, 1), (2, 2). Similarly p, q can have powers of t as factors in $(2\times 5-1)$ ways, namely (0, 4), (1, 4), (2, 4), (3, 4), (4, 0), (4, 1), (4, 2), (4, 3), (4, 4). Also, p, q can have powers of s as factors in $(2\times 3-1)$ ways. Hence,

the number of ordered pairs $(p, q)=(2\times 3-1)(2\times 5-1)(2\times 3-1)$.

This includes the case when both p, q are equal to $r^2t^4s^2$. Excluding this, the required number $=5\times 9\times 5-1=224$.

34. The vertex $=(\alpha,\alpha)$. Then $(\alpha-0)^2+(\alpha-0)^2=(\sqrt{2})^2$

$\therefore \quad \alpha^2=1 \quad \Rightarrow \alpha=1$ ($\because$ vertex is in the first quadrant).

The focus $=(\beta,\beta)$. Then $(\beta-0)^2+(\beta-0)^2=(2\sqrt{2})^2 \quad \Rightarrow \beta=2$

$\therefore$ focus = (2, 2).

$\therefore$ the directrix passes through O ($\because$ V is the midpoint of OS).

and it is perpendicular to $y=x$.

$\therefore$ the directrix is $x+y=0$.

$\therefore$ the equation of the parabola is

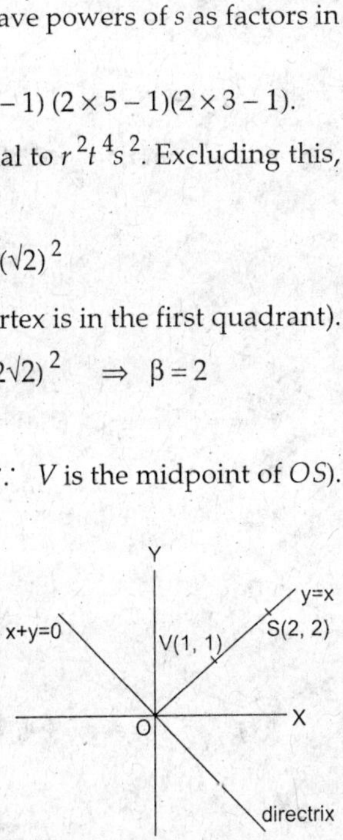

$$\frac{x+y}{\sqrt{2}}=\sqrt{(x-2)^2+(y-2)^2}$$

or $\quad (x+y)^2=2(x^2+y^2-4x-4y+8)$

or $\quad (x-y)^2=8(x+y-2)$.

35. The equation of the tangent at (x, y) is $Y-y=\dfrac{dy}{dx}(X-x)$

$\therefore \quad A=\left(x-y\dfrac{dx}{dy}, 0\right)$ and $B=\left(0, y-x\dfrac{dy}{dx}\right)$.

Now, $\frac{BP}{AP} = \frac{3}{1}$. So, $P = \left(\frac{3}{4}\left(x - y\frac{dx}{dy}\right), \frac{1}{4}\left(y - x\frac{dy}{dx}\right)\right) = (x, y)$

$\Rightarrow \quad \frac{1}{4}\left(y - x\frac{dy}{dx}\right) = y \quad \Rightarrow \quad 3y + x\frac{dy}{dx} = 0$

$\Rightarrow \quad 3\frac{dx}{x} + \frac{dy}{y} = 0 \quad \Rightarrow \quad 3\log x + \log y = \log c$

$\Rightarrow \quad x^3 y = c$. Now, $f(1) = 1 \quad \Rightarrow \quad (1, 1)$ is on it.

$\therefore \quad c = 1$. Hence, $x^3 y = 1$. $\quad \therefore \quad \left(2, \frac{1}{8}\right)$ is on it.

The normal to $x^3 y = 1$ at $(1, 1)$ has the equation

$$y - 1 = \frac{-1}{\left.\frac{dy}{dx}\right]_{(1,1)}}(x - 1) \quad \text{or} \quad y - 1 = \frac{-1}{-3}(x - 1) \quad \text{or} \quad x - 3y + 2 = 0.$$

36. $\text{ar}(\Delta ABC) = \text{ar}(\Delta ABD) + \text{ar}(\Delta ACD)$

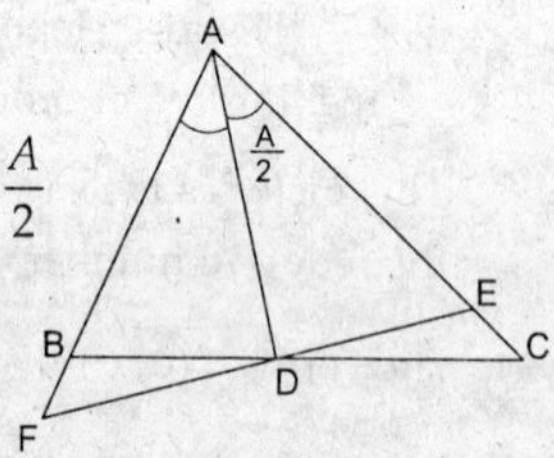

$\Rightarrow \quad \frac{1}{2}bc\sin A = \frac{1}{2}c \cdot AD \cdot \sin\frac{A}{2} + \frac{1}{2}b \cdot AD \cdot \sin\frac{A}{2}$

$\Rightarrow \quad AD = \frac{bc\sin A}{(b + c)\sin\frac{A}{2}} = \frac{2bc}{b + c}\cos\frac{A}{2}$.

ΔAEF is isosceles because the bisector of $\angle FAE$ is perpendicular to EF.

$EF = 2ED = 2AD \cdot \tan\frac{A}{2} = 2 \cdot \frac{2bc}{b + c}\cos\frac{A}{2} \cdot \tan\frac{A}{2} = \frac{4bc}{b + c}\sin\frac{A}{2}$.

$AE = AD \cdot \sec\frac{A}{2} = \frac{2bc}{b + c}\cos\frac{A}{2} \cdot \sec\frac{A}{2} = \frac{2bc}{b + c} =$ HM of b, c.

37. Let $f(x) = ax^3 + bx^2 + cx + d$.

$f(1) = -1 \quad \Rightarrow \quad a + b + c + d = -1$.

$f(2) = 18 \quad \Rightarrow \quad 8a + 4b + 2c + d = 18$.

$f''(0) = 0 \quad \Rightarrow \quad b = 0 \quad$ and $\quad f'(-1) = 0 \quad \Rightarrow \quad 3a - 2b + c = 0$.

Solving, $a = \frac{19}{4}, b = 0, c = -\frac{57}{4}, d = \frac{17}{2}$.

$\therefore \quad f(x) = \frac{1}{4}(19x^3 - 57x + 34)$.

$f'(x) = 0 \quad \Rightarrow \quad 57x^2 - 57 = 0 \quad \Rightarrow \quad x = 1, -1$.

$f''(x) = \frac{57}{2}x$. So, $f''(-1) < 0$ and $f''(1) > 0$.

$\therefore$ $f(x)$ has a local minimum at $x = 1$.

$f(x)$ is increasing $\Rightarrow f'(x) \geq 0 \Rightarrow x^2 - 1 \geq 0 \Rightarrow x \leq -1$ or $x \geq 1$.

$\therefore$ in $[1, 2\sqrt{5}]$, $f(x)$ is increasing.

The distance of $(-1, 2)$ from $(1, -1)$ is not $2\sqrt{5}$.

38. (i) Using $\int_a^b f(x)\,dx = \frac{b-a}{2}\{f(a) + f(b)\}$,

$$\int_0^{\pi/2} \sin x\,dx = \frac{\frac{\pi}{2} - 0}{2}\left(\sin\frac{\pi}{2} + \sin 0\right) = \frac{\pi}{4}.$$

Using $\int_a^b f(x)\,dx = \frac{b-a}{4}\{f(a) + f(b) + 2f(c)\}$, where $c = \frac{a+b}{2}$,

$$\int_0^{\pi/2} \sin x\,dx = \frac{\frac{\pi}{2} - 0}{4}\left(\sin\frac{\pi}{2} + \sin 0 + 2\sin\frac{\pi}{4}\right) = \frac{\pi}{8}(1 + \sqrt{2}).$$

(ii) Putting $c = \frac{a+b}{2}$ in $F(c) = \frac{c-a}{2}\{f(a) + f(c)\} + \frac{b-c}{2}\{f(b) + f(c)\}$,

$$F\left(\frac{a+b}{2}\right) = \frac{\frac{a+b}{2} - a}{2}\left\{f(a) + f\left(\frac{a+b}{2}\right)\right\} + \frac{b - \frac{a+b}{2}}{2}\left\{f(b) + f\left(\frac{a+b}{2}\right)\right\}$$

$$= \frac{b-a}{4}\left\{f(a) + f(b) + 2f\left(\frac{a+b}{2}\right)\right\}$$

$$= \frac{b-a}{4}\{f(a) + f(b) + 2f(c)\}, \text{ where } c = \frac{a+b}{2}$$

$$= \int_a^b f(x)\,dx.$$

(iii) $F(c) = \frac{c-a}{2}f(a) + \frac{b-c}{2}f(b) + \frac{b-a}{2}f(c)$

$\therefore$ $2F'(c) = f(a) - f(b) + (b-a)f'(c)$ ... (1)

$\Rightarrow$ $2F''(c) = (b-a)f''(c) < 0$ (because $f''(c) < 0$).

$\therefore$ $F''(c) < 0$. Also, for having a maximum, $F'(c) = 0$.

$\therefore$ from (1), $0 = f(a) - f(b) + (b-a)f'(c) \quad \Rightarrow f'(c) = \dfrac{f(b)-f(a)}{b-a}$.

39. (i) $\tan^{-1}\dfrac{1}{2n^2} = \tan^{-1}\dfrac{2}{4n^2} = \tan^{-1}\dfrac{2}{1+(4n^2-1)}$

$$= \tan^{-1}\frac{(2n+1)-(2n-1)}{1+(2n+1)(2n-1)} = \tan^{-1}(2n+1) - \tan^{-1}(2n-1)$$

$$\therefore \sum_{n=1}^{n} \tan^{-1}\frac{1}{2n^2} = (\tan^{-1}3 - \tan^{-1}1) + (\tan^{-1}5 - \tan^{-1}3)$$

$$+ \dots \dots + \{\tan^{-1}(2n+1) - \tan^{-1}(2n-1)\}$$

$$= \tan^{-1}(2n+1) - \tan^{-1}1.$$

When $n \to \infty$, $\displaystyle\sum_{n=1}^{\infty} \tan^{-1}\frac{1}{2n^2} = \frac{\pi}{2} - \frac{\pi}{4} = \frac{\pi}{4} = t$. So, $\tan t = \tan\dfrac{\pi}{4} = 1$.

(ii) $\cos\theta_1 = \dfrac{a}{b+c} \quad \Rightarrow \quad \dfrac{1-\tan^2\dfrac{\theta_1}{2}}{1+\tan^2\dfrac{\theta_1}{2}} = \dfrac{a}{b+c}$.

By componendo–dividendo, $\tan^2\dfrac{\theta_1}{2} = \dfrac{b+c-a}{a+b+c}$.

Similarly, $\tan^2\dfrac{\theta_3}{2} = \dfrac{a+b-c}{a+b+c}$.

$\therefore \tan^2\dfrac{\theta_1}{2} + \tan^2\dfrac{\theta_3}{2} = \dfrac{2b}{a+b+c} = \dfrac{2b}{2b+b}$ ($\because$ a, b, c are in AP)

$$= \frac{2}{3}.$$

(iii) The equation of the line is $\dfrac{x-0}{1} = \dfrac{y-1}{2} = \dfrac{z-0}{2}$.

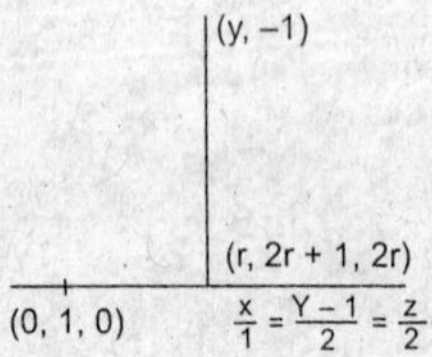

Any point on it is $(r, 2r+1, 2r)$.

It is the foot of the perpendicular if

$$r \cdot 1 + (2r+1) \cdot 2 + 2r \cdot 2 = 0$$

$$\Rightarrow r = -\frac{2}{9}.$$

$\therefore$ the foot of the perpendicular $= \left(-\dfrac{2}{9}, \dfrac{5}{9}, -\dfrac{4}{9}\right)$

$\therefore$ the length of the perpendicular $= \sqrt{\left(\frac{2}{9}\right)^2 + \left(-\frac{5}{9}\right)^2 + \left(\frac{4}{9}\right)^2}$

$$= \frac{1}{9}\sqrt{2^2+5^2+4^2} = \frac{\sqrt{5}}{3}.$$

(iv) Putting $\log x = z^2$,

$$I_1 = \int_1^e (-1+\sqrt{\log x})\,dx = -\int_1^e dx + \int_0^1 z \cdot 2ze^{z^2}\,dz.$$

$$\therefore\ I_1 = -e + 1 + z \cdot e^{z^2}\Big]_0^1 - \int_0^1 e^{z^2}\,dz = 1 - \int_0^1 e^{x^2}\,dx.$$

$$\therefore\ I = \int_0^1 e^{x^2}\,dx + 1 - \int_0^1 e^{x^2}\,dx = 1.$$

40. Here, $g(x) = \{f'(x)\}^2 + f''(x) \cdot f(x) = \frac{d}{dx}\{f'(x) \cdot f(x)\}$.

Let $\phi(x) = f'(x) \cdot f(x)$. Then $g(x) = \phi'(x)$.

We know that between two real roots of $\phi(x) = 0$ there is a real root of $\phi'(x) = 0$, that is, $g(x) = 0$.

Now, $\phi(x) = 0 \Rightarrow f'(x) \cdot f(x) = 0 \Rightarrow f'(x) = 0$ or $f(x) = 0$.

Now, $f(a) = 0$, $f(e) = 0$, $f(b) > 0$ and $f(c) < 0$, $f(c) < 0$ and $f(d) > 0$.

$\therefore$ $f(x)$ has four real roots at least, namely $x = a$, $x = e$, $x = \lambda$, $x = \mu$, where $b < \lambda < c$ and $c < \mu < d$ in the interval $[a, e]$.

$\therefore$ $f'(x) = 0$ has three real roots at least—one in (a, λ), one in (λ, μ) and one in (μ, e) in the interval $[a, e]$.

$\therefore$ $\phi(x) = 0$ has at least seven real roots in the interval $[a, e]$.

$\therefore$ $\phi'(x) = g(x) = 0$ has at least six real roots in $[a, e]$.

$\therefore$ the minimum number of zeros of $g(x)$ in $[a, e]$ is 6.

□

IIT Questions—4

Straight Objective Type (One-Option-Correct Questions)

1. The number of solutions of the pair of equations

$$2\sin^2\theta - \cos 2\theta = 0$$
$$2\cos^2\theta - 3\sin\theta = 0$$

in the interval $[0, 2\pi]$ is

(a) zero (b) one (c) two (d) four

2. Let α, β be the roots of the equation $x^2 - px + r = 0$ and $\frac{\alpha}{2}, 2\beta$ be the roots of the equation $x^2 - qx + r = 0$. Then the value of r is

(a) $\frac{2}{9}(p-q)(2q-p)$ (b) $\frac{2}{9}(q-p)(2p-q)$

(c) $\frac{2}{9}(q-2p)(2q-p)$ (c) $\frac{2}{9}(2p-q)(2q-p)$

3. The number of distinct real values of λ, for which the vectors $-\lambda^2\hat{i}+\hat{j}+\hat{k}$, $\hat{i}-\lambda^2\hat{j}+\hat{k}$ and $\hat{i}+\hat{j}-\lambda^2\hat{k}$ are coplanar, is

(a) zero (b) one (c) two (d) three

4. One Indian and four American men and their wives are to be seated randomly around a circular table. Then the conditional probability that the Indian man is seated adjacent to his wife given that each American man is seated adjacent to his wife is

(a) $\frac{1}{2}$ (b) $\frac{1}{3}$ (c) $\frac{2}{5}$ (d) $\frac{1}{5}$

5. $\lim\limits_{x \to \frac{\pi}{4}} \dfrac{\int_2^{\sec^2 x} f(t)\, dt}{x^2 - \frac{\pi^2}{16}}$ equals

(a) $\frac{8}{\pi}f(2)$ (b) $\frac{2}{\pi}f(2)$ (c) $\frac{2}{\pi}f\left(\frac{1}{2}\right)$ (d) $4f(2)$

6. Let $f(x) = \dfrac{x}{(1+x^n)^{1/n}}$ for $n \geq 2$ and $g(x) = \underbrace{(f \circ f \circ \ldots \circ f)}_{f \text{ occurs } n \text{ times}}(x)$. Then $\int x^{n-2} g(x)\, dx$ equals

(a) $\dfrac{1}{n(n-1)}(1+n\,x^n)^{1-\frac{1}{n}} + K$ (b) $\dfrac{1}{n-1}(1+n\,x^n)^{1-\frac{1}{n}} + K$

(c) $\dfrac{1}{n(n+1)}(1+n\,x^n)^{1+\frac{1}{n}} + K$ (d) $\dfrac{1}{n+1}(1+n\,x^n)^{1+\frac{1}{n}} + K$

7. The letters of the word COCHIN are permuted and all the permutations are arranged in an alphabetical order as in an English dictionary. The number of words that appear before the word COCHIN is

(a) 360 (b) 192 (c) 96 (d) 48

8. Let $ABCD$ be a quadrilateral with area 18, with side AB parallel to the side CD and $AB = 2CD$. Let AD be perpendicular to AB and CD. If a circle is drawn inside the quadrilateral $ABCD$ touching all the sides, then its radius is

(a) 3 (b) 2 (c) $\dfrac{3}{2}$ (d) 1

9. A hyperbola, having the transverse axis of length $2\sin\theta$, is confocal with the ellipse $3x^2 + 4y^2 = 12$. Then its equation is

(a) $x^2\text{cosec}^2\theta - y^2\sec^2\theta = 1$ (b) $x^2\sec^2\theta - y^2\text{cosec}^2\theta = 1$

(c) $x^2\sin^2\theta - y^2\cos^2\theta = 1$ (d) $x^2\cos^2\theta - y^2\sin^2\theta = 1$

10. The tangent to the curve $y = e^x$ drawn at the point (c, e^c) intersects the line joining the points $(c-1, e^{c-1})$ and $(c+1, e^{c+1})$

(a) on the left of $x = c$ (b) on the right of $x = c$

(c) at no point (d) at all points

11. A man walks a distance of 3 units from the origin towards the north-east ($N\,45°\,E$) direction. From there, he walks a distance of 4 units towards the north-west ($N\,45°\,W$) direction to reach a point P. Then the position of P in the Argand plane is

(a) $3e^{i\pi/4} + 4i$ (b) $(3-4i)e^{i\pi/4}$

(c) $(4+3i)e^{i\pi/4}$ (d) $(3+4i)e^{i\pi/4}$

12. Let $f(x)$ be differentiable on the interval $(0, \infty)$ such that $f(1) = 1$, and

$$\lim_{t \to x} \frac{t^2 f(x) - x^2 f(t)}{t - x} = 1$$

for each $x > 0$. Then $f(x)$ is

(a) $\frac{1}{3x} + \frac{2x^2}{3}$ (b) $\frac{-1}{3x} + \frac{4x^2}{3}$

(c) $\frac{-1}{x} + \frac{2}{x^2}$ (d) $\frac{1}{x}$

13. $\frac{d^2x}{dy^2}$ equals

(a) $\left(\frac{d^2y}{dx^2}\right)^{-1}$ (b) $-\left(\frac{d^2y}{dx^2}\right)^{-1}\left(\frac{dy}{dx}\right)^{-3}$

(c) $\left(\frac{d^2y}{dx^2}\right)\left(\frac{dy}{dx}\right)^{-2}$ (d) $-\left(\frac{d^2y}{dx^2}\right)\left(\frac{dy}{dx}\right)^{-3}$

14. The differential equation $\frac{dy}{dx} = \frac{\sqrt{1 - y^2}}{y}$ determines a family of circles with

(a) variable radii and a fixed centre at (0, 1)
(b) variable radii and a fixed centre at (0, –1)
(c) fixed radius 1 and variable centres along the x-axis
(d) fixed radius 1 and variable centres along the y-axis

15. If $|z| = 1$ and $z \neq \pm 1$ then all the values of $\frac{z}{1 - z^2}$ lie on

(a) a line not passing through the origin
(b) $|z| = \sqrt{2}$
(c) the x-axis
(d) the y-axis

16. Let $O(0, 0)$, $P(3, 4)$, $Q(6, 0)$ be the vertices of the triangle OPQ. The point R inside the triangle OPQ is such that the triangles OPR, PQR, OQR are of equal area. The coordinates of R are

(a) $\left(\frac{4}{3}, 3\right)$ (b) $\left(3, \frac{2}{3}\right)$

(c) $\left(3, \frac{4}{3}\right)$ (d) $\left(\frac{4}{3}, \frac{2}{3}\right)$

Assertion-Reason Type (One-Option-Correct Questions)

17. Let $F(x)$ be an indefinite integral of $\sin^2 x$.

STATEMENT-1: The function $F(x)$ satisfies $F(x+\pi) = F(x)$ for all real x.

because

STATEMENT-2: $\sin^2(x+\pi) = \sin^2 x$ for all real x.

(a) Statement-1 is True, Statement-2 is True; Statement-2 is a correct explanation for Statement-1

(b) Statement-1 is True, Statement-2 is True; Statement-2 is *Not* a correct explanation for Statement-1

(c) Statement-1 is True, Statement-2 is False

(d) Statement-1 is False, Statement-2 is True

18. Consider the planes $3x - 6y - 2z = 15$ and $2x + y - 2z = 5$.

STATEMENT-1: The parametric equations of the line of intersection of the given planes are $x = 3 + 14t, y = 1 + 2t, z = 15t$.

because

STATEMENT-2: The vector $14\hat{i} + 2\hat{j} + 15\hat{k}$ is parallel to the line of intersection of given planes.

(a) Statement-1 is True, Statement-2 is True; Statement-2 is a correct explanation for Statement-1

(b) Statement-1 is True, Statement-2 is True; Statement-2 is *Not* a correct explanation for Statement-1

(c) Statement-1 is True, Statement-2 is False

(d) Statement-1 is False, Statement-2 is True

19. STATEMENT-1: The curve $y = \frac{-x^2}{2} + x + 1$ is symmetric with respect to the line $x = 1$.

because

STATEMENT-2: A parabola is symmetric about its axis.

(a) Statement-1 is True, Statement-2 is True; Statement-2 is a correct explanation for Statement-1

(b) Statement-1 is True, Statement-2 is True; Statement-2 is *Not* a correct explanation for Statement-1

(c) Statement-1 is True, Statement-2 is False

(d) Statement-1 is False, Statement-2 is True

20. Let $H_1, H_2, H_3, \ldots, H_n$ be mutually exclusive and exhaustive events with $P(H_i) > 0; i = 1, 2, 3, \ldots, n$. Let E be any other event with $0 < P(E) < 1$.

STATEMENT-1: $P(H_i \mid E) > P(E \mid H_i) \cdot P(H_i)$ for $i = 1, 2, 3, \ldots, n$.

because

STATEMENT-2: $\sum_{i=1}^{n} P(H_i) = 1$.

(a) Statement-1 is True, Statement-2 is True; Statement-2 is a correct explanation for Statement-1

(b) Statement-1 is True, Statement-2 is True; Statement-2 is *Not* a correct explanation for Statement-1

(c) Statement-1 is True, Statement-2 is False

(d) Statement-1 is False, Statement-2 is True

21. Tangents are drawn from the point (17, 7) to the circle $x^2 + y^2 = 169$.

STATEMENT-1: The tangents are mutually perpendicular.

because

STATEMENT-2: The locus of the points from which mutually perpendicular tangents can be drawn to the given circle is $x^2 + y^2 = 338$.

(a) Statement-1 is True, Statement-2 is True; Statement-2 is a correct explanation for Statement-1

(b) Statement-1 is True, Statement-2 is True; Statement-2 is *Not* a correct explanation for Statement-1

(c) Statement-1 is True, Statement-2 is False

(d) Statement-1 is False, Statement-2 is True

22. Let $f(x) = 2 + \cos x$ for all real x.

STATEMENT-1: For each real t, there exists a point c in $[t, t + \pi]$ such that $f'(c) = 0$.

because

STATEMENT-2: $f(t) = f(t + 2\pi)$ for each real t.

(a) Statement-1 is True, Statement-2 is True; Statement-2 is a correct explanation for Statement-1

(b) Statement-1 is True, Statement-2 is True; Statement-2 is *Not* a correct explanation for Statement-1

(c) Statement-1 is True, Statement-2 is False

(d) Statement-1 is False, Statement-2 is True

23. Lines $L_1 \equiv y - x = 0$ and $L_2 \equiv 2x + y = 0$ intersect the line $L_3 : y + 2 = 0$ at P and Q respectively. The bisector of the acute angle between L_1 and L_2 intersects L_3 at R.

STATEMENT-1: The ratio $PR : RQ = 2\sqrt{2} : \sqrt{5}$.

because

STATEMENT-2: In any triangle, bisector of an angle divides the triangle into two similar triangles.

(a) Statement-1 is True, Statement-2 is True; Statement-2 is a correct explanation for Statement-1

(b) Statement-1 is True, Statement-2 is True; Statement-2 is *Not* a correct explanation for Statement-1

(c) Statement-1 is True, Statement-2 is False

(d) Statement-1 is False, Statement-2 is True

Linked Comprehension Type (One-Option-Correct Questions)
Paragraph for Question Nos. 24 to 26

Let A_1, G_1, H_1 denote the arithmetic, geometric and harmonic means, respectively, of two distinct positive numbers. For $n \geq 2$, let A_{n-1} and H_{n-1} have arithmetic, geometric and harmonic means as A_n, G_n, H_n respectively.

24. Which one of the following statements is correct?

(a) $G_1 > G_2 > G_3 > \ldots$

(b) $G_1 < G_2 < G_3 < \ldots$

(c) $G_1 = G_2 = G_3 = \ldots$

(d) $G_1 < G_3 < G_5 < \ldots$ and $G_2 > G_4 > G_6 > \ldots$

25. Which one of the following statements is correct?

(a) $A_1 > A_2 > A_3 > \ldots$

(b) $A_1 < A_2 < A_3 < \ldots$

(c) $A_1 > A_3 > A_5 > \ldots$ and $A_2 < A_4 < A_6 < \ldots$

(d) $A_1 < A_3 < A_5 < \ldots$ and $A_2 > A_4 > A_6 > \ldots$

26. Which one of the following statements is correct?

(a) $H_1 > H_2 > H_3 > \ldots$

(b) $H_1 < H_2 < H_3 < \ldots$

(c) $H_1 > H_3 > H_5 > \ldots$ and $H_2 < H_4 < H_6 < \ldots$

(d) $H_1 < H_3 < H_5 < \ldots$ and $H_2 > H_4 > H_6 > \ldots$

Paragraph for Question Nos. 27 to 29

Consider the circle $x^2 + y^2 = 9$ and the parabola $y^2 = 8x$. They intersect at P and Q in the first and the fourth quadrants, respectively. Tangents to the circle at P and Q intersect the x-axis at R and tangents to the parabola at P and Q intersect the x-axis at S.

27. The ratio of the areas of the triangles PQS and PQR is

(a) $1 : \sqrt{2}$ (b) $1 : 2$ (c) $1 : 4$ (d) $1 : 8$

28. The radius of the circumcircle of the triangle PRS is

(a) 5 (b) $3\sqrt{3}$ (c) $3\sqrt{2}$ (d) $2\sqrt{3}$

29. The radius of the incircle of the triangle PQR is

(a) 4 (b) 3 (c) $\frac{8}{3}$ (d) 2

Paragraph for Question Nos. 30 to 32

Let V_r denote the sum of the first r terms of an arithmetic progression (AP) whose first term is r and the common difference is $(2r - 1)$. Let

$$T_r = V_{r+1} - V_r - 2 \text{ and } Q_r = T_{r+1} - T_r \text{ for } r = 1, 2, \ldots$$

30. The sum $V_1 + V_2 + \ldots + V_n$ is

(a) $\frac{1}{12}n(n+1)(3n^2 - n + 1)$ (b) $\frac{1}{12}n(n+1)(3n^2 + n + 2)$

(c) $\frac{1}{2}n(2n^2 - n + 1)$ (d) $\frac{1}{3}(2n^3 - 2n + 3)$

31. T_r is always

(a) an odd number (b) an even number

(c) a prime number (d) a composite number

32. Which one of the following is the correct statement?

(a) $Q_1, Q_2, Q_3, \ldots$ are in AP with common difference 5

(b) $Q_1, Q_2, Q_3, \ldots$ are in AP with common difference 6

(c) $Q_1, Q_2, Q_3, \ldots$ are in AP with common difference 11

(d) $Q_1 = Q_2 = Q_3 = \ldots$

Matrix-Match Type

Statements (A, B, C, D) in **Column I** have to be matched with statements (p, q, r, s) in **Column II.** The answers to these questions have to be appropriately bubbled as illustrated in the following example.

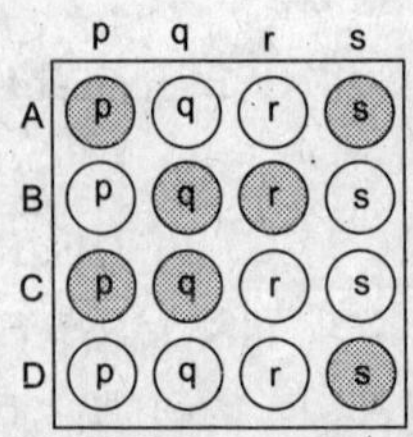

If the correct matches are A-p, A-s, B-q, B-r, C-p, C-q and D-s then the correctly bubbled 4×4 matrix should be as given above.

33. Match the integrals in **Column I** with the values in **Column II** and indicate your answer by darkening the appropriate bubbles in the 4×4 matrix

Column I	Column II
(A) $\int_{-1}^{1} \frac{dx}{1+x^2}$	(p) $\frac{1}{2} \log\left(\frac{2}{3}\right)$
(B) $\int_{0}^{1} \frac{dx}{\sqrt{1-x^2}}$	(q) $2 \log\left(\frac{2}{3}\right)$
(C) $\int_{2}^{3} \frac{dx}{1-x^2}$	(r) $\frac{\pi}{3}$
(D) $\int_{1}^{2} \frac{dx}{x\sqrt{x^2-1}}$	(s) $\frac{\pi}{2}$

34. In the following $[x]$ denotes the greatest integer less than for equal to x. Match the functions in **Column I** with the properties in **Column II** and indicate your answer by darkening the appropriate bubles in the 4×4 matrix.

Column I	Column II
(A) $x\|x\|$	(p) continuous in (–1, 1)
(B) $\sqrt{\|x\|}$	(q) differentiable in (–1, 1)
(C) $x + [x]$	(r) strictly increasing in (–1, 1)
(D) $\|x-1\| + \|x+1\|$	(s) not differentiable at least at one point in (–1, 1)

35. Match the statement in **Column I** with the properties in **Column II** and indicate your answer by darkening the appropriate bubbles in the 4×4 matrix.

Column I	Column II
(A) Two intersecting circles	(p) have a common tangent
(B) Two mutually external circles	(q) have a common normal
(C) Two circles, one strictly inside the other	(r) do not have a common tangent
(D) Two branches of a hyperbola	(s) do not have a common normal

36. Consider the following linear equations

$$ax + by + cz = 0$$
$$bx + cy + az = 0$$
$$cx + ay + bz = 0$$

Match the conditions/expressions in **Column I** with statements in **Column II** and indicate your answer by darkening the appropriate bubbles in the 4×4 matrix given in the ORS.

Column I	Column II
(A) $a + b + c \neq 0$ and $a^2 + b^2 + c^2 = ab + bc + ca$	(p) the equations represent planes meeting only at a single point.
(B) $a + b + c = 0$ and $a^2 + b^2 + c^2 \neq ab + bc + ca$	(q) the equations represent the line $x = y = z$.
(C) $a + b + c \neq 0$ and $a^2 + b^2 + c^2 \neq ab + bc + ca$	(r) the equations represent identical planes.
(D) $a + b + c = 0$ and $a^2 + b^2 + c^2 = ab + bc + ca$	(s) the equations represent the whole of the three dimensional space.

37. Let (x, y) be such that

$$\sin^{-1}(ax) + \cos^{-1}(y) + \cos^{-1}(bxy) = \frac{\pi}{2}.$$

Match the statements in **Column I** with statements in **Column II** and indicate your answer by darkening appropriate bubbles in the 4×4 matrix.

Column I	Column II
(A) If $a = 1$ and $b = 0$ then (x, y)	(p) lies on the circle $x^2 + y^2 = 1$
(B) If $a = 1$ and $b = 1$ then (x, y)	(q) lies on $(x^2 - 1)(y^2 - 1) = 0$
(C) If $a = 1$ and $b = 2$ then (x, y)	(r) lies on $y = x$
(D) If $a = 2$ and $b = 2$ then (x, y)	(s) lies on $(4x^2 - 1)(y^2 - 1) = 0$

38. Let $f(x) = \dfrac{x^2 - 6x + 5}{x^2 - 5x + 6}$

Match the expression/statements in **Column I** with expression/statements in **Column II** and indicate your answer by darkening the appropriate bubbles in the 4×4 matrix.

Column I	Column II
(A) If $-1 < x < 1$ then $f(x)$ satisfies	(p) $0 < f(x) < 1$
(B) If $1 < x < 2$ then $f(x)$ satisfies	(q) $f(x) < 0$

(C) If $3 < x < 5$ then $f(x)$ satisfies (r) $f(x) > 0$

(D) If $x > 5$ then $f(x)$ satisfies (s) $f(x) < 1$

Answers

1. c **2.** d **3.** c **4.** c **5.** a
6. a **7.** c **8.** b **9.** a **10.** a
11. d **12.** a **13.** d **14.** c **15.** d
16. c **17.** d **18.** d **19.** a **20.** b
21. a **22.** a **23.** c **24.** c **25.** a
26. b **27.** c **28.** b **29.** d **30.** b
31. d **32.** b

33.

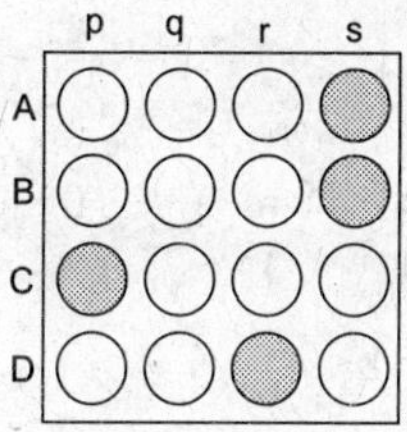

(i.e., $A \to s, B \to s, C \to p, D \to r$)

34.

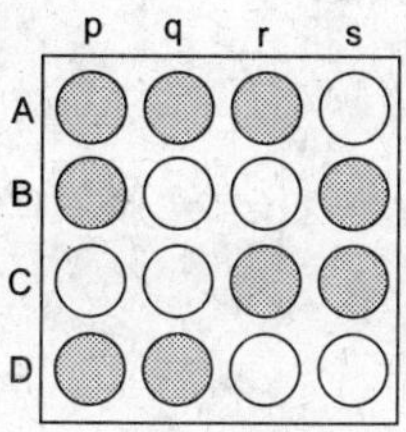

(i.e., $A \to p, A \to q, A \to r, B \to p, B \to s, C \to r, C \to s, D \to p, D \to q$)

35.

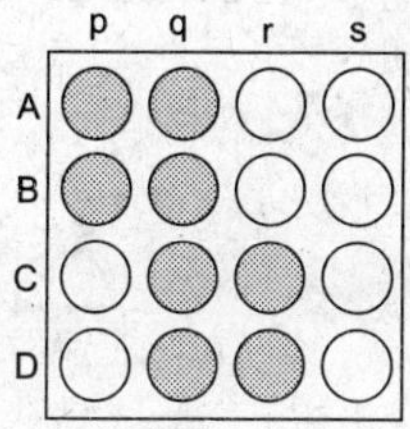

(i.e., $A \to p, A \to q, B \to p, B \to q, C \to q, C \to r, D \to q, D \to r$)

36.

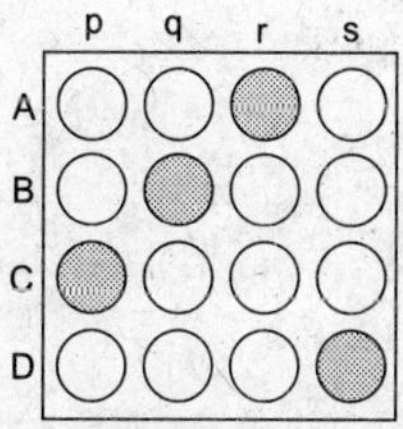

(i.e., $A \to r, B \to q, C \to p, D \to s$)

37.

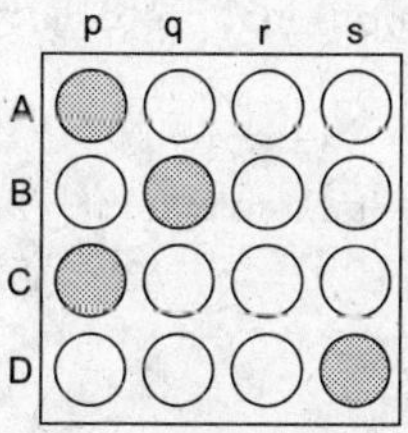

(i.e., $A \to p, B \to q, C \to p, D \to s$)

38.

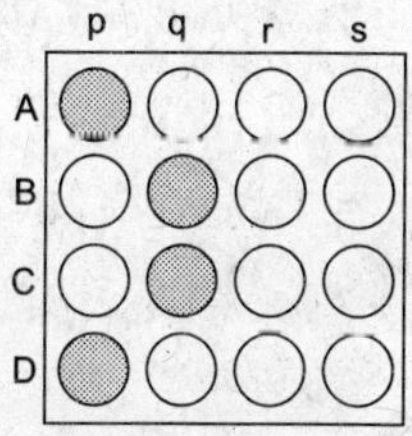

(i.e., $A \to p, B \to q, C \to q, D \to p$)

Hints and Solutions of Selected Questions

1. $2\sin^2\theta - \cos 2\theta = 0 \Rightarrow 4\sin^2\theta = 1 \Rightarrow \sin\theta = \pm\frac{1}{2}$

$2\cos^2\theta - 3\sin\theta = 0 \Rightarrow 2\sin^2\theta + 3\sin\theta - 2 = 0 \Rightarrow (2\sin\theta - 1)(\sin\theta + 2) = 0$

$\Rightarrow \sin\theta = \frac{1}{2}, -2(\text{absurd}).$

$\therefore$ satisfying both, $\sin\theta = \frac{1}{2}$. $\therefore\ \theta = \frac{\pi}{6}, \frac{5\pi}{6}$ in $[0, 2\pi]$.

2. $\alpha + \beta = p,\ \frac{\alpha}{2} + 2\beta = q \Rightarrow \alpha = \frac{2(2p-q)}{3},\ \beta = \frac{1}{3}(2q - p)$

$\therefore\ \gamma = \alpha\beta = \frac{2}{3}(2p - q)\cdot\frac{1}{3}(2q - p) = \frac{2}{9}(2p - q)(2q - p).$

3. The scalar triple product $= 0 \Rightarrow \begin{vmatrix} -\lambda^2 & 1 & 1 \\ 1 & -\lambda^2 & 1 \\ 1 & 1 & -\lambda^2 \end{vmatrix} = 0$

$\Rightarrow\ (2-\lambda^2)\begin{vmatrix} 1 & 1 & 1 \\ 1 & -\lambda^2 & 1 \\ 1 & 1 & -\lambda^2 \end{vmatrix} = 0 \Rightarrow\ (2-\lambda^2)\begin{vmatrix} 1 & 0 & 0 \\ 1 & -(1+\lambda^2) & 0 \\ 1 & 0 & -(1+\lambda^2) \end{vmatrix} = 0$

$\Rightarrow\ (2-\lambda^2)(1+\lambda^2)^2 = 0 \Rightarrow \lambda = \pm\sqrt{2}$ (real values).

4. The number of ways in which the American man and wife are together

$= (6-1)!\times 2!\times 2!\times 2!\times 2!$

The number of ways in which all men and wives are together

$= (5-1)!\times 2!\times 2!\times 2!\times 2!\times 2!$

$\therefore$ the required conditional probability $= \frac{4!\times(2!)^5}{5!\times(2!)^4} = \frac{2}{5}$.

5. Form is $\frac{0}{0}$. By L'Hospital's rule, limit $= \lim\limits_{x\to\pi/4} \frac{f(\sec^2 x)\cdot 2\sec x\cdot\sec x\cdot\tan x}{2x}$

$= f(2)\cdot\frac{2^2\cdot 1}{2\cdot\frac{\pi}{4}} = \frac{8}{\pi}f(2).$

6. $f(f(x)) = \dfrac{f(x)}{\{1+f(x)^n\}^{1/n}} = \dfrac{\dfrac{x}{(1+x^n)^{1/n}}}{\left(1+\dfrac{x^n}{1+x^n}\right)^{1/n}} = \dfrac{x}{(1+2x^n)^{1/n}}$.

Similarly, $f\{f(f(x))\} = \dfrac{x}{(1+3x^n)^{1/n}}$, etc. So, $g(x) - \dfrac{x}{(1+nx^n)^{1/n}}$

$$\therefore \quad \int x^{n-2}\, g(x)\, dx = \int \frac{x^{n-1}}{(1+nx^n)^{1/n}}\, dx = \frac{1}{n(n-1)}\,(1+nx^n)^{-\frac{1}{n}+1} + k,$$

(putting $1 + nx^n = z$)

7. Order of letters—CCHINO.

C | C × × × × — 4 !

C | H × × × × — 4 !

C | I × × × × — 4 !

C | N × × × × — 4 !

C | O | C | H | I | N

∴ the required number of words = 4 ! + 4 ! + 4 ! + 4 ! = 96.

8. $ar(\text{ABCD}) = ar(\text{rectangle AICD}) + ar(\text{CIB})$

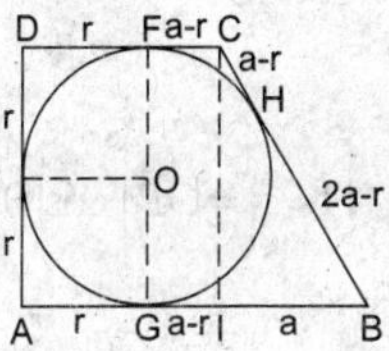

$\therefore \quad 18 = 2r(r + a - r) + \frac{1}{2} a \cdot CI = 2ra + \frac{1}{2} \cdot a \cdot 2r$

$\therefore \quad ar = 6$ ... (1)

In Δ CIB, $CI^2 + IB^2 = BC^2$

$\Rightarrow \quad (2r)^2 + a^2 = (a - r + 2a - r)^2$

or $8a^2 - 12ar = 0$ or $2a = 3r$ ... (2)

Solving (1) and (2), $2 \cdot \frac{6}{r} = 3r \quad \therefore \quad r = 2$

9. As the hyperbola is confocal with the ellipse $\frac{x^2}{4} + \frac{y^2}{3} = 1$, the equation of the hyperbola will be of the form $\frac{x^2}{a^2} - \frac{y^2}{b^2} = 1$.

Transverse axis $= 2a = 2\sin\theta \Rightarrow a = \sin\theta$.

Now, $b^2 = a^2(e^2 - 1) = \sin^2\theta(e^2 - 1)$ ... (1)

But for the ellipse, foci $= (\pm 2e', 0)$, where $3 = 4(1 - e'^2)$

$\therefore$ foci $= (\pm 1, 0)$

For the ellipse, the distance between foci = 2 and for the hyperbola it is $2ae$. So, $2 = 2ae \quad \therefore e = \frac{1}{a} = \frac{1}{\sin\theta}$

$\therefore$ (1) $\Rightarrow b^2 = \sin^2\theta\left(\frac{1}{\sin^2\theta - 1}\right) = 1 - \sin^2\theta = \cos^2\theta.$

$\therefore$ the hyperbola has the equation $\frac{x^2}{\sin^2\theta} - \frac{y^2}{\cos^2\theta} = 1.$

10. All the points are on the curve $y = e^x$.

The slope of the tangent at $(c, e^c) = \left.\frac{dy}{dx}\right)_{x=c} = e^c = m_1.$

The slope of the line joining the points

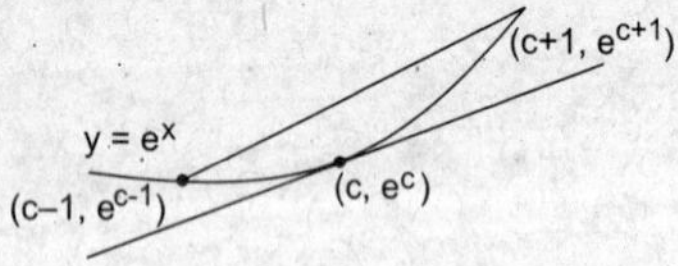

$$= \frac{e^{c+1} - e^{c-1}}{(c+1)-(c-1)}$$

$$= \frac{e^{c-1}}{2}(e^2 - 1) = m_2.$$

Now $m_2 - m_1 = \frac{e^{c-1}}{2}(e^2-1) - e^c = e^{c-1}\left(\frac{e^2-1}{2} - e\right) = \frac{e^{c-1}}{2}\{(e-1)^2 - 2\} > 0,$

because $e = 2.72$

$\therefore \quad m_2 > m_1$. So, the point of intersection is on the left of $x = c$.

11. $z = OP\left[\cos\left(\theta + \frac{\pi}{4}\right) + i\sin\left(\theta + \frac{\pi}{4}\right)\right]$

$$= 5 \cdot e^{i(\theta + \frac{\pi}{4})} = 5 \cdot e^{i\frac{\pi}{4}} \cdot e^{i\theta}$$

$$= 5\, e^{i\frac{\pi}{4}}(\cos\theta + i\sin\theta) = 5\, e^{i\frac{\pi}{4}}\left(\frac{3}{5} + i\cdot\frac{4}{5}\right)$$

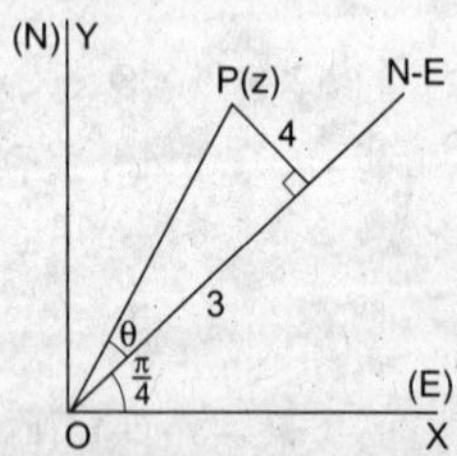

12. $\lim\limits_{t\to x} \frac{2t\, f(x) - x^2 f'(t)}{1} = 1 \quad$ or $\quad 2x\, f(x) - x^2 f'(x) = 1$

or $\quad x^2 f'(x) - 2x\, f(x) + 1 = 0 \quad$ or $\quad \frac{x^2 f'(x) - 2x\, f(x)}{x^4} + \frac{1}{x^4} = 0$

or $\quad \frac{d}{dx}\left(\frac{f(x)}{x^2}\right) + \frac{1}{x^4} = 0 \quad$ or $\quad \frac{f(x)}{x^2} + \frac{x^{-3}}{-3} = c$

But $f(1) = 1 \Rightarrow c = \frac{2}{3}$. $\quad \therefore \quad \frac{f(x)}{x^2} - \frac{1}{3x^3} = \frac{2}{3} \quad \therefore \quad f(x) = x^2\left(\frac{2}{3} + \frac{1}{3x^3}\right)$

13. $\dfrac{d^2x}{dy^2}=\dfrac{d}{dy}\left(\dfrac{dx}{dy}\right)=\dfrac{d}{dx}\left(\dfrac{1}{\frac{dy}{dx}}\right)\cdot\dfrac{dx}{dy}=-\left(\dfrac{dy}{dx}\right)^{-2}\cdot\dfrac{d^2y}{dx^2}\cdot\left(\dfrac{dy}{dx}\right)^{-1}$

$$=-\frac{d^2y}{dx^2}\cdot\left(\frac{dy}{dx}\right)^{-3}.$$

14. $dx=\dfrac{y}{\sqrt{1-y^2}}dy \Rightarrow -2dx=\dfrac{-2y}{\sqrt{1-y^2}}dy$

$\therefore \quad -2x+2x=\int\dfrac{dz}{\sqrt{z}},\ (1-y^2=z) \qquad \therefore \quad -2x+2c=2z^{\frac{1}{2}}$

or $(-x+c)^2=1-y^2$ or $x^2+y^2-2cx+c^2-1=0.$

It is a circle whose centre $=(c,0)$, radius $=\sqrt{c^2-(c^2-1)}=1.$

15. $|z|=1 \Rightarrow z=\cos\theta+i\sin\theta.$

$\therefore \quad z_1=\dfrac{z}{1-z^2}=\dfrac{\cos\theta+i\sin\theta}{1-(\cos\theta+i\sin\theta)^2}=\dfrac{\cos\theta+i\sin\theta}{1-\cos 2\theta-i\sin 2\theta}$

$$=\frac{\cos\theta+i\sin\theta}{-2i\sin\theta\,(\cos\theta+i\sin\theta)^2}=\frac{1}{-2i\sin\theta}=\frac{i}{2\sin\theta}=0+i\cdot\frac{1}{2\sin\theta}$$

$\therefore \quad z_1$ is on the y-axis for different θ.

16. $ar(\Delta\,\text{OPR})=ar(\Delta\,\text{PQR})=ar(\Delta\,\text{OQR})=\dfrac{1}{3}\cdot ar(\Delta\,\text{OPQ})$

$\therefore \quad$ distance of R from $OP=\dfrac{1}{3}\times$ distance of Q from OP, etc.

So, R is the centroid of Δ OPQ.

$\therefore \quad R=\left(\dfrac{0+3+6}{3},\dfrac{0+4+0}{3}\right)=\left(3,\dfrac{4}{3}\right).$

17. $\int\sin^2x\,dx=\dfrac{1}{2}\int(1-\cos 2x)\,dx=\dfrac{1}{2}\left\{x-\dfrac{\sin 2x}{2}\right\}+c.$

So, $F(x)=\dfrac{1}{4}(2x-\sin 2x)+c.$

$\therefore \quad F(x+\pi)=\dfrac{1}{4}\{2(x+\pi)-\sin 2(x+\pi)\}+c=\dfrac{1}{4}(2x-\sin 2x)+\dfrac{\pi}{2}+c$

$\therefore \quad F(x+\pi)-F(x)=\dfrac{\pi}{2}\neq 0.$ So, Statement-1 is False.

But $\sin^2(x+\pi)=(-\sin x)^2=\sin^2x.$ So, Statement-2 is True.

18. The symmetric equations of the line of intersection of the two planes are

$$\frac{x-\dfrac{(-6)(-5)-(-15)}{3+12}}{12+2}=\frac{y-\dfrac{(-15)2-3(-5)}{3+12}}{-4+6}=\frac{z-0}{3+12}=t$$

or $\quad \dfrac{x-3}{14}=\dfrac{y+1}{2}=\dfrac{z}{15}=t$

$\therefore \quad x=3+14t,\ y=-1+2t,\ z=15t.$

$\therefore \quad$ Statement-1 is False. The line of intersection has the direction ratios 14, 2, 15. So, the line is parallel to the vector $14\vec{i}+2\vec{j}+15\vec{k}$. So, Statement-2 is True.

19. Here $2y=-x^2+2x+2$ or $x^2-2x+1=3-2y$ or $(x-1)^2=2\left(\dfrac{3}{2}-y\right)$.

Put $x-1=Y$, $\dfrac{3}{2}-y=X$. So, the equation is $Y^2=2X$, which is a parabola whose axis is $Y=0$, i.e., $x-1=0$.

A parabola is always symmetrical about its axis. So, Statement-2 is correct and as a consequence Statement-1 is correct.

20. Statement-2 is True because $H_1, H_2, \ldots, H_n$ are exhaustive events.

$P(H_i\cap E)=P(E)\cdot P(H_i\mid E)$. Also, $P(H_i\cap E)=P(H_i)\cdot P(E\mid H_i)$.

$\therefore \quad P(H_i)\cdot P(E\mid H_i)=P(E)\cdot P(H_i\mid E)<P(H_i\mid E)$. So, Statement-1 is True.

But Statement-2 is not a correct explanation for Statement-1.

21. The locus of the points from which mutually perpendicular tangents can be drawn to the circle $x^2+y^2=13^2$ is the circle (director circle) $x^2+y^2=(13\sqrt{2})^2$ or $x^2+y^2=338$. So statement-2 is True.

(17, 7) satisfies $x^2+y^2=338$ because $17^2+7^2=289+49=338$.

$\therefore$ (17, 7) is on the director circle. So, the tangents from (17, 7) to the given circle are mutually perpendicular. So, Statement-1 is True using the Statement-2.

22. $f(x)=2+\cos x$ is differentiable everywhere and $f'(x)=-\sin x$.

$\therefore \quad f'(c)=0 \Rightarrow c=0, \pm\pi, \pm 2\pi, \pm 3\pi, \ldots$

$\therefore \quad$ for each real t, $[t, t+\pi]$ contains a value from $\{0, \pm\pi, \pm 2\pi, \pm 3\pi, \ldots\}$

$\therefore \quad$ Statement-1 is True.

$f(t+2\pi)=2+\cos(t+2\pi)=2+\cos t=f(t)$.

So, Statement-2 is True.

Therefore $f(x)$ is a periodic function of period 2π.

$\therefore$ whatever is true for $f(x)$ in some $[t, t+\pi]$ is true for $f(x)$ in $[t+2n\pi, t+(2n+1)\pi]$.

23. $P: \left.\begin{array}{l} y-x=0 \\ y+2=0 \end{array}\right\} \Rightarrow P=(-2,-2)$

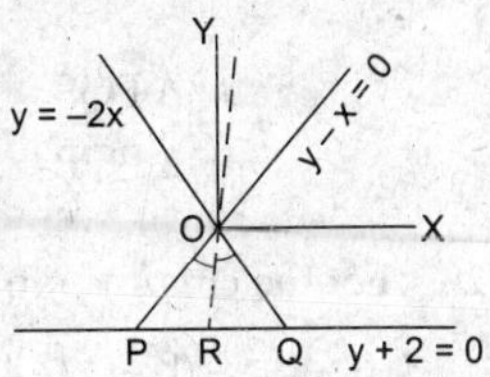

$Q: \left.\begin{array}{l} y=-2x \\ y+2=0 \end{array}\right\} \Rightarrow Q=(1,-2)$

$\therefore \quad OP=\sqrt{2^2+2^2}=2\sqrt{2}$

$OQ=\sqrt{1^2+2^2}=\sqrt{5}$

OR bisects $\angle POQ$.

So, $PR:RQ=OP:OQ=2\sqrt{2}:\sqrt{5}$.

$\triangle\, OPR$ and $\triangle\, OQR$ are not similar.

So, Statement-1 is True but Statement-2 is False.

24. Let the positive numbers be $a, b\ (a<b)$.

Then $A_1=\dfrac{a+b}{2}, G_1=\sqrt{ab}, H_1=\dfrac{2ab}{a+b}$.

From question, A_1, G_2, H_1 are in GP. So, $G_2^2=A_1H_1=ab=G_1^2$. So, $G_1=G_2$

A_2, G_3, H_2 are in GP. So, $G_3^2=A_2H_2=G_2^2$. So, $G_2=G_3$, etc.

25. & **26.** Also, $a<H_1<G_1<A_1<b, H_1<H_2<G_2<A_2<A_1$,

$H_2<H_3<G_3<A_3<A_2$, etc.

$\therefore \quad H_1<H_2<H_3<\ldots$ and $A_1>A_2>A_3>\ldots$.

27. Solving $x^2+y^2=9, y^2=8x$

we get $x=1, -9$ (not possible)

$y=2\sqrt{2}, -2\sqrt{2}$

$\therefore \quad P=(1, 2\sqrt{2}), Q=(1, -2\sqrt{2})$

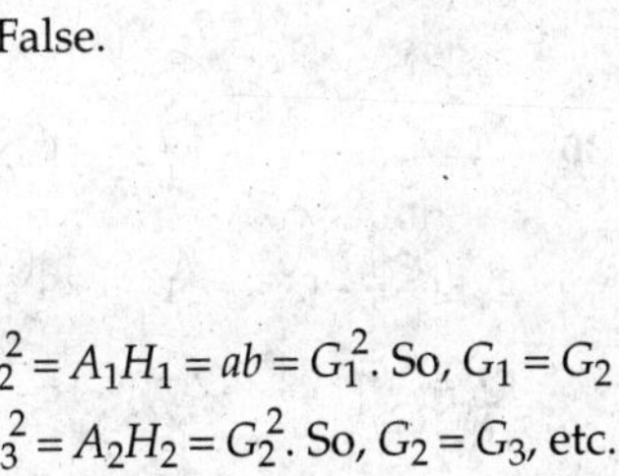

Let $R=(a, 0)$. The chord of contact of $(a, 0)$ w.r.t. $x^2+y^2=9$ is the line PQ whose equation is $x=1$.

$\therefore \quad ax=9$ and $x=1$ are the same line. So, $a=9$.

Let $S=(b, 0)$. The chord of contact of $(b, 0)$ w.r.t. $y^2=8x$ is the line PQ whose equation is $x=1$.

$\therefore \quad y\cdot 0-4(x+b)=0$ and $x=1$ are the same line. So, $b=-1$. $\therefore \quad S=(-1, 0)$.

Now, $ar(\Delta\,PQS) = \frac{1}{2} \cdot PQ \cdot TS = \frac{1}{2} \cdot 4\sqrt{2} \cdot (1+1) = 4\sqrt{2}$

$ar(\Delta\,PQR) = \frac{1}{2} \cdot PQ \cdot TR = \frac{1}{2} \cdot 4\sqrt{2} \cdot (9-1) = 16\sqrt{2}$

$\therefore \quad \cdot \; \frac{ar(\Delta\,PQS)}{ar(\Delta\,PQR)} = \frac{4\sqrt{2}}{16\sqrt{2}} = 1:4.$

28. Let the circumcentre of Δ PRS be (α, β). Then

$(\alpha-1)^2 + (\beta - 2\sqrt{2})^2 = (\alpha+1)^2 + \beta^2 = (\alpha-9)^2 + \beta^2 = (\text{radius of circumcircle})^2$

or $\quad -2\alpha - 4\sqrt{2}\beta + 9 = 2\alpha + 1 = -18\alpha + 81$. Solving, $\alpha = 4,\ \beta = -\sqrt{2}$.

$\therefore \quad$ radius of circumcircle $= \sqrt{(\alpha+1)^2 + \beta^2} = \sqrt{5^2+2} = 3\sqrt{3}$.

29. Radius of incircle of Δ PQR $= \frac{\Delta}{s} = \frac{16\sqrt{2}}{\frac{1}{2}(4\sqrt{2} + 6\sqrt{2} + 6\sqrt{2})} = \frac{16\sqrt{2}}{8\sqrt{2}} = 2.$

30. $V_r = \frac{r}{2}\{2r + (r-1)(2r-1)\} = \frac{r}{2}(2r^2 - r + 1)$

$T_r = \frac{r+1}{2}\{2(r+1)^2 - (r+1) + 1\} - \frac{r}{2}(2r^2 - r + 1) - 2 = 3r^2 + 2r - 1$

$Q_r = 3(r+1)^2 + 2(r+1) - 1 - (3r^2 + 2r - 1) = 3(2r+1) + 2 = 6r + 5.$

$$V_1 + V_2 + \ldots + V_n = \sum_1^n \left(r^3 - \frac{1}{2}r^2 + \frac{1}{2}r\right)$$

$$= \left\{\frac{n(n+1)}{2}\right\}^2 - \frac{1}{2} \cdot \frac{n(n+1)(2n+1)}{6} + \frac{1}{2} \cdot \frac{n(n+1)}{2}$$

$$= \frac{n(n+1)}{12}(3n^2 + n + 2).$$

31. $T_r = 3r^2 + 2r - 1 = (3r-1)(r+1)$. So, T_r is a composite number.

32. $Q_r = 6r + 5$. So, 11, 17, 23, … are in AP whose c.d. = 6.

33. $\int_{-1}^{1} \frac{dx}{1+x^2} = \tan^{-1}x \Big]_{-1}^{1} = \tan^{-1}1 - \tan^{-1}(-1) = \frac{\pi}{2}$.

$\int_0^1 \frac{dx}{\sqrt{1-x^2}} = \sin^{-1}x \Big]_0^1 = \sin^{-1}1 = \frac{\pi}{2}$.

$\int_2^3 \frac{dx}{1-x^2} = -\int_2^3 \frac{dx}{x^2-1} = \frac{-1}{2}\log\frac{x-1}{x+1}\Big]_2^3 = \frac{-1}{2}\left(\log\frac{1}{2} - \log\frac{1}{3}\right)$

$$= -\frac{1}{2}\log\frac{3}{2} = \frac{1}{2}\log\frac{2}{3}.$$

$$\int_1^2 \frac{dx}{x\sqrt{x^2-1}} = \sec^{-1}x\Big]_1^2 = \sec^{-1}2 - \sec^{-1}1 = \frac{\pi}{3}.$$

34. $f(x) = x\,|x| = x^2, x \geq 0$

$-x^2, x < 0$. So, $f(x)$ is continuous everywhere.

Also, $f(x)$ is differentiable everywhere except perhaps at $x = 0$.

$$f'(0^+) = \lim_{h \to 0} \frac{f(0+h) - f(0)}{h} = \lim_{h \to 0} \frac{f(0+h)^2 - 0}{h} = 0$$

$$f'(0^-) = \lim_{h \to 0} \frac{f(0-h) - f(0)}{-h} = \lim_{h \to 0} \frac{-(0-h)^2 - 0}{-h} = 0$$

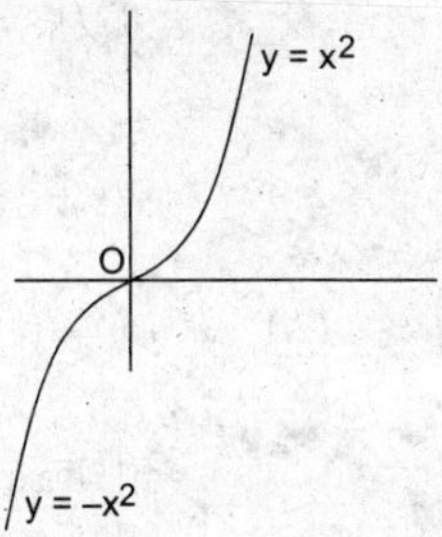

$\therefore$ $f(x)$ is differentiable in $(-1, 1)$.

Also from the graph, $f(x)$ is increasing in $(-1, 1)$.

So, $A \to p, A \to q, A \to r$.

$$y = \sqrt{|x|} \Rightarrow y = \sqrt{x}, x \geq 0 \Rightarrow y^2 = x, x \geq 0$$
$$\sqrt{-x}, x < 0 \qquad -x, x < 0$$

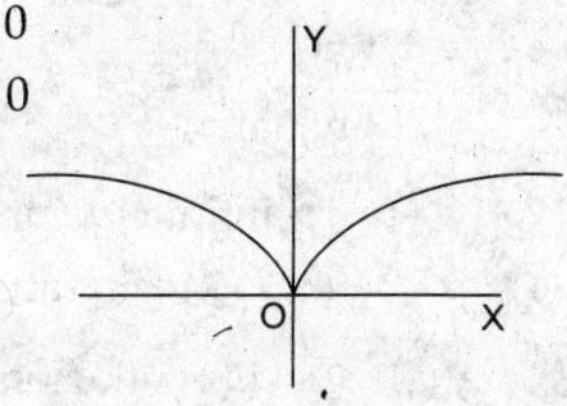

Clearly from the graph, the function is continuous everywhere, differentiable everywhere except at $x = 0$. So, $B \to p, B \to s$.

$g(x) = x + [x] = x - 1, \quad -1 \leq x < 0$

$x, \quad 0 \leq x < 1.$

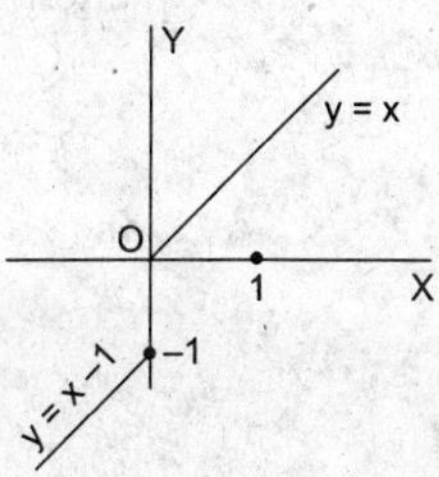

$g(x)$ is not continuous at $x = 0$, hence not differentiable at $x = 0$, Clearly from the graph, $g(x)$ is increasing in $(-1, 1)$. So, $C \to s, C \to r$.

$h(x) = -(x-1) + (x+1), \ -1 \leq x \leq 1.$

So, $h(x) = 2$ in $(-1, 1)$. So $D \to p, D \to q$.

35.

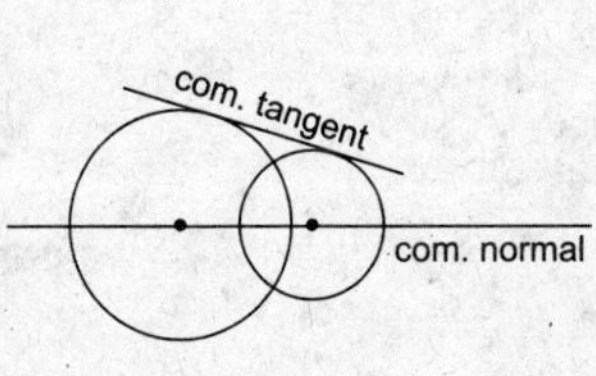

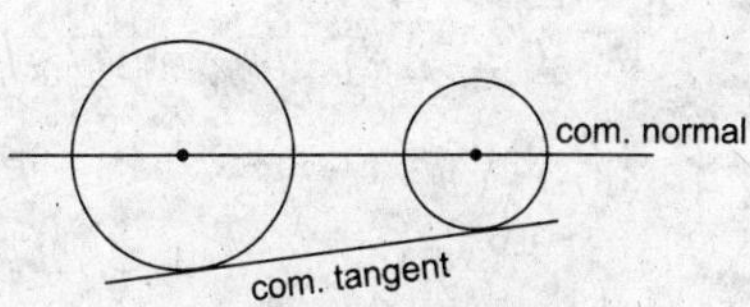

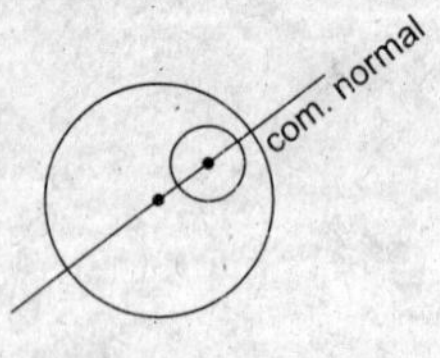

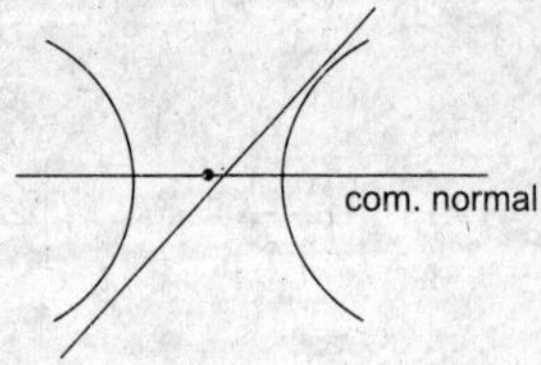

$\therefore \quad A \to p, A \to q;\ B \to p, B \to q;\ C \to q, C \to r;\ D \to q, D \to r.$

36. The system will have the trivial solution (0, 0, 0). It will have infinite number of nontrivial solutions if

$$\Delta = \begin{vmatrix} a & b & c \\ b & c & a \\ c & a & b \end{vmatrix} = 0.$$

Now, $\Delta = (a+b+c)\begin{vmatrix} 1 & 1 & 1 \\ b & c & a \\ c & a & b \end{vmatrix} = (a+b+c)(ab+bc+ca-a^2-b^2-c^2)$

(A) $a^2+b^2+c^2 = ab+bc+ca \Rightarrow (a-b)^2+(b-c)^2+(c-a)^2 = 0$

$\Rightarrow a = b = c \neq 0\ (\because a+b+c \neq 0)$

$\therefore$ all the equations are identical and infinite number of points (x, y, z) satisfy the equations. So, $A \to r$.

(B) $a+b+c = 0 \Rightarrow \Delta = 0;\ a^2+b^2+c^2 \neq ab+bc+ca \Rightarrow a, b, c$ are not equal.

$\therefore$ infinite points (x, y, z) and all points with equal coordinates $(\lambda, \lambda, \lambda)$ will satisfy the three equation. So, $B \to q$.

(C) $\Delta \neq 0$. So, no nontrivial solution. So, only (0, 0, 0) satisfied the three equation. So, $C \to p$.

(D) $a^2+b^2+c^2 = ab+bc+ca \Rightarrow a = b = c$. Also $a+b+c = 0$.

So, $a = b = c = 0$.

$\therefore$ all (x, y, z), will satisfy the three equations. So, $D \to s$.

37. $\cos^{-1}y + \cos^{-1}(bxy) = \frac{\pi}{2} - \sin^{-1}(ax) = \cos^{-1}(ax)$

$\Rightarrow \quad y \cdot bxy - \sqrt{1-y^2} \cdot \sqrt{1-b^2x^2y^2} = ax.$

(A) $\Rightarrow -\sqrt{1-y^2} = x$ or $1-y^2 = x^2$ or $x^2+y^2 = 1$. So, $A \to p$.

(B) $\Rightarrow xy^2 - \sqrt{1-y^2} \cdot \sqrt{1-x^2y^2} = x$

or $(1-y^2)(1-x^2y^2) = x^2(1-y^2)^2$ or $(1-y^2)\{1-x^2y^2-x^2(1-y^2)\} = 0$

or $(1-y^2)(1-x^2) = 0$. So, $B \to q$.

(C) $\Rightarrow 2xy^2 - \sqrt{1-y^2} \cdot \sqrt{1-4x^2y^2} = x$

or $(1-y^2)(1-4x^2y^2) = x^2(1-2y^2)^2$ or $1-y^2 = x^2$. So, $C \to p$.

(D) $\Rightarrow 2xy^2 - \sqrt{1-y^2}\cdot\sqrt{1-4x^2y^2} = 2x$

or $(1-y^2)(1-4x^2y^2) = 4x^2(1-y^2)^2$

or $(1-y^2)\{1-4x^2y^2-4x^2(1-y^2)\} = 0$

or $(1-y^2)(1-4x^2) = 0$. So, $D \to s$.

38. $f(x) = \dfrac{(x-5)(x-1)}{(x-3)(x-2)}$.

If $-1 < x < 1,\ f(x) = \dfrac{(-)(-)}{(-)(-)} = (+) > 0.$

If $1 < x < 2,\ f(x) = \dfrac{(-)(+)}{(-)(-)} = (-) < 0.$ So, $B \to q$.

If $3 < x < 5,\ f(x) = \dfrac{(-)(+)}{(+)(+)} = (-) < 0.$ So, $C \to q$.

If $x > 5,\ f(x) = \dfrac{(+)(+)}{(+)(+)} = (+) > 0.$

Again $f(x) = 1 - \dfrac{x+1}{(x-2)(x-3)}$.

If $-1 < x < 1,\ f(x) = 1 - \dfrac{(+)}{(-)(-)} = 1-(+) < 1.$ So, $A \to p$.

If $x > 5,\ f(x) = 1 - \dfrac{(+)}{(+)(+)} = 1-(+) < 1.$ So, $D \to p$.

□

IIT Questions—5

Straight Objective Type (One-Option-Correct Questions)

1. The total number of local maxima and local minima of the function

$$f(x) = \begin{cases} (2+x)^3, & -3 < x \le -1 \\ x^{2/3}, & -1 < x < 2 \end{cases}$$

is

(a) 0 (b) 1
(c) 2 (d) 3

2. Let $g(x) = \dfrac{(x-1)^n}{\log \cos^m (x-1)}$; $0 < x < 2$, m and n are integers, $m \ne 0$, $n > 0$, and let p be the left hand derivative of $|x - 1|$ at $x = 1$.
If $\lim\limits_{x \to 1+} g(x) = p$ then

(a) $n = 1, m = 1$ (b) $n = 1, m = -1$
(c) $n = 2, m = 2$ (d) $n > 2, m = n$

3. Let $I = \displaystyle\int \frac{e^x}{e^{4x} + e^{2x} + 1}\,dx$, $J = \displaystyle\int \frac{e^{-x}}{e^{-4x} + e^{-2x} + 1}\,dx$.
Then, for an arbitrary constant c, the value of $J - I$ equals

(a) $\dfrac{1}{2}\log\left(\dfrac{e^{4x} - e^{2x} + 1}{e^{4x} + e^{2x} + 1}\right) + c$ (b) $\dfrac{1}{2}\log\left(\dfrac{e^{2x} + e^{x} + 1}{e^{2x} - e^{x} + 1}\right) + c$

(c) $\dfrac{1}{2}\log\left(\dfrac{e^{2x} - e^{x} + 1}{e^{2x} + e^{x} + 1}\right) + c$ (d) $\dfrac{1}{2}\log\left(\dfrac{e^{4x} + e^{2x} + 1}{e^{4x} - e^{2x} + 1}\right) + c$

4. Let $g(x) = \log f(x)$, where $f(x)$ is twice differentiable positive function on $(0, \infty)$ such that $f(x + 1) = x f(x)$. Then for $N = 1, 2, 3, \ldots$

$$g''\left(N + \frac{1}{2}\right) - g''\left(\frac{1}{2}\right) =$$

(a) $-4\left\{1 + \dfrac{1}{9} + \dfrac{1}{25} + \ldots + \dfrac{1}{(2N-1)^2}\right\}$ (b) $4\left\{1 + \dfrac{1}{9} + \dfrac{1}{25} + \ldots + \dfrac{1}{(2N-1)^2}\right\}$

(c) $-4\left\{1 + \dfrac{1}{9} + \dfrac{1}{25} + \ldots + \dfrac{1}{(2N+1)^2}\right\}$ (d) $4\left\{1 + \dfrac{1}{9} + \dfrac{1}{25} + \ldots + \dfrac{1}{(2N+1)^2}\right\}$

5. Let the function $g : (-\infty, \infty) \to \left(-\frac{\pi}{2}, \frac{\pi}{2}\right)$ be given by $g(u) = 2\tan^{-1}(e^u) - \frac{\pi}{2}$. Then g is

(a) even and is strictly increasing in $(0, \infty)$

(b) odd and is strictly decreasing in $(-\infty, \infty)$

(c) odd and is strictly increasing in $(-\infty, \infty)$

(d) neither even nor odd but is strictly increasing in $(-\infty, \infty)$

6. The area of the region between the curves $y = \sqrt{\sec x + \tan x}$ and $y = \sqrt{\sec x - \tan x}$ bounded by the lines $x = 0$ and $x = \frac{\pi}{4}$ is

(a) $\int_0^{\sqrt{2}-1} \frac{t}{(1+t^2)\sqrt{1-t^2}}\,dt$ (b) $\int_0^{\sqrt{2}-1} \frac{4t}{(1+t^2)\sqrt{1-t^2}}\,dt$

(c) $\int_0^{\sqrt{2}+1} \frac{4t}{(1+t^2)\sqrt{1-t^2}}\,dt$ (d) $\int_0^{\sqrt{2}+1} \frac{t}{(1+t^2)\sqrt{1-t^2}}\,dt$

7. Consider a branch of the hyperbola

$$x^2 - 2y^2 - 2\sqrt{2}x - 4\sqrt{2}y - 6 = 0$$

with vertex at the point A. Let B be one end of its latus rectum. If S is the focus of the hyperbola nearest to the point A then ar(ΔASB) is

(a) $1 - \sqrt{2/3}$ (b) $\sqrt{3/2} - 1$

(c) $1 + \sqrt{2/3}$ (d) $\sqrt{3/2} + 1$

8. Consider the two curves $C_1 : y^2 = 4x$

$C_2 : x^2 + y^2 - 6x + 1 = 0.$

Then

(a) C_1 and C_2 touch each other only at one point

(b) C_1 and C_2 touch each other exactly at two points

(c) C_1 and C_2 intersect (but do not touch) at exactly at two points

(d) C_1 and C_2 neither intersect nor touch each other

9. The edges of a parallelopiped are of unit length and are parallel to noncoplanar unit vectors $\hat{a}, \hat{b}, \hat{c}$ such that $\hat{a} \cdot \hat{b} = \hat{b} \cdot \hat{c} = \hat{c} \cdot \hat{a} = \frac{1}{2}$.

Then the volume of the parallelopiped is

(a) $\frac{1}{\sqrt{2}}$ (b) $\frac{1}{2\sqrt{2}}$ (c) $\frac{\sqrt{3}}{2}$ (d) $\frac{1}{\sqrt{3}}$

10. An experiment has 10 equally likely outcomes. Let A and B be two non-empty events of the experiment. If A consists of 4 outcomes, the number of outcomes that B must have so that A and B are independent, is

(a) 2, 4 or 8 (b) 3, 6 or 9 (c) 4 or 8 (d) 5 or 10

11. If $0 < x < 1$, then $\sqrt{1+x^2}\left[\{x \cos(\cot^{-1} x) + \sin(\cot^{-1} x)\}^2 - 1\right]^{1/2} =$

(a) $\dfrac{x}{\sqrt{1+x^2}}$ (b) x

(c) $x\sqrt{1+x^2}$ (d) $\sqrt{1+x^2}$

12. Let two non-collinear unit vectors $\hat{a}$ and $\hat{b}$ form an acute angle. A point P moves so that at any time t the position vector $\overrightarrow{OP}$ (where O is the origin) is given by $\hat{a}\cos t + \hat{b}\sin t$. When P is the farthest from origin O, let M be the length of $\overrightarrow{OP}$ and $\hat{u}$ be the unit vector along $\overrightarrow{OP}$. Then

(a) $\hat{u} = \dfrac{\hat{a}+\hat{b}}{|\hat{a}+\hat{b}|}$ and $M = (1 + \hat{a}\cdot\hat{b})^{1/2}$

(b) $\hat{u} = \dfrac{\hat{a}-\hat{b}}{|\hat{a}-\hat{b}|}$ and $M = (1 + \hat{a}\cdot\hat{b})^{1/2}$

(c) $\hat{u} = \dfrac{\hat{a}+\hat{b}}{|\hat{a}+\hat{b}|}$ and $M = (1 + 2\hat{a}\cdot\hat{b})^{1/2}$

(d) $\hat{u} = \dfrac{\hat{a}-\hat{b}}{|\hat{a}-\hat{b}|}$ and $M = (1 + 2\hat{a}\cdot\hat{b})^{1/2}$

13. A particle P starts from the point $z_0 = 1 + 2i$, where $i = \sqrt{-1}$. It moves first horizontally away from origin by 5 units and then vertically away from origin by 3 units to reach a point z_1. From z_1 the particle moves $\sqrt{2}$ units in the direction of the vector $\hat{i} + \hat{j}$ and then moves through an angle $\pi/2$ in anticlockwise direction on a circle with centre at origin, to reach a point z_2. The point z_2 is given by

(a) $6 + 7i$ (b) $-7 + 6i$

(c) $7 + 6i$ (d) $-6 + 7i$

14. Consider three points $P = [-\sin(\beta - \alpha), -\cos\beta]$, $Q = [\cos(\beta - \alpha), \sin\beta]$ and $R = [\cos(\beta - \alpha + \theta), \sin(\beta - \theta)]$, where $0 < \alpha, \beta, \theta < \frac{\pi}{4}$. Then

(a) P lies on the line segment RQ (b) Q lies on the line segment PR

(c) R lies on the line segment QP (d) P, Q, R are non-collinear

15. Let a, b be non-zero real numbers. The equation

$$(ax^2 + by^2 + c)(x^2 - 5xy + 6y^2) = 0$$

represents

(a) four straight lines, when $c = 0$ and a, b are of the same sign

(b) two straight lines and a circle, when $a = b$ and c is of sign opposite to that of a

(c) two straight lines and a hyperbola, when a and b are of the same sign and c is of sign opposite to that of a

(d) a circle and an ellipse, when a and b are of the same sign

Multiple Correct Answers Type (One or More-Options-Correct Questions)

16. Let $f(x)$ be a non-constant twice differentiable function defined on $(-\infty, \infty)$ such that $f(x) = f(1-x)$ and $f'\left(\frac{1}{4}\right) = 0$. Then,

(a) $f''(x)$ vanishes at least twice on $[0, 1]$

(b) $f'\left(\frac{1}{2}\right) = 0$

(c) $\int_{-1/2}^{1/2} f\left(x + \frac{1}{2}\right) \cdot \sin x \, dx = 0$

(d) $\int_0^{1/2} f(t) \cdot e^{\sin \pi t} \, dt = \int_{1/2}^{1} f(1-t) \cdot e^{\sin \pi t} \, dt$

17. Let $S_n = \sum_{k=1}^{n} \frac{n}{n^2 + kn + k^2}$ and $T_n = \sum_{k=0}^{n-1} \frac{n}{n^2 + kn + k^2}$, for $n = 1, 2, 3, \ldots$. Then,

(a) $S_n < \frac{\pi}{3\sqrt{3}}$

(b) $S_n > \frac{\pi}{3\sqrt{3}}$

(c) $T_n < \frac{\pi}{3\sqrt{3}}$

(d) $T_n > \frac{\pi}{3\sqrt{3}}$

18. A straight line through the vertex P of the $\Delta\, PQR$ intersects the side QR at S and the circumcircle of the ΔPQR at T. If S is not the centre of the circumcircle, then

(a) $\frac{1}{PS} + \frac{1}{ST} < \frac{2}{\sqrt{QS \cdot SR}}$

(b) $\frac{1}{PS} + \frac{1}{ST} > \frac{2}{\sqrt{QS \cdot SR}}$

(c) $\frac{1}{PS} + \frac{1}{ST} < \frac{4}{QR}$

(d) $\frac{1}{PS} + \frac{1}{ST} > \frac{4}{QR}$

19. Let $P(x_1, y_1)$ and $Q(x_2, y_2)$; where $y_1 < 0, y < 0$ be the end points of the latus rectum of the ellipse $x^2 + 4y^2 = 4$. The equation of parabola with latus rectum PQ are

(a) $x^2 + 2\sqrt{3}y = 3 + \sqrt{3}$ (b) $x^2 - 2\sqrt{3}y = 3 + \sqrt{3}$
(c) $x^2 + 2\sqrt{3}y = 3 - \sqrt{3}$ (d) $x^2 - 2\sqrt{3}y = 3 - \sqrt{3}$

Linked Comprehension Type

Paragraph for Question Nos. 20 to 22

Let A, B, C be three sets of complex numbers as given below.

$$A = \{z : \text{Im}(z) \geq 1\},\ B = \{z : |z - 2 - i| = 3\},\ C = \{z : \text{Re}((1 - i)z) = \sqrt{2}\}$$

20. The number of elements in the set $A \cap B \cap C$ is

(a) 0 (b) 1 (c) 2 (d) ∞

21. Let z be any point in $A \cap B \cap C$. Then, $|z + 1 - i|^2 + |z - 5 - i|^2$ lies between

(a) 25 and 29 (b) 30 and 34 (c) 35 and 39 (d) 40 and 44

22. Let z be any point in $A \cap B \cap C$ and let w be any point satisfying $|w - 2 - i| < 3$. Then, $|z| - |w| + 3$ lies between

(a) –6 and 3 (b) –3 and 6 (c) –6 and 6 (d) –3 and 9

Paragraph for Question Nos. 23 to 25

Consider the function $f : R \to R$ defined by $f(x) = \dfrac{x^2 - ax + 1}{x^2 + ax + 1}$, $a \in (0, 2)$

23. Which of the following is true?

(a) $(2 + a)^2 \cdot f''(1) + (2 - a)^2 \cdot f''(-1) = 0$

(b) $(2 - a)^2 f''(1) - (2 + a)^2 f''(-1) = 0$

(c) $f'(1)f'(-1) = (2 - a)^2$

(d) $f'(1)f'(-1) = -(2 + a)^2$

24. Which of the following is true?

(a) $f(x)$ is decreasing in (–1, 1) and has a local minimum at $x = 1$

(b) $f(x)$ is increasing in (–1, 1) and has a local maximum at $x = 1$

(c) $f(x)$ is increasing in (–1, 1) but has neither a local maximum nor a local minimum at $x = 1$

(d) $f(x)$ is decreasing in $(-1, 1)$ but has neither a local maximum nor a local minimum at $x = 1$

25. Let $g(x) = \int_0^{e^x} \frac{f'(t)}{1+t^2}\,dt$, which of the following is true?

(a) $g'(x)$ is positive on $(-\infty, 0)$ and negative on $(0, \infty)$
(b) $g'(x)$ is negative on $(-\infty, 0)$ and positive on $(0, \infty)$
(c) $g'(x)$ changes sign on both $(-\infty, 0)$ and $(0, \infty)$
(d) $g'(x)$ does not change sign on $(-\infty, \infty)$

Paragraph for Question Nos. 26 to 28

A circle C of radius 1 is inscribed in an equilateral triangle PQR. The points of contact of C with the sides PQ, QR, RP are D, E, F respectively. The line PQ is given by the equation $\sqrt{3}x + y - 6 = 0$ and the point D is $\left(\frac{3\sqrt{3}}{2}, \frac{3}{2}\right)$. Further, it is given that the origin and the centre of C are on the same side of the line PQ.

26. The equation of the circle C is

(a) $(x - 2\sqrt{3})^2 + (y - 1)^2 = 1$ (b) $(x - 2\sqrt{3})^2 + \left(y + \frac{1}{2}\right)^2 = 1$
(c) $(x - \sqrt{3})^2 + (y + 1)^2 = 1$ (d) $(x - \sqrt{3})^2 + (y - 1)^2 = 1$

27. Points E and F are given by

(a) $\left(\frac{\sqrt{3}}{2}, \frac{3}{2}\right), (\sqrt{3}, 0)$ (b) $\left(\frac{\sqrt{3}}{2}, \frac{1}{2}\right), (\sqrt{3}, 0)$
(c) $\left(\frac{\sqrt{3}}{2}, \frac{3}{2}\right), \left(\frac{\sqrt{3}}{2}, \frac{1}{2}\right)$ (d) $\left(\frac{3}{2}, \frac{\sqrt{3}}{2}\right), \left(\frac{\sqrt{3}}{2}, \frac{1}{2}\right)$

28. Equations of the sides QR, RP are

(a) $y = \frac{2}{\sqrt{3}}x + 1, y = -\frac{2}{\sqrt{3}}x - 1$ (b) $y = \frac{x}{\sqrt{3}}, y = 0$
(c) $y = \frac{\sqrt{3}}{2}x + 1, y = -\frac{\sqrt{3}}{2}x - 1$ (d) $y = \sqrt{3}x, y = 0$

Paragraph for Question Nos. 29 to 31

Consider the lines: $L_1 : \frac{x+1}{3} = \frac{y+2}{1} = \frac{z+1}{2}$

$L_2 : \frac{x-2}{1} = \frac{y+2}{2} = \frac{z-3}{3}$

29. The unit vector perpendicular to both L_1 and L_2 is

(a) $\frac{1}{\sqrt{99}}(-\hat{i}+7\hat{j}+7\hat{k})$ (b) $\frac{1}{5\sqrt{3}}(-\hat{i}-7\hat{j}+5\hat{k})$

(c) $\frac{1}{5\sqrt{3}}(-\hat{i}+7\hat{j}+5\hat{k})$ (d) $\frac{1}{\sqrt{99}}(7\hat{i}-7\hat{j}-\hat{k})$

30. The shortest distance between L_1 and L_2 is

(a) 0 (b) $\frac{17}{\sqrt{3}}$ (c) $\frac{41}{5\sqrt{3}}$ (d) $\frac{17}{5\sqrt{3}}$

31. The distance of the point (1, 1, 1) from the plane passing through the point (–1, –2, –1) and whose normal is perpendicular to both the lines, L_1 and L_2 is

(a) $\frac{2}{\sqrt{75}}$ (b) $\frac{7}{\sqrt{75}}$ (c) $\frac{13}{\sqrt{75}}$ (d) $\frac{23}{\sqrt{75}}$

Paragraph for Question Nos. 32 to 34

Consider the functions defined implicitly by the equation $y^3 - 3y + x = 0$ on various intervals in the real line.

If $x \in (-\infty, -2) \cup (2, \infty)$, the equation implicitly defines a unique real valued differentiable function $y = f(x)$.

If $x \in (-2, 2)$, the equation implicitly defines a unique real valued differentiable function $y = g(x)$ satisfying $g(0) = 0$.

32. If $f(-10\sqrt{2}) = 2\sqrt{2}$ then $f''(-10\sqrt{2}) =$

(a) $\frac{4\sqrt{2}}{7^3 \cdot 3^2}$ (b) $-\frac{4\sqrt{2}}{7^3 \cdot 3^2}$ (c) $\frac{4\sqrt{2}}{7^3 \cdot 3}$ (d) $-\frac{4\sqrt{2}}{7^3 \cdot 3}$

33. The area of the region bounded by the curve $y = f(x)$, x-axis, and the lines $x = a$ and $x = b$, where $-\infty < a < b < -2$, is

(a) $\int_a^b \frac{x}{3\{(f(x))^2 - 1\}} dx + bf(b) - af(a)$

(b) $-\int_a^b \frac{x}{3\{(f(x))^2 - 1\}} dx + bf(b) - af(a)$

(c) $\int_a^b \frac{x}{3\{(f(x))^2 - 1\}} dx - bf(b) + af(a)$

(d) $-\int_a^b \frac{x}{3\{(f(x))^2 - 1\}} dx - bf(b) + af(a)$

34. $\int_{-1}^{1} g'(x)\, dx =$

(a) $2g(-1)$ (b) 0 (c) $-2g(1)$ (d) $2g(1)$

Assertion-Reasoning Type

35. Suppose four distinct positive numbers a_1, a_2, a_3, a_4 are in GP. Let $b_1 = a_1, b_2 = b_1 + a_2, b_3 = b_2 + a_3$ and $b_4 = b_3 + a_4$.

STATEMENT-1: The numbers b_1, b_2, b_3, b_4 are neither in AP nor in GP, *and*

STATEMENT-2 : The numbers b_1, b_2, b_3, b_4 are in HP.

(a) STATEMENT-1 is True, STATEMENT-2 is True; STATEMENT-2 is a correct explanation for STATEMENT-1.

(b) STATEMENT-1 is True, STATEMENT-2 is True; STATEMENT-2 is NOT a correct explanation for STATEMENT-1.

(c) STATEMENT-1 is True, STATEMENT-2 is False.

(d) STATEMENT-1 is False, STATEMENT-2 is True.

36. Consider the three planes $P_1 : x - y + z = 1$

$$P_2 : x + y - z = -1$$
$$P_3 : x - 3y + 3z = 2$$

Let L_1, L_2, L_3 be the lines of intersection of the planes P_2 and P_3, P_3 and P_1, and P_1 and P_2 respectively.

STATEMENT-1: At least two of the lines L_1, L_2 and L_3 are non-parallel, *and*

STATEMENT-2: The three planes do not have a common point.

(a) STATEMENT-1 is True, STATEMENT-2 is True; STATEMENT-2 is a correct explanation for STATEMENT-1.

(b) STATEMENT-1 is True, STATEMENT-2 is True; STATEMENT-2 is NOT a correct explanation for STATEMENT- 1.

(c) STATEMENT-1 is True, STATEMENT-2 is False.

(d) STATEMENT-1 is False, STATEMENT-2 is True.

37. Consider the system of equations

$$x \quad 2y + 3z = -1$$
$$-x + y - 2z = k$$
$$x - 3y + 4z = 1$$

STATEMENT-1: The system of equations has no solution for $k \neq 3$, *and*

STATEMENT-2: The determinant $\begin{vmatrix} 1 & 3 & -1 \\ -1 & -2 & k \\ 1 & 4 & 1 \end{vmatrix} \neq 0$, for $k \neq 3$.

(a) STATEMENT-1 is True, STATEMENT-2 is True; STATEMENT-2 is a correct explanation for STATEMENT-1.

(b) STATEMENT-1 is True, STATEMENT-2 is True; STATEMENT-2 is NOT a correct explanation for STATEMENT-1.

(c) STATEMENT-1 is True, STATEMENT-2 is False.

(d) STATEMENT-1 is False, STATEMENT-2 is True.

38. Consider the system of equations

$$ax + by = 0, cx + dy = 0; \text{ where } a, b, c, d \in \{0, 1\}.$$

STATEMENT-1: The probability that the system of equations has a unique solution is $\frac{3}{8}$, *and*

STATEMENT-2: The probability that the system of equations has a solution is 1.

(a) STATEMENT-1 is True, STATEMENT-2 is True; STATEMENT-2 is a correct explanation for STATEMENT-1.

(b) STATEMENT-1 is True, STATEMENT-2 is True; STATEMENT-2 is NOT a correct explanation for STATEMENT-1.

(c) STATEMENT-1 is True, STATEMENT-2 is False.

(d) STATEMENT-1 is False, STATEMENT-2 is True.

39. Let f and g be real-valued functions defined on interval (–1, 1) such that $g''(x)$ is continuous, $g(0) \neq 0$, $g'(0) = 0$, $g''(0) \neq 0$ and $f(x) = g(x).\sin x$.

STATEMENT-1: $\lim_{x \to 0} \{g(x) \cdot \cot x - g(0) \cdot \operatorname{cosec} x\} = f''(0)$, *and*

STATEMENT-2: $f'(0) = g(0)$.

(a) STATEMENT-1 is True, STATEMENT-2 is True; STATEMENT-2 is a correct explanation for STATEMENT-1.

(b) STATEMENT-1 is True, STATEMENT-2 is True; STATEMENT-2 is NOT a correct explanation for STATEMENT-1.

(c) STATEMENT-1 is True, STATEMENT-2 is False.

(d) STATEMENT-1 is False, STATEMENT-2 is True.

40. Consider $L_1 : 2x + 3y + p - 3 = 0$

$$L_2 : 2x + 3y + p + 3 = 0.$$

where p is a real number and $C : x^2 + y^2 + 6x - 10y + 30 = 0$.

STATEMENT-1: If the line L_1 is a chord of the circle C then the line L_2 is not always a diameter of the circle C *and*

STATEMENT-2: If the line L_1 is a diameter of the circle C then the line L_2 is not a chord of the circle C.

(a) STATEMENT-1 is True, STATEMENT-2 is True; STATEMENT-2 is a correct explanation for STATEMENT-1.

(b) STATEMENT-1 is True, STATEMENT-2 is True; STATEMENT 2 is NOT a correct explanation for STATEMENT-1.

(c) STATEMENT-1 is True, STATEMENT-2 is False.

(d) STATEMENT-1 is False, STATEMENT-2 is True.

41. Let a, b, c, p, q be real numbers. Suppose α, β are the roots of the equation $x^2 + 2px + q = 0$ and $\alpha, \frac{1}{\beta}$ are the roots of the equation $ax^2 + 2bx + c = 0$ where $\beta^2 \notin \{-1, 0, 1\}$.

STATEMENT-1: $(p^2 - q)(b^2 - ac) \geq 0$, *and*

STATEMENT-2: $b \neq pa$ or $c \neq qa$

(a) STATEMENT-1 is True, STATEMENT-2 is True; STATEMENT-2 is a correct explanation for STATEMENT-1.

(b) STATEMENT-1 is True, STATEMENT-2 is True; STATEMENT-2 is NOT a correct explanation for STATEMENT-1.

(c) STATEMENT-1 is True, STATEMENT-2 is False.

(d) STATEMENT-1 is False, STATEMENT-2 is True.

42. Let a solution $y = y(x)$ of the differential equation

$$x\sqrt{x^2 - 1}\, dy - y\sqrt{y^2 - 1}\, dx = 0$$

satisfy $y(2) = \frac{2}{\sqrt{3}}$.

STATEMENT-1: $y(x) = \sec\left(\sec^{-1} x - \frac{\pi}{6}\right)$, *and*

STATEMENT-2: $y(x)$ is given by $\frac{1}{y} = \frac{2\sqrt{3}}{x} - \sqrt{1 - \frac{1}{x^2}}$

(a) STATEMENT-1 is True, STATEMENT-2 is True; STATEMENT-2 is a correct explanation for STATEMENT-1.

(b) STATEMENT-1 is True, STATEMENT-2 is True; STATEMENT-2 is NOT a correct explanation for STATEMENT-1.

(c) STATEMENT-1 is True, STATEMENT-2 is False.

(d) STATEMENT-1 is False, STATEMENT-2 is True.

Matrix Match Type Questions

Statements A, B, C, D in Column I have to be matched with statements p, q, r, s in Column II. The answers to these questions have to be appropriately bubbled as illustrated below.

If the correct matches are A–q, A–r, B–p, B–s, C–r, C–s, D–q then the correctly bubbled matrix will be the following.

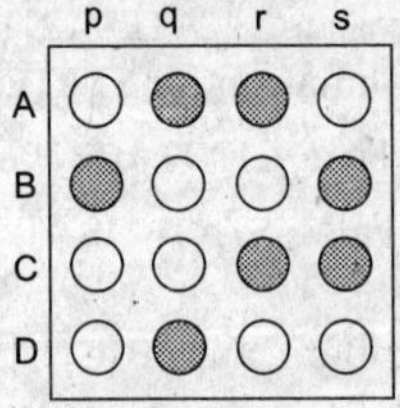

43. Consider the lines given by $L_1 : x + 3y - 5 = 0$

$$L_2 : 3x - ky - 1 = 0$$

$$L_3 : 5x + 2y - 12 = 0$$

Column I	**Column II**
(A) L_1, L_2, L_3 are concurrent, if	(p) $k = -9$
(B) One of L_1, L_2, L_3 is parallel to at least one of the other two, if	(q) $k = -\frac{6}{5}$
(C) L_1, L_2, L_3 form a triangle, if	(r) $k = \frac{5}{6}$
(D) L_1, L_2, L_3 do not form a triangle, if	(s) $k = 5$

44. Consider all possible permutations of letters of the word ENDEANOEL.

Column I	**Column II**
(A) The number of permutations containing the word ENDEA is	(p) $\underline{\lvert 5}$
(B) The number of permutations in which the letter *E* occurs in the first and the last positions is	(q) $2\,\underline{\lvert 5}$
(C) The number of permutations in which none of the letters *D*, *L*, *N* occurs in the last five positions	(r) $7\,\underline{\lvert 5}$
(D) The number of permutations in which the letters *A*, *E*, *O* occur only in odd positions is	(s) $21\,\underline{\lvert 5}$

45. **Column I** | **Column II**

Column I	Column II
(A) The minimum positive value of $\frac{x^2+2x+4}{x+2}$, $x \in R$ is	(p) 0
(B) Let A and B be 3×3 matrices of real numbers, where A is symmetric, B is skew-symmetric and $(A+B)(A-B)=(A-B)(A+B)$. If $(AB)^T = (-1)^k AB$, where $(AB)^T$ is the transpose of AB then the possible values of k are	(q) 1
(C) Let $a = \log_3 \log_3 2$. An integer k satisfying $1 < 2^{-k+3^{-a}} < 2$, must be less than	(r) 2
(D) If $\sin\theta = \cos\phi$, then the possible values of $\frac{1}{\pi}\left(\theta \pm \phi - \frac{\pi}{2}\right)$ are	(s) 3

Answers

1. c	**2.** c	**3.** c	**4.** a	**5.** c
6. b	**7.** b	**8.** b	**9.** a	**10.** d
11. c	**12.** a	**13.** d	**14.** d	**15.** b
16. a, b, c, d	**17.** a, d	**18.** b, d	**19.** b, c	**20.** b
21. c	**22.** d	**23.** a	**24.** a	**25.** b
26. d	**27.** a	**28.** d	**29.** b	**30.** d
31. c	**32.** b	**33.** a	**34.** d	**35.** c
36. d	**37.** a	**38.** b	**39.** b	**40.** c
41. b	**42.** c			

43.

	p	q	r	s
A	○	○	○	●
B	●	●	○	○
C	○	○	●	○
D	●	●	○	●

44.

	p	q	r	s
A	●	○	○	○
B	○	○	○	●
C	○	●	○	○
D	○	●	○	○

45.

	p	q	r	s
A	○	○	●	○
B	○	●	○	●
C	○	○	●	●
D	●	○	●	○

Hints and Solutions to Selected Questions

1. $\lim_{h\to 0} f(-1-h) = \lim_{h\to 0} (2-1-h)^3 = 1;$

$\lim_{h\to 0} f(-1+h) = \lim_{h\to 0} (-1+h)^{2/3} = 1,\ f(-1) = 1$

$\therefore$ $f(x)$ is continuous in $(-3, 2)$. Draw a rough sketch of $f(x)$.

The function has a local minimum at $x = 0$ and a local maximum at $x = -1$.

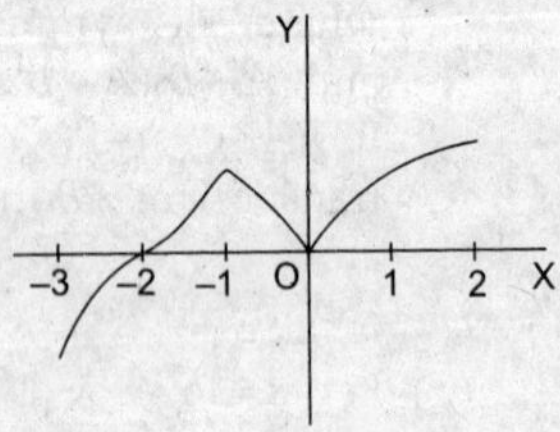

2. LH derivative at $x = 1$ is

$$\lim_{h\to 0} \frac{f(1-h)-f(1)}{-h} = \lim_{h\to 0} \frac{|1-h-1|-0}{-h} = \lim_{h\to 0} \frac{h}{-h} = -1.$$

$$\therefore \quad \lim_{x\to 1+} g(x) = p = -1 \qquad \Rightarrow \quad \lim_{h\to 0} \frac{(1+h-1)^n}{\log \cos^m(1+h-1)} = -1$$

$$\Rightarrow \quad \lim_{h\to 0} \frac{h^n}{m \log \cos h} = -1 \qquad \Rightarrow \quad \lim_{h\to 0} \frac{nh^{n-1}}{m(-\tan h)} = -1$$

$$\Rightarrow \quad \lim_{h\to 0} \frac{-n}{m} \cdot \left(\frac{h}{\tan h}\right) \cdot h^{n-2} = -1 \Rightarrow \quad n = 2, m = n.$$

3. $$J - I = \int \left(\frac{e^{3x}}{1+e^{2x}+e^{4x}} - \frac{e^x}{e^{4x}+e^{2x}+1} \right) dx = \int \frac{(e^{2x}-1)\, e^x}{e^{4x}+e^{2x}+1}\, dx$$

$$= \int \frac{z^2-1}{z^4+z^2+1}\, dz, \text{ (putting } e^x = z)$$

$$= \int \frac{1-\frac{1}{z^2}}{z^2+\frac{1}{z^2}+1}\, dz = \int \frac{d\left(z+\frac{1}{z}\right)}{\left(z+\frac{1}{z}\right)^2 - 1} = \frac{1}{2} \log \frac{\left(z+\frac{1}{z}\right)-1}{\left(z+\frac{1}{z}\right)+1} + c$$

$$= \frac{1}{2} \log \frac{e^x+e^{-x}-1}{e^x+e^{-x}+1} + c = \frac{1}{2} \log \frac{e^{2x}-e^x+1}{e^{2x}+e^x+1} + c.$$

4. $g(x+1) = \log f(x+1) = \log \{x f(x)\} = \log x + \log f(x) = \log x + g(x)$

$\therefore \quad g(x+1) - g(x) = \log x$

$$\therefore \quad g''(x+1) - g''(x) = -\frac{1}{x^2}, \text{ (differentiating twice)}$$

$$\therefore \quad g''\left(1+\frac{1}{2}\right) - g''\left(\frac{1}{2}\right) = -\frac{1}{\left(\frac{1}{2}\right)^2}$$

$$g''\left(2+\frac{1}{2}\right) - g''\left(1+\frac{1}{2}\right) = -\frac{1}{\left(\frac{3}{2}\right)^2}$$

....

$$g''\left(N+\frac{1}{2}\right) - g''\left(N-\frac{1}{2}\right) = -\frac{1}{\left(N-\frac{1}{2}\right)^2}$$

Adding, $g''\left(N+\frac{1}{2}\right) - g''\left(\frac{1}{2}\right) = -4\left\{1+\frac{1}{3^2}+\frac{1}{5^2}+\ldots+\frac{1}{(2N-1)^2}\right\}$.

5. $g(-u) = 2\tan^{-1} e^{-u} - \frac{\pi}{2} = 2\cot^{-1} e^{u} - \frac{\pi}{2} = 2\left(\frac{\pi}{2} - \tan^{-1} e^{u}\right) - \frac{\pi}{2}$

$$= -\left(2\tan^{-1} e^{u} - \frac{\pi}{2}\right) = -g(u)$$

$g'(u) = 2 \cdot \frac{1}{1+e^{2u}} \cdot e^{u} > 0$. So, $g(u)$ is odd and strictly increasing.

6. Area $= \int_0^{\pi/4} (\sqrt{\sec x + \tan x} - \sqrt{\sec x - \tan x})\, dx$

$$= \int_0^{\pi/4} \left(\sqrt{\frac{1+\sin x}{\cos x}} - \sqrt{\frac{1-\sin x}{\cos x}}\right) dx$$

$$= \int_0^{\pi/4} \left\{\sqrt{\frac{\left(\cos\frac{x}{2}+\sin\frac{x}{2}\right)^2}{\left(\cos^2\frac{x}{2}-\sin^2\frac{x}{2}\right)}} - \sqrt{\frac{\left(\cos\frac{x}{2}-\sin\frac{x}{2}\right)^2}{\left(\cos^2\frac{x}{2}-\sin^2\frac{x}{2}\right)}}\right\} dx$$

$$= \int_0^{\pi/4} \left\{\sqrt{\frac{1+\tan\frac{x}{2}}{1-\tan\frac{x}{2}}} - \sqrt{\frac{1-\tan\frac{x}{2}}{1+\tan\frac{x}{2}}}\right\} dx$$

$$= \int_0^{\pi/4} \frac{2\tan\frac{x}{2}}{\sqrt{1-\tan^2\frac{x}{2}}}\,dx$$

$$= \int_0^{\tan\frac{\pi}{8}} \frac{2t}{\sqrt{1-t^2}} \cdot \frac{2dt}{1+t^2}, \qquad \left(\text{putting } \tan\frac{x}{2} = t\right)$$

$$= 4\int_0^{\sqrt{2}-1} \frac{t\,dt}{(1+t^2)\sqrt{1-t^2}}, \quad \left\{\because \tan\frac{\pi}{8} = \frac{1-\cos\frac{\pi}{4}}{\sin\frac{\pi}{4}} = \frac{1-\frac{1}{\sqrt{2}}}{\frac{1}{\sqrt{2}}} = \sqrt{2}-1\right\}$$

7. Here, $(x^2 - 2\sqrt{2}x + 2) - 2(y^2 + 2\sqrt{2}y + 2) = 4$ or $\frac{(x-\sqrt{2})^2}{2^2} - \frac{(y+\sqrt{2})^2}{(\sqrt{2})^2} = 1.$

Putting $x - \sqrt{2} = X, y + \sqrt{2} = Y$ we get $\frac{X^2}{2^2} - \frac{Y^2}{(\sqrt{2})^2} = 1,$ the standard equation.

$\therefore$ vertex $A = (a, 0)_{X,Y} = (2, 0)_{X,Y} = (2+\sqrt{2}, -\sqrt{2});$

also $b^2 = a^2(e^2 - 1) \Rightarrow 2 = 4(e^2 - 1) \Rightarrow e = \sqrt{\frac{3}{2}}.$

The focus $S = (ae, 0)_{X,Y} = \left(2 \cdot \sqrt{\frac{3}{2}}, 0\right)_{X,Y} = (\sqrt{6}, 0)_{X,Y} = (\sqrt{6}+\sqrt{2}, -\sqrt{2})$

An end B of the latus rectum $= \left(ae, \frac{b^2}{a}\right)_{X,Y} = \left(2 \cdot \sqrt{\frac{3}{2}}, \frac{2}{2}\right)_{X,Y}$

$$= (\sqrt{6}, 1)_{X,Y} = (\sqrt{6}+\sqrt{2}, 1-\sqrt{2})$$

$$\therefore \quad \text{ar}(ASB) = \frac{1}{2}\begin{vmatrix} 2+\sqrt{2} & -\sqrt{2} & 1 \\ \sqrt{6}+\sqrt{2} & -\sqrt{2} & 1 \\ \sqrt{6}+\sqrt{2} & 1-\sqrt{2} & 1 \end{vmatrix}$$

$$= \frac{1}{2}\begin{vmatrix} 2+\sqrt{2} & -\sqrt{2} & 1 \\ \sqrt{6}-2 & 0 & 0 \\ 0 & 1 & 0 \end{vmatrix} = \frac{1}{2}(2-\sqrt{6})(-1)$$

$$= \frac{1}{2}(\sqrt{6}-2) = \sqrt{\frac{3}{2}} - 1.$$

Alternatively,

$CS = ae,\ CA = a,\ BS = \dfrac{b^2}{a}$ and $\triangle ASB$ is right angled at S.

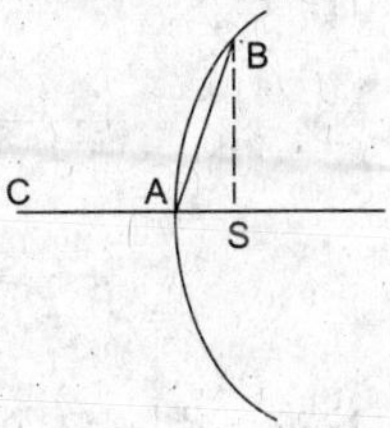

$$\therefore \quad \text{ar}(\triangle ASB) = \frac{1}{2} \cdot AS \cdot BS = \frac{1}{2}(CS - CA)\, BS.$$

$$= \frac{1}{2}(ae - a)\frac{b^2}{a} = \frac{1}{2}(e-1)b^2$$

$$= \frac{1}{2}\left(\sqrt{\frac{3}{2}} - 1\right)2 = \sqrt{\frac{3}{2}} - 1.$$

8. The centre of the circle C_2 is $(3, 0)$ and radius $= \sqrt{3^2 + 0^2 - 1} = 2\sqrt{2} < 3$.

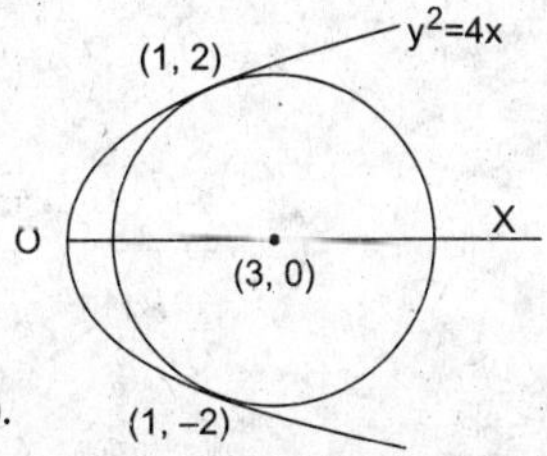

Solving equations of C_1, C_2—

$$x^2 + 4x - 6x + 1 = 0$$

or $\quad (x-1)^2 = 0$, having equal roots.

$\therefore$ they touch at $x = 1$, i.e., the points $(1, \pm 2)$.

9. $V = [\hat{a}\ \hat{b}\ \hat{c}]$

$$\Rightarrow \quad V^2 = [\hat{a}\ \hat{b}\ \hat{c}]^2$$

$$= \begin{vmatrix} a_1 & a_2 & a_3 \\ b_1 & b_2 & b_3 \\ c_1 & c_2 & c_3 \end{vmatrix} \begin{vmatrix} a_1 & b_1 & c_1 \\ a_2 & b_2 & c_2 \\ a_3 & b_3 & c_3 \end{vmatrix}, \text{ where } \hat{a} = a_1\hat{i} + a_2\hat{j} + a_3\hat{k}, \text{ etc.}$$

$$= \begin{vmatrix} \hat{a}\cdot\hat{a} & \hat{a}\cdot\hat{b} & \hat{a}\cdot\hat{c} \\ \hat{b}\cdot\hat{a} & \hat{b}\cdot\hat{b} & \hat{b}\cdot\hat{c} \\ \hat{c}\cdot\hat{a} & \hat{c}\cdot\hat{b} & \hat{c}\cdot\hat{c} \end{vmatrix} = \begin{vmatrix} 1 & \frac{1}{2} & \frac{1}{2} \\ \frac{1}{2} & 1 & \frac{1}{2} \\ \frac{1}{2} & \frac{1}{2} & 1 \end{vmatrix} = 2\begin{vmatrix} 1 & \frac{1}{2} & \frac{1}{2} \\ 1 & 1 & \frac{1}{2} \\ 1 & \frac{1}{2} & 1 \end{vmatrix}$$

$$\therefore \quad V^2 = 2\begin{vmatrix} 1 & \frac{1}{2} & \frac{1}{2} \\ 0 & \frac{1}{2} & 0 \\ 0 & 0 & \frac{1}{2} \end{vmatrix} = 2 \cdot \frac{1}{2} \cdot \frac{1}{2} = \frac{1}{2}; \qquad \therefore \quad V = \frac{1}{\sqrt{2}}.$$

10. $P(A) = \dfrac{4}{10},\ P(B) = \dfrac{x}{10}$. A and B are independent if $P(A \cap B) = P(A) \cdot P(B)$

$$\therefore \quad P(A \cap B) = \frac{4x}{100} = \frac{\frac{2}{5}x}{10}.$$

The numerator must be an integer lying between 1 and 10. So x must be a multiple of 5 in [1, 10]. $\therefore$ $x = 5$ or 10.

11. $$\text{Exp.} = \sqrt{1+x^2}\left[\left\{x\cdot\cos\left(\cos^{-1}\frac{x}{\sqrt{1+x^2}}\right)+\sin\left(\sin^{-1}\frac{1}{\sqrt{1+x^2}}\right)\right\}^2-1\right]^{1/2}$$

$$=\sqrt{1+x^2}\left[\left\{x\cdot\frac{x}{\sqrt{1+x^2}}+\frac{1}{\sqrt{1+x^2}}\right\}^2-1\right]^{1/2}$$

$$=\sqrt{1+x^2}\,[1+x^2-1]^{1/2}=x\sqrt{1+x^2}.$$

12. $$|\overrightarrow{OP}| = (\hat{a}\cos t+\hat{b}\sin t)^2=\cos^2 t+\sin^2 t+\sin 2t\,\hat{a}\cdot\hat{b}$$

$$=1+\sin 2t\,\hat{a}\cdot\hat{b} \qquad \therefore\ |\overrightarrow{OP}| = (1+\sin 2t\,\hat{a}\cdot\hat{b})^{1/2}$$

$$\therefore\quad M=\max|\overrightarrow{OP}| = (1+\hat{a}\cdot\hat{b})^{1/2}\text{ when } t=\frac{\pi}{4}$$

$$\therefore\quad \hat{u}=\frac{\overrightarrow{OP}}{|\overrightarrow{OP}|}=\frac{\hat{a}\cos\frac{\pi}{4}+\hat{b}\sin\frac{\pi}{4}}{\left|\hat{a}\cos\frac{\pi}{4}+\hat{b}\sin\frac{\pi}{4}\right|}=\frac{\hat{a}+\hat{b}}{|\hat{a}+\hat{b}|}.$$

13. $$P=\left(6+\sqrt{2}\cos\frac{\pi}{4},\,5+\sqrt{2}\cos\frac{\pi}{4}\right)=(7,6)$$

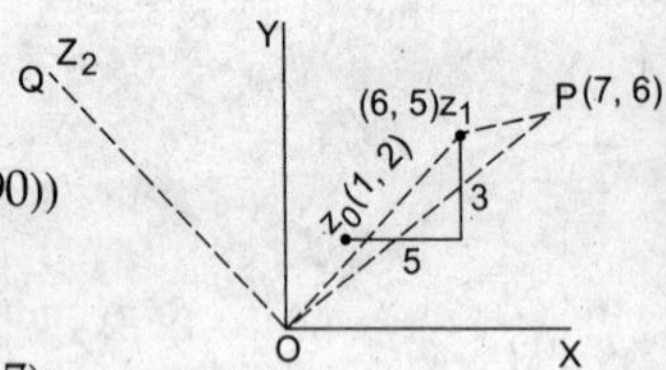

$$\therefore\quad Q=(OP\cos(\alpha+90^\circ),\,OP\sin(\alpha+90))$$

$$=(\sqrt{85}\cdot(-\sin\alpha),\,\sqrt{85}\cos\alpha)$$

$$=\left(-\sqrt{85}\cdot\frac{6}{\sqrt{85}},\,\sqrt{85}\cdot\frac{7}{\sqrt{85}}\right)=(-6,7)$$

$$\therefore\quad z_2=-6+7i.$$

14. $R=(\cos(\beta-\alpha)\cos\theta-\sin(\beta-\alpha)\sin\theta,\ \sin\beta\cdot\cos\theta-\cos\beta\cdot\sin\theta)$

Let $P=(x_1,y_1)$ and $Q=(x_2,y_2)$.

Then $R=(x_2\cos\theta+x_1\sin\theta,\ y_2\cos\theta+y_1\sin\theta)$

Clearly, $\left(\dfrac{x_2\cos\theta+x_1\sin\theta}{\cos\theta+\sin\theta},\dfrac{y_2\cos\theta+y_1\sin\theta}{\cos\theta+\sin\theta}\right)$ lies on the line segment PQ.

So, (c) can be true, if $\cos\theta+\sin\theta=1$ or $1+\sin 2\theta=1$ or $\sin 2\theta=0$, which is impossible as $0<\theta<\dfrac{\pi}{4}$.

Hence, P, Q and R are noncollinear.

Alternatively, $$2\text{ar}(\Delta PQR)=\begin{vmatrix}-\sin(\beta-\alpha) & -\cos\beta & 1\\ \cos(\beta-\alpha) & \sin\beta & 1\\ \cos(\beta-\alpha+\theta) & \sin(\beta-\theta) & 1\end{vmatrix}$$

$$= \begin{vmatrix} -\sin(\beta-\alpha) & -\cos\beta & 1 \\ \cos(\beta-\alpha) & \sin\beta & 1 \\ 0 & 0 & 1-\sin\theta-\cos\theta \end{vmatrix}$$

$$= (1-\sin\theta-\cos\theta)\cos(2\beta-\alpha) = 0$$

only if $\cos\theta + \sin\theta = 1$ $(\because \ 2\beta - \alpha < 90°)$

But $\cos\theta + \sin\theta \neq 1$ (as above). So, $ar(\Delta PQR) \neq 0$.

15. Here, $(ax^2 + by^2 + c)(x - 2y)(x - 3y) = 0$

$\Rightarrow$ it represents two straight lines and the locus $ax^2 + by^2 + c = 0$, when $c = 0$ and a, b have the same sign $ax^2 + by^2 + c = 0$ does not represent two straight lines. If $a = b$ and c is of opposite sign to that of a, $ax^2 + by^2 + c = 0 \Rightarrow x^2 + y^2 = \frac{-c}{a} = k^2 \left(\because \ \frac{-c}{a} > 0\right)$ which is a circle.

16. $f(x) = f(1-x) \Rightarrow f'(x) = -f'(1-x)$ ... (1)

Putting $x = \frac{1}{2}$, $f'\left(\frac{1}{2}\right) = -f'\left(\frac{1}{2}\right) \Rightarrow f'\left(\frac{1}{2}\right) = 0$

Putting $\frac{1}{2} + x$ for x, $f\left(\frac{1}{2}+x\right) = f\left\{1 - \left(\frac{1}{2}+x\right)\right\}$ or $f\left(\frac{1}{2}+x\right) = f\left(\frac{1}{2}-x\right)$

$\therefore$ $f\left(\frac{1}{2}+x\right)$ is an even function. But $\sin x$ is an odd function.

$\therefore$ $f\left(\frac{1}{2}+x\right)\cdot\sin x$ is an odd function; $\therefore \displaystyle\int_{-1/2}^{1/2} f\left(\frac{1}{2}+x\right)\sin x\,dx = 0.$

Putting $t = 1 - z$, $\displaystyle\int_0^{1/2} f(t)\,e^{\sin \pi t}\,dt = \int_1^{1/2} f(1-z)\,e^{\sin \pi(1-z)}\,(-dz)$

$$= \int_{1/2}^{1} f(1-z)\,e^{\sin \pi z}\,dz = \int_{1/2}^{1} f(1-t)\,e^{\sin \pi t}\,dt.$$

Now, from (1), $f'\left(\frac{1}{4}\right) = -f'\left(\frac{3}{4}\right)$. But $f'\left(\frac{1}{4}\right) = 0$.

$\therefore$ $f'\left(\frac{1}{4}\right) = 0 = f'\left(\frac{3}{4}\right)$. Also $f'\left(\frac{1}{2}\right) = 0$ and $\frac{1}{4} < \frac{1}{2} < \frac{3}{4}$

$\therefore$ by Rolle's theorem, $f''(x) = 0$ when $\frac{1}{4} < x < \frac{1}{2}$

0 when $\frac{1}{2} < x < \frac{3}{4}$

17. Clearly, $f(x)=\dfrac{1}{1+x+x^2}$, [0, 1] is a decreasing function because

$$f'(x)=\frac{-(1+2x)}{(1+x+x^2)^2}<0.$$

$\therefore \quad S_n < \lim\limits_{n\to\infty} S_n$ and $T_n > \lim\limits_{n\to\infty} T_n$ and

$$\lim_{n\to\infty} S_n = \lim_{n\to\infty} T_n = \int_0^1 \frac{dx}{1+x+x^2},$$

by definition of definite integration.

Now, $\displaystyle\int_0^1 \frac{dx}{1+x+x^2} = \int_0^1 \frac{dx}{\left(x+\frac{1}{2}\right)^2+\left(\frac{\sqrt{3}}{2}\right)^2}$

$$=\frac{2}{\sqrt{3}}\tan^{-1}\frac{x+\frac{1}{2}}{\frac{\sqrt{3}}{2}}\Bigg]_0^1=\frac{2}{\sqrt{3}}\left(\tan^{-1}\sqrt{3}-\tan^{-1}\frac{1}{\sqrt{3}}\right)=\frac{\pi}{3\sqrt{3}}.$$

$\therefore \quad S_n < \dfrac{\pi}{3\sqrt{3}}, T_n > \dfrac{\pi}{3\sqrt{3}}.$

18. Taking AM, GM of $\dfrac{1}{PS}$ and $\dfrac{1}{ST}$ and using $A > G$, we get

$$\frac{\frac{1}{PS}+\frac{1}{ST}}{2} > \sqrt{\frac{1}{PS}\cdot\frac{1}{ST}}.$$

But $PS \cdot ST = QS \cdot SR$, from geometry.

$\therefore \quad \dfrac{1}{PS}+\dfrac{1}{ST} > 2\cdot\dfrac{1}{\sqrt{QS\cdot SR}}.$

Similarly, $\dfrac{QS+SR}{2} > \sqrt{QS\cdot SR}$ or $QR > 2\sqrt{QS\cdot SR}$

$\therefore \quad \dfrac{1}{\sqrt{QS\cdot SR}} > \dfrac{2}{QR}$

$\therefore \quad \dfrac{1}{PS}+\dfrac{1}{ST} > 2\cdot\dfrac{1}{\sqrt{QS\cdot SR}} > 2\cdot\dfrac{2}{QR}.$

19. Here $\dfrac{x^2}{2^2}+\dfrac{y^2}{1^2}=1$. So, $b^2=a^2(1-e^2) \Rightarrow 1=4(1-e^2) \Rightarrow e=\dfrac{\sqrt{3}}{2}$

$\therefore \quad ae = 2\cdot\dfrac{\sqrt{3}}{2}=\sqrt{3}, \dfrac{b^2}{a}=\dfrac{1}{2}$

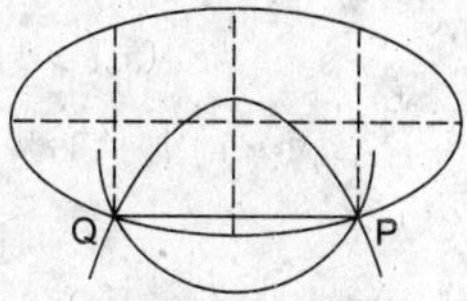

$\therefore$ P, Q are $\left(\sqrt{3}, -\frac{1}{2}\right), \left(-\sqrt{3}, -\frac{1}{2}\right)$

$\therefore$ focus of the parabola $= \left(0, -\frac{1}{2}\right)$.

Vertex is on $x = 0$ at a distance $\frac{\sqrt{3}}{2}$ (= one-fourth of latus rectum = $2\sqrt{3}$.)

$\therefore$ two possible vertices $= \left(0, -\frac{\sqrt{3}}{2} - \frac{1}{2}\right), \left(0, \frac{\sqrt{3}}{2} - \frac{1}{2}\right)$

$$= \left(0, -\frac{\sqrt{3}+1}{2}\right), \left(0, \frac{\sqrt{3}-1}{2}\right)$$

$\therefore$ parabolas are $x^2 = 2\sqrt{3}\left\{y - \left(-\frac{\sqrt{3}+1}{2}\right)\right\}$ or $x^2 - 2\sqrt{3}y = 3 + \sqrt{3}$

$$x^2 = -2\sqrt{3}\left\{y - \frac{\sqrt{3}-1}{2}\right\} \text{ or } x^2 + 2\sqrt{3}y = 3 - \sqrt{3}$$

Hints for question numbers 20 to 22

20. $\text{Im}(z) \geq 1 \Rightarrow y \geq 1$; $|z - 2 - i| = 3 \Rightarrow (x-2)^2 + (y-1)^2 = 3^2$
$\text{Re}((1-i)z) = \sqrt{2} \Rightarrow x + y = \sqrt{2}$.

$\therefore$ the set $A \cap B \cap C$ has only one point namely the point P.

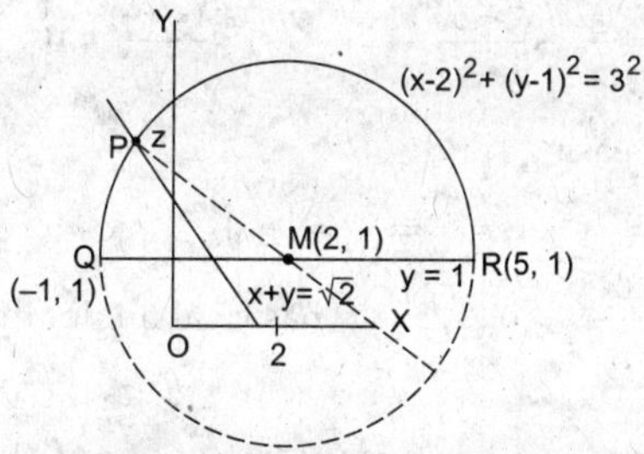

21. Clearly, from the diagram $z = -1 + i$ and $z = 5 + i$ are the ends of a diameter of the circle $(x-2)^2 + (y-1)^2 = 3^2$.

$\therefore$ $PQ = |z - (-1 + i)|$, $PR = |z - (5 + i)|$

$\therefore$ $PQ^2 + PR^2 = QR^2$

$\Rightarrow$ $|z + 1 - i|^2 + |z - 5 - i|^2 = QR^2 = (5+1)^2 + (1-1)^2 = 36$.

22. Any point w satisfying $|w - 2 - i| < 3$ is a point in the circle with centre at M and radius 3, i.e., the circle in the figure.

$\therefore$ $|z - w| < 6$

$$|z-w| > ||z|-|w||$$

$$\therefore \quad ||z|-|w|| < 6 \Rightarrow -6 < |z|-|w| < 6$$

$$\Rightarrow -6+3 < |z|-|w|+3 < 6+3$$

$$\Rightarrow -3 < |z|-|w|+3 < 9$$

Hint for question numbers 23 to 25

23. $f(x) = 1 - \dfrac{2ax}{x^2+ax+1}$

$$\Rightarrow \quad f'(x) = -\frac{2a(x^2+ax+1) - 2ax(2x+a)}{(x^2+ax+1)^2} = \frac{2a(x^2-1)}{(x^2+ax+1)}$$

$$\therefore \quad f''(x) = \frac{4ax(x^2+ax+1)^2 - 2a(x^2-1)\cdot 2(x^2+ax+1)\cdot(2x+a)}{(x^2+ax+1)^4}$$

$$= \frac{4ax(x^2+ax+1) - 4a(x^2-1)(2x+a)}{(x^2+ax+1)^3}$$

$$\therefore \quad f''(1) = \frac{4a(2+a)}{(2+a)^3} = \frac{4a}{(2+a)^2};\ f''(-1) = \frac{-4a(2-a)}{(2-a)^3} = \frac{-4a}{(2-a)^2}$$

$$\therefore \quad (2+a)^2\cdot f''(1) + (2-a)^2\cdot f''(-1) = 4a - 4a = 0.$$

24. In $(-1, 1), f'(x) = \dfrac{2a(x^2-1)}{(x^2+ax+1)^2} \Rightarrow f'(x) = \dfrac{(+)(-)}{(+)} < 0$

$\therefore \quad f(x)$ is decreasing

$f'(x) = 0 \Rightarrow x^2 - 1 = 0 \Rightarrow x = 1, -1$

$\therefore \quad f''(1) = \dfrac{4a}{(2+a)^2} > 0 \quad \therefore\ f(x)$ has local minimum at $x = 1$.

25. Here, $g'(x) = \dfrac{dg(x)}{d(e^x)}\cdot\dfrac{d(e^x)}{dx} = \dfrac{f'(e^x)}{1+e^{2x}}\cdot e^x$

$$= \frac{e^x}{1+e^{2x}}\cdot\frac{2a(e^{2x}-1)}{(e^{2x}+ae^x+1)^2} > 0, \text{ when } x \in (0, \infty)$$

$$< 0, \text{ when } x \in (-\infty, 0)$$

Hint for question numbers 26 to 28

26. The equation of DI (perpendicular to $\sqrt{3}x + y = 6$):

$$y - \frac{3}{2} = \frac{1}{\sqrt{3}}\left(x - \frac{3\sqrt{3}}{2}\right)$$

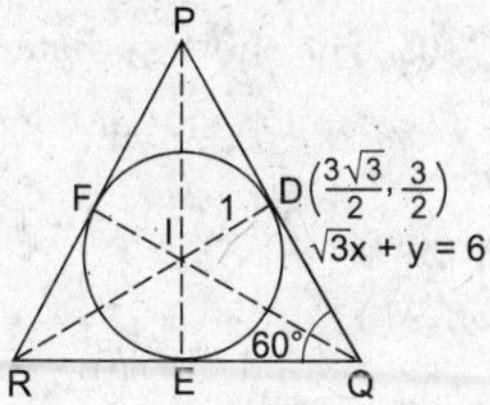

or $\quad y - \dfrac{3}{2} = \tan 30° \cdot \left(x - \dfrac{3\sqrt{3}}{2}\right)$

$$\therefore \quad \frac{x - \frac{3\sqrt{3}}{2}}{\cos 30°} = \frac{y - \frac{3}{2}}{\sin 30°} = \pm 1 (= r)$$

$$\therefore \quad I = \left(\frac{3\sqrt{3}}{2} \pm \cos 30°, \ \frac{3}{2} \pm \sin 30°\right)$$

$$= \left(\frac{3\sqrt{3}}{2} \pm \frac{\sqrt{3}}{2}, \ \frac{3}{2} \pm \frac{1}{2}\right)$$

$$= (2\sqrt{3}, 2) \text{ or } (\sqrt{3}, 1).$$

The origin and I are on the same side of PQ. So $I = (\sqrt{3}, 1)$.

$\therefore$ the equation of the circle C is $(x - \sqrt{3})^2 + (y - 1)^2 = 1$.

27. PQ makes 120° with the x-axis ($\because\ m = -\sqrt{3}$). So, RQ is parallel to the x-axis

$\therefore$ IE is parallel to y-axis passing through $(\sqrt{3}, 1)$.

$\therefore$ the equation of EI(and so EP) is $x = \sqrt{3}$

Solving $x = \sqrt{3}$ with $\sqrt{3}x + y = 6$, we get $P = (\sqrt{3}, 3)$.

I divides PE in the ratio 2 : 1. So, $E = (\sqrt{3}, 0)$

$\therefore$ QR has the equation $y = 0$. Solving with $\sqrt{3}x + y = 6$, $Q = (2\sqrt{3}, 0)$.

I divides FQ in the ratio 1 : 2. So, $F = \left(\dfrac{\sqrt{3}}{2}, \dfrac{3}{2}\right)$.

28. Slope of PR is tan 60° ($\because\ QR \parallel x$-axis, its equation being $y = 0$) and it passes through F.

$\therefore$ the equation of PR is $y - \dfrac{3}{2} = \tan 60°. \left(x - \dfrac{\sqrt{3}}{2}\right)$ or $y = \sqrt{3}x$.

Hint for question numbers 29 to 31

29. Direction ratios are 3, 1, 2 and 1, 2, 3.

$\therefore$ vectors along the lines $= 3\hat{i} + \hat{j} + 2\hat{k}$ and $\hat{i} - 2\hat{j} + 3\hat{k}$

$\therefore$ vector perpendicular to both L_1 and L_2

$= (3\hat{i} + \hat{j} + 2\hat{k}) \times (\hat{i} + 2\hat{j} + 3\hat{k}) = -\hat{i} - 7\hat{j} + 5\hat{k}$

$\therefore$ unit vector perpendicular to both L_1 and L_2

$$= \frac{-\hat{i} - 7\hat{j} + 5\hat{k}}{\sqrt{(-1)^2 + (-7)^2 + 5^2}} = \frac{1}{5\sqrt{3}}(-\hat{i} - 7\hat{j} + 5\hat{k}).$$

30. The shortest distance between the lines $= \overrightarrow{PQ}$.

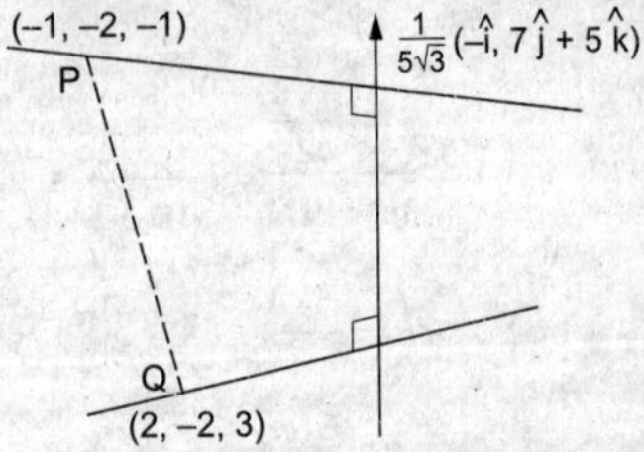

Unit vector along s.d.

$$= \{(2+1)\vec{i} + (3+1)\vec{k}\}.\ \frac{1}{5\sqrt{3}}(-\hat{i} - 7\hat{j} + 5\hat{k})$$

$$= \frac{1}{5\sqrt{3}}(-3+20)$$

$$= \frac{17}{5\sqrt{3}};$$

31. The equation of the plane is

$$-1(x+1) - 7(y+2) + 5(z+1) = 0$$

or $\quad x + 7y - 5z + 10 = 0.$

$\therefore \quad$ distance $= \left|\dfrac{1 + 7\cdot 1 - 5\cdot 1 + 10}{\sqrt{1^2 + 7^2 + (-5)^2}}\right| = \dfrac{13}{\sqrt{75}}.$

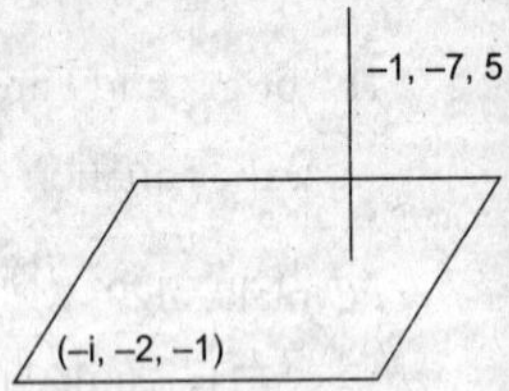

Hint for question numbers 32 to 34

32. Differentiating $y^3 - 3y + x = 0$ w.r.t. x,

$$3y^2\frac{dy}{dx} - 3\frac{dy}{dx} + 1 = 0 \quad \text{or} \quad (3y^2 - 3)\frac{dy}{dx} + 1 = 0 \quad \ldots (1)$$

When $x = -10\sqrt{2},\ y = f(-10\sqrt{2}) = 2\sqrt{2}$ and $\frac{dy}{dx} = f'(-10\sqrt{2})$.

$$\therefore \quad \{3(2\sqrt{2})^2 - 3\}f'(-10\sqrt{2}) + 1 = 0 \quad \therefore \quad f'(-10\sqrt{2}) = \frac{-1}{21}$$

Again, differentiating (1) w.r.t. x,

$$6y\left(\frac{dy}{dx}\right)^2 + (3y^2 - 3)\frac{d^2y}{dx^2} = 0$$

When

$$x = -10\sqrt{2},\ y = f(-10\sqrt{2}) = 2\sqrt{2},\ \frac{dy}{dx} = f'(-10\sqrt{2}) = \frac{-1}{21},\ \frac{d^2y}{dx^2} = f''(-10\sqrt{2}).$$

$$\therefore \quad 6\cdot 2\sqrt{2}\cdot\left(\frac{-1}{21}\right)^2 + 3((2\sqrt{2})^2 -)f''(-10\sqrt{2}) = 0.$$

or $\quad \dfrac{12\sqrt{2}}{21^2} + 21\cdot f''(-10\sqrt{2}) = 0 \quad \therefore \quad f''(-10\sqrt{2}) = -\dfrac{12\sqrt{2}}{(21)^3} = \dfrac{-4\sqrt{2}}{7^3\cdot 3^2}.$

33. Area $= \displaystyle\int_a^b f(x)dx = [f(x)\cdot x]_a^b - \int_a^b x\cdot f'(x)dx$

$$= bf(b) - af(a) - \int_a^b x \cdot \frac{-1}{3y^2 - 3}\,dx, \text{ by (1) above}$$

$$= \int_a^b \frac{x}{3\{(f(x))^2 - 1\}}\,dx + b\,f(b) - a\,f(a).$$

34. $g'(x) = \frac{dy}{dx} = \frac{-1}{3\{(f(x))^2 - 1\}}$; $\therefore$ $g'(x)$ is an even function.

$$\therefore \quad \int_{-1}^{1} g'(x)dx = 2\int_0^1 g'(x)dx = 2[g(x)]_0^1 = 2g(1) - 2g(0) = 2g(1).$$

35. Here, $b_1 = a_1,\ b_2 = a_1 + a_2 = a_1(1 + r),\ b_3 = a_1(1 + r) + a_1r^2 = a_1(1 + r + r^2)$

$b_4 = a_1(1 + r + r^2) + ar^3 = a_1(1 + r + r^2 + r^3),$

r being the common ratio of the GP.

Clearly, b_1, b_2, b_3, b_4 are neither in AP nor in GP nor in HP.

$\therefore$ STATEMENT-1 is true but STATEMENT-2 is false.

36. Solving the equations $x = 0$ (adding the first two)

$\therefore$ $-y + z = 1$ and $-3y + 3z = 2 \Rightarrow -y + z = 1 = \frac{2}{3}$ (absurd)

$\therefore$ STATEMENT-2 is true.

Direction ratios of L_1 are given by 0, –4, –4

L_2 are given by 0, –2, –2

L_3 are given by 0, 2, 2

$\therefore$ each of L_1, L_2, L_3 has the direction ratios 0, 1, 1.

So, $L_1 \parallel L_2 \parallel L_3$. $\therefore$ STATEMENT-1 is false.

37. $$\Delta = \begin{vmatrix} 1 & -2 & 3 \\ -1 & 1 & -2 \\ 1 & -3 & 4 \end{vmatrix} = \begin{vmatrix} 1 & -2 & 3 \\ 0 & -1 & 1 \\ 0 & -1 & 1 \end{vmatrix} = 0, \ (\because R_2 \equiv R_3)$$

$$\Delta_x = \begin{vmatrix} -2 & 3 & 1 \\ 1 & -2 & -k \\ -3 & 4 & -1 \end{vmatrix} = \begin{vmatrix} 0 & -1 & 1-2k \\ 1 & -2 & -k \\ 0 & -2 & -1-3k \end{vmatrix} = -(1 + 3k + 2 - 4k) = k - 3$$

$$\Delta_y = \begin{vmatrix} 1 & 3 & 1 \\ -1 & -2 & -k \\ 1 & 4 & -1 \end{vmatrix} = k - 3,\ \Delta_z = \begin{vmatrix} 1 & -2 & 1 \\ -1 & 1 & -k \\ 1 & -3 & -1 \end{vmatrix} = 3 - k$$

$\therefore$ if $k \neq 3$ then $\Delta = 0, \Delta_x \neq 0, \Delta_y \neq 0, \Delta_z \neq 0$

$\therefore$ the system has no solution if $\begin{vmatrix} 1 & 3 & -1 \\ -1 & -2 & k \\ 1 & 4 & 1 \end{vmatrix} = -\Delta_y = 3 - k \neq 0$, i.e., $k \neq 3$

$\therefore$ STATEMENT-1 is true and STATEMENT-2 is true;

Also STATEMENT-2 is a correct explanation for STATEMENT-1.

38. For unique solution $\Delta = \begin{vmatrix} a & b \\ c & d \end{vmatrix} \neq 0$

$\therefore$ $ad - bc \neq 0$. As $a, b, c, d \in \{0, 1\}$, we have

$ad = 1, bc = 0$ or $ad = 0, bc = 1$

$\therefore$ $(a, d, b, c) = (1, 1, 0, 0), (1, 1, 0, 1), (1, 1, 1, 0),$

$(0, 0, 1, 1), (0, 1, 1, 1), (1, 0, 1, 1)$

$\therefore$ $n(E) = 6, n(S) = 2 \times 2 \times 2 \times 2 = 16.$ $\therefore$ $P(E) = \frac{6}{16} = \frac{3}{8}$.

If $\Delta \neq 0$, the system has unique solution.

If $\Delta = 0$ then $ad = bc$ or $\frac{a}{c} = \frac{b}{d}$, i.e., only one equation in x, y.

So, there will be infinite solutions.

$\therefore$ the system has a unique solution or infinite solution.

$\therefore$ existence of solution is certain, i.e., probability = 1.

$\therefore$ STATEMENT-1 is true, STATEMENT-2 is true; but STATEMENT-2 is Not a correct explanation for STATEMENT-1

39. $\lim_{x \to 0} \{g(x) \cdot \cot x - g(0) \cdot \operatorname{cosec} x\} = \lim_{x \to 0} \frac{g(x) \cdot \cos x - g(0)}{\sin x}, \left(\frac{0}{0}\right)$

$$= \lim_{x \to 0} \frac{g'(x)\cos x - g(x)\sin x}{\cos x} = g'(0) = 0.$$

$f(x) = g(x) \cdot \sin x \Rightarrow f'(x) = g'(x) \sin x + g(x)\cos x, [f'(0) = g(0)]$

$\Rightarrow f''(x) = g''(x) \cdot \sin x + 2g'(x) \cos x - g(x) \cdot \sin x$

$\therefore f''(0) = 2g'(0) = 0$

$\therefore$ STATEMENT-1 is true, STATEMENT-2 is true; but STATEMENT-2 is NOT an explanation for STATEMENT-1.

40. For the circle C, centre = (–3, 5) and radius = $\sqrt{(-3)^2 + 5^2 - 30} = 2$.

L_1 is a chord of the circle C $\Rightarrow$ the length of the perpendicular for (–3, 5) to the line L_1 < radius

$$\Rightarrow \quad \left|\frac{2(-3)+3\cdot 5+p-3}{\sqrt{2^2+3^2}}\right| < 2 \text{ or } \left|\frac{6+p}{\sqrt{13}}\right| < 2 \text{ or } |6+p| < 2\cdot\sqrt{13}$$

or $\quad -2\sqrt{13} < 6+p < 2\sqrt{13}$ or $-6-2\sqrt{13} < p < 2\sqrt{13}-6$ ... (i)

L_2 is a diameter of the circle C $\Rightarrow$ (–3, 5) is on $2x+3y+p+3=0$

$\Rightarrow 2(-3)+3\cdot 5+p+3=0 \Rightarrow p=-12$ ... (ii)

Clearly (i) does not necessarily mean (ii). So, STATEMENT-1 is true.

L_1 is a diameter of the circle C $\Rightarrow$ (–3, 5) is on $2x+3y+p-3=0$

$\Rightarrow 2(-3)+3.5+p-3=0 \Rightarrow p=-6$ ... (iii)

L_2 is a chord of the circle C $\Rightarrow$ the length of the perpendicular for (–3, 5) to the line L_2 < radius

$$\Rightarrow \quad \left|\frac{2(-3)+3.5+p+3}{\sqrt{2^2+3^2}}\right| < 2 \text{ or } \left|\frac{12+p}{\sqrt{13}}\right| < 2 \text{ or } |12+p| < 2\sqrt{13}$$

or $\quad -2\sqrt{13} < 12+p < 2\sqrt{13}$ or $-2\sqrt{3}-12 < p < 2\sqrt{13}-12$ ... (iv)

$\therefore$ When (iii) is satisfied (iv) holds. So, STATEMENT-2 is false.

41. α, β are both real or both imaginary.

If α, β are both imaginary, $\beta=\overline{\alpha}$ and $\frac{1}{\beta}=\overline{\alpha}$ ($\because$ imaginary roots occur in conjugate pair)

$\therefore \quad \beta=\frac{1}{\beta} \quad \therefore \quad \beta^2=1.$ Given, $\beta^2 \neq 1.$

So, α and β are both real. So, both equations have real roots.

$\therefore \quad 4p^2-4q \geq 0$ and $4b^2-4ac \geq 0 \quad \Rightarrow \quad (p^2-4q)(b^2-ac) \geq 0.$

$\therefore$ STATEMENT-1 is true.

Now, $\alpha+\beta=-2p, \; \alpha\beta=q$

$$\alpha+\frac{1}{\beta}=\frac{-2b}{a}, \; \alpha\cdot\frac{1}{\beta}=\frac{c}{a}$$

$$\therefore \quad \frac{\alpha\beta}{\frac{\alpha}{\beta}}=\frac{q}{\frac{c}{a}} \text{ or } \beta^2=\frac{qa}{c}\neq 1 \text{ (given, } \beta^2\neq 1)$$

Also, $\beta^2 \neq 1 \Rightarrow \beta \neq \frac{1}{\beta} \Rightarrow \alpha + \beta \neq \alpha + \frac{1}{\beta}$

$$\Rightarrow -2p \neq \frac{-2b}{a} \Rightarrow b \neq pa.$$

$\therefore$ STATEMENT-2 is true. But STATEMENT-2 is not a correct explanation for STATEMENT-1.

42. Here $\frac{dy}{y\sqrt{y^2-1}} - \frac{dx}{x\sqrt{x^2-1}} = 0$ or $\sec^{-1}y - \sec^{-1}x = k$ ... (i)

$\therefore$ $y(2) = \frac{2}{\sqrt{3}} \Rightarrow \sec^{-1}\frac{2}{\sqrt{3}} - \sec^{-1}2 = k \Rightarrow \frac{\pi}{6} - \frac{\pi}{3} = k$

$\therefore$ (1) gives, $\sec^{-1}y = \sec^{-1}x - \frac{\pi}{6}$ or $y = \sec\left(\sec^{-1}x - \frac{\pi}{6}\right)$

$\therefore$ STATEMENT-1 is true.

$\therefore$ $\frac{1}{y} = \cos\left(\sec^{-1}x - \frac{\pi}{6}\right) = \cos(\sec^{-1}x)\cdot\cos\frac{\pi}{6} + \sin(\sec^{-1}x)\cdot\sin\frac{\pi}{6}$

$$= \cos\left(\cos^{-1}\frac{1}{x}\right)\cdot\frac{\sqrt{3}}{2} + \sqrt{1-\cos^2\left(\cos^{-1}\frac{1}{x}\right)}\cdot\frac{1}{2}$$

$\therefore$ $\frac{1}{y} = \frac{\sqrt{3}}{2x} + \frac{1}{2}\cdot\sqrt{1-\frac{1}{x^2}}$ $\therefore$ STATEMENT-2 is false.

43. L_1, L_2, L_3 are concurrent if $\begin{vmatrix} 1 & 3 & -5 \\ 3 & -k & -1 \\ 5 & 2 & -12 \end{vmatrix} = 0 \Rightarrow k = \frac{65}{13} = 5.$ $\therefore$ A – s

$L_1 \parallel L_2$ if $\frac{1}{3} = \frac{3}{-k} \Rightarrow k = -9.$

$L_2 \parallel L_3$ if $\frac{3}{5} = \frac{-k}{2} \Rightarrow k = \frac{-6}{5}$ $\therefore$ B – p, B – q

L_1, L_2, L_3 form a triangle if *A*, *B* are not true. $\therefore$ C – r

L_1, L_2, L_3 do not form a triangle if *A* or *B* are true. $\therefore$ D – p, D – q, D – s.

44. (ENDEA) NOEL Number of arrangements $= {}^5P_5 = 5!$ $\therefore$ A – p

E × × × × × × × E
NDANOEL Number of arrangements $= \frac{7!}{2!} = 21\ 5!$ $\therefore$ B – s

E, E, A, O, L
× × × × × × × × ×
D, L, N, N Number of arrangements $= \frac{4!}{2!} \times \frac{5!}{3!} = 2 \times 5!$ $\therefore$ C – q

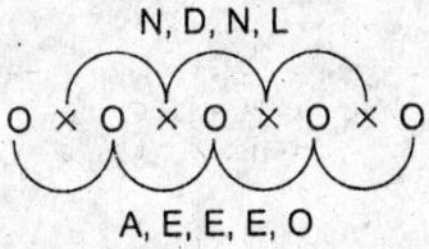

Number of arrangement $= \frac{5!}{3!} \times \frac{4!}{2!} = 2 \times 5!$ $\therefore$ D – q

45. $\frac{x^2 + 2x + 4}{x+2} = y$ or $x^2 + (2-y)x + 4 - 2y = 0$

As x is real, $D \geq 0 \Rightarrow (2-y)^2 - 4(4-2y) \geq 0$

$$\Rightarrow y^2 + 4y - 12 \geq 0 \Rightarrow (y+6)(y-2) \geq 0$$

$\therefore$ by sign-scheme, $y \leq -6$ or $y \geq 2$. But y is +ve. So, A – r

$(A+B)(A-B) = (A-B)(A+B) \Rightarrow A^2 - AB + BA - B^2 = A^2 + AB - BA - B^2$

$$\Rightarrow 2BA = 2AB \Rightarrow AB = BA$$

As A is symmetric and B is skew symmetric, $(AB)^T = -AB$

But $(AB)^T = (-1)^k AB$, given. So, $(-1)^k AB = -AB$ $\therefore$ $(-1)^k = -1 \Rightarrow k = 1, 3$

$\therefore$ B – q, B – s

$$a = \log_3 \log_3 2 \Rightarrow 3^a = \log_3 2 \Rightarrow 3^{-a} = \frac{1}{\log_3 2} = \log_2 3.$$

$$2^0 < 2^{-k+3^{-a}} < 2^1 \Rightarrow 0 < -k + 3^{-a} < 1 \Rightarrow k < 3^{-a} < 1 + k$$

$$\Rightarrow k < \log_2 3 < 1 + k \Rightarrow k = 1 \text{ when } k \text{ is integer}$$

$\therefore$ $k < 2, k < 3.$ $\therefore$ C – r, C – s

$$\sin\theta = \cos\varphi = \sin\left(\frac{\pi}{2} - \varphi\right)$$

$$\Rightarrow \theta = \frac{\pi}{2} - \varphi, \pi - \left(\frac{\pi}{2} - \varphi\right), 2\pi + \left(\frac{\pi}{2} - \varphi\right), 3\pi - \left(\frac{\pi}{2} - \varphi\right), 4\pi + \left(\frac{\pi}{2} - \varphi\right), \ldots$$

$$\Rightarrow \theta \pm \varphi - \frac{\pi}{2} = 0, 2\pi, 4\pi, \ldots$$

$$\Rightarrow \frac{1}{\pi}\left(\theta \pm \varphi - \frac{\pi}{2}\right) = 0, 2, 4, \ldots$$

$\therefore$ D – p, D – r.

□

IIT Questions—6

SECTION I

Single Correct Choice Type

Each question in this section has 4 choices (a), (b), (c) and (d) for its answer, out of which **ONLY ONE** is correct.

1. If the sum of the first n terms of an AP is cn^2 then the sum of squares of these n terms is

(a) $\dfrac{n(4n^2-1)c^2}{6}$ (b) $\dfrac{n(4n^2+1)c^2}{3}$

(c) $\dfrac{n(4n^2-1)c^2}{3}$ (d) $\dfrac{n(4n^2+1)c^2}{6}$

2. Let $z = x + iy$ be a complex number where x and y are integers. Then the area of the rectangle whose vertices are the roots of the equation $z\bar{z}^3 + \bar{z}z^3 = 350$ is

(a) 48 (b) 32

(c) 40 (d) 80

3. Let $z = \cos\theta + i\sin\theta$. Then the value of $\sum_{m=1}^{15} \text{Im}(z^{2m-1})$ at $\theta = 2°$ is

(a) $\dfrac{1}{\sin 2°}$ (b) $\dfrac{1}{3\sin 2°}$

(c) $\dfrac{1}{2\sin 2°}$ (d) $\dfrac{1}{4\sin 2°}$

4. The number of seven digit integers, with sum of the digits equal to 10 and formed by using the digits 1, 2 and 3 only, is

(a) 55 (b) 66

(c) 77 (d) 88

5. The locus of orthocentre of the triangle formed by the lines $(1+p)x - py + p(1+p) = 0$, $(1+q)x - qy + q(1+q) = 0$ and $y = 0$ where $p \neq q$, is

(a) a hyperbola (b) a parabola
(c) an ellipse (d) a straight line

6. Tangents drawn from the point $P(1, 8)$ to the circle $x^2 + y^2 - 6x - 4y - 11 = 0$ touch the circle at the points A and B. The equation of the circumcircle of the $\triangle PAB$ is

(a) $x^2 + y^2 + 4x - 6y + 19 = 0$ (b) $x^2 + y^2 - 4x - 10y + 19 = 0$
(c) $x^2 + y^2 - 2x + 6y - 29 = 0$ (d) $x^2 + y^2 - 6x - 4y + 19 = 0$

7. The line passing through the extremity A of the major axis and extremity B of the minor axis of the ellipse $x^2 + 9y^2 = 9$ meets its auxiliary circle at the point M. Then the area of the triangle with vertices at A, M and the origin O is

(a) $\frac{31}{10}$ (b) $\frac{29}{10}$ (c) $\frac{21}{10}$ (d) $\frac{27}{10}$

8. The normal at a point P on the ellipse $x^2 + 4y^2 = 16$ meets the x-axis at Q. If M is the mid-point of the line segment PQ then the locus of M intersects the latus rectum of the given ellipse at the points

(a) $\left(\pm\frac{3\sqrt{5}}{2}, \pm\frac{2}{7}\right)$ (b) $\left(\pm\frac{3\sqrt{5}}{2}, \pm\frac{\sqrt{19}}{4}\right)$
(c) $\left(\pm 2\sqrt{3}, \pm\frac{1}{7}\right)$ (d) $\left(\pm 2\sqrt{3}, \pm\frac{4\sqrt{3}}{7}\right)$

9. If $\vec{a}, \vec{b}, \vec{c}$ and $\vec{d}$ are unit vectors such that $(\vec{a} \times \vec{b}) \cdot (\vec{c} \times \vec{d}) = 1$ and $\vec{a} \cdot \vec{c} = \frac{1}{2}$ then

(a) $\vec{a}, \vec{b}, \vec{c}$ are non-coplanar
(b) $\vec{b}, \vec{c}, \vec{d}$ are non-coplanar
(c) $\vec{b}, \vec{d}$ are non-parallel
(d) $\vec{a}, \vec{d}$ are parallel and $\vec{b}, \vec{c}$ are parallel

10. Let $P(3, 2, 6)$ be a point in space and Q be a point on the line $\vec{r} = (\hat{i} - \hat{j} + 2\hat{k}) + \mu(-3\hat{i} + \hat{j} + 5\hat{k})$. Then the value of μ for which the vector $\overrightarrow{PQ}$ is parallel to the plane $x - 4y + 3z = 1$ is

(a) $\frac{1}{4}$ (b) $-\frac{1}{4}$
(c) $\frac{1}{8}$ (d) $-\frac{1}{8}$

11. A line with positive direction cosines passes through the point $P(2, -1, 2)$ and makes equal angles with the coordinate axes. The line meets the plane $2x + y + z = 9$ at point Q. The length of the line segment PQ equals

(a) 1 (b) $\sqrt{2}$ (c) $\sqrt{3}$ (d) 2

12. Let f be a non-negative function defined on the interval $[0, 1]$. If $\int_0^x \sqrt{1 - \{f'(t)\}^2}\, dt = \int_0^x f(t)\, dt$, $0 \le x \le 1$ and $f(0) = 0$ then

(a) $f\left(\frac{1}{2}\right) < \frac{1}{2}$ and $f\left(\frac{1}{3}\right) > \frac{1}{3}$ (b) $f\left(\frac{1}{2}\right) > \frac{1}{2}$ and $f\left(\frac{1}{3}\right) > \frac{1}{3}$

(c) $f\left(\frac{1}{2}\right) < \frac{1}{2}$ and $f\left(\frac{1}{3}\right) < \frac{1}{3}$ (d) $f\left(\frac{1}{2}\right) > \frac{1}{2}$ and $f\left(\frac{1}{3}\right) < \frac{1}{3}$

SECTION II

Multiple Correct Choice Type

Each question in this section has four choices (a), (b), (c) and (d) for its answer, out of which **ONE OR MORE** is/are correct.

13. If $\frac{\sin^4 x}{2} + \frac{\cos^4 x}{3} = \frac{1}{5}$, then

(a) $\tan^2 x = \frac{3}{2}$ (b) $\frac{\sin^8 x}{8} + \frac{\cos^8 x}{27} = \frac{1}{125}$

(c) $\tan^2 x = \frac{2}{3}$ (d) $\frac{\sin^8 x}{8} + \frac{\cos^8 x}{27} = \frac{2}{125}$

14. For $0 < \theta < \frac{\pi}{2}$, the solution(s) of

$$\sum_{m=1}^{6} \operatorname{cosec}\left(\theta + \frac{m-1}{4}\pi\right) \cdot \operatorname{cosec}\left(\theta + \frac{m}{4}\pi\right) = 4\sqrt{2} \text{ is/are}$$

(a) $\frac{\pi}{4}$ (b) $\frac{\pi}{6}$

(c) $\frac{\pi}{12}$ (d) $\frac{5\pi}{12}$

15. In a triangle ABC with fixed base BC, the vertex A moves such that $\cos B + \cos C = 4\sin^2 \frac{A}{2}$. If a, b and c denote the lengths of the sides of the triangle opposite to the angles A, B and C respectively then

(a) $b + c = 4a$

(b) $b + c = 2a$

(c) locus of A is an ellipse

(d) locus of A is a pair of straight lines

16. The tangent PT and the normal PN to the parabola $y^2 = 4ax$ at a point P on it meet its axis at points T and N respectively. The locus of the centroid of the $\triangle PTN$ is a parabola whose

(a) vertex $= \left(\frac{2a}{3}, 0\right)$

(b) directrix is $x = 0$

(c) latus rectum $= \frac{2a}{3}$

(d) focus $= (a, 0)$

17. An ellipse intersects the hyperbola $2x^2 - 2y^2 = 1$ orthogonally. The eccentricity of the ellipse is reciprocal of that of the hyperbola. If the axes of the ellipse are along the coordinate axes, then the

(a) equation of the ellipse is $x^2 + 2y^2 = 2$

(b) foci of the ellipse are $(\pm 1, 0)$

(c) equation of the ellipse is $x^2 + 2y^2 = 4$

(d) foci of the ellipse are $(\pm\sqrt{2}, 0)$

18. For the function $f(x) = x \cos \frac{1}{x}$, $x \geq 1$

(a) for at least one $x \in [1, \infty)$, $f(x+2) - f(x) < 2$

(b) $\lim_{x \to \infty} f'(x) = 1$

(c) for all $x \in [1, \infty)$, $f(x+2) - f(x) > 2$

(d) $f'(x)$ is strictly decreasing in $[1, \infty)$

19. If $I_n = \int_{-\pi}^{\pi} \frac{\sin nx}{(1+\pi^x) \cdot \sin x} dx$, $n = 0, 1, 2, \ldots$, then

(a) $I_n = I_{n+2}$

(b) $\sum_{m=1}^{10} I_{2m+1} = 10\pi$

(c) $\sum_{m=1}^{10} I_{2m} = 0$

(d) $I_n = I_{n+1}$

20. The area of the region bounded by the curve $y = e^x$ and lines $x = 0$ and $y = e$ is

(a) $e - 1$

(b) $\int_1^e \log_e (e + 1 - y)\, dy$

(c) $e - \int_0^1 e^x\, dx$

(d) $\int_1^e \log_e y\, dy$

21. Let $L = \lim_{x \to 0} \dfrac{a - \sqrt{a^2 - x^2} - \dfrac{x^2}{4}}{x^4}$, $a > 0$. If L is finite, then

(a) $a = 2$

(b) $a = 1$

(c) $L = \dfrac{1}{64}$

(d) $L = \dfrac{1}{32}$

SECTION III

Comprehension Type

In this section, each group has 3 multiple choice questions based on a paragraph. Each question has 4 choices (a), (b), (c) and (d) for its answer, out of which **ONLY ONE** is correct.

Paragraph for Q. N. 22 to 24

Let $\mathscr{A}$ be the set of all 3×3 symmetric matrices all of whose elements are either 0 or 1. Five of the elements are 1 and four are 0.

22. The number of matrices in $\mathscr{A}$ is

(a) 12 (b) 6 (c) 9 (d) 3

23. The number of matrices A in $\mathscr{A}$ for which the system of linear equations $A \begin{bmatrix} x \\ y \\ z \end{bmatrix} = \begin{bmatrix} 1 \\ 0 \\ 0 \end{bmatrix}$ has a unique solution, is

(a) less than 4

(b) at least 4 but less than 7

(c) at least 7 but less than 10

(d) at least 10

24. The number of matrices A in $\mathscr{A}$ for which the system of linear equations $A \begin{bmatrix} x \\ y \\ z \end{bmatrix} = \begin{bmatrix} 1 \\ 0 \\ 0 \end{bmatrix}$ is inconsistent, is

(a) 0

(b) more than 2

(c) 2

(d) 1

Paragraph for Q. N. 25 to 27

A fair die is tossed repeatedly until a six is obtained. Let X denote the number of tosses required.

25. The probability that $X = 3$ equals

(a) $\frac{25}{216}$ (b) $\frac{25}{36}$

(c) $\frac{5}{36}$ (d) $\frac{125}{216}$

26. The probability that $X \geq 3$ equals

(a) $\frac{125}{216}$ (b) $\frac{25}{216}$

(c) $\frac{5}{36}$ (d) $\frac{25}{36}$

27. The conditional probability that $X \geq 6$ given $X > 3$ equals

(a) $\frac{125}{216}$ (b) $\frac{25}{216}$

(c) $\frac{5}{36}$ (d) $\frac{25}{36}$

SECTION IV

Matrix-Match Type

In this section each question contains statements given in two columns, which are to be matched. The statements in **Column I** are labelled (A), (B), (C) and (D), while the statements in **Column II** are labelled (p), (q), (r), (s) and (t). Any given statement in **Column I** can have correct matching with **ONE OR MORE** statement(s) in **Column II.** The appropriate bubbles corresponding to the answers to these questions have to be darkened as illustrated in the following example.

If the correct matches are A—p, s and t; B—q and r; C—p and q; D—s then the correct darkening of bubbles will look like the following.

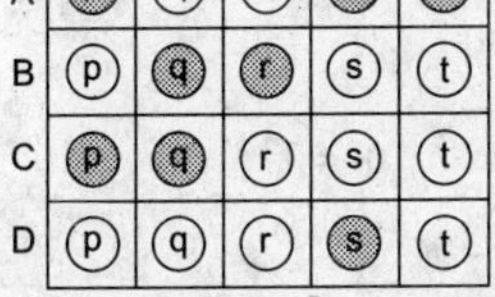

28. Match the statements/experessions given in **Column I** with the values given in **Column II.**

Column I	Column II
(A) Root(s) of the equation $2\sin^2\theta + \sin^2 2\theta = 2$	(p) $\pi/6$
(B) Points of discontinuity of $f(x) = \left[\frac{6x}{\pi}\right]\cos\left[\frac{3x}{\pi}\right]$, where $[y]$ = greatest integer less than or equal to y	(q) $\pi/4$
	(r) $\pi/3$
(C) The volume of the parallelopiped with edges represented by the vectors $\vec{i}+\vec{j}$, $\vec{i}+2\vec{j}$ and $\vec{i}+\vec{j}+\pi\vec{k}$	(s) $\pi/2$
(D) The angle between vectors $\vec{a}$ and $\vec{b}$ where $\vec{a}, \vec{b}$ and $\vec{c}$ are unit vectors satisfying $\vec{a}+\vec{b}+\sqrt{3}\,\vec{c}=\vec{0}$	(t) π

29. Match the statements/expressions given in **Column I** with the values given in **Column II.**

Column I	Column II
(A) The number of solutions of the equation $x\,e^{\sin x} - \cos x = 0$ in the interval $\left(0, \frac{\pi}{2}\right)$	(p) 1
	(q) 2
(B) The value(s) of k for which the planes $kx + 4y + z = 0$, $4x + ky + 2z = 0$ and $2x + 2y + z = 0$ intersect in a straight line	(r) 3
(C) The value(s) of k for which $\lvert x-1\rvert + \lvert x-2\rvert + \lvert x+1\rvert + \lvert x+2\rvert = 4k$ has integral solution(s)	(s) 4
(D) If $y' = y + 1$ and $y(0) = 1$ then value(s) of $y(\log_e 2)$	(t) 5

30. Match the statements/expressions in **Column I** with the open intervals in **Column II.**

Column I	Column II
(A) The interval contained in the domain of definition of non-zero solutions of the differential equation $(x-3)^2 \cdot y' + y = 0$	(p) $\left(-\frac{\pi}{2}, \frac{\pi}{2}\right)$
(B) The interval containing the value of the integral $\int_1^5 (x-1)(x-2)(x-3)(x-4)(x-5)\,dx$	(q) $\left(0, \frac{\pi}{2}\right)$
	(r) $\left(\frac{\pi}{8}, \frac{5\pi}{4}\right)$
(C) The interval in which at least one of the points of local maximum of $\cos^2 x + \sin x$ lies	(s) $\left(0, \frac{\pi}{8}\right)$
(D) The interval in which $\tan^{-1}(\sin x + \cos x)$ is increasing	(t) $(-\pi, \pi)$

31. Match the conic sections in **Column I** with the statements/expressions in **Column II.**

Column I	Column II
(A) Circle	(p) The locus of the point (h, k) for which the line $hx + ky = 1$ touches the circle $x^2 + y^2 = 4$
	(q) Points z in the complex plane satisfying $\lvert z+2 \rvert - \lvert z-2 \rvert = \pm 3$
(B) Parabola	(r) Points of the conic have parametric representation $x = \sqrt{3}\left(\frac{1-t^2}{1+t^2}\right), \; y = \frac{2t}{1+t^2}$
(C) Ellipse	(s) The eccentricity of the conic lies in the interval $1 \le x < \infty$
(D) Hyperbola	(t) Points z in the complex plane satisfying $\text{Re}\,(z+1)^2 = \lvert z \rvert^2 + 1$

SECTION V

Integer Answer Type

In this section the answer to each of the questions is a single-digit integer, ranging from 0 to 9. If the correct answers to question numbers X, Y, Z and W (say) are 6, 0, 9 and 2 respectively, then the correct darkening of bubbles will look like the following:

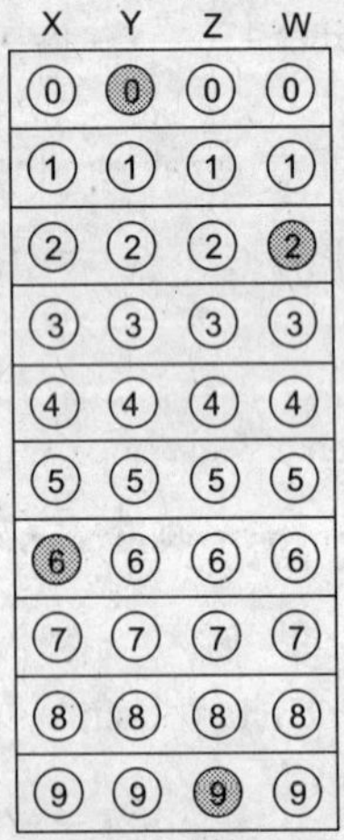

32. The centres of two circles C_1 and C_2 each of unit radius are at a distance of 6 units from each other. Let P be the mid-point of the line segment joining the centres of C_1 and C_2, and C be a circle touching circles C_1 and C_2 externally. If a common tangent to C_1 and C passing through P is also a common tangent to C_2 and C then the radius of the circle C is

33. Let ABC and ABC' be two non-congruent triangles with sides $AB = 4$, $AC = AC' = 2\sqrt{2}$ and angle $B = 30°$. The absolute value of the difference between the areas of these triangles is

34. Let (x, y, z) be points with integral coordinates satisfying the system of homogeneous equations

$$3x - y - z = 0, \quad -3x + z = 0, \quad -3x + 2y + z = 0.$$

Then the number of such points for which $x^2 + y^2 + z^2 \leq 100$ is

35. The smallest value of k for which both the roots of the equation $x^2 - 8kx + 16(k^2 - k + 1) = 0$ are real, distinct and have values at least 4, is

36. The maximum value of the function $f(x) = 2x^3 - 15x^2 + 36x - 48$ on the set $A = \{x \mid x^2 + 20 \leq 9x\}$ is

37. If the function $f(x) = x^3 + e^{x/2}$ and $g(x) = f^{-1}(x)$ then the value of $g'(1)$ is

38. Let $p(x)$ be a polynomial of degree 4 having extremum at $x = 1, 2$ and $\lim\limits_{x \to 0}\left(1 + \dfrac{p(x)}{x^2}\right) = 2$. Then the value of $p(2)$ is

39. Let $f : R \to R$ be a continuous function which satisfies $f(x) = \int_0^x f(t)\, dt$. Then the value of $f(\log_e 5)$ is

ANSWERS

1. c	**2.** a	**3.** d	**4.** c	**5.** d
6. b	**7.** d	**8.** c	**9.** c	**10.** a
11. c	**12.** c	**13.** b, c	**14.** c, d	**15.** b, c
16. a, d	**17.** a, b	**18.** b, c, d	**19.** a, b, c	**20.** b, c, d
21. a, c	**22.** a	**23.** b	**24.** b	**25.** a
26. b	**27.** d			

28.

	p	q	r	s	t
A	p	**q**	r	**s**	t
B	**p**	q	**r**	**s**	**t**
C	p	q	r	s	**t**
D	p	q	**r**	s	t

29.

	p	q	r	s	t
A	**p**	q	r	s	t
B	p	**q**	r	s	t
C	p	**q**	**r**	**s**	**t**
D	p	q	**r**	s	t

30.

	p	q	r	s	t
A	**p**	**q**	r	**s**	t
B	**p**	q	r	s	**t**
C	**p**	**q**	**r**	s	**t**
D	p	q	r	**s**	t

31.

	p	q	r	s	t
A	**p**	q	r	s	t
B	p	q	r	**s**	**t**
C	p	q	**r**	s	t
D	p	**q**	r	**s**	t

32	33	34	35	36	37	38	39
0	0	0	0	0	0	**0**	**0**
1	1	1	1	1	1	1	1
2	2	2	**2**	2	**2**	2	2
3	3	3	3	3	3	3	3
4	**4**	4	4	4	4	4	4
5	5	5	5	5	5	5	5
6	6	6	6	6	6	6	6
7	7	**7**	7	**7**	7	7	7
8	8	8	8	8	8	8	8
9	9	9	9	9	9	9	9

Hints and Solutions

1. We know, $1 + 3 + 5 + \ldots$ upto n terms $= n^2$

$\therefore \quad c + 3c + 5c + \ldots$ upto n terms $= cn^2$

$\therefore \quad c^2 + (3c)^2 + (5c)^2 + \ldots$ upto n terms $= c^2\,\{1^2 + 3^2 + 5^2 + \ldots$ to n terms$\}$

$$= c^2 \sum_{n=1}^{n} (2n-1)^2 = c^2 \sum_{n=1}^{n} (4n^2 - 4n + 1) = c^2 \left\{4 \sum_{n=1}^{n} n^2 - 4 \sum_{n=1}^{n} n + \sum_{n=1}^{n} 1\right\}$$

$$= c^2 \left\{4 \cdot \frac{n\,(n+1)\,(2n+1)}{6} - 4 \cdot \frac{n\,(n+1)}{2} + n\right\}$$

$$= \frac{nc^2}{3}\,\{2(n+1)\,(2n+1) - 6(n+1) + 3\} = \frac{nc^2}{3}\,\{4n^2 - 1\}.$$

2. Here $z = x + iy$ and $z\,\bar{z}^3 + \bar{z}\,z^3 = z\,\bar{z}\,(\bar{z}^2 + z^2)$

$\therefore \quad (x^2 + y^2)\,\{(x + iy)^2 + (x - iy)^2\} = 350$

$\Rightarrow \quad (x^2 + y^2) \cdot 2(x^2 - y^2) = 350$

$\Rightarrow \quad (x^2 + y^2)\,(x^2 - y^2) = 175 = 35 \times 5$ or 25×7 or 175×1

$\therefore \quad x^2 + y^2 = 35$ or $x^2 + y^2 = 25$ or $x^2 + y^2 = 175$ x and y being integers

$x^2 - y^2 = 5 \qquad x^2 - y^2 = 7 \qquad x^2 - y^2 = 1$

$\Rightarrow \quad x^2 = 20 \qquad \Rightarrow x^2 = 16 \qquad \Rightarrow x^2 = 88$

$y^2 = 15 \qquad y^2 = 9 \qquad y^2 = 87$

$\Rightarrow \quad x = \sqrt{20} \qquad \Rightarrow x = 4, -4 \qquad \Rightarrow x = \sqrt{88}$

$y = \sqrt{15}$ (absurd) $\qquad y = 3, -3 \qquad y = \sqrt{87}$ (abused)

$\therefore$ vertices of the rectangle are

(4, 3), (4, –3), (–4, 3), (–4, –3)

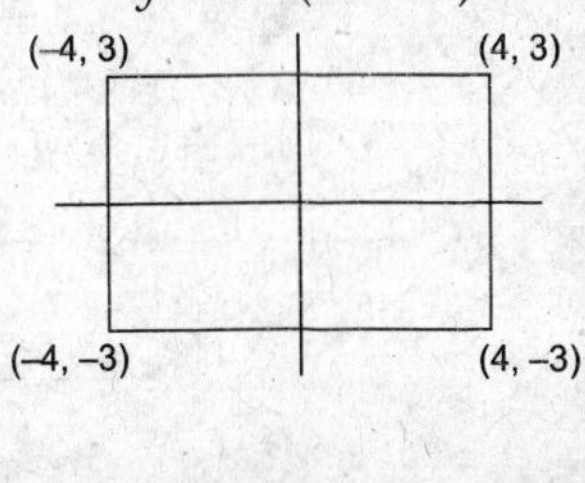

$\therefore$ area of the reactangle $= 8 \times 6$

$= 48.$

3. $\sum_{m=1}^{15} \text{Im}(z^{2m-1}) = \sum_{m=1}^{15} \text{Im}\{(\cos\theta + i\sin\theta)^{2m-1}\}$

$$= \sum_{m=1}^{15} \text{Im}\{\cos(2m-1)\theta + i\sin(2m-1)\theta\}$$

$$= \sum_{m=1}^{15} \sin(2m-1)\theta = \sin\theta + \sin 3\theta + \sin 5\theta + \ldots \text{ to 15 terms}$$

$$= \frac{\sin\frac{15\cdot 2\theta}{2}}{\sin\frac{2\theta}{2}} \cdot \sin\frac{\theta + 29\theta}{2}$$

$$= \frac{\sin 15\theta}{\sin\theta} \cdot \sin 15\theta = \frac{\sin^2 30°}{\sin 2°}, \text{ when } \theta = 2°$$

$$= \frac{\left(\frac{1}{2}\right)^2}{\sin 2°} = \frac{1}{4\sin 2°}.$$

4. Numbers must have 1, 1, 1, 1, 1, 2, 3 or 1, 1, 1, 1, 2, 2, 2

$\therefore$ the required number of numbers $= \frac{\lfloor 7}{\lfloor 5} + \frac{\lfloor 7}{\lfloor 4 \, \lfloor 3} = 42 + 35 = 77.$

5. Solving $(1+p)x - py + p(1+p) = 0$, $y = 0$; we get a vertex $= (-p, 0)$

The altitude through the vertex $(-p, 0)$ is

$$qx + (1+q)y = q(-p) + (1+q)\cdot 0$$

or $$qx + (1+q)y = -pq \quad \ldots (1)$$

Similarly, the altitude through the vertex $(-q, 0)$ is

$$px + (1+p)y = -pq \quad \ldots (2)$$

Eliminating p, q from (1) and (2), we get the locus of orthocentre

(1) – (2) $\Rightarrow (q-p)x + (q-p)y = 0$ or $x + y = 0$, which is a straight line.

6. The equation of AB (the chord of contact of tangents from the point P) is

$$x \cdot 1 + y \cdot 8 - 3(x+1) - 2(y+8) - 11 = 0$$

or $-2x + 6y - 30 = 0$ or $x - 3y + 15 = 0$

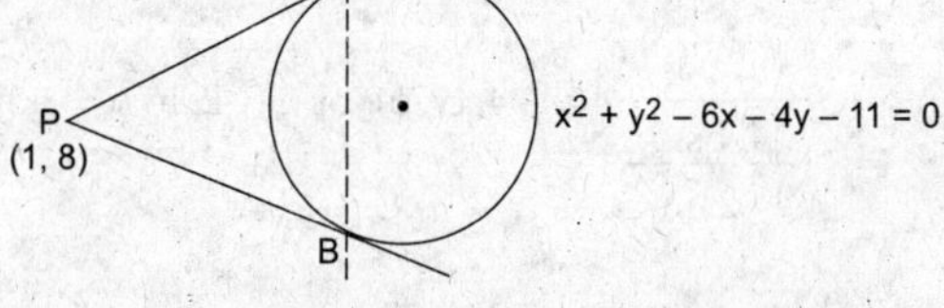

$\therefore$ the circumcircle of $\triangle PAB$ is

$x^2 + y^2 - 6x - 4y - 11 + \lambda(x - 3y + 15) = 0$, which passes through $(1, 8)$

$\therefore \quad 1 + 64 - 6 - 32 - 11 + \lambda(1 - 24 + 15) = 0$

or $\quad 16 - 8\lambda = 0; \quad \therefore \lambda = 2.$

$\therefore$ the required equation is

$x^2 + y^2 - 6x - 4y - 11 + 2(x - 3y + 15) = 0$

or $\quad x^2 + y^2 - 4x - 10y + 19 = 0.$

7. Here $\dfrac{x^2}{3^2} + \dfrac{y^2}{1^2} = 1; \quad \therefore A = (-3, 0), B = (0, 1).$

The equation of AB is $\dfrac{x}{-3} + \dfrac{y}{1} = 1$ or $x - 3y + 3 = 0.$

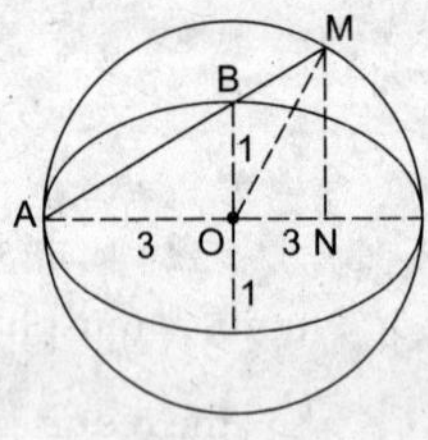

Auxiliary circle is $x^2 + y^2 = 9.$

Solving, $(3y - 3)^2 + y^2 = 9$

or $\quad 10y^2 - 18y = 0; \quad \therefore y = 0, \dfrac{9}{5}.$

When $y = \dfrac{9}{5}$, $x - 3 \times \dfrac{9}{5} + 3 = 0$ or $x = \dfrac{27}{5} - 3 = \dfrac{12}{5}.$

$\therefore \quad M = \left(\dfrac{12}{5}, \dfrac{9}{5}\right).$

$$\therefore \quad \text{ar}(\triangle OAM) = \frac{1}{2}\left|x_1(y_2 - y_3) + x_2(y_3 - y_1) + x_3(y_1 - y_2)\right|$$

$$= \frac{1}{2}\left|-3\left(\frac{9}{5} - 0\right) + \frac{12}{5}(0 - 0) + 0\left(0 - \frac{9}{5}\right)\right| = \frac{27}{10}.$$

Alternatively, ar $(OAM) = \dfrac{1}{2} AO \times MN = \dfrac{1}{2} \times 3 \times \dfrac{9}{5} = \dfrac{27}{10}.$

8. The ellipse is $\dfrac{x^2}{4^2} + \dfrac{y^2}{2^2} = 1.$ Let $P = (4\cos\phi, 2\sin\phi)$

Now, $\dfrac{2x}{16} + \dfrac{2y}{4}\dfrac{dy}{dx} = 0$ or $\dfrac{dy}{dx} = -\dfrac{x}{4y}$

$$\therefore \quad \text{the normal at } P \text{ is} \quad y - 2\sin\phi = \frac{-1}{\left(\dfrac{dy}{dx}\right)_P} \cdot (x - 4\cos\phi)$$

$$\text{or } y - 2\sin\phi = \frac{-1}{-\dfrac{4\cos\phi}{4 \cdot 2\sin\phi}} \cdot (x - 4\cos\phi) \text{ or } y - 2\sin\phi = 2\tan\phi \cdot x - 8\sin\phi$$

or $\quad y = 2\tan\phi \cdot x - 6\sin\phi$; its interesection with x-axis is $\left(\dfrac{3\sin\phi}{\tan\phi}, 0\right)$

i.e., $\quad Q = (3\cos\phi, 0)$.

$\therefore \quad M = \left(\dfrac{4\cos\phi + 3\cos\phi}{2}, \dfrac{2\sin\phi + 0}{2}\right) = \left(\dfrac{7\cos\phi}{2}, \sin\phi\right)$

$\therefore$ the locus of $M : x = \dfrac{7\cos\phi}{2}, y = \sin\phi$

$$\Rightarrow \left(\frac{2}{7}x\right)^2 + y^2 = \cos^2\phi + \sin^2\phi = 1.$$

$\therefore$ the equation of the locus of M is $\dfrac{x^2}{49/4} + y^2 = 1$.

The latus rectum of the given ellipse :

$$x = \pm ae = \pm 4 \cdot \sqrt{\frac{3}{4}} = \pm 2\sqrt{3}, \qquad [\because 2^2 = 4^2(1 - e^2)]$$

Required points of intersection are found by solving

$$\frac{x^2}{49/4} + y^2 = 1 \text{ and } x = \pm 2\sqrt{3}.$$

Solving, $y^2 = 1 - \dfrac{48}{49} = \dfrac{1}{49}$; $\quad \therefore\ y = \pm\dfrac{1}{7}\cdot$

$\therefore$ required points are $\left(\pm 2\sqrt{3}, \pm\dfrac{1}{7}\right)$.

9. $\left(|\vec{a}|\,|\vec{b}|\sin\theta_1\,\vec{n_1}\right)\cdot\left(|\vec{c}|\,|\vec{d}|\sin\theta_2\,\vec{n_2}\right) = 1$

$\therefore \quad \sin\theta_1 \sin\theta_2\, \vec{n_1}\cdot\vec{n_2} = 1$ or $\sin\theta_1\sin\theta_2\cos\theta = 1$

$\Rightarrow \quad \sin\theta_1 = 1,\ \sin\theta_2 = 1, \cos\theta = 1 \quad \Rightarrow \vec{a}\perp\vec{b},\ \vec{c}\perp\vec{d},\ \vec{n_1} \parallel \vec{n_2}$

$\vec{n_1} \parallel \vec{n_2} \Rightarrow$ plane of $\vec{a}, \vec{b}$ is parallel to the plane of $\vec{c}, \vec{d}$

i.e., $\vec{a}, \vec{b}, \vec{c}, \vec{d}$ arc coplanar vectors (being free vectors)

such that $\vec{a}\perp\vec{b},\ \vec{c}\perp\vec{d},\ \vec{a}$ and $\vec{c}$ are inclined at 60° $\quad \left(\because \vec{a}\cdot\vec{c} = \dfrac{1}{2}\right)$

$\therefore$

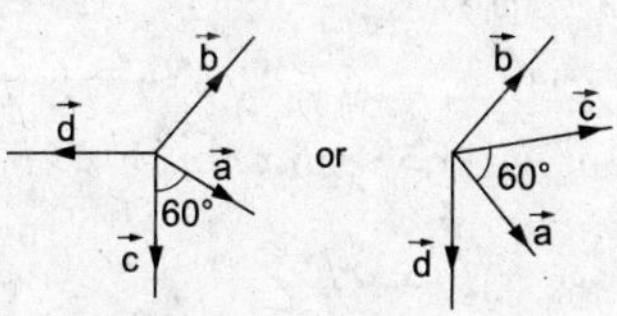

In both the cases $\vec{b}, \vec{d}$ are not parallel.

10. Q is a point on the line $\dfrac{x-1}{-3}=\dfrac{y+1}{1}=\dfrac{z-2}{5}=\mu$

$\therefore \quad Q=(1-3\mu, -1+\mu, 2+5\mu)$

$\therefore \quad \overrightarrow{PQ}=(1-3\mu-3, -1+\mu-2, 2+5\mu-6)$

i.e., $(-2-3\mu)\vec{i}+(-3+\mu)\vec{j}+(-4+5\mu)\vec{k}.$

Perpendicular to the plane $x-4y+3z=1$ has the d.r. $=1, -4, 3$.

$\therefore$ from the question, $1\cdot(-2-3\mu)+(-4)(-3+\mu)+3(-4+5\mu)=0$

or $-2+8\mu=0; \quad \therefore \mu=\dfrac{1}{4}\cdot$

11. As the line makes equal angles with the axis, its d.c. are $\dfrac{1}{\sqrt{3}}, \dfrac{1}{\sqrt{3}}, \dfrac{1}{\sqrt{3}}\cdot$

The equation of the line is $\dfrac{x-2}{1/\sqrt{3}}=\dfrac{y+1}{1/\sqrt{3}}=\dfrac{z-2}{1/\sqrt{3}}=r.$

If $PQ=r, \quad Q=\left(2+\dfrac{r}{\sqrt{3}}, -1+\dfrac{r}{\sqrt{3}}, 2+\dfrac{r}{\sqrt{3}}\right).$

It is on $2x+y+z=9$.

$\therefore \quad 2\left(2+\dfrac{r}{\sqrt{3}}\right)+\left(-1+\dfrac{r}{\sqrt{3}}\right)+\left(2+\dfrac{r}{\sqrt{3}}\right)=9$

or $\dfrac{4r}{\sqrt{3}}+5=9; \quad \therefore r=\sqrt{3}.$

12. $\displaystyle\int_0^x \left\{\sqrt{1-\{f'(t)\}^2}-f(t)\right\}dt=0 \quad$ for all $x\in[0,1]$

$\therefore \quad \sqrt{1-\{f'(x)\}^2}-f(x)=0, \quad x\in[0,1]$

or $1-\{f'(x)\}^2=\{f(x)\}^2$ or $\dfrac{dy}{dx}=\pm\sqrt{1-y^2}, \quad$ where $y=f(x)$

or $\dfrac{dy}{\sqrt{1-y^2}}=\pm\,dx$

or $\sin^{-1}y=\pm x+c.$ When $x=0$, $f(0)=y=0$

$\therefore \quad \sin^{-1}0=0+c; \ \therefore \sin^{-1}y=\pm x; \ \therefore y=\sin x$, [$\because f(x)=y$ is non-negative]

$\therefore \quad f(x)=\sin x.$ But we know $\sin x<x$ for $x>0$.

$\therefore \quad f(x)<x$ for $x>0$.

So, $f\left(\dfrac{1}{2}\right)<\dfrac{1}{2}$ and $f\left(\dfrac{1}{3}\right)<\dfrac{1}{3}\cdot$

13. $15\tan^4 x + 10 = 6\sec^4 x$ or $15\tan^4 x + 10 = 6(1 + \tan^2 x)^2$

or $9\tan^4 x - 12\tan^2 x + 4 = 0$ or $(3\tan^2 x - 2)^2 = 0$; $\therefore \tan^2 x = \frac{2}{3}\cdot$

$$\therefore \frac{\sin^2 x}{2} = \frac{\cos^2 x}{3} = \frac{\sin^2 x + \cos^2 x}{2+3} = \frac{1}{5}; \quad \therefore \sin^2 x = \frac{2}{5}, \cos^2 x = \frac{3}{5}$$

$$\therefore \frac{\sin^8 x}{8} + \frac{\cos^8 x}{27} = \frac{\left(\frac{2}{5}\right)^4}{8} + \frac{\left(\frac{3}{5}\right)^4}{27} = \frac{2}{625} + \frac{3}{625} = \frac{5}{625} = \frac{1}{125}\cdot$$

14. $\operatorname{cosec}\left(\theta + \frac{m-1}{4}\pi\right)\operatorname{cosec}\left(\theta + \frac{m}{4}\pi\right) = \frac{1}{\sin\left(\theta + \frac{m-1}{4}\pi\right)\sin\left(\theta + \frac{m}{4}\pi\right)}$

$$= \frac{1}{\sin\frac{\pi}{4}} \cdot \frac{\sin\left\{\left(\theta + \frac{m}{4}\pi\right) - \left(\theta + \frac{m-1}{4}\pi\right)\right\}}{\sin\left(\theta + \frac{m-1}{4}\pi\right)\sin\left(\theta + \frac{m}{4}\pi\right)}$$

$$= \sqrt{2}\cdot\left\{\cot\left(\theta + \frac{m-1}{4}\pi\right) - \cot\left(\theta + \frac{m}{4}\pi\right)\right\}$$

$\therefore$ the equation is $\sqrt{2}\left\{\cot\theta - \cot\left(\theta + \frac{6}{4}\pi\right)\right\} = 4\sqrt{2}$

or $\cot\theta - \cot\left(\theta + \frac{3\pi}{2}\right) = 4$ or $\cot\theta + \tan\theta = 4$ or $\frac{\cos^2\theta + \sin^2\theta}{\sin\theta\cos\theta} = 4$

or $\sin 2\theta = \frac{1}{2}\cdot$ $\therefore 2\theta = \frac{\pi}{6}, \frac{5\pi}{6}$; $\therefore \theta = \frac{\pi}{12}, \frac{5\pi}{12}\cdot$

15. $\cos B + \cos C = 4\sin^2\frac{A}{2}$ or $2\cos\frac{B+C}{2}\cos\frac{B-C}{2} = 2\cdot 2\sin^2\frac{A}{2}$

or $\cos\frac{B-C}{2} = 2\sin\frac{A}{2}$ $\left(\because \text{ in a } \Delta ABC, \sin\frac{A}{2} \neq 0\right)$

or $2\cos\frac{A}{2}\cdot\cos\frac{B-C}{2} = 4\sin\frac{A}{2}\cdot\cos\frac{A}{2}$

or $2\sin\frac{B+C}{2}\cdot\cos\frac{B-C}{2} = 2\sin A$

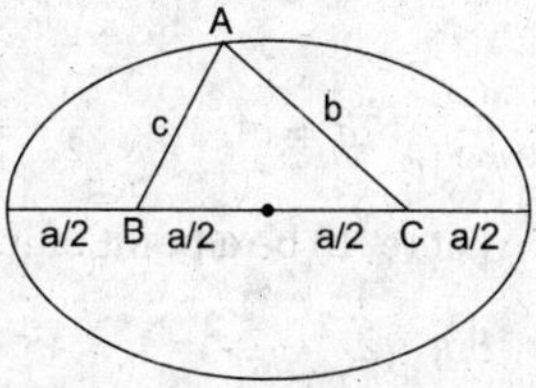

or $\sin B + \sin C = 2\sin A$

or $\frac{b}{2R} + \frac{c}{2R} = 2\cdot\frac{a}{2R}$

or $b + c = 2a$.

$\therefore$ A moves such that $AC + AB = 2a$ = constant, where B and C are fixed points. So, the locus of A is an ellipse.

16. The equation of PT is $ty = x + at^2$ and
the equation of PN is $y = -tx + at^3 + 2at$.
Solving $ty = x + at^2$ and $y = 0$, $T = (-at^2, 0)$.
Solving $y = -tx + at^3 + 2at$ and $y = 0$, $N = (at^2 + 2a, 0)$.

$\therefore$ the centroid of $\triangle PTN$ is $\left(\dfrac{at^2 - at^2 + at^2 + 2a}{3}, \dfrac{2at + 0 + 0}{3}\right)$,

i.e., $\left(\dfrac{at^2 + 2a}{3}, \dfrac{2at}{3}\right)$.

$\therefore$ the locus of the centroid of $\triangle PTN$ is obtained by eliminating t from

$$x = \frac{at^2 + 2a}{3},\ y = \frac{2at}{3}$$

$$\therefore \quad \frac{3x - 2a}{a} = t^2,\ \frac{3y}{2a} = t \quad \Rightarrow \quad \frac{3x - 2a}{a} = \left(\frac{3y}{2a}\right)^2$$

$\therefore$ $y^2 = \dfrac{4a^2}{9} \cdot \dfrac{3x - 2a}{a}$ or $y^2 = 4 \cdot \dfrac{a}{3} \cdot \left(x - \dfrac{2a}{3}\right)$ which is a parabola.

$\therefore$ vertex $= \left(\dfrac{2a}{3}, 0\right)$; directrix is $\left(x - \dfrac{2a}{3}\right) + \dfrac{a}{3} = 0$, i.e., $x = \dfrac{a}{3}$

latus rectum $= \dfrac{4a}{3}$; focus $= \left(\dfrac{a}{3} + \dfrac{2a}{3}, 0\right) = (a, 0)$.

17. Hyperbola is $\dfrac{x^2}{1/2} - \dfrac{y^2}{1/2} = 1$.

$\therefore$ eccentricity of the hyperbola is given by $\left(\dfrac{1}{2}\right)^2 = \left(\dfrac{1}{2}\right)^2 \cdot (e^2 - 1)$

or $e^2 - 1 = 1$; $\quad \therefore\ e = \sqrt{2}$.

Let the equation of the ellipse be (because axes are along coordinate axes)

$$\frac{x^2}{a^2} + \frac{y^2}{b^2} = 1, \quad \text{where } e = \frac{1}{\sqrt{2}}$$

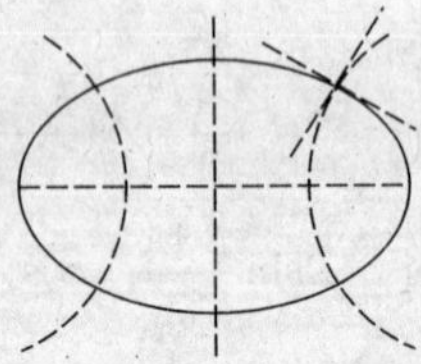

$\therefore$ $b^2 = a^2(1 - e^2)$ or $b^2 = a^2\left(1 - \dfrac{1}{2}\right)$

or $2b^2 = a^2$; $\quad \therefore\ \dfrac{b^2}{a^2} = \dfrac{1}{2}$.

Let (α, β) be a point of intersection.

For $2x^2 - 2y^2 = 1$, $\quad 4x - 4y\dfrac{dy}{dx} = 0$; $\quad \therefore\ \left(\dfrac{dy}{dx}\right)_{\alpha, \beta} = \left(\dfrac{x}{y}\right)_{\alpha, \beta} = \dfrac{\alpha}{\beta}$.

For $\dfrac{x^2}{a^2} + \dfrac{y^2}{b^2} = 1$, $\quad \dfrac{2x}{a^2} + \dfrac{2y}{b^2}\dfrac{dy}{dx} = 0$; $\quad \therefore\ \left(\dfrac{dy}{dx}\right)_{\alpha, \beta} = -\left(\dfrac{b^2x}{a^2y}\right)_{\alpha, \beta} = -\dfrac{b^2\alpha}{a^2\beta}$

They cut orthogonally. So, $\frac{\alpha}{\beta}\cdot\left(-\frac{b^2\alpha}{a^2\beta}\right)=-1$ or $\frac{\alpha^2}{\beta^2}\cdot\frac{1}{2}=1$; $\therefore\ \frac{\alpha^2}{\beta^2}=2.$

$(\alpha, \beta)\in$ hyperbola. So, $2\alpha^2-2\beta^2=1$ or $2\alpha^2-\alpha^2=1$; $\therefore\ \alpha^2=1$; $\therefore\ \beta^2=\frac{1}{2}\cdot$

$(\alpha, \beta)\in$ ellipse. So, $\frac{\alpha^2}{a^2}+\frac{\beta^2}{b^2}=1$; $\therefore\ \frac{1}{a^2}+\frac{1}{2b^2}=1$ or $\frac{1}{a^2}+\frac{1}{a^2}=1$

$\therefore\quad a^2=2;\quad \therefore\ b^2=\frac{a^2}{2}=\frac{2}{2}=1.$

$\therefore\quad$ ellipse is $\frac{x^2}{2}+\frac{y^2}{1}=1$ or $x^2+2y^2=2;$

foci $=(\pm ae, 0)=\left(\pm\sqrt{2}\cdot\frac{1}{\sqrt{2}}, 0\right)=(\pm 1, 0).$

18. In $x\ge 1$, $f'(x)=\cos\frac{1}{x}+x\cdot\left(-\sin\frac{1}{x}\right)\cdot\frac{-1}{x^2}=\cos\frac{1}{x}+\frac{1}{x}\sin\frac{1}{x}$

$\therefore\quad \lim_{x\to\infty} f'(x)=\lim_{x\to\infty}\left\{\cos\frac{1}{x}+\frac{1}{x}\sin\frac{1}{x}\right\}=1+0=1.$

In $x\ge 1$, $f''(x)=\frac{d}{dx}\left\{\cos\frac{1}{x}+\frac{1}{x}\sin\frac{1}{x}\right\}=\frac{1}{x^2}\sin\frac{1}{x}-\frac{1}{x^2}\sin\frac{1}{x}+\frac{1}{x}\cdot\cos\frac{1}{x}\cdot\frac{-1}{x^2}$

$=-\frac{1}{x^3}\cos\frac{1}{x}<0;$ $\therefore$ $f'(x)$ is strictly decreasing in $[1, \infty)$.

In $x\ge 1$, $f(x)$ is continuous and differentiable.

$\therefore\quad$ by Lagrange's theorem, $\frac{f(x+2)-f(x)}{(x+2)-x}=f'(\xi),\ x<\xi<x+2$

$\therefore\quad f(x+2)-f(x)=2f'(\xi).$

But $f'(x)$ is strictly decreasing and $f'(\xi)\to 1$, when $\xi\to\infty$

$\therefore\quad$ In $x\ge 1$, $f'(\xi)>1$. So, $f(x+2)-f(x)=2f'(\xi)>2\times 1=2.$

19. $I_n=\int_{-\pi}^{0}\frac{\sin nx}{(1+\pi^x)\sin x}\,dy+\int_0^{\pi}\frac{\sin nx}{(1+\pi^x)\sin x}\,dx$

$=\int_{\pi}^{0}\frac{\sin(-nx)}{(1+\pi^{-x})\sin(-x)}\,d(-x)+\int_0^{\pi}\frac{\sin nx}{(1+\pi^x)\sin x}\,dx$

$=\int_0^{\pi}\frac{\sin nx}{\sin x}\left\{\frac{1}{1+\pi^{-x}}+\frac{1}{1+\pi^x}\right\}dx=\int_0^{\pi}\frac{\sin nx}{\sin x}\left\{\frac{\pi^x}{\pi^x+1}+\frac{1}{1+\pi^x}\right\}dx$

$\therefore\quad I_n=\int_0^{\pi}\frac{\sin nx}{\sin x}\,dx.$

$$\therefore \quad I_{n+2} - I_n = \int_0^{\pi} \frac{\sin(n+2)x - \sin nx}{\sin x}\,dx = \int_0^{\pi} \frac{2\cos(n+1)x \cdot \sin x}{\sin x}\,dx$$

$$= 2\int_0^{\pi} \cos(n+1)x\,dx = 2 \cdot \left.\frac{\sin(n+1)x}{n+1}\right]_0^{\pi} = \frac{2}{n+1}\{\sin(n+1)\pi - 0\} = 0.$$

$$\therefore \quad I_n = I_{n+2}; \quad \therefore\ I_3 = I_1,\ I_5 = I_3 = I_1, \text{ etc., and } I_2 = I_0,\ I_4 = I_2 = I_0, \text{ etc.}$$

$$\therefore \quad \sum_{m=1}^{10} I_{2m+1} = I_3 + I_5 + \ldots + I_{21} = I_1 + I_1 + \ldots \text{ 10 times}$$

$$= 10I_1 = 10\int_0^{\pi} \frac{\sin x}{\sin x}\,dx = 10\int_0^{\pi} dx = 10\pi.$$

$$\sum_{m=1}^{10} I_{2m} = I_2 + I_4 + I_6 + \ldots + I_{20} = I_0 + I_0 + \ldots \text{ 10 times}$$

$$= 10I_0 = 10\int_0^{\pi} \frac{\sin 0x}{\sin x}\,dx = 10\int_0^{\pi} 0\,dx = 0.$$

20. Area $= \int_1^e x\,dy = \int_1^e \log_e y\,dy.$

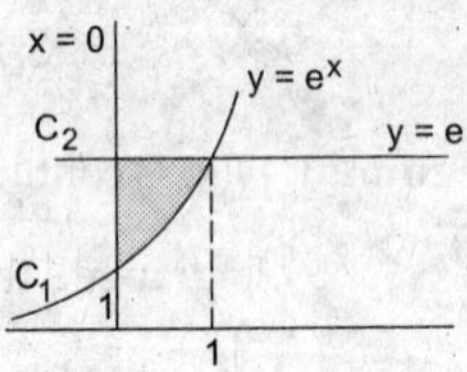

Also, area $= \int_0^1 (y)\,C_2\,dx - \int_0^1 (y)\,C_1\,dx$

$$= \int_0^1 e\,dx - \int_0^1 e^x\,dx = e - \int_0^1 e^x\,dx.$$

Again, $\int_1^e \log_e y\,dy = \int_e^1 \log_e(e+1-z)\,(-dz)$, putting $y = e + 1 - z$

$$= \int_1^e \log(e+1-z)\,dz = \int_1^e \log(e+1-y)\,dy.$$

21. $L = \lim\limits_{x \to 0} \dfrac{0 - \dfrac{-x}{\sqrt{a^2 - x^2}} - \dfrac{2x}{4}}{4x^3}$, by L'Hospital's Rule

$$= \lim_{x \to 0} \frac{\dfrac{1}{\sqrt{a^2 - x^2}} - \dfrac{1}{2}}{4x^2} = \infty \text{ unless } a = 2$$

$$\therefore \quad \text{finite } L = \lim_{x \to 0} \frac{\frac{1}{\sqrt{4-x^2}} - \frac{1}{2}}{4x^2} = \lim_{x \to 0} \frac{2 - \sqrt{4-x^2}}{8x^2\sqrt{4-x^2}}$$

$$= \lim_{x \to 0} \frac{4-(4-x^2)}{8x^2\sqrt{4-x^2}\,(2+\sqrt{4-x^2})} = \lim_{x \to 0} \frac{1}{8\sqrt{4-x^2}\,(2+\sqrt{4-x^2})}$$

$$= \frac{1}{8 \cdot 2 \cdot 4} = \frac{1}{64}.$$

22. There are two possibilities for diagonal elements:

(i) 1, 1, 1 (ii) 1, 0, 0

(i)			(ii)		
1	(1, 0, 0)		1	(1, 1, 0)	
	1			0	
(1, 0, 0)		1	(1, 1, 0)		0

In case of (i), the number of symmetric matrices

= arrangement in leading diagonal elements 1, 1, 1

× arrangements of elements 1, 0, 0 in three places.

$$= \frac{\lfloor 3}{\lfloor 3} \times \frac{\lfloor 3}{\lfloor 2} = 3.$$

In case of (ii), the number of symmetric matrices

= arrangement in leading diagonal elements 1, 0, 0

× arrangement of elements 1, 1, 0 in three places

$$= \frac{\lfloor 3}{\lfloor 2} \times \frac{\lfloor 3}{\lfloor 2} = 3 \times 3 = 9.$$

$\therefore$ the number of matrices in $\mathcal{A} = 3 + 9 = 12$.

23. Here $\begin{bmatrix} x \\ y \\ z \end{bmatrix} = A^{-1}\begin{bmatrix} 1 \\ 0 \\ 0 \end{bmatrix}$. For unique solution, $|A| \neq 0$ for otherwise A^{-1} will not exist.

Let the symmetric matrix $A = \begin{bmatrix} a & h & g \\ h & b & f \\ g & f & c \end{bmatrix}$;

then $|A| = abc + 2fgh - (af^2 + bg^2 + ch^2)$.

As in QN 22 *I*, ($a = b = c = 1$) or (a, b or $c = 1$ and other two zero).

When $a = b = c = 1$ then two of f, g, h are '0' and the third is '1'

$\therefore \quad |A| = 1 + 0 - 1 = 0.$

When a, b or $c = 1$ and other two zero then $|A| = 0$

when $abc = 0, fgh = 0, a = 1, f = 0$

or $abc = 0, fgh = 0, b = 1, g = 0$

or $abc = 0, fgh = 0, c = 1, h = 0.$

$\therefore$ there are $3 + 3$ matrices for which $|A| = 0.$

$\therefore$ total number of matrices with nonzero $|A| = 12 - (3+3) = 6.$

24. If $A = \begin{bmatrix} a & h & g \\ h & b & f \\ g & f & c \end{bmatrix}$, equations are $ax + hy + gz = 1,$

$$hx + by + fz = 0,$$

$$gx + fy + cz = 0.$$

Inconsistent if $\begin{vmatrix} a & h & g \\ h & b & f \\ g & f & c \end{vmatrix} = 0$ and at least one of

$\begin{vmatrix} h & g & -1 \\ b & f & 0 \\ f & c & 0 \end{vmatrix}, \begin{vmatrix} a & g & -1 \\ h & f & 0 \\ g & c & 0 \end{vmatrix}, \begin{vmatrix} a & h & -1 \\ h & b & 0 \\ g & f & 0 \end{vmatrix}$ is nonzero

i.e., $abc + 2fgh - af^2 - bg^2 - ch^2 = 0$ and

$bc - f^2 \neq 0$ or $hc - fg \neq 0$ or $hf - bg \neq 0.$

When $a = b = c = 1, f = 0, g = 1, h = 0;$ we get $|A| = 1 + 0 - 0 - 1 - 0 = 0$

and $bc - f^2 = 1 - 0 \neq 0.$

When $a = b = c = 1, f = 0, g = 0, h = 1;$ we get $|A| = 1 + 0 - 0 - 0 - 1 = 0$

and $hc - fg = 1 - 0 \neq 0.$

When $a = 0, b = 0, c = 1, h = 0, f = 1, g = 1;$ we get $|A| = 0 + 0 - 0 - 0 - 0 = 0$

and $bc - f^2 = 0 - 1^2 \neq 0.$

Thus we have detected three cases for inconsistency. Hence more than 2 matrices.

25. The probability of getting a six in a throw (event E) $= \frac{1}{6}\cdot$

The probability of not getting a six in a throw (event $\bar{E}$) $= \frac{5}{6}\cdot$

$\therefore$ required probability $= P(\bar{E}) \cdot P(\bar{E}) \cdot P(E) = \frac{5}{6} \cdot \frac{5}{6} \cdot \frac{1}{6} = \frac{25}{216}\cdot$

26. Required probability $= P(\bar{E}) \cdot P(\bar{E}) \cdot P(E) + P(\bar{E}) \cdot P(\bar{E}) \cdot P(\bar{E}) \cdot P(E) + \ldots$

$$= \left(\frac{5}{6}\right)^2 \cdot \frac{1}{6} + \left(\frac{5}{6}\right)^3 \cdot \frac{1}{6} + \left(\frac{5}{6}\right)^4 \cdot \frac{1}{6} + \ldots \text{ to } \infty$$

$$= \left(\frac{5}{6}\right)^2 \cdot \frac{1}{6}\left\{1 + \frac{5}{6} + \left(\frac{5}{6}\right)^2 + \ldots \text{ to } \infty\right\} = \left(\frac{5}{6}\right)^2 \cdot \frac{1}{6} \cdot \frac{1}{1 - 5/6} = \frac{25}{36}.$$

27. Required probability $P\left(\frac{X \geq 6}{X > 3}\right) = P\left(\frac{X \geq 6}{X \geq 4}\right)$

$$= \frac{\left(\frac{5}{6}\right)^5 \cdot \frac{1}{6}\left\{1 + \frac{5}{6} + \left(\frac{5}{6}\right)^2 + \ldots \text{ to } \infty\right\}}{\left(\frac{5}{6}\right)^3 \cdot \frac{1}{6}\left\{1 + \frac{5}{6} + \left(\frac{5}{6}\right)^2 + \ldots \text{ to } \infty\right\}} = \frac{25}{36}.$$

28. (A) $4\sin^2\theta \cdot \cos^2\theta - 2\cos^2\theta = 0$ or $\cos^2\theta\,(1 - 2\sin^2\theta) = 0$

or $\cos^2\theta \cdot \cos 2\theta = 0$ $\Rightarrow \cos\theta = 0,\ \cos 2\theta = 0$

$$\Rightarrow \theta = \frac{\pi}{2}, \frac{3\pi}{2}, \ldots;\quad 2\theta = \frac{\pi}{2}, \frac{3\pi}{2}, \ldots$$

$$\therefore \theta = \frac{\pi}{2}, \frac{\pi}{4}. \qquad \therefore \text{A—q, s.}$$

(B) $f(x)$ will be discontinuous when at least one of $\left[\frac{6x}{\pi}\right]$ or $\cos\left[\frac{3x}{\pi}\right]$ is discontinuous, i.e., either $\frac{6x}{\pi}$ or $\frac{3x}{\pi}$ is integer.

This will happen if $x = \frac{\pi}{6}, \frac{\pi}{3}, \frac{\pi}{2}, \pi.$ $\therefore$ B—p, r, s, t.

(C) Volume $= \begin{vmatrix} 1 & 1 & 0 \\ 1 & 2 & 0 \\ 1 & 1 & \pi \end{vmatrix} = \pi$

because volume $= [\vec{i} + \vec{j}\ \ \vec{i} + 2\vec{j}\ \ \vec{i} + \vec{j} + \pi\vec{k}]$. $\therefore$ C—t.

(D) $\vec{a} + \vec{b} = -\sqrt{3}\,\vec{c}$ or $(\vec{a} + \vec{b}) \cdot (\vec{a} + \vec{b}) = 3\,\vec{c} \cdot \vec{c}$

or $\vec{a}^2 + 2\vec{a} \cdot \vec{b} + \vec{b}^2 = 3\vec{c}^2$ or $1 + 2\,|\vec{a}|\,|\vec{b}|\cos\theta + 1 = 3$

because $\vec{a} \cdot \vec{a} = \vec{b} \cdot \vec{b} = \vec{c} \cdot \vec{c} = 1.$

$\therefore$ $2\cos\theta = 1$; $\therefore \theta = \frac{\pi}{3}.$ $\therefore$ D—r.

29. (A) Here the equation is $e^{\sin x} = \frac{\cos x}{x}.$

Consider $f(x) = e^{\sin x}$; $f(0) = e^0 = 1$

$f'(x) = e^{\sin x} \cdot \cos x$; $\therefore$ in $(0, \pi/2)$, $f'(x) > 0$.

Hence $f(x)$ is increasing in $(0, \pi/2)$.

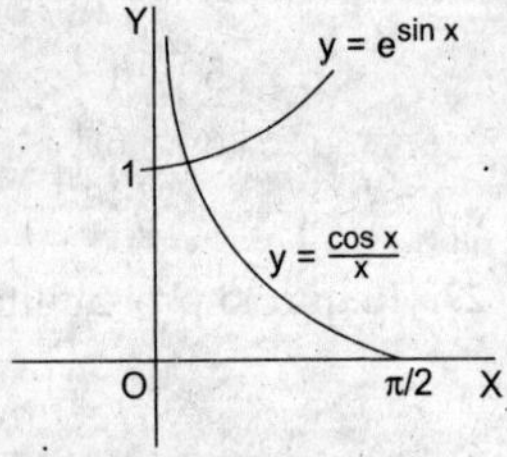

Consider, $g(x) = \dfrac{\cos x}{x}$; $g(0) = \dfrac{1}{0} = \infty$

$$g'(x) = \frac{-x\sin x - \cos x}{x^2} < 0 \text{ in } \left(0, \frac{\pi}{2}\right).$$

Hence, $g(x)$ is decreasing in $(0, \pi/2)$ and $g(\pi/2) = 0$.

Two curves cut at only one point in $(0, \pi/2)$.

So, only one solution. A—p.

(B) They interesect along a line if the normal vectors to the planes are coplanar, i.e., vectors $k\vec{i} + 4\vec{j} + \vec{k}$, $4\vec{i} + k\vec{j} + 2\vec{k}$ and $2\vec{i} + 2\vec{j} + \vec{k}$ are coplanar.

$\therefore \quad [k\vec{i} + 4\vec{j} + \vec{k} \;\; 4\vec{i} + k\vec{j} + 2\vec{k} \;\; 2\vec{i} + 2\vec{j} + \vec{k}] = 0$

or $\begin{vmatrix} k & 4 & 1 \\ 4 & k & 2 \\ 2 & 2 & 1 \end{vmatrix} = 0$ or $\begin{vmatrix} k-2 & 2 & 0 \\ 0 & k-4 & 0 \\ 2 & 2 & 1 \end{vmatrix} = 0$ or $(k-2)(k-4) = 0$

$\therefore \quad k = 2, 4$. But when $k = 4$, the last two planes become the same plane. So, $k = 2$. B—q.

(C) For $x < -2$, $-(x-1) - (x-2) - (x+1) - (x+2) = 4k$

or $-4x = 4k$ or $k = -x$ $\Rightarrow k = 3, 4, 5, 6, \ldots$

For $x = -2$, $3 + 4 + 1 = 4k$ or $k = 2$.

For $x = -1$, $2 + 3 + 1 = 4k$ or $k = \dfrac{3}{2} \notin I$.

For $x = 0$, $1 + 2 + 1 + 2 = 4k$ or $k = \dfrac{3}{2} \notin I$.

For $x = 1$, $1 + 2 + 3 = 4k$ or $k = \dfrac{3}{2} \notin I$.

For $x \geq 2$, $x - 1 + x - 2 + x + 1 + x + 2 = 4k$.

or $4x = 4k$ or $k = x$ $\Rightarrow k = 2, 3, 4, 5, 6, \ldots$.

Hence, $k = 2, 3, 4, 5$. C—q, r, s, t.

(D) $\dfrac{dy}{dx} = y + 1$ or $\dfrac{dy}{y+1} = dx$; $\therefore \log(y+1) = x + k$

$y(0) = 1$ (given); $\therefore \log(1+1) = 0 + k$; $\therefore k = \log 2$.

$\therefore \quad \log(y+1) = x + \log 2$; $\therefore y + 1 = e^{x + \log 2} = e^x \cdot e^{\log 2} = 2e^x$.

When $x = \log_e 2$, $y + 1 = 2e^{\log_e 2} = 2 \times 2$; $\therefore y = 4 - 1 = 3$. D—r.

30. (A) $(x-3)^2 \cdot \dfrac{dy}{dx} + y = 0$ or $\dfrac{dy}{y} + \dfrac{dx}{(x-3)^2} = 0$, $x \neq 3$

$\therefore \quad \log y - \dfrac{1}{x-3} = \log c$ or $\log \dfrac{y}{c} = \dfrac{1}{x-3}$; $\therefore y = ce^{\frac{1}{x-3}}$, $x \neq 3$

$\therefore$ the domain of definition of nonzero solutions is $R - \{3\}$.
So, A—p, q, s.

(B) Put $x - 3 = t$. Then $I = \int_{-2}^{2} (t+2)(t+1)\,t(t-1)(t-2)\,dt$

$$= \int_{-2}^{2} (t^2 - 1)(t^2 - 4)\,t\,dt = 0,$$

because integrand is an odd-function.

$0 \in \left(-\frac{\pi}{2}, \frac{\pi}{2}\right)$, $0 \in (-\pi, \pi)$. B—p, t.

(C) $\cos^2 x + \sin x = 1 - \sin^2 x + \sin x = 1 - \left(\sin^2 x - \sin x + \frac{1}{4}\right) + \frac{1}{4}$

$$= 1 + \frac{1}{4} - \left(\sin x - \frac{1}{2}\right)^2 = \frac{5}{4} - \left(\sin x - \frac{1}{2}\right)^2$$

$\therefore$ $\cos^2 x + \sin x$ is maximum, when $\sin x - \frac{1}{2} = 0$.

i.e., $\sin x = \frac{1}{2}$; $x = \frac{\pi}{6}, \frac{5\pi}{6}$.

Clearly, $\frac{\pi}{6} \in \left(-\frac{\pi}{2}, \frac{\pi}{2}\right)$, $\frac{\pi}{6} \in \left(0, \frac{\pi}{2}\right)$, $\frac{\pi}{6}$ and $\frac{5\pi}{6} \in \left(\frac{\pi}{8}, \frac{5\pi}{4}\right)$,

$\frac{\pi}{6}$ and $\frac{5\pi}{6} \in (-\pi, \pi)$. C—p, q, r, t.

(D) $f(x) = \tan^{-1}(\sin x + \cos x)$; $\therefore$ $f'(x) = \dfrac{1 \cdot (\cos x - \sin x)}{1 + (\sin x + \cos x)^2}$

$f(x)$ is increasing if $f'(x) > 0$ $\Rightarrow$ $\cos x > \sin x$

$$\Rightarrow x \in \left(-\frac{3\pi}{4}, \frac{\pi}{4}\right) \supset \left(0, \frac{\pi}{8}\right).$$ D—s.

31. (p) Condition for tangency gives $\left|\dfrac{-1}{\sqrt{h^2 + k^2}}\right| = 2$ or $h^2 + k^2 = \frac{1}{4}$

$\therefore$ the locus of the point (h, k) has the equation $x^2 + y^2 = \frac{1}{4}$. $\therefore$ p—A.

(q) Here $|PS - PS'| = 3$

$\therefore$ the locus of P is a hyperbola with S, S' as foci.

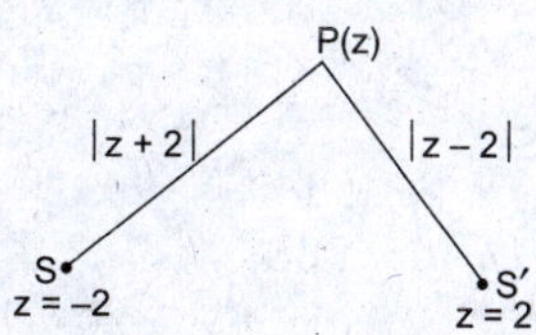

$\therefore$ q—D.

(r) Let $t = \tan\theta$. Then, $x = \sqrt{3} \cdot \dfrac{1 - \tan^2\theta}{1 + \tan^2\theta} = \sqrt{3}\cos 2\theta$

$$y = \frac{2t}{1+t^2} = \frac{2\tan\theta}{1+\tan^2\theta} = \sin 2\theta$$

$\therefore \quad \left(\dfrac{x}{\sqrt{3}}\right)^2 + y^2 = \cos^2 2\theta + \sin^2 2\theta = 1$

or $\quad \dfrac{x^2}{3} + \dfrac{y^2}{1} = 1$ which is an ellipse. r—C.

(s) Here the eccentricity $e = 1$ or $1 < e < \infty$.

So, the conic is a parabola or a hyperbola. s—B, D.

(t) $(x+1)^2 - y^2 = x^2 + y^2 + 1$ or $2x = 2y^2$ or $y^2 = x$ which is a parabola.

$\therefore$ t—B.

$\therefore$ A—p; B—s, t; C—r; D—q, s.

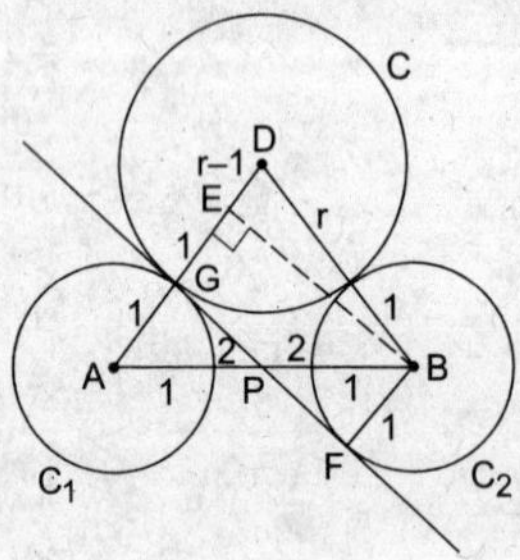

32. In $\triangle BFP$, $PF^2 = PB^2 - BF^2 = 3^2 - 1^2 = 8$

$\therefore \quad PF = 2\sqrt{2}$.

In $\triangle AGP$, $PG^2 = PA^2 - AG^2 = 3^2 - 1^2 = 8$

$\therefore \quad PG = 2\sqrt{2}$.

$\therefore \quad FG = 2\sqrt{2} + 2\sqrt{2} = 4\sqrt{2}$.

In right-angled triangle BED,

$BE^2 + ED^2 = BD^2$ or $FG^2 + (DG - EG)^2 = BD^2$

or $\quad (4\sqrt{2})^2 + (r-1)^2 = (r+1)^2$ or $4r = 32$; $\therefore$ $r = 8$.

33. In $\triangle ADB$, $\dfrac{AD}{AB} = \sin 30°$ or $\dfrac{AD}{4} = \dfrac{1}{2}$; $\therefore$ $AD = 2$.

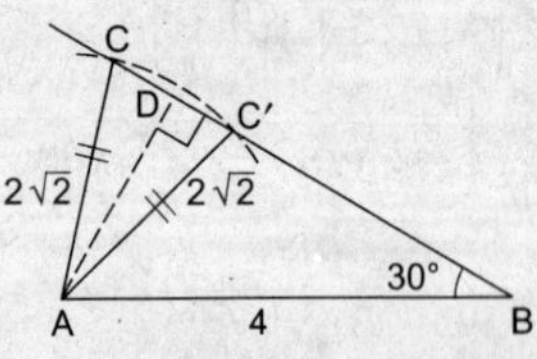

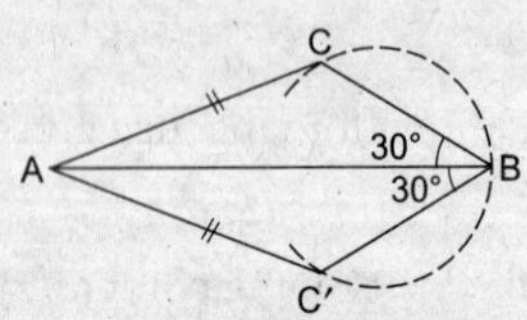

Note : This possibility is rejected because in that case $\triangle ABC \cong \triangle ABC'$

In $\triangle ADC'$, $\sin \angle AC'D = \dfrac{AD}{AC'} = \dfrac{2}{2\sqrt{2}} = \dfrac{1}{\sqrt{2}}$

$\therefore \quad \angle AC'D = 45°$. Also, $\angle BAC' = 45° - 30° = 15°$.

$AC = AC' \quad \Rightarrow \angle ACD = 45°$. So, $\angle CAC' = 180° - 45° - 45° = 90°$.

So, $\angle BAC = 90° + 15° = 105°$.

$$\therefore \quad \text{ar}(\Delta BAC) = \frac{1}{2} CA \cdot BA \cdot \sin CAB$$
$$= \frac{1}{2} \cdot 2\sqrt{2} \cdot 4 \cdot \sin 105°$$
$$= 4\sqrt{2} \cos 15°$$
$$= 4\sqrt{2} \cos (45° - 30°)$$
$$= 4\sqrt{2} \{\cos 45° \cdot \cos 30° + \sin 45° \cdot \sin 30°\}$$
$$= 4\sqrt{2}\left(\frac{1}{\sqrt{2}} \cdot \frac{\sqrt{3}}{2} + \frac{1}{\sqrt{2}} \cdot \frac{1}{2}\right) = 2(\sqrt{3}+1).$$

$$\text{ar}(\Delta BAC') = \frac{1}{2} \cdot 2\sqrt{2} \cdot 4 \cdot \sin 15° = 4\sqrt{2} \sin (45° - 30°)$$
$$= 4\sqrt{2}\left(\frac{1}{\sqrt{2}} \cdot \frac{\sqrt{3}}{2} - \frac{1}{\sqrt{2}} \cdot \frac{1}{2}\right) = 2(\sqrt{3}-1).$$

$\therefore \quad |\text{ar}(\Delta BAC) - \text{ar}(\Delta BAC')| = |2(\sqrt{3}+1) - 2(\sqrt{3}-1)| = 4.$

34. Here $z = 3x$. Therefore, $3x - y - z = 0 \quad \Rightarrow y = 0$

$z = 3x$ and $y = 0$ satisfies the equation $-3x + 2y + z = 0$.

$\therefore \quad (x, y, z)$ is such that $z = 3x, . y = 0$.

$\therefore \quad x = t, \; y = 0, \; z = 3t,$ where t is an integer.

$\therefore \quad x^2 + y^2 + z^2 \le 100 \quad \Rightarrow \quad t^2 + 0^2 + (3t)^2 \le 100 \quad \Rightarrow \quad t^2 \le 10$

$\therefore \quad t = 0, \pm 1, \pm 2, \pm 3$. So, we get 7 points.

35. Roots are real and distinct $\Rightarrow 64k^2 - 4 \cdot 1 \cdot 16(k^2 - k + 1) > 0$ or $k > 1$.

Again $\quad \alpha \ge 4, \; \beta \ge 4 \quad \Rightarrow \alpha - 4 \ge 0, \; \beta - 4 \ge 0$

$\Rightarrow \quad (\alpha - 4) + (\beta - 4) \ge 0, \quad (\alpha - 4)(\beta - 4) \ge 0$

$\Rightarrow \quad \alpha + \beta - 8 \ge 0, \quad \alpha\beta - 4(\alpha + \beta) + 16 \ge 0$

$\Rightarrow \quad 8k - 8 \ge 0, \; 16(k^2 - k + 1) - 4 \cdot 8k + 16 \ge 0$

$\Rightarrow \quad k \ge 1, \quad k^2 \quad 3k + 2 \ge 0$, i.e., 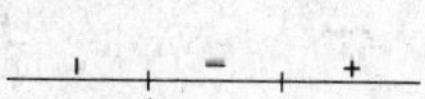 $\Rightarrow k \ge 2$.

$\therefore \quad$ the smallest value of k is 2.

36. $x^2 + 20 \le 9x \quad$ or $x^2 - 9x + 20 \le 0 \quad$ or $(x - 4)(x - 5) \le 0$

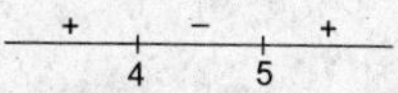 $\therefore \; x \in [4, 5] = A$.

Now, $\quad f'(x) = 6x^2 - 30x + 36 = 6(x^2 - 5x + 6) = 6(x - 2)(x - 3)$

$\therefore$ $f'(x) > 0$ in the interval [4, 5]. So, $f(x)$ is increasing in [4, 5]

$\therefore$ maximum value of $f(x)$ on the set A is at $x = 5$.

$\therefore$ maximum value $= f(5) = 2 \cdot 5^3 - 15 \cdot 5^2 + 36 \cdot 5 - 48 = 430 - 423 = 7$.

37. $f(0) = e^0 = 1;\quad \therefore f^{-1}(1) = 0;\quad \therefore g(1) = 0$.

Also $f(g(x)) = x;\quad \therefore f'(g(x)) \cdot g'(x) = 1;\quad \therefore\quad g'(x) = \dfrac{1}{f'(g(x))}$.

$$\therefore\quad g'(1) = \frac{1}{f'(g(1))} = \frac{1}{f'(0)}.$$

Again, $f'(x) = 3x^2 + \dfrac{1}{2} e^{x/2};\qquad \therefore f'(0) = \dfrac{1}{2} e^0 = \dfrac{1}{2}$.

$$\therefore\quad g'(1) = \frac{1}{1/2} = 2.$$

38. Let $p(x) = ax^4 + bx^3 + cx^2 + dx + e$.

Then $p'(x) = 4ax^3 + 3bx^2 + 2cx + d$. For extremum $p'(x) = 0$.

As $p(x)$ has extremum at $x = 1, 2$.

$$4a + 3b + 2c + d = 0,\quad 32a + 12b + 4c + d = 0.$$

Again, $\lim\limits_{x \to 0} \left(1 + \dfrac{p(x)}{x^2}\right) = 2 \;\Rightarrow\; \lim\limits_{x \to 0} \left\{1 + ax^2 + bx + c + \dfrac{d}{x} + \dfrac{e}{x^2}\right\} = 2$.

This is possible only when $d = 0, e = 0$ and then $1 + c = 2$, i.e., $c = 1$

$\therefore\quad c = 1,\ d = 0, e = 0$ and $4a + 3b + 2 = 0,\ 32a + 12b + 4 = 0$.

Solving, $a = \dfrac{1}{4}$ and $b = -1$.

$\therefore\quad p(x) = \dfrac{1}{4}x^4 - x^3 + x^2;\qquad \therefore p(2) = \dfrac{1}{4} \cdot 2^4 - 2^3 + 2^2 = 4 - 8 + 4 = 0$.

39. Here, $f'(x) = f(t)\,]_{t=x}$ or $f'(x) = f(x)$ or $\dfrac{f'(x)}{f(x)} = 1$

$\therefore\quad \log f(x) = x + \log c\quad$ or $\log \dfrac{f(x)}{c} = x\quad$ or $f(x) = ce^x$ ... (1)

As $f(x) = \int_0^x f(t)\,dt,\quad f(0) = \int_0^0 f(t)\,dt = 0$.

$\therefore\quad (1) \Rightarrow 0 = c;\qquad \therefore f(x) = 0 \cdot e^x = 0;\qquad \therefore f(\log_e 5) = 0$.

□

IIT Questions—7

SECTION I

Single Correct Choice Type

This section contains 14 multiple-choice questions. Each question has 4 choices (a), (b), (c) and (d) for its answer, out of which *only one* is correct.

1. Let p and q be real numbers such that $p \neq 0$, $p^3 \neq q$ and $p^3 \neq -q$. If α and β are nonzero complex numbers satisfying $\alpha + \beta = -p$ and $\alpha^3 + \beta^3 = q$ then a quadratic equation having $\frac{\alpha}{\beta}$ and $\frac{\beta}{\alpha}$ as its roots is

   (a) $(p^3 + q)x^2 - (p^3 + 2q)x + (p^3 + q) = 0$

   (b) $(p^3 + q)x^2 - (p^3 - 2q)x + (p^3 + q) = 0$

   (c) $(p^3 - q)x^2 - (5p^3 - 2q)x + (p^3 - q) = 0$

   (d) $(p^3 - q)x^2 - (5p^3 + 2q)x + (p^3 - q) = 0$

2. Let $S = \{1, 2, 3, 4\}$. The total number of unordered pairs of disjoint subsets of S is equal to

   (a) 25 (b) 34

   (c) 42 (d) 41

3. For $r = 0, 1, 2, 3, \ldots, 10$ let A_r, B_r, C_r denote respectively the coefficient of x^r in the expansions of $(1+x)^{10}$, $(1+x)^{20}$ and $(1+x)^{30}$. Then

   $\sum_{r=1}^{10} A_r(B_{10}B_r - C_{10}A_r)$ is equal to

   (a) $B_{10} - C_{10}$ (b) $A_{10}(B_{10}^2 - C_{10}A_{10})$

   (c) 0 (d) $C_{10} - B_{10}$

4. The number of 3×3 matrices A whose entries are either 0 or 1 and for which the system $A\begin{bmatrix} x \\ y \\ z \end{bmatrix} = \begin{bmatrix} 1 \\ 0 \\ 0 \end{bmatrix}$ has exactly two distinct solutions, is

   (a) 0 (b) $2^9 - 1$ (c) 168 (d) 2

5. If the angles A, B and C of a triangle are in an AP and if a, b and c denote the lengths of the sides opposite to A, B and C respectively then the value of the expression $\frac{a}{c}\sin 2C + \frac{c}{a}\sin 2A$ is

(a) $\frac{1}{2}$ (b) $\frac{\sqrt{3}}{2}$

(c) 1 (d) $\sqrt{3}$

6. Let P, Q, R and S be points on the plane with position vectors $-2\hat{i} - \hat{j}$, $4\hat{i}$, $3\hat{i} + 3\hat{j}$ and $-3\hat{i} + 2\hat{j}$ respectively. The quadrilateral $PQRS$ must be a

(a) parallelogram, which is neither a rhombus nor a rectangle
(b) square
(c) rectangle but not a square
(d) rhombus but not a square

7. Two adjacent sides of a parallelogram $ABCD$ are given by $\vec{AB} = 2\hat{i} + 10\hat{j} + 11\hat{k}$ and $\vec{AD} = -\hat{i} + 2\hat{j} + 2\hat{k}$. The side AD is rotated by an acute angle α in the plane of the parallelogram so that AD becomes AD'. If AD' makes a right angle with the side AB then the cosine of the angle α is given by

(a) $\frac{8}{9}$ (b) $\frac{\sqrt{17}}{9}$ (c) $\frac{1}{9}$ (d) $\frac{4\sqrt{5}}{9}$

8. The equation of the plane containing the straight line $\frac{x}{2} = \frac{y}{3} = \frac{z}{4}$ and perpendicular to the plane containing the straight lines $\frac{x}{3} = \frac{y}{4} = \frac{z}{2}$ and $\frac{x}{4} = \frac{y}{2} = \frac{z}{3}$ is

(a) $x + 2y - 2z = 0$ (b) $3x + 2y - 2z = 0$
(c) $x - 2y + z = 0$ (d) $5x + 2y - 4z = 0$

9. If the distance of the point $P(1, -2, 1)$ from the plane $x + 2y - 2z = \alpha$, where $\alpha > 0$, is 5 then the foot of the perpendicular from P to the plane is

(a) $\left(\frac{8}{3}, \frac{4}{3}, \frac{-7}{3}\right)$ (b) $\left(\frac{4}{3}, -\frac{4}{3}, \frac{1}{3}\right)$

(c) $\left(\frac{1}{3}, \frac{2}{3}, \frac{10}{3}\right)$ (d) $\left(\frac{2}{3}, -\frac{1}{3}, \frac{5}{2}\right)$

10. Let f, g and h be real-valued functions defined on the interval [0, 1] by $f(x) = e^{x^2} + e^{-x^2}$, $g(x) = xe^{x^2} + e^{-x^2}$ and $h(x) = x^2e^{x^2} + e^{-x^2}$. If a, b and c denote the absolute maximum of f, g and h on [0, 1] then

(a) $a = b$ and $b \neq c$ (b) $c = a$ and $a \neq b$

(c) $a \neq b$ and $b \neq c$ (d) $a = b = c$

11. The value of $\lim_{x \to 0} \frac{1}{x^3} \int_0^x \frac{t \log(1+t)}{t^4 + 4} dt$ is

(a) 0 (b) $\frac{1}{12}$ (c) $\frac{1}{24}$ (d) $\frac{1}{64}$

12. Let f be a real-valued function defined on the interval (–1, 1) such that

$e^{-x} \cdot f(x) = 2 + \int_0^x \sqrt{t^4 + 1}\, dt \;\forall\, x \in (-1, 1)$. Let f^{-1} be the inverse function of

f. Then $(f^{-1})'(2)$ is equal to

(a) 1 (b) $\frac{1}{3}$

(c) $\frac{1}{2}$ (d) $\frac{1}{e}$

13. Let ω be a complex cube root of unity with $\omega \neq 1$. A fair die is thrown three times. If r_1, r_2 and r_3 are the numbers obtained on the die then the probability that $\omega^{r_1} + \omega^{r_2} + \omega^{r_3} = 0$ is

(a) $\frac{1}{18}$ (b) $\frac{1}{9}$

(c) $\frac{2}{9}$ (d) $\frac{1}{36}$

14. A signal which can be green or red with probability $\frac{4}{5}$ and $\frac{1}{5}$ respectively is received by station A and then transmitted to station B. The probability of each station receiving the signal correctly is $\frac{3}{4}$. If the signal received at station B is green then the probability that the original signal was green is

(a) 3 (b) $\frac{6}{7}$

(c) $\frac{20}{23}$ (d) $\frac{9}{20}$

SECTION II

Multiple Correct Choice Type

This section contains 5 multiple-choice questions. Each question has four choices (a), (b), (c) and (d) out of which ***one* or *more*** may be correct.

15. Let z_1 and z_2 be two distinct complex numbers and let $z = (1-t)z_1 + tz_2$ for some real number t, where $0 < t < 1$. If $\arg(w)$ denotes the principal argument of a nonzero complex number w then

(a) $|z - z_1| + |z - z_2| = |z_1 - z_2|$

(b) $\arg(z - z_1) = \arg(z - z_2)$

(c) $\begin{vmatrix} z - z_1 & \bar{z} - \bar{z}_1 \\ z_2 - z_1 & \bar{z}_2 - \bar{z}_1 \end{vmatrix} = 0$

(d) $\arg(z - z_1) = \arg(z_2 - z_1)$

16. Let ABC be a triangle such that $\angle ACB = \frac{\pi}{6}$ and let a, b, c denote the lengths of the sides opposite to A, B, C respectively. The value(s) of x for which $a = x^2 + x + 1$, $b = x^2 - 1$ and $c = 2x + 1$ is (are)

(a) $-(2 + \sqrt{3})$ (b) $\sqrt{3} + 1$

(c) $2 + \sqrt{3}$ (d) $4\sqrt{3}$

17. Let A and B be two distinct points on the parabola $y^2 = 4x$. If the axis of the parabola touches a circle of radius r having AB as its diameter then the slope of the line AB can be

(a) $-\frac{1}{r}$ (b) $\frac{1}{r}$

(c) $\frac{2}{r}$ (d) $-\frac{2}{r}$

18. The value(s) of $\int_0^1 \frac{x^4(1-x)^4}{1+x^2}\,dx$ is (are)

(a) $\frac{22}{7} - \pi$ (b) $\frac{2}{105}$ (c) 0 (d) $\frac{71}{15} - \frac{3\pi}{2}$

19. Let f be a real-valued function defined on the interval $(0, \infty)$ by

$f(x) = \log_e x + \int_0^x \sqrt{1 + \sin t}\,dt$. Which of the following statement(s) is (are) true?

(a) $f''(x)$ exists for all $x \in (0, \infty)$

(b) $f'(x)$ exists for all $x \in (0, \infty)$ and $f'(x)$ is continuous on $(0, \infty)$ but not differentiable on $(0, \infty)$

(c) there exists $\alpha > 1$ such that $|f'(x)| < |f(x)|$ for all $x \in (\alpha, \infty)$

(d) there exists $\beta > 0$ such that $|f(x)| + |f'(x)| \leq \beta$ for all $x \in (\alpha, \infty)$

SECTION III

Paragraph Type

Paragraph for Question Nos. 20 to 21

The circle $x^2 + y^2 - 8x = 0$ and hyperbola $\frac{x^2}{9} - \frac{y^2}{4} = 1$ intersect at the points A and B.

20. The equation of a common tangent with positive slope to the circle as well as to the hyperbola is

(a) $2x - \sqrt{5}y - 20 = 0$ (b) $2x - \sqrt{5}y + 4 = 0$

(c) $3x - 4y + 8 = 0$ (d) $4x - 3y + 4 = 0$

21. The equation of the circle with AB as its diameter is

(a) $x^2 + y^2 - 12x + 24 = 0$ (b) $x^2 + y^2 + 12x + 24 = 0$

(c) $x^2 + y^2 + 24x - 12 = 0$ (d) $x^2 + y^2 - 24x - 12 = 0$

Paragraph for Question Nos. 22 to 24

Consider the polynomial $f(x) = 1 + 2x + 3x^2 + 4x^3$. Let s be the sum of all distinct real roots of $f(x) = 0$ and let $|s| = t$.

22. The real number s lies in the interval

(a) $\left(-\frac{1}{4}, 0\right)$ (b) $\left(-11, -\frac{3}{4}\right)$

(c) $\left(-\frac{3}{4}, -\frac{1}{2}\right)$ (d) $\left(0, \frac{1}{4}\right)$

23. The area bounded by the curve $y = f(x)$ and the lines $x = 0, y = 0$ and $x = t$, lies in the interval

(a) $\left(\frac{3}{4}, 3\right)$ (b) $\left(\frac{21}{64}, \frac{11}{16}\right)$

(c) $(9, 10)$ (d) $\left(0, \frac{21}{64}\right)$

24. The function $f'(x)$ is

(a) increasing in $\left(-t, -\frac{1}{4}\right)$ and decreasing in $\left(-\frac{1}{4}, t\right)$

(b) decreasing in $\left(-t, -\frac{1}{4}\right)$ and increasing in $\left(-\frac{1}{4}, t\right)$

(c) increasing in $(-t, t)$

(d) decreasing in $(-t, t)$

Paragraph for Question Nos. 25 to 27

Tangents are drawn from the point $P(3, 4)$ to the ellipse $\frac{x^2}{9} + \frac{y^2}{4} = 1$ touching the ellipse at the points A and B.

25. The coordinates of A and B are

(a) $(3, 0)$ and $(0, 2)$

(b) $\left(-\frac{8}{5} \cdot \frac{2\sqrt{161}}{15}\right)$ and $\left(-\frac{9}{5}, \frac{8}{5}\right)$

(c) $\left(-\frac{8}{5} \cdot \frac{2\sqrt{161}}{15}\right)$ and $(0, 2)$

(d) $(3, 0)$ and $\left(-\frac{9}{5}, \frac{8}{5}\right)$

26. The orthocentre of the triangle PAB is

(a) $\left(5, \frac{8}{7}\right)$

(b) $\left(\frac{7}{5}, \frac{25}{8}\right)$

(c) $\left(\frac{11}{5}, \frac{8}{5}\right)$

(d) $\left(\frac{8}{25}, \frac{7}{5}\right)$

27. The equation of the locus of the point whose distances from the point P and the line AB are equal, is

(a) $9x^2 + y^2 - 6xy - 54x - 62y + 241 = 0$

(b) $x^2 + 9y^2 + 6xy - 54x + 62y - 241 = 0$

(c) $9x^2 + 9y^2 - 6xy - 54x - 62y - 241 = 0$

(d) $x^2 + y^2 - 2xy + 27x + 31y - 120 = 0$

SECTION IV

Matrix-Match Type

28. Match the statements in **Column I** with those in **Column II.**

[**Note:** Here z takes values in the complex plane and Im(z) and Re(z) denote the imaginary part and real part of z respectively.]

Column I	Column II
(A) The set of points z satisfying $\lvert z - i\lvert z\rvert\rvert = \lvert z + i\lvert z\rvert\rvert$ is contained in or equal to	(p) an ellipse with eccentricity $\frac{4}{5}$
	(q) the set of points z satisfying $\text{Im}(z) = 0$
(B) The set of points z satisfying $\lvert z+4\rvert + \lvert z-4\rvert = 10$ is contained in or equal to	(r) the set of points z satisfying $\lvert\text{Im}(z)\rvert \le 1$
(C) If $\lvert\omega\rvert = 2$ then the set of points $z = \omega - 1/\omega$ is contained in or equal to	(s) the set of points z satisfying $\lvert\text{Re}(z)\rvert \le 2$
(D) If $\lvert\omega\rvert = 1$ then the set of points $z = \omega + 1/\omega$ is contained in or equal to	(t) the set of points z satisfying $\lvert z\rvert \le 3$

29. Match the statements in **Column I** with the values in **Column II.**

Column I	Column II
(A) A line from the origin meets the lines $\frac{x-2}{1} = \frac{y-1}{-2} = \frac{z+1}{1}$ and $\frac{x - 8/3}{2} = \frac{y+3}{-1} = \frac{z-1}{1}$ at P and Q respectively. If the length $PQ = d$ then d^2 is	(p) –4
	(q) 0
(B) The values of x satisfying $\tan^{-1}(x+3) - \tan^{-1}(x-3) = \sin^{-1}(3/5)$ are	(r) 4
(C) Nonzero vectors $\vec{a}, \vec{b}$ and $\vec{c}$ satisfy $\vec{a} \cdot \vec{b} = 0$, $(\vec{b} - \vec{a}) \cdot (\vec{b} + \vec{c}) = 0$ and $2\lvert\vec{b} + \vec{c}\rvert = \lvert\vec{b} - \vec{a}\rvert$. If $\vec{a} = \mu\vec{b} + 4\vec{c}$ then the possible values of μ are	(s) 5

(D) Let f be the function on $[-\pi, \pi]$ given by $f(0) = 9$ and $f(x) = \dfrac{\sin \frac{9}{2} x}{\sin \frac{1}{2} x}$ for $x \neq 0$.

Then value of $2\pi \int_{-\pi}^{\pi} f(x)\, dx$ is

(t) 6

SECTION V

Integer Answer Type

30. Let $\omega = \cos \frac{2\pi}{3} + i \sin \frac{2\pi}{3}$. Then the number of distinct complex number z satisfying $\begin{vmatrix} z+1 & \omega & \omega^2 \\ \omega & z+\omega^2 & 1 \\ \omega^2 & 1 & z+\omega \end{vmatrix} = 0$ is equal to

31. Let S_k, $k = 1, 2, 3, \ldots, 100$ denote the sum of the infinite geometric series whose first term is $\frac{k-1}{\lfloor k}$ and the common ratio is $\frac{1}{k}$. Then the value of

$$\frac{100^2}{\lfloor 100} + \sum_{k=2}^{100} |(k^2 - 3k + 1)s_k| \text{ is}$$

32. The number of all possible values of θ, where $0 < \theta < \pi$, for which the system of equations

$$(y + z) \cos 3\theta = (xyz) \sin 3\theta$$

$$x \sin 3\theta = \frac{2\cos 3\theta}{y} + \frac{2\sin 3\theta}{z}$$

$$(xyz) \sin 3\theta = (y + 2z) \cos 3\theta + y \sin 3\theta$$

have a solution (x_0, y_0, z_0) with $y_0 z_0 \neq 0$, is

33. The number of values of θ in the interval $\left(-\frac{\pi}{2}, \frac{\pi}{2}\right)$ such that $\theta \neq \frac{1}{5} n\pi$, where $n = 0, \pm 1, \pm 2$ and $\tan \theta = \cot 5\theta$ as well as $\sin 2\theta = \cos 4\theta$ is

34. The maximum value of the expression $\dfrac{1}{\sin^2\theta + 3\sin\theta\cos\theta + 5\cos^2\theta}$ is

35. The line $2x + y = 1$ is tangent to the hyperbola $\frac{x^2}{a^2} - \frac{y^2}{b^2} = 1$. If this line passes through the point of intersection of the nearest directrix and the x-axis then the eccentricity of the hyperbola is

36. If $\vec{a}$ and $\vec{b}$ are vectors in space given by $\vec{a} = \frac{\hat{i} - 2\hat{j}}{\sqrt{5}}$ and $\vec{b} = \frac{2\hat{i} + \hat{j} + 3\hat{k}}{\sqrt{14}}$ then the value of $(2\vec{a} + \vec{b}) \cdot \{(\vec{a} \times \vec{b}) \times (\vec{a} - 2\vec{b})\}$ is

37. If the distance between the plane $Ax - 2y + z = d$ and the plane containing the lines $\frac{x-1}{2} = \frac{y-2}{3} = \frac{z-3}{4}$ and $\frac{x-2}{3} = \frac{y-3}{4} = \frac{z-4}{5}$ is $\sqrt{6}$ then $|\ d\ |$ is

38. Let f be a real-valued differentiable function on R, the set of real numbers, such that $f(1) = 1$. If the y-intercept of the tangent at any point $P(x, y)$ on the curve $y = f(x)$ is equal to the cube of the abscissa of P then the value of $f(-3)$ is equal to

39. For any real number x, let $[x]$ = largest integer less than or equal to x. Let f be a real valued function defined on the interval $[-10, 10]$ by

$$f(x) = \begin{cases} x - [x] & \text{if } [x] \text{ is odd} \\ 1 + [x] - x & \text{if } [x] \text{ is even.} \end{cases}$$

Then the value of $\frac{\pi^2}{10} \int_{-10}^{10} f(x) \cos \pi x \, dx$ is

40. Let $a_1, a_2, a_3, \ldots, a_{11}$ be real numbers satisfying $a_1 = 15$, $27 - 2a_2 > 0$ and $a_k = 2a_{k-1} - a_{k-2}$ for $k = 3, 4, \ldots, 11$. If $\frac{a_1^2 + a_2^2 + \ldots + a_{11}^2}{11} = 90$ then the value of $\frac{a_1 + a_2 + \ldots + a_{11}}{11}$ is equal to

41. Two parallel chords of a circle of radius 2 are at a distance $\sqrt{3} + 1$ apart. If the chords subtend at the centre, angles of magnitude $\frac{\pi}{k}$ and $\frac{2\pi}{k}$, where $k > 0$ then the value of $[k]$, the largest integer less than or equal to k, is

42. Consider a $\Delta\, ABC$. Suppose $BC = 6$, $CA = 10$ and the area of the triangle is $15\sqrt{3}$. If $\angle ACB > 90°$ and r denotes the radius of the incircle of the triangle then r^2 is equal to

43. Let k be a positive real number and

$$A = \begin{bmatrix} 2k-1 & 2\sqrt{k} & 2\sqrt{k} \\ 2\sqrt{k} & 1 & -2k \\ -2\sqrt{k} & 2k & -1 \end{bmatrix} \text{ and } B = \begin{bmatrix} 0 & 2k-1 & \sqrt{k} \\ 1-2k & 0 & 2\sqrt{k} \\ -\sqrt{k} & -2\sqrt{k} & 0 \end{bmatrix}.$$

If det (adj A) + det (adj B) = 10^6 then $[k]$ is equal to

[**Note:** Adj M denotes the adjoint of the matrix M and $[k]$ denotes the largest integer $\leq k$.]

44. Let f be a function defined on R, the set of real numbers, such that

$$f'(x) = 2010(x-2009)(x-2010)^2(x-2011)^3(x-2012)^4 \text{ for all } x \in R.$$

If g is a function defined on R with values in $(0, \infty)$ such that

$$f(x) = \log_e g(x) \text{ for all } x \in R$$

then the number of points in R at which g has a local maximum is

ANSWERS

1. b	**2.** a	**3.** d	**4.** a	**5.** d
6. a	**7.** b	**8.** c	**9.** a	**10.** d
11. b	**12.** b	**13.** c	**14.** c	**15.** a, c, d
16. b	**17.** c, d	**18.** a	**19.** b, c	**20.** b
21. a	**22.** c	**23.** a	**24.** b	**25.** d
26. c	**27.** a			

28.

	p	q	r	s	t
A	○	●	●	○	○
B	●	○	○	○	○
C	●	○	○	●	●
D	○	●	●	●	●

29.

	p	q	r	s	t
A	○	○	○	○	●
B	●	○	●	○	○
C	○	●	○	○	○
D	○	○	●	○	○

30.	31.	32.	33.	34.	35.	36.	37.	38.	39.	40.	41.	42.	43.	44.
0	0	0	0	0	0	0	0	0	0	**0**	0	0	0	0
1	1	1	1	1	1	1	1	1	1	1	1	1	1	**1**
2	2	2	2	**2**	**2**	2	2	2	2	2	2	2	2	2
3	**3**	**3**	**3**	3	3	3	3	3	3	3	**3**	**3**	3	3
4	4	4	4	4	4	4	4	4	**4**	4	4	4	**4**	4
5	5	5	5	5	5	**5**	5	5	5	5	5	5	5	5
6	6	6	6	6	6	6	**6**	6	6	6	6	6	6	6
7	7	7	7	7	7	7	7	7	7	7	7	7	7	7
8	8	8	8	8	8	8	8	8	8	8	8	8	8	8
9	9	9	9	9	9	9	9	**9**	9	9	9	9	9	9

Hints and Solutions to Selected Questions

1. The quadratic equation is $x^2 - \left(\frac{\alpha}{\beta} + \frac{\beta}{\alpha}\right)x + \frac{\alpha}{\beta} \cdot \frac{\beta}{\alpha} = 0$ or $x^2 - \frac{\alpha^2 + \beta^2}{\alpha\beta}x + 1 = 0.$

Now $(\alpha + \beta)^3 = \alpha^3 + \beta^3 + 3\alpha\beta(\alpha + \beta) \Rightarrow -p^3 = q + 3\alpha\beta(-p) \quad \therefore \ \alpha\beta = \frac{p^3 + q}{3p}$

$$\therefore \quad \alpha^2 + \beta^2 = (\alpha + \beta)^2 - 2\alpha\beta = p^2 - \frac{2(p^3 + q)}{3p} = \frac{p^3 - 2q}{3p}$$

$$\therefore \quad \text{the equation is } x^2 - \frac{p^3 - 2q}{3p} \cdot \frac{3p}{p^3 + q}x + 1 = 0$$

or $\quad (p^3 + q)x^2 - (p^3 - 2q)x + (p^3 + q) = 0.$

2. No. of unordered pairs of disjoint subsets, one having one element and the other having elements from the remaining three elements = ${}^4C_1 \times (2^3 - 1)$.

No. of unordered pairs of disjoint subsets, one having two elements and the other having elements from the remaining two elements = ${}^4C_2 \times (2^2 - 1)$.

No. of unordered pairs of disjoint subsets, one having three elements and the other having elements from the remaining one element = ${}^4C_3 \times 1$.

Each way will appear twice.

So, the required total number $= \frac{{}^4C_1 \times (2^3 - 1) + {}^4C_2 \times (2^2 - 1) + {}^4C_3 \times 1}{2}$

$$= \frac{4 \times 7 + 6 \times 3 + 4}{2} = \frac{50}{2} = 25.$$

3. $A_r = {}^{10}C_r, B_r = {}^{20}C_r, C_r = {}^{30}C_r$

$A_r(B_{10}B_r - C_{10}A_r)$

$$= {}^{10}C_r({}^{20}C_{10}B_r - {}^{30}C_{10}A_r)$$

$$= {}^{10}C_r({}^{20}C_{10} \cdot {}^{20}C_r - {}^{30}C_{10} \cdot {}^{10}C_r)$$

$$= {}^{20}C_{10}({}^{10}C_r \cdot {}^{20}C_r) - {}^{30}C_{10}({}^{10}C_r)^2$$

$$\therefore \quad \text{value} = {}^{20}C_{10} \cdot \sum_{1}^{10} {}^{10}C_r \cdot {}^{20}C_r - {}^{30}C_{10} \cdot \sum_{1}^{10} ({}^{10}C_r)^2$$

Now $(1 + x)^{10} \cdot (x + 1)^{20} = (1 + {}^{10}C_1x + \ldots + {}^{10}C_rx^r + \ldots)$
$\times (x^{20} + {}^{20}C_1x^{19} + \ldots + {}^{20}C_rx^{20-r} + \ldots)$

Equating coefficients of x^{20},

$${}^{30}C_{20} = 1 + {}^{10}C_1 \cdot {}^{20}C_1 + {}^{10}C_2 \cdot {}^{20}C_2 + \ldots + {}^{10}C_{10} \cdot {}^{20}C_{10}.$$

Again $(1+x)^{10}(x+1)^{10} = (1 + {}^{10}C_1 x + {}^{10}C_2 x^2 + \ldots)\,(x^{10} + {}^{10}C_1 x^9 + \ldots)$

Equating coefficients of x^{10},

$$^{20}C_{10} = 1 + {}^{10}C_1 \cdot {}^{10}C_1 + \ldots + {}^{10}C_{10} \cdot {}^{10}C_{10}.$$

$$\therefore \quad \text{value} = {}^{20}C_{10}({}^{30}C_{20} - 1) - {}^{30}C_{10}({}^{20}C_{10} - 1)$$

$$= {}^{30}C_{10} - {}^{20}C_{10} = C_{10} - B_{10}.$$

4. Three linear equations in three unknown will have either no sol. or unique sol. or infinite sol. So for exactly two solutions A does not exist.

5. Here $B = \dfrac{A+C}{2}$ and $A + B + C = 180°$ $\quad \therefore \quad B = \dfrac{180°}{3} = 60°$

$$\text{Value} = \frac{\sin A}{\sin C} \cdot 2\sin C \cdot \cos C + \frac{\sin C}{\sin A} \cdot 2\sin A \cdot \cos A$$

$$= 2(\sin A \cdot \cos C + \sin C \cdot \cos A)$$

$$= 2 \cdot \sin(A+C) = 2 \cdot \sin 120° = 2 \cdot \frac{\sqrt{3}}{2} = \sqrt{3}.$$

6. $\vec{PQ} = \vec{OQ} - \vec{OP} = 4\hat{i} - (-2\hat{i} - \hat{j}) = 6\hat{i} + \hat{j}$

$\vec{SR} = \vec{OR} - \vec{OS} = 3\hat{i} + 3\hat{j} - (-3\hat{i} + 2\hat{j}) = 6\hat{i} + \hat{j}$

$\therefore \quad PQ \parallel = SR$. So, $PQRS$ is a parallelogram.

$\vec{QR} = \vec{OR} - \vec{OQ} = 3\hat{i} + 3\hat{j} - 4\hat{i} = -\hat{i} + 3\hat{j}.$

Clearly $PQ \neq QR$ and $\vec{PQ}$ is not perpendicular to $\vec{QR}$.

So, $PQRS$ is neither a rhombus nor a rectangle.

7. $\cos(90° - \alpha) = \dfrac{2\hat{i} + 10\hat{j} + 11\hat{k}}{|2\hat{i} + 10\hat{j} + 11\hat{k}|} \cdot \dfrac{-\hat{i} + 2\hat{j} + 2\hat{k}}{|\hat{i} + 2\hat{j} + 2\hat{k}|}$

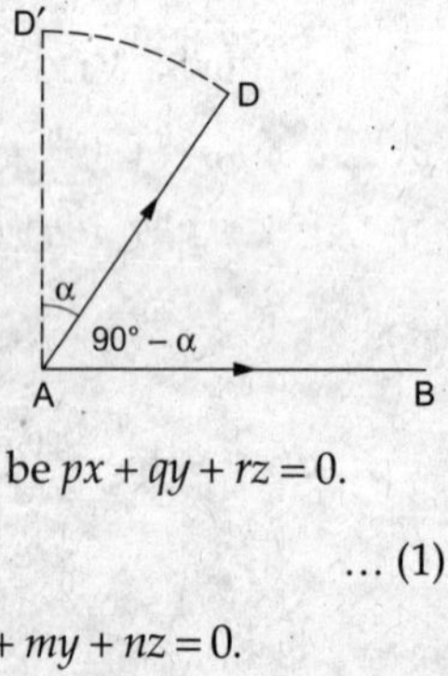

$$\therefore \quad \sin\alpha = \frac{-2 + 20 + 22}{15 \times 3} = \frac{40}{45} = \frac{8}{9}$$

$$\therefore \quad \cos\alpha = \frac{\sqrt{81 - 64}}{9} = \frac{\sqrt{17}}{9}.$$

8. Let the equation of the plane containing $\frac{x}{2} = \frac{y}{3} = \frac{z}{4}$ be $px + qy + rz = 0$.

$\therefore \quad 2p + 3q + 4r = 0 \quad \ldots (1)$

Let the plane containing the other two lines be $lx + my + nz = 0$.

Then $3l + 4m + 2n = 0$ and $4l + 2m + 3n = 0 \Rightarrow \dfrac{l}{8} = \dfrac{m}{-1} = \dfrac{n}{-10}$.

$\therefore$ the plane containing the other two lines is $8x - y - 10z = 0$.

The two planes are perpendicular to each other. So, $8p - q - 10r = 0$ ... (2)

Solving (1), (2) we get $\frac{p}{-26} = \frac{q}{52} = \frac{r}{-26}$ or $\frac{p}{1} = \frac{q}{-2} = \frac{r}{1}$.

$\therefore$ the equation of the plane containing the first line is $x - 2y + z = 0$.

9. Here $5 = \left| \frac{1 + 2(-2) - 2 \cdot 1 - \alpha}{\sqrt{1^2 + 2^2 + (-2)^2}} \right| = \frac{|5 + \alpha|}{3}$

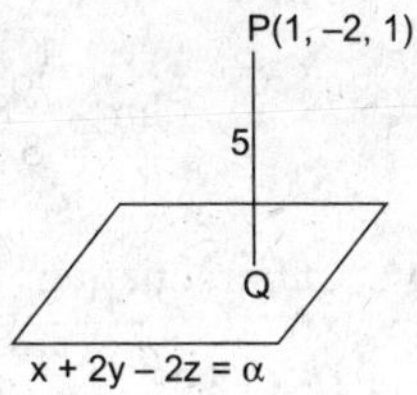

$\therefore$ $5 + \alpha = \pm 15$

$\therefore$ $\alpha = 10, -20$. But $\alpha > 0$. So, $\alpha = 10$.

d.r. of the normal = 1, 2, –2

$\therefore$ equation of PQ : $\frac{x-1}{1} = \frac{y+2}{2} = \frac{z-1}{-2} = r$ (say)

$\therefore$ $Q = (r + 1, 2r - 2, 1 - 2r)$

$\therefore$ $r + 1 + 2(2r - 2) - 2(1 - 2r) = \alpha = 10$

$\therefore$ $r = \frac{5}{3}$

$\therefore$ $Q = \left(\frac{5}{3} + 1, \frac{10}{3} - 2, 1 - \frac{10}{3}\right) = \left(\frac{8}{3}, \frac{4}{3}, \frac{-7}{3}\right)$.

10. $f'(x) = e^{x^2} \cdot 2x + e^{-x^2} \cdot (-2x)$

$= 2x(e^{x^2} - e^{-x^2}) \geq 0$, where $0 \leq x \leq 1$

$\therefore$ $f(x)$ is *m. i.* in [0, 1]. So, $f(1)$ is max.

$\therefore$ $a = f(1) = e + e^{-1}$

$g(x) = (x - 1)e^{x^2} + f(x)$

$\therefore$ $g'(x) = e^{x^2} + (x - 1)\, 2xe^{x^2} + f'(x)$

$= e^{x^2}\{1 + 2x^2 - 2x\} + f'(x)$

$= e^{x^2}\{x^2 + (x - 1)^2\} + f'(x)$

≥ 0 when $0 \leq x \leq 1$ because $f'(x) \geq 0$

$\therefore$ $g(x)$ is *m. i* in [0, 1]. So, $g(1)$ is max.

$\therefore$ $b = g(1) = f(1) = e + e^{-1}$

$h(x) = (x^2 - 1)\, e^{x^2} + f(x)$

$\therefore$ $h'(x) = 2xe^{x^2} + (x^2 - 1) \cdot 2xe^{x^2} + f'(x)$

$= 2xe^{x^2} \cdot (1 + x^2 - 1) + f'(x) = 2x^3 e^{x^2} + f'(x)$

≥ 0 when $0 \leq x \leq 1$ because $f'(x) \geq 0$

$\therefore$ $h(x)$ is $m.\ i.$ in $[0, 1]$. So, $h(1)$ is max

$\therefore$ $c = h(1) = f(1) = e + \dfrac{1}{e}$ $\therefore$ $a = b = c.$

11. Limit $= \lim\limits_{x \to 0} \dfrac{\left\{\dfrac{x\log(1+x)}{x^4+4}\right\}}{3x^2}$, (using L 'Hospital's rule)

$$= \lim_{x \to 0} \frac{\log(1+x)}{3x(x^4+4)} = \lim_{x \to 0} \frac{\dfrac{1}{3(1+x)}}{5x^4+4} = \frac{1}{3 \cdot 4} = \frac{1}{12}.$$

12. Differentiating,

$$-e^{-x} \cdot f(x) + e^{-x} \cdot f'(x) = \sqrt{x^4+1}$$

or $f'(x) - f(x) = e^x \cdot \sqrt{x^4+1}$

$\therefore$ $f'(0) - f(0) = 1$

But $f(0) = e^0 \cdot \left\{2 + \int_0^0 \sqrt{t^4+1}\, dt\right\} = 2$

$\therefore$ $f^{-1}(2) = 0$

$\therefore$ $f'(0) = 1 + f(0) = 1 + 2 = 3.$

Now, $f(f^{-1}(x)) = x.$

Differentiating w.r.t., x, $f'(f^{-1}(x)) \cdot (f^{-1}(x))' = 1$

$\therefore$ $(f^{-1}(x))' = \dfrac{1}{f'(f^{-1}(x))}$

$\therefore$ $(f^{-1}(2))' = (f^{-1})'(2) = \dfrac{1}{f'(f^{-1}(2))} = \dfrac{1}{f'(0)} = \dfrac{1}{3}.$

13. $n(S) = 6 \times 6 \times 6$

For favourable situations r_1, r_2, r_3 must appear from $\{1, 4\}, \{2, 5\}, \{3, 6\}$ in any order to make $w^{r_1} + w^{r_2} + w^{r_3} = 0.$

$\therefore$ $n(E) = {}^2C_1 \times {}^2C_1 \times {}^2C_1 \times {}^3P_3 = 2 \times 2 \times 2 \times 6$

$\therefore$ $P(E) = \dfrac{n(E)}{n(S)} = \dfrac{2 \times 2 \times 2 \times 6}{6 \times 6 \times 6} = \dfrac{2}{9}.$

14. Let G = the event of original signal is green

E_1 = the event of A receives the signal correctly

E_2 = the event of B receives the signal correctly

E = the event of the signal received by B is correct.

Here $P(G) = \frac{4}{5}$, $P(E_1) = \frac{3}{4}$, $P(E_2) = \frac{3}{4}$.

Now, $P(E) = P(GE_1E_2) + P(G\overline{E}_1\overline{E}_2) + P(\overline{G}E_1\overline{E}_2) + P(\overline{G}\,\overline{E}_1E_2)$

$$= P(G)\cdot P(E_1)\cdot P(E_2) + \ldots\ldots + P(\overline{G})\,P(\overline{E}_1)\,P(E_2)$$

$$= \frac{4}{5}\cdot\frac{3}{4}\cdot\frac{3}{4} + \frac{4}{5}\cdot\frac{1}{4}\cdot\frac{1}{4} + \frac{1}{5}\cdot\frac{3}{4}\cdot\frac{1}{4} + \frac{1}{5}\cdot\frac{1}{4}\cdot\frac{3}{4} = \frac{46}{80}$$

$$\text{Required probability} = P(G/E) = \frac{P(GE_1E_2) + P(G\overline{E}_1\overline{E}_2)}{P(E)} = \frac{\frac{40}{80}}{\frac{46}{80}} = \frac{20}{23}.$$

15. As $z = (1-t)z_1 + tz_2$, we get z_1, z, z_2 are colliner.

$\therefore \quad |z - z_1| + |z - z_2| = |z_1 - z_2| \quad (\because PA + PB = AB)$

$\arg(z - z_1) - \arg(z_2 - z_1) = \angle PAB = 0$

$\therefore \quad \arg(z - z_1) = \arg(z_2 - z_1)$

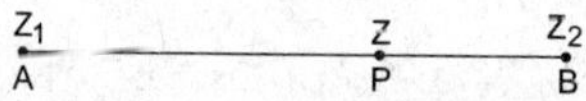

Also, $\arg\dfrac{z - z_1}{z - z_1} = 0 = \arg\dfrac{\overline{z} - \overline{z}_1}{\overline{z}_2 - \overline{z}_1}$

$\therefore \quad \dfrac{z - z_1}{z_2 - z_1} = \dfrac{\overline{z} - \overline{z}_1}{\overline{z}_2 - \overline{z}_1} \quad$ or $\quad (z - z_1)(\overline{z}_2 - \overline{z}_1) - (z_2 - z_1)(\overline{z} - \overline{z}_1) = 0$

or $\quad \begin{vmatrix} z - z_1 & \overline{z} - \overline{z}_1 \\ z_2 - z_1 & \overline{z}_2 - \overline{z}_1 \end{vmatrix} = 0.$

16. $\cos C = \dfrac{a^2 + b^2 - c^2}{2ab} = \dfrac{(x^2 + x + 1)^2 + (x^2 - 1)^2 - (2x+1)^2}{2(x^2 + x + 1)(x^2 - 1)}$

$\therefore \quad \cos\dfrac{\pi}{6} = \dfrac{\sqrt{3}}{2} = \dfrac{(x^2 + 3x + 2)(x^2 - x) + (x^2 - 1)^2}{2(x^2 + x + 1)(x^2 - 1)}$

or $\quad \sqrt{3} = \dfrac{(x+2)\,x(x^2 - 1) + (x^2 - 1)^2}{(x^2 + x + 1)(x^2 - 1)} = \dfrac{x(x+2) + x^2 - 1}{x^2 + x + 1}$

$$= \frac{2(x^2 + x + 1) - 3}{x^2 + x + 1} = 2 - \frac{3}{x^2 + x + 1}$$

$\therefore \quad \dfrac{3}{x^2 + x + 1} = 2 - \sqrt{3} \quad$ or $\quad x^2 + x + 1 = \dfrac{3}{2 - \sqrt{3}} = 3(2 + \sqrt{3})$

or $\quad \left(x + \dfrac{1}{2}\right)^2 + \dfrac{3}{4} = 6 + 3\sqrt{3}$

or $\quad \left(x + \dfrac{1}{2}\right)^2 = 6 - \dfrac{3}{4} + 3\sqrt{3} = \dfrac{21 + 12\sqrt{3}}{4} = \dfrac{3}{4}(7 + 4\sqrt{3})$

$\therefore \quad x+\frac{1}{2}=\pm\frac{\sqrt{3}}{2}(2+\sqrt{3})=\pm\left(\sqrt{3}+\frac{3}{2}\right)$

or $\quad x=\frac{-1}{2}+\sqrt{3}+\frac{3}{2}, -\frac{1}{2}-\sqrt{3}-\frac{3}{2}$

$\therefore \quad x=1+\sqrt{3}, -\sqrt{3}-2$. But $c=2x+1>0$

$\therefore \quad x>-\frac{1}{2}$ but $-\sqrt{3}-2<-\frac{1}{2}$.

$\therefore \quad x=1+\sqrt{3}$.

17. 'm' of $AB=\frac{2t_1-2t_2}{t_1^2-t_2^2}=\frac{2}{t_1+t_2}$

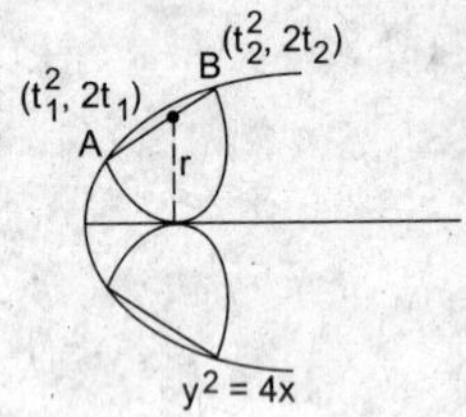

As per circle touches the axis, the perpendicular to the axis passes through the centre.

$\therefore \quad r=\pm$(y-coordinate of the centre) $=\pm(t_1+t_2)$

$\therefore \quad$ 'm' of $AB=\pm\frac{2}{r}$.

18. Value $=\int_0^1 \frac{x^4(x^4-4x^3+6x^2-4x+1)}{1+x^2}dx=\int_0^1 \frac{x^8-4x^7+6x^6-4x^5+x^4}{x^2+1}dx$

$=\int_0^1\left(x^6-4x^5+5x^4-4x^2+4-\frac{4}{1+x^2}\right)dx$

$=\left[\frac{x^7}{7}-\frac{2}{3}x^6+x^5-\frac{4}{3}x^3+4x-4\tan^{-1}x\right]_0^1$

$=\frac{1}{7}-\frac{2}{3}+1-\frac{4}{3}+4-4\cdot\frac{\pi}{4}=\frac{22}{7}-\pi.$

19. $f(x)=\log_e x+\int_0^x \sqrt{1+\sin t}\,dt \qquad \therefore \qquad f'(x)=\frac{1}{x}+\sqrt{1+\sin x}$

$\therefore \quad f''(x)=-\frac{1}{x^2}+\frac{1}{2\sqrt{1+\sin x}}\cdot\cos x$

which does not exist when $x=0$ or $\sin x=-1$.

$\therefore \quad f'(x)$ is not differentiable on $(0, \infty)$. But $f'(x)$ exists in $(0, \infty)$ and $f'(x)$ is continuous in $(0, \infty)$ because $y=\frac{1}{x}$ and $y=\sqrt{1+\sin x}$ are continuous.

Clearly, $|f'(x)|\le\left|\frac{1}{x}\right|+\sqrt{1+\sin x}\le 3$ if $x>1$.

If $x>e^3$, $|f(x)|>3$. $\quad\therefore\quad$ for $\alpha=e^3$, $|f'(x)|<|f(x)|$ for all $x\in(\alpha,\infty)$.

20. Any tangent to the hyperbola is $y = mx + \sqrt{9m^2 - 4}$. Here $m > 0$.

It touches the circle if the length of perpendicular from the centre (4, 0) to the line is equal to radius of the circle, that is

$$\frac{4m + \sqrt{9m^2 - 4} - 0}{\sqrt{m^2 + 1}} = 4 \text{ or } 16m^2 + 9m^2 - 4 + 8m\sqrt{9m^2 - 4} = 16(1 + m^2)$$

or $\quad 8m\sqrt{9m^2 - 4} = 20 - 9m^2 \quad$ or $\quad 64m^2(9m^2 - 4) = (20 - 9m^2)^2$

or $\quad 495m^4 + 104m^2 - 400 = 0 \quad$ or $\quad (5m^2 - 4)(99m^2 + 100) = 0$

$\therefore \quad 5m^2 - 4 = 0$ because m is real.

$\therefore \quad m = \frac{2}{\sqrt{5}} \; (\because m > 0) \quad \therefore$ the common tangent is $y = \frac{2}{\sqrt{5}}x + \sqrt{9 \cdot \frac{4}{5} - 4}$

or $\quad y = \frac{2}{\sqrt{5}}x + \frac{4}{\sqrt{5}} \quad$ or $\quad 2x - \sqrt{5}y + 4 = 0.$

21. Any point on the hyperbola $\frac{x^2}{9} - \frac{y^2}{4} = 1$ is $(3\sec\theta, 2\tan\theta)$. It is on the circle

if $\quad (3\sec\theta)^2 + (2\tan\theta)^2 - 8 \cdot 3\sec\theta = 0$

or $\quad 9\sec^2\theta + 4(\sec^2\theta - 1) - 24\sec\theta = 0$

or $\quad 13\sec^2\theta - 24\sec\theta - 4 = 0$

or $\quad (13\sec\theta + 2)(\sec\theta - 2) = 0 \quad \therefore \quad \sec\theta = 2, \frac{-2}{13}$ (absurd)

When $\sec\theta = 2$, $\tan\theta = \pm\sqrt{3}$. $\quad \therefore \quad A, B = (6, 2\sqrt{3}), (6, -2\sqrt{3})$

$\therefore$ the required circle is $(x - 6)(x - 6) + (y - 2\sqrt{3})(y + 2\sqrt{3}) = 0$

or $\quad x^2 - 12x + 36 + y^2 - 12 = 0. \quad$ or $\quad x^2 + y^2 - 12x + 24 = 0.$

22. $f'(x) = 2 + 6x + 12x^2$ for $f'(x) = 0$, discriminant $= 6^2 - 4 \cdot 12 \cdot 2 < 0$.

So, $f'(x) = 0$ has imaginary roots. So, $f(x) = 0$ must have a pair of imaginary roots. $f(x) = 0$ being a cubic equation, $f(x) = 0$ has only one real root which is s.

Now, $f\left(-\frac{1}{2}\right) = 1 + 2\left(-\frac{1}{2}\right) + 3\left(-\frac{1}{2}\right)^2 + 4\left(-\frac{1}{2}\right)^3 = 1 - 1 + \frac{3}{4} - \frac{1}{2} > 0$

$$f\left(-\frac{3}{4}\right) = 1 + 2\left(-\frac{3}{4}\right) + 3\left(-\frac{3}{4}\right)^2 + 4\left(-\frac{3}{4}\right)^3 = 1 - \frac{3}{2} + \frac{27}{16} - \frac{27}{16} = -\frac{1}{2} < 0.$$

$\therefore f(x)$ changes its sign in $\left(-\frac{3}{4}, -\frac{1}{2}\right)$. So, $f(x) = 0$ has its root s in $\left(-\frac{3}{4}, -\frac{1}{2}\right)$.

23. $|s| = t$. As $-\frac{3}{4} < s < -\frac{1}{2}$ we get $\frac{1}{2} < 1 < \frac{3}{4}$.

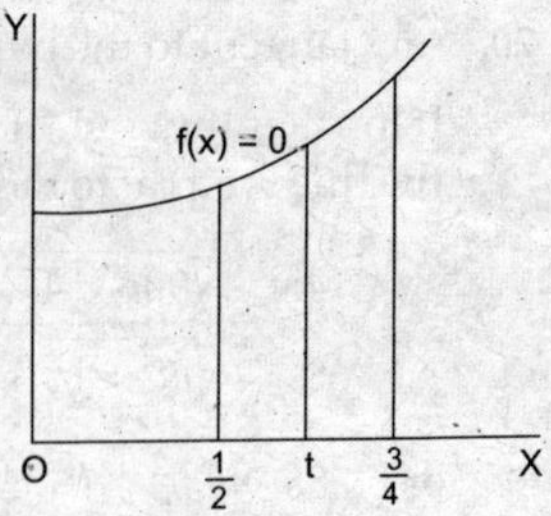

Clearly, $\int_0^{1/2} f(x)\,dx < \int_0^t f(x)\,dx < \int_0^{3/4} f(x)\,dx$ ... (1)

As $f(x) = 1 + 2x + 3x^2 + 4x^3$,

$\int_0^t f(x)\,dx = [x + x^2 + x^3 + x^4]_0^t$

$\therefore$ (1) gives, $\frac{1}{2} + \frac{1}{4} + \frac{1}{8} + \frac{1}{16} < \int_0^t f(x)\,dx < \frac{3}{4} + \frac{9}{16} + \frac{27}{64} + \frac{81}{256}$

or $\frac{15}{16} < \int_0^t f(x)\,dx < \frac{525}{256}$

$\therefore$ $\frac{3}{4} < \frac{15}{16} < \int_0^t f(x)\,dx < \frac{525}{256} < 3.$

24. $f'(x) = 2 + 6x + 12x^2$

$\therefore$ $f''(x) = 6 + 24x = 24\left(x - \frac{-1}{4}\right) \Rightarrow f''(x) < 0$ if $x < \frac{-1}{4}$

> 0 if $x > -\frac{1}{4}$.

$\therefore$ $f'(x)$ is decreasing if $x < -\frac{1}{4}$ and so in $\left(-t, -\frac{1}{4}\right)$

$\therefore$ $f'(x)$ is increasing if $x > -\frac{1}{4}$ and so in $\left(-\frac{1}{4}, t\right)$.

25. The tangent at (x_1, y_1) is $\frac{xx_1}{9} + \frac{yy_1}{4} = 1$. It passes through (3, 4)

$\therefore$ $\frac{x_1}{3} + y_1 = 1$. Also $\frac{x_1^2}{9} + \frac{y_1^2}{4} = 1 \Rightarrow (1 - y_1)^2 + \frac{y_1^2}{4} = 1$ or $y_1\left(\frac{5y_1}{4} - 2\right) = 0$

$\therefore$ $y_1 = 0, \frac{8}{5}$. So, $x_1 = (1-0)3, \left(1 - \frac{8}{5}\right)3 = 3, \frac{-9}{5}$

$\therefore$ $A, B = (3, 0), \left(-\frac{9}{5}, \frac{8}{5}\right)$.

26. Here the vertices of the $\Delta\, PAB$ are

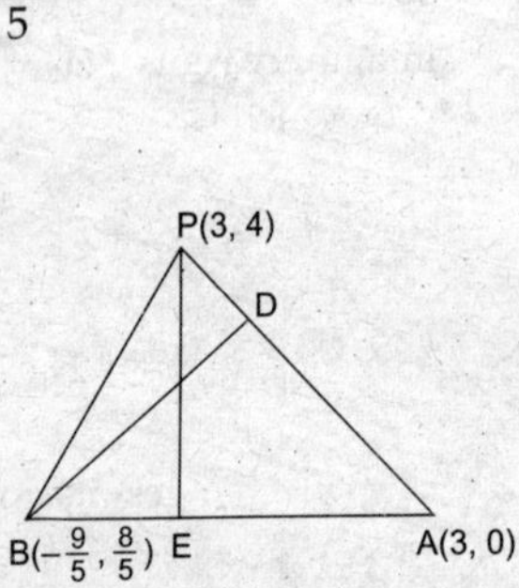

$P = (3, 4), A = (3, 0), B = \left(-\frac{9}{5}, \frac{8}{5}\right)$.

Equation of AB is $y - 0 = \frac{\frac{8}{5}}{-\frac{9}{5} - 3}(x - 3)$

or $\quad y = -\frac{1}{3}(x-3).$

$\therefore$ equation of PE is $y - 4 = 3(x-3)$ or $3x - y = 5$ ... (1)

AP has 'm' $= \frac{4-0}{3-3} = \infty$

$\therefore$ BD has $m = 0$ (being $\perp^{r}$ to AP)

$\therefore$ equation of BD is $y - \frac{8}{5} = 0$ ... (2)

Solving (1) and (2), $3x = 5 + \frac{8}{5}$ or $x = \frac{11}{5}$

$\therefore$ orthocentre $= \left(\frac{11}{5}, \frac{8}{5}\right)$.

27. Equation of AB is $x + 3y = 3$ and $P = (3, 4)$ [See Q. N. 26]

$\therefore$ the equation of the locus is

$$\left|\frac{x+3y-3}{\sqrt{1^2+3^2}}\right| = \sqrt{(x-3)^2+(y-4)^2}$$

or $\quad 10(x^2 + y^2 - 6x - 8y + 25) = (x + 3y - 3)^2$

or $\quad 9x^2 + y^2 - 6xy - 54x - 62y + 241 = 0.$

28. (A) $\left|\frac{z}{|z|} - i\right| = \left|\frac{z}{|z|} + i\right| \Rightarrow |\cos\theta + i\sin\theta - i| = |\cos\theta + i\sin\theta + i|,$

$\because \frac{z}{|z|}$ = unimodular complex number

$\therefore \cos^2\theta + (\sin\theta - 1)^2 = \cos\theta + (\sin\theta + 1)^2$ or $\sin\theta = 0$

$\therefore \text{Im}\left(\frac{z}{|z|}\right) = 0 \qquad \therefore \text{Im}(z) = 0.$

This satisfies $|\text{Im}(z)| \le 1$ also. $\therefore$ A $\to$ q, r.

(B) The locus of z is an ellipse with foci at $S'(z=-4)$, $S(z=4)$ and focal distance $2ae = SS' = 8$ and major axis $= 2a = 10$.

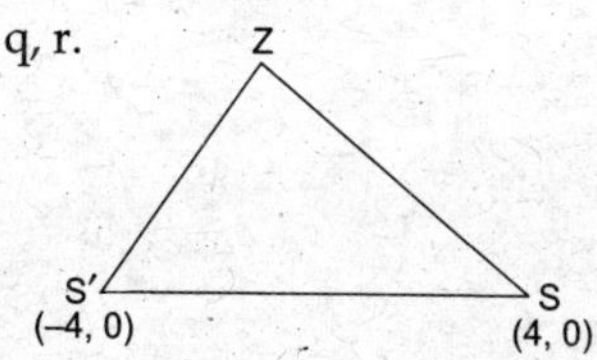

$\therefore ae = 4, a = 5 \quad \therefore e = \frac{4}{5} \quad \therefore$ B $\to$ p.

(C) $\omega = 2(\cos\theta + i\sin\theta).$

So, $z = 2(\cos\theta + i\sin\theta) - \frac{1}{2(\cos\theta + i\sin\theta)}$

$= 2(\cos\theta + i\sin\theta) - \frac{1}{2}(\cos\theta - i\sin\theta)$

$= \frac{3}{2}\cos\theta + i\frac{5}{2}\sin\theta.$

$$\therefore \quad |z| = \sqrt{\frac{9}{4}\cos^2\theta + \frac{25}{4}\sin^2\theta} = \sqrt{\frac{9}{4} + 4\sin^2\theta} \le \sqrt{\frac{9}{4} + 4} = \frac{5}{2} < 3$$

$$|\text{Re}(z)| = \frac{3}{2}\cos\theta \le \frac{3}{2} < 2.$$

Again, $z = x + iy$, where $x = \frac{3}{2}\cos\theta,\ y = \frac{5}{2}\sin\theta$

$$\therefore \quad \left(\frac{2x}{3}\right)^2 + \left(\frac{2y}{5}\right)^2 = 1 \text{ or } \frac{x^2}{\left(\frac{3}{2}\right)^2} + \frac{y^2}{\left(\frac{5}{2}\right)^2} = 1.$$ Here $b = \frac{5}{2}, a = \frac{2}{3}$ in the ellipse.

$$\therefore \quad a^2 = b^2(1 - e^2) \text{ or } \frac{9}{4} = \frac{25}{4}(1 - e^2) \text{ or } \frac{9}{25} = 1 - e^2$$

$$\therefore \quad e^2 = \frac{16}{25} \qquad \therefore \quad e = \frac{4}{5} \qquad \therefore \quad C \to t, s, p.$$

(D) $|\omega| = \Rightarrow \omega = \cos\theta + i\sin\theta \qquad \therefore \quad \frac{1}{\omega} = \cos\theta - i\sin\theta$

$$\therefore \quad z = \omega + \frac{1}{\omega} = 2\cos\theta \qquad \therefore \quad \text{Im}(z) = 0; \text{ so, Im}(z) < 1.$$

Also, $\text{Re}(z) = 2\cos\theta \le 2$ and $|z| = |2\cos\theta| \le 2 < 3 \quad \therefore \quad D \to q, r, s, t.$

29. (A) Here $P = (2 + r, 1 - 2r, -1 + r)$ and $Q = \left(\frac{8}{3} + 2s, -3 - s, 1 + s\right)$, and O, P, Q are collinear.

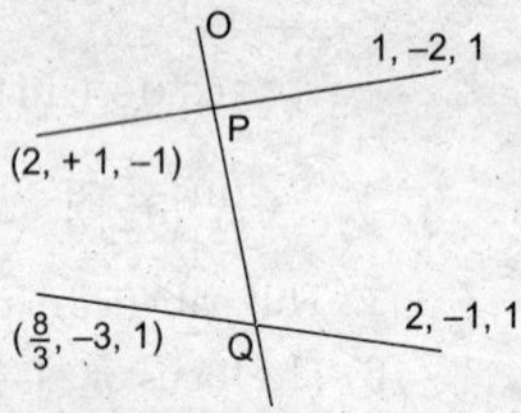

$$\therefore \quad \frac{2 + r}{\frac{8}{3} + 2s} = \frac{1 - 2r}{-3 - s} = \frac{-1 + r}{1 + s}$$

$$= \frac{-r}{-2} = \frac{3}{\frac{5}{3} + s} = \frac{-1}{-1 + s} = \frac{4}{\frac{8}{3}} = \frac{3}{2}$$

$$\therefore \quad r = 3, -3 + 3s = -2, \text{ i.e., } s = \frac{1}{3}$$

$$\therefore \quad P = (5, -5, 2) \text{ and } Q = \left(\frac{10}{3}, -\frac{10}{3}, \frac{4}{3}\right)$$

$$\therefore \quad PQ^2 = \left(\frac{5}{3}\right)^2 + \left(-\frac{5}{3}\right)^2 + \left(\frac{2}{3}\right)^2 = \frac{54}{9} = 6. \qquad \therefore \quad A \to t.$$

(B) $\tan^{-1}\frac{x + 3 - (x - 3)}{1 + (x + 3)(x - 3)} = \tan^{-1}\frac{3}{4}$

or $\tan^{-1}\frac{6}{x^2 - 8} = \tan^{-1}\frac{3}{4} \qquad \therefore \quad \frac{6}{x^2 - 8} = \frac{3}{4}$

$$\therefore \quad x^2 = 16 \qquad \therefore \quad x = \pm 4 \qquad \therefore \quad B \to p, r$$

(C) Here $\vec{c}=\dfrac{\vec{a}-\mu\vec{b}}{4}$ and $\vec{a}\cdot\vec{b}=0$

$\therefore \quad (\vec{b}-\vec{a})\cdot\left(\vec{b}+\dfrac{\vec{a}-\mu\vec{b}}{4}\right)=0$ and $2\left|\vec{b}+\dfrac{\vec{a}-\mu\vec{b}}{4}\right|=|\vec{b}-\vec{a}|$

$\Rightarrow \quad (\vec{b}-\vec{a})\cdot(4\vec{b}+\vec{a}-\mu\vec{b})=0$ and $2|(4-\mu)\vec{b}+\vec{a}|=4|\vec{b}-\vec{a}|$

$\Rightarrow \quad (4-\mu)\vec{b}^2-\vec{a}^2=0$ and $(4-\mu)^2\vec{b}^2+\vec{a}^2=4(\vec{b}^2+\vec{a}^2)$

$\therefore \quad (4-\mu)^2\vec{b}^2-4\vec{b}^2=3\vec{a}^2=3(4-\mu)\vec{b}^2 \Rightarrow (4-\mu)^2-4=3(4-\mu)$

$\therefore \quad \mu^2-5\mu=0 \quad \therefore \quad \mu=0, 5.$

But $\mu\neq 5$ because $(4-\mu)\vec{b}^2-\vec{a}^2=0$ will give $\vec{a}^2+\vec{b}^2=0\Rightarrow\vec{a}=\vec{0},\ \vec{b}=\vec{0}$

$\therefore \quad C\to q.$

(D) Here $f(x)$ is an even function because $f(x)=f(-x)$.

$$\therefore \quad \frac{2}{\pi}\int_{-\pi}^{\pi} f(x)\,dx$$

$$=\frac{2}{\pi}\cdot 2\int_0^{\pi} f(x)\,dx=\frac{4}{\pi}\int_0^{\pi}\frac{\sin\frac{9x}{2}}{\sin\frac{9x}{2}}\,dx$$

$$=\frac{4}{\pi}\cdot 2\int_0^{\pi/2}\frac{\sin 9\theta}{\sin\theta}\,d\theta, \quad \left(\text{taking } \frac{x}{2}=\theta\right)$$

$$=\frac{8}{\pi}\int_0^{\pi/2}\frac{(\sin 9\theta-\sin 7\theta)+(\sin 7\theta-\sin 5\theta)+(\sin 5\theta-\sin 3\theta)+(\sin 3\theta-\sin\theta)+\sin\theta}{\sin\theta}\,d\theta$$

$$=\frac{8}{\pi}\int_0^{\pi/2}(\cos 8\theta+\cos 6\theta+\cos 4\theta+\cos 2\theta+1)\,d\theta$$

$$=\frac{8}{\pi}\left[\frac{\sin 8\theta}{8}+\frac{\sin 6\theta}{6}+\frac{\sin 4\theta}{4}+\frac{\sin 2\theta}{2}+\theta\right]_0^{\pi/2}$$

$$=\frac{8}{\pi}\left[0+\frac{\pi}{2}-0\right]=4. \qquad D\to r.$$

30. $$\text{Det.}=z\begin{vmatrix}1 & w & \omega^2\\ 1 & z+\omega^2 & 1\\ 1 & 1 & z+\omega\end{vmatrix}=z\begin{vmatrix}1 & \omega & \omega^2\\ 0 & z+\omega^2-\omega & 1-\omega^2\\ 0 & 1-\omega & z+\omega-\omega^2\end{vmatrix}$$

$$=(z+\omega^2-\omega)(z+\omega-\omega^2)-(1-\omega)(1-\omega^2)$$

(Clearly $\omega=-\dfrac{1}{2}+i\dfrac{\sqrt{3}}{2}=$ imaginary cube root of 1)

$$=z^2+(0)z+(\omega^2-\omega)(\omega-\omega^2)-(1-\omega)(1-\omega^2)$$

$$= z^2 + 2 - \omega - \omega^2 - (2 - \omega - \omega^2) = z^2$$

$\therefore$ Det. $= 0 \Rightarrow z^2 = 0 \Rightarrow$ the number of distinct complex number z is 1.

31. Here $S_k = \dfrac{\dfrac{k-1}{k!}}{1-\dfrac{1}{k}} = \dfrac{1}{(k-1)!}$.

So, $\displaystyle\sum_{k=2}^{100} \left| (k^2 - 3k + 1)\frac{1}{(k-1)!} \right| = \sum_{k=2}^{100} \left| \frac{(k-1)^2 - k}{(k-1)!} \right| = \sum_{k=2}^{100} \left| \frac{k-1}{(k-2)!} - \frac{k}{(k-1)!} \right|$

$$= \left| \frac{1}{0!} - \frac{2}{1!} \right| + \left| \frac{2}{1!} - \frac{3}{2!} \right| + \left| \frac{3}{2!} - \frac{4}{3!} \right| + \ldots + \left| \frac{99}{98!} - \frac{100}{99!} \right|$$

$$= \left(\frac{2}{1!} - \frac{1}{0!} + \frac{2}{1!} \right) - \frac{3}{2!} + \frac{3}{2!} - \frac{4}{3!} + \ldots + \frac{99}{98!} - \frac{100}{99!}$$

$$= 2 - 1 + 2 - \frac{100}{99!} = 3 - \frac{100^2}{100!}$$

$\therefore$ the value of the sum $= \dfrac{100^2}{100!} + \left(3 - \dfrac{100^2}{100!} \right) = 3.$

32. From the first and third equations, $(y + z)\cos 3\theta = (y + 2z)\cos 3\theta + y \sin 3\theta$

or $z\cos 3\theta = 2z\cos 3\theta + y\sin 3\theta$ or $y\sin 3\theta + z\sin 3\theta = 0$.

From the second and third equations,

$$yz\left(\frac{2\cos 3\theta}{y} + \frac{2\cos 3\theta}{z} \right) = (y + 2z)\cos 3\theta + y\sin 3\theta$$

or $y\sin 3\theta = y\cos 3\theta$ or $\sin 3\theta = \cos 3\theta$ $(\because\ y \neq 0)$

or $\cos 3\theta = \cos\left(\dfrac{\pi}{2} - 3\theta \right)$ or $3\theta = 2n\pi \pm \left(\dfrac{\pi}{2} - 3\theta \right)$

$\therefore$ $6\theta = 2n\pi + \dfrac{\pi}{2}$ $\therefore$ $\theta = \dfrac{n\pi}{3} + \dfrac{\pi}{12}$ $\therefore$ $\theta = \dfrac{\pi}{12}, \dfrac{\pi}{3} + \dfrac{\pi}{12}, \dfrac{2\pi}{3} + \dfrac{\pi}{12}$.

33. $\tan\theta = \cot 5\theta = \tan\left(\dfrac{\pi}{2} - 5\theta \right) \Rightarrow \theta = n\pi + \dfrac{\pi}{2} - 5\theta \Rightarrow \theta = \dfrac{n\pi}{6} + \dfrac{\pi}{12}$

$\therefore$ $\theta = -\dfrac{\pi}{2} + \dfrac{\pi}{12}, -\dfrac{\pi}{3} + \dfrac{\pi}{12}, -\dfrac{\pi}{6} + \dfrac{\pi}{12}, \dfrac{\pi}{12}, \dfrac{\pi}{6} + \dfrac{\pi}{12}, \dfrac{\pi}{3} + \dfrac{\pi}{12} \left(\because\ -\dfrac{\pi}{2} < \theta < \dfrac{\pi}{2} \right)$

$$= -\frac{5\pi}{12}, -\frac{\pi}{4}, -\frac{\pi}{12}, \frac{\pi}{12}, \frac{\pi}{4}, \frac{5\pi}{12}$$

$\cos 4\theta = \sin 2\theta = \cos\left(\dfrac{\pi}{2} - 2\theta \right) \Rightarrow 4\theta = 2n\pi \pm \left(\dfrac{\pi}{2} - 2\theta \right)$

$$\Rightarrow 6\theta = 2n\pi + \frac{\pi}{2} \text{ and } 2\theta = 2n\pi - \frac{\pi}{2}$$

$$\Rightarrow \theta = \frac{n\pi}{3} + \frac{\pi}{12} \text{ and } \theta = n\pi - \frac{\pi}{4}$$

$$\therefore \quad \theta = -\frac{\pi}{3} + \frac{\pi}{12}, \frac{\pi}{12}, \frac{\pi}{3} + \frac{\pi}{12} \text{ and } -\frac{\pi}{4} \quad \left(\because \quad -\frac{\pi}{2} < \theta < \frac{\pi}{2}\right)$$

$$\therefore \quad \text{satisfying both, } \theta = -\frac{\pi}{4}, \frac{\pi}{12}, \frac{5\pi}{12}.$$

34. Expression $= \dfrac{1}{1 + 4\cos^2\theta + \frac{3}{2}\sin 2\theta} = \dfrac{1}{1 + 2(1 + \cos 2\theta) + \frac{3}{2}\sin 2\theta}$

$$= \frac{2}{6 + (4\cos 2\theta + 3\sin 2\theta)}$$

Also, $-\sqrt{4^2 + 3^2} \le 4\cos 2\theta + 3\sin 2\theta \le \sqrt{4^2 + 3^2}$

or $\quad -5 \le 4\cos 2\theta + 3\sin 2\theta \le 5$

$\therefore \quad \dfrac{2}{6+5} \le \text{expression} \le \dfrac{2}{6-5} \qquad \therefore$ maximum value of expression = 2.

35. $2x + y = 1$ passes through $\left(\dfrac{a}{e}, 0\right)$.

$\therefore \quad 2\dfrac{a}{e} = 1 \qquad \therefore \quad a = \dfrac{1}{2}e$

Using condition of tangency,

$1 = \sqrt{a^2m^2 - b^2} \quad$ (here $m = -2$)

$\therefore \quad 1 = 4a^2 - b^2 \qquad$ or $\quad 1 = e^2 - b^2$

But $b^2 = a^2(e^2 - 1)$. So, $e^2 - 1 = \left(\dfrac{1}{2}e\right)^2 \cdot (e^2 - 1)$ or $e^2 = 4$ ($\because \; e \neq \pm 1$)

$\therefore \quad e = 2$ ($\because$ e is positive).

36. $(2\vec{a} + \vec{b}) \cdot \{(\vec{a} \times \vec{b}) \times (\vec{a} - 2\vec{b})\} = (2\vec{a} + \vec{b}) \cdot \left[\{\vec{a} \cdot (\vec{a} - 2\vec{b})\}\,\vec{b} - \{\vec{b} \cdot (\vec{a} - 2\vec{b})\}\,\vec{a}\right]$

$= (2\vec{a} + \vec{b}) \cdot \{(\vec{a}^2 - 2\vec{a} \cdot \vec{b})\,\vec{b} - (\vec{b} \cdot \vec{a} - 2\vec{b}^2)\,\vec{a}\}$

$= (2\vec{a} + \vec{b}) \cdot (\vec{b} + 2\vec{a}) \quad (\because \; \vec{a} \cdot \vec{a} = 1,\ \vec{b} \cdot \vec{b} = 1,\ \vec{a} \cdot \vec{b} = 0)$

$= 4\vec{a}^2 + 4\vec{a} \cdot \vec{b} + \vec{b}^2 = 4 + 0 + 1 = 5.$

37. Direction ratios l, m, n of the normal to the plane of lines are given by

$l \cdot 2 + m \cdot 3 + \cdot\, n \cdot 4 = 0$

$l \cdot 3 + m \cdot 4 + n \cdot 5 = 0$

or $\frac{l}{-1} = \frac{m}{2} = \frac{n}{-1}$

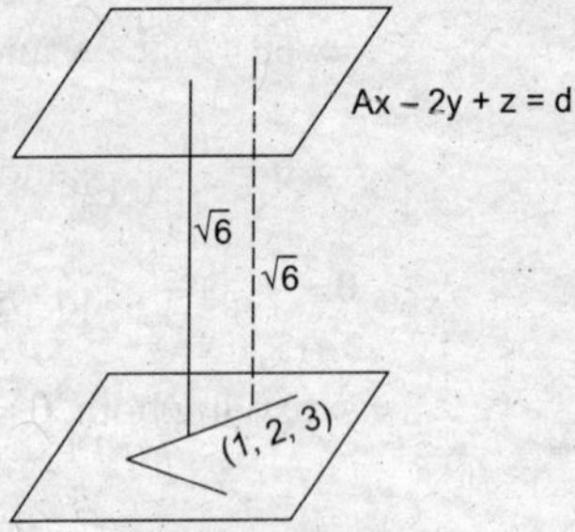

As the planes must be parallel,

$$\frac{A}{-1} = \frac{-2}{2} = \frac{1}{-1} \quad \therefore \quad A = 1$$

$\therefore$ distance between the planes

$= \sqrt{6} =$ distance of (1, 2, 3) from the first plane

$$\therefore \quad \sqrt{6} = \left| \frac{1 \cdot 1 - 2 \cdot 2 + 3 - d}{\sqrt{1^2 + (-2)^2 + 1^2}} \right| = \left| \frac{d}{\sqrt{6}} \right|$$

$\therefore \quad |d| = 6.$

38. The equation of tangent at (x, y) is $Y - y = \frac{dy}{dx}(X - x)$

or $\frac{dy}{dx} X - Y = x\frac{dy}{dx} - y$; $\therefore$ y-intercept of the tangent $= y - x\frac{dy}{dx}$

Given $y - x\frac{dy}{dx} = x^3$ or $\frac{dy}{dx} - \frac{1}{x}y = -x^2$. Now I.F. $= e^{\int -1/x\, dx} = e^{-\log x} = \frac{1}{x}$

$$\therefore \quad y \cdot \frac{1}{x} = \int -x^2 \cdot \frac{1}{x} dx = -\frac{x^2}{2} + c. \text{ When } x = 1, y = 1 \quad \therefore \quad c = \frac{3}{2}$$

$$\therefore \quad \frac{y}{x} = -\frac{x^2}{2} + \frac{3}{2} \qquad \therefore \quad f(x) = -\frac{1}{2}x^3 + \frac{3}{2}x$$

$$\therefore \quad f(-3) = -\frac{1}{2}(-3)^3 + \frac{3}{2}(-3) = \frac{27}{2} - \frac{9}{2} = 9.$$

39. Here $1 \le x < 2 \Rightarrow f(x) = x - 1$, $\quad 0 \le x < 1 \Rightarrow f(x) = 1 + 0 - x = 1 - x$,

$3 \le x < 4 \Rightarrow f(x) = x - 3$, $\quad 2 \le x < 3 \Rightarrow f(x) = 1 + 2 - x = 3 - x$,

etc. $\quad 4 \le x < 5 \Rightarrow f(x) = 1 + 4 - x = 5 - x$, etc.

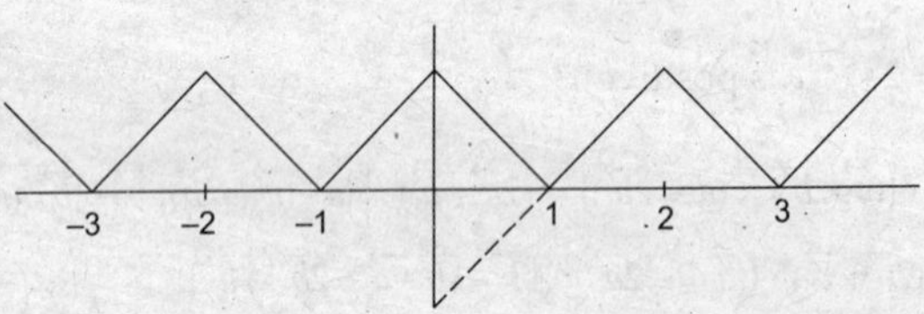

Clearly $f(x)$ is periodic having a period 2. Also $\cos \pi x = \cos \pi(x + 2)$. So $\cos \pi x$ is also periodic having a period 2. So $f(x) \cos \pi x$ is periodic having a period 2. Also $f(x)$, $\cos \pi x$ are even functions and so $f(x) \cos \pi x$ is even.

$$\therefore \quad \int_{-10}^{10} f(x) \cos \pi x \, dx = 2\int_0^{10} f(x) \cos \pi x \, dx \quad \text{(being even function)}$$

$$= 2 \times 5 \int_0^2 f(x) \cos \pi x \, dx, \text{ (being periodic of period 2)}$$

$$\int_0^2 f(x) \cos \pi x \, dx = \int_0^1 f(x) \cos \pi x \, dx + \int_1^2 f(x) \cos \pi x \, dx = I_1 + I_2$$

$$I_1 = \int_0^1 f(x) \cos \pi x \, dx = \int_0^1 (1 - x) \cos \pi x \quad [\because \text{ in } (0, 1), f(x) = 1 - x]$$

$$= \int_0^1 \{1 - (1 - x)\} \cos \pi(1 - x) \, dx = -\int_0^1 x \cos \pi x \, dx$$

$$I_2 = \int_1^2 f(x) \cos \pi x \, dx = \int_1^2 (x - 1) \cos \pi x \, dx \quad [\because \text{ in } (1, 2), f(x) = x - 1]$$

$$= \int_0^1 z \cos \pi (1 + z) \, dz \quad \text{(putting } x - 1 = z\text{)}$$

$$= -\int_0^1 z \cos \pi z \, dz = -\int_0^1 x \cos \pi x \, dx$$

$$\therefore \quad \int_{-10}^{10} f(x) \cos \pi x \, dx = 10(I_1 + I_2) = 10 \times (-2) \int_0^1 x \cos \pi x \, dx$$

$$= -20 \left[x \frac{\sin \pi x}{x} - \int \frac{\sin \pi x}{x} dx \right]_0^1$$

$$= -20 \cdot \frac{1}{\pi} \left[\frac{-\cos \pi x}{x} \right]_0^1 = \frac{20}{\pi^2} [1 - (-1)]$$

$$\therefore \quad \text{value} = \frac{\pi^2}{10} \cdot \frac{40}{\pi^2} = 4.$$

40. $a_k + a_{k-2} = 2a_{k-1}$ $\quad \therefore \quad$ a_{k-2}, a_{k-1}, a_k are in AP for $k = 3, 4, \ldots, 11$.

$\therefore$ $a_1, a_2, a_3, \ldots, a_{11}$ are in AP. Let d be the common difference.

$$\therefore \quad 90 = \frac{a_1^2 + (a_1 + d)^2 + (a_2 + 2d)^2 + \ldots + (a_1 + 10d)^2}{11}$$

$$= a_1^2 + 2a_1 d \frac{1 + 2 + 3 + \ldots + 10}{11} + d^2 \cdot \frac{1^2 + 2^2 + 3^2 + \ldots + 10^2}{11}$$

or $\quad 90 = a_1^2 + 2a_1 d \cdot \dfrac{10 \cdot 11}{2 \cdot 11} + d^2 \cdot \dfrac{10 \cdot 11 \cdot 21}{6} \cdot \dfrac{1}{11} \Rightarrow a_1^2 + 10a_1 d + 35d^2 = 90$

or $\quad 15^2 + 150d + 35d^2 = 90$ $\qquad$ or $\quad 7d^2 + 30d + 27 = 0$

or $\quad 7d^2 + 21d + 9d + 27 = 0$ $\qquad$ or $\quad (7d + 9)(d + 3) = 0.$

But $a_2 = a_1 + d = 15 - \dfrac{9}{7}, 15 - 3$. As $a_2 < \dfrac{27}{2}, a_2 \neq 15 - \dfrac{9}{7}$

$\therefore \quad a_2 = 12$ and $d = -3$.

$$\text{Now, value} = \frac{\frac{11}{2}\{2a_1 + 10d\}}{11} = a_1 + 5d = 15 + 5(-3) = 0.$$

41. Let $\frac{\pi}{2k} = \theta$. Then $\cos\theta = \frac{x}{2}$ and $\cos 2\theta = \frac{\sqrt{3}+1-x}{2}$

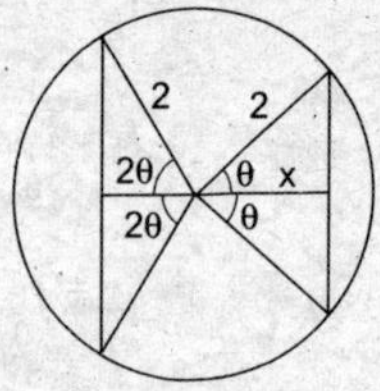

$\therefore \quad 2\cos^2\theta - 1 = \frac{\sqrt{3}+1-x}{2}$

or $\quad 2\cdot\frac{x^2}{4} - 1 = \frac{\sqrt{3}+1}{2} - \frac{x}{2}$ or $\frac{x^2}{2} + \frac{x}{2} - \frac{3+\sqrt{3}}{2} = 0$

or $\quad x^2 + x - (3+\sqrt{3}) = 0$ or $x^2 - 3 + x - \sqrt{3} = 0$

or $\quad (x-\sqrt{3})(x+\sqrt{3}+1) = 0$

$\therefore \quad x = \sqrt{3}, (-\sqrt{3}-1)$. But $x > 0$. So, $x = \sqrt{3}$.

$\therefore \quad \cos\theta = \frac{x}{2} = \frac{\sqrt{3}}{2} \quad \therefore \quad \theta = \frac{\pi}{6} \quad \therefore \quad \frac{\pi}{2k} = \frac{\pi}{6} \quad \therefore \quad k = 3.$

42. $\sin C = \frac{2\Delta}{ab} = \frac{2\times 15\sqrt{3}}{6\times 10} = \frac{\sqrt{3}}{2} \qquad \therefore \quad C = 120^\circ \;(\because\; C \text{ is obtuse})$

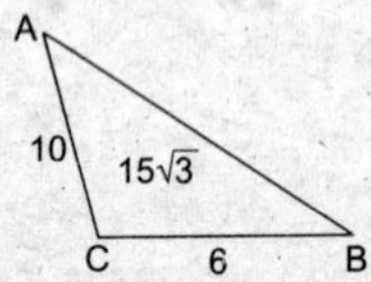

$\therefore \quad c^2 = a^2 + b^2 - 2ab\cos C$

$= 36 + 100 - 2\times 6\times 10\times\left(-\frac{1}{2}\right)$

$= 136 + 60 = 196; \quad \therefore \quad c = 14$

$\therefore \quad r = \frac{\Delta}{s} = \frac{15\sqrt{3}}{\frac{10+6+14}{2}} = \sqrt{3}; \quad \therefore \quad r^2 = 3.$

43. $|A| = \begin{vmatrix} 2k-1 & 2\sqrt{k} & 2\sqrt{k} \\ 2\sqrt{k} & 1 & -2k \\ -2\sqrt{k} & 2k & -1 \end{vmatrix} = \begin{vmatrix} 2k-1 & 2\sqrt{k} & 2\sqrt{k} \\ 2\sqrt{k} & 1 & -2k \\ 0 & 2k+1 & -(2k+1) \end{vmatrix}$

$= \begin{vmatrix} 2k-1 & 4\sqrt{k} & 2\sqrt{k} \\ 2\sqrt{k} & 1-2k & -2k \\ 0 & 0 & -(2k+1) \end{vmatrix}$

$= -(2k+1)\cdot\{-(2k-1)^2 - 8k\} = (2k+1)(4k^2 - 4k + 1 + 8k)$

$= (2k+1)^3$

$|B| = \begin{vmatrix} 0 & 2k-1 & \sqrt{k} \\ 1-2k & 0 & 2\sqrt{k} \\ -\sqrt{k} & -2\sqrt{k} & 0 \end{vmatrix} = \begin{vmatrix} 0 & 2k-1 & \sqrt{k} \\ 1-2k & 2-4k & 0 \\ -\sqrt{k} & -2\sqrt{k} & 0 \end{vmatrix}$

$$= \sqrt{k}\,\{-2k(1-2k) + \sqrt{k}(2-4k)\} = 0.$$

$\therefore$ $\det(\text{adj. } A) = |A|^{n-1} = |A|^2 = (2k+1)^6;$

$\det(\text{adj. } B) = |B|^{n-1} = |B|^2 = 0$

$\therefore$ $\det(\text{adj. } A) + \det(\text{adj. } B) = 10^6 \Rightarrow (2k+1)^6 + 0 = 10^6$

$$\Rightarrow 2k + 1 = 10$$

$$\Rightarrow k = 4.5 \Rightarrow [k] = 4.$$

44. $g(x) = e^{f(x)}$

$\therefore$ $g'(x) = e^{f(x)} \cdot f'(x) = (+)\,2010(x-2009)(x-2010)^2(x-2011)^3(x-2012)^4$

$\therefore$ $g'(x) = 0$, when $x = 2009, 2010, 2011, 2012.$

Now, $\left.\begin{array}{l} g'(2009-t) = (+)(-t)(-t-1)^2(-t-2)^3(-t-3)^4 = +\text{ve} \\ g'(2009+t) = (+)(t)(t-1)^2(t-2)^3(t-3)^4 = -\text{ve} \end{array}\right\}$

$\therefore$ $g(x)$ has a maximum at $x = 2009$.

$\left.\begin{array}{l} g'(2010-t) = (+)(1-t)(-t)^2(-t-1)^3(-t-2)^4 = -\text{ve} \\ g'(2010+t) = (+)(1+t)(t)^2(t-1)^3(t-2)^4 = -\text{ve} \end{array}\right\}$

no maximum at $x = 2010$

$\left.\begin{array}{l} g'(2011-t) = (+)(2-t)(1-t)^2(-t)^3(-t-1)^4 = -\text{ve} \\ g'(2011+t) = (+)(2+t)(1+t)^2(t)^3(t-1)^4 = +\text{ve} \end{array}\right\}$

no maximum at $x = 2011$

$\left.\begin{array}{l} g'(2012-t) = (+)(3-t)(2-t)^2(1-t)^3(-t)^4 = +\text{ve} \\ g'(2012+t) = (+)(3+t)(2+t)^2(1+t)^3(t^4) = +\text{ve} \end{array}\right\}$

no maximum at $x = 2012$

$\therefore$ $g(x)$ has local maximum only at $x = 2009$.

□

IIT Questions—8

SECTION I

Single Correct Choice Type

This section contains 7 multiple-choice questions. Each question has 4 choices (a), (b), (c) and (d) for its answer, out of which *only one* is correct.

1. Let (x_0, y_0) be the solution of the following equations

$$(2x)^{\ln 2} = (3y)^{\ln 3}$$
$$3^{\ln x} = 2^{\ln y}$$

Then x_0 is

(a) $\frac{1}{6}$ (b) $\frac{1}{3}$ (c) $\frac{1}{2}$ (d) 6

2. Let $P = \{\theta : \sin\theta - \cos\theta = \sqrt{2}\cos\theta\}$ and $Q = \{\theta : \sin\theta + \cos\theta = \sqrt{2}\sin\theta\}$ be two sets. Then

(a) $P \subset Q$ and $Q - P \neq \phi$ (b) $Q \not\subset P$

(c) $P \not\subset Q$ (d) $P = Q$

3. Let $\vec{a} = \hat{i} + \hat{j} + \hat{k}$, $\vec{b} = \hat{i} - \hat{j} + \hat{k}$ and $\vec{c} = \hat{i} - \hat{j} - \hat{k}$. A vector $\vec{v}$ in the plane of $\vec{a}$ and $\vec{b}$, whose projection on $\vec{c}$ is $\frac{1}{\sqrt{3}}$, is given by

(a) $\hat{i} - 3\hat{j} + 3\hat{k}$ (b) $-3\hat{i} - 3\hat{j} + \hat{k}$

(c) $3\hat{i} - \hat{j} + 3\hat{k}$ (d) $\hat{i} + 3\hat{j} - 3\hat{k}$

4. Let α and β be the roots of $x^2 - 6x - 2 = 0$, $\alpha > \beta$. If $a_n = \alpha^n - \beta^n$, $n \geq 1$ then the value of $\frac{a_{10} - 2a_8}{2a_9}$ is equal to

(a) 1 (b) 2 (c) 3 (d) 4

5. A straight line L through the point $(3, -2)$ is inclined at an angle $60°$ to the line $\sqrt{3}x + y = 1$. If L also intersects the x-axis then the equation of the line L is

(a) $y + \sqrt{3}x + 2 - 3\sqrt{3} = 0$ (b) $y - \sqrt{3}x + 2 + 3\sqrt{3} = 0$

(c) $\sqrt{3}y - x + 3 + 2\sqrt{3} = 0$ (d) $\sqrt{3}y + x - 3 + 2\sqrt{3} = 0$

6. The value of $\int_{\sqrt{\ln 2}}^{\sqrt{\ln 3}} \frac{x\sin x^2}{\sin x^2 + \sin(\ln 6 - x^2)} dx$ is

(a) $\frac{1}{4}\ln\frac{3}{2}$ (b) $\frac{1}{2}\ln\frac{3}{2}$ (c) $\ln\frac{3}{2}$ (d) $\frac{1}{6}\ln\frac{3}{2}$

7. Let the straight line $x = b$ divide the area enclosed by $y = (1-x)^2$, $y = 0$ and $x = 0$ into two parts $R_1 (0 \le x \le b)$ and $R_2 (b \le x \le 1)$ such that $R_1 - R_2 = \frac{1}{4}$. Then b equals

(a) $\frac{3}{4}$ (b) $\frac{1}{2}$ (c) $\frac{1}{3}$ (d) $\frac{1}{4}$

SECTION II

Multiple Correct Choice Type

This section contains 4 multiple-choice questions. Each question has four choices (a), (b), (c) and (d) out of which ***one* or *more*** may be correct.

8. The vector(s) which is/are coplanar with vectors $\hat{i} + \hat{j} + 2\hat{k}$ and $\hat{i} + 2\hat{j} + \hat{k}$, and perpendicular to the vector $\hat{i} + \hat{j} + \hat{k}$ is/are

(a) $\hat{j} - \hat{k}$ (b) $-\hat{i} + \hat{j}$ (c) $\hat{i} - \hat{j}$ (d) $-\hat{j} + \hat{k}$

9. Let the eccentricity of the hyperbola $\frac{x^2}{a^2} - \frac{y^2}{b^2} = 1$ be reciprocal to that of the ellipse $x^2 + 4y^2 = 4$. If the hyperbola passes through a focus of the ellipse, then

(a) the hyperbola has the equation $\frac{x^2}{3} - \frac{y^2}{2} = 1$

(b) a focus of the hyperbola is (2, 0)

(c) the eccentricity of the hyperbola is $\sqrt{5/3}$

(d) the hyperbola has the equation $x^2 - 3y^2 = 3$

10. Let M and N be two 3×3 non-singular skew symmetric matrices such that $MN = NM$. If P^T denotes the transpose of P then $M^2N^2(M^TN)^{-1}(MN^{-1})^T$ is equal to

(a) M^2 (b) $-N^2$ (c) $-M^2$ (d) MN

11. Let $f: R \to R$ be a function such that $f(x+y) = f(x) + f(y), \forall\, x, y \in R$. If $f(x)$ is differentiable at $x = 0$ then

(a) $f(x)$ is differentiable only in a finite interval containing zero

(b) $f(x)$ is continuous $\forall\, x \in R$

(c) $f'(x)$ is constant $\forall\, x \in R$

(d) $f(x)$ is differentiable except at finitely many points

SECTION III

Paragraph Type

This section contains 2 paragraphs. Based upon one of paragraphs 2 multiple-choice questions and based on the other paragraph 3 multiple-choice questions are to be answered. Each of these questions has four choices (a), (b), (c) and (d) out of which *only one* is correct.

Paragraph for Question Nos. 12 to 13

Let U_1 and U_2 be two urns such that U_1 contains 3 white and 2 red balls and U_2 contains only one white ball. A fair coin is tossed. If head appears then 1 ball is drawn from U_1 and put in U_2. However, if tail appears then 2 balls are drawn at random from U_1 and put in U_2. Now 1 ball is drawn at random from U_2.

12. The probability of the drawn ball from U_2 being white is

(a) $\frac{13}{30}$ (b) $\frac{23}{30}$ (c) $\frac{19}{30}$ (d) $\frac{11}{30}$

13. Given that the drawn ball from U_2 is white, the probability that head appeared on the coin is

(a) $\frac{17}{23}$ (b) $\frac{11}{23}$ (c) $\frac{15}{23}$ (d) $\frac{12}{23}$

Paragraph for Question Nos. 14 to 16

Let a, b and c be three real numbers satisfying $[a \;\; b \;\; c]\begin{bmatrix} 1 & 9 & 7 \\ 8 & 2 & 7 \\ 7 & 3 & 7 \end{bmatrix} = [0 \;\; 0 \;\; 0] \ldots (E)$

14. If the point $P(a, b, c)$, with reference to (E), lies on the plane $2x + y + z = 1$ then the value of $7a + b + c$ is

(a) 0 (b) 12

(c) 7 (d) 6

15. Let ω be a solution of $x^3 - 1 = 0$ with $\text{Im}(\omega) > 0$. If $a = 2$ with b and c satisfying (E) then the value of $\frac{3}{\omega^a} + \frac{1}{\omega^b} + \frac{3}{\omega^c}$ is equal to

(a) -2 (b) 2 (c) 3 (d) -3

16. Let $b = 6$, with a and c satisfying (E). If α and β are the roots of the quadratic equation $ax^2 + bx + c = 0$ then

$$\sum_{n=0}^{\infty}\left(\frac{1}{\alpha}+\frac{1}{\beta}\right)^n \text{ is equal to}$$

(a) 6 (b) 7 (c) $\frac{6}{7}$ (d) ∞

SECTION IV

Integer Answer Type

This section contains 7 questions. The answer to each question is a single digit integer. The bubble corresponding to the correct answer is to be darkened in the ORS.

17. Consider the parabola $y^2 = 8x$. Let Δ_1 be the area of the triangle formed by the end points of its latus rectum and the point $P\left(\frac{1}{2}, 2\right)$ on the parabola and Δ_2 be the area of the triangle formed by drawing tangents at P and the end points of the latus rectum. Then $\frac{\Delta_1}{\Delta_2}$ is

18. Let $f(\theta) = \sin\left\{\tan^{-1}\left(\frac{\sin\theta}{\sqrt{\cos 2\theta}}\right)\right\}$ where $-\frac{\pi}{4} < \theta < \frac{\pi}{4}\cdot$ Then the value of $\frac{d}{d(\tan\theta)}f(\theta)$ is

19. Let $f : [1, \infty) \to [2, \infty)$ be a differentiable function such that $f(1) = 2$. If $6\int_1^x f(t)dt = 3x(f(x) - x^3$ for all $x \geq 1$ then the value of $f(2)$ is

20. The positive integral value of $n > 3$ satisfying the equation

$$\frac{1}{\sin\frac{\pi}{n}} = \frac{1}{\sin\frac{2\pi}{n}} + \frac{1}{\sin\frac{3\pi}{n}}$$

is

21. Let $a_1, a_2, a_3, \ldots, a_{100}$ be in AP with $a_1 = 3$ and $S_p = \sum_{i=1}^{p} a_i$, $1 \le p \le 100$. For any integer n with $1 \le n \le 20$, let $m = 5n$. If $\frac{S_m}{S_n}$ does not depend on n then a_2 is

22. The minimum value of the sum of the real numbers $a^{-5}, a^{-4}, 3a^{-3}, 1, a^8$ and a^{10} with $a > 0$ is

23. If z is a complex number satisfying $|z - 3 - 2i| \le 2$ then the minimum value of $|2z - 6 + 5i|$ is

ANSWERS

1. c **2.** d **3.** c **4.** c **5.** b
6. a **7.** b **8.** a, d **9.** b, d **10.** c
11. b, c **12.** b **13.** d **14.** d **15.** a
16. b

17.	0	1	**(2)**	3	4	5	6	7	8	9
18.	0	**(1)**	2	3	4	5	6	7	8	9
19.	0	1	2	3	4	5	**(6)**	7	8	9
20.	0	1	2	3	4	5	6	**(7)**	8	9
21.	0	1	2	3	4	5	6	7	8	**(9)**
22.	0	1	2	3	4	5	6	7	**(8)**	9
23.	0	1	2	3	4	**(5)**	6	7	8	9

Hints and Solutions to Selected Questions

1. $(2x)^{\log 2} = (3y)^{\log 3} \Rightarrow \log 2 \cdot (\log 2 + \log x) = \log 3 \cdot (\log 3 + \log y)$

or $(\log 3)^2 - (\log 2)^2 = \log 2 \cdot \log x - \log 3 \cdot \log y$ ... (1)

$3^{\log x} = 2^{\log y} \Rightarrow \log x \cdot \log 3 = \log y \cdot \log 2$ ... (2)

From (1) and (2), $\log y = \dfrac{-(\log 3)^2 + (\log 2)^2 + \log 2 \cdot \log x}{\log 3}$

or $\dfrac{\log x \cdot \log 3}{\log 2} = \dfrac{\log 2 \cdot \log x - (\log 3)^2 + (\log 2)^2}{\log 3}$

or $\log x \{(\log 3)^2 - (\log 2)^2\} = \log 2 \{-(\log 3)^2 + (\log 2)^2\}$

$\therefore$ $\log x = -\log 2 = \log \frac{1}{2} \Rightarrow x = \frac{1}{2}$.

2. $\sin\theta - \cos\theta = \sqrt{2}\cos\theta$ or $\sin\theta = (\sqrt{2}+1)\cos\theta$

$\therefore \quad \tan\theta = \sqrt{2}+1 \left(\because \ \theta \neq 0, \frac{\pi}{2}\right)$

$\sin\theta + \cos\theta = \sqrt{2}\sin\theta$ or $\cos\theta = (\sqrt{2}-1)\sin\theta$

$\therefore \quad \tan\theta = \frac{1}{\sqrt{2}-1} = \frac{\sqrt{2}+1}{2-1} = \sqrt{2}+1.$

$\therefore$ the two sets P and Q are equal.

3. Let $\vec{v} = x\vec{a} + y\vec{b}$

$= x(\hat{i}+\hat{j}+\hat{k}) + y(\hat{i}-\hat{j}+\hat{k}).$

Here, $\vec{v}\cdot\frac{\vec{c}}{|\vec{c}|} = \frac{1}{\sqrt{3}} \quad \therefore \quad \frac{\hat{i}-\hat{j}-\hat{k}}{\sqrt{3}}\cdot\{(x+y)\hat{i}+(x-y)\hat{j}+(x+y)\hat{k}\} = \frac{1}{\sqrt{3}}$

or $\quad x+y-(x-y)-(x+y) = 1 \quad \therefore \quad y-x=1$

$\therefore \quad \vec{v} = x(\hat{i}+\hat{j}+\hat{k}) + (x+1)(\hat{i}-\hat{j}+\hat{k})$

$= (2x+1)\hat{i} - \hat{j} + (2x+1)\hat{k}$

Taking $x = 1$, $\vec{v} = 3\hat{i} - \hat{j} + 3\hat{k}.$

4. Here, $\alpha^2 - 6\alpha - 2 = 0 \quad \therefore \quad \alpha^{10} - 6\alpha^9 - 2\alpha^8 = 0$ ... (1)

$\beta^2 - 6\beta - 2 = 0 \quad \therefore \quad \beta^{10} - 6\beta^9 - 2\beta^8 = 0$ ... (2)

From (1) – (2), $\alpha^{10} - \beta^{10} - 6(\alpha^9 - \beta^9) - 2(\alpha^8 - \beta^8) = 0$

$\therefore \quad a_{10} - 6a_9 - 2a_8 = 0 \quad \therefore \quad \frac{a_{10} - 2a_8}{2a_9} = \frac{6a_9}{2a_9} = 3.$

5. Let L be $y + 2 = m(x-3)$, $x \neq 0$.

Given, $\left|\frac{m-(-\sqrt{3})}{1+m(-\sqrt{3})}\right| = \sqrt{3}; \quad \therefore \quad m + \sqrt{3} = \pm\sqrt{3}(1-\sqrt{3}m)$

or $\quad m + \sqrt{3} = \pm(\sqrt{3} + 3m) \quad \therefore \quad m = 0, \sqrt{3}$. But $m \neq 0$.

$\therefore$ the equation of L is $y + 2 = \sqrt{3}(x-3) \Rightarrow y - \sqrt{3}x + 2 + 3\sqrt{3} = 0.$

6. Putting $x^2 = z$, $I = \int_{\log 2}^{\log 3} \frac{\frac{1}{2}\sin z}{\sin z + \sin(\log 6 - z)}\,dz$

$= \frac{1}{2}\int_{\log 2}^{\log 3} \frac{\sin z}{\sin z + \sin(\log 6 - z)}\,dz$

Putting $\log 6 - z = t$, $I = \frac{1}{2}\int_{\log 3}^{\log 2} \frac{\sin(\log 6 - t)}{\sin(\log 6 - t) + \sin t}(-dt)$

Adding, $2I = \frac{1}{2}\int_{\log 2}^{\log 3} \frac{\sin\theta + \sin(\log 6 - \theta)}{\sin\theta + \sin(\log 6 - \theta)}\, d\theta$

$= \frac{1}{2}\int_{\log 2}^{\log 3} d\theta = \frac{1}{2}(\log 3 - \log 2)$

$\therefore \quad I = \frac{1}{4}\cdot \log\frac{3}{2}$.

7. $R_1 = \int_0^b y\,dx = \int_0^b (1-x)^2\,dx$

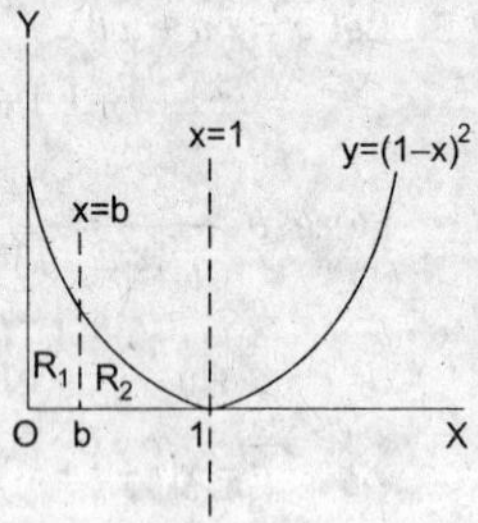

$R_2 = \int_b^1 y\,dx = \int_b^1 (1-x)^2 dx$

$\therefore \quad R_1 - R_2 = \int_0^b (1-x)^2\,dx - \int_b^1 (1-x)^2 dx$

$= \left[-\frac{(1-x)^3}{3}\right]_0^b - \left[-\frac{(1-x)^3}{3}\right]_b^1$

$= -\frac{(1-b)^3}{3} + \frac{1}{3} - \frac{(1-b)^3}{3}$

$\therefore \quad \frac{1}{4} = -\frac{2}{3}(1-b)^3 + \frac{1}{3}; \qquad \therefore \quad \frac{2}{3}(1-b)^3 = \frac{1}{3} - \frac{1}{4} = \frac{1}{12};$

$\therefore \quad (1-b)^3 = \frac{1}{8} \qquad \therefore \quad 1 - b = \frac{1}{2}; \qquad \therefore \quad b = \frac{1}{2}$.

8. Here, $\vec{r} = x(\hat{i} + \hat{j} + 2\hat{k}) + y(\hat{i} + 2\hat{j} + \hat{k})$

$\therefore \quad \vec{r}\cdot(\hat{i} + \hat{j} + \hat{k}) = 0 \Rightarrow x(1+1+2) + y(1+2+1) = 0$

$\therefore \quad y = -x$. So, $\vec{r} = x(\hat{i} + \hat{j} + 2\hat{k}) - x(\hat{i} + 2\hat{j} + \hat{k})$

$= x(-\hat{j} + \hat{k})$.

Taking $x = \pm 1$, $\vec{r} = \pm(-\hat{j} + \hat{k})$.

9. For the ellipse, $x^2 + 4y^2 = 4$ or $\frac{x^2}{2^2} + \frac{y^2}{1^2} = 1$.

$\therefore \quad 1^2 = 2^2\cdot(1-e^2); \qquad \therefore \quad 1 - e^2 = \frac{1}{4}; \qquad \therefore \quad e = \sqrt{\frac{3}{4}} = \frac{\sqrt{3}}{2}$.

$\therefore$ the eccentricity of the hyperbola $= \frac{1}{\sqrt{3}/2} = \frac{2}{\sqrt{3}}$.

The foci of the ellipse are $\left(\pm 2\cdot\frac{\sqrt{3}}{2}, 0\right)$, i.e., $(\pm\sqrt{3}, 0)$.

The hyperbola passes through a focus of the ellipse $\Rightarrow \frac{3}{a^2} = 1 \Rightarrow a^2 = 3$.

$\therefore \quad b^2 = a^2(e^2-1) \Rightarrow b^2 = 3\left\{\left(\frac{2}{\sqrt{3}}\right)^2 - 1\right\} \Rightarrow b^2 = 3\left(\frac{4}{3}-1\right) = 1.$

$\therefore$ the equation of the hyperbola is $\frac{x^2}{3} - \frac{y^2}{1} = 1$, i.e., $x^2 - 3y^2 = 3$.

A focus of the hyperbola is $(ae, 0) = \left(\sqrt{3}\cdot\frac{2}{\sqrt{3}}, 0\right) = (2, 0)$.

10. $M^2N^2(M^TN)^{-1}(MN^{-1})^T = M^2NNN^{-1}(M^T)^{-1}(N^{-1})^TM^T$

$[\because (AB)^{-1} = B^{-1}A^{-1}, (AB)^T = B^TA^T]$

$= M^2N(-M)^{-1}(N^T)^{-1}(-M)$

$[\because (A^T)' = (A')^T, AA^{-1} = I, M^T = -M$, being skew symmetric$]$

$= M^2NM^{-1}(-1)^2(-N)^{-1}M \quad [\because N^T = -N$, being skew symmetric$]$

$= -M^2NM^{-1}N^{-1}M = -M(MN)(NM)^{-1}M$

$= -M(NM)(NM)^{-1}M \quad [\because MN = NM$, given$]$

$= -MIM = -MM = -M^2.$

11. $f'(x) = \lim_{h\to 0}\frac{f(x+h)-f(x)}{h} = \lim_{h\to 0}\frac{f(h)}{h} \quad [\because f(x+h) = f(x)+f(h)$, given$]$

$= \lim_{h\to 0}\frac{f(0+h)-f(0)}{h} \quad [\because f(x+0) = f(x)+f(0) \Rightarrow f(0) = 0]$

$= f'(0) = \text{constant}.$

$\therefore \quad f(x) = f'(0)\cdot x + c; \quad \therefore \quad f(0) = c = 0. \quad \therefore \quad f(x) = f'(0)\cdot x.$

12. By the theorem of total probability, required prob.

= prob. of head appearing {prob. of drawing white from U_1 × prob. of drawing white from 2 whites from U_2 + prob. of drawing red from U_1 × prob. of drawing white from 1 white from U_2}
+ prob. of tail appearing {prob. of drawing 2 whites from U_1 × prob. of drawing white from 3 whites from U_2 + prob. of drawing 2 reds from U_1 × prob. of drawing white from 1 white and 2 reds from U_2 + prob. of drawing 1 white and 1 red from U_1 × prob. of drawing white from 2 whites and 1 red from U_2}

$= \frac{1}{2}\left\{\frac{3}{5}\times 1 + \frac{2}{5}\times\frac{1}{2}\right\} + \frac{1}{2}\left\{\frac{{}^3C_2}{{}^5C_2}\times 1 + \frac{{}^2C_2}{{}^5C_2}\times\frac{1}{3} + \frac{{}^3C_1\cdot{}^2C_1}{{}^5C_2}\times\frac{{}^2C_1}{{}^3C_1}\right\}$

$= \frac{2}{5} + \frac{1}{2}\left\{\frac{3}{10} + \frac{1}{10}\times\frac{1}{3} + \frac{6}{10}\times\frac{2}{3}\right\} = \frac{2}{5} + \frac{11}{30} = \frac{23}{30}.$

13. By Bayes's theorem on inverse probability,

$$\text{required probability} = \frac{\begin{array}{c}\text{prob. of head appearing \{prob. of drawing white from } U_1 \times \text{prob. of drawing white}\\ \text{from 2 whites from } U_2 + \text{prob. of drawing red from } U_1 \times \text{prob. of drawing white from 1 white from } U_2\}\end{array}}{\text{total probability}}$$

$$= \frac{\frac{1}{2}\left\{\frac{3}{5} \times 1 + \frac{2}{5} \times \frac{1}{2}\right\}}{\frac{23}{30}} \text{ (see Q.N. 12)}$$

$$= \frac{12}{23}.$$

14. Here $[a + 8b + 7c \quad 9a + 2b + 3c \quad 7a + 7b + 7c] = [0 \ \ 0 \ \ 0]$

$$\Rightarrow \left.\begin{array}{r} a + 8b + 7c = 0 \\ 9a + 2b + 3c = 0 \\ a + b + c = 0 \end{array}\right\} \Rightarrow \frac{a}{1} = \frac{b}{6} = \frac{c}{-7} = t \text{ (say)}.$$

$P(a, b, c)$ is on the plane $2x + y + z = 1 \Rightarrow 2t + 6t - 7t = 1 \Rightarrow t = 1$.

$\therefore \quad a = 1, b = 6, c = -7; \quad \therefore \quad 7a + b + c = 7 + 6 - 7 = 6.$

15. $x = 1, \dfrac{-1 \pm \sqrt{3}i}{2}$; so $\omega = \dfrac{-1 + \sqrt{3}i}{2}$ and $\omega^3 = 1, 1 + \omega + \omega^2 = 0$

As in Q.N. 14, $\dfrac{a}{1} = \dfrac{b}{6} = \dfrac{c}{-7}$; so, $\dfrac{2}{1} = \dfrac{b}{6} = \dfrac{c}{-7}$.

$\therefore \quad b = 12, c = -14$ when $a = 2$.

So, required value $= \dfrac{3}{\omega^2} + \dfrac{1}{\omega^{12}} + \dfrac{3}{\omega^{-14}}$

$$= \frac{3\omega^3}{\omega^2} + \frac{1}{(\omega^3)^4} + \frac{3}{\omega^{-14} \cdot \omega^{15}} = 3\omega + 1 + \frac{3}{\omega}$$

$$= 3\omega + \frac{3\omega^3}{\omega} + 1 = 3(\omega + \omega^2) + 1 = 3(-1) + 1 = -2.$$

16. As in Q.N. 14, $\dfrac{a}{1} = \dfrac{b}{6} = \dfrac{c}{-7}$ · So, $\dfrac{a}{1} = \dfrac{6}{6} = \dfrac{c}{-7} \quad \therefore \quad a = 1, c = -7.$

$\therefore \quad \alpha, \beta$ are roots of $x^2 + 6x - 7 = 0$ or $(x + 7)(x - 1) = 0$; $\therefore \ \alpha, \beta = -7, 1.$

$$\therefore \quad \sum_{n=0}^{\infty} \left(\frac{1}{\alpha} + \frac{1}{\beta}\right)^n = \sum_{n=0}^{\infty} \left(\frac{1}{-7} + 1\right)^n = \sum_{n=0}^{\infty} \left(\frac{6}{7}\right)^n$$

$$= 1 + \frac{6}{7} + \left(\frac{6}{7}\right)^2 + \ldots \text{ to } \infty = \frac{1}{1 - \frac{6}{7}} = 7.$$

17. The parabola is $y^2 = 8x$ or $y^2 = 4 \times 2 \times x$ $\therefore\ a = 2.$

$\therefore$ end points of the latus rectum are $(a, \pm 2a) = (2, \pm 4)$.

$\therefore\ \Delta_1 = \text{ar}\,(\Delta PLM) = \frac{1}{2} \cdot \left(2 - \frac{1}{2}\right) \cdot 8 = 6.$

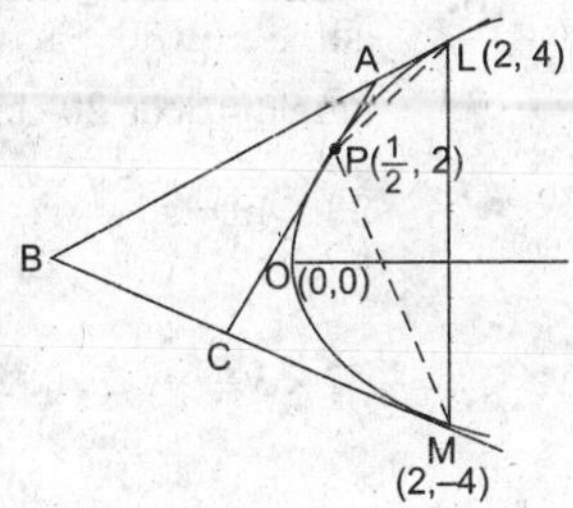

Tangents at $L(2, 4)$, $M(2, -4)$ are

$$y = x + 2 \text{ and } -y = x + 2.$$

Tangent at $P\left(\frac{1}{2}, 2\right)$ is $y = 2x + 1$.

$\therefore\quad B = (0, -2), A = (1, 3)$ and $C = (-1, -1)$

$$\therefore\quad \Delta_2 = \text{ar}\,(\Delta BCA) = \left|\frac{1}{2}\begin{vmatrix} 0 & -2 & 1 \\ -1 & -1 & 1 \\ 1 & 3 & 1 \end{vmatrix}\right| = 3.$$

$$\therefore\quad \frac{\Delta_1}{\Delta_2} = \frac{6}{3} = 2.$$

18. Let $\tan^{-1}\left(\frac{\sin\theta}{\sqrt{\cos 2\theta}}\right) = \lambda$. Then $\tan\lambda = \frac{\sin\theta}{\sqrt{\cos 2\theta}}$.

$$\therefore\quad f(\theta) = \sin\lambda = \frac{\sin\theta}{\sqrt{\sin^2\theta + (\sqrt{\cos 2\theta})^2}}$$

$$= \frac{\sin\theta}{\sqrt{\sin^2\theta + \cos^2\theta - \sin^2\theta}} = \frac{\sin\theta}{\cos\theta} = \tan\theta.$$

$$\therefore\quad \frac{d}{d(\tan\theta)} f(\theta) = \frac{d}{d(\tan\theta)} \tan\theta = 1.$$

19. Differentiating w.r.t. x, $6f(x) = 3f(x) + 3xf'(x) - 3x^2$

or $\quad xf'(x) - f(x) = x^2 \qquad$ or $\quad \frac{dy}{dx} - \frac{1}{x}y = x$ where $y = f(x)$

$\therefore$ integrating factor $= e^{\int -\frac{1}{x}dx} = e^{-\log x} = e^{\log x^{-1}} = x^{-1}$

$\therefore\quad \frac{d}{dx}\{y \cdot x^{-1}\} = x \cdot x^{-1} = 1; \qquad \therefore\quad \frac{y}{x} = x + c.$

From the question, when $x = 1$, $y = f(1) = 2$.

So, $2 = 1 + c$; $\qquad \therefore\ c = 1.$

$\therefore\quad \frac{y}{x} = x + 1$ or $y = f(x) = x^2 + x$

$\therefore\quad f(2) = 2^2 + 2 = 6.$

20. Let $\frac{\pi}{n} = \theta$. Then the equation is $\frac{1}{\sin\theta} = \frac{1}{\sin 2\theta} + \frac{1}{\sin 3\theta}$

or $\quad \frac{\sin 3\theta - \sin\theta}{\sin\theta \cdot \sin 3\theta} = \frac{1}{\sin 2\theta} \quad$ or $\quad \frac{2\cos 2\theta \cdot \sin\theta}{\sin\theta \cdot \sin 3\theta} = \frac{1}{\sin 2\theta}$

$\therefore \quad \sin 3\theta = 2\cos 2\theta \cdot \sin 2\theta = \sin 4\theta;$

$\therefore \quad 3\theta + 4\theta = \pi; \quad \therefore \quad 7 \cdot \frac{\pi}{n} = \pi \Rightarrow n = 7.$

21. $S_p = \frac{p}{2}\{2 \times 3 + (p-1)d\}$. Then $\frac{S_m}{S_n} = \frac{S_{5n}}{S_n} = \frac{\frac{5n}{2}\{6 + (5n-1)d\}}{\frac{n}{2}\{6 + (n-1)d\}}$

$\therefore \quad \frac{S_m}{S_n} = \frac{5(6 - d + 5nd)}{6 - d + nd}.$

This will be independent of n if $6 - d = 0 \Rightarrow d = 6$.

$\therefore \quad a_2 = a_1 + d = 3 + 6 = 9.$

$\{d = 6$ can also be obtained by equating the values of $\frac{S_m}{S_n}$ when $n = 1, 2.\}$

22. AM $\geq$ GM for positive numbers. So,

$$\frac{a^{-5} + a^{-4} + a^{-3} + a^{-3} + a^{-3} + 1 + a^8 + a^{10}}{8}$$

$$\geq \sqrt[8]{a^{-5} \cdot a^{-4} \cdot a^{-3} \cdot a^{-3} \cdot a^{-3} \cdot 1 \cdot a^8 \cdot a^{10}} = \sqrt[8]{1 \cdot a^0} = 1$$

$\therefore \quad a^{-5} + a^{-4} + 3a^{-3} + 1 + a^8 + a^{10} \geq 8.$

23. $|2z - 6 + 5i| = 2\left|z - \left(3 - \frac{5}{2}i\right)\right|$

$= 2 \times \left[\text{distance between } z \text{ and the point } \left(3, -\frac{5}{2}\right)\right]$

$|z - 3 - 2i| \leq 2 \Rightarrow |z - (3 + 2i)| \leq 2$

$\Rightarrow z$ is a point on or inside a circle whose radius is 2 and the centre is (3, 2).

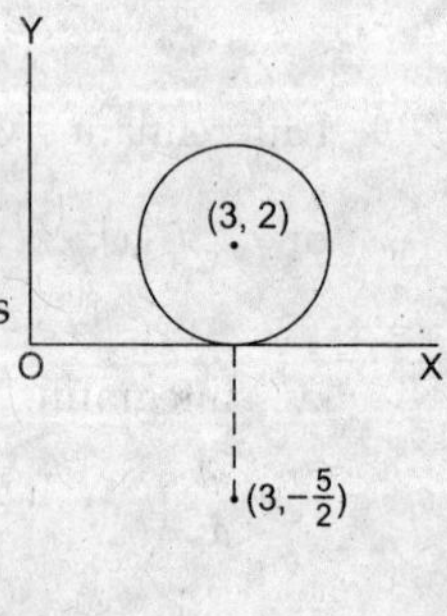

$\therefore$ minimum value of $|2z - 6 + 5i| = 2 \times \frac{5}{2} = 5.$

□